Family-group names in Coleoptera (Insecta)

Patrice Bouchard, Yves Bousquet, Anthony E. Davies,
Miguel A. Alonso-Zarazaga, John F. Lawrence, Chris H. C. Lyal,
Alfred F. Newton, Chris A. M. Reid, Michael Schmitt,
S. Adam Ślipiński and Andrew B. T. Smith

PENSOFT.
Sofia–Moscow
2011

ZooKeys 88 (Special Issue)

Family-group names in Coleoptera (Insecta)

Patrice Bouchard, Yves Bousquet, Anthony E. Davies, Miguel A. Alonso-Zarazaga, John F. Lawrence, Chris H. C. Lyal, Alfred F. Newton, Chris A. M. Reid, Michael Schmitt, S. Adam Ślipiński, Andrew B. T. Smith

First published 2011
ISBN 978-954-642-583-6 (hardback)
ISBN 978-954-642-584-3 (paperback)

Pensoft Publishers
Geo Milev Str. 13a, Sofia 1111, Bulgaria
Fax: +359-2-870-42-82
info@pensoft.net
www.pensoft.net

Printed in Bulgaria, April 2011

Contents

ZooKeys 88: 1–972 (2011)
doi: 10.3897/zookeys.88.807
www.zookeys.org

CATALOGUE

A peer-reviewed open-access journal
ZooKeys
Launched to accelerate biodiversity research

Family-group names in Coleoptera (Insecta)

Patrice Bouchard[1], Yves Bousquet[1], Anthony E. Davies[1], Miguel A. Alonso-Zarazaga[2], John F. Lawrence[3], Chris H. C. Lyal[4], Alfred F. Newton[5], Chris A. M. Reid[6], Michael Schmitt[7], S. Adam Ślipiński[3], Andrew B. T. Smith[8]

1 *Canadian National Collection of Insects, Arachnids and Nematodes, Agriculture and Agri-Food Canada, 960 Carling Avenue, Ottawa, Ontario, K1A 0C6, Canada* **2** *Departamento de Biodiversidad y Biología Evolutiva, Museo Nacional de Ciencias Naturales, Jose Gutierrez Abascal, 2; E-28006, Madrid, Spain* **3** *Australian National Insect Collection, CSIRO Entomology, GPO Box 1700, Canberra, ACT 2601, Australia* **4** *Department of Entomology, The Natural History Museum, Cromwell Road, London SW7 5BD, United Kingdom* **5** *Zoology Department, Field Museum of Natural History, 1400 South Lake Shore Drive Chicago, IL 60605, USA* **6** *Australian Museum, 6 College Street, Sydney, NSW 2010, Australia* **7** *Ernst-Moritz-Arndt-Universitaet, Allgemeine & Systematische Zoologie, Anklamer Str. 20, D-17489 Greifswald, Germany* **8** *Canadian Museum of Nature, P. O. Box 3443, Station D, Ottawa, Ontario, K1P 6P4, Canada*

Corresponding author: *Patrice Bouchard* (patrice.bouchard@agr.gc.ca)

Academic editor: *Terry Erwin* | Received 17 December 2010 | Accepted 17 February 2011 | Published 4 April 2011

Citation: Bouchard P, Bousquet Y, Davies AE, Alonso-Zarazaga MA, Lawrence JF, Lyal CHC, Newton AF, Reid CAM, Schmitt M, Ślipiński SA, Smith ABT (2011) Family-group names in Coleoptera (Insecta). ZooKeys 88: 1–972. doi: 10.3897/zookeys.88.807

Abstract

We synthesize data on all known extant and fossil Coleoptera family-group names for the first time. A catalogue of 4887 family-group names (124 fossil, 4763 extant) based on 4707 distinct genera in Coleoptera is given. A total of 4492 names are available, 183 of which are permanently invalid because they are based on a preoccupied or a suppressed type genus. Names are listed in a classification framework. We recognize as valid 24 superfamilies, 211 families, 541 subfamilies, 1663 tribes and 740 subtribes. For each name, the original spelling, author, year of publication, page number, correct stem and type genus are included. The original spelling and availability of each name were checked from primary literature. A list of necessary changes due to Priority and Homonymy problems, and actions taken, is given. Current usage of names was conserved, whenever possible, to promote stability of the classification.

New synonymies (family-group names followed by genus-group names): Agronomina Gistel, 1848 **syn. n.** of Amarina Zimmermann, 1832 (Carabidae), Hylepnigalioini Gistel, 1856 **syn. n.** of Melandryini Leach, 1815 (Melandryidae), Polycystophoridae Gistel, 1856 **syn. n.** of Malachiinae Fleming, 1821 (Melyridae), Sclerasteinae Gistel, 1856 **syn. n.** of Ptilininae Shuckard, 1839 (Ptinidae), Phloeonomini Ádám, 2001 **syn. n.** of Omaliini MacLeay, 1825 (Staphylinidae),

Sepedophilini Ádám, 2001 **syn. n.** of Tachyporini MacLeay, 1825 (Staphylinidae), Phibalini Gistel, 1856 **syn. n.** of Ctenioopodini Solier, 1835 (Tenebrionidae); *Agronoma* Gistel 1848 (type species *Carabus familiaris* Duftschmid, 1812, designated herein) **syn. n.** of *Amara* Bonelli, 1810 (Carabidae), *Hylepnigalio* Gistel, 1856 (type species *Chrysomela caraboides* Linnaeus, 1760, by monotypy) **syn. n.** of *Melandrya* Fabricius, 1801 (Melandryidae), *Polycystophorus* Gistel, 1856 (type species *Cantharis aeneus* Linnaeus, 1758, designated herein) **syn. n.** of *Malachius* Fabricius, 1775 (Melyridae), *Sclerastes* Gistel, 1856 (type species *Ptilinus costatus* Gyllenhal, 1827, designated herein) **syn. n.** of *Ptilinus* Geoffroy, 1762 (Ptinidae), *Paniscus* Gistel, 1848 (type species *Scarabaeus fasciatus* Linnaeus, 1758, designated herein) **syn. n.** of *Trichius* Fabricius, 1775 (Scarabaeidae), *Phibalus* Gistel, 1856 (type species *Chrysomela pubescens* Linnaeus, 1758, by monotypy) **syn. n.** of *Omophlus* Dejean, 1834 (Tenebrionidae). The following new replacement name is proposed: Gompeliina Bouchard, 2011 **nom. n.** for Olotelina Báguena Corella, 1948 (Aderidae).

Reversal of Precedence (Article 23.9) is used to conserve usage of the following names (family-group names followed by genus-group names): Perigonini Horn, 1881 nom. protectum over Trechicini Bates, 1873 nom. oblitum (Carabidae), Anisodactylina Lacordaire, 1854 nom. protectum over Eurytrichina LeConte, 1848 nom. oblitum (Carabidae), Smicronychini Seidlitz, 1891 nom. protectum over Desmorini LeConte, 1876 nom. oblitum (Curculionidae), Bagoinae Thomson, 1859 nom. protectum over Lyprinae Gistel 1848 nom. oblitum (Curculionidae), Aterpina Lacordaire, 1863 nom. protectum over Heliomenina Gistel, 1848 nom. oblitum (Curculionidae), Naupactini Gistel, 1848 nom. protectum over Iphiini Schönherr, 1823 nom. oblitum (Curculionidae), Cleonini Schönherr, 1826 nom. protectum over Geomorini Schönherr, 1823 nom. oblitum (Curculionidae), Magdalidini Pascoe, 1870 nom. protectum over Scardamyctini Gistel, 1848 nom. oblitum (Curculionidae), Agrypninae/-ini Candèze, 1857 nom. protecta over Adelocerinae/-ini Gistel, 1848 nom. oblita and Pangaurinae/-ini Gistel, 1856 nom. oblita (Elateridae), Prosternini Gistel, 1856 nom. protectum over Diacanthini Gistel, 1848 nom. oblitum (Elateridae), Calopodinae Costa, 1852 nom. protectum over Sparedrinae Gistel, 1848 nom. oblitum (Oedemeridae), Adesmiini Lacordaire, 1859 nom. protectum over Macropodini Agassiz, 1846 nom. oblitum (Tenebrionidae), Bolitophagini Kirby, 1837 nom. protectum over Eledonini Billberg, 1820 nom. oblitum (Tenebrionidae), Throscidae Laporte, 1840 nom. protectum over Stereolidae Rafinesque, 1815 nom. oblitum (Throscidae) and Lophocaterini Crowson, 1964 over Lycoptini Casey, 1890 nom. oblitum (Trogossitidae); *Monotoma* Herbst, 1799 nom. protectum over *Monotoma* Panzer, 1792 nom. oblitum (Monotomidae); *Pediacus* Shuckard, 1839 nom. protectum over *Biophloeus* Dejean, 1835 nom. oblitum (Cucujidae), *Pachypus* Dejean, 1821 nom. protectum over *Pachypus* Billberg, 1820 nom. oblitum (Scarabaeidae), *Sparrmannia* Laporte, 1840 nom. protectum over *Leocaeta* Dejean, 1833 nom. oblitum and *Cephalotrichia* Hope, 1837 nom. oblitum (Scarabaeidae).

Keywords

Beetles, nomenclature, classification, world fauna, family-group names, type genera, stem

Zoological nomenclature affects the work of all zoologists, yet only a minuscule fraction of one percent of [the craziest] zoologists deal directly with problems associated with scientific names of animals (Bock 1994).

Introduction

The accurate use of scientific names in zoology is necessary to minimize nomenclatural instability and allow the maximal retrieval of scientific information from the ever increasing body of literature. Akin to other scientific names such as species- and genus-group names, family-group names are extremely important in the exchange of information about the world we live in. Rules about how to treat family-group names, as determined by the *International Commission on Zoological Nomenclature* [henceforth the Commission], have been an integral part of zoological nomenclature since the early 1900s (for a historical review of family-group names refer to Bock 1994).

The first family-group names based on the stem of their type genus appeared in zoological literature in the early 19th Century (see Sabrosky 1999). The first authors inconsistently used a variety of endings (–IDES, –ITES, –IDA, –I, –IDE, etc.) for supra-generic divisions at various levels. The ending –IDAE was apparently first suggested by Kirby (1813) as a characteristic ending for names at the rank of family. Although many ranks have been used in the past for suprageneric names, only the following suffixes are currently regulated by the *International Code of Zoological Nomenclature* [henceforth the Code] (ICZN 1999a): superfamily (–OIDEA), family (–IDAE), subfamily (–INAE), tribe (–INI) and subtribe (–INA).

The last twenty five years has seen an increase in the number of studies on family-group names of entire groups in zoology (e.g., Štys and Jansson 1988 for Hemiptera: Nepomorpha; Bock 1994 for Aves; McKenna and Bell 1997 for Mammalia; Ferraris and de Pinna 1999 for Pisces: Siluriformes; Sabrosky 1999 for Diptera; Engel and Krishna 2004 for Isoptera; Speidel and Naumann 2004 for Lepidoptera: NOCTUOIDEA; Bouchet and Rocroi 2005 for Gastropoda; Engel 2005 for Hymenoptera: APOIDEA; Engel and Haas 2007 for Dermaptera; Bouchet and Rocroi 2010 for Bivalvia). The contents (i.e., whether they include fossils or not) and presentation of the data (i.e., in a classification scheme, in chronological order or in alphabetic order) in those works vary greatly. The most extensive treatments of animal family-group names to date are those of Diptera (Sabrosky 1999) and Gastropoda (Bouchet and Rocroi 2005) which covered names based on approximately 2000 and 2400 distinct type genera respectively. Some of the reasons given by authors of these works for providing comprehensive lists of family-group names include avoiding the unnecessary proposal of new names, facilitating decisions on priority and promoting long-term stability of the classification. We believe that such catalogues are especially important as the number of studies dealing with the higher relationships of major clades will continue to increase as new algorithms and data sets (e.g., molecular data) become available.

Given the importance of family-group names in the scientific literature, it is rather surprising that still many authors do not cite the author or year of publication of those names correctly, or at all. According to Recommendation 51A of the Code (ICZN 1999a) "The original author and date of a name should be cited at least once in each work dealing with the taxon denoted by that name." In addition to satisfying Recommendation 51A of the Code, citation of authorities of scientific names can also have a positive impact on modern world taxonomists (Werner 2006; Agnarson and Kuntner 2007). Adding to further confusion and errors in the literature is the fact that the Principle of Coordination (Art. 36.1) is still overlooked by some authors.

Coleoptera are currently the most species-rich group of organisms on this planet with approximately 360 000 described species (Bouchard et al. 2009). The great morphological diversity of beetles has led to the proliferation of suprageneric taxa at various ranks. Latreille (1797) was apparently the first to introduce the concept of family-level taxa (see Bock 1994: 244) but it is only a few years later (Latreille 1802) that he proposed available names for these groupings, including several in the order Coleoptera. More recent evidence suggests that available family-group names in some groups of animals other than Coleoptera (e.g., Chordata: Sauropsida) appeared in the late 18th Century (see Dubois and Bour 2010), even earlier than in Latreille (1802).

The nomenclature of family-group names in Coleoptera did not receive much attention until the treatment of Geadephaga names by Madge (1989). Similar lists, based on rules from two different editions of the Code (ICZN 1985g, 1999a), have now been published in Dytiscidae (Nilsson et al. 1989), Staphyliniformia (Newton and Thayer 1992), Cucujoidea (Pakaluk et al. 1994), Curculionoidea (Alonso-Zarazaga and Lyal 1999), Buprestoidea (Bellamy 2003, 2008a-d, 2009), Tenebrionidae (Bouchard et al. 2005), Scarabaeoidea (Smith 2006) and Cerambycidae (Bousquet et al. 2009). Another important contribution was that of Lawrence and Newton (1995) which included a review of the nomenclature of all beetle family-group names for the rank of subfamily and above.

The vast body of scientific literature dealing with beetles has been a deterrent to producing a complete review of all Coleoptera family-group names in the past. This publication was only made possible by the collaboration of several coleopterists.

The specific objectives of this study are to: 1) establish the first comprehensive list of all Coleoptera family-group names with information on type genus, author(s), year of publication and complete bibliographical references; 2) assess the availability and validity of each name using rules laid out in the most recent *Code of Zoological Nomenclature* (ICZN 1999a), 3) summarize Priority and Homonymy problems with currently used names and, 4) propose or implement solutions to these problems in order to promote stability. We include family-group names that were published on or before December 31, 2010.

This publication should be seen as the starting point of what could eventually serve as the basis for a submission of Part of a *List of Available names in Zoology* (Art. 79).

With this goal in mind, we ask all coleopterists to either send us, or to publish, corrections or differences of opinions in the months and years to come. It is our intention to update our list of family-group names as new data are published and to provide a second edition of this work in approximately five years.

Methods

Criteria of availability

Decisions about the availability and validity of each Coleoptera family-group name in our catalogue were made following the process outline in Figure 1. First we established whether the name was available or unavailable based on the criteria of availability summarized below. If a name first proposed on a particular type genus was determined to be unavailable then searches were conducted to establish if a family-group name based on the same type genus was made available subsequently. If so then this available name was entered in our catalogue. From the pool of available names, we removed those that are permanently invalid (Art. 39). All the remaining available names were then separated into those that are valid based on the classification used here (at any rank from subtribe to superfamily) and those that are invalid (i.e., synonyms).

In order to be available, a family-group name proposed before 1931 must to be a scientific name (i.e., in latinized form) in the nominative plural based on the stem of an available genus name then used as valid in the new suprageneric taxon (Art. 11.7).

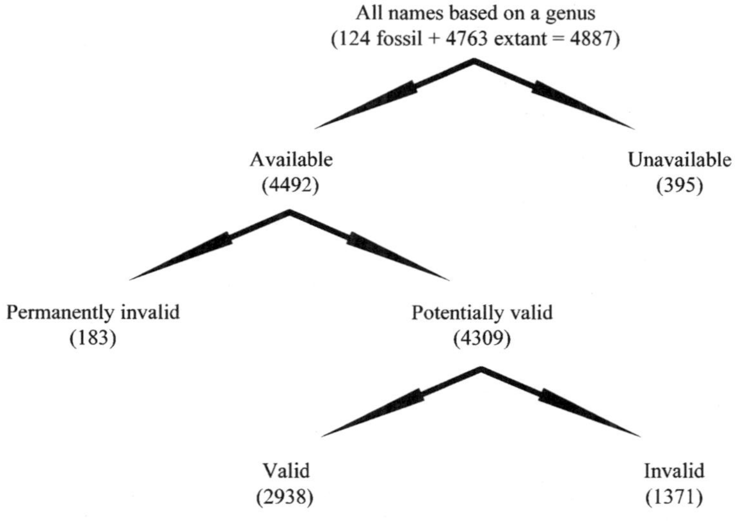

Figure 1. Overview of the process used to determine the availability and validity of family-group names in Coleoptera (modified from Bouchet and Rocroi 2005). The number of names for each category is given in parentheses.

In addition to the criteria of availability mentioned above, new family-group names proposed between 1931 and 1999 had to be described in words, or be associated with a bibliographic reference to such a description, in order to be considered available (Arts 13, 15). In works published after 1930 which contained more than one use of a new family-group name, we selected a page where a description was clearly associated with the new taxon for the catalogue. It should be noted that replacement names proposed in that time period are available without description (Art. 13.1.3). Additionally, "a family-group name first published after 1930 and before 1961 which does not satisfy the provisions of Article 13.1 is available from its original publication only if it was used as valid before 2000, and also was not rejected by an author who, after 1960 and before 2000, expressly applied Article 13 of the then current editions of the Code" (Art. 13.2.1). Names proposed between 1931 and 1960 can also be considered available if the description of the new family-group name and a single new genus-group name is combined (Art. 13.5). Finally, family-group names proposed since 2000 have to be explicitly indicated as new and the name of the type genus has to be clearly cited in order to be available (Art. 16). Based on our interpretation of Article 16.2, any new family-group name proposed after 1999 in a paper in which the name of the type genus is clearly cited in connection with it (although maybe not explicitly with a formula such as "Type genus = *Aus* Doe, 2010") is available.

One of the most difficult tasks while working on this review was to decide on the most consistent and objective way to apply Article 11.7.2, which deals with the availability of names that were originally proposed in a vernacular form. Vernacular names are generally not treated as scientific names in zoological nomenclature (see Recommendation 11A). However, the International Commission on Zoological Nomenclature has made a single, but rather restricted, exception that applies to family-group names. According to the Code, a vernacular name "…is available with its original author and date only if it has been latinized by later authors and has been generally accepted as valid by authors interested in the group concerned and as dating from that first publication in vernacular form" (Art. 11.7.2).

We have interpreted names proposed in the following languages to be vernacular: all German names with the suffix "–EN", all Spanish names with the suffix "–OS" or "–AS" and all French names with the suffix "–IENS". The most important issue was to determine the correct status of names originally proposed with the suffix "–ES." We have used the principle that all family-group names proposed by non-French writers with the suffix "–ES" were in fact latinized names with an ending that differs from those regulated by the Commission (these are mostly older names proposed before rules of zoological nomenclature became well-established). For each name with the suffix "–ES" first proposed by French workers (e.g., Latreille, Lacordaire, Lameere) we went through the entire work containing those names and established if these authors consistently used either vernacular or latinized names in their work. French vernacular names often have accents in them ("é" or "è") while latinized names do not. We have found that determining whether an author used vernacular or latinized names in a particular work was fairly straightforward (Bousquet et al. 2009). Latreille proposed new family-group names with the suffix "–ES" in several of his works. We noticed that he most often used both the vernacular (listed first, with accents when required by French language) and latinized (listed second, always in italics) forms of each name together in the same heading. We have listed the latinized form of the name in each of those cases.

As in Bousquet et al. (2009), we have interpreted the requirements listed in Art. 11.7.2. as three separate conditions to be met. For each vernacular name we determined if it had 1) been subsequently latinized, 2) been generally accepted as valid by authors interested in the group and 3) been attributed to the author and date of original publication. Failure to fulfil any of the requirements resulted in the treatment of that name as unavailable. In those cases, we provide a comment explaining why this taxon was treated as unavailable (e.g., "original vernacular name unavailable (Art. 11.7.2): not subsequently latinized"). Many recent authors have treated family-group names first proposed in vernacular form as available if they were latinized by later authors but we do not believe that this practice is in line with the requirements of the Code. We have accepted as available all vernacular names published before 1900 that have subsequently been used in latinized form, while being used as valid, and credited to the publication in their vernacular form. For every vernacular name that meets the requirement of availability, we have added a comment in the format of the following example "original vernacular name available (Art. 11.7.2): first used in latinized form by J. L. LeConte (1862: 240, as MERACANTHINI), generally accepted as in Gebien (1911: 567, as MERACANTHINAE)".

Formation and treatment of family-group names

Article 29.2 includes a list of suffixes to be used for groups at the superfamily (-OIDEA), family (-IDAE), subfamily (-INAE), tribe (-INI) and subtribe (-INA) names. The suffix of other categories is not regulated by the Code. The rank of supertribe (-ITAE) was used here for a small number of groups in which recent changes in classification required an additional rank between subfamily and tribe (e.g., STAPHYLINOIDEA).

The correct spelling of family-group names depends on the stem of its type genus (Art. 29.3). As pointed out by Newton and Thayer (1992) the family-group name stems of most generic names that can be regarded as Latin nouns are determined easily by dropping the following nominative case ending *–us*, *–um*, *–es* and *–a*. Examples of such straightforward case in Coleoptera include CARABIDAE (type genus *Carabus*), ADELIINI (type genus *Adelium*), TRECHODINA (type genus *Trechodes*) and ANTHIINI (type genus *Anthia*). A summary of other commonly encountered generic endings, along with their corresponding stems, is given in Table 1. Note that generic names ending with *-gaster* (Greek for stomach) can have either *-gaster-* or *-gastr-* as their correct stem and therefore we have accepted the stem formation of the first author of the family-group name based on such genera as correct. Stems were reviewed for all type genera included in our catalogue.

Table 1. List of common Coleoptera generic suffixes with their associated family-group name endings (based partly on Bouchet and Rocroi 2005, with additions by YB, AFN, MAAZ and M. K. Thayer).

Generic ending	Meaning	Derived family-group name ending	Type genus example	Family-group name example
-apion	pear (Greek)	-API-	*Aspidapion*	ASPIDAPIINA
-arthron	joint (Greek)	-ARTHR-	*Decarthron*	DECARTHRINA
-aspis	shield (Greek)	-ASPID-	*Anaspis*	ANASPIDINAE
-baris	flat-bottomed boat (Greek)	-BARID-	*Baris*	BARIDINI
-chlamys	mantle (Greek)	-CLAMYD-	*Spodochlamys*	SPODOCHLAMYDINI
-celis	spot (Greek)	-CELID-	*Xiphoscelis*	XIPHOSCELIDINI
-ceras	horn (Greek)	-CERAT-	*Megaceras*	MEGACERATINI
-cnema	shin or tibia (Greek)	-CNEM-	*Pachycnema*	PACHYCNEMINA
-cupes	dainty (Latin)	-CUPED-	*Cupes*	CUPEDIDAE
-dacne	bite (Greek)	-DACN-	*Dacne*	DACNINI
-deres	neck, throat (Greek)	-DER-	*Aglycyderes*	AGLYCYDERINI
-derma	skin (Greek)	-DERMAT-	*Cryptoderma*	CRYPTODERMATINAE
-dytes	diver (Greek)	-DYT-	*Aspidytes*	ASPIDYTIDAE
-genys	jaw (Greek)	-GENY-	*Chaetogenys*	CHAETOGENYINI
-hospes	guest (Latin)	-HOSPIT-	*Termitohospes*	TERMITOHOSPITINI
-ides	similar to, derived from (Greek)	-ID-	*Anaides*	ANAIDINAE
-ifer	carrier of (Latin)	-IFER-	*Undulifer*	UNDULIFERINAE
-iger	carrier of (Latin)	-IGER-	*Apoderiger*	APODERIGERINA
-ites	like (Latin, Greek)	-IT-	*Aegialites*	AEGIALITINAE
-loma	edge or fringe (Greek)	-LOMAT-	*Discoloma*	DISCOLOMATINAE
-macer	thin (Latin)	-MACR-	*Rhynchitomacer*	RHYNCHITOMACRINI
-mycter	nose (Greek)	-MYCTER-	*Eurymycter*	EURYMYCTERINI
-odes	similar to (Greek)	-OD-	*Agyrtodes*	AGYRTODINI
-odon	tooth (Greek)	-ODONT-	*Pentodon*	PENTODONTINI
-oides	like (Greek)	-OID-	*Acmaeoderoides*	ACMAEODEROIDINA
-omma	eye (Greek)	-OMMAT-	*Omma*	OMMATIDAE
-onyx	yellow gem stone (Greek)	-ONYCH-	*Trichonyx*	TRICHONYCHINI
-ops	eye (Greek)	-OP-	*Achaenops*	ACHAENOPINA
-opsis	appearance (Greek)	-OPSE-	*Brachyceropsis*	BRACHYCEROPSEINI
-otes	quality, nature (Greek)	-OT-	*Agriotes*	AGRIOTINI
-pholis	horny scale (Greek)	-PHOLID-	*Trachypholis*	TRACHYPHOLIDINI
-pteryx	wing (Greek)	-PTERYG-	*Trichopteryx*	TRICHOPTERYGINI
-pus	foot (Greek)	-POD-	*Baripus*	BARIPODINA
-rhinus	snout (Greek)	-RHIN-	*Platyrhinus*	PLATYRHININI
-rhynchus	snout (Greek)	-RHYNCH-	*Doydirhynchus*	DOYDIRHYNCHINI
-rhipis	fan (Greek)	-RHIPID-	*Xenorhipis*	XENORHIPIDINI
-soma	body (Greek)	-SOMAT-	*Platysoma*	PLATYSOMATINI
-stoma	mouth (Greek)	-STOMAT-	*Stenostoma*	STENOSTOMATINI
-teles	perfect (Greek)	-TEL-	*Abroteles*	ABROTELINA
-termes	wood-worm (Latin)	-TERMIT-	*Philotermes*	PHILOTERMITINI
-thorax	chest (Greek)	-THORAC-	*Mecyclothorax*	MECYCLOTHORACINI
-trox	gnawer (Greek)	-TROG-	*Trox*	TROGIDAE
-trupes	borer (Greek)	-TRUP-	*Ceratotrupes*	CERATOTRUPINI
-typus	shape (Greek)	-TYP-	*Amarotypus*	AMAROTYPINI

It should be noted that if a family-group name was not formed in accordance with Art. 29.3 but its original spelling is in prevailing usage then the current spelling is to be maintained (Art. 29.5). We have conserved the spelling of several family-group names currently used as valid however we did not do so for names that are listed as synonyms. For names based on incorrect stems proposed after 1999, we have considered that prevailing usage cannot be used to conserve the original spellings because too few references using these names could be found. We have therefore corrected the stems of such names unless the name of the type genus was an arbitrary combination of letters (Art. 29.4).

As stated in Art. 35.4.1 "A family-group name based upon an unjustified emendation ... or an incorrect spelling of the name of the type genus must be corrected, unless it is preserved under Article 29.5 or unless the spelling of the genus-group name used to form the family-group name is preserved under Articles 33.2.3.1 or 33.3.1." When an unjustified emendation or an incorrect subsequent spelling of the type genus is in prevailing usage and is attributed to the original author and date (Art. 33.2.3.1; 33.3.1), the correct spelling of the type genus is that in current usage.

In the glossary of the Code, a name in prevailing usage is defined as a "name which is adopted by at least a substantial majority of the most recent authors concerned with the relevant taxon, irrespective of how long ago their work was published". The unfortunate subjectivity in this definition, as pointed out by Ferraris (2000), left us with no choice but determine prevailing usage in an *ad hoc* fashion throughout.

Principle of Coordination

The Principle of Coordination (Art. 36.1) is, unfortunately, still overlooked by some authors. In some instances authors who propose new ranks for previously established suprageneric names are sometimes treated as the author of those names when in fact only a change of rank was presented. Based on the Code "A name established at any rank in the family-group is deemed to have been simultaneously established for nominal taxa at all ranks in the family-group; all these taxa have the same type genus, and their names are formed from the stem of the name of the type genus [Art. 29.3] with appropriate change of suffix [Art. 34.1]. The name has the same authorship and date at every rank."

Principle of Priority

As for species- and genus-group names, the oldest available name for a family-group taxon should be considered as valid (Art. 23). However four important exceptions need further discussion. Firstly, when a little-known family-group name was discovered to be older than a name currently used as valid for a particular taxon, we used the Reversal of Precedence to conserve usage of the younger name if the conditions of

Art. 23.9.2 could be fulfilled. Younger names conserved using Reversal of Precedence are listed in Appendix 1. In some instances we could not fulfill all conditions of Art. 23.9.2 although we considered that using the newly discovered older name as valid would threaten stability or cause confusion. In those cases we maintained usage of the younger name as valid and either submitted an application to the Commission to conserve the younger name or made a recommendation that such submission should be submitted in the near future (Art. 23.9.3).

According to Art. 35.5 "If after 1999 a name in use for a family-group taxon (e.g., for a subfamily) is found to be older than a name in prevailing usage for a taxon at higher rank in the same family-group taxon (e.g., for the family within which the older name is the name of a subfamily) the older name is not to displace the younger name." Indeed, we encountered a small number of cases in which the replacement of a name at the higher rank (e.g., Lymexyloidea Fleming, 1821) by the discovery of an older name for a taxon at a lower rank (e.g., Hylecoetoidea Germar, 1818) would not have served the stability of well-established names. In such cases usage of the younger name at the higher rank was conserved.

In cases where a family-group name was replaced before 1961 because of the synonymy of its type genus (e.g., Lepiceridae Hinton, 1936 instead of Cyathoceridae Sharp, 1882), the substitute name is to be maintained if it is in prevailing usage (Art. 40.2). In such cases, the valid family-group name retains its own author but takes the priority of the replaced name. Based on Recommendation 40A, we have cited those names with their original author and date, followed by the date of its priority enclosed in parentheses, as determined by Art. 40.2.1 (e.g., Lepiceridae Hinton, 1936 (1882)).

Lastly, if the stability and continuity of the meaning of a family-group name is threatened by the discovery that the type genus was originally misidentified, or that the type genus was based on a misidentified type species, or that an older type species of the type genus had been overlooked, then the case is to be referred to the Commission for a ruling (Art. 65.2.1). In cases where the oldest available name for a family-group taxon is based on misidentified type genus, or an altered concept of the type genus (e.g., see Scydmaenini Reitter, 1882), we have preferred to consider this name as invalid and to use instead the family-group name which is in prevailing usage (e.g., Cyrtoscydmini L. W. Schaufuss, 1889) until an application is submitted and a ruling is rendered by the Commission.

A list of problem cases based on the Principle of Priority, with comments on implementation of solutions or necessary actions to be taken in the future, is given in Appendix 2.

Principle of Homonymy

Based on the Principle of Homonymy "when two or more names are homonyms, only the senior, as determined by the Principle of Priority…, may be used as the valid name..." (Art. 55.2). Here we report several instances in which family-group names in Coleoptera are identical to other family-group names in zoological nomenclature. These names are either based on identical type genera or on type genera that are similar but not identical. In the first instance the family-group name based on the preoccupied type genus is permanently invalid (Art. 39) but available.

When family-group names are homonyms because their type genera are similar but not identical, the case must be referred to the Commission for a ruling to remove homonymy (Art. 55.3.1). Such is the case with ADELIINI Kirby, 1825 (type genus *Adelium* Kirby, 1819) and the hymenopteran name ADELIINI Viereck, 1918 (type genus *Adelius* Haliday, 1833) which are both correctly formed from the stem of their type genus. More than thirty such cases were encountered during our research on Coleoptera family-group names.

Junior homonyms conserved using Reversal of Precedence are listed in Appendix 1. A list of problem cases based on the Principle of Homonymy, with comments on implementation of solutions or necessary actions to be taken in the future, is given in Appendix 3.

Submissions to the Commission

As mentioned above, we have encountered several cases which require an application to the Commission because of problems with priority and/or homonymy. Some of these cases have been submitted recently (Engel and Bouchard 2009, Bousquet et al. 2010, Bousquet and Bouchard 2010) and others will be submitted in the near future (see summary in Appendices 2 and 3). However, it is not our intention to submit applications for all cases. Since we intend to submit this work as Part of the *List of Available Names in Zoology* in approximately five years, we hope the coleopterist community will take this opportunity to submit applications to the Commission in order to resolve some of the remaining problems outlined here.

Bibliographic notes

Von Hayek (1973: 282, 290) and Stibick (1979: 157) reported that the important works on ELATERIDAE classification by Candèze (1857) and Lacordaire (1857) were published in May and June of 1857 respectively. Since then, when the same new name appeared in both of these works, Candèze's names have been given priority over Lacordaire's. We have discovered that Lacordaire's work was in fact published before 25 May 1857 (as recorded by the Académie des Sciences de France) which would make the names in his work the oldest based on strict adherence to the Code (Art. 21.3). In order to avoid unnecessary changes in authorship in the future and maintain prevailing usage, we have opted to continue attributing elaterid family-group names to Candèze, 1857.

Authorship of the three volumes on Coleoptera of the series *Encyclopédie d'histoire naturelle* is debated. The title pages on the volumes list Jean Charles Chenu as the author of the *Encyclopédie d'histoire naturelle* with the assistance of Eugène Desmarest for the Coleoptera section. However, the *Société Entomologique de France* recorded the livraisons received for 1851 in their *Bulletin* (p. cxxv) as "*Encyclopédie d'histoire naturelle, ou traité complet de cette science, sous la direction de M. le docteur Chenu. Coléoptères, par M.E. Desmarest.*" This was likely the title on the wrapper and clearly suggests that Desmarest was responsible alone for the three volume series on Coleoptera. This is also substained by Desmarest himself who in the *Bulletin de la Société Entomologique de France* for 1860 (p. lxiii) speaks of the three volumes series as being his work, published under the direction of Chenu. Based on the above fact, we credit the three-volume series to Desmarest alone (Desmarest 1851, 1857, 1860).

The articles presented by Mulsant (and Rey) in the "Opuscules Entomologiques" are considered here to be reprints (as a compilation) of original publications in various Annales, mostly printed at Lyon, for the following reasons: In the dedication to the first volume Mulsant wrote "ces Opuscules, publiés déjà ça et là dans nos Recueils académiques". The dates of the dedications in the Opuscules are always later than the dates on which they were presented to the various Sociétés, frequently in the following year. Although published at Paris, the Opuscules (see verso of title pages) were printed at Lyon by F. Dumoulin and others, the same printers which published the *Mémoires de l'Académie... des Sciences, Belles-Lettres et Arts de Lyon*, etc. Rey himself, in later publications such as the continuation of the "Histoire Naturelle des Coléoptères de France" (e.g., *Annales de la Société Linnéenne de Lyon* (N. S.) 32[1885]: 1-186 + [4], pls. 1-2 [see p. 106]) cited the *Annales de la Société Linnéenne de Lyon* article before the *Opuscules* version. E. A. Fitch (1881: 46-47), in his obituary of Mulsant, indicated that both the Histoire Naturelle des Coléoptères de France and the Opuscules Entomologiques «appeared originally... in the Annals» and «later republished in separate form at Paris». These concepts were reiterated by Tottenham (1949: 352). Marseul (1882: 20) wrote, "Tous les ouvrages de Mulsant, sauf quelques-uns, ont été publiés dans les Annales des trois sociétés... et l'Académie des Sciences. Tous ont été reproduits séparément soit dans l'*Histoire Naturelle des Coléoptères et des Punaises de France*, soit dans les *Opuscules.*»

However, the first fascicles of the *Histoire Naturelle* series up to the "*Pectinipèdes*" were not reprinted in journals. It seems that most of the succeeding fascicles were submitted to the various Sociétés, were printed by Dumoulin, Barret, Pinier, etc. at Lyon, the dedications were then added and they were sent to Paris for separate publication with the same typesetting, and with changes only to the pagination and some titles. Since the issuing of the journal volumes was frequently delayed by the compilation of many papers (usually in the following year), the separate edition often appeared first, as shown by records of other journals and, in particular, the *Bibliographie de la France*. Therefore we have given date priority to the *Histoire Naturelle* versions in most instances, only those page numbers are cited with the scientific names in the catalogue unless evidence was found to the contrary, and the alternate versions are cited only in the bibliography, following each reference.

Format of the catalogue and conventions used

This paper is organized into two main parts: 1) a synoptic classification containing only the names which we consider as valid herein and 2) a catalogue of all family-group names organized in the same order as in the synoptic classification. It should be noted that our publication is first and foremost a nomenclatural treatment of family-group names. We want to emphasize that the classification used here is not based on newly generated phylogenetic data. Furthermore, we do not necessarily endorse all parts of the classification presented, particularly those which were not based on phylogenetic approaches. For the most part, we follow currently accepted concepts in recent taxon-specific catalogues as well as comprehensive syntheses such as the *Handbook of Zoology* (Beutel and Leschen 2005, Leschen et al. 2010), *American Beetles* (Arnett and Thomas 2000; Arnett et al. 2002) and the *Catalogue of Palaearctic Coleoptera* (Löbl and Smetana 2003, 2004, 2006, 2007, 2008, 2010). Although we recognize that the classification we use will likely become outdated in the near future, we believe that this is the best way to present the assembled data.

Only family-group names that are based on genera are included in our catalogue. For each family-group name the original spelling, author, year of publication, page number, correct stem and type genus are given. Complete data and comments regarding a particular family-group name are presented with the lowest-rank name when the same stem is used in more than one rank, since the same criteria apply in accordance with the Principle of Coordination (Art. 36.1). The correct stem to be used, which is given in square brackets (e.g., [stem: *Uralocole-*]), is especially important for synonyms, since the status of a name may be changed from synonym to valid in the future. Subsequent alternative spellings of family-group names (i.e., simple changes in suffix depending on change in rank) are not given here. Both available and unavailable names are listed together in the catalogue in order to enhance information retrieval. Unavailable names are preceded by an asterisk "*" and are identified as such in the "Comments" section along with the reason for this status. As is common practice in publications of the International Commission on Zoological Nomenclature, we use small capital letters for family-group names. Names in the catalogue are organized in a "phylogenetic" framework, as accepted in recent publications on the group, down to subfamily level. Valid tribes and subtribes are listed in alphabetical order within each subfamily. Synonyms are listed in chronological order under each valid taxon.

Known works of Johannes Gistel (who also published under the name Johannes Gistl and as G. Tilesius; see Evenhuis 1997a) were included in this paper. Although Gistel's works were largely ignored by contemporaries (e.g., Gemminger and Harold 1868a) and continue to be ignored by some workers (e.g., Cate 2007) they are nevertheless broadly available and represent a significant contribution to the nomenclature of family-group and genus-group names in Coleoptera. In order to promote stability, some names proposed for the first time by Gistel (but ignored until now) that threaten names in current usage are treated here as *nomina oblita* (when conditions of Art. 23.9 could be met) or are subject to appeals to the Commission. According to the glossary of the Code, a *nomen oblitum* is a "Latin term applied after 1 January 2000 to a name, unused since 1899, which as a result of an action taken under Article 23.9.2 does not take precedence over a younger synonym or homonym in prevailing usage. The term *nomen oblitum* was also applied to a disused senior synonym rejected between 6 November 1961 and 1 January 1973 under Art. 23b of the Code editions then in force (see Art. 23.12.2). *Nomina oblita* are available names; see Articles 23.9 and 23.12 for conditions controlling their use as valid names."

Because of the importance of the Principle of Priority, we have tried to find the most accurate date of publication (given in square brackets in the References section) for works cited in the manuscript. The date of publication was determined either from the original publication itself, from the date of "stamps" when received in libraries of natural history institutions or from secondary literature sources. A list of natural history institutions and secondary literature sources is given at the beginning of the References section. Data on dates of publication were provided by AFN, PB, AED, MAAZ, CHCL, ABTS and YB. References for type genera are not included here to conserve space and because they are for the most part available in other recent publications (e.g., Löbl and Smetana 2003, 2004, 2006, 2007, 2008, 2010).

Names on the *Official List of Family-Group Names in Zoology* and type genera on the *Official List of Generic Names in Zoology* are included in our catalogue and are summarized in Appendix 4 and 5 respectively. Cases involving family-group names and/or their type genera that are awaiting a ruling by the Commission are summarized in Appendix 6.

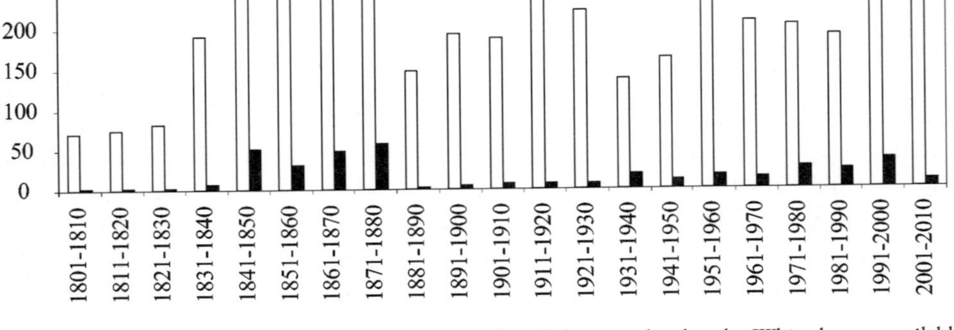

Figure 2. Number of family-group names proposed in Coleoptera by decade. White bars = available names. Black bars = unavailable names.

Results

Number of family-group names

A total of 4887 family-group names are included in our catalogue (see Figure 1). Of the names recorded, 4492 are available, of which 183 are permanently invalid because they are based on a preoccupied type genus or a genus which has been suppressed by the Commission. The majority of names were proposed in the suborder Polyphaga (4314) followed by Adephaga (531), Myxophaga (16), Archostemata (17) and Protocoleoptera (9). Within Adephaga the family CARABIDAE contains the highest number of names by far (441) followed by DYTISCIDAE (47). The five superfamilies with the highest number of names in POLYPHAGA are the CURCULIONOIDEA (862), CHRYSOMELOIDEA (794), STAPHYLINOIDEA (594), TENEBRIONOIDEA (579) and SCARABAEOIDEA (436). Overall, the five families with the highest number of family-group names proposed are CURCULIONIDAE (555), STAPHYLINIDAE (493), CERAMBYCIDAE (468), CARABIDAE (441) and TENEBRIONIDAE (323).

The number of Coleoptera family-group names that appeared in the 19th and 20th Centuries are almost identical (2331 and 2556 respectively) while the last decade saw an increase of 5% (246 names) in the total number of names in the literature (Figure 2). A large number of names proposed in the middle of the 19th Century are unavailable. Those names were generally proposed in vernacular form and were not made available subsequently (Art. 11.7.2). Unavailable names proposed after 1930 generally lacked a description or bibliographic reference to such a description (Arts 13, 15). Several names proposed in the last decade were either not explicitly introduced as new taxa or did not include necessary information about the type genus and are therefore unavailable (Art. 16).

Significant contributions

A summary of the most significant contributions, in terms of the total number of names proposed by author, is shown in Table 2. Eighteen authors proposed 50 or more family-group names in Coleoptera. Lacordaire was the most prolific with 353 new names. Although the majority of Lacordaire's names were proposed in vernacular form, a large proportion of them (90%) were subsequently made available and many are used as valid today.

The authors with the lowest percentage of available family-group names (Mulsant, Chapuis, Rey and Blanchard) originally proposed their names in vernacular form and a significant proportion of those names were not made available subsequently. All names proposed by LeConte, Reitter, Casey, Horn, Legalov, C. G. Thomson and Laporte are considered available.

Thirteen of the authors in Table 2 were active exclusively during the 19th Century while two authors, Reitter and Casey, published their works containing new family-group names in both the 19th and 20th Centuries. Two authors published family-group names exclusively during the 20th Century (Jeannel and Voss). Legalov is our only contemporary colleague in this list; he proposed more than 50 new family-group names in the last decade. It should be noted that the recent proliferation of new scientific names proposed by Legalov is treated by some authors as "extreme splitting" (Riedel 2006) and "based on numerous spurious characters of doubtful phylogenetic value" (Oberprieler et al. 2007).

The broad taxonomic interests and expertise of Lacordaire, Gistel, LeConte, Mulsant, Latreille and C. G. Thomson are exemplified by the fact that they introduced new family-group names in 25 or more families. This was done either in catalogues (e.g., Gistel, C. G. Thomson) and/or in large scale taxonomic treatments (e.g., Lacordaire, LeConte, Latreille, Mulsant). On the other hand, the taxonomic expertise of J. Thomson, Chapuis, Jeannel, Voss, Burmeister and Legalov is much more targeted towards a small number of families (fewer than 10).

Table 2. Summary of number of family-group names in Coleoptera by author. Only authors that have proposed fifty or more names are included. Authors listed in decreasing order by total number of names proposed. For each author we give the total number of name proposed, the percentage of those names that are available, the range of years in which those names were proposed and the number of families in which they proposed new family-group names in.

Rank	Author	Country of origin	Total family-group names	Percent available	Publication (year range)	Number of families
1	Lacordaire, Jean Théodore	France	353	90	1848–1872	34
2	Gistel, Johannes von Nepomuk Franz Xaver	Germany	135	99	1848–1856	42
3	LeConte, John Lawrence	USA	131	100	1847–1883	45
4	Thomson, James	USA	124	99	1857–1877	6
5	Mulsant, Étienne	France	122	76	1839–1880	25
6	Chapuis, Félicien	France	120	72	1869–1876	7
7	Jeannel, René	France	109	99	1910–1967	6
8	Reitter, Edmund	Germany	87	100	1875–1926	24
9	Legalov, Andrei Aleksandrovich	Russia	73	100	2001–2009	9
10	Voss, Eduard	Germany	66	97	1922–1972	6
11	Burmeister, Hermann Carl Conrad	Germany	66	98	1840–1873	2
12	Casey, Thomas Lincoln	USA	64	100	1884–1922	14
13	Horn, George Henry	USA	63	100	1867–1893	20
14	Latreille, Pierre André	France	62	98	1802–1834	30
15	Rey, Claudius	France	57	53	1853–1886	12
16	Thomson, Carl Gustaf	Sweden	55	100	1857–1867	27
17	Blanchard, Charles Émile	France	55	71	1845–1853	20
18	Laporte, François Louis Nompar de Caumont (Comte Castelnau)	France	51	100	1834–1840	21

New nomenclatural acts

The following synonymies are recorded here for the first time: AGRONOMINA Gistel, 1848 syn. nov. of AMARINA Zimmermann, 1832 (CARABIDAE), HYLEPNIGALIOINI Gistel, 1856 syn. nov. of MELANDRYINI Leach, 1815 (MELANDRYIDAE), POLYCYSTOPHORIDAE Gistel, 1856 syn. nov. of MALACHIINAE Fleming, 1821 (MELYRIDAE), SCLERASTEINAE Gistel, 1856 syn. nov. of PTILININAE Shuckard, 1839 (PTINIDAE), PHLOEONOMINI Ádám, 2001 syn. nov. of OMALIINI MacLeay, 1825 (STAPHYLINIDAE), SEPEDOPHILINI Ádám, 2001 syn. nov. of TACHYPORINI MacLeay, 1825 (STAPHYLINIDAE), PHIBALINI Gistel, 1856 syn. nov. of CTENIOPODINI Solier, 1835 (TENEBRIONIDAE); *Agronoma* Gistel 1848 (type species *Carabus familiaris* Duftschmid, 1812, designated herein) syn. nov. of *Amara* Bonelli, 1810 (CARABIDAE), *Hylepnigalio* Gistel, 1856 (type species *Chrysomela caraboides* Linnaeus, 1760, by monotypy) syn. nov. of *Melandrya* Fabricius, 1801 (MELANDRYIDAE), *Polycystophorus* Gistel, 1856 (type species *Cantharis aeneus* Linnaeus, 1758, designated herein) syn. nov. of *Malachius* Fabricius, 1775 (MELYRIDAE), *Sclerastes* Gistel, 1856 (type species *Ptilinus costatus* Gyllenhal, 1827, designated herein) syn. nov. of *Ptilinus* Geoffroy, 1762 (PTINIDAE), *Paniscus* Gistel, 1848 (type species *Scarabaeus fasciatus* Linnaeus, 1758, designated herein) syn. nov. of *Trichius* Fabricius, 1775 (SCARABAEIDAE), *Phibalus* Gistel, 1856 (type species *Chrysomela pubescens* Linnaeus, 1758, by monotypy) syn. nov. of *Omophlus* Dejean, 1834 (TENEBRIONIDAE). The following replacement name is used here for the first time: GOMPELIINA Bouchard, 2011 nom. nov. for OLOTELINA Báguena Corella, 1948 (ADERIDAE).

Reversal of Precedence (Article 23.9) is used to conserve usage of the following names (family-group names followed by genus-group names): PERIGONINI Horn, 1881 nom. protectum over TRECHICINI Bates, 1873 nom. oblitum (CARABIDAE), ANISODACTYLINA Lacordaire, 1854 nom. protectum over EURYTRICHINA LeConte, 1848 nom. oblitum (CARABIDAE), SMICRONYCHINI Seidlitz, 1891 nom. protectum over DESMORINI LeConte, 1876 nom. oblitum (CURCULIONIDAE), BAGOINAE Thomson, 1859 nom. protectum over LYPRINAE Gistel 1848 nom. oblitum (CURCULIONIDAE), ATERPINA Lacordaire, 1863 nom. protectum over HELIOMENINA Gistel, 1848 nom. oblitum (CURCULIONIDAE), NAUPACTINI Gistel, 1848 nom. protectum over IPHIINI Schönherr, 1823 nom. oblitum (CURCULIONIDAE), CLEONINI Schönherr, 1826 nom. protectum over GEOMORINI Schönherr, 1823 nom. oblitum (CURCULIONIDAE), MAGDALIDINI Pascoe, 1870 nom. protectum over SCARDAMYCTINI Gistel, 1848 nom. oblitum (CURCULIONIDAE), AGRYPNINAE/-INI Candèze, 1857 nom. protecta over ADELOCERINAE/-INI Gistel, 1848 nom. oblita and PANGAURINAE/-INI Gistel, 1856 nom. oblita (ELATERIDAE), PROSTERNINI Gistel, 1856 nom. protectum over DIACANTHINI Gistel, 1848 nom. oblitum (ELATERIDAE), CALOPODINAE Costa, 1852 nom. protectum over SPAREDRINAE Gistel, 1848 nom. oblitum (OEDEMERIDAE), ADESMIINI Lacordaire, 1859 nom. protectum over MACROPODINI Agassiz, 1846 nom. oblitum (TENEBRIONIDAE), BOLITOPHAGINI Kirby, 1837 nom. protectum over ELEDONINI Billberg, 1820 nom. oblitum (TENEBRIONIDAE), THROSCIDAE Laporte, 1840 nom. protectum over STEREOLIDAE Rafinesque, 1815 nom. oblitum (THROSCIDAE) and LOPHOcaterini Crowson, 1964 over Lycoptini Casey, 1890 nom. oblitum (Trogossitidae); *Monotoma* Herbst, 1799 nom. protectum over *Monotoma* Panzer, 1792 nom. oblitum (Monotomidae); *Pediacus* Shuckard, 1839 nom. protectum over *Biophloeus* Dejean, 1835 nom. oblitum (Cucujidae), *Pachypus* Dejean, 1821 nom. protectum over *Pachypus* Billberg, 1820 nom. oblitum (Scarabaeidae), *Sparrmannia* Laporte, 1840 nom. protectum over *Leocaeta* Dejean, 1833 nom. oblitum and *Cephalotrichia* Hope, 1837 nom. oblitum (Scarabaeidae).

Synoptic classification of the world Coleoptera

Order COLEOPTERA

†Suborder PROTOCOLEOPTERA

†Superfamily Tshekardocoleoidea Rohdendorf, 1944

†Family Tshekardocoleidae Rohdendorf, 1944
†Family Labradorocoleidae Ponomarenko, 1969
†Family Oborocoleidae Kukalová, 1969

†Superfamily Permocupedoidea Martynov, 1933

†Family Permocupedidae Martynov, 1933
†Family Taldycupedidae Rohdendorf, 1961

†Superfamily Permosynoidea Tillyard, 1924

†Family Ademosynidae Ponomarenko, 1968
†Family Permosynidae Tillyard, 1924

Suborder ARCHOSTEMATA

Family Crowsoniellidae Iablokoff-Khnzorian, 1983
Family Cupedidae Laporte, 1836
  Subfamily Priacminae Crowson, 1962
  †Subfamily Mesocupedinae Ponomarenko, 1969
  Subfamily Cupedinae Laporte, 1836
Family Micromalthidae Barber, 1913
Family Ommatidae Sharp and Muir, 1912
  †Subfamily Brochocoleinae Hong, 1982
  Subfamily Tetraphalerinae Crowson, 1962
  Subfamily Ommatinae Sharp and Muir, 1912
    †Tribe Lithocupedini Ponomarenko, 1969
    †Tribe Notocupedini Ponomarenko, 1966

Tribe OMMATINI Sharp and Muir, 1912
Family JURODIDAE Ponomarenko, 1985
†Family TRIADOCUPEDIDAE Ponomarenko, 1966
†Family MAGNOCOLEIDAE Hong, 1998
†Family OBRIENIIDAE Zherikhin and Gratshev, 1994
†Subfamily KARARHYNCHINAE Zherikhin and Gratshev, 1994
†Tribe KARARHYNCHINI Zherikhin and Gratshev, 1994
†Tribe KENDERLYKAINI Legalov, 2009
†Subfamily OBRIENIINAE Zherikhin and Gratshev, 1994

Suborder MYXOPHAGA

†Superfamily ASIOCOLEOIDEA Rohdendorf, 1961
†Family ASIOCOLEIDAE Rohdendorf, 1961
†Family TRICOLEIDAE Ponomarenko, 1969

†Superfamily RHOMBOCOLEOIDEA Rohdendorf, 1961
†Family RHOMBOCOLEIDAE Rohdendorf, 1961

†Superfamily SCHIZOPHOROIDEA Ponomarenko, 1968
†Family SCHIZOPHORIDAE Ponomarenko, 1968
†Family CATINIIDAE Ponomarenko, 1968
†Family SCHIZOCOLEIDAE Rohdendorf, 1961

Superfamily LEPICEROIDEA Hinton, 1936 (1882)
Family LEPICERIDAE Hinton, 1936 (1882)

Superfamily SPHAERIUSOIDEA Erichson, 1845
Family TORRIDINCOLIDAE Steffan, 1964
Subfamily TORRIDINCOLINAE Steffan, 1964
Subfamily DELEVEINAE Endrödy-Younga, 1997
Family HYDROSCAPHIDAE LeConte, 1874
Family SPHAERIUSIDAE Erichson, 1845

Suborder ADEPHAGA

†Family TRITARSIDAE Hong, 2002
Family GYRINIDAE Latreille, 1810
Subfamily SPANGLEROGYRINAE Folkerts, 1979
Subfamily GYRININAE Latreille, 1810
Tribe ENHYDRINI Régimbart, 1882
Subtribe DINEUTINA Desmarest, 1851
Subtribe ENHYDRINA Régimbart, 1882

Tribe Gyrinini Latreille, 1810
Subtribe Gyrinina Latreille, 1810
Subtribe Heterogyrina Brinck, 1956
Tribe Orectochilini Régimbart, 1882
Family Trachypachidae Thomson, 1857
†Subfamily Eodromeinae Ponomarenko, 1977
Subfamily Trachypachinae Thomson, 1857
Family Rhysodidae Laporte, 1840
Tribe Leoglymmiini Bell and Bell, 1978
Tribe Dhysorini Bell and Bell, 1978
Tribe Medisorini Bell and Bell, 1987
Tribe Rhysodini Laporte, 1840
Tribe Clinidiini Bell and Bell, 1978
Tribe Omoglymmiini Bell and Bell, 1978
Tribe Sloanoglymmiini Bell and Bell, 1991
Family Carabidae Latreille, 1802
†Subfamily Protorabinae Ponomarenko, 1977
†Subfamily Conjunctiinae Ponomarenko, 1977
Subfamily Nebriinae Laporte, 1834
Tribe Nebriini Laporte, 1834
Tribe Notiokasiini Kavanaugh and Nègre, 1983
Tribe Notiophilini Motschulsky, 1850
Tribe Opisthiini Dupuis, 1912
Tribe Pelophilini Kavanaugh, 1996
Subfamily Cicindinae Csiki, 1927
Subfamily Cicindelinae Latreille, 1802
Tribe Amblycheilini Csiki, 1903
Tribe Cicindelini Latreille, 1802
Subtribe Apteroessina Rivalier, 1971
Subtribe Cicindelina Latreille, 1802
Subtribe Dromicina Thomson, 1859
Subtribe Iresiina Rivalier, 1971
Subtribe Theratina Horn, 1893
Tribe Collyridini Brullé, 1834
Subtribe Collyridina Brullé, 1834
Subtribe Tricondylina Naviaux, 1991
Tribe Ctenostomatini Laporte, 1834
Tribe Manticorini Laporte, 1834
Tribe Megacephalini Laporte, 1834
Subfamily Carabinae Latreille, 1802
Tribe Carabini Latreille, 1802
Tribe Ceroglossini Lapouge, 1927
Tribe Cychrini Perty, 1830

Tribe Pamborini Hope, 1838
Subfamily Loricerinae Bonelli, 1810
Subfamily Omophroninae Bonelli, 1810
Subfamily Elaphrinae Latreille, 1802
Subfamily Migadopinae Chaudoir, 1861
Tribe Amarotypini Erwin, 1985
Tribe Migadopini Chaudoir, 1861
Subtribe Aquilicina Moret, 2005
Subtribe Migadopina Chaudoir, 1861
Subfamily Hiletinae Schiødte, 1848
Subfamily Scaritinae Bonelli, 1810
Tribe Carenini MacLeay, 1887
Tribe Clivinini Rafinesque, 1815
Subtribe Ardistomina Putzeys, 1867
Subtribe Clivinina Rafinesque, 1815
Subtribe Forcipatorina Bänninger, 1938
Tribe Dalyatini Mateu, 2002
Tribe Dyschiriini Kolbe, 1880
†Tribe Palaeoaxinidiini McKay, 1991
Tribe Pasimachini Putzeys, 1867
Tribe Promecognathini LeConte, 1853
Tribe Salcediini Alluaud, 1930 (1929)
Subtribe Androzelmina Bell, 1998
Subtribe Salcediina Alluaud, 1930 (1929)
Subtribe Solenogenyina Bell, 1998
Tribe Scaritini Bonelli, 1810
Subtribe Acanthoscelina Csiki, 1927
Subtribe Corintascarina Basilewsky, 1973
Subtribe Dyscherina Basilewsky, 1973
Subtribe Ochyropina Basilewsky, 1973
Subtribe Oxylobina Andrewes, 1929
Subtribe Scapterina Putzeys, 1867
Subtribe Scaritina Bonelli, 1810
Subtribe Storthodontina Jeannel, 1946
Subfamily Broscinae Hope, 1838
Tribe Broscini Hope, 1838
Subtribe Axonyina Roig-Juñent, 2000
Subtribe Baripodina Jeannel, 1941
Subtribe Broscina Hope, 1838
Subtribe Creobiina Jeannel, 1941
Subtribe Nothobroscina Roig-Juñent, 2000
Subfamily Apotominae LeConte, 1853
Subfamily Siagoninae Bonelli, 1813

Tribe Enceladini Horn, 1881
Tribe Lupercini Lecordier, 1977
Tribe Siagonini Bonelli, 1813
Subfamily Melaeninae Csiki, 1933
Subfamily Gehringiinae Darlington, 1933
Tribe Gehringiini Darlington, 1933
Subtribe Gehringiina Darlington, 1933
Subtribe Helenaeina Deuve, 2007
Subfamily Trechinae Bonelli, 1810
Tribe Bembidiini Stephens, 1827
Subtribe Anillina Jeannel, 1937
Subtribe Bembidiina Stephens, 1827
Subtribe Tachyina Motschulsky, 1862
Subtribe Xystosomina Erwin, 1994
Tribe Horologionini Jeannel, 1949
Tribe Pogonini Laporte, 1834
Tribe Trechini Bonelli, 1810
Subtribe Aepina Fowler, 1887
Subtribe Cnidina Jeannel, 1958
Subtribe Perileptina Sloane, 1903
Subtribe Plocamotrechina Jeannel, 1960
Subtribe Trechina Bonelli, 1810
Subtribe Trechodina Jeannel, 1926
Tribe Zolini Sharp, 1886
Subtribe Chalteniina Roig-Juñent and Cicchino, 2001
Subtribe Sinozolina Deuve, 1997
Subtribe Zolina Sharp, 1886
Subfamily Patrobinae Kirby, 1837
Tribe Lissopogonini Zamotajlov, 2000
Tribe Patrobini Kirby, 1837
Subtribe Deltomerina Chaudoir, 1871
Subtribe Deltomerodina Zamotajlov, 2002
Subtribe Patrobina Kirby, 1837
Subtribe Platidiolina Zamotajlov and Lafer, 2001
Subfamily Psydrinae LeConte, 1853
Tribe Amblytelini Blackburn, 1892
Tribe Mecyclothoracini Jeannel, 1940
Tribe Meonini Sloane, 1898
Tribe Moriomorphini Sloane, 1890
Tribe Psydrini LeConte, 1853
Tribe Tropopterini Sloane, 1898
Subfamily Nototylinae Bänninger, 1927
Subfamily Paussinae Latreille, 1806

Tribe Metriini LeConte, 1853
Tribe Mystropomini Horn, 1881
Tribe Ozaenini Hope, 1838
Tribe Paussini Latreille, 1806
†Subtribe Arthropteritina Luna de Carvalho, 1961
Subtribe Carabidomemnina Wasmann, 1928
Subtribe Cerapterina Billberg, 1820
†Subtribe Eopaussina Luna de Carvalho, 1951
Subtribe Heteropaussina Janssens, 1950
Subtribe Homopterina Wasmann, 1920
Subtribe Paussina Latreille, 1806
Subtribe Pentaplatarthrina Jeannel, 1946
Tribe Protopaussini Gestro, 1892
Subfamily Brachininae Bonelli, 1810
Tribe Brachinini Bonelli, 1810
Subtribe Aptinina Gistel, 1848
Subtribe Brachinina Bonelli, 1810
Subtribe Mastacina Erwin, 1970
Subtribe Pheropsophina Jeannel, 1949
Tribe Crepidogastrini Jeannel, 1949
Subfamily Harpalinae Bonelli, 1810
Tribe Abacetini Chaudoir, 1873
Tribe Amorphomerini Sloane, 1923
Tribe Anthiini Bonelli, 1813
Tribe Atranini Horn, 1881
Tribe Bascanini Basilewsky, 1953
Tribe Calophaenini Jeannel, 1948
Tribe Catapieseini Bates, 1882
Tribe Chaetodactylini Tschitschérine, 1903
Tribe Chaetogenyini Emden, 1958
Tribe Chlaeniini Brullé, 1834
Subtribe Callistina Laporte, 1834
Subtribe Chlaeniina Brullé, 1834
Tribe Cnemalobini Germain, 1911
Tribe Cratocerini Lacordaire, 1854
Tribe Ctenodactylini Laporte, 1834
Tribe Cuneipectini Sloane, 1907
Tribe Cyclosomini Laporte, 1834
Subtribe Cyclosomina Laporte, 1834
Subtribe Masoreina Chaudoir, 1871
Tribe Dercylini Sloane, 1923
Tribe Drimostomatini Chaudoir, 1872
Tribe Dryptini Bonelli, 1810

Tribe Enoicini Basilewsky, 1985
Tribe Galeritini Kirby, 1825
  Subtribe Galeritina Kirby, 1825
  Subtribe Planetina Jedlička, 1941
Tribe Geobaenini Péringuey, 1896
Tribe Ginemini Ball and Shpeley, 2002
Tribe Glyptini Horn, 1881
Tribe Graphipterini Latreille, 1802
Tribe Harpalini Bonelli, 1810
  Subtribe Anisodactylina Lacordaire, 1854 ***nomen protectum***
  Subtribe Harpalina Bonelli, 1810
  Subtribe Pelmatellina Bates, 1882
  Subtribe Stenolophina Kirby, 1837
Tribe Helluonini Hope, 1838
  Subtribe Helluonina Hope, 1838
  Subtribe Omphrina Jedlička, 1941
Tribe Hexagoniini Horn, 1881 (1834)
Tribe Idiomorphini Bates, 1891
Tribe Lachnophorini LeConte, 1853
  Subtribe Lachnophorina LeConte, 1853
  Subtribe Selinina Jeannel, 1948
Tribe Lebiini Bonelli, 1810
  Subtribe Actenonycina Bates, 1871
  Subtribe Agrina Kirby, 1837
  Subtribe Apenina Ball, 1983
  Subtribe Calleidina Chaudoir, 1873
  Subtribe Celaenephina Habu, 1982
  Subtribe Cymindidina Laporte, 1834
  Subtribe Demetriadina Bates, 1886
  Subtribe Dromiusina Bonelli, 1810
  Subtribe Gallerucidiina Chaudoir, 1872
  Subtribe Lebiina Bonelli, 1810
  Subtribe Metallicina Basilewsky, 1984
  Subtribe Nemotarsina Bates, 1883
  Subtribe Pericalina Hope, 1838
  Subtribe Pseudotrechina Basilewsky, 1984
  Subtribe Sugimotoina Habu, 1975
  Subtribe Trichina Basilewsky, 1984
Tribe Licinini Bonelli, 1810
  Subtribe Dicaelina Laporte, 1834
  Subtribe Dicrochilina Ball, 1992
  Subtribe Lestignathina Ball, 1992
  Subtribe Licinina Bonelli, 1810

Tribe Melanchitonini Jeannel, 1948
Tribe Microcheilini Jeannel, 1948
Tribe Morionini Brullé, 1835
Tribe Odacanthini Laporte, 1834
Tribe Omphreini Ganglbauer, 1891
Tribe Oodini LaFerté-Sénectère, 1851
Tribe Orthogoniini Schaum, 1857
Tribe Panagaeini Bonelli, 1810
  Subtribe Brachygnathina Basilewsky, 1946
  Subtribe Panagaeina Bonelli, 1810
  Subtribe Tefflina Basilewsky, 1946
Tribe Peleciini Chaudoir, 1880
  Subtribe Agonicina Sloane, 1920
  Subtribe Peleciina Chaudoir, 1880
Tribe Pentagonicini Bates, 1873
Tribe Perigonini Horn, 1881 ***nomen protectum***
Tribe Physocrotaphini Chaudoir, 1863
Tribe Platynini Bonelli, 1810
Tribe Pseudomorphini Hope, 1838
Tribe Pterostichini Bonelli, 1810
  Subtribe Abacomorphina Tschitschérine, 1902
  Subtribe Euchroina Chaudoir, 1874
  Subtribe Metiina Straneo, 1951
  Subtribe Microcephalina Tschitschérine, 1898
  Subtribe Pterostichina Bonelli, 1810
Tribe Sphodrini Laporte, 1834
  Subtribe Atranopsina Baehr, 1982
  Subtribe Calathina Laporte, 1834
  Subtribe Dolichina Brullé, 1834
  Subtribe Pristosiina Lindroth, 1956
  Subtribe Sphodrina Laporte, 1834
  Subtribe Synuchina Lindroth, 1956
Tribe Xenaroswellianini Erwin, 2007
Tribe Zabrini Bonelli, 1810
  Subtribe Amarina Zimmermann, 1832
  Subtribe Zabrina Bonelli, 1810
Tribe Zuphiini Bonelli, 1810
  Subtribe Dicrodontina Machado, 1992
  Subtribe Leleupidiina Basilewsky, 1951
  Subtribe Metazuphiina Mateu, 1992
  Subtribe Mischocephalina Mateu, 1992
  Subtribe Patriziina Basilewsky, 1953
  Subtribe Zuphiina Bonelli, 1810

Family Haliplidae Aubé, 1836
†Family Triaplidae Ponomarenko, 1977
†Family Colymbotethidae Ponomarenko, 1994
†Family Parahygrobiidae Ponomarenko, 1977
†Family Coptoclavidae Ponomarenko, 1961
  †Subfamily Necronectinae Ponomarenko, 1977
  †Subfamily Charonoscaphinae Ponomarenko, 1977
  †Subfamily Coptoclavinae Ponomarenko, 1961
  †Subfamily Coptoclaviscinae Soriano, Ponomarenko and Delclos, 2007
  †Subfamily Hispanoclavinae Soriano, Ponomarenko and Delclos, 2007
†Family Liadytidae Ponomarenko, 1977
Family Meruidae Spangler and Steiner, 2005
Family Noteridae Thomson, 1860
  Subfamily Noterinae Thomson, 1860
    Tribe Neohydrocoptini Zalat, Saleh, Angus and Kaschef, 2000
    Tribe Noterini Thomson, 1860
    Tribe Pronoterini Nilsson, 2005
    Tribe Tonerini Miller, 2009
  Subfamily Notomicrinae Zimmermann, 1919
  Subfamily Phreatodytinae Uéno, 1957
Family Amphizoidae LeConte, 1853
Family Aspidytidae Ribera, Beutel, Balke and Vogler, 2002
Family Hygrobiidae Régimbart, 1879 (1837)
Family Dytiscidae Leach, 1815
  Subfamily Agabinae Thomson, 1867
  Subfamily Colymbetinae Erichson, 1837
    Tribe Anisomeriini Brinck, 1948
    Tribe Carabdytini Pederzani, 1995
    Tribe Colymbetini Erichson, 1837
  Subfamily Copelatinae Branden, 1885
  Subfamily Coptotominae Branden, 1885
  Subfamily Dytiscinae Leach, 1815
    Tribe Aciliini Thomson, 1867
    Tribe Aubehydrini Guignot, 1942
    Tribe Cybisterini Sharp, 1880
    Tribe Dytiscini Leach, 1815
    Tribe Eretini Crotch, 1873
    Tribe Hydaticini Sharp, 1880
    Tribe Hyderodini Miller, 2000
  Subfamily Hydrodytinae Miller, 2001
  Subfamily Hydroporinae Aubé, 1836
    Tribe Bidessini Sharp, 1880
    Tribe Carabhydrini Watts, 1978

Tribe Hydroporini Aubé, 1836
Tribe Hydrovatini Sharp, 1880
Tribe Hygrotini Portevin, 1929
Tribe Hyphydrini Gistel, 1848
Tribe Laccornini Wolfe and Roughley, 1990
Tribe Methlini Branden, 1885
†Tribe Schistomerini Palmer, 1957
Tribe Vatellini Sharp, 1880
Subfamily Laccophilinae Gistel, 1848
Tribe Agabetini Branden, 1885
Tribe Laccophilini Gistel, 1848
Subfamily Lancetinae Branden, 1885
Subfamily Matinae Branden, 1885
†Subfamily Palaeogyrininae Schlechtendal, 1894
†Subfamily Liadytiscinae Prokin and Ren, 2010

Suborder POLYPHAGA

Series STAPHYLINIFORMIA

Superfamily Hydrophiloidea Latreille, 1802
Family Hydrophilidae Latreille, 1802
Subfamily Helophorinae Leach, 1815
Subfamily Epimetopinae Zaitzev, 1908
Subfamily Georissinae Laporte, 1840
Subfamily Hydrochinae Thomson, 1859
Subfamily Spercheinae Erichson, 1837
Subfamily Horelophinae Hansen, 1991
Subfamily Horelophopsinae Hansen, 1997
Subfamily Hydrophilinae Latreille, 1802
Tribe Anacaenini Hansen, 1991
Tribe Berosini Mulsant, 1844
Tribe Chaetarthriini Bedel, 1881
Tribe Hydrophilini Latreille, 1802
Subtribe Acidocerina Zaitzev, 1908
Subtribe Globuloseina García, 2001
Subtribe Hydrobiusina Mulsant, 1844
Subtribe Hydrophilina Latreille, 1802
Tribe Laccobiini Houlbert, 1922
Tribe Sperchopsini Hansen, 1991
Subfamily Sphaeridiinae Latreille, 1802
Tribe Andotypini Hansen, 1991
Tribe Borborophorini Hansen, 1991

Tribe Coelostomatini Heyden, 1891 (1890)
Tribe Megasternini Mulsant, 1844
Tribe Omicrini Smetana, 1975
Tribe Protosternini Hansen, 1991
Tribe Rygmodini Orchymont, 1916
Tribe Sphaeridiini Latreille, 1802
Tribe Tormissini Hansen, 1991
Family Sphaeritidae Shuckard, 1839
Family Synteliidae Lewis, 1882
Family Histeridae Gyllenhal, 1808
Subfamily Niponiinae Fowler, 1912
Subfamily Abraeinae MacLeay, 1819
Tribe Abraeini MacLeay, 1819
Tribe Acritini Wenzel, 1944
Tribe Acritomorphini Wenzel, 1944
Tribe Plegaderini Portevin, 1929
Tribe Teretriini Bickhardt, 1914
Subfamily Trypeticinae Bickhardt, 1913
Subfamily Trypanaeinae Marseul, 1857
Subfamily Saprininae Blanchard, 1845
Subfamily Dendrophilinae Reitter, 1909
Tribe Anapleini Olexa, 1982
Tribe Bacaniini Kryzhanovskij, 1976
Tribe Dendrophilini Reitter, 1909
Tribe Paromalini Reitter, 1909
Subfamily Onthophilinae MacLeay, 1819
Subfamily Tribalinae Bickhardt, 1914
Subfamily Histerinae Gyllenhal, 1808
Tribe Exosternini Bickhardt, 1914
Tribe Histerini Gyllenhal, 1808
Tribe Hololeptini Hope, 1840
Tribe Omalodini Kryzhanovskij, 1972
Tribe Platysomatini Bickhardt, 1914
Subfamily Haeteriinae Marseul, 1857
Tribe Haeteriini Marseul, 1857
Tribe Nymphistrini Tishechkin, 2007
Tribe Synoditulini Tishechkin, 2007
Subfamily Chlamydopsinae Bickhardt, 1914

Superfamily Staphylinoidea Latreille, 1802

Family Hydraenidae Mulsant, 1844
Subfamily Orchymontiinae Perkins, 1997
Subfamily Prosthetopinae Perkins, 1994

Tribe Coelometoponini Perkins, 2005
Tribe Nucleotopini Perkins, 1994
Tribe Parasthetopini Perkins, 1994
Tribe Prosthetopini Perkins, 1994
Tribe Protosthetopini Perkins, 1994
Tribe Pterosthetopini Perkins, 1994
Subfamily Hydraeninae Mulsant, 1844
Tribe Hydraenidini Perkins, 1980
Tribe Hydraenini Mulsant, 1844
Tribe Limnebiini Mulsant, 1844
Tribe Madagastrini Perkins, 1997
Tribe Parhydraenini Perkins, 1997
Subfamily Ochthebiinae Thomson, 1859
Tribe Ochthebiini Thomson, 1859
Subtribe Enicocerina Perkins, 1997
Subtribe Meropathina Perkins, 1997
Subtribe Neochthebiina Perkins, 1997
Subtribe Ochthebiina Thomson, 1859
Subtribe Protochthebiina Perkins, 1997
Tribe Ochtheosini Perkins, 1997
Family Ptiliidae Erichson, 1845
Subfamily Ptiliinae Erichson, 1845
Tribe Discheramocephalini Grebennikov, 2009
Tribe Nanosellini Barber, 1924
Tribe Ptenidiini Flach, 1889
Tribe Ptiliini Erichson, 1845
Tribe Ptinellini Reitter, 1906 (1891)
Subfamily Cephaloplectinae Sharp, 1883
Subfamily Acrotrichinae Reitter, 1909 (1856)
Family Agyrtidae Thomson, 1859
Subfamily Agyrtinae Thomson, 1859
Subfamily Necrophilinae Newton, 1997
Subfamily Pterolomatinae Thomson, 1862
Family Leiodidae Fleming, 1821
Subfamily Camiarinae Jeannel, 1911
Tribe Agyrtodini Jeannel, 1936
Tribe Camiarini Jeannel, 1911
Tribe Neopelatopini Jeannel, 1962
Subfamily Catopocerinae Hatch, 1927 (1880)
Tribe Catopocerini Hatch, 1927 (1880)
Tribe Glacicavicolini Westcott, 1968
Subfamily Leiodinae Fleming, 1821
Tribe Agathidiini Westwood, 1838

Tribe Estadiini Portevin, 1914
Tribe Leiodini Fleming, 1821
Tribe Pseudoliodini Portevin, 1926
Tribe Scotocryptini Reitter, 1884
Tribe Sogdini Lopatin, 1961
Subfamily Coloninae Horn, 1880 (1859)
Subfamily Cholevinae Kirby, 1837
Tribe Anemadini Hatch, 1928
Subtribe Anemadina Hatch, 1928
Subtribe Eocatopina Jeannel, 1936
Subtribe Eunemadina Newton, 1998
Subtribe Nemadina Jeannel, 1936
Subtribe Paracatopina Jeannel, 1936
Tribe Cholevini Kirby, 1837
Subtribe Catopina Chaudoir, 1845
Subtribe Cholevina Kirby, 1837
Tribe Eucatopini Jeannel, 1921
Tribe Leptodirini Lacordaire, 1854 (1849)
Subtribe Anthroherponina Jeannel, 1910
Subtribe Bathysciina Horn, 1880
Subtribe Bathysciotina Guéorguiev, 1974
Subtribe Leptodirina Lacordaire, 1854 (1849)
Subtribe Pholeuina Reitter, 1886
Subtribe Platycholeina Horn, 1880
Subtribe Spelaeobatina Guéorguiev, 1974
Tribe Oritocatopini Jeannel, 1936
Tribe Ptomaphagini Jeannel, 1911
Subtribe Baryodirina Perreau, 2000
Subtribe Ptomaphagina Jeannel, 1911
Subtribe Ptomaphaginina Szymczakowski, 1964
Tribe Sciaphyini Perreau, 2000
Subfamily Platypsyllinae Ritsema, 1869
Family Silphidae Latreille, 1806
Subfamily Silphinae Latreille, 1806
Subfamily Nicrophorinae Kirby, 1837
Family Staphylinidae Latreille, 1802
Subfamily Glypholomatinae Jeannel, 1962
Subfamily Microsilphinae Crowson, 1950
Subfamily Omaliinae MacLeay, 1825
Tribe Anthophagini Thomson, 1859
Tribe Aphaenostemmini Peyerimhoff, 1914
Tribe Corneolabiini Steel, 1950
Tribe Coryphiini Jakobson, 1908

Subtribe Boreaphilina Zerche, 1990
Subtribe Coryphiina Jakobson, 1908
Tribe Eusphalerini Hatch, 1957
Tribe Hadrognathini Portevin, 1929
Tribe Omaliini MacLeay, 1825
Subfamily Empelinae Newton and Thayer, 1992
Subfamily Proteininae Erichson, 1839
Tribe Anepiini Steel, 1966
Tribe Austrorhysini Newton and Thayer, 1995
Tribe Nesoneini Steel, 1966
Tribe Proteinini Erichson, 1839
Tribe Silphotelini Newton and Thayer, 1995
Subfamily Micropeplinae Leach, 1815
Subfamily Neophoninae Fauvel, 1905
Subfamily Dasycerinae Reitter, 1887
Subfamily Protopselaphinae Newton and Thayer, 1995
Subfamily Pselaphinae Latreille, 1802
Supertribe Batrisitae Reitter, 1882
Tribe Amauropini Jeannel, 1948
Tribe Batrisini Reitter, 1882
Subtribe Ambicocerina Leleup, 1970
Subtribe Batrisina Reitter, 1882
Subtribe Leupeliina Jeannel, 1954
Subtribe Stilipalpina Jeannel, 1954
Tribe Thaumastocephalini Poggi, Nonveiller, Colla, Pavićević and Rada, 2001
Supertribe Clavigeritae Leach, 1815
Tribe Clavigerini Leach, 1815
Subtribe Apoderigerina Jeannel, 1954
Subtribe Clavigerina Leach, 1815
Subtribe Clavigerodina Schaufuss, 1882
Subtribe Disarthricerina Jeannel, 1949
Subtribe Hoplitoxenina Célis, 1969
Subtribe Lunillina Célis, 1969
Subtribe Mastigerina Jeannel, 1954
Subtribe Miroclavigerina Jeannel, 1949
Subtribe Neocerina Jeannel, 1954
Subtribe Radamina Jeannel, 1954
Subtribe Thysdariina Jeannel, 1954
Tribe Colilodionini Besuchet, 1991
Tribe Tiracerini Besuchet, 1986
Supertribe Euplectitae Streubel, 1839
Tribe Bythinoplectini Schaufuss, 1890
Subtribe Bythinoplectina Schaufuss, 1890

Subtribe Pyxidicerina Raffray, 1904
Tribe Dimerini Raffray, 1908
Tribe Euplectini Streubel, 1839
Tribe Jubini Raffray, 1904
Tribe Mayetiini Winkler, 1925
Tribe Metopiasini Raffray, 1904
Subtribe Metopiasina Raffray, 1904
Subtribe Rhinoscepsina Bowman, 1934
Tribe Trichonychini Reitter, 1882
Subtribe Bibloporina Park, 1951
Subtribe Panaphantina Jeannel, 1950
Subtribe Trichonychina Reitter, 1882
Subtribe Trimiina Bowman, 1934
Tribe Trogastrini Jeannel, 1949
Subtribe Phtegnomina Park, 1951
Subtribe Rhexiina Park, 1951
Subtribe Trogastrina Jeannel, 1949
Supertribe Faronitae Reitter, 1882
Supertribe Goniaceritae Reitter, 1882 (1872)
Tribe Arnylliini Jeannel, 1952
Tribe Barrosellini Leleup, 1973
Tribe Brachyglutini Raffray, 1904
Subtribe Baradina Park, 1951
Subtribe Brachyglutina Raffray, 1904
Subtribe Decarthrina Park, 1951
Subtribe Eupseniina Park, 1951
Tribe Bythinini Raffray, 1890
Subtribe Bythinina Raffray, 1890
Subtribe Machaeritina Jeannel, 1950
Subtribe Xenobythina Jeannel, 1950
Tribe Cyathigerini Schaufuss, 1872
Tribe Goniacerini Reitter, 1882 (1872)
Tribe Imirini Jeannel, 1949
Tribe Iniocyphini Park, 1951
Subtribe Iniocyphina Park, 1951
Subtribe Natypleurina Newton and Thayer, 1992
Tribe Machadoini Jeannel, 1951
Tribe Proterini Jeannel, 1949
Tribe Pygoxyini Reitter, 1909
Tribe Speleobamini Park, 1951
Tribe Tychini Raffray, 1904
Tribe Valdini Park, 1953
Supertribe Pselaphitae Latreille, 1802

Tribe Arhytodini Raffray, 1890
Tribe Attapseniini Bruch, 1933
Tribe Ctenistini Blanchard, 1845
Tribe Hybocephalini Raffray, 1890
Tribe Odontalgini Jeannel, 1949
Tribe Pachygastrodini Leleup, 1969
Tribe Phalepsini Jeannel, 1949
Tribe Pselaphini Latreille, 1802
Tribe Schistodactylini Raffray, 1890
Tribe Tmesiphorini Jeannel, 1949
Tribe Tyrini Reitter, 1882
Subtribe Centrophthalmina Jeannel, 1949
Subtribe Janusculina Cerruti, 1970
Subtribe Somatipionina Jeannel, 1949
Subtribe Tyrina Reitter, 1882
Subfamily Phloeocharinae Erichson, 1839
Subfamily Olisthaerinae Thomson, 1858
Subfamily Tachyporinae MacLeay, 1825
Tribe Deropini Smetana, 1983
Tribe Megarthropsini Cameron, 1919
Tribe Mycetoporini Thomson, 1859
Tribe Tachyporini MacLeay, 1825
Tribe Vatesini Seevers, 1958
Subfamily Trichophyinae Thomson, 1858
Subfamily Habrocerinae Mulsant and Rey, 1876
Subfamily Aleocharinae Fleming, 1821
Tribe Actocharini Bernhauer and Schubert, 1911
Tribe Aenictoteratini Kistner, 1993
Tribe Akatastopsisini Pace, 2000
Tribe Aleocharini Fleming, 1821
Subtribe Aleocharina Fleming, 1821
Subtribe Compactopediina Kistner, 1970
Subtribe Hodoxenina Kistner, 1970
Tribe Athetini Casey, 1910
Subtribe Athetina Casey, 1910
Subtribe Coptotermoeciina Kistner and Pasteels, 1970
Subtribe Microceroxenina Kistner, 1970
Subtribe Nasutiphilina Kistner, 1970
Subtribe Schistogeniina Fenyes, 1918
Subtribe Taxicerina Lohse, 1989
Subtribe Termitotelina Kistner, 1970
Subtribe Thamiaraeina Fenyes, 1921
Tribe Autaliini Thomson, 1859

Tribe Cordobanini Bernhauer, 1910
Tribe Corotocini Fenyes, 1918
  Subtribe Abrotelina Seevers, 1957
  Subtribe Corotocina Fenyes, 1918
  Subtribe Eburniogastrina Jacobson, Kistner and Pasteels, 1986
  Subtribe Nasutitellina Jacobson, Kistner and Pasteels, 1986
  Subtribe Sphuridaethina Pace, 1988
  Subtribe Termitocharina Seevers, 1957
  Subtribe Termitocupidina Jacobson, Kistner and Pasteels, 1986
  Subtribe Termitogastrina Bernhauer and Scheerpeltz, 1926
  Subtribe Termitoiceina Jacobson, Kistner and Pasteels, 1986
  Subtribe Termitopithina Jacobson, Kistner and Pasteels, 1986
  Subtribe Termitoptochina Fenyes, 1921
  Subtribe Timeparthenina Fenyes, 1921
Tribe Crematoxenini Mann, 1921
Tribe Cryptonotopseini Pace, 2003
Tribe Deinopsini Sharp, 1883
Tribe Diestotini Mulsant and Rey, 1871
Tribe Diglottini Jakobson, 1909
Tribe Digrammini Fauvel, 1900
Tribe Dorylogastrini Wasmann, 1916
Tribe Dorylomimini Wasmann, 1916
Tribe Drepanoxenini Kistner and Watson, 1972
Tribe Ecitocharini Seevers, 1965
Tribe Ecitogastrini Fenyes, 1918
Tribe Eusteniamorphini Bernhauer and Scheerpeltz, 1926
Tribe Falagriini Mulsant and Rey, 1873
Tribe Feldini Kistner, 1972
Tribe Gymnusini Heer, 1839
Tribe Himalusini Klimaszewski, Pace and Center, 2010
Tribe Homalotini Heer, 1839
  Subtribe Bolitocharina Thomson, 1859
  Subtribe Dinardopsina Bernhauer and Scheerpeltz, 1926
  Subtribe Gyrophaenina Kraatz, 1856
  Subtribe Homalotina Heer, 1839
  Subtribe Silusina Fenyes, 1918
Tribe Hoplandriini Casey, 1910
  Subtribe Hoplandriina Casey, 1910
  Subtribe Platandriina Hanley, 2002
  Subtribe Pseudoplandriina Hanley, 2002
Tribe Hygronomini Thomson, 1859
  Subtribe Hygronomina Thomson, 1859
  Subtribe Saphoglossina Bernhauer and Scheerpeltz, 1926

Tribe Hypocyphtini Laporte, 1835
Tribe Leucocraspedini Fenyes, 1921
Tribe Liparocephalini Fenyes, 1918
Tribe Lomechusini Fleming, 1821
 Subtribe Aenictobiina Kistner, 1997
 Subtribe Lomechusina Fleming, 1821
 Subtribe Myrmedoniina Thomson, 1867
 Subtribe Termitozyrina Seevers, 1957
Tribe Masuriini Cameron, 1939
Tribe Mesoporini Cameron, 1959
Tribe Mimanommatini Wasmann, 1912
 Subtribe Dorylophilina Fenyes, 1921
 Subtribe Mimanommatina Wasmann, 1912
Tribe Mimecitini Wasmann, 1917
 Subtribe Labidopullina Jacobson and Kistner, 1991
 Subtribe Leptanillophilina Fenyes, 1918
 Subtribe Mimecitina Wasmann, 1917
 Subtribe Mimonillina Bernhauer and Scheerpeltz, 1926
Tribe Myllaenini Ganglbauer, 1895
Tribe Oxypodini Thomson, 1859
 Subtribe Aphytopodina Bernhauer and Scheerpeltz, 1926
 Subtribe Blepharhymenina Klimaszewski and Peck, 1986
 Subtribe Dinardina Mulsant and Rey, 1873
 Subtribe Meoticina Seevers, 1978
 Subtribe Oxypodina Thomson, 1859
 Subtribe Tachyusina Thomson, 1859
Tribe Oxypodinini Fenyes, 1921
Tribe Paglini Newton and Thayer, 1992
Tribe Paradoxenusini Bruch, 1937
Tribe Pediculotini Ádám, 1987
Tribe Philotermitini Seevers, 1957
Tribe Phyllodinardini Wasmann, 1916
Tribe Phytosini Thomson, 1867
Tribe Placusini Mulsant and Rey, 1871
Tribe Pronomaeini Mulsant and Rey, 1873
Tribe Pseudoperinthini Cameron, 1939
Tribe Pygostenini Fauvel, 1899
Tribe Sahlbergiini Kistner, 1993
Tribe Sceptobiini Seevers, 1978
Tribe Skatitoxenini Kistner and Pasteels, 1969
Tribe Termitodiscini Wasmann, 1904
 Subtribe Athexeniina Pace, 2000
 Subtribe Termitodiscina Wasmann, 1904

Tribe Termitohospitini Seevers, 1941
Subtribe Hetairotermitina Seevers, 1957
Subtribe Termitohospitina Seevers, 1941
Tribe Termitonannini Fenyes, 1918
Subtribe Perinthina Bernhauer and Scheerpeltz, 1926
Subtribe Termitonannina Fenyes, 1918
Tribe Termitopaediini Seevers, 1957
Tribe Termitusini Fenyes, 1918
Subtribe Termitospectrina Seevers, 1957
Subtribe Termitusina Fenyes, 1918
Tribe Trichopseniini LeConte and Horn, 1883
Tribe Trilobitideini Fauvel, 1899
Subfamily Trigonurinae Reiche, 1866
Subfamily Apateticinae Fauvel, 1895
Subfamily Scaphidiinae Latreille, 1806
Tribe Cypariini Achard, 1924
Tribe Scaphidiini Latreille, 1806
Tribe Scaphiini Achard, 1924
Tribe Scaphisomatini Casey, 1893
Subfamily Piestinae Erichson, 1839
Subfamily Osoriinae Erichson, 1839
Tribe Eleusinini Sharp, 1887
Tribe Leptochirini Sharp, 1887
Tribe Osoriini Erichson, 1839
Subtribe Osoriina Erichson, 1839
Subtribe Parosoriina Bernhauer and Schubert, 1911
Tribe Thoracophorini Reitter, 1909
Subtribe Clavilispinina Newton and Thayer, 1992
Subtribe Glyptomina Newton and Thayer, 1992
Subtribe Lispinina Bernhauer and Schubert, 1910
Subtribe Thoracophorina Reitter, 1909
Subfamily Oxytelinae Fleming, 1821
Tribe Blediini Ádám, 2001
Tribe Coprophilini Heer, 1839
Tribe Euphaniini Reitter, 1909
Tribe Oxytelini Fleming, 1821
Tribe Planeustomini Jacquelin du Val, 1857
Subfamily Oxyporinae Fleming, 1821
Subfamily Megalopsidiinae Leng, 1920
Subfamily Scydmaeninae Leach, 1815
†Supertribe Hapsomelitae Poinar and Brown, 2004
Supertribe Mastigitae Fleming, 1821
Tribe Clidicini Casey, 1897

Tribe Leptomastacini Casey, 1897
Tribe Mastigini Fleming, 1821
Supertribe Scydmaenitae Leach, 1815
Tribe Cephenniini Reitter, 1882
Tribe Chevrolatiini Reitter, 1882
Tribe Cyrtoscydmini Schaufuss, 1889
Tribe Eutheiini Casey, 1897
Tribe Leptoscydmini Casey, 1897
Tribe Plaumanniolini Costa Lima, 1962
Tribe Scydmaenini Leach, 1815
Subfamily Steninae MacLeay, 1825
Subfamily Euaesthetinae Thomson, 1859
Tribe Alzadaesthetini Scheerpeltz, 1974
Tribe Austroesthetini Cameron, 1944
Tribe Euaesthetini Thomson, 1859
Tribe Fenderiini Scheerpeltz, 1974
Tribe Nordenskioldiini Bernhauer and Schubert, 1911
Tribe Stenaesthetini Bernhauer and Schubert, 1911
Subfamily Solieriinae Newton and Thayer, 1992
Subfamily Leptotyphlinae Fauvel, 1874
Tribe Cephalotyphlini Coiffait, 1963
Tribe Entomoculiini Coiffait, 1957
Tribe Leptotyphlini Fauvel, 1874
Tribe Metrotyphlini Coiffait, 1963
Tribe Neotyphlini Coiffait, 1963
Subfamily Pseudopsinae Ganglbauer, 1895
Subfamily Paederinae Fleming, 1821
Tribe Paederini Fleming, 1821
Subtribe Astenina Hatch, 1957
Subtribe Cryptobiina Casey, 1905
Subtribe Cylindroxystina Bierig, 1943
Subtribe Dolicaonina Casey, 1905
Subtribe Echiasterina Casey, 1905
Subtribe Lathrobiina Laporte, 1835
Subtribe Lithocharina Casey, 1905
Subtribe Medonina Casey, 1905
Subtribe Paederina Fleming, 1821
Subtribe Scopaeina Mulsant and Rey, 1878
Subtribe Stilicina Casey, 1905
Subtribe Stilicopsina Casey, 1905
Tribe Pinophilini Nordmann, 1837
Subtribe Pinophilina Nordmann, 1837
Subtribe Procirrina Bernhauer and Schubert, 1912

Subfamily Staphylininae Latreille, 1802
Tribe Arrowinini Solodovnikov and Newton, 2005
Tribe Diochini Casey, 1906
Tribe Maorothiini Assing, 2000
Tribe Othiini Thomson, 1859
Tribe Platyprosopini Lynch Arribálzaga, 1884
Tribe Staphylinini Latreille, 1802
Subtribe Amblyopinina Seevers, 1944
Subtribe Anisolinina Hayashi, 1993
Subtribe Eucibdelina Sharp, 1889
Subtribe Hyptiomina Casey, 1906
Subtribe Philonthina Kirby, 1837
Subtribe Quediina Kraatz, 1857
Subtribe Staphylinina Latreille, 1802
Subtribe Tanygnathinina Reitter, 1909
Subtribe Xanthopygina Sharp, 1884
Tribe Xantholinini Erichson, 1839
†Subfamily Protactinae Heer, 1847

Series SCARABAEIFORMIA

Superfamily Scarabaeoidea Latreille, 1802

Family Pleocomidae LeConte, 1861
Subfamily Pleocominae LeConte, 1861
†Subfamily Cretocominae Nikolajev, 2002
†Subfamily Archescarabaeinae Nikolajev, 2010
Family Geotrupidae Latreille, 1802
Subfamily Taurocerastinae Germain, 1897
Subfamily Bolboceratinae Mulsant, 1842
Tribe Athyreini Lynch Arribálzaga, 1878
Tribe Bolbelasmini Nikolajev, 1996
Tribe Bolboceratini Mulsant, 1842
Tribe Bolbochromini Nikolajev, 1970
Tribe Eubolbitini Nikolajev, 1970
Tribe Eucanthini Nikolajev, 2003
Tribe Gilletinini Nikolajev, 1990
Tribe Odonteini Shokhin, 2007
Tribe Stenaspidiini Nikolajev, 2003
Subfamily Geotrupinae Latreille, 1802
Tribe Ceratotrupini Zunino, 1984
Tribe Enoplotrupini Paulian, 1945
†Tribe Cretogeotrupini Nikolajev, 1996
Tribe Geotrupini Latreille, 1802

Tribe Lethrini Oken, 1843
Family Belohinidae Paulian, 1959
Family Passalidae Leach, 1815
Subfamily Aulacocyclinae Kaup, 1868
Tribe Aulacocyclini Kaup, 1868
Tribe Ceracupedini Boucher, 2006
Subfamily Passalinae Leach, 1815
Tribe Leptaulacini Kaup, 1871
Tribe Macrolinini Kaup, 1871
Tribe Passalini Leach, 1815
Tribe Proculini Kaup, 1868
Tribe Solenocyclini Kaup, 1871
Family Trogidae MacLeay, 1819
†Subfamily Avitortorinae Nikolajev, 2007
Subfamily Troginae MacLeay, 1819
Subfamily Omorginae Nikolajev, 2005
Family Glaresidae Kolbe, 1905
Family Diphyllostomatidae Holloway, 1972
Family Lucanidae Latreille, 1804
†Subfamily Protolucaninae Nikolajev, 2007
Subfamily Aesalinae MacLeay, 1819
Tribe Aesalini MacLeay, 1819
Tribe Ceratognathini Sharp, 1899
Tribe Nicagini LeConte, 1861
†Subfamily Ceruchitinae Nikolajev, 2006
Subfamily Syndesinae MacLeay, 1819
Subfamily Lampriminae MacLeay, 1819
Tribe Lamprimini MacLeay, 1819
Tribe Streptocerini Kikuta, 1986
Subfamily Lucaninae Latreille, 1804
Tribe Chiasognathini Burmeister, 1847
Tribe Lucanini Latreille, 1804
Tribe Platycerini Mulsant, 1842
Tribe Platyceroidini Paulsen and Hawks, 2008
†Subfamily Paralucaninae Nikolajev, 2000
Family Ochodaeidae Mulsant and Rey, 1871
†Subfamily Cretochodaeinae Nikolajev, 1995
Subfamily Ochodaeinae Mulsant and Rey, 1871
Tribe Enodognathini Scholtz, 1988
Tribe Ochodaeini Mulsant and Rey, 1871
Subfamily Chaetocanthinae Scholtz, 1988
Tribe Chaetocanthini Scholtz, 1988
Tribe Pseudochodaeini Scholtz, 1988

Tribe Synochodaeini Scholtz, 1988
Family Hybosoridae Erichson, 1847
†Subfamily Mimaphodiinae Nikolajev, 2007
Subfamily Anaidinae Nikolajev, 1996
Subfamily Ceratocanthinae Martínez, 1968
Tribe Ceratocanthini Martínez, 1968
Tribe Ivieolini Howden and Gill, 2000
Tribe Scarabatermitini Nikolajev, 1999
Subfamily Hybosorinae Erichson, 1847
Subfamily Liparochrinae Ocampo, 2006
Subfamily Pachyplectrinae Ocampo, 2006
Family Glaphyridae MacLeay, 1819
Subfamily Glaphyrinae MacLeay, 1819
Subfamily Amphicominae Blanchard, 1845
†Subfamily Cretoglaphyrinae Nikolajev, 2005
Family Scarabaeidae Latreille, 1802
†Subfamily Lithoscarabaeinae Nikolajev, 1992
Subfamily Chironinae Blanchard, 1845
Subfamily Aegialiinae Laporte, 1840
Subfamily Eremazinae Iablokoff-Khnzorian, 1977
Subfamily Aphodiinae Leach, 1815
Tribe Aphodiini Leach, 1815
Subtribe Aphodiina Leach, 1815
Subtribe Didactyliina Pittino, 1985
Subtribe Proctophanina Stebnicka and Howden, 1995
Tribe Corythoderini Schmidt, 1910
Tribe Eupariini Schmidt, 1910 ***nomen protectum***
Tribe Odontolochini Stebnicka and Howden, 1996
Tribe Odochilini Rakovič, 1987
Tribe Psammodiini Mulsant, 1842
Subtribe Phycocina Landin, 1960
Subtribe Psammodiina Mulsant, 1842
Subtribe Rhyssemina Pittino and Mariani, 1986
Tribe Rhyparini Schmidt, 1910
Tribe Stereomerini Howden and Storey, 1992
Tribe Termitoderini Tangelder and Krikken, 1982
Subfamily Aulonocneminae Janssens, 1946
Subfamily Termitotroginae Wasmann, 1918
Subfamily Scarabaeinae Latreille, 1802
Tribe Ateuchini Perty, 1830
Subtribe Ateuchina Perty, 1830
Subtribe Scatimina Vaz-de-Mello, 2008
Tribe Coprini Leach, 1815

Tribe Deltochilini Lacordaire, 1856
Tribe Eucraniini Burmeister, 1873
Tribe Gymnopleurini Lacordaire, 1856
Tribe Oniticellini Kolbe, 1905
Subtribe Drepanocerina van Lansberge, 1875
Subtribe Eurysternina Vulcano, Martínez and Pereira, 1961
Subtribe Helictopleurina Janssens, 1946
Subtribe Oniticellina Kolbe, 1905
Tribe Onitini Laporte, 1840
Tribe Onthophagini Burmeister, 1846
Tribe Phanaeini Hope, 1838
Tribe Scarabaeini Latreille, 1802
Tribe Sisyphini Mulsant, 1842
†Subfamily Prototroginae Nikolajev, 2000
†Subfamily Cretoscarabaeinae Nikolajev, 1995
Subfamily Dynamopodinae Arrow, 1911
Tribe Dynamopodini Arrow, 1911
Tribe Thinorycterini Semenov and Reichardt, 1925
Subfamily Phaenomeridinae Erichson, 1847
Subfamily Orphninae Erichson, 1847
Tribe Aegidiini Paulian, 1984
Tribe Orphnini Erichson, 1847
Subfamily Allidiostomatinae Arrow, 1940
Subfamily Aclopinae Blanchard, 1850
Tribe Aclopini Blanchard, 1850
†Tribe Holcorobeini Nikolajev, 1992
Tribe Phaenognathini Iablokoff-Khnzorian, 1977
Subfamily Melolonthinae Leach, 1819
Tribe Ablaberini Blanchard, 1850
Tribe Automoliini Britton, 1978
Tribe Chasmatopterini Lacordaire, 1856
Tribe Colymbomorphini Blanchard, 1850
Tribe Comophorinini Britton, 1957
†Tribe Cretomelolonthini Nikolajev, 1998
Tribe Dichelonychini Burmeister, 1855
Tribe Diphucephalini Laporte, 1840
Tribe Diphycerini Medvedev, 1952
Tribe Diplotaxini Kirby, 1837
Tribe Euchirini Hope, 1840
Tribe Heteronychini Lacordaire, 1856
Tribe Hopliini Latreille, 1829
Subtribe Hopliina Latreille, 1829
Subtribe Pachycnemina Laporte, 1840

Tribe Lichniini Burmeister, 1844
Tribe Liparetrini Burmeister, 1855
Tribe Macrodactylini Kirby, 1837
Tribe Maechidiini Burmeister, 1855
Tribe Melolonthini Leach, 1819
Subtribe Enariina Dewailly, 1950
Subtribe Heptophyllina Medvedev, 1951
Subtribe Leucopholina Burmeister, 1855
Subtribe Melolonthina Leach, 1819
Subtribe Pegylina Lacroix, 1989
Subtribe Rhizotrogina Burmeister, 1855
Subtribe Schizonychina Burmeister, 1855
Tribe Oncerini LeConte, 1861
Tribe Pachypodini Erichson, 1840
Tribe Pachytrichini Burmeister, 1855
Tribe Phyllotocidiini Britton, 1957
Tribe Podolasiini Howden, 1997
Tribe Scitalini Britton, 1957
Tribe Sericini Kirby, 1837
Subtribe Phyllotocina Burmeister, 1855
Subtribe Sericina Kirby, 1837
Subtribe Trochalina Brenske, 1898
Tribe Sericoidini Erichson, 1847
Tribe Systellopini Sharp, 1877
Tribe Tanyproctini Erichson, 1847
Subtribe Macrophyllina Burmeister, 1855
Subtribe Tanyproctina Erichson, 1847

Subfamily Rutelinae MacLeay, 1819
Tribe Adoretini Burmeister, 1844
Subtribe Adoretina Burmeister, 1844
Subtribe Adorrhinyptiina Arrow, 1917
Subtribe Pachyrhinadoretina Ohaus, 1912
Subtribe Prodoretina Ohaus, 1912
Subtribe Trigonostomusina Ohaus, 1912
Tribe Alvarengiini Frey, 1975
Tribe Anatistini Lacordaire, 1856
Tribe Anomalini Streubel, 1839 ***nomen protectum***
Subtribe Anisopliina Burmeister, 1844
Subtribe Anomalina Streubel, 1839 ***nomen protectum***
Subtribe Isopliina Péringuey, 1902
Subtribe Leptohopliina Potts, 1974
Subtribe Popilliina Ohaus, 1918
Tribe Anoplognathini MacLeay, 1819

Subtribe Anoplognathina MacLeay, 1819
Subtribe Brachysternina Burmeister, 1844
Subtribe Phalangogoniina Ohaus, 1918
Subtribe Platycoeliina Burmeister, 1844
Subtribe Schizognathina Ohaus, 1918
Tribe Geniatini Burmeister, 1844
Tribe Rutelini MacLeay, 1819
Subtribe Areodina Burmeister, 1844
Subtribe Desmonychina Arrow, 1917
Subtribe Didrepanephorina Ohaus, 1918
Subtribe Heterosternina Bates, 1888 ***nomen protectum***
Subtribe Lasiocalina Ohaus, 1918
Subtribe Oryctomorphina Burmeister, 1847
Subtribe Parastasiina Burmeister, 1844
Subtribe Rutelina MacLeay, 1819
Subfamily Dynastinae MacLeay, 1819
Tribe Agaocephalini Burmeister, 1847
Tribe Cyclocephalini Laporte, 1840
Tribe Dynastini MacLeay, 1819
Tribe Hexodontini Lacordaire, 1856
Tribe Oryctini Mulsant, 1842
Tribe Oryctoderini Endrödi, 1966
Tribe Pentodontini Mulsant, 1842
Subtribe Cheiroplatina Carne, 1957
Subtribe Dipelicina Carne, 1957
Subtribe Pentodontina Mulsant, 1842
Subtribe Pseudoryctina Carne, 1957
Tribe Phileurini Burmeister, 1847
Subtribe Cryptodina Burmeister and Schaum, 1840
Subtribe Phileurina Burmeister, 1847
Subfamily Cetoniinae Leach, 1815
Tribe Cetoniini Leach, 1815
Subtribe Cetoniina Leach, 1815
Subtribe Euphoriina Horn, 1880
Subtribe Leucocelina Kraatz, 1882
Tribe Cremastocheilini Burmeister and Schaum, 1841
Subtribe Aspilina Krikken, 1984
Subtribe Coenochilina Burmeister, 1842
Subtribe Cremastocheilina Burmeister and Schaum, 1841
Subtribe Cymophorina Krikken, 1984
Subtribe Genuchina Krikken, 1984
Subtribe Goliathopsidina Krikken, 1984
Subtribe Heterogeniina Krikken, 1984

Subtribe Lissogeniina Krikken, 1984
Subtribe Macromina Burmeister and Schaum, 1840
Subtribe Nyassinina Krikken, 1984
Subtribe Oplostomina Krikken, 1984
Subtribe Pilinurgina Krikken, 1984
Subtribe Spilophorina Krikken, 1984
Subtribe Telochilina Krikken, 1984
Subtribe Trichoplina Krikken, 1984
Subtribe Trogodina Krikken, 1984
Tribe Diplognathini Burmeister, 1842
Tribe Goliathini Latreille, 1829
Subtribe Coryphocerina Burmeister, 1842
Subtribe Dicronocephalina Krikken, 1984
Subtribe Goliathina Latreille, 1829
Subtribe Ichnestomatina Burmeister, 1842
Tribe Gymnetini Kirby, 1827
Subtribe Blaesiina Schoch, 1895
Subtribe Gymnetina Kirby, 1827
Tribe Phaedimini Schoch, 1894
Tribe Schizorhinini Burmeister, 1842
Subtribe Lomapterina Burmeister, 1842
Subtribe Schizorhinina Burmeister, 1842
Tribe Stenotarsiini Kraatz, 1880
Subtribe Anochiliina Krikken, 1984
Subtribe Coptomiina Schenkling, 1921
Subtribe Chromoptiliina Krikken, 1984
Subtribe Doryscelina Schenkling, 1921
Subtribe Euchroeina Paulian and Descarpentries, 1982
Subtribe Heterophanina Schoch, 1894
Subtribe Heterosomatina Krikken, 1984
Subtribe Pantoliina Krikken, 1984
Subtribe Parachiliina Krikken, 1984
Subtribe Stenotarsiina Kraatz, 1880
Tribe Taenioderini Mikšić, 1976
Subtribe Chalcotheina Mikšić, 1976
Subtribe Taenioderina Mikšić, 1976
Tribe Trichiini Fleming, 1821
Subtribe Cryptodontina Lacordaire, 1856
Subtribe Incina Burmeister, 1842
Subtribe Osmodermatina Schenkling, 1922
Subtribe Platygeniina Krikken, 1984
Subtribe Trichiina Fleming, 1821
Tribe Valgini Mulsant, 1842

Subtribe Microvalgina Kolbe, 1904
Subtribe Valgina Mulsant, 1842
Tribe Xiphoscelidini Burmeister, 1842
†Family Coprinisphaeridae Genise, 2004
†Family Pallichnidae Genise, 2004

Series ELATERIFORMIA

Superfamily Scirtoidea Fleming, 1821

Family Decliniidae Nikitsky, Lawrence, Kirejtshuk and Gratshev, 1994
Family Eucinetidae Lacordaire, 1857
Family Clambidae Fischer von Waldheim, 1821
Subfamily Calyptomerinae Crowson, 1955
Subfamily Acalyptomerinae Crowson, 1979
Subfamily Clambinae Fischer von Waldheim, 1821
Family Scirtidae Fleming, 1821
Subfamily Scirtinae Fleming, 1821
Subfamily Nipponocyphoninae Lawrence and Yoshitomi, 2007
Subfamily Stenocyphoninae Lawrence and Yoshitomi, 2007
†Family Elodophthalmidae Kirejtshuk and Azar, 2008
†Family Mesocinetidae Kirejtshuk and Ponomarenko, 2010

Superfamily Dascilloidea Guérin-Méneville, 1843 (1834)

Family Dascillidae Guérin-Méneville, 1843 (1834)
Subfamily Dascillinae Guérin-Méneville, 1843 (1834)
Tribe Cinnabariini Pic, 1914
Tribe Dascillini Guérin-Méneville, 1843 (1834)
Subfamily Karumiinae Escalera, 1913
Tribe Emmitini Escalera, 1914
Tribe Escalerinini Paulus, 1972
Tribe Genecerini Pic, 1914
Tribe Karumiini Escalera, 1913
Family Rhipiceridae Latreille, 1834

Superfamily Buprestoidea Leach, 1815

Family Schizopodidae LeConte, 1859
Subfamily Schizopodinae LeConte, 1859
Tribe Dystaxiini Théry, 1929
†Tribe Electrapatini Iablokoff-Khnzorian, 1962
Tribe Schizopodini LeConte, 1859
Family Buprestidae Leach, 1815
Subfamily Julodinae Lacordaire, 1857
Subfamily Polycestinae Lacordaire, 1857

Tribe Acmaeoderini Kerremans, 1893
Subtribe Acmaeoderina Kerremans, 1893
Subtribe Acmaeoderoidina Cobos, 1955
Subtribe Nothomorphina Cobos, 1955
Tribe Astraeini Cobos, 1980
Tribe Bulini Bellamy, 1995
Tribe Haplostethini LeConte, 1861
Tribe Paratracheini Cobos, 1980
Tribe Perucolini Cobos, 1980
Tribe Polycestini Lacordaire, 1857
Subtribe Polycestina Lacordaire, 1857
Subtribe Xenopseina Volkovitsh, 2008
Tribe Polyctesini Cobos, 1955
Tribe Prospherini Cobos, 1980
Tribe Ptosimini Kerremans, 1903
Tribe Thrincopygini LeConte, 1861
Tribe Tyndaridini Cobos, 1955
Subtribe Mimicoclytrinina Bellamy, 2003
Subtribe Pseudacherusiina Cobos, 1980
Subtribe Tylaucheniina Cobos, 1959
Subtribe Tyndaridina Cobos, 1955
Tribe Xyroscelidini Cobos, 1955
Subfamily Galbellinae Reitter, 1911
Subfamily Chrysochroinae Laporte, 1835
Tribe Chrysochroini Laporte, 1835
Subtribe Chalcophorina Lacordaire, 1857 (1848)
Subtribe Chrysochroina Laporte, 1835
Subtribe Eucallopistina Bellamy, 2003
Tribe Dicercini Gistel, 1848
Subtribe Dicercina Gistel, 1848
Subtribe Haplotrinchina Holyński, 1993
Subtribe Hippomelanina Holyński, 1993
Subtribe Pseudoperotina Tôyama, 1987
Tribe Evidini Tôyama, 1987
Tribe Paraleptodemini Cobos, 1975
Subtribe Euchromatina Holyński, 1993
Subtribe Euplectaleciina Holyński, 1993
Subtribe Hypoprasina Holyński, 1993
Subtribe Paraleptodemina Cobos, 1975
Subtribe Pristipterina Holyński, 1993
Tribe Paratassini Bílý and Volkovitsh, 1996
Tribe Poecilonotini Jakobson, 1913
Subtribe Poecilonotina Jakobson, 1913

Subtribe Nesotrinchina Bílý, Kubáň and Volkovitsh, 2009
Tribe Sphenopterini Lacordaire, 1857
Tribe Vadonaxiini Descarpentries, 1970
Subfamily Buprestinae Leach, 1815
Tribe Actenodini Gistel, 1848
Tribe Anthaxiini Gory and Laporte, 1839
Tribe Bubastini Obenberger, 1920
Tribe Buprestini Leach, 1815
Subtribe Agaeocerina Nelson, 1982
Subtribe Buprestina Leach, 1815
Subtribe Lamprocheilina Holyński, 1993
Subtribe Trachykelina Holyński, 1988
Tribe Chrysobothrini Gory and Laporte, 1836
Tribe Coomaniellini Bílý, 1974
Tribe Curidini Holyński, 1988
Subtribe Anilarina Bílý, 2000
Subtribe Curidina Holyński, 1988
Subtribe Neocuridina Holyński, 1988
Tribe Epistomentini Levey, 1978
Tribe Exagistini Tôyama, 1987
Tribe Julodimorphini Kerremans, 1903
Tribe Kisanthobiini Richter, 1949
Tribe Maoraxiini Holyński, 1984
Tribe Melanophilini Bedel, 1921
Tribe Melobaseini Bílý, 2000
Tribe Mendizabaliini Cobos, 1968
Tribe Nascionini Holyński, 1988
Tribe Phrixiini Cobos, 1975
Tribe Pterobothrini Volkovitsh, 2001
Tribe Stigmoderini Lacordaire, 1857
Tribe Thomassetiini Bellamy, 1987
Subtribe Philanthaxiina Holyński, 1988
Subtribe Thomassetiina Bellamy, 1987
Tribe Trigonogeniini Cobos, 1956
Tribe Xenorhipidini Cobos, 1986
Subtribe Trichinorhipidina Bellamy, 2006
Subtribe Xenorhipidina Cobos, 1986
Subfamily Agrilinae Laporte, 1835
Tribe Agrilini Laporte, 1835
Subtribe Agrilina Laporte, 1835
Subtribe Amorphosternina Cobos, 1974
Subtribe Amyiina Holyński, 1993
Subtribe Rhaeboscelidina Cobos, 1976

Tribe Aphanisticini Jacquelin du Val, 1859
Subtribe Anthaxomorphina Holyński, 1993
Subtribe Aphanisticina Jacquelin du Val, 1859
Subtribe Cylindromorphina Portevin, 1931
Subtribe Cylindromorphoidina Cobos, 1979
Subtribe Germaricina Cobos, 1979
Tribe Coraebini Bedel, 1921
Subtribe Amorphosomatina Majer, 2000
Subtribe Cisseina Majer, 2000
Subtribe Clematina Majer, 2000
Subtribe Coraebina Bedel, 1921
Subtribe Dismorphina Cobos, 1990
Subtribe Ethoniina Majer, 2000
Subtribe Geraliina Cobos, 1988
Subtribe Meliboeina Majer, 2000
Subtribe Synechocerina Majer, 2000
Subtribe Toxoscelina Majer, 2000
Tribe Tracheini Laporte, 1835
Subtribe Brachina LeConte, 1861
Subtribe Leiopleurina Holyński, 1993
Subtribe Pachyschelina Böving and Craighead, 1931
Subtribe Tracheina Laporte, 1835
†Subfamily Parathyreinae Alexeev, 1994

Superfamily Byrrhoidea Latreille, 1804

Family Byrrhidae Latreille, 1804
Subfamily Byrrhinae Latreille, 1804
Tribe Byrrhini Latreille, 1804
Tribe Exomellini Casey, 1914
Tribe Morychini El Moursy, 1961
Tribe Pedilophorini Casey, 1912
Tribe Simplocariini Mulsant and Rey, 1869
Subfamily Syncalyptinae Mulsant and Rey, 1869
Tribe Microchaetini Paulus, 1973
Tribe Syncalyptini Mulsant and Rey, 1869
Subfamily Amphicyrtinae LeConte, 1861
Family Elmidae Curtis, 1830
Subfamily Larainae LeConte, 1861
Tribe Laraini LeConte, 1861
Tribe Potamophilini Mulsant and Rey, 1872
Subfamily Elminae Curtis, 1830
Tribe Ancyronychini Ganglbauer, 1904
Tribe Elmini Curtis, 1830

Subtribe Elmina Curtis, 1830
Subtribe Stenelmina Mulsant and Rey, 1872
Tribe Macronychini Gistel, 1848
Family Dryopidae Billberg, 1820 (1817)
Family Lutrochidae Kasap and Crowson, 1975
Family Limnichidae Erichson, 1846
Subfamily Hyphalinae Britton, 1971
Subfamily Limnichinae Erichson, 1846
Tribe Limnichini Erichson, 1846
Tribe Wooldridgeini Spangler, 1999
Subfamily Cephalobyrrhinae Champion, 1925
Subfamily Thaumastodinae Champion, 1924
Family Heteroceridae MacLeay, 1825
Subfamily Elythomerinae Pacheco, 1964
Subfamily Heterocerinae MacLeay, 1825
Tribe Augylini Pacheco, 1964
Tribe Heterocerini MacLeay, 1825
Tribe Micilini Pacheco, 1964
Tribe Tropicini Pacheco, 1964
Family Psephenidae Lacordaire, 1854
Subfamily Afroeubriinae Lee, Satô, Shepard and Jäch, 2007
Subfamily Eubriinae Lacordaire, 1857
Subfamily Eubrianacinae Jakobson, 1913
Subfamily Psephenoidinae Bollow, 1938
Subfamily Psepheninae Lacordaire, 1854
Family Cneoglossidae Champion, 1897
Family Ptilodactylidae Laporte, 1836
Subfamily Anchytarsinae Champion, 1897
Subfamily Cladotominae Pic, 1914
Subfamily Aploglossinae Champion, 1897
Subfamily Araeopidiinae Lawrence, 1991
Subfamily Ptilodactylinae Laporte, 1836
Family Podabrocephalidae Pic, 1930
Family Chelonariidae Blanchard, 1845
Family Eulichadidae Crowson, 1973
Family Callirhipidae Emden, 1924

Superfamily Elateroidea Leach, 1815

Family Rhinorhipidae Lawrence, 1988
Family Artematopodidae Lacordaire, 1857
Subfamily Electribiinae Crowson, 1975
Subfamily Allopogoniinae Crowson, 1973
Subfamily Artematopodinae Lacordaire, 1857

Tribe Artematopodini Lacordaire, 1857
Tribe Ctesibiini Crowson, 1973
Tribe Macropogonini LeConte, 1861
Family Brachypsectridae LeConte and Horn, 1883
Family Cerophytidae Latreille, 1834
Family Eucnemidae Eschscholtz, 1829
Subfamily Perothopinae Lacordaire, 1857
Subfamily Phyllocerinae Reitter, 1905
Tribe Anelastini Reitter, 1911
Tribe Phyllocerini Reitter, 1905
Subfamily Pseudomeninae Muona, 1993
Tribe Pseudomenini Muona, 1993
Tribe Schizophilini Muona, 1993
Subfamily Palaeoxeninae Muona, 1993
Subfamily Phleogninae Muona, 1993
Subfamily Anischiinae Fleutiaux, 1936
Subfamily Melasinae Fleming, 1821
Tribe Calyptocerini Muona, 1993
Tribe Ceballosmelasini Muona, 1993
Tribe Dirhagini Reitter, 1911
Tribe Epiphanini Muona, 1993
Tribe Hylocharini Jacquelin du Val, 1859
Tribe Melasini Fleming, 1821
Subtribe Compsocnemina Muona, 1993
Subtribe Melasina Fleming, 1821
Tribe Neocharini Muona, 1993
Tribe Xylobiini Reitter, 1911
Subfamily Eucneminae Eschscholtz, 1829
Tribe Dendrocharini Fleutiaux, 1920
Tribe Dyscharachthini Muona, 1993
Tribe Entomosatopini Muona, 1993
Tribe Eucnemini Eschscholtz, 1829
Tribe Galbitini Muona, 1991
Tribe Mesogenini Muona, 1993
Tribe Muonajini Özdikmen, 2008
Tribe Perrotiini Muona, 1993
Tribe Phaenocerini Muona, 1993
Tribe Proutianini Muona, 1993
Subfamily Macraulacinae Fleutiaux, 1923
Tribe Anelastidini Muona, 1993
Tribe Echthrogasterini Cobos, 1965
Tribe Euryptychini Mamaev, 1976

Tribe Jenibuntorini Muona, 1993
Tribe Macraulacini Fleutiaux, 1923
Tribe Nematodini Leiler, 1976
Tribe Oisocerini Muona, 1993
Tribe Orodotini Muona, 1993
†Tribe Throscogeniini Iablokoff-Khnzorian, 1962
Family Throscidae Laporte, 1840 *nomen protectum*
†Family Praelateriidae Dolin, 1973
Family Elateridae Leach, 1815
Subfamily Cebrioninae Latreille, 1802
Subfamily Agrypninae Candèze, 1857 *nomen protectum*
Tribe Agrypnini Candèze, 1857 ***nomen protectum***
Tribe Anaissini Golbach, 1984
†Tribe Cryptocardiini Dolin, 1980
Tribe Euplinthini Costa, 1975
Subtribe Cleidecostina Johnson, 2002
Subtribe Compsoplinthina Costa, 1975
Subtribe Euplinthina Costa, 1975
Tribe Hemirhipini Candèze, 1857
Tribe Oophorini Gistel, 1848
Tribe Platycrepidiini Costa and Casari-Chen, 1993
Tribe Pseudomelanactini Arnett, 1967
Tribe Pyrophorini Candèze, 1863
Subtribe Hapsodrilina Costa, 1975
Subtribe Nyctophyxina Costa, 1975
Subtribe Pyrophorina Candèze, 1863
Tribe Tetralobini Laporte, 1840
Subfamily Thylacosterninae Fleutiaux, 1920
Subfamily Lissominae Laporte, 1835
Subfamily Semiotinae Jakobson, 1913
Subfamily Campyloxeninae Costa, 1975
Subfamily Pityobiinae Hyslop, 1917
Subfamily Oxynopterinae Candèze, 1857
Subfamily Dendrometrinae Gistel, 1848
Tribe Crepidomenini Candèze, 1863
Tribe Dendrometrini Gistel, 1848
Subtribe Dendrometrina Gistel, 1848
Subtribe Denticollina Stein and Weise, 1877 (1848)
Subtribe Hemicrepidiina Champion, 1896
Tribe Dimini Candèze, 1863
Tribe Hypnoidini Schwarz, 1906 (1860)
Tribe Pleonomini Semenov and Pjatakova, 1936

Tribe Prosternini Gistel, 1856 ***nomen protectum***
Tribe Senodoniini Schenkling, 1927
Subfamily Negastriinae Nakane and Kishii, 1956
Tribe Negastriini Nakane and Kishii, 1956
Tribe Quasimusini Schimmel and Tarnawski, 2009
Subtribe Loebliquasimusina Schimmel and Tarnawski, 2009
Subtribe Quasimusina Schimmel and Tarnawski, 2009
Subtribe Striatoquasimusina Schimmel and Tarnawski, 2009
Subtribe Wittmeroquasimusina Schimmel and Tarnawski, 2009
Subfamily Elaterinae Leach, 1815
Tribe Agriotini Laporte, 1840
Subtribe Agriotina Laporte, 1840
Subtribe Cardiorhinina Candèze, 1863
Tribe Ampedini Gistel, 1848
Tribe Dicrepidiini Thomson, 1858
Tribe Elaterini Leach, 1815
Tribe Megapenthini Gurjeva, 1973
Tribe Melanotini Candèze, 1859 (1848)
Tribe Odontonychini Girard, 1973
Tribe Physorhinini Candèze, 1859
Tribe Pomachiliini Candèze, 1859
Tribe Synaptini Gistel, 1856
Subfamily Cardiophorinae Candèze, 1859
Subfamily Hemiopinae Fleutiaux, 1941
Subfamily Physodactylinae Lacordaire, 1857
Subfamily Eudicronychinae Girard, 1971
Subfamily Subprotelaterinae Fleutiaux, 1920
Subfamily Morostomatinae Dolin, 2000
†Subfamily Protagrypninae Dolin, 1973
†Tribe Desmatini Dolin, 1975
†Tribe Hypnomorphini Dolin, 1975
†Tribe Protagrypnini Dolin, 1973
Family Plastoceridae Crowson, 1972
Family Drilidae Blanchard, 1845
Subfamily Drilinae Blanchard, 1845
Subfamily Thilmaninae Kazantsev, 2004
Tribe Euanomini Kazantsev, 2010
Tribe Thilmanini Kazantsev, 2004
Family Omalisidae Lacordaire, 1857
†Family Berendtimiridae Winkler, 1987
Family Lycidae Laporte, 1836
Subfamily Libnetinae Bocák and Bocáková, 1990

Subfamily Dictyopterinae Houlbert, 1922
Tribe Dictyopterini Houlbert, 1922
Tribe Lycoprogenthini Bocák and Bocáková, 2008
Tribe Taphini Bocák and Bocáková, 1990
Subfamily Lyropaeinae Bocák and Bocáková, 1989
Tribe Alyculini Bocák and Bocáková, 2008
Tribe Antennolycini Bocák and Bocáková, 2008
Tribe Lyropaeini Bocák and Bocáková, 1989
Tribe Miniduliticolini Kazantsev, 2003
Tribe Platerodrilini Kazantsev, 2004
Subfamily Ateliinae Kleine, 1929
Tribe Ateliini Kleine, 1929
Tribe Dilophotini Kleine, 1929
Subfamily Lycinae Laporte, 1836
Tribe Calochromini Lacordaire, 1857
Tribe Calopterini Green, 1949
Subtribe Acroleptina Bocáková, 2005
Subtribe Calopterina Green, 1949
Tribe Conderini Bocák and Bocáková, 1990
Tribe Dihammatini Bocák and Bocáková, 2008
Tribe Erotini LeConte, 1881
Tribe Eurrhacini Bocáková, 2005
Tribe Leptolycini Leng and Mutchler, 1922
Tribe Lycini Laporte, 1836
Tribe Lyponiini Bocák and Bocáková, 1990
Tribe Macrolycini Kleine, 1929
Tribe Melanerotini Kazantsev, 2010
Tribe Metriorrhynchini Kleine, 1926
Subtribe Hemiconderinina Bocák and Bocáková, 1990
Subtribe Metriorrhynchina Kleine, 1926
Subtribe Trichalina Kleine, 1929
Tribe Platerodini Kleine, 1929
Tribe Slipinskiini Bocák and Bocáková, 1992
Tribe Thonalmini Kleine, 1933
Subfamily Dexorinae Bocák and Bocáková, 1989
Family Telegeusidae Leng, 1920
Family Phengodidae LeConte, 1861
Subfamily Phengodinae LeConte, 1861
Subfamily Mastinocerinae LeConte, 1881
Subfamily Penicillophorinae Paulus, 1975
Family Rhagophthalmidae Olivier, 1907
Family Lampyridae Rafinesque, 1815

Subfamily Psilocladinae McDermott, 1964
Subfamily Amydetinae Olivier, 1907
Tribe Amydetini Olivier, 1907
Tribe Vestini McDermott, 1964
Subfamily Lampyrinae Rafinesque, 1815
Tribe Cratomorphini Green, 1948
Tribe Lamprocerini Olivier, 1907
Tribe Lamprohizini Kazantsev, 2010
Tribe Lampyrini Rafinesque, 1815
Tribe Lucidotini Lacordaire, 1857
Subtribe Dadophorina Olivier, 1907
Subtribe Lamprigerina McDermott, 1964
Subtribe Lucidotina Lacordaire, 1857
Subtribe Photinina LeConte, 1881
Tribe Pleotomini Summers, 1874
Subfamily Luciolinae Lacordaire, 1857
Tribe Curtosini McDermott, 1964
Tribe Luciolini Lacordaire, 1857
Subfamily Photurinae Lacordaire, 1857
Family Omethidae LeConte, 1861
Subfamily Omethinae LeConte, 1861
Subfamily Matheteinae LeConte, 1881
Subfamily Driloniinae Crowson, 1972
Family Cantharidae Imhoff, 1856 (1815)
Subfamily Cantharinae Imhoff, 1856 (1815)
Tribe Cantharini Imhoff, 1856 (1815)
Tribe Podabrini Gistel, 1856
Subfamily Silinae Mulsant, 1862
Tribe Silini Mulsant, 1862
Tribe Tytthonyxini Arnett, 1962
Subfamily Dysmorphocerinae Brancucci, 1980
Subfamily Malthininae Kiesenwetter, 1852
Tribe Malchinini Brancucci, 1980
Tribe Malthinini Kiesenwetter, 1852
Tribe Malthodini Böving and Craighead, 1931
Subfamily Chauliognathinae LeConte, 1861
Tribe Chauliognathini LeConte, 1861
Tribe Ichthyurini Champion, 1915
Subfamily Cydistinae Paulus, 1972
Subfamily Pterotinae LeConte, 1861
Subfamily Ototretinae McDermott, 1964
Subfamily Ototretadrilinae Crowson, 1972
†Subfamily Lasiosynidae Kirejtshuk, Chang, Ren and Kun, 2010

Series DERODONTIFORMIA

Superfamily Derodontoidea LeConte, 1861

Family Derodontidae LeConte, 1861
- **Subfamily Peltasticinae LeConte, 1861**
- **Subfamily Derodontinae LeConte, 1861**
- **Subfamily Laricobiinae Mulsant and Rey, 1864**

Family Nosodendridae Erichson, 1846

Family Jacobsoniidae Heller, 1926

Series BOSTRICHIFORMIA

Superfamily Bostrichoidea Latreille, 1802

Family Dermestidae Latreille, 1804
- **Subfamily Dermestinae Latreille, 1804**
  - Tribe Dermestini Latreille, 1804
  - Tribe Marioutini Jakobson, 1913
- **Subfamily Thorictinae Agassiz, 1846**
  - Tribe Thaumaphrastini Anderson, 1949
  - Tribe Thorictini Agassiz, 1846
- **Subfamily Orphilinae LeConte, 1861**
- **Subfamily Trinodinae Casey, 1900**
  - †Tribe Cretonodini Kirejtshuk and Azar, 2009
  - Tribe Thylodriini Semenov, 1909
  - Tribe Trinodini Casey, 1900
  - Tribe Trinoparvini Háva, 2010
- **Subfamily Attageninae Laporte, 1840**
  - Tribe Attagenini Laporte, 1840
  - Tribe Egidyellini Semenov, 1914
- **Subfamily Megatominae Leach, 1815**
  - Tribe Anthrenini Gistel, 1848
  - Tribe Megatomini Leach, 1815

Family Endecatomidae LeConte, 1861

Family Bostrichidae Latreille, 1802
- **Subfamily Dysidinae Lesne, 1921**
- **Subfamily Polycaoninae Lesne, 1896**
- **Subfamily Bostrichinae Latreille, 1802**
  - Tribe Apatini Billberg, 1820
  - Tribe Bostrichini Latreille, 1802
  - Tribe Dinapatini Lesne, 1910
  - Tribe Sinoxylini Marseul, 1857
  - Tribe Xyloperthini Lesne, 1921
- **Subfamily Psoinae Blanchard, 1851**

Subfamily Dinoderinae Thomson, 1863
Subfamily Lyctinae Billberg, 1820
Tribe Lyctini Billberg, 1820
Tribe Trogoxylini Lesne, 1921
Subfamily Euderiinae Lesne, 1934
Family Ptinidae Latreille, 1802
Subfamily Eucradinae LeConte, 1861
Tribe Eucradini LeConte, 1861
Tribe Hedobiini Mulsant and Rey, 1868
Subfamily Ptininae Latreille, 1802
Tribe Gibbiini Jacquelin du Val, 1860
Tribe Meziini Bellés, 1985
Tribe Ptinini Latreille, 1802
Tribe Sphaericini Portevin, 1931
Subfamily Dryophilinae Gistel, 1848
Tribe Dryophilini Gistel, 1848
Tribe Ptilineurini Böving, 1927
Subfamily Ernobiinae Pic, 1912
Subfamily Anobiinae Fleming, 1821
Subfamily Ptilininae Shuckard, 1839
Subfamily Alvarenganiellinae Viana and Martínez, 1971
Subfamily Xyletininae Gistel, 1848
Tribe Lasiodermini Böving, 1927
Tribe Metholcini Zahradník, 2009
Tribe Xyletinini Gistel, 1848
Subfamily Dorcatominae Thomson, 1859
Subfamily Mesocoelopodinae Mulsant and Rey, 1864
Tribe Tricorynini White, 1971
Tribe Mesocoelopodini Mulsant and Rey, 1864

Series CUCUJIFORMIA

Superfamily Lymexyloidea Fleming, 1821

Family Lymexylidae Fleming, 1821
Subfamily Hylecoetinae Germar, 1818
Subfamily Lymexylinae Fleming, 1821
Subfamily Atractocerinae Laporte, 1840
Subfamily Melittommatinae Wheeler, 1986

Superfamily Cleroidea Latreille, 1802

Family Phloiophilidae Kiesenwetter, 1863
Family Trogossitidae Latreille, 1802
Subfamily Peltinae Latreille, 1806

Tribe Ancyronini Kolibáč, 2006
Tribe Colydiopeltini Kolibáč, 2006
Tribe Decamerini Crowson, 1964
Tribe Lophocaterini Crowson, 1964 ***nomen protectum***
Tribe Peltini Latreille, 1806
Tribe Thymalini Léveillé, 1888
Subfamily Trogossitinae Latreille, 1802
Tribe Calityini Reitter, 1922
Tribe Egoliini Lacordaire, 1854
Tribe Gymnochilini Lacordaire, 1854
Tribe Larinotini Ślipiński, 1992
†Tribe Lithostomatini Kolibáč and Huang, 2008
Tribe Trogossitini Latreille, 1802
Family Chaetosomatidae Crowson, 1952
Family Metaxinidae Kolibáč, 2004
Family Thanerocleridae Chapin, 1924
Subfamily Zenodosinae Kolibáč, 1992
Subfamily Thaneroclerinae Chapin, 1924
Tribe Isoclerini Kolibáč, 1992
Tribe Thaneroclerini Chapin, 1924
Tribe Viticlerini Winkler, 1982
Family Cleridae Latreille, 1802
Subfamily Tillinae Fischer von Waldheim, 1813
Subfamily Hydnocerinae Spinola, 1844
Tribe Callimerini Kolibáč, 1998
Tribe Hydnocerini Spinola, 1844
Tribe Lemidiini Kolibáč, 1998
Subfamily Clerinae Latreille, 1802
Subfamily Korynetinae Laporte, 1836
Family Acanthocnemidae Crowson, 1964
Family Phycosecidae Crowson, 1952
Family Prionoceridae Lacordaire, 1857
Tribe Lobonychini Majer, 1987
Tribe Prionocerini Lacordaire, 1857
Family Mauroniscidae Majer, 1995
Family Melyridae Leach, 1815
Subfamily Rhadalinae LeConte, 1861
Subfamily Melyrinae Leach, 1815
Tribe Arthrobrachini Majer, 1987
Tribe Astylini Pic, 1929
Tribe Cerallini Pic, 1929
Tribe Melyrini Leach, 1815
Subfamily Dasytinae Laporte, 1840

Tribe Chaetomalachiini Majer, 1987
Tribe Danaceini Thomson, 1859
Tribe Dasytini Laporte, 1840
Tribe Gietellini Constantin and Menier, 1987
Tribe Listrini Majer, 1990
Subfamily Malachiinae Fleming, 1821
Tribe Amalthocini Majer, 2002
Tribe Attalomimini Majer, 1995
Tribe Carphurini Champion, 1923
Tribe Lemphini Wittmer, 1976
Tribe Malachiini Fleming, 1821
Tribe Pagurodactylini Constantin, 2001

Superfamily Cucujoidea Latreille, 1802
†Family Parandrexidae Kirejtshuk, 1994
†Family Sinisilvanidae Hong, 2002
Family Boganiidae Sen Gupta and Crowson, 1966
Subfamily Paracucujinae Endrödy-Younga and Crowson, 1986
Subfamily Boganiinae Sen Gupta and Crowson, 1966
Family Byturidae Gistel, 1848
Subfamily Platydascillinae Pic, 1914
Subfamily Byturinae Gistel, 1848
Family Helotidae Chapuis, 1876
Family Protocucujidae Crowson, 1954
Family Sphindidae Jacquelin du Val, 1860
Subfamily Protosphindinae Sen Gupta and Crowson, 1979
Subfamily Odontosphindinae Sen Gupta and Crowson, 1979
Subfamily Sphindiphorinae Sen Gupta and Crowson, 1979
Subfamily Sphindinae Jacquelin du Val, 1860
Family Biphyllidae LeConte, 1861
Family Erotylidae Latreille, 1802
Subfamily Xenoscelinae Ganglbauer, 1899
Subfamily Pharaxonothinae Crowson, 1952
Subfamily Loberinae Bruce, 1951
Subfamily Languriinae Hope, 1840
Tribe Hapalipini Leschen, 2003
Tribe Languriini Hope, 1840
Tribe Thallisellini Sen Gupta, 1968
Subfamily Cryptophilinae Casey, 1900
Tribe Cryptophilini Casey, 1900
Tribe Empocryptini Leschen, 2003
Tribe Toramini Sen Gupta, 1967
Subfamily Erotylinae Latreille, 1802

Tribe Dacnini Gistel, 1848
Tribe Encaustini Crotch, 1876
Tribe Erotylini Latreille, 1802
Tribe Megalodacnini Sen Gupta, 1970
Tribe Tritomini Curtis, 1834
Family Monotomidae Laporte, 1840
Subfamily Rhizophaginae Redtenbacher, 1845
Subfamily Monotominae Laporte, 1840
Tribe Europini Sen Gupta, 1988
Tribe Lenacini Crowson, 1952
Tribe Monotomini Laporte, 1840
†Tribe Rhizophtomini Kirejtshuk and Azar, 2009
Tribe Thionini Crowson, 1952
Family Hobartiidae Sen Gupta and Crowson, 1966
Family Cryptophagidae Kirby, 1826
Subfamily Cryptophaginae Kirby, 1826
Tribe Caenoscelini Casey, 1900
Tribe Cryptophagini Kirby, 1826
Tribe Picrotini Crowson, 1980
Subfamily Atomariinae LeConte, 1861
Tribe Atomariini LeConte, 1861
Tribe Cryptafricini Leschen, 1996
Tribe Hypocoprini Reitter, 1879
Family Agapythidae Sen Gupta and Crowson, 1969
Family Priasilphidae Crowson, 1973
Family Phloeostichidae Reitter, 1911
Family Silvanidae Kirby, 1837
Subfamily Brontinae Blanchard, 1845
Tribe Brontini Blanchard, 1845
Tribe Telephanini LeConte, 1861
Subfamily Silvaninae Kirby, 1837
Family Cucujidae Latreille, 1802
Family Myraboliidae Lawrence and Britton, 1991
Family Cavognathidae Sen Gupta and Crowson, 1966
Family Lamingtoniidae Sen Gupta and Crowson, 1969
Family Passandridae Blanchard, 1845
Family Phalacridae Leach, 1815
Subfamily Phaenocephalinae Matthews, 1899
Subfamily Phalacrinae Leach, 1815
Family Propalticidae Crowson, 1952
Family Laemophloeidae Ganglbauer, 1899
Family Tasmosalpingidae Lawrence and Britton, 1991
Family Cyclaxyridae Gimmel, Leschen and Ślipiński, 2009

Family Kateretidae Kirby, 1837
Family Nitidulidae Latreille, 1802
Subfamily Calonecrinae Kirejtshuk, 1982
Subfamily Maynipeplinae Kirejtshuk, 1998
Subfamily Epuraeinae Kirejtshuk, 1986
Tribe Epuraeini Kirejtshuk, 1986
Tribe Taenioncini Kirejtshuk, 1998
Subfamily Carpophilinae Erichson, 1842
Subfamily Amphicrossinae Kirejtshuk, 1986
Subfamily Meligethinae Thomson, 1859
Subfamily Nitidulinae Latreille, 1802
Tribe Cychramini Gistel, 1848
Tribe Cychramptodini Kirejtshuk and Lawrence, 1992
Tribe Cyllodini Everts, 1898
Tribe Lawrencerosini Kirejtshuk, 1991
Tribe Mystropini Murray, 1864
Tribe Nitidulini Latreille, 1802
Subfamily Cillaeinae Kirejtshuk and Audisio, 1986
Subfamily Cryptarchinae Thomson, 1859
Tribe Arhinini Kirejtshuk, 1987
Tribe Cryptarchini Thomson, 1859
Tribe Eucalosphaerini Kirejtshuk, 1987
Tribe Platyarchini Kirejtshuk, 1998
Subfamily Cybocephalinae Jacquelin du Val, 1858
Family Smicripidae Horn, 1880
Family Bothrideridae Erichson, 1845
Subfamily Teredinae Seidlitz, 1888
Tribe Sosylopsini Dajoz, 1980
Tribe Sysolini Ślipiński and Pal, 1985
Tribe Teredini Seidlitz, 1888
Subfamily Xylariophilinae Pal and Lawrence, 1986
Subfamily Anommatinae Ganglbauer, 1899
Subfamily Bothriderinae Erichson, 1845
Family Cerylonidae Billberg, 1820
Subfamily Euxestinae Grouvelle, 1908
Subfamily Loebliorylоninae Ślipiński, 1990
Subfamily Ostomopsinae Sen Gupta and Crowson, 1973
Subfamily Murmidiinae Jacquelin du Val, 1858
Subfamily Ceryloninae Billberg, 1820
Family Alexiidae Imhoff, 1856
Family Discolomatidae Horn, 1878
Subfamily Notiophyginae Jakobson, 1915
Tribe Dystheamonini John, 1954

Tribe Notiophygini Jakobson, 1915
Tribe Pachyplacini John, 1954
Subfamily Discolomatinae Horn, 1878
Subfamily Aphanocephalinae Jakobson, 1904
Subfamily Cephalophaninae John, 1954
Subfamily Pondonatinae John, 1954
Family Endomychidae Leach, 1815
Subfamily Merophysiinae Seidlitz, 1872
Subfamily Pleganophorinae Jacquelin du Val, 1858
Subfamily Anamorphinae Strohecker, 1953
Subfamily Leiestinae Thomson, 1863
Subfamily Mycetaeinae Jacquelin du Val, 1857
Subfamily Eupsilobiinae Casey, 1895
Subfamily Xenomycetinae Strohecker, 1962
Subfamily Danascelinae Tomaszewska, 2000
Subfamily Endomychinae Leach, 1815
Subfamily Epipocinae Gorham, 1873
Subfamily Stenotarsinae Chapuis, 1876
Subfamily Lycoperdininae Bromhead, 1838
Family Coccinellidae Latreille, 1807
Subfamily Microweiseinae Leng, 1920
Tribe Microweiseini Leng, 1920
Tribe Serangiini Pope, 1962
Tribe Sukunahikonini Kamiya, 1960
Subfamily Coccinellinae Latreille, 1807
Tribe Argentipilosini Gordon and de Almeida, 1991
Tribe Aspidimerini Mulsant, 1850
Tribe Azyini Mulsant, 1850
Tribe Brachiacanthini Mulsant, 1850
Tribe Carinodulini Gordon, Pakaluk and Ślipiński, 1989
Tribe Cephaloscymnini Gordon, 1985
Tribe Chilocorini Mulsant, 1846
Tribe Chnoodini Mulsant, 1850
Tribe Coccidulini Mulsant, 1846
Tribe Coccinellini Latreille, 1807
Tribe Cranophorini Mulsant, 1850
Tribe Cryptognathini Mulsant, 1850
Tribe Cynegetini Thomson, 1866
Tribe Diomini Gordon, 1999
Tribe Epilachnini Mulsant, 1846
Tribe Epivertini Pang and Mao, 1979
Tribe Eremochilini Gordon and Vanderberg, 1987
Tribe Hyperaspidini Mulsant, 1846

Tribe Limnichopharini Miyatake, 1994
Tribe Monocorynini Miyatake, 1988
Tribe Noviini Mulsant, 1846
Tribe Ortaliini Mulsant, 1850
Tribe Oryssomini Gordon, 1974
Tribe Platynaspini Mulsant, 1846
Tribe Plotinini Miyatake, 1994
Tribe Poriini Mulsant, 1850
Tribe Scymnillini Casey, 1899
Tribe Scymnini Mulsant, 1846
Tribe Selvadiini Gordon, 1985
Tribe Shirozuellini Sasaji, 1967
Tribe Stethorini Dobzhansky, 1924
Tribe Sticholotidini Weise, 1901
Tribe Telsimiini Casey, 1899
Tribe Tetrabrachini Kapur, 1948
Family Corylophidae LeConte, 1852
Subfamily Periptyctinae Ślipiński, Lawrence and Tomaszewska, 2001
Subfamily Corylophinae LeConte, 1852
Tribe Aenigmaticini Casey, 1900
Tribe Cleidostethini Bowestead, Booth, Ślipiński and Lawrence, 2001
Tribe Corylophini LeConte, 1852
Tribe Foadiini Ślipiński, Tomaszewska and Lawrence, 2009
Tribe Orthoperini Jacquelin du Val, 1857
Tribe Parmulini Poey, 1854
Tribe Peltinodini Paulian, 1950
Tribe Rypobiini Paulian, 1950
Tribe Sericoderini Matthews, 1888
Tribe Teplinini Pakaluk, Ślipiński and Lawrence, 1994
Family Akalyptoischiidae Lord, Hartley, Lawrence, McHugh and Miller, 2010
Family Latridiidae Erichson, 1842
Subfamily Latridiinae Erichson, 1842
Subfamily Corticariinae Curtis, 1829
†Subfamily Tetrameropseinae Kirejtshuk and Azar, 2008

Superfamily Tenebrionoidea Latreille, 1802

Family Mycetophagidae Leach, 1815
Subfamily Esarcinae Reitter, 1882
Subfamily Mycetophaginae Leach, 1815
Tribe Mycetophagini Leach, 1815
Tribe Typhaeini Thomson, 1863
Subfamily Bergininae Leng, 1920
Family Archeocrypticidae Kaszab, 1964

Family Pterogeniidae Crowson, 1953
Family Ciidae Leach, 1819
Subfamily Sphindociinae Lawrence, 1974
Subfamily Ciinae Leach, 1819
Tribe Ciini Leach, 1819
Tribe Orophiini Thomson, 1863
Tribe Xylographellini Kawanabe and Miyatake, 1996
Subtribe Syncosmetina Lopes-Andrade, 2008
Subtribe Xylographellina Kawanabe and Miyatake, 1996
Family Tetratomidae Billberg, 1820
Subfamily Tetratominae Billberg, 1820
Subfamily Piseninae Miyatake, 1960
Subfamily Penthinae Lacordaire, 1859
Subfamily Hallomeninae Gistel, 1848
Subfamily Eustrophinae Gistel, 1848
Tribe Eustrophini Gistel, 1848
Tribe Holostrophini Nikitsky, 1998
Family Melandryidae Leach, 1815
Subfamily Melandryinae Leach, 1815
Tribe Anisoxiellini Nikitsky, 2007
Tribe Dircaeini Kirby, 1837
Tribe Hypulini Gistel, 1848
Tribe Melandryini Leach, 1815
Tribe Orchesiini Mulsant, 1856
Tribe Serropalpini Latreille, 1829
Tribe Xylitini Thomson, 1864
Tribe Zilorini Desbrochers des Loges, 1900
Subfamily Osphyinae Mulsant, 1856 (1839)
Family Mordellidae Latreille, 1802
Subfamily Ctenidiinae Franciscolo, 1951
Subfamily Mordellinae Latreille, 1802
Tribe Conaliini Ermisch, 1956
Tribe Mordellini Latreille, 1802
Tribe Mordellistenini Ermisch, 1941
Tribe Reynoldsiellini Franciscolo, 1957
Tribe Stenaliini Franciscolo, 1955
Family Ripiphoridae Gemminger, 1870 (1855)
Subfamily Ptilophorinae Gerstaecker, 1855
Subfamily Pelecotominae Seidlitz, 1875
Subfamily Hemirhipidiinae Heller, 1921
Subfamily Ripidiinae Gerstaecker, 1855
Tribe Eorhipidiini Iablokoff-Khnzorian, 1986
Tribe Ripidiini Gerstaecker, 1855

Subfamily Ripiphorinae Gemminger, 1870 (1855)
Tribe Macrosiagonini Heyden, 1908
Tribe Ripiphorini Gemminger, 1870 (1855)
Family Zopheridae Solier, 1834
Subfamily Colydiinae Billberg, 1820
Tribe Acropini Sharp, 1894
Tribe Adimerini Sharp, 1894
Tribe Colydiini Billberg, 1820
Tribe Gempylodini Sharp, 1893
Tribe Nematidiini Horn, 1878
Tribe Orthocerini Blanchard, 1845
Tribe Rhagoderini Horn, 1878
Tribe Rhopalocerini Reitter, 1911
Tribe Synchitini Erichson, 1845
Subfamily Zopherinae Solier, 1834
Tribe Latometini Ślipiński and Lawrence, 1999
Tribe Monommatini Blanchard, 1845
Tribe Phellopsini Ślipiński and Lawrence, 1999
Tribe Pycnomerini Erichson, 1845
Tribe Usechini Horn, 1867
Tribe Zopherini Solier, 1834
Family Ulodidae Pascoe, 1869
Family Promecheilidae Lacordaire, 1859
Family Chalcodryidae Watt, 1974
Family Trachelostenidae Lacordaire, 1859
Family Tenebrionidae Latreille, 1802
Subfamily Zolodininae Watt, 1975
Subfamily Lagriinae Latreille, 1825 (1820)
Tribe Adeliini Kirby, 1828
Tribe Belopini Reitter, 1917
Tribe Chaerodini Doyen, Matthews and Lawrence, 1990
Tribe Cossyphini Latreille, 1802
Tribe Goniaderini Lacordaire, 1859
Tribe Laenini Seidlitz, 1895
Tribe Lagriini Latreille, 1825 (1820)
Subtribe Lagriina Latreille, 1825 (1820)
Subtribe Statirina Blanchard, 1845
Tribe Lupropini Ardoin, 1958
Tribe Pycnocerini Lacordaire, 1859 ***nomen protectum***
Subfamily Nilioninae Oken, 1843
Subfamily Phrenapatinae Solier, 1834
Tribe Archaeoglenini Watt, 1975
Tribe Penetini Lacordaire, 1859

Tribe Phrenapatini Solier, 1834

Subfamily Pimeliinae Latreille, 1802

Tribe Adelostomini Solier, 1834
Tribe Adesmiini Lacordaire, 1859 ***nomen protectum***
Tribe Akidini Billberg, 1820
Tribe Anepsiini LeConte, 1862
Tribe Asidini Fleming, 1821
Tribe Boromorphini Skopin, 1978
Tribe Branchini LeConte, 1862
Tribe Caenocrypticini Koch, 1958
Tribe Ceratanisini Gebien, 1937
Tribe Cnemeplatiini Jacquelin du Val, 1861
Subtribe Actizetina Watt, 1992
Subtribe Cnemeplatiina Jacquelin du Val, 1861
Subtribe Rondoniellina Ferrer and Moragues, 2000
Subtribe Thorictosomatina Watt, 1992
Tribe Cnemodinini Gebien, 1910
Tribe Coniontini Waterhouse, 1858
Tribe Cossyphodini Wasmann, 1899
Subtribe Cossyphodina Wasmann, 1899
Subtribe Cossyphoditina Basilewsky, 1950
Subtribe Esemephina Steiner, 1980
Subtribe Paramellonina Andreae, 1961
Tribe Cryptochilini Solier, 1841
Subtribe Calognathina Lacordaire, 1859
Subtribe Cryptochilina Solier, 1841
Subtribe Homebiina Endrödy-Younga, 1989
Subtribe Horatomina Koch, 1955
Subtribe Vansoniina Koch, 1955
Tribe Cryptoglossini LeConte, 1862 ***nomen protectum***
Tribe Edrotini Lacordaire, 1859
Tribe Elenophorini Solier, 1837
Tribe Epitragini Blanchard, 1845 ***nomen protectum***
Tribe Erodiini Billberg, 1820 ***nomen protectum***
Tribe Evaniosomini Lacordaire, 1859
Tribe Falsomycterini Gebien, 1910
Tribe Idisiini Medvedev, 1973
Tribe Klewariini Gebien, 1910
Tribe Kuhitangiini Medvedev, 1962
Tribe Lachnogyini Seidlitz, 1894
Subtribe Lachnodactylina Reitter, 1904
Subtribe Lachnogyina Seidlitz, 1894
Subtribe Netuschiliina Ferrer and Yvinec, 2004

Tribe Leptodini Lacordaire, 1859
Tribe Nyctellini Solier, 1834
Tribe Nyctoporini Lacordaire, 1859
Tribe Phrynocarenini Gebien, 1928
Tribe Physogasterini Lacordaire, 1859
Tribe Pimeliini Latreille, 1802
Tribe Praociini Eschscholtz, 1829
Tribe Sepidiini Eschscholtz, 1829
Subtribe Hypomelina Koch, 1955
Subtribe Molurina Solier, 1834
Subtribe Oxurina Koch, 1955
Subtribe Phanerotomeina Koch, 1958
Subtribe Sepidiina Eschscholtz, 1829
Subtribe Trachynotina Koch, 1955
Tribe Stenosini Schaum, 1859 (1834)
Tribe Tentyriini Eschscholtz, 1831
Tribe Thinobatini Lacordaire, 1859
Tribe Trilobocarini Lacordaire, 1859
Tribe Vacronini Gebien, 1910
Tribe Zophosini Solier, 1834
Subfamily Tenebrioninae Latreille, 1802
Tribe Acropteronini Doyen, 1989
Tribe Alphitobiini Reitter, 1917
Tribe Amarygmini Gistel, 1848
Tribe Amphidorini LeConte, 1862
Tribe Apocryphini Lacordaire, 1859
Tribe Blaptini Leach, 1815
Subtribe Blaptina Leach, 1815
Subtribe Gnaptorina Medvedev, 2001
Subtribe Gnaptorinina Medvedev, 2001
Subtribe Prosodina Skopin, 1960
Subtribe Remipedellina Semenov, 1907
Tribe Bolitophagini Kirby, 1837 ***nomen protectum***
Tribe Centronopini Doyen, 1989
Tribe Cerenopini Horn, 1870
Tribe Dissonomini Medvedev, 1968
Tribe Eulabini Horn, 1870
Tribe Falsocossyphini Ferrer, 2006
Tribe Heleini Fleming, 1821
Subtribe Asphalina Matthews and Lawrence, 2005
Subtribe Cyphaleina Lacordaire, 1859
Subtribe Heleina Fleming, 1821
Tribe Helopini Latreille, 1802

Subtribe Helopina Latreille, 1802
Subtribe Cylindrinotina Español, 1956
Tribe Helopinini Lacordaire, 1859
Subtribe Aptilina Koch, 1958
Subtribe Helopinina Lacordaire, 1859
Subtribe Micrantereina Reitter, 1917
Subtribe Oncosomina Koch, 1958
Tribe Melanimonini Seidlitz, 1894 (1854)
Tribe Opatrini Brullé, 1832
Subtribe Heterocheirina Koch, 1956
Subtribe Heterotarsina Blanchard, 1845
Subtribe Opatrina Brullé, 1832
Subtribe Neopachypterina Bouchard, Löbl and Merkl, 2007
Tribe Palorini Matthews, 2003
Tribe Pedinini Eschscholtz, 1829
Subtribe Dendarina Mulsant and Rey, 1854
Subtribe Eurynotina Mulsant and Rey, 1854
Subtribe Leichenina Mulsant, 1854
Subtribe Loensina Koch, 1956
Subtribe Melambiina Mulsant and Rey, 1854
Subtribe Pedinina Eschscholtz, 1829
Subtribe Platynotina Mulsant and Rey, 1853
Subtribe Pythiopina Koch, 1953
Tribe Platyscelidini Lacordaire, 1859
Tribe Praeugenini De Moor, 1970
Tribe Rhysopaussini Wasmann, 1896
Tribe Scaurini Billberg, 1820
Tribe Scotobiini Solier, 1838
Tribe Tenebrionini Latreille, 1802
Tribe Titaenini Fauvel, 1905
Tribe Toxicini Oken, 1843
Subtribe Eudysantina Bouchard, Lawrence, Davies and Newton, 2005
Subtribe Nycteropina Lacordaire, 1859
Subtribe Toxicina Oken, 1843
Tribe Triboliini Gistel, 1848
Tribe Ulomini Blanchard, 1845
Subfamily Alleculinae Laporte, 1840
Tribe Alleculini Laporte, 1840
Subtribe Alleculina Laporte, 1840
Subtribe Gonoderina Seidlitz, 1896
Subtribe Mycetocharina Gistel, 1848
Subtribe Xystropodina Solier, 1835
Tribe Cteniopodini Solier, 1835

Subfamily Diaperinae Latreille, 1802
Tribe Crypticini Brullé, 1832
Tribe Diaperini Latreille, 1802
Subtribe Adelinina LeConte, 1862
Subtribe Diaperina Latreille, 1802
Tribe Ectychini Doyen, Matthews and Lawrence, 1990
Tribe Gnathidiini Gebien, 1921
Subtribe Anopidiina Jeannel and Paulian, 1945
Subtribe Gnathidiina Gebien, 1921
Tribe Hyociini Medvedev and Lawrence, 1982
Subtribe Brittonina Medvedev and Lawrence, 1986
Subtribe Hyociina Medvedev and Lawrence, 1982
Subtribe Uptonina Medvedev and Lawrence, 1986
Tribe Hypophlaeini Billberg, 1820
Tribe Leiochrinini Lewis, 1894
Tribe Myrmechixenini Jacquelin du Val, 1858
Tribe Phaleriini Blanchard, 1845
Tribe Scaphidemini Reitter, 1922
Tribe Trachyscelini Blanchard, 1845
Subfamily Stenochiinae Kirby, 1837
Tribe Cnodalonini Oken, 1843
Tribe Stenochiini Kirby, 1837
Tribe Talanini Champion, 1887 (1883)
Family Prostomidae Thomson, 1859
Family Synchroidae Lacordaire, 1859
Family Stenotrachelidae Thomson, 1859
Subfamily Stenotrachelinae Thomson, 1859
Subfamily Cephaloinae LeConte, 1862
Subfamily Nematoplinae LeConte, 1862
Subfamily Stoliinae Nikitsky, 1985
Family Oedemeridae Latreille, 1810
Subfamily Polypriinae Lawrence, 2005
Subfamily Calopodinae Costa, 1852 *nomen protectum*
Subfamily Oedemerinae Latreille, 1810
Tribe Asclerini Gistel, 1848
Tribe Ditylini Mulsant, 1858
Tribe Nacerdini Mulsant, 1858
Tribe Oedemerini Latreille, 1810
Tribe Stenostomatini Mulsant, 1858
Family Meloidae Gyllenhal, 1810
Subfamily Eleticinae Wellman, 1910
Tribe Derideini Wellman, 1910
Tribe Eleticini Wellman, 1910

Subtribe Eleticina Wellman, 1910
Subtribe Eospastina Selander, 1966
Tribe Ertlianini Selander, 1966
Tribe Spasticini Kaszab, 1959
Subtribe Anthicoxenina Selander, 1966
Subtribe Protomeloina Abdullah, 1965
Subtribe Spasticina Kaszab, 1959
Subtribe Xenospastina Selander, 1966
Subfamily Meloinae Gyllenhal, 1810
Tribe Cerocomini Leach, 1815
Tribe Epicautini Parker and Böving, 1924
Tribe Eupomphini LeConte, 1862
Tribe Lyttini Solier, 1851
Tribe Meloini Gyllenhal, 1810
Tribe Mylabrini Rafinesque, 1815
Tribe Pyrotini MacSwain, 1956
Subfamily Tetraonycinae Böving and Craighead, 1931
Subfamily Nemognathinae Laporte, 1840
Tribe Horiini Latreille, 1802
Tribe Nemognathini Laporte, 1840
Subtribe Nemognathina Laporte, 1840
Subtribe Zonitidina Mulsant, 1857
Subtribe Sitarina Mulsant, 1857
Tribe Stenoderini Selander, 1991
Family Mycteridae Oken, 1843
Subfamily Mycterinae Oken, 1843
Subfamily Eurypinae Thomson, 1860
Subfamily Hemipeplinae Lacordaire, 1854
Family Boridae Thomson, 1859
Subfamily Borinae Thomson, 1859
Subfamily Synercticinae Lawrence and Pollock, 1994
Family Trictenotomidae Blanchard, 1845
Family Pythidae Solier, 1834
Family Pyrochroidae Latreille, 1806
Subfamily Tydessinae Nikitsky, 1986
Subfamily Pilipalpinae Abdullah, 1964
Subfamily Pedilinae Lacordaire, 1859
Subfamily Pyrochroinae Latreille, 1806
Subfamily Agnathinae Lacordaire, 1859
Family Salpingidae Leach, 1815
Subfamily Othniinae LeConte, 1861
Subfamily Prostominiinae Grouvelle, 1914
Subfamily Agleninae Horn, 1878

Subfamily Inopeplinae Grouvelle, 1908
Subfamily Dacoderinae LeConte, 1862
Subfamily Aegialitinae LeConte, 1862
Subfamily Salpinginae Leach, 1815
Family Anthicidae Latreille, 1819
Subfamily Eurygeniinae LeConte, 1862
Tribe Eurygeniini LeConte, 1862
Tribe Ictistygnini Borchmann, 1936
Tribe Mitraelabrini Abdullah, 1969
Subfamily Macratriinae LeConte, 1862
Tribe Macratriini LeConte, 1862
†Tribe Camelomorphini Kirejtshuk and Azar, 2008
Subfamily Steropinae Jacquelin du Val, 1863
Subfamily Copobaeninae Abdullah, 1969
Subfamily Lemodinae Lawrence and Britton, 1991
Subfamily Tomoderinae Bonadona, 1961
Subfamily Anthicinae Latreille, 1819
Tribe Anthicini Latreille, 1819
Tribe Endomiini Kaszab, 1956
Tribe Formicomini Bonadona, 1974
Tribe Microhoriini Bonadona, 1974
Subfamily Notoxinae Stephens, 1829
Family Aderidae Csiki, 1909
Tribe Aderini Csiki, 1909
Subtribe Aderina Csiki, 1909
Subtribe Cnopina Báguena Corella, 1948
Subtribe Gompeliina Bouchard, 2011
Subtribe Syzetoninina Báguena Corella, 1948
Tribe Emelinini Báguena Corella, 1948
Tribe Euglenesini Seidlitz, 1875
Subtribe Euglenesina Seidlitz, 1875
Subtribe Pseudolotelina Báguena Corella, 1948
Tribe Phytobaenini Báguena Corella, 1948
Family Scraptiidae Gistel, 1848
Subfamily Scraptiinae Gistel, 1848
Tribe Allopodini Franciscolo, 1964
Tribe Scraptiini Gistel, 1848
Subfamily Anaspidinae Mulsant, 1856
Tribe Anaspidini Mulsant, 1856
Tribe Anaspimordini Franciscolo, 1954
Tribe Menuthianaspidini Franciscolo, 1972
Tribe Pentariini Franciscolo, 1954
Subfamily Lagrioidinae Abdullah and Abdullah, 1968

Subfamily Afreminae Levey, 1985
Subfamily Ischaliinae Blair, 1920

Superfamily Chrysomeloidea Latreille, 1802
Family Oxypeltidae Lacordaire, 1868
Family Vesperidae Mulsant, 1839
Subfamily Philinae Thomson, 1861
Subfamily Vesperinae Mulsant, 1839
Subfamily Anoplodermatinae Guérin-Méneville, 1840
Tribe Anoplodermatini Guérin-Méneville, 1840
Tribe Hypocephalini Blanchard, 1845
Tribe Mysteriini Prosen, 1960
Family Disteniidae Thomson, 1861
Tribe Cyrtonopini Gressitt, 1940
Tribe Disteniini Thomson, 1861
Tribe Dynamostini Lacordaire, 1868
Tribe Heteropalpini Villiers, 1961
Family Cerambycidae Latreille, 1802
Subfamily Parandrinae Blanchard, 1845
Tribe Erichsoniini Thomson, 1861
Tribe Parandrini Blanchard, 1845
Subfamily Prioninae Latreille, 1802
Tribe Acanthophorini Thomson, 1864
Tribe Aegosomatini Thomson, 1861
Tribe Anacolini Thomson, 1857
Tribe Cacoscelini Thomson, 1861
Tribe Callipogonini Thomson, 1861
Tribe Calocomini Galileo and Martins, 1993
Tribe Cantharocnemini Thomson, 1861
Tribe Ergatini Fairmaire, 1864
Tribe Eurypodini Gahan, 1906 (1868)
Tribe Hopliderini Thomson, 1864
Tribe Macrodontiini Thomson, 1861
Tribe Macrotomini Thomson, 1861
Subtribe Archetypina Lameere, 1912
Subtribe Basitoxina Lameere, 1912
Subtribe Macrotomina Thomson, 1861
Subtribe Platygnathina Gilmour, 1954
Subtribe Xixuthrina Lameere, 1912
Tribe Mallaspini Thomson, 1861
Tribe Mallodonini Thomson, 1861
Tribe Meroscelisini Thomson, 1861
Tribe Prionini Latreille, 1802

Tribe Remphanini Lacordaire, 1868
Tribe Solenopterini Lacordaire, 1868
Tribe Tereticini Lameere, 1913
Tribe Vesperoctenini Vives, 2005
Subfamily Lepturinae Latreille, 1802
Tribe Desmocerini Blanchard, 1845
Tribe Encyclopini LeConte, 1873
Tribe Lepturini Latreille, 1802
Tribe Oxymirini Danilevsky, 1997
Tribe Rhagiini Kirby, 1837
Tribe Rhamnusiini Sama, 2009
Tribe Teledapini Pascoe, 1871
Tribe Sachalinobiini Danilevsky, 2010
Tribe Xylosteini Reitter, 1913
Subfamily Spondylidinae Audinet-Serville, 1832
Tribe Anisarthrini Mamaev and Danilevsky, 1973
Tribe Asemini Thomson, 1861
Tribe Atimiini LeConte, 1873
Tribe Saphanini Gistel, 1848
Tribe Spondylidini Audinet-Serville, 1832
Subfamily Necydalinae Latreille, 1825
Subfamily Dorcasominae Lacordaire, 1868
Subfamily Apatophyseinae Lacordaire, 1869
Subfamily Cerambycinae Latreille, 1802
Tribe Acangassuini Galileo and Martins, 2001
Tribe Achrysonini Lacordaire, 1868
Tribe Agallissini LeConte, 1873
Tribe Alanizini Di Iorio, 2003
Tribe Anaglyptini Lacordaire, 1868 ***nomen protectum***
Tribe Aphanasiini Lacordaire, 1868
Tribe Aphneopini Lacordaire, 1868
Tribe Auxesini Lepesme and Breuning, 1952
Tribe Basipterini Fragoso, Monné and Campos Seabra, 1987
Tribe Bimiini Lacordaire, 1868
Tribe Bothriospilini Lane, 1950
Tribe Brachypteromatini Sama, 2008
Tribe Callichromatini Swainson, 1840
Tribe Callidiini Kirby, 1837
Tribe Callidiopini Lacordaire, 1868
Tribe Cerambycini Latreille, 1802
Subtribe Cerambycina Latreille, 1802
Subtribe Sphallotrichina Martins and Monné, 2005
Tribe Certallini Fairmaire, 1864

Tribe Chlidonini Waterhouse, 1879
Tribe Cleomenini Lacordaire, 1868
Tribe Clytini Mulsant, 1839
Tribe Compsocerini Thomson, 1864
Tribe Coptommatini Lacordaire, 1869
Tribe Curiini LeConte, 1873
Tribe Deilini Fairmaire, 1864
Tribe Dejanirini Lacordaire, 1868
Tribe Diorini Lane, 1950
Tribe Distichocerini Pascoe, 1867
Tribe Dodecosini Aurivillius, 1912
Tribe Dryobiini Arnett, 1962 ***nomen protectum***
Tribe Eburiini Blanchard, 1845
Tribe Ectenessini Martins, 1998
Tribe Elaphidiini Thomson, 1864
Tribe Eligmodermini Lacordaire, 1868
Tribe Erlandiini Aurivillius, 1912
Tribe Eroschemini Lacordaire, 1868
Tribe Eumichthini Linsley, 1940
Tribe Gahaniini Quentin and Villiers, 1969
Tribe Glaucytini Lacordaire, 1868
Tribe Graciliini Mulsant, 1839
Tribe Hesperophanini Mulsant, 1839
  Subtribe Daramina Sama, 2008
  Subtribe Hesperophanina Mulsant, 1839
Tribe Hesthesini Pascoe, 1867
Tribe Heteropsini Lacordaire, 1868 ***nomen protectum***
Tribe Hexoplini Martins, 2006
Tribe Holopleurini Chemsak and Linsley, 1974
Tribe Holopterini Lacordaire, 1868
Tribe Hyboderini Linsley, 1940
Tribe Hylotrupini Zagajkevich, 1991
Tribe Ibidionini Thomson, 1861
  Subtribe Compsina Martins and Galileo, 2007
  Subtribe Ibidionina Thomson, 1861
  Subtribe Tropidina Martins and Galileo, 2007
Tribe Ideratini Martins and Napp, 2009
Tribe Lissonotini Swainson, 1840
Tribe Luscosmodicini Martins, 2003
Tribe Lygrini Sama, 2008
Tribe Macronini Lacordaire, 1868
Tribe Megacoelini Quentin and Villiers, 1969
Tribe Methiini Thomson, 1860

Tribe Molorchini Gistel, 1848
Tribe Mythodini Lacordaire, 1868
Tribe Necydalopsini Lacordaire, 1868
Tribe Neocorini Martins, 2005
Tribe Neostenini Lacordaire, 1868
Tribe Obriini Mulsant, 1839
Tribe Ochyrini Pascoe, 1871
Tribe Oedenoderini Aurivillius, 1912
Tribe Oemini Lacordaire, 1868
  Subtribe Methioidina Martins, 1997
  Subtribe Oemina Lacordaire, 1868
Tribe Opsimini LeConte, 1873
Tribe Oxycoleini Martins and Galileo, 2003
Tribe Paraholopterini Martins, 1997
Tribe Phalotini Lacordaire, 1868
Tribe Phlyctaenodini Lacordaire, 1868
Tribe Phoracanthini Newman, 1840
Tribe Phyllarthriini Lepesme and Breuning, 1956
Tribe Piesarthriini McKeown, 1947
Tribe Piezocerini Lacordaire, 1868
  Subtribe Haruspicina Martins, 1976
  Subtribe Piezocerina Lacordaire, 1868
Tribe Platyarthrini Bates, 1870
Tribe Plectogasterini Quentin and Villiers, 1969
Tribe Plectromerini Nearns and Braham, 2008
Tribe Pleiarthrocerini Lane, 1950
Tribe Protaxini Gahan, 1906
Tribe Prothemini Lacordaire, 1868
Tribe Psebiini Lacordaire, 1868
Tribe Pseudocephalini Aurivillius, 1912 (1861)
Tribe Pseudolepturini Thomson, 1861
Tribe Psilomorphini Lacordaire, 1868
Tribe Pteroplatini Thomson, 1861
Tribe Pyrestini Lacordaire, 1868
Tribe Rhagiomorphini Newman, 1841
Tribe Rhinotragini Thomson, 1861
Tribe Rhopalophorini Blanchard, 1845
Tribe Rosaliini Fairmaire, 1864
Tribe Sestyrini Lacordaire, 1868
Tribe Smodicini Lacordaire, 1868
Tribe Spintheriini Lacordaire, 1869
Tribe Stenhomalini Miroshnikov, 1989
Tribe Stenoderini Pascoe, 1867

Tribe Stenopterini Gistel, 1848
Tribe Strongylurini Lacordaire, 1868
Tribe Tessarommatini Lacordaire, 1868
Tribe Thraniini Gahan, 1906
Tribe Thyrsiini Marinoni and Napp, 1984
Tribe Tillomorphini Lacordaire, 1868
Tribe Torneutini Thomson, 1861
Tribe Trachyderini Dupont, 1836
Subtribe Ancylocerina Thomson, 1864
Subtribe Trachyderina Dupont, 1836
Tribe Tragocerini Pascoe, 1867
Tribe Trichomesiini Aurivillius, 1912
Tribe Tropocalymmatini Lacordaire, 1868
Tribe Typhocesini Lacordaire, 1868
Tribe Unxiini Napp, 2007
Tribe Uracanthini Blanchard, 1853
Tribe Vesperellini Sama, 2008
Tribe Xystrocerini Blanchard, 1845

Subfamily Lamiinae Latreille, 1825

Tribe Acanthocinini Blanchard, 1845
Tribe Acanthoderini Thomson, 1860
Tribe Acmocerini Thomson, 1864
Tribe Acridocephalini Dillon and Dillon, 1959
Tribe Acrocinini Swainson, 1840
Tribe Aderpasini Breuning and Teocchi, 1978
Tribe Aerenicini Lacordaire, 1872
Tribe Agapanthiini Mulsant, 1839
Tribe Amphoecini Breuning, 1951
Tribe Ancitini Aurivillius, 1917
Tribe Ancylonotini Lacordaire, 1869
Tribe Anisocerini Thomson, 1860
Tribe Apomecynini Thomson, 1860
Tribe Astathini Thomson, 1864
Tribe Batocerini Thomson, 1864
Tribe Calliini Thomson, 1864
Tribe Ceroplesini Thomson, 1860
Subtribe Ceroplesina Thomson, 1860
Subtribe Crossotina Thomson, 1864
Tribe Cloniocerini Lacordaire, 1872
Tribe Colobotheini Thomson, 1860
Tribe Compsosomatini Thomson, 1857
Tribe Cyrtinini Thomson, 1864
Tribe Desmiphorini Thomson, 1860

Tribe Dorcadionini Swainson, 1840
Tribe Dorcaschematini Thomson, 1860
Tribe Elytracanthinini Bousquet, 2009
Tribe Enicodini Thomson, 1864
Tribe Eupromerini Galileo and Martins, 1995
Tribe Forsteriini Tippmann, 1960
Tribe Gnomini Thomson, 1860
Tribe Gyaritini Breuning, 1950
Tribe Heliolini Breuning, 1951
Tribe Hemilophini Thomson, 1868 ***nomen protectum***
Tribe Homonoeini Thomson, 1864
Tribe Hyborhabdini Aurivillius, 1911
Tribe Lamiini Latreille, 1825
Tribe Laticraniini Lane, 1959
Tribe Mauesiini Lane, 1956
Tribe Megabasini Thomson, 1860
Tribe Mesosini Mulsant, 1839
Tribe Microcymaturini Breuning and Teocchi, 1985
Tribe Moneilemini Thomson, 1864
Tribe Monochamini Gistel, 1848
Tribe Morimonellini Lobanov, Danilevsky and Murzin, 1981
Tribe Morimopsini Lacordaire, 1869
Tribe Nyctimeniini Gressitt, 1951
Tribe Obereini Thomson, 1864
Tribe Oculariini Breuning, 1950
Tribe Onciderini Thomson, 1860
Tribe Oncideropsidini Aurivillius, 1922
Tribe Onocephalini Thomson, 1860
Tribe Onychogleneini Aurivillius, 1923
Tribe Parmenini Mulsant, 1839
Tribe Petrognathini Blanchard, 1845
Tribe Phacellini Lacordaire, 1872
Tribe Phantasini Kolbe, 1897
Tribe Phrynetini Thomson, 1864
Tribe Phymasternini Teocchi, 1989
Tribe Phytoeciini Mulsant, 1839
Tribe Pogonocherini Mulsant, 1839
Tribe Polyrhaphidini Thomson, 1860
Tribe Pretiliini Martins and Galileo, 1990
Tribe Proctocerini Aurivillius, 1922
Tribe Prosopocerini Thomson, 1864
Tribe Pteropliini Thomson, 1860
Tribe Rhodopinini Gressitt, 1951

Tribe Saperdini Mulsant, 1839
Tribe Stenobiini Breuning, 1950
Tribe Sternotomini Thomson, 1860
Tribe Tapeinini Thomson, 1857
Tribe Tetraopini Thomson, 1860
Tribe Tetraulaxini Breuning and Teocchi, 1977
Tribe Tetropini Portevin, 1927
Tribe Theocrini Lacordaire, 1872
Tribe Tmesisternini Blanchard, 1853
Tribe Tragocephalini Thomson, 1857
Tribe Xenicotelini Matsushita, 1933
Tribe Xenofreini Aurivillius, 1923
Tribe Xenoleini Lacordaire, 1872
Tribe Xylorhizini Lacordaire, 1872
Tribe Zygocerini Thomson, 1864
Family Megalopodidae Latreille, 1802
Subfamily Megalopodinae Latreille, 1802
Subfamily Palophaginae Kuschel and May, 1990
Subfamily Zeugophorinae Böving and Craighead, 1931
Family Orsodacnidae Thomson, 1859
Subfamily Orsodacninae Thomson, 1859
Subfamily Aulacoscelidinae Chapuis, 1874
Family Chrysomelidae Latreille, 1802
Subfamily Sagrinae Leach, 1815
Tribe Carpophagini Chapuis, 1874
Tribe Diaphanopsidini Monrós, 1958
Tribe Megamerini Chapuis, 1874
Tribe Sagrini Leach, 1815
Subfamily Bruchinae Latreille, 1802
Tribe Amblycerini Bridwell, 1932
Subtribe Amblycerina Bridwell, 1932
Subtribe Spermophagina Borowiec, 1987
Tribe Bruchini Latreille, 1802
Subtribe Acanthoscelidina Bridwell, 1946
Subtribe Bruchina Latreille, 1802
Subtribe Megacerina Bridwell, 1946
Tribe Eubaptini Bridwell, 1932
Tribe Kytorhinini Bridwell, 1932
Tribe Pachymerini Bridwell, 1929
Subtribe Caryedontina Bridwell, 1929
Subtribe Caryopemina Bridwell, 1929
Subtribe Pachymerina Bridwell, 1929
Tribe Rhaebini Blanchard, 1845

Subfamily Donaciinae Kirby, 1837
- Tribe Donaciini Kirby, 1837
- Tribe Haemoniini Chen, 1941
- Tribe Plateumarini Böving, 1922

Subfamily Criocerinae Latreille, 1804
- Tribe Criocerini Latreille, 1804
- Tribe Lemini Gyllenhal, 1813
- Tribe Pseudocriocerini Heinze, 1962

Subfamily Cassidinae Gyllenhal, 1813
- Tribe Alurnini Chapuis, 1875
- Tribe Anisoderini Chapuis, 1875
- Tribe Aproidini Weise, 1911
- Tribe Arescini Chapuis, 1875
- Tribe Aspidimorphini Chapuis, 1875
- Tribe Basiprionotini Gressitt, 1952 (1929)
- Tribe Botryonopini Chapuis, 1875
- Tribe Callispini Chapuis, 1875
- Tribe Callohispini Uhmann, 1960
- Tribe Cassidini Gyllenhal, 1813
- Tribe Cephaloleiini Chapuis, 1875
- Tribe Chalepini Weise, 1910
- Tribe Coelaenomenoderini Weise, 1911
- Tribe Cryptonychini Chapuis, 1875
- Tribe Delocraniini Spaeth, 1929
- Tribe Dorynotini Monrós and Viana, 1949 (1923)
- Tribe Eugenysini Hincks, 1952
- Tribe Eurispini Chapuis, 1875
- Tribe Exothispini Weise, 1911
- Tribe Goniocheniini Spaeth, 1942
- Tribe Gonophorini Chapuis, 1875
- Tribe Hemisphaerotini Monrós and Viana, 1951 (1929)
- Tribe Hispini Gyllenhal, 1813
- Tribe Hispoleptini Chapuis, 1875
- Tribe Hybosispini Weise, 1910
- Tribe Imatidiini Hope, 1840
- Tribe Ischyrosonychini Chapuis, 1875
- Tribe Leptispini Fairmaire, 1868
- Tribe Mesomphaliini Hope, 1840
- Tribe Nothosacanthini Gressitt, 1952 (1929)
- Tribe Oediopalpini Monrós and Viana, 1947 (1910)
- Tribe Omocerini Hincks, 1952 (1923)
- Tribe Oncocephalini Chapuis, 1875
- Tribe Promecothecini Chapuis, 1875

Tribe Prosopodontini Weise, 1910
Tribe Sceloenoplini Uhmann, 1930
Tribe Spilophorini Chapuis, 1875
Tribe Uroplatini Weise, 1910
Subfamily Chrysomelinae Latreille, 1802
Tribe Chrysomelini Latreille, 1802
Tribe Timarchini Motschulsky, 1860
Subfamily Galerucinae Latreille, 1802
Tribe Alticini Newman, 1834
Tribe Decarthrocerini Laboissière, 1937
Tribe Galerucini Latreille, 1802
Tribe Hylaspini Chapuis, 1875
Tribe Luperini Gistel, 1848
Tribe Metacyclini Chapuis, 1875
Tribe Oidini Laboissière, 1921 (1875)
Subfamily Lamprosomatinae Lacordaire, 1848
Tribe Lamprosomatini Lacordaire, 1848
Tribe Neochlamysini Monrós, 1959
Tribe Sphaerocharini Chapuis, 1874
Subfamily Cryptocephalinae Gyllenhal, 1813
Tribe Clytrini Kirby, 1837
Tribe Cryptocephalini Gyllenhal, 1813
Subtribe Achaenopina Chapuis, 1874
Subtribe Cryptocephalina Gyllenhal, 1813
Subtribe Monachulina Leng, 1920
Subtribe Pachybrachina Chapuis, 1874
Subtribe Stylosomina Chapuis, 1874
Tribe Fulcidacini Jakobson, 1924
Subfamily Eumolpinae Hope, 1840
Tribe Bromiini Baly, 1865 (1863)
Tribe Caryonodini Bechyné, 1951
Tribe Cubispini Monrós, 1954
Tribe Eumolpini Hope, 1840
Tribe Euryopini Chapuis, 1874
Tribe Habrophorini Bechyné and Špringlová de Bechyné, 1969
Tribe Hemydacnini Bechyné, 1951
Tribe Megascelidini Chapuis, 1874
Tribe Merodini Chapuis, 1874
Tribe Pygomolpini Bechyné, 1949
Tribe Rosiroiini Bechyné, 1950
Tribe Typophorini Baly, 1865
Subfamily Spilopyrinae Chapuis, 1874
Subfamily Synetinae LeConte and Horn, 1883

†Subfamily Protoscelidinae Medvedev, 1968

Superfamily Curculionoidea Latreille, 1802

Family Nemonychidae Bedel, 1882
Subfamily Nemonychinae Bedel, 1882
Subfamily Cimberidinae Gozis, 1882
Tribe Cimberidini Gozis, 1882
Tribe Doydirhynchini Pierce, 1916
†Tribe Kuschelomacrini Riedel, 2010
Subfamily Rhinorhynchinae Voss, 1922
Tribe Mecomacerini Kuschel, 1994
Subtribe Brarina Legalov, 2009
Subtribe Mecomacerina Kuschel, 1994
Tribe Rhinorhynchini Voss, 1922
†Subfamily Slonikinae Zherikhin, 1977
†Tribe Slonikini Zherikhin, 1977
†Tribe Ulyaniscini Legalov, 2009
†Subfamily Eccoptarthrinae Arnoldi, 1977
†Subfamily Brenthorrhininae Arnoldi, 1977
†Tribe Brenthorrhinini Arnoldi, 1977
†Tribe Brenthorrhinoidini Legalov, 2003
†Subfamily Distenorrhininae Arnoldi, 1977
†Subfamily Eobelinae Arnoldi, 1977
†Tribe Eobelini Arnoldi, 1977
†Subtribe Eobelina Arnoldi, 1977
†Subtribe Procurculionina Arnoldi, 1977
†Tribe Karataucarini Legalov, 2009
†Tribe Nanophydini Arnoldi, 1977
†Tribe Oxycorynoidini Arnoldi, 1977
†Tribe Probelini Legalov, 2009
†Subfamily Paleocartinae Legalov, 2003
†Tribe Nebrenthorrhinini Legalov, 2007
†Tribe Paleocartini Legalov, 2003
†Subfamily Metrioxenoidinae Legalov, 2009
†Subfamily Cretonemonychinae Gratshev and Legalov, 2009
†Subfamily Selengarhynchinae Gratshev and Legalov, 2009
Family Anthribidae Billberg, 1820
Subfamily Anthribinae Billberg, 1820
Tribe Anthribini Billberg, 1820
Tribe Basitropini Lacordaire, 1865
Tribe Corrhecerini Lacordaire, 1865
Tribe Cratoparini LeConte, 1876
†Tribe Cretanthribini Legalov, 2009

Tribe Decataphanini Lacordaire, 1865
Tribe Discotenini Lacordaire, 1865
Tribe Ecelonerini Lacordaire, 1865
Tribe Ischnocerini Lacordaire, 1865
Tribe Gymnognathini Valentine, 1960
Tribe Jordanthribini Morimoto, 1980
Tribe Mauiini Valentine, 1990
Tribe Mecocerini Lacordaire, 1865
Tribe Mycteini Morimoto, 1972
Tribe Ozotomerini Morimoto, 1972
Tribe Piesocorynini Valentine, 1960
Tribe Platyrhinini Bedel, 1882
Tribe Platystomini Pierce, 1916
Tribe Proscoporhinini Lacordaire, 1865
Tribe Ptychoderini Jekel, 1855
Tribe Sintorini Lacordaire, 1865
Tribe Stenocerini Kolbe, 1895
Tribe Tophoderini Lacordaire, 1865
Tribe Trigonorhinini Valentine, 1999
Tribe Tropiderini Lacordaire, 1865
Tribe Xenocerini Lacordaire, 1865
Tribe Xylinadini Lacordaire, 1865
Tribe Zygaenodini Lacordaire, 1865
Subfamily Choraginae Kirby, 1819
Tribe Apolectini Lacordaire, 1865
Tribe Araecerini Lacordaire, 1865
Tribe Cisanthribini Zimmerman, 1994
Tribe Choragini Kirby, 1819
Tribe Valenfriesiini Alonso-Zarazaga and Lyal, 1999
Tribe Xenorchestini Lacordaire, 1865
Subfamily Urodontinae Thomson, 1859
†Family Ulyanidae Zherikhin, 1993
Family Belidae Schönherr, 1826
Subfamily Belinae Schönherr, 1826
Tribe Agnesiotidini Zimmerman, 1994
Tribe Belini Schönherr, 1826
Subtribe Belina Schönherr, 1826
Subtribe Homalocerina Legalov, 2009
Tribe Pachyurini Kuschel, 1959
Subfamily Oxycoryninae Schönherr, 1840
Tribe Aglycyderini Wollaston, 1864
Tribe Alloxycorynini Legalov, 2009
Tribe Distenorrhinoidini Legalov, 2009

Tribe Metrioxenini Voss, 1953
Subtribe Afrocorynina Voss, 1957
Subtribe Metrioxenina Voss, 1953
Subtribe Zherichinixenina Legalov, 2009
Tribe Oxycorynini Schönherr, 1840
Subtribe Allocorynina Sharp, 1890
Subtribe Oxycorynina Schönherr, 1840
Subtribe Oxycraspedina Marvaldi and Oberprieler, 2006
Family Caridae Thompson, 1992
Subfamily Carinae Thompson, 1992
Subfamily Chilecarinae Legalov, 2009
Tribe Carodini Legalov, 2009
Tribe Chilecarini Legalov, 2009
†Subfamily Baissorhynchinae Zherikhin, 1993
Family Attelabidae Billberg, 1820
Subfamily Attelabinae Billberg, 1820
Tribe Attelabini Billberg, 1820
Subtribe Attelabina Billberg, 1820
Subtribe Euscelina Voss, 1925
Subtribe Euscelophilina Voss, 1925
Subtribe Henicolabina Legalov, 2007
Subtribe Himatolabina Legalov, 2003
Subtribe Hybolabina Voss, 1925
Subtribe Isolabina Legalov, 2007
Subtribe Lagenoderina Voss, 1925
Subtribe Lamprolabina Voss, 1925
Subtribe Metocalolabina Legalov, 2003
Subtribe Omolabina Legalov, 2003
Subtribe Paramecolabina Legalov, 2003
Subtribe Phialodina Legalov, 2003
Subtribe Phymatolabina Voss, 1925
Subtribe Phymatopsinina Legalov, 2003
Subtribe Pleurolabina Legalov, 2003
Tribe Euopini Voss, 1925
Tribe Pilolabini Voss, 1925
Subfamily Apoderinae Jekel, 1860
Tribe Apoderini Jekel, 1860
Tribe Clitostylini Voss, 1929
Subtribe Allapoderina Legalov, 2003
Subtribe Clitostylina Voss, 1929
Subtribe Pseudophrysina Legalov, 2003
Tribe Hoplapoderini Voss, 1926
Subtribe Afroapoderina Legalov, 2003

Subtribe Hoplapoderina Voss, 1926
Subtribe Paratomapoderina Legalov, 2003
Tribe Trachelophorini Voss, 1926
Subfamily Rhynchitinae Gistel, 1848
Tribe Auletini Desbrochers des Loges, 1908
Subtribe Auletina Desbrochers des Loges, 1908
Subtribe Auletobiina Legalov, 2001
Subtribe Guineauletina Legalov, 2003
Subtribe Mandelschtamiina Legalov, 2003
Subtribe Pseudauletina Voss, 1933
Subtribe Pseudomesauletina Legalov, 2003
Tribe Auletorhinini Voss, 1935
Tribe Byctiscini Voss, 1923
Subtribe Byctiscina Voss, 1923
Subtribe Listrobyctiscina Legalov, 2003
Subtribe Svetlanaebyctiscina Legalov, 2003
Tribe Cesauletini Legalov, 2003
Tribe Deporaini Voss, 1929
Subtribe Chonostropheina Morimoto, 1962
Subtribe Deporaina Voss, 1929
Tribe Minurini Legalov, 2003
Tribe Rhinocartini Voss, 1931
Tribe Rhynchitini Gistel, 1848
Subtribe Acritorrhynchitina Legalov, 2007
Subtribe Anisomerinina Legalov, 2003
Subtribe Eugnamptina Voss, 1930
Subtribe Lasiorhynchitina Legalov, 2003
Subtribe Perrhynchitina Legalov, 2003
Subtribe Rhynchitallina Legalov, 2003
Subtribe Rhynchitina Gistel, 1848
Subtribe Temnocerina Legalov, 2003
Subfamily Isotheinae Scudder, 1893
Tribe Isotheini Scudder, 1893
Subtribe Depasophilina Legalov, 2003
†Subtribe Isotheina Scudder, 1893
†Tribe Toxorhynchini Scudder, 1893
Subfamily Pterocolinae Lacordaire, 1865
Family Brentidae Billberg, 1820
Subfamily Brentinae Billberg, 1820
Tribe Brentini Billberg, 1820
Subtribe Arrhenodina Lacordaire, 1865
Subtribe Brentina Billberg, 1820
Subtribe Eremoxenina Semenov, 1892

Tribe Cyladini Schönherr, 1823
Tribe Cyphagogini Kolbe, 1892
Subtribe Atopobrentina Damoiseau, 1965
Subtribe Cyphagogina Kolbe, 1892
†Subtribe Dominibrentina Poinar, 2009
Subtribe Hoplopisthiina Senna and Calabresi, 1919
Subtribe Stereodermina Sharp, 1895
Tribe Pholidochlamydini Damoiseau, 1962
Tribe Taphroderini Lacordaire, 1865
Tribe Trachelizini Lacordaire, 1865
Subtribe Acratina Alonso-Zarazaga, Lyal, Bartolozzi and Sforzi, 1999
Subtribe Ithystenina Lacordaire, 1865
Subtribe Microtrachelizina Zimmerman, 1994
Subtribe Pseudoceocephalina Kleine, 1922
Subtribe Rhyticephalina Kleine, 1922
Subtribe Trachelizina Lacordaire, 1865
Subtribe Tychaeina Schoenfeldt, 1908
Tribe Ulocerini Schönherr, 1823
Subfamily Eurhynchinae Lacordaire, 1863
†Tribe Axelrodiellini Legalov, 2009
Tribe Eurhynchini Lacordaire, 1863
Subfamily Apioninae Schönherr, 1823
Supertribe Apionitae Schönherr, 1823
Tribe Apionini Schönherr, 1823
Subtribe Apionina Schönherr, 1823
Subtribe Aplemonina Kissinger, 1968
Subtribe Aspidapiina Alonso-Zarazaga, 1990
Subtribe Catapiina Alonso-Zarazaga, 1990
Subtribe Ceratapiina Alonso-Zarazaga, 1990
Subtribe Exapiina Alonso-Zarazaga, 1990
Subtribe Ixapiina Alonso-Zarazaga, 1990
Subtribe Kalcapiina Alonso-Zarazaga, 1990
Subtribe Malvapiina Alonso-Zarazaga, 1990
Subtribe Metapiina Alonso-Zarazaga, 1990
Subtribe Oxystomatina Alonso-Zarazaga, 1990
Subtribe Piezotrachelina Voss, 1959
Subtribe Prototrichapiina Wanat, 1995
Subtribe Synapiina Alonso-Zarazaga, 1990
Subtribe Trichapiina Alonso-Zarazaga, 1990
Tribe Chilapiini Wanat, 2001
Tribe Noterapiini Kissinger, 2004
Tribe Podapiini Wanat, 2001
Tribe Rhinorhynchidiini Zimmerman, 1994

Supertribe Antliarhinitae Schönherr, 1823
Supertribe Cybebitae Lacordaire, 1863
Supertribe Mecolenitae Wanat, 2001
Supertribe Myrmacicelitae Zimmerman, 1994
Tribe Lispotheriini Wanat, 2001
Tribe Myrmacicelini Zimmerman, 1994
Supertribe Rhadinocybitae Alonso-Zarazaga, 1992
Tribe Notapionini Zimmerman, 1994
Tribe Rhadinocybini Alonso-Zarazaga, 1992
Supertribe Tanaitae Schönherr, 1839
Subfamily Ithycerinae Schönherr, 1823
Subfamily Microcerinae Lacordaire, 1863
Subfamily Nanophyinae Gistel, 1848
Tribe Corimaliini Alonso-Zarazaga, 1989
Tribe Nanophyini Gistel, 1848
Family Dryophthoridae Schönherr, 1825
Subfamily Dryophthorinae Schönherr, 1825
Subfamily Cryptodermatinae Bovie, 1908
Subfamily Orthognathinae Lacordaire, 1865
Tribe Orthognathini Lacordaire, 1865
Tribe Rhinostomini LeConte, 1874
Subfamily Rhynchophorinae Schönherr, 1833
Tribe Diocalandrini Zimmerman, 1993
Tribe Litosomini Lacordaire, 1865
Tribe Ommatolampini Lacordaire, 1865
Tribe Polytini Zimmerman, 1993
Tribe Rhynchophorini Schönherr, 1833
Tribe Sphenophorini Lacordaire, 1865
Subfamily Stromboscerinae Lacordaire, 1865
Family Brachyceridae Billberg, 1820
Subfamily Brachycerinae Billberg, 1820
Tribe Brachycerini Billberg, 1820
Tribe Byrsopini Germar, 1829
Subfamily Cryptolarynginae Schalkwyk, 1966
Subfamily Erirhininae Schönherr, 1825
Tribe Aonychini Zimmerman, 1993
Tribe Arthrostenini Reitter, 1913
†Tribe Cretuliini Legalov, 2009
Tribe Erirhinini Schönherr, 1825
Tribe Himasthlophallini Zherikhin, 1991
Tribe Stenopelmini LeConte, 1876
Tribe Tadiini Zimmerman, 1993
Tribe Tanysphyrini Gistel, 1848

Subfamily Ocladiinae Lacordaire, 1865
Tribe Desmidophorini Morimoto, 1962
Tribe Ocladiini Lacordaire, 1865
Subfamily Raymondionyminae Reitter, 1913
Tribe Myrtonymini Kuschel, 1990
Tribe Raymondionymini Reitter, 1913
Family Curculionidae Latreille, 1802
Subfamily Curculioninae Latreille, 1802
Tribe Acalyptini Thomson, 1859
Subtribe Acalyptina Thomson, 1859
Subtribe Derelomina Lacordaire, 1865
Subtribe Notolomina Franz, 2006
Subtribe Phyllotrogina Franz, 2006
Subtribe Staminodeina Franz, 2006
Tribe Acentrusini Alonso-Zarazaga, 2005
Tribe Ancylocnemidini Voss, 1962
Tribe Anthonomini Thomson, 1859
Tribe Camarotini Schönherr, 1833
Subtribe Camarotina Schönherr, 1833
Subtribe Prionomerina Lacordaire, 1863
Tribe Ceratopodini Lacordaire, 1863
Tribe Cionini Schönherr, 1825
Tribe Cranopoeini Kuschel, 2009
Tribe Cryptoplini Lacordaire, 1863
Tribe Curculionini Latreille, 1802
Subtribe Curculionina Latreille, 1802
Subtribe Pseudobalaninina Heller, 1925
Subtribe Timolina Heller, 1925
Tribe Diabathrariini Lacordaire, 1863
Tribe Ellescini Thomson, 1859
Subtribe Dorytomina Bedel, 1886
Subtribe Ellescina Thomson, 1859
Tribe Erodiscini Lacordaire, 1863
Tribe Eugnomini Lacordaire, 1863
Subtribe Eugnomina Lacordaire, 1863
Subtribe Meriphina Marshall, 1937
Tribe Gonipterini Lacordaire, 1863
Tribe Mecinini Gistel, 1848
Tribe Nerthopini Lacordaire, 1865
Tribe Otidocephalini Lacordaire, 1863
Tribe Piazorhinini Lacordaire, 1863
Tribe Prionobrachiini Hustache, 1938
Tribe Pyropini Lacordaire, 1865

Tribe Rhamphini Rafinesque, 1815
Subtribe Dinorhopalina Voss, 1936
Subtribe Ixalmina Voss, 1936
Subtribe Rhamphina Rafinesque, 1815
Subtribe Tachygonina Lacordaire, 1865
Tribe Smicronychini Seidlitz, 1891 ***nomen protectum***
Tribe Sphaeriopoeini Kuschel, 2003
Tribe Storeini Lacordaire, 1863
Tribe Styphlini Jekel, 1861
Tribe Tychiini Gistel, 1848
Subtribe Demimaeina Voss, 1937
Subtribe Lignyodina Bedel, 1883
Subtribe Ochyromerina Voss, 1935
Subtribe Tychiina Gistel, 1848
Tribe Ulomascini Lacordaire, 1865
Tribe Viticiini Morimoto, 1983
Subfamily Bagoinae Thomson, 1859 ***nomen protectum***
Subfamily Baridinae Schönherr, 1836
Tribe Ambatini Lacordaire, 1863
Tribe Anopsilini Bondar, 1942
Tribe Apostasimerini Schönherr, 1844
Subtribe Apostasimerina Schönherr, 1844
Subtribe Madopterina Lacordaire, 1865
Subtribe Thaliabaridina Bondar, 1943
Subtribe Torcina Bondar, 1943
Subtribe Zygobaridina Pierce, 1907
Tribe Baridini Schönherr, 1836
Subtribe Baridina Schönherr, 1836
Subtribe Coelonertina Casey, 1922
Subtribe Coleomerina Casey, 1922
Subtribe Diorymerina Jekel, 1865
Subtribe Eurhinina Lacordaire, 1865
Tribe Madarini Jekel, 1865
Subtribe Barymerina Lacordaire, 1865
Subtribe Eutoxina Champion, 1908
Subtribe Leptoschoinina Lacordaire, 1865
Subtribe Madarina Jekel, 1865
Subtribe Tonesiina Alonso-Zarazaga and Lyal, 1999
Tribe Neosharpiini Hoffmann, 1956
Tribe Nertinini Voss, 1954
Tribe Optatini Champion, 1907
Tribe Pantotelini Lacordaire, 1865
Subtribe Cyrionychina Casey, 1922

Subtribe Pantotelina Lacordaire, 1865
Tribe Peridinetini Lacordaire, 1865
Subfamily Ceutorhynchinae Gistel, 1848
Tribe Ceutorhynchini Gistel, 1848
Tribe Cnemogonini Colonnelli, 1979
Tribe Egriini Pajni and Kohli, 1982
Tribe Hypohypurini Colonnelli, 2004
Tribe Hypurini Schultze, 1902
Tribe Lioxyonychini Colonnelli, 1984
Tribe Mecysmoderini Wagner, 1938
Tribe Mononychini LeConte, 1876
Tribe Phytobiini Gistel, 1848
Tribe Scleropterini Schultze, 1902
Subfamily Conoderinae Schönherr, 1833
Tribe Arachnopodini Lacordaire, 1865
Tribe Campyloscelini Schönherr, 1845
Subtribe Campyloscelina Schönherr, 1845
Subtribe Corynemerina Hustache, 1929
Subtribe Phaenomerina Faust, 1898
Tribe Conoderini Schönherr, 1833
Tribe Coryssomerini Thomson, 1859
Tribe Coryssopodini Lacordaire, 1865
Tribe Lechriopini Lacordaire, 1865
Tribe Lobotrachelini Lacordaire, 1865
Tribe Mecopini Lacordaire, 1865
Tribe Menemachini Lacordaire, 1865
Tribe Othippiini Morimoto, 1962
Tribe Peloropodini Hustache, 1932
Tribe Piazurini Lacordaire, 1865
Tribe Sphadasmini Lacordaire, 1865
Tribe Trichodocerini Champion, 1906
Tribe Zygopini Lacordaire, 1865
Subfamily Cossoninae Schönherr, 1825
Tribe Acamptini LeConte, 1876
Tribe Acanthinomerini Voss, 1972
Tribe Allomorphini Folwaczny, 1973
Tribe Aphyllurini Voss, 1955
Tribe Araucariini Kuschel, 1966
Tribe Choerorhinini Folwaczny, 1973
Tribe Cossonini Schönherr, 1825
Tribe Cryptommatini Voss, 1972
Tribe Dryotribini LeConte, 1876
Tribe Microxylobiini Voss, 1972

Tribe Nesiobiini Alonso-Zarazaga and Lyal, 1999
Tribe Neumatorini Folwaczny, 1973
Tribe Onychiini Chapuis, 1869
Tribe Onycholipini Wollaston, 1873
Tribe Pentarthrini Lacordaire, 1865
Tribe Proecini Voss, 1956
Tribe Pseudapotrepini Champion, 1909
Tribe Rhyncolini Gistel, 1848
Subtribe Phloeophagina Voss, 1955
Subtribe Pseudomimina Voss, 1939
Subtribe Rhyncolina Gistel, 1848
Tribe Tapiromimini Voss, 1972
Subfamily Cryptorhynchinae Schönherr, 1825
Tribe Aedemonini Faust, 1898
Tribe Camptorhinini Lacordaire, 1865
Tribe Cryptorhynchini Schönherr, 1825
Subtribe Cryptorhynchina Schönherr, 1825
Subtribe Mecistostylina Lacordaire, 1865
Subtribe Tylodina Lacordaire, 1865
Tribe Gasterocercini Zherikhin, 1991
Tribe Psepholacini Lacordaire, 1865
Tribe Sophrorhinini Lacordaire, 1865
Tribe Torneumatini Bedel, 1884
Subfamily Cyclominae Schönherr, 1826
Tribe Amycterini Waterhouse, 1854
Tribe Aterpini Lacordaire, 1863 ***nomen protectum***
Subtribe Aterpina Lacordaire, 1863 ***nomen protectum***
Subtribe Rhadinosomina Lacordaire, 1863
Tribe Cyclomini Schönherr, 1826
Tribe Dichotrachelini Hoffmann, 1957
Tribe Hipporhinini Lacordaire, 1863
Tribe Listroderini LeConte, 1876
Tribe Notiomimetini Wollaston, 1873
Tribe Rhythirrinini Lacordaire, 1863
Subfamily Entiminae Schönherr, 1823
Tribe Agraphini Horn, 1876
Tribe Alophini LeConte, 1874
Tribe Anomophthalmini Morrone, 1998
Tribe Anypotactini Champion, 1911
Tribe Blosyrini Lacordaire, 1863
Tribe Brachyderini Schönherr, 1826
Tribe Celeuthetini Lacordaire, 1863
Subtribe Celeuthetina Lacordaire, 1863

Subtribe Isopterina Morimoto and Kojima, 2001
Tribe Cneorhinini Lacordaire, 1863
Tribe Cratopodini Hustache, 1919
Tribe Cylydrorhinini Lacordaire, 1863
Tribe Cyphicerini Lacordaire, 1863
Subtribe Acanthotrachelina Marshall, 1944
Subtribe Cyphicerina Lacordaire, 1863
Subtribe Mylacorrhinina Reitter, 1913
Subtribe Myllocerina Pierce, 1913
Subtribe Phytoscaphina Lacordaire, 1863
Tribe Ectemnorhinini Lacordaire, 1863
Tribe Elytrurini Marshall, 1956
Tribe Embrithini Marshall, 1942
Tribe Entimini Schönherr, 1823
Tribe Episomini Lacordaire, 1863
Tribe Eudiagogini LeConte, 1874
Tribe Eupholini Günther, 1943
Tribe Eustylini Lacordaire, 1863
Tribe Geonemini Gistel, 1848
Tribe Holcorhinini Desbrochers des Loges, 1898
Tribe Hormorini Horn, 1876
Tribe Laparocerini Lacordaire, 1863
Tribe Leptostethini Lacordaire, 1863
Tribe Lordopini Schönherr, 1823
Tribe Mesostylini Reitter, 1913
Tribe Myorhinini Marseul, 1863
Tribe Nastini Reitter, 1913
Tribe Naupactini Gistel, 1848 ***nomen protectum***
Tribe Nothognathini Marshall, 1916
Tribe Omiini Shuckard, 1839
Tribe Oosomini Lacordaire, 1863
Tribe Ophryastini Lacordaire, 1863
Tribe Ophtalmorrhynchini Hoffmann, 1965
Tribe Otiorhynchini Schönherr, 1826
Tribe Ottistirini Heller, 1925
Tribe Pachyrhynchini Schönherr, 1826
Tribe Peritelini Lacordaire, 1863
Tribe Phyllobiini Schönherr, 1826
Tribe Polycatini Marshall, 1956
Tribe Polydrusini Schönherr, 1823
Tribe Premnotrypini Kuschel, 1956
†Tribe Pristorhynchini Heer, 1847
Tribe Prypnini Lacordaire, 1863

Tribe Psallidiini Lacordaire, 1863
Tribe Rhyncogonini Sharp, 1919
Tribe Sciaphilini Sharp, 1891
Tribe Sitonini Gistel, 1848
Tribe Tanymecini Lacordaire, 1863
Subtribe Piazomiina Reitter, 1913
Subtribe Tainophthalmina Desbrochers des Loges, 1873
Subtribe Tanymecina Lacordaire, 1863
Tribe Tanyrhynchini Schönherr, 1826
Tribe Thecesternini Lacordaire, 1863
Tribe Trachyphloeini Gistel, 1848
Subtribe Trachyphilina Voss, 1948
Subtribe Trachyphloeina Gistel, 1848
Tribe Tropiphorini Marseul, 1863
Tribe Typhlorhinini Kuschel, 1954
Subfamily Hyperinae Marseul, 1863 (1848)
Tribe Cepurini Capiomont, 1867
Tribe Hyperini Marseul, 1863 (1848)
Subfamily Lixinae Schönherr, 1823
Tribe Cleonini Schönherr, 1826 ***nomen protectum***
Tribe Lixini Schönherr, 1823
Tribe Rhinocyllini Lacordaire, 1863
Subfamily Mesoptiliinae Lacordaire, 1863
Tribe Carciliini Pierce, 1916
Tribe Laemosaccini Lacordaire, 1865
Tribe Magdalidini Pascoe, 1870 ***nomen protectum***
Tribe Mesoptiliini Lacordaire, 1863
Subfamily Molytinae Schönherr, 1823
Tribe Anoplini Bedel, 1883
Tribe Amalactini Lacordaire, 1863
Tribe Aminyopini Voss, 1956
Tribe Amorphocerini Voss, 1939
Tribe Anchonini Imhoff, 1856
Tribe Brachyceropseini Aurivillius, 1926
Tribe Cholini Schönherr, 1825
Subtribe Cholina Schönherr, 1825
Subtribe Cholomina Vaurie, 1974
Subtribe Rhinastina Vaurie, 1973
Tribe Cleogonini Gistel, 1848
Tribe Conotrachelini Jekel, 1865
Tribe Cycloterini Lacordaire, 1863
Subtribe Cycloterina Lacordaire, 1863
Subtribe Thrombosternina Voss, 1965

Tribe Dinomorphini Lacordaire, 1863
Tribe Emphyastini Lacordaire, 1863
Tribe Euderini Lacordaire, 1865
Tribe Galloisiini Morimoto, 1962
Tribe Guioperini Lacordaire, 1865
Tribe Hylobiini Kirby, 1837
Subtribe Epistrophina Marshall, 1932
Subtribe Hylobiina Kirby, 1837
Tribe Ithyporini Lacordaire, 1865
Subtribe Colobodina Voss, 1958
Subtribe Ithyporina Lacordaire, 1865
Subtribe Sclerocardiina Lacordaire, 1865
Tribe Itini Reitter, 1913
Tribe Juanorhinini Aurivillius, 1931
Tribe Lepyrini Kirby, 1837
Tribe Lithinini Lacordaire, 1863
Subtribe Lithinina Lacordaire, 1863
Subtribe Rhytidophloeina Voss, 1963
Tribe Lymantini Lacordaire, 1865
Tribe Mecysolobini Reitter, 1913
Tribe Metatygini Pascoe, 1888
Tribe Molytini Schönherr, 1823
Subtribe Leiosomatina Reitter, 1913
Subtribe Molytina Schönherr, 1823
Subtribe Plinthina Lacordaire, 1863
Subtribe Typoderina Voss, 1965
Tribe Nettarhinini Lacordaire, 1865
Tribe Pacholenini Lacordaire, 1863
Tribe Paipalesomini Marshall, 1932
Tribe Petalochilini Lacordaire, 1863
Tribe Phoenicobatini Champion, 1914
Tribe Phrynixini Kuschel, 1964
Tribe Pissodini Gistel, 1848
Subtribe Cotasteromimina Morimoto, 1962
Subtribe Orthorhinina Jekel, 1865
Subtribe Pissodina Gistel, 1848
Tribe Sternechini Lacordaire, 1863
Tribe Styanacini Chûjô and Voss, 1960
Tribe Trachodini Gistel, 1848
Tribe Trigonocolini Lacordaire, 1863
Tribe Trypetidini Lacordaire, 1865
Subfamily Orobitidinae Thomson, 1859
Subfamily Xiphaspidinae Marshall, 1920

Subfamily Scolytinae Latreille, 1804

Tribe Amphiscolytini Mandelshtam and Beaver, 2003
Tribe Bothrosternini Blandford, 1896
Tribe Cactopinini Chamberlin, 1939
Tribe Carphodicticini Wood, 1971
Tribe Coptonotini Chapuis, 1869
Tribe Corthylini LeConte, 1876
  Subtribe Corthylina LeConte, 1876
  Subtribe Pityophthorina Eichhoff, 1878
Tribe Cryphalini Lindemann, 1877
Tribe Crypturgini LeConte, 1876
†Tribe Cylindrobrotini Kirejtshuk, Azar, Beaver, Mandelshtam and Nel, 2009
Tribe Diamerini Hagedorn, 1909
Tribe Dryocoetini Lindemann, 1877
Tribe Hexacolini Eichhoff, 1878
Tribe Hylastini LeConte, 1876
Tribe Hylesinini Erichson, 1836
Tribe Hylurgini Gistel, 1848
Tribe Hyorrhynchini Hopkins, 1915
Tribe Hypoborini Nüsslin, 1911
Tribe Ipini Bedel, 1888
Tribe Micracidini LeConte, 1876
Tribe Phloeosinini Nüsslin, 1912
Tribe Phloeotribini Chapuis, 1869
Tribe Phrixosomatini Wood, 1978
Tribe Polygraphini Chapuis, 1869
Tribe Premnobiini Browne, 1962
Tribe Scolytini Latreille, 1804
Tribe Scolytoplatypodini Blandford, 1893
Tribe Xyleborini LeConte, 1876
Tribe Xyloctonini Eichhoff, 1878
Tribe Xyloterini LeConte, 1876

Subfamily Platypodinae Shuckard, 1839

Tribe Mecopelmini Thompson, 1992
Tribe Platypodini Shuckard, 1839
Tribe Schedlariini Wood and Bright, 1992
Tribe Tesserocerini Strohmeyer, 1914
  Subtribe Diapodina Strohmeyer, 1914
  Subtribe Tesserocerina Strohmeyer, 1914

Catalogue of Coleoptera family-group names

Order COLEOPTERA

†Suborder PROTOCOLEOPTERA

†Superfamily Tshekardocoleoidea Rohdendorf, 1944

Tshekardocoleidae Rohdendorf, 1944: 252 [stem: *Tshekardocole-*]. Type genus: *Tshekardocoleus* Rohdendorf, 1944.r

†Family Tshekardocoleidae Rohdendorf, 1944

Tshekardocoleidae Rohdendorf, 1944: 252 [stem: *Tshekardocole-*]. Type genus: *Tshekardocoleus* Rohdendorf, 1944.

Uralocoleidae Zalesskiy, 1947: 857 [stem: *Uralocole-*]. Type genus: *Uralocoleus* Zalessky, 1947.

†Family Labradorocoleidae Ponomarenko, 1969

Labradorocoleidae Ponomarenko, 1969b: 307 [stem: *Labradorocole-*]. Type genus: *Labradorocoleus* Ponomarenko, 1969.

†Family Oborocoleidae Kukalová, 1969

Oborocoleidae Kukalová, 1969: 155 [stem: *Oborocole-*]. Type genus: *Oborocoleus* Kukalová, 1969.

†Superfamily Permocupedoidea Martynov, 1933

Permocupidae Martynov, 1933: 85 [stem: *Permocuped-*]. Type genus: *Permocupes* Martynov, 1933.

†Family Permocupedidae Martynov, 1933

Permocupidae Martynov, 1933: 85 [stem: *Permocuped-*]. Type genus: *Permocupes* Martynov, 1933. Comment: incorrect original stem formation, not in prevailing usage.

Kaltanocoleidae Rohdendorf, 1961: 397 [stem: *Kaltanocole-*]. Type genus: *Kaltanocoleus* Rohdendorf, 1961.

†Family Taldycupedidae Rohdendorf, 1961

Taldycupidae Rohdendorf, 1961: 412 [stem: *Taldycuped-*]. Type genus: *Taldycupes* Rohdendorf, 1961. Comment: usage of Trachycupidae by Fujiyama (1973: 375) was in error for Taldycupidae (Ponomarenko pers. comm. August 2009); incorrect original stem formation, not in prevailing usage.

†Superfamily Permosynoidea Tillyard, 1924

Permosynidae Tillyard, 1924: 431 [stem: *Permosyn-*]. Type genus: *Permosyne* Tillyard, 1924.

†Family Ademosynidae Ponomarenko, 1968

Ademosynidae Ponomarenko, 1968: 128 [stem: *Ademosyn-*]. Type genus: *Ademosyne* Handlirsch, 1906.

†Family Permosynidae Tillyard, 1924

Permosynidae Tillyard, 1924: 431 [stem: *Permosyn-*]. Type genus: *Permosyne* Tillyard, 1924.

Suborder ARCHOSTEMATA

Family Crowsoniellidae Iablokoff-Khnzorian, 1983

Crowsoniellidae Iablokoff-Khnzorian, 1983: 65 [stem: *Crowsoniell-*]. Type genus: *Crowsoniella* Pace, 1975.

Family Cupedidae Laporte, 1836

Cupesidae Laporte, 1836: 56 [stem: *Cuped-*]. Type genus: *Cupes* Fabricius, 1801.

Subfamily Priacminae Crowson, 1962

Priacmini Crowson, 1962: 152 [stem: *Priacm-*]. Type genus: *Priacma* J. L. LeConte, 1874.

†Subfamily Mesocupedinae Ponomarenko, 1969

Mesocupedini Ponomarenko, 1969a: 105 [stem: *Mesocuped-*]. Type genus: *Mesocupes* Martynov, 1926.

Subfamily Cupedinae Laporte, 1836

Cupesidae Laporte, 1836: 56 [stem: *Cuped-*]. Type genus: *Cupes* Fabricius, 1802. Comment: incorrect original stem formation, not in prevailing usage.

Family Micromalthidae Barber, 1913

Micromalthidae Barber, 1913: 185 [stem: *Micromalth-*]. Type genus: *Micromalthus* J. L. LeConte, 1878.

Family Ommatidae Sharp and Muir, 1912

Ommadidae Sharp and Muir, 1912: 521 [stem: *Ommat-*]. Type genus: *Omma* Newman, 1839.

†Subfamily Brochocoleinae Hong, 1982

Brochocoleidae Hong, 1982: 100 [stem: *Brochocole-*]. Type genus: *Brochocoleus* Hong, 1982.

Subfamily Tetraphalerinae Crowson, 1962

Tetraphalerini Crowson, 1962: 152 [stem: *Tetraphaler-*]. Type genus: *Tetraphalerus* C. O. Waterhouse, 1901.

Subfamily Ommatinae Sharp and Muir, 1912

Ommadidae Sharp and Muir, 1912: 521 [stem: *Ommat-*]. Type genus: *Omma* Newman, 1839.

†Tribe Lithocupedini Ponomarenko, 1969

Lithocupedini Ponomarenko, 1969a: 82 [stem: *Lithocuped-*]. Type genus: *Lithocupes* Ponomarenko, 1966.

†Tribe Notocupedini Ponomarenko, 1966

Notocupedini Ponomarenko, 1966: 60 [stem: *Notocup-*]. Type genus: *Notocupes* Ponomarenko, 1966.

Tribe Ommatini Sharp and Muir, 1912

Ommadidae Sharp and Muir, 1912: 521 [stem: *Ommat-*]. Type genus: *Omma* Newman, 1839. Comment: incorrect original stem formation, not in prevailing usage.

Family Jurodidae Ponomarenko, 1985

Jurodidae Ponomarenko, 1985: 53 [stem: *Jurod-*]. Type genus: *Jurodes* Ponomarenko, 1985.

Sikhotealiniidae Lafer, 1996: 390 [stem: *Sikhotealini-*]. Type genus: *Sikhotealinia* Lafer, 1996.

†Family Triadocupedidae Ponomarenko, 1966

Triadocupedinae Ponomarenko, 1966: 48 [stem: *Triadocuped-*]. Type genus: *Triadocupes* Ponomarenko, 1966.

†Family Magnocoleidae Hong, 1998

Magnocoleidae Hong, 1998: 41 [stem: *Magnocole-*]. Type genus: *Magnocoleus* Hong, 1998.

†Family Obrieniidae Zherikhin and Gratshev, 1994

Obrieniidae Zherikhin and Gratshev, 1994: 51 [stem: *Obrieni-*]. Type genus: *Obrienia* Zherikhin and Gratshev, 1994. Comment: precedence (Obrieniidae Zherikhin and Gratshev, 1994 vs Kararhynchidae Zherikhin and Gratshev, 1994) given to taxon originally proposed at the higher rank (Art. 24.1).

†Subfamily Kararhynchinae Zherikhin and Gratshev, 1994

Kararhynchinae Zherikhin and Gratshev, 1994: 58 [stem: *Kararhynch-*]. Type genus: *Kararhynchus* Zherikhin and Gratshev, 1994.

†Tribe Kararhynchini Zherikhin and Gratshev, 1994

Kararhynchinae Zherikhin and Gratshev, 1994: 58 [stem: *Kararhynch-*]. Type genus: *Kararhynchus* Zherikhin and Gratshev, 1994.

†Tribe Kenderlykaini Legalov, 2009

Kenderlykanini Legalov, 2009c: 285 [stem: *Kenderlyka-*]. Type genus: *Kenderlyka* Zherikhin and Gratshev, 1994. Comment: incorrect original stem formation, not in prevailing usage.

†Subfamily Obrieniinae Zherikhin and Gratshev, 1994

Obrieniidae Zherikhin and Gratshev, 1994: 51 [stem: *Obrieni-*]. Type genus: *Obrienia* Zherikhin and Gratshev, 1994.

Suborder MYXOPHAGA

†Superfamily Asiocoleoidea Rohdendorf, 1961

Asiocoleidae Rohdendorf, 1961: 396 [stem: *Asiocole-*]. Type genus: *Asiocoleus* Rohdendorf, 1961.

†Family Asiocoleidae Rohdendorf, 1961

Asiocoleidae Rohdendorf, 1961: 396 [stem: *Asiocole-*]. Type genus: *Asiocoleus* Rohdendorf, 1961.

†Family Tricoleidae Ponomarenko, 1969

Tricoleidae Ponomarenko, 1969a: 138 [stem: *Tricole-*]. Type genus: *Tricoleus* Ponomarenko, 1969.

†Superfamily Rhombocoleoidea Rohdendorf, 1961

Rhombocoleidae Rohdendorf, 1961: 432 [stem: *Rhombocole-*]. Type genus: *Rhombocoleus* Rohdendorf, 1961.

†Family Rhombocoleidae Rohdendorf, 1961

Rhombocoleidae Rohdendorf, 1961: 432 [stem: *Rhombocole-*]. Type genus: *Rhombocoleus* Rohdendorf, 1961.

†Superfamily Schizophoroidea Ponomarenko, 1968

Schizophoridae Ponomarenko, 1968: 130 [stem: *Schizophor-*]. Type genus: *Schizophorus* Ponomarenko, 1968. Comment: usage of this name conserved over Schizocoleoidea Rohdendorf, 1961 (Art. 35.5).

†Family Schizophoridae Ponomarenko, 1968

Schizophoridae Ponomarenko, 1968: 130 [stem: *Schizophor-*]. Type genus: *Schizophorus* Ponomarenko, 1968.

†Family Catiniidae Ponomarenko, 1968

Catiniidae Ponomarenko, 1968: 137 [stem: *Catini-*]. Type genus: *Catinius* Ponomarenko, 1968.

†Family Schizocoleidae Rohdendorf, 1961

Curculiopsidae Martynov, 1937: 39 [stem: *Curculiopse-*]. Type genus: *Curculiopsis* Martynov, 1937 [preoccupied genus name, not *Curculiopsis* Handlirsch, 1907 [fossil Curculionoidea]; syn. of *Aenigmocoleus* Rohdendorf, 1961 or *Rossocoleus* Rohdendorf, 1961]. Comment: permanently invalid (Art. 39): based on preoccupied type genus; incorrect original stem formation, not in prevailing usage.

Schizocoleidae Rohdendorf, 1961: 438 [stem: *Schizocole-*]. Type genus: *Schizocoleus* Rohdendorf, 1961.

Superfamily Lepiceroidea Hinton, 1936 (1882)

Lepiceridae Hinton, 1936: 473 [stem: *Lepicer-*]. Type genus: *Lepicerus* Motschulsky, 1855. Comment: use of family-group name conserved over Cyathoceroidea Sharp, 1882 (Art. 40.2) (see Lawrence and Newton 1995: 804).

Family Lepiceridae Hinton, 1936 (1882)

Cyathoceridae Sharp, 1882: 141 [stem: *Cyathocer-*]. Type genus: *Cyathocerus* Sharp, 1882 [syn. of *Lepicerus* Motschulsky, 1855]. Comment: use of younger name Lepiceridae Hinton, 1936 conserved over this name (Art. 40.2) (see Lawrence and Newton 1995: 804).

Lepiceridae Hinton, 1936: 473 [stem: *Lepicer-*]. Type genus: *Lepicerus* Motschulsky, 1855. Comment: use of family-group name conserved over Cyathoceridae Sharp, 1882 (Art. 40.2) (see Lawrence and Newton 1995: 804).

†Haplochelidae Kirejtshuk and Poinar, 2006: 156 [stem: *Haplochel-*]. Type genus: *Haplochelus* Kirejtshuk and Poinar, 2006. Comment: synonymy with Lepiceridae by Ge et al. (2010: 336).

Superfamily Sphaeriusoidea Erichson, 1845

Sphaerina Erichson, 1845: 38 [stem: *Sphaerius-*]. Type genus: *Sphaerius* Waltl, 1838 [originally placed on the Official Index of Rejected and Invalid Generic Names in Zoology by the Commission (ICZN 1985d) but later placed on the Official List of Generic Names in Zoology (ICZN 2000)]. Comment: correct stem determined to be *Sphaerius-* and Sphaeriusidae Erichson, 1845 placed on the Official List of Family-Group Names in Zoology (ICZN 2000).

Family Torridincolidae Steffan, 1964

Torridincolidae Steffan, 1964: 199 [stem: *Torridincol-*]. Type genus: *Torridincola* Steffan, 1964.

Subfamily Torridincolinae Steffan, 1964

Torridincolidae Steffan, 1964: 199 [stem: *Torridincol-*]. Type genus: *Torridincola* Steffan, 1964.

PTYOPTERINAE M. Abdullah, 1974: 961 [stem: *Ptyopteryg-*]. Type genus: *Ptyopteryx* H. Reichardt and C. Costa, 1967 [preoccupied genus name, not *Ptyopteryx* Kolenati, 1848 [Trichoptera]; syn. of *Iapir* Py-Daniel et al., 1993]. Comment: permanently invalid (Art. 39): based on preoccupied type genus; incorrect original stem formation, not in prevailing usage.

Subfamily DELEVEINAE Endrödy-Younga, 1997

DELEVEINAE Endrödy-Younga, 1997: 317 [stem: *Deleve-*]. Type genus: *Delevea* H. Reichardt, 1976.

Family HYDROSCAPHIDAE LeConte, 1874

HYDROSCAPHIDAE J. L. LeConte, 1874a: 45 [stem: *Hydroscaph-*]. Type genus: *Hydroscapha* J. L. LeConte, 1874.

Family SPHAERIUSIDAE Erichson, 1845

SPHAERINA Erichson, 1845: 38 [stem: *Sphaerius-*]. Type genus: *Sphaerius* Waltl, 1838 [originally placed on the Official Index of Rejected and Invalid Generic Names in Zoology by the Commission (ICZN 1985d) but later placed on the Official List of Generic Names in Zoology (ICZN 2000)]. Comment: family-group name and its type genus were rejected by the Commission (ICZN 1985d) but rescinded afterwards (ICZN 2000); family-group name placed on the Official List of Family-Group Names in Zoology and correct stem ruled to be *Sphaerius-* (ICZN 2000).

MICROSPORIDAE Crotch, 1873a: 78 [stem: *Microspor-*]. Type genus: *Microsporus* Kolenati, 1846 [placed on the Official List of Generic Names in Zoology (ICZN 1985d)]. Comment: name originally placed on the Official List of Family-Group Names in Zoology as "MICROSPORIDAE H. Reichardt, 1976" (ICZN 1985d) but author and year subsequently corrected to Crotch, 1873 (ICZN 2000).

Suborder ADEPHAGA

†Family TRITARSIDAE Hong, 2002

TRITARSUSIDAE Hong, 2002: 102 [stem: *Tritars-*]. Type genus: *Tritarsus* Hong, 2002. Comment: incorrect original stem formation, not in prevailing usage.

Family GYRINIDAE Latreille, 1810

GYRINITES Latreille, 1810: 141 [stem: *Gyrin-*]. Type genus: *Gyrinus* Geoffroy, 1762 [placed on the Official List of Generic Names in Zoology (ICZN 1994a)].

Subfamily SPANGLEROGYRINAE Folkerts, 1979

SPANGLEROGYRINAE Folkerts, 1979: 7 [stem: *Spanglerogyr-*]. Type genus: *Spanglerogyrus* Folkerts, 1979.

Subfamily Gyrininae Latreille, 1810

Gyrinites Latreille, 1810: 141 [stem: *Gyrin-*]. Type genus: *Gyrinus* Geoffroy, 1762 [placed on the Official List of Generic Names in Zoology (ICZN 1994a)].

Tribe Enhydrini Régimbart, 1882

Enhydrini Régimbart, 1882: 392 [stem: *Enhydr-*]. Type genus: *Enhydrus* Laporte, 1834 [placed on the Official List of Generic Names in Zoology (ICZN 1964)]. Comment: usage of this name conserved over Dineutini Desmarest, 1851 (Art. 35.5).

Subtribe Dineutina Desmarest, 1851

Dineutides Desmarest, 1851: 223 [stem: *Dineut-*]. Type genus: *Dineutes* W. S. MacLeay, 1825.

Subtribe Enhydrina Régimbart, 1882

Enhydrini Régimbart, 1882: 392 [stem: *Enhydr-*]. Type genus: *Enhydrus* Laporte, 1834 [placed on the Official List of Generic Names in Zoology (ICZN 1964)]. Comment: junior homonym of Enhydrini Gray, 1825 (type genus *Enhydra* Fleming, 1822) in Mammalia; an application was recently submitted to the Commission by Özdikmen and Darilmaz (2010; see Appendix 6) to remove the homonymy (Art. 55.3.1).

Tribe Gyrinini Latreille, 1810

Gyrinites Latreille, 1810: 141 [stem: *Gyrin-*]. Type genus: *Gyrinus* Geoffroy, 1762 [placed on the Official List of Generic Names in Zoology (ICZN 1994a)].

Subtribe Gyrinina Latreille, 1810

Gyrinites Latreille, 1810: 141 [stem: *Gyrin-*]. Type genus: *Gyrinus* Geoffroy, 1762 [placed on the Official List of Generic Names in Zoology (ICZN 1994a)].

Subtribe Heterogyrina Brinck, 1956

Heterogyrini P. Brinck, 1956: 37, in key [stem: *Heterogyr-*]. Type genus: *Heterogyrus* Legros, 1953.

Tribe Orectochilini Régimbart, 1882

Orectochilini Régimbart, 1882: 391 [stem: *Orectochil-*]. Type genus: *Orectochilus* Dejean, 1833.

Family Trachypachidae Thomson, 1857

Trachypachini C. G. Thomson, 1857: 5 [stem: *Trachypach-*]. Type genus: *Trachypachus* Motschulsky, 1844.

†Subfamily Eodromeinae Ponomarenko, 1977

Eodromeinae Ponomarenko, 1977: 46 [stem: *Eodrome-*]. Type genus: *Eodromeus* Ponomarenko, 1977. Comment: the type genus originally included four species: *antiquus, stenalis, dissectus* and *major*; Ponomarenko (1977: 66) designated "*fasciatus*" as the type species of the genus which was a *lapsus calami* for *dissectus* (see Carpenter 1992: 291); this genus is considered here as being available in Ponomarenko (1977: 66) and not Ponomarenko (in Carpenter 1992: 291); this action enables us to treat the family-group name Eodromeinae as available from the original description in Ponomarenko (1977: 46).

Leptopodocoleidae Hong, 1982: 118 [stem: *Leptopodocole-*]. Type genus: *Leptopodocoleus* Hong, 1982.

Subfamily Trachypachinae Thomson, 1857

Trachypachini C. G. Thomson, 1857: 5 [stem: *Trachypach-*]. Type genus: *Trachypachus* Motschulsky, 1844.

*Systolosomini Erwin, 1985: 467 [stem: *Systolosomat-*]. Type genus: *Systolosoma* Solier, 1849. Comment: unavailable family-group name, proposed after 1930 without description or bibliographic reference to such a description (Art. 13.1); incorrect original stem formation, not in prevailing usage.

Systolosomini Erwin, 1991: 4, in key [stem: *Systolosomat-*]. Type genus: *Systolosoma* Solier, 1849. Comment: incorrect original stem formation, not in prevailing usage.

Family Rhysodidae Laporte, 1840

Rhysodites Laporte, 1840a: 291 [stem: *Rhysod-*]. Type genus: *Rhysodes* Germar, 1822.

Tribe Leoglymmiini Bell and Bell, 1978

Leoglymmiina R. T. Bell and J. R. Bell, 1978: 53 [stem: *Leoglymmi-*]. Type genus: *Leoglymmius* Bell and Bell, 1978.

Tribe Dhysorini Bell and Bell, 1978

Dhysorina R. T. Bell and J. R. Bell, 1978: 53 [stem: *Dhysor-*]. Type genus: *Dhysores* Grouvelle, 1903.

Tribe Medisorini Bell and Bell, 1987

Medisorina R. T. Bell and J. R. Bell, 1987: 287 [stem: *Medisor-*]. Type genus: *Medisores* Bell and Bell, 1987.

Tribe Rhysodini Laporte, 1840

Rhysodites Laporte, 1840a: 291 [stem: *Rhysod-*]. Type genus: *Rhysodes* Germar, 1822. Comment: original vernacular name available (Art. 11.7.2): first used in latinized form by Agassiz (1846b: 326, as Rhyssodeoidae [incorrect stem formation]), generally accepted as in R. T. Bell (2003: 78, as Rhysodidae).

Tribe Clinidiini Bell and Bell, 1978

Clinidiina R. T. Bell and J. R. Bell, 1978: 59 [stem: *Clinidi-*]. Type genus: *Clinidium* Kirby, 1830.

Tribe Omoglymmiini Bell and Bell, 1978

Omoglymmiina R. T. Bell and J. R. Bell, 1978: 66 [stem: *Omoglymmi-*]. Type genus: *Omoglymmius* Ganglbauer, 1891.

Tribe Sloanoglymmiini Bell and Bell, 1991

Sloanoglymmiina R. T. Bell and J. R. Bell, 1991: 183 [stem: *Sloanoglymmi-*]. Type genus: *Sloanoglymmius* Bell and Bell, 1991.

Family Carabidae Latreille, 1802

Carabici Latreille, 1802: 80 [stem: *Carab-*]. Type genus: *Carabus* Linnaeus, 1758 [placed on the Official List of Generic Names in Zoology (ICZN 1950)]. Comment: First Reviser (Carabidae Latreille, 1802 vs Cicindelidae Latreille, 1802 vs Elaphridae Latreille, 1802) not determined, current usage maintained.

†Subfamily Protorabinae Ponomarenko, 1977

Protorabinae Ponomarenko, 1977: 71 [stem: *Protorab-*]. Type genus: *Protorabus* Ponomarenko, 1977.

†Subfamily Conjunctiinae Ponomarenko, 1977

Conjunctiini Ponomarenko, 1977: 85 [stem: *Conjuncti-*]. Type genus: *Conjunctia* Ponomarenko, 1977.

Subfamily Nebriinae Laporte, 1834

Nebriidae Laporte, 1834b: 90 [stem: *Nebri-*]. Type genus: *Nebria* Latreille, 1802.

Tribe Nebriini Laporte, 1834

Nebriidae Laporte, 1834b: 90 [stem: *Nebri-*]. Type genus: *Nebria* Latreille, 1802.

Tribe Notiokasiini Kavanaugh and Nègre, 1983

*Notiokasiini Erwin, 1979: 578 [stem: *Notiokasi-*]. Type genus: *Notiokasis* Kavanaugh and Nègre, 1983. Comment: family-group name unavailable (Art. 11.7.1.1): not based on an available genus name at the time.

Notiokasiini Kavanaugh and Nègre, 1983: 551 [stem: *Notiokasi-*]. Type genus: *Notiokasis* Kavanaugh and Nègre, 1983.

Tribe Notiophilini Motschulsky, 1850

*Notiophiles Motschulsky, 1849: 54 [stem: *Notiophil-*]. Type genus: *Notiophilus* Duméril, 1805. Comment: original vernacular name unavailable (Art.

11.7.2): subsequently used in latinized form but not generally attributed to Motschulsky (1849).

NOTIOPHILI Motschulsky, 1850: 16 [stem: *Notiophil-*]. Type genus: *Notiophilus* Duméril, 1805.

Tribe OPISTHIINI Dupuis, 1912

OPISTHIINAE Dupuis, 1912: 1 [stem: *Opisthi-*]. Type genus: *Opisthius* Kirby, 1837.

Tribe PELOPHILINI Kavanaugh, 1996

PELOPHILINI Kavanaugh, 1996: 35 [stem: *Pelophil-*]. Type genus: *Pelophila* Dejean, 1821.

Subfamily CICINDINAE Csiki, 1927

CICINDINI Csiki, 1927: 425 [stem: *Cicind-*]. Type genus: *Cicindis* Bruch, 1908.

Subfamily CICINDELINAE Latreille, 1802

CICINDELETAE Latreille, 1802: 77 [stem: *Cicindel-*]. Type genus: *Cicindela* Linné, 1758.

Tribe AMBLYCHEILINI Csiki, 1903

AMBLYCHILINAE Csiki, 1903: 124 [stem: *Amblycheil-*]. Type genus: *Amblycheila* Say, 1830 [as *Amblychila*, unjustified emendation of type genus name by Agassiz (1846b), not in prevailing usage]. Comment: incorrect original stem formation, not in prevailing usage.

OMITES W. Horn, 1907: 466 [stem: *Om-*]. Type genus: *Omus* Eschscholtz, 1829.

Tribe CICINDELINI Latreille, 1802

CICINDELETAE Latreille, 1802: 77 [stem: *Cicindel-*]. Type genus: *Cicindela* Linnaeus, 1758.

Subtribe APTEROESSINA Rivalier, 1971

APTEROESSINA Rivalier, 1971: 143 [stem: *Apteroess-*]. Type genus: *Apteroessa* Hope, 1838.

Subtribe CICINDELINA Latreille, 1802

CICINDELETAE Latreille, 1802: 77 [stem: *Cicindel-*]. Type genus: *Cicindela* Linnaeus, 1758.

Subtribe DROMICINA Thomson, 1859

DROMICITAE J. Thomson, 1859: 89 [stem: *Dromic-*]. Type genus: *Dromica* Dejean, 1826.

CALEDONICINI W. Horn, 1893: 324 [stem: *Caledonic-*]. Type genus: *Caledonica* Chaudoir, 1861.

Odontochilini W. Horn, 1899: 41 [stem: *Odontocheil-*]. Type genus: *Odontocheila* Laporte, 1834 [as *Odontochila*, unjustified emendation of type genus name by Agassiz (1846b), not in prevailing usage]. Comment: incorrect original stem formation, not in prevailing usage.

Prepusini W. Horn, 1899: 44 [stem: *Prepus-*]. Type genus: *Prepusa* Chaudoir, 1850 [syn. of *Eulampra* Chaudoir, 1848].

Euryodini W. Horn, 1899: 37 [stem: *Euryod-*]. Type genus: *Euryoda* Lacordaire, 1842 [syn. of *Prothyma* Hope, 1838].

Prothymini W. Horn, 1906: 86 [stem: *Prothym-*]. Type genus: *Prothyma* Hope, 1838.

Subtribe Iresiina Rivalier, 1971

Euprosopini W. Horn, 1893: 324 [stem: *Euprosop-*]. Type genus: *Euprosopus* Dejean, 1825.

Eucalliini W. Horn, 1893: 324 [stem: *Eucalli-*]. Type genus: *Eucallia* Guérin-Méneville, 1844 [syn. of *Callidema* Guérin-Méneville, 1843].

Distypsiderini W. Horn, 1893: 324 [stem: *Distipsider-*]. Type genus: *Distipsidera* Westwood, 1837 [as *Distypsidera*, unjustified emendation of type genus name by Gemminger and Harold (1868a: 32), not in prevailing usage]. Comment: incorrect original stem formation, not in prevailing usage.

Iresina Rivalier, 1971: 138 [stem: *Iresi-*]. Type genus: *Iresia* Dejean, 1829. Comment: although this is not the oldest name for the subtribe, we recommend that an application be sent to the Commission in order to conserve usage of Iresiina Rivalier, 1971 over the three older names proposed by W. Horn in 1893 which have not been used as valid recently to our knowledge; incorrect original stem formation, not in prevailing usage.

Subtribe Theratina Horn, 1893

Theratidae W. Horn, 1893: 324 [stem: *Therat-*]. Type genus: *Therates* Latreille, 1816.

Tribe Collyridini Brullé, 1834

Collyriens Brullé, 1834: 96 [stem: *Collyrid-*]. Type genus: *Collyris* Fabricius, 1801.

Subtribe Collyridina Brullé, 1834

Collyriens Brullé, 1834: 96 [stem: *Collyrid-*]. Type genus: *Collyris* Fabricius, 1801. Comment: original vernacular name available (Art. 11.7.2): first used in latinized form by Hope (1838a: 30, as Collyridae [incorrect stem formation]), generally accepted as in Puchkov and Matalin (2003: 116, as Collyridina); incorrect original stem formation, not in prevailing usage.

TRACHELONIADAE Gistel, 1850: 75 [stem: *Tracheloni-*]. Type genus: *Trachelonia* Gistel, 1850 [syn. of *Collyris* Fabricius, 1801].

*COLLIURIDES Motschulsky, 1855: 34 [stem: *Colliur-*]. Type genus: *Colliuris* sensu Latreille, 1802 [not *Colliuris* DeGeer, 1774; syn. of *Collyris* Fabricius, 1801]. Comment: original vernacular name unavailable (Art. 11.7.2): not subsequently latinized.

Subtribe TRICONDYLINA Naviaux, 1991

TRICONDYLINA Naviaux, 1991: 219 [stem: *Tricondyl-*]. Type genus: *Tricondyla* Latreille, 1822.

Tribe CTENOSTOMATINI Laporte, 1834

CTENOSTOMIDAE Laporte, 1834b: 38 [stem: *Ctenostomat-*]. Type genus: *Ctenostoma* Klug, 1821. Comment: incorrect original stem formation, not in prevailing usage.

Tribe MANTICORINI Laporte, 1834

MANTICORIDAE Laporte, 1834b: 33 [stem: *Manticor-*]. Type genus: *Manticora* Fabricius, 1781.

Tribe MEGACEPHALINI Laporte, 1834

MEGACEPHALIDAE Laporte, 1834b: 33 [stem: *Megacephal-*]. Type genus: *Megacephala* Latreille, 1802.

OXYCHEILITES J. Thomson, 1857a: 17 [stem: *Oxycheil-*]. Type genus: *Oxycheila* Dejean, 1825. Comment: original vernacular name available (Art. 11.7.2): first used in latinized form by Chaudoir (1861a: 326, as OXYCHILINI [incorrect stem formation]), generally accepted as in Fleutiaux (1892a: 13, as OXYCHILINI [incorrect stem formation]); the stem *Oxychil-* should not be used for this taxon in order to avoid homonymy with OXYCHILINI Hesse, 1927 (type genus *Oxychilus* Fitzinger 1833), currently being used as valid in Mollusca.

PLATYCHILIDAE W. Horn, 1893: 325 [stem: *Platychil-*]. Type genus: *Platychile* W. S. MacLeay, 1825.

TETRACHAE Leng and Mutchler, 1916: 683 [stem: *Tetrach-*]. Type genus: *Tetracha* Hope, 1838.

Subfamily CARABINAE Latreille, 1802

CARABICI Latreille, 1802: 80 [stem: *Carab-*]. Type genus: *Carabus* Linnaeus, 1758.

Tribe CARABINI Latreille, 1802

CARABICI Latreille, 1802: 80 [stem: *Carab-*]. Type genus: *Carabus* Linnaeus, 1758.

CALOSOMII Bonelli, 1810: Tabula Synoptica [stem: *Calosomat-*]. Type genus: *Calosoma* Weber, 1801. Comment: incorrect original stem formation, not in prevailing usage.

Proceridae Laporte, 1834b: 87 [stem: *Procer-*]. Type genus: *Procerus* Dejean, 1821.

Callisthenisidae Gistel, 1848: [2] [stem: *Callisthen-*]. Type genus: *Callisthenes* Fischer von Waldheim, 1820. Comment: incorrect original stem formation, not in prevailing usage.

*Cechenogenici Morawitz, 1889: 40 [stem: *Cechen-*]. Type genus: *Cechenus* Fischer von Waldheim, 1822. Comment: family-group name unavailable (Art. 11.7.1.1): original name not proposed as a noun.

*Tribacogenici Morawitz, 1889: 40 [stem: *Tribac-*]. Type genus: *Tribax* Fischer von Waldheim, 1817. Comment: family-group name unavailable (Art. 11.7.1.1): original name not proposed as a noun.

*Procrustogenici Morawitz, 1889: 40 [stem: *Procrust-*]. Type genus: *Procrustes* Bonelli, 1810. Comment: family-group name unavailable (Art. 11.7.1.1): original name not proposed as a noun.

Procrustogenici Roeschke, 1898: 285 [stem: *Procrust-*]. Type genus: *Procrustes* Bonelli, 1810.

Aplothoracina Lapouge, 1927: 45 [stem: *Aplothorac-*]. Type genus: *Aplothorax* G. R. Waterhouse, 1842.

*Callistheniens Lapouge, 1927: 47 [stem: *Callisthen-*]. Type genus: *Callisthenes* Fischer von Waldheim, 1820. Comment: original vernacular name unavailable (Art. 11.7.2): proposed after 1899; family-group name proposed as new without reference to Callisthenisidae Gistel, 1848.

Cechenogenici Csiki, 1927: 110 [stem: *Cechen-*]. Type genus: *Cechenus* Fischer von Waldheim, 1822.

Tribacogenici Csiki, 1927: 121 [stem: *Tribac-*]. Type genus: *Tribax* Fischer von Waldheim, 1817.

Megadontici Csiki, 1927: 60 [stem: *Megodont-*]. Type genus: *Megodontus* Solier, 1848 [as *Megadontus*, incorrect subsequent spelling of type genus name, not in prevailing usage]. Comment: incorrect original stem formation, not in prevailing usage.

Tribe Ceroglossini Lapouge, 1927

Ceroglossina Lapouge, 1927: 45 [stem: *Cerogloss-*]. Type genus: *Ceroglossus* Solier, 1848.

Tribe Cychrini Perty, 1830

Cychrii Perty, 1830: 6 [stem: *Cychr-*]. Type genus: *Cychrus* Fabricius, 1794. Comment: name previously attributed to Laporte (1834b: 86).

Tribe Pamborini Hope, 1838

Pamboridae Hope, 1838a: 47 [stem: *Pambor-*]. Type genus: *Pamborus* Latreille, 1812.

Subfamily Loricerinae Bonelli, 1810

Loricerides Bonelli, 1810: Tabula Synoptica [stem: *Loricer-*]. Type genus: *Loricera* Latreille, 1802.

Subfamily Omophroninae Bonelli, 1810

Omophronii Bonelli, 1810: Tabula Synoptica [stem: *Omophron-*]. Type genus: *Omophron* Latreille, 1802.

Scolyti Motschulsky, 1850: 91 [stem: *Scolyt-*]. Type genus: *Scolytus* Fabricius, 1790 [preoccupied genus name, not *Scolytus* Geoffroy, 1762 [Coleoptera: Curculionidae]; syn. of *Omophron* Latreille, 1802]. Comment: permanently invalid (Art. 39): based on preoccupied type genus; Scolytinae/-ini Latreille, 1804 are currently used as valid in Curculionidae.

Epactiini Fauvel, 1888: 1 [stem: *Epacti-*]. Type genus: *Epactius* Schneider, 1791 [syn. of *Omophron* Latreille, 1802].

Subfamily Elaphrinae Latreille, 1802

Elaphrii Latreille, 1802: 81 [stem: *Elaphr-*]. Type genus: *Elaphrus* Fabricius, 1775.

Subfamily Migadopinae Chaudoir, 1861

Migadopidae Chaudoir, 1861b: 510 [stem: *Migadop-*]. Type genus: *Migadops* G. R. Waterhouse, 1842.

Tribe Amarotypini Erwin, 1985

Amarotypini Erwin, 1985: 468 [stem: *Amarotyp-*]. Type genus: *Amarotypus* H. W. Bates, 1872.

Tribe Migadopini Chaudoir, 1861

Migadopidae Chaudoir, 1861b: 510 [stem: *Migadop-*]. Type genus: *Migadops* G. R. Waterhouse, 1842.

Subtribe Aquilicina Moret, 2005

Aquilicina Moret, 2005: 30 [stem: *Aquilic-*]. Type genus: *Aquilex* Moret, 1989.

Subtribe Migadopina Chaudoir, 1861

Migadopidae Chaudoir, 1861b: 510 [stem: *Migadop-*]. Type genus: *Migadops* G. R. Waterhouse, 1842.

Monolobinae Jeannel, 1938: 10, in key [stem: *Monolob-*]. Type genus: *Monolobus* Solier, 1849.

Loxomeriformes Erwin, 1985: 446 [stem: *Loxomer-*]. Type genus: *Loxomerus* Chaudoir, 1842.

Subfamily HILETINAE Schiødte, 1848

HILETINI Schiødte, 1848: 69 [stem: *Hilet-*]. Type genus: *Hiletus* Schiødte, 1848.

CAMARAGNATHINI Csiki, 1927: 341 [stem: *Camaragnath-*]. Type genus: *Camaragnathus* Bertrand-Bocandé, 1849 [syn. of *Hiletus* Schiødte, 1847].

Subfamily SCARITINAE Bonelli, 1810

SCARITIDES Bonelli, 1810: Tabula Synoptica [stem: *Scarit-*]. Type genus: *Scarites* Fabricius, 1775.

Tribe CARENINI MacLeay, 1887

CARENIDES W. J. MacLeay, 1887: 117 [stem: *Caren-*]. Type genus: *Carenum* Bonelli, 1813.

Tribe CLIVININI Rafinesque, 1815

CLIVINIDIA Rafinesque, 1815: 109 [stem: *Clivin-*]. Type genus: *Clivina* Latreille, 1802.

Subtribe ARDISTOMINA Putzeys, 1867

ARDISTOMIDES Putzeys, 1867: 200 [stem: *Ardistom-*]. Type genus: *Ardistomis* Putzeys, 1846. Comment: original vernacular name available (Art. 11.7.2): first used in latinized form by Csiki (1927: 547, as ARDISTOMINA), generally accepted as in Balkenohl (2003: 219, as ARDISTOMINA); current spelling maintained (Art. 29.3.1.1): incorrect stem formation in prevailing usage (should be *Ardistomid-*).

Subtribe CLIVININA Rafinesque, 1815

CLIVINIDIA Rafinesque, 1815: 109 [stem: *Clivin-*]. Type genus: *Clivina* Latreille, 1802.

REICHEINA Jeannel, 1957: 141, in key [stem: *Reichei-*]. Type genus: *Reicheia* Saulcy, 1862. Comment: incorrect original stem formation, not in prevailing usage.

ITALODYTINA Jeannel, 1957: 141, in key [stem: *Italodyt-*]. Type genus: *Italodytes* Müller, 1938.

REICHEIINA Basilewsky, 1980: 293 [stem: *Reichei-*]. Type genus: *Reicheia* Saulcy, 1862. Comment: family-group name proposed as new without reference to REICHEIINA Jeannel, 1957.

Subtribe FORCIPATORINA Bänninger, 1938

*OXYSTOMIDES Putzeys, 1867: 12 [stem: *Oxystom-*]. Type genus: *Oxystomus* Dejean, 1825 [preoccupied genus name, not *Oxystomus* Fischer von Waldheim, 1803 [Mammalia]; syn. of *Forcipator* Maindron, 1904]. Comment: original vernacular name unavailable (Art. 11.7.2): subsequently used in latinized form but not generally attributed to Putzeys (1867).

Oxystomina Csiki, 1927: 491. Type genus: *Oxystomus* Dejean, 1825 [preoccupied genus name, not *Oxystomus* Fischer von Waldheim, 1803 [Mammalia]; syn. of *Forcipator* Maindron, 1904]. Comment: permanently invalid (Art. 39): based on preoccupied type genus.

Forcipatorina Bänninger, 1938: 83 [stem: *Forcipator-*]. Type genus: *Forcipator* Maindron, 1904.

Tribe Dalyatini Mateu, 2002

Dalyatinae Mateu, 2002: 67 [stem: *Dalyat-*]. Type genus: *Dalyat* Mateu, 2002.

Tribe Dyschiriini Kolbe, 1880

Dyschiriini Kolbe, 1880: 266 [stem: *Dyschiri-*]. Type genus: *Dyschirius* Bonelli, 1810.

†Tribe Palaeoaxinidiini McKay, 1991

Palaeoaxinidini McKay, 1991: 10 [stem: *Palaeoaxinidi-*]. Type genus: *Palaeoaxinidium* McKay, 1991. Comment: incorrect original stem formation, not in prevailing usage.

Tribe Pasimachini Putzeys, 1867

Pasimachides Putzeys, 1867: 3 [stem: *Pasimach-*]. Type genus: *Pasimachus* Bonelli, 1813. Comment: original vernacular name available (Art. 11.7.2): first used in latinized form and generally accepted as in Csiki (1927: 444, as Pasimachina).

Tribe Promecognathini LeConte, 1853

Promecognathi J. L. LeConte, 1853b: 394 [stem: *Promecognath-*]. Type genus: *Promecognathus* Chaudoir, 1846.

Axinidiini Basilewsky, 1963: 307 [stem: *Axinidi-*]. Type genus: *Axinidium* Sturm, 1843.

Tribe Salcediini Alluaud, 1930

Salcediini Alluaud, 1930a: 21 [stem: *Salcedi-*]. Type genus: *Salcedia* Fairmaire, 1899.

Subtribe Androzelmina Bell, 1998

Androzelmina R. T. Bell, 1998: 271 [stem: *Androzelm-*]. Type genus: *Androzelma* Dostal, 1993.

Subtribe Salcediina Alluaud, 1930 (1929)

Zelmides Andrewes, 1929: 209, 416 [stem: *Zelm-*]. Type genus: *Zelma* Andrewes, 1920 [syn. of *Salcedia* Fairmaire, 1899]. Comment: use of younger name Salcediina Alluaud, 1930 conserved over this name (Art. 40.2).

Salcediini Alluaud, 1930a: 21 [stem: *Salcedi-*]. Type genus: *Salcedia* Fairmaire, 1899. Comment: name proposed to replace Zelmini Andrewes, 1929 because of the synonymy of the type genus; use of Salcediina conserved over Zelmina Andrewes, 1929 (Art. 40.2).

Subtribe Solenogenyina Bell, 1998

Solenogenyina R. T. Bell, 1998: 271 [stem: *Solenogeny-*]. Type genus: *Solenogenys* Westwood, 1859.

Tribe Scaritini Bonelli, 1810

Scaritides Bonelli, 1810: Tabula Synoptica [stem: *Scarit-*]. Type genus: *Scarites* Fabricius, 1775.

Subtribe Acanthoscelina Csiki, 1927

Acanthoscelina Csiki, 1927: 489 [stem: *Acanthoscel-*]. Type genus: *Acanthoscelis* Dejean, 1825. Comment: the incorrect stem formation is in prevailing usage (should be *Acanthoscelid-*) and is maintained (Art. 29.3.1.1); usage of the stem *Acanthoscel-* avoids homonymy with Acanthoscelidina Bridwell, 1946 (type genus *Acanthoscelides* Schilsky, 1905) in Chrysomelidae.

Subtribe Corintascarina Basilewsky, 1973

Corintascarini Basilewsky, 1973a: 10 [stem: *Corintascar-*]. Type genus: *Corintascaris* Basilewsky, 1952. Comment: current spelling maintained (Art. 29.3.1.1): incorrect stem formation in prevailing usage (should be *Corintascarid-*).

Subtribe Dyscherina Basilewsky, 1973

Dyscherina Basilewsky, 1973a: 101 [stem: *Dyscher-*]. Type genus: *Dyscherus* Chaudoir, 1855.

Subtribe Ochyropina Basilewsky, 1973

Ochyropini Basilewsky, 1973a: 9 [stem: *Ochyrop-*]. Type genus: *Ochyropus* Schiødte, 1847. Comment: current spelling maintained (Art. 29.5): incorrect stem formation in prevailing usage (should be *Ochyropod-*).

Subtribe Oxylobina Andrewes, 1929

Oxylobides Andrewes, 1929: 209, 292 [stem: *Oxylob-*]. Type genus: *Oxylobus* Chaudoir, 1855.

Subtribe Scapterina Putzeys, 1867

Scaptérides Putzeys, 1867: 7 [stem: *Scapter-*]. Type genus: *Scapterus* Dejean, 1826. Comment: original vernacular name available (Art. 11.7.2): first used in latinized form by Csiki (1927: 490, as SCAPTERINA), generally accepted as in Balkenohl (2003: 231, as SCAPTERINA).

PASSALIDIINA Csiki, 1927: 489 [stem: *Passalidi-*]. Type genus: *Passalidius* Chaudoir, 1863.

Subtribe SCARITINA Bonelli, 1810

SCARITIDES Bonelli, 1810: Tabula Synoptica [stem: *Scarit-*]. Type genus: *Scarites* Fabricius, 1775.

Subtribe STORTHODONTINA Jeannel, 1946

STORTHODONTINI Jeannel, 1946: 237, in key [stem: *Storthodont-*]. Type genus: *Storthodontus* Chaudoir, 1855.

Subfamily BROSCINAE Hope, 1838

BROSCHIDAE Hope, 1838a: 80 [stem: *Brosc-*]. Type genus: *Broscus* Panzer, 1813.

Tribe BROSCINI Hope, 1838

BROSCHIDAE Hope, 1838a: 80 [stem: *Brosc-*]. Type genus: *Broscus* Panzer, 1813.

Subtribe AXONYINA Roig-Juñent, 2000

AXONYINA Roig-Juñent, 2000: 18 [stem: *Axony-*]. Type genus: *Axonya* Andrewes, 1923.

Subtribe BARIPODINA Jeannel, 1941

BARYPITAE Jeannel, 1941: 287, in key [stem: *Baripod-*]. Type genus: *Baripus* Dejean, 1828 [as *Barypus*, unjustified emendation of type genus name by Agassiz (1846b: 43), not in prevailing usage]. Comment: incorrect original stem formation, not in prevailing usage.

Subtribe BROSCINA Hope, 1838

BROSCHIDAE Hope, 1838a: 80 [stem: *Brosc-*]. Type genus: *Broscus* Panzer, 1813 [as *Broschus*, incorrect subsequent spelling of type genus name, not in prevailing usage]. Comment: incorrect original stem formation, not in prevailing usage.

CEPHALOTIDA Heer, 1838: 12 [stem: *Cephalot-*]. Type genus: *Cephalotes* Bonelli, 1810 [preoccupied genus name, not *Cephalotes* Latreille, 1802 [Hymenoptera]; syn. of *Broscus* Panzer, 1813]. Comment: permanently invalid (Art. 39): based on preoccupied type genus.

ZACOTINI G. H. Horn, 1881: 169 [stem: *Zacot-*]. Type genus: *Zacotus* J. L. LeConte, 1869.

Subtribe Creobiina Jeannel, 1941

Creobitae Jeannel, 1941: 287, in key [stem: *Creobi-*]. Type genus: *Creobius* Guérin-Méneville, 1838. Comment: incorrect original stem formation, not in prevailing usage.

Subtribe Nothobroscina Roig-Juñent, 2000

Nothobroscina Roig-Juñent, 2000: 32 [stem: *Nothobrosc-*]. Type genus: *Nothobroscus* Roig-Juñent and Ball, 1995.

Subfamily Apotominae LeConte, 1853

Apotomi J. L. LeConte, 1853b: 370 [stem: *Apotom-*]. Type genus: *Apotomus* Illiger, 1807.

Subfamily Siagoninae Bonelli, 1813

Siagones Bonelli, 1813: 24 [stem: *Siagon-*]. Type genus: *Siagona* Latreille, 1804.

Tribe Enceladini Horn, 1881

Enceladini G. H. Horn, 1881: 118 [stem: *Encelad-*]. Type genus: *Enceladus* Bonelli, 1813.

Tribe Lupercini Lecordier, 1977

Lupercini Lecordier, 1977: 628 [stem: *Luperc-*]. Type genus: *Luperca* Laporte, 1840.

Tribe Siagonini Bonelli, 1813

Siagones Bonelli, 1813: 24 [stem: *Siagon-*]. Type genus: *Siagona* Latreille, 1804.

Subfamily Melaeninae Csiki, 1933

Coscinides Chaudoir, 1876b: 115 [stem: *Coscini-*]. Type genus: *Coscinia* Dejean, 1831 [preoccupied genus name, not *Coscinia* Hübner, 1819 [Lepidoptera]; syn. of *Cymbionotum* Baudi, 1864]. Comment: permanently invalid (Art. 39): based on preoccupied type genus; incorrect original stem formation, not in prevailing usage.

Granigerini Bedel, 1900: 24 [stem: *Graniger-*]. Type genus: *Graniger* sensu Chaudoir, 1876 [not *Graniger* Motschulsky, 1864; syn. of *Cymbionotum* Baudi, 1864]. Comment: based on a misidentified type genus; an application will need to be submitted to the Commission to suppress this name for the Principles of Priority and Homonymy (Art. 65.2.1) if Granigerini Antoine, 1959 in Harpalinae: Harpalini: Harpalina is to be used as valid in the future.

Melaeninae Csiki, 1933: 1650 [stem: *Melaen-*]. Type genus: *Melaenus* Dejean, 1831. Comment: published 26 May 1933; name proposed after 1930 without description or bibliographic reference to such a description (Art. 13.1), however available because it was used as valid before 2000 as in Lorenz (1998b: 149) and was not rejected by an author who, between 1961 and 1999, applied Article 13 of the then current edition of the Code (see Art. 13.2.1).

CYMBIONOTINI Andrewes, 1933: 3 [stem: *Cymbionot-*]. Type genus: *Cymbionotum* Baudi di Selve, 1864. Comment: published 30 June 1933.

Subfamily GEHRINGIINAE Darlington, 1933

GEHRINGIINI Darlington, 1933: 110 [stem: *Gehringi-*]. Type genus: *Gehringia* Darlington, 1933.

Tribe GEHRINGIINI Darlington, 1933

GEHRINGIINI Darlington, 1933: 110 [stem: *Gehringi-*]. Type genus: *Gehringia* Darlington, 1933.

Subtribe GEHRINGIINA Darlington, 1933

GEHRINGIINI Darlington, 1933: 110 [stem: *Gehringi-*]. Type genus: *Gehringia* Darlington, 1933.

Subtribe HELENAEINA Deuve, 2007

HELENAEINA Deuve, 2007: 217 [stem: *Helenae-*]. Type genus: *Helenaea* Schatzmayr and Koch, 1934.

Subfamily TRECHINAE Bonelli, 1810

TRECHII Bonelli, 1810: Tabula Synoptica [stem: *Trech-*]. Type genus: *Trechus* Clairville, 1806.

Tribe BEMBIDIINI Stephens, 1827

BEMBIDIIDAE Stephens, 1827: 5 [stem: *Bembidi-*]. Type genus: *Bembidion* Latreille, 1802 [as *Bembidium*, unjustified emendation of type genus name by Gyllenhal (1810: 12), not in prevailing usage].

Subtribe ANILLINA Jeannel, 1937

ANILLINI Jeannel, 1937: 244, in key [stem: *Anill-*]. Type genus: *Anillus* Jacquelin du Val, 1851. Comment: precedence (ANILLINA Jeannel, 1937 vs SCOTODIPNINA Jeannel, 1937) given to taxon originally proposed at the higher rank (Art. 24.1).

SCOTODIPNINA Jeannel, 1937: 265, in key [stem: *Scotodipn-*]. Type genus: *Scotodipnus* Schaum, 1860.

TYPHLOCHARINA Jeanne, 1973: 95, in key [stem: *Typhlocharit-*]. Type genus: *Typhlocharis* Dieck, 1869. Comment: incorrect original stem formation, not in prevailing usage.

Subtribe BEMBIDIINA Stephens, 1827

BEMBIDIIDAE Stephens, 1827: 5 [stem: *Bembidi-*]. Type genus: *Bembidion* Latreille, 1802 [as *Bembidium*, unjustified emendation of type genus name by Gyllenhal (1810), not in prevailing usage].

PERYPHIDAE Kirby, 1837: 52 [stem: *Peryph-*]. Type genus: *Peryphus* Dejean, 1821.

Subtribe TACHYINA Motschulsky, 1862

TACHYAIRES Motschulsky, 1862: 24 [stem: *Tachy-*]. Type genus: *Tachys* Dejean, 1821 [placed on the Official List of Generic Names in Zoology (ICZN 1990c)]. Comment: original vernacular name available (Art. 11.7.2): first used in latinized form by Jeannel (1941: 401, 422, as TACHYINI), generally accepted as in J. K. Park et al. (2006: 91, as TACHYINI); current spelling maintained (Art. 29.5): incorrect stem formation in prevailing usage (should be *Tache-*, see Alonso-Zarazaga 2007).

MICRATOPINI Casey, 1914a: 42 [stem: *Micratop-*]. Type genus: *Micratopus* Casey, 1914.

LIMNASTINI Jeannel, 1937: 245, in key [stem: *Lymnast-*]. Type genus: *Lymnastis* Motschulsky, 1862 [as *Limnastis*, unjustified emendation of type genus name by Ganglbauer (1891a: 181), not in prevailing usage]. Comment: incorrect original stem formation, not in prevailing usage.

Subtribe XYSTOSOMINA Erwin, 1994

XYSTOSOMINA Erwin, 1994: 560 [stem: *Xystosom-*]. Type genus: *Xystosomus* Schaum, 1863.

Tribe HOROLOGIONINI Jeannel, 1949

HOROLOGIONIDAE Jeannel, 1949b: 91 [stem: *Horologion-*]. Type genus: *Horologion* J. M. Valentine, 1932. Comment: current spelling maintained (Art. 29.5): incorrect stem formation in prevailing usage (should be *Horologi-*).

Tribe POGONINI Laporte, 1834

POGONIDAE Laporte, 1834b: 70 [stem: *Pogon-*]. Type genus: *Pogonus* Dejean, 1821.

POGONOPSINI Bedel, 1900: 20 [stem: *Pogonopse-*]. Type genus: *Pogonopsis* Bedel, 1898. Comment: incorrect original stem formation, not in prevailing usage.

Tribe TRECHINI Bonelli, 1810

TRECHII Bonelli, 1810: Tabula Synoptica [stem: *Trech-*]. Type genus: *Trechus* Clairville, 1806.

Subtribe AEPINA Fowler, 1887

AËPYES Fowler, 1887: 123 [stem: *Aep-*]. Type genus: *Aepus* Samouelle, 1819.

TEMNOSTEGINI Enderlein, 1909: 376 [stem: *Temnosteg-*]. Type genus: *Temnostega* Enderlein, 1905.

Subtribe CNIDINA Jeannel, 1958

CNIDINA Jeannel, 1958b: 733 [stem: *Cnid-*]. Type genus: *Cnides* Motschulsky, 1862. Comment: name proposed after 1930 without description or bibliographic reference to such a description (Art. 13.1), however available because it was used as valid before 2000 as in Casale and Laneyrie (1982: 9, as CNIDINI) and was not rejected by an author who, between 1961 and 1999, applied Article 13 of the then current edition of the Code (see Art. 13.2.1).

Subtribe PERILEPTINA Sloane, 1903

PERILEPTIDES Sloane, 1903: 583 [stem: *Perilept-*]. Type genus: *Perileptus* Schaum, 1860. Comment: PERILEPTIDES Sloane, 1903 was treated as Latin and available in Madge (1989: 466) but the valid name for this taxon was listed as PERILEPTI Jeannel, 1922 by Csiki (1928) and PERILEPTINA Jeannel, 1922 by Lorenz (2005); we follow the usage of Madge (1989).

OCHTHEPHILINI Jeannel, 1922: 165 [stem: *Ochthephil-*]. Type genus: *Ochthephilus* Nietner, 1857 [preoccupied genus name, not *Ochthephilus* Mulsant and Rey, 1856 [Coleoptera: STAPHYLINIDAE]; syn. of *Perileptus* Schaum, 1860]. Comment: permanently invalid (Art. 39): based on preoccupied type genus.

Subtribe PLOCAMOTRECHINA Jeannel, 1960

*PLOCAMOTRECHINA Jeannel, 1958b: 733 [stem: *Plocamotrech-*]. Type genus: *Plocamotrechus* Jeannel, 1926 [syn. of *Pachydesus* Motschulsky, 1864]. Comment: unavailable family-group name, proposed after 1930 without description or bibliographic reference to such a description (Art. 13.1).

PLOCAMOTRECHINA Jeannel, 1960: 53, in key [stem: *Plocamotrech-*]. Type genus: *Plocamotrechus* Jeannel, 1926 [syn. of *Pachydesus* Motschulsky, 1864].

Subtribe TRECHINA Bonelli, 1810

TRECHII Bonelli, 1810: Tabula Synoptica [stem: *Trech-*]. Type genus: *Trechus* Clairville, 1806.

*APHAENOPSES Fauvel, 1888: 7 [stem: *Aphoenop-*]. Type genus: *Aphoenops* Bonvouloir, 1861 [as *Aphaenops*, incorrect subsequent spelling of type genus name, not in prevailing usage]. Comment: original vernacular name unavailable (Art. 11.7.2): not subsequently latinized; incorrect original stem formation, not in prevailing usage.

HOMALODERINI Jeannel, 1926: 397, in key [stem: *Omaloder-*]. Type genus: *Omalodera* Blanchard, 1842 [as *Homalodera*, unjustified emendation of type genus name by Gemminger and Harold (1868a: 389), not in prevailing usage]. Comment: incorrect original stem formation, not in prevailing usage.

Subtribe TRECHODINA Jeannel, 1926

TRECHODINI Jeannel, 1926: 469 [stem: *Trechod-*]. Type genus: *Trechodes* Blackburn, 1901.

THALASSOPHILI Csiki, 1928: 233 [stem: *Thalassophil-*]. Type genus: *Thalassophilus* Wollaston, 1854.

Tribe ZOLINI Sharp, 1886

ZOLINI Sharp, 1886a: 371 [stem: *Zol-*]. Type genus: *Zolus* Sharp, 1886.

Subtribe CHALTENIINA Roig-Juñent and Cicchino, 2001

CHALTENIINA Roig-Juñent and Cicchino, 2001: 658 [stem: *Chalteni-*]. Type genus: *Chaltenia* Roig-Juñent and Cicchino, 2001.

Subtribe SINOZOLINA Deuve, 1997

SINOZOLINI Deuve, 1997: 35 [stem: *Sinozol-*]. Type genus: *Sinozolus* Bedel, 1898.

Subtribe ZOLINA Sharp, 1886

ZOLINI Sharp, 1886a: 371 [stem: *Zol-*]. Type genus: *Zolus* Sharp, 1886.

MERIZODINI Sloane, 1920: 139 [stem: *Merizodont-*]. Type genus: *Merizodus* Solier, 1849. Comment: incorrect original stem formation, not in prevailing usage.

*OOPTERINI Jeannel, 1938: 45 [stem: *Oopter-*]. Type genus: *Oopterus* Guérin-Méneville, 1841. Comment: unavailable family-group name, proposed after 1930 without description or bibliographic reference to such a description (Art. 13.1).

OOPTERINI Jeannel, 1940: 93 [stem: *Oopter-*]. Type genus: *Oopterus* Guérin-Méneville, 1841.

Subfamily PATROBINAE Kirby, 1837

PATROBIDAE Kirby, 1837: 50 [stem: *Patrob-*]. Type genus: *Patrobus* Dejean, 1821.

Tribe LISSOPOGONINI Zamotajlov, 2000

LISSOPOGONINI Zamotajlov, 2000: 266 [stem: *Lissopogon-*]. Type genus: *Lissopogonus* Andrewes, 1923.

ZOLINOPATROBINA Deuve and Tian, 2001: 421 [stem: *Zolinopatrob-*]. Type genus: *Zolinopatrobus* Deuve and Tian, 2001 [syn. of *Lissopogonus* Andrewes, 1923].

Tribe PATROBINI Kirby, 1837

PATROBIDAE Kirby, 1837: 50 [stem: *Patrob-*]. Type genus: *Patrobus* Dejean, 1821.

Subtribe DELTOMERINA Chaudoir, 1871

DELTOMERIDAE Chaudoir, 1871a: 51 [stem: *Deltomer-*]. Type genus: *Deltomerus* Motschulsky, 1850.

Subtribe Deltomerodina Zamotajlov, 2002

Deltomerodina Zamotajlov, 2002: 83 [stem: *Deltomerod-*]. Type genus: *Deltomerodes* Deuve, 1992.

Subtribe Patrobina Kirby, 1837

Patrobidae Kirby, 1837: 50 [stem: *Patrob-*]. Type genus: *Patrobus* Dejean, 1821.

Subtribe Platidiolina Zamotajlov and Lafer, 2001

Platidiolini Zamotajlov and Lafer, 2001: 411 [stem: *Platidiol-*]. Type genus: *Platidiolus* Chaudoir, 1878.

Subfamily Psydrinae LeConte, 1853

Psydri J. L. LeConte, 1853b: 393 [stem: *Psydr-*]. Type genus: *Psydrus* J. L. LeConte, 1846.

Tribe Amblytelini Blackburn, 1892

Amblytelides Blackburn, 1892: 85 [stem: *Amblytel-*]. Type genus: *Amblytelus* Erichson, 1842. Comment: the junior homonym Amblytelinae Viereck, 1918 (type genus *Amblyteles* Wesmael, 1845) is available in Hymenoptera: Ichneumonidae, it is presently considered a synonym of Ichneumonini Latreille, 1802; both family-group names are based on the same stem; this case is to be referred to the Commission to remove the homonymy (Art. 55.3.1).

Tribe Mecyclothoracini Jeannel, 1940

Mecyclothoracitae Jeannel, 1940: 97 [stem: *Mecyclothorac-*]. Type genus: *Mecyclothorax* Sharp, 1903. Comment: name proposed after 1930 without description or bibliographic reference to such a description (Art. 13.1), however available because it was used as valid before 2000 as in Lorenz (1998a: 222) and was not rejected by an author who, between 1961 and 1999, applied Article 13 of the then current edition of the Code (see Art. 13.2.1).

Tribe Meonini Sloane, 1898

Meonides Sloane, 1898: 470 [stem: *Meon-*]. Type genus: *Meonis* Laporte, 1867. Comment: current spelling maintained (Art. 29.3.1.1): incorrect stem formation in prevailing usage (should be *Meonid-*).

Tribe Moriomorphini Sloane, 1890

Moriomorphini Sloane, 1890: 646 [stem: *Moriomorph-*]. Type genus: *Moriomorpha* Laporte, 1867.

Melisoderides Sloane, 1898: 470 [stem: *Melisoder-*]. Type genus: *Melisodera* Westwood, 1835.

Tribe Psydrini LeConte, 1853

Psydri J. L. LeConte, 1853b: 393 [stem: *Psydr-*]. Type genus: *Psydrus* J. L. LeConte, 1846.

Nomiidae Gozis, 1875: 3 [stem: *Nomi-*]. Type genus: *Nomius* Laporte, 1835. Comment: the younger name Nomiidae Robertson, 1904 (type genus *Nomia* Latreille, 1804) has been used in Hymenoptera and a Case was recently submitted to the Commission to emend the beetle family-group name to Nomiusidae (Art. 55.3) in order to remove the homonymy (Engel and Bouchard 2009).

Tribe Tropopterini Sloane, 1898

Tropopterides Sloane, 1898: 470 [stem: *Tropopter-*]. Type genus: *Tropopterus* Solier, 1849.

Subfamily Nototylinae Bänninger, 1927

Tylonotini Schaum, 1863: 74 [stem: *Tylonot-*]. Type genus: *Tylonotus* Schaum, 1863 [preoccupied genus name, not *Tylonotus* Haldeman, 1847 [Coleoptera: Cerambycidae], not *Tylonotus* Fieber, 1858 [Hemiptera]; syn. of *Nototylus* Gemminger and Harold, 1868]. Comment: permanently invalid (Art. 39): based on preoccupied type genus.

Nototylini Bänninger, 1927: 177 [stem: *Nototyl-*]. Type genus: *Nototylus* Gemminger and Harold, 1868.

Subfamily Paussinae Latreille, 1806

Paussili Latreille, 1806: 234 [stem: *Pauss-*]. Type genus: *Paussus* Linnaeus, 1775.

Tribe Metriini LeConte, 1853

Metrii J. L. LeConte, 1853b: 394 [stem: *Metri-*]. Type genus: *Metrius* Eschscholtz, 1829.

Tribe Mystropomini Horn, 1881

Mystropomini G. H. Horn, 1881: 116 [stem: *Mystropom-*]. Type genus: *Mystropomus* Chaudoir, 1848.

Tribe Ozaenini Hope, 1838

Ozaenidae Hope, 1838a: 107 [stem: *Ozaen-*]. Type genus: *Ozaena* A. G. Olivier, 1812.

*Tropopsitos Solier, 1849: 179 [stem: *Tropopse-*]. Type genus: *Tropopsis* Solier, 1849. Comment: original vernacular name unavailable (Art. 11.7.2): not subsequently latinized; incorrect original stem formation, not in prevailing usage.

Pseudozaenini Sloane, 1905: 704 [stem: *Pseudozaen-*]. Type genus: *Pseudozaena* Laporte, 1834.

Eustrini Jeannel, 1946: 48 [stem: *Eustr-*]. Type genus: *Eustra* Schmidt-Göbel, 1846.

PACHYTELINI Jeannel, 1946: 47, in key [stem: *Pachytel-*]. Type genus: *Pachyteles* Perty, 1830.

PHYSEITAE Jeannel, 1946: 47, in key [stem: *Physe-*]. Type genus: *Physea* Brullé, 1834.

Tribe PAUSSINI Latreille, 1806

PAUSSILI Latreille, 1806: 234 [stem: *Pauss-*]. Type genus: *Paussus* Linnaeus, 1775.

†Subtribe ARTHROPTERITINA Luna de Carvalho, 1961

ARTHROPTERITINA Luna de Carvalho, 1961: 3 [stem: *Arthropterit-*]. Type genus: *Arthropterites* Wasmann, 1926.

Subtribe CARABIDOMEMNINA Wasmann, 1928

*CARABIDOMEMNINEN Kolbe, 1927a: 178 [stem: *Carabidomemn-*]. Type genus: *Carabidomemnus* Kolbe, 1924. Comment: original vernacular name unavailable (Art. 11.7.2): proposed after 1899.

*EOHOMOPTERINEN Kolbe, 1927b: 214 [stem: *Eohomopter-*]. Type genus: *Eohomopterus* Wasmann, 1920. Comment: original vernacular name unavailable (Art. 11.7.2): proposed after 1899; the latinized form EOHOMOPTERINI was used by Nagel (1987: 33, 54, 55, 56) but it is unavailable because it was proposed after 1930 without description or bibliographic reference to such a description (Art. 13.1).

CARABIDOMEMNINAE Wasmann, 1928: 271 [stem: *Carabidomemn-*]. Type genus: *Carabidomemnus* Kolbe, 1924.

Subtribe CERAPTERINA Billberg, 1820

CERAPTERIDES Billberg, 1820a: 47 [stem: *Cerapter-*]. Type genus: *Cerapterus* Swederus, 1788.

MEGALOPAUSSINAE Wasmann, 1920: 111 [stem: *Megalopauss-*]. Type genus: *Megalopaussus* Lea, 1906.

*ARTHROPTERINEN Kolbe, 1927b: 211 [stem: *Arthropter-*]. Type genus: *Arthropterus* W. S. MacLeay, 1838. Comment: original vernacular name unavailable (Art. 11.7.2): proposed after 1899.

ARTHROPTERINI Wasmann, 1928: 274 [stem: *Arthropter-*]. Type genus: *Arthropterus* W. S. MacLeay, 1838. Comment: ARTHROPTERIDAE Fieber, 1861 has been used in Hemiptera but it is not based on a genus and is therefore unavailable; ARTHROPTERINI Wasmann, 1928 is a junior homonym of ARTHROPTERIDAE Jordan, 1923 proposed in Pisces (type genus *Arthropterus* Agassiz, 1843, preoccupied genus name replaced by *Arthrobatis* Whitley 1940); the Pisces name is permanently invalid since its type genus is a junior homonym; an application to the Commission is needed to conserve the Coleoptera name if it is to be used as valid.

*MESARTHROPTERINA Nagel, 1987: 60 [stem: *Mesarthropter-*]. Type genus: *Mesarthropterus* Wasmann, 1926. Comment: unavailable family-group name, proposed after 1930 without description or bibliographic reference to such a description (Art. 13.1).

†Subtribe Eopaussina Luna de Carvalho, 1951

*Eopaussina Darlington, 1950: 84 [stem: *Eopauss-*]. Type genus: *Eopaussus* Wasmann, 1926. Comment: unavailable family-group name, proposed after 1930 without description or bibliographic reference to such a description (Art. 13.1).

Eopaussinae Luna de Carvalho, 1951: 22 [stem: *Eopauss-*]. Type genus: *Eopaussus* Wasmann, 1926.

Subtribe Heteropaussina Janssens, 1950

Pleuropterini Wasmann, 1920: 111 [stem: *Pleuropter-*]. Type genus: *Pleuropterus* Westwood, 1841 [preoccupied genus name, not *Pleuropterus* Burnett, 1829 [Mammalia]; syn. of *Janssenius* Luna de Carvalho, 1951]. Comment: permanently invalid (Art. 39): based on preoccupied type genus.

Heteropaussines Janssens, 1950: 5 (footnote) [stem: *Heteropauss-*]. Type genus: *Heteropaussus* J. Thomson, 1860.

Janssenini Luna de Carvalho, 1951: 18 [stem: *Jansseni-*]. Type genus: *Janssenius* Luna de Carvalho, 1951. Comment: incorrect original stem formation, not in prevailing usage.

Subtribe Homopterina Wasmann, 1920

Homopterinae Wasmann, 1920: 111 [stem: *Homopter-*]. Type genus: *Homopterus* Westwood, 1841.

Subtribe Paussina Latreille, 1806

Paussili Latreille, 1806: 234 [stem: *Pauss-*]. Type genus: *Paussus* Linnaeus, 1775.

Platyrhopalini Jeannel, 1946: 65 [stem: *Platyrhopal-*]. Type genus: *Platyrhopalus* Westwood, 1833.

Enneapaussini Jeannel, 1946: 62 [stem: *Enneapauss-*]. Type genus: *Enneapaussus* Jeannel, 1946.

Ceratoderina Darlington, 1950: 107 [stem: *Ceratoder-*]. Type genus: *Ceratoderus* Westwood, 1841.

Hylotorini Luna de Carvalho, 1951: 50 [stem: *Hylotor-*]. Type genus: *Hylotorus* Dalman, 1823.

Leleupaussina Luna de Carvalho, 1989: 430 [stem: *Leleupauss-*]. Type genus: *Leleupaussus* Luna de Carvalho, 1962.

Subtribe Pentaplatarthrina Jeannel, 1946

*Pentaplatarthrinen Kolbe, 1927b: 214 [stem: *Pentaplatarthr-*]. Type genus: *Pentaplatarthrus* Westwood, 1833. Comment: original vernacular name unavailable (Art. 11.7.2): proposed after 1899.

PENTAPLATARTHRINI Jeannel, 1946: 65 [stem: *Pentaplatarthr-*]. Type genus: *Pentaplatarthrus* Westwood, 1833.

HEXAPLATARTHRINA Luna de Carvalho, 1961: 3 [stem: *Hexaplatarthr-*]. Type genus: *Hexaplatarthrus* Jeannel, 1955.

Tribe PROTOPAUSSINI Gestro, 1892

PROTOPAUSSINI Gestro, 1892: 707 [stem: *Protopauss-*]. Type genus: *Protopaussus* Gestro, 1892.

Subfamily BRACHININAE Bonelli, 1810

BRACHINII Bonelli, 1810: Tabula Synoptica [stem: *Brachin-*]. Type genus: *Brachinus* Weber, 1801.

Tribe BRACHININI Bonelli, 1810

BRACHINII Bonelli, 1810: Tabula Synoptica [stem: *Brachin-*]. Type genus: *Brachinus* Weber, 1801.

Subtribe APTININA Gistel, 1848

APTINIDAE Gistel, 1848: [2] [stem: *Aptin-*]. Type genus: *Aptinus* Bonelli, 1810. Comment: APTEMIDAE Gistel (1856a: 356) is probably a misspelling of APTINIDAE (see Madge 1989: 460).

STYPHLOMERINI Habu, 1984: 133 [stem: *Styphlomer-*]. Type genus: *Styphlomerus* Chaudoir, 1875.

Subtribe BRACHININA Bonelli, 1810

BRACHINII Bonelli, 1810: Tabula Synoptica [stem: *Brachin-*]. Type genus: *Brachinus* Weber, 1801.

Subtribe MASTACINA Erwin, 1970

MASTACINA Erwin, 1970: 28, 32 [stem: *Mastac-*]. Type genus: *Mastax* Fischer von Waldheim, 1828.

Subtribe PHEROPSOPHINA Jeannel, 1949

PHEROPSOPHINI Jeannel, 1949c: 1083, in key [stem: *Pheropsoph-*]. Type genus: *Pheropsophus* Solier, 1833.

Tribe CREPIDOGASTRINI Jeannel, 1949

CREPIDOGASTRITAE Jeannel, 1949c: 1080 [stem: *Crepidogastr-*]. Type genus: *Crepidogaster* Boheman, 1848.

Subfamily HARPALINAE Bonelli, 1810

HARPALII Bonelli, 1810: Tabula Synoptica [stem: *Harpal-*]. Type genus: *Harpalus* Latreille, 1802. Comment: First Reviser (HARPALINAE Bonelli, 1810 vs

Dryptinae Bonelli, 1810 vs Lebiinae Bonelli, 1810 vs Licininae Bonelli, 1810 vs Panagaeinae Bonelli, 1810 vs Platyninae Bonelli, 1810 vs Pterostichinae Bonelli, 1810 vs Zabrinae Bonelli, 1810 vs Zuphiinae Bonelli, 1810) not determined, current usage maintained; Graphipterinae Latreille, 1802 has precedence over this taxon, however, Harpalinae Bonelli, 1810 is in prevailing usage and is maintained (Art. 35.5).

Tribe Abacetini Chaudoir, 1873

Abacétides Chaudoir, 1873a: 5 [stem: *Abacet-*]. Type genus: *Abacetus* Dejean, 1828. Comment: original vernacular name available (Art. 11.7.2): first used in latinized form by H. W. Bates (1873: 224, as Abacetinae), generally accepted as in Bousquet (2003: 346, as Abacetini).

Celioschesini Jeannel, 1948b: 442 [stem: *Celioschese-*]. Type genus: *Celioschesis* Tschitschérine, 1898 [syn. of *Aristopus* LaFerté-Sénectère, 1853]. Comment: incorrect original stem formation, not in prevailing usage.

Loxandrina Erwin and Sims, 1984: 383, in key [stem: *Loxandr-*]. Type genus: *Loxandrus* J. L. LeConte, 1852.

Loxandrini Bousquet and Larochelle, 1993: 31 [stem: *Loxandr-*]. Type genus: *Loxandrus* J. L. LeConte, 1852. Comment: family-group name proposed as new without reference to Loxandrina Erwin and Sims, 1984.

Tribe Amorphomerini Sloane, 1923

Trimerinae Alluaud, 1922: 500 [stem: *Trimer-*]. Type genus: *Trimerus* Chaudoir, 1878 [preoccupied genus name, not *Trimerus* Green, 1832 [Trilobita]; syn. of *Amorphomerus* Sloane, 1923]. Comment: permanently invalid (Art. 39): based on preoccupied type genus.

Amorphomerini Sloane, 1923a: 249 [stem: *Amorphomer-*]. Type genus: *Amorphomerus* Sloane, 1923.

Tribe Anthiini Bonelli, 1813

Anthies Bonelli, 1813: 18 [stem: *Anthi-*]. Type genus: *Anthia* Weber, 1801.

Polyhirmi Rousseau, 1905: 3, in key [stem: *Polyhirm-*]. Type genus: *Polyhirma* Chaudoir, 1850.

Cypholobini G. Strohmeyer, 1928: 287 [stem: *Cypholob-*]. Type genus: *Cypholoba* Chaudoir, 1850.

Tribe Atranini Horn, 1881

Atrani G. H. Horn, 1881: 145 [stem: *Atran-*]. Type genus: *Atranus* J. L. LeConte, 1848.

Tribe Bascanini Basilewsky, 1953

Bascanini Basilewsky, 1953a: 165 [stem: *Bascan-*]. Type genus: *Bascanus* Péringuey, 1896.

Tribe Calophaenini Jeannel, 1948

*Calophaenidae Jeannel, 1942: 1017 [stem: *Calophaen-*]. Type genus: *Calophaena* Klug, 1821. Comment: unavailable family-group name, proposed after 1930 without description or bibliographic reference to such a description (Art. 13.1).

Calophaenidae Jeannel, 1948b: 378 [stem: *Calophaen-*]. Type genus: *Calophaena* Klug, 1821.

Tribe Catapieseini Bates, 1882

Catapiesinae H. W. Bates, 1882: 90 [stem: *Catapiese-*]. Type genus: *Catapiesis* Solier, 1835. Comment: incorrect original stem formation, not in prevailing usage.

Tribe Chaetodactylini Tschitschérine, 1903

Chaetodactylini Tschitschérine, 1903: 157 [stem: *Chaetodactyl-*]. Type genus: *Chaetodactyla* Tschitschérine, 1897.

Tribe Chaetogenyini Emden, 1958

Chaetogenyina Emden, 1958: 24 [stem: *Chaetogeny-*]. Type genus: *Chaetogenys* Emden, 1958.

Tribe Chlaeniini Brullé, 1834

Chlaenides Brullé, 1834: 123 [stem: *Chlaeni-*]. Type genus: *Chlaenius* Bonelli, 1810.

Subtribe Callistina Laporte, 1834

Callistidae Laporte, 1834b: 80 [stem: *Callist-*]. Type genus: *Callistus* Bonelli, 1810. Comment: published before 9 August 1834.

Eusynetadae Gistel, 1856a: 356 [stem: *Eusynet-*]. Type genus: *Eusyneta* Gistel, 1856 [syn. of *Callistus* Bonelli, 1810]. Comment: incorrect original stem formation, not in prevailing usage.

Subtribe Chlaeniina Brullé, 1834

Chlaenides Brullé, 1834: 123 [stem: *Chlaeni-*]. Type genus: *Chlaenius* Bonelli, 1810. Comment: published before 2 August 1834; original vernacular name available (Art. 11.7.2): first used in latinized form by Erichson (1837: 96, as Chlaeniini), generally accepted as in Lorenz (2005: 328, as Chlaeniini); incorrect original stem formation, not in prevailing usage.

Lissaucheniidae Gistel, 1848: [2] [stem: *Lissaucheni-*]. Type genus: *Lissauchenius* W. S. MacLeay, 1825.

Rhopalomelini Alluaud, 1930b: 105 [stem: *Rhopalomel-*]. Type genus: *Rhopalomelus* Boheman, 1848.

Chlaeniodini Jeannel, 1949c: 777 [stem: *Chlaeniod-*]. Type genus: *Chlaeniodus* Jeannel, 1949.

Eccoptomenini Jeannel, 1949c: 821 [stem: *Eccoptomen-*]. Type genus: *Eccoptomenus* Chaudoir, 1850.

Chlaenionini Jeannel, 1949c: 776, in key [stem: *Chlaenion-*]. Type genus: *Chlaenionus* Kuntzen, 1913.

Procletini Basilewsky, 1950b: 49 [stem: *Proclet-*]. Type genus: *Procletus* Péringuey, 1896.

Pleroticini Basilewsky, 1950b: 50 [stem: *Plerotic-*]. Type genus: *Pleroticus* Péringuey, 1896.

Callistoidini Basilewsky, 1950b: 51 [stem: *Callistoid-*]. Type genus: *Callistoides* Motschulsky, 1865.

Harpaglossini Basilewsky, 1950b: 52 [stem: *Harpagloss-*]. Type genus: *Harpaglossus* Motschulsky, 1858.

Leptodinodini Basilewsky and Grundmann, 1955: 205 [stem: *Leptodinod-*]. Type genus: *Leptodinodes* Jeannel, 1949.

Chlaenioctenini Basilewsky and Grundmann, 1955: 204 [stem: *Chlaenioctenen-*]. Type genus: *Chlaenioctenus* H. W. Bates, 1892.

Brachylobini Basilewsky and Grundmann, 1955: 204 [stem: *Brachylob-*]. Type genus: *Brachylobus* Chaudoir, 1876.

Tribe Cnemalobini Germain, 1911

Cnémacanthides Lacordaire, 1854a: 237 [stem: *Cnemacanth-*]. Type genus: *Cnemacanthus* sensu Brullé, 1834 [not *Cnemacanthus* Gray, 1832; syn. of *Cnemalobus* Guérin-Méneville, 1838]. Comment: original vernacular name available (Art. 11.7.2): first used in latinized form by Schaum (1860: 353, as Cnemacanthidae), generally accepted as in Broun (1880: 7, as Cnemacanthidae); based on a misidentified type genus, name treated here as invalid until an application is submitted to the Commission to suppress it for the Principle of Priority (Art. 65.2.1).

Cnemalobini Germain, 1911: 53 [stem: *Cnemalob-*]. Type genus: *Cnemalobus* Guérin-Méneville, 1838.

Cnemalobini Bousquet and Larochelle, 1993: 27 [stem: *Cnemalob-*]. Type genus: *Cnemalobus* Guérin-Méneville, 1838. Comment: proposed as a new without reference to Cnemalobini Germain, 1911.

Tribe Cratocerini Lacordaire, 1854

Cratocérides Lacordaire, 1854a: 257 [stem: *Cratocer-*]. Type genus: *Cratocerus* Dejean, 1829. Comment: original vernacular name available (Art. 11.7.2): first used in latinized form by Murray (1858: 343, as Cratoceridae), generally accepted as in Csiki (1929: 493, as Cratoceri).

Tribe Ctenodactylini Laporte, 1834

Ctenodactylidae Laporte, 1834b: 45 [stem: *Ctenodactyl-*]. Type genus: *Ctenodactyla* Dejean, 1825.

*Leptotrachélides Chaudoir, 1848: 52 [stem: *Leptotrachel-*]. Type genus: *Leptotrachelus* Latreille, 1829. Comment: original vernacular name unavailable (Art. 11.7.2): not subsequently latinized, also originally proposed in synonymy with Rhagocrepides (see Madge 1989: 464).

Rhagocrepides Chaudoir, 1848: 52 [stem: *Rhagocrepid-*]. Type genus: *Rhagocrepis* Eschscholtz, 1829 [syn. of *Leptotrachelus* Latreille, 1829]. Comment: original vernacular name available (Art. 11.7.2): first used in latinized form by and generally accepted as in Chaudoir (1861b: 528, as Rhagocrepidae); incorrect original stem formation, not in prevailing usage.

Tribe Cuneipectini Sloane, 1907

Cuneipectini Sloane, 1907a: 358 [stem: *Cuneipect-*]. Type genus: *Cuneipectus* Sloane, 1907. Comment: current spelling maintained (Art. 29.5): incorrect stem formation in prevailing usage (should be *Cuneipector-*).

Tribe Cyclosomini Laporte, 1834

Cyclosomidae Laporte, 1834b: 69 [stem: *Cyclosom-*]. Type genus: *Cyclosomus* Latreille, 1829.

Subtribe Cyclosomina Laporte, 1834

Cyclosomidae Laporte, 1834b: 69 [stem: *Cyclosom-*]. Type genus: *Cyclosomus* Latreille, 1829.

Tétragonodérides Chaudoir, 1871b: 111 [stem: *Tetragonoder-*]. Type genus: *Tetragonoderus* Dejean, 1829. Comment: original vernacular name available (Art. 11.7.2): first used in latinized form by and generally accepted as in Chaudoir (1876a: 2, as Tetragonoderidae).

Sarothrocrepidae Chaudoir, 1876a: 83 [stem: *Sarothrocrepid-*]. Type genus: *Sarothrocrepis* Chaudoir, 1850. Comment: incorrect original stem formation, not in prevailing usage; Chaudoir also used the spelling Sarothrocrépides in his original publication on page 80.

Subtribe Masoreina Chaudoir, 1871

*Somoplatides Chaudoir, 1846: 511 [stem: *Somoplat-*]. Type genus: *Somoplatus* Dejean, 1829. Comment: original vernacular name unavailable (Art. 11.7.2): subsequently used in latinized form but not generally attributed to Chaudoir (1846) and generally accepted as valid; this name was first used in latinized form by Carus and Engelmann (1861: 1806 [index], as Somoplatidae) referring to Chaudoir's paper, but the name was not used as valid; this name was treated as "vernacular, not scientific" by Ball (1979: 77); Basilewsky (1984: 527, as Somoplatini) used this taxon as valid but did not refer to Chaudoir's original vernacular name, Somoplatini Basilewsky, 1984 is also unavailable since it was proposed after 1930 without description or bibliographic reference to such a description (Art. 13.1); Lorenz (2005: 451) listed this name as as "Somoplatides Chaudoir 1846 [suppr.]".

Mazoréides Chaudoir, 1871b: 111 [stem: *Masore-*]. Type genus: *Masoreus* Dejean, 1821 [implicit use of *Mazoreus* as the type genus, which is an incorrect subsequent spelling of the type genus name, not in prevailing usage]. Comment: original vernacular name available (Art. 11.7.2): first used in latinized form by Chaudoir (1876: 2, as Masoreidae), generally accepted as in Lorenz (2005: 451, as Masoreina); incorrect stem formation, not in prevailing usage.

Corsyrini Ganglbauer, 1891b: 53 [stem: *Corsyr-*]. Type genus: *Corsyra* Dejean, 1825.

Aephnidiina Jakobson, 1907: 390 [stem: *Aephnidi-*]. Type genus: *Aephnidius* W. S. MacLeay, 1825.

Anaulacini Csiki, 1932b: 1287 [stem: *Anaulac-*]. Type genus: *Anaulacus* W. S. MacLeay, 1825. Comment: name proposed after 1930 without description or bibliographic reference to such a description (Art. 13.1), however available because it was used as valid before 2000 as in Blackwelder (1944: 52) and was not rejected by an author who, between 1961 and 1999, applied Article 13 of the then current edition of the Code (see Art. 13.2.1).

Discopterini Jedlička, 1941: 6, in key [stem: *Discopter-*]. Type genus: *Discoptera* Semenov, 1889.

Tribe Dercylini Sloane, 1923

Dercylini Sloane, 1923b: 250a [stem: *Dercyl-*]. Type genus: *Dercylus* Laporte, 1832.

Dercylidae Jeannel, 1948b: 626 [stem: *Dercyl-*]. Type genus: *Dercylus* Laporte, 1832. Comment: family-group name proposed as new without reference to Dercylini Sloane, 1923.

Tribe Drimostomatini Chaudoir, 1872

Drimostomidas Chaudoir, 1872a: 283 [stem: *Drimostomat-*]. Type genus: *Drimostoma* Dejean, 1830 [syn. of *Caelostomus* W. S. MacLeay, 1825]. Comment: the original spelling Drimostomidas is the accusative (Latin) form of Drimostomidae and was recognized as such by Madge (1989: 463); original incorrect original stem formation, not in prevailing usage.

Caelostomina Burgeon, 1935: 194 [stem: *Caelostom-*]. Type genus: *Caelostomus* W. S. MacLeay, 1825.

Cyrtolaina Whitehead and Ball, 1975: 595 [stem: *Cyrtola-*]. Type genus: *Cyrtolaus* H. W. Bates, 1882.

Tribe Dryptini Bonelli, 1810

Dryptinae Bonelli, 1810: Tabula Synoptica [stem: *Drypt-*]. Type genus: *Drypta* Latreille, 1797.

Tribe Enoicini Basilewsky, 1985

Enoicini Basilewsky, 1985: 16 [stem: *Enoic-*]. Type genus: *Enoicus* Péringuey, 1896.

Tribe Galeritini Kirby, 1825

Galeritidae Kirby, 1825: 564 [stem: *Galerit-*]. Type genus: *Galerita* Fabricius, 1801 [*Galerita* Gouan, 1770 [Aves] has been suppressed both for the Principle of Priority and the Principle of Homonymy (ICZN 1968b); *Galerita* Fabricius, 1801 placed on the Official List of Generic Names in Zoology (ICZN 1968b)].

Subtribe Galeritina Kirby, 1825

Galeritidae Kirby, 1825: 564 [stem: *Galerit-*]. Type genus: *Galerita* Fabricius, 1801 [*Galerita* Gouan, 1770 [Aves] has been suppressed both for the Principle of Priority and the Principle of Homonymy (ICZN 1968b); *Galerita* Fabricius, 1801 placed on the Official List of Generic Names in Zoology (ICZN 1968b)]. Comment: family-group name previously attributed to J. L. LeConte (1853b) or Lacordaire (1854a); name placed on the Official List of Family-Group Names in Zoology (ICZN 1968b, as Galeritini Lacordaire, 1854).

Galeritinini Jeannel, 1949c: 1057 [stem: *Galeritin-*]. Type genus: *Galeritina* Jeannel, 1949 [syn. of *Galerita* Fabricius, 1801]. Comment: replacement name for "Galeritini Lacordaire, 1854" because of the homonymy of the type genus; name placed on the Official Index of Rejected and Invalid Family-Group Names in Zoology (ICZN 1968b).

Galeritulini Jedlička, 1963: 279, in key [stem: *Galeritul-*]. Type genus: *Galeritula* Strand, 1936 [syn. of *Galerita* Fabricius, 1801]. Comment: replacement name for "Galeritini Lacordaire, 1854" because of the homonymy of the type genus; name placed on the Official Index of Rejected and Invalid Family-Group Names in Zoology (ICZN 1968b).

Subtribe Planetina Jedlička, 1941

Planetini Jedlička, 1941: 7, in key [stem: *Planet-*]. Type genus: *Planetes* W. S. MacLeay, 1825.

Tribe Geobaenini Péringuey, 1896

Geobaenides Péringuey, 1896: 469 [stem: *Geobaen-*]. Type genus: *Geobaenus* Dejean, 1829.

Tribe Ginemini Ball and Shpeley, 2002

Ginemini Ball and Shpeley, 2002: 77 [stem: *Ginem-*]. Type genus: *Ginema* Ball and Shpeley, 2002.

Tribe Glyptini Horn, 1881

Glypti G. H. Horn, 1881: 179 [stem: *Glypt-*]. Type genus: *Glyptus* Brullé, 1835 [*Glyptus* Hoffmannsegg, 1818 placed on the Official Index of Rejected and Invalid Generic Names in Zoology and *Glyptus* Brullé, 1835 placed on the Official List of Generic Names in Zoology (ICZN 1985b)].

Tribe Graphipterini Latreille, 1802

Graphipterides Latreille, 1802: 83 [stem: *Graphipter-*]. Type genus: *Graphipterus* Latreille, 1802. Comment: usage of Harpalinae Bonelli, 1810 is maintained over Graphipterinae Latreille, 1802 (Art. 35.5).

Tribe Harpalini Bonelli, 1810

Harpalii Bonelli, 1810: Tabula Synoptica [stem: *Harpal-*]. Type genus: *Harpalus* Latreille, 1802.

Subtribe Anisodactylina Lacordaire, 1854 *nomen protectum*

Eurytrichini J. L. LeConte, 1847: 376 [stem: *Eurytrich-*]. Type genus: *Eurytrichus* J. L. LeConte, 1847 [syn. of *Anisotarsus* Chaudoir, 1837]. Comment: *nomen oblitum* (Appendix 1).

Anisodactylides Lacordaire, 1854a: 268 [stem: *Anisodactyl-*]. Type genus: *Anisodactylus* Dejean, 1829. Comment: *nomen protectum* (see Appendix 1); original vernacular name available (Art. 11.7.2): first used in latinized form by Murray (1858: 346, as Anisodactylidae), generally accepted as in Lorenz (2005: 348, as Anisodactylini).

Anisotarsi Csiki, 1932a: 1039 [stem: *Anisotars-*]. Type genus: *Anisotarsus* Chaudoir, 1837. Comment: name proposed after 1930 without description or bibliographic reference to such a description (Art. 13.1), however available because it was used as valid before 2000 as in Blackwelder (1944: 46) and was not rejected by an author who, between 1961 and 1999, applied Article 13 of the then current edition of the Code (see Art. 13.2.1).

*Geopini Csiki, 1932a: 1026 [stem: *Geopin-*]. Type genus: *Geopinus* J. L. LeConte, 1847. Comment: unavailable family-group name, proposed after 1930 without description or bibliographic reference to such a description (Art. 13.1).

*Gnathaphani Schauberger, 1934: 104, 106 [stem: *Gnathaphan-*]. Type genus: *Gnathaphanus* W. S. MacLeay, 1825. Comment: unavailable family-group name, proposed after 1930 without description or bibliographic reference to such a description (Art. 13.1).

Subtribe Harpalina Bonelli, 1810

Harpalii Bonelli, 1810: Tabula Synoptica [stem: *Harpal-*]. Type genus: *Harpalus* Latreille, 1802. Comment: First Reviser found (Harpalini Bonelli, 1810 vs Ditomini Bonelli, 1810) is Noonan (1976: 28).

Ditomici Bonelli, 1810: Tabula Synoptica [stem: *Ditom-*]. Type genus: *Ditomus* Bonelli, 1810.

Acinopidae Laporte, 1834b: 67 [stem: *Acinopod-*]. Type genus: *Acinopus* Dejean, 1821. Comment: incorrect original stem formation, not in prevailing usage.

Ophonidae Laporte, 1834b: 68 [stem: *Ophon-*]. Type genus: *Ophonus* Dejean, 1821 [placed on the Official List of Generic Names in Zoology (ICZN 1990c)].

Cratognathidae Laporte, 1834b: 70 [stem: *Cratognath-*]. Type genus: *Cratognathus* Dejean, 1829.

Stenomorphidae Laporte, 1834b: 71 [stem: *Stenomorph-*]. Type genus: *Stenomorphus* Dejean, 1831.

Daptini J. L. LeConte, 1847: 371 [stem: *Dapt-*]. Type genus: *Daptus* Fischer von Waldheim, 1823.

Amblystomini Fauvel, 1889: 17 [stem: *Amblystom-*]. Type genus: *Amblystomus* Erichson, 1837.

Trichopselaphini Tschitschérine, 1900: 351 [stem: *Trichopselaph-*]. Type genus: *Trichopselaphus* Chaudoir, 1843.

Selenophorini Casey, 1914b: 134 [stem: *Selenophor-*]. Type genus: *Selenophorus* Dejean, 1829.

Pachycarina Stichel, 1923: 81 [stem: *Pachycar-*]. Type genus: *Pachycarus* Solier, 1835.

*Dioctini Csiki, 1932a: 1023a [stem: *Dioct-*]. Type genus: *Dioctes* Ménétriés, 1849 [syn. of *Machozetus* Chaudoir, 1850]. Comment: family-group name unavailable (Art. 11.6): originally published as synonym and not made available subsequently.

Diorychi Csiki, 1932a: 1193 [stem: *Diorych-*]. Type genus: *Dioryche* W. S. MacLeay, 1825. Comment: name proposed after 1930 without description or bibliographic reference to such a description (Art. 13.1), however available because it was used as valid before 2000 as in Blackwelder (1944: 48) and was not rejected by an author who, between 1961 and 1999, applied Article 13 of the then current edition of the Code (see Art. 13.2.1).

*Machozeti Csiki, 1932a: 1023 [stem: *Machozet-*]. Type genus: *Machozetus* Chaudoir, 1850. Comment: unavailable family-group name, proposed after 1930 without description or bibliographic reference to such a description (Art. 13.1).

Bradybaeni Csiki, 1932a: 1187 [stem: *Bradybaen-*]. Type genus: *Bradybaenus* Dejean, 1829. Comment: name proposed after 1930 without description or bibliographic reference to such a description (Art. 13.1), however available because it was used as valid before 2000 as in Noonan (1976: 28) and was not rejected by an author who, between 1961 and 1999, applied Article 13 of the then current edition of the Code (see Art. 13.2.1); Bradybaenidae Pilsbry, 1939 (type genus *Bradybaena* Beck, 1837) is currently used as valid in Mollusca, both family-group names have a correct stem; this case is to be referred to the Commission to remove the homonymy (Art. 55.3.1).

*Euryderi Csiki, 1932a: 1081 [stem: *Euryder-*]. Type genus: *Euryderus* J. L. LeConte, 1846. Comment: unavailable family-group name, proposed after 1930 without description or bibliographic reference to such a description (Art. 13.1).

*Heteracanthi Csiki, 1932a: 1085 [stem: *Heteracanth-*]. Type genus: *Heteracantha* Brullé, 1834. Comment: unavailable family-group name, proposed after 1930 without description or bibliographic reference to such a description (Art. 13.1).

*Barysomi Csiki, 1932a: 1192 [stem: *Barysom-*]. Type genus: *Barysomus* Dejean, 1829. Comment: unavailable family-group name, proposed after 1930 without description or bibliographic reference to such a description (Art. 13.1).

Machozeti Schauberger, 1934: 99 [stem: *Machozet-*]. Type genus: *Machozetus* Chaudoir, 1850.

Trichotichnini Jeannel, 1942: 624 [stem: *Trichotichn-*]. Type genus: *Trichotichnus* Morawitz, 1863.

Bleusei Antoine, 1959: 386 [stem: *Bleuse-*]. Type genus: *Bleusea* Bedel, 1896.

Eriotomi Antoine, 1959: 354 [stem: *Eriotom-*]. Type genus: *Eriotomus* Piochard de la Brûlerie, 1873 [syn. of *Oedesis* Motschulsky, 1850].

Graniger Antoine, 1959: 357 [stem: *Graniger-*]. Type genus: *Graniger* Motschulsky, 1864. Comment: an application will need to be submitted to the Commission to suppress Granigerini Bedel, 1900 (based on the misidentified type genus *Graniger* sensu Chaudoir, 1876) for the Principles of Priority and Homonymy (Art. 65.2.1) if this name is to be used as valid.

Cratacanthi Lindroth, 1968: 742 [stem: *Cratacanth-*]. Type genus: *Cratacanthus* Dejean, 1829.

*Carterophonini Jeanne, 1971: 11 [stem: *Carterophon-*]. Type genus: *Carterophonus* Ganglbauer, 1891 [syn. of *Graniger* Motschulsky, 1864]. Comment: unavailable family-group name, proposed after 1930 without description or bibliographic reference to such a description (Art. 13.1).

Subtribe Pelmatellina Bates, 1882

Pelmatellinae H. W. Bates, 1882: 67 [stem: *Pelmatell-*]. Type genus: *Pelmatellus* H. W. Bates, 1882.

Subtribe Stenolophina Kirby, 1837

Stenolophidae Kirby, 1837: 46 [stem: *Stelonoph-*]. Type genus: *Stenolophus* Dejean, 1821.

Polpochilinae H. W. Bates, 1891a: 10 [stem: *Polpochil-*]. Type genus: *Polpochila* Solier, 1849.

Acupalpini Tschitschérine, 1900: 351 [stem: *Acupalp-*]. Type genus: *Acupalpus* Latreille, 1829.

Cratocarini Casey, 1914b: 48 [stem: *Cratocar-*]. Type genus: *Cratocara* J. L. LeConte, 1863 [syn. of *Polpochila* Solier, 1849].

Pachytracheli Csiki, 1932a: 1082 [stem: *Pachytrachel-*]. Type genus: *Pachytrachelus* Chaudoir, 1852. Comment: name proposed after 1930 without description or bibliographic reference to such a description (Art. 13.1), however available because it was used as valid before 2000 as in Jeannel (1948b: 720, as Pachytrachelini) and was not rejected by an author who, between 1961 and 1999, applied Article 13 of the then current edition of the Code (see Art. 13.2.1).

*Agonoderi Csiki, 1932a: 1188 [stem: *Agonoder-*]. Type genus: *Agonoderus* Dejean, 1829. Comment: unavailable family-group name, proposed after 1930 without description or bibliographic reference to such a description (Art. 13.1).

*Bradycelli Csiki, 1932a: 1222 [stem: *Bradycell-*]. Type genus: *Bradycellus* Erichson, 1837. Comment: unavailable family-group name, proposed after 1930 without description or bibliographic reference to such a description (Art. 13.1).

Anoplogenii Schauberger, 1937: 272 [stem: *Anoplogeni-*]. Type genus: *Anoplogenius* Chaudoir, 1852 [syn. of *Loxoncus* Schmidt-Göbel, 1846].

*Dichirotrichi F. Burmeister, 1939: 186 [stem: *Dicheirotrich-*]. Type genus: *Dicheirotrichus* Jacquelin du Val, 1855 [as *Dichirotrichus*, unjustified emendation of type genus name by Gemminger and Harold (1868a: 262), not in prevailing usage]. Comment: unavailable family-group name, proposed after 1930 without description or bibliographic reference to such a description (Art. 13.1); incorrect original stem formation, not in prevailing usage (see Madge 1989: 463).

Bradycellini Jeannel, 1942: 700 [stem: *Bradycell-*]. Type genus: *Bradycellus* Erichson, 1837.

Hippolaetina Basilewsky, 1951b: 258 [stem: *Hippoloet-*]. Type genus: *Hippoloetis* Laporte, 1835 [as *Hippolaetis*, incorrect subsequent spelling of type genus name, not in prevailing usage]. Comment: incorrect original stem formation, not in prevailing usage.

Anthracini Schuler, 1970: 114 [stem: *Anthrac-*]. Type genus: *Anthracus* Motschulsky, 1850.

Tribe Helluonini Hope, 1838

Helluonidae Hope, 1838a: 110 [stem: *Helluon-*]. Type genus: *Helluo* Bonelli, 1813.

Subtribe Helluonina Hope, 1838

Helluonidae Hope, 1838a: 110 [stem: *Helluon-*]. Type genus: *Helluo* Bonelli, 1813.

Subtribe Omphrina Jedlička, 1941

Omphrini Jedlička, 1941: 6, in key [stem: *Omphr-*]. Type genus: *Omphra* Dejean, 1825.

Helluomorphina H. Reichardt, 1974: 226 [stem: *Helluomorph-*]. Type genus: *Helluomorpha* Laporte, 1834.

Tribe Hexagoniini Horn, 1881 (1834)

Trigonodactyliens Brullé, 1834: 127 [stem: *Trigonodactyl-*]. Type genus: *Trigonodactyla* Dejean, 1831 [syn. of *Hexagonia* Kirby, 1825]. Comment: original vernacular name available (Art. 11.7.2): first used in latinized form and generally accepted as in Desmarest (1851: 56, as Trigonodactylidae); use of the younger name Hexagoniini G. H. Horn, 1881 conserved over this name (Art. 40.2).

Hexagoniae G. H. Horn, 1881: 146 [stem: *Hexagoni-*]. Type genus: *Hexagonia* Kirby, 1825. Comment: use of family-group name conserved over Trigonodactylini Brullé, 1834 (Art. 40.2).

Tribe Idiomorphini Bates, 1891

Idiomorphinae H. W. Bates, 1891b: cccxxxiii [stem: *Idiomorph-*]. Type genus: *Idiomorphus* Chaudoir, 1846.

Perochnoristhinae Basilewsky, 1973b: 224 [stem: *Perochnoristh-*]. Type genus: *Perochnoristhus* Basilewsky, 1973.

Tribe Lachnophorini LeConte, 1853

Lachnophori J. L. LeConte, 1853b: 375 [stem: *Lachnophor-*]. Type genus: *Lachnophorus* Dejean, 1831.

Subtribe Lachnophorina LeConte, 1853

Lachnophori J. L. LeConte, 1853b: 375 [stem: *Lachnophor-*]. Type genus: *Lachnophorus* Dejean, 1831.

Anchonodérides Lacordaire, 1854a: 373 [stem: *Anchonoder-*]. Type genus: *Anchonoderus* Reiche, 1843. Comment: original vernacular name available (Art. 11.7.2): first used in latinized form and generally accepted as in H. W. Bates (1871: 30, as Anchonoderinae).

Eucaeri J. L. LeConte, 1861: 22 [stem: *Eucaer-*]. Type genus: *Eucaerus* J. L. LeConte, 1853.

EGINI G. H. Horn, 1881: 152 [stem: *Eg-*]. Type genus: *Ega* Laporte, 1834.

Subtribe SELININA Jeannel, 1948

SELININI Jeannel, 1948b: 743, in key [stem: *Selin-*]. Type genus: *Selina* Motschulsky, 1858. Comment: the junior homonym SELININA Koch, 1956 (type genus *Selinus* Mulsant and Rey, 1853) is available in TENEBRIONIDAE; this case is to be referred to the Commission to remove the homonymy (Art. 55.3.1).

Tribe LEBIINI Bonelli, 1810

LEBIOTAE Bonelli, 1810: Tabula Synoptica [stem: *Lebi-*]. Type genus: *Lebia* Latreille, 1802. Comment: First Reviser (LEBIINI Bonelli, 1810 vs DROMIUSINI Bonelli, 1810) not determined, current usage maintained.

Subtribe ACTENONYCINA Bates, 1871

ACTENONYCINAE H. W. Bates, 1871: 30 [stem: *Actenonyc-*]. Type genus: *Actenonyx* A. White, 1846. Comment: current spelling maintained (Art. 29.5): incorrect stem formation in prevailing usage (should be *Actenonych-*).

Subtribe AGRINA Kirby, 1837

AGRIDAE Kirby, 1837: 13 [stem: *Agr-*]. Type genus: *Agra* Fabricius, 1801.

Subtribe APENINA Ball, 1983

APENINA Ball, 1983: 516 [stem: *Apen-*]. Type genus: *Apenes* J. L. LeConte, 1851.

*TRYMOSTERNINI Zaballos and Jeanne, 1994: 116 [stem: *Trymostern-*]. Type genus: *Trymosternus* Chaudoir, 1873. Comment: unavailable family-group name, proposed after 1930 without description or bibliographic reference to such a description (Art. 13.1).

*PLATYTARINI Zaballos and Jeanne, 1994: 116 [stem: *Platytar-*]. Type genus: *Platytarus* Fairmaire, 1850. Comment: unavailable family-group name, proposed after 1930 without description or bibliographic reference to such a description (Art. 13.1).

Subtribe CALLEIDINA Chaudoir, 1873

CALLIDIDES Chaudoir, 1873b: 97 [stem: *Calleid-*]. Type genus: *Calleida* Dejean, 1824 [as *Callida*, unjustified emendation of type genus name by Agassiz (1846b: 58), not in prevailing usage]. Comment: original vernacular name available (Art. 11.7.2): first used in latinized form by H. W. Bates (1873: 311, as CALLEIDINAE), generally accepted as in Csiki (1932b: 1436, as CALLIDI [incorrect stem formation]); incorrect stem formation, not in prevailing usage.

Plochionidae Gozis, 1875: 3 [stem: *Plochion-*]. Type genus: *Plochionus* Dejean, 1821.

Anomotarina Habu, 1967: 117 [stem: *Anomotar-*]. Type genus: *Anomotarus* Chaudoir, 1875.

Subtribe Celaenephina Habu, 1982

Celaenephina Habu, 1982: 111 [stem: *Celaeneph-*]. Type genus: *Celaenephes* Schmidt-Göbel, 1846.

Subtribe Cymindidina Laporte, 1834

Cymindidae Laporte, 1834b: 46 [stem: *Cymindid-*]. Type genus: *Cymindis* Latreille, 1806. Comment: incorrect original stem formation, not in prevailing usage.

Taridae Gistel, 1848: [2] [stem: *Tar-*]. Type genus: *Tarus* Clairville, 1806 [syn. of *Cymindis* Latreille, 1806].

Pseudomasoreini Jeannel, 1942: 1039, in key [stem: *Pseudomasore-*]. Type genus: *Pseudomasoreus* Desbrochers des Loges, 1904.

Subtribe Demetriadina Bates, 1886

Demetriinae H. W. Bates, 1886: 207 [stem: *Demetriad-*]. Type genus: *Demetrias* Bonelli, 1810. Comment: incorrect original stem formation, not in prevailing usage.

Peliocypini Basilewsky, 1984: 553 [stem: *Peliocypad-*]. Type genus: *Peliocypas* Schmidt-Göbel, 1846. Comment: incorrect original stem formation, not in prevailing usage.

Subtribe Dromiusina Bonelli, 1810

Dromiei Bonelli, 1810: Tabula Synoptica [stem: *Dromius-*]. Type genus: *Dromius* Bonelli, 1810 [placed on the Official List of Generic Names in Zoology (ICZN 2006b)]. Comment: Dromiidae Bonelli, 1810 placed on the Official Index of Rejected and Invalid Family-Group Names in Zoology, correct stem ruled to be *Dromius-* and Dromiusidae Bonelli, 1810 placed on the Official List of Family-Group Names in Zoology (ICZN 2006b).

Lichnasthenitae J. Thomson, 1858: 35 [stem: *Lichnasthen-*]. Type genus: *Lichnasthenus* J. Thomson, 1858. Comment: J. Thomson (1858: 35) used the spelling Lichnastenitae but corrected the name to Lichnasthenitae in the same publication in the index (page 458) and in the errata (page [472, unn.]), the corrected spelling is considered a justified emendation (see Madge 1989: 464).

Lionychidae Jeannel, 1948b: 378, in key [stem: *Lionych-*]. Type genus: *Lionychus* Wissmann, 1846.

SINGILINI Jeannel, 1949c: 915 [stem: *Singil-*]. Type genus: *Singilis* Rambur, 1837.

*SYNTOMINI Jeanne, 1972: 101 [stem: *Syntom-*]. Type genus: *Syntomus* Hope, 1838. Comment: unavailable family-group name, proposed after 1930 without description or bibliographic reference to such a description (Art. 13.1).

METADROMIINA Basilewsky, 1984: 545 [stem: *Metadromi-*]. Type genus: *Metadromius* Bedel, 1907.

METAXYMORPHINA Basilewsky, 1984: 551 [stem: *Metaxymorph-*]. Type genus: *Metaxymorphus* Chaudoir, 1850.

SINGILIOMIMINA Basilewsky, 1984: 552 [stem: *Singiliomim-*]. Type genus: *Singiliomimus* Péringuey, 1896.

Subtribe GALLERUCIDIINA Chaudoir, 1872

GALLERUCIDIAE Chaudoir, 1872b: 416 [stem: *Gallerucidi-*]. Type genus: *Gallerucidia* Chaudoir, 1872.

LEBIDIINA Jakobson, 1907: 394 [stem: *Lebidi-*]. Type genus: *Lebidia* Morawitz, 1862.

Subtribe LEBIINA Bonelli, 1810

LEBIOTAE Bonelli, 1810: Tabula Synoptica [stem: *Lebi-*]. Type genus: *Lebia* Latreille, 1802.

ENCRATIDAE Gistel, 1856a: 355 [stem: *Encrat-*]. Type genus: *Encrates* Gistel, 1848 [syn. of *Lebia* Latreille, 1802].

LAMPRIADAE Chaudoir, 1871b: 115 [stem: *Lampri-*]. Type genus: *Lamprias* Bonelli, 1810. Comment: incorrect original stem formation, not in prevailing usage.

PHYSODÉRIDES Chaudoir, 1877: 213 [stem: *Physoder-*]. Type genus: *Physodera* Eschscholtz, 1829. Comment: original vernacular name available (Art. 11.7.2): first used in latinized form by H. W. Bates (1883: 207, as PHYSODERINAE), generally accepted as in Lorenz (2005: 489, as PHYSODERINA).

Subtribe METALLICINA Basilewsky, 1984

METALLICINI Basilewsky, 1984: 542 [stem: *Metallic-*]. Type genus: *Metallica* Chaudoir, 1872.

Subtribe NEMOTARSINA Bates, 1883

NEMOTARSINAE H. W. Bates, 1883: 173 [stem: *Nemotars-*]. Type genus: *Nemotarsus* J. L. LeConte, 1853.

Subtribe PERICALINA Hope, 1838

PERICALLIDAE Hope, 1838a: 105 [stem: *Perical-*]. Type genus: *Pericalus* W. S. MacLeay, 1825. Comment: incorrect original stem formation, not in prevailing usage.

*Mormolycites Blanchard, 1845a: 366 [stem: *Mormolyc-*]. Type genus: *Mormolyce* Hagenbach, 1825. Comment: original vernacular name unavailable (Art. 11.7.2): subsequently used in latinized form but not generally attributed to Blanchard (1845a).

Costodérides Chaudoir, 1848: 116 [stem: *Coptoder-*]. Type genus: *Coptodera* Dejean, 1825. Comment: original vernacular name available (Art. 11.7.2): first used in latinized form by H. W. Bates (1871: 30, as Coptoderinae), generally accepted as in Handlirsch (1925: 547, as Coptoderini); incorrect stem formation, not in prevailing usage.

Mormolycidae Desmarest, 1851: 62 [stem: *Mormolyc-*]. Type genus: *Mormolyce* Hagenbach, 1825.

Thyréoptérides Chaudoir, 1869: 113 [stem: *Thyreopter-*]. Type genus: *Thyreopterus* Dejean, 1831. Comment: original vernacular name available (Art. 11.7.2): first used in latinized form by Bertkau (1878: 417, as Thyreopterini), generally accepted as in Jeannel (1950b: 165, as Thyreopteridae).

Eucheilinae H. W. Bates, 1883: 168 [stem: *Eucheil-*]. Type genus: *Eucheila* Dejean, 1829 [*Eucheila* is an incorrect subsequent spelling of the original spelling *Eucheyla*, in prevailing usage and so deemed to be the correct original spelling (Art. 33.3.1)].

Miscelini Sloane, 1907b: 473 [stem: *Miscel-*]. Type genus: *Miscelus* Klug, 1834.

Periglossiinae Liebke, 1929: 247 [stem: *Periglossi-*]. Type genus: *Periglossium* Liebke, 1929 [syn. of *Inna* Putzeys, 1861].

Catascopi Csiki, 1932b: 1352 [stem: *Catascop-*]. Type genus: *Catascopus* Kirby, 1823. Comment: name proposed after 1930 without description or bibliographic reference to such a description (Art. 13.1), however available because it was used as valid before 2000 as in Blackwelder (1944: 57) and was not rejected by an author who, between 1961 and 1999, applied Article 13 of the then current edition of the Code (see Art. 13.2.1).

Miscelini Jeannel, 1949c: 1006 [stem: *Miscel-*]. Type genus: *Miscelus* Klug, 1834. Comment: family-group name proposed as new without reference to Miscelini Sloane, 1907.

Lobodontini Jeannel, 1949c: 1007, in key [stem: *Lobodont-*]. Type genus: *Lobodontus* Chaudoir, 1842.

Thysanotini Jeannel, 1949c: 947, in key [stem: *Thysanot-*]. Type genus: *Thysanotus* Chaudoir, 1848.

Somotrichini Mateu, 1963: 125, in key [stem: *Somotrich-*]. Type genus: *Somotrichus* Seidlitz, 1887.

Subtribe Pseudotrechina Basilewsky, 1984

Pseudotrechini Basilewsky, 1984: 555 [stem: *Pseudotrech-*]. Type genus: *Pseudotrechus* Rosenhauer, 1856.

Subtribe Sugimotoina Habu, 1975

Sugimotoina Habu, 1975: 77 [stem: *Sugimoto-*]. Type genus: *Sugimotoa* Habu, 1975.

Subtribe Trichina Basilewsky, 1984

Trichini Basilewsky, 1984: 549 [stem: *Trich-*]. Type genus: *Trichis* Klug, 1832.

Tribe Licinini Bonelli, 1810

Licinii Bonelli, 1810: Tabula Synoptica [stem: *Licin-*]. Type genus: *Licinus* Latreille, 1802.

Subtribe Dicaelina Laporte, 1834

Dicoelidae Laporte, 1834b: 83 [stem: *Dicael-*]. Type genus: *Dicaelus* Bonelli, 1813 [as *Dicoelus*, incorrect subsequent spelling of type genus name, not in prevailing usage]. Comment: incorrect original stem formation, not in prevailing usage.

Rembidae Gistel, 1848: [2] [stem: *Remb-*]. Type genus: *Rembus* W. S. MacLeay, 1825 [preoccupied genus name, not *Rembus* Germar, 1824 [Coleoptera: Curculionidae]; syn. of *Diplocheila* Brullé, 1834]. Comment: permanently invalid (Art. 39): based on preoccupied type genus.

*Submerini Lafer, 1989: 205 [stem: *Submer-*]. Type genus: *Submera* Habu, 1956. Comment: unavailable family-group name, proposed after 1930 without description or bibliographic reference to such a description (Art. 13.1).

Subtribe Dicrochilina Ball, 1992

Dicrochilina Ball, 1992: 347 [stem: *Dicrochil-*]. Type genus: *Dicrochile* Guérin-Méneville, 1846.

Subtribe Lestignathina Ball, 1992

Lestignathina Ball, 1992: 347 [stem: *Lestignath-*]. Type genus: *Lestignathus* Erichson, 1842.

Subtribe Licinina Bonelli, 1810

Licinii Bonelli, 1810: Tabula Synoptica [stem: *Licin-*]. Type genus: *Licinus* Latreille, 1802.

Badistidae Gistel, 1856a: 357 [stem: *Badister-*]. Type genus: *Badister* Clairville, 1806 [as *Badistes*, unjustified emendation of type genus name by Agassiz (1846b: 42), not in prevailing usage]. Comment: incorrect original stem formation, not in prevailing usage.

Badistritae Jeannel, 1942: 999 [stem: *Badister-*]. Type genus: *Badister* Clairville, 1806 [as *Badistes*, unjustified emendation of type genus name by Agassiz (1846b: 42), not in prevailing usage]. Comment: family-group name proposed as new without reference to Badistidae Gistel, 1856; incorrect original stem formation, not in prevailing usage.

Tribe Melanchitonini Jeannel, 1948

Melanodini Alluaud, 1916: 228 [stem: *Melanod-*]. Type genus: *Melanodes* Chaudoir, 1876 [preoccupied type genus, not *Melanodes* Guenée, 1857 [Lepidoptera]; syn. of *Melanchiton* Basilewsky, 1946]. Comment: permanently invalid (Art. 39): based on preoccupied type genus.

Melanchitonitae Jeannel, 1948b: 627 [stem: *Melanchiton-*]. Type genus: *Melanchiton* Basilewsky, 1946. Comment: replacement name for Melanodini Alluaud, 1916.

Tribe Microcheilini Jeannel, 1948

Microchilitae Jeannel, 1948b: 616 [stem: *Microcheil-*]. Type genus: *Microcheila* Brullé, 1834 [as *Microchila*, unjustified emendation of type genus name by Agassiz (1846b: 233), not in prevailing usage]. Comment: incorrect original stem formation, not in prevailing usage.

Tribe Morionini Brullé, 1835

Morioniens Brullé, 1835: 36 [stem: *Morion-*]. Type genus: *Morion* Latreille, 1810. Comment: original vernacular name available (Art. 11.7.2): first used in latinized form by Hope (1838a: 109, as Morionidae [based on "*Morio* Latreille", an incorrect subsequent spellling of the type genus name, see Madge (1989: 465)], generally accepted as in Lorenz (2005: 247, as Morionini).

Tribe Odacanthini Laporte, 1834

Odacanthidae Laporte, 1834b: 40 [stem: *Odacanth-*]. Type genus: *Odacantha* Paykull, 1798.

Casnoniae J. L. LeConte, 1861: 21 [stem: *Cosnani-*]. Type genus: *Cosnania* Dejean, 1821 [as *Casnonia*, incorrect subsequent spelling of type genus name, not in prevailing usage]. Comment: incorrect original stem formation, not in prevailing usage.

Colliurini Bedel, 1910: 72 [stem: *Colliur-*]. Type genus: *Colliuris* DeGeer, 1774.

Lasiocerini Jeannel, 1948b: 747, in key [stem: *Lasiocer-*]. Type genus: *Lasiocera* Dejean, 1831.

Tribe Omphreini Ganglbauer, 1891

Omphreini Ganglbauer, 1891b: 26 [stem: *Omphre-*]. Type genus: *Omphreus* Dejean, 1828.

Tribe Oodini LaFerté-Sénectère, 1851

Oodites LaFerté-Sénectère, 1851: 266 [stem: *Ood-*]. Type genus: *Oodes* Bonelli, 1810. Comment: original vernacular name available (Art. 11.7.2): first used in latinized form by Bedel (1879: 52, as OODINI), generally accepted as in Lorenz (2005: 324, as OODINI).

THRYPTOCERINI Jeannel, 1949c: 841 [stem: *Thryptocer-*]. Type genus: *Thryptocerus* Chaudoir, 1878.

SPHAERODINI Jeannel, 1949c: 829, in key [stem: *Sphoerod-*]. Type genus: *Sphoerodes* Chaudoir, 1883 [as *Sphaerodes*, incorrect subsequent spelling of type genus name, not in prevailaing usage]. Comment: incorrect original stem formation, not in prevailing usage.

SIMOINI Basilewsky, 1953a: 153 [stem: *Simo-*]. Type genus: *Simous* Chaudoir, 1882.

Tribe ORTHOGONIINI Schaum, 1857

ORTHOGONIDEN Schaum, 1857b: 308 [stem: *Orthogoni-*]. Type genus: *Orthogonius* W. S. MacLeay, 1825. Comment: original vernacular name available (Art. 11.7.2): first used in latinized form by Schaum (1860: 774, as ORTHOGONINI [incorrect stem formation]), generally accepted as in Lorenz (2005: 391, as ORTHOGONIINAE); incorrect original stem formation, not in prevailing usage.

Tribe PANAGAEINI Bonelli, 1810

PANAGAEIDES Bonelli, 1810: Tabula Synoptica [stem: *Panagae-*]. Type genus: *Panagaeus* Latreille, 1802.

Subtribe BRACHYGNATHINA Basilewsky, 1946

BRACHYGNATHINI Basilewsky, 1946: 7 [stem: *Brachygnath-*]. Type genus: *Brachygnathus* Perty, 1830.

Subtribe PANAGAEINA Bonelli, 1810

PANAGAEIDES Bonelli, 1810: Tabula Synoptica [stem: *Panagae-*]. Type genus: *Panagaeus* Latreille, 1802.

Subtribe TEFFLINA Basilewsky, 1946

TEFFLINI Basilewsky, 1946: 7 [stem: *Teffl-*]. Type genus: *Tefflus* Leach, 1819.

Tribe PELECIINI Chaudoir, 1880

PÉLÉCIDES Chaudoir, 1880: 317 [stem: *Peleci-*]. Type genus: *Pelecium* Kirby, 1819.

Subtribe AGONICINA Sloane, 1920

AGONICINI Sloane, 1920: 129 [stem: *Agonic-*]. Type genus: *Agonica* Sloane, 1920.

Subtribe PELECIINA Chaudoir, 1880

PÉLÉCIDES Chaudoir, 1880: 317 [stem: *Peleci-*]. Type genus: *Pelecium* Kirby, 1819. Comment: original vernacular name available (Art. 11.7.2): first used in latinized form by G. H. Horn (1881: 170, as Peleciini), generally accepted as in Csiki (1932b: 1285, as Peleciini); incorrect original stem formation, not in prevailing usage.

Disphaericini Sloane, 1923a: 248 [stem: *Dispheric-*]. Type genus: *Dsiphericus* G. R. Waterhouse, 1842 [as *Disphaericus*, unjustified emendation of type genus name by Agassiz (1846b: 127), not in prevailing usage]. Comment: incorrect original stem formation, not in prevailing usage.

Tribe Pentagonicini Bates, 1873

Pentagonicinae H. W. Bates, 1873: 320 [stem: *Pentagonic-*]. Type genus: *Pentagonica* Schmidt-Göbel, 1846.

Scopodinae H. W. Bates, 1874b: 275 [stem: *Scopod-*]. Type genus: *Scopodes* Erichson, 1842.

Tribe Perigonini Horn, 1881 *nomen protectum*

Trechichinae H. W. Bates, 1873: 281 [stem: *Trechic-*]. Type genus: *Trechicus* J. L. LeConte, 1853 [as *Trechichus*, incorrect subsequent spelling of type genus name, not in prevailing usage; subgenus of *Perigona* Laporte, 1835]. Comment: *nomen oblitum* (see Appendix 1); incorrect original stem formation, not in prevailing usage.

Perigonae G. H. Horn, 1881: 143 [stem: *Perigon-*]. Type genus: *Perigona* Laporte, 1835. Comment: *nomen protectum* (see Appendix 1).

Tribe Physocrotaphini Chaudoir, 1863

Physocrotaphides Chaudoir, 1863: 303 [stem: *Physocrotaph-*]. Type genus: *Physocrotaphus* Parry, 1849. Comment: original vernacular name available (Art. 11.7.2): first used in latinized form by H. W. Bates (1892: 388, as Physocrotaphinae), generally accepted as in B. P. Moore (1998: 369, as Physocrotaphini).

Helluodini Csiki, 1932b: 1571 [stem: *Helluod-*]. Type genus: *Helluodes* Westwood, 1847. Comment: name proposed after 1930 without description or bibliographic reference to such a description (Art. 13.1), however available because it was used as valid before 2000 as in Darlington (1968: 222) and was not rejected by an author who, between 1961 and 1999, applied Article 13 of the then current edition of the Code (see Art. 13.2.1).

Tribe Platynini Bonelli, 1810

Platynii Bonelli, 1810: Tabula Synoptica [stem: *Platyn-*]. Type genus: *Platynus* Bonelli, 1810. Comment: First Reviser found (Platynina Bonelli, 1810 vs Anchomenina Bonelli, 1810) is Jakobson (1906: 317).

Anchomenii Bonelli, 1810: Tabula Synoptica [stem: *Anchomen-*]. Type genus: *Anchomenus* Bonelli, 1810.

Sericodiadae Kirby, 1837: 14 [stem: *Sericod-*]. Type genus: *Sericoda* Kirby, 1837.

AGONIDAE Kirby, 1837: 23 [stem: *Agonum-*]. Type genus: *Agonum* Bonelli, 1810 [placed on the Official List of Generic Names in Zoology (ICZN 1996d)]. Comment: senior homonym of AGONIDAE Swainson, 1839 (type genus *Agonus* Bloch and Schneider, 1801) in Pisces; AGONIDAE Kirby, 1837 placed on the Official Index of Rejected and Invalid Family-Group Names in Zoology (ICZN 1996d), stem emended to *Agonum-* and AGONUMIDAE Kirby, 1837 placed on the Official List of Family-Group Names in Zoology (ICZN 1996d).

COLPODIDAS Chaudoir, 1872a: 285 [stem: *Colpod-*]. Type genus: *Colpodes* W. S. MacLeay, 1825. Comment: the original spelling COLPODIDAS is the accusative (Latin) form of and COLPODIDAE was recognized as such by Madge (1989: 463); COLPODIDAE Poche, 1913 (type genus *Colpoda* Müller, 1773) has been used as valid in Protista as recently as in 2006; this case is to be referred to the Commission to remove the homonymy (Art. 55.3.1).

AGELAEINA Jakobson, 1907: 334 [stem: *Agelae-*]. Type genus: *Agelaea* Gené, 1839.

PROSPHODRINI J. M. Valentine, 1987: 74 [stem: *Prosphodr-*]. Type genus: *Prosphodrus* Britton, 1959.

MELEAGROSINI Morvan, 2004: 2 [stem: *Meleagr-*]. Type genus: *Meleagros* Kirschenhofer, 1999. Comment: incorrect original stem formation, not in prevailing usage.

Tribe PSEUDOMORPHINI Hope, 1838

HETEROMORPHIDAE Hope, 1838a: 108 [stem: *Pseudomorph-*]. Type genus: *Pseudomorpha* Kirby, 1823 [as *Heteromorpha*, incorrect original spelling of type genus name, not in prevailing usage]. Comment: this family-group name is based on an incorrect original spelling of the name of its type genus (*Heteromorpha*) and must be corrected (ICZN 1999a: Article 35.4.1); Kirby (1823a) originally used two different names for the type genus, *Pseudomorpha* (in the description on p. 98) and *Heteromorpha* (in the explanation of the figures on p. 109). Both names are considered different original spellings of the same name. Kirby reissued the paper later the same year in another journal, without the plate, and so only *Pseudomorpha* was used; as such, Kirby (1823b) is the "First Reviser" and *Pseudomorpha* is the correct original spelling (ICZN 1999a: Article 24.2.4), therefore HETEROMORPHIDAE Hope, 1838 is corrected to PSEUDOMORPHIDAE Hope, 1838.

Tribe PTEROSTICHINI Bonelli, 1810

PTEROSTICHII Bonelli, 1810: Tabula Synoptica [stem: *Pterostich-*]. Type genus: *Pterostichus* Bonelli, 1810.

Subtribe ABACOMORPHINA Tschitschérine, 1902

ABACOMORPHINI Tschitschérine, 1902: 507 [stem: *Abacomorph-*]. Type genus: *Abacomorphus* Chaudoir, 1878. Comment: First Reviser (ABACOMORPHINA Tschitschérine, 1902 vs SPHODROSOMINA Tschitschérine, 1902) not determined, current usage maintained.

SPHODROSOMINI Tschitschérine, 1902: 507 [stem: *Sphodrosom-*]. Type genus: *Sphodrosomus* Perroud and Montrouzier, 1864.

Subtribe EUCHROINA Chaudoir, 1874

EUCHROIDES Chaudoir, 1874: 16 [stem: *Euchro-*]. Type genus: *Euchroa* Brullé, 1834. Comment: original vernacular name available (Art. 11.7.2): first used in latinized form by Tschitschérine (1899: 84, as EUCHROINI), generally accepted as in Frania and Ball (2007: 12, as EUCHROINA).

Subtribe METIINA Straneo, 1951

ANTARCTIIDES Lacordaire, 1854a: 336 [stem: *Antarcti-*]. Type genus: *Antarctia* Dejean, 1828 [preoccupied genus name, not *Antarctia* Hübner, 1820 [Lepidoptera]; syn. of *Metius* Curtis, 1838]. Comment: original vernacular name available (Art. 11.7.2): first used in latinized form by Reed (1874: 58, as ANTARCTIINAE), generally accepted as in Jeannel (1942: 734, as ANTARCTIITAE); permanently invalid (Art. 39): based on preoccupied type genus.

METIINI Straneo, 1951: 56 [stem: *Meti-*]. Type genus: *Metius* Curtis, 1838.

KUSCHELININI Jeannel, 1967: 437 [stem: *Kuschelin-*]. Type genus: *Kuschelinus* Straneo, 1963.

Subtribe MICROCEPHALINA Tschitschérine, 1898

MICROCEPHALIDES Tschitschérine, 1898: 46 [stem: *Microcephal-*]. Type genus: *Microcephalus* Dejean, 1828. Comment: the older name MICROCEPHALI Latreille, 1825: 345 is unavailable (Art. 11.7.1.1) because it is not based on an available generic name [descriptive name for a group of STAPHYLINIDAE including *Tachyporus* and other genera with relatively small heads].

TICHONIINI Semenov, 1904a: 201 [stem: *Tichoni-*]. Type genus: *Tichonia* Semenov, 1904 [preoccupied genus name, not *Tichonia* Hübner, 1826 [Lepidoptera]; syn. of *Microcephalus* Dejean, 1828]. Comment: replacement name for MICROCEPHALINI Tschitschérine, 1898 because of the homonymy of the type genus; permanently invalid (Art. 39): based on preoccupied type genus.

TICHONILLINA Emden, 1958: 24 [stem: *Tichonill-*]. Type genus: *Tichonilla* Strand, 1942 [syn. of *Microcephalus* Dejean, 1828].

Subtribe PTEROSTICHINA Bonelli, 1810

PTEROSTICHII Bonelli, 1810: Tabula Synoptica [stem: *Pterostich-*]. Type genus: *Pterostichus* Bonelli, 1810. Comment: First Reviser (PTEROSTICHINA Bonelli, 1810 vs POECILINA Bonelli, 1810 vs MOLOPINA Bonelli, 1810) not determined, current usage maintained.

POECILII Bonelli, 1810: Tabula Synoptica [stem: *Poecil-*]. Type genus: *Poecilus* Bonelli, 1810.

Molopides Bonelli, 1810: Tabula Synoptica [stem: *Molop-*]. Type genus: *Molops* Bonelli, 1810.

Féroniens Dejean, 1825: 3 [stem: *Feroni-*]. Type genus: *Feronia* Latreille, 1816 [syn. of *Poecilus* Bonelli, 1810]. Comment: original vernacular name available (Art. 11.7.2): first used in latinized form by Thon (1827: 15, as Feroniae), generally accepted as in Schaum (1857a: 136, as Feronidae).

Trogonotomidae Laporte, 1834b: 75 [stem: *Trigonotom-*]. Type genus: *Trigonotoma* Dejean, 1828. Comment: the original spelling was Trogonotomidae, but this was apparently a typographical error since both the vernacular spelling (Trigonotomites) and the genus name (*Trigonotoma*) were spelled with the stem "*Trigonotom-*" on the same page.

Catadromiens Brullé, 1834: 277, 328 [stem: *Catadrom-*]. Type genus: *Catadromus* W. S. MacLeay, 1825. Comment: original vernacular name available (Art. 11.7.2): first used in latinized form and generally accepted as in Desmarest (1851: 143, as Catadromidae).

Thaliadae Hope, 1838a: 71 [stem: *Thali-*]. Type genus: *Thalia* Hope, 1838 [preoccupied genus name, not *Thalia* Blumenbach, 1827 [Tunicata]; syn. of *Poecilus* Bonelli, 1810]. Comment: permanently invalid (Art. 39): based on preoccupied type genus.

Stomidae Chaudoir, 1846: 514 [stem: *Stomid-*]. Type genus: *Stomis* Clairville, 1806. Comment: incorrect original stem formation, not in prevailing usage.

Rhathyminae H. W. Bates, 1891b: cccxxxiv [stem: *Rathym-*]. Type genus: *Rathymus* Dejean, 1831 [as *Rhathymus*, unjustified emendation of type genus name by Agassiz (1846b: 321), not in prevailing usage]. Comment: incorrect original stem formation, not in prevailing usage.

Trigonognathides Tschitschérine, 1898: 66 [stem: *Trigonognath-*]. Type genus: *Trigonognatha* Motschulsky, 1858.

Platysmatini Tschitschérine, 1899: 84 [stem: *Platysmat-*]. Type genus: *Platysma* Bonelli, 1810.

Cyrtoderini Tschitschérine, 1902: 506 [stem: *Cyrtoder-*]. Type genus: *Cyrtoderus* Hope, 1842.

Darodiliini Tschitschérine, 1902: 506 [stem: *Darodili-*]. Type genus: *Darodilia* Laporte, 1867.

Deliniini Tschitschérine, 1902: 506 [stem: *Delini-*]. Type genus: *Delinius* Westwood, 1864.

Cyphosomatini Tschitschérine, 1902: 507 [stem: *Cyphosomat-*]. Type genus: *Cyphosoma* Hope, 1842 [preoccupied genus name, not *Cyphosoma* Mannerheim, 1837 [Coleoptera: Buprestidae]; syn. of *Cratogaster* Blanchard, 1853]. Comment: permanently invalid (Art. 39): based on preoccupied type genus.

Myadina Jakobson, 1907: 334 [stem: *Myad-*]. Type genus: *Myas* Sturm, 1826.

Cratogastri Csiki, 1929: 524 [stem: *Cratogastr-*]. Type genus: *Cratogaster* Blanchard, 1843.

Abaxini Schuler, 1970: 115 [stem: *Abac-*]. Type genus: *Abax* Bonelli, 1810. Comment: incorrect original stem formation, not in prevailing usage.

Aristochroodini Sciaky, 1996: 437 [stem: *Aristochrood-*]. Type genus: *Aristochroodes* Marcilhac, 1993.

Tribe Sphodrini Laporte, 1834

Sphodridae Laporte, 1834b: 78 [stem: *Sphodr-*]. Type genus: *Sphodrus* Clairville, 1806. Comment: published before 9 August 1834; First Reviser found (Sphodrini Laporte, 1834 vs Calathini Laporte, 1834) is Habu (1978: 296).

Subtribe Atranopsina Baehr, 1982

Atranopsina Baehr, 1982: 265 [stem: *Atranops-*]. Type genus: *Atranopsis* Baehr, 1982. Comment: current spelling maintained (Art. 29.5): incorrect stem formation in prevailing usage (should be *Atranopse-*).

*Platyderina Baehr, 1982: 265 [stem: *Platyder-*]. Type genus: *Platyderes* Stephens, 1827. Comment: unavailable family-group name, proposed after 1930 without description or bibliographic reference to such a description (Art. 13.1).

Subtribe Calathina Laporte, 1834

Calathidae Laporte, 1834b: 71 [stem: *Calath-*]. Type genus: *Calathus* Bonelli, 1810.

Subtribe Dolichina Brullé, 1834

Dolichiens Brullé, 1834: 295 [stem: *Dolich-*]. Type genus: *Dolichus* Bonelli, 1810. Comment: original vernacular name available (Art. 11.7.2): first used in latinized form by Hope (1838a: 72, as Dolichidae), generally accepted as in Lorenz (2005: 399, as Dolichina); published "31 Dec" 1834, therefore Sphodrini Laporte, 1834 and Calathini Laporte, 1834 have priority over this name for the tribe name.

Subtribe Pristosiina Lindroth, 1956

Pristosiae Lindroth, 1956: 489, in key [stem: *Pristosi-*]. Type genus: *Pristosia* Motschulsky, 1865.

*Eucalathi Lindroth, 1956: 489, in key [stem: *Eucalath-*]. Type genus: *Eucalathus* H. W. Bates, 1883 [syn. of *Pristosia* Motschulsky, 1865]. Comment: family-group name unavailable (Art. 11.6): originally published as synonym and not made available subsequently.

Subtribe Sphodrina Laporte, 1834

Sphodridae Laporte, 1834b: 78 [stem: *Sphodr-*]. Type genus: *Sphodrus* Clairville, 1806.

Pristonychidae Gistel, 1848: [2] [stem: *Pristonych-*]. Type genus: *Pristonychus* Dejean, 1828.

*Laemostenina Baehr, 1982: 265 [stem: *Laemosten-*]. Type genus: *Laemostenus* Bonelli, 1810. Comment: unavailable family-group name, proposed after 1930 without description or bibliographic reference to such a description (Art. 13.1).

Subtribe Synuchina Lindroth, 1956

Synuchi Lindroth, 1956: 489, in key [stem: *Synuch-*]. Type genus: *Synuchus* Gyllenhal, 1810.

Tribe Xenaroswellianini Erwin, 2007

Xenaroswellanini Erwin, 2007: 563 [stem: *Xenaroswellian-*]. Type genus: *Xenaroswelliana* Erwin, 2007. Comment: the original spelling Xenaroswellanini is considered a *lapsus calami* and the stem is corrected to *Xenaroswellian-* here.

Tribe Zabrini Bonelli, 1810

Zabrides Bonelli, 1810: Tabula Synoptica [stem: *Zabr-*]. Type genus: *Zabrus* Clairville, 1806.

Subtribe Amarina Zimmermann, 1832

Amaroiden C. Zimmermann, 1832: 6 [stem: *Amar-*]. Type genus: *Amara* Bonelli, 1810. Comment: original vernacular name available (Art. 11.7.2): first used in latinized form by Laporte (1834b: 78, as Amaridae), generally accepted as in Hieke (2003: 547, as Amarina).

Isopleuridae Kirby, 1837: 49 [stem: *Isopleur-*]. Type genus: *Isopleurus* Kirby, 1837 [syn. of *Celia* C. Zimmermann, 1832].

Agronomaeidae Gistel, 1848: [2] [stem: *Agronom-*]. Type genus: *Agronoma* Gistel, 1848 [*Carabus familiaris* Duftschmid, 1812, one of the species originally included, is here chosen as type species of *Agronoma* Gistel, 1848; **syn. nov.** of *Amara* Bonelli, 1810]. Comment: **syn. nov.**; incorrect original stem formation, not in prevailing usage.

Pangeteidae Gistel, 1856a: 358 [stem: *Panget-*]. Type genus: *Pangetes* Gistel, 1856 [syn. of *Amara* Bonelli, 1810]. Comment: incorrect original stem formation, not in prevailing usage.

Subtribe Zabrina Bonelli, 1810

Zabrides Bonelli, 1810: Tabula Synoptica [stem: *Zabr-*]. Type genus: *Zabrus* Clairville, 1806.

Tribe Zuphiini Bonelli, 1810

Zuphietae Bonelli, 1810: Tabula Synoptica [stem: *Zuphi-*]. Type genus: *Zuphium* Latreille, 1806.

Subtribe Dicrodontina Machado, 1992

Dicrodontini Machado, 1992: 569 [stem: *Dicrodont-*]. Type genus: *Dicrodontus* Chaudoir, 1872.

Subtribe Leleupidiina Basilewsky, 1951

Leleupidiini Basilewsky, 1951a: 178 [stem: *Leleupidi-*]. Type genus: *Leleupidia* Basilewsky, 1951.

Subtribe Metazuphiina Mateu, 1992

Metazuphiina Mateu, 1992: 198, in key [stem: *Metazuphi-*]. Type genus: *Metazuphium* Mateu, 1992.

Subtribe Mischocephalina Mateu, 1992

Mischocephalina Mateu, 1992: 198, in key [stem: *Mischocephal-*]. Type genus: *Mischocephalus* Chaudoir, 1863. Comment: this name was incorrectly spelled Mischocephaliina in the key on page 198 but correctly spelled Mischocephalina on p. 203 of the same publication.

Subtribe Patriziina Basilewsky, 1953

Patriziini Basilewsky, 1953b: 266 [stem: *Patrizi-*]. Type genus: *Patrizia* Alluaud, 1931.

Subtribe Zuphiina Bonelli, 1810

Zuphietae Bonelli, 1810: Tabula Synoptica [stem: *Zuphi-*]. Type genus: *Zuphium* Latreille, 1806.

*Polystichides Chaudoir, 1863: 308 [stem: *Polistich-*]. Type genus: *Polistichus* Bonelli, 1810 [as *Polystichus*, unjustified emendation of type genus name by Agassiz (1846b: 301), not in prevailing usage]. Comment: original vernacular name unavailable (Art. 11.7.2): subsequently used in latinized form but not generally attributed to Chaudoir (1863); incorrect original spelling of family-group name, not in prevailing usage.

Polystichinae H. W. Bates, 1886: 199 [stem: *Polistich-*]. Type genus: *Polistichus* Bonelli, 1810 [as *Polystichus*, unjustified emendation of type genus name by Agassiz (1846b: 301), not in prevailing usage]. Comment: incorrect original stem formation, not in prevailing usage.

Family Haliplidae Aubé, 1836

Haliplides Aubé, 1836: 15 [stem: *Halipl-*]. Type genus: *Haliplus* Latreille, 1802. Comment: original vernacular name available (Art. 11.7.2): first used in latinized form by Erichson (1837: 183, as Haliplini), generally accepted as in Lawrence and Newton (1995: 807, as Haliplidae).

Peltodytinae Böving and Craighead, 1931: 17 [stem: *Peltodyt-*]. Type genus: *Peltodytes* Régimbart, 1879.

Brychiini Ádám, 1996: 15 [stem: *Brychi-*]. Type genus: *Brychius* C. G. Thomson, 1859.

†Family Triaplidae Ponomarenko, 1977

Triaplidae Ponomarenko, 1977: 17 [stem: *Triapl-*]. Type genus: *Triaplus* Ponomarenko, 1977.

†Family Colymbotethidae Ponomarenko, 1994

Colymbotethidae Ponomarenko, 1994: 188 [stem: *Colymboteth-*]. Type genus: *Colymbotethis* Ponomarenko, 1994.

†Family Parahygrobiidae Ponomarenko, 1977

Parahygrobiidae Ponomarenko, 1977: 19 [stem: *Parahygrobi-*]. Type genus: *Parahygrobia* Ponomarenko, 1977.

†Family Coptoclavidae Ponomarenko, 1961

Coptoclavidae Ponomarenko, 1961: 68 [stem: *Coptoclav-*]. Type genus: *Coptoclava* Ping, 1928.

†Subfamily Necronectinae Ponomarenko, 1977

Necronectinae Ponomarenko, 1977: 22 [stem: *Necronect-*]. Type genus: *Necronectes* Ponomarenko, 1977 [syn. of *Timarchopsis* Brauer, Redtenbacher and Ganglbauer, 1889].

†Subfamily Charonoscaphinae Ponomarenko, 1977

Charonoscaphinae Ponomarenko, 1977: 32 [stem: *Charonoscaph-*]. Type genus: *Charonoscapha* Ponomarenko, 1977.

†Subfamily Coptoclavinae Ponomarenko, 1961

Coptoclavidae Ponomarenko, 1961: 68 [stem: *Coptoclav-*]. Type genus: *Coptoclava* Ping, 1928.

†Subfamily Coptoclaviscinae Soriano, Ponomarenko and Delclos, 2007

Coptoclaviscinae Soriano et al., 2007: 532 [stem: *Coptoclavisc-*]. Type genus: *Coptoclavisca* Ponomarenko, 1987.

†Subfamily Hispanoclavinae Soriano, Ponomarenko and Delclos, 2007

Hispanoclavinae Soriano et al., 2007: 527 [stem: *Hispanoclav-*]. Type genus: *Hispanoclava* Soriano, Ponomarenko and Delclos, 2007.

†Family Liadytidae Ponomarenko, 1977

Liadytidae Ponomarenko, 1977: 37 [stem: *Liadyt-*]. Type genus: *Liadytes* Ponomarenko, 1977.

Family Meruidae Spangler and Steiner, 2005

Meruidae Spangler and Steiner, 2005: 351 [stem: *Meru-*]. Type genus: *Meru* Spangler and Steiner, 2005.

Family Noteridae Thomson, 1860

Noterides C. G. Thomson, 1860: 34 [stem: *Noter-*]. Type genus: *Noterus* Clairville, 1806.

Subfamily Noterinae Thomson, 1860

Noterides C. G. Thomson, 1860: 34 [stem: *Noter-*]. Type genus: *Noterus* Clairville, 1806.

Tribe Neohydrocoptini Zalat, Saleh, Angus and Kaschef, 2000

Neohydrocoptini Zalat et al., 2000: 11 [stem: *Neohydrocopt-*]. Type genus: *Neohydrocoptus* Satô, 1972.

Tribe Noterini Thomson, 1860

Noterides C. G. Thomson, 1860: 34 [stem: *Noter-*]. Type genus: *Noterus* Clairville, 1806.

Hydrocanthini Sharp, 1880: cxlviii [stem: *Hydrocanth-*]. Type genus: *Hydrocanthus* Say, 1823.

Suphisini Sharp, 1880: cxlviii [stem: *Suphid-*]. Type genus: *Suphis* Aubé, 1836. Comment: incorrect original stem formation, not in prevailing usage.

Tribe Pronoterini Nilsson, 2005

Pronoterini Nilsson, 2005: 90 [stem: *Pronoter-*]. Type genus: *Pronoterus* Sharp, 1882.

Tribe Tonerini Miller, 2009

Tonerini K. B. Miller, 2009: 201 [stem: *Toner-*]. Type genus: *Tonerus* K. B. Miller, 2009.

Subfamily Notomicrinae Zimmermann, 1919

Notomicrini A. Zimmermann, 1919: 110 [stem: *Notomicr-*]. Type genus: *Notomicrus* Sharp, 1882.

Subfamily Phreatodytinae Uéno, 1957

Phreatodytidae Uéno, 1957: 251 [stem: *Phreatodyt-*]. Type genus: *Phreatodytes* Uéno, 1957.

Family Amphizoidae LeConte, 1853

Amphizoidae J. L. LeConte, 1853a: 227 [stem: *Amphizo-*]. Type genus: *Amphizoa* J. L. LeConte, 1853.

Family Aspidytidae Ribera, Beutel, Balke and Vogler, 2002

Aspidytidae Ribera et al., 2002: 2354 [stem: *Aspidyt-*]. Type genus: *Aspidytes* Ribera, Beutel, Balke and Vogler, 2002.

Family Hygrobiidae Régimbart, 1879 (1837)

Pelobiinae Erichson, 1837: 182 [stem: *Paelobi-*]. Type genus: *Paelobius* Schönherr, 1808 [as *Pelobius*, unjustified emendation of type genus name by Erichson (1832: 45); syn. of *Hygrobia* Latreille, 1804]. Comment: use of younger family-group name Hygrobiidae Régimbart, 1879 conserved (Art. 40.2) (see Lawrence and Newton 1995: 808); incorrect original stem formation, not in prevailing usage.

Hygrobiinae Régimbart, 1879: 449 [stem: *Hygrobi-*]. Type genus: *Hygrobia* Latreille, 1804 [the original spelling of the type genus, *Hygriobia*, has been corrected to *Hygrobia* and *Hygrobia* Latreille, 1804 placed on the Official List of Generic Names in Zoology (ICZN 1954)]. Comment: use of this name conserved over the older name Paelobiidae Erichson, 1837 (Art. 40.2) (see Lawrence and Newton 1995: 808).

Family Dytiscidae Leach, 1815

Dyticides Leach, 1815: 84 [stem: *Dytisc-*]. Type genus: *Dytiscus* Linnaeus, 1758 [placed on the Official List of Generic Names in Zoology (ICZN 1961e)]. Comment: name placed on the Official List of Family-Group Names in Zoology (ICZN 1961e, as Dytiscidae Leach, 1817).

Subfamily Agabinae Thomson, 1867

Agabides C. G. Thomson, 1867: 84 [stem: *Agab-*]. Type genus: *Agabus* Leach, 1817.

Agabinini Crotch, 1873d: 385 [stem: *Agabin-*]. Type genus: *Agabinus* Crotch, 1873 [syn. of *Platambus* C. G. Thomson, 1859].

Ilybii Acloque, 1896: 81 [stem: *Ilybi-*]. Type genus: *Ilybius* Erichson, 1832.

Hydronebriini Guignot, 1948: 168, in key [stem: *Hydronebri-*]. Type genus: *Hydronebrius* Jakovlev, 1897. Comment: printed in the issue of November 1948; this family-group name was also used in the same year by C. Brinck (1948 [December]: 112, as Hydronebriini).

Hydrotrupinae Roughley, 2000: 52 [stem: *Hydrotrup-*]. Type genus: *Hydrotrupes* Sharp, 1882.

Subfamily Colymbetinae Erichson, 1837

Colymbetini Erichson, 1837: 149 [stem: *Colymbet-*]. Type genus: *Colymbetes* Clairville, 1806.

Tribe Anisomeriini Brinck, 1948

Anisomeriini C. Brinck, 1948: 112 [stem: *Anisomeri-*]. Type genus: *Anisomeria* C. Brinck, 1943.

Tribe Carabdytini Pederzani, 1995

Carabdytini Pederzani, 1995: 45, in key [stem: *Carabdyt-*]. Type genus: *Carabdytes* Balke, Hendrich and Wewalka, 1992.

Tribe Colymbetini Erichson, 1837

Colymbetini Erichson, 1837: 149 [stem: *Colymbet-*]. Type genus: *Colymbetes* Clairville, 1806.

Cymatopterini Portevin, 1929: 198, in key [stem: *Cymatopter-*]. Type genus: *Cymatopterus* Dejean, 1833 [syn. of *Colymbetes* Clairville, 1806].

Subfamily Copelatinae Branden, 1885

Copelatini Branden, 1885: 82 [stem: *Copelat-*]. Type genus: *Copelatus* Erichson, 1832. Comment: First Reviser (Copelatini Branden, 1885 vs Aglymbini Branden, 1885 vs Lacconectini Branden, 1885) not determined, current usage maintained.

Aglymbini Branden, 1885: 87 [stem: *Aglymb-*]. Type genus: *Aglymbus* Sharp, 1882.

Lacconectini Branden, 1885: 87 [stem: *Lacconect-*]. Type genus: *Lacconectus* Motschulsky, 1855.

Subfamily Coptotominae Branden, 1885

Coptotomini Branden, 1885: 88 [stem: *Coptotom-*]. Type genus: *Coptotomus* Say, 1830.

Coptotomini C. Brinck, 1948: 112, 116 [stem: *Coptotom-*]. Type genus: *Coptotomus* Say, 1830. Comment: family-group name proposed as new without reference to Coptotomini Branden, 1885.

Subfamily Dytiscinae Leach, 1815

Dyticides Leach, 1815: 84 [stem: *Dytisc-*]. Type genus: *Dytiscus* Linnaeus, 1758 [placed on the Official List of Generic Names in Zoology (ICZN 1961e)]. Comment: name placed on the Official List of Family-Group Names in Zoology (ICZN 1961e, as Dytiscidae Leach, 1817).

Tribe Aciliini Thomson, 1867

Aciliides C. G. Thomson, 1867: 84 [stem: *Acili-*]. Type genus: *Acilius* Leach, 1817 [placed on the Official List of Generic Names in Zoology (ICZN 1961e)].

Thermonectini Sharp, 1880: cl [stem: *Thermonect-*]. Type genus: *Thermonectus* Dejean, 1833.

Graphoderini Houlbert, 1934: 126 [stem: *Graphoder-*]. Type genus: *Graphoderus* Dejean, 1833 [placed on the Official List of Generic Names in Zoology (ICZN 1961d)].

Tribe Aubehydrini Guignot, 1942

Aubehydrinae Guignot, 1942: 11 [stem: *Aubehydr-*]. Type genus: *Aubehydrus* Guignot, 1942 [syn. of *Notaticus* A. Zimmermann, 1928].

Tribe Cybisterini Sharp, 1880

Cybistrini Sharp, 1880: cl [stem: *Cybister-*]. Type genus: *Cybister* Curtis, 1827. Comment: incorrect original stem formation, not in prevailing usage.

Tribe Dytiscini Leach, 1815

Dyticides Leach, 1815: 84 [stem: *Dytisc-*]. Type genus: *Dytiscus* Linnaeus, 1758 [placed on the Official List of Generic Names in Zoology (ICZN 1961e)]. Comment: incorrect original stem formation, not in prevailing usage; name placed on the Official List of Family-Group Names in Zoology (ICZN 1961e, as Dytiscidae Leach, 1817).

Tribe Eretini Crotch, 1873

Eretini Crotch, 1873d: 385 [stem: *Eret-*]. Type genus: *Eretes* Laporte, 1833.

Eunectini Portevin, 1929: 208 [stem: *Eunect-*]. Type genus: *Eunectes* Erichson, 1832 [preoccupied genus name, not *Eunectes* Wagler, 1830 [Reptilia]; syn. of *Eretes* Laporte, 1833]. Comment: permanently invalid (Art. 39): based on preoccupied type genus.

Tribe Hydaticini Sharp, 1880

Hydaticini Sharp, 1880: cl [stem: *Hydatic-*]. Type genus: *Hydaticus* Leach, 1817.

Tribe Hyderodini Miller, 2000

Hyderodini K. B. Miller, 2000: 175 [stem: *Hyderod-*]. Type genus: *Hyderodes* Hope, 1838.

Subfamily Hydrodytinae Miller, 2001

Hydrodytinae K. B. Miller, 2001: 76 [stem: *Hydrodyt-*]. Type genus: *Hydrodytes* K. B. Miller, 2001.

Subfamily Hydroporinae Aubé, 1836

Hydroporides Aubé, 1836: 15 [stem: *Hydropor-*]. Type genus: *Hydroporus* Clairville, 1806.

Tribe Bidessini Sharp, 1880

Bidessini Sharp, 1880: cxlviii [stem: *Bidess-*]. Type genus: *Bidessus* Sharp, 1882.

Tribe Carabhydrini Watts, 1978

Carabhydrini Watts, 1978: 26 [stem: *Carabhydr-*]. Type genus: *Carabhydrus* Watts, 1978.

Tribe Hydroporini Aubé, 1836

Hydroporides Aubé, 1836: 15 [stem: *Hydropor-*]. Type genus: *Hydroporus* Clairville, 1806. Comment: original vernacular name available (Art. 11.7.2): first used in latinized form by Erichson (1837: 166, as Hydroporini), generally accepted as in Nilsson (2003: 54, as Hydroporinae).

Hydrocoptini Branden, 1885: 13 [stem: *Hydrocopt-*]. Type genus: *Hydrocoptus* Motschulsky, 1853 [syn. of *Hydroporus* Clairville, 1806].

Sternopriscini Branden, 1885: 38 [stem: *Sternoprisc-*]. Type genus: *Sternopriscus* Sharp, 1882.

Siettitiini Smrz, 1982: 289 [stem: *Siettiti-*]. Type genus: *Siettitia* Abeille de Perrin, 1904.

Deronectini Galewski, 1994: 98, in key [stem: *Deronect-*]. Type genus: *Deronectes* Sharp, 1882.

Tribe Hydrovatini Sharp, 1880

Hydrovatini Sharp, 1880: cxlviii [stem: *Hydrovat-*]. Type genus: *Hydrovatus* Motschulsky, 1853.

Tribe Hygrotini Portevin, 1929

Hygrotini Portevin, 1929: 180 [stem: *Hygrot-*]. Type genus: *Hygrotus* Stephens, 1828.

Tribe Hyphydrini Gistel, 1848

Hyphydriidae Gistel, 1848: [2] [stem: *Hyphydr-*]. Type genus: *Hyphydrus* Illiger, 1802. Comment: incorrect original stem formation, not in prevailing usage.

Actobaenidae Gistel, 1856a: 355 [stem: *Actobaen-*]. Type genus: *Actobaena* Gistel, 1856 [syn. of *Hyphydrus* Illiger, 1802].

Pachydrini Biström et al., 1997: 66 [stem: *Pachydr-*]. Type genus: *Pachydrus* Sharp, 1882.

Tribe Laccornini Wolfe and Roughley, 1990

Laccornini Wolfe and Roughley, 1990: 302 [stem: *Laccorn-*]. Type genus: *Laccornis* Gozis, 1914. Comment: current spelling maintained (Art. 29.5): incorrect stem formation in prevailing usage (should be *Laccornith-*).

Tribe Methlini Branden, 1885

Methlidae Branden, 1885: 65 [stem: *Methl-*]. Type genus: *Methles* Sharp, 1882. Comment: precedence (Methlini Branden, 1885 vs Celinini Branden, 1885) given to taxon originally proposed at the higher rank (Art. 24.1).

Celinini Branden, 1885: 65 [stem: *Celin-*]. Type genus: *Celina* Aubé, 1837.

†Tribe Schistomerini Palmer, 1957

Schistomerini Palmer, 1957: 259 [stem: *Schistomer-*]. Type genus: *Schistomerus* Palmer, 1957.

Tribe Vatellini Sharp, 1880

Vatellini Sharp, 1880: cxlviii [stem: *Vatell-*]. Type genus: *Vatellus* Aubé, 1837 [placed on the Official List of Generic Names in Zoology (ICZN 1992)].

Subfamily Laccophilinae Gistel, 1848

Laccophilidae Gistel, 1848: [1] [stem: *Laccophil-*]. Type genus: *Laccophilus* Leach, 1815.

Tribe Agabetini Branden, 1885

Agabetini Branden, 1885: 87 [stem: *Agabet-*]. Type genus: *Agabetes* Crotch, 1873.

Tribe Laccophilini Gistel, 1848

Laccophilidae Gistel, 1848: [1] [stem: *Laccophil-*]. Type genus: *Laccophilus* Leach, 1817.

Subfamily Lancetinae Branden, 1885

Lancetini Branden, 1885: 88 [stem: *Lancet-*]. Type genus: *Lancetes* Sharp, 1882.

Subfamily Matinae Branden, 1885

Matini Branden, 1885: 88 [stem: *Mat-*]. Type genus: *Matus* Aubé, 1836.

†Subfamily Palaeogyrininae Schlechtendal, 1894

Palaeogyrinidae Schlechtendal, 1894: 200 [stem: *Palaeogyrin-*]. Type genus: *Palaeogyrinus* Schlechtendal, 1894.

†Subfamily Liadytiscinae Prokin and Ren, 2010

Liadytiscinae Prokin and Ren, 2010: 48 [stem: *Liadytisc-*]. Type genus: *Liadytiscus* Prokin and Ren, 2010.

Suborder POLYPHAGA

Series STAPHYLINIFORMIA

Superfamily Hydrophiloidea Latreille, 1802

Hydrophilii Latreille, 1802: 136 [stem: *Hydrophil-*]. Type genus: *Hydrophilus* Geoffroy, 1762 [placed on the Official List of Generic Names in Zoology (ICZN 1994a)]. Comment: First Reviser (Hydrophiloidea Latreille, 1802 vs Sphaeridioidea Latreille, 1802) not determined, current usage maintained.

Family Hydrophilidae Latreille, 1802

Hydrophilii Latreille, 1802: 136 [stem: *Hydrophil-*]. Type genus: *Hydrophilus* Geoffroy, 1762 [placed on the Official List of Generic Names in Zoology (ICZN 1994a)]. Comment: First Reviser (Hydrophilidae Latreille, 1802 vs Sphaeridiidae Latreille, 1802) not determined, current usage maintained.

Subfamily Helophorinae Leach, 1815

Helopherida Leach, 1815: 95 [stem: *Helophor-*]. Type genus: *Helophorus* Fabricius, 1775 [the original spelling *Elophorus* Fabricius, 1775 was considered an incorrect original spelling and placed on the Official Index of Rejected and Invalid Generic Names in Zoology (ICZN 1993b); *Helophorus* Fabricius, 1775 was placed on the Official List of Generic Names in Zoology (ICZN 1993b)]. Comment: published in April 1815; incorrect original stem formation, not in prevailing usage; this family-group name was also used in the same year by Rafinesque (1815 [between April 1815 and 21 July 1815]: 112, as Elophoria).

Subfamily Epimetopinae Zaitzev, 1908

Epimetopina Zaitzev, 1908: 353 [stem: *Epimetop-*]. Type genus: *Epimetopus* Lacordaire, 1854.

Subfamily Georissinae Laporte, 1840

Géorissites Laporte, 1840b: 44 [stem: *Georiss-*]. Type genus: *Georissus* Latreille, 1809 [placed on the Official List of Generic Names in Zoology (ICZN 1998)]. Comment: original vernacular name available (Art. 11.7.2): first used in latinized form by Heer (1841a: 471, as Georissida), generally accepted as in Hansen (2004: 42, as Georissidae).

Subfamily Hydrochinae Thomson, 1859

Hydrochidae C. G. Thomson, 1859: 15 [stem: *Hydroch-*]. Type genus: *Hydrochus* Leach, 1817.

Subfamily Spercheinae Erichson, 1837

Spercheini Erichson, 1837: 193 [stem: *Sperche-*]. Type genus: *Spercheus* Kugelann, 1798.

Subfamily Horelophinae Hansen, 1991

Horelophinae Hansen, 1991: 104 [stem: *Horeloph-*]. Type genus: *Horelophus* Orchymont, 1913.

Subfamily Horelophopsinae Hansen, 1997

Horelophopsinae Hansen, 1997: 108 [stem: *Horelophops-*]. Type genus: *Horelophopsis* Hansen, 1997. Comment: current spelling maintained (Art. 29.5): incorrect stem formation in prevailing usage (should be *Horelophopse-*).

Subfamily Hydrophilinae Latreille, 1802

Hydrophilii Latreille, 1802: 136 [stem: *Hydrophil-*]. Type genus: *Hydrophilus* Geoffroy, 1762 [placed on the Official List of Generic Names in Zoology (ICZN 1994a)].

Tribe Anacaenini Hansen, 1991

Anacaenini Hansen, 1991: 129 [stem: *Anacaen-*]. Type genus: *Anacaena* C. G. Thomson, 1859.

Tribe Berosini Mulsant, 1844

Bérosaires Mulsant, 1844: 97 [stem: *Beros-*]. Type genus: *Berosus* Leach, 1817 [placed on the Official List of Generic Names in Zoology (ICZN 1990a)]. Comment: original vernacular name available (Art. 11.7.2): first used in latinized form by C. G. Thomson (1859: 17, as Berosina), generally accepted as in Hansen (2004: 46, as Berosini).

Derallini Bertrand, 1967: 86, in key [stem: *Derall-*]. Type genus: *Derallus* Sharp, 1882.

Tribe Chaetarthriini Bedel, 1881

*Cyllidiaires Mulsant, 1844: 143 [stem: *Cyllidi-*]. Type genus: *Cyllidium* Erichson, 1837 [syn. of *Chaetarthria* Stephens, 1835]. Comment: original vernacular name unavailable (Art. 11.7.2): not subsequently latinized.

Chaetarthriini Bedel, 1881: 314 [stem: *Chaetarthri-*]. Type genus: *Chaetarthria* Stephens, 1835.

Amphiopitae Kuwert, 1890: 120 [stem: *Amphiop-*]. Type genus: *Amphiops* Erichson, 1843.

Tribe Hydrophilini Latreille, 1802

Hydrophilii Latreille, 1802: 136 [stem: *Hydrophil-*]. Type genus: *Hydrophilus* Geoffroy, 1762 [placed on the Official List of Generic Names in Zoology (ICZN 1994a)].

Subtribe Acidocerina Zaitzev, 1908

*Philhydrates Mulsant, 1844: 131 [stem: *Philydr-*]. Type genus: *Philydrus* Solier, 1834 [as *Philhydrus*, incorrect subsequent spelling of type genus name, not in prevailing usage; preoccupied genus name, not *Philydrus* Duftschmid, 1805 [Coleoptera: Elmidae]; syn. of *Enochrus* C. G. Thomson, 1859]. Comment: original vernacular name unavailable (Art. 11.7.2): subsequently used in latinized form but not generally attributed to Mulsant (1844); incorrect original stem formation, not in prevailing usage.

Philhydrida J. E. LeConte, 1849: 27 [stem: *Philydr-*]. Type genus: *Philydrus* Solier, 1834 [as *Philhydrus*, incorrect subsequent spelling of type genus name, not in prevailing usage; preoccupied genus name, not *Philydrus* Duftschmid, 1805 [Coleoptera: Elmidae]; syn. of *Enochrus* C. G. Thomson, 1859]. Comment: incorrect original stem formation, not in prevailing usage; permanently invalid (Art. 39): based on preoccupied type genus.

Helopeltini G. H. Horn, 1873: 118 [stem: *Helopelt-*]. Type genus: *Helopeltis* G. H. Horn, 1873 [preoccupied genus name, not *Helopeltis* Signoret, 1858

[Hemiptera]; syn. of *Helobata* Bergroth, 1888]. Comment: permanently invalid (Art. 39): based on preoccupied type genus.

Acidocerini Zaitzev, 1908: 353 [stem: *Acidocer-*]. Type genus: *Acidocerus* Klug, 1855.

Helocharae Orchymont, 1919: 147 [stem: *Helochar-*]. Type genus: *Helochares* Mulsant, 1844 [placed on the Official List of Generic Names in Zoology (ICZN 1964)]. Comment: senior homonym of Helocharini Metcalf, 1965 in Homoptera (type genus *Helochara* Fitch, 1851), the homopteran name does not appear to be available (Art. 13.1).

Subtribe Globuloseina García, 2001

Globulina García, 2001: 153 [stem: *Globulose-*]. Type genus: *Globulosis* García, 2001. Comment: incorrect original stem formation, not in prevailing usage.

Subtribe Hydrobiusina Mulsant, 1844

Hydrobiaires Mulsant, 1844: 116 [stem: *Hydrobius-*]. Type genus: *Hydrobius* Leach, 1815 [placed on the Official List of Generic Names in Zoology (ICZN 1990a)]. Comment: spelling emended to *Hydrobius-* to avoid homonymy with Hydrobiidae Troschel, 1857 in Mollusca (type genus *Hydrobia* Hartmann, 1821) (ICZN 2003a); Hydrobiusina Mulsant, 1844 placed on the Official List of Family-Group Names in Zoology (ICZN 2003a); Hydrobiina Mulsant, 1844 placed on the Official Index of Rejected and Invalid Family-Group Names in Zoology (ICZN 2003a); the family-group name Hydrobia was used earlier by Rafinesque (1815: 110) but since the list of genera did not include *Hydrobius*, it was probably not based on that genus name.

Subtribe Hydrophilina Latreille, 1802

Hydrophilii Latreille, 1802: 136 [stem: *Hydrophil-*]. Type genus: *Hydrophilus* Geoffroy, 1762 [placed on the Official List of Generic Names in Zoology (ICZN 1994a)].

Hydatophilidae Gistel, 1856a: 353 [stem: *Hydatophil-*]. Type genus: *Hydatophilus* Gistel, 1856 [syn. of *Hydrophilus* Geoffroy, 1762]. Comment: published by 18 February 1856; this family-group name was also used in the same year by Gistel (1856b ["31 December"]: 95, as Hydatophilida).

Tribe Laccobiini Houlbert, 1922

Laccobiini Houlbert, 1922a: 154, in key [stem: *Laccobi-*]. Type genus: *Laccobius* Erichson, 1837. Comment: family-group name previously attributed to Bertrand (1954: 439; see Hansen 2004).

Oocyclini Hansen, 1991: 136 [stem: *Oocycl-*]. Type genus: *Oocyclus* Sharp, 1882.

Tribe Sperchopsini Hansen, 1991

Sperchopsini Hansen, 1991: 108 [stem: *Sperchops-*]. Type genus: *Sperchopsis* J. L. LeConte, 1861. Comment: current spelling maintained (Art. 29.5): incorrect stem formation in prevailing usage (should be *Sperchopse-*).

Subfamily Sphaeridiinae Latreille, 1802

Sphaeridiota Latreille, 1802: 135 [stem: *Sphaeridi-*]. Type genus: *Sphaeridium* Fabricius, 1775.

Tribe Andotypini Hansen, 1991

Andotypini Hansen, 1991: 186 [stem: *Andotyp-*]. Type genus: *Andotypus* Spangler, 1979.

Tribe Borborophorini Hansen, 1991

Borborophorini Hansen, 1991: 190 [stem: *Borborophor-*]. Type genus: *Borborophorus* Hansen, 1990.

Tribe Coelostomatini Heyden, 1891 (1890)

*Cyclonotides Motschulsky, 1857: 74 [stem: *Cyclonot-*]. Type genus: *Cyclonotum* Erichson, 1837 [syn. of *Coelostoma* Brullé, 1835]. Comment: original vernacular name unavailable (Art. 11.7.2): subsequently used in latinized form but not generally attributed to Motschulsky (1857).

Cyclonotides G. H. Horn, 1890: 281 [stem: *Cyclonot-*]. Type genus: *Cyclonotum* Erichson, 1837 [syn. of *Coelostoma* Brullé, 1835]. Comment: use of younger family-group name Coelostomatini L. Heyden, 1891 conserved (Art. 40.2).

Coelostomitae L. Heyden, 1891: 71 [stem: *Coelostomat-*]. Type genus: *Coelostoma* Brullé, 1835. Comment: incorrect original stem formation, not in prevailing usage; use of family-group name conserved over Cyclonotini G. H. Horn, 1890 (Art. 40.2) (see Hansen 1991); Hansen (1991: 282) mentions that "Cyclonotaires Rey, 1886" is the oldest name for this tribe but he used Art. 40b of the then Code to conserve usage of Coelostomatini L. Heyden, 1891; we could not find any instance where the requirements of Art. 11.7.2 were met and therefore treat the vernacular name Cyclonotaires as unavailable.

Cyllomina Zaitzev, 1908: 400 [stem: *Cylomat-*]. Type genus: *Cyloma* Sharp, 1872 [as *Cylloma*, unjustified emendation of type genus name, not in prevailing usage]. Comment: incorrect original stem formation, not in prevailing usage.

Tribe Megasternini Mulsant, 1844

Mégasternaires Mulsant, 1844: 186 [stem: *Megastern-*]. Type genus: *Megasternum* Mulsant, 1844 [placed on the Official List of Generic Names in Zoology (ICZN 1981b)]. Comment: original vernacular name available (Art. 11.7.2):

first used in latinized form by G. H. Horn (1890: 307, as Megasterni), generally accepted as in Hansen (2004: 61, as Megasternini).

Cercyones G. H. Horn, 1890: 287 [stem: *Cercyon-*]. Type genus: *Cercyon* Leach, 1817.

Tribe Omicrini Smetana, 1975

Omicrini Smetana, 1975: 155 [stem: *Omicr-*]. Type genus: *Omicrus* Sharp, 1879.

Tribe Protosternini Hansen, 1991

Protosternini Hansen, 1991: 212 [stem: *Protostern-*]. Type genus: *Protosternum* Sharp, 1890.

Tribe Rygmodini Orchymont, 1916

Rygmodini Orchymont, 1916: 238 [stem: *Rygmod-*]. Type genus: *Rygmodus* A. White, 1846.

Rygmodini Orchymont, 1919: 105 [stem: *Rygmod-*]. Type genus: *Rygmodus* A. White, 1846. Comment: family-group name proposed as new without reference to Rygmodini Orchymont, 1916.

Tribe Sphaeridiini Latreille, 1802

Sphaeridiota Latreille, 1802: 135 [stem: *Sphaeridi-*]. Type genus: *Sphaeridium* Fabricius, 1775. Comment: Latreille also used the incorrect original spelling Sphéridiote on page 138 of the same work.

Tribe Tormissini Hansen, 1991

Tormissini Hansen, 1991: 181 [stem: *Tormiss-*]. Type genus: *Tormissus* Broun, 1893.

Family Sphaeritidae Shuckard, 1839

Sphaeritidae Shuckard, 1839a: 149 [stem: *Sphaerit-*]. Type genus: *Sphaerites* Duftschmid, 1805.

Family Synteliidae Lewis, 1882

Synteliidae Lewis, 1882: 137 [stem: *Synteli-*]. Type genus: *Syntelia* Westwood, 1864.

Family Histeridae Gyllenhal, 1808

Histeroides Gyllenhal, 1808: 74 [stem: *Hister-*]. Type genus: *Hister* Linnaeus, 1758.

Subfamily Niponiinae Fowler, 1912

Niponiidae Fowler, 1912: 93 [stem: *Niponi-*]. Type genus: *Niponius* Lewis, 1885.

Subfamily Abraeinae MacLeay, 1819

Abreidae W. S. MacLeay, 1819: 25 [stem: *Abrae-*]. Type genus: *Abraeus* Leach, 1817.

Tribe ABRAEINI MacLeay, 1819

ABREIDAE W. S. MacLeay, 1819: 25 [stem: *Abrae-*]. Type genus: *Abraeus* Leach, 1817.

Tribe ACRITINI Wenzel, 1944

ACRITINI Wenzel, 1944: 55, in key [stem: *Acrit-*]. Type genus: *Acritus* J. L. LeConte, 1853.

Tribe ACRITOMORPHINI Wenzel, 1944

ACRITOMORPHINI Wenzel, 1944: 55, in key [stem: *Acritomorph-*]. Type genus: *Acritomorphus* Wenzel, 1944.

Tribe PLEGADERINI Portevin, 1929

PLEGADERINI Portevin, 1929: 601, in key [stem: *Plegader-*]. Type genus: *Plegaderus* Erichson, 1834.

Tribe TERETRIINI Bickhardt, 1914

TERETRIINAE Bickhardt, 1914: 306 [stem: *Teretri-*]. Type genus: *Teretrius* Erichson, 1834.

Subfamily TRYPETICINAE Bickhardt, 1913

TRYPETICINAE Bickhardt, 1913: 166 [stem: *Trypetic-*]. Type genus: *Trypeticus* Marseul, 1864.

Subfamily TRYPANAEINAE Marseul, 1857

TRYPANÉENS Marseul, 1857b: 148 [stem: *Trypanae-*]. Type genus: *Trypanaeus* Eschscholtz, 1829. Comment: original vernacular name available (Art. 11.7.2): first used in latinized form by Jakobson (1910: 638, as TRYPANAEINA), generally accepted as in Newton and Thayer (1992: 25, as TRYPANAEINAE); incorrect stem formation, not in prevailing usage.

Subfamily SAPRININAE Blanchard, 1845

SAPRINITES Blanchard, 1845a: 276 [stem: *Saprin-*]. Type genus: *Saprinus* Erichson, 1834. Comment: original vernacular name available (Art. 11.7.2): first used in latinized form by Fairmaire and Laboulbène (1855: 273, as SAPRINII), generally accepted as in Mazur (2004: 90, as SAPRININAE).

MYRMETINI Portevin, 1929: 593 [stem: *Myrmet-*]. Type genus: *Myrmetes* Marseul, 1862.

Subfamily DENDROPHILINAE Reitter, 1909

DENDROPHILINI Reitter, 1909: 288 [stem: *Dendrophil-*]. Type genus: *Dendrophilus* Leach, 1817. Comment: First Reviser (DENDROPHILINAE Reitter, 1909 vs PAROMALINAE Reitter, 1909) not determined, current usage maintained.

Tribe Anapleini Olexa, 1982

Anapleini Olexa, 1982: 38 [stem: *Anaple-*]. Type genus: *Anapleus* G. H. Horn, 1873.

Tribe Bacaniini Kryzhanovskij, 1976

*Bacaniini Vienna, 1974: 273 [stem: *Bacani-*]. Type genus: *Bacanius* J. L. LeConte, 1853. Comment: unavailable name (Art. 13.1): proposed after 1930 without description or bibliographic reference to such a description.

Bacaniini Kryzhanovskij, 1976: 266 [stem: *Bacani-*]. Type genus: *Bacanius* J. L. LeConte, 1853.

Tribe Dendrophilini Reitter, 1909

Dendrophilini Reitter, 1909: 288 [stem: *Dendrophil-*]. Type genus: *Dendrophilus* Leach, 1817.

Tribe Paromalini Reitter, 1909

Paromalini Reitter, 1909: 287 [stem: *Paromal-*]. Type genus: *Paromalus* Erichson, 1834.

Subfamily Onthophilinae MacLeay, 1819

Onthophilidae W. S. MacLeay, 1819: 25 [stem: *Onthophil-*]. Type genus: *Onthophilus* Leach, 1817.

Scolytini Jakobson, 1911a: 652 [stem: *Scolyt-*]. Type genus: *Scolytus* Müller, 1764 [preoccupied genus name, not *Scolytus* Geoffroy, 1762 [Coleoptera: Curculionidae]; syn. of *Onthophilus* Leach, 1817]. Comment: permanently invalid (Art. 39): based on preoccupied type genus.

Subfamily Tribalinae Bickhardt, 1914

Tribalini Bickhardt, 1914: 307 [stem: *Tribal-*]. Type genus: *Tribalus* Erichson, 1834.

Subfamily Histerinae Gyllenhal, 1808

Histeroides Gyllenhal, 1808: 74 [stem: *Hister-*]. Type genus: *Hister* Linnaeus, 1758.

Tribe Exosternini Bickhardt, 1914

Exosternini Bickhardt, 1914: 308 [stem: *Exostern-*]. Type genus: *Exosternus* Lewis, 1902.

Tribe Histerini Gyllenhal, 1808

Histeroides Gyllenhal, 1808: 74 [stem: *Hister-*]. Type genus: *Hister* Linnaeus, 1758. Comment: current spelling maintained (Art. 29.5): incorrect stem formation in prevailing usage (should be *Histr-*).

Tribe Hololeptini Hope, 1840

Hololeptidae Hope, 1840a: 106 [stem: *Hololept-*]. Type genus: *Hololepta* Paykull, 1811. Comment: W. S. MacLeay (1819: 25) published the name Omoleptidae, presumably as a *lapsus calami* for Hololeptidae, but he did not cite the type genus, nor did he use the family-group name subsequently; in the absence of validating evidence we treat Omoleptidae as a *nomen dubium*.

Tribe Omalodini Kryzhanovskij, 1972

*Omalodini A. N. Reichardt, 1941: 37 [stem: *Omalod-*]. Type genus: *Omalodes* Dejean, 1833. Comment: unavailable family-group name, proposed after 1930 without description or bibliographic reference to such a description (Art. 13.1).

Omalodini Kryzhanovskij, 1972: 19 [stem: *Omalod-*]. Type genus: *Omalodes* Dejean, 1833.

Tribe Platysomatini Bickhardt, 1914

Platysomini Bickhardt, 1914: 307 [stem: *Platysomat-*]. Type genus: *Platysoma* Leach, 1817. Comment: incorrect original stem formation, not in prevailing usage; correction of original spelling by Mazur (1973); the corrected spelling should be used in order to avoid homonymy with Platysomidae Young, 1866 (type genus *Platysomus* Agassiz, 1833) in Pisces.

Althanini Cooman, 1939: 138 [stem: *Althan-*]. Type genus: *Althanus* Lewis, 1903.

Subfamily Haeteriinae Marseul, 1857

Hétériens Marseul, 1857b: 148 [stem: *Haeteri-*]. Type genus: *Haeterius* Dejean, 1833 [as *Hetaerius*, incorrect subsequent spelling of type genus name, not in prevailing usage].

Tribe Haeteriini Marseul, 1857

Hétériens Marseul, 1857b: 148 [stem: *Haeteri-*]. Type genus: *Haeterius* Dejean, 1833 [as *Hetaerius*, incorrect subsequent spelling of type genus name, not in prevailing usage]. Comment: original vernacular name available (Art. 11.7.2): first used in latinized form by J. Schmidt (1885: 283, as Hetaeriini [incorrect stem formation]), generally accepted as in Mazur (2004: 75, as Haeteriinae); incorrect stem formation, not in prevailing usage, correction of original spelling by Bousquet and Laplante (1999: 104).

Tribe Nymphistrini Tishechkin, 2007

Nymphisterini Tishechkin, 2007: 51 [stem: *Nymphistr-*]. Type genus: *Nymphister* Reichensperger, 1933. Comment: incorrect original stem formation, not in prevailing usage.

Tribe Synoditulini Tishechkin, 2007

Synoditulini Tishechkin, 2007: 50 [stem: *Synoditul-*]. Type genus: *Synoditulus* Reichensperger, 1924.

Subfamily Chlamydopsinae Bickhardt, 1914

Chlamydopsini Bickhardt, 1914: 308 [stem: *Chlamydops-*]. Type genus: *Chlamydopsis* Westwood, 1869. Comment: current spelling maintained (Art. 29.5): incorrect stem formation in prevailing usage (should be *Chlamydopse-*).

Superfamily Staphylinoidea Latreille, 1802

Staphyliniae Latreille, 1802: 124 [stem: *Staphylin-*]. Type genus: *Staphylinus* Linnaeus, 1758 [placed on the Official List of Generic Names in Zoology (ICZN 1959a)]. Comment: First Reviser (Staphylinoidea Latreille, 1802 vs Pselaphoidea Latreille, 1802) not determined, current usage maintained.

Family Hydraenidae Mulsant, 1844

Hydraenaires Mulsant, 1844: 50 [stem: *Hydraen-*]. Type genus: *Hydraena* Kugelann, 1794.

Subfamily Orchymontiinae Perkins, 1997

Orchymontinae Perkins, 1997: 189 [stem: *Orchymonti-*]. Type genus: *Orchymontia* Broun, 1919. Comment: incorrect original stem formation, not in prevailing usage.

Subfamily Prosthetopinae Perkins, 1994

Prosthetopinae Perkins, 1994: 7 [stem: *Prosthetop-*]. Type genus: *Prosthetops* C. O. Waterhouse, 1879. Comment: precedence (Nucleotopinae Perkins, 1994 vs Parasthetopinae Perkins, 1994 vs Prosthetopinae Perkins, 1994 vs Protosthetopinae Perkins, 1994 vs Pterosthetopinae Perkins, 1994) given to taxon originally proposed at the higher rank (Art. 24.1).

Tribe Coelometoponini Perkins, 2005

Coelometoponini Perkins, 2005: 9 [stem: *Coelometopon-*]. Type genus: *Coelometopon* Janssens, 1972. Comment: current spelling maintained (Art. 29.5): incorrect stem formation in prevailing usage (should be *Coelometop-*).

Tribe Nucleotopini Perkins, 1994

Nucleotopini Perkins, 1994: 8, in key [stem: *Nucleotop-*]. Type genus: *Nucleotops* Perkins and Balfour-Browne, 1994.

Tribe Parasthetopini Perkins, 1994

Parasthetopini Perkins, 1994: 8, in key [stem: *Parasthetop-*]. Type genus: *Parasthetops* Perkins and Balfour-Browne, 1994.

Tribe PROSTHETOPINI Perkins, 1994

PROSTHETOPINAE Perkins, 1994: 7 [stem: *Prosthetop-*]. Type genus: *Prosthetops* C. O. Waterhouse, 1879.

Tribe PROTOSTHETOPINI Perkins, 1994

PROTOSTHETOPINI Perkins, 1994: 8, in key [stem: *Protosthetop-*]. Type genus: *Protosthetops* Perkins, 1994.

Tribe PTEROSTHETOPINI Perkins, 1994

PTEROSTHETOPINI Perkins, 1994: 9, in key [stem: *Pterosthetop-*]. Type genus: *Pterosthetops* Perkins, 1994.

Subfamily HYDRAENINAE Mulsant, 1844

HYDRAENAIRES Mulsant, 1844: 50 [stem: *Hydraen-*]. Type genus: *Hydraena* Kugelann, 1794. Comment: First Reviser (HYDRAENINAE Mulsant, 1844 vs LIMNEBIINAE Mulsant, 1844) not determined, current usage maintained.

Tribe HYDRAENIDINI Perkins, 1980

HYDRAENIDINI Perkins, 1980: 414 [stem: *Hydraenid-*]. Type genus: *Hydraenida* Germain, 1901.

Tribe HYDRAENINI Mulsant, 1844

HYDRAENAIRES Mulsant, 1844: 50 [stem: *Hydraen-*]. Type genus: *Hydraena* Kugelann, 1794. Comment: original vernacular name available (Art. 11.7.2): first used in latinized form by Gistel (1848: [1], as HYDRAENAEIDAE [incorrect stem formation]), generally accepted as in Jäch (2004: 102, as HYDRAENINI).

Tribe LIMNEBIINI Mulsant, 1844

LIMNÉBIAIRES Mulsant, 1844: 88 [stem: *Limnebi-*]. Type genus: *Limnebius* Leach, 1815. Comment: original vernacular name available (Art. 11.7.2): first used in latinized form by C. G. Thomson (1859: 14, as LIMNEBIIDAE), generally accepted as in Jäch (2004: 110, as LIMNEBIINI).

Tribe MADAGASTRINI Perkins, 1997

MADAGASTRINI Perkins, 1997: 176 [stem: *Madagastr-*]. Type genus: *Madagaster* Perkins, 1997.

Tribe PARHYDRAENINI Perkins, 1997

PARHYDRAENINI Perkins, 1997: 170 [stem: *Parhydraen-*]. Type genus: *Parhydraena* Orchymont, 1937.

Subfamily Ochthebiinae Thomson, 1859

Ochtebiidae C. G. Thomson, 1859: 15 [stem: *Ochthebi-*]. Type genus: *Ochthebius* Leach, 1815 [placed on the Official List of Generic Names in Zoology (ICZN 1991a)].

Tribe Ochthebiini Thomson, 1859

Ochtebiidae C. G. Thomson, 1859: 15 [stem: *Ochthebi-*]. Type genus: *Ochthebius* Leach, 1815 [placed on the Official List of Generic Names in Zoology (ICZN 1991a)].

Subtribe Enicocerina Perkins, 1997

Enicocerina Perkins, 1997: 157 [stem: *Enicocer-*]. Type genus: *Enicocerus* Stephens, 1829.

Subtribe Meropathina Perkins, 1997

Meropathina Perkins, 1997: 143 [stem: *Meropath-*]. Type genus: *Meropathus* Enderlein, 1901.

Subtribe Neochthebiina Perkins, 1997

Neochthebina Perkins, 1997: 152 [stem: *Neochthebi-*]. Type genus: *Neochthebius* Orchymont, 1932. Comment: incorrect original stem formation, not in prevailing usage.

Subtribe Ochthebiina Thomson, 1859

Ochtebiidae C. G. Thomson, 1859: 15 [stem: *Ochthebi-*]. Type genus: *Ochthebius* Leach, 1815 [as *Ochtebius*, incorrect subsequent spelling of type genus name, not in prevailing usage; *Ochthebius* placed on the Official List of Generic Names in Zoology (ICZN 1991a)]. Comment: incorrect original stem formation, not in prevailing usage.

Ochthebiinae Perkins, 1980: 414 [stem: *Ochthebi-*]. Type genus: *Ochthebius* Leach, 1815 [placed on the Official List of Generic Names in Zoology (ICZN 1991a)]. Comment: family-group name proposed as new without reference to Ochtebiidae C. G. Thomson, 1859.

Subtribe Protochthebiina Perkins, 1997

Protochthebiina Perkins, 1997: 152 [stem: *Protochthebi-*]. Type genus: *Protochthebius* Perkins, 1997.

Tribe Ochtheosini Perkins, 1997

Ochtheosini Perkins, 1997: 123 [stem: *Ochtheos-*]. Type genus: *Ochtheosus* Perkins, 1997.

Family PTILIIDAE Erichson, 1845

PTILINA Erichson, 1845: 15 [stem: *Ptili-*]. Type genus: *Ptilium* Gyllenhal, 1827 [placed on the Official List of Generic Names in Zoology (ICZN 1984b)].

Subfamily PTILIINAE Erichson, 1845

PTILINA Erichson, 1845: 15 [stem: *Ptili-*]. Type genus: *Ptilium* Gyllenhal, 1827 [placed on the Official List of Generic Names in Zoology (ICZN 1984b)].

Tribe DISCHERAMOCEPHALINI Grebennikov, 2009

DISCHERAMOCEPHALINI Grebennikov, 2009: 115 [stem: *Discheramocephal-*]. Type genus: *Discheramocephalus* C. Johnson, 2007.

Tribe NANOSELLINI Barber, 1924

NANOSELLINAE Barber, 1924: 170 [stem: *Nanosell-*]. Type genus: *Nanosella* Motschulsky, 1869.

Tribe PTENIDIINI Flach, 1889

PTENIDIINI Flach, 1889: 489 [stem: *Ptenidi-*]. Type genus: *Ptenidium* Erichson, 1845 [placed on the Official List of Generic Names in Zoology and given precedence over the generic name *Anisarthria* Stephens, 1830 whenever the two names are considered synonyms (ICZN 1984b)].

Tribe PTILIINI Erichson, 1845

*PTILINA Heer, 1843: 60 [stem: *Ptili-*]. Type genus: *Ptilium* Gyllenhal, 1827 [as a synonym of *Trichopteryx* Kirby, 1826; *Ptilium* Gyllenhal, 1827 placed on the Official List of Generic Names in Zoology (ICZN 1984b)]. Comment: family-group name unavailable (Art. 11.7.1.1): not based on a genus used as valid at the time; incorrect original stem formation, not in prevailing usage.

PTILINA Erichson, 1845: 15 [stem: *Ptili-*]. Type genus: *Ptilium* Gyllenhal, 1827 [placed on the Official List of Generic Names in Zoology (ICZN 1984b)]. Comment: published 28 May 1845; incorrect original stem formation, not in prevailing usage; this family-group name was also used in the same year by Motschulsky (1845 [after 6 August]: 504, as PTILIEN); Motschulsky's name, originally proposed in a vernacular form, was treated as available by Lawrence and Newton (1995: 818).

PTILOPTERIIDAE Gistel, 1848: [4] [stem: *Ptilopteri-*]. Type genus: *Ptilopterium* Gistel, 1848 [syn. of *Ptilium* Gyllenhal, 1827].

ACTIDIINI Portevin, 1929: 568, in key [stem: *Actidi-*]. Type genus: *Actidium* A. Matthews, 1868.

Tribe PTINELLINI Reitter, 1906 (1891)

NEUGLENINI Reitter, 1891: 146 [stem: *Neuglen-*]. Type genus: *Neuglenes* C. G. Thomson, 1859 [syn. of *Ptinella* Motschulsky, 1844].

Ptinellini Reitter, 1906: 259 [stem: *Ptinell-*]. Type genus: *Ptinella* Motschulsky, 1844 [placed on the Official List of Generic Names in Zoology (ICZN 1985b)]. Comment: name proposed to replace Neuglenini Reitter, 1891 because of the synonymy of the type genus; use of younger family-group name Ptinellini Reitter, 1906 conserved (Art. 40.2).

*Pterycini Dybas, 1966: 16, 44 [stem: *Pteryg-*]. Type genus: *Pteryx* A. Matthews, 1858. Comment: incorrect original stem formation, not in prevailing usage; the earlier usage of "pterycine group" by Dybas (1955: 562) is unavailable because it is not a noun (Art. 11.7.1.1); also this name has been used subsequently by Hall (2003: 85) but the name Pterycini has not been made available yet (see Hall 2005: 257).

Subfamily Cephaloplectinae Sharp, 1883

Cephaloplectinae Sharp, 1883: 295 [stem: *Cephaloplect-*]. Type genus: *Cephaloplectus* Sharp, 1883.

Limulodinae Ganglbauer, 1898: 297 [stem: *Limulod-*]. Type genus: *Limulodes* A. Matthews, 1866.

Subfamily Acrotrichinae Reitter, 1909 (1856)

Trichopterygia Erichson, 1845: 13 [stem: *Trichopteryg-*]. Type genus: *Trichopteryx* Kirby, 1826 [preoccupied genus name, not *Trichopteryx* Hübner, 1825 [Lepidoptera]; syn. of *Acrotrichis* Motschulsky, 1848]. Comment: permanently invalid (Art. 39): based on preoccupied type genus.

Cleopteriidae Gistel, 1856a: 360 [stem: *Cleopteri-*]. Type genus: *Cleopterium* Gistel, 1856 [syn. of *Acrotrichis* Motschulsky, 1848]. Comment: use of younger family-group name Acrotrichinae Reitter, 1909 conserved (Art. 40.2) (see Newton and Thayer 1992: 44).

Acrotrichini Reitter, 1909: 272 [stem: *Acrotrich-*]. Type genus: *Acrotrichis* Motschulsky, 1848. Comment: use of family-group name conserved over Cleopteriinae Gistel, 1856 (Art. 40.2) (see Newton and Thayer 1992: 44).

Nephanini Portevin, 1929: 573, in key [stem: *Nephan-*]. Type genus: *Nephanes* C. G. Thomson, 1859 [placed on the Official List of Generic Names in Zoology (ICZN 1985b)].

Family Agyrtidae Thomson, 1859

Agyrtidae C. G. Thomson, 1859: 57 [stem: *Agyrt-*]. Type genus: *Agyrtes* Frölich, 1799.

Subfamily Agyrtinae Thomson, 1859

*Agyrtes Motschulsky, 1849: 57 [stem: *Agyrt-*]. Type genus: *Agyrtes* Frölich, 1799. Comment: original vernacular name unavailable (Art. 11.7.2): subsequently used in latinized form but not generally attributed to Motschulsky (1849).

Agyrtidae C. G. Thomson, 1859: 57 [stem: *Agyrt-*]. Type genus: *Agyrtes* Frölich, 1799.

LYROSOMINI G. H. Horn, 1880b: 247 [stem: *Lyrosomat-*]. Type genus: *Lyrosoma* Mannerheim, 1853. Comment: incorrect original stem formation, not in prevailing usage; correction of stem by Newton and Thayer (1992: 33).

LENDOMINI Bradley, 1930: 159, in key [stem: *Lendom-*]. Type genus: *Lendomus* Casey, 1924 [syn. of *Agyrtes* Frölich, 1799].

Subfamily NECROPHILINAE Newton, 1997

*NECROPHILINI Jeannel, 1936: 10 [stem: *Necrophil-*]. Type genus: *Necrophilus* Latreille, 1829. Comment: unavailable family-group name, proposed after 1930 without description or bibliographic reference to such a description (Art. 13.1).

NECROPHILINAE Newton, 1997: 127 [stem: *Necrophil-*]. Type genus: *Necrophilus* Latreille, 1829. Comment: the type genus of the older names NECROPHILIDAE Gistel, 1848: [4] and NECROPHILIDAE Gistel, 1856a: 362 cannot be determined (either *Necrophilus* Latreille, 1829 [AGYRTIDAE] or *Necrophila* Kirby and Spence, 1828 [SILPHIDAE]) and therefore those names are considered unavailable (also see comments in Newton and Thayer (1992: 23) and Newton (1997: 147)).

Subfamily PTEROLOMATINAE Thomson, 1862

PTEROLOMINA C. G. Thomson, 1862: 20 [stem: *Pterolomat-*]. Type genus: *Pteroloma* Gyllenhal, 1827. Comment: incorrect original stem formation, not in prevailing usage; correction of stem by Newton and Thayer (1992: 33).

Family LEIODIDAE Fleming, 1821

LEIODESIDAE Fleming, 1821: 51 [stem: *Leiod-*]. Type genus: *Leiodes* Latreille, 1797. Comment: incorrect original stem formation, not in prevailing usage.

Subfamily CAMIARINAE Jeannel, 1911

CAMIARINAE Jeannel, 1911: 192, in key [stem: *Camiar-*]. Type genus: *Camiarus* Sharp, 1878.

Tribe AGYRTODINI Jeannel, 1936

AGYRTODINI Jeannel, 1936: 99 [stem: *Agyrtod-*]. Type genus: *Agyrtodes* Portevin, 1907.

*†GUVANOCOLEINA Perkovsky, 2002: 16 [stem: *Gurvanocole-*]. Type genus: *Gurvanocoleus* Ponomarenko, 1986. Comment: unavailable family-group name, proposed after 1930 without description or bibliographic reference to such a description (Art. 13.1); incorrect original stem formation, not in prevailing usage.

Tribe CAMIARINI Jeannel, 1911

CAMIARINAE Jeannel, 1911: 192, in key [stem: *Camiar-*]. Type genus: *Camiarus* Sharp, 1878.

*†MESECANINA Perkovsky, 2002: 15 [stem: *Mesecan-*]. Type genus: *Mesecanus* Newton, 1982. Comment: unavailable family-group name, proposed after 1930 without description or bibliographic reference to such a description (Art. 13.1).

Tribe NEOPELATOPINI Jeannel, 1962

NEOPELATOPINI Jeannel, 1962b: 487 [stem: *Neopelatop-*]. Type genus: *Neopelatops* Jeannel, 1936.

Subfamily CATOPOCERINAE Hatch, 1927 (1880)

CATOPOCERINI Hatch, 1927: 4, in key [stem: *Catopocer-*]. Type genus: *Catopocerus* Motschulsky, 1869. Comment: use of family-group name conserved over PINODYTINAE G. H. Horn, 1880 (Art. 40.2) (see Newton and Thayer 1992: 34).

Tribe CATOPOCERINI Hatch, 1927 (1880)

PINODYTINI G. H. Horn, 1880b: 248 [stem: *Pinodyt-*]. Type genus: *Pinodytes* G. H. Horn, 1880 [syn. of *Catopocerus* Motschulsky, 1869]. Comment: use of younger family-group name CATOPOCERINI Hatch, 1927 conserved (Art. 40.2) (see Newton and Thayer 1992: 34).

CATOPOCERINI Hatch, 1927: 4, in key [stem: *Catopocer-*]. Type genus: *Catopocerus* Motschulsky, 1869. Comment: new name for PINODYTINI G. H. Horn, 1880 because of synonymy of type genus; use of family-group name conserved over PINODYTINI G. H. Horn, 1880 (Art. 40.2) (see Newton and Thayer 1992: 34).

Tribe GLACICAVICOLINI Westcott, 1968

GLACICAVICOLINAE Westcott, 1968: 1 [stem: *Glacicavicol-*]. Type genus: *Glacicavicola* Westcott, 1968.

Subfamily LEIODINAE Fleming, 1821

LEIODESIDAE Fleming, 1821: 51 [stem: *Leiod-*]. Type genus: *Leiodes* Latreille, 1797.

Tribe AGATHIDIINI Westwood, 1838

*ANISOTOMIDAE Stephens, 1828: 99 [stem: *Anisotom-*]. Type genus: *Anisotoma* Panzer, 1797. Comment: family-group name unavailable (Art. 11.7.1.1): not based on a genus used as valid at the time; see Newton and Thayer (1992: 20).

AGATHIDIIDAE Westwood, 1838: 10 [stem: *Agathidi-*]. Type genus: *Agathidium* Illiger, 1798.

ANISOTOMIDAE Reitter, 1884: 6 [stem: *Anisotom-*]. Type genus: *Anisotoma* Panzer, 1797. Comment: an application will need to be submitted to the Commission to suppress ANISOTOMIDAE Erichson, 1845 (based on the misidentified type genus *Anisotoma* sensu Schmidt, 1841) for the Principles of Priority and Homonymy (Art. 65.2.1) if this name is to be used as valid.

Tribe ESTADIINI Portevin, 1914

ESTADIINI Portevin, 1914: 199 [stem: *Estadi-*]. Type genus: *Estadia* Fairmaire, 1903 [syn. of *Dietta* Sharp, 1876].

Tribe Leiodini Fleming, 1821

Leiodesidae Fleming, 1821: 51 [stem: *Leiod-*]. Type genus: *Leiodes* Latreille, 1797. Comment: incorrect original stem formation, not in prevailing usage; correction of stem by Newton and Thayer (1992: 34).

Anisotomidae Erichson, 1845: 41 [stem: *Anisotom-*]. Type genus: *Anisotoma* sensu Schmidt, 1841 [not *Anisotoma* Panzer, 1797; syn. of *Leiodes* Latreille, 1797]. Comment: based on a misidentified type genus; an application will need to be submitted to the Commission to suppress this name for the Principles of Priority and Homonymy (Art. 65.2.1) if Anisotomidae Reitter, 1884 in Leiodinae: Agathidiini is to be used as valid in the future.

*Cyrtusina Perkovsky, 1997b: 168 [stem: *Cyrtus-*]. Type genus: *Cyrtusa* Erichson, 1842. Comment: unavailable family-group name, proposed after 1930 without description or bibliographic reference to such a description (Art. 13.1).

*Hypoliodina Perkovsky, 2002: 15 [stem: *Hypoliod-*]. Type genus: *Hypoliodes* Portevin, 1908. Comment: unavailable family-group name, proposed after 1930 without description or bibliographic reference to such a description (Art. 13.1).

Tribe Pseudoliodini Portevin, 1926

Pseudoliodini Portevin, 1926: 75 [stem: *Pseudoliod-*]. Type genus: *Pseudoliodes* Portevin, 1926 [syn. of *Pseudcolenis* Reitter, 1884].

Dermatohomoeini Hlisnikovský, 1963: 311 [stem: *Dermatohomoe-*]. Type genus: *Dermatohomoeus* Hlisnikovský, 1963.

Tribe Scotocryptini Reitter, 1884

Scotocryptini Reitter, 1884: 91 [stem: *Scotocrypt-*]. Type genus: *Scotocryptus* Girard, 1874.

Tribe Sogdini Lopatin, 1961

Sogdiidae Lopatin, 1961: 121 [stem: *Sogd-*]. Type genus: *Sogda* Lopatin, 1961. Comment: incorrect original stem formation, not in prevailing usage; correction of stem by Newton and Thayer (1992: 36).

Triarthrini Jeannel, 1962b: 486, in key [stem: *Triarthr-*]. Type genus: *Triarthron* Märkel, 1840. Comment: junior homonym of Triarthridae Ulrich, 1930 (type genus *Triarthrus* Green, 1832) in Trilobita; this case is to be referred to the Commission to remove the homonymy (Art. 55.3.1).

Hydnobiini Jeannel, 1962b: 492 [stem: *Hydnobi-*]. Type genus: *Hydnobius* Schmidt, 1841.

*Stereina Perkovsky, 2002: 15 [stem: *Stere-*]. Type genus: *Stereus* Wollaston, 1857 [preoccupied genus name, not *Stereus* Mannerheim, 1846 [Coleoptera: Curculionidae]; syn. of *Deltocnemis* Sahlberg, 1886]. Comment: unavailable family-group name, proposed after 1930 without description or bibliographic reference to such a description (Art. 13.1).

Subfamily Coloninae Horn, 1880 (1859)

Myloechina C. G. Thomson, 1859: 60 [stem: *Myloech-*]. Type genus: *Myloechus* Latreille, 1807 [syn. of *Colon* Herbst, 1797]. Comment: use of younger family-group name Coloninae G. H. Horn, 1880 conserved (Art. 40.2); see Newton and Thayer (1992: 36).

Colones G. H. Horn, 1880b: 266 [stem: *Colon-*]. Type genus: *Colon* Herbst, 1797 [the original spelling *Kolon* was placed on the Official Index of Rejected and Suppressed Generic Names in Zoology, *Colon* was considered the correct original spelling of the genus and placed on the Official List of Generic Names in Zoology (ICZN 1995b)]. Comment: use of family-group name conserved over Myloechinae C. G. Thomson, 1859 (Art. 40.2); see Newton and Thayer (1992: 36).

Subfamily Cholevinae Kirby, 1837

Cholevidae Kirby, 1837: 108 [stem: *Cholev-*]. Type genus: *Choleva* Latreille, 1797.

Tribe Anemadini Hatch, 1928

Anemadina Hatch, 1928: 159 [stem: *Anemad-*]. Type genus: *Anemadus* Reitter, 1884.

Subtribe Anemadina Hatch, 1928

Anemadina Hatch, 1928: 159 [stem: *Anemad-*]. Type genus: *Anemadus* Reitter, 1884.

Anemadinae Jeannel, 1936: 179 [stem: *Anemad-*]. Type genus: *Anemadus* Reitter, 1884. Comment: family-group name proposed as new without reference to Anemadina Hatch, 1928.

Subtribe Eocatopina Jeannel, 1936

Eocatopina Jeannel, 1936: 124, in key [stem: *Eocatop-*]. Type genus: *Eocatops* Peyerimhoff, 1924.

Subtribe Eunemadina Newton, 1998

Eunemadina Newton, 1998: 103 [stem: *Eunemad-*]. Type genus: *Eunemadus* Portevin, 1914.

Subtribe Nemadina Jeannel, 1936

Nemadinae Jeannel, 1936: 96 [stem: *Nemad-*]. Type genus: *Nemadus* C. G. Thomson, 1867.

Subtribe Paracatopina Jeannel, 1936

Paracatopini Jeannel, 1936: 181 [stem: *Paracatop-*]. Type genus: *Paracatops* Portevin, 1907.

Tribe Cholevini Kirby, 1837

Cholevidae Kirby, 1837: 108 [stem: *Cholev-*]. Type genus: *Choleva* Latreille, 1797.

Subtribe Catopina Chaudoir, 1845

Catopides Chaudoir, 1845: 195 [stem: *Catop-*]. Type genus: *Catops* Paykull, 1798. Comment: original vernacular name available (Art. 11.7.2): first used in latinized form by Agassiz (1846b: 68, as Catopidae), generally accepted as in Perreau (2004: 136, as Catopina).

Subtribe Cholevina Kirby, 1837

Cholevidae Kirby, 1837: 108 [stem: *Cholev-*]. Type genus: *Choleva* Latreille, 1797.

Tribe Eucatopini Jeannel, 1921

Eucatopini Jeannel, 1921: 233 [stem: *Eucatop-*]. Type genus: *Eucatops* Portevin, 1903.

Tribe Leptodirini Lacordaire, 1854 (1849)

Leptodérides Lacordaire, 1854b: 195 [stem: *Leptodir-*]. Type genus: *Leptodirus* Schmidt, 1832 [as *Leptoderus*, unjustified emendation of type genus name by F. Schmidt (1852), not in prevailing usage]. Comment: use of family-group name conserved over Stagobiini Schiødte, 1849 (Art. 40.2); see Newton and Thayer (1992: 35).

Subtribe Anthroherponina Jeannel, 1910

Antroherpona Jeannel, 1910: 25 [stem: *Antroherpon-*]. Type genus: *Anthroherpon* Reitter, 1889 [as *Antroherpon*, incorrect subsequent spelling of type genus name, not in prevailing usage]. Comment: incorrect original stem formation, not in prevailing usage.

Hadesiini Absolon, 1913: 108 [stem: *Hadesi-*]. Type genus: *Hadesia* J. Müller, 1911.

Antroherponina Guéorguiev, 1974: 841, in key [stem: *Antroherpon-*]. Type genus: *Anthroherpon* Reitter, 1889 [as *Antroherpon*, incorrect subsequent spelling of type genus name, not in prevailing usage]. Comment: family-group name proposed as new without reference to Antroherpona Jeannel, 1910.

Subtribe Bathysciina Horn, 1880

Bathysciae G. H. Horn, 1880b: 251 [stem: *Bathysci-*]. Type genus: *Bathyscia* Schiødte, 1847.

Oriotini Reitter, 1889b: 296 [stem: *Oryot-*]. Type genus: *Oryotus* Miller, 1856 [as *Oriotus*, incorrect subsequent spelling of type genus name, not in prevailing usage]. Comment: incorrect original stem formation, not in prevailing usage.

BATHYSCIINA Guéorguiev, 1974: 841, in key [stem: *Bathysci-*]. Type genus: *Bathyscia* Schiødte, 1847. Comment: family-group name proposed as new without reference to BATHYSCIAE G. H. Horn, 1880.

NEOTROPOSPEONELLINA Perkovsky, 1997a: 59 [stem: *Neotropospeonell-*]. Type genus: *Neotropospeonella* Pace, 1985 [syn. of *Oryotus* Miller, 1856].

*APHAOBIINA Perkovsky, 2002: 16 [stem: *Aphaobi-*]. Type genus: *Aphaobius* Abeille de Perrin, 1878. Comment: unavailable family-group name, proposed after 1930 without description or bibliographic reference to such a description (Art. 13.1).

Subtribe BATHYSCIOTINA Guéorguiev, 1974

BATHYSCIOTINA Guéorguiev, 1974: 841, in key [stem: *Bathysciot-*]. Type genus: *Bathysciotes* Jeannel, 1910.

Subtribe LEPTODIRINA Lacordaire, 1854 (1849)

STAGOBIINAE Schiødte, 1849: 16 [stem: *Stagobi-*]. Type genus: *Stagobius* Schiødte, 1847 [syn. of *Leptodirus* Schmidt, 1832]. Comment: use of younger family-group name LEPTODIRINA Lacordaire, 1854 conserved (Art. 40.2); see Newton and Thayer (1992: 35).

LEPTODÉRIDES Lacordaire, 1854b: 195 [stem: *Leptodir-*]. Type genus: *Leptodirus* Schmidt, 1832 [as *Leptoderus*, unjustified emendation of type genus name by Schmidt (1852), not in prevailing usage]. Comment: original vernacular name available (Art. 11.7.2): first used in latinized form by Kraatz (1859a: 35, as LEPTODERIDAE), generally accepted as in Perreau (2004: 147, as LEPTODIRINI); use of family-group name conserved over STAGOBIINA Schiødte, 1849 (Art. 40.2); see Newton and Thayer (1992: 35); incorrect original stem formation, not in prevailing usage.

LEPTODIRINA Guéorguiev, 1974: 841, in key [stem: *Leptodir-*]. Type genus: *Leptodirus* Schmidt, 1832. Comment: family-group name proposed as new without reference to LEPTODÉRIDES Lacordaire, 1854.

*LANEYRIELLINA Perkovsky, 2002: 16 [stem: *Laneyriell-*]. Type genus: *Laneyriella* Guéorguiev, 1976. Comment: unavailable family-group name, proposed after 1930 without description or bibliographic reference to such a description (Art. 13.1).

Subtribe PHOLEUINA Reitter, 1886

PHOLEUONES Reitter, 1886: 314 [stem: *Pholeu-*]. Type genus: *Pholeuon* Hampe, 1856. Comment: incorrect original stem formation, not in prevailing usage; correction of stem by Newton and Thayer (1992: 35); current spelling maintained (Art. 29.5): corrected stem formation in prevailing usage (should be *Pholeuont-*).

GHIDINIINA Guéorguiev, 1974: 841, in key [stem: *Ghidini-*]. Type genus: *Ghidinia* Pavan, 1938 [syn. of *Boldoria* Jeannel, 1924].

*COIFFAITIOLINA Perkovsky, 2002: 16 [stem: *Coiffaitiol-*]. Type genus: *Coiffaitiola* Jeannel, 1955. Comment: unavailable family-group name, proposed after 1930 without description or bibliographic reference to such a description (Art. 13.1).

Subtribe PLATYCHOLEINA Horn, 1880

PLATYCHOLEI G. H. Horn, 1880b: 251 [stem: *Platychole-*]. Type genus: *Platycholeus* G. H. Horn, 1880.

Subtribe SPELAEOBATINA Guéorguiev, 1974

SPELAEOBATINA Guéorguiev, 1974: 841, in key [stem: *Spelaeobat-*]. Type genus: *Spelaeobates* Müller, 1901.

Tribe ORITOCATOPINI Jeannel, 1936

ORITOCATOPINI Jeannel, 1936: 116 [stem: *Oritocatop-*]. Type genus: *Oritocatops* Jeannel, 1921.

Tribe PTOMAPHAGINI Jeannel, 1911

PTOMAPHAGINI Jeannel, 1911: 193 [stem: *Ptomaphag-*]. Type genus: *Ptomaphagus* Hellwig, 1795.

Subtribe BARYODIRINA Perreau, 2000

BARYODIRINA Perreau, 2000: 24 [stem: *Baryodir-*]. Type genus: *Baryodirus* Perreau, 2000.

Subtribe PTOMAPHAGINA Jeannel, 1911

PTOMAPHAGINI Jeannel, 1911: 193 [stem: *Ptomaphag-*]. Type genus: *Ptomaphagus* Hellwig, 1795.

Subtribe PTOMAPHAGININA Szymczakowski, 1964

PTOMAPHAGININI Szymczakowski, 1964: 66 [stem: *Ptomaphagin-*]. Type genus: *Ptomaphaginus* Portevin, 1914.

Tribe SCIAPHYINI Perreau, 2000

SCIAPHYINI Perreau, 2000: 23 [stem: *Sciaphy-*]. Type genus: *Sciaphyes* Jeannel, 1910.

Subfamily PLATYPSYLLINAE Ritsema, 1869

PLATYPSYLLIDAE Ritsema, 1869: 38 [stem: *Platypsyll-*]. Type genus: *Platypsyllus* Ritsema, 1869.

LEPTINIDAE J. L. LeConte, 1872: 802 [stem: *Leptin-*]. Type genus: *Leptinus* Müller, 1817. Comment: this family-group name has been attributed to J. L. LeConte (1866 [error for 1867]) in the literature, however J. L. LeConte (1867: 368) did not give a name to the new taxon and therefore it is not available; see Newton and Thayer (1992: 37).

Family SILPHIDAE Latreille, 1806

SILPHALES Latreille, 1806: 1 [stem: *Silph-*]. Type genus: *Silpha* Linnaeus, 1758.

Subfamily SILPHINAE Latreille, 1806

SILPHALES Latreille, 1806: 1 [stem: *Silph-*]. Type genus: *Silpha* Linnaeus, 1758.

NECRODISIDAE Gistel, 1848: [4] [stem: *Necrod-*]. Type genus: *Necrodes* Leach, 1815. Comment: incorrect original stem formation, not in prevailing usage.

Subfamily NICROPHORINAE Kirby, 1837

NECROPHORIDAE Kirby, 1837: 95 [stem: *Nicrophor-*]. Type genus: *Nicrophorus* Fabricius, 1775 [as *Necrophorus*, unjustified emendation of type genus name by Thunberg (1789), not in prevailing usage]. Comment: incorrect original stem formation, not in prevailing usage.

Family STAPHYLINIDAE Latreille, 1802

STAPHYLINIAE Latreille, 1802: 124 [stem: *Staphylin-*]. Type genus: *Staphylinus* Linnaeus, 1758 [placed on the Official List of Generic Names in Zoology (ICZN 1959a)]. Comment: placed on the Official List of Family-Group Names in Zoology (as STAPHYLINIDAE Latreille, [1803-1804]) and "STAPHYLINII Latreille, [1803-1804]" placed on the Official Index of Rejected and Invalid Family-Group Names in Zoology (ICZN 1959a); First Reviser (STAPHYLINIDAE Latreille, 1802 vs PSELAPHIDAE Latreille, 1802) not determined, current usage maintained.

Subfamily GLYPHOLOMATINAE Jeannel, 1962

GLYPHOLOMINI Jeannel, 1962b: 482 [stem: *Glypholomat-*]. Type genus: *Glypholoma* Jeannel, 1962. Comment: originally proposed as a tribe of SILPHIDAE; incorrect original stem formation, not in prevailing usage; correction of stem by Newton and Thayer (1992: 58).

Subfamily MICROSILPHINAE Crowson, 1950

*MICRAGYRTINI Blackwelder, 1944: 84 [stem: *Micragyrt-*]. Type genus: *Micragyrtes* Champion, 1918 [syn. of *Microsilpha* Broun, 1886]. Comment: unavailable family-group name, proposed after 1930 without description or bibliographic reference to such a description (Art. 13.1); originally proposed as a tribe of LEIODIDAE.

MICROSILPHINAE Crowson, 1950: 279, in key [stem: *Microsilph-*]. Type genus: *Microsilpha* Broun, 1886. Comment: originally proposed as a subfamily of SILPHIDAE.

MICRAGYRTINI Jeannel, 1962b: 484 [stem: *Micragyrt-*]. Type genus: *Micragyrtes* Champion, 1918 [syn. of *Microsilpha* Broun, 1886]. Comment: originally proposed as a tribe of SILPHIDAE.

Subfamily OMALIINAE MacLeay, 1825

OMALIDAE W. S. MacLeay, 1825: 49 [stem: *Omali-*]. Type genus: *Omalium* Gravenhorst, 1802.

Tribe ANTHOPHAGINI Thomson, 1859

ANTHOPHAGIDES C. G. Thomson, 1859: 48 [stem: *Anthophag-*]. Type genus: *Anthophagus* Gravenhorst, 1802 [placed on the Official List of Generic Names in Zoology (ICZN 2004d)].

BRATHINIDAE J. L. LeConte, 1861: 48 [stem: *Brathin-*]. Type genus: *Brathinus* J. L. LeConte, 1852. Comment: originally proposed as a subfamily of SILPHIDAE.

*LESTÉVATES Mulsant and Rey, 1880: 8 [stem: *Lestev-*]. Type genus: *Lesteva* Latreille, 1797 [placed on the Official List of Generic Names in Zoology (ICZN 2004d)]. Comment: original vernacular name unavailable (Art. 11.7.2): subsequently used in latinized form but not generally attributed to Mulsant and Rey (1880).

LESTEVINA Jakobson, 1908: 450 [stem: *Lestev-*]. Type genus: *Lesteva* Latreille, 1797 [placed on the Official List of Generic Names in Zoology (ICZN 2004d)].

Tribe APHAENOSTEMMINI Peyerimhoff, 1914

APHAENOSTEMMINI Peyerimhoff, 1914: 248, in key [stem: *Aphaenostemm-*]. Type genus: *Aphaenostemmus* Peyerimhoff, 1914.

Tribe CORNEOLABIINI Steel, 1950

CORNEOLABIINI Steel, 1950: 54 [stem: *Corneolabi-*]. Type genus: *Corneolabium* Steel, 1950.

Tribe CORYPHIINI Jakobson, 1908

CORYPHIINA Jakobson, 1908: 452 [stem: *Coryphi-*]. Type genus: *Coryphium* Stephens, 1834 [placed on the Official List of Generic Names in Zoology (ICZN 1990b)].

Subtribe BOREAPHILINA Zerche, 1990

*BORÉAPHILAIRES Mulsant and Rey, 1880: 391 [stem: *Boreaphil-*]. Type genus: *Boreaphilus* Sahlberg, 1832. Comment: original vernacular name unavailable (Art. 11.7.2): subsequently used in latinized form but not generally attributed to Mulsant and Rey (1880).

BOREAPHILINA Zerche, 1990: 22, in key [stem: *Boreaphil-*]. Type genus: *Boreaphilus* Sahlberg, 1832. Comment: family-group name proposed as new without reference to BORÉAPHILAIRES Mulsant and Rey, 1880.

Subtribe Coryphiina Jakobson, 1908

Coryphiina Jakobson, 1908: 452 [stem: *Coryphi-*]. Type genus: *Coryphium* Stephens, 1834 [placed on the Official List of Generic Names in Zoology (ICZN 1990b)].

Tribe Eusphalerini Hatch, 1957

*Anthobiates Mulsant and Rey, 1880: 290 [stem: *Anthobi-*]. Type genus: *Anthobium* sensu Erichson, 1840 [not *Anthobium* Leach, 1819; syn. of *Eusphalerum* Kraatz, 1857]. Comment: original vernacular name unavailable (Art. 11.7.2): subsequently used in latinized form but not generally attributed to Mulsant and Rey (1880).

Anthobiini Portevin, 1929: 450 [stem: *Anthobi-*]. Type genus: *Anthobium* sensu Erichson, 1840 [not *Anthobium* Leach, 1819; syn. of *Eusphalerum* Kraatz, 1857]. Comment: based on a misidentified type genus, name treated here as invalid until an application is submitted to the Commission to suppress it for the Principle of Priority (Art. 65.2.1).

Eusphalerini Hatch, 1957: 82 [stem: *Eusphaler-*]. Type genus: *Eusphalerum* Kraatz, 1857. Comment: new name for Anthobiini; although this is not the oldest name for the tribe, we recommend that an application be submitted to the Commission to suppress the older name Anthobiini Poertevin, 1929 because it is based on a misidentified type genus (Art. 65.2.1).

Tribe Hadrognathini Portevin, 1929

*Eugnathates Mulsant and Rey, 1880: 386 [stem: *Eugnath-*]. Type genus: *Eugnathus* Mulsant and Rey, 1851 [preoccupied genus name, not *Eugnathus* Schönherr, 1833 [Coleoptera: Curculionidae] or *Eugnathus* Agassiz, 1836 [Pisces]; syn. of *Hadrognathus* Schaum, 1852]. Comment: original vernacular name unavailable (Art. 11.7.2): not subsequently latinized; if found to be available then permanently invalid (Art. 39): based on preoccupied type genus.

Hadrognathini Portevin, 1929: 431 [stem: *Hadrognath-*]. Type genus: *Hadrognathus* Schaum, 1852.

Tribe Omaliini MacLeay, 1825

Omalidae W. S. MacLeay, 1825: 49 [stem: *Omali-*]. Type genus: *Omalium* Gravenhorst, 1802. Comment: incorrect original stem formation, not in prevailing usage.

Micralymmates Mulsant and Rey, 1880: 3 [stem: *Micralymmat-*]. Type genus: *Micralymma* Westwood, 1838. Comment: original vernacular name available (Art. 11.7.2): first used in latinized form by Portevin (1929: 443, as Micralymmini), generally accepted as in Newton and Thayer (1992: 58, as Micralymmini); incorrect original stem formation, not in prevailing usage.

Tetradeli Fauvel, 1904: 90 [stem: *Tetradel-*]. Type genus: *Tetradelus* Fauvel, 1904.

Arpediopsini Cameron, 1917a: 123 [stem: *Arpediopse-*]. Type genus: *Arpediopsis* Cameron, 1917 [preoccupied genus name, not *Arpediopsis* Ganglbauer, 1895 [Coleoptera: Staphylinidae: Omaliinae: Anthophagini]; syn. of *Crymus* Fauvel, 1904]. Comment: permanently invalid (Art. 39): based on preoccupied type genus; incorrect original stem formation, not in prevailing usage.

Arpediomimi Cameron, 1917b: 277 [stem: *Arpediomim-*]. Type genus: *Arpediomimus* Cameron, 1917 [syn. of *Crymus* Fauvel, 1904]. Comment: replacement name for Arpediopsini Cameron, 1917; incorrect original stem formation, not in prevailing usage.

Phloeonomini Ádám, 2001: 231 [stem: *Phloeonom-*]. Type genus: *Phloeonomus* Heer, 1839. Comment: **syn. nov.**

Subfamily Empelinae Newton and Thayer, 1992

*Empelidae M. Abdullah, 1969b: 683 [stem: *Empel-*]. Type genus: *Empelus* J. L. LeConte, 1861. Comment: unavailable family-group name, proposed after 1930 without description or bibliographic reference to such a description (Art. 13.1).

Empelinae Newton and Thayer, 1992: 25 [stem: *Empel-*]. Type genus: *Empelus* J. L. LeConte, 1861.

Subfamily Proteininae Erichson, 1839

Proteinini Erichson, 1839a: 641 [stem: *Protein-*]. Type genus: *Proteinus* Latreille, 1797 [placed on the Official List of Generic Names in Zoology (ICZN 1969a)].

Tribe Anepiini Steel, 1966

Anepiini Steel, 1966: 306 [stem: *Anepi-*]. Type genus: *Anepius* Blackburn, 1902.

Tribe Austrorhysini Newton and Thayer, 1995

Austrorhysini Newton and Thayer, 1995: 298 [stem: *Austrorhys-*]. Type genus: *Austrorhysus* Steel, 1966.

Tribe Nesoneini Steel, 1966

Nesoneini Steel, 1966: 292 [stem: *Nesone-*]. Type genus: *Nesoneus* Bernhauer, 1939.

Tribe Proteinini Erichson, 1839

Proteinini Erichson, 1839a: 641 [stem: *Protein-*]. Type genus: *Proteinus* Latreille, 1797 [placed on the Official List of Generic Names in Zoology (ICZN 1969a)]. Comment: this family-group name was also proposed in the same year by Erichson (1839b: 31, as Proteinini) and Heer (1839b: 4, as Proteinina); for comments about the priority of these works see Newton and Thayer (1992: 24).

*Phléobiens Mulsant and Rey, 1876: 209 [stem: *Phloeobi-*]. Type genus: *Phloeobium* sensu Erichson, 1840 [not *Phloeobium* Dejean, 1833; syn. of

Metopsia Wollaston, 1854]. Comment: original vernacular name unavailable (Art. 11.7.2): subsequently used in latinized form but not generally attributed to Mulsant and Rey, 1876; incorrect original stem formation, not in prevailing usage.

PHLOEOBIINAE Fowler, 1888: 431 [stem: *Phloeobi-*]. Type genus: *Phloeobium* sensu Erichson, 1840 [not *Phloeobium* Dejean, 1833; syn. of *Metopsia* Wollaston, 1854]. Comment: based on a misidentified type genus.

MEGARTHRINI Joy, 1932: 93, in key [stem: *Megarthr-*]. Type genus: *Megarthrus* Curtis, 1829.

METOPSIINAE Tottenham, 1954: 8, in key [stem: *Metopsi-*]. Type genus: *Metopsia* Wollaston, 1854.

PTERONINI Arnett, 1961: 238, in key [stem: *Pteroni-*]. Type genus: *Pteronius* Blackwelder, 1952 [syn. of *Proteinus* Latreille, 1797]. Comment: incorrect original stem formation, not in prevailing usage.

PTERONIINAE I. Moore, 1964: 85, in key [stem: *Pteroni-*]. Type genus: *Pteronius* Blackwelder, 1952 [syn. of *Proteinus* Latreille, 1797]. Comment: proposed as new without reference to PTERONINI Arnett, 1961.

Tribe SILPHOTELINI Newton and Thayer, 1995

SILPHOTELINI Newton and Thayer, 1995: 297 [stem: *Silphotel-*]. Type genus: *Silphotelus* Broun, 1895.

Subfamily MICROPEPLINAE Leach, 1815

MICROPEPLIDA Leach, 1815: 90 [stem: *Micropepl-*]. Type genus: *Micropeplus* Latreille, 1809.

MICROPEPLIDA Heer, 1839a: 169 [stem: *Micropepl-*]. Type genus: *Micropeplus* Latreille, 1809. Comment: family-group name proposed as new without reference to MICROPEPLIDA Leach, 1815; this family-group name was also used in the same year by Heer (1839b: 4, as MICROPEPLIDA); for comments about the priority of these works see Newton and Thayer (1992: 24).

Subfamily NEOPHONINAE Fauvel, 1905

NEOPHONI Fauvel, 1905a: 98 [stem: *Neophon-*]. Type genus: *Neophonus* Fauvel, 1905.

Subfamily DASYCERINAE Reitter, 1887

DASYCERINI Reitter, 1887: 8, in key [stem: *Dasycer-*]. Type genus: *Dasycerus* Brongniart, 1800. Comment: originally proposed as a new tribe of LATRIDIIDAE; the earlier usage of the DASYCERINAE by Swainson (1840: 293) was not based on a type genus and is therefore unavailable.

Subfamily PROTOPSELAPHINAE Newton and Thayer, 1995

PROTOPSELAPHINAE Newton and Thayer, 1995: 227 [stem: *Protopselaph-*]. Type genus: *Protopselaphus* Newton and Thayer, 1995.

Subfamily Pselaphinae Latreille, 1802

Pselaphii Latreille, 1802: 239 [stem: *Pselaph-*]. Type genus: *Pselaphus* Herbst, 1791.

Supertribe Batrisitae Reitter, 1882

Batrisini Reitter, 1882b: 183 [stem: *Batris-*]. Type genus: *Batrisus* Aubé, 1833.

Tribe Amauropini Jeannel, 1948

Amauropsini Jeannel, 1948a: 1 [stem: *Amaurop-*]. Type genus: *Amaurops* Fairmaire, 1851. Comment: incorrect original stem formation, not in prevailing usage; correction of stem by Newton and Thayer (1992: 37).

Tribe Batrisini Reitter, 1882

Batrisini Reitter, 1882b: 183 [stem: *Batris-*]. Type genus: *Batrisus* Aubé, 1833.

Subtribe Ambicocerina Leleup, 1970

Ambicocerina Leleup, 1970: 309 [stem: *Ambicocer-*]. Type genus: *Ambicocerus* Leleup, 1970.

Subtribe Batrisina Reitter, 1882

Batrisini Reitter, 1882b: 183 [stem: *Batris-*]. Type genus: *Batrisus* Aubé, 1833.

Oropygiina Jeannel, 1949a: 113, in key [stem: *Orropygi-*]. Type genus: *Orropygia* Raffray, 1910 [as *Oropygia*, incorrect subsequent spelling of type genus name, not in prevailing usage]. Comment: incorrect original stem formation, not in prevailing usage; correction of stem by Newton and Thayer (1992: 37).

Trabisina Jeannel, 1949a: 114, in key [stem: *Trabis-*]. Type genus: *Trabisus* Raffray, 1890 [syn. of *Atheropterus* Raffray, 1882].

Subtribe Leupeliina Jeannel, 1954

Leupeliina Jeannel, 1954c: 106 [stem: *Leupeli-*]. Type genus: *Leupelia* Jeannel, 1954.

Subtribe Stilipalpina Jeannel, 1954

Stilipalpina Jeannel, 1954c: 118 [stem: *Stilipalp-*]. Type genus: *Stilipalpus* Jeannel, 1951.

Tribe Thaumastocephalini Poggi, Nonveiller, Colla, Pavićević and Rada, 2001

Thaumastocephalini Poggi et al., 2001: 5, in key [stem: *Thaumastocephal-*]. Type genus: *Thaumastocephalus* Poggi et al., 2001.

Supertribe Clavigeritae Leach, 1815

Clavigerides Leach, 1815: 117 [stem: *Claviger-*]. Type genus: *Claviger* Preyssler, 1790.

Tribe Clavigerini Leach, 1815

Clavigerides Leach, 1815: 117 [stem: *Claviger-*]. Type genus: *Claviger* Preyssler, 1790.

Subtribe Apoderigerina Jeannel, 1954

Apoderigerini Jeannel, 1954a: 310 [stem: *Apoderiger-*]. Type genus: *Apoderiger* Wasmann, 1897.

Subtribe Clavigerina Leach, 1815

Clavigerides Leach, 1815: 117 [stem: *Claviger-*]. Type genus: *Claviger* Preyssler, 1790.

Adranites Desmarest, 1857: 144 [stem: *Adran-*]. Type genus: *Adranes* J. L. LeConte, 1849.

Adraniini O. Park, 1951: 58, in key [stem: *Adran-*]. Type genus: *Adranes* J. L. LeConte, 1849. Comment: family-group name proposed as new without reference to Adranites Desmarest, 1857; incorrect original stem formation, not in prevailing usage.

Subtribe Clavigerodina Schaufuss, 1882

Clavigerodini L. W. Schaufuss, 1882a: 205 [stem: *Clavigerod-*]. Type genus: *Clavigerodes* Raffray, 1877.

Commatocerini L. W. Schaufuss, 1882b: 349 [stem: *Commatocer-*]. Type genus: *Commatocerus* Raffray, 1882 [syn. of *Fustiger* J. L. LeConte, 1866].

Clavigeropsini L. W. Schaufuss, 1890: in table [stem: *Clavigeropse-*]. Type genus: *Clavigeropsis* Raffray, 1882. Comment: incorrect original stem formation, not in prevailing usage.

Fustigerini Jeannel, 1949a: 31 [stem: *Fustiger-*]. Type genus: *Fustiger* J. L. LeConte, 1866.

Neoceratopsini Célis, 1970: 260 [stem: *Neoceratopse-*]. Type genus: *Neoceratopsis* Jeannel, 1956. Comment: incorrect original stem formation, not in prevailing usage.

Subtribe Disarthricerina Jeannel, 1949

Diarthricerini Jeannel, 1949a: 29, in key [stem: *Disarthricer-*]. Type genus: *Disarthricerus* Raffray, 1895 [as *Diarthricerus*, incorrect subsequent spelling of type genus name, not in prevailing usage]. Comment: incorrect original stem formation, not in prevailing usage.

Subtribe Hoplitoxenina Célis, 1969

Hoplitoxenini Célis, 1969: 418 [stem: *Hoplitoxen-*]. Type genus: *Hoplitoxenus* Jeannel, 1960.

Dimerometopini Célis, 1970: 244 [stem: *Dimerometop-*]. Type genus: *Dimerometopus* Célis, 1970.

Subtribe Lunillina Célis, 1969

Lunillini Célis, 1969: 416 [stem: *Lunill-*]. Type genus: *Lunilla* Jeannel, 1957.

Subtribe Mastigerina Jeannel, 1954

Mastigerini Jeannel, 1954a: 291, in key [stem: *Mastiger-*]. Type genus: *Mastiger* Motschulsky, 1851.

Subtribe Miroclavigerina Jeannel, 1949

Miroclavigerini Jeannel, 1949a: 29, in key [stem: *Miroclaviger-*]. Type genus: *Miroclaviger* Wasmann, 1893.

Subtribe Neocerina Jeannel, 1954

Neocerini Jeannel, 1954a: 316 [stem: *Neocer-*]. Type genus: *Neocerus* Wasmann, 1893. Comment: First Reviser (Neocerina Jeannel, 1954 vs Theocerina Jeannel, 1954) not determined, current usage maintained.

Theocerini Jeannel, 1954a: 314 [stem: *Theocer-*]. Type genus: *Theocerus* Raffray, 1897.

Subtribe Radamina Jeannel, 1954

Radamini Jeannel, 1954a: 319 [stem: *Radam-*]. Type genus: *Radama* Raffray, 1883.

Subtribe Thysdariina Jeannel, 1954

Thysdariini Jeannel, 1954a: 332 [stem: *Thysdari-*]. Type genus: *Thysdarius* Fairmaire, 1904.

Tribe Colilodionini Besuchet, 1991

Colilodionini Besuchet, 1991: 514 [stem: *Colilodion-*]. Type genus: *Colilodion* Besuchet, 1991.

Tribe Tiracerini Besuchet, 1986

Articerides Desmarest, 1857: 145 [stem: *Articer-*]. Type genus: *Articerus* sensu Hope, 1845 [not *Articerus* Dalman, 1826; syn. of *Tiracerus* Besuchet, 1986]. Comment: based on a misidentified type genus, name treated here as invalid until an application is submitted to the Commission to suppress it for the Principle of Priority (Art. 65.2.1).

Tiracerini Besuchet, 1986: 263 [stem: *Tiracer-*]. Type genus: *Tiracerus* Besuchet, 1986.

Supertribe Euplectitae Streubel, 1839

Euplectidae Streubel, 1839: 135 [stem: *Euplect-*]. Type genus: *Euplectus* Leach, 1817.

Tribe Bythinoplectini Schaufuss, 1890

Bythinoplectini L. W. Schaufuss, 1890: in table [stem: *Bythinoplect-*]. Type genus: *Bythinoplectus* Reitter, 1882.

Subtribe Bythinoplectina Schaufuss, 1890

Bythinoplectini L. W. Schaufuss, 1890: in table [stem: *Bythinoplect-*]. Type genus: *Bythinoplectus* Reitter, 1882.

Zethini L. W. Schaufuss, 1890: in table [stem: *Zeth-*]. Type genus: *Zethus* L. W. Schaufuss, 1872 [preoccupied genus name, not *Zethus* Fabricius, 1805 [Hymenoptera], not *Zethus* Pander, 1830 [Trilobita]; syn. of *Zethopsus* Reitter, 1880]. Comment: permanently invalid (Art. 39): based on preoccupied type genus.

Zethopsina Jeannel, 1952a: 51, in key [stem: *Zethops-*]. Type genus: *Zethopsus* Reitter, 1880.

Subtribe Pyxidicerina Raffray, 1904

Pyxidicerini Raffray, 1904: 489, in key [stem: *Pyxidicer-*]. Type genus: *Pyxidicerus* Motschulsky, 1863.

Tribe Dimerini Raffray, 1908

Dimerini Raffray, 1908: 412 [stem: *Dimer-*]. Type genus: *Dimerus* Fiori, 1899 [syn. of *Octomicrus* L. W. Schaufuss, 1877]. Comment: published before 17 March 1908; this family-group name was also used in the same year by Bernhauer (1908 [23 March]: 327, as Dimerini); use of family-group name Dimerini Raffray, 1908 (based on synonym) conserved (Art. 40.2); see Newton and Thayer (1992: 39).

Octomicrini Jeannel, 1952a: 43 [stem: *Octomicr-*]. Type genus: *Octomicrus* L. W. Schaufuss, 1877.

Tribe Euplectini Streubel, 1839

Euplectidae Streubel, 1839: 135 [stem: *Euplect-*]. Type genus: *Euplectus* Leach, 1817.

Mitracephalini O. Park, 1951: 64, in key [stem: *Mitracephal-*]. Type genus: *Mitracephala* Raffray, 1890 [preoccupied genus name, not *Mitracephala* J. Thomson, 1859 [Coleoptera: Scarabaeidae]; syn. of *Mitrametopus* Raffray, 1911]. Comment: permanently invalid (Art. 39): based on preoccupied type genus.

Mitrametopina O. Park, 1952: 87 [stem: *Mitrametop-*]. Type genus: *Mitrametopus* Raffray, 1911. Comment: replacement name for Mitracephalini O. Park, 1951 because of the homonymy of the type genus.

Tribe Jubini Raffray, 1904

Jubini Raffray, 1904: 489, in key [stem: *Jub-*]. Type genus: *Jubus* L. W. Schaufuss, 1872.

AUXENOCERINI Jeannel, 1962a: 319 [stem: *Auxenocer-*]. Type genus: *Auxenocerus* Jeannel, 1962.

Tribe MAYETIINI Winkler, 1925

MAYETIINI A. Winkler, 1925: 348 [stem: *Mayeti-*]. Type genus: *Mayetia* Mulsant and Rey, 1875.

Tribe METOPIASINI Raffray, 1904

METOPIINI Raffray, 1904: 490, in key [stem: *Metopias-*]. Type genus: *Metopias* Gory, 1832 [placed on the Official List of Generic Names in Zoology (ICZN 1994d)]. Comment: *Metopias-* established as the correct stem and METOPIASINI Raffray, 1904 placed on the Official List of Family-Group Names in Zoology (ICZN 1994d).

Subtribe METOPIASINA Raffray, 1904

METOPIINI Raffray, 1904: 490, in key [stem: *Metopias-*]. Type genus: *Metopias* Gory, 1832 [placed on the Official List of Generic Names in Zoology (ICZN 1994d)]. Comment: correct original stem *Metopi-* modified to *Metopias-* and METOPIASINI Raffray, 1904 placed on the Official List of Family-Group Names in Zoology (ICZN 1994d); METOPIINI Raffray, 1904 placed on the Official Index of Rejected and Invalid Family-Group Names in Zoology (ICZN 1994d).

Subtribe RHINOSCEPSINA Bowman, 1934

RHINOSCEPSII Bowman, 1934: 8 [stem: *Rhinosceps-*]. Type genus: *Rhinoscepsis* J. L. LeConte, 1878. Comment: current spelling maintained (Art. 29.5): incorrect stem formation in prevailing usage (should be *Rhinoscepse-*).

Tribe TRICHONYCHINI Reitter, 1882

TRICHONYIDES Reitter, 1882b: 194 [stem: *Trichonych-*]. Type genus: *Trichonyx* Chaudoir, 1845.

Subtribe BIBLOPORINA Park, 1951

BIBLOPORINI O. Park, 1951: 64, in key [stem: *Biblopor-*]. Type genus: *Bibloporus* C. G. Thomson, 1859.

BIBLOPORELLINA Jeannel, 1952a: 92 [stem: *Bibloporell-*]. Type genus: *Bibloporellus* Jeannel, 1949.

Subtribe PANAPHANTINA Jeannel, 1950

PANAPHANTINA Jeannel, 1950a: 76 [stem: *Panaphant-*]. Type genus: *Panaphantus* Kiesenwetter, 1858.

TRISIGNINA O. Park and Schuster, 1955: 1 [stem: *Trisign-*]. Type genus: *Trisignis* O. Park and Schuster, 1955.

Acetaliini Jeannel, 1958a: 81 [stem: *Acetali-*]. Type genus: *Acetalius* Sharp, 1883.

Bibloplectina Jeannel, 1959: 96, in key [stem: *Bibloplect-*]. Type genus: *Bibloplectus* Reitter, 1881.

Subtribe Trichonychina Reitter, 1882

Trichonyides Reitter, 1882b: 194 [stem: *Trichonych-*]. Type genus: *Trichonyx* Chaudoir, 1845. Comment: incorrect original stem formation, not in prevailing usage.

Raffrayina Jeannel, 1949a: 76 [stem: *Raffrayi-*]. Type genus: *Raffrayia* Reitter, 1882. Comment: incorrect original stem formation, not in prevailing usage; correction of stem by Jeannel (1964: 65, as Raffrayiini).

Ranavalini Jeannel, 1954a: 183 [stem: *Ranaval-*]. Type genus: *Ranavala* Raffray, 1898.

Chrestomerina Jeannel, 1962a: 344 [stem: *Chrestomer-*]. Type genus: *Chrestomera* Jeannel, 1962.

Pteracmini Jeannel, 1962a: 347 [stem: *Pteracmet-*]. Type genus: *Pteracmes* Raffray, 1890. Comment: incorrect original stem formation, not in prevailing usage.

Trimiodytina Jeannel, 1964: 39 [stem: *Trimiodyt-*]. Type genus: *Trimiodytes* Raffray, 1897.

Subtribe Trimiina Bowman, 1934

Trimii Bowman, 1934: 8 [stem: *Trimi-*]. Type genus: *Trimium* Aubé, 1833.

Trimiina Jeannel, 1950a: 139 [stem: *Trimi-*]. Type genus: *Trimium* Aubé, 1833. Comment: family-group name proposed as new without reference to Trimii Bowman, 1934.

Tribe Trogastrini Jeannel, 1949

Trogastrini Jeannel, 1949a: 41, in key [stem: *Trogastr-*]. Type genus: *Trogaster* Sharp, 1874.

Subtribe Phtegnomina Park, 1951

Phtegnomini O. Park, 1951: 64, in key [stem: *Phtegnom-*]. Type genus: *Phtegnomus* Raffray, 1890.

Subtribe Rhexiina Park, 1951

Rhexini O. Park, 1951: 63, in key [stem: *Rhexi-*]. Type genus: *Rhexius* J. L. LeConte, 1849. Comment: incorrect original stem formation, not in prevailing usage; correction of stem by Newton and Thayer (1992: 39).

Subtribe Trogastrina Jeannel, 1949

Trogastrini Jeannel, 1949a: 41, in key [stem: *Trogastr-*]. Type genus: *Trogaster* Sharp, 1874.

Supertribe FARONITAE Reitter, 1882

FARONIDES Reitter, 1882b: 194 [stem: *Faron-*]. Type genus: *Faronus* Aubé, 1844.

Supertribe GONIACERITAE Reitter, 1882 (1872)

GONIACERIDES Reitter, 1882b: 188 [stem: *Goniacer-*]. Type genus: *Goniacerus* Motschulsky, 1855. Comment: use of family-group name conserved over GONIASTITAE L. W. Schaufuss, 1872 (Art. 40.2) (see Newton and Thayer 1992: 39).

Tribe ARNYLLIINI Jeannel, 1952

ARNYLLIINI Jeannel, 1952b: 100 [stem: *Arnylli-*]. Type genus: *Arnyllium* Reitter, 1884.

Tribe BARROSELLINI Leleup, 1973

BARROSELLINI Leleup, 1973: 81 [stem: *Barrosell-*]. Type genus: *Barrosellus* Jeannel, 1951.

Tribe BRACHYGLUTINI Raffray, 1904

BRACHYGLUTINI Raffray, 1904: 490, in key [stem: *Brachyglut-*]. Type genus: *Brachygluta* C. G. Thomson, 1859.

Subtribe BARADINA Park, 1951

BARADIINI O. Park, 1951: 62, in key [stem: *Barad-*]. Type genus: *Barada* Raffray, 1891 [syn. of *Euphalepsus* Reitter, 1883]. Comment: incorrect original stem formation, not in prevailing usage; correction of stem by Newton and Thayer (1992: 40).

Subtribe BRACHYGLUTINA Raffray, 1904

BRYAXES J. L. LeConte, 1861: 57 [stem: *Bryaxe-*]. Type genus: *Bryaxis* Leach, 1817 [preoccupied genus name, not *Bryaxis* Kugelann, 1794 [Coleoptera: STAPHYLINIDAE: PSELAPHINAE: BYTHININI]; placed on the Official Index of Rejected and Invalid Generic Names in Zoology (ICZN 1969b); syn. of *Rybaxis* Saulcy, 1876]. Comment: permanently invalid (Art. 39): based on preoccupied and suppressed type genus; incorrect original stem formation, not in prevailing usage.

BRACHYGLUTINI Raffray, 1904: 490, in key [stem: *Brachyglut-*]. Type genus: *Brachygluta* C. G. Thomson, 1859. Comment: replacement name for BRYAXES J. L. LeConte, 1861.

REICHENBACHIINA Jakobson, 1910: 577 [stem: *Reichenbachi-*]. Type genus: *Reichenbachia* Leach, 1826.

GLOBINA Jeannel, 1959: 462, in key [stem: *Glob-*]. Type genus: *Globa* Raffray, 1887.

HALORABYXINA Leleup, 1969a: 138 [stem: *Halorabyx-*]. Type genus: *Halorabyxis* Jeannel, 1954 [syn. of *Physoplectus* Reitter, 1882].

Pselaptina O. Park, 1976: 48 [stem: *Pselapt-*]. Type genus: *Pselaptus* J. L. LeConte, 1880.

Subtribe Decarthrina Park, 1951

Decarthronini O. Park, 1951: 61, in key [stem: *Decarthr-*]. Type genus: *Decarthron* Brendel, 1865. Comment: incorrect original stem formation, not in prevailing usage; correction of stem by Newton and Thayer (1992: 40).

Subtribe Eupseniina Park, 1951

Eupseniini O. Park, 1951: 61, in key [stem: *Eupseni-*]. Type genus: *Eupsenius* J. L. LeConte, 1849.

Tribe Bythinini Raffray, 1890

Bythinini Raffray, 1890: 83, in key [stem: *Bythin-*]. Type genus: *Bythinus* Leach, 1817 [placed on the Official List of Generic Names in Zoology (ICZN 1969b)].

Subtribe Bythinina Raffray, 1890

Bythinini Raffray, 1890: 83, in key [stem: *Bythin-*]. Type genus: *Bythinus* Leach, 1817 [placed on the Official List of Generic Names in Zoology (ICZN 1969b)].

Bryaxina Jakobson, 1910: 579 [stem: *Bryaxe-*]. Type genus: *Bryaxis* Kugelann, 1794 [placed on the Official List of Generic Names in Zoology (ICZN 1969b)]. Comment: incorrect original stem formation, not in prevailing usage.

Subtribe Machaeritina Jeannel, 1950

Machaeritina Jeannel, 1950a: 168 [stem: *Machaerit-*]. Type genus: *Machaerites* Miller, 1855.

Subtribe Xenobythina Jeannel, 1950

Xenobythina Jeannel, 1950a: 201 [stem: *Xenobyth-*]. Type genus: *Xenobythus* Peyerimhoff, 1901.

Tribe Cyathigerini Schaufuss, 1872

Cyathigerini L. W. Schaufuss, 1872: 245 [stem: *Cyathiger-*]. Type genus: *Cyathiger* King, 1865 [syn. of *Plagiophorus* Motschulsky, 1851].

Tribe Goniacerini Reitter, 1882 (1872)

Goniastini L. W. Schaufuss, 1872: 245 [stem: *Goniast-*]. Type genus: *Goniastes* Westwood, 1870. Comment: use of younger family-group name Goniacerini Reitter, 1882 conserved (Art. 40.2) (see Newton and Thayer 1992: 40).

GONIACERIDES Reitter, 1882b: 188 [stem: *Goniacer-*]. Type genus: *Goniacerus* Motschulsky, 1855. Comment: use of family-group name conserved over GONIASTINI L. W. Schaufuss, 1872 (Art. 40.2) (see Newton and Thayer 1992: 40).

SIMINI L. W. Schaufuss, 1890: in table [stem: *Sim-*]. Type genus: *Simus* Raffray, 1882 [preoccupied genus name, not *Simus* Bonaparte, 1838 [Reptilia] or *Simus* Hodgson, 1841 [Aves]; syn. of *Ipsimus* Reitter, 1885]. Comment: permanently invalid (Art. 39): based on preoccupied type genus.

LISTRIOPHORINI L. W. Schaufuss, 1890: in table [stem: *Listriophor-*]. Type genus: *Listriophorus* L. W. Schaufuss, 1872.

Tribe IMIRINI Jeannel, 1949

MIRINI Raffray, 1917: 110 [stem: *Mir-*]. Type genus: *Mirus* Saulcy, 1877 [preoccupied genus name, not *Mirus* Albers, 1850 [Mollusca]; syn. of *Imirus* Reitter, 1885]. Comment: permanently invalid (Art. 39): based on preoccupied type genus; junior homonym of MIRIDAE Hahn, 1833 (type genus *Miris* Fabricius, 1794) in Hemiptera.

IMIRINI Jeannel, 1949a: 41, in key [stem: *Imir-*]. Type genus: *Imirus* Reitter, 1885.

Tribe INIOCYPHINI Park, 1951

INIOCYPHINI O. Park, 1951: 60, in key [stem: *Iniocyph-*]. Type genus: *Iniocyphus* Raffray, 1912.

Subtribe INIOCYPHINA Park, 1951

INIOCYPHINI O. Park, 1951: 60, in key [stem: *Iniocyph-*]. Type genus: *Iniocyphus* Raffray, 1912. Comment: First Reviser found (INIOCYPHINA Park, 1951 vs DALMODIINA Park, 1951) is Chandler (2001: 377).

DALMODIINI O. Park, 1951: 61, in key [stem: *Dalmod-*]. Type genus: *Dalmodes* Reitter, 1882. Comment: incorrect original stem formation, not in prevailing usage; correction of stem by Newton and Thayer (1992: 41).

Subtribe NATYPLEURINA Newton and Thayer, 1992

TANYPLEURINI Jeannel, 1949a: 79 [stem: *Tanypleur-*]. Type genus: *Tanypleurus* Raffray, 1890 [preoccupied genus name, not *Tanypleurus* Steenstrup and Luetken, 1861 [Crustacea]; syn. of *Natypleurus* Newton and Thayer, 1992]. Comment: permanently invalid (Art. 39): based on preoccupied type genus.

NATYPLEURINA Newton and Thayer, 1992: 41 [stem: *Natypleur-*]. Type genus: *Natypleurus* Newton and Thayer, 1992. Comment: replacement name for TANYPLEURINI Jeannel, 1949 because of the homonymy of the type genus.

Tribe MACHADOINI Jeannel, 1951

MACHADOINI Jeannel, 1951: 105 [stem: *Machado-*]. Type genus: *Machadous* Jeannel, 1951.

Tribe Proterini Jeannel, 1949

Proterini Jeannel, 1949a: 41, in key [stem: *Proter-*]. Type genus: *Proterus* Raffray, 1897.

Tribe Pygoxyini Reitter, 1909

Pygoxyini Reitter, 1909: 202 [stem: *Pygoxy-*]. Type genus: *Pygoxyon* Reitter, 1880.

Tribe Speleobamini Park, 1951

Speleobamini O. Park, 1951: 51 [stem: *Speleobam-*]. Type genus: *Speleobama* O. Park, 1951.

Tribe Tychini Raffray, 1904

Tychini Raffray, 1904: 490, in key [stem: *Tych-*]. Type genus: *Tychus* Leach, 1817.

Tribe Valdini Park, 1953

Valdiini O. Park, 1953: 261 [stem: *Vald-*]. Type genus: *Valda* Casey, 1894. Comment: incorrect original stem formation, not in prevailing usage; correction of stem by Newton and Thayer (1992: 41).

Supertribe Pselaphitae Latreille, 1802

Pselaphii Latreille, 1802: 239 [stem: *Pselaph-*]. Type genus: *Pselaphus* Herbst, 1791.

Tribe Arhytodini Raffray, 1890

Arhytodini Raffray, 1890: 84, in key [stem: *Arhytod-*]. Type genus: *Arhytodes* Reitter, 1882 [syn. of *Rhytus* Westwood, 1870].

Holozodini Raffray, 1900: 518 [stem: *Holozod-*]. Type genus: *Holozodus* Fairmaire, 1898.

Tribe Attapseniini Bruch, 1933

Attapseniini Bruch, 1933a: 26 [stem: *Attapseni-*]. Type genus: *Attapsenius* Bruch, 1933.

Tribe Ctenistini Blanchard, 1845

Cténistites Blanchard, 1845a: 306 [stem: *Ctenist-*]. Type genus: *Ctenistes* Reichenbach, 1816. Comment: original vernacular name available (Art. 11.7.2): first used in latinized form by Reitter (1882b: 183, as Ctenistini), generally accepted as in Chandler (2001: 484, as Ctenistini).

Chenniides Reitter, 1882b: 183 [stem: *Chenni-*]. Type genus: *Chennium* Latreille, 1807.

Tribe Hybocephalini Raffray, 1890

Hybocephalini Raffray, 1890: 83, in key [stem: *Hybocephal-*]. Type genus: *Hybocephalus* Motschulsky, 1851.

Mestogastrina Jakobson, 1910: 585 [stem: *Mestogastr-*]. Type genus: *Mestogaster* Schmidt-Göbel, 1838.

Tribe Odontalgini Jeannel, 1949

Odontalgini Jeannel, 1949a: 177 [stem: *Odontalg-*]. Type genus: *Odontalgus* Raffray, 1877.

Tribe Pachygastrodini Leleup, 1969

Pachygastrodini Leleup, 1969b: 282 [stem: *Pachygastrod-*]. Type genus: *Pachygastrodes* Leleup, 1969.

Tribe Phalepsini Jeannel, 1949

Phalepsini Jeannel, 1949a: 208 [stem: *Phaleps-*]. Type genus: *Phalepsus* Westwood, 1870.

Tribe Pselaphini Latreille, 1802

Pselaphii Latreille, 1802: 239 [stem: *Pselaph-*]. Type genus: *Pselaphus* Herbst, 1791.

Tribe Schistodactylini Raffray, 1890

Schistodactylini Raffray, 1890: 84, in key [stem: *Schistodactyl-*]. Type genus: *Schistodactylus* Raffray, 1883.

Tribe Tmesiphorini Jeannel, 1949

Tmesiphorini Jeannel, 1949a: 202 [stem: *Tmesiphor-*]. Type genus: *Tmesiphorus* J. L. LeConte, 1849.

Tribe Tyrini Reitter, 1882

Tyrides Reitter, 1882b: 184 [stem: *Tyr-*]. Type genus: *Tyrus* Aubé, 1833.

Subtribe Centrophthalmina Jeannel, 1949

Centrophthalmina Jeannel, 1949a: 209, in key [stem: *Centrophthalm-*]. Type genus: *Centrophthalmus* Schmidt-Göbel, 1838.

Petanopini Jeannel, 1954b: 102, in key [stem: *Petanop-*]. Type genus: *Petanops* Jeannel, 1954 [syn. of *Daveyia* Lea, 1912].

Subtribe Janusculina Cerruti, 1970

Janusculina Cerruti, 1970: 123 [stem: *Januscul-*]. Type genus: *Janusculus* Cerruti, 1970.

Subtribe Somatipionina Jeannel, 1949

Somatipionini Jeannel, 1949a: 208 [stem: *Somatipion-*]. Type genus: *Somatipion* L. W. Schaufuss, 1877.

Hamotini O. Park, 1951: 67, in key [stem: *Hamot-*]. Type genus: *Hamotus* Aubé, 1844.

Subtribe Tyrina Reitter, 1882

Tyrides Reitter, 1882b: 184 [stem: *Tyr-*]. Type genus: *Tyrus* Aubé, 1833.

Chalcoplectini Oke, 1925: 13 [stem: *Chalcoplect-*]. Type genus: *Chalcoplectus* Oke, 1925.

Ceophyllini O. Park, 1951: 67, in key [stem: *Ceophyll-*]. Type genus: *Ceophyllus* J. L. LeConte, 1849.

Subfamily Phloeocharinae Erichson, 1839

Phloeocharini Erichson, 1839a: 612 [stem: *Phloeochar-*]. Type genus: *Phloeocharis* Mannerheim, 1830. Comment: this family-group name was also used in the same year by Erichson (1839b: 31, as Phloeocharini) and Heer (1839b: 5, as Phloeocharina); for comments about the priority of these works see Newton and Thayer (1992: 24); current spelling maintained (Art. 29.5): incorrect stem formation in prevailing usage (should be *Phloeocharit-*).

Scotodytidae Saulcy, 1870: 90 [stem: *Scotodyt-*]. Type genus: *Scotodytes* Saulcy, 1865 [syn. of *Phloeocharis* Mannerheim, 1830].

Subfamily Olisthaerinae Thomson, 1858

Olisthaerini C. G. Thomson, 1858: 38 [stem: *Olisthaer-*]. Type genus: *Olisthaerus* Dejean, 1833.

Subfamily Tachyporinae MacLeay, 1825

Tachyporidae W. S. MacLeay, 1825: 49 [stem: *Tachypor-*]. Type genus: *Tachyporus* Gravenhorst, 1802.

Tribe Deropini Smetana, 1983

Deropsini Smetana, 1983: 272 [stem: *Derop-*]. Type genus: *Derops* Sharp, 1889. Comment: incorrect original stem formation, not in prevailing usage; correction of stem by Newton and Thayer (1992: 66).

Tribe Megarthropsini Cameron, 1919

Megarthropsini Cameron, 1919: 231 [stem: *Megarthrops-*]. Type genus: *Megarthropsis* Cameron, 1919. Comment: current spelling maintained (Art. 29.5): incorrect stem formation in prevailing usage (should be *Megarthropse-*).

Tribe Mycetoporini Thomson, 1859

Mycetoporides C. G. Thomson, 1859: 46 [stem: *Mycetopor-*]. Type genus: *Mycetoporus* Mannerheim, 1830 [placed on the Official List of Generic Names in Zoology (ICZN 1993c)].

Bolitobii G. H. Horn, 1877: 83 [stem: *Bolitobi-*]. Type genus: *Bolitobius* Leach, 1819.

Tribe Tachyporini MacLeay, 1825

Tachinidae Fleming, 1821: 49 [stem: *Tachinus-*]. Type genus: *Tachinus* Gravenhorst, 1802 [placed on the Official List of Generic Names in Zoology (ICZN 1993f)]. Comment: senior homonym of Tachinariae Robineau-Desvoidy, 1830 [Diptera: Tachinidae], stem emended to *Tachinus-* and Tachinusidae Fleming, 1821 placed on the Official List of Family-Group Names in Zoology (ICZN 1993f); Tachinidae Fleming, 1821 placed on the Official Index of Rejected and Invalid Family-Group Names in Zoology (ICZN 1993f).

Tachyporidae W. S. MacLeay, 1825: 49 [stem: *Tachypor-*]. Type genus: *Tachyporus* Gravenhorst, 1802 [placed on the Official List of Generic Names in Zoology (ICZN 1993f)]. Comment: family-group name given precedence over Tachinusidae Fleming, 1821 and placed on the Official List of Family-Group Names in Zoology (ICZN 1993f).

Symmixini Bernhauer, 1915: 56 [stem: *Symmix-*]. Type genus: *Symmixus* Bernhauer, 1915.

Euconosomini Cameron, 1918: 216 [stem: *Euconosomat-*]. Type genus: *Euconosoma* Cameron, 1918. Comment: incorrect original stem formation, not in prevailing usage.

Conosomini Jeannel and Jarrige, 1949: 335 [stem: *Conosomat-*]. Type genus: *Conosoma* Kraatz, 1857 [preoccupied genus name, not *Conosoma* Lenz, 1794 [Diptera]; syn. of *Tachinus* Gravenhorst, 1802]. Comment: permanently invalid (Art. 39): based on preoccupied type genus; incorrect original stem formation, not in prevailing usage.

Sepedophilini Ádám, 2001: 199 [stem: *Sepedophil-*]. Type genus: *Sepedophilus* Gistel, 1856. Comment: **syn. nov.**

Tribe Vatesini Seevers, 1958

Xenocephalini Wasmann, 1887: 411 [stem: *Xenocephal-*]. Type genus: *Xenocephalus* Wasmann, 1887 [preoccupied genus name, not *Xenocephalus* Kaup, 1858 [Pisces]; replaced by *Wasmannotherium* Bernhauer, 1921; syn. of *Vatesus* Sharp, 1876]. Comment: permanently invalid (Art. 39): based on preoccupied type genus.

Vatesini Seevers, 1958: 183 [stem: *Vates-*]. Type genus: *Vatesus* Sharp, 1876. Comment: replacement name for Xenocephalini Wasmann, 1887 because of the homonymy of the type genus.

Subfamily Trichophyinae Thomson, 1858

Trichophyini C. G. Thomson, 1858: 30 [stem: *Trichophy-*]. Type genus: *Trichophya* Mannerheim, 1830.

Subfamily Habrocerinae Mulsant and Rey, 1876

Habrocériens Mulsant and Rey, 1876: 210 [stem: *Habrocer-*]. Type genus: *Habrocerus* Erichson, 1839. Comment: original vernacular name available (Art. 11.7.2): first used in latinized form by G. H. Horn (1877: 83, as Habroceri), generally accepted as in Newton and Thayer (1992: 56, as Habrocerinae).

Subfamily Aleocharinae Fleming, 1821

Aleocharidae Fleming, 1821: 49 [stem: *Aleochar-*]. Type genus: *Aleochara* Gravenhorst, 1802. Comment: First Reviser (Aleocharinae Fleming, 1821 vs Lomechusinae Fleming, 1821) not determined, current usage maintained.

Tribe Actocharini Bernhauer and Schubert, 1911

Actochari Bernhauer and Schubert, 1911: 91 [stem: *Actochar-*]. Type genus: *Actocharis* Fauvel, 1869. Comment: originally proposed as a subtribe of Oxytelini Fleming, 1821; current spelling maintained (Art. 29.5): incorrect stem formation in prevailing usage (should be *Actocharit-*).

Tribe Aenictoteratini Kistner, 1993

Aenictoteratini Kistner, 1993: 242 [stem: *Aenictoterat-*]. Type genus: *Aenictoteras* W. M. Wheeler, 1932.

Tribe Akatastopsisini Pace, 2000

Akatastopsisini Pace, 2000a: 112 [stem: *Akatastopsis-*]. Type genus: *Akatastopsis* Pace, 2000.

Tribe Aleocharini Fleming, 1821

Aleocharidae Fleming, 1821: 49 [stem: *Aleochar-*]. Type genus: *Aleochara* Gravenhorst, 1802.

Subtribe Aleocharina Fleming, 1821

Aleocharidae Fleming, 1821: 49 [stem: *Aleochar-*]. Type genus: *Aleochara* Gravenhorst, 1802.

Piochardiae Fenyes, 1918: 20 [stem: *Piochardi-*]. Type genus: *Piochardia* Heyden, 1870.

Subtribe Compactopediina Kistner, 1970

Compactopedina Kistner, 1970b: 18 [stem: *Compactopedi-*]. Type genus: *Compactopedia* Kistner, 1970. Comment: incorrect original stem formation, not in prevailing usage; correction of stem by Newton and Thayer (1992: 47).

Subtribe Hodoxenina Kistner, 1970

Hodoxenina Kistner, 1970a: 12 [stem: *Hodoxen-*]. Type genus: *Hodoxenus* Kistner, 1970.

Tribe Athetini Casey, 1910

Athetae Casey, 1910: 2 [stem: *Athet-*]. Type genus: *Atheta* C. G. Thomson, 1858 [placed on the Official List of Generic Names in Zoology (ICZN 1961b)]. Comment: notice of a new application for the conservation of this name over the older name Callicerini Jakobson, 1908 by Gusarov was recently published in the Bulletin of Zoological Nomenclature (2010: 270; also see Appendix 6).

Subtribe Athetina Casey, 1910

Callicerina Jakobson, 1908: 448 [stem: *Callicer-*]. Type genus: *Callicerus* Gravenhorst, 1802. Comment: junior homonym of Callicerina Rondani, 1856 (type genus *Callicera* Panzer, 1806) in Diptera: Syrphidae; notice of a new application for the suppression of this name by Gusarov was recently published in the Bulletin of Zoological Nomenclature (2010: 270; also see Appendix 6).

Athetae Casey, 1910: 2 [stem: *Athet-*]. Type genus: *Atheta* C. G. Thomson, 1858 [placed on the Official List of Generic Names in Zoology (ICZN 1961b)]. Comment: notice of a new application for the conservation of this name over the older name Callicerina Jakobson, 1908 by Gusarov was recently published in the Bulletin of Zoological Nomenclature (2010: 270; also see Appendix 6).

Strigotae Casey, 1910: 176 [stem: *Strigot-*]. Type genus: *Strigota* Casey, 1910.

Plagiarthrini Cameron, 1926: 184 [stem: *Plagiarthrin-*]. Type genus: *Plagiarthrina* Keys, 1920 [syn. of *Atheta* C. G. Thomson, 1858]. Comment: incorrect original stem formation, not in prevailing usage.

Ischnopodini Hatch, 1957: 141 [stem: *Ischnopod-*]. Type genus: *Ischnopoda* sensu Westwood, 1838 [not *Ischnopoda* Stephens, 1835 (see ICZN 1961b); syn. of *Acrotona* C. G. Thomson, 1859]. Comment: based on misidentified type genus.

*Xenotae Seevers, 1978: 113 [stem: *Xenot-*]. Type genus: *Xenota* Mulsant and Rey, 1873 [subgenus of *Atheta* C. G. Thomson, 1858]. Comment: unavailable family-group name, proposed after 1930 without description or bibliographic reference to such a description (Art. 13.1).

Acrotonae Seevers, 1978: 97 [stem: *Acroton-*]. Type genus: *Acrotona* C. G. Thomson, 1859 [placed on the Official List of Generic Names in Zoology (ICZN 1961b)].

Geostibae Seevers, 1978: 126 [stem: *Geostib-*]. Type genus: *Geostiba* C. G. Thomson, 1858 [placed on the Official List of Generic Names in Zoology (ICZN 2005a)].

Dimetrotae Seevers, 1978: 102 [stem: *Dimetrot-*]. Type genus: *Dimetrota* Mulsant and Rey, 1873 [subgenus of *Atheta* C. G. Thomson, 1858].

*Hydrosmectina Muona, 1979: 23 [stem: *Hydrosmect-*]. Type genus: *Hydrosmecta* C. G. Thomson, 1858. Comment: unavailable family-group name, proposed after 1930 without description or bibliographic reference to such a description (Art. 13.1).

*Trichomicrina Muona, 1979: 23 [stem: *Trichomicr-*]. Type genus: *Trichomicra* Brundin, 1945. Comment: unavailable family-group name, proposed after 1930 without description or bibliographic reference to such a description (Art. 13.1).

*Dadobiina Muona, 1979: 23 [stem: *Dadobi-*]. Type genus: *Dadobia* C. G. Thomson, 1858. Comment: unavailable family-group name, proposed after 1930 without description or bibliographic reference to such a description (Art. 13.1).

*Amischina Muona, 1979: 25 [stem: *Amisch-*]. Type genus: *Amischa* C. G. Thomson, 1858. Comment: unavailable family-group name, proposed after 1930 without description or bibliographic reference to such a description (Art. 13.1).

Subtribe Coptotermoeciina Kistner and Pasteels, 1970

*Coptotermoeciina Seevers, 1957: 248 [stem: *Coptotermoeci-*]. Type genus: *Coptotermoecia* Oke, 1933. Comment: unavailable family-group name, proposed after 1930 without description or bibliographic reference to such a description (Art. 13.1).

Coptotermoeciina Kistner and Pasteels, 1970: 86 [stem: *Coptotermoeci-*]. Type genus: *Coptotermoecia* Oke, 1933.

Subtribe Microceroxenina Kistner, 1970

Microceroxenina Kistner, 1970c: 10 [stem: *Microceroxen-*]. Type genus: *Microceroxenus* Kistner, 1970.

Subtribe Nasutiphilina Kistner, 1970

Nasutiphilina Kistner, 1970d: 500 [stem: *Nasutiphil-*]. Type genus: *Nasutiphilus* Kistner, 1970.

Subtribe Schistogeniina Fenyes, 1918

Schistogeniae Fenyes, 1918: 18 [stem: *Schistogeni-*]. Type genus: *Schistogenia* Kraatz, 1857.

Subtribe Taxicerina Lohse, 1989

Taxicerina Lohse, 1989: 210 [stem: *Taxicer-*]. Type genus: *Taxicera* Mulsant and Rey, 1873.

Subtribe Termitotelina Kistner, 1970

*Termitotelina Seevers, 1957: 250 [stem: *Termitotel-*]. Type genus: *Termitotelus* Wasmann, 1908. Comment: unavailable family-group name, proposed after 1930 without description or bibliographic reference to such a description (Art. 13.1).

Termitotelina Kistner, 1970a: 4 [stem: *Termitotel-*]. Type genus: *Termitotelus* Wasmann, 1908.

Subtribe Thamiaraeina Fenyes, 1921

Thamiaraeini Fenyes, 1921: 34 [stem: *Thamiarae-*]. Type genus: *Thamiaraea* C. G. Thomson, 1858.

Tribe Autaliini Thomson, 1859

Autaliides C. G. Thomson, 1859: 30 [stem: *Autali-*]. Type genus: *Autalia* Leach, 1819.

Ophioglossae Fenyes, 1918: 18 [stem: *Ophiogloss-*]. Type genus: *Ophioglossa* Fauvel, 1866.

Rhopalogastra Fenyes, 1918: 17 [stem: *Rhopalogastr-*]. Type genus: *Rhopalogastrum* Bernhauer, 1912.

Tribe Cordobanini Bernhauer, 1910

Cordobanini Bernhauer, 1910: 386 [stem: *Cordoban-*]. Type genus: *Cordobanus* Bernhauer, 1910.

Tribe Corotocini Fenyes, 1918

Corotocini Fenyes, 1918: 61 [stem: *Corotoc-*]. Type genus: *Corotoca* Schiødte, 1847.

Subtribe Abrotelina Seevers, 1957

Abrotelina Seevers, 1957: 121 [stem: *Abrotel-*]. Type genus: *Abroteles* Casey, 1889.

Subtribe Corotocina Fenyes, 1918

Corotocini Fenyes, 1918: 61 [stem: *Corotoc-*]. Type genus: *Corotoca* Schiødte, 1847.

Termitomimini Fenyes, 1921: 33 [stem: *Termitomim-*]. Type genus: *Termitomimus* Trägårdh, 1907.

Subtribe Eburniogastrina Jacobson, Kistner and Pasteels, 1986

Eburniogastrina Jacobson et al., 1986: 27 [stem: *Eburniogastr-*]. Type genus: *Eburniogaster* Seevers, 1938.

Subtribe Nasutitellina Jacobson, Kistner and Pasteels, 1986

Nasutitellina Jacobson et al., 1986: 95 [stem: *Nasutitell-*]. Type genus: *Nasutitella* Pasteels, 1967.

Subtribe Sphuridaethina Pace, 1988

Sphuridaethina Pace, 1988: 980 [stem: *Sphuridaeth-*]. Type genus: *Sphuridaethes* Pace, 1988.

Subtribe Termitocharina Seevers, 1957

Termitocharina Seevers, 1957: 64, in key [stem: *Termitochar-*]. Type genus: *Termitochara* Wasmann, 1893.

Subtribe Termitocupidina Jacobson, Kistner and Pasteels, 1986

Termitocupidina Jacobson et al., 1986: 35 [stem: *Termitocupid-*]. Type genus: *Termitocupidus* Jacobson et al., 1986.

Subtribe Termitogastrina Bernhauer and Scheerpeltz, 1926

Termitogastri Bernhauer and Scheerpeltz, 1926: 734 [stem: *Termitogastr-*]. Type genus: *Termitogaster* Casey, 1889.

Termitellici Jacobson et al., 1986: 47 [stem: *Termitell-*]. Type genus: *Termitella* Wasmann, 1911. Comment: proposed as an infratribe, a rank not used here.

Termitogastrici Jacobson et al., 1986: 58 [stem: *Termitogastr-*]. Type genus: *Termitogaster* Casey, 1889. Comment: family-group name proposed as a new taxon, without reference to Termitogastri Bernhauer and Scheerpeltz, 1926; proposed as an infratribe, a rank not used here.

Subtribe Termitoiceina Jacobson, Kistner and Pasteels, 1986

Termitoiceina Jacobson et al., 1986: 84 [stem: *Termitoice-*]. Type genus: *Termitoiceus* Silvestri, 1901.

Subtribe Termitopithina Jacobson, Kistner and Pasteels, 1986

Termitopithina Jacobson et al., 1986: 80 [stem: *Termitopith-*]. Type genus: *Termitopithus* Seevers, 1957.

Subtribe Termitoptochina Fenyes, 1921

Termitoptochini Fenyes, 1921: 33 [stem: *Termitoptoch-*]. Type genus: *Termitoptochus* Silvestri, 1911.

TERMITOPTOCHINA Jacobson et al., 1986: 96 [stem: *Termitoptoch-*]. Type genus: *Termitoptochus* Silvestri, 1911. Comment: family-group name proposed as new without reference to TERMITOPTOCHINI Fenyes, 1921.

Subtribe TIMEPARTHENINA Fenyes, 1921

TIMEPARTHENINI Fenyes, 1921: 34 [stem: *Timeparthen-*]. Type genus: *Timeparthenus* Silvestri, 1901.

Tribe CREMATOXENINI Mann, 1921

CREMATOXENINI Mann, 1921: 547 [stem: *Crematoxen-*]. Type genus: *Crematoxenus* Mann, 1921.

PULICOMORPHINI Mann, 1924: 87 [stem: *Pulicomorph-*]. Type genus: *Pulicomorpha* Mann, 1924.

PHILACAMATINI Bruch, 1933b: 206 [stem: *Philacamat-*]. Type genus: *Philacamatus* Bruch, 1933.

Tribe CRYPTONOTOPSEINI Pace, 2003

CRYPTONOTOPSISINI Pace, 2003: 38 [stem: *Cryptonotopse-*]. Type genus: *Cryptonotopsis* Pace, 2003. Comment: incorrect original stem formation, not in prevailing usage.

Tribe DEINOPSINI Sharp, 1883

DEINOPSINI Sharp, 1883: 294 [stem: *Deinops-*]. Type genus: *Deinopsis* A. Matthews, 1838. Comment: current spelling maintained (Art. 29.5): incorrect original stem formation in prevailing usage (should be *Deinopse-*).

ADINOPSINI Cameron, 1919: 242 [stem: *Adinopse-*]. Type genus: *Adinopsis* Cameron, 1919. Comment: incorrect original stem formation, not in prevailing usage.

Tribe DIESTOTINI Mulsant and Rey, 1871

DIESTOTATES Mulsant and Rey, 1871c: 96 [stem: *Diestot-*]. Type genus: *Diestota* Mulsant and Rey, 1870. Comment: original vernacular name available (Art. 11.7.2): first used in latinized form by Lohse (1989: 186, as DIESTOTINI), generally accepted as in Newton and Thayer (1992: 49, as DIESTOTINI).

ELACHISTARTHRONINI Notman, 1920: 714 [stem: *Elachistarthr-*]. Type genus: *Elachistarthron* Notman, 1920 [syn. of *Diestota* Mulsant and Rey, 1870]. Comment: incorrect original stem formation, not in prevailing usage; correction of stem by Newton and Thayer (1992: 49).

Tribe DIGLOTTINI Jakobson, 1909

DIGLOSSAIRES Mulsant and Rey, 1873a: 73 [stem: *Digloss-*]. Type genus: *Diglossa* Haliday, 1837 [preoccupied genus name, not *Diglossa* Wagler, 1832 [Aves]; syn. of *Diglotta* Champion, 1899]. Comment: original vernacular name available (Art. 11.7.2): first used in latinized form and generally accepted as in Ganglbauer (1895: 313, as Diglossini); permanently invalid (Art. 39): based on preoccupied type genus.

Diglottina Jakobson, 1909: 529 [stem: *Diglott-*]. Type genus: *Diglotta* Champion, 1899. Comment: published 4 March 1909; this family-group name was also used in the same year by Eichelbaum (1909 [before 26 December]: 204, as Diglottini).

Tribe Digrammini Fauvel, 1900

Digrammini Fauvel, 1900: 123 [stem: *Digramm-*]. Type genus: *Digrammus* Fauvel, 1900.

Tribe Dorylogastrini Wasmann, 1916

Dorylogastrini Wasmann, 1916a: 103 [stem: *Dorylogastr-*]. Type genus: *Dorylogaster* Wasmann, 1904.

Tribe Dorylomimini Wasmann, 1916

Dorylomimini Wasmann, 1916a: 99 [stem: *Dorylomim-*]. Type genus: *Dorylomimus* Wasmann, 1902.

Tribe Drepanoxenini Kistner and Watson, 1972

Drepanoxenini Kistner and Watson, 1972: 2 [stem: *Drepanoxen-*]. Type genus: *Drepanoxenus* Kistner and Watson, 1972.

Tribe Ecitocharini Seevers, 1965

Ecitocharini Seevers, 1965: 287 [stem: *Ecitochar-*]. Type genus: *Ecitochara* Wasmann, 1887.

Tribe Ecitogastrini Fenyes, 1918

Ecitogastrini Fenyes, 1918: 74 [stem: *Ecitogastr-*]. Type genus: *Ecitogaster* Wasmann, 1899.

Tribe Eusteniamorphini Bernhauer and Scheerpeltz, 1926

Eusteniamorphini Bernhauer and Scheerpeltz, 1926: 517 [stem: *Eusteniamorph-*]. Type genus: *Eusteniamorpha* Cameron, 1920.

Tribe Falagriini Mulsant and Rey, 1873

Falagriates Mulsant and Rey, 1873b: 8 [stem: *Falagri-*]. Type genus: *Falagria* Leach, 1819. Comment: original vernacular name available (Art. 11.7.2): first used in latinized form by Seidlitz (1874 [Gatt.]: 71, as Falagriina), generally accepted as in Newton and Thayer (1992: 50, as Falagriini).

Tribe Feldini Kistner, 1972

*Feldina Seevers, 1957: 236 [stem: *Feld-*]. Type genus: *Felda* Blackwelder, 1952. Comment: unavailable family-group name, proposed after 1930 without description or bibliographic reference to such a description (Art. 13.1).

Feldini Kistner, 1972: 2 [stem: *Feld-*]. Type genus: *Felda* Blackwelder, 1952.

Tribe Gymnusini Heer, 1839

Gymnusida Heer, 1839a: 302 [stem: *Gymnus-*]. Type genus: *Gymnusa* Gravenhorst, 1806. Comment: this family-group name was also used in the same year by Heer (1839b: 49, as Gymnusida); for comments about the priority of these works see Newton and Thayer (1992: 24).

Tribe Himalusini Klimaszewski, Pace and Center, 2010

Himalusini Klimaszewski et al., 2010: 3 [stem: *Himalus-*]. Type genus: *Himalusa* Pace, 2006.

Tribe Homalotini Heer, 1839

Homalotida Heer, 1839a: 305 [stem: *Homalot-*]. Type genus: *Homalota* Mannerheim, 1830.

Subtribe Bolitocharina Thomson, 1859

Bolitocharides C. G. Thomson, 1859: 31 [stem: *Bolitochar-*]. Type genus: *Bolitochara* Mannerheim, 1830 [placed on the Official List of Generic Names in Zoology (ICZN 1961a)]. Comment: First Revisers found (Bolitocharini C. G. Thomson, 1859 vs Euryusini C. G. Thomson, 1859) are Newton and Thayer (1992: 50); Bolitocharini C. G. Thomson, 1859 placed on the Official List of Family-Group Names in Zoology (ICZN 1961a); the original spelling Bolitocharides C. G. Thomson, 1859 and several subsequent spellings of this name placed on the Official Index of Rejected and Invalid Family-Group Names in Zoology (ICZN 1961a).

Euryusides C. G. Thomson, 1859: 40 [stem: *Euryus-*]. Type genus: *Euryusa* Erichson, 1837.

Sipaliae Casey, 1910: 167 [stem: *Sipali-*]. Type genus: *Sipalia* Mulsant and Rey, 1853 [placed on the Official Index of Rejected and Invalid Generic Names in Zoology (ICZN 2005d)]. Comment: permanently invalid (Art. 39): based on suppressed type genus.

Heterotae Fenyes, 1918: 18 [stem: *Heterot-*]. Type genus: *Heterota* Mulsant and Rey, 1873.

Leptusae Fenyes, 1918: 18 [stem: *Leptus-*]. Type genus: *Leptusa* Kraatz, 1856 [placed on the Official List of Generic Names in Zoology (ICZN 2005d)].

Nanoglossae Fenyes, 1918: 20 [stem: *Nanogloss-*]. Type genus: *Nanoglossa* Fauvel, 1868 [subgenus of *Leptusa* Kraatz, 1856].

DITROPALIINI Hatch, 1957: 134, in key, 147 [stem: *Ditropali-*]. Type genus: *Ditropalia* Casey, 1906 [syn. of *Bolitochara* Mannerheim, 1830]. Comment: this taxon was named correctly DITROPALIINI in the text on p. 147 but incorrect as BOLITOCHARINI in the key on p. 134.

Subtribe DINARDOPSINA Bernhauer and Scheerpeltz, 1926

DINARDOPSES Bernhauer and Scheerpeltz, 1926: 525, 804 [stem: *Dinardops-*]. Type genus: *Dinardopsis* Bruch, 1917. Comment: the erroneous spelling DINARDOPSIS was used in the catalogue on p. 525 but this was corrected to DINARDOPSES on page 804 of the same work; current spelling maintained (Art. 29.5): incorrect stem formation in prevailing usage (should be *Dinardopse-*).

Subtribe GYROPHAENINA Kraatz, 1856

GYROPHAENINI Kraatz, 1856: 351 [stem: *Gyrophaen-*]. Type genus: *Gyrophaena* Mannerheim, 1830.

Subtribe HOMALOTINA Heer, 1839

HOMALOTIDA Heer, 1839a: 305 [stem: *Homalot-*]. Type genus: *Homalota* Mannerheim, 1830. Comment: this family-group name was also used in the same year by Heer (1839b: 50, as HOMALOTIDA); for comments about the priority of these works see Newton and Thayer (1992: 24).

THECTUROTAE Fenyes, 1918: 18 [stem: *Thecturot-*]. Type genus: *Thecturota* Casey, 1893.

*CYPHEAE Seevers, 1978: 272 [stem: *Cyphe-*]. Type genus: *Cyphea* Fauvel, 1863. Comment: unavailable family-group name, proposed after 1930 without description or bibliographic reference to such a description (Art. 13.1).

Subtribe SILUSINA Fenyes, 1918

SILUSAE Fenyes, 1918: 17 [stem: *Silus-*]. Type genus: *Silusa* Erichson, 1837.

Tribe HOPLANDRIINI Casey, 1910

HOPLANDRIAE Casey, 1910: 170 [stem: *Hoplandri-*]. Type genus: *Hoplandria* Kraatz, 1857.

Subtribe HOPLANDRIINA Casey, 1910

HOPLANDRIAE Casey, 1910: 170 [stem: *Hoplandri-*]. Type genus: *Hoplandria* Kraatz, 1857.

Subtribe PLATANDRIINA Hanley, 2002

PLATANDRIINA Hanley, 2002: 317 [stem: *Platandri-*]. Type genus: *Platandria* Casey, 1893.

Subtribe Pseudoplandriina Hanley, 2002

Pseudoplandriina Hanley, 2002: 317 [stem: *Pseudoplandri-*]. Type genus: *Pseudoplandria* Fenyes, 1921.

Tribe Hygronomini Thomson, 1859

Hygronomides C. G. Thomson, 1859: 31 [stem: *Hygronom-*]. Type genus: *Hygronoma* Erichson, 1837.

Subtribe Hygronomina Thomson, 1859

Hygronomides C. G. Thomson, 1859: 31 [stem: *Hygronom-*]. Type genus: *Hygronoma* Erichson, 1837.

Subtribe Saphoglossina Bernhauer and Scheerpeltz, 1926

Saphoglossae Bernhauer and Scheerpeltz, 1926: 521 [stem: *Saphogloss-*]. Type genus: *Saphoglossa* Sharp, 1883.

Tribe Hypocyphtini Laporte, 1835

Hypocyphtidae Laporte, 1835a: 135 [stem: *Hypocypht-*]. Type genus: *Hypocyphtus* Gyllenhal, 1827 [syn. of *Cypha* Leach, 1819].

Oligotides C. G. Thomson, 1859: 30 [stem: *Oligot-*]. Type genus: *Oligota* Mannerheim, 1830.

Nematoscelini Fenyes, 1921: 33 [stem: *Nematoscelid-*]. Type genus: *Nematoscelis* Wollaston, 1867. Comment: incorrect original stem formation, not in prevailing usage.

*Cyphinae Lohse, 1974: 7 [stem: *Cyph-*]. Type genus: *Cypha* Leach, 1819. Comment: family-group name unavailable (Art. 11.6): originally published as synonym and not made available subsequently; also see Cyphini Lacordaire, 1863 (type genus *Cyphus* Germar, 1824) in Coleoptera: Curculionidae.

Tribe Leucocraspedini Fenyes, 1921

Leucocraspedini Fenyes, 1921: 34 [stem: *Leucocrasped-*]. Type genus: *Leucocraspedum* Kraatz, 1859.

Tribe Liparocephalini Fenyes, 1918

Liparocephali Fenyes, 1918: 18 [stem: *Liparocephal-*]. Type genus: *Liparocephalus* Mäklin, 1853.

Tribe Lomechusini Fleming, 1821

Lomechusidae Fleming, 1821: 49 [stem: *Lomechus-*]. Type genus: *Lomechusa* Gravenhorst, 1806.

Subtribe Aenictobiina Kistner, 1997

Aenictobiina Kistner, 1997: 174 [stem: *Aenictobi-*]. Type genus: *Aenictobia* Seevers, 1953.

Subtribe Lomechusina Fleming, 1821

Lomechusidae Fleming, 1821: 49 [stem: *Lomechus-*]. Type genus: *Lomechusa* Gravenhorst, 1806.

Xenodusae Seevers, 1978: 155 [stem: *Xenodus-*]. Type genus: *Xenodusa* Wasmann, 1894.

Subtribe Myrmedoniina Thomson, 1867

Myrmedoniides C. G. Thomson, 1867: 209 [stem: *Myrmedoni-*]. Type genus: *Myrmedonia* Erichson, 1837 [syn. of *Zyras* Stephens, 1835].

*Myrméciates Mulsant and Rey, 1873b: 98 [stem: *Myrmoeci-*]. Type genus: *Myrmoecia* Mulsant and Rey, 1873. Comment: original vernacular name unavailable (Art. 11.7.2): not subsequently latinized; incorrect original stem formation, not in prevailing usage.

Zyrini Bradley, 1930: 83 [stem: *Zyr-*]. Type genus: *Zyras* Stephens, 1835 [placed on the Official List of Generic Names in Zoology (ICZN 1961a)].

*Ecitoporae Seevers, 1978: 13 [stem: *Ecitopor-*]. Type genus: *Ecitopora* Wasmann, 1887. Comment: unavailable family-group name, proposed after 1930 without description or bibliographic reference to such a description (Art. 13.1).

*Tetradoniae Seevers, 1978: 13 [stem: *Tetradoni-*]. Type genus: *Tetradonia* Wasmann, 1894. Comment: unavailable family-group name, proposed after 1930 without description or bibliographic reference to such a description (Art. 13.1).

*Dinocorynae Seevers, 1978: 13 [stem: *Dinocoryn-*]. Type genus: *Dinocoryna* Casey, 1893. Comment: unavailable family-group name, proposed after 1930 without description or bibliographic reference to such a description (Art. 13.1).

Subtribe Termitozyrina Seevers, 1957

Termitozyrina Seevers, 1957: 62, in key [stem: *Termitozyr-*]. Type genus: *Termitozyras* Seevers, 1957.

Tribe Masuriini Cameron, 1939

Masuriini Cameron, 1939: 24 [stem: *Masuri-*]. Type genus: *Masuria* Cameron, 1928.

Tribe Mesoporini Cameron, 1959

Mesoporinae Cameron, 1959: 119 [stem: *Mesopor-*]. Type genus: *Mesoporus* Cameron, 1959.

Tribe Mimanommatini Wasmann, 1912

Mimanommatinae Wasmann, 1912: 478 [stem: *Mimanommat-*]. Type genus: *Mimanomma* Wasmann, 1912.

Subtribe Dorylophilina Fenyes, 1921

Dorylophilini Fenyes, 1921: 34 [stem: *Dorylophil-*]. Type genus: *Dorylophila* Wasmann, 1904.

Deremini Seevers, 1965: 294 [stem: *Derem-*]. Type genus: *Derema* Fauvel, 1899.

Subtribe Mimanommatina Wasmann, 1912

Mimanommatinae Wasmann, 1912: 478 [stem: *Mimanommat-*]. Type genus: *Mimanomma* Wasmann, 1912.

Tribe Mimecitini Wasmann, 1917

Mimecitonini Wasmann, 1917: 325 [stem: *Mimecit-*]. Type genus: *Mimeciton* Wasmann, 1893.

Subtribe Labidopullina Jacobson and Kistner, 1991

Labidopullina Jacobson and Kistner, 1991: 7 [stem: *Labidopull-*]. Type genus: *Labidopullus* Borgmeier, 1958.

Subtribe Leptanillophilina Fenyes, 1918

Leptanillophilini Fenyes, 1918: 59 [stem: *Leptanillophil-*]. Type genus: *Leptanillophilus* Holmgren, 1908.

Subtribe Mimecitina Wasmann, 1917

Mimecitonini Wasmann, 1917: 325 [stem: *Mimecit-*]. Type genus: *Mimeciton* Wasmann, 1893. Comment: incorrect original stem formation, not in prevailing usage; correction of stem by Newton and Thayer (1992: 52).

Subtribe Mimonillina Bernhauer and Scheerpeltz, 1926

Mimonillae Bernhauer and Scheerpeltz, 1926: 518 [stem: *Mimonill-*]. Type genus: *Mimonilla* Wasmann, 1913.

Tribe Myllaenini Ganglbauer, 1895

Myllaenini Ganglbauer, 1895: 317 [stem: *Myllaen-*]. Type genus: *Myllaena* Erichson, 1837.

Dimonomerini Cameron, 1933: 103 [stem: *Dimonomer-*]. Type genus: *Dimonomera* Cameron, 1933.

Tribe Oxypodini Thomson, 1859

Oxypodides C. G. Thomson, 1859: 36 [stem: *Oxypod-*]. Type genus: *Oxypoda* Mannerheim, 1830 [placed on the Official List of Generic Names in Zoology (ICZN 1957)]. Comment: First Revisers found (Oxypodini C. G. Thomson, 1859 vs Tachyusini C. G. Thomson, 1859 vs Ocaleini C. G. Thomson, 1859 vs Phloeoporini C. G. Thomson, 1859) are Newton and Thayer (1992:

53); name placed on the Official List of Family-Group Names in Zoology (ICZN 1957).

Subtribe Aphytopodina Bernhauer and Scheerpeltz, 1926

Aphytopi Bernhauer and Scheerpeltz, 1926: 740 [stem: *Aphytopod-*]. Type genus: *Aphytopus* Sharp, 1886. Comment: incorrect original stem formation, not in prevailing usage; correction of stem by Newton and Thayer (1992: 53).

Subtribe Blepharhymenina Klimaszewski and Peck, 1986

*Blepharhymeni Seevers, 1978: 82 [stem: *Blepharhymen-*]. Type genus: *Blepharhymenus* Solier, 1849. Comment: unavailable family-group name, proposed after 1930 without description or bibliographic reference to such a description (Art. 13.1).

Blepharrhymeni Klimaszewski and Peck, 1986: 58 [stem: *Blepharhymen-*]. Type genus: *Blepharhymenus* Solier, 1849 [as *Blepharrhymenus*, unjustified emendation of genus name by Gemminger and Harold (1868b: 505), not in prevailing usage]. Comment: incorrect original stem formation, not in prevailing usage; correction of stem by Newton and Thayer (1992: 53).

Subtribe Dinardina Mulsant and Rey, 1873

Dinardaires Mulsant and Rey, 1873a: 6 [stem: *Dinard-*]. Type genus: *Dinarda* Leach, 1819. Comment: original vernacular name available (Art. 11.7.2): first used in latinized form by Wasmann (1904a: 218, as Dinardini), generally accepted as in Newton and Thayer (1992: 53, as Dinardina).

*Homéusates Mulsant and Rey, 1874: 286 [stem: *Homoeus-*]. Type genus: *Homoeusa* Kraatz, 1856. Comment: original vernacular name unavailable (Art. 11.7.2): not subsequently latinized; incorrect original stem formation, not in prevailing usage.

Decusini Fenyes, 1918: 19 [stem: *Decus-*]. Type genus: *Decusa* Casey, 1900.

Subtribe Meoticina Seevers, 1978

Meoticae Seevers, 1978: 78 [stem: *Meotic-*]. Type genus: *Meotica* Mulsant and Rey, 1873.

Subtribe Oxypodina Thomson, 1859

Oxypodides C. G. Thomson, 1859: 36 [stem: *Oxypod-*]. Type genus: *Oxypoda* Mannerheim, 1830 [placed on the Official List of Generic Names in Zoology (ICZN 1957)]. Comment: First Reviser (Oxypodina C. G. Thomson, 1859 vs Ocaleina C. G. Thomson, 1859 vs Phloeoporina C.

G. Thomson, 1859) not determined, current usage maintained; placed on the Official List of Family-Group Names in Zoology (ICZN 1957).

Ocaleides C. G. Thomson, 1859: 38 [stem: *Ocale-*]. Type genus: *Ocalea* Erichson, 1837.

Phloeoporides C. G. Thomson, 1859: 33 [stem: *Phloeopor-*]. Type genus: *Phloeopora* Erichson, 1837.

Ocyusates Mulsant and Rey, 1874: 286 [stem: *Ocyus-*]. Type genus: *Ocyusa* Kraatz, 1856. Comment: original vernacular name available (Art. 11.7.2): first used in latinized form by Fenyes (1918: 20, as Ocyusae), generally accepted as in Newton and Thayer (1992: 53, as Ocyusina).

Calodérates Mulsant and Rey, 1874: 286 [stem: *Caloder-*]. Type genus: *Calodera* Mannerheim, 1830. Comment: original vernacular name available (Art. 11.7.2): first used in latinized form by Fenyes (1918: 20, as Caloderae), generally accepted as in Ádám (2001:142, as Caloderini).

Microglottae Fenyes, 1918: 20 [stem: *Microglott-*]. Type genus: *Microglotta* Kraatz, 1862 [syn. of *Haploglossa* Kraatz, 1856].

Phloeoporini Cameron, 1939: 562 [stem: *Phloeopor-*]. Type genus: *Phloeopora* Erichson, 1837. Comment: proposed as new without reference to Phloeoporides C. G. Thomson, 1859.

Subtribe Tachyusina Thomson, 1859

Tachyusides C. G. Thomson, 1859: 34 [stem: *Tachyus-*]. Type genus: *Tachyusa* Erichson, 1837 [placed on the Official List of Generic Names in Zoology (ICZN 1961b)].

Tribe Oxypodinini Fenyes, 1921

Oxypodinini Fenyes, 1918: 18 [stem: *Oxypodin-*]. Type genus: *Oxypodinus* Bernhauer, 1901.

Heterotaxini Fenyes, 1921: 33 [stem: *Heterotax-*]. Type genus: *Heterotaxus* Bernhauer, 1915.

Tribe Paglini Newton and Thayer, 1992

Pachyglossini Fenyes, 1918: 60 [stem: *Pachygloss-*]. Type genus: *Pachyglossa* Fauvel, 1868 [preoccupied genus name, not *Pachyglossa* Hodgson, 1843 [Aves]; syn. of *Pagla* Blackwelder, 1952]. Comment: permanently invalid (Art. 39): based on preoccupied type genus.

Paglini Newton and Thayer, 1992: 54 [stem: *Pagl-*]. Type genus: *Pagla* Blackwelder, 1952. Comment: replacement name for Pachyglossini Fenyes, 1918 because of the homonymy of the type genus.

Tribe Paradoxenusini Bruch, 1937

Paradoxenusini Bruch, 1937: 354 [stem: *Paradoxenus-*]. Type genus: *Paradoxenusa* Bruch, 1937.

Tribe Pediculotini Ádám, 1987

Pediculotini Ádám, 1987: 156 [stem: *Pediculot-*]. Type genus: *Pediculota* Ádám, 1987.

Tribe Philotermitini Seevers, 1957

Philotermitini Seevers, 1957: 63, in key [stem: *Philotermit-*]. Type genus: *Philotermes* Kraatz, 1857.

Tribe Phyllodinardini Wasmann, 1916

Phyllodinardini Wasmann, 1916a: 105 [stem: *Phyllodinard-*]. Type genus: *Phyllodinarda* Wasmann, 1916.

Tribe Phytosini Thomson, 1867

Phytosides C. G. Thomson, 1867: 206 [stem: *Phytos-*]. Type genus: *Phytosus* Curtis, 1838.

Tribe Placusini Mulsant and Rey, 1871

Placusates Mulsant and Rey, 1871c: 102 [stem: *Placus-*]. Type genus: *Placusa* Erichson, 1837. Comment: original vernacular name available (Art. 11.7.2): first used in latinized form by Fenyes (1918: 17, as Placusae), generally accepted as in Newton and Thayer (1992: 54, as Placusini).

*Euvirae Seevers, 1978: 272 [stem: *Euvir-*]. Type genus: *Euvira* Sharp, 1883. Comment: unavailable family-group name, proposed after 1930 without description or bibliographic reference to such a description (Art. 13.1).

Tribe Pronomaeini Mulsant and Rey, 1873

Pronoméates Mulsant and Rey, 1873b: 8 [stem: *Pronomae-*]. Type genus: *Pronomaea* Erichson, 1837. Comment: original vernacular name available (Art. 11.7.2): first used in latinized form by Ganglbauer (1895: 315, as Pronomaeini), generally accepted as in Newton and Thayer (1992: 54, as Pronomaeini); incorrect original stem formation, not in prevailing usage.

Tribe Pseudoperinthini Cameron, 1939

Pseudoperinthinae Cameron, 1939: 1 [stem: *Pseudoperinth-*]. Type genus: *Pseudoperinthus* Wasmann, 1916.

Tribe Pygostenini Fauvel, 1899

Pygostenini Fauvel, 1899: 5 [stem: *Pygosten-*]. Type genus: *Pygostenus* Kraatz, 1858.

Sympolemonini Fenyes, 1918: 51 [stem: *Sympolemont-*]. Type genus: *Sympolemon* Wasmann, 1900. Comment: incorrect original stem formation, not in prevailing usage.

Tribe Sahlbergiini Kistner, 1993

Sahlbergiini Kistner, 1993: 315 [stem: *Sahlbergi-*]. Type genus: *Sahlbergius* Bernhauer, 1927.

Tribe Sceptobiini Seevers, 1978

Sceptobiini Seevers, 1978: 148 [stem: *Sceptobi-*]. Type genus: *Sceptobius* Sharp, 1883.

Tribe Skatitoxenini Kistner and Pasteels, 1969

Skatitoxenini Kistner and Pasteels, 1969: 1190 [stem: *Skatitoxen-*]. Type genus: *Skatitoxenus* Kistner and Pasteels, 1969.

Tribe Termitodiscini Wasmann, 1904

Termitodiscini Wasmann, 1904b: 656 [stem: *Termitodisc-*]. Type genus: *Termitodiscus* Wasmann, 1899.

Subtribe Athexeniina Pace, 2000

Athexenina Pace, 2000b: 336, in key [stem: *Athexeni-*]. Type genus: *Athexenia* Pace, 1999. Comment: incorrect original stem formation, not in prevailing usage.

Subtribe Termitodiscina Wasmann, 1904

Termitodiscini Wasmann, 1904b: 656 [stem: *Termitodisc-*]. Type genus: *Termitodiscus* Wasmann, 1899.

Tribe Termitohospitini Seevers, 1941

Termitohospini Seevers, 1941: 331 [stem: *Termitohospit-*]. Type genus: *Termitohospes* Seevers, 1941.

Subtribe Hetairotermitina Seevers, 1957

Hetairotermitina Seevers, 1957: 191 [stem: *Hetairotermit-*]. Type genus: *Hetairotermes* Cameron, 1920.

Subtribe Termitohospitina Seevers, 1941

Termitohospini Seevers, 1941: 331 [stem: *Termitohospit-*]. Type genus: *Termitohospes* Seevers, 1941. Comment: incorrect original stem formation, not in prevailing usage; correction of stem by Seevers (1957: 191).

Tribe Termitonannini Fenyes, 1918

Termitonannini Fenyes, 1918: 75 [stem: *Termitonann-*]. Type genus: *Termitonannus* Wasmann, 1902.

Subtribe Perinthina Bernhauer and Scheerpeltz, 1926

Perinthi Bernhauer and Scheerpeltz, 1926: 521 [stem: *Perinth-*]. Type genus: *Perinthus* Casey, 1889.

*Poduroideae Scheerpeltz, 1934: 1537 [stem: *Poduroid-*]. Type genus: *Poduroides* Mann, 1926. Comment: unavailable family-group name, proposed after 1930 without description or bibliographic reference to such a description (Art. 13.1).

Subtribe Termitonannina Fenyes, 1918

Termitonannini Fenyes, 1918: 75 [stem: *Termitonann-*]. Type genus: *Termitonannus* Wasmann, 1902.

Tribe Termitopaediini Seevers, 1957

Termitopaediini Seevers, 1957: 214 [stem: *Termitopaedi-*]. Type genus: *Termitopaedia* Wasmann, 1911.

Termitondina Seevers, 1957: 63, in key [stem: *Termitond-*]. Type genus: *Termitonda* Seevers, 1957. Comment: as Termitendina, name spelled correctly on page 238.

Tribe Termitusini Fenyes, 1918

Termitusae Fenyes, 1918: 18 [stem: *Termitus-*]. Type genus: *Termitusa* Wasmann, 1905.

Subtribe Termitospectrina Seevers, 1957

Termitospectrina Seevers, 1957: 191 [stem: *Termitospectr-*]. Type genus: *Termitospectrum* Mann, 1926.

Subtribe Termitusina Fenyes, 1918

Termitusae Fenyes, 1918: 18 [stem: *Termitus-*]. Type genus: *Termitusa* Wasmann, 1905.

Tribe Trichopseniini LeConte and Horn, 1883

Trichopsenii J. L. LeConte and G. H. Horn, 1883: 100 [stem: *Trichopseni-*]. Type genus: *Trichopsenius* G. H. Horn, 1877.

Termitopsenini Wasmann, 1916b: 196 [stem: *Termitopseni-*]. Type genus: *Termitopsenius* Wasmann, 1902. Comment: incorrect original stem formation, not in prevailing usage.

Schizelythrinae Kemner, 1925: 122 [stem: *Schizelythr-*]. Type genus: *Schizelythron* Kemner, 1925.

Tribe Trilobitideini Fauvel, 1899

Trilobitideidae Fauvel, 1899: 3 [stem: *Trilobitide-*]. Type genus: *Trilobitideus* Raffray, 1898.

Subfamily Trigonurinae Reiche, 1866

Trigonurides Reiche, 1866: 642 [stem: *Trigonur-*]. Type genus: *Trigonurus* Mulsant, 1847.

Subfamily Apateticinae Fauvel, 1895

Apateticae Fauvel, 1895: 190 [stem: *Apatetic-*]. Type genus: *Apatetica* Westwood, 1848.

Subfamily Scaphidiinae Latreille, 1806

Scaphidilia Latreille, 1806: 3 [stem: *Scaphidi-*]. Type genus: *Scaphidium* A. G. Olivier, 1790.

Tribe Cypariini Achard, 1924

Cypariini Achard, 1924: 28 [stem: *Cypari-*]. Type genus: *Cyparium* Erichson, 1845.

Tribe Scaphidiini Latreille, 1806

Scaphidilia Latreille, 1806: 3 [stem: *Scaphidi-*]. Type genus: *Scaphidium* A. G. Olivier, 1790. Comment: incorrect original stem formation, not in prevailing usage.

Cerambyciscaphini Pic, 1915: 30 [stem: *Cerambyciscaph-*]. Type genus: *Cerambyciscapha* Pic, 1915.

Diateliitae Achard, 1924: 28 [stem: *Diateli-*]. Type genus: *Diatelium* Pascoe, 1863.

Tribe Scaphiini Achard, 1924

Scaphiitae Achard, 1924: 27 [stem: *Scaphi-*]. Type genus: *Scaphium* Kirby, 1837.

Tribe Scaphisomatini Casey, 1893

Scaphisomini Casey, 1893: 511 [stem: *Scaphisomat-*]. Type genus: *Scaphisoma* Leach, 1815. Comment: incorrect original stem formation, not in prevailing usage; correction of stem by Newton and Thayer (1992: 64).

Heteroscaphini Achard, 1914: 395 [stem: *Heteroscaph-*]. Type genus: *Heteroscapha* Achard, 1914 [syn. of *Bironium* Csiki, 1909].

Cyparellini Achard, 1924: 28 [stem: *Cyparell-*]. Type genus: *Cyparella* Achard, 1924 [syn. of *Baeocera* Erichson, 1845].

Baeoceritae Achard, 1924: 30 [stem: *Baeocer-*]. Type genus: *Baeocera* Erichson, 1845 [placed on the Official List of Generic Names in Zoology (ICZN 1982)].

Sciatrophitae Achard, 1924: 30 [stem: *Sciatroph-*]. Type genus: *Sciatrophes* Blackburn, 1903 [syn. of *Baeocera* Erichson, 1845].

Baeoceridiitae Achard, 1924: 30 [stem: *Baeoceridi-*]. Type genus: *Baeoceridium* Reitter, 1889.

Scaphicomitae Achard, 1924: 31 [stem: *Scaphicom-*]. Type genus: *Scaphicoma* Motschulsky, 1863.

Toxidiini Achard, 1924: 31 [stem: *Toxidi-*]. Type genus: *Toxidium* J. L. LeConte, 1860.

Subfamily Piestinae Erichson, 1839

Piestini Erichson, 1839b: 31 [stem: *Piest-*]. Type genus: *Piestus* Gravenhorst, 1806.

Prognathites Blanchard, 1845a: 290 [stem: *Prognath-*]. Type genus: *Prognathus* Berthold, 1827 [as *Prognata*, incorrect subsequent spelling of type genus name not in prevailing usage; syn. of *Siagonium* Kirby and Spence, 1815]. Comment: original vernacular name available (Art. 11.7.2): first used in latinized form and generally accepted as in Blanchard (1853: 54, as Prognathitae).

*Siagoniini Crowson, 1980: 289 [stem: *Siagoni-*]. Type genus: *Siagonium* Kirby and Spence, 1815. Comment: unavailable family-group name, proposed after 1930 without description or bibliographic reference to such a description (Art. 13.1).

Subfamily Osoriinae Erichson, 1839

Osorini Erichson, 1839b: 30 [stem: *Osori-*]. Type genus: *Osorius* Latreille, 1829.

Tribe Eleusinini Sharp, 1887

Eleusinina Sharp, 1887: 728 [stem: *Eleusin-*]. Type genus: *Eleusis* Laporte, 1835.

Tribe Leptochirini Sharp, 1887

Leptochirina Sharp, 1887: 733 [stem: *Leptochir-*]. Type genus: *Leptochirus* Germar, 1824.

Tribe Osoriini Erichson, 1839

Osorini Erichson, 1839b: 30 [stem: *Osori-*]. Type genus: *Osorius* Latreille, 1829.

Subtribe Osoriina Erichson, 1839

Osorini Erichson, 1839b: 30 [stem: *Osori-*]. Type genus: *Osorius* Latreille, 1829. Comment: incorrect original stem formation, not in prevailing usage; correction of stem by J. L. LeConte (1861: 68).

Subtribe Parosoriina Bernhauer and Schubert, 1911

Parosorii Bernhauer and Schubert, 1911: 146 [stem: *Parosori-*]. Type genus: *Parosorius* Bernhauer, 1904.

Tribe Thoracophorini Reitter, 1909

Thoracophorinae Reitter, 1909: 199 [stem: *Thoracophor-*]. Type genus: *Thoracophorus* Motschulsky, 1837 [unjustified emendation of original type genus name *Thoraxophorus* by Erichson (1840a: 908); unjustified emendation in prevailing usage, treated as justified emendation (Art. 33.2.3.1)].

Subtribe CLAVILISPININA Newton and Thayer, 1992

PARALISPINI Blackwelder, 1942: 79 [stem: *Paralispin-*]. Type genus: *Paralispinus* Bernhauer, 1921 [preoccupied genus name, not *Paralispinus* Eichelbaum, 1913 [Coleoptera: STAPHYLINIDAE]; syn. of *Clavilispinus* Bernhauer, 1926]. Comment: permanently invalid (Art. 39): based on preoccupied type genus.

CLAVILISPININA Newton and Thayer, 1992: 59 [stem: *Clavilispin-*]. Type genus: *Clavilispinus* Bernhauer, 1926. Comment: replacement name for PARALISPINI Blackwelder, 1942 because of the homonymy of the type genus.

Subtribe GLYPTOMINA Newton and Thayer, 1992

CALOCERI Blackwelder, 1942: 78 [stem: *Calocer-*]. Type genus: *Calocerus* Fauvel, 1891 [preoccupied genus name, not *Calocerus* J. L. LeConte, 1853 [Coleoptera: ELATERIDAE]; syn. of *Glyptoma* Erichson, 1839]. Comment: permanently invalid (Art. 39): based on preoccupied type genus.

GLYPTOMINA Newton and Thayer, 1992: 59 [stem: *Glyptom-*]. Type genus: *Glyptoma* Erichson, 1839. Comment: replacement name for CALOCERI Blackwelder, 1942 because of the homonymy of the type genus; current spelling maintained (Art. 29.5): incorrect stem formation in prevailing usage (should be *Glyptomat-*).

Subtribe LISPININA Bernhauer and Schubert, 1910

LISPINI Bernhauer and Schubert, 1910: 19 [stem: *Lispin-*]. Type genus: *Lispinus* Erichson, 1839.

Subtribe THORACOPHORINA Reitter, 1909

THORACOPHORINAE Reitter, 1909: 199 [stem: *Thoracophor-*]. Type genus: *Thoracophorus* Motschulsky, 1837 [unjustified emendation of original type genus name *Thoraxophorus* by Erichson (1840a: 908); unjustified emendation in prevailing usage, treated as justified emendation (Art. 33.2.3.1)]. Comment: based on corrected spelling of type genus.

Subfamily OXYTELINAE Fleming, 1821

OXYTELIDAE Fleming, 1821: 49 [stem: *Oxytel-*]. Type genus: *Oxytelus* Gravenhorst, 1802.

Tribe BLEDIINI Ádám, 2001

BLEDIINI Ádám, 2001: 216 [stem: *Bledi-*]. Type genus: *Bledius* Leach, 1819.

Tribe COPROPHILINI Heer, 1839

COPROPHILINA Heer, 1839a: 198 [stem: *Coprophil-*]. Type genus: *Coprophilus* Latreille, 1829 [placed on the Official List of Generic Names in Zoology (ICZN 1993a)]. Comment: this family-group name was also used in the same year by

Erichson (1839b: 30, as COPROPHILINI) and Heer (1839b: 13, as COPROPHILIDA); for comments about the priority of these works see Newton and Thayer (1992: 24).

*HOMALOTRIQUITOS Solier, 1849: 321 [stem: *Homalotrich-*]. Type genus: *Homalotrichus* Solier, 1849. Comment: original vernacular name unavailable (Art. 11.7.2): not subsequently latinized.

TOXODERI Bernhauer and Schubert, 1911: 91 [stem: *Toxoder-*]. Type genus: *Toxoderus* Fauvel, 1900 [syn. of *Homalotrichus* Solier, 1849]. Comment: junior homonym of TOXODERINI Saussure, 1869 (type genus *Toxodera* Audinet-Serville, 1837) in Mantodea: MANTIDAE; this case is to be referred to the Commission to remove the homonymy (Art. 55.3.1).

Tribe EUPHANIINI Reitter, 1909

*PHOLIDIENS Mulsant and Rey, 1876: 209 [stem: *Pholid-*]. Type genus: *Pholidus* Mulsant and Rey, 1856 [preoccupied genus name, not *Pholidus* Rafinesque, 1815 [Pisces], or *Pholidus* Gray, 1840 [Aves]; syn. of *Euphanias* Fairmaire and Laboulbène, 1856]. Comment: original vernacular name unavailable (Art. 11.7.2): subsequently used in latinized form but not generally attributed to Mulsant and Rey (1876).

PHOLIDINI Acloque, 1896: 145 [stem: *Pholid-*]. Type genus: *Pholidus* Mulsant and Rey, 1856 [preoccupied genus name, not *Pholidus* Rafinesque, 1815 [Pisces], or *Pholidus* Gray, 1840 [Aves]; syn. of *Euphanias* Fairmaire and Laboulbène, 1856]. Comment: permanently invalid (Art. 39): based on preoccupied type genus.

EUPHANIAE Reitter, 1909: 16 [stem: *Euphani-*]. Type genus: *Euphanias* Fairmaire and Laboulbène, 1856. Comment: precedence (EUPHANIINI Reitter, 1909 vs DELEASTERINI Reitter, 1909) given to taxon originally proposed at the higher rank (Art. 24.1).

DELEASTERINI Reitter, 1909: 164 [stem: *Deleaster-*]. Type genus: *Deleaster* Erichson, 1839.

SYNTOMIINAE Böving and Craighead, 1931: 28 [stem: *Syntomi-*]. Type genus: *Syntomium* Curtis, 1828.

Tribe OXYTELINI Fleming, 1821

OXYTELIDAE Fleming, 1821: 49 [stem: *Oxytel-*]. Type genus: *Oxytelus* Gravenhorst, 1802.

THINOBIIDES J. Sahlberg, 1876: 242 [stem: *Thinobi-*]. Type genus: *Thinobius* Kiesenwetter, 1844.

TROGOPHLÉAIRES Mulsant and Rey, 1878b: 688 [stem: *Trogophloe-*]. Type genus: *Trogophloeus* Mannerheim, 1830 [syn. of *Carpelimus* Leach, 1819]. Comment: original vernacular name available (Art. 11.7.2): first used in latinized form by Seidlitz (1889 [Gatt.]: 90, as TROGOPHLOEINA), generally accepted as in

Johansen (1914: 533, as TROGOPHLOEINA); incorrect original stem formation, not in prevailing usage.

APOCELLARIA Lynch Arribálzaga, 1884: 344 [stem: *Apocell-*]. Type genus: *Apocellus* Erichson, 1839.

ECITOCLIMACINI Borgmeier, 1934: 452 [stem: *Ecitoclimac-*]. Type genus: *Ecitoclimax* Borgmeier, 1934. Comment: originally proposed as a tribe of ALEOCHARINAE.

TORRENTOMI Bierig, 1934: 213 [stem: *Torrentom-*]. Type genus: *Torrentomus* Bierig, 1934 [syn. of *Thinobius* Kiesenwetter, 1844].

TRIGONOBREGMINI Scheerpeltz, 1944: 170, in key [stem: *Trigonobregmat-*]. Type genus: *Trigonobregma* Scheerpeltz, 1944. Comment: incorrect original stem formation, not in prevailing usage; placement following Herman (2001).

CARPELIMINI Hatch, 1957: 85, in key [stem: *Carpelim-*]. Type genus: *Carpelimus* Leach, 1819.

*THINODROMINI Gildenkov, 2000: 56 [stem: *Thinodrom-*]. Type genus: *Thinodromus* Kraatz, 1858. Comment: name unavailable (Art. 16.1): name not indicated as intentionally new.

APLODERINI Ádám, 2001: 218 [stem: *Aploder-*]. Type genus: *Aploderus* Stephens, 1834.

Tribe PLANEUSTOMINI Jacquelin du Val, 1857

PLANEUSTOMITES Jacquelin du Val, 1857a: 58 [stem: *Planeustom-*]. Type genus: *Planeustomus* Jacquelin du Val, 1857. Comment: original vernacular name available (Art. 11.7.2): first used in latinized form and generally accepted as in Ádám (2001: 208, as PLANEUSTOMINI).

ACROGNATHINI Reitter, 1909: 164 [stem: *Acrognath-*]. Type genus: *Acrognathus* Erichson, 1839 [preoccupied genus name, not *Acrognathus* Agassiz, 1836 [Pisces]; syn. of *Manda* Blackwelder, 1952]. Comment: permanently invalid (Art. 39): based on preoccupied type genus.

MANDINI Gildenkov, 2003: 35 [stem: *Mand-*]. Type genus: *Manda* Blackwelder, 1952. Comment: replacement name for ACROGNATHINI Reitter, 1909 because of the homonymy of the type genus.

Subfamily OXYPORINAE Fleming, 1821

OXYPORIDAE Fleming, 1821: 49 [stem: *Oxypor-*]. Type genus: *Oxyporus* Fabricius, 1775.

Subfamily MEGALOPSIDIINAE Leng, 1920

MEGALOPINI Erichson, 1839b: 30 [stem: *Megalop-*]. Type genus: *Megalops* Erichson, 1839 [preoccupied genus name, not *Megalops* Lacepède, 1803 [Pisces], not *Megalops* Rafinesque, 1815 [Pisces]; syn. of *Megalopinus* Eichelbaum, 1915]. Comment: permanently invalid (Art. 39): based on preoccupied type genus.

MEGALOPSIDIINI Leng, 1920: 98 [stem: *Megalopsidi-*]. Type genus: *Megalopsidia* Leng, 1918 [syn. of *Megalopinus* Eichelbaum, 1915].

AULACOTRACHELINAE L. Benick, 1920: 1 [stem: *Aulacotrachel-*]. Type genus: *Aulacotrachelus* Benick, 1920 [syn. of *Megalopinus* Eichelbaum, 1915]. Comment: replacement name for MEGALOPINAE Erichson, 1839 because of the homonymy of the type genus.

STYLOPODINAE Blackwelder, 1943: 202 [stem: *Stylopod-*]. Type genus: *Stylopodus* Benick, 1917 [syn. of *Megalopinus* Eichelbaum, 1915]. Comment: name proposed to replace MEGALOPINAE Erichson, 1839 and MEGALOPSIDIINAE Leng, 1920 because of the synonymy of the type genus.

*MEGALOPININAE Puthz, 1967: 192 [stem: *Megalopin-*]. Type genus: *Megalopinus* Eichelbaum, 1915. Comment: unavailable family-group name, proposed after 1930 without description or bibliographic reference to such a description (Art. 13.1).

MEGALOPININAE Naomi, 1986: 344 [stem: *Megalopin-*]. Type genus: *Megalopinus* Eichelbaum, 1915. Comment: name proposed to replace MEGALOPINAE Erichson, 1839 and MEGALOPSIDIINAE Leng, 1920 because of the synonymy of the type genus.

Subfamily SCYDMAENINAE Leach, 1815

SCYDMAENIDES Leach, 1815: 92 [stem: *Scydmaen-*]. Type genus: *Scydmaenus* Latreille, 1802. Comment: placement follows Grebennikov and Newton (2009).

†Supertribe HAPSOMELITAE Poinar and Brown, 2004

HAPSOMELINAE Poinar and Brown, 2004: 790 [stem: *Hapsomel-*]. Type genus: *Hapsomela* Poinar and Brown, 2004.

Supertribe MASTIGITAE Fleming, 1821

MASTIGOIDAE Fleming, 1821: 49 [stem: *Mastig-*]. Type genus: *Mastigus* Latreille, 1802. Comment: incorrect original stem formation, not in prevailing usage.

Tribe CLIDICINI Casey, 1897

CLIDICINI Casey, 1897: 541 [stem: *Clidic-*]. Type genus: *Clidicus* Laporte, 1833.

Tribe LEPTOMASTACINI Casey, 1897

LEPTOMASTACINI Casey, 1897: 541 [stem: *Leptomastac-*]. Type genus: *Leptomastax* Pirazzoli, 1855.

Tribe MASTIGINI Fleming, 1821

MASTIGOIDAE Fleming, 1821: 49 [stem: *Mastig-*]. Type genus: *Mastigus* Latreille, 1802. Comment: incorrect original stem formation, not in prevailing usage.

Supertribe SCYDMAENITAE Leach, 1815

SCYDMAENIDES Leach, 1815: 92 [stem: *Scydmaen-*]. Type genus: *Scydmaenus* Latreille, 1802.

Tribe Cephenniini Reitter, 1882

Cephenniini Reitter, 1882c: 142 [stem: *Cephenni-*]. Type genus: *Cephennium* Müller and Kunze, 1822.

Anisosphaeridae Tömösváry, 1883: 128 [stem: *Anisosphaer-*]. Type genus: *Anisosphaera* Tömösváry, 1883 [syn. of *Cephennium* Müller and Kunze, 1822]. Comment: family-group name originally based on larva only.

Tribe Chevrolatiini Reitter, 1882

Chevrolatini Reitter, 1882c: 142 [stem: *Chevrolati-*]. Type genus: *Chevrolatia* Jacquelin du Val, 1859. Comment: incorrect original stem formation, not in prevailing usage.

Tribe Cyrtoscydmini Schaufuss, 1889

Cyrtoscydmini L. W. Schaufuss, 1889: 2 [stem: *Cyrtoscydm-*]. Type genus: *Cyrtoscydmus* Motschulsky, 1869 [syn. of *Stenichnus* C. G. Thomson, 1859].

Glandulariidae L. W. Schaufuss, 1889: 3 [stem: *Glandulari-*]. Type genus: *Glandularia* L. W. Schaufuss, 1889 [syn. of *Euconnus* (*Napochus*) C. G. Thomson, 1859].

Euconnini Casey, 1897: 354 [stem: *Euconn-*]. Type genus: *Euconnus* C. G. Thomson, 1859.

Opresini Casey, 1897: 354 [stem: *Opres-*]. Type genus: *Opresus* Casey, 1897 [syn. of *Microscydmus* Saulcy and Croissandeau, 1893].

Lophioderini Casey, 1897: 356 [stem: *Lophioder-*]. Type genus: *Lophioderus* Casey, 1897.

Stenichnini Ganglbauer, 1898: 25 [stem: *Stenichn-*]. Type genus: *Stenichnus* C. G. Thomson, 1859.

Neuraphini Csiki, 1909b: 18 [stem: *Neuraph-*]. Type genus: *Neuraphes* C. G. Thomson, 1859 [this name is an incorrect subsequent spelling of *Nevraphes*, in prevailing usage and so deemed to be the correct original spelling (Art. 33.3.1)].

Syndicini Csiki, 1919: 17 [stem: *Syndic-*]. Type genus: *Syndicus* Motschulsky, 1851.

Sciacharini Csiki, 1919: 69 [stem: *Sciacharit-*]. Type genus: *Sciacharis* Broun, 1893. Comment: incorrect original stem formation, not in prevailing usage.

Siamitini H. Franz, 1989: 44 [stem: *Siamit-*]. Type genus: *Siamites* Franz, 1989.

Tribe Eutheiini Casey, 1897

Eutheiini Casey, 1897: 507 [stem: *Euthei-*]. Type genus: *Eutheia* Stephens, 1830.

Ascydmini Casey, 1897: 355 [stem: *Ascydm-*]. Type genus: *Ascydmus* Casey, 1897 [syn. of *Euthiconus* Reitter, 1882].

Tribe Leptoscydmini Casey, 1897

Leptoscydmini Casey, 1897: 355 [stem: *Leptoscydm-*]. Type genus: *Leptoscydmus* Casey, 1897.

Tribe Plaumanniolini Costa Lima, 1962

Plaumanniolinae Costa Lima, 1962: 415 [stem: *Plaumanniol-*]. Type genus: *Plaumanniola* Costa Lima, 1962. Comment: originally proposed as a subfamily of Ptinidae.

Tribe Scydmaenini Leach, 1815

Scydmaenides Leach, 1815: 92 [stem: *Scydmaen-*]. Type genus: *Scydmaenus* Latreille, 1802.

Eumicrini Reitter, 1882c: 192 [stem: *Eumicr-*]. Type genus: *Eumicrus* Laporte, 1833 [syn. of *Scydmaenus* Latreille, 1802].

Subfamily Steninae MacLeay, 1825

Stenidae W. S. MacLeay, 1825: 49 [stem: *Sten-*]. Type genus: *Stenus* Latreille, 1797. Comment: the younger name Steninae Fraser and Purves, 1960 (type genus *Steno* Gray, 1846) in Mammalia: Delphinidae is unavailable according to Newton and Thayer (1992: 66).

Subfamily Euaesthetinae Thomson, 1859

Euaesthetina C. G. Thomson, 1859: 42 [stem: *Euaesthet-*]. Type genus: *Euaesthetus* Gravenhorst, 1806.

Tribe Alzadaesthetini Scheerpeltz, 1974

Alzadaesthetini Scheerpeltz, 1974: 102, in key [stem: *Alzadaesthet-*]. Type genus: *Alzadaesthetus* Kistner, 1961.

Tribe Austroesthetini Cameron, 1944

Austroaesthetini Cameron, 1944: 69 [stem: *Austroesthet-*]. Type genus: *Austroesthetus* Oke, 1933 [as *Austroaesthetus*, unjustified emendation of genus name not in prevailing usage]. Comment: incorrect original stem formation, not in prevailing usage.

Tribe Euaesthetini Thomson, 1859

Euaesthetina C. G. Thomson, 1859: 42 [stem: *Euaesthet-*]. Type genus: *Euaesthetus* Gravenhorst, 1806.

Tamotini Coiffait, 1984: 353 [stem: *Tamot-*]. Type genus: *Tamotus* L. W. Schaufuss, 1872.

Tribe Fenderiini Scheerpeltz, 1974

Fenderiini Scheerpeltz, 1974: 103, in key [stem: *Fenderi-*]. Type genus: *Fenderia* Hatch, 1957.

Tribe NORDENSKIOLDIINI Bernhauer and Schubert, 1911

NORDENSKIOELDIINI Bernhauer and Schubert, 1911: 186 [stem: *Nordenskioldi-*]. Type genus: *Nordenskioldia* Sahlberg, 1880 [as *Nordenskioeldia*, incorrect subsequent spelling of genus name not in prevailing usage]. Comment: incorrect original stem formation, not in prevailing usage (see Newton and Thayer 1992: 56).

Tribe STENAESTHETINI Bernhauer and Schubert, 1911

STENAESTHETINI Bernhauer and Schubert, 1911: 186 [stem: *Stenaesthet-*]. Type genus: *Stenaesthetus* Sharp, 1874.

Subfamily SOLIERIINAE Newton and Thayer, 1992

FISOGNATITOS Solier, 1849: 303 [stem: *Physognath-*]. Type genus: *Physognathus* Solier, 1849 [preoccupied genus name, not *Physognathus* Agassiz, 1846 [Reptilia]; syn. of *Solierius* Bernhauer, 1921]. Comment: original vernacular name available (Art. 11.7.2): first used in latinized form and generally accepted as in Kraatz (1859b: 3, as PHYSOGNATHITES [treated as Latin]); permanently invalid (Art. 39): based on preoccupied type genus.

SOLIERIINAE Newton and Thayer, 1992: 27 [stem: *Solieri-*]. Type genus: *Solierius* Bernhauer, 1921. Comment: replacement name for PHYSOGNATHINAE Solier, 1849 because of the homonymy of the type genus.

Subfamily LEPTOTYPHLINAE Fauvel, 1874

LEPTOTYPHLI Fauvel, 1874: 35 [stem: *Leptotyphl-*]. Type genus: *Leptotyphlus* Fauvel, 1874.

Tribe CEPHALOTYPHLINI Coiffait, 1963

CEPHALOTYPHLINI Coiffait, 1963: 380, in key [stem: *Cephalotyphl-*]. Type genus: *Cephalotyphlus* Coiffait, 1955.

Tribe ENTOMOCULIINI Coiffait, 1957

ENTOMOCULINI Coiffait, 1957: 61 [stem: *Entomoculi-*]. Type genus: *Entomoculia* Croissandeau, 1891. Comment: incorrect original stem formation, not in prevailing usage; correction of stem by Newton and Thayer (1992: 56).

Tribe LEPTOTYPHLINI Fauvel, 1874

LEPTOTYPHLI Fauvel, 1874: 35 [stem: *Leptotyphl-*]. Type genus: *Leptotyphlus* Fauvel, 1874.

LEPTOTYPHLINI Coiffait, 1957: 61 [stem: *Leptotyphl-*]. Type genus: *Leptotyphlus* Fauvel, 1874. Comment: family-group name proposed as new without reference to LEPTOTYPHLI Fauvel, 1874.

Tribe Metrotyphlini Coiffait, 1963

Metrotyphlini Coiffait, 1963: 381 [stem: *Metrotyphl-*]. Type genus: *Metrotyphlus* Coiffait, 1959.

Tribe Neotyphlini Coiffait, 1963

Neotyphlini Coiffait, 1963: 381, in key [stem: *Neotyphl-*]. Type genus: *Neotyphlus* Coiffait, 1959.

Subfamily Pseudopsinae Ganglbauer, 1895

Pseudopsini Ganglbauer, 1895: 690 [stem: *Pseudops-*]. Type genus: *Pseudopsis* Newman, 1834. Comment: current spelling maintained (Art. 29.5): incorrect stem formation in prevailing usage (should be *Pseudopse-*).

Subfamily Paederinae Fleming, 1821

Poederidae Fleming, 1821: 49 [stem: *Paeder-*]. Type genus: *Paederus* Fabricius, 1775.

Tribe Paederini Fleming, 1821

Poederidae Fleming, 1821: 49 [stem: *Paeder-*]. Type genus: *Paederus* Fabricius, 1775.

Subtribe Astenina Hatch, 1957

Suniina Sharp, 1886b: 591 [stem: *Suni-*]. Type genus: *Sunius* sensu Erichson, 1839 [not *Sunius* Stephens, 1829; syn. of *Astenus* Dejean, 1833]. Comment: based on a misidentified type genus, name treated here as invalid until an application is submitted to the Commission to suppress it for the Principle of Priority (Art. 65.2.1).

Astenina Hatch, 1957: 151, in key [stem: *Asten-*]. Type genus: *Astenus* Dejean, 1833. Comment: although this is not the oldest name for the subtribe, we recommend that an application be submitted to the Commission to suppress Suniina Sharp, 1886 because it is based on a misidentified type genus (Art. 65.2.1).

Subtribe Cryptobiina Casey, 1905

Cryptobia Casey, 1905: 21 [stem: *Cryptobi-*]. Type genus: *Cryptobium* Mannerheim, 1830 [syn. of *Ochthephilum* Stephens, 1829]. Comment: the younger name Cryptobiinae Hollande, 1952 (type genus *Cryptobia* Leidy, 1846) in Protozoa: Bodonidae is unavailable (see Newton and Thayer 1992: 61).

Cryptobiina Bordoni, 1975: 420 [stem: *Cryptobi-*]. Type genus: *Cryptobium* Mannerheim, 1830 [syn. of *Ochthephilum* Stephens, 1829]. Comment: family-group name proposed as new without reference to Cryptobia Casey, 1905.

Subtribe Cylindroxystina Bierig, 1943

Cylindroxystini Bierig, 1943: 158 [stem: *Cylindroxyst-*]. Type genus: *Cylindroxystus* Bierig, 1943.

Subtribe Dolicaonina Casey, 1905

Gnatimenitos Solier, 1849: 326 [stem: *Gnathymen-*]. Type genus: *Gnathymenus* Solier, 1849. Comment: original vernacular name available (Art. 11.7.2): first used in latinized form and generally accepted as in Ádám (2001: 115, as Gnathymenini); incorrect original stem formation, not in prevailing usage; this name was treated as unavailable by Newton and Thayer (1992: 61) which led to their use of Dolicaonina Casey, 1905 as the valid name for this subtribe; all subsequent authors have followed Newton and Thayer (1992: 61) except for the subsequent use of "Gnathymenini (Solier, 1849)" as valid by Ádám (2001: 115) which made Solier's name available according to our criteria of availability for names originally proposed in vernacular form; here we continue to use Dolicaonina Casey, 1905 as the valid name for this subtribe and recommend that an application be submitted to the Commission to suppress the name Gnathymenina Solier, 1849.

Dolicaones Casey, 1905: 56 [stem: *Dolicaon-*]. Type genus: *Dolicaon* Laporte, 1835. Comment: see comments under the name Gnatimenitos Solier, 1849 above.

Leptobii Bordoni, 1980: 170 [stem: *Leptobi-*]. Type genus: *Leptobium* Casey, 1905. Comment: unnecessary replacement name for "Dolicaina Bordoni, 1975".

Subtribe Echiasterina Casey, 1905

Echiasteres Casey, 1905: 245 [stem: *Echiaster-*]. Type genus: *Echiaster* Erichson, 1839.

Subtribe Lathrobiina Laporte, 1835

Lathrobidae Laporte, 1835a: 117 [stem: *Lathrobi-*]. Type genus: *Lathrobium* Gravenhorst, 1802. Comment: incorrect original stem formation, not in prevailing usage.

Sphaeronia Casey, 1905: 54 [stem: *Sphaeron-*]. Type genus: *Sphaeronum* Sharp, 1876 [as *Sphaeronium*, incorrect subsequent spelling of type genus name, not in prevailing usage]. Comment: incorrect original stem formation, not in prevailing usage.

Subtribe Lithocharina Casey, 1905

Lithochares Casey, 1905: 146 [stem: *Lithochar-*]. Type genus: *Lithocharis* Dejean, 1833. Comment: current spelling maintained (Art. 29.5): incorrect stem formation in prevailing usage (should be *Lithocharit-*).

Lithocharina Bordoni, 1974: 324 [stem: *Lithochar-*]. Type genus: *Lithocharis* Dejean, 1833. Comment: family-group name proposed as new without reference to Lithochares Casey, 1905.

Subtribe Medonina Casey, 1905

Medones Casey, 1905: 20 [stem: *Medon-*]. Type genus: *Medon* Stephens, 1833. Comment: current spelling maintained (Art. 29.5): incorrect stem formation in prevailing usage (should be *Medont-*).

Medina Bordoni, 1975: 420 [stem: *Medon-*]. Type genus: *Medon* Stephens, 1833. Comment: family-group name proposed as new without reference to Medones Casey, 1905; incorrect original stem formation, not in prevailing usage (should be *Medont-*).

Acanthoglossi Coiffait, 1982: 10 [stem: *Acanthogloss-*]. Type genus: *Acanthoglossa* Kraatz, 1859.

Subtribe Paederina Fleming, 1821

Poederidae Fleming, 1821: 49 [stem: *Paeder-*]. Type genus: *Paederus* Fabricius, 1775 [as *Poederus*, incorrect subsequent spelling of type genus name, not in prevailing usage]. Comment: incorrect original stem formation, not in prevailing usage, see comments in Newton and Thayer (1992: 60).

Geopaederidae Gistel, 1848: [13] [stem: *Geopaeder-*]. Type genus: *Geopaederus* Gistel, 1848 [syn. of *Paederus* Fabricius, 1775].

Subtribe Scopaeina Mulsant and Rey, 1878

*Poliodontidos Solier, 1849: 303 [stem: *Polyodont-*]. Type genus: *Polyodontus* Solier, 1849 [preoccupied genus name, not *Polyodontus* Eysenhardt, 1818 [Vermes]; syn. of *Scopaeus* Erichson, 1839]. Comment: original vernacular name unavailable (Art. 11.7.2): not subsequently latinized; if found to be available then permanently invalid (Art. 39): based on preoccupied type genus; Polyodontidae Bonaparte, 1838 (type genus *Polyodon* Lacepède, 1797) is available in Pisces; incorrect original stem formation, not in prevailing usage.

Scopéates Mulsant and Rey, 1878a: 178 [stem: *Scopae-*]. Type genus: *Scopaeus* Erichson, 1839. Comment: original vernacular name available (Art. 11.7.2): first used in latinized form by Seidlitz (1889 [Gatt.]: 92, as Scopaeina), generally accepted as in Newton and Thayer (1992: 62, as Scopaeina); incorrect original stem formation, not in prevailing usage.

Subtribe Stilicina Casey, 1905

Stilici Casey, 1905: 218 [stem: *Stilic-*]. Type genus: *Stilicus* Berthold, 1827 [syn. of *Rugilus* Leach, 1819].

RUGILINA Hatch, 1957: 151, in key [stem: *Rugil-*]. Type genus: *Rugilus* Leach, 1819.

Subtribe STILICOPSINA Casey, 1905

STILICOPSES Casey, 1905: 230 [stem: *Stilicops-*]. Type genus: *Stilicopsis* Sachse, 1852. Comment: current spelling maintained (Art. 29.5): incorrect stem formation in prevailing usage (should be *Stilicopse-*).

XENASTERES Bierig, 1939: 179 [stem: *Xenaster-*]. Type genus: *Xenaster* Bierig, 1939 [preoccupied genus name, not *Xenaster* Simonwitsch, 1871 [Echinodermata]; the nomenclatural status of *Xenaster* will be addressed in the near future by L. Herman (pers. comm. 2010) therefore we refrain from proposing a new replacement name here]. Comment: permanently invalid (Art. 39): based on preoccupied type genus.

*STAMNODERES Blackwelder, 1944: 126 [stem: *Stamnoder-*]. Type genus: *Stamnoderus* Sharp, 1886. Comment: unavailable family-group name, proposed after 1930 without description or bibliographic reference to such a description (Art. 13.1).

Tribe PINOPHILINI Nordmann, 1837

PINOPHILINIFORMES Nordmann, 1837: 6 [stem: *Pinophil-*]. Type genus: *Pinophilus* Gravenhorst, 1802.

Subtribe PINOPHILINA Nordmann, 1837

PINOPHILINIFORMES Nordmann, 1837: 6 [stem: *Pinophil-*]. Type genus: *Pinophilus* Gravenhorst, 1802.

Subtribe PROCIRRINA Bernhauer and Schubert, 1912

PROCIRRI Bernhauer and Schubert, 1912: 197 [stem: *Procirr-*]. Type genus: *Procirrus* Latreille, 1829.

Subfamily STAPHYLININAE Latreille, 1802

STAPHYLINIAE Latreille, 1802: 124 [stem: *Staphylin-*]. Type genus: *Staphylinus* Linnaeus, 1758 [placed on the Official List of Generic Names in Zoology (ICZN 1959a)].

Tribe ARROWININI Solodovnikov and Newton, 2005

ARROWININI Solodovnikov and Newton, 2005: 420 [stem: *Arrowin-*]. Type genus: *Arrowinus* Bernhauer, 1935.

Tribe DIOCHINI Casey, 1906

DIOCHI Casey, 1906: 429 [stem: *Dioch-*]. Type genus: *Diochus* Erichson, 1839.

DIOCHINAE I. Moore, 1964: 86, in key [stem: *Dioch-*]. Type genus: *Diochus* Erichson, 1839. Comment: family-group name proposed as new without reference to DIOCHI Casey, 1906.

Tribe Maorothiini Assing, 2000

Maorothiini Assing, 2000: 16 [stem: *Maorothi-*]. Type genus: *Maorothius* Assing, 2000.

Tribe Othiini Thomson, 1859

Othiides C. G. Thomson, 1859: 26 [stem: *Othi-*]. Type genus: *Othius* Stephens, 1829 [placed on the Official List of Generic Names in Zoology (ICZN 1983c)].

Atrecini Hatch, 1957: 172, in key [stem: *Atrec-*]. Type genus: *Atrecus* Jacquelin du Val, 1856.

Tribe Platyprosopini Lynch Arribálzaga, 1884

Platyprosoparia Lynch Arribálzaga, 1884: 165 [stem: *Platyprosop-*]. Type genus: *Platyprosopus* Mannerheim, 1830.

Platyprosopinae I. Moore, 1964: 86, in key [stem: *Platyprosop-*]. Type genus: *Platyprosopus* Mannerheim, 1830. Comment: family-group name proposed as new without reference to Platyprosopinae Lynch Arribálzaga, 1884.

Tribe Staphylinini Latreille, 1802

Staphyliniae Latreille, 1802: 124 [stem: *Staphylin-*]. Type genus: *Staphylinus* Linnaeus, 1758 [placed on the Official List of Generic Names in Zoology (ICZN 1959a)]. Comment: placed on the Official List of Family-Group Names in Zoology (as Staphylinidae Latreille, [1803-1804]) (ICZN 1959a).

Subtribe Amblyopinina Seevers, 1944

Amblyopininae Seevers, 1944: 157 [stem: *Amblyopin-*]. Type genus: *Amblyopinus* Solsky, 1875.

Heterothopsi Coiffait, 1978: 300 [stem: *Heterothop-*]. Type genus: *Heterothops* Stephens, 1829. Comment: incorrect original stem formation, not in prevailing usage; correction of stem by Newton and Thayer (1992: 65); placement based on Chatzimanolis et al. (2010).

Subtribe Anisolinina Hayashi, 1993

Anisolinina Hayashi, 1993: 288 [stem: *Anisolin-*]. Type genus: *Anisolinus* Sharp, 1889. Comment: originally proposed as a subtribe of Philonthini Kirby, 1837.

Subtribe Eucibdelina Sharp, 1889

Eucibdelini Sharp, 1889: 112 [stem: *Eucibdel-*]. Type genus: *Eucibdelus* Kraatz, 1859.

Subtribe Hyptiomina Casey, 1906

Hyptiomae Casey, 1906: 361 [stem: *Hyptiom-*]. Type genus: *Hyptioma* Casey, 1906 [syn. of *Holisus* Erichson, 1839]. Comment: current spelling maintained (Art. 29.5): incorrect stem formation in prevailing usage (should be *Hyptiomat-*).

*Holisi Blackwelder, 1944: 143 [stem: *Holis-*]. Type genus: *Holisus* Erichson, 1839. Comment: unavailable family-group name, proposed after 1930 without description or bibliographic reference to such a description (Art. 13.1).

Holisina Newton, 1988: 259 [stem: *Holis-*]. Type genus: *Holisus* Erichson, 1839.

Subtribe Philonthina Kirby, 1837

Philonthidae Kirby, 1837: 91 [stem: *Philonth-*]. Type genus: *Philonthus* Curtis, 1829.

*Rémates Mulsant and Rey, 1876: 596 [stem: *Rem-*]. Type genus: *Remus* sensu C. G. Thomson, 1859 [not *Remus* Holme, 1837; syn. of *Erichsonius* Fauvel, 1874]. Comment: original vernacular name unavailable (Art. 11.7.2): not subsequently latinized; also based on a misidentified type genus.

Craspedomeri Bernhauer, 1911: 88 [stem: *Craspedomer-*]. Type genus: *Craspedomerus* Bernhauer, 1911.

Subtribe Quediina Kraatz, 1857

Quediiformes Kraatz, 1857: 473 [stem: *Quedi-*]. Type genus: *Quedius* Stephens, 1829 [placed on the Official List of Generic Names in Zoology (ICZN 1996c)]. Comment: use of family-group name given precedence over Platycnemini Nordmann 1837 and "Quediini Kraatz, [1857]" placed on the Official List of Family-Group Names in Zoology (ICZN 1996c).

*Acylophorini Scheerpeltz, 1968: 97 [stem: *Acylophor-*]. Type genus: *Acylophorus* Nordmann, 1837. Comment: unavailable family-group name, proposed after 1930 without description or bibliographic reference to such a description (Art. 13.1).

Acylophorini Outerelo and Gamarra, 1985: 48, in key [stem: *Acylophor-*]. Type genus: *Acylophorus* Nordmann, 1837.

Subtribe Staphylinina Latreille, 1802

Staphyliniae Latreille, 1802: 124 [stem: *Staphylin-*]. Type genus: *Staphylinus* Linnaeus, 1758 [placed on the Official List of Generic Names in Zoology (ICZN 1959a)]. Comment: placed on the Official List of Family-Group Names in Zoology (as Staphylinidae Latreille, [1803-1804]) and "Staphylinii Latreille, [1803-1804]" placed on the Official Index of Rejected and Invalid Family-Group Names in Zoology (ICZN 1959a).

Creophilidae Kirby, 1837: 95 [stem: *Creophil-*]. Type genus: *Creophilus* Leach, 1819 [*Creophilus* (attributed to Samouelle, 1819) placed on the Official List of Generic Names in Zoology (ICZN 1959a)].

THINOPININAE Böving and Craighead, 1931: 30 [stem: *Thinopin-*]. Type genus: *Thinopinus* J. L. LeConte, 1852.

OCYPINA Hatch, 1957: 173, in key [stem: *Ocypod-*]. Type genus: *Ocypus* Leach, 1819. Comment: incorrect original stem formation, not in prevailing usage; correction of stem by Newton and Thayer (1992: 65).

Subtribe TANYGNATHININA Reitter, 1909

TANYGNATHININI Reitter, 1909: 105 [stem: *Tanygnathin-*]. Type genus: *Tanygnathinus* Reitter, 1909 [syn. of *Atanygnathus* Jakobson, 1909].

TANYGNATHINI Casey, 1915b: 424 [stem: *Tanygnath-*]. Type genus: *Tanygnathus* Erichson, 1839 [preoccupied genus name, not *Tanygnathus* Wagler, 1832 [Aves]; syn. of *Atanygnathus* Jakobson, 1909]. Comment: permanently invalid (Art. 39): based on preoccupied type genus.

ATANYGNATHINI Lohse, 1964: 220 [stem: *Atanygnath-*]. Type genus: *Atanygnathus* Jakobson, 1909.

Subtribe XANTHOPYGINA Sharp, 1884

PLATYCNEMIDIFORMES Nordmann, 1837: 6 [stem: *Platycnem-*]. Type genus: *Platycnemus* Nordmann, 1837 [placed on the Official List of Generic Names in Zoology (ICZN 1996c); syn. of *Haematodes* Laporte, 1835]. Comment: younger name XANTHOPYGINA Sharp, 1884 given precedence over this name (ICZN 1996c); placed on the Official List of Family-Group Names in Zoology (ICZN 1996c, as PLATYCNEMINI Nordmann, 1837).

XANTHOPYGINA Sharp, 1884: 342 [stem: *Xanthopyg-*]. Type genus: *Xanthopygus* Kraatz, 1857. Comment: family-group name given precedence over PLATYCNEMINI Nordmann, 1837 and placed on the Official List of Family-Group Names in Zoology (ICZN 1996c).

TRIACRI Bernhauer, 1931: 84 [stem: *Triacr-*]. Type genus: *Triacrus* Nordmann, 1837.

Tribe XANTHOLININI Erichson, 1839

AGRAEFORMES Nordmann, 1837: 7 [stem: *Agrod-*]. Type genus: *Agrodes* Nordmann, 1837 [placed on the Official List of Generic Names in Zoology (ICZN 1996c); syn. of *Plochionocerus* Dejean, 1833]. Comment: name placed on the Official List of Family-Group Names in Zoology (ICZN 1996c, as AGRODINI Nordmann, 1837); younger name XANTHOLININI Erichson, 1839 given precedence over this name (ICZN 1996c).

GYROHYPNIDAE Kirby, 1837: 88 [stem: *Gyrohypn-*]. Type genus: *Gyrohypnus* Leach, 1819 [placed on the Official List of Generic Names in Zoology (ICZN 1983c, as *Gyrohypnus* Samouelle, 1819)]. Comment: placed on the Official List of Family-Group Names in Zoology (ICZN 1996c, as GYROHYPNINI Kirby, 1837); younger name XANTHOLININI Erichson, 1839 given precedence over this name (ICZN 1996c).

Xantholinini Erichson, 1839b: 28 [stem: *Xantholin-*]. Type genus: *Xantholinus* Dejean, 1821 [placed on the Official List of Generic Names in Zoology (ICZN 1983c)]. Comment: family-group name given precedence over Agrodini Nordmann, 1837 and Gyrohypnini Kirby, 1837 and placed on the Official List of Family-Group Names in Zoology (ICZN 1996c).

Araeocnemes Casey, 1906: 359 [stem: *Araeocnem-*]. Type genus: *Araeocnemus* Nordmann, 1837 [as *Araeocnemis*, incorrect subsequent spelling of type genus name, not in prevailing usage; syn. of *Plochionocerus* Dejean, 1833].

Metoponci Casey, 1906: 360 [stem: *Metoponc-*]. Type genus: *Metoponcus* Kraatz, 1857 [syn. of *Zeteotomus* Jacquelin du Val, 1856].

†Subfamily Protactinae Heer, 1847

Protactiden Heer, 1847: 28 [stem: *Protact-*]. Type genus: *Protactus* Heer, 1847. Comment: original vernacular name available (Art. 11.7.2): first used in latinized form by Bronn (1848: 1045, as Protactidae), generally accepted as in Herman (2001: 3839, as Protactinae).

Staphylinoidea *incertae sedis*

Hameedini M. Abdullah and Quadri, 1968: 310 [stem: *Hameedi-*]. Type genus: *Hameedia* M. Abdullah and Quadri, 1968. Comment: incorrect original stem formation, not in prevailing usage; originally proposed as a tribe of Staphylinidae: Oxyporinae; correct placement uncertain (see Newton and Thayer 1992: 67).

Series SCARABAEIFORMIA

Superfamily Scarabaeoidea Latreille, 1802

Scarabaeïdes Latreille, 1802: 144 [stem: *Scarabae-*]. Type genus: *Scarabaeus* Linnaeus, 1758. Comment: First Reviser (Scarabaeoidea Latreille, 1802 vs Geotrupoidea Latreille, 1802) not determined, current usage maintained.

Family Pleocomidae LeConte, 1861

Pleocomini J. L. LeConte, 1861: 128 [stem: *Pleocom-*]. Type genus: *Pleocoma* J. L. LeConte, 1856.

Subfamily Pleocominae LeConte, 1861

Pleocomini J. L. LeConte, 1861: 128 [stem: *Pleocom-*]. Type genus: *Pleocoma* J. L. LeConte, 1856.

†Subfamily Cretocominae Nikolajev, 2002

Cretocomini Nikolajev, 2002: 53 [stem: *Cretocom-*]. Type genus: *Cretocoma* Nikolajev, 2002.

†Subfamily Archescarabaeinae Nikolajev, 2010

Archescarabaeinae Nikolajev, 2010: 69 [stem: *Archescarabae-*]. Type genus: *Archescarabaeus* Nikolajev, 2010.

Family Geotrupidae Latreille, 1802

Geotrupini Latreille, 1802: 142 [stem: *Geotrup-*]. Type genus: *Geotrupes* Latreille, 1797 [placed on the Official List of Generic Names in Zoology (ICZN 1955b)]. Comment: name placed on the Official List of Family-Group Names in Zoology (ICZN 1955c, as Geotrupini Latreille, 1806).

Subfamily Taurocerastinae Germain, 1897

Taurocerastidae Germain, 1897: 288 [stem: *Taurocerast-*]. Type genus: *Taurocerastes* Philippi, 1866.

Subfamily Bolboceratinae Mulsant, 1842

Bolbocéraires Mulsant, 1842: 347 [stem: *Bolbocerat-*]. Type genus: *Bolboceras* Kirby, 1819 [placed on the Official List of Generic Names in Zoology (ICZN 2006a)].

Tribe Athyreini Lynch Arribálzaga, 1878

*Athyréites Blanchard, 1845a: 221 [stem: *Athyre-*]. Type genus: *Athyreus* W. S. MacLeay, 1819 [placed on the Official List of Generic Names in Zoology (ICZN 1985a)]. Comment: original vernacular name unavailable (Art. 11.7.2): subsequently used in latinized form but not generally attributed to Blanchard (1845a) (see A. B. T. Smith 2006: 148).

Athyreitae Lynch Arribálzaga, 1878: 145 [stem: *Athyre-*]. Type genus: *Athyreus* W. S. MacLeay, 1819 [placed on the Official List of Generic Names in Zoology (ICZN 1985a)]. Comment: name previously attributed to Howden and Martínez (1963).

Athyreini Howden and Martínez, 1963: 346 [stem: *Athyre-*]. Type genus: *Athyreus* W. S. MacLeay, 1819 [placed on the Official List of Generic Names in Zoology (ICZN 1985a)]. Comment: family-group name proposed as new without reference to Athyréides Blanchard, 1845 or Athyreitae Lynch Arribálzaga, 1878.

Tribe Bolbelasmini Nikolajev, 1996

Bolbelasmini Nikolajev, 1996: 96 [stem: *Bolbelasm-*]. Type genus: *Bolbelasmus* Boucomont, 1911.

Tribe Bolboceratini Mulsant, 1842

Bolbocéraires Mulsant, 1842: 347 [stem: *Bolbocerat-*]. Type genus: *Bolboceras* Kirby, 1819 [placed on the Official List of Generic Names in Zoology (ICZN 2006a)]. Comment: original vernacular name available (Art. 11.7.2): first used in latinized form by Boucomont (1911: 333, as Bolbocerinae [incorrect stem formation]), generally accepted as in A. B. T. Smith (2006: 148, as Bolboceratini); concept of *Bolboceras* Kirby, 1819 fixed by the Commission (ICZN 2006a); incorrect original stem formation, not in prevailing usage.

Australobolbini Nikolajev, 1996: 96 [stem: *Australobolb-*]. Type genus: *Australobolbus* Howden and Cooper, 1977.

Tribe Bolbochromini Nikolajev, 1970

Bolbochromini Nikolajev, 1970: 34 [stem: *Bolbochrom-*]. Type genus: *Bolbochromus* Boucomont, 1909.

Tribe Eubolbitini Nikolajev, 1970

Eubolbitini Nikolajev, 1970: 33 [stem: *Eubolbit-*]. Type genus: *Eubolbitus* Reitter, 1892.

Tribe Eucanthini Nikolajev, 2003

Eucanthini Nikolajev, 2003b: 209 [stem: *Eucanth-*]. Type genus: *Eucanthus* Westwood, 1848.

Tribe Gilletinini Nikolajev, 1990

Gilletinini Nikolajev, 1990: 99 [stem: *Gilletin-*]. Type genus: *Gilletinus* Boucomont, 1932.

Tribe Odonteini Shokhin, 2007

Odonteini Shokhin, 2007: 111 [stem: *Odonte-*]. Type genus: *Odonteus* Samouelle, 1819 [placed on the Official List of Generic Names in Zoology (ICZN 2006a)]. Comment: replacement name for Bolboceratini of authors because the concept of *Bolboceras* Kirby, 1819 was fixed differently by the Commission (ICZN 2006a).

Tribe Stenaspidiini Nikolajev, 2003

Stenaspidiini Nikolajev, 2003a: 190, in key [stem: *Stenaspidi-*]. Type genus: *Stenaspidius* Westwood, 1848.

Subfamily Geotrupinae Latreille, 1802

Geotrupini Latreille, 1802: 142 [stem: *Geotrup-*]. Type genus: *Geotrupes* Latreille, 1797 [placed on the Official List of Generic Names in Zoology (ICZN 1955b)]. Comment: name placed on the Official List of Family-Group Names in Zoology (ICZN 1955c, as Geotrupini Latreille, 1806).

Tribe Ceratotrupini Zunino, 1984

Ceratotrupini Zunino, 1984: 89 [stem: *Ceratotrup-*]. Type genus: *Ceratotrupes* Jekel, 1866.

Tribe Enoplotrupini Paulian, 1945

Enoplotrupini Paulian, 1945: 40, in key [stem: *Enoplotrup-*]. Type genus: *Enoplotrupes* Lucas, 1869.

Chromogeotrupini Zunino, 1984: 30 [stem: *Chromogeotrup-*]. Type genus: *Chromogeotrupes* Bovo and Zunino, 1983.

†Tribe Cretogeotrupini Nikolajev, 1996

Cretogeotrupini Nikolajev, 1996: 97 [stem: *Cretogeotrup-*]. Type genus: *Cretogeotrupes* Nikolajev, 1992.

Tribe Geotrupini Latreille, 1802

Geotrupini Latreille, 1802: 142 [stem: *Geotrup-*]. Type genus: *Geotrupes* Latreille, 1797 [placed on the Official List of Generic Names in Zoology (ICZN 1955b)]. Comment: name placed on the Official List of Family-Group Names in Zoology (ICZN 1955c, as Geotrupini Latreille, 1806).

Tribe Lethrini Oken, 1843

Lethren Oken, 1843: 484 [stem: *Lethr-*]. Type genus: *Lethrus* Scopoli, 1777. Comment: original vernacular name available (Art. 11.7.2): first used in latinized form and generally accepted as in Tulk (1847: 614, as Lethri); name previously attributed to Mulsant and Rey (1871a).

Family Belohinidae Paulian, 1959

Belohininae Paulian, 1959: 40 [stem: *Belohin-*]. Type genus: *Belohina* Paulian, 1959.

Family Passalidae Leach, 1815

Passalida Leach, 1815: 100 [stem: *Passal-*]. Type genus: *Passalus* Fabricius, 1792.

Subfamily Aulacocyclinae Kaup, 1868

Aulacocyclinae Kaup, 1868a: 4 [stem: *Aulacocycl-*]. Type genus: *Aulacocyclus* Kaup, 1868.

Tribe Aulacocyclini Kaup, 1868

Aulacocyclinae Kaup, 1868a: 4 [stem: *Aulacocycl-*]. Type genus: *Aulacocyclus* Kaup, 1868.

Tribe Ceracupedini Boucher, 2006

Ceracupini Boucher, 2006: 319 [stem: *Ceracuped-*]. Type genus: *Ceracupes* Kaup, 1871. Comment: incorrect original stem formation, not in prevailing usage.

Subfamily Passalinae Leach, 1815

Passalida Leach, 1815: 100 [stem: *Passal-*]. Type genus: *Passalus* Fabricius, 1792.

Tribe Leptaulacini Kaup, 1871

Leptaulaceae Kaup, 1871: 28 [stem: *Leptaulac-*]. Type genus: *Leptaulax* Kaup, 1868.

Tribe Macrolinini Kaup, 1871

Macrolineae Kaup, 1871: 42 [stem: *Macrolin-*]. Type genus: *Macrolinus* Kaup, 1868. Comment: First Reviser found (Macrolinini Kaup, 1871 vs Aceraiini Kaup, 1871) is Gravely (1918: 76).

Aceraiae Kaup, 1871: 47 [stem: *Acerai-*]. Type genus: *Aceraius* Kaup, 1868.

Eriocnemiae Kaup, 1871: 35 [stem: *Eriocnemid-*]. Type genus: *Eriocnemis* Kaup, 1868 [preoccupied genus name, not *Eriocnemis* Reichenbach, 1849 [Aves]; syn. of *Pelopides* Kuwert, 1896]. Comment: permanently invalid (Art. 39): based on preoccupied type genus; incorrect original stem formation, not in prevailing usage.

Gonatinae Kuwert, 1891: 169 [stem: *Gonat-*]. Type genus: *Gonatas* Kaup, 1871.

Mastachilinae Kuwert, 1891: 167 [stem: *Mastachil-*]. Type genus: *Mastachilus* Kaup, 1868.

Pharochilinae Kuwert, 1891: 166 [stem: *Pharochil-*]. Type genus: *Pharochilus* Kaup, 1868.

Tarquininae Kuwert, 1891: 164 [stem: *Tarquini-*]. Type genus: *Tarquinius* Kuwert, 1891. Comment: incorrect original stem formation, not in prevailing usage.

Vellejinae Kuwert, 1891: 167 [stem: *Vellej-*]. Type genus: *Vellejus* Kaup, 1871 [preoccupied genus name, not *Vellejus* Mannerheim, 1830 [Coleoptera: Staphylinidae], not *Vellejus* Stål, 1865 [Hemiptera]; syn. of *Labienus* Kaup, 1871]. Comment: permanently invalid (Art. 39): based on preoccupied type genus.

Aureliinae Kuwert, 1896: 230 [stem: *Aureli-*]. Type genus: *Aurelius* Kuwert, 1891.

Lachinae Kuwert, 1896: 230 [stem: *Lach-*]. Type genus: *Laches* Kaup, 1871 [preoccupied genus name, not *Laches* Gistel, 1848 [Hymenoptera], not *Laches* Thorell, 1869 [Arachnida]; syn. of *Aceraius* Kaup, 1868]. Comment: permanently invalid (Art. 39): based on preoccupied type genus.

Pelopinae Kuwert, 1896: 229 [stem: *Pelopid-*]. Type genus: *Pelopides* Kuwert, 1896. Comment: incorrect original stem formation, not in prevailing usage.

Pleurariinae Kuwert, 1896: 224 [stem: *Pleurari-*]. Type genus: *Pleurarius* Kaup, 1868.

Gnaphalocneminae Gravely, 1914: 194 [stem: *Gnaphalocnemid-*]. Type genus: *Gnaphalocnemis* Heller, 1900 [syn. of *Pelopides* Kuwert, 1896]. Comment: incorrect original stem formation, not in prevailing usage.

Austropassalinae Mjöberg, 1917: 11 [stem: *Austropassal-*]. Type genus: *Austropassalus* Mjöberg, 1917.

Tribe Passalini Leach, 1815

Passalida Leach, 1815: 100 [stem: *Passal-*]. Type genus: *Passalus* Fabricius, 1792.

Neleinae Kaup, 1869: 28 [stem: *Nele-*]. Type genus: *Neleus* Kaup, 1869 [preoccupied genus name, not *Neleus* Desbonne and Schramm, 1867 [Crustacea]; syn. of *Passalus* Fabricius, 1792]. Comment: permanently invalid (Art. 39): based on preoccupied type genus.

Pertinaceae Kaup, 1871: 89 [stem: *Pertinac-*]. Type genus: *Pertinax* Kaup, 1869.

Phoroneae Kaup, 1871: 97 [stem: *Phorone-*]. Type genus: *Phoroneus* Kaup, 1869 [preoccupied genus name, not *Phoroneus* Stål, 1865 [Hemiptera]; syn. of *Passalus* Fabricius, 1792]. Comment: permanently invalid (Art. 39): based on preoccupied type genus.

Mitrorhinae Kuwert, 1891: 190 [stem: *Mitrorhin-*]. Type genus: *Mitrorhinus* Kaup, 1871.

Paxillinae Kuwert, 1891: 182 [stem: *Paxill-*]. Type genus: *Paxillus* W. S. MacLeay, 1819.

Petrejinae Kuwert, 1891: 176 [stem: *Petrej-*]. Type genus: *Petrejus* Kaup, 1869.

Nelidinae Kuwert, 1896: 222 [stem: *Neleid-*]. Type genus: *Neleides* Kaup, 1869. Comment: incorrect original stem formation, not in prevailing usage.

Ptichopinae Kuwert, 1896: 224 [stem: *Ptichopod-*]. Type genus: *Ptichopus* Kaup, 1869. Comment: incorrect original stem formation, not in prevailing usage.

Rhodacanthopinae Kuwert, 1896: 222 [stem: *Rhodocanthopod-*]. Type genus: *Rhodocanthopus* Kaup, 1871. Comment: incorrect original stem formation, not in prevailing usage.

Vatiniinae Kuwert, 1896: 226 [stem: *Vatini-*]. Type genus: *Vatinius* Kaup, 1869 [preoccupied genus name, not *Vatinius* Stål, 1865 [Hemiptera]; syn. of *Passalus* Fabricius, 1792]. Comment: permanently invalid (Art. 39): based on preoccupied type genus.

Tribe Proculini Kaup, 1868

Proculinae Kaup, 1868b: 8 [stem: *Procul-*]. Type genus: *Proculus* Kaup, 1868.

Pseudacantheae Kaup, 1871: 73 [stem: *Pseudacanth-*]. Type genus: *Pseudacanthus* Kaup, 1869.

Oileinae Kuwert, 1891: 192 [stem: *Oile-*]. Type genus: *Oileus* Kaup, 1869.

Sertorinae Kuwert, 1891: 175 [stem: *Sertori-*]. Type genus: *Sertorius* Kaup, 1871 [preoccupied genus name, not *Sertorius* Stål, 1866 [Hemiptera]; syn. of *Arrox* Zang, 1905]. Comment: permanently invalid (Art. 39): based on preoccupied type genus; incorrect original stem formation, not in prevailing usage.

Unduliferinae Kuwert, 1891: 176 [stem: *Undulifer-*]. Type genus: *Undulifer* Kaup, 1869.

Veturinae Kuwert, 1891: 173 [stem: *Veturi-*]. Type genus: *Veturius* Kaup, 1871. Comment: incorrect original stem formation, not in prevailing usage.

Popiliinae Kuwert, 1896: 221 [stem: *Popili-*]. Type genus: *Popilius* Kaup, 1871.

PROCULEJINAE Kuwert, 1896: 221 [stem: *Proculej-*]. Type genus: *Proculejus* Kaup, 1868.

SPURIINAE Kuwert, 1896: 221 [stem: *Spuri-*]. Type genus: *Spurius* Kaup, 1871.

VINDICINAE Kuwert, 1896: 227 [stem: *Vindic-*]. Type genus: *Vindex* Kaup, 1871.

Tribe SOLENOCYCLINI Kaup, 1871

SOLENOCYCLEAE Kaup, 1871: 24 [stem: *Solenocycl-*]. Type genus: *Solenocyclus* Kaup, 1868. Comment: First Reviser found (SOLENOCYCLINI Kaup, 1871 vs STEPHANOCEPHALINI Kaup, 1871) is Boucher (2006).

STEPHANOCEPHALEAE Kaup, 1871: 78 [stem: *Stephanocephal-*]. Type genus: *Stephanocephalus* Kaup, 1868.

CICERONINAE Kuwert, 1891: 183 [stem: *Ciceroni-*]. Type genus: *Ciceronius* Kaup, 1871. Comment: incorrect original stem formation, not in prevailing usage.

ERIONOMINAE Kuwert, 1891: 176 [stem: *Erionom-*]. Type genus: *Erionomus* Kaup, 1868.

FLAMININAE Kuwert, 1891: 185 [stem: *Flamini-*]. Type genus: *Flaminius* Kuwert, 1891. Comment: incorrect original stem formation, not in prevailing usage.

SEMICYCLINAE Kuwert, 1891: 177 [stem: *Semicycl-*]. Type genus: *Semicyclus* Kaup, 1871.

Family TROGIDAE MacLeay, 1819

TROGIDAE W. S. MacLeay, 1819: 59 [stem: *Trog-*]. Type genus: *Trox* Fabricius, 1775.

†Subfamily AVITORTORINAE Nikolajev, 2007

AVITORTORINAE Nikolajev, 2007a: 110 [stem: *Avitortor-*]. Type genus: *Avitortor* Ponomarenko, 1977.

Subfamily TROGINAE MacLeay, 1819

TROGIDAE W. S. MacLeay, 1819: 59 [stem: *Trog-*]. Type genus: *Trox* Fabricius, 1775.

PHOBERIDAE Gistel, 1848: [5] [stem: *Phober-*]. Type genus: *Phoberus* W. S. MacLeay, 1819.

Subfamily OMORGINAE Nikolajev, 2005

OMORGINI Nikolajev, 2005a: 322 [stem: *Omorg-*]. Type genus: *Omorgus* Erichson, 1847.

Family GLARESIDAE Kolbe, 1905

GLARESINI Kolbe, 1905: 543 [stem: *Glares-*]. Type genus: *Glaresis* Erichson, 1848.

Family DIPHYLLOSTOMATIDAE Holloway, 1972

DIPHYLLOSTOMATIDAE Holloway, 1972: 38 [stem: *Diphyllostomat-*]. Type genus: *Diphyllostoma* Fall, 1901.

Family Lucanidae Latreille, 1804

Lucanides Latreille, 1804c: 149 [stem: *Lucan-*]. Type genus: *Lucanus* Scopoli, 1763.

†Subfamily Protolucaninae Nikolajev, 2007

Protolucaninae Nikolajev, 2007a: 18 [stem: *Protolucan-*]. Type genus: *Protolucanus* Nikolajev, 2007.

Subfamily Aesalinae MacLeay, 1819

Aesalidae W. S. MacLeay, 1819: 102 [stem: *Aesal-*]. Type genus: *Aesalus* Fabricius, 1801.

Tribe Aesalini MacLeay, 1819

Aesalidae W. S. MacLeay, 1819: 102 [stem: *Aesal-*]. Type genus: *Aesalus* Fabricius, 1801.

Tribe Ceratognathini Sharp, 1899

Ceratognathini Sharp, 1899a: 194 [stem: *Ceratognath-*]. Type genus: *Ceratognathus* Westwood, 1838.

Tribe Nicagini LeConte, 1861

Nicagini J. L. LeConte, 1861: 130 [stem: *Nicag-*]. Type genus: *Nicagus* J. L. LeConte, 1861.

†Subfamily Ceruchitinae Nikolajev, 2006

Ceruchitinae Nikolajev, 2006: 133 [stem: *Ceruchit-*]. Type genus: *Ceruchites* Statz, 1952.

Subfamily Syndesinae MacLeay, 1819

Syndesidae W. S. MacLeay, 1819: 103 [stem: *Syndes-*]. Type genus: *Syndesus* W. S. MacLeay, 1819.

Sinodendriens Mulsant, 1842: 600 [stem: *Sinodendr-*]. Type genus: *Sinodendron* Hellwig, 1791. Comment: original vernacular name available (Art. 11.7.2): first used in latinized form by Imhoff (1856: xii, as Sinodendridae), generally accepted as in Ratcliffe (2002: 8, as Sinodendrini).

Ceruchites Jacquelin du Val, 1859: 4 [stem: *Ceruch-*]. Type genus: *Ceruchus* W. S. MacLeay, 1819. Comment: original vernacular name available (Art. 11.7.2): first used in latinized form by J. L. LeConte (1861: 121, as Ceruchini), generally accepted as in Ratcliffe (2002: 8, as Ceruchini).

Subfamily Lampriminae MacLeay, 1819

Lamprimidae W. S. MacLeay, 1819: 97 [stem: *Lamprim-*]. Type genus: *Lamprima* Latreille, 1804.

Tribe Lamprimini MacLeay, 1819

Lamprimidae W. S. MacLeay, 1819: 97 [stem: *Lamprim-*]. Type genus: *Lamprima* Latreille, 1804.

Tribe Streptocerini Kikuta, 1986

Streptocerini Kikuta, 1986: 131 [stem: *Streptocer-*]. Type genus: *Streptocerus* Fairmaire, 1850.

Subfamily Lucaninae Latreille, 1804

Lucanides Latreille, 1804c: 149 [stem: *Lucan-*]. Type genus: *Lucanus* Scopoli, 1763.

Tribe Chiasognathini Burmeister, 1847

Chiasognathidae H. C. C. Burmeister, 1847: 334 [stem: *Chiasognath-*]. Type genus: *Chiasognathus* Stephens, 1831.

Tribe Lucanini Latreille, 1804

Lucanides Latreille, 1804c: 149 [stem: *Lucan-*]. Type genus: *Lucanus* Scopoli, 1763. Comment: published 7 March 1804; this family-group name was also used in the same year by Latreille (1804b [between 19 August and 17 September]: 234, as Lucanides).

Figulidae H. C. C. Burmeister, 1847: 428 [stem: *Figul-*]. Type genus: *Figulus* W. S. MacLeay, 1819.

Corypticidae Gistel, 1848: [5] [stem: *Coryptic-*]. Type genus: *Corypticus* Sturm, 1843 [the genus was spelled *Coryptius* (Sturm 1843: 136) and *Corypticus* (Sturm 1843: 347) in the paper making the genus name available but the latter spelling is in prevailing usage and is the correct spelling].

Dorcidae Parry, 1864: 86 [stem: *Dorc-*]. Type genus: *Dorcus* W. S. MacLeay, 1819.

Cladognathidae Parry, 1870: 75 [stem: *Cladognath-*]. Type genus: *Cladognathus* H. C. C. Burmeister, 1847 [syn. of *Prosopocoilus* Hope, 1845].

Odontolabidae Parry, 1870: 106 [stem: *Odontolabid-*]. Type genus: *Odontolabis* Hope, 1842. Comment: incorrect original stem formation, not in prevailing usage.

Nigidiini Jakobson, 1911b: 142 [stem: *Nigidi-*]. Type genus: *Nigidius* W. S. MacLeay, 1819. Comment: Nigidiini was attributed to Benesh (1960) and treated as unavailable by A. B. T. Smith (2006: 154).

Leptinopterini Jakobson, 1911b: 142 [stem: *Leptinopter-*]. Type genus: *Leptinopterus* Hope, 1838 [unjustified emendation of *Leptynopterus* by Hope (1845: 5), in prevailing usage and so deemed to be a justified emendation (Article 33.2.3.1)].

Rhaetulinae Miwa, 1931: 323 [stem: *Rhaetul-*]. Type genus: *Rhaetulus* Westwood, 1871.

Penichrolucaninae Arrow, 1950: 233 [stem: *Penichrolucan-*]. Type genus: *Penichrolucanus* Deyrolle, 1863.

*Chalcodinae Didier and Séguy, 1953: 91 [stem: *Chalcod-*]. Type genus: *Chalcodes* H. C. C. Burmeister, 1847. Comment: unavailable family-group name, proposed after 1930 without description or bibliographic reference to such a description (Art. 13.1).

Dendeziini Benesh, 1955b: 72 [stem: *Dendezi-*]. Type genus: *Dendezia* Basilewsky, 1952.

Lissotini Benesh, 1955b: 73 [stem: *Lissotet-*]. Type genus: *Lissotes* Westwood, 1855. Comment: incorrect original stem formation, not in prevailing usage.

Sclerostomini Benesh, 1955a: 97 [stem: *Sclerostom-*]. Type genus: *Sclerostomus* H. C. C. Burmeister, 1847.

Scortizini Benesh, 1955a: 103 [stem: *Scortiz-*]. Type genus: *Scortizus* Westwood, 1834.

*Prosopocoilini Benesh, 1960: 50 [stem: *Prosopocoil-*]. Type genus: *Prosopocoilus* Hope, 1845. Comment: unavailable family-group name, proposed after 1930 without description or bibliographic reference to such a description (Art. 13.1).

*Pseudodorcini Benesh, 1960: 97 [stem: *Pseudodorc-*]. Type genus: *Pseudodorcus* Parry, 1870. Comment: unavailable family-group name, proposed after 1930 without description or bibliographic reference to such a description (Art. 13.1).

*Rhyssonotini Benesh, 1960: 148 [stem: *Ryssonot-*]. Type genus: *Ryssonotus* W. S. MacLeay, 1819 [as *Rhyssonotus*, unjustified emendation of type genus name by Agassiz (1846b: 329), not in prevailing usage]. Comment: unavailable family-group name, proposed after 1930 without description or bibliographic reference to such a description (Art. 13.1); incorrect original stem formation, not in prevailing usage.

*Chalcodinae J. P. Lacroix, 1979: 258 [stem: *Chalcod-*]. Type genus: *Chalcodes* H. C. C. Burmeister, 1847. Comment: unavailable family-group name, proposed after 1930 without description or bibliographic reference to such a description (Art. 13.1).

Pholidotini Kikuta, 1986: 131 [stem: *Pholidot-*]. Type genus: *Pholidotus* W. S. MacLeay, 1819 [preoccupied genus name, not *Pholidotus* Brisson, 1762 [Mammalia]; syn. of *Casignetus* W. S. MacLeay, 1819]. Comment: permanently invalid (Art. 39): based on preoccupied type genus.

*Aegini Maes, 1992b: 97 [stem: *Aeg-*]. Type genus: *Aegus* W. S. MacLeay, 1819. Comment: unavailable family-group name, proposed after 1930 without description or bibliographic reference to such a description (Art. 13.1); Aegidae A. White, 1850 (type genus *Aega* Leach, 1815) is currently used as valid in Isopoda and therefore anyone wishing to base a new family-group name on the genus *Aegus* W. S. MacLeay should change the stem in order to avoid homonymy with the isopod name.

*Allotopini Maes, 1992a: 56 [stem: *Allotop-*]. Type genus: *Allotopus* Albers, 1894. Comment: unavailable family-group name, proposed after 1930 without description or bibliographic reference to such a description (Art. 13.1).

*Casignetini Maes, 1992b: 61 [stem: *Casignet-*]. Type genus: *Casignetus* W. S. MacLeay, 1819. Comment: unavailable family-group name, proposed after 1930 without description or bibliographic reference to such a description (Art. 13.1); Casignetini Falkovitsh, 1972 (type genus *Casigneta* Wallengren, 1881) proposed in Lepidoptera is permanently invalid since it is based on a preoccupied type genus name.

*Colophonini Maes, 1992b: 107 [stem: *Colophon-*]. Type genus: *Colophon* Gray, 1832. Comment: unavailable family-group name, proposed after 1930 without description or bibliographic reference to such a description (Art. 13.1).

*Cyclommatini Maes, 1992a: 56 [stem: *Cyclommat-*]. Type genus: *Cyclommatus* Parry, 1863. Comment: unavailable family-group name, proposed after 1930 without description or bibliographic reference to such a description (Art. 13.1).

*Homoderini Maes, 1992b: 85 [stem: *Homoder-*]. Type genus: *Homoderus* Parry, 1863. Comment: unavailable family-group name, proposed after 1930 without description or bibliographic reference to such a description (Art. 13.1).

Casignetini Reid, 1999: 175 [stem: *Casignet-*]. Type genus: *Casignetus* W. S. MacLeay, 1819. Comment: replacement name for Pholidotini Kikuta, 1986 because of the homonymy of the type genus.

Brasilucanini Nikolajev, 1999a: 171 [stem: *Brasilucan-*]. Type genus: *Brasilucanus* Vulcano and Pereira, 1961.

Tribe Platycerini Mulsant, 1842

Platycéraires Mulsant, 1842: 593 [stem: *Platycer-*]. Type genus: *Platycerus* Geoffroy, 1762 [placed on the Official List of Generic Names in Zoology (ICZN 1994a)]. Comment: original vernacular name available (Art. 11.7.2): first used in latinized form by Gistel (1856a: 365, as Platyceridae), generally accepted as in Paulsen and Hawks (2008: 1, as Platycerini).

Systenocerini Portevin, 1931: 2, in key [stem: *Systenocer-*]. Type genus: *Systenocerus* Weise, 1883 [syn. of *Platycerus* Geoffroy, 1762].

Tribe Platyceroidini Paulsen and Hawks, 2008

Platyceroidini Paulsen and Hawks, 2008: 2 [stem: *Platyceroid-*]. Type genus: *Platyceroides* Benesh, 1946.

†Subfamily Paralucaninae Nikolajev, 2000

Paralucaninae Nikolajev, 2000b: S328 [stem: *Paralucan-*]. Type genus: *Paralucanus* Nikolajev, 2000.

Family Ochodaeidae Mulsant and Rey, 1871

Ochodéens Mulsant and Rey, 1871b: 493 [stem: *Ochodae-*]. Type genus: *Ochodaeus* Dejean, 1821.

†Subfamily CRETOCHODAEINAE Nikolajev, 1995

CRETOCHODAEINI Nikolajev, 1995a: 78 [stem: *Cretochodae-*]. Type genus: *Cretochodaeus* Nikolajev, 1995.

Subfamily OCHODAEINAE Mulsant and Rey, 1871

OCHODÉENS Mulsant and Rey, 1871b: 493 [stem: *Ochodae-*]. Type genus: *Ochodaeus* Dejean, 1821.

Tribe ENODOGNATHINI Scholtz, 1988

ENDOGNATHINI Scholtz, 1988: 228 [stem: *Enodognath-*]. Type genus: *Enodognathus* Benderitter, 1920 [as *Endognathus*, incorrect subsequent spelling of type genus name, not in prevailing usage]. Comment: incorrect original stem formation, not in prevailing usage; correction of stem by Scholtz (1991: 30).

Tribe OCHODAEINI Mulsant and Rey, 1871

OCHODÉENS Mulsant and Rey, 1871b: 493 [stem: *Ochodae-*]. Type genus: *Ochodaeus* Dejean, 1821. Comment: original vernacular name available (Art. 11.7.2): first used in latinized form by Arrow (1904: 747, as OCHODAEINAE), generally accepted as in A. B. T. Smith (2006: 155, as OCHODAEIDAE); incorrect original stem formation, not in prevailing usage.

Subfamily CHAETOCANTHINAE Scholtz, 1988

CHAETOCANTHINAE Scholtz, 1988: 231 [stem: *Chaetocanth-*]. Type genus: *Chaetocanthus* Péringuey, 1901. Comment: precedence (CHAETOCANTHINAE Scholtz, 1988 vs PSEUDOCHODAEINAE Scholtz, 1988 vs SYNOCHODAEINAE Scholtz, 1988) given to taxon originally proposed at the higher rank (Art. 24.1).

Tribe CHAETOCANTHINI Scholtz, 1988

CHAETOCANTHINAE Scholtz, 1988: 231 [stem: *Chaetocanth-*]. Type genus: *Chaetocanthus* Péringuey, 1901.

Tribe PSEUDOCHODAEINI Scholtz, 1988

PSEUDOCHODAEINI Scholtz, 1988: 235 [stem: *Pseudochodae-*]. Type genus: *Pseudochodaeus* Carlson and Ritcher, 1974.

Tribe SYNOCHODAEINI Scholtz, 1988

SYNOCHODAEINI Scholtz, 1988: 237 [stem: *Synochodae-*]. Type genus: *Synochodaeus* Kolbe, 1907.

Family HYBOSORIDAE Erichson, 1847

HYBOSORIDAE Erichson, 1847a: 104 [stem: *Hybosor-*]. Type genus: *Hybosorus* W. S. MacLeay, 1819.

†Subfamily Mimaphodiinae Nikolajev, 2007

Mimaphodiinae Nikolajev, 2007b: 47 [stem: *Mimaphodi-*]. Type genus: *Mimaphodius* Nikolajev, 2007.

Subfamily Anaidinae Nikolajev, 1996

Anaidini Nikolajev, 1996: 94 [stem: *Anaid-*]. Type genus: *Anaides* Westwood, 1842.

Cryptogeniini Howden, 2001: 199 [stem: *Cryptogeni-*]. Type genus: *Cryptogenius* Westwood, 1842.

Subfamily Ceratocanthinae Martínez, 1968

Ceratocanthini Martínez, 1968: 14 [stem: *Ceratocanth-*]. Type genus: *Ceratocanthus* A. White, 1842.

Tribe Ceratocanthini Martínez, 1968

Acanthocérides Lacordaire, 1856: 155 [stem: *Acanthocer-*]. Type genus: *Acanthocerus* W. S. MacLeay, 1819 [preoccupied genus name, not *Acanthocerus* Palisot de Beauvois, 1818 [Hemiptera]; syn. of *Ceratocanthus* A. White, 1842]. Comment: original vernacular name available (Art. 11.7.2): first used in latinized form and generally accepted as in J. L. LeConte (1861: 129, as Acanthocerini); permanently invalid (Art. 39): based on preoccupied type genus.

Ceratocanthini Martínez, 1968: 14 [stem: *Ceratocanth-*]. Type genus: *Ceratocanthus* A. White, 1842. Comment: replacement name for Acanthocerini Lacordaire, 1856 because of the homonymy of the type genus.

Tribe Ivieolini Howden and Gill, 2000

Ivieolini Howden and Gill, 2000: 315 [stem: *Ivieol-*]. Type genus: *Ivieolus* Howden and Gill, 1988.

Tribe Scarabatermitini Nikolajev, 1999

Scarabatermitini Nikolajev, 1999b: 175 [stem: *Scarabatermit-*]. Type genus: *Scarabatermes* Howden, 1973.

Subfamily Hybosorinae Erichson, 1847

Hybosoridae Erichson, 1847a: 104 [stem: *Hybosor-*]. Type genus: *Hybosorus* W. S. MacLeay, 1819.

Subfamily Liparochrinae Ocampo, 2006

Liparochrinae Ocampo, 2006: 29 [stem: *Liparochr-*]. Type genus: *Liparochrus* Erichson, 1848.

Subfamily Pachyplectrinae Ocampo, 2006

Pachyplectrinae Ocampo, 2006: 30 [stem: *Pachyplectr-*]. Type genus: *Pachyplectrus* J. L. LeConte, 1874.

Family Glaphyridae MacLeay, 1819

Glaphyridae W. S. MacLeay, 1819: 76 [stem: *Glaphyr-*]. Type genus: *Glaphyrus* Latreille, 1802.

Subfamily Glaphyrinae MacLeay, 1819

Glaphyridae W. S. MacLeay, 1819: 76 [stem: *Glaphyr-*]. Type genus: *Glaphyrus* Latreille, 1802.

Subfamily Amphicominae Blanchard, 1845

Amphicomites Blanchard, 1845a: 211 [stem: *Amphicom-*]. Type genus: *Amphicoma* Latreille, 1807. Comment: original vernacular name available (Art. 11.7.2): first used in latinized form and generally accepted as in Blanchard (1850: 52, as Amphicomitae); name treated as unavailable by A. B. T. Smith (2006: 157).

†Subfamily Cretoglaphyrinae Nikolajev, 2005

Cretoglaphyrini Nikolajev, 2005b: 70 [stem: *Cretoglaphyr-*]. Type genus: *Cretoglaphyrus* Nikolajev, 2005.

Family Scarabaeidae Latreille, 1802

Scarabaeïdes Latreille, 1802: 144 [stem: *Scarabae-*]. Type genus: *Scarabaeus* Linnaeus, 1758.

†Subfamily Lithoscarabaeinae Nikolajev, 1992

Lithoscarabaeinae Nikolajev, 1992: 76 [stem: *Lithoscarabae-*]. Type genus: *Lithoscarabaeus* Nikolajev, 1992.

Subfamily Chironinae Blanchard, 1845

Chironites Blanchard, 1845a: 225 [stem: *Chiron-*]. Type genus: *Chiron* W. S. MacLeay, 1819. Comment: original vernacular name available (Art. 11.7.2): first used in latinized form by Harold (1868: 278, as Chironidae), generally accepted as in A. B. T. Smith (2006: 157, as Chironinae).

Subfamily Aegialiinae Laporte, 1840

Aegialites Laporte, 1840b: 99 [stem: *Aegiali-*]. Type genus: *Aegialia* Latreille, 1807. Comment: original vernacular name available (Art. 11.7.2): first used in latinized form by Harold (1868: 278, as Aegialidae [incorrect stem formation]), generally accepted as in A. B. T. Smith (2006: 157, as Aegialiinae); incorrect original stem formation, not in prevailing usage.

Silluviinae Landin, 1950: 3 [stem: *Silluvi-*]. Type genus: *Silluvia* Landin, 1950.

Saprini Nikolajev, 2008: 149 [stem: *Sapr-*]. Type genus: *Saprus* Blackburn, 1904 [preoccupied genus name, not *Saprus* Gistel, 1856 [Coleoptera: Sphaeriusidae]; syn. of *Sapriniana* Strand, 1917]. Comment: permanently invalid (Art. 39): based on preoccupied type genus.

Subfamily Eremazinae Iablokoff-Khnzorian, 1977

Eremazini Iablokoff-Khnzorian, 1977 [3 October]: 168 [stem: *Eremaz-*]. Type genus: *Eremazus* Mulsant, 1851.

Eremazina Stebnicka, 1977 ["31 December"]: 412 [stem: *Eremaz-*]. Type genus: *Eremazus* Mulsant, 1851. Comment: proposed as new, without reference to Eremazini Iablokoff-Khnzorian, 1977.

Subfamily Aphodiinae Leach, 1815

Aphodida Leach, 1815: 97 [stem: *Aphodi-*]. Type genus: *Aphodius* Illiger, 1798.

Tribe Aphodiini Leach, 1815

Aphodida Leach, 1815: 97 [stem: *Aphodi-*]. Type genus: *Aphodius* Illiger, 1798.

Subtribe Aphodiina Leach, 1815

Aphodida Leach, 1815: 97 [stem: *Aphodi-*]. Type genus: *Aphodius* Illiger, 1798.

*Ammoeciates Mulsant, 1842: 301 [stem: *Ammoeci-*]. Type genus: *Ammoecius* Mulsant, 1842. Comment: original vernacular name unavailable (Art. 11.7.2): not subsequently latinized.

*Plagiogonates Mulsant and Rey, 1871a: 609 [stem: *Plagiogon-*]. Type genus: *Plagiogonus* Mulsant, 1842. Comment: original vernacular name unavailable (Art. 11.7.2): not subsequently latinized.

*Heptaulacates Mulsant and Rey, 1871a: 585 [stem: *Heptaulac-*]. Type genus: *Heptaulacus* Mulsant, 1842. Comment: original vernacular name unavailable (Art. 11.7.2): not subsequently latinized.

Subtribe Didactyliina Pittino, 1985

Didactyliina Pittino, 1985: 270, in key [stem: *Didactyli-*]. Type genus: *Didactylia* d'Orbigny, 1896.

Subtribe Proctophanina Stebnicka and Howden, 1995

Proctophanini Stebnicka and Howden, 1995: 742 [stem: *Proctophan-*]. Type genus: *Proctophanes* Harold, 1861.

Tribe Corythoderini Schmidt, 1910

Corythoderina A. Schmidt, 1910a: 137 [stem: *Corythoder-*]. Type genus: *Corythoderus* Klug, 1845. Comment: the earliest known publication date of this work is 31 December 1910 (see Evenhuis, 1994: 55); although Corythoderina A. Schmidt (1910b: 93) was published on 30 September 1910, we use A. Schmidt, 1910a as the correct original work since "Corythoderina Schmidt 1910: 137" was used as the correct citation for this name by A. Schmidt himself (1910b: 93).

Tribe Eupariini Schmidt, 1910 *nomen protectum*

Ataenidae Harold, 1868: 278 [stem: *Ataeni-*]. Type genus: *Ataenius* Harold, 1867 [placed on the Official List of Generic Names in Zoology (ICZN 2010a)]. Comment: *nomen oblitum* (see A. B. T. Smith 2006: 158).

*Hexalates Mulsant and Rey, 1871a: 605 [stem: *Hexal-*]. Type genus: *Hexalus* Mulsant and Rey, 1871 [syn. of *Ataenius* Harold, 1867]. Comment: original vernacular name unavailable (Art. 11.7.2): not subsequently latinized.

Eupariina A. Schmidt, 1910a: 102 [stem: *Eupari-*]. Type genus: *Euparia* Lepeletier and Audinet-Serville, 1828. Comment: *nomen protectum* (see A. B. T. Smith 2006: 158); senior homonym of Eupariini B. D. Valentine, 1960 (type genus *Euparius* Schönherr, 1823) in Anthribidae; this case is to be referred to the Commission to remove the homonymy (Art. 55.3.1); the earliest known publication date of this work is 31 December 1910 (see Evenhuis, 1994: 55); although Eupariina A. Schmidt (1910b: 71) was published on 30 September 1910, we use A. Schmidt, 1910a as the correct original work since "Eupariina Schmidt 1910: 102" was used as the correct citation for this name by A. Schmidt himself (1910b: 71).

Lomanoxiini Stebnicka, 1999: 280 [stem: *Lomanoxi-*]. Type genus: *Lomanoxia* Martínez, 1951.

Tribe Odontolochini Stebnicka and Howden, 1996

Odontolochini Stebnicka and Howden, 1996: 99 [stem: *Odontoloch-*]. Type genus: *Odontolochus* Schmidt, 1916.

Tribe Odochilini Rakovič, 1987

Odochilini Rakovič, 1987: 29 [stem: *Odochil-*]. Type genus: *Odochilus* Harold, 1877.

Tribe Psammodiini Mulsant, 1842

Psammodiaires Mulsant, 1842: 317 [stem: *Psammodi-*]. Type genus: *Psammodius* Fallén, 1807.

Subtribe Phycocina Landin, 1960

Phycochi Landin, 1960: 59 [stem: *Phycoc-*]. Type genus: *Phycocus* Broun, 1886 [as *Phycochus*, incorrect subsequent spelling of type genus name, not in prevailing usage (see Stebnicka 2001)]. Comment: incorrect original stem formation, not in prevailing usage.

Phycochini Rakovič and Král, 1997: 59 [stem: *Phycoc-*]. Type genus: *Phycocus* Broun, 1886 [as *Phycochus*, incorrect subsequent spelling of type genus name, not in prevailing usage (see Stebnicka 2001)]. Comment: incorrect original stem formation, not in prevailing usage; family-group name proposed as new without reference to Phycochi Landin, 1960.

Subtribe PSAMMODIINA Mulsant, 1842

PSAMMODIAIRES Mulsant, 1842: 317 [stem: *Psammodi-*]. Type genus: *Psammodius* Fallén, 1807. Comment: original vernacular name available (Art. 11.7.2): first used in latinized form by Harold (1868: 278, as PSAMMODIDAE [incorrect stem formation]), generally accepted as in A. B. T. Smith (2006: 159, as PSAMMODIINI).

PLEUROPHORATES Mulsant, 1842: 304 [stem: *Pleurophor-*]. Type genus: *Pleurophorus* Mulsant, 1842. Comment: original vernacular name available (Art. 11.7.2): first used in latinized form and generally accepted as in Ádám (1994: 15, as PLEUROPHORINI).

PSAMMOBIINA Reitter, 1909: 303 [stem: *Psammobi-*]. Type genus: *Psammobius* Heer, 1841 [syn. of *Psammodius* Fallén, 1807].

Subtribe RHYSSEMINA Pittino and Mariani, 1986

RHYSSEMINA Pittino and Mariani, 1986: 17, in key [stem: *Rhyssem-*]. Type genus: *Rhyssemus* Mulsant, 1842.

Tribe RHYPARINI Schmidt, 1910

RHYPARINA A. Schmidt, 1910a: 130 [stem: *Rhypar-*]. Type genus: *Rhyparus* Westwood, 1845 [*Rhyparus* is an unjustified emendation of *Ryparus* Westwood, 1845 by Agassiz (1846b: 328), in prevailing usage, and so deemed to be a justified emendation (Article 33.2.3.1); the emended spelling avoids homonymy with *Ryparus* Spinola, 1844 [Coleoptera: CLERIDAE] (see A. B. T. Smith 2006: 159)]. Comment: the earliest known publication date of this work is 31 December 1910 (see Evenhuis, 1994: 55); although RHYPARINA A. Schmidt (1910b: 91) was published on 30 September 1910, we use A. Schmidt, 1910a as the correct original work since "RHYPARINA Schmidt 1910: 130" was used as the correct citation for this name by A. Schmidt himself (1910b: 91).

Tribe STEREOMERINI Howden and Storey, 1992

STEREOMERINI Howden and Storey, 1992: 1811 [stem: *Stereomer-*]. Type genus: *Stereomera* Arrow, 1905.

Tribe TERMITODERINI Tangelder and Krikken, 1982

TERMITODERINI Tangelder and Krikken, 1982: 10, in key [stem: *Termitoder-*]. Type genus: *Termitoderus* Mateu, 1966.

Subfamily AULONOCNEMINAE Janssens, 1946

AULONOCNEMINAE Janssens, 1946: 7, in key [stem: *Aulonocnem-*]. Type genus: *Aulonocnemis* Klug, 1837. Comment: current spelling maintained (Art. 29.3.1.1): incorrect stem formation in prevailing usage (should be *Aulonocnemid-*).

Subfamily Termitotroginae Wasmann, 1918

Termitotrogini Wasmann, 1918: 4 [stem: *Termitotrog-*]. Type genus: *Termitotrox* Reichensperger, 1915.

Subfamily Scarabaeinae Latreille, 1802

Scarabaeïdes Latreille, 1802: 144 [stem: *Scarabae-*]. Type genus: *Scarabaeus* Linnaeus, 1758.

Tribe Ateuchini Perty, 1830

Ateuchidae Perty, 1830: 38 [stem: *Ateuch-*]. Type genus: *Ateuchus* Weber, 1801.

Subtribe Ateuchina Perty, 1830

Ateuchidae Perty, 1830: 38 [stem: *Ateuch-*]. Type genus: *Ateuchus* Weber, 1801. Comment: name previously attributed to Laporte (1840b: 63); use of subtribes follow Vaz-de-Mello (2008).

Choerididae Harold, 1867: 9 [stem: *Choeridi-*]. Type genus: *Choeridium* Lepeletier and Audinet-Serville, 1828.

Pinotinae Kolbe, 1905: 548 [stem: *Pinot-*]. Type genus: *Pinotus* Erichson, 1847.

Dichotomiini Pereira, 1954: 55 [stem: *Dichotomi-*]. Type genus: *Dichotomius* Hope, 1838.

Subtribe Scatimina Vaz-de-Mello, 2008

Scatimina Vaz-de-Mello, 2008: 10 [stem: *Scatim-*]. Type genus: *Scatimus* Erichson, 1847.

Ateuchini *incertae sedis*

Demarziellini Balthasar, 1961: 178 [stem: *Demarziell-*]. Type genus: *Demarziella* Balthasar, 1961. Comment: family-group taxon not originally described but available (Art. 13.5) (see Matthews and Stebnicka 1986); current placement follows Vaz-de-Mello (2008).

Tribe Coprini Leach, 1815

*Coprides Baudet-Lafarge, 1809: 44 [stem: *Copr-*]. Type genus: *Copris* Geoffroy, 1762 [placed on the Official List of Generic Names in Zoology (ICZN 1994a)]. Comment: original vernacular name unavailable (Art. 11.7.2): subsequently used in latinized form but not generally attributed to Baudet-Lafarge (1809).

Coprides Leach, 1815: 96 [stem: *Copr-*]. Type genus: *Copris* Geoffroy, 1762 [placed on the Official List of Generic Names in Zoology (ICZN 1994a)]. Comment: current spelling maintained (Art. 29.3.1.1): incorrect stem formation in prevailing usage (should be *Coprid-*).

Coptodactylini Paulian, 1933: 67 [stem: *Coptodactyl-*]. Type genus: *Coptodactyla* H. C. C. Burmeister, 1846.

Tribe Deltochilini Lacordaire, 1856

Deltochilides Lacordaire, 1856: 78 [stem: *Deltochil-*]. Type genus: *Deltochilum* Eschscholtz, 1822. Comment: original vernacular name available (Art. 11.7.2): first used in latinized form by J. L. LeConte (1861: 125, as Deltochila), generally accepted as in Germain (1903: 354, as Deltochilidae); First Reviser (Deltochilini Lacordaire, 1856 vs Mentophilini Lacordaire, 1856 vs Scatonomini Lacordaire, 1856) not determined, current usage maintained.

Minthophilides Lacordaire, 1856: 80 [stem: *Mentophil-*]. Type genus: *Mentophilus* Laporte, 1840 [as *Minthophilus*, both *Mentophilus* (p. 74) and *Minthophilus* (p. 63) were used in the original publication by Laporte (1840b), although *Mentophilus* has been used as the correct spelling in recent literature, we could not determine the First Reviser]. Comment: original vernacular name available (Art. 11.7.2): first used in latinized form and generally accepted as in Gerstaecker (1861: 464, as Minthophilidae); name treated as unavailable by A. B. T. Smith (2006: 160); incorrect original stem formation, not in prevailing usage.

Scatonomides Lacordaire, 1856: 87 [stem: *Scatonom-*]. Type genus: *Scatonomus* Erichson, 1835. Comment: original vernacular name available (Art. 11.7.2): first used in latinized form and generally accepted as in J. L. LeConte (1861: 125, as Scatonomi); transferred from Ateuchini by Vaz-de-Mello (2008).

Coprobiadae H. C. C. Burmeister, 1873: 407 [stem: *Coprobi-*]. Type genus: *Coprobius* Latreille, 1829.

Canthonides van Lansberge, 1875a: 184 [stem: *Canthon-*]. Type genus: *Canthon* Hoffmannsegg, 1817. Comment: original vernacular name available (Art. 11.7.2): first used in latinized form by Kolbe (1905: 530, as Canthoninae), generally accepted as in A. B. T. Smith (2006: 160, as Canthonini).

Epilissides van Lansberge, 1875a: 188 [stem: *Epiliss-*]. Type genus: *Epilissus* Dejean, 1836. Comment: original vernacular name available (Art. 11.7.2): first used in latinized form and generally accepted as in Shipp (1894: 255, as Epilissini).

Epirinides van Lansberge, 1875a: 189 [stem: *Epirin-*]. Type genus: *Epirinus* Dejean, 1833. Comment: original vernacular name available (Art. 11.7.2): first used in latinized form and generally accepted as in Bertkau (1875: 335, as Epirini); name treated as unavailable by A. B. T. Smith (2006: 160).

Panelini Arrow, 1931: 404 [stem: *Panel-*]. Type genus: *Panelus* Lewis, 1895.

Tribe Eucraniini Burmeister, 1873

Eucraniadae H. C. C. Burmeister, 1873: 405 [stem: *Eucrani-*]. Type genus: *Eucranium* Brullé, 1837.

Ennearabdini Pereira and Martínez, 1956: 238 [stem: *Ennearabd-*]. Type genus: *Ennearabdus* van Lansberge, 1874.

Tribe Gymnopleurini Lacordaire, 1856

Gymnopleurides Lacordaire, 1856: 72 [stem: *Gymnopleur-*]. Type genus: *Gymnopleurus* Illiger, 1803. Comment: original vernacular name available (Art.

11.7.2): first used in latinized form by J. L. LeConte (1861: 124, as Gymnopleuri), generally accepted as in A. B. T. Smith (2006: 160, as Gymnopleurini).

Tribe Oniticellini Kolbe, 1905

Oniticellini Kolbe, 1905: 547 [stem: *Oniticell-*]. Type genus: *Oniticellus* Dejean, 1821. Comment: usage of this name over Drepanocerini conserved (Art. 35.5) (see A. B. T. Smith 2006).

Subtribe Drepanocerina van Lansberge, 1875

Drèpanocérides van Lansberge, 1875b: 14 [stem: *Drepanocer-*]. Type genus: *Drepanocerus* Kirby, 1828. Comment: original vernacular name available (Art. 11.7.2): first used in latinized form by Kolbe (1905: 531, as Drepanocerini), generally accepted as in A. B. T. Smith (2006: 161, as Drepanocerini); although Drepanocerini has priority over Oniticellini, the latter is in prevailing usage at the tribal level and must not be displaced by the older name (Art. 35.5) (see A. B. T. Smith 2006).

Subtribe Eurysternina Vulcano, Martínez and Pereira, 1961

Eurysternini Vulcano et al., 1961: 268 [stem: *Eurystern-*]. Type genus: *Eurysternus* Dalman, 1824. Comment: current placement based on Génier (2009).

Subtribe Helictopleurina Janssens, 1946

Helictopleurides Janssens, 1946: 11, in key [stem: *Helictopleur-*]. Type genus: *Helictopleurus* d'Orbigny, 1915.

Subtribe Oniticellina Kolbe, 1905

Oniticellini Kolbe, 1905: 547 [stem: *Oniticell-*]. Type genus: *Oniticellus* Dejean, 1821.

Tribe Onitini Laporte, 1840

Onitides Laporte, 1840b: 88 [stem: *Onit-*]. Type genus: *Onitis* Fabricius, 1798. Comment: original vernacular name available (Art. 11.7.2): first used in latinized form by Shipp (1894: 255, as Onitidae), generally accepted as in A. B. T. Smith (2006: 161, as Onitini); although Fabricius originally treated the type genus name as masculine, it appears in Greek dictionaries as a feminine noun (with stem *Onitid-*); we recommend that an application be submitted to the Commission to establish the correct gender of *Onitis* and the correct stem based on that genus.

Tribe Onthophagini Burmeister, 1846

Onthophagidae H. C. C. Burmeister, 1846: [1] [stem: *Onthophag-*]. Type genus: *Onthophagus* Latreille, 1802.

Alloscelides Janssens, 1946: 10, in key [stem: *Alloscel-*]. Type genus: *Alloscelus* Boucomont, 1923.

Tribe Phanaeini Hope, 1838

Phanaeidae Hope, 1838a: 321 [stem: *Phanae-*]. Type genus: *Phanaeus* W. S. MacLeay, 1819.

Gromphina Zunino, 1985a: 22 [stem: *Gromphad-*]. Type genus: *Gromphas* Brullé, 1837. Comment: published 20 November 1985; this family-group name was also used in the same year by Zunino (1985b [20 December]: 107, as Gromphina); incorrect original stem formation, not in prevailing usage.

Tribe Scarabaeini Latreille, 1802

Scarabaeïdes Latreille, 1802: 144 [stem: *Scarabae-*]. Type genus: *Scarabaeus* Linnaeus, 1758.

Pachysomides Ferreira, 1953: 5, in key [stem: *Pachysomat-*]. Type genus: *Pachysoma* W. S. MacLeay, 1821. Comment: incorrect original stem formation, not in prevailing usage.

Actinophorini Ádám, 2003: 130 [stem: *Actinophor-*]. Type genus: *Actinophorus* Creutzer, 1799 [syn. of *Scarabaeus* Linnaeus, 1758].

Tribe Sisyphini Mulsant, 1842

Sisyphaires Mulsant, 1842: 41 [stem: *Sisyph-*]. Type genus: *Sisyphus* Latreille, 1807 [the original spelling of the type genus is *Sisyphe*, however, the incorrect subsequent spelling *Sisyphus* is in prevailing usage and should be considered the correct original spelling (Art. 33.3.1) (see A. B. T. Smith 2006)]. Comment: original vernacular name available (Art. 11.7.2): first used in latinized form by Shipp (1894: 255, as Sisyphinae), generally accepted as in A. B. T. Smith (2006: 162, as Sisyphini).

†Subfamily Prototroginae Nikolajev, 2000

Prototroginae Nikolajev, 2000a: 63 [stem: *Prototrog-*]. Type genus: *Prototrox* Nikolajev, 2000.

†Subfamily Cretoscarabaeinae Nikolajev, 1995

Cretoscarabaeinae Nikolajev, 1995b: 147 [stem: *Cretoscarabae-*]. Type genus: *Cretoscarabaeus* Nikolajev, 1995.

Subfamily Dynamopodinae Arrow, 1911

Dynamopinae Arrow, 1911: 611 [stem: *Dynamopod-*]. Type genus: *Dynamopus* Semenov, 1895.

Tribe Dynamopodini Arrow, 1911

Dynamopinae Arrow, 1911: 611 [stem: *Dynamopod-*]. Type genus: *Dynamopus* Semenov, 1895. Comment: incorrect original stem formation, not in prevailing usage; correction of stem by Lawrence and Newton (1995).

Tribe Thinorycterini Semenov and A. N. Reichardt, 1925

Thinorycterina Semenov and A. N. Reichardt, 1925: 86 [stem: *Thinorycter-*]. Type genus: *Thinorycter* Semenov and A. N. Reichardt, 1925.

Subfamily Phaenomeridinae Erichson, 1847

Phaenomerini Erichson, 1847b: 655 [stem: *Phaenomerid-*]. Type genus: *Phaenomeris* Hope, 1833 [placed on the Official List of Generic Names in Zoology (ICZN 1962)]. Comment: stem emended from *Phaenomer-* to *Phaenomerid-* thus removing it from homonymy with another family-group name in Curculionidae and placed on the Official List of Family-Group Names in Zoology (ICZN 1962, as Phaenomerididae Ohaus, 1913); in this ruling (ICZN 1962) "Ohaus, 1913" was erroneously given as the original author of this family-group name (see A. B. T. Smith 2006).

Subfamily Orphninae Erichson, 1847

Orphnidae Erichson, 1847a: 111 [stem: *Orphn-*]. Type genus: *Orphnus* W. S. MacLeay, 1819.

Tribe Aegidiini Paulian, 1984

Aegidiinae Paulian, 1984: 68 [stem: *Aegidi-*]. Type genus: *Aegidium* Westwood, 1845.

Tribe Orphnini Erichson, 1847

Orphnidae Erichson, 1847a: 111 [stem: *Orphn-*]. Type genus: *Orphnus* W. S. MacLeay, 1819.

Hybalidae Marseul, 1857a: 83 [stem: *Hybal-*]. Type genus: *Hybalus* Dejean, 1833. Comment: name previously attributed to Jacquelin du Val (1859: 31, as Hybalites) and treated as unavailable by A. B. T. Smith (2006: 163).

Subfamily Allidiostomatinae Arrow, 1940

Idiostominae Arrow, 1904: 747 [stem: *Idiostomat-*]. Type genus: *Idiostoma* Arrow, 1904 [preoccupied genus name, not *Idiostoma* Walsingham, 1882 [Lepidoptera]; syn. of *Allidiostoma* Arrow, 1940]. Comment: permanently invalid (Art. 39): based on preoccupied type genus; incorrect original stem formation, not in prevailing usage.

Allidiostomidae Arrow, 1940: 16 [stem: *Allidiostomat-*]. Type genus: *Allidiostoma* Arrow, 1940. Comment: replacement name for Idiostominae Arrow, 1904 because of the homonymy of the type genus; incorrect original stem formation, not in prevailing usage; correction of stem by Lawrence and Newton (1995).

Subfamily ACLOPINAE Blanchard, 1850

ACLOPITAE Blanchard, 1850: 96 [stem: *Aclop-*]. Type genus: *Aclopus* Erichson, 1835.

Tribe ACLOPINI Blanchard, 1850

ACLOPITAE Blanchard, 1850: 96 [stem: *Aclop-*]. Type genus: *Aclopus* Erichson, 1835.

†Tribe HOLCOROBEINI Nikolajev, 1992

HOLCOROBEINI Nikolajev, 1992: 81 [stem: *Holcorobe-*]. Type genus: *Holcorobeus* Nikritin, 1977.

Tribe PHAENOGNATHINI Iablokoff-Khnzorian, 1977

PHAENOGNATHINI Iablokoff-Khnzorian, 1977: 137, 172 [stem: *Phaenognath-*]. Type genus: *Phaenognatha* Hope, 1842.

Subfamily MELOLONTHINAE Leach, 1819

MELOLONTHIDAE Leach, 1819: 189 [stem: *Melolonth-*]. Type genus: *Melolontha* Fabricius, 1775 [placed on the Official List of Generic Names in Zoology (ICZN 1994a)].

Tribe ABLABERINI Blanchard, 1850

ABLABERITAE Blanchard, 1850: 100 [stem: *Ablaber-*]. Type genus: *Ablabera* Dejean, 1833.

CAMENTINI Machatschke, 1959: 738 [stem: *Cament-*]. Type genus: *Camenta* Erichson, 1847.

Tribe AUTOMOLIINI Britton, 1978

CAULOBIINA H. C. C. Burmeister, 1855: 204 [stem: *Caulobi-*]. Type genus: *Caulobius* Le Guillou, 1844 [preoccupied genus name, not *Caulobius* Duponchel, 1838 [Lepidoptera]; syn. of *Deuterocaulobius* Dalla Torre, 1912]. Comment: permanently invalid (Art. 39): based on preoccupied type genus.

AUTOMOLINI Britton, 1957: 72 [stem: *Automol-*]. Type genus: *Automolus* H. C. C. Burmeister, 1855 [preoccupied genus name, not *Automolus* Reichenbach, 1853 [Aves]; syn. of *Automolius* Britton, 1978]. Comment: permanently invalid (Art. 39): based on preoccupied type genus.

AUTOMOLIINI Britton, 1978: 7 [stem: *Automoli-*]. Type genus: *Automolius* Britton, 1978. Comment: replacement name for AUTOMOLINI Britton, 1957 because of the homonymy of the type genus.

Tribe Chasmatopterini Lacordaire, 1856

Chasmatoptérides Lacordaire, 1856: 220 [stem: *Chasmatopter-*]. Type genus: *Chasmatopterus* Dejean, 1821. Comment: original vernacular name available (Art. 11.7.2): first used in latinized form by H. W. Bates (1887: 130, as Chasmatopteridae), generally accepted as in A. B. T. Smith (2006: 164, as Chasmatopterini).

Tribe Colymbomorphini Blanchard, 1850

Colymbomorphitae Blanchard, 1850: 97 [stem: *Colymbomorph-*]. Type genus: *Colymbomorpha* Blanchard, 1850.

Stethaspididae H. C. C. Burmeister, 1855: 218 [stem: *Stethaspid-*]. Type genus: *Stethaspis* Hope, 1837 [placed on the Official List of Generic Names in Zoology (ICZN 1983b)]. Comment: A. B. T. Smith (2006) used "Stethaspini" as the valid name for this tribe, however, the fact that the genus *Colymbomorpha* is considered to be in this tribe, e.g., Houston and Weir (1992), brings the name Colymbomorphini into synonymy with Stethaspini and Xylonichini; Colymbomorphini has nomenclatural priority and should be considered the valid name for this tribe.

Xylonychini Britton, 1957: 9 [stem: *Xylonich-*]. Type genus: *Xylonichus* Boisduval, 1835 [as *Xylonychus*, incorrect subsequent spelling of type genus name, not in prevailing usage (see A. B. T. Smith 2006: 166)]. Comment: incorrect original stem formation, not in prevailing usage.

Tribe Comophorinini Britton, 1957

Comophorini Britton, 1957: 10 [stem: *Comophorin-*]. Type genus: *Comophorina* Strand, 1928. Comment: incorrect original stem formation, not in prevailing usage.

†Tribe Cretomelolonthini Nikolajev, 1998

Cretomelolonthini Nikolajev, 1998: 80 [stem: *Cretomelolonth-*]. Type genus: *Cretomelolontha* Nikolajev, 1998.

Tribe Dichelonychini Burmeister, 1855

Dichelonychidae H. C. C. Burmeister, 1855: 70 [stem: *Dichelonych-*]. Type genus: *Dichelonyx* Harris, 1827.

Tribe Diphucephalini Laporte, 1840

Diphucéphalites Laporte, 1840b: 145 [stem: *Diphucephal-*]. Type genus: *Diphucephala* Dejean, 1821. Comment: original vernacular name available (Art. 11.7.2): first used in latinized form by Agassiz (1846b: 125, as Diphycocephaloidae [incorrect stem formation]), generally accepted as in A. B. T. Smith (2006: 165, as Diphucephalini).

Tribe Diphycerini Medvedev, 1952

Diphycerini S. I. Medvedev, 1952: 186 [stem: *Diphycer-*]. Type genus: *Diphycerus* Deyrolle and Fairmaire, 1878.

Tribe Diplotaxini Kirby, 1837

Diplotaxidae Kirby, 1837: 129 [stem: *Diplotax-*]. Type genus: *Diplotaxis* Kirby, 1837. Comment: current spelling maintained (Art. 29.5): incorrect stem formation in prevailing usage (should be *Diplotaxe-*).

Apogoniitae Blanchard, 1851a: 228 [stem: *Apogoni-*]. Type genus: *Apogonia* Kirby, 1819.

Liogenitae Blanchard, 1851a: 166 [stem: *Liogeny-*]. Type genus: *Liogenys* Guérin-Méneville, 1831. Comment: incorrect original stem formation, not in prevailing usage.

Tribe Euchirini Hope, 1840

Eucheiridae Hope, 1840b: 300 [stem: *Euchir-*]. Type genus: *Euchirus* Kirby, 1828 [as *Eucheirus*, incorrect subsequent spelling of type genus name, not in prevailing usage]. Comment: incorrect original stem formation, not in prevailing usage.

Tribe Heteronychini Lacordaire, 1856

Hétéronycides Lacordaire, 1856: 225 [stem: *Heteronych-*]. Type genus: *Heteronyx* Guérin-Méneville, 1831. Comment: original vernacular name available (Art. 11.7.2): first used in latinized form by Bertkau (1890: 277, as Heteronychini), generally accepted as in A. B. T. Smith (2006: 167, as Heteronychini); incorrect original stem formation, not in prevailing usage.

Tribe Hopliini Latreille, 1829

Hoplides Latreille, 1829a: 563 [stem: *Hopli-*]. Type genus: *Hoplia* Illiger, 1803.

Subtribe Hopliina Latreille, 1829

Hoplides Latreille, 1829a: 563 [stem: *Hopli-*]. Type genus: *Hoplia* Illiger, 1803. Comment: incorrect original stem formation, not in prevailing usage.

Dicheliden Oken, 1843: 483 [stem: *Dichel-*]. Type genus: *Dichelus* Lepeletier and Audinet-Serville, 1828. Comment: original vernacular name available (Art. 11.7.2): first used in latinized form and generally accepted as in Tulk (1847: 614, as Dichelidae).

Gymnolomidae H. C. C. Burmeister, 1844: 138 [stem: *Gymnolomat-*]. Type genus: *Gymnoloma* Dejean, 1833. Comment: incorrect original stem formation, not in prevailing usage.

Heterochelidae H. C. C. Burmeister, 1844: 86 [stem: *Heterochel-*]. Type genus: *Heterochelus* H. C. C. Burmeister, 1844.

Lepisiidae H. C. C. Burmeister, 1844: 166 [stem: *Lepisi-*]. Type genus: *Lepisia* Lepeletier and Audinet-Serville, 1828.

Scelophysides Péringuey, 1902: 624 [stem: *Scelophys-*]. Type genus: *Scelophysa* H. C. C. Burmeister, 1844.

*Madahopliini M. Lacroix, 1997: 21 [stem: *Madahopli-*]. Type genus: *Madahoplia* Lacroix, 1998. Comment: unavailable family-group name, not based on available genus name at the time (see A. B. T. Smith 2006).

Subtribe Pachycnemina Laporte, 1840

Pachycnémides Laporte, 1840b: 155 [stem: *Pachycnem-*]. Type genus: *Pachycnema* Lepeletier and Audinet-Serville, 1828. Comment: original vernacular name available (Art. 11.7.2): first used in latinized form by Burmeister (1844: 53, as Pachycnemidae), generally accepted as in A. B. T. Smith (2006: 170, as Pachycnemina).

Lepitrichiden Oken, 1843: 483 [stem: *Lepitrich-*]. Type genus: *Lepitrix* Lepeletier and Audinet-Serville, 1828. Comment: original vernacular name available (Art. 11.7.2): first used in latinized form and generally accepted as in Tulk (1847: 614, as Lepitrichidae).

Anisonychidae H. C. C. Burmeister, 1844: 35 [stem: *Anisonych-*]. Type genus: *Anisonyx* Latreille, 1807. Comment: Lucas (1920: 3) listed "Anisochelidae H. C. C. Burmeister" in error for "Anisonychidae H. C. C. Burmeister"; this name is a senior homonym of Anisonychini Legalov, 2003 (type genus *Anisonychus* Voss, 1927) in Attelabidae; this case is to be referred to the Commission to remove the homonymy (Art. 55.3.1).

Tribe Lichniini Burmeister, 1844

Lichniadae H. C. C. Burmeister, 1844: 8 [stem: *Lichni-*]. Type genus: *Lichnia* Erichson, 1835.

Tribe Liparetrini Burmeister, 1855

Liparetridae H. C. C. Burmeister, 1855: 187 [stem: *Liparetr-*]. Type genus: *Liparetrus* Guérin-Méneville, 1831.

Haplonychidae H. C. C. Burmeister, 1855: 224 [stem: *Haplonych-*]. Type genus: *Haplonycha* Dejean, 1836 [unjustified emendation of *Aplonycha* by Agassiz (1846b: 29), in prevailing usage and so deemed to be a justified emendation (Article 33.2.3.1) (see A. B. T. Smith 2006: 166); syn. of *Colpochila* Erichson, 1843]. Comment: this name is a senior homonym of Haplonychini Lacordaire, 1865 (type genus *Haplonyx* Schönherr, 1836) proposed in Curculionidae and now a synonym of Cryptoplini Lacordaire, 1863; this case is to be referred to the Commission to remove the homonymy (Art. 55.3.1).

Allarini Britton, 1955: 125 [stem: *Allar-*]. Type genus: *Allara* Britton, 1955.

Colpochilini Britton, 1957: 10 [stem: *Colpochil-*]. Type genus: *Colpochila* Erichson, 1843.

Tribe Macrodactylini Kirby, 1837

Macrodactylidae Kirby, 1837: 133 [stem: *Macrodactyl-*]. Type genus: *Macrodactylus* Dejean, 1821.

Ceraspididae H. C. C. Burmeister, 1855: 91 [stem: *Ceraspid-*]. Type genus: *Ceraspis* Lepeletier and Audinet-Serville, 1828.

Dicraniadae H. C. C. Burmeister, 1855: 65 [stem: *Dicrani-*]. Type genus: *Dicrania* Lepeletier and Audinet-Serville, 1828.

Isonychidae H. C. C. Burmeister, 1855: 22 [stem: *Isonych-*]. Type genus: *Isonychus* Mannerheim, 1829.

Microcraniadae H. C. C. Burmeister, 1855: 75 [stem: *Microcrani-*]. Type genus: *Microcrania* H. C. C. Burmeister, 1855.

Plectridae H. C. C. Burmeister, 1855: 80 [stem: *Plectr-*]. Type genus: *Plectris* Lepeletier and Audinet-Serville, 1828.

*Philochlénides Lacordaire, 1856: 256 [stem: *Philochloeni-*]. Type genus: *Philochloenia* Dejean, 1833 [as *Philochlaenia*, incorrect subsequent spelling of type genus name, not in prevailing usage]. Comment: original vernacular name unavailable (Art. 11.7.2): not subsequently latinized; incorrect original stem formation, not in prevailing usage.

Clavipalpides Lacordaire, 1856: 267 [stem: *Clavipalp-*]. Type genus: *Clavipalpus* Laporte, 1832. Comment: published before 29 March 1856; original vernacular name available (Art. 11.7.2): first used in latinized form by Imhoff (1856 [before 25 December]: xi, as Clavipalpidae), generally accepted as in G. H. Horn (1880e: 147, as Clavipalpides [treated as Latin]).

Tribe Maechidiini Burmeister, 1855

Maechidiina H. C. C. Burmeister, 1855: 208 [stem: *Maechidi-*]. Type genus: *Maechidius* W. S. MacLeay, 1819.

Tribe Melolonthini Leach, 1819

Melolonthidae Leach, 1819: 189 [stem: *Melolonth-*]. Type genus: *Melolontha* Fabricius, 1775 [placed on the Official List of Generic Names in Zoology (ICZN 1994a)].

Subtribe Enariina Dewailly, 1950

Enarina Dewailly, 1950: 323 [stem: *Enari-*]. Type genus: *Enaria* Erichson, 1847. Comment: incorrect original stem formation, not in prevailing usage.

Subtribe Heptophyllina Medvedev, 1951

Heptophyllini S. I. Medvedev, 1951: 197 [stem: *Heptophyll-*]. Type genus: *Heptophylla* Motschulsky, 1858.

Subtribe Leucopholina Burmeister, 1855

Leucopholidae H. C. C. Burmeister, 1855: 285 [stem: *Leucophol-*]. Type genus: *Leucopholis* Dejean, 1833. Comment: current spelling maintained (Art. 29.3.1.1): incorrect stem formation in prevailing usage (should be *Leucopholid-*).

Subtribe Melolonthina Leach, 1819

*Mélolonthides Baudet-Lafarge, 1809: 12 [stem: *Melolonth-*]. Type genus: *Melolontha* Fabricius, 1775 [placed on the Official List of Generic Names in Zoology (ICZN 1994a)]. Comment: original vernacular name unavailable (Art. 11.7.2): subsequently used in latinized form but not generally attributed to Baudet-Lafarge (1809).

Melolonthidae Leach, 1819: 189 [stem: *Melolonth-*]. Type genus: *Melolontha* Fabricius, 1775 [placed on the Official List of Generic Names in Zoology (ICZN 1994a)]. Comment: published in June 1819; this family-group name was also used in the same year by W. S. MacLeay (1819 [November]: 79, as Melolonthidae).

Polyphyllidae H. C. C. Burmeister, 1855: 397 [stem: *Polyphyll-*]. Type genus: *Polyphylla* Harris, 1841.

Psilonychides Péringuey, 1904: 184 [stem: *Psilonych-*]. Type genus: *Psilonychus* H. C. C. Burmeister, 1855.

Subtribe Pegylina Lacroix, 1989

Pegylini M. Lacroix, 1989: 115 [stem: *Pegyl-*]. Type genus: *Pegylis* Erichson, 1847. Comment: current spelling maintained (Art. 29.3.1.1): incorrect stem formation in prevailing usage (should be *Pegylid-*).

Subtribe Rhizotrogina Burmeister, 1855

Rhizotrogidae H. C. C. Burmeister, 1855: 308 [stem: *Rhizotrog-*]. Type genus: *Rhizotrogus* Latreille, 1825.

Tostegopterae J. L. LeConte, 1861: 139 [stem: *Tostegopter-*]. Type genus: *Tostegoptera* Blanchard, 1851.

Subtribe Schizonychina Burmeister, 1855

Schizonychidae H. C. C. Burmeister, 1855: 265 [stem: *Schizonych-*]. Type genus: *Schizonycha* Dejean, 1833.

Tribe Oncerini LeConte, 1861

Oncerini J. L. LeConte, 1861: 133 [stem: *Oncer-*]. Type genus: *Oncerus* J. L. LeConte, 1856.

Tribe Pachypodini Erichson, 1840

Pachypoden Erichson, 1840b: 29 [stem: *Pachypod-*]. Type genus: *Pachypus* Dejean, 1821 [*nomen protectum*; this genus name is a junior homonym of *Pachypus* Billberg, 1820 *nomen oblitum*; we provide references to support the conservation of *Pachypus* Dejean, 1821 as the valid name for this genus (Art. 23.9.1) (see Appendix 1)]. Comment: original vernacular name available (Art. 11.7.2): first used in latinized form by Erichson (1847a: 100, as Pachypoda), generally accepted as in A. B. T. Smith (2006: 164, as Pachypodini).

Tribe Pachytrichini Burmeister, 1855

Pachytrichiadae H. C. C. Burmeister, 1855: 241 [stem: *Pachytrich-*]. Type genus: *Pachytricha* Hope, 1841.

Tribe Phyllotocidiini Britton, 1957

Phyllotocidiini Britton, 1957: 58 [stem: *Phyllotocidi-*]. Type genus: *Phyllotocidium* Blackburn, 1898.

Tribe Podolasiini Howden, 1997

Lasiopodes J. L. LeConte, 1856: 282 [stem: *Lasiopod-*]. Type genus: *Lasiopus* J. L. LeConte, 1856 [preoccupied genus name, not *Lasiopus* Schönherr, 1823 [Coleoptera: Curculionidae], not *Lasiopus* Geoffroy, 1835 [Mammalia]; syn. of *Podolasia* Harold, 1869]. Comment: permanently invalid (Art. 39): based on preoccupied type genus.

Podolasiini Howden, 1997: 224 [stem: *Podolasi-*]. Type genus: *Podolasia* Harold, 1869.

Tribe Scitalini Britton, 1957

Scitalini Britton, 1957: 10 [stem: *Scital-*]. Type genus: *Scitala* Erichson, 1842.

Tribe Sericini Kirby, 1837

Sericidae Kirby, 1837: 128 [stem: *Seric-*]. Type genus: *Serica* W. S. MacLeay, 1819.

Subtribe Phyllotocina Burmeister, 1855

Phyllotocidae H. C. C. Burmeister, 1855: 182 [stem: *Phyllotoc-*]. Type genus: *Phyllotocus* Fisher von Waldheim, 1823.

Subtribe Sericina Kirby, 1837

Sericidae Kirby, 1837: 128 [stem: *Seric-*]. Type genus: *Serica* W. S. MacLeay, 1819. Comment: published 23 October 1837; this family-group name was also proposed in the same year by Hope (1837 ["31 December"]: 73, 107, as SERICIDAE); Hope (1840a: 112) refers to the publication by Kirby (1837) which is further evidence that the publication by Kirby was published first (see A. B. T. Smith 2006: 165).

OMALOPLIITES Blanchard, 1845a: 212 [stem: *Omalopli-*]. Type genus: *Omaloplia* Schönherr, 1817. Comment: original vernacular name available (Art. 11.7.2): first used in latinized form by Gistel (1848: [5], as OMALOPLIAEIDAE [incorrect stem formation]), generally accepted as in Blanchard (1853: 129, as OMALOPLIITAE); OMALOPLIITES Blanchard, 1845 treated as unavailable and the first available name based on this type genus listed as HOMALOPLIINA H. C. C. Burmeister, 1855 by A. B. T. Smith (2006: 165).

ASTAENIDAE H. C. C. Burmeister, 1855: 123 [stem: *Astaen-*]. Type genus: *Astaena* Erichson, 1847.

Subtribe TROCHALINA Brenske, 1898

TROCHALINAE Brenske, 1898: 354 [stem: *Trochal-*]. Type genus: *Trochalus* Laporte, 1832.

Tribe SERICOIDINI Erichson, 1847

SERICOIDEAE Erichson, 1847a: 102 [stem: *Sericoid-*]. Type genus: *Sericoides* Guérin-Méneville, 1840.

Tribe SYSTELLOPINI Sharp, 1877

SYSTELLOPIDES Sharp, 1877: 311 [stem: *Systellop-*]. Type genus: *Systellopus* Sharp, 1877. Comment: current spelling maintained (Art. 29.5): incorrect stem formation in prevailing usage (should be *Systellopod-*).

Tribe TANYPROCTINI Erichson, 1847

TANYPROCTINI Erichson, 1847b: 653 [stem: *Tanyproct-*]. Type genus: *Tanyproctus* Ménétriés, 1832.

Subtribe MACROPHYLLINA Burmeister, 1855

MACROPHYLLIDAE H. C. C. Burmeister, 1855: 447 [stem: *Macrophyll-*]. Type genus: *Macrophylla* Hope, 1837 [syn. of *Aegostheta* Dejean, 1833]. Comment: senior homonym of MACROPHYLLINA Gray, 1866 (type genus *Macrophyllum* Gray, 1838) in Mammalia; this case is to be referred to the Commission to remove the homonymy (Art. 55.3.1).

AEGOSTHETINI Lacroix, 2007: 203 [stem: *Aegosthet-*]. Type genus: *Aegostheta* Dejean, 1833.

Subtribe TANYPROCTINA Erichson, 1847

TANYPROCTINI Erichson, 1847b: 653 [stem: *Tanyproct-*]. Type genus: *Tanyproctus* Ménétriés, 1832.

ELAPHOCERITAE Blanchard, 1851a: 164 [stem: *Elaphocer-*]. Type genus: *Elaphocera* Géné, 1836. Comment: ELAPHOCERINI no longer has priority over PACHYDEMINI through a reversal of precedence (A. B. T. Smith 2006: 168).

PACHYDEMIDAE H. C. C. Burmeister, 1855: 437 [stem: *Pachydem-*]. Type genus: *Pachydema* Laporte, 1832. Comment: name conserved over ELAPHOCERINI Blanchard, 1851 as a *nomen protectum* and treated as valid by A. B. T. Smith (2006: 168).

ACHLOIDAE H. C. C. Burmeister, 1855: 465 [stem: *Achlo-*]. Type genus: *Achloa* Erichson, 1840.

CEPHALOTRICHIADAE H. C. C. Burmeister, 1855: 433 [stem: *Cephalotrichi-*]. Type genus: *Cephalotrichia* Hope, 1837.

*LEPTOPODIDAE H. C. C. Burmeister, 1855: 428 [stem: *Leptopod-*]. Type genus: *Leptopus* Waltl, 1838 [preoccupied genus name, not *Leptopus* Latreille, 1809 [Hemiptera], not *Leptopus* Rafinesque, 1814 [Pisces], not *Leptopus* Lamarck, 1818 [Crustacea], and not *Leptopus* Fallén, 1823 [Diptera]; syn. of *Elaphocera* Gené, 1836]. Comment: family-group name unavailable (Art. 11.7.1.1): not based on a genus used as valid at the time.

SPARRMANNINI Péringuey, 1904: 170 [stem: *Sparrmanni-*]. Type genus: *Sparrmannia* Laporte, 1840 [*nomen protectum*; this genus name is a junior synonym of *Leocaeta* Dejean, 1833 *nomen oblitum* and *Cephalotrichia* Hope, 1837 *nomen oblitum*; we provide references to support the conservation of *Sparrmannia* as the valid name for this genus (Art. 23.9.1) (see Appendix 1); *Sparrmannia* is an incorrect subsequent spelling of *Sparmannia* Laporte, 1840, in prevailing usage and so deemed to be the correct original spelling (see Evans 1989)]. Comment: incorrect original stem formation, not in prevailing usage.

Subfamily RUTELINAE MacLeay, 1819

RUTELIDAE W. S. MacLeay, 1819: 69 [stem: *Rutel-*]. Type genus: *Rutela* Latreille, 1802. Comment: First Reviser (RUTELINAE MacLeay, 1819 vs ANOPLOGNATHINAE MacLeay, 1819) not determined, current usage maintained.

Tribe ADORETINI Burmeister, 1844

ADORETIDAE H. C. C. Burmeister, 1844: 466 [stem: *Adoret-*]. Type genus: *Adoretus* Dejean, 1833.

Subtribe ADORETINA Burmeister, 1844

ADORETIDAE H. C. C. Burmeister, 1844: 466 [stem: *Adoret-*]. Type genus: *Adoretus* Dejean, 1833.

ADORODOCIINA Ohaus, 1912: 151 [stem: *Adorodoci-*]. Type genus: *Adorodocia* Brenske, 1893.

ADOROLEPTINA Ohaus, 1912: 151 [stem: *Adorolept-*]. Type genus: *Adoroleptus* Brenske, 1893.

Pseudadoretina Ohaus, 1912: 151 [stem: *Pseudadoret-*]. Type genus: *Pseudadoretus* Semenov, 1889.

Scaphorhinadoretina Ohaus, 1912: 151 [stem: *Scaphorhinadoret-*]. Type genus: *Scaphorhinadoretus* Ohaus, 1912. Comment: erroneously cited as a new genus originally but correctly listed as a subtribe on p. 426 of the same work.

Subtribe Adorrhinyptiina Arrow, 1917

Adorrhinyptiini Arrow, 1917: 273 [stem: *Adorrhinypti-*]. Type genus: *Adorrhinyptia* Arrow, 1917.

Subtribe Pachyrhinadoretina Ohaus, 1912

Pachyrhinadoretina Ohaus, 1912: 151 [stem: *Pachyrhinadoret-*]. Type genus: *Pachyrhinadoretus* Ohaus, 1912.

Subtribe Prodoretina Ohaus, 1912

Prodoretina Ohaus, 1912: 151 [stem: *Prodoret-*]. Type genus: *Prodoretus* Brenske, 1893.

Subtribe Trigonostomusina Ohaus, 1912

Trigonostomina Ohaus, 1912: 151 [stem: *Trigonostomus-*]. Type genus: *Trigonostomus* Brenske, 1893 [placed on the Official List of Generic Names in Zoology (ICZN 2009b)]. Comment: *Trigonostomus-* determined to be the correct stem of this name to avoid homonymy with Trigonostomidae Graff, 1905 (type genus *Trigonostomum* Schmidt, 1852) in Platyhelminthes and Trigonostomusina Ohaus, 1912 placed on the Official List of Family-group Names in Zoology (ICZN 2009b).

Tribe Alvarengiini Frey, 1975

*Pachylides Lacordaire, 1856: 394 [stem: *Pachyl-*]. Type genus: *Pachylus* H. C. C. Burmeister, 1847 [preoccupied genus name, not *Pachylus* C. L. Koch, 1839 [Arachnida]; syn. of *Ottokelleria* d'Andretta and Martínez, 1957]. Comment: original vernacular name unavailable (Art. 11.7.2): not subsequently latinized; if discovered to be available then permanently invalid (Art. 39): based on preoccupied type genus; Pachylinae Sørensen, 1884 (type genus *Pachylus* C. L. Koch, 1839) is used as valid in Arachnida.

Alvarengiini Frey, 1975: 84 [stem: *Alvarengi-*]. Type genus: *Alvarengius* Frey, 1975.

Tribe Anatistini Lacordaire, 1856

Anatistides Lacordaire, 1856: 321 [stem: *Anatist-*]. Type genus: *Anatista* Brême, 1844. Comment: published before 29 March 1856; original vernacular name available (Art. 11.7.2): first used in latinized form and generally accepted as in Imhoff (1856 [before 25 December]: xi, as ANATISTIDAE).

SPODOCHLAMYDINI Ohaus, 1918: 166 [stem: *Spodochlamyd-*]. Type genus: *Spodochlamys* H. C. C. Burmeister, 1855.

Tribe ANOMALINI Streubel, 1839 *nomen protectum*

ANOMALIDAE Streubel, 1839: 136 [stem: *Anomal-*]. Type genus: *Anomala* Samouelle, 1819 [placed on the Official List of Generic Names in Zoology (ICZN 1989d); *Anomala* von Block, 1799 was suppressed for the purposes of the Principle of Priority and the Principle of Homonymy (ICZN 1989d)]. Comment: *nomen protectum* (see A. B. T. Smith 2006: 173).

Subtribe ANISOPLIINA Burmeister, 1844

ANISOPLIADAE H. C. C. Burmeister, 1844: 208 [stem: *Anisopli-*]. Type genus: *Anisoplia* Schönherr, 1817. Comment: incorrect original stem formation, not in prevailing usage.

Subtribe ANOMALINA Streubel, 1839 *nomen protectum*

EUCHLORIDAE Hope, 1839: 67 [stem: *Euchlor-*]. Type genus: *Euchlora* W. S. MacLeay, 1819. Comment: *nomen oblitum* (see A. B. T. Smith 2006: 173).

ANOMALIDAE Streubel, 1839: 136 [stem: *Anomal-*]. Type genus: *Anomala* Samouelle, 1819 [placed on the Official List of Generic Names in Zoology (ICZN 1989d); *Anomala* von Block, 1799 was suppressed for the purposes of the Principle of Priority and the Principle of Homonymy (ICZN 1989d)]. Comment: *nomen protectum* (see A. B. T. Smith 2006: 173).

PHYLLURGAEIDAE Gistel, 1848: [5] [stem: *Phyllurg-*]. Type genus: *Phyllurga* Gistel, 1848 [syn. of *Euchlora* W. S. MacLeay, 1819].

DILOPHOCHILINA Ohaus, 1918: 166 [stem: *Dilophochil-*]. Type genus: *Dilophochila* H. W. Bates, 1888.

Subtribe ISOPLIINA Péringuey, 1902

ISOPLIINI Péringuey, 1902: 564 [stem: *Isopli-*]. Type genus: *Isoplia* H. C. C. Burmeister, 1855.

Subtribe LEPTOHOPLIINA Potts, 1974

LEPOTHOPLIINI Potts, 1974: 152 [stem: *Leptohopli-*]. Type genus: *Leptohoplia* Saylor, 1935. Comment: incorrect original stem formation, not in prevailing usage.

Subtribe POPILLIINA Ohaus, 1918

POPILLIINA Ohaus, 1918: 133 [stem: *Popilli-*]. Type genus: *Popillia* Dejean, 1821.

Tribe Anoplognathini MacLeay, 1819

Anoplognathidae W. S. MacLeay, 1819: 81 [stem: *Anoplognath-*]. Type genus: *Anoplognathus* Leach, 1815.

Subtribe Anoplognathina MacLeay, 1819

Anoplognathidae W. S. MacLeay, 1819: 81 [stem: *Anoplognath-*]. Type genus: *Anoplognathus* Leach, 1815.

Subtribe Brachysternina Burmeister, 1844

Brachysternidae H. C. C. Burmeister, 1844: 455 [stem: *Brachystern-*]. Type genus: *Brachysternus* Guérin-Méneville, 1831.

Subtribe Phalangogoniina Ohaus, 1918

Phalangogoniina Ohaus, 1918: 176 [stem: *Phalangogoni-*]. Type genus: *Phalangogonia* H. C. C. Burmeister, 1844.

Subtribe Platycoeliina Burmeister, 1844

Platycoeliidae H. C. C. Burmeister, 1844: 451 [stem: *Platycoeli-*]. Type genus: *Platycoelia* Dejean, 1833.

Subtribe Schizognathina Ohaus, 1918

Schizognathina Ohaus, 1918: 174 [stem: *Schizognath-*]. Type genus: *Schizognathus* Fischer von Waldheim, 1823.

Tribe Geniatini Burmeister, 1844

Geniatidae H. C. C. Burmeister, 1844: 478 [stem: *Geniat-*]. Type genus: *Geniates* Kirby, 1819. Comment: First Reviser (Geniatini H. C. C. Burmeister, 1844 vs Leucothyreini H. C. C. Burmeister, 1844) not determined, current usage maintained.

Leucothyreidae H. C. C. Burmeister, 1844: 485 [stem: *Leucothyre-*]. Type genus: *Leucothyreus* W. S. MacLeay, 1819.

Tribe Rutelini MacLeay, 1819

Rutelidae W. S. MacLeay, 1819: 69 [stem: *Rutel-*]. Type genus: *Rutela* Latreille, 1802.

Subtribe Areodina Burmeister, 1844

Areodidae H. C. C. Burmeister, 1844: 423 [stem: *Areod-*]. Type genus: *Areoda* W. S. MacLeay, 1819.

Subtribe Desmonychina Arrow, 1917

Desmonychinae Arrow, 1917: 359 [stem: *Desmonych-*]. Type genus: *Desmonyx* Arrow, 1907. Comment: spelled Desmonycinae pp. xiii, 359, 382, but spelled correctly Desmonychinae p. 3 [in key].

Subtribe Didrepanephorina Ohaus, 1918

Didrepanophorina Ohaus, 1918: 14 [stem: *Didrepanephor-*]. Type genus: *Didrepanephorus* Wood-Mason, 1878 [as *Didrepanophorus*, incorrect subsequent spelling of type genus name, not in prevailing usage]. Comment: incorrect original stem formation, not in prevailing usage.

Subtribe Heterosternina Bates, 1888 *nomen protectum*

Macropni G. H. Horn, 1867b: 398 [stem: *Macropn-*]. Type genus: *Macropnus* G. H. Horn, 1867. Comment: *nomen oblitum* (see A. B. T. Smith 2006: 171).

Heterosterninae H. W. Bates, 1888: 286 [stem: *Heterostern-*]. Type genus: *Heterosternus* Dupont, 1832. Comment: *nomen protectum* (see A. B. T. Smith 2006: 171).

Subtribe Lasiocalina Ohaus, 1918

Lasiocalina Ohaus, 1918: 30 [stem: *Lasiocal-*]. Type genus: *Lasiocala* Blanchard, 1851.

Subtribe Oryctomorphina Burmeister, 1847

Oryctomorphidae H. C. C. Burmeister, 1847: 28 [stem: *Oryctomorph-*]. Type genus: *Oryctomorphus* Guérin-Méneville, 1831.

Subtribe Parastasiina Burmeister, 1844

Parastasiidae H. C. C. Burmeister, 1844: 368 [stem: *Parastasi-*]. Type genus: *Parastasia* Westwood, 1841.

Subtribe Rutelina MacLeay, 1819

Rutelidae W. S. MacLeay, 1819: 69 [stem: *Rutel-*]. Type genus: *Rutela* Latreille, 1802.

Chasmodiidae H. C. C. Burmeister, 1844: 333 [stem: *Chasmodi-*]. Type genus: *Chasmodia* W. S. MacLeay, 1819.

Chrysophoridae H. C. C. Burmeister, 1844: 412 [stem: *Chrysophor-*]. Type genus: *Chrysophora* Dejean, 1821.

Macraspididae H. C. C. Burmeister, 1844: 343 [stem: *Macraspid-*]. Type genus: *Macraspis* W. S. MacLeay, 1819.

Pelidnotidae H. C. C. Burmeister, 1844: 388 [stem: *Pelidnot-*]. Type genus: *Pelidnota* W. S. MacLeay, 1819 [placed on the Official List of Generic Names in Zoology (ICZN 2003b)].

Antichirides Lacordaire, 1856: 341 [stem: *Anticheir-*]. Type genus: *Anticheira* Eschscholtz, 1818 [as *Antichira*, unjustified emendation of type genus name by Agassiz (1846b: 27), not in prevailing usage]. Comment: original vernacular name available (Art. 11.7.2): first used in latinized form and generally accepted as in H. W. Bates (1888: 262, as Antichirina); incorrect original stem formation, not in prevailing usage.

Plusiotina H. W. Bates, 1888: 276 [stem: *Plusiotid-*]. Type genus: *Plusiotis* H. C. C. Burmeister, 1844. Comment: incorrect original stem formation, not in prevailing usage.

Fruhstorferiina Ohaus, 1918: 43 [stem: *Fruhstorferi-*]. Type genus: *Fruhstorferia* Kolbe, 1894.

Subfamily Dynastinae MacLeay, 1819

Dynastidae W. S. MacLeay, 1819: 64 [stem: *Dynast-*]. Type genus: *Dynastes* W. S. MacLeay, 1819.

Tribe Agaocephalini Burmeister, 1847

Agaocephalidae H. C. C. Burmeister, 1847: 280 [stem: *Agaocephal-*]. Type genus: *Agaocephala* Lepeletier and Audinet-Serville, 1828 [as *Agacephala*, incorrect subsequent spelling for *Agacephala*, in prevailing usage and so deemed to be the correct original spelling (Art. 33.3.1) (see A. B. T. Smith 2006)].

Tribe Cyclocephalini Laporte, 1840

Cyclocephalites Laporte, 1840b: 124 [stem: *Cyclocephal-*]. Type genus: *Cyclocephala* Dejean, 1821. Comment: original vernacular name available (Art. 11.7.2): first used in latinized form by Imhoff (1856: xi, as Cyclocephalidae), generally accepted as in A. B. T. Smith (2006: 175, as Cyclocephalini).

Chalepidae H. C. C. Burmeister, 1847: 71 [stem: *Chalep-*]. Type genus: *Chalepus* W. S. MacLeay, 1819 [preoccupied genus name, not *Chalepus* Thunberg, 1805 [Coleoptera: Chrysomelidae]; syn. of *Dyscinetus* Harold, 1869]. Comment: permanently invalid (Art. 39): based on preoccupied type genus; this family-group name is a senior homonym of Chalepini Weise, 1910 (type genus *Chalepus* Thunberg, 1805) currently used as valid in Chrysomelidae.

Peltonotini Arrow, 1917: 27 [stem: *Peltonot-*]. Type genus: *Peltonotus* H. C. C. Burmeister, 1847.

Acrobolbiina Ohaus, 1918: 13 [stem: *Acrobolbi-*]. Type genus: *Acrobolbia* Ohaus, 1912.

Tribe Dynastini MacLeay, 1819

Dynastidae W. S. MacLeay, 1819: 64 [stem: *Dynast-*]. Type genus: *Dynastes* W. S. MacLeay, 1819.

Xylotrupidae Hope, 1838b: 319 [stem: *Xylotrup-*]. Type genus: *Xylotrupes* Hope, 1837.

Megasominae Swainson, 1840: 210 [stem: *Megasomat-*]. Type genus: *Megasoma* Kirby, 1825. Comment: family-group name attributed to Imhoff (1856) in A. B. T. Smith (2006: 175); incorrect original stem formation, not in prevailing usage.

Tribe Hexodontini Lacordaire, 1856

Hexodontides Lacordaire, 1856: 391 [stem: *Hexodont-*]. Type genus: *Hexodon* A. G. Olivier, 1789. Comment: published before 29 March 1856; original vernacular name available (Art. 11.7.2): first used in latinized form by Imhoff (1856 [before 25 December]: xi, as Hexodontidae), generally accepted as in Arrow (1937: 4, as Hexodontini).

Tribe Oryctini Mulsant, 1842

Oryctésaires Mulsant, 1842: 372 [stem: *Oryct-*]. Type genus: *Oryctes* Illiger, 1798. Comment: original vernacular name available (Art. 11.7.2): first used in latinized form by MacDonald (1845: 185, as Oryctesiae [incorrect stem formation]), generally accepted as in A. B. T. Smith (2006: 175, as Oryctini); incorrect original stem formation, not in prevailing usage.

Megaceridae H. C. C. Burmeister, 1847: 212 [stem: *Megacerat-*]. Type genus: *Megaceras* Hope, 1837. Comment: incorrect original stem formation, not in prevailing usage.

Strategidae H. C. C. Burmeister, 1847: 87 [stem: *Strateg-*]. Type genus: *Strategus* Kirby, 1828.

Tribe Oryctoderini Endrödi, 1966

Oryctoderini Endrödi, 1966: 25 [stem: *Oryctoder-*]. Type genus: *Oryctoderus* Boisduval, 1835.

Tribe Pentodontini Mulsant, 1842

Pentodonaires Mulsant, 1842: 381 [stem: *Pentodont-*]. Type genus: *Pentodon* Hope, 1837 [placed on the Official List of Generic Names in Zoology (ICZN 2003b)].

Subtribe Cheiroplatina Carne, 1957

Cheiroplatina Carne, 1957: 61 [stem: *Cheiroplat-*]. Type genus: *Cheiroplatys* Hope, 1837. Comment: Casey (1915a) cites "Cheiroplatids" several times within the Pentodontini without formally erecting a subtribe; current spelling maintained (Art. 29.5): incorrect stem formation in prevailing usage (should be *Cheiroplate-*).

Subtribe Dipelicina Carne, 1957

Dipelicina Carne, 1957: 117 [stem: *Dipelic-*]. Type genus: *Dipelicus* Hope, 1843.

Subtribe Pentodontina Mulsant, 1842

Pentodonaires Mulsant, 1842: 381 [stem: *Pentodont-*]. Type genus: *Pentodon* Hope, 1837 [placed on the Official List of Generic Names in Zoology (ICZN 2003b)]. Comment: original vernacular name available (Art. 11.7.2): first used in latinized form by H. W. Bates (1888: 314, as Pentodontinae), generally accepted as in A. B. T. Smith (2006: 176, as Pentodontina); incorrect original stem formation, not in prevailing usage.

Calicnémiens Mulsant, 1842: 386 [stem: *Calicnemid-*]. Type genus: *Calicnemis* Laporte, 1832. Comment: original vernacular name available (Art. 11.7.2): first used in latinized form by MacDonald (1845: 185, as Callicnemiae [incorrect stem formation]) and generally accepted as in Acloque (1896: 267, as Calicnemisii) and Houlbert (1922b: 225, as Callicnemini [incorrect stem formation]); name attributed to Blanchard (1845a) and treated as unavailable by A. B. T. Smith (2006: 168); incorrect original stem formation, not in prevailing usage.

Bothynidae H. C. C. Burmeister, 1847: 90 [stem: *Bothyn-*]. Type genus: *Bothynus* Hope, 1837 [placed on the Official List of Generic Names in Zoology (ICZN 2008a)].

Pimelopodea H. C. C. Burmeister, 1847: 172 [stem: *Pimelopod-*]. Type genus: *Pimelopus* Erichson, 1842.

Podalgidae H. C. C. Burmeister, 1847: 90 [stem: *Podalg-*]. Type genus: *Podalgus* H. C. C. Burmeister, 1847 [placed on the Official List of Generic Names in Zoology (ICZN 2004a)].

Metanastina Carne, 1957: 32 [stem: *Metanast-*]. Type genus: *Metanastes* Arrow, 1911.

Subtribe Pseudoryctina Carne, 1957

Pseudoryctina Carne, 1957: 121 [stem: *Pseudoryct-*]. Type genus: *Pseudoryctes* Sharp, 1873.

Tribe Phileurini Burmeister, 1847

Phileuridae H. C. C. Burmeister, 1847: 138 [stem: *Phileur-*]. Type genus: *Phileurus* Latreille, 1807. Comment: usage of this name conserved over Cryptodini H. C. C. Burmeister and Schaum, 1840 (Art. 35.5).

Subtribe Cryptodina Burmeister and Schaum, 1840

Cryptodinae H. C. C. Burmeister and Schaum, 1840: 360 [stem: *Cryptod-*]. Type genus: *Cryptodus* MacLeay, 1819. Comment: current spelling maintained (Art. 29.5): incorrect stem formation in prevailing usage (should be *Cryptodont-*); conservation of original stem avoids homonymy problems with Cryptodontidae Dall, 1895 (type genus *Cryptodon* Turton, 1822) in Mollusca: Bivalvia and Cryptodontina Lacordaire, 1856 (type genus *Cryptodontes* H. C. C. Burmeister, 1847) in Scarabaeidae: Cetoniinae: Trichiini.

Subtribe Phileurina Burmeister, 1847

Phileuridae H. C. C. Burmeister, 1847: 138 [stem: *Phileur-*]. Type genus: *Phileurus* Latreille, 1807.

Subfamily Cetoniinae Leach, 1815

Cetonida Leach, 1815: 99 [stem: *Cetoni-*]. Type genus: *Cetonia* Fabricius, 1775. Comment: A. B. T. Smith (2006: 177) noted that although Cetoniinae has priority over Melolonthinae, the latter is in prevailing usage at the family level and must not be displaced by the older name under Article 35.5 (only relevant in cases where authors consider Melolonthidae to be a family containing the subfamily Cetoniinae).

Tribe Cetoniini Leach, 1815

Cetonida Leach, 1815: 99 [stem: *Cetoni-*]. Type genus: *Cetonia* Fabricius, 1775.

Subtribe Cetoniina Leach, 1815

Cetonida Leach, 1815: 99 [stem: *Cetoni-*]. Type genus: *Cetonia* Fabricius, 1775.

Elaphini Schoch, 1894: 175 [stem: *Elaphin-*]. Type genus: *Elaphinis* H. C. C. Burmeister, 1842.

Glycyphanae Schoch, 1894: 175 [stem: *Glycyphan-*]. Type genus: *Glycyphana* H. C. C. Burmeister, 1842.

Pachnodii Péringuey, 1907: 371 [stem: *Pachnod-*]. Type genus: *Pachnoda* H. C. C. Burmeister, 1842.

Tephraeides Schenkling, 1921: 313 [stem: *Tephrae-*]. Type genus: *Tephraea* H. C. C. Burmeister, 1842.

Subtribe Euphoriina Horn, 1880

Euphoriae G. H. Horn, 1880c: 397 [stem: *Euphori-*]. Type genus: *Euphoria* H. C. C. Burmeister, 1842. Comment: name attributed to Schoch (1894) by A. B. T. Smith (2006: 182).

Subtribe Leucocelina Kraatz, 1882

Leucoceliden Kraatz, 1882b: 65 [stem: *Leucocel-*]. Type genus: *Leucocelis* H. C. C. Burmeister, 1842. Comment: original vernacular name available (Art. 11.7.2): first used in latinized form by Schoch (1894: 213, as Leucocelidae), generally accepted as in A. B. T. Smith (2006: 182, as Leucocelina); current spelling maintained (Art. 29.3.1.1): incorrect stem formation in prevailing usage (should be *Leucocelid-*).

Tribe Cremastocheilini Burmeister and Schaum, 1841

Cremastochilidae H. C. C. Burmeister and Schaum, 1841: 243 [stem: *Cremastocheil-*]. Type genus: *Cremastocheilus* Knoch, 1801. Comment: usage of this name conserved over Macromini H. C. C. Burmeister and Schaum, 1840 (Art. 35.5).

Subtribe Aspilina Krikken, 1984

Aspilina Krikken, 1984: 25, in key [stem: *Aspil-*]. Type genus: *Aspilus* Westwood in Schaum, 1848.

Subtribe Coenochilina Burmeister, 1842

Coenochilidae H. C. C. Burmeister, 1842: 148 [stem: *Coenochil-*]. Type genus: *Coenochilus* Schaum, 1841.

Subtribe Cremastocheilina Burmeister and Schaum, 1841

Cremastochilidae H. C. C. Burmeister and Schaum, 1841: 243 [stem: *Cremastocheil-*]. Type genus: *Cremastocheilus* Knoch, 1801. Comment: incorrect original stem formation, not in prevailing usage.

Subtribe Cymophorina Krikken, 1984

Cymophorina Krikken, 1984: 23, in key [stem: *Cymophor-*]. Type genus: *Cymophorus* Kirby, 1827.

Subtribe Genuchina Krikken, 1984

Genuchina Krikken, 1984: 21, in key [stem: *Genuch-*]. Type genus: *Genuchus* Kirby, 1825.

Subtribe Goliathopsidina Krikken, 1984

Goliathopsidina Krikken, 1984: 25, in key [stem: *Goliathopsid-*]. Type genus: *Goliathopsis* Janson, 1881. Comment: current spelling maintained (Art. 29.5): incorrect stem formation in prevailing usage (should be *Goliathopse-*).

Subtribe Heterogeniina Krikken, 1984

Heterogeniina Krikken, 1984: 25, in key [stem: *Heterogeni-*]. Type genus: *Heterogenius* Moser, 1911.

Subtribe Lissogeniina Krikken, 1984

Lissogeniina Krikken, 1984: 25, in key [stem: *Lissogeni-*]. Type genus: *Lissogenius* Schaum, 1845.

Subtribe Macromina Burmeister and Schaum, 1840

Macrominae H. C. C. Burmeister and Schaum, 1840: 360 [stem: *Macrom-*]. Type genus: *Macroma* Gory and Percheron, 1833 [syn. of *Campsiura* Hope, 1831]. Comment: current spelling maintained (Art. 29.5): incorrect stem formation in prevailing usage (should be *Macromat-*).

Subtribe Nyassinina Krikken, 1984

Nyassinina Krikken, 1984: 25, in key [stem: *Nyassin-*]. Type genus: *Nyassinus* Westwood, 1879.

Subtribe Oplostomina Krikken, 1984

Oplostomina Krikken, 1984: 23, in key [stem: *Oplostom-*]. Type genus: *Oplostomus* W. S. MacLeay, 1838.

Subtribe Pilinurgina Krikken, 1984

Pilinurgina Krikken, 1984: 25, in key [stem: *Pilinurg-*]. Type genus: *Pilinurgus* H. C. C. Burmeister, 1842.

Subtribe Spilophorina Krikken, 1984

Spilophorina Krikken, 1984: 25, in key [stem: *Spilophor-*]. Type genus: *Spilophorus* Westwood in Schaum, 1848. Comment: this family-group name is a junior homonym of Spilophorini Chapuis, 1875 (type genus *Spilophora* Boheman, 1850) currently used as valid in Chrysomelidae; this case is to be referred to the Commission to remove the homonymy (Art. 55.3.1).

Subtribe Telochilina Krikken, 1984

Telochilina Krikken, 1984: 21, in key [stem: *Telochil-*]. Type genus: *Telochilus* Krikken, 1975.

Subtribe Trichoplina Krikken, 1984

Trichoplina Krikken, 1984: 23, in key [stem: *Trichopl-*]. Type genus: *Trichoplus* H. C. C. Burmeister, 1842.

Subtribe Trogodina Krikken, 1984

Trogodina Krikken, 1984: 27, in key [stem: *Trogod-*]. Type genus: *Trogodes* Westwood, 1874.

Tribe Diplognathini Burmeister, 1842

Diplognathidae H. C. C. Burmeister, 1842: 617 [stem: *Diplognath-*]. Type genus: *Diplognatha* Gory and Percheron, 1833.

Porphyronotii Péringuey, 1907: 371 [stem: *Porphyronot-*]. Type genus: *Porphyronota* H. C. C. Burmeister, 1842.

Tribe Goliathini Latreille, 1829

Goliathides Latreille, 1829a: 571 [stem: *Goliath-*]. Type genus: *Goliathus* Lamarck, 1801. Comment: name attributed to Griffith and Pidgeon (1832: 492) in A. B. T. Smith (2006: 181).

Subtribe Coryphocerina Burmeister, 1842

Coryphoceridae H. C. C. Burmeister, 1842: 215 [stem: *Coryphocer-*]. Type genus: *Coryphocera* H. C. C. Burmeister, 1842.

Heterorrhinidae Kraatz, 1880a: 21 [stem: *Heterorhin-*]. Type genus: *Heterorhina* Westwood, 1842 [as *Heterorrhina*, incorrect subsequent spelling of type genus name, not in prevailing usage]. Comment: incorrect original stem formation, not in prevailing usage.

Ceratorrhinidae Kraatz, 1880a: 18 [stem: *Ceratorhin-*]. Type genus: *Ceratorhina* Westwood, 1843 [as *Ceratorrhina*, incorrect subsequent spelling of type genus name, not in prevailing usage]. Comment: incorrect original stem formation, not in prevailing usage.

Bothrorrhinae Schoch, 1894: 173 [stem: *Bothrorrhin-*]. Type genus: *Bothrorrhina* H. C. C. Burmeister, 1842.

Gnathoceridae Schoch, 1894: 170 [stem: *Gnathocer-*]. Type genus: *Gnathocera* Kirby, 1825.

Ischnosceli Schoch, 1894: 170 [stem: *Ischnoscelid-*]. Type genus: *Ischnoscelis* H. C. C. Burmeister, 1842. Comment: incorrect original stem formation, not in prevailing usage.

Rhomborrhinae Schoch, 1894: 171 [stem: *Rhomborhin-*]. Type genus: *Rhomborhina* Hope, 1837 [as *Rhomborrhina*, incorrect subsequent spelling of type genus name, not in prevailing usage]. Comment: incorrect original stem formation, not in prevailing usage.

Tmesorrhinae Schoch, 1894: 170 [stem: *Tmesorrhin-*]. Type genus: *Tmesorrhina* Westwood, 1842.

Coelorrhinae Schoch, 1895: iii [stem: *Coelorrhin-*]. Type genus: *Coelorrhina* Hope, 1841 [*Coelorrhina* is an incorrect subsequent spelling for *Caelorrhina* in prevailing usage, and so deemed to be the correct original spelling (Article 33.3.1) (see A. B. T. Smith 2006: 181)].

Mecynorrhinina Schenkling, 1921: 15 [stem: *Mecynorhin-*]. Type genus: *Mecynorhina* Hope, 1837 [as *Mecynorrhina*, incorrect subsequent spelling of type genus name, not in prevailing usage]. Comment: incorrect original stem formation, not in prevailing usage.

Stephanorrhinina Schenkling, 1921: 35 [stem: *Stephanorrhin-*]. Type genus: *Stephanorrhina* H. C. C. Burmeister, 1842.

Subtribe Dicronocephalina Krikken, 1984

Dicronocephalina Krikken, 1984: 37, in key [stem: *Dicronocephal-*]. Type genus: *Dicronocephalus* Hope, 1831 [*Dicronocephalus* is an incorrect subsequent spelling for *Dicranocephalus*, since the incorrect subsequent spelling is in prevailing usage, it is now considered the correct original spelling of the name under Article 33.3.1; the original spelling placed the name in homonymy with *Dicranocephalus* Hahn, 1826 (Hemiptera) but usage of the subsequent spelling as the correct spelling avoids this homonymy problem (see A. B. T. Smith 2006)].

Subtribe Goliathina Latreille, 1829

Goliathides Latreille, 1829a: 571 [stem: *Goliath-*]. Type genus: *Goliathus* Lamarck, 1801. Comment: name attributed to Griffith and Pidgeon (1832) in A. B. T. Smith (2006: 181).

Hypselogeniae Schoch, 1894: 169 [stem: *Hypselogeni-*]. Type genus: *Hypselogenia* H. C. C. Burmeister, 1840.

Subtribe Ichnestomatina Burmeister, 1842

Ischnostomidae H. C. C. Burmeister, 1842: 600 [stem: *Ichnestomat-*]. Type genus: *Ichnestoma* Gory and Percheron, 1833 [as *Ichnostoma*, incorrect subsequent spelling of type genus name, not in prevailing usage]. Comment: incorrect original stem formation, not in prevailing usage.

Tribe Gymnetini Kirby, 1827

Gymnetidae Kirby, 1827: 150 [stem: *Gymnet-*]. Type genus: *Gymnetis* W. S. MacLeay, 1819 [placed on the Official List of Generic Names in Zoology (ICZN 1967a)].

Subtribe Blaesiina Schoch, 1895

Blaesiae Schoch, 1895: iii [stem: *Blaesi-*]. Type genus: *Blaesia* H. C. C. Burmeister, 1842.

Subtribe Gymnetina Kirby, 1827

Gymnetidae Kirby, 1827: 150 [stem: *Gymnet-*]. Type genus: *Gymnetis* W. S. MacLeay, 1819. Comment: current spelling maintained (Art. 29.3.1.1): incorrect stem formation in prevailing usage (should be *Gymnetid-*).

Clinteriidae Kraatz, 1882a: 49 [stem: *Clinteri-*]. Type genus: *Clinteria* H. C. C. Burmeister, 1842.

Stethodesmae Schoch, 1894: 172 [stem: *Stethodesmat-*]. Type genus: *Stethodesma* Bainbridge, 1841. Comment: incorrect original stem formation, not in prevailing usage.

Tribe Phaedimini Schoch, 1894

Phaedimi Schoch, 1894: 169 [stem: *Phaedim-*]. Type genus: *Phaedimus* Westwood, 1841.

Tribe Schizorhinini Burmeister, 1842

Schizorrhinidae H. C. C. Burmeister, 1842: 530 [stem: *Schizorhin-*]. Type genus: *Schizorhina* Kirby, 1825. Comment: First Reviser (Lomapterini H. C. C. Burmeister, 1842 vs Macronotini H. C. C. Burmeister, 1842 vs Schizorrhinini H. C. C. Burmeister, 1842) not determined, current usage maintained.

Subtribe Lomapterina Burmeister, 1842

Lomapteridae H. C. C. Burmeister, 1842: 310 [stem: *Lomapter-*]. Type genus: *Lomaptera* Gory and Percheron, 1833. Comment: First Reviser (Lomapterina H. C. C. Burmeister, 1842 vs Macronotina H. C. C. Burmeister, 1842) not determined, current usage maintained.

Macronotidae H. C. C. Burmeister, 1842: 318 [stem: *Macronot-*]. Type genus: *Macronota* Hoffmannsegg, 1817.

Subtribe Schizorhinina Burmeister, 1842

Schizorrhinidae H. C. C. Burmeister, 1842: 530 [stem: *Schizorhin-*]. Type genus: *Schizorhina* Kirby, 1825 [as *Schizorrhina*, incorrect subsequent spelling of type genus name, not in prevailing usage]. Comment: incorrect original stem formation, not in prevailing usage.

Diaphoniadae Kraatz, 1880b: 195 [stem: *Diaphoni-*]. Type genus: *Diaphonia* Newman, 1840.

Eupoecilidae Kraatz, 1880b: 188 [stem: *Eupoecil-*]. Type genus: *Eupoecila* H. C. C. Burmeister, 1842.

Hemipharidae Kraatz, 1880b: 182 [stem: *Hemiphar-*]. Type genus: *Hemipharis* H. C. C. Burmeister, 1842.

Tribe Stenotarsiini Kraatz, 1880

Stenotarsiden Kraatz, 1880b: 182 [stem: *Stenotarsi-*]. Type genus: *Stenotarsia* H. C. C. Burmeister, 1842.

Subtribe Anochiliina Krikken, 1984

*Anochiliens Pouillaude, 1917: 54 [stem: *Anochili-*]. Type genus: *Anochilia* H. C. C. Burmeister, 1842. Comment: original vernacular name unavailable (Art. 11.7.2): proposed after 1899.

Anochiliina Krikken, 1984: 31, in key [stem: *Anochili-*]. Type genus: *Anochilia* H. C. C. Burmeister, 1842.

Subtribe Coptomiina Schenkling, 1921

Coptomiini Schenkling, 1921: 147 [stem: *Coptomi-*]. Type genus: *Coptomia* H. C. C. Burmeister, 1842.

Subtribe Chromoptiliina Krikken, 1984

Chromoptiliina Krikken, 1984: 31, in key [stem: *Chromoptili-*]. Type genus: *Chromoptilia* Westwood, 1842.

Subtribe Doryscelina Schenkling, 1921

*Dorysceliens Pouillaude, 1917: 53, in key [stem: *Doryscel-*]. Type genus: *Doryscelis* Dejean, 1836. Comment: original vernacular name unavailable (Art. 11.7.2): proposed after 1899.

Doryscelina Schenkling, 1921: 148 [stem: *Doryscel-*]. Type genus: *Doryscelis* Dejean, 1836. Comment: name attributed to Krikken (1984) by A. B. T. Smith (2006: 180); current spelling maintained (Art. 29.3.1.1): incorrect stem formation in prevailing usage (should be *Doryscelid-*).

Subtribe Euchroeina Paulian and Descarpentries, 1982

*Euchroeens Pouillaude, 1917: 64 [stem: *Euchroe-*]. Type genus: *Euchroea* H. C. C. Burmeister, 1842. Comment: original vernacular name unavailable (Art. 11.7.2): proposed after 1899; the author also used the incorrect spelling *Euchaeeus* in the key on page 54.

Euchroeina Paulian and Descarpentries, 1982: 5 [stem: *Euchroe-*]. Type genus: *Euchroea* H. C. C. Burmeister, 1842. Comment: the name Euchroeidae Dahlbom, 1854 (type genus *Euchroeus* Latreille, 1809) is available in Hymenoptera; this case is to be referred to the Commission to remove the homonymy (Art. 55.3.1).

Subtribe Heterophanina Schoch, 1894

Heterophanae Schoch, 1894: 173 [stem: *Heterophan-*]. Type genus: *Heterophana* H. C. C. Burmeister, 1842.

Subtribe Heterosomatina Krikken, 1984

Heterosomatina Krikken, 1984: 29, in key [stem: *Heterosomat-*]. Type genus: *Heterosoma* Schaum, 1845.

Subtribe Pantoliina Krikken, 1984

*Pantoliens Pouillaude, 1917: 54 [stem: *Pantoli-*]. Type genus: *Pantolia* H. C. C. Burmeister, 1842. Comment: original vernacular name unavailable (Art. 11.7.2): proposed after 1899.

Pantoliina Krikken, 1984: 33, in key [stem: *Pantoli-*]. Type genus: *Pantolia* H. C. C. Burmeister, 1842.

Subtribe Parachiliina Krikken, 1984

Parachiliina Krikken, 1984: 33, in key [stem: *Parachili-*]. Type genus: *Parachilia* H. C. C. Burmeister, 1842.

Subtribe Stenotarsiina Kraatz, 1880

Stenotarsiden Kraatz, 1880b: 182 [stem: *Stenotarsi-*]. Type genus: *Stenotarsia* H. C. C. Burmeister, 1842. Comment: original vernacular name available (Art. 11.7.2): first used in latinized form by Jakobson (1915: 961, as

STENOTARSINA [incorrect stem formation]), generally accepted as in A. B. T. Smith (2006: 179, as STENOTARSIINI); incorrect original stem formation, not in prevailing usage.

Tribe TAENIODERINI Mikšić, 1976

TAENIODERINA Mikšić, 1976: 29 [stem: *Taenioder-*]. Type genus: *Taeniodera* H. C. C. Burmeister, 1842. Comment: First Reviser found (TAENIODERINI Mikšić, 1976 vs CHALCOTHEINI Mikšić, 1976) is Krikken (1984: 63).

Subtribe CHALCOTHEINA Mikšić, 1976

CHALCOTHEINA Mikšić, 1976: 357 [stem: *Chalcothe-*]. Type genus: *Chalcothea* H. C. C. Burmeister, 1842.

Subtribe TAENIODERINA Mikšić, 1976

TAENIODERINA Mikšić, 1976: 29 [stem: *Taenioder-*]. Type genus: *Taeniodera* H. C. C. Burmeister, 1842.

Tribe TRICHIINI Fleming, 1821

TRICHIADAE Fleming, 1821: 50 [stem: *Trichi-*]. Type genus: *Trichius* Fabricius, 1775 [placed on the Official List of Generic Names in Zoology (ICZN 2004c)]. Comment: TRICHIINAE Lozek, 1956 (type genus *Trichia* Hartmann, 1840) proposed in Mollusca: Gastropoda is a junior homonym of this family-group name; TRICHIINAE Lozek, 1956 was recently placed on the Official index of Rejected and Invalid Family-Group Names in Zoology (ICZN 2004c).

Subtribe CRYPTODONTINA Lacordaire, 1856

CRYPTODONTIDES Lacordaire, 1856: 462 [stem: *Cryptodont-*]. Type genus: *Cryptodontes* H. C. C. Burmeister, 1847. Comment: published before 29 March 1856; original vernacular name available (Art. 11.7.2): first used in latinized form by Imhoff (1856 [before 25 December]: xii, as CRYPTODONTIDAE), generally accepted as in A. B. T. Smith (2006: 177, as CRYPTODONTINA); the name CRYPTODONTIDAE Dall, 1895 (type genus *Cryptodon* Turton, 1822) is available in Mollusca: Bivalvia while the name CRYPTODONTIDAE Owen, 1860 used in Therapsida is unavailable because it is not based on an available genus-group name; this case is to be referred to the Commission to remove the homonymy (Art. 55.3.1).

Subtribe INCINA Burmeister, 1842

INCADAE H. C. C. Burmeister, 1842: 704 [stem: *Inc-*]. Type genus: *Inca* Lepeletier and Audinet-Serville, 1828.

Subtribe Osmodermatina Schenkling, 1922

Osmodermini Schenkling, 1922: 3 [stem: *Osmodermat-*]. Type genus: *Osmoderma* Lepeletier and Audinet-Serville, 1828 [placed on the Official List of Generic Names in Zoology (ICZN 2007)]. Comment: incorrect original stem formation, not in prevailing usage.

Subtribe Platygeniina Krikken, 1984

Platygeniini Krikken, 1984: 18, in key [stem: *Platygeni-*]. Type genus: *Platygenia* W. S. MacLeay, 1819.

Subtribe Trichiina Fleming, 1821

*Trichides Baudet-Lafarge, 1809: 34 [stem: *Trichi-*]. Type genus: *Trichius* Fabricius, 1775 [placed on the Official List of Generic Names in Zoology (ICZN 2004c)]. Comment: original vernacular name unavailable (Art. 11.7.2): subsequently used in latinized form but not generally attributed to Baudet-Lafarge (1809); incorrect original stem formation, not in prevailing usage.

Trichiadae Fleming, 1821: 50 [stem: *Trichi-*]. Type genus: *Trichius* Fabricius, 1775 [placed on the Official List of Generic Names in Zoology (ICZN 2004c)].

Paniscidae Gistel, 1848: [5] [stem: *Panisc-*]. Type genus: *Paniscus* Gistel, 1848 [preoccupied genus name, not *Paniscus* Schrank, 1802 [Hymenoptera]; Gistel (1848: [5]) originally included *Scarabaeus fasciatus* Linnaeus, 1758 and *Trichius zonatus* Germar, 1831 in his genus *Paniscus*, we hereby select *Scarabaeus fasciatus* Linnaeus, 1758 as the type species of *Paniscus* Gistel, 1848; **syn. nov.** of *Trichius* Fabricius, 1775]. Comment: permanently invalid (Art. 39): based on preoccupied type genus.

Elpidides Péringuey, 1907: 314 [stem: *Elpid-*]. Type genus: *Elpidus* Péringuey, 1907.

Myodermini Péringuey, 1907: 313 [stem: *Myoderm-*]. Type genus: *Myodermum* H. C. C. Burmeister and Schaum, 1840.

Tribe Valgini Mulsant, 1842

Valguaires Mulsant, 1842: 519 [stem: *Valg-*]. Type genus: *Valgus* Scriba, 1790.

Subtribe Microvalgina Kolbe, 1904

Microvalginae Kolbe, 1904: 10 [stem: *Microvalg-*]. Type genus: *Microvalgus* Kraatz, 1883.

Subtribe Valgina Mulsant, 1842

Valguaires Mulsant, 1842: 519 [stem: *Valg-*]. Type genus: *Valgus* Scriba, 1790. Comment: published before 6 August 1842; original vernacular name available (Art. 11.7.2): used in latinized form by several authors, generally accepted as in A. B. T. Smith (2006: 177, as VALGINI); this family-group name was also used in the same year by Burmeister (1842 [before 28 December]: 718, as VALGIDAE); incorrect original stem formation, not in prevailing usage.

ACANTHOVALGINAE Kolbe, 1904: 11 [stem: *Acanthovalg-*]. Type genus: *Acanthovalgus* Kraatz, 1895.

COSMOVALGINAE Kolbe, 1904: 11 [stem: *Cosmovalg-*]. Type genus: *Cosmovalgus* Kolbe, 1897.

DASYVALGINAE Kolbe, 1904: 11 [stem: *Dasyvalg-*]. Type genus: *Dasyvalgus* Kolbe, 1904.

ISCHNOVALGINAE Kolbe, 1904: 9 [stem: *Ischnovalg-*]. Type genus: *Ischnovalgus* Kolbe, 1897.

SPHINCTOVALGINAE Kolbe, 1904: 9 [stem: *Sphinctovalg-*]. Type genus: *Sphinctovalgus* Kolbe, 1904.

Tribe XIPHOSCELIDINI Burmeister, 1842

XIPHOSCELIDEAE H. C. C. Burmeister, 1842: 613 [stem: *Xiphoscelid-*]. Type genus: *Xiphoscelis* H. C. C. Burmeister, 1842. Comment: incorrect original stem formation, not in prevailing usage.

†Family COPRINISPHAERIDAE Genise, 2004

COPRINISPHAERIDAE Genise, 2004: 426 [stem: *Coprinisphaer-*]. Type genus: *Coprinisphaera* Sauer, 1955 [placed on the Official List of Generic Names in Zoology and given precedence over *Fontanai* Roselli, 1939 (ICZN 2008b)]. Comment: ichnotaxon based on dung beetle burrows.

†Family PALLICHNIDAE Genise, 2004

PALLICHNIDAE Genise, 2004: 432 [stem: *Pallichn-*]. Type genus: *Pallichnus* Retallack, 1984. Comment: ichnotaxon based on dung beetle burrows.

Series ELATERIFORMIA

Superfamily SCIRTOIDEA Fleming, 1821

SCIRTESIDAE Fleming, 1821: 50 [stem: *Scirt-*]. Type genus: *Scirtes* Illiger, 1807. Comment: as pointed out by Lawrence and Newton (1995: 838) the name SCIRTOIDEA Fleming, 1821 has priority over CLAMBOIDEA Fischer von Waldheim, 1821.

Family DECLINIIDAE Nikitsky, Lawrence, Kirejtshuk and Gratshev, 1994

DECLINIIDAE Nikitsky et al., 1994: 7 [stem: *Declini-*]. Type genus: *Declinia* Nikitsky et al., 1994.

Family Eucinetidae Lacordaire, 1857

Eucinétides Lacordaire, 1857: 281 [stem: *Eucinet-*]. Type genus: *Eucinetus* Germar, 1818. Comment: original vernacular name available (Art. 11.7.2): first used in latinized form by J. L. LeConte (1861: 181, as Eucinetini), generally accepted as in Hansen (1996: 135, as Eucinetidae).

Cryptomeridae Broun, 1893: 1358 [stem: *Cryptomer-*]. Type genus: *Cryptomera* Broun, 1893 [syn. of *Eucinetus* Germar, 1818].

Family Clambidae Fischer von Waldheim, 1821

Clambini Fischer von Waldheim, 1821: 52 [stem: *Clamb-*]. Type genus: *Clambus* Fischer von Waldheim, 1821.

Subfamily Calyptomerinae Crowson, 1955

Calyptomeridae Crowson, 1955: 11 [stem: *Calyptomer-*]. Type genus: *Calyptomerus* Redtenbacher, 1849.

Subfamily Acalyptomerinae Crowson, 1979

Acalyptomerinae Crowson, 1979: 612, in key [stem: *Acalyptomer-*]. Type genus: *Acalyptomerus* Crowson, 1979.

Subfamily Clambinae Fischer von Waldheim, 1821

Clambini Fischer von Waldheim, 1821: 52 [stem: *Clamb-*]. Type genus: *Clambus* Fischer von Waldheim, 1821.

Family Scirtidae Fleming, 1821

Scirtesidae Fleming, 1821: 50 [stem: *Scirt-*]. Type genus: *Scirtes* Illiger, 1807.

†Sinodryopitidae Hong, 2002: 107 [stem: *Sinodryopit-*]. Type genus: *Sinodryopites* Hong, 2002. Comment: originally described in Byrrhoidea; synonymy with Scirtidae by Kirejtshuk and Azar (2008: 29).

Subfamily Scirtinae Fleming, 1821

Scirtesidae Fleming, 1821: 50 [stem: *Scirt-*]. Type genus: *Scirtes* Illiger, 1807. Comment: incorrect original stem formation, not in prevailing usage.

Cyphonidae Stephens, 1829a: 11 [stem: *Cyphon-*]. Type genus: *Cyphon* Paykull, 1799.

Elodiidae Shuckard, 1839b: 41 [stem: *Elod-*]. Type genus: *Elodes* Latreille, 1797. Comment: incorrect original stem formation, not in prevailing usage.

Atopidini Pic, 1914: 16 [stem: *Atopid-*]. Type genus: *Atopida* A. White, 1846.

Subfamily Nipponocyphoninae Lawrence and Yoshitomi, 2007

Nipponocyphoninae Lawrence and Yoshitomi, 2007: 522 [stem: *Nipponocyphon-*]. Type genus: *Nipponocyphon* Lawrence and Yoshitomi, 2007.

Subfamily Stenocyphoninae Lawrence and Yoshitomi, 2007

Stenocyphoninae Lawrence and Yoshitomi, 2007: 522 [stem: *Stenocyphon-*]. Type genus: *Stenocyphon* Lawrence, 2001.

†Family Elodophthalmidae Kirejtshuk and Azar, 2008

Elodophthalmidae Kirejtshuk and Azar, 2008: 24 [stem: *Elodophthalm-*]. Type genus: *Elodophthalmus* Kirejtshuk and Azar, 2008.

†Family Mesocinetidae Kirejtshuk and Ponomarenko, 2010

Mesocinetidae Kirejtshuk and Ponomarenko, 2010: 304 [stem: *Mesocinet-*]. Type genus: *Mesocinetus* Ponomarenko, 1986.

Superfamily Dascilloidea Guérin-Méneville, 1843 (1834)

Dascillidae Guérin-Méneville, 1843: 193 [stem: *Dascill-*]. Type genus: *Dascillus* Latreille, 1797. Comment: usage of younger name conserved over Atopoidea Laporte, 1834 (Art. 40.2); although Rhipiceroidea is also an older name for this superfamily, its use for a taxon including the families Dascilidae (including Karumiinae) and Rhipiceridae is likely to cause some confusion and we therefore continue to use Dascilloidea as valid; the concept of Rhipiceroidea has varied among authors, Crowson (1953) used Rhipiceroidea for the families Rhipiceridae and Callirhipidae, which are not now considered to form a monophyletic group, and Dascilloidea for Dascillidae plus those families now included in Scirtoidea; the family group name Sandalidae (which is a junior synonym of Rhipiceridae) was used by Craighead (1921) and Böving and Craighead (1931) for Rhipiceridae in the strict sense, while Emden (1924, 1931, 1933) used the same name for a family which also included those genera now placed in Callirhipidae; Crowson (1971, 1973b) proposed a reconstituted Dascilloidea for Dascillidae plus Rhipiceridae (sensu stricto) and excluded Callirhipidae from Rhipiceroidea and placed it in another superfamily Artematopoidea (along with Artematopidae (now Artematopodidae) and Brachypsectridae); finally, the evidence for placing Dascillidae and Rhipiceridae in the same superfamily is not convincing (especially when the larvae are taken into account) so it is quite possible that Dascilloidea in the sense of Crowson (1971) may cease to exist when more evidence is presented.

Family Dascillidae Guérin-Méneville, 1843 (1834)

Dascillidae Guérin-Méneville, 1843: 193 [stem: *Dascill-*]. Type genus: *Dascillus* Latreille, 1797.

Subfamily Dascillinae Guérin-Méneville, 1843 (1834)

Dascillidae Guérin-Méneville, 1843: 193 [stem: *Dascill-*]. Type genus: *Dascillus* Latreille, 1797.

Tribe Cinnabariini Pic, 1914

Cinnabariini Pic, 1914: 15 [stem: *Cinnabari-*]. Type genus: *Cinnabarium* Fairmaire, 1895 [syn. of *Coptocera* Murray, 1868].

Tribe Dascillini Guérin-Méneville, 1843 (1834)

Atopites Laporte, 1834a: 227 [stem: *Atop-*]. Type genus: *Atopa* Paykull, 1799 [syn. of *Dascillus* Latreille, 1797]. Comment: original vernacular name available (Art. 11.7.2): first used in latinized form and generally accepted as in Laporte (1836: 21, as Atopidae).

Dascillidae Guérin-Méneville, 1843: 193 [stem: *Dascill-*]. Type genus: *Dascillus* Latreille, 1797. Comment: name proposed to replace Atopidae Laporte, 1834 because of the synonymy of the type genus; usage of younger name conserved over Atopini Laporte, 1834 (Art. 40.2) (see Lawrence and Newton 1995).

Subfamily Karumiinae Escalera, 1913

Karuminae Escalera, 1913: 320 [stem: *Karumi-*]. Type genus: *Karumia* Escalera, 1913.

Tribe Emmitini Escalera, 1914

Emminae Escalera, 1913: 318 [stem: *Emm-*]. Type genus: *Emma* Escalera, 1913 [preoccupied genus name, not *Emma* Gray, 1843 [Bryozoa]; syn. of *Emmita* Escalera, 1914]. Comment: permanently invalid (Art. 39): based on preoccupied type genus.

Emmitinae Escalera, 1914a: 349 [stem: *Emmit-*]. Type genus: *Emmita* Escalera, 1914. Comment: replacement name for Emminae Escalera, 1913 because of the homonymy of the type genus.

Tribe Escalerinini Paulus, 1972

Escalerini Paulus, 1972a: 49, in key [stem: *Escalerin-*]. Type genus: *Escalerina* Bolívar, 1926. Comment: incorrect original stem formation, not in prevailing usage.

Tribe Genecerini Pic, 1914

Genecerini Pic, 1914: 16 [stem: *Genecer-*]. Type genus: *Genecerus* Walker, 1871.

Tribe Karumiini Escalera, 1913

Karuminae Escalera, 1913: 320 [stem: *Karumi-*]. Type genus: *Karumia* Escalera, 1913. Comment: incorrect original stem formation, not in prevailing usage.

Zarudniolidae Semenov and Martynov, 1925: 74 [stem: *Zarudniol-*]. Type genus: *Zarudniola* Semenov and Martynov, 1925 [syn. of *Karumia* Escalera, 1913].

Family Rhipiceridae Latreille, 1834

Rhipicerides Latreille, 1834: 167 [stem: *Rhipicer-*]. Type genus: *Rhipicera* Latreille, 1816. Comment: published in issue 1 of volume 3 of the Annales de la Société

Entomologique de France; this family-group name was also used in the same year by Laporte (1834a: 226, as Rhipicérites) in issue 2 of volume 3 of the same journal.
Sandalidae Jakobson, 1913: 729 [stem: *Sandal-*]. Type genus: *Sandalus* Knoch, 1801.

Superfamily Buprestoidea Leach, 1815

Buprestides Leach, 1815: 85 [stem: *Buprest-*]. Type genus: *Buprestis* Linnaeus, 1758 [placed on the Official List of Generic Names in Zoology (ICZN 1994e)].

Family Schizopodidae LeConte, 1859

Schizopodidae J. L. LeConte, 1859b: 122 [stem: *Schizopod-*]. Type genus: *Schizopus* J. L. LeConte, 1858 [placed on the Official List of Generic Names in Zoology (ICZN 1993d)].

Subfamily Schizopodinae LeConte, 1859

Schizopodidae J. L. LeConte, 1859b: 122 [stem: *Schizopod-*]. Type genus: *Schizopus* J. L. LeConte, 1858 [placed on the Official List of Generic Names in Zoology (ICZN 1993d)].

Tribe Dystaxiini Théry, 1929

Dystaxini Théry, 1929: 60 [stem: *Dystaxi-*]. Type genus: *Dystaxia* J. L. LeConte, 1866. Comment: incorrect original stem formation, not in prevailing usage.
Glyptoscelimorphini Cobos, 1963: 354 [stem: *Glyptoscelimorph-*]. Type genus: *Glyptoscelimorpha* G. H. Horn, 1893.

†Tribe Electrapatini Iablokoff-Khnzorian, 1962

Electrapatidae Iablokoff-Khnzorian, 1962: 87 [stem: *Electrapat-*]. Type genus: *Electrapate* Iablokoff-Khnzorian, 1962.

Tribe Schizopodini LeConte, 1859

Schizopodidae J. L. LeConte, 1859b: 122 [stem: *Schizopod-*]. Type genus: *Schizopus* J. L. LeConte, 1858 [*Schizopus* J. L. LeConte, 1858 has precedence over *Schizopus* Claparéde et Lachman, 1858 and was placed on the Official List of Generic Names in Zoology (ICZN 1993d)].

Family Buprestidae Leach, 1815

Buprestides Leach, 1815: 85 [stem: *Buprest-*]. Type genus: *Buprestis* Linnaeus, 1758 [placed on the Official List of Generic Names in Zoology (ICZN 1994e)].

Subfamily Julodinae Lacordaire, 1857

Julodides Lacordaire, 1857: 10 [stem: *Julod-*]. Type genus: *Julodis* Eschscholtz, 1829. Comment: original vernacular name available (Art. 11.7.2): first used in latinized form by J. L. LeConte (1861: 154, as JULODINI), generally accepted as in Bellamy (2008a: 23, as JULODINAE).

STERNOCERINI Csiki, 1904: 132 [stem: *Sternocer-*]. Type genus: *Sternocera* Eschscholtz, 1829.

AMBLYSTERNINI Cobos, 1955: 22, in key [stem: *Amblystern-*]. Type genus: *Amblysterna* Saunders, 1871.

Subfamily POLYCESTINAE Lacordaire, 1857

POLYCESTIDES Lacordaire, 1857: 61 [stem: *Polycest-*]. Type genus: *Polycesta* Dejean, 1833.

Tribe ACMAEODERINI Kerremans, 1893

ACMAEODERINI Kerremans, 1893: 112 [stem: *Acmaeoder-*]. Type genus: *Acmaeodera* Eschscholtz, 1829 [placed on the Official List of Generic Names in Zoology (ICZN 2005b)].

Subtribe ACMAEODERINA Kerremans, 1893

ACMAEODERINI Kerremans, 1893: 112 [stem: *Acmaeoder-*]. Type genus: *Acmaeodera* Eschscholtz, 1829 [placed on the Official List of Generic Names in Zoology (ICZN 2005b)].

Subtribe ACMAEODEROIDINA Cobos, 1955

ACMAEODEROIDINI Cobos, 1955: 15 [stem: *Acmaeoderoid-*]. Type genus: *Acmaeoderoides* Van Dyke, 1942.

Subtribe NOTHOMORPHINA Cobos, 1955

NOTOMORPHINI Cobos, 1955: 17 [stem: *Nothomorph-*]. Type genus: *Nothomorpha* Saunders, 1871 [as *Notomorpha*, incorrect subsequent spelling of type genus name, not in prevailing usage]. Comment: Cobos (1955: 23) also used the spelling NOTHOMORPHINI in his original paper, this is considered a misspelling since he listed the type genus name as *Notomorpha*; incorrect original stem formation, not in prevailing usage.

ACMAEODERINI *incertae sedis*

*ODETTEINA Volkovitsh, 2001: 52, 91 [stem: *Odette-*]. Type genus: *Odettea* Baudon, 1966. Comment: unavailable family-group name, proposed after 1930 without description or bibliographic reference to such a description (Art. 13.1).

Tribe ASTRAEINI Cobos, 1980

ASTRAEUSINI Cobos, 1980: 28 [stem: *Astrae-*]. Type genus: *Astraeus* Laporte and Gory, 1838 [placed on the Official List of Generic Names in Zoology (ICZN 1966)]. Comment: incorrect original stem formation, not in prevailing usage.

Tribe Bulini Bellamy, 1995

Bulisina Bellamy, 1995: 173 [stem: *Bul-*]. Type genus: *Bulis* Laporte and Gory, 1838. Comment: incorrect original stem formation, not in prevailing usage; stem correction by Bellamy (1996: 222).

Tribe Haplostethini LeConte, 1861

Haplostethini J. L. LeConte, 1861: 155 [stem: *Haplosteth-*]. Type genus: *Haplostethus* J. L. LeConte, 1860.

Mastogenini J. L. LeConte and G. H. Horn, 1883: 199 [stem: *Mastogeni-*]. Type genus: *Mastogenius* Solier, 1849. Comment: incorrect original stem formation, not in prevailing usage.

Tribe Paratracheini Cobos, 1980

Paratrachysae Cobos, 1980: 47 [stem: *Paratrache-*]. Type genus: *Paratrachys* Saunders, 1873. Comment: here we adopt the stem *Paratrache-* since the correct stem for the genus *Trachys*, with the same ending, was recently determined to be *Trache-* (ICZN 2009a).

Tribe Perucolini Cobos, 1980

Perucolini Cobos, 1980: 81 [stem: *Perucol-*]. Type genus: *Perucola* Théry, 1925.

Tribe Polycestini Lacordaire, 1857

Polycestides Lacordaire, 1857: 61 [stem: *Polycest-*]. Type genus: *Polycesta* Dejean, 1833.

Subtribe Polycestina Lacordaire, 1857

Polycestides Lacordaire, 1857: 61 [stem: *Polycest-*]. Type genus: *Polycesta* Dejean, 1833. Comment: original vernacular name available (Art. 11.7.2): first used in latinized form by Stein (1868: 62, as Polycestini), generally accepted as in Bellamy (2008a: 23, as Polycestini).

Subtribe Xenopseina Volkovitsh, 2008

Xenopsina Volkovitsh, 2008: 628 [stem: *Xenopse-*]. Type genus: *Xenopsis* Saunders, 1867. Comment: incorrect original stem formation, not in prevailing usage.

Tribe Polyctesini Cobos, 1955

Polyctesini Cobos, 1955: 6 [stem: *Polyctes-*]. Type genus: *Polyctesis* Marseul, 1865. Comment: current spelling maintained (Art. 29.5): incorrect stem formation in prevailing usage (should be *Polyctese-*).

Tribe Prospherini Cobos, 1980

Prospheresini Cobos, 1980: 84 [stem: *Prospher-*]. Type genus: *Prospheres* Saunders, 1868. Comment: incorrect original stem formation, not in prevailing usage.

Tribe Ptosimini Kerremans, 1903

Ptosimites Kerremans, 1903: 37 [stem: *Ptosim-*]. Type genus: *Ptosima* Dejean, 1833.

Tribe Thrincopygini LeConte, 1861

Thrincopygini J. L. LeConte, 1861: 154 [stem: *Thrincopyg-*]. Type genus: *Thrincopyge* J. L. LeConte, 1858.

Tribe Tyndaridini Cobos, 1955

Tyndarini Cobos, 1955: 11 [stem: *Tyndarid-*]. Type genus: *Tyndaris* J. Thomson, 1857.

Subtribe Mimicoclytrinina Bellamy, 2003

Acherusini Cobos, 1955: 24, in key [stem: *Acherusi-*]. Type genus: *Acherusia* Laporte and Gory, 1838 [preoccupied genus name, not *Acherusia* Costa, 1834 [Crustacea]; syn. of *Mimicoclytrina* Bellamy, 2003]. Comment: permanently invalid (Art. 39): based on preoccupied type genus.

Mimicoclytrinina Bellamy, 2003: 25 [stem: *Mimicoclytrin-*]. Type genus: *Mimicoclytrina* Bellamy, 2003. Comment: replacement name for Acherusiina Cobos, 1955 because of the homonymy of the type genus.

Subtribe Pseudacherusiina Cobos, 1980

Pseudoacherusini Cobos, 1980: 78 [stem: *Pseudacherusi-*]. Type genus: *Pseudacherusia* Kerremans, 1905. Comment: incorrect original stem formation, not in prevailing usage.

Subtribe Tylaucheniina Cobos, 1959

Tylacheniae Cobos, 1959: 4 [stem: *Tylaucheni-*]. Type genus: *Tylauchenia* H. C. C. Burmeister, 1872 [syn. of *Ocypetes* Saunders, 1871]. Comment: incorrect original stem formation, not in prevailing usage.

Subtribe Tyndaridina Cobos, 1955

Tyndarini Cobos, 1955: 11 [stem: *Tyndarid-*]. Type genus: *Tyndaris* J. Thomson, 1857. Comment: incorrect original stem formation, not in prevailing usage.

Tribe Xyroscelidini Cobos, 1955

Xiroscelini Cobos, 1955: 19 [stem: *Xyroscelid-*]. Type genus: *Xyroscelis* Saunders, 1868 [as *Xiroscelis*, incorrect subsequent spelling of type genus name, not in prevailing usage]. Comment: incorrect original stem formation, not in prevailing usage.

Subfamily Galbellinae Reitter, 1911

Galbellinae Reitter, 1911: 178 [stem: *Galbell-*]. Type genus: *Galbella* Westwood, 1848.

Subfamily Chrysochroinae Laporte, 1835

Chrysochroidae Laporte, 1835b: 158 [stem: *Chrysochro-*]. Type genus: *Chrysochroa* Dejean, 1833.

Tribe Chrysochroini Laporte, 1835

Chrysochroidae Laporte, 1835b: 158 [stem: *Chrysochro-*]. Type genus: *Chrysochroa* Dejean, 1833.

Subtribe Chalcophorina Lacordaire, 1857 (1848)

Anaglyptisidae Gistel, 1848: [5] [stem: *Anaglypt-*]. Type genus: *Anaglyptes* Gistel, 1848 [syn. of *Chalcophora* Dejean, 1833]. Comment: senior homonym of Anaglyptini Lacordaire, 1868 (type genus *Anaglyptus* Mulsant, 1839) in Cerambycidae; incorrect original stem formation, not in prevailing usage; the younger name Chalcophorina Lacordaire, 1857 is conserved over this name (Art. 40.2).

Chalcophorides Lacordaire, 1857: 14 [stem: *Chalcophor-*]. Type genus: *Chalcophora* Dejean, 1833. Comment: original vernacular name available (Art. 11.7.2): first used in latinized form by J. L. LeConte (1861: 151, as Chalcophorae), generally accepted as in Bellamy (2008a: 23, as Chalcophorina); name conserved over the older name Anaglyptina Gistel, 1848 (Art. 40.2).

*Chrysodemides H. Deyrolle, 1865: 11 [stem: *Chrysodem-*]. Type genus: *Chrysodema* Laporte and Gory, 1837 [placed on the Official List of Generic Names in Zoology (ICZN 2004b)]. Comment: original vernacular name unavailable (Art. 11.7.2): subsequently used in latinized form but not generally attributed to H. Deyrolle (1865).

Chrysodémides Kerremans, 1892: 49 [stem: *Chrysodem-*]. Type genus: *Chrysodema* Laporte and Gory, 1837 [placed on the Official List of Generic Names in Zoology (ICZN 2004b)]. Comment: original vernacular name available (Art. 11.7.2): first used in latinized form and generally accepted as in Skinner (1892: 124, as Chrysodemidae).

Chalcophorellini Tôyama, 1986: 189 [stem: *Chalcophorell-*]. Type genus: *Chalcophorella* Kerremans, 1903.

Iridotaenini Tôyama, 1987: 5 [stem: *Iridotaeni-*]. Type genus: *Iridotaenia* Deyrolle, 1864 [placed on the Official List of Generic Names in Zoology (ICZN 2004b)]. Comment: incorrect original stem formation, not in prevailing usage.

Lampropeplina Holyński, 1993: 22 [stem: *Lampropepl-*]. Type genus: *Lampropepla* Fairmaire, 1904 [syn. of *Madecassia* Kerremans, 1903].

Subtribe Chrysochroina Laporte, 1835

Chrysochroidae Laporte, 1835b: 158 [stem: *Chrysochro-*]. Type genus: *Chrysochroa* Dejean, 1833.

CATOXANTHINA Jakobson, 1913: 772 [stem: *Catoxanth-*]. Type genus: *Catoxantha* Solier, 1833.

Subtribe EUCALLOPISTINA Bellamy, 2003

CALLOPISTINA Kurosawa, 1990: 63 [stem: *Callopist-*]. Type genus: *Callopistus* Deyrolle, 1864 [preoccupied genus name, not *Callopistus* Say, 1831 [Coleoptera: CURCULIONIDAE]; syn. of *Eucallopistus* Bellamy, 2003]. Comment: permanently invalid (Art. 39): based on preoccupied type genus.

EUCALLOPISTINA Bellamy, 2003: 31 [stem: *Eucallopist-*]. Type genus: *Eucallopistus* Bellamy, 2003. Comment: replacement name for CALLOPISTINA Kurosawa, 1990 because of the homonymy of the type genus.

Tribe DICERCINI Gistel, 1848

DICERCAEIDAE Gistel, 1848: [5] [stem: *Dicerc-*]. Type genus: *Dicerca* Eschscholtz, 1829 [placed on the Official List of Generic Names in Zoology (ICZN 1994e)]. Comment: First Reviser found (DICERCINI Gistel, 1848 vs POLYBOTHRISINI Gistel, 1848) is Bellamy (2008b: 819).

Subtribe DICERCINA Gistel, 1848

DICERCAEIDAE Gistel, 1848: [5] [stem: *Dicerc-*]. Type genus: *Dicerca* Eschscholtz, 1829 [placed on the Official List of Generic Names in Zoology (ICZN 1994e)]. Comment: incorrect original stem formation, not in prevailing usage.

POLYBOTHRISIDAE Gistel, 1848: [5] [stem: *Polybothrid-*]. Type genus: *Polybothris* Spinola, 1837. Comment: incorrect original stem formation, not in prevailing usage.

PSILOPTÉRIDES Lacordaire, 1857: 26 [stem: *Psilopter-*]. Type genus: *Psiloptera* Dejean, 1833. Comment: original vernacular name available (Art. 11.7.2): first used in latinized form by J. Thomson (1858: 72, as PSILOPTERITAE), generally accepted as in Bellamy (2002: 61, as PSILOPTERINI).

CAPNODINA Jakobson, 1913: 779 [stem: *Capnod-*]. Type genus: *Capnodis* Eschscholtz, 1829.

Subtribe HAPLOTRINCHINA Holyński, 1993

HAPLOTRINCHINA Holyński, 1993: 27 [stem: *Haplotrinch-*]. Type genus: *Haplotrinchus* Kerremans, 1903.

Subtribe HIPPOMELANINA Holyński, 1993

HIPPOMELANINA Holyński, 1993: 24 [stem: *Hippomelan-*]. Type genus: *Hippomelas* Gory and Laporte, 1837.

Subtribe Pseudoperotina Tôyama, 1987

Pseudoperotina Tôyama, 1987: 4 [stem: *Pseudoperot-*]. Type genus: *Pseudoperotis* Obenberger, 1936.

Tribe Evidini Tôyama, 1987

Evidini Tôyama, 1987: 6 [stem: *Evid-*]. Type genus: *Evides* Dejean, 1833.

Tribe Paraleptodemini Cobos, 1975

Paraleptodemini Cobos, 1975: 88 [stem: *Paraleptodem-*]. Type genus: *Paraleptodema* Obenberger, 1936 [syn. of *Cinyra* Laporte and Gory, 1837].

Subtribe Euchromatina Holyński, 1993

Euchromatina Holyński, 1993: 23 [stem: *Euchromat-*]. Type genus: *Euchroma* Dejean, 1833.

Subtribe Euplectaleciina Holyński, 1993

Euplectaleciina Holyński, 1993: 24 [stem: *Euplectaleci-*]. Type genus: *Euplectalecia* Obenberger, 1924.

Subtribe Hypoprasina Holyński, 1993

Hypoprasina Holyński, 1993: 23 [stem: *Hypopras-*]. Type genus: *Hypoprasis* Fairmaire and Germain, 1864. Comment: current spelling maintained (Art. 29.5): incorrect stem formation in prevailing usage (should be *Hypoprase-*).

Subtribe Paraleptodemina Cobos, 1975

Paraleptodemini Cobos, 1975: 88 [stem: *Paraleptodem-*]. Type genus: *Paraleptodema* Obenberger, 1936 [syn. of *Cinyra* Laporte and Gory, 1837].

Cinyrini Cobos, 1979a: 226 [stem: *Cinyr-*]. Type genus: *Cinyra* Gory and Laporte, 1837.

Subtribe Pristipterina Holyński, 1993

Pristipterina Holyński, 1993: 23 [stem: *Pristipter-*]. Type genus: *Pristiptera* Dejean, 1833 [syn. of *Pelecopselaphus* Solier, 1833].

Tribe Paratassini Bílý and Volkovitsh, 1996

Paratassini Bílý and Volkovitsh, 1996: 329 [stem: *Paratass-*]. Type genus: *Paratassa* Marseul, 1882.

Tribe Poecilonotini Jakobson, 1913

Poecilonotina Jakobson, 1913: 773 [stem: *Poecilonot-*]. Type genus: *Poecilonota* Eschscholtz, 1829 [placed on the Official List of Generic Names in Zoology (ICZN 1996a)].

Subtribe Poecilonotina Jakobson, 1913

Poecilonotina Jakobson, 1913: 773 [stem: *Poecilonot-*]. Type genus: *Poecilonota* Eschscholtz, 1829 [placed on the Official List of Generic Names in Zoology (ICZN 1996a)].

Subtribe Nesotrinchina Bílý, Kubáň and Volkovitsh, 2009

Nesotrinchina Bílý et al., 2009: 750 [stem: *Nesotrinch-*]. Type genus: *Nesotrinchus* Obenberger, 1924.

Tribe Sphenopterini Lacordaire, 1857

Sphénoptérides Lacordaire, 1857: 68 [stem: *Sphenopter-*]. Type genus: *Sphenoptera* Dejean, 1833. Comment: original vernacular name available (Art. 11.7.2): first used in latinized form by Stein (1868: 62, as Sphenopterini), generally accepted as in Bellamy (2008a: 23, as Sphenopterini).

Tribe Vadonaxiini Descarpentries, 1970

Vadonaxiini Descarpentries, 1970: 188 [stem: *Vadonaxi-*]. Type genus: *Vadonaxia* Descarpentries, 1970.

Subfamily Buprestinae Leach, 1815

Buprestides Leach, 1815: 85 [stem: *Buprest-*]. Type genus: *Buprestis* Linnaeus, 1758 [placed on the Official List of Generic Names in Zoology (ICZN 1994e)].

Tribe Actenodini Gistel, 1848

Actenodeidae Gistel, 1848: [5] [stem: *Actenod-*]. Type genus: *Actenodes* Dejean, 1833 [placed on the Official List of Generic Names in Zoology (ICZN 2002b)]. Comment: name previously attributed to Kerremans (1890) in the literature; incorrect original stem formation, not in prevailing usage.

Belionotina Jakobson, 1913: 793 [stem: *Belionot-*]. Type genus: *Belionota* Eschscholtz, 1829.

Tribe Anthaxiini Gory and Laporte, 1839

Anthaxidae Gory and Laporte, 1839: unnumbered page [stem: *Anthaxi-*]. Type genus: *Anthaxia* Eschscholtz, 1829 [placed on the Official List of Generic Names in Zoology (ICZN 2002c)]. Comment: incorrect original stem formation, not in prevailing usage.

Tribe Bubastini Obenberger, 1920

Bubastini Obenberger, 1920: 89 [stem: *Bubast-*]. Type genus: *Bubastes* Laporte and Gory, 1838.

Tribe Buprestini Leach, 1815

Buprestides Leach, 1815: 85 [stem: *Buprest-*]. Type genus: *Buprestis* Linnaeus, 1758 [placed on the Official List of Generic Names in Zoology (ICZN 1994e)].

Subtribe Agaeocerina Nelson, 1982

Agaeocerini Nelson, 1982: 440 [stem: *Agaeocer-*]. Type genus: *Agaeocera* Saunders, 1871.

Subtribe Buprestina Leach, 1815

Buprestides Leach, 1815: 85 [stem: *Buprest-*]. Type genus: *Buprestis* Linnaeus, 1758 [placed on the Official List of Generic Names in Zoology (ICZN 1994e)]. Comment: current spelling maintained (Art. 29.3.1.1): incorrect stem formation in prevailing usage (should be *Buprestid-*).

Ancylochirina Jakobson, 1913: 787 [stem: *Ancylochir-*]. Type genus: *Ancylochira* Eschscholtz, 1829 [subgenus of *Buprestis* Linnaeus, 1758].

Subtribe Lamprocheilina Holyński, 1993

Lamprocheilina Holyński, 1993: 13 [stem: *Lamprocheil-*]. Type genus: *Lamprocheila* Saunders, 1871.

Subtribe Trachykelina Holyński, 1988

Trachykelina Holyński, 1988: 51, in key [stem: *Trachykel-*]. Type genus: *Trachykele* Marseul, 1865.

Tribe Chrysobothrini Gory and Laporte, 1836

Chrysobothridae Gory and Laporte, 1836: [1] [stem: *Chrysobothr-*]. Type genus: *Chrysobothris* Eschscholtz, 1829 [placed on the Official List of Generic Names in Zoology (ICZN 1994e)]. Comment: current spelling maintained (Art. 29.3.1.1): incorrect stem formation in prevailing usage (should be *Chrysobothrid-*).

Tribe Coomaniellini Bílý, 1974

Coomaniellini Bílý, 1974: 41 [stem: *Coomaniell-*]. Type genus: *Coomaniella* Bourgoin, 1924.

Tribe Curidini Holyński, 1988

Curidina Holyński, 1988: 52, in key [stem: *Curid-*]. Type genus: *Curis* Gory and Laporte, 1837 [syn. of *Selagis* Dejean, 1836 (also see ICZN 2008c)]. Comment: First Reviser (Curidini Holyński, 1988 vs Neocuridini Holyński, 1988) not determined, current usage maintained.

Subtribe Anilarina Bílý, 2000

Anilarini Bílý, 2000: 113 [stem: *Anilar-*]. Type genus: *Anilara* Saunders, 1868.

Subtribe Curidina Holyński, 1988

Curidina Holyński, 1988: 52, in key [stem: *Curid-*]. Type genus: *Curis* Gory and Laporte, 1837 [syn. of *Selagis* Dejean, 1836 (also see ICZN 2008c)].

Subtribe Neocuridina Holyński, 1988

Neocuridina Holyński, 1988: 52, in key [stem: *Neocurid-*]. Type genus: *Neocuris* Saunders, 1868.

Tribe Epistomentini Levey, 1978

Epistomentini Levey, 1978: 155 [stem: *Epistoment-*]. Type genus: *Epistomentis* Solier, 1849.

Tribe Exagistini Tôyama, 1987

Exagistini Tôyama, 1987: 2 [stem: *Exagist-*]. Type genus: *Exagistus* Deyrolle, 1864.

Tribe Julodimorphini Kerremans, 1903

Julodimorphites Kerremans, 1903: 16 [stem: *Julodimorph-*]. Type genus: *Julodimorpha* Harold, 1869.

Tribe Kisanthobiini Richter, 1949

Kisanthobiini Richter, 1949: 215 [stem: *Kisanthobi-*]. Type genus: *Kisanthobia* Marseul, 1865.

Tribe Maoraxiini Holyński, 1984

Maoraxiini Holyński, 1984: 106 [stem: *Maoraxi-*]. Type genus: *Maoraxia* Obenberger, 1937.

Tribe Melanophilini Bedel, 1921

Melanophilini Bedel, 1921: 171 [stem: *Melanophil-*]. Type genus: *Melanophila* Eschscholtz, 1829 [placed on the Official List of Generic Names in Zoology (ICZN 1996b)].

Tribe Melobaseini Bílý, 2000

Melobasini Bílý, 2000: 113 [stem: *Melobase-*]. Type genus: *Melobasis* Gory and Laporte, 1837. Comment: incorrect original stem formation, not in prevailing usage.

Tribe Mendizabaliini Cobos, 1968

Mendizabaliini Cobos, 1968: 19 [stem: *Mendizabali-*]. Type genus: *Mendizabalia* Cobos, 1957.

Tribe Nascionini Holyński, 1988

Nascionina Holyński, 1988: 51, in key [stem: *Nascion-*]. Type genus: *Nascio* Laporte and Gory, 1838 [placed on the Official List of Generic Names in Zoology (ICZN 2002b)].

Tribe Phrixiini Cobos, 1975

Phrixiini Cobos, 1975: 102 [stem: *Phrixi-*]. Type genus: *Phrixia* Deyrolle, 1864.

Tribe Pterobothrini Volkovitsh, 2001

Pterobothrini Volkovitsh, 2001: 86, 103 [stem: *Pterobothr-*]. Type genus: *Pterobothris* Fairmaire and Germain, 1858. Comment: current spelling maintained (Art. 29.3.1.1): incorrect original stem formation in prevailing usage (should be *Pterobothrid-*).

Tribe Stigmoderini Lacordaire, 1857

Stigmodérides Lacordaire, 1857: 52 [stem: *Stigmoder-*]. Type genus: *Stigmodera* Eschscholtz, 1829. Comment: original vernacular name available (Art. 11.7.2): first used in latinized form by Wallace (1860: 183, as Stigmoderidae), generally accepted as in Bellamy (2008a: 23, as Stigmoderini).

Tribe Thomassetiini Bellamy, 1987

Thomassetiini Bellamy, 1987: 223 [stem: *Thomasseti-*]. Type genus: *Thomassetia* Théry, 1928.

Subtribe Philanthaxiina Holyński, 1988

Philanthaxiina Holyński, 1988: 51, in key [stem: *Philanthaxi-*]. Type genus: *Philanthaxia* Deyrolle, 1864.

Subtribe Thomassetiina Bellamy, 1987

Thomassetiini Bellamy, 1987: 223 [stem: *Thomasseti-*]. Type genus: *Thomassetia* Théry, 1928.

Tribe Trigonogeniini Cobos, 1956

Trigonogeniini Cobos, 1956: 72 [stem: *Trigonogeni-*]. Type genus: *Trigonogenium* Harold, 1869.

Tribe Xenorhipidini Cobos, 1986

Xenorhipini Cobos, 1986: 136 [stem: *Xenorhipid-*]. Type genus: *Xenorhipis* J. L. LeConte, 1866. Comment: incorrect original stem formation, not in prevailing usage.

Subtribe Trichinorhipidina Bellamy, 2006

Trichinorhipidina Bellamy, 2006: 139 [stem: *Trichinorhipid-*]. Type genus: *Trichinorhipis* Barr, 1948.

Subtribe Xenorhipidina Cobos, 1986

Xenorhipini Cobos, 1986: 136 [stem: *Xenorhipid-*]. Type genus: *Xenorhipis* J. L. LeConte, 1866. Comment: incorrect original stem formation, not in prevailing usage.

Buprestinae *incertae sedis*

†Glaphyropteridae Pongrácz, 1935: 541 [stem: *Glaphyropter-*]. Type genus: *Glaphyroptera* Heer, 1852. Comment: the older name Glaphyropteridae Brauer, 1852 is a collective name for most Neuroptera in the present sense and it is not based on a genus name (Ponomarenko 2009 pers. comm.).

Subfamily Agrilinae Laporte, 1835

Agrilidae Laporte, 1835b: 165 [stem: *Agril-*]. Type genus: *Agrilus* Curtis, 1825. Comment: First Reviser (Agrilinae Laporte, 1835 vs Tracheinae Laporte, 1835) not determined, current usage maintained.

Tribe Agrilini Laporte, 1835

Agrilidae Laporte, 1835b: 165 [stem: *Agril-*]. Type genus: *Agrilus* Curtis, 1825.

Subtribe Agrilina Laporte, 1835

Agrilidae Laporte, 1835b: 165 [stem: *Agril-*]. Type genus: *Agrilus* Curtis, 1825.

Subtribe Amorphosternina Cobos, 1974

Amorphosternae Cobos, 1974: 69 [stem: *Amorphostern-*]. Type genus: *Amorphosternus* Deyrolle, 1864.

Subtribe Amyiina Holyński, 1993

Amyiina Holyński, 1993: 14 [stem: *Amyi-*]. Type genus: *Amyia* Saunders, 1871.

Subtribe Rhaeboscelidina Cobos, 1976

*Rhaeboscelini Böving and Craighead, 1931: 77 [stem: *Rhaeboscelid-*]. Type genus: *Rhaeboscelis* Chevrolat, 1838. Comment: unavailable family-group name, proposed after 1930 without description or bibliographic reference to such a description (Art. 13.1); incorrect original stem formation, not in prevailing usage.

Rhaeboscelidi Cobos, 1976: 20 [stem: *Rhaeboscelid-*]. Type genus: *Rhaeboscelis* Chevrolat, 1838.

Tribe Aphanisticini Jacquelin du Val, 1859

Aphanisticites Jacquelin du Val, 1859: 104 [stem: *Aphanistic-*]. Type genus: *Aphanisticus* Latreille, 1810.

Subtribe Anthaxomorphina Holyński, 1993

Anthaxomorphina Holyński, 1993: 32 [stem: *Anthaxomorph-*]. Type genus: *Anthaxomorphus* Deyrolle, 1864.

Subtribe Aphanisticina Jacquelin du Val, 1859

Aphanisticites Jacquelin du Val, 1859: 104 [stem: *Aphanistic-*]. Type genus: *Aphanisticus* Latreille, 1829. Comment: original vernacular name available (Art. 11.7.2): first used in latinized form by Acloque (1896: 280, as Aphanisticii), generally accepted as in Hansen (1996: 137, as Aphanisticini).

Subtribe Cylindromorphina Portevin, 1931

Cylindromorphini Portevin, 1931: 335 [stem: *Cylindromorph-*]. Type genus: *Cylindromorphus* Kiesenwetter, 1857.

Subtribe Cylindromorphoidina Cobos, 1979

Cylindromorphoidini Cobos, 1979b: 420, in key [stem: *Cylindromorphoid-*]. Type genus: *Cylindromorphoides* Kerremans, 1903.

Subtribe Germaricina Cobos, 1979

Germaricini Cobos, 1979b: 420, in key [stem: *Germaric-*]. Type genus: *Germarica* Blackburn, 1887.

Tribe Coraebini Bedel, 1921

Coraebini Bedel, 1921: 170 [stem: *Coraeb-*]. Type genus: *Coraebus* Gory and Laporte, 1839.

Subtribe Amorphosomatina Majer, 2000

Amorphosomina Majer, 2000: 210 [stem: *Amorphosomat-*]. Type genus: *Amorphosoma* Laporte, 1835. Comment: incorrect original stem formation, not in prevailing usage.

Subtribe Cisseina Majer, 2000

Cisseina Majer, 2000: 203 [stem: *Cisse-*]. Type genus: *Cisseis* Gory and Laporte, 1839 [syn. of *Diphucrania* Dejean, 1833 (see ICZN 2008c)].

Subtribe Clematina Majer, 2000

Clemina Majer, 2000: 215 [stem: *Clemat-*]. Type genus: *Clema* Semenov, 1900. Comment: incorrect original stem formation, not in prevailing usage.

Subtribe Coraebina Bedel, 1921

Coraebini Bedel, 1921: 170 [stem: *Coraeb-*]. Type genus: *Coraebus* Gory and Laporte, 1839.

Subtribe Dismorphina Cobos, 1990

Dismorphina Cobos, 1990: 542 [stem: *Dismorph-*]. Type genus: *Dismorpha* Gistel, 1848.

Subtribe Ethoniina Majer, 2000

Ethoniina Majer, 2000: 201 [stem: *Ethoni-*]. Type genus: *Ethonion* Kubáň, 2000.

Subtribe Geraliina Cobos, 1988

Geraliina Cobos, 1988: 10 [stem: *Gerali-*]. Type genus: *Geralius* Harold, 1869.

Subtribe Meliboeina Majer, 2000

Meliboeina Majer, 2000: 213 [stem: *Meliboe-*]. Type genus: *Meliboeus* Deyrolle, 1864.

Subtribe Synechocerina Majer, 2000

Synechocerina Majer, 2000: 214 [stem: *Synechocer-*]. Type genus: *Synechocera* Deyrolle, 1864.

Subtribe Toxoscelina Majer, 2000

Toxoscelina Majer, 2000: 207 [stem: *Toxoscel-*]. Type genus: *Toxoscelus* Deyrolle, 1864.

Tribe Tracheini Laporte, 1835

Trachisidae Laporte, 1835b: 166 [stem: *Trache-*]. Type genus: *Trachys* Fabricius, 1801 [placed on the Official List of Generic Names in Zoology (ICZN 2009a)]. Comment: correct stem ruled to be *Trache-* and Tracheidae Laporte, 1835 placed on the Official List of Family-Group Names in Zoology (ICZN 2009a).

Subtribe Brachina LeConte, 1861

Braches J. L. LeConte, 1861: 156 [stem: *Brach-*]. Type genus: *Brachys* Dejean, 1833. Comment: current spelling maintained (Art. 29.5): incorrect stem formation in prevailing usage (should be *Brache-*).

Brachyini Cobos, 1979b: 417, in key [stem: *Brach-*]. Type genus: *Brachys* Dejean, 1833. Comment: family-group name proposed as a new taxon, without reference to Braches J. L. LeConte, 1861; incorrect original stem formation, not in prevailing usage.

Subtribe Leiopleurina Holyński, 1993

Leiopleurina Holyński, 1993: 32 [stem: *Leiopleur-*]. Type genus: *Leiopleura* Deyrolle, 1864.

Subtribe Pachyschelina Böving and Craighead, 1931

Pachyschelinae Böving and Craighead, 1931: 49, in key [stem: *Pachyschel-*]. Type genus: *Pachyschelus* Solier, 1833.

Subtribe Tracheina Laporte, 1835

Trachisidae Laporte, 1835b: 166 [stem: *Trache-*]. Type genus: *Trachys* Fabricius, 1801 [placed on the Official List of Generic Names in Zoology (ICZN 2009a)]. Comment: published before 6 May 1835; correct stem ruled to be *Trache-* and Tracheidae Laporte, 1835 placed on the Official List of Family-Group Names in Zoology (ICZN 2009a); Trachisidae Laporte, 1835 deemed to be an incorrect original spelling and placed on the Official Index of Rejected and Invalid Family-Group Names in Zoology along with subsequent family-group names based on *Trachys* Fabricius (ICZN 2009a); the correct spelling Tracheidae was first used by Gistel (1848: [5]); this family-group name was also used in the same year by Solier (1835c [before 4 May]: c, as Trachysides).

Phytoteradae Gistel, 1856a: 366 [stem: *Phytoter-*]. Type genus: *Phytotera* Gistel, 1856 [syn. of *Trachys* Fabricius, 1801]. Comment: incorrect original stem formation, not in prevailing usage.

†Subfamily Parathyreinae Alexeev, 1994

Parathyreinae Alexeev, 1994: 10 [stem: *Parathyre-*]. Type genus: *Parathyrea* Alexeev, 1994.

Superfamily Byrrhoidea Latreille, 1804

Byrrhii Latreille, 1804c: 146 [stem: *Byrrh-*]. Type genus: *Byrrhus* Linnaeus, 1767 [placed on the Official List of Generic Names in Zoology (ICZN 1994a)].

Family Byrrhidae Latreille, 1804

Byrrhii Latreille, 1804c: 146 [stem: *Byrrh-*]. Type genus: *Byrrhus* Linnaeus, 1767 [placed on the Official List of Generic Names in Zoology (ICZN 1994a)].

Subfamily Byrrhinae Latreille, 1804

Byrrhii Latreille, 1804c: 146 [stem: *Byrrh-*]. Type genus: *Byrrhus* Linnaeus, 1767 [placed on the Official List of Generic Names in Zoology (ICZN 1994a)].

Tribe Byrrhini Latreille, 1804

Byrrhii Latreille, 1804c: 146 [stem: *Byrrh-*]. Type genus: *Byrrhus* Linnaeus, 1767 [placed on the Official List of Generic Names in Zoology (ICZN 1994a)]. Comment: published 7 March 1804; this family-group name was also used in the same year by Latreille (1804a [between 19 August and 17 September]: 190, as Byrrhii).

*Tylicini Johnson, 1991: 160 [stem: *Tylic-*]. Type genus: *Tylicus* Casey, 1912 [syn. of *Arctobyrrhus* Munster, 1902]. Comment: unavailable family-group name, proposed after 1930 without description or bibliographic reference to such a description (Art. 13.1).

Tribe Exomellini Casey, 1914

Exomellini Casey, 1914a: 378 [stem: *Exomell-*]. Type genus: *Exomella* Casey, 1914.

Tribe Morychini El Moursy, 1961

Morychini El Moursy, 1961: 11 [stem: *Morych-*]. Type genus: *Morychus* Erichson, 1847.

Tribe Pedilophorini Casey, 1912

Pedilophorini Casey, 1912: 4 [stem: *Pedilophor-*]. Type genus: *Pedilophorus* Steffahny, 1843.

Tribe Simplocariini Mulsant and Rey, 1869

Simplocariates Mulsant and Rey, 1869: 151 [stem: *Simplocari-*]. Type genus: *Simplocaria* Stephens, 1829 [placed on the Official List of Generic Names in Zoology (ICZN 1985c)]. Comment: original vernacular name available (Art. 11.7.2): first used in latinized form by Casey (1912: 14, as Simplocariini), generally accepted as in P. J. Johnson (2002b: 115, as Simplocariini).

Lioonini Leng, 1920: 193 [stem: *Lio-*]. Type genus: *Lioon* Casey, 1912. Comment: incorrect original stem formation, not in prevailing usage.

Subfamily Syncalyptinae Mulsant and Rey, 1869

Syncalyptaires Mulsant and Rey, 1869: 31 [stem: *Syncalypt-*]. Type genus: *Syncalypta* Dillwyn, 1829 [syn. of *Chaetophora* Kirby and Spence, 1817].

Tribe Microchaetini Paulus, 1973

Microchaetini Paulus, 1973: 353, in key [stem: *Microchaet-*]. Type genus: *Microchaetes* Hope, 1834.

Tribe Syncalyptini Mulsant and Rey, 1869

Syncalyptaires Mulsant and Rey, 1869: 31 [stem: *Syncalypt-*]. Type genus: *Syncalypta* Dillwyn, 1829 [syn. of *Chaetophora* Kirby and Spence, 1817]. Comment: original vernacular name available (Art. 11.7.2): first used in latinized form by Portevin (1931: 290, as Syncalyptini), generally accepted as in Hansen (1996: 138, as Syncalyptinae).

Subfamily Amphicyrtinae LeConte, 1861

Amphicyrtini J. L. LeConte, 1861: 111 [stem: *Amphicyrt-*]. Type genus: *Amphicyrta* Erichson, 1843.

Family Elmidae Curtis, 1830

Elmidae Curtis, 1830: pl. 294 [stem: *Elm-*]. Type genus: *Elmis* Latreille, 1802 [placed on the Official List of Generic Names in Zoology (ICZN 1995d)]. Comment: name placed on the Official List of Family-Group Names in Zoology and correct stem ruled to be *Elm-* (ICZN 1995d).

Subfamily Larainae LeConte, 1861

Larini J. L. LeConte, 1861: 116 [stem: *Lara-*]. Type genus: *Lara* J. L. LeConte, 1852 [placed on the Official List of Generic Names in Zoology (ICZN 1988g)].

Tribe Laraini LeConte, 1861

Larini J. L. LeConte, 1861: 116 [stem: *Lara-*]. Type genus: *Lara* J. L. LeConte, 1852 [placed on the Official List of Generic Names in Zoology (ICZN 1988g)]. Comment: Laraini J. L. LeConte, 1861 placed on the Official List of Family-Group Names in Zoology and correct stem ruled to be *Lara-* to avoid homonymy with Laridae Rafinesque-Schmaltz, 1815 (type genus *Larus* Linnaeus, 1758) in Aves (ICZN 1988g); Larini J. L. LeConte, 1861 placed on the Official Index of Rejected and Invalid Family-Group Names in Zoology (ICZN 1988g).

Tribe Potamophilini Mulsant and Rey, 1872

*Potamophiles Motschulsky, 1849: 54 [stem: *Potamophil-*]. Type genus: *Potamophilus* Germar, 1811. Comment: original vernacular name unavailable (Art. 11.7.2): subsequently used in latinized form but not generally attributed to Motschulsky (1849).

Potamophilaires Mulsant and Rey, 1872b: 11 [stem: *Potamophil-*]. Type genus: *Potamophilus* Germar, 1811. Comment: original vernacular name available (Art. 11.7.2): first used in latinized form by Ganglbauer (1904: 99, as Potamophilini), generally accepted as in Jäch et al. (2006: 432, as Potamophilini).

Subfamily Elminae Curtis, 1830

Elmidae Curtis, 1830: pl. 294 [stem: *Elm-*]. Type genus: *Elmis* Latreille, 1802 [placed on the Official List of Generic Names in Zoology (ICZN 1995d)]. Comment:

placed on the Official List of Family-Group Names in Zoology and correct stem ruled to be *Elm-* (ICZN 1995d).

Tribe ANCYRONYCHINI Ganglbauer, 1904

ANCYRONYCHINI Ganglbauer, 1904: 108 [stem: *Ancyronych-*]. Type genus: *Ancyronyx* Erichson, 1847.

Tribe ELMINI Curtis, 1830

ELMIDAE Curtis, 1830: pl. 294 [stem: *Elm-*]. Type genus: *Elmis* Latreille, 1802 [placed on the Official List of Generic Names in Zoology (ICZN 1995d)]. Comment: ELMIDAE Curtis, 1830 placed on the Official List of Family-Group Names in Zoology and correct stem ruled to be *Elm-* (ICZN 1995d).

Subtribe ELMINA Curtis, 1830

*LIMNIIDAE Stephens, 1828: 104 [stem: *Limni-*]. Type genus: *Limnius* Illiger, 1802. Comment: unavailable family-group name, not based on genus used as valid at the time (see Jäch 1994).

*LIMNIIDAE Stephens, 1829a: 5 [stem: *Limni-*]. Type genus: *Limnius* Illiger, 1802. Comment: unavailable family-group name, not based on genus used as valid at the time.

ELMIDAE Curtis, 1830: pl. 294 [stem: *Elm-*]. Type genus: *Elmis* Latreille, 1802 [placed on the Official List of Generic Names in Zoology (ICZN 1995d)]. Comment: ELMIDAE Curtis, 1830 placed on the Official List of Family-Group Names in Zoology and correct stem ruled to be *Elm-* (ICZN 1995d).

LIMNIIDAE Hope, 1838a: 153 [stem: *Limni-*]. Type genus: *Limnius* Illiger, 1802. Comment: family-group name previously attributed to C. G. Thomson (1859) in the literature.

Subtribe STENELMINA Mulsant and Rey, 1872

STENELMISATES Mulsant and Rey, 1872a: 49 [stem: *Stenelm-*]. Type genus: *Stenelmis* Dufour, 1835. Comment: original vernacular name available (Art. 11.7.2): first used in latinized form by Bollow (1941: 3, as STENELMINI), generally accepted as in Jäch et al. (2006: 437, as STENELMINA); incorrect original stem formation, not in prevailing usage.

Tribe MACRONYCHINI Gistel, 1848

MACRONYCHIDAE Gistel, 1848: [1] [stem: *Macronych-*]. Type genus: *Macronychus* P. W. J. Müller, 1806. Comment: family-group name previously attributed to Mulsant and Rey (1872) in the literature.

Family Dryopidae Billberg, 1820 (1817)

Parnidea Leach, 1817: 88 [stem: *Parn-*]. Type genus: *Parnus* Fabricius, 1792 [syn. of *Dryops* A. G. Olivier, 1791]. Comment: younger name Dryopidae Billberg, 1820 conserved (Art. 40.2) (see Lawrence and Newton 1995).

Dryopides Billberg, 1820a: 38 [stem: *Dryop-*]. Type genus: *Dryops* A. G. Olivier, 1791. Comment: younger name conserved over Parnidae Leach, 1817 (Art. 40.2) (see Lawrence and Newton 1995).

Pelonominae Böving and Craighead, 1931: 45, in key [stem: *Pelonom-*]. Type genus: *Pelonomus* Erichson, 1847.

Chiloeidae Dajoz, 1973: 179 [stem: *Chiloe-*]. Type genus: *Chiloea* Dajoz, 1973 [syn. of *Sosteamorphus* Hinton, 1936].

Family Lutrochidae Kasap and Crowson, 1975

*Lutrochidae Hinton, 1971: 297 [stem: *Lutroch-*]. Type genus: *Lutrochus* Erichson, 1847. Comment: unavailable family-group name, proposed after 1930 without description or bibliographic reference to such a description (Art. 13.1).

Lutrochidae Kasap and Crowson, 1975: 442 [stem: *Lutroch-*]. Type genus: *Lutrochus* Erichson, 1847.

Family Limnichidae Erichson, 1846

Limnichini Erichson, 1846: 465 [stem: *Limnich-*]. Type genus: *Limnichus* Dejean, 1821.

Subfamily Hyphalinae Britton, 1971

Hyphalinae Britton, 1971: 88 [stem: *Hyphal-*]. Type genus: *Hyphalus* Britton, 1971.

Subfamily Limnichinae Erichson, 1846

Limnichini Erichson, 1846: 465 [stem: *Limnich-*]. Type genus: *Limnichus* Dejean, 1821.

Tribe Limnichini Erichson, 1846

Limnichini Erichson, 1846: 465 [stem: *Limnich-*]. Type genus: *Limnichus* Dejean, 1821.

Botriophorates Mulsant and Rey, 1869: 160 [stem: *Bothriophor-*]. Type genus: *Bothriophorus* Mulsant and Rey, 1852 [as *Botriophorus*, incorrect subsequent spelling of type genus name, not in prevailing usage]. Comment: family-group name also spelled Botriaphorates in the original publication (p. 173); original vernacular name available (Art. 11.7.2): first used in latinized form by Ganglbauer (1904: 51, as Bothriophorini), generally accepted as in Shepard (2002: 125, as Bothriophorini); incorrect original stem formation, not in prevailing usage.

Tribe Wooldridgeini Spangler, 1999

Wooldridgeini Spangler, 1999: 181 [stem: *Wooldridge-*]. Type genus: *Wooldridgeus* Spangler, 1998.

Subfamily Cephalobyrrhinae Champion, 1925

Cephalobyrrhinae Champion, 1925: 174 [stem: *Cephalobyrrh-*]. Type genus: *Cephalobyrrhus* Pic, 1923.

Subfamily Thaumastodinae Champion, 1924

Thaumastodinae Champion, 1924: 25 [stem: *Thaumastod-*]. Type genus: *Thaumastodus* Champion, 1924 [syn. of *Pseudeucinetus* Heller, 1921].

Family Heteroceridae MacLeay, 1825

Heteroceridae W. S. MacLeay, 1825: 34 [stem: *Heterocer-*]. Type genus: *Heterocerus* Fabricius, 1792.

Subfamily Elythomerinae Pacheco, 1964

Elythomerini Pacheco, 1964: 120 [stem: *Elythomer-*]. Type genus: *Elythomerus* C. O. Waterhouse, 1874.

Subfamily Heterocerinae MacLeay, 1825

Heteroceridae W. S. MacLeay, 1825: 34 [stem: *Heterocer-*]. Type genus: *Heterocerus* Fabricius, 1792.

Tribe Augylini Pacheco, 1964

Augyliini Pacheco, 1964: 19 [stem: *Augyl-*]. Type genus: *Augyles* Schiødte, 1866. Comment: incorrect original stem formation, not in prevailing usage.

Tribe Heterocerini MacLeay, 1825

Heteroceridae W. S. MacLeay, 1825: 34 [stem: *Heterocer-*]. Type genus: *Heterocerus* Fabricius, 1792.

Tribe Micilini Pacheco, 1964

Micilini Pacheco, 1964: 32 [stem: *Micil-*]. Type genus: *Micilus* Mulsant and Rey, 1872.

Tribe Tropicini Pacheco, 1964

Tropicini Pacheco, 1964: 101 [stem: *Tropic-*]. Type genus: *Tropicus* Pacheco, 1964.

Family Psephenidae Lacordaire, 1854

Pséphénides Lacordaire, 1854b: 497 [stem: *Psephen-*]. Type genus: *Psephenus* Haldeman, 1853.

Subfamily Afroeubriinae Lee, Satô, Shepard and Jäch, 2007

Afroeubriinae Lee, Satô, Shepard and Jäch, 2007: 532 [stem: *Afroeubri-*]. Type genus: *Afroeubria* Villiers, 1961.

Subfamily Eubriinae Lacordaire, 1857

Eubriades Lacordaire, 1857: 283 [stem: *Eubri-*]. Type genus: *Eubria* Germar, 1818. Comment: original vernacular name available (Art. 11.7.2): first used in latinized form by J. L. LeConte (1861: 180, as Eubriini), generally accepted as in Hansen (1996: 139, as Eubriinae).

Subfamily Eubrianacinae Jakobson, 1913

Placonychini G. H. Horn, 1880a: 110 [stem: *Placonych-*]. Type genus: *Placonycha* G. H. Horn, 1880 [syn. of *Eubrianax* Kiesenwetter, 1874]. Comment: usage of the younger name Eubrianacinae Jakobson, 1913 conserved over this name (Art. 40.2) (see Lawrence and Newton 1995: 846).

Eubrianacini Jakobson, 1913: 723 [stem: *Eubrianac-*]. Type genus: *Eubrianax* Kiesenwetter, 1874. Comment: current spelling maintained (Art. 29.5): incorrect stem formation in prevailing usage (should be *Eubrianact-*); use of this name conserved over Placonychinae G. H. Horn, 1880 (Art. 40.2) (see Lawrence and Newton 1995: 846).

Subfamily Psephenoidinae Bollow, 1938

Psephenoidini Bollow, 1938: 156 [stem: *Psephenoid-*]. Type genus: *Psephenoides* Gahan, 1914.

Subfamily Psepheninae Lacordaire, 1854

Eurypalpini J. L. LeConte, 1852a: 41 [stem: *Eurypalp-*]. Type genus: *Eurypalpus* J. L. LeConte, 1852 [preoccupied genus name, not *Eurypalpus* Macquart, 1835 [Diptera]; syn. of *Psephenus* Haldeman, 1853]. Comment: permanently invalid (Art. 39): based on preoccupied type genus.

Pséphénides Lacordaire, 1854b: 497 [stem: *Psephen-*]. Type genus: *Psephenus* Haldeman, 1853. Comment: original vernacular name available (Art. 11.7.2): first used in latinized form by Imhoff (1856: xv, as Psephenidae), generally accepted as in Hansen (1996: 139, as Psephenidae).

Family Cneoglossidae Champion, 1897

Cneoglossini Champion, 1897: 594 [stem: *Cneogloss-*]. Type genus: *Cneoglossa* Guérin-Méneville, 1843.

Family Ptilodactylidae Laporte, 1836

Ptilodactylidae Laporte, 1836: 21 [stem: *Ptilodactyl-*]. Type genus: *Ptilodactyla* Illiger, 1807.

Subfamily Anchytarsinae Champion, 1897

Coloboderides Erichson, 1847a: 174 [stem: *Coloboder-*]. Type genus: *Colobodera* Klug, 1838 [syn. of *Daemon* Laporte, 1836].

ANCHYTARSINI Champion, 1897: 593 [stem: *Anchytars-*]. Type genus: *Anchytarsus* Guérin-Méneville, 1843. Comment: COLOBODERINAE Erichson, 1847 is the oldest name for this tribe, however since COLOBODERINAE has not been used as valid after 1899 to our knowledge, we believe that usage of ANCHYTARSINAE Champion, 1897 should be conserved and an application be submitted to the Commission to suppress the older name.

*EPILICHADINAE Lawrence and Stribling, 1992: 19 [stem: *Epilichad-*]. Type genus: *Epilichas* A. White, 1859. Comment: although Ivie (2002: 136) treated this name as available we do not believe that requirements of Art. 13.1 were met and we therefore treat this name here as unavailable.

Subfamily CLADOTOMINAE Pic, 1914

CLADOTOMINI Pic, 1914: 45 [stem: *Cladotom-*]. Type genus: *Cladotoma* Westwood, 1836.

Subfamily APLOGLOSSINAE Champion, 1897

APLOGLOSSINAE Champion, 1897: 623 [stem: *Aplogloss-*]. Type genus: *Aploglossa* Guérin-Méneville, 1849.

Subfamily ARAEOPIDIINAE Lawrence, 1991

ARAEOPIDIINAE Lawrence, 1991: 250, in key [stem: *Araeopidi-*]. Type genus: *Araeopidius* Cockerell, 1906.

Subfamily PTILODACTYLINAE Laporte, 1836

PTILODACTYLIDAE Laporte, 1836: 21 [stem: *Ptilodactyl-*]. Type genus: *Ptilodactyla* Illiger, 1807.

Family PODABROCEPHALIDAE Pic, 1930

PODABROCEPHALIDAE Pic, 1930: 314 [stem: *Podabrocephal-*]. Type genus: *Podabrocephalus* Pic, 1913.

Family CHELONARIIDAE Blanchard, 1845

CHELONARIITES Blanchard, 1845b: 70 [stem: *Chelonari-*]. Type genus: *Chelonarium* Fabricius, 1801. Comment: original vernacular name available (Art. 11.7.2): first used in latinized form by Imhoff (1856: 140, as CHELONARII), generally accepted as in Satô (2006: 454, as CHELONARIIDAE).

Family EULICHADIDAE Crowson, 1973

*LICHADIDEN Kolbe, 1908: 249 [stem: *Lichad-*]. Type genus: *Lichas* Westwood, 1853 [preoccupied genus name, not *Lichas* Dalman, 1827 [Trilobita]; syn. of *Eulichas* Jakobson, 1913]. Comment: original vernacular name unavailable (Art. 11.7.2): proposed after 1899.

Lichadidae Forbes, 1926: 102 [stem: *Lichad-*]. Type genus: *Lichas* Westwood, 1853 [preoccupied genus name, not *Lichas* Dalman, 1827 [Trilobita]; syn. of *Eulichas* Jakobson, 1913]. Comment: permanently invalid (Art. 39): based on preoccupied type genus.

Eulichadidae Crowson, 1973b: 237 [stem: *Eulichad-*]. Type genus: *Eulichas* Jakobson, 1913.

Family Callirhipidae Emden, 1924

Zenoini J. L. LeConte, 1866a: 50 [stem: *Zeno-*]. Type genus: *Zenoa* Say, 1835. Comment: this name is older than Callirhipidae Emden, 1924 however we recommend that an application be submitted to conserve the younger name.

Callirhipini Emden, 1924: 87 [stem: *Callirhip-*]. Type genus: *Callirhipis* Latreille, 1829. Comment: the oldest name for this family is Zenoidae LeConte, 1866, however, as pointed out by Lawrence and Newton (1995: 849) an application to the Commission is needed in order to preserve the broadly accepted younger name Callirhipidae Emden, 1924; current spelling maintained (Art. 29.3.1.1): incorrect stem formation in prevailing usage (should be *Callirhipid-*).

Superfamily Elateroidea Leach, 1815

Elaterides Leach, 1815: 85 [stem: *Elater-*]. Type genus: *Elater* Linnaeus, 1758. Comment: the oldest available name for this superfamily is Cebrionoidea Latreille, 1802, however Elateroidea Leach, 1815 is maintained here pending resolution of an application to the ICZN being prepared by P. J. Johnson (see Lawrence and Newton 1995: 849; P. J. Johnson pers. comm. 2009).

Family Rhinorhipidae Lawrence, 1988

Rhinorhipidae Lawrence, 1988: 3 [stem: *Rhinorhip-*]. Type genus: *Rhinorhipus* Lawrence, 1988.

Family Artematopodidae Lacordaire, 1857

Artématopides Lacordaire, 1857: 260 [stem: *Artematopod-*]. Type genus: *Artematopus* Perty, 1832.

Subfamily Electribiinae Crowson, 1975

Electropogonini Crowson, 1975: 77 [stem: *Electribi-*]. Type genus: *Electribius* Crowson, 1973. Comment: incorrect original stem formation, not in prevailing usage; Electropogonini (based on *Electropogon*) in Crowson's work has been treated as a *lapsus calami* for Electribiidae based on *Electribius* (see Lawrence and Newton 1995: 850).

Subfamily Allopogoniinae Crowson, 1973

Allopogonini Crowson, 1973b: 231, in key [stem: *Allopogoni-*]. Type genus: *Allopogonia* Cockerell, 1906. Comment: incorrect original stem formation, not in prevailing usage.

Subfamily Artematopodinae Lacordaire, 1857

Artématopides Lacordaire, 1857: 260 [stem: *Artematopod-*]. Type genus: *Artematopus* Perty, 1832.

Tribe Artematopodini Lacordaire, 1857

Artématopides Lacordaire, 1857: 260 [stem: *Artematopod-*]. Type genus: *Artematopus* Perty, 1832. Comment: original vernacular name available (Art. 11.7.2): first used in latinized form by Champion (1897: 586, as Artematopinae [incorrect stem formation]), generally accepted as in Lawrence and Newton (1995: 849, as Artematopodidae); incorrect original stem formation, not in prevailing usage.

Tribe Ctesibiini Crowson, 1973

Ctesibiinae Crowson, 1973b: 228, in key [stem: *Ctesibi-*]. Type genus: *Ctesibius* Champion, 1897.

Tribe Macropogonini LeConte, 1861

Macropogonini J. L. LeConte, 1861: 178 [stem: *Macropogon-*]. Type genus: *Macropogon* Motschulsky, 1845.

Eurypogonidae Böving and Craighead, 1931: 45, in key [stem: *Eurypogon-*]. Type genus: *Eurypogon* Motschulsky, 1859.

Family Brachypsectridae LeConte and Horn, 1883

Brachypsectrini J. L. LeConte and G. H. Horn, 1883: 170 [stem: *Brachypsectr-*]. Type genus: *Brachypsectra* J. L. LeConte, 1874.

Family Cerophytidae Latreille, 1834

Cerophytides Latreille, 1834: 133 [stem: *Cerophyt-*]. Type genus: *Cerophytum* Latreille, 1809.

Family Eucnemidae Eschscholtz, 1829

Eucnemides Eschscholtz, 1829a: 10 [stem: *Eucnem-*]. Type genus: *Eucnemis* Ahrens, 1812. Comment: the name Melasidae Fleming, 1821 has priority over this name, however Muona and Alaruikka (2007: 32) mentioned that an application has been sent to the Commission to conserve usage of Eucnemidae over the older name Melasidae for reasons of stability, we follow current usage until the case is resolved (also see Appendix 6).

Subfamily Perothopinae Lacordaire, 1857

Pérothopides Lacordaire, 1857: 128 [stem: *Perothop-*]. Type genus: *Perothops* Eschscholtz, 1836. Comment: original vernacular name available (Art. 11.7.2): first used in latinized form by J. L. LeConte (1861: 162, as Perothopini), generally accepted as in Lawrence and Newton (1995: 851, as Perothopinae).

Subfamily Phyllocerinae Reitter, 1905

Phylloceridae Reitter, 1905: 4 [stem: *Phyllocer-*]. Type genus: *Phyllocerus* Lepeletier and Audinet-Serville, 1825.

Tribe Anelastini Reitter, 1911

Anelastini Reitter, 1911: 202 [stem: *Anelast-*]. Type genus: *Anelastes* Kirby, 1819.

Tribe Phyllocerini Reitter, 1905

Phylloceridae Reitter, 1905: 4 [stem: *Phyllocer-*]. Type genus: *Phyllocerus* Lepeletier and Audinet-Serville, 1825.

Subfamily Pseudomeninae Muona, 1993

Pseudomeninae Muona, 1993: 41 [stem: *Pseudomen-*]. Type genus: *Pseudomenes* Fleutiaux, 1902. Comment: precedence (Pseudomeninae Muona, 1993 vs Schizophilinae Muona, 1993) given to taxon originally proposed at the higher rank (Art. 24.1).

Tribe Pseudomenini Muona, 1993

* Pseudomeninae Muona, 1991a: 167 [stem: *Pseudomen-*]. Type genus: *Pseudomenes* Fleutiaux, 1902. Comment: unavailable family-group name, proposed after 1930 without description or bibliographic reference to such a description (Art. 13.1).

Pseudomeninae Muona, 1993: 41 [stem: *Pseudomen-*]. Type genus: *Pseudomenes* Fleutiaux, 1902.

Tribe Schizophilini Muona, 1993

*Schizophilini Muona, 1991a: 167 [stem: *Schizophil-*]. Type genus: *Schizophilus* Bonvouloir, 1871. Comment: unavailable family-group name, proposed after 1930 without description or bibliographic reference to such a description (Art. 13.1).

Schizophilini Muona, 1993: 42 [stem: *Schizophil-*]. Type genus: *Schizophilus* Bonvouloir, 1871.

Subfamily Palaeoxeninae Muona, 1993

Palaeoxeninae Muona, 1993: 42 [stem: *Palaeoxen-*]. Type genus: *Palaeoxenus* G. H. Horn, 1891.

Subfamily PHLEGONINAE Muona, 1993

PHLEGONINAE Muona, 1993: 42 [stem: *Phlegon-*]. Type genus: *Phlegon* Laporte, 1840. Comment: current spelling maintained (Art. 29.5): incorrect stem formation in prevailing usage (should be *Phlegont-*).

Subfamily ANISCHIINAE Fleutiaux, 1936

ANISCHINAE Fleutiaux, 1936: 292 [stem: *Anischi-*]. Type genus: *Anischia* Fleutiaux, 1896. Comment: incorrect original stem formation, not in prevailing usage.

Subfamily MELASINAE Fleming, 1821

MELASIDAE Fleming, 1821: 49 [stem: *Melas-*]. Type genus: *Melasis* A. G. Olivier, 1790.

Tribe CALYPTOCERINI Muona, 1993

*CALYPTOCERINI Muona, 1991a: 167 [stem: *Calyptocer-*]. Type genus: *Calyptocerus* Guérin-Méneville, 1843. Comment: unavailable family-group name, proposed after 1930 without description or bibliographic reference to such a description (Art. 13.1).

CALYPTOCERINI Muona, 1993: 43 [stem: *Calyptocer-*]. Type genus: *Calyptocerus* Guérin-Méneville, 1843.

Tribe CEBALLOSMELASINI Muona, 1993

*CEBALLOSMELASINI Muona, 1991a: 167 [stem: *Ceballosmelas-*]. Type genus: *Ceballosmelasis* Cobos, 1964. Comment: unavailable family-group name, proposed after 1930 without description or bibliographic reference to such a description (Art. 13.1).

CEBALLOSMELASINI Muona, 1993: 42 [stem: *Ceballosmelas-*]. Type genus: *Ceballosmelasis* Cobos, 1964.

Tribe DIRHAGINI Reitter, 1911

DIRRHAGINI Reitter, 1911: 202 [stem: *Dirhag-*]. Type genus: *Dirhagus* Latreille, 1834 [as *Dirrhagus*, incorrect subsequent spelling of type genus name, not in prevailing usage; syn. of *Microrhagus* Dejean, 1833]. Comment: incorrect original stem formation, not in prevailing usage.

MICRORHAGINAE Fleutiaux, 1919: 112 [stem: *Microrhag-*]. Type genus: *Microrhagus* Dejean, 1833.

ARHIPINI Cobos, 1965: 396 [stem: *Arrhipid-*]. Type genus: *Arrhipis* Bonvouloir, 1871 [as *Arhipis*, incorrect subsequent spelling type genus name, not in prevailing usage]. Comment: incorrect original stem formation, not in prevailing usage.

Tribe EPIPHANINI Muona, 1993

*EPIPHANINI Muona, 1991a: 167 [stem: *Epiphan-*]. Type genus: *Epiphanis* Eschscholtz, 1829. Comment: unavailable family-group name, proposed after

1930 without description or bibliographic reference to such a description (Art. 13.1).

EPIPHANINI Muona, 1993: 45 [stem: *Epiphan-*]. Type genus: *Epiphanis* Eschscholtz, 1829.

Tribe HYLOCHARINI Jacquelin du Val, 1859

HYLOCHARITES Jacquelin du Val, 1859: 119 [stem: *Hylochar-*]. Type genus: *Hylochares* Latreille, 1834. Comment: original vernacular name available (Art. 11.7.2): generally accepted as in Muona (2007: 84, as HYLOCHARINI); family-group names with the incorrectly formed stem *Hylochar-* have been used in Aves (type genus *Hylocharis* Boie, 1831), the correct stem for the bird family-group name is *Hylocharit-* (see Bock 1994: 143).

HYLOCHARINI Cobos, 1965: 369 [stem: *Hylochar-*]. Type genus: *Hylochares* Latreille, 1834. Comment: proposed as new without reference to HYLOCHARINI Jacquelin du Val, 1859.

Tribe MELASINI Fleming, 1821

MELASIDAE Fleming, 1821: 49 [stem: *Melas-*]. Type genus: *Melasis* A. G. Olivier, 1790.

Subtribe COMPSOCNEMINA Muona, 1993

COMPSOCNEMINA Muona, 1993: 43 [stem: *Compsocnem-*]. Type genus: *Compsocnemis* Bonvouloir, 1871. Comment: current spelling maintained (Art. 29.3.1.1): incorrect stem formation in prevailing usage (should be *Compsocnemid-*).

Subtribe MELASINA Fleming, 1821

MELASIDAE Fleming, 1821: 49 [stem: *Melas-*]. Type genus: *Melasis* A. G. Olivier, 1790.

Tribe NEOCHARINI Muona, 1993

*NEOCHARINI Muona, 1991a: 167 [stem: *Neochar-*]. Type genus: *Neocharis* Sharp, 1887. Comment: unavailable family-group name, proposed after 1930 without description or bibliographic reference to such a description (Art. 13.1).

NEOCHARINI Muona, 1993: 44 [stem: *Neochar-*]. Type genus: *Neocharis* Sharp, 1887. Comment: current spelling maintained (Art. 29.5): incorrect stem formation in prevailing usage (should be *Neocharit-*).

Tribe XYLOBIINI Reitter, 1911

XYLOBIINI Reitter, 1911: 203 [stem: *Xylobi-*]. Type genus: *Xylobius* Latreille, 1834 [syn. of *Xylophilus* Mannerheim, 1823].

Subfamily Eucneminae Eschscholtz, 1829

Eucnemides Eschscholtz, 1829a: 10 [stem: *Eucnem-*]. Type genus: *Eucnemis* Ahrens, 1812.

Tribe Dendrocharini Fleutiaux, 1920

Dendrocharini Fleutiaux, 1920: 100, in key [stem: *Dendrochar-*]. Type genus: *Dendrocharis* Guérin-Méneville, 1843. Comment: current spelling maintained (Art. 29.5): incorrect stem formation in prevailing usage (should be *Dendrocharit-*).

Tribe Dyscharachthini Muona, 1993

*Dyscharachthini Muona, 1991a: 167 [stem: *Dyscharachth-*]. Type genus: *Dyscharachthis* Blackburn, 1900. Comment: unavailable family-group name, proposed after 1930 without description or bibliographic reference to such a description (Art. 13.1).

Dyscharachthini Muona, 1993: 49 [stem: *Dyscharachth-*]. Type genus: *Dyscharachthis* Blackburn, 1900. Comment: current spelling maintained (Art. 29.5): incorrect stem formation in prevailing usage (should be *Dyscharachthent-*).

Tribe Entomosatopini Muona, 1993

*Entomosatopini Muona, 1991a: 167 [stem: *Entomosatop-*]. Type genus: *Entomosatopus* Bonvouloir, 1871. Comment: unavailable family-group name, proposed after 1930 without description or bibliographic reference to such a description (Art. 13.1).

Entomosatopini Muona, 1993: 48 [stem: *Entomosatop-*]. Type genus: *Entomosatopus* Bonvouloir, 1871.

Tribe Eucnemini Eschscholtz, 1829

Eucnemides Eschscholtz, 1829a: 10 [stem: *Eucnem-*]. Type genus: *Eucnemis* Ahrens, 1812. Comment: current spelling maintained (Art. 29.3.1.1): incorrect stem formation in prevailing usage (should be *Eucnemid-*).

Gastraulaci Fleutiaux, 1902: 648 [stem: *Gastraulac-*]. Type genus: *Gastraulacus* Guérin-Méneville, 1843.

Tribe Galbitini Muona, 1991

*Galbites Blanchard, 1845b: 71 [stem: *Galb-*]. Type genus: *Galba* Latreille, 1829 [preoccupied genus name, not *Galba* Schrank, 1803 [Mollusca]; syn. of *Galbites* Fleutiaux, 1918]. Comment: original vernacular name unavailable (Art. 11.7.2): not subsequently latinized; if found to be availabe in the future then permanently invalid (Art. 39): based on preoccupied type genus.

Pterotarsini Cobos, 1965: 294 [stem: *Pterotars-*]. Type genus: *Pterotarsus* sensu Guérin-Méneville, 1838 [syn. of *Galbites* Fleutiaux, 1918]. Comment:

proposed as new without reference to Pterotarsini Fleutiaux, 1902 in Elateridae; based on misidentified type genus; name treated here as invalid until an application is submitted to the Commission to suppress it for the Principle of Priority (Art. 65.2.1).

*Galbitini Muona, 1991a: 167 [stem: *Galbit-*]. Type genus: *Galbites* Fleutiaux, 1918. Comment: unavailable family-group name, proposed after 1930 without description or bibliographic reference to such a description (Art. 13.1).

Galbitini Muona, 1991b: 19 [stem: *Galbit-*]. Type genus: *Galbites* Fleutiaux, 1918.

Tribe Mesogenini Muona, 1993

*Mesogenini Muona, 1991a: 167 [stem: *Mesogen-*]. Type genus: *Mesogenus* Bonvouloir, 1871. Comment: unavailable family-group name, proposed after 1930 without description or bibliographic reference to such a description (Art. 13.1).

Mesogenini Muona, 1993: 49 [stem: *Mesogen-*]. Type genus: *Mesogenus* Bonvouloir, 1871.

Tribe Muonajini Özdikmen, 2008

*Yangini Muona, 1991a: 167 [stem: *Yang-*]. Type genus: *Yanga* Muona, 1993 [preoccupied genus name, not *Yanga* Distant, 1904 [Hemiptera]; syn. of *Muonaja* Özdikmen, 2008]. Comment: unavailable family-group name, proposed after 1930 without description or bibliographic reference to such a description (Art. 13.1).

Yangini Muona, 1993: 48 [stem: *Yang-*]. Type genus: *Yanga* Muona, 1993 [preoccupied genus name, not *Yanga* Distant, 1904 [Hemiptera]; syn. of *Muonaja* Özdikmen, 2008]. Comment: permanently invalid (Art. 39): based on preoccupied type genus.

Muonajini Özdikmen, 2008: 675 [stem: *Muonaj-*]. Type genus: *Muonaja* Özdikmen, 2008. Comment: replacement name for Yangini Muona, 1993 because of the homonymy of the type genus.

Tribe Perrotiini Muona, 1993

*Perrotini Muona, 1991a: 167 [stem: *Perroti-*]. Type genus: *Perrotius* Fleutiaux, 1938. Comment: unavailable family-group name, proposed after 1930 without description or bibliographic reference to such a description (Art. 13.1).

Perrotini Muona, 1993: 47 [stem: *Perroti-*]. Type genus: *Perrotius* Fleutiaux, 1938. Comment: incorrect original stem formation, not in prevailing usage.

Tribe Phaenocerini Muona, 1993

* Phaenocerini Muona, 1991a: 167 [stem: *Phaenocer-*]. Type genus: *Phaenocerus* Bonvouloir, 1871. Comment: unavailable family-group name, proposed after 1930 without description or bibliographic reference to such a description (Art. 13.1).

Phaenocerini Muona, 1993: 47 [stem: *Phaenocer-*]. Type genus: *Phaenocerus* Bonvouloir, 1871.

Tribe Proutianini Muona, 1993

*Proutianini Muona, 1991a: 167 [stem: *Proutian-*]. Type genus: *Proutianus* Muona, 1993. Comment: unavailable family-group name, proposed after 1930 without description or bibliographic reference to such a description (Art. 13.1).

Proutianini Muona, 1993: 47 [stem: *Proutian-*]. Type genus: *Proutianus* Muona, 1993.

Subfamily Macraulacinae Fleutiaux, 1923

Macraulacinae Fleutiaux, 1923: 304 [stem: *Micraulac-*]. Type genus: *Macraulacus* Bonvouloir, 1872.

Tribe Anelastidini Muona, 1993

Anelastidini Muona, 1993: 52 [stem: *Anelastid-*]. Type genus: *Anelastidius* Jacquelin du Val, 1863. Comment: current spelling maintained (Art. 29.5): incorrect original stem formation in prevailing usage (should be *Anelastidi-*).

Tribe Echthrogasterini Cobos, 1965

Echthrogasterini Cobos, 1965: 369 [stem: *Echthrogaster-*]. Type genus: *Echthrogaster* Blackburn, 1900.

Tribe Euryptychini Mamaev, 1976

Euryptychini Mamaev, 1976: 154 [stem: *Euryptych-*]. Type genus: *Euryptychus* J. L. LeConte, 1852.

Tribe Jenibuntorini Muona, 1993

*Jenibuntorini Muona, 1991a: 167 [stem: *Jenibuntor-*]. Type genus: *Jenibuntor* Muona, 1993. Comment: unavailable family-group name, proposed after 1930 without description or bibliographic reference to such a description (Art. 13.1).

Jenibuntorini Muona, 1993: 51 [stem: *Jenibuntor-*]. Type genus: *Jenibuntor* Muona, 1993.

Tribe Macraulacini Fleutiaux, 1923

Macraulacinae Fleutiaux, 1923: 304 [stem: *Macraulac-*]. Type genus: *Macraulacus* Bonvouloir, 1872.

Fornaxini Cobos, 1965: 294 [stem: *Fornac-*]. Type genus: *Fornax* Laporte, 1835. Comment: incorrect original stem formation, not in prevailing usage.

Dromaeolini Leiler, 1976: 48 [stem: *Dromaeol-*]. Type genus: *Dromaeolus* Kiesenwetter, 1858.

Tribe Nematodini Leiler, 1976

Nematodini Leiler, 1976: 48 [stem: *Nematod-*]. Type genus: *Nematodes* Berthold, 1827.

Tribe Oisocerini Muona, 1993

Oisocerini Muona, 1993: 51 [stem: *Oisocer-*]. Type genus: *Oisocerus* Bonvouloir, 1868.

Tribe Orodotini Muona, 1993

Cryptostomidae Laporte, 1835b: 181 [stem: *Cryptostomat-*]. Type genus: *Cryptostoma* Dejean, 1821 [preoccupied genus name, not *Cryptostoma* Blainville, 1818 [Mollusca]; syn. of *Ceratogonys* Perty, 1830]. Comment: permanently invalid (Art. 39): based on preoccupied type genus; incorrect original stem formation, not in prevailing usage.

*Orodotini Muona, 1991a: 167 [stem: *Orodot-*]. Type genus: *Orodotes* Bonvouloir, 1871. Comment: unavailable family-group name, proposed after 1930 without description or bibliographic reference to such a description (Art. 13.1).

Orodotini Muona, 1993: 52 [stem: *Orodot-*]. Type genus: *Orodotes* Bonvouloir, 1871.

†Tribe Throscogeniini Iablokoff-Khnzorian, 1962

Throscogeniinae Iablokoff-Khnzorian, 1962: 81 [stem: *Throscogeni-*]. Type genus: *Throscogenius* Iablokoff-Khnzorian, 1962.

Family Throscidae Laporte, 1840 *nomen protectum*

Stereolia Rafinesque, 1815: 112 [stem: *Stereol-*]. Type genus: *Stereolus* Rafinesque, 1815 [unjustified emendation of *Throscus* Latreille, 1797 not in prevailing usage; syn. of *Trixagus* Kugelann, 1794]. Comment: *nomen oblitum* (see Appendix 1).

Throscites Laporte, 1840a: 228 [stem: *Throsc-*]. Type genus: *Throscus* Latreille, 1797 [syn. of *Trixagus* Kugelann, 1794]. Comment: *nomen protectum* (see Appendix 1); original vernacular name available (Art. 11.7.2): first used in latinized form by Agassiz (1846b: 369, as Throscoidae), generally accepted as in P. J. Johnson (2002b: 158, as Throscidae).

Trixagidae Gistel, 1848: [4] [stem: *Trixag-*]. Type genus: *Trixagus* Kugelann, 1794.

Potergini Cobos, 1961: 5 [stem: *Poterg-*]. Type genus: *Potergus* Bonvouloir, 1871.

†Family Praelateriidae Dolin, 1973

Praelateriidae Dolin, 1973: 78 [stem: *Praelateri-*]. Type genus: *Praelaterium* Dolin, 1973.

Family Elateridae Leach, 1815

Elaterides Leach, 1815: 85 [stem: *Elater-*]. Type genus: *Elater* Linnaeus, 1758. Comment: the oldest available name for this family is Cebrionidae Latreille,

1802, however ELATERIDAE Leach, 1815 is maintained here pending resolution of an application to the ICZN being prepared by P. J. Johnson (see Lawrence and Newton 1995: 852; P. J. Johnson pers. comm. 2009).

Subfamily CEBRIONINAE Latreille, 1802

CEBRIONATES Latreille, 1802: 97 [stem: *Cebrion-*]. Type genus: *Cebrio* A. G. Olivier, 1790.

PLASTOCERINI J. L. LeConte, 1861: 172 [stem: *Plastocer-*]. Type genus: *Plastocerus* sensu J. L. LeConte, 1853 [not *Plastocerus* Schaum, 1852; syn. of *Octinodes* Candèze, 1863]. Comment: based on a misidentified type genus; an application should be submitted to the Commission to suppress this name for the Principles of Priority and Homonymy (Art. 65.2.1) since PLASTOCERIDAE Crowson, 1972 is currently used as valid in ELATEROIDEA.

APLASTINAE Stibick, 1979: 175 [stem: *Aplast-*]. Type genus: *Aplastus* J. L. LeConte, 1859.

CEBRIOGNATHINAE Paulus, 1981: 264, in key [stem: *Cebriognath-*]. Type genus: *Cebriognathus* Chobaut, 1899.

Subfamily AGRYPNINAE Candèze, 1857 *nomen protectum*

AGRYPNIDES Candèze, 1857: 17 [stem: *Agrypn-*]. Type genus: *Agrypnus* Eschscholtz, 1829. Comment: *nomen protectum* (see Appendix 1); AGRYPNINAE Candèze, 1857 given precedence for subfamily name over OOPHORINAE Gistel, 1848 (Art. 35.5).

Tribe AGRYPNINI Candèze, 1857 *nomen protectum*

ADELOCERAEIDAE Gistel, 1848: [5] [stem: *Adelocer-*]. Type genus: *Adelocera* Latreille, 1829. Comment: *nomen oblitum* (see Appendix 1); incorrect original stem formation, not in prevailing usage.

PANGAURADAE Gistel, 1856a: 366 [stem: *Pangaur-*]. Type genus: *Pangaura* Gistel, 1856 [syn. of *Lacon* Laporte, 1838]. Comment: *nomen oblitum* (see Appendix 1); incorrect original stem formation, not in prevailing usage.

AGRYPNIDES Candèze, 1857: 17 [stem: *Agrypn-*]. Type genus: *Agrypnus* Eschscholtz, 1829. Comment: published before 29 June 1857; *nomen protectum* (see Appendix 1); original vernacular name available (Art. 11.7.2): first used in latinized form by Kiesenwetter (1858: 230, as AGRYPNINI), generally accepted as in Lawrence and Newton (1995: 854, as AGRYPNINAE); this family-group name was also proposed in the same year by Lacordaire (1857 [before 25 May]: 138, as AGRYPNIDES) (see *Bibliographic notes* in Introduction).

*OCTOCRYPTITES Candèze, 1892: 486 [stem: *Octocrypt-*]. Type genus: *Octocryptus* Candèze, 1892. Comment: original vernacular name unavailable (Art. 11.7.2): subsequently used in latinized form but not generally attributed to Candèze (1892).

ADELOCERINI Buysson, 1893: 15 [stem: *Adelocer-*]. Type genus: *Adelocera* Latreille, 1829. Comment: family-group name proposed as new without reference to ADELOCERIDAE Gistel, 1848.

Octocryptini Schwarz, 1906: 31 [stem: *Octocrypt-*]. Type genus: *Octocryptus* Candèze, 1892.

Cavicoxumidae Pic, 1928: 21 [stem: *Cavicox-*]. Type genus: *Cavicoxum* Pic, 1928 [syn. of *Agraeus* Candèze, 1857]. Comment: incorrect original stem formation, not in prevailing usage.

Laconini Dajoz, 1964: 60 [stem: *Lacon-*]. Type genus: *Lacon* Laporte, 1838.

Tribe Anaissini Golbach, 1984

Alampina C. Costa, 1975: 53 [stem: *Alamp-*]. Type genus: *Alampes* Champion, 1895 [preoccupied genus name, not *Alampes* Horváth, 1884 [Hemiptera]; syn. of *Peralampes* P. J. Johnson, 2002]. Comment: permanently invalid (Art. 39): based on preoccupied type genus.

Anaissini Golbach, 1984: 81 [stem: *Anaiss-*]. Type genus: *Anaissus* Candèze, 1857.

†Tribe Cryptocardiini Dolin, 1980

Cryptocardiini Dolin, 1980: 74 [stem: *Cryptocardi-*]. Type genus: *Cryptocardius* Dolin, 1980.

Tribe Euplinthini Costa, 1975

Euplinthina C. Costa, 1975: 66 [stem: *Euplinth-*]. Type genus: *Euplinthus* C. Costa, 1975. Comment: we act as First Revisers (Compsoplinthini C. Costa, 1975 vs Euplinthini C. Costa, 1975).

Subtribe Cleidecostina Johnson, 2002

Heligmini C. Costa, 1975: 53 [stem: *Heligm-*]. Type genus: *Heligmus* Candèze, 1865 [preoccupied genus name, not *Heligmus* Dujardin, 1845 [Nematoda]; syn. of *Cleidecosta* P. J. Johnson, 2002]. Comment: permanently invalid (Art. 39): based on preoccupied type genus.

Cleidecostini P. J. Johnson, 2002a: 16 [stem: *Cleidecost-*]. Type genus: *Cleidecosta* P. J. Johnson, 2002. Comment: replacement name for Heligmini C. Costa, 1975 because of the homonymy of the type genus.

Subtribe Compsoplinthina Costa, 1975

Compsoplinthina C. Costa, 1975: 71 [stem: *Compsoplinth-*]. Type genus: *Compsoplinthus* C. Costa, 1975.

Subtribe Euplinthina Costa, 1975

Euplinthina C. Costa, 1975: 66 [stem: *Euplinth-*]. Type genus: *Euplinthus* C. Costa, 1975.

Tribe Hemirhipini Candèze, 1857

Hémirhipides Candèze, 1857: 199 [stem: *Hemirhip-*]. Type genus: *Hemirhipus* Berthold, 1827. Comment: published before 29 June; original vernacular name available (Art. 11.7.2): first used in latinized form by J. Thomson (1858: 75, as Hemirhipitae), generally accepted as in P. J. Johnson (2002b: 169, as Hemirhipini); this family-group name was also proposed in the same year by Lacordaire (1857 [before 25 May]: 148, as Hémirhipides) (see *Bibliographic notes* in Introduction); First Reviser (Hemirhipini Candèze, 1857 vs Chalcolepidiini Candèze, 1857) not determined, current usage maintained.

Chalcolépidiides Candèze, 1857: 257 [stem: *Chalcolepidi-*]. Type genus: *Chalcolepidius* Eschscholtz, 1829. Comment: published before 29 June 1857; original vernacular name available (Art. 11.7.2): first used in latinized form by J. L. LeConte (1861: 164, as Chalcolepidiini), generally accepted as in Hyslop (1917: 252, as Chalcolepidina [incorrect stem formation]); this family-group name was also proposed in the same year by Lacordaire (1857 [before 25 May]: 153, as Chalcolépidiides) (see *Bibliographic notes* in Introduction).

Alaites Candèze, 1874: 112 [stem: *Ala-*]. Type genus: *Alaus* Eschscholtz, 1829. Comment: original vernacular name available (Art. 11.7.2): first used in latinized form by Champion (1894: 269, as Alaini), generally accepted as in Heyne and Taschenberg (1905: 152, as Alaini).

Ludioctenina Jakobson, 1913: 755 [stem: *Ludiocten-*]. Type genus: *Ludioctenus* Fairmaire, 1893.

Alauinae Laurent, 1974: 16 [stem: *Ala-*]. Type genus: *Alaus* Eschscholtz, 1829. Comment: family-group name proposed as new without reference to Alaites Candèze, 1874; incorrect original stem formation, not in prevailing usage.

Tribe Oophorini Gistel, 1848

Oophoridae Gistel, 1848: [5] [stem: *Oophor-*]. Type genus: *Oophorus* Eschscholtz, 1833 [syn. of *Aeolus* Eschscholtz, 1829]. Comment: family-group name attributed to Gistel (1856a: 367) in recent literature.

Monocrépidiites Candèze, 1859: 176 [stem: *Monocrepidi-*]. Type genus: *Monocrepidius* Eschscholtz, 1829 [syn. of *Conoderus* Eschscholtz, 1829]. Comment: original vernacular name available (Art. 11.7.2): first used in latinized form by J. L. LeConte (1861: 167, as Monocrepidii), generally accepted as in Burakowski et al. (1985: 147, as Monocrepidiinae).

Drasteriini Houlbert, 1912: 184 [stem: *Drasteri-*]. Type genus: *Drasterius* Eschscholtz, 1829 [placed on the Official List of Generic Names in Zoology (ICZN 1987b)]. Comment: the junior homonym Drasteriini Wiltshire, 1976 (type genus *Drasteria* Hübner, 1818) is available in Lepidoptera: Noctuidae; this case is to be referred to the Commission to remove the homonymy (Art. 55.3.1).

*Aeolina Jakobson, 1913: 747 [stem: *Aeol-*]. Type genus: *Aeolus* Eschscholtz, 1829. Comment: family-group name unavailable (Art. 11.6): originally published as synonym and not made available subsequently.

Conoderinae Fleutiaux, 1919: 58 [stem: *Conoder-*]. Type genus: *Conoderus* Eschscholtz, 1829. Comment: the older name Conoderinae Schönherr,

1833 (type genus *Conoderes* Schönherr, 1833) is curently used as valid in Coleoptera: Curculionidae; this case is to be referred to the Commission to remove the homonymy (Art. 55.3.1).

Pachyderinae Fleutiaux, 1919: 57 [stem: *Pachyder-*]. Type genus: *Pachyderes* Guérin-Méneville, 1830.

Tribe Platycrepidiini Costa and Casari-Chen, 1993

Eudactylites Candèze, 1859: 153 [stem: *Eudactyl-*]. Type genus: *Eudactylus* Sallé, 1855 [preoccupied genus name, not *Eudactylus* Fitzinger, 1843 [Reptilia]; syn. of *Platycrepidius* Candèze, 1859]. Comment: original vernacular name available (Art. 11.7.2): first used in latinized form by Champion (1895: 337, as Eudactylini), generally accepted as in Hyslop (1917: 259, as Eudactylini); permanently invalid (Art. 39): based on preoccupied type genus.

Platycrepidiini C. Costa and Casari-Chen, 1993: 62 [stem: *Platycrepidi-*]. Type genus: *Platycrepidius* Candèze, 1859.

Tribe Pseudomelanactini Arnett, 1967

Pseudomelanactini Arnett, 1967: 111 [stem: *Pseudomelanact-*]. Type genus: *Pseudomelanactes* Mathieu, 1961 [syn. of *Anthracalaus* Fairmaire, 1889].

Tribe Pyrophorini Candèze, 1863

Pyrophorites Candèze, 1863: 3 [stem: *Pyrophor-*]. Type genus: *Pyrophorus* Billberg, 1820.

Subtribe Hapsodrilina Costa, 1975

Hapsodrilina C. Costa, 1975: 88 [stem: *Hapsodril-*]. Type genus: *Hapsodrilus* C. Costa, 1975.

Subtribe Nyctophyxina Costa, 1975

Nyctophyxina C. Costa, 1975: 85 [stem: *Nyctophyx-*]. Type genus: *Nyctophysis* C. Costa, 1975. Comment: current spelling maintained (Art. 29.5): incorrect stem formation in prevailing usage (should be *Nyctophyse-*).

Subtribe Pyrophorina Candèze, 1863

Pyrophorites Candèze, 1863: 3 [stem: *Pyrophor-*]. Type genus: *Pyrophorus* Billberg, 1820. Comment: original vernacular name available (Art. 11.7.2): first used in latinized form by Champion (1896: 463, as Pyrophorini), generally accepted as in P. J. Johnson (2002b: 169, as Pyrophorini).

Tribe Tetralobini Laporte, 1840

Tetralobites Laporte, 1840a: 230 [stem: *Tetralob-*]. Type genus: *Tetralobus* Lepeletier and Audinet-Serville, 1828. Comment: original vernacular name available (Art. 11.7.2): first used in latinized form by Blanchard (1853: 84, as

TETRALOBITAE), generally accepted as in Lawrence and Newton (1995: 853, as TETRALOBINAE).

PHYLLOPHORIDAE Hope, 1842: 73 [stem: *Phyllophor-*]. Type genus: *Phyllophorus* Hope, 1842 [preoccupied genus name, not *Phyllophorus* Grube, 1840 [Echinodermata]; syn. of *Tetralobus* Lepeletier and Audinet-Serville, 1828]. Comment: permanently invalid (Art. 39): based on preoccupied type genus.

PIEZOPHYLLINI Laurent, 1967: 85 [stem: *Piezophyll-*]. Type genus: *Piezophyllus* Hope, 1842.

Subfamily THYLACOSTERNINAE Fleutiaux, 1920

SOLENISCINAE Lameere, 1900: 377 [stem: *Solenisc-*]. Type genus: *Soleniscus* Bonvouloir, 1875 [preoccupied genus name, not *Soleniscus* Meek and Worthen, 1860 [Mollusca]; syn. of *Cussolenis* Fleutiaux, 1918]. Comment: permanently invalid (Art. 39): based on preoccupied type genus.

PTEROTARSINI Fleutiaux, 1902: 648 [stem: *Pterotars-*]. Type genus: *Pterotarsus* Guérin-Méneville, 1829. Comment: although this name has priority over THYLACOSTERNINAE Fleutiaux, 1920, an application will be submitted to the Commission by J. Muona and H. Silfverberg to conserve the well-established younger name.

THYLACOSTERNINAE Fleutiaux, 1920: 94 [stem: *Thylacostern-*]. Type genus: *Thylacosternus* Bonvouloir, 1875. Comment: an application will be submitted to the Commission by J. Muona and H. Silfverberg (pers. comm. 2010) to conserve prevailing usage of THYLACOSTERNINAE Fleutiaux, 1920 over the older name PTEROTARSINAE Fleutiaux, 1902.

BALGINAE Fleutiaux, 1926: 30 [stem: *Balg-*]. Type genus: *Balgus* Fleutiaux, 1920.

CUSSOLENITAE Cobos, 1961: 3 [stem: *Cussolen-*]. Type genus: *Cussolenis* Fleutiaux, 1918.

Subfamily LISSOMINAE Laporte, 1835

LISSOMIDAE Laporte, 1835b: 178 [stem: *Lissom-*]. Type genus: *Lissomus* Dalman, 1824.

DRAPETINI J. L. LeConte, 1863: 44 [stem: *Drapet-*]. Type genus: *Drapetes* Dejean, 1821. Comment: DRAPETINI Collin, 1961 (type genus *Drapetis* Meigen, 1822) has been used as valid to this day in Diptera although Sabrosky (1999: 118) pointed out that the correct stem for the Diptera name is *Drapetid-*.

PROTELATERIDAE Schwarz, 1902: 365 [stem: *Protelater-*]. Type genus: *Protelater* Sharp, 1877.

OESTODINI Hyslop, 1917: 251 [stem: *Oestod-*]. Type genus: *Oestodes* J. L. LeConte, 1853.

ATHOOMORPHINAE Laurent, 1966: 818 [stem: *Athoomorph-*]. Type genus: *Athoomorphus* Schwarz, 1898.

DRAPETINI Dolin, 1975b: 1627, in key [stem: *Drapet-*]. Type genus: *Drapetes* Dejean, 1821. Comment: proposed as new without reference to DRAPETINI J. L. LeConte, 1863.

Sphaenelaterini Stibick, 1979: 179 [stem: *Sphaenelater-*]. Type genus: *Sphaenelater* Schwarz, 1902.

Subfamily Semiotinae Jakobson, 1913

Semiotina Jakobson, 1913: 736 [stem: *Semiot-*]. Type genus: *Semiotus* Eschscholtz, 1829.

Semiotinae Golbach, 1970: 320 [stem: *Semiot-*]. Type genus: *Semiotus* Eschscholtz, 1829. Comment: family-group name proposed as new without reference to Semiotina Jakobson, 1913.

Subfamily Campyloxeninae Costa, 1975

Campyloxeninae C. Costa, 1975: 114 [stem: *Campyloxen-*]. Type genus: *Campyloxenus* Fairmaire, 1860.

Subfamily Pityobiinae Hyslop, 1917

Pityobini Hyslop, 1917: 249 [stem: *Pityobi-*]. Type genus: *Pityobius* J. L. LeConte, 1853. Comment: incorrect original stem formation, not in prevailing usage.

Rostricephalinae Fleutiaux, 1947: 238, in key [stem: *Rostricephal-*]. Type genus: *Rostricephalus* Fleutiaux, 1918.

Adzusini Kishii, 1989: 2 [stem: *Adzus-*]. Type genus: *Adzusa* Kishii, 1957.

Subfamily Oxynopterinae Candèze, 1857

Oxynoptérides Candèze, 1857: 355 [stem: *Oxynopter-*]. Type genus: *Oxynopterus* Hope, 1842. Comment: published before 29 June 1857; original vernacular name available (Art. 11.7.2): first used in latinized form by Heyne and Taschenberg (1905: 155, as Oxynopterini), generally accepted as in Johnson (2002b: 169, as Oxynopterini); this family-group name was also proposed in the same year by Lacordaire (1857 [before 25 May]: 158, as Oxynoptérides) (see *Bibliographic notes* in Introduction); First Revisor (Melanactinae Candèze, 1857 vs Oxynopterinae Candèze, 1857) not determined, current usage maintained.

Mélanactides Candèze, 1857: 182 [stem: *Melanact-*]. Type genus: *Melanactes* J. L. LeConte, 1853. Comment: published before 29 June 1857; original vernacular name available (Art. 11.7.2): first used in latinized form by Holmgren (1899: 199, as Melanactidae), generally accepted as in Stibick (1979: 164, as Melanactinae); this family-group name was also proposed in the same year by Lacordaire (1857 [before 25 May]: 144, as Mélanactides) (see *Bibliographic notes* in Introduction).

Asaphites Candèze, 1863: 207 [stem: *Asaph-*]. Type genus: *Asaphes* Kirby, 1837 [preoccupied genus name, not *Asaphes* Walker, 1834 [Hymenoptera]; syn. of *Hemicrepidius* Germar, 1839]. Comment: original vernacular name available (Art. 11.7.2): first used in latinized form and generally accepted as in Heyne and Taschenberg (1905: 163, as Asaphini); permanently invalid (Art. 39): based on preoccupied type genus; the older name Asaphidae H. C. C. Burmeister, 1843 (type genus *Asaphus* Brongniart 1822) is currently used as valid in Trilobita; the younger name Asaphinae Ashmead, 1904 (type genus *Asaphes* Walker, 1834) is currently used as valid in Hymenoptera: Pteromalidae.

Campsosterninae Fleutiaux, 1927a: 104 [stem: *Campsostern-*]. Type genus: *Campsosternus* Latreille, 1834. Comment: published 25 April 1927; this family-group name was also used in the same year by Fleutiaux (1927b ["31 December"]: 108, as Campsosterninae).

Pectocerini Gurjeva, 1974: 107, in key [stem: *Pectocer-*]. Type genus: *Pectocera* Hope, 1842.

Subfamily Dendrometrinae Gistel, 1848

Dendrometridae Gistel, 1848: [5] [stem: *Dendrometr-*]. Type genus: *Dendrometrus* Gistel, 1848 [syn. of *Limonius* Eschscholtz, 1829]. Comment: First Reviser found (Denticollinae Stein and Weise, 1877 (1848) vs Dendrometrinae Gistel, 1848) is Sánchez-Ruiz (1996: 74).

Tribe Crepidomenini Candèze, 1863

Crépidoménites Candèze, 1863: 190 [stem: *Crepidomen-*]. Type genus: *Crepidomenus* Erichson, 1842. Comment: original vernacular name available (Art. 11.7.2): first used in latinized form by Heyne and Taschenberg (1905: 163, as Crepidomenini), generally accepted as in Calder (1978: 295, as Crepidomeninae).

Tribe Dendrometrini Gistel, 1848

Dendrometridae Gistel, 1848: [5] [stem: *Dendrometr-*]. Type genus: *Dendrometrus* Gistel, 1848 [syn. of *Limonius* Eschscholtz, 1829]. Comment: First Reviser found (Denticollini Stein and Weise, 1877 (1848) vs Dendrometrini Gistel, 1848) is Sánchez-Ruiz (1996: 75).

Subtribe Dendrometrina Gistel, 1848

Dendrometridae Gistel, 1848: [5] [stem: *Dendrometr-*]. Type genus: *Dendrometrus* Gistel, 1848 [syn. of *Limonius* Eschscholtz, 1829].

Athoites Candèze, 1859: 4 [stem: *Atho-*]. Type genus: *Athous* Eschscholtz, 1829. Comment: original vernacular name available (Art. 11.7.2): first used in latinized form by J. L. LeConte (1861: 170, as Athoi), generally accepted as in Burakowski et al. (1985: 189, as Athoinae).

Limoniina Jakobson, 1913: 755 [stem: *Limoni-*]. Type genus: *Limonius* Eschscholtz, 1829.

Subtribe Denticollina Stein and Weise, 1877 (1848)

Campylidae Gistel, 1848: [5] [stem: *Campyl-*]. Type genus: *Campylus* Fischer von Waldheim, 1824 [syn. of *Denticollis* Piller and Mitterpacher, 1783]. Comment: usage of the younger name Denticollina Stein and Weise, 1877 conserved over this name (Art. 40.2).

DENTICOLLINI Stein and Weise, 1877: 96 [stem: *Denticoll-*]. Type genus: *Denticollis* Piller and Mitterpacher, 1783. Comment: younger name conserved over CAMPYLINA Gistel, 1848 (Art. 40.2).

LEPTUROIDINI Schwarz, 1906: 3, in key [stem: *Lepturoid-*]. Type genus: *Lepturoides* Herbst, 1784 [syn. of *Denticollis* Piller and Mitterpacher, 1783].

Subtribe HEMICREPIDIINA Champion, 1896

HEMICREPIDIINI Champion, 1896: 477 [stem: *Hemicrepidi-*]. Type genus: *Hemicrepidius* Germar, 1839.

Tribe DIMINI Candèze, 1863

DIMITES Candèze, 1863: 237 [stem: *Dim-*]. Type genus: *Dima* Charpentier, 1825. Comment: original vernacular name available (Art. 11.7.2): first used in latinized form by Champion (1896: 476, as DIMINI), generally accepted as in Gurjeva (1974: 107, as DIMINAE); current spelling maintained (Art. 29.5): incorrect stem formation in prevailing usage (should be *Dimat-*).

BELIOPHORINA Jakobson, 1913: 737 [stem: *Beliophor-*]. Type genus: *Beliophorus* Eschscholtz, 1829.

PENIINI Dolin, 1990: 17 [stem: *Peni-*]. Type genus: *Penia* Laporte, 1836.

Tribe HYPNOIDINI Schwarz, 1906 (1860)

CRYPTOHYPNITES Candèze, 1860: 50 [stem: *Cryptohypn-*]. Type genus: *Cryptohypnus* Eschscholtz, 1830 [syn. of *Hypnoidus* Dillwyn, 1829]. Comment: original vernacular name available (Art. 11.7.2): first used in latinized form by J. L. LeConte (1861: 166, as CRYPTOHYPNI), generally accepted as in Burakowski et al. (1985: 214, as CRYPTOHYPNINAE); HYPNOIDINI Schwarz, 1906 conserved over this name (Art. 40.2).

HYPNOIDINI Schwarz, 1906: 150 [stem: *Hypnoid-*]. Type genus: *Hypnoidus* Dillwyn, 1829. Comment: name proposed to replace CRYPTOHYPNITES Candèze, 1860 because of the synonymy of the type genus; younger name conserved over CRYPTOHYPNINI Candèze, 1860 (Art. 40.2).

HYPOLITHINAE Fleutiaux, 1928: 252 [stem: *Hypolith-*]. Type genus: *Hypolithus* Eschscholtz, 1829. Comment: name proposed to replace CRYPTOHYPNITES Candèze, 1860 because of the synonymy of the type genus.

*PRISAHYPINI Stibick, 1976: 197 [stem: *Prisahypn-*]. Type genus: *Prisahypnus* Stibick, 1979. Comment: unavailable family-group name, not based on available genus name; incorrect original stem formation, not in prevailing usage.

PRISAHYPINI Stibick, 1979: 166 [stem: *Prisahypn-*]. Type genus: *Prisahypnus* Stibick, 1979. Comment: incorrect original stem formation, not in prevailing usage.

Tribe PLEONOMINI Semenov and Pjatakova, 1936

PLEONOMINI Semenov and Pjatakova, 1936: 103 [stem: *Pleonom-*]. Type genus: *Pleonomus* Ménétriés, 1849. Comment: name proposed after 1930 without description or bibliographic reference to such a description (Art. 13.1), however available because it was used as valid before 2000 as in Stibick (1979: 176) and was not rejected by an author who, between 1961 and 1999, applied Article 13 of the then current edition of the Code (see Art. 13.2.1).

Tribe PROSTERNINI Gistel, 1856 *nomen protectum*

DIACANTHIDAE Gistel, 1848: [5] [stem: *Diacanth-*]. Type genus: *Diacanthus* Latreille, 1834 [syn. of *Selatosomus* Stephens, 1830]. Comment: *nomen oblitum* (see Appendix 1).

PROSTERNIDAE Gistel, 1856a: 367 [stem: *Prostern-*]. Type genus: *Prosternon* Latreille, 1834. Comment: *nomen protectum* (see Appendix 1).

CORYMBITINI J. L. LeConte, 1861: 169 [stem: *Corymbit-*]. Type genus: *Corymbites* Latreille, 1834 [syn. of *Ctenicera* Latreille, 1829].

CTENICERINA Jakobson, 1913: 736 [stem: *Ctenicer-*]. Type genus: *Ctenicera* Latreille, 1829. Comment: family-group name previously attributed to Fleutiaux (1936) in the literature; "CTENICERINI Fleutiaux, 1936" used as valid instead of PROSTERNINI Gistel, 1856 by Cate (2007: 33, 173).

CTENICERINAE Neboiss, 1956: 47 [stem: *Ctenicer-*]. Type genus: *Ctenicera* Latreille, 1829. Comment: name proposed to replace CORYMBITINAE J. L. LeConte, 1861 because of the synonymy of the type genus; family-group name proposed as new without reference to CTENICERINA Jakobson, 1913 or CTENICERINAE Fleutiaux, 1936.

Tribe SENODONIINI Schenkling, 1927

ALLOTRIITES Candèze, 1863: 225 [stem: *Allotri-*]. Type genus: *Allotrius* Laporte, 1840 [preoccupied genus name, not *Allotrius* Temminck, 1835 [Aves]; syn. of *Senodonia* Laporte, 1838]. Comment: original vernacular name available (Art. 11.7.2): first used in latinized form by Champion (1896: 489, as ALLOTRIINI), generally accepted as in Heyne and Taschenberg (1905: 163, as ALLOTRIINI); permanently invalid (Art. 39): based on preoccupied type genus.

SENODONIINAE Schenkling, 1927: 417 [stem: *Senodoni-*]. Type genus: *Senodonia* Laporte, 1838.

SENODONIINI Dolin, 2000: 19 [stem: *Senodoni-*]. Type genus: *Senodonia* Laporte, 1838. Comment: family-group name proposed as new without reference to SENODONIINAE Schenkling, 1927.

Subfamily NEGASTRIINAE Nakane and Kishii, 1956

NEGASTRIINAE Nakane and Kishii, 1956: 203 [stem: *Negastri-*]. Type genus: *Negastrius* C. G. Thomson, 1859.

Tribe NEGASTRIINI Nakane and Kishii, 1956

NEGASTRIINAE Nakane and Kishii, 1956: 203 [stem: *Negastri-*]. Type genus: *Negastrius* C. G. Thomson, 1859.

Tribe Quasimusini Schimmel and Tarnawski, 2009

Quasimusini Schimmel and Tarnawski, 2009: 17 [Stem: *Quasimus-*]. Type genus: *Quasimus* Gozis, 1886.

Subtribe Loebliquasimusina Schimmel and Tarnawski, 2009

Loebliquasina Schimmel and Tarnawski, 2009: 18 [Stem: *Loebliquas-*]. Type genus: *Loebliquasis* Dolin, 1997. Comment: incorrect original stem formation, not in prevailing usage.

Subtribe Quasimusina Schimmel and Tarnawski, 2009

Quasimusina Schimmel and Tarnawski, 2009: 19 [Stem: *Quasimus-*]. Type genus: *Quasimus* Gozis, 1886.

Subtribe Striatoquasimusina Schimmel and Tarnawski, 2009

Striatoquasina Schimmel and Tarnawski, 2009: 20 [Stem: *Striatoquasimus-*]. Type genus: *Striatoquasimus* Schimmel and Tarnawski, 2009. Comment: incorrect original stem formation, not in prevailing usage.

Subtribe Wittmeroquasimusina Schimmel and Tarnawski, 2009

Wittmeroquasina Schimmel and Tarnawski, 2009: 20 [Stem: *Wittmeroquasimus-*]. Type genus: *Wittmeroquasimus* Dolin, 1997. Comment: incorrect original stem formation, not in prevailing usage.

Subfamily Elaterinae Leach, 1815

Elaterides Leach, 1815: 85 [stem: *Elater-*]. Type genus: *Elater* Linnaeus, 1758.

Tribe Agriotini Laporte, 1840

Agriotites Laporte, 1840a: 233 [stem: *Agriot-*]. Type genus: *Agriotes* Eschscholtz, 1829.

Subtribe Agriotina Laporte, 1840

Agriotites Laporte, 1840a: 233 [stem: *Agriot-*]. Type genus: *Agriotes* Eschscholtz, 1829. Comment: original vernacular name available (Art. 11.7.2): first used in latinized form by Champion (1896: 511, as Agriotini), generally accepted as in Hansen (1996: 143, as Agriotini); current spelling maintained (Art. 29.5): incorrect stem formation in prevailing usage (should be *Agriotet-*).

Subtribe Cardiorhinina Candèze, 1863

Cardiorhinites Candèze, 1863: 247 [stem: *Cardiorhin-*]. Type genus: *Cardiorhinus* Eschscholtz, 1829. Comment: original vernacular name available (Art. 11.7.2): first used in latinized form by Champion (1896: 495, as Cardiorhini), generally accepted as in Heyne and Taschenberg (1905: 164, as Cardiorhinini).

Tribe Ampedini Gistel, 1848

Ampedidae Gistel, 1848: [5] [stem: *Amped-*]. Type genus: *Ampedus* Dejean, 1833. Comment: family-group name previously attributed to Gistel (1856a: 367), e.g., P. J. Johnson (2002b: 168).

Tribe Dicrepidiini Thomson, 1858

Dicrepidiitae J. Thomson, 1858: 75 [stem: *Dicrepidi-*]. Type genus: *Dicrepidius* Eschscholtz, 1829.

Tribe Elaterini Leach, 1815

Elaterides Leach, 1815: 85 [stem: *Elater-*]. Type genus: *Elater* Linnaeus, 1758.

Arneidae Gistel, 1848: [6] [stem: *Arne-*]. Type genus: *Arneus* Gistel, 1848 [syn. of *Sericus* Eschscholtz, 1829].

Steatoderidae Gistel, 1848: [5] [stem: *Steatoder-*]. Type genus: *Steatoderus* Dejean, 1833 [syn. of *Elater* Linnaeus, 1758].

Amphilabridae Gistel, 1856a: 367 [stem: *Amphilabr-*]. Type genus: *Amphilabris* Gistel, 1834 [syn. of *Sericus* Eschscholtz, 1829].

Ludiides Lacordaire, 1857: 197 [stem: *Ludi-*]. Type genus: *Ludius* Berthold, 1827 [syn. of *Elater* Linnaeus, 1758]. Comment: original vernacular name available (Art. 11.7.2): first used in latinized form and generally accepted as in J. L. LeConte (1861: 168, as Ludii); the junior homonym Ludiini Aurivillius, 1904 (type genus *Ludia* Wallengren, 1865) in Lepidoptera: Saturniidae was replaced by Micragonini Cockerell, 1914 (see Oberprieler 1997: 146).

Hypodésites Candèze, 1863: 242 [stem: *Hypodese-*]. Type genus: *Hypodesis* Latreille, 1834. Comment: original vernacular name available (Art. 11.7.2): first used in latinized form by Champion (1896: 490, as Hypodesini), generally accepted as in Heyne and Taschenberg (1905: 164, as Hypodesini); incorrect original stem formation, not in prevailing usage.

Sericosomina Hyslop, 1917: 258 [stem: *Sericosom-*]. Type genus: *Sericosomus* Dejean, 1833 [syn. of *Sericus* Eschscholtz, 1829].

*Dolerosomini Dolin, 1975b: 1632 [stem: *Dolerosom-*]. Type genus: *Dolerosomus* Motschulsky, 1859. Comment: unavailable family-group name, proposed after 1930 without description or bibliographic reference to such a description (Art. 13.1).

Tribe Megapenthini Gurjeva, 1973

Megapenthini Gurjeva, 1973: 448 [stem: *Megapenth-*]. Type genus: *Megapenthes* Kiesenwetter, 1858.

Tribe Melanotini Candèze, 1859 (1848)

Cratonychidae Gistel, 1848: [5] [stem: *Cratonych-*]. Type genus: *Cratonychus* Dejean, 1833 [syn. of *Melanotus* Eschscholtz, 1829]. Comment: use of Melanotini Candèze, 1859 converved over this name (Art. 40.2).

Mélanotites Candèze, 1859: 4 [stem: *Melanot-*]. Type genus: *Melanotus* Eschscholtz, 1829. Comment: original vernacular name available (Art. 11.7.2): first used in latinized form by J. L. LeConte (1861: 169, as Melanoti), generally accepted as in Hansen (1996: 143, as Melanotini); name conserved over Cratonychini Gistel, 1848 (Art. 40.2) (see Sánchez-Ruiz 1996: 167).

Tribe Odontonychini Girard, 1973

Odontonychini Girard, 1973: 276 [stem: *Odontonych-*]. Type genus: *Odontonychus* Candèze, 1897.

Tribe Physorhinini Candèze, 1859

Physorhinites Candèze, 1859: 384 [stem: *Physorhin-*]. Type genus: *Physorhinus* Germar, 1840. Comment: original vernacular name available (Art. 11.7.2): first used in latinized form by J. L. LeConte (1861: 167, as Physorhini), generally accepted as in Burakowski et al. (1985: 114, as Physorhininae).

Tribe Pomachiliini Candèze, 1859

Pomachiliites Candèze, 1859: 4 [stem: *Pomachili-*]. Type genus: *Pomachilius* Eschscholtz, 1829. Comment: original vernacular name available (Art. 11.7.2): first used in latinized form by Champion (1895: 402, as Pomachiliini), generally accepted as in Burakowski et al. (1985: 131, as Pomachiliinae).

Tribe Synaptini Gistel, 1856

Dairaeidae Gistel, 1848: [5] [stem: *Dair-*]. Type genus: *Daira* Gistel, 1848 [preoccupied genus name, not *Daira* Milne-Edwards, 1830 [Crustacea]; syn. of *Synaptus* Eschscholtz, 1829]. Comment: permanently invalid (Art. 39): based on preoccupied type genus; incorrect original stem formation, not in prevailing usage.

Synaptidae Gistel, 1856a: 366 [stem: *Synapt-*]. Type genus: *Synaptus* Eschscholtz, 1829.

Adrastites Candèze, 1863: 448 [stem: *Adrast-*]. Type genus: *Adrastus* Eschscholtz, 1829. Comment: original vernacular name available (Art. 11.7.2): first used in latinized form by Champion (1896: 533, as Adrastini), generally accepted as in Burakowski et al. (1985: 133, as Adrastinae).

Subfamily Cardiophorinae Candèze, 1859

Cardiophorites Candèze, 1859: 4 [stem: *Cardiophor-*]. Type genus: *Cardiophorus* Eschscholtz, 1829. Comment: original vernacular name available (Art. 11.7.2): first used in latinized form by J. L. LeConte (1861: 166, as Cardiophori), generally accepted as in Burakowski et al. (1985: 227, as Cardiophorinae).

Aphrici J. L. LeConte, 1861: 173 [stem: *Aphric-*]. Type genus: *Aphricus* J. L. LeConte, 1854.

APTOPINA Jakobson, 1913: 760 [stem: *Aptopod-*]. Type genus: *Aptopus* Eschscholtz, 1829. Comment: incorrect original stem formation, not in prevailing usage.

ESTHESOPINAE Fleutiaux, 1919: 76 [stem: *Esthesopod-*]. Type genus: *Esthesopus* Eschscholtz, 1829. Comment: incorrect original stem formation, not in prevailing usage.

NYCTORINI Semenov and Pjatakova, 1936: 102 [stem: *Nyctor-*]. Type genus: *Nyctor* Semenov and Pjatakova, 1936 [syn. of *Cardiophorus* Eschscholtz, 1829].

Subfamily HEMIOPINAE Fleutiaux, 1941

HEMIOPINAE Fleutiaux, 1941: 31 [stem: *Hemiop-*]. Type genus: *Hemiops* Laporte, 1838.

Subfamily PHYSODACTYLINAE Lacordaire, 1857

PHYSODACTYLIDES Lacordaire, 1857: 236 [stem: *Physodactyl-*]. Type genus: *Physodactylus* Fischer von Waldheim, 1823. Comment: original vernacular name available (Art. 11.7.2): first used in latinized form by Fleutiaux (1892b: 404, as PHYSODACTYLINI), generally accepted as in Lawrence and Newton (1995: 855, as PHYSODACTYLINAE).

TOXOGNATHINAE Fleutiaux, 1941: 34 [stem: *Toxognath-*]. Type genus: *Toxognathus* Fairmaire, 1878.

Subfamily EUDICRONYCHINAE Girard, 1971

DICRONYCHIDAE Schwarz, 1897: 11 [stem: *Dicronych-*]. Type genus: *Dicronychus* Laporte, 1840 [preoccupied genus name, not *Dicronychus* Brullé, 1832 [Coleoptera: ELATERIDAE: CARDIOPHORINAE]; syn. of *Eudicronychus* Méquignon, 1931]. Comment: permanently invalid (Art. 39): based on preoccupied type genus.

EUDICRONYCHINAE Girard, 1971: 645 [stem: *Eudicronych-*]. Type genus: *Eudicronychus* Méquignon, 1931. Comment: description by indication (distinguishing characters given in Schwarz (1907: 2, as DICRONYCHIDAE)).

Subfamily SUBPROTELATERINAE Fleutiaux, 1920

SUBPROTELATERINAE Fleutiaux, 1920: 99 [stem: *Subprotelater-*]. Type genus: *Subprotelater* Fleutiaux, 1916.

Subfamily MOROSTOMATINAE Dolin, 2000

MOROSTOMINAE Dolin, 2000: 18 [stem: *Morostomat-*]. Type genus: *Morostoma* Candèze, 1879. Comment: incorrect original stem formation, not in prevailing usage.

†Subfamily PROTAGRYPNINAE Dolin, 1973

PROTAGRYPNINI Dolin, 1973: 74 [stem: *Protagrypn-*]. Type genus: *Protagrypnus* Dolin, 1973.

†Tribe Desmatini Dolin, 1975

Desmatini Dolin, 1975a: 60 [stem: *Desmat-*]. Type genus: *Desmatus* Dolin, 1975.

†Tribe Hypnomorphini Dolin, 1975

Hypnomorphini Dolin, 1975a: 54 [stem: *Hypnomorph-*]. Type genus: *Hypnomorphus* Dolin, 1975.

†Tribe Protagrypnini Dolin, 1973

Protagrypnini Dolin, 1973: 74 [stem: *Protagrypn-*]. Type genus: *Protagrypnus* Dolin, 1973.

Family Plastoceridae Crowson, 1972

Plastoceridae Crowson, 1972: 37, in key [stem: *Plastocer-*]. Type genus: *Plastocerus* Schaum, 1852. Comment: an application needs to be submitted to the Commission to suppress Plastoceri J. L. LeConte, 1861 (based on the misidentified type genus *Plastocerus* sensu J. L. LeConte, 1853) for the Principles of Priority and Homonymy (Art. 65.2.1) to conserve this name as valid.

Family Drilidae Blanchard, 1845

Drilites Blanchard, 1845b: 53 [stem: *Dril-*]. Type genus: *Drilus* A. G. Olivier, 1790.

Subfamily Drilinae Blanchard, 1845

Drilites Blanchard, 1845b: 53 [stem: *Dril-*]. Type genus: *Drilus* A. G. Olivier, 1790. Comment: original vernacular name available (Art. 11.7.2): first used in latinized form by Gistel (1848: [11], as Drilidae), generally accepted as in Hansen (1996: 144, as Drilidae).

Subfamily Thilmaninae Kazantsev, 2004

Thilmaninae Kazantsev, 2004b: 240 [stem: *Thilman-*]. Type genus: *Thilmanus* Baudi di Selve, 1871. Comment: transferred from Lycidae by Bocák and Brlik (2008).

Tribe Euanomini Kazantsev, 2010

Euanomini Kazantsev, 2010a: 55 [stem: *Euanom-*]. Type genus: *Euanoma* Reitter, 1889.

Tribe Thilmanini Kazantsev, 2004

Thilmaninae Kazantsev, 2004b: 240 [stem: *Thilman-*]. Type genus: *Thilmanus* Baudi di Selve, 1871.

Family Omalisidae Lacordaire, 1857

*Homalises Motschulsky, 1849: 55 [stem: *Omalis-*]. Type genus: *Omalisus* Geoffroy, 1762 [as *Homalisus*, unjustified emendation of type genus name by Illiger (1801: 139), not in prevailing usage; placed on the Official List of Generic Names in

Zoology (ICZN 1994a)]. Comment: original vernacular name unavailable (Art. 11.7.2): subsequently used in latinized form but not generally attributed to Motschulsky (1849).

HOMALISIDES Lacordaire, 1857: 303 [stem: *Omalis-*]. Type genus: *Omalisus* Geoffroy, 1762 [as *Homalisus*, unjustified emendation of type genus name by Illiger (1801: 139), not in prevailing usage; placed on the Official List of Generic Names in Zoology (ICZN 1994a)]. Comment: original vernacular name available (Art. 11.7.2): first used in latinized form by Kiesenwetter (1860: 442, as HOMALISIDAE [incorrect stem formation]), generally accepted as in Lawrence and Newton (1995: 856, as OMALISIDAE); incorrect original stem formation, not in prevailing usage.

†Family BERENDTIMIRIDAE Winkler, 1987

BERENDTIMIRIDAE J. R. Winkler, 1987: 52 [stem: *Berendtimir-*]. Type genus: *Berendtimirus* J. R. Winkler, 1987.

Family LYCIDAE Laporte, 1836

LYCUSIDAE Laporte, 1836: 25 [stem: *Lyc-*]. Type genus: *Lycus* Fabricius, 1787.

Subfamily LIBNETINAE Bocák and Bocáková, 1990

LIBNETININA Bocák and Bocáková, 1990: 652 [stem: *Libnet-*]. Type genus: *Libnetis* C. O. Waterhouse, 1878. Comment: incorrect original stem formation, not in prevailing usage.

Subfamily DICTYOPTERINAE Houlbert, 1922

DICTYOPTERINI Houlbert, 1922a: 338 [stem: *Dictyopter-*]. Type genus: *Dictyoptera* Latreille, 1829.

Tribe DICTYOPTERINI Houlbert, 1922

DICTYOPTERINI Houlbert, 1922a: 338 [stem: *Dictyopter-*]. Type genus: *Dictyoptera* Latreille, 1829. Comment: family-group name previously attributed to Kleine (1929: 226); although Houlbert (1922a) uses the tribe name LYCINI for a group of genera that includes *Dictyoptera* Latreille in the main text (page 240), this was certainly a mistake since the tribe name DICTYOPTERINI is correctly used in the "Index alphabétique" (page 319) and in the "Table systématique" (page 338).

Tribe LYCOPROGENTHINI Bocák and Bocáková, 2008

LYCOPROGENTHINI Bocák and Bocáková, 2008: 709 [stem: *Lycoprogenth-*]. Type genus: *Lycoprogenthes* Pic, 1915.

Tribe TAPHINI Bocák and Bocáková, 1990

TAPHININA Bocák and Bocáková, 1990: 650 [stem: *Taph-*]. Type genus: *Taphes* C. O. Waterhouse, 1878. Comment: incorrect original stem formation, not in prevailing usage.

Subfamily Lyropaeinae Bocák and Bocáková, 1989

Lyropaeini Bocák and Bocáková, 1989: 718 [stem: *Lyropae-*]. Type genus: *Lyropaeus* C. O. Waterhouse, 1878.

Tribe Alyculini Bocák and Bocáková, 2008

Alyculini Bocák and Bocáková, 2008: 710 [stem: *Alycul-*]. Type genus: *Alyculus* Kazantsev, 1999.

Tribe Antennolycini Bocák and Bocáková, 2008

Antennolycini Bocák and Bocáková, 2008: 710 [stem: *Antennolyc-*]. Type genus: *Antennolycus* Bocák and Bocáková, 1999.

Tribe Lyropaeini Bocák and Bocáková, 1989

Lyropaeini Bocák and Bocáková, 1989: 718 [stem: *Lyropae-*]. Type genus: *Lyropaeus* C. O. Waterhouse, 1878.

Paralycinae L. N. Medvedev and Kazantsev, 1992: 59 [stem: *Paralyc-*]. Type genus: *Paralycus* L. N. Medvedev and Kazantsev, 1992 [syn. of *Lyropaeus* C. O. Waterhouse, 1878].

Tribe Miniduliticolini Kazantsev, 2003

Miniduliticolini Kazantsev, 2003: 20 [stem: *Miniduliticol-*]. Type genus: *Miniduliticola* Kazantsev, 2003.

Tribe Platerodrilini Kazantsev, 2004

*Duliticolinae Mjöberg, 1925: 140 [stem: *Duliticol-*]. Type genus: *Duliticola* Mjöberg, 1925 [syn. of *Platerodrilus* Pic, 1921]. Comment: unavailable family-group name (Art. 8.3): nomenclatural act disclaimed.

*Duliticolinae Kazantsev, 2003: 19 [stem: *Duliticol-*]. Type genus: *Duliticola* Mjöberg, 1925 [syn. of *Platerodrilus* Pic, 1921]. Comment: family-group name proposed as a new taxon but unavailable (Art. 11.7.1.1) because it was based on a type genus which was considered a synonym of *Platerodrilus* Pic, 1921 by the author.

Platerodrilini Kazantsev, 2004b: 241 [stem: *Platerodril-*]. Type genus: *Platerodrilus* Pic, 1921.

Subfamily Ateliinae Kleine, 1929

Atelinae Kleine, 1929: 222 [stem: *Ateli-*]. Type genus: *Atelius* C. O. Waterhouse, 1878. Comment: First Revisers found (Ateliinae Kleine, 1928 vs Dilophotinae Kleine, 1928) are Bocák and Bocáková (2008: 712).

Tribe Ateliini Kleine, 1929

Atelinae Kleine, 1929: 222 [stem: *Ateli-*]. Type genus: *Atelius* C. O. Waterhouse, 1878. Comment: incorrect original stem formation, not in prevailing usage.

Tribe Dilophotini Kleine, 1929

Dilophotinae Kleine, 1929: 222 [stem: *Dilophot-*]. Type genus: *Dilophotes* C. O. Waterhouse, 1879. Comment: current spelling maintained (Art. 29.5): incorrect stem formation in prevailing usage (should be *Dilophotet-*).

Subfamily Lycinae Laporte, 1836

Lycusidae Laporte, 1836: 25 [stem: *Lyc-*]. Type genus: *Lycus* Fabricius, 1787.

Tribe Calochromini Lacordaire, 1857

Calochromides Lacordaire, 1857: 301 [stem: *Calochrom-*]. Type genus: *Calochromus* Guérin-Méneville, 1833. Comment: original vernacular name available (Art. 11.7.2): first used in latinized form by Gorham (1880: 1, as Calochrominae), generally accepted as in Hansen (1996: 144, as Calochrominae).

Lygistopteri J. L. LeConte, 1881: 27 [stem: *Lygistopter-*]. Type genus: *Lygistopterus* Dejean, 1833.

Tribe Calopterini Green, 1949

Calopterini Green, 1949: 56, in key [stem: *Calopter-*]. Type genus: *Calopteron* Laporte, 1838.

Subtribe Acroleptina Bocáková, 2005

Acroleptina Bocáková, 2005: 445 [stem: *Acrolept-*]. Type genus: *Acroleptus* Bourgeois, 1886.

Subtribe Calopterina Green, 1949

*Calopterini Kleine, 1933: 20 [stem: *Calopter-*]. Type genus: *Calopteron* Laporte, 1838. Comment: unavailable family-group name, proposed after 1930 without description or bibliographic reference to such a description (Art. 13.1).

Calopterini Green, 1949: 56, in key [stem: *Calopter-*]. Type genus: *Calopteron* Laporte, 1838.

Tribe Conderini Bocák and Bocáková, 1990

Conderini Bocák and Bocáková, 1990: 643 [stem: *Conder-*]. Type genus: *Conderis* C. O. Waterhouse, 1879. Comment: current spelling maintained (Art. 29.5): incorrect stem formation in prevailing usage (should be *Conderid-*).

Tribe Dihammatini Bocák and Bocáková, 2008

Dihammatini Bocák and Bocáková, 2008: 716 [stem: *Dihammat-*]. Type genus: *Dihammatus* C. O. Waterhouse, 1879.

Tribe Erotini LeConte, 1881

Erotes J. L. LeConte, 1881: 23 [stem: *Erot-*]. Type genus: *Eros* Newman, 1838.

Flagraxina Kazantsev, 2004a: 35, in key [stem: *Flagrax-*]. Type genus: *Flagrax* Kazantsev, 1992.

Aferotini Kazantsev, 2004a: 4 [stem: *Aferot-*]. Type genus: *Aferos* Kazantsev, 1992.

Tribe Eurrhacini Bocáková, 2005

Eurrhacina Bocáková, 2005: 444 [stem: *Eurrhac-*]. Type genus: *Eurrhacus* C. O. Waterhouse, 1879.

Tribe Leptolycini Leng and Mutchler, 1922

Leptolycini Leng and Mutchler, 1922: 430 [stem: *Leptolyc-*]. Type genus: *Leptolycus* Leng and Mutchler, 1922.

Tribe Lycini Laporte, 1836

Lycusidae Laporte, 1836: 25 [stem: *Lyc-*]. Type genus: *Lycus* Fabricius, 1787. Comment: incorrect original stem formation, not in prevailing usage.

Tribe Lyponiini Bocák and Bocáková, 1990

Lyponiinina Bocák and Bocáková, 1990: 652 [stem: *Lyponi-*]. Type genus: *Lyponia* C. O. Waterhouse, 1878. Comment: incorrect original stem formation, not in prevailing usage.

Tribe Macrolycini Kleine, 1929

Macrolycinae Kleine, 1929: 222 [stem: *Macrolyc-*]. Type genus: *Macrolycus* C. O. Waterhouse, 1878.

Tribe Melanerotini Kazantsev, 2010

Melanerotini Kazantsev, 2010b: 195 [stem: *Melanerot-*]. Type genus: *Melaneros* Fairmaire, 1877.

Tribe Metriorrhynchini Kleine, 1926

Metriorrhynchinae Kleine, 1926: 97 [stem: *Metriorrhynch-*]. Type genus: *Metriorrhynchus* Guérin-Méneville, 1838.

Subtribe Hemiconderinina Bocák and Bocáková, 1990

Hemiconderinina Bocák and Bocáková, 1990: 645 [stem: *Hemiconderin-*]. Type genus: *Hemiconderis* Kleine, 1926. Comment: current spelling maintained (Art. 29.5): incorrect stem formation in prevailing usage (should be *Hemiconderid-*).

Subtribe Metriorrhynchina Kleine, 1926

Metriorrhynchinae Kleine, 1926: 97 [stem: *Metriorrhynch-*]. Type genus: *Metriorrhynchus* Guérin-Méneville, 1838 [*Metriorrhynchus* is an incorrect subsequent spelling by Gemminger (1869: 1629) of the original name

Metriorhynchus in prevailing usage and so deemed to be the correct original spelling (Art. 33.3.1)]. Comment: First Reviser (METRIORRHYNCHINA Kleine, 1926 vs DILOLYCINA Kleine, 1926) not determined, current usage maintained.

DILOLYCINAE Kleine, 1926: 186 [stem: *Dilolyc-*]. Type genus: *Dilolycus* Kleine, 1926. Comment: one of the original spellings of this family-group name was HAPLOTHORACINAE on page 95 (based on the new genus *Haplothorax* Kleine), the spelling of the family-group and genus-group names were changed to DILOLYCINAE and *Dilolycus* on page 186 and in the "Corrigenda" on page 188 of the same work because *Haplothorax* Agassiz, 1846 (unjustified emendation of *Aplothorax* G. R. Waterhouse, 1841) was already available in CARABIDAE; DILOLYCINAE is therefore treated as the correct original spelling of this family-group name (Art. 19.2).

CLADOPHORINAE Kleine, 1929: 222 [stem: *Cladophor-*]. Type genus: *Cladophorus* Guérin-Méneville, 1830.

Subtribe TRICHALINA Kleine, 1929

TRICHALINAE Kleine, 1929: 222 [stem: *Trichal-*]. Type genus: *Trichalus* C. O. Waterhouse, 1877.

Tribe PLATERODINI Kleine, 1929

PLATERODINAE Kleine, 1929: 222 [stem: *Platerod-*]. Type genus: *Plateros* Bourgeois, 1879. Comment: current spelling maintained (Art. 29.5): incorrect stem formation in prevailing usage (should be *Platerot-*) (see Kazantsev 2010c: 278).

Tribe SLIPINSKIINI Bocák and Bocáková, 1992

SLIPINSKIININA Bocák and Bocáková, 1992: 256 [stem: *Slipinski-*]. Type genus: *Slipinskia* Bocák and Bocáková, 1992. Comment: incorrect original stem formation, not in prevailing usage.

Tribe THONALMINI Kleine, 1933

THONALMINI Kleine, 1933: 18 [stem: *Thonalm-*]. Type genus: *Thonalmus* Bourgeois, 1883. Comment: name proposed after 1930 without description or bibliographic reference to such a description (Art. 13.1), however available because it was used as valid before 2000 as in Blackwelder (1945: 343, as THONALMINI) and was not rejected by an author who, between 1961 and 1999, applied Article 13 of the then current edition of the Code (see Art. 13.2.1) (see Bocák and Bocáková 2008: 717).

Subfamily DEXORINAE Bocák and Bocáková, 1989

*DEXORINI Kleine, 1933: 113 [stem: *Dexor-*]. Type genus: *Dexoris* C. O. Waterhouse, 1878. Comment: unavailable family-group name, proposed after 1930 without description or bibliographic reference to such a description (Art. 13.1).

Dexorinae Bocák and Bocáková, 1989: 718 [stem: *Dexor-*]. Type genus: *Dexoris* C. O. Waterhouse, 1878.

Family Telegeusidae Leng, 1920

Telegeusidae Leng, 1920: 152 [stem: *Telegeus-*]. Type genus: *Telegeusis* G. H. Horn, 1895. Comment: current spelling maintained (Art. 29.5): incorrect stem formation in prevailing usage (should be *Telegeuse-*).

Family Phengodidae LeConte, 1861

Phengodini J. L. LeConte, 1861: 185 [stem: *Phengod-*]. Type genus: *Phengodes* Illiger, 1807.

Subfamily Phengodinae LeConte, 1861

Phengodini J. L. LeConte, 1861: 185 [stem: *Phengod-*]. Type genus: *Phengodes* Illiger, 1807.

Pseudophengodidae Pic, 1930: 319 [stem: *Pseudophengod-*]. Type genus: *Pseudophengodes* Pic, 1930.

Subfamily Mastinocerinae LeConte, 1881

Mastinocerini J. L. LeConte, 1881: 40 [stem: *Mastinocer-*]. Type genus: *Mastinocerus* Solier, 1849.

Subfamily Penicillophorinae Paulus, 1975

Penicillophorini Paulus, 1975: 80 [stem: *Penicillophor-*]. Type genus: *Penicillophorus* Paulus, 1975.

Family Rhagophthalmidae Olivier, 1907

Rhagophthalmidae E. Olivier, 1907: 63 [stem: *Rhagophthalm-*]. Type genus: *Rhagophthalmus* Motschulsky, 1853.

Family Lampyridae Rafinesque, 1815

Lampyria Rafinesque, 1815: 110 [stem: *Lampyr-*]. Type genus: *Lampyris* Geoffroy, 1762 [placed on the Official List of Generic Names in Zoology (ICZN 1984a)].

Subfamily Psilocladinae McDermott, 1964

Psilocladina McDermott, 1964: 12, in key [stem: *Psiloclad-*]. Type genus: *Psilocladus* Blanchard, 1846.

Cyphonocerinae Crowson, 1972: 55, in key [stem: *Cyphonocer-*]. Type genus: *Cyphonocerus* Kiesenwetter, 1879.

Subfamily Amydetinae Olivier, 1907

Amydetini E. Olivier, 1907: 48 [stem: *Amydet-*]. Type genus: *Amydetes* Hoffmansegg, 1807.

Tribe Amydetini Olivier, 1907

Amydetini E. Olivier, 1907: 48 [stem: *Amydet-*]. Type genus: *Amydetes* Hoffmansegg, 1807.

Megalophthalmini E. Olivier, 1907: 46 [stem: *Megalophthalm-*]. Type genus: *Megalophthalmus* Gray, 1832 [preoccupied genus name, not *Megalophthalmus* Leach, 1830 [Crustacea]; syn. of *Magnoculus* McDermott, 1964]. Comment: permanently invalid (Art. 39): based on preoccupied type genus.

Tribe Vestini McDermott, 1964

Vestina McDermott, 1964: 12, in key [stem: *Vest-*]. Type genus: *Vesta* Laporte, 1833.

Subfamily Lampyrinae Rafinesque, 1815

Lampyria Rafinesque, 1815: 110 [stem: *Lampyr-*]. Type genus: *Lampyris* Geoffroy, 1762 [placed on the Official List of Generic Names in Zoology (ICZN 1984a)].

Tribe Cratomorphini Green, 1948

Cratomorphi Green, 1948: 68, in key [stem: *Cratomorph-*]. Type genus: *Cratomorphus* Motschulsky, 1853.

Tribe Lamprocerini Olivier, 1907

Lamprocerini E. Olivier, 1907: 7 [stem: *Lamprocer-*]. Type genus: *Lamprocera* Laporte, 1833.

Tribe Lamprohizini Kazantsev, 2010

Lamprohizini Kazantsev, 2010d: 189 [stem: *Lamprohiz-*]. Type genus: *Lamprohiza* Motschulsky, 1853.

Tribe Lampyrini Rafinesque, 1815

Lampyria Rafinesque, 1815: 110 [stem: *Lampyr-*]. Type genus: *Lampyris* Geoffroy, 1762 [placed on the Official List of Generic Names in Zoology (ICZN 1984a)]. Comment: family-group name previously attributed to Latreille (1816: 236); current spelling maintained (Art. 29.3.1.1): incorrect stem formation in prevailing usage (should be *Lampyrid-*).

Tribe Lucidotini Lacordaire, 1857

Lucidotides Lacordaire, 1857: 310 [stem: *Lucidot-*]. Type genus: *Lucidota* Laporte, 1833.

Subtribe Dadophorina Olivier, 1907

Dadophorini E. Olivier, 1907: 26 [stem: *Dadophor-*]. Type genus: *Dadophora* E. Olivier, 1907.

Subtribe Lamprigerina McDermott, 1964

Lamprigerina McDermott, 1964: 12, in key [stem: *Lampriger-*]. Type genus: *Lamprigera* Motschulsky, 1853.

Subtribe Lucidotina Lacordaire, 1857

Lucidotides Lacordaire, 1857: 310 [stem: *Lucidot-*]. Type genus: *Lucidota* Laporte, 1833. Comment: original vernacular name available (Art. 11.7.2): first used in latinized form by J. L. LeConte (1861: 184, as Lucidotae), generally accepted as in Gorham (1881: 29, as Lucidotides [treated as Latin]).

Pristolycini J. R. Winkler, 1953: 409 [stem: *Pristolyc-*]. Type genus: *Pristolycus* Gorham, 1883.

Pristolycini Kazantsev, 2010d: 205 [stem: *Pristolyc-*]. Type genus: *Pristolycus* Gorham, 1883. Comment: name proposed as new without reference to Pristolycini J. R. Winkler, 1953.

Subtribe Photinina LeConte, 1881

*Phosphaenaires Mulsant, 1862: 116 [stem: *Phosphaen-*]. Type genus: *Phosphaenus* Laporte, 1833. Comment: original vernacular name unavailable (Art. 11.7.2): subsequently used in latinized form but not generally attributed to Mulsant (1862).

Photini J. L. LeConte, 1881: 30 [stem: *Photin-*]. Type genus: *Photinus* Laporte, 1833. Comment: incorrect original stem formation, not in prevailing usage; this name is a senior homonym of Photininae Giglio-Tos, 1915 (type genus *Photina* Burmeister, 1838) in Mantodea; this case was submitted to the Commission to remove the homonymy by emending the stem of the mantid name (see Svenson and Branham 2007: 243).

Phosphaenides E. Olivier, 1884: 4 [stem: *Phosphaen-*]. Type genus: *Phosphaenus* Laporte, 1833. Comment: original vernacular name available (Art. 11.7.2): first used in latinized form and generally accepted as in Ganglbauer (1886: 280, as Phosphaenini).

Tribe Pleotomini Summers, 1874

Pleotomini Summers, 1874: 91 [stem: *Pleotom-*]. Type genus: *Pleotomus* J. L. LeConte, 1861.

Calyptocephalina Jakobson, 1911a: 667 [stem: *Calyptocephal-*]. Type genus: *Calyptocephalus* Gray, 1832.

Pleotomi Green, 1948: 68, in key [stem: *Pleotom-*]. Type genus: *Pleotomus* J. L. LeConte, 1861. Comment: family-group name proposed as new without reference to Pleotomini Summers, 1874.

Subfamily Luciolinae Lacordaire, 1857

Luciolides Lacordaire, 1857: 333 [stem: *Luciol-*]. Type genus: *Luciola* Laporte, 1833.

Tribe CURTOSINI McDermott, 1964

CURTOSINI McDermott, 1964: 47 [stem: *Curtos-*]. Type genus: *Curtos* Motschulsky, 1845. Comment: current spelling maintained (Art. 29.5): incorrect stem formation in prevailing usage (should be *Curt-*).

Tribe LUCIOLINI Lacordaire, 1857

LUCIOLIDES Lacordaire, 1857: 333 [stem: *Luciol-*]. Type genus: *Luciola* Laporte, 1833. Comment: original vernacular name available (Art. 11.7.2): first used in latinized form by J. L. LeConte (1861: 184, as LUCIOLINI), generally accepted as in Lawrence and Newton (1995: 859, as LUCIOLINAE).

Subfamily PHOTURINAE Lacordaire, 1857

PHOTURIDES Lacordaire, 1857: 338 [stem: *Photur-*]. Type genus: *Photuris* J. L. LeConte, 1851. Comment: original vernacular name available (Art. 11.7.2): first used in latinized form by Heyne and Taschenberg (1906: 177, as PHOTURINI), generally accepted as in Lawrence and Newton (1995: 859, as PHOTURINAE); current spelling maintained (Art. 29.3.1.1): incorrect stem formation in prevailing usage (should be *Photurid-*).

LAMPYRIDAE *incertae sedis*

CHEGUEVARIINI Kazantsev, 2007: 370 [stem: *Cheguevari-*]. Type genus: *Cheguevaria* Kazantsev, 2007.

Family OMETHIDAE LeConte, 1861

OMETHES J. L. LeConte, 1861: 187 [stem: *Ometh-*]. Type genus: *Omethes* J. L. LeConte, 1861.

Subfamily OMETHINAE LeConte, 1861

OMETHES J. L. LeConte, 1861: 187 [stem: *Ometh-*]. Type genus: *Omethes* J. L. LeConte, 1861.

Subfamily MATHETEINAE LeConte, 1881

MATHETEI J. L. LeConte, 1881: 29 [stem: *Mathete-*]. Type genus: *Matheteus* J. L. LeConte, 1874.

Subfamily DRILONIINAE Crowson, 1972

DRILONIINAE Crowson, 1972: 58, in key [stem: *Driloni-*]. Type genus: *Drilonius* Kiesenwetter, 1874.

Family CANTHARIDAE Imhoff, 1856 (1815)

CANTHARIDAE Imhoff, 1856: [2] 69 [stem: *Canthar-*]. Type genus: *Cantharis* Linnaeus, 1758. Comment: this family-group name was used by many authors prior to Imhoff's usage, these were based on a the misidentified type genus *Cantharis* (syn.

of *Lytta* Fabricius, 1775) and are therefore not used as valid for this group (see Lawrence and Newton 1995:860); this case should be referred to the Commission to suppress any use of Cantharidae prior to Imhoff (1856); usage of the younger name Cantharidae Imhoff, 1856 over Telephoridae Leach, 1815 is conserved (Art. 40.2) (see Lawrence and Newton 1995: 860).

Subfamily Cantharinae Imhoff, 1856 (1815)

Cantharidae Imhoff, 1856: [2] 69 [stem: *Canthar-*]. Type genus: *Cantharis* Linnaeus, 1758. Comment: usage of the younger name over Telephorinae Leach, 1815 is conserved (Art. 40.2) (see Lawrence and Newton 1995: 860).

Tribe Cantharini Imhoff, 1856 (1815)

Telephorides Leach, 1815: 85 [stem: *Telephor-*]. Type genus: *Telephorus* Schaeffer, 1766 [syn. of *Cantharis* Linnaeus, 1758]. Comment: usage of Cantharidae/-inae/-ini over this name is conserved (Art. 40.2) (see Lawrence and Newton 1995: 860).

Cantharidae Imhoff, 1856: [2] 69 [stem: *Canthar-*]. Type genus: *Cantharis* Linnaeus, 1758. Comment: usage of the younger name over Telephorini Leach, 1815 is conserved (Art. 40.2) (see Lawrence and Newton 1995: 860); current spelling maintained (Art. 29.3.1.1): incorrect stem formation in prevailing usage (should be *Cantharid-*).

Tribe Podabrini Gistel, 1856

Podabridae Gistel, 1856a: 385 [stem: *Podabr-*]. Type genus: *Podabrus* Dejean, 1833. Comment: family-group name attributed to J. L. LeConte (1881: 45) in recent literature although the same author had used the family-group name Podabri twenty years earlier (J. L. LeConte, 1861: 188).

Subfamily Silinae Mulsant, 1862

Siliaires Mulsant, 1862: 342 [stem: *Sil-*]. Type genus: *Silis* Charpentier, 1825.

Tribe Silini Mulsant, 1862

Siliaires Mulsant, 1862: 342 [stem: *Sil-*]. Type genus: *Silis* Charpentier, 1825. Comment: original vernacular name available (Art. 11.7.2): first used in latinized form by Gorham (1881: 91, as Silini), generally accepted as in Hansen (1996: 145, as Silinae).

Tribe Tytthonyxini Arnett, 1962

Tytthonyini Arnett, 1962a: 537 [stem: *Tytthonyx-*]. Type genus: *Tytthonyx* J. L. LeConte, 1851. Comment: spelling changed to Tytthonyxini by Wittmer (1970: 42); current spelling maintained (Art. 29.5): incorrect stem formation in prevailing usage (should be *Tytthonych-*).

Subfamily Dysmorphocerinae Brancucci, 1980

Dysmorphocerinae Brancucci, 1980: 292 [stem: *Dysmorphocer-*]. Type genus: *Dysmorphocerus* Solier, 1849.

Subfamily Malthininae Kiesenwetter, 1852

Malthinen Kiesenwetter, 1852: 239 [stem: *Malthin-*]. Type genus: *Malthinus* Latreille, 1806.

Tribe Malchinini Brancucci, 1980

Malchinini Brancucci, 1980: 313 [stem: *Malchin-*]. Type genus: *Malchinus* Kiesenwetter, 1863.

Tribe Malthinini Kiesenwetter, 1852

Malthinen Kiesenwetter, 1852: 239 [stem: *Malthin-*]. Type genus: *Malthinus* Latreille, 1806. Comment: original vernacular name available (Art. 11.7.2): first used in latinized form by J. L. LeConte (1861: 187, as Malthini), generally accepted as in Lawrence and Newton (1995: 860, as Malthininae); this family-group name was also used in the same year by Motschulsky (1852: 1, as Malthinides) but this vernacular name is unavailable (Art. 11.7.2) because it is not generally attributed to Motschulsky (1852).

Tribe Malthodini Böving and Craighead, 1931

Malthodinae Böving and Craighead, 1931: 48, in key [stem: *Malthod-*]. Type genus: *Malthodes* Kiesenwetter, 1852.

Malthodini Brancucci, 1980: 307 [stem: *Malthod-*]. Type genus: *Malthodes* Kiesenwetter, 1852. Comment: family-group name proposed as new without reference to Malthodinae Böving and Craighead, 1931.

Subfamily Chauliognathinae LeConte, 1861

Chauliognathini J. L. LeConte, 1861: 186 [stem: *Chauliognath-*]. Type genus: *Chauliognathus* Hentz, 1829.

Tribe Chauliognathini LeConte, 1861

Chauliognathini J. L. LeConte, 1861: 186 [stem: *Chauliognath-*]. Type genus: *Chauliognathus* Hentz, 1829.

Tribe Ichthyurini Champion, 1915

Ichthyurini Champion, 1915: 128 [stem: *Ichthyur-*]. Type genus: *Ichthyurus* Westwood, 1848.

Subfamily Cydistinae Paulus, 1972

Cydistinae Paulus, 1972a: 48, in key [stem: *Cydist-*]. Type genus: *Cydistus* Bourgeois, 1885.

Subfamily Pterotinae LeConte, 1861

Pterotini J. L. LeConte, 1861: 185 [stem: *Pterot-*]. Type genus: *Pterotus* J. L. LeConte, 1859.

Subfamily Ototretinae McDermott, 1964

Ototretinae McDermott, 1964: 11, in key [stem: *Ototret-*]. Type genus: *Ototreta* E. Olivier, 1900 [syn. of *Drilaster* Kiesenwetter, 1879].

Subfamily Ototretadrilinae Crowson, 1972

Ototretadrilinae Crowson, 1972: 55, in key [stem: *Ototretadril-*]. Type genus: *Ototretadrilus* Pic, 1921.

†Subfamily Lasiosynidae Kirejtshuk, Chang, Ren and Kun, 2010

Lasiosynidae Kirejtshuk et al., 2010: 68 [stem: *Lasiosyn-*]. Type genus: *Lasiosyne* Tan et al., 2007.

Series DERODONTIFORMIA

Superfamily Derodontoidea LeConte, 1861

Derodontidae J. L. LeConte, 1861: 100 [stem: *Derodont-*]. Type genus: *Derodontus* J. L. LeConte, 1861. Comment: name conserved over Nosodendroidea Erichson, 1846 (Art. 35.5).

Family Derodontidae LeConte, 1861

Derodontidae J. L. LeConte, 1861: 100 [stem: *Derodont-*]. Type genus: *Derodontus* J. L. LeConte, 1861. Comment: First Reviser found (Derodontidae J. L. LeConte, 1861 vs Peltasticidae J. L. LeConte, 1861) is Schenkling (1915: 3).

Subfamily Peltasticinae LeConte, 1861

Peltasticidae J. L. LeConte, 1861: 88 [stem: *Peltastic-*]. Type genus: *Peltastica* Mannerheim, 1852.

Subfamily Derodontinae LeConte, 1861

Derodontidae J. L. LeConte, 1861: 100 [stem: *Derodont-*]. Type genus: *Derodontus* J. L. LeConte, 1861.

Subfamily Laricobiinae Mulsant and Rey, 1864

Laricobiens Mulsant and Rey, 1864a: 374 [stem: *Laricobi-*]. Type genus: *Laricobius* Rosenhauer, 1846. Comment: original vernacular name available (Art. 11.7.2): first used in latinized form by Ganglbauer (1899: 766, as Laricobiidae), generally accepted as in Hansen (1996: 146, as Laricobiinae).

Family NOSODENDRIDAE Erichson, 1846

NOSODENDRINI Erichson, 1846: 465 [stem: *Nosodendr-*]. Type genus: *Nosodendron* Latreille, 1804.

Family JACOBSONIIDAE Heller, 1926

JACOBSONIIDAE Heller, 1926: 127 [stem: *Jacobsoni-*]. Type genus: *Jacobsonium* Heller, 1926 [syn. of *Sarothrias* Grouvelle, 1918].

SAROTHRIIDAE Crowson, 1955: 75 [stem: *Sarothri-*]. Type genus: *Sarothrias* Grouvelle, 1918. Comment: replacement name for JACOBSONIIDAE Heller, 1926 because of synonymy of the type genus.

DEROLATHRIINAE Sen Gupta, 1979: 692 [stem: *Derolathr-*]. Type genus: *Derolathrus* Sharp, 1900. Comment: incorrect original stem formation, not in prevailing usage.

Series BOSTRICHIFORMIA

Superfamily BOSTRICHOIDEA Latreille, 1802

BOSTRICHINI Latreille, 1802: 202 [stem: *Bostrich-*]. Type genus: *Bostrichus* Geoffroy, 1762 [placed on the Official List of Generic Names in Zoology (ICZN 1994a)]. Comment: First Reviser (BOSTRICHOIDEA Latreille, 1802 vs PTINOIDEA Latreille, 1802) not determined, current usage maintained.

Family DERMESTIDAE Latreille, 1804

DERMESTINI Latreille, 1804c: 142 [stem: *Dermest-*]. Type genus: *Dermestes* Linnaeus, 1758.

Subfamily DERMESTINAE Latreille, 1804

DERMESTINI Latreille, 1804c: 142 [stem: *Dermest-*]. Type genus: *Dermestes* Linnaeus, 1758.

Tribe DERMESTINI Latreille, 1804

DERMESTINI Latreille, 1804c: 142 [stem: *Dermest-*]. Type genus: *Dermestes* Linnaeus, 1758. Comment: published 7 March 1804; this family-group name was also used in the same year by Latreille (1804a [between 19 August and 17 September]: 233, as DERMESTINI).

Tribe MARIOUTINI Jakobson, 1913

MARIOUTINI Jakobson, 1913: 826 [stem: *Mariout-*]. Type genus: *Mariouta* Pic, 1899.

RHOPALOSILPHINAE Arrow, 1929: 98 [stem: *Rhopalosilph-*]. Type genus: *Rhopalosilpha* Arrow, 1929. Comment: originally proposed as a subfamily of SILPHIDAE.

Subfamily Thorictinae Agassiz, 1846

Thorictides Agassiz, 1846a: 162 [stem: *Thorict-*]. Type genus: *Thorictus* Germar, 1834.

Tribe Thaumaphrastini Anderson, 1949

Thaumaphrastinae W. H. Anderson, 1949: 127 [stem: *Thaumaphrast-*]. Type genus: *Thaumaphrastus* Blaisdell, 1927 [syn. of *Thorictodes* Reitter, 1875].

Tribe Thorictini Agassiz, 1846

Thorictides Agassiz, 1846a: 162 [stem: *Thorict-*]. Type genus: *Thorictus* Germar, 1834.

Subfamily Orphilinae LeConte, 1861

Orphili J. L. LeConte, 1861: 109 [stem: *Orphil-*]. Type genus: *Orphilus* Erichson, 1846.

Subfamily Trinodinae Casey, 1900

Trinodini Casey, 1900: 139 [stem: *Trinod-*]. Type genus: *Trinodes* Dejean, 1821.

†Tribe Cretonodini Kirejtshuk and Azar, 2009

Cretonodini Kirejtshuk and Azar, 2009: 121 [stem: *Cretonod-*]. Type genus: *Cretonodes* Kirejtshuk and Azar, 2009.

Tribe Thylodriini Semenov, 1909

Thelydriini Semenov, 1909: xxv [stem: *Thylodri-*]. Type genus: *Thylodrias* Motschulsky, 1839 [as *Thelydrias*, unjustified emendation of type genus name by Agassiz (1846b: 370), not in prevailing usage]. Comment: incorrect original stem formation, not in prevailing usage.

Trichelodini Peacock, 1978: 344 [stem: *Trichelod-*]. Type genus: *Trichelodes* Carter, 1935.

Tribe Trinodini Casey, 1900

Trinodini Casey, 1900: 139 [stem: *Trinod-*]. Type genus: *Trinodes* Dejean, 1821.

Tribe Trinoparvini Háva, 2010

Trinoparvini Háva, 2010: 56 [stem: *Trinoparv-*]. Type genus: *Trinoparvus* Háva, 2004.

Subfamily Attageninae Laporte, 1840

Attagénites Laporte, 1840b: 35 [stem: *Attagen-*]. Type genus: *Attagenus* Latreille, 1802.

Tribe Attagenini Laporte, 1840

Attagénites Laporte, 1840b: 35 [stem: *Attagen-*]. Type genus: *Attagenus* Latreille, 1802. Comment: original vernacular name available (Art. 11.7.2): first used in latinized form by J. L. LeConte (1861: 108, as Attageni), generally accepted as in Hansen (1996: 147, as Attageninae).

Tribe Egidyellini Semenov, 1914

Epidyellini Semenov, 1914: 15 [stem: *Egidyell-*]. Type genus: *Egidyella* Reitter, 1899 [as *Epidyella*, incorrect subsequent spelling of type genus name, not in prevailing usage]. Comment: incorrect original stem formation, not in prevailing usage.

Subfamily Megatominae Leach, 1815

Megatomida Leach, 1815: 94 [stem: *Megatom-*]. Type genus: *Megatoma* Herbst, 1791.

Tribe Anthrenini Gistel, 1848

Anthrenidae Gistel, 1848: [5] [stem: *Anthren-*]. Type genus: *Anthrenus* Geoffroy, 1762 [placed on the Official List of Generic Names in Zoology (ICZN 1994a)].

*Trogodermates Mulsant and Rey, 1867a: 120 [stem: *Trogodermat-*]. Type genus: *Trogoderma* Dejean, 1821. Comment: original vernacular name unavailable (Art. 11.7.2): not subsequently latinized; incorrect original stem formation, not in prevailing usage.

Tribe Megatomini Leach, 1815

Megatomida Leach, 1815: 94 [stem: *Megatom-*]. Type genus: *Megatoma* Herbst, 1791.

Ctesiini Rees, 1943: 12 [stem: *Ctesi-*]. Type genus: *Ctesias* Stephens, 1830.

Family Endecatomidae LeConte, 1861

Endecatomini J. L. LeConte, 1861: 207 [stem: *Endecatom-*]. Type genus: *Endecatomus* Mellié, 1847.

Family Bostrichidae Latreille, 1802

Bostrichini Latreille, 1802: 202 [stem: *Bostrich-*]. Type genus: *Bostrichus* Geoffroy, 1762 [placed on the Official List of Generic Names in Zoology (ICZN 1994a)].

Subfamily Dysidinae Lesne, 1921

Dysididae Lesne, 1921b: 286 [stem: *Dysid-*]. Type genus: *Dysides* Perty, 1832.

Apoleoninae Gardner, 1933: 3 [stem: *Apoleont-*]. Type genus: *Apoleon* Gorham, 1885. Comment: incorrect original stem formation, not in prevailing usage.

Subfamily Polycaoninae Lesne, 1896

Polycaoninae Lesne, 1896: 96 [stem: *Polycaon-*]. Type genus: *Polycaon* Laporte, 1836.

Subfamily Bostrichinae Latreille, 1802

Bostrichini Latreille, 1802: 202 [stem: *Bostrich-*]. Type genus: *Bostrichus* Geoffroy, 1762 [placed on the Official List of Generic Names in Zoology (ICZN 1994a)].

Tribe Apatini Billberg, 1820

Apatides Billberg, 1820a: 47 [stem: *Apat-*]. Type genus: *Apate* Fabricius, 1775 [*nomen protectum* (see Borowski and Węgrzynowicz 2009)]. Comment: this family-group name was also used in the same year by Billberg (1820b: 394, as Apatides).

Ligniperdidae Jacobi, 1906: 139 [stem: *Ligniperd-*]. Type genus: *Ligniperda* Pallas, 1772 [*nomen oblitum* (see Borowski and Węgrzynowicz 2009); syn. of *Apate* Fabricius, 1775].

Chileniidae Lesne, 1921b: 287 [stem: *Chileni-*]. Type genus: *Chilenius* Lesne, 1921.

Bostrychopsini Lesne, 1921b: 288 [stem: *Bostrychopse-*]. Type genus: *Bostrychopsis* Lesne, 1899 [syn. of *Apate* Fabricius, 1775]. Comment: incorrect original stem formation, not in prevailing usage.

Tribe Bostrichini Latreille, 1802

Bostrichini Latreille, 1802: 202 [stem: *Bostrich-*]. Type genus: *Bostrichus* Geoffroy, 1762 [placed on the Official List of Generic Names in Zoology (ICZN 1994a)].

Apatidini Bradley, 1930: 207, in key [stem: *Apatid-*]. Type genus: *Apatides* Casey, 1898.

Lichenophanini Portevin, 1931: 470, in key [stem: *Lichenophan-*]. Type genus: *Lichenophanes* Lesne, 1899.

Tribe Dinapatini Lesne, 1910

Dinapatinae Lesne, 1910: 471 [stem: *Dinapat-*]. Type genus: *Dinapate* G. H. Horn, 1886.

Tribe Sinoxylini Marseul, 1857

Sinoxylidae Marseul, 1857a: 107 [stem: *Sinoxyl-*]. Type genus: *Sinoxylon* Duftschmid, 1825.

Tribe Xyloperthini Lesne, 1921

Xyloperthini Lesne, 1921b: 288 [stem: *Xyloperth-*]. Type genus: *Xylopertha* Guérin-Méneville, 1845.

Subfamily Psoinae Blanchard, 1851

Psoitas Blanchard, 1851b: 434 [stem: *Pso-*]. Type genus: *Psoa* Herbst, 1797. Comment: original vernacular name available (Art. 11.7.2): first used in latinized form by J.

L. LeConte (1861: 208, as Psoini), generally accepted as in Ivie (2002: 241, as Psoinae).

Subfamily Dinoderinae Thomson, 1863

Dinoderina C. G. Thomson, 1863: 201 [stem: *Dinoder-*]. Type genus: *Dinoderus* Stephens, 1830.

Subfamily Lyctinae Billberg, 1820

Lyctides Billberg, 1820a: 48 [stem: *Lyct-*]. Type genus: *Lyctus* Fabricius, 1792.

Tribe Lyctini Billberg, 1820

Lyctides Billberg, 1820a: 48 [stem: *Lyct-*]. Type genus: *Lyctus* Fabricius, 1792.

Tribe Trogoxylini Lesne, 1921

Trogoxylini Lesne, 1921a: 231 [stem: *Trogoxyl-*]. Type genus: *Trogoxylon* J. L. LeConte, 1862.

Tristariini Lesne, 1921b: 287 [stem: *Tristari-*]. Type genus: *Tristaria* Reitter, 1878.

Subfamily Euderiinae Lesne, 1934

Euderiitae Lesne, 1934: 392 [stem: *Euderi-*]. Type genus: *Euderia* Broun, 1880.

Family Ptinidae Latreille, 1802

Ptiniores Latreille, 1802: 112 [stem: *Ptin-*]. Type genus: *Ptinus* Linnaeus, 1767 [placed on the Official List of Generic Names in Zoology (ICZN 1995a)]. Comment: Ptinidae Latreille, 1802 placed on the Official List of Family-Group Names in Zoology (ICZN 1995a); several subfamilies in Ptinidae contain tribes introduced in the literature by R. E. White (1982), those tribal names were not described originally and have not been made available subsequently; although some of White's tribes have been used as available and valid names recently, e.g., Zahradník (2007), they are not available; in subfamilies containing tribes first proposed by R. E. White (1982), we have decided not to use tribal names at all in order to avoid adding to the confusion, the necessary phylogenetic work needed to establish a meaningful tribal classification in those subfamilies is currently under way (K. Philips pers. comm. 2009).

Subfamily Eucradinae LeConte, 1861

Eucradini J. L. LeConte, 1861: 202 [stem: *Eucrad-*]. Type genus: *Eucrada* J. L. LeConte, 1861.

Tribe Eucradini LeConte, 1861

Eucradini J. L. LeConte, 1861: 202 [stem: *Eucrad-*]. Type genus: *Eucrada* J. L. LeConte, 1861.

Tribe Hedobiini Mulsant and Rey, 1868

Hédobiaires Mulsant and Rey, 1868b: 23 [stem: *Hedobi-*]. Type genus: *Hedobia* Dejean, 1821. Comment: original vernacular name available (Art. 11.7.2): first used in latinized form by Seidlitz (1889 [Gatt.]: 116, as Hedobiini), generally accepted as in Philips (2002: 254, as Hedobiini).

Subfamily Ptininae Latreille, 1802

Ptiniores Latreille, 1802: 112 [stem: *Ptin-*]. Type genus: *Ptinus* Linnaeus, 1767 [placed on the Official List of Generic Names in Zoology (ICZN 1995a)]. Comment: Ptinidae Latreille, 1802 placed on the Official List of Family-Group Names in Zoology (ICZN 1995a).

Tribe Gibbiini Jacquelin du Val, 1860

Gibbiites Jacquelin du Val, 1860: 211 [stem: *Gibbi-*]. Type genus: *Gibbium* Scopoli, 1777. Comment: original vernacular name available (Art. 11.7.2): first used in latinized form by Kiesenwetter (1877: 44, as Gibbiini), generally accepted as in Bellés (1985: 13, as Gibbiinae).

Tribe Meziini Bellés, 1985

Meziini Bellés, 1985: 37, in key [stem: *Mezi-*]. Type genus: *Mezium* Curtis, 1828.

Tribe Ptinini Latreille, 1802

Ptiniores Latreille, 1802: 112 [stem: *Ptin-*]. Type genus: *Ptinus* Linnaeus, 1767 [placed on the Official List of Generic Names in Zoology (ICZN 1995a)]. Comment: Ptinidae Latreille, 1802 placed on the Official List of Family-Group Names in Zoology (ICZN 1995a).

Tribe Sphaericini Portevin, 1931

Sphaericini Portevin, 1931: 494 [stem: *Sphaeric-*]. Type genus: *Sphaericus* Wollaston, 1854.

Ptininae *incertae sedis*

Gnostidae Gemminger and Harold, 1868: 700 [stem: *Gnost-*]. Type genus: *Gnostus* Westwood, 1855.

Ectrephidae Wasmann, 1894: 121 [stem: *Ectreph-*]. Type genus: *Ectrephes* Pascoe, 1866.

Subfamily Dryophilinae Gistel, 1848

Dryophilidae Gistel, 1848: [6] [stem: *Dryophil-*]. Type genus: *Dryophilus* Chevrolat, 1832.

Tribe Dryophilini Gistel, 1848

Dryophilidae Gistel, 1848: [6] [stem: *Dryophil-*]. Type genus: *Dryophilus* Chevrolat, 1832. Comment: family-group name previously attributed to J. L. LeConte (1861: 205) in the literature.

Dryobiadae Gistel, 1856a: 368 [stem: *Dryobi-*]. Type genus: *Dryobia* Gistel, 1856 [syn. of *Dryophilus* Chevrolat, 1832]. Comment: senior homonym of Dryobiini Arnett, 1962 (type genus *Dryobius* J. L. LeConte, 1850) used as valid in Cerambycidae; *nomen oblitum* (see Bousquet et al. 2009: 45); incorrect original stem formation, not in prevailing usage.

Tribe Ptilineurini Böving, 1927

Ptilineurini Böving, 1927a: 56 [stem: *Ptilineur-*]. Type genus: *Ptilineurus* Reitter, 1901.

Subfamily Ernobiinae Pic, 1912

Cosmoceroideos Solier, 1849: 476 [stem: *Cosmocer-*]. Type genus: *Cosmocerus* Solier, 1849 [preoccupied genus name, not *Cosmocerus* Guérin-Méneville, 1844 [Coleoptera: Cerambycidae]; syn. of *Cerocosmus* Gemminger, 1873]. Comment: permanently invalid (Art. 39): based on preoccupied type genus; Lawrence and Newton (1995: 864) used the latinized form Cosmocerinae and treated Solier's name as available therefore we treat this name as available but permanently invalid.

Ernobiinae Pic, 1912: 12 [stem: *Ernobi-*]. Type genus: *Ernobius* C. G. Thomson, 1859. Comment: First Reviser found (Ernobiinae Pic, 1912 vs Cerocosminae Pic, 1912) is White (1974: 419).

Cerocosminae Pic, 1912: 45 [stem: *Cerocosm-*]. Type genus: *Cerocosmus* Gemminger, 1873.

Xestobiini Böving, 1927a: 56 [stem: *Xestobi-*]. Type genus: *Xestobium* Motschulsky, 1845.

*Ozognathini R. E. White, 1982: 2 [stem: *Ozognath-*]. Type genus: *Ozognathus* J. L. LeConte, 1861. Comment: unavailable family-group name, proposed after 1930 without description or bibliographic reference to such a description (Art. 13.1).

Subfamily Anobiinae Fleming, 1821

Anobiumedae Fleming, 1821: 50 [stem: *Anobi-*]. Type genus: *Anobium* Fabricius, 1775 [placed on the Official List of Generic Names in Zoology (ICZN 1976)]. Comment: incorrect original stem formation, not in prevailing usage.

*Euceratocerini R. E. White, 1982: 7 [stem: *Euceratocer-*]. Type genus: *Euceratocerus* J. L. LeConte, 1874. Comment: unavailable family-group name, proposed after 1930 without description or bibliographic reference to such a description (Art. 13.1).

*Colposternini R. E. White, 1982: 9 [stem: *Colpostern-*]. Type genus: *Colposternus* Fall, 1905. Comment: unavailable family-group name, proposed after 1930 without description or bibliographic reference to such a description (Art. 13.1).

*Gastrallini R. E. White, 1982: 9 [stem: *Gastrall-*]. Type genus: *Gastrallus* Jacquelin du Val, 1860. Comment: unavailable family-group name, proposed after 1930 without description or bibliographic reference to such a description (Art. 13.1).

*Hadrobregmini R. E. White, 1982: 15 [stem: *Hadrobregm-*]. Type genus: *Hadrobregmus* C. G. Thomson, 1859. Comment: unavailable family-group name, proposed after 1930 without description or bibliographic reference to such a description (Art. 13.1).

*NICOBIINI R. E. White, 1982: 10 [stem: *Nicobi-*]. Type genus: *Nicobium* J. L. LeConte, 1861. Comment: unavailable family-group name, proposed after 1930 without description or bibliographic reference to such a description (Art. 13.1).

*STEGOBIINI R. E. White, 1982: 11 [stem: *Stegobi-*]. Type genus: *Stegobium* Motschulsky, 1860. Comment: unavailable family-group name, proposed after 1930 without description or bibliographic reference to such a description (Art. 13.1).

Subfamily PTILININAE Shuckard, 1839

PTILINIDAE Shuckard, 1839b: 45 [stem: *Ptilin-*]. Type genus: *Ptilinus* Geoffroy, 1762 [placed on the Official List of Generic Names in Zoology (ICZN 1994a)].

SCLERASTEIDAE Gistel, 1856a: 368 [stem: *Sclerast-*]. Type genus: *Sclerastes* Gistel, 1856 [this genus originally included "*pectinicornis*" and "*costatus* Gy", we here chose *Ptilinus costatus* Gyllenhal, 1827 as the type of *Sclerastes*; **syn. nov.** of *Ptilinus* Geoffroy, 1762]. Comment: **syn. nov.**; incorrect original stem formation, not in prevailing usage.

Subfamily ALVARENGANIELLINAE Viana and Martínez, 1971

ALVARENGANIELLINAE Viana and Martínez, 1971: 121 [stem: *Alvarenganiell-*]. Type genus: *Alvarenganiella* Viana and Martínez, 1971.

Subfamily XYLETININAE Gistel, 1848

XYLETINIDAE Gistel, 1848: [6] [stem: *Xyletin-*]. Type genus: *Xyletinus* Latreille, 1810 [placed on the Official List of Generic Names in Zoology (ICZN 1971)].

Tribe LASIODERMINI Böving, 1927

LASIODERMINI Böving, 1927a: 56 [stem: *Lasioderm-*]. Type genus: *Lasioderma* Stephens, 1835. Comment: current spelling maintained (Art. 29.5): incorrect stem formation in prevailing usage (should be *Lasiodermat-*).

Tribe METHOLCINI Zahradník, 2009

METHOLCINI Zahradník, 2009: 180 [stem: *Metholc-*]. Type genus: *Metholcus* Jacquelin du Val, 1860.

Tribe XYLETININI Gistel, 1848

XYLETINIDAE Gistel, 1848: [6] [stem: *Xyletin-*]. Type genus: *Xyletinus* Latreille, 1810 [placed on the Official List of Generic Names in Zoology (ICZN 1971)].

VRILETTINI Böving, 1927a: 56 [stem: *Vrillett-*]. Type genus: *Vrilletta* J. L. LeConte, 1874. Comment: incorrect original stem formation, not in prevailing usage.

Subfamily DORCATOMINAE Thomson, 1859

DORCATOMINA C. G. Thomson, 1859: 90 [stem: *Dorcatom-*]. Type genus: *Dorcatoma* Herbst, 1791 [the original spelling *Dorkatoma* was placed on the Official Index of Rejected and Suppressed Generic Names in Zoology, *Dorcatoma* was chosen as correct original spelling and placed on the Official List of Generic Names in Zoology (ICZN 1995b)].

CAENOCARINI Böving, 1927a: 57 [stem: *Caenocar-*]. Type genus: *Caenocara* C. G. Thomson, 1859. Comment: also misspelled as COENECARINI (page 56) in the original publication.

*CALYMMADERINI R. E. White, 1982: 22 [stem: *Calymmader-*]. Type genus: *Calymmaderus* Solier, 1849. Comment: unavailable family-group name, proposed after 1930 without description or bibliographic reference to such a description (Art. 13.1).

*CRYPTORAMORPHINI R. E. White, 1982: 25 [stem: *Cryptoramorph-*]. Type genus: *Cryptoramorphus* R. E. White, 1966. Comment: unavailable family-group name, proposed after 1930 without description or bibliographic reference to such a description (Art. 13.1).

*PETALIINI R. E. White, 1982: 23 [stem: *Petali-*]. Type genus: *Petalium* J. L. LeConte, 1861. Comment: unavailable family-group name, proposed after 1930 without description or bibliographic reference to such a description (Art. 13.1); the tribe PETALIINI Tillyard, 1917 in Odonata (type genus *Petalia* Hagen, 1854, a junior homonym of *Petalia* Gray, 1838 [Mammalia]) is available but permanently invalid because it is based on a preoccupied type genus.

*PROTHECINI R. E. White, 1982: 25 [stem: *Prothec-*]. Type genus: *Protheca* J. L. LeConte, 1865. Comment: unavailable family-group name, proposed after 1930 without description or bibliographic reference to such a description (Art. 13.1).

Subfamily MESOCOELOPODINAE Mulsant and Rey, 1864

MÉSOCOELOPAIRES Mulsant and Rey, 1864b: 311 [stem: *Mesocoelopod-*]. Type genus: *Mesocoelopus* Jacquelin du Val, 1860.

Tribe TRICORYNINI White, 1971

TRICORYNINAE R. E. White, 1971: 1301 [stem: *Tricoryn-*]. Type genus: *Tricorynus* G. R. Waterhouse, 1849.

Tribe MESOCOELOPODINI Mulsant and Rey, 1864

MÉSOCOELOPAIRES Mulsant and Rey, 1864b: 311 [stem: *Mesocoelopod-*]. Type genus: *Mesocoelopus* Jacquelin du Val, 1860. Comment: original vernacular name available (Art. 11.7.2): first used in latinized form by Kiesenwetter (1877: 153, as MESOCOELOPINI [incorrect stem formation]), generally accepted as in Philips (2002: 258, as MESOCOELOPODINAE); incorrect original stem formation, not in prevailing usage.

MESOTHINI Portevin, 1931: 489 [stem: *Mesothet-*]. Type genus: *Mesothes* Mulsant and Rey, 1864. Comment: incorrect original stem formation, not in prevailing usage.

PTINIDAE ***incertae sedis***

FABIINAE Martínez and Viana, 1964: 7 [stem: *Fabi-*]. Type genus: *Fabia* Martínez and Viana, 1964.

Series CUCUJIFORMIA

Superfamily LYMEXYLOIDEA Fleming, 1821

LYMOXYLONIDAE Fleming, 1821: 49 [stem: *Lymexyl-*]. Type genus: *Lymexylon* Fabricius, 1775. Comment: LYMEXYLOIDEA given precedence for superfamily name over HYLECOETOIDEA Germar, 1818 (Art. 35.5).

Family LYMEXYLIDAE Fleming, 1821

LYMOXYLONIDAE Fleming, 1821: 49 [stem: *Lymexyl-*]. Type genus: *Lymexylon* Fabricius, 1775. Comment: LYMEXYLIDAE given precedence for family name over HYLECOETIDAE Germar, 1818 (Art. 35.5).

Subfamily HYLECOETINAE Germar, 1818

HYLECOETI Germar, 1818: 344 [stem: *Hylecoet-*]. Type genus: *Hylecoetus* Latreille, 1806. Comment: name previously attributed to Gistel (1856a).

Subfamily LYMEXYLINAE Fleming, 1821

LYMOXYLONIDAE Fleming, 1821: 49 [stem: *Lymexyl-*]. Type genus: *Lymexylon* Fabricius, 1775 [as *Lymoxylon*, incorrect subsequent spelling of type genus name, not in prevailing usage]. Comment: published in July 1821; incorrect original stem formation, not in prevailing usage; this family-group name was also used in the same year by Fischer von Waldheim (1821 ["31 December"]: 37, as LYMEXYLA).

Subfamily ATRACTOCERINAE Laporte, 1840

ATRACTOCÉRITES Laporte, 1840a: 290 [stem: *Atractocer-*]. Type genus: *Atractocerus* Palisot de Beauvois, 1802. Comment: original vernacular name available (Art. 11.7.2): first used in latinized form by Brues and Melander (1932: 23, as ATRACTOCERIDAE), generally accepted as in Cuccodoro (2007: 363, as ATRACTOCERINAE).

Subfamily MELITTOMMATINAE Wheeler, 1986

MELITTOMMINAE Q. D. Wheeler, 1986: 160 [stem: *Melittommat-*]. Type genus: *Melittomma* Murray, 1867. Comment: incorrect original stem formation, not in prevailing usage.

Superfamily CLEROIDEA Latreille, 1802

CLERII Latreille, 1802: 110 [stem: *Cler-*]. Type genus: *Clerus* Geoffroy, 1762 [placed on the Official List of Generic Names in Zoology (ICZN 1984a)]. Comment:

First Reviser (Cleroidea Latreille, 1802 vs Trogossitoidea Latreille, 1802) not determined, current usage maintained.

Family Phloiophilidae Kiesenwetter, 1863

Phloeophilini Kiesenwetter, 1863: 626, in key [stem: *Phloiophil-*]. Type genus: *Phloiophilus* Stephens, 1830 [as *Phloeophilus*, unjustified emendation of type genus name by Agassiz (1846b: 286), not in prevailing usage; *Phloeophilus* Agassiz, 1846 is a junior homonym of *Phloeophilus* Schönherr, 1833 [Coleoptera: Anthribidae]]. Comment: incorrect original stem formation, not in prevailing usage; using the original spelling would place Phloeophilini Lacordaire, 1865 [Coleoptera: Anthribidae] into homonymy with this name.

Family Trogossitidae Latreille, 1802

Trogossitarii Latreille, 1802: 159 [stem: *Trogossit-*]. Type genus: *Trogossita* A. G. Olivier, 1790 [syn. of *Tenebroides* Piller and Mitterpacher, 1783].

Subfamily Peltinae Latreille, 1806

Peltides Latreille, 1806: 8 [stem: *Pelt-*]. Type genus: *Peltis* Kugelann, 1792 [placed on the Official List of Generic Names in Zoology (ICZN 1994a)].

Tribe Ancyronini Kolibáč, 2006

Ancyronini Kolibáč, 2006: 127 [stem: *Ancyron-*]. Type genus: *Ancyrona* Reitter, 1876.

Tribe Colydiopeltini Kolibáč, 2006

Colydiopeltini Kolibáč, 2006: 126 [stem: *Colydiopelt-*]. Type genus: *Colydiopeltis* Ślipiński, 1992.

Tribe Decamerini Crowson, 1964

Decamerinae Crowson, 1964: 287 [stem: *Decamer-*]. Type genus: *Decamerus* Solier, 1849.

Tribe Lophocaterini Crowson, 1964 *nomen protectum*

Lycoptini Casey, 1890: 311 [stem: *Lycopt-*]. Type genus: *Lycoptis* Casey, 1890. Comment: *nomen oblitum* (see Appendix 1).

Lophocaterinae Crowson, 1964: 297 [stem: *Lophocater-*]. Type genus: *Lophocateres* Olliff, 1883. Comment: *nomen protectum* (see Appendix 1).

Tribe Peltini Latreille, 1806

Peltides Latreille, 1806: 8 [stem: *Pelt-*]. Type genus: *Peltis* Kugelann, 1792 [placed on the Official List of Generic Names in Zoology (ICZN 1994a)]. Comment: name previously attributed to Kirby (1837).

OSTOMINI Harold, 1876: 169 [stem: *Ostomat-*]. Type genus: *Ostoma* Laicharting, 1781. Comment: incorrect original stem formation, not in prevailing usage.

Tribe THYMALINI Léveillé, 1888

*THYMALITES Blanchard, 1845a: 277 [stem: *Thymal-*]. Type genus: *Thymalus* Latreille, 1802. Comment: original vernacular name unavailable (Art. 11.7.2): subsequently used in latinized form but not generally attributed to Blanchard (1845a).

THYMALINI Léveillé, 1888: 444 [stem: *Thymal-*]. Type genus: *Thymalus* Latreille, 1802.

RENTONIINAE Crowson, 1966: 120 [stem: *Rentoni-*]. Type genus: *Rentonium* Crowson, 1966.

PROTOPELTINI Crowson, 1966: 120 [stem: *Protopelt-*]. Type genus: *Protopeltis* Crowson, 1964.

†MELIGETHIELLINAE Kirejtshuk and Ponomarenko, 1990: 79 [stem: *Meligethiell-*]. Type genus: *Meligethiella* L. N. Medvedev, 1969.

Subfamily TROGOSSITINAE Latreille, 1802

TROGOSSITARII Latreille, 1802: 159 [stem: *Trogossit-*]. Type genus: *Trogossita* A. G. Olivier, 1790 [syn. of *Tenebroides* Piller and Mitterpacher, 1783].

Tribe CALITYINI Reitter, 1922

CALITYNI Reitter, 1922a: 66 [stem: *Cality-*]. Type genus: *Calitys* C. G. Thomson, 1859. Comment: incorrect original stem formation, not in prevailing usage.

Tribe EGOLIINI Lacordaire, 1854

ÉGOLIIDES Lacordaire, 1854b: 334 [stem: *Egoli-*]. Type genus: *Egolia* Erichson, 1842. Comment: original vernacular name available (Art. 11.7.2): first used in latinized form by Ganglbauer (1899: 419, as EGOLIIDES [treated as Latin]), generally accepted as in Kolibáč (2006: 119, as EGOLIINI).

Tribe GYMNOCHILINI Lacordaire, 1854

GYMNOCHILIDES Lacordaire, 1854b: 344 [stem: *Gymnochil-*]. Type genus: *Gymnochila* Klug, 1844. Comment: original vernacular name available (Art. 11.7.2): first used in latinized form by J. Thomson (1858: 43, as GYMNOCHILITAE), generally accepted as in Burakowski et al. (1986: 118, as GYMNOCHILINAE).

LEPERINI Reitter, 1876a: 55 [stem: *Leperin-*]. Type genus: *Leperina* Erichson, 1844. Comment: incorrect original stem formation, not in prevailing usage.

Tribe LARINOTINI Ślipiński, 1992

LARINOTINAE Ślipiński, 1992: 443 [stem: *Larinot-*]. Type genus: *Larinotus* Carter and Zeck, 1937.

†Tribe Lithostomatini Kolibáč and Huang, 2008

Lithostomini Kolibáč and Huang, 2008: 142 [stem: *Lithostomat-*]. Type genus: *Lithostoma* Martynov, 1926. Comment: incorrect original stem formation, not in prevailing usage.

Tribe Trogossitini Latreille, 1802

Trogossitarii Latreille, 1802: 159 [stem: *Trogossit-*]. Type genus: *Trogossita* A. G. Olivier, 1790 [syn. of *Tenebroides* Piller and Mitterpacher, 1783].

Nemosomida Leach, 1815: 110 [stem: *Nemozomat-*]. Type genus: *Nemozoma* Latreille, 1804 [as *Nemosoma*, incorrect subsequent spelling of type genus name, not in prevailing usage]. Comment: incorrect original stem formation, not in prevailing usage.

Temnochilini Léveillé, 1888: 432 [stem: *Temnoscheil-*]. Type genus: *Temnoscheila* Westwood, 1830 [as *Temnochila*, unjustified emendation of type genus name by Erichson (1844: 449), not in prevailing usage]. Comment: incorrect original stem formation, not in prevailing usage.

Tenebrioidini Ganglbauer, 1899: 420 [stem: *Tenebroid-*]. Type genus: *Tenebroides* Piller and Mitterpacher, 1783 [as *Tenebrioides*, incorrect subsequent spelling of type genus name, not in prevailing usage]. Comment: incorrect original stem formation, not in prevailing usage.

Family Chaetosomatidae Crowson, 1952

Chaetosomatidae Crowson, 1952: 66 [stem: *Chaetosomat-*]. Type genus: *Chaetosoma* Westwood, 1851 [this genus is a junior homonym of *Chaetosoma* Chevrolat, 1843 in Cerambycidae; *Chaetosoma* Chevrolat, 1843 is also a senior objective synonym of the well-established name *Apodasya* Pascoe, 1863]. Comment: this name is a junior homonym of Chaetosomatidae Claus, 1872 (type genus *Chaetosoma* Claparède, 1863, which is a junior homonym of *Chaetosoma* Westwood, 1851) in Nematoda; the nematode family-group name is permanently invalid (Art. 39) because it is based on a junior homonym, the valid name for this nematode family is Draconematidae Filipjev, 1918; furthermore, the type genus of Chaetosomatidae Crowson is a junior homonym of *Chaetosoma* Chevrolat, 1843 in Cerambycidae; an application was recently submitted by Bousquet and Bouchard (2010) to suppress *Chaetosoma* Chevrolat, 1843 for the Principles of Priority and Homonymy and Chaetosomatidae Claus, 1872 for the Principle of Priority, therefore conserving the names Chaetosomatidae Crowson and *Chaetosoma* Westwood (see Appendix 6).

Family Metaxinidae Kolibáč, 2004

Metaxinidae Kolibáč, 2004: 247 [stem: *Metaxin-*]. Type genus: *Metaxina* Broun, 1909.

Family Thaneroclerdae Chapin, 1924

Thanerocleriae Chapin, 1924: 251 [stem: *Thanerocler-*]. Type genus: *Thaneroclerus* Westwood, 1838.

Subfamily Zenodosinae Kolibáč, 1992

Zenodosini Kolibáč, 1992: 307 [stem: *Zenodos-*]. Type genus: *Zenodosus* Wolcott, 1910.

Subfamily Thaneroclerinae Chapin, 1924

Thaneroclerinae Chapin, 1924: 251 [stem: *Thanerocler-*]. Type genus: *Thaneroclerus* Westwood, 1838.

Tribe Isoclerini Kolibáč, 1992

Isoclerina Kolibáč, 1992: 315 [stem: *Isocler-*]. Type genus: *Isoclerus* Lewis, 1892.

Tribe Thaneroclerini Chapin, 1924

Thaneroclerinae Chapin, 1924: 251 [stem: *Thanerocler-*]. Type genus: *Thaneroclerus* Westwood, 1838.

Tribe Viticlerini Winkler, 1982

Viticlerini J. R. Winkler, 1982: 529, in key [stem: *Viticler-*]. Type genus: *Viticlerus* Miyatake, 1977.

Family Cleridae Latreille, 1802

Clerii Latreille, 1802: 110 [stem: *Cler-*]. Type genus: *Clerus* Geoffroy, 1762 [placed on the Official List of Generic Names in Zoology (ICZN 1984a)]. Comment: see Opitz (2010) for an alternative subfamilial classification within this family.

Subfamily Tillinae Fischer von Waldheim, 1813

Tillii Fischer von Waldheim, 1813: 232 [stem: *Till-*]. Type genus: *Tillus* A. G. Olivier, 1790. Comment: family-group name previously attributed to Leach (1815).

Cylidrina Reitter, 1894: 38 [stem: *Cylidr-*]. Type genus: *Cylidrus* Latreille, 1829.

*Monophyllini Böving and Craighead, 1931: 78 [stem: *Monophyll-*]. Type genus: *Monophylla* Spinola, 1841. Comment: unavailable family-group name, proposed after 1930 without description or bibliographic reference to such a description (Art. 13.1).

Subfamily Hydnocerinae Spinola, 1844

Hydnocéroïdes Spinola, 1844: Tableau générique (3) [stem: *Hydnocer-*]. Type genus: *Hydnocera* Newman, 1838 [syn. of *Phyllobaenus* Dejean, 1833].

Tribe Callimerini Kolibáč, 1998

Callimerini Kolibáč, 1998: 165 [stem: *Callimer-*]. Type genus: *Callimerus* Gorham, 1876.

Tribe Hydnocerini Spinola, 1844

Hydnocéroïdes Spinola, 1844: Tableau générique (3) [stem: *Hydnocer-*]. Type genus: *Hydnocera* Newman, 1838 [syn. of *Phyllobaenus* Dejean, 1833]. Comment: original vernacular name available (Art. 11.7.2): first used in latinized form by Desmarest (1857: 263, as Hydnoceridae), generally accepted as in Kolibáč (1998: 127, as Hydnocerinae).

Phyllobénides Lacordaire, 1857: 466 [stem: *Phyllobaen-*]. Type genus: *Phyllobaenus* Dejean, 1833. Comment: original vernacular name available (Art. 11.7.2): first used in latinized form by and generally accepted as in Gorham (1876: 59, as Phyllobenides [treated as Latin]); incorrect original stem formation, not in prevailing usage.

Tribe Lemidiini Kolibáč, 1998

Lemidiini Kolibáč, 1998: 183 [stem: *Lemidi-*]. Type genus: *Lemidia* Spinola, 1841.

Subfamily Clerinae Latreille, 1802

Clerii Latreille, 1802: 110 [stem: *Cler-*]. Type genus: *Clerus* Geoffroy, 1762 [placed on the Official List of Generic Names in Zoology (ICZN 1984a)].

Notoxii Sturm, 1826: 38 [stem: *Notox-*]. Type genus: *Notoxus* sensu Fabricius, 1775 [not *Notoxus* Geoffroy, 1762; syn. of *Opilo* Latreille, 1802]. Comment: based on a misidentified type genus; an application should be submitted to the Commission to suppress this name for the Principles of Priority and Homonymy (Art. 65.2.1) since Notoxinae Stephens, 1829 is currently used as valid in Anthicidae.

Prioceridae Laporte, 1836: 33 [stem: *Priocer-*]. Type genus: *Priocera* Kirby, 1819.

*Trichodites Blanchard, 1845b: 84 [stem: *Trichod-*]. Type genus: *Trichodes* Herbst, 1792 [placed on the Official List of Generic Names in Zoology (ICZN 1984a)]. Comment: original vernacular name unavailable (Art. 11.7.2): subsequently used in latinized form but not generally attributed to Blanchard (1845b).

Opilonidae Gistel, 1848: [6] [stem: *Opilon-*]. Type genus: *Opilo* Latreille, 1802.

Thanasimiidae Gistel, 1848: [6] [stem: *Thanasim-*]. Type genus: *Thanasimus* Latreille, 1806. Comment: incorrect original stem formation, not in prevailing usage.

Dendroplaneteidae Gistel, 1856a: 395 [stem: *Dendroplanet-*]. Type genus: *Dendroplanetes* Gistel, 1856 [syn. of *Opilo* Latreille, 1802]. Comment: also spelled Dondroplaneteidae in the same publication on page 368; incorrect original stem formation, not in prevailing usage.

Trichodini Portevin, 1931: 457, in key [stem: *Trichod-*]. Type genus: *Trichodes* Herbst, 1792 [placed on the Official List of Generic Names in Zoology (ICZN 1984a)]. Comment: the older name originally proposed as Trichodina Maitland, 1851 (type genus *Trichoda* Müller, 1773) is available in Protozoa: Ciliophora; this case is to be referred to the Commission to remove the homonymy (Art. 55.3.1).

Dieropsinae J. R. Winkler, 1964: 317 [stem: *Dieropse-*]. Type genus: *Dieropsis* Gahan, 1908. Comment: incorrect original stem formation, not in prevailing usage.

Cleropiestinae J. R. Winkler, 1980: 437 [stem: *Cleropiest-*]. Type genus: *Cleropiestus* Fairmaire, 1889.

Anthicoclerinae Opitz, 2010: 58 [stem: *Anthicocler-*]. Type genus: *Anthicoclerus* Schenkling, 1906.

Subfamily Korynetinae Laporte, 1836

Corynetidae Laporte, 1836: 34 [stem: *Korynet-*]. Type genus: *Korynetes* Herbst, 1792 [as *Corynetes*, unjustified emendation of type genus name by Paykull (1798: 274), not in prevailing usage; *Korynetes* Herbst, 1792 placed on the Official List of Generic Names in Zoology and *Corynetes* Paykull, 1798 placed on the Official Index of Rejected and Invalid Generic Names in Zoology (ICZN 1961c)]. Comment: incorrect original stem formation, not in prevailing usage.

Ichnoïdes Spinola, 1841: 71 [stem: *Ichne-*]. Type genus: *Ichnea* Laporte, 1836. Comment: original vernacular name available (Art. 11.7.2): first used in latinized form by Agassiz (1846b: 193, as Ichneoidae), treated as available by Opitz and Herman (2009: 183, as Ichneinae); *nomen oblitum*, Epiphloeinae Kuwert, 1893 conserved as valid subfamily over this name by Opitz and Herman (2009).

Platynoptéroïdes Spinola, 1844: Tableau générique (4) [stem: *Platynopter-*]. Type genus: *Platynoptera* Chevrolat, 1834. Comment: original vernacular name available (Art. 11.7.2): first used in latinized form and generally accepted as in Desmarest (1857: 269, as Platynopteridae).

Necrobiaeidae Gistel, 1848: [6] [stem: *Necrobi-*]. Type genus: *Necrobia* A. G. Olivier, 1795 [placed on the Official List of Generic Names in Zoology (ICZN 1961c)]. Comment: incorrect original stem formation, not in prevailing usage.

Enopliidae Gistel, 1848: [6] [stem: *Enopli-*]. Type genus: *Enoplium* Latreille, 1802.

Tarsosténites Jacquelin du Val, 1860: 198 [stem: *Tarsosten-*]. Type genus: *Tarsostenus* Spinola, 1844. Comment: original vernacular name available (Art. 11.7.2): first used in latinized form by Böving and Craighead (1931: 57, as Tarsosteninae), generally accepted as in Opitz (2002: 279, as Tarsostenninae [incorrect stem formation]).

Epiphlöinen Kuwert, 1893: 492 [stem: *Epiphloe-*]. Type genus: *Epiphloeus* Spinola, 1841. Comment: original vernacular name available (Art. 11.7.2): first used in latinized form by Schenkling (1910: 113, as Epiphloeini), generally accepted as in Opitz and Herman (2009: 183, as Epiphloeinae); *nomen protectum* (see Opitz and Herman (2009: 184)); although Kuwert apparently originally published his name as a synonym of Phyllobaenides Lacordaire, 1857, the name Epiphloeini was treated as valid by several authors, e.g., Chapin (1924: 165), before 1961 which made the name available (Art. 11.6.1) (see Opitz and Herman 2009: 184).

Dermestoidini Jakobson, 1911a: 719 [stem: *Dermestoid-*]. Type genus: *Dermestoides* Schaeffer, 1771.

Orthopleurinae Böving and Craighead, 1931: 56, in key [stem: *Orthopleur-*]. Type genus: *Orthopleura* Spinola, 1844.

NEORTHOPLEURINAE Opitz, 2009: 138 [stem: *Neorthopleur-*]. Type genus: *Neorthopleura* Barr, 1976.

PELONIINAE Opitz, 2010: 97 [stem: *Peloni-*]. Type genus: *Pelonium* Spinola, 1844.

Family ACANTHOCNEMIDAE Crowson, 1964

ACANTHOCNEMINAE Crowson, 1964: 317, in key [stem: *Acanthocnem-*]. Type genus: *Acanthocnemus* Perris, 1866.

Family PHYCOSECIDAE Crowson, 1952

PHYCOSECIDAE Crowson, 1952: 117 [stem: *Phycosec-*]. Type genus: *Phycosecis* Pascoe, 1875. Comment: current spelling maintained (Art. 29.3.1.1): incorrect stem formation in prevailing usage (should be *Phycosecid-*).

Family PRIONOCERIDAE Lacordaire, 1857

PRIONOCÉRIDES Lacordaire, 1857: 411 [stem: *Prionocer-*]. Type genus: *Prionocerus* Perty, 1831. Comment: PRIONOCÉRITES was used earlier by Laporte (1840a: 275) but this name was treated as a *lapsus calami* for PRIOCÉRITES Laporte (1834b: 33) by Lawrence and Newton (1995: 871).

Tribe LOBONYCHINI Majer, 1987

LOBONYCHINI Majer, 1987: 790, in key [stem: *Lobonych-*]. Type genus: *Lobonyx* Jacquelin du Val, 1859.

Tribe PRIONOCERINI Lacordaire, 1857

PRIONOCÉRIDES Lacordaire, 1857: 411 [stem: *Prionocer-*]. Type genus: *Prionocerus* Perty, 1831. Comment: original vernacular name available (Art. 11.7.2): first used in latinized form by Ritsema (1884: 494, as PRIONOCERINI), generally accepted as in M. C. Thomas (2008: 445, as PRIONOCERIDAE); the earlier usage of PRIONOCERITES by Laporte (1840a: 285) was in error for PRIOCERITES in CLERIDAE: CLERINAE (see Lawrence and Newton 1995: 871); PRIONOCERINI Savchenko, 1966 (type genus *Prionocera* Loew, 1844) is available in Diptera; this case is to be referred to the Commission to remove the homonymy (Art. 55.3.1).

Family MAURONISCIDAE Majer, 1995

MAURONISCIDAE Majer, 1995a: 340 [stem: *Mauronisc-*]. Type genus: *Mauroniscus* Bourgeois, 1911. Comment: published 30 April 1995; this family-group name was also used in the same year by Majer (1995b [1 December]: 57, as MAURONISCIDAE).

Family MELYRIDAE Leach, 1815

MELYRIDES Leach, 1815: 87 [stem: *Melyr-*]. Type genus: *Melyris* Fabricius, 1775.

Subfamily Rhadalinae LeConte, 1861

Rhadalini J. L. LeConte, 1861: 194 [stem: *Rhadal-*]. Type genus: *Rhadalus* J. L. LeConte, 1852.

Haplocnémates Mulsant and Rey, 1868a: 181 [stem: *Aplocnem-*]. Type genus: *Aplocnemus* Stephens, 1830 [as *Haplocnemus*, unjustified emendation of type genus name by Agassiz (1846b: 29), not in prevailing usage]. Comment: original vernacular name available (Art. 11.7.2): first used in latinized form by Crowson (1964: 316, as Haplocneminae [incorrect stem formation]), generally accepted as in Majer (1987: 800, as Aplocnemini); incorrect original stem formation, not in prevailing usage.

Microjulistini Majer, 1987: 797, in key [stem: *Microjulist-*]. Type genus: *Microjulistus* Reitter, 1889.

Pelecophorini Majer, 1987: 797, in key [stem: *Pelecophor-*]. Type genus: *Pelecophora* Lepeletier and Audinet-Serville, 1825.

Subfamily Melyrinae Leach, 1815

Melyrides Leach, 1815: 87 [stem: *Melyr-*]. Type genus: *Melyris* Fabricius, 1775.

Tribe Arthrobrachini Majer, 1987

Arthrobrachini Majer, 1987: 793, in key [stem: *Arthrobrach-*]. Type genus: *Arthrobrachus* Solier, 1849.

Tribe Astylini Pic, 1929

Astylini Pic, 1929: 1 [stem: *Astyl-*]. Type genus: *Astylus* Laporte, 1836.

Tribe Cerallini Pic, 1929

Cerallini Pic, 1929: 13 [stem: *Cerall-*]. Type genus: *Cerallus* Jacquelin du Val, 1859.

Tribe Melyrini Leach, 1815

Melyrides Leach, 1815: 87 [stem: *Melyr-*]. Type genus: *Melyris* Fabricius, 1775. Comment: published April 1815; this family-group name was also used in the same year by Rafinesque (1815 [between April and 21 July]: 110, as Melyria).

Zygiidae Jakobson, 1911a: 687 [stem: *Zygi-*]. Type genus: *Zygia* Fabricius, 1775 [syn. of *Melyris* Fabricius, 1775].

Subfamily Dasytinae Laporte, 1840

Dasydites Laporte, 1840a: 280 [stem: *Dasyt-*]. Type genus: *Dasytes* Paykull, 1799.

Tribe Chaetomalachiini Majer, 1987

Chaetomalachiini Majer, 1987: 799, in key [stem: *Chaetomalachi-*]. Type genus: *Chaetomalachius* Kraatz, 1882.

Tribe DANACEINI Thomson, 1859

DANACAEINA C. G. Thomson, 1859: 108 [stem: *Danace-*]. Type genus: *Danacea* Laporte, 1838 [as *Danacaea*, incorrect subsequent spelling of type genus name, not in prevailing usage]. Comment: incorrect original stem formation, not in prevailing usage.

AMAURONOININi Majer, 1987: 799, in key [stem: *Amauronioid-*]. Type genus: *Amauronioides* Champion, 1923 [syn. of *Pseudamauronia* Pic, 1915]. Comment: incorrect original stem formation, not in prevailing usage.

DANACAEOMIMINI Majer, 1987: 799, in key [stem: *Danacaeomim-*]. Type genus: *Danacaeomimus* Champion, 1922.

Tribe DASYTINI Laporte, 1840

DASYDITES Laporte, 1840a: 280 [stem: *Dasyt-*]. Type genus: *Dasytes* Paykull, 1799. Comment: original vernacular name available (Art. 11.7.2): first used in latinized form by Agassiz (1846b: 117, as DASYTETOIDAE [incorrect stem formation]), generally accepted as in Pic (1937: 1, as DASYTINAE); current spelling maintained (Art. 29.5): incorrect stem formation in prevailing usage (should be *Dasytet-*).

COLPOTHISIDAE Gistel, 1848: [11] [stem: *Colpoth-*]. Type genus: *Colpothis* Gistel, 1848 [syn. of *Dasytes* Paykull, 1799]. Comment: incorrect original stem formation, not in prevailing usage.

*HÉNICOPAIRES Mulsant and Rey, 1868a: 264 [stem: *Enicopod-*]. Type genus: *Enicopus* Stephens, 1830 [as *Henicopus*, unjustified emendation of type genus name by Agassiz (1846b: 138), not in prevailing usage]. Comment: original vernacular name unavailable (Art. 11.7.2): subsequently used in latinized form but not generally attributed to Mulsant and Rey (1868a); incorrect original stem formation, not in prevailing usage.

HENICOPINI Escalera, 1927: 7 [stem: *Enicopod-*]. Type genus: *Enicopus* Stephens, 1830 [as *Henicopus*, unjustified emendation of type genus name by Agassiz (1846b: 138), not in prevailing usage]. Comment: incorrect original stem formation, not in prevailing usage.

Tribe GIETELLINI Constantin and Menier, 1987

GIETELLINAE Constantin and Menier, 1987: 62 [stem: *Gietell-*]. Type genus: *Gietella* Constantin and Menier, 1987.

Tribe LISTRINI Majer, 1990

LISTRINI Majer, 1990: 371 [stem: *Listr-*]. Type genus: *Listrus* Motschulsky, 1859.

Subfamily MALACHIINAE Fleming, 1821

MALACHIUSIDAE Fleming, 1821: 50 [stem: *Malachi-*]. Type genus: *Malachius* Fabricius, 1775.

Tribe Amalthocini Majer, 2002

Amalthocinae Majer, 2002: 186 [stem: *Amalthoc-*]. Type genus: *Amalthocus* Fairmaire, 1886.

Tribe Attalomimini Majer, 1995

Attalomimidae Majer, 1995a: 378 [stem: *Attalomim-*]. Type genus: *Attalomimus* Wittmer, 1976.

Tribe Carphurini Champion, 1923

Carphurinae Champion, 1923: 2 [stem: *Carphur-*]. Type genus: *Carphurus* Erichson, 1840.

Tribe Lemphini Wittmer, 1976

Lemphini Wittmer, 1976: 427 [stem: *Lemph-*]. Type genus: *Lemphus* Erichson, 1840.

Tribe Malachiini Fleming, 1821

Malachiusidae Fleming, 1821: 50 [stem: *Malachi-*]. Type genus: *Malachius* Fabricius, 1775. Comment: incorrect original stem formation, not in prevailing usage.

Tamulidae Gistel, 1848: [11] [stem: *Tamul-*]. Type genus: *Tamulus* Gistel, 1848 [syn. of *Malachius* Fabricius, 1775].

Polycystophoridae Gistel, 1856a: 385 [stem: *Polycystophor-*]. Type genus: *Polycystophorus* Gistel, 1856 [this genus was included by Gistel in his new family Polycystophoridae along with one genus now classified in Cantharidae (*Malthinus* Latreille) and four genera of Malachiinae (*Anthocomus* Erichson, *Ebaeus* Erichson, *Charopus* Erichson and *Troglops* Erichson); several available species names were originally included in *Polycystophorus* (Gistel, 1856a: 385), we hereby select *Cantharis aeneus* Linnaeus, 1758 as the type species; **syn. nov.** of *Malachius* Fabricius, 1775]. Comment: **syn. nov.**

Apalochraires Mulsant and Rey, 1867b: 19 [stem: *Apalochr-*]. Type genus: *Apalochrus* Erichson, 1840. Comment: original vernacular name available (Art. 11.7.2): first used in latinized form and generally accepted as in Houlbert (1922a: 250, as Apalochrini).

*Anthocomates Mulsant and Rey, 1867b: 127 [stem: *Anthocom-*]. Type genus: *Anthocomus* Erichson, 1840. Comment: original vernacular name unavailable (Art. 11.7.2): not subsequently latinized.

*Troglopates Mulsant and Rey, 1867b: 281 [stem: *Troglop-*]. Type genus: *Troglops* Erichson, 1840. Comment: original vernacular name unavailable (Art. 11.7.2): not subsequently latinized.

Caulautaires Abeille de Perrin, 1890: 244 [stem: *Colot-*]. Type genus: *Colotes* Erichson, 1840 [as *Caulautes*, unjustified emendation of type genus name by Abeille de Perrin (1890), not in prevailing usage]. Comment: original vernacular name available (Art. 11.7.2): first used in latinized form by Bertkau (1891: 304, as COLOTINI), generally accepted as in Hansen (1996: 152, as COLOTINI); incorrect original stem formation, not in prevailing usage; COLOTINI Larsen, 1983 (type genus *Colotis* Hübner, 1819) does not appear to be available in Lepidoptera: PIERIDAE; if the Lepidoptera name is found to be available then this case is to be referred to the Commission to remove the homonymy (Art. 55.3.1).

ATTALAIRES Abeille de Perrin, 1891: 364 [stem: *Attal-*]. Type genus: *Attalus* Erichson, 1840. Comment: original vernacular name available (Art. 11.7.2): first used in latinized form by Hatch (1962: 87, as ATTALINA), generally accepted as in Mayor (2007: 419, as ATTALINI).

LAIINA Jakobson, 1911a: 688 [stem: *Lai-*]. Type genus: *Laius* Guérin-Méneville, 1838.

ILLOPINA Jakobson, 1911a: 688 [stem: *Illop-*]. Type genus: *Illops* Erichson, 1840.

PARATINI Portevin, 1931: 426, in key [stem: *Paratin-*]. Type genus: *Paratinus* Abeille de Perrin, 1891 [syn. of *Apalochrus* Erichson, 1840]. Comment: the correct spelling PARATININI was used in the same work on page 442.

EBAEINI Portevin, 1931: 426, in key [stem: *Ebae-*]. Type genus: *Ebaeus* Erichson, 1840.

COLLOPINA Hatch, 1962: 88, in key [stem: *Collop-*]. Type genus: *Collops* Erichson, 1840.

Tribe PAGURODACTYLINI Constantin, 2001

PAGURODACTYLINAE Constantin, 2001: 6 [stem: *Pagurodactyl-*]. Type genus: *Pagurodactylus* Gorham, 1900.

MELYRIDAE *incertae sedis*

ELOSOMATIDAE Jakobson, 1915: 993 [stem: *Elosomat-*]. Type genus: *Elosoma* Motschulsky, 1845. Comment: transferred from SALPINGIDAE by Zerche (2004), correct placement within the family uncertain.

Superfamily CUCUJOIDEA Latreille, 1802

CUCUJIPES Latreille, 1802: 210 [stem: *Cucuj-*]. Type genus: *Cucujus* Fabricius, 1775 [placed on the Official List of Generic Names in Zoology (ICZN 1994a)]. Comment: First Reviser (CUCUJOIDEA Latreille, 1802 vs EROTYLOIDEA Latreille, 1802 vs NITIDULOIDEA Latreille, 1802) not determined, current usage maintained.

†Family PARANDREXIDAE Kirejtshuk, 1994

PARANDREXIDAE Kirejtshuk, 1994: 57 [stem: *Parandrex-*]. Type genus: *Parandrexis* Martynov, 1926. Comment: current spelling maintained (Art. 29.5): incorrect stem formation in prevailing usage (should be *Parandrexe-*).

†Family SINISILVANIDAE Hong, 2002

SINISILVANIDAE Hong, 2002: 132 [stem: *Sinisilvan-*]. Type genus: *Sinisilvana* Hong, 2002.

Family Boganiidae Sen Gupta and Crowson, 1966

Boganiidae Sen Gupta and Crowson, 1966: 63 [stem: *Bogani-*]. Type genus: *Boganium* Sen Gupta and Crowson, 1966.

Subfamily Paracucujinae Endrödy-Younga and Crowson, 1986

Paracucujinae Endrödy-Younga and Crowson, 1986: 255 [stem: *Paracucuj-*]. Type genus: *Paracucujus* Sen Gupta and Crowson, 1966.

Athertoniini Crowson, 1990: 91 [stem: *Athertoni-*]. Type genus: *Athertonium* Crowson, 1990.

Subfamily Boganiinae Sen Gupta and Crowson, 1966

Boganiidae Sen Gupta and Crowson, 1966: 63 [stem: *Bogani-*]. Type genus: *Boganium* Sen Gupta and Crowson, 1966.

Family Byturidae Gistel, 1848

Byturidae Gistel, 1848: [3] [stem: *Bytur-*]. Type genus: *Byturus* Latreille, 1797.

Subfamily Platydascillinae Pic, 1914

Platydascillidae Pic, 1914: 20 [stem: *Platydascill-*]. Type genus: *Platydascillus* Everts, 1909.

Subfamily Byturinae Gistel, 1848

Byturidae Gistel, 1848: [3] [stem: *Bytur-*]. Type genus: *Byturus* Latreille, 1797. Comment: family-group name previously attributed to Jacquelin du Val (1858: 211).

Family Helotidae Chapuis, 1876

Hélotides Chapuis, 1876: 15 [stem: *Helot-*]. Type genus: *Helota* W. S. MacLeay, 1825. Comment: published before 1 May 1876; original vernacular name available (Art. 11.7.2): used in latinized form by several authors subsequently, generally accepted as in Lawrence and Newton (1995: 875, as "Helotidae Reitter, 1876/Chapuis, 1876"); this family-group name was also used in the same year by Reitter (1876a [before 15 August]: 5, as Helotidae); this is a junior homonym of Helotidae Adams et al., 1854 (type genus *Helotes* Cuvier, 1829) in Pisces, this case is to be referred to the Commission to remove the homonymy (Art. 55.3.1).

Family Protocucujidae Crowson, 1954

Protocucujidae Crowson, 1954: 60 [stem: *Protocucuj-*]. Type genus: *Protocucujus* Crowson, 1954.

Family Sphindidae Jacquelin du Val, 1860

Sphindides Jacquelin du Val, 1860: 224 [stem: *Sphind-*]. Type genus: *Sphindus* Dejean, 1821 [placed on the Official List of Generic Names in Zoology (ICZN 1997)].

Comment: this name (as Sphindidae Jacquelin du Val, [1861]) was placed on the Official List of Family-Group Names in Zoology and ruled under the plenary power that this and other family-group names based on *Sphindus* Dejean, 1821 are to be given precedence over Aspidiphoridae Kiesenwetter, 1877 (1859) and other family-group names based on *Aspidiphorus* Dejean, 1821 whenever their type genera are placed in the same family-group taxon (ICZN 1997).

Subfamily Protosphindinae Sen Gupta and Crowson, 1979

Protosphindinae Sen Gupta and Crowson, 1979: 179, in key [stem: *Protosphind-*]. Type genus: *Protosphindus* Sen Gupta and Crowson, 1979.

Subfamily Odontosphindinae Sen Gupta and Crowson, 1979

Odontosphindini Sen Gupta and Crowson, 1979: 180, in key [stem: *Odontosphind-*]. Type genus: *Odontosphindus* Sen Gupta and Crowson, 1979.

Subfamily Sphindiphorinae Sen Gupta and Crowson, 1979

Sphindiphorini Sen Gupta and Crowson, 1979: 180, in key [stem: *Sphindiphor-*]. Type genus: *Sphindiphorus* Sen Gupta and Crowson, 1979.

Subfamily Sphindinae Jacquelin du Val, 1860

Coniporina C. G. Thomson, 1859: 90 [stem: *Conipor-*]. Type genus: *Coniporus* C. G. Thomson, 1859 [placed on the Official Index of Rejected and Invalid Generic Names in Zoology (ICZN 1997); syn. of *Aspidiphorus* Dejean, 1821]. Comment: permanently invalid (Art. 39): based on suppressed type genus.

Sphindides Jacquelin du Val, 1860: 224 [stem: *Sphind-*]. Type genus: *Sphindus* Dejean, 1821 [placed on the Official List of Generic Names in Zoology (ICZN 1997)]. Comment: original vernacular name available (Art. 11.7.2): first used in latinized form by Kiesenwetter (1877: 7, as Sphindini), generally accepted as in Lawrence and Newton (1995: 872, as Sphindidae).

Aspidiphoridae Kiesenwetter, 1877: 198 [stem: *Aspidiphor-*]. Type genus: *Aspidiphorus* Dejean, 1821 [*Arpidiphorus*, the original spelling of the type genus, as well as *Aspidophorus* Agassiz, 1846 were placed on the Official Index of Rejected and Invalid Generic Names in Zoology and *Aspidiphorus* was placed on the Official List of Generic Names in Zoology (ICZN 1997)]. Comment: this name (as Aspidiphoridae Kiesenwetter, 1877 (1859)) was placed on the Official List of Family-Group Names in Zoology and ruled under the plenary power that this and other family-group names based on *Aspidiphorus* are not to be given priority over Sphindidae and other family-group names based on *Sphindus* Dejean, 1821 whenever their type genera are placed in the same family-group taxon (ICZN 1997).

Erysphindinae Sen Gupta and Crowson, 1979: 180, in key [stem: *Eurysphind-*]. Type genus: *Eurysphindus* J. L. LeConte, 1878.

Family Biphyllidae LeConte, 1861

Diphyllidae J. L. LeConte, 1861: 105 [stem: *Biphyll-*]. Type genus: *Biphyllus* Dejean, 1821 [as *Diphyllus*, unjustified emendation of type genus name by Agassiz (1846b: 47), not in prevailing usage]. Comment: incorrect original stem formation, not in prevailing usage (see Cline and Shockley 2010).

Family Erotylidae Latreille, 1802

Erotilenae Latreille, 1802: 233 [stem: *Erotyl-*]. Type genus: *Erotylus* Fabricius, 1775.

Subfamily Xenoscelinae Ganglbauer, 1899

Xenoscelini Ganglbauer, 1899: 649 [stem: *Xenoscel-*]. Type genus: *Xenoscelis* Wollaston, 1864. Comment: current spelling maintained (Art. 29.3.1.1): incorrect stem formation in prevailing usage (should be *Xenoscelid-*).

Eicolyctini Vogt, 1967: 103 [stem: *Eicolyct-*]. Type genus: *Eicolyctus* J. R. Sahlberg, 1919 [syn. of *Zavaljus* Reitter, 1880].

Loberonothini Sen Gupta and Crowson, 1969a: 127 [stem: *Loberonoth-*]. Type genus: *Loberonotha* Sen Gupta and Crowson, 1969.

Subfamily Pharaxonothinae Crowson, 1952

Setariini Casey, 1900: 77 [stem: *Setari-*]. Type genus: *Setaria* Mulsant and Rey, 1863 [preoccupied genus name, not *Setaria* Viborg, 1795 [Vermes], not *Setaria* Oken, 1815 [Vermes], not *Setaria* Blyth, 1844 [Aves]; syn. of *Setariola* Jakobson, 1915]. Comment: permanently invalid (Art. 39): based on preoccupied type genus.

Pharaxonothinae Crowson, 1952: 127 [stem: *Pharaxonoth-*]. Type genus: *Pharaxonotha* Reitter, 1875. Comment: First Reviser found (Pharaxonothinae Crowson, 1952 vs Setariolinae Crowson, 1952) is Leschen (2003: 35).

Setariolinae Crowson, 1952: 127 [stem: *Setariol-*]. Type genus: *Setariola* Jakobson, 1915.

Xenoscelinini Sen Gupta and Crowson, 1971: 5, in key [stem: *Xenoscelin-*]. Type genus: *Xenoscelinus* Grouvelle, 1910 [syn. of *Cathartocryptus* Sharp, 1886].

Subfamily Loberinae Bruce, 1951

Loberinae Bruce, 1951: 4 [stem: *Lober-*]. Type genus: *Loberus* J. L. LeConte, 1861. Comment: name proposed after 1930 without description or bibliographic reference to such a description (Art. 13.1), however available because it was used as valid before 2000 as in Sen Gupta (1968a: 1, as Loberini) and was not rejected by an author who, between 1961 and 1999, applied Article 13 of the then current edition of the Code (see Art. 13.2.1).

Subfamily Languriinae Hope, 1840

Languiridae Hope, 1840a: 190 [stem: *Languri-*]. Type genus: *Languria* Latreille, 1802 [as *Languiria*, incorrect subsequent spelling of type genus name, not in prevailing usage].

Tribe Hapalipini Leschen, 2003

Hapalipini Leschen, 2003: 38 [stem: *Hapalip-*]. Type genus: *Hapalips* Reitter, 1877.

Tribe Languriini Hope, 1840

Languiridae Hope, 1840a: 190 [stem: *Languri-*]. Type genus: *Languria* Latreille, 1802 [as *Languiria*, incorrect subsequent spelling of type genus name, not in prevailing usage]. Comment: incorrect original stem formation, not in prevailing usage; family-group name previously attributed to Crotch (1873c) by Pakaluk et al. (1994) and subsequent authors; usage of «Wiedeman, 1823» as the correct author and year for this family-group name in the literature is incorrect (see Pakaluk et al. 1994).

Cladoxeninae Arrow, 1925: 166 [stem: *Cladoxen-*]. Type genus: *Cladoxena* Motschulsky, 1866.

Tribe Thallisellini Sen Gupta, 1968

Thallisellini Sen Gupta, 1968b: 470 [stem: *Thallisell-*]. Type genus: *Thallisella* Crotch, 1876.

Subfamily Cryptophilinae Casey, 1900

Cryptophilini Casey, 1900: 77 [stem: *Cryptophil-*]. Type genus: *Cryptophilus* Reitter, 1874.

Tribe Cryptophilini Casey, 1900

Cryptophilini Casey, 1900: 77 [stem: *Cryptophil-*]. Type genus: *Cryptophilus* Reitter, 1874.

Tribe Empocryptini Leschen, 2003

Empocryptini Leschen, 2003: 41 [stem: *Empocrypt-*]. Type genus: *Empocryptus* Sharp, 1900.

Tribe Toramini Sen Gupta, 1967

Toraminae Sen Gupta, 1967: 168 [stem: *Toram-*]. Type genus: *Toramus* Grouvelle, 1916.

Subfamily Erotylinae Latreille, 1802

Erotilenae Latreille, 1802: 233 [stem: *Erotyl-*]. Type genus: *Erotylus* Fabricius, 1775.

Tribe Dacnini Gistel, 1848

*Engidites Latreille, 1829a: 506 [stem: *Eng-*]. Type genus: *Engis* Paykull, 1800 [syn. of *Dacne* Latreille, 1797]. Comment: family-group name unavailable (Art. 11.7.1.1): not based on a genus used as valid at the time.

DACNEIDAE Gistel, 1848: [3] [stem: *Dacn-*]. Type genus: *Dacne* Latreille, 1797. Comment: incorrect original stem formation, not in prevailing usage.

CRYPTODACNINI Sen Gupta, 1970: 101, in key [stem: *Cryptodacn-*]. Type genus: *Cryptodacne* Sharp, 1878.

Tribe ENCAUSTINI Crotch, 1876

ENGIDAE W. S. MacLeay, 1825: 40 [stem: *Engid-*]. Type genus: *Engis* sensu W. S. MacLeay, 1825 [not *Engis* Paykull, 1800; syn. of *Encaustes* Lacordaire, 1842]. Comment: based on a misidentified type genus, name treated here as invalid until an application is submitted to the Commission to suppress it for the Principle of Priority (Art. 65.2.1).

ENCAUSTINI Crotch, 1876: 476 [stem: *Encaust-*]. Type genus: *Encaustes* Lacordaire, 1842. Comment: published after February 1876; this family-group name was also used in the same year by Chapuis (1876 [before 1 May]: 46, as ENCAUSTITES).

Tribe EROTYLINI Latreille, 1802

EROTILENAE Latreille, 1802: 233 [stem: *Erotyl-*]. Type genus: *Erotylus* Fabricius, 1775. Comment: incorrect original stem formation, not in prevailing usage.

Tribe MEGALODACNINI Sen Gupta, 1970

MEGALODACNINI Sen Gupta, 1970: 100, in key [stem: *Megalodacn-*]. Type genus: *Megalodacne* Crotch, 1873.

Tribe TRITOMINI Curtis, 1834

TRITOMIDAE Curtis, 1834: plate 498 [stem: *Tritom-*]. Type genus: *Tritoma* Fabricius, 1775 [placed on the Official List of Generic Names in Zoology (ICZN 1994a)].

TRIPLACINAE Erichson, 1847a: 179 [stem: *Triplac-*]. Type genus: *Triplax* Herbst, 1793.

RENANIINAE Chûjô, 1941: 10 [stem: *Renani-*]. Type genus: *Renania* Lewis, 1887.

CYRTOTRIPLACINA Chûjô, 1969: 201 [stem: *Cyrtotriplac-*]. Type genus: *Cyrtotriplax* Crotch, 1873 [syn. of *Tritoma* Fabricius, 1775].

Family MONOTOMIDAE Laporte, 1840

MONOTOMITES Laporte, 1840b: 377 [stem: *Monotom-*]. Type genus: *Monotoma* Herbst, 1793.

Subfamily RHIZOPHAGINAE Redtenbacher, 1845

RHIZOPHAGI L. Redtenbacher, 1845: 125 [stem: *Rhizophag-*]. Type genus: *Rhizophagus* Herbst, 1793 [the original spelling *Ryzophagus* was placed on the Official Index of Rejected and Suppressed Generic Names in Zoology, *Rhizophagus* was considered the correct original spelling of the genus and placed on the Official List of Generic Names in Zoology (ICZN 1995b)].

Subfamily Monotominae Laporte, 1840

Monotomites Laporte, 1840b: 377 [stem: *Monotom-*]. Type genus: *Monotoma* Herbst, 1793.

Tribe Europini Sen Gupta, 1988

Europini Sen Gupta, 1988: 14, in key [stem: *Europ-*]. Type genus: *Europs* Wollaston, 1854.

Tribe Lenacini Crowson, 1952

Lenacinae Crowson, 1952: 121 [stem: *Lenac-*]. Type genus: *Lenax* Sharp, 1877.

Tribe Monotomini Laporte, 1840

Monotomites Laporte, 1840b: 377 [stem: *Monotom-*]. Type genus: *Monotoma* Herbst, 1793 [*nomen protectum*; this genus name is a junior homonym of *Monotoma* Panzer, 1792 *nomen oblitum*; we provide references to support the conservation of *Monotoma* Herbst, 1793 as the valid name for this genus (Art. 23.9.1) (see Appendix 1)]. Comment: original vernacular name available (Art. 11.7.2): first used in latinized form by Agassiz (1846b: 238, as Monotomoidae), generally accepted as in Pakaluk et al. (1994: 230, as Monotomidae).

†Tribe Rhizophtomini Kirejtshuk and Azar, 2009

Rhizophtominae Kirejtshuk and Azar, 2009: 123 [stem: *Rhizophtom-*]. Type genus: *Rhizophtoma* Kirejtshuk and Azar, 2009.

Tribe Thionini Crowson, 1952

Thioninae Crowson, 1952: 121 [stem: *Thion-*]. Type genus: *Thione* Sharp, 1899.

Family Hobartiidae Sen Gupta and Crowson, 1966

Hobartiini Sen Gupta and Crowson, 1966: 65, in key [stem: *Hobarti-*]. Type genus: *Hobartius* Sen Gupta and Crowson, 1966.

Family Cryptophagidae Kirby, 1826

Cryptophagidae Kirby, 1826: 504 [stem: *Cryptophag-*]. Type genus: *Cryptophagus* Herbst, 1792.

Subfamily Cryptophaginae Kirby, 1826

Cryptophagidae Kirby, 1826: 504 [stem: *Cryptophag-*]. Type genus: *Cryptophagus* Herbst, 1792.

Tribe Caenoscelini Casey, 1900

Caenoscelini Casey, 1900: 103 [stem: *Caenoscel-*]. Type genus: *Caenoscelis* C. G. Thomson, 1863. Comment: First Reviser found (Caenoscelini Casey, 1900 vs Sternodeini Casey, 1900) is Leschen (1996: 607); current spelling maintained (Art. 29.3.1.1): incorrect stem formation in prevailing usage (should be *Caenoscelid-*).

Sternodeini Casey, 1900: 103 [stem: *Sternode-*]. Type genus: *Sternodea* Reitter, 1875.

Tribe Cryptophagini Kirby, 1826

Ips Latreille, 1802: 129 [stem: *Ip-*]. Type genus: *Ips* sensu Latreille, 1802 [not *Ips* DeGeer, 1775; syn. of *Cryptophagus* Herbst, 1792]. Comment: although the spelling of the family-group names is identical to the spelling of the type genus, it is clear, based on other evidence from the same work, that Latreille was using it as a family-group name (see Lawrence and Newton 1995: 913); based on a misidentified type genus; name treated here as invalid until an application is submitted to the Commission to suppress it for the Principles of Priority and Homonymy (Art. 65.2.1); also see Ipini Bedel, 1888 in Curculionidae: Scolytinae.

Cryptophagidae Kirby, 1826: 504 [stem: *Cryptophag-*]. Type genus: *Cryptophagus* Herbst, 1792 [the original spelling *Kryptophagus* was placed on the Official Index of Rejected and Suppressed Generic Names in Zoology, *Cryptophagus* was considered the correct original spelling of the genus and placed on the Official List of Generic Names in Zoology (ICZN 1995b)].

Telmatophilides Jacquelin du Val, 1858: 209 [stem: *Telmatophil-*]. Type genus: *Telmatophilus* Heer, 1841. Comment: original vernacular name available (Art. 11.7.2): first used in latinized form by J. L. LeConte (1861: 98, as Telmatophilini), generally accepted as in Telnov (2004: 73, as Telmatophilini).

Antherophagi J. L. LeConte, 1861: 98 [stem: *Antherophag-*]. Type genus: *Antherophagus* Dejean, 1821.

Paramecosomina Reitter, 1875: 4 [stem: *Paramecosomat-*]. Type genus: *Paramecosoma* Curtis, 1833. Comment: incorrect original stem formation, not in prevailing usage.

Catopochrotidae Reitter, 1889a: 289 [stem: *Catopochrot-*]. Type genus: *Catopochrotus* Reitter, 1889.

Emphyli Casey, 1900: 86 [stem: *Emphyl-*]. Type genus: *Emphylus* Erichson, 1846 [syn. of *Spavius* Motschulsky, 1844].

Spaniophaeni Casey, 1900: 86 [stem: *Spaniophaen-*]. Type genus: *Spaniophaenus* Reitter, 1875.

Tribe Picrotini Crowson, 1980

Picrotini Crowson, 1980: 283 [stem: *Picrot-*]. Type genus: *Picrotus* Sharp, 1886.

*Cryptosomatulini Crowson, 1980: 284 [stem: *Cryptosomatul-*]. Type genus: *Cryptosomatula* Bruce, 1940. Comment: unavailable family-group name, proposed after 1930 without description or bibliographic reference to such a description (Art. 13.1).

Cryptosomatulini Leschen, 1996: 605 [stem: *Cryptosomatul-*]. Type genus: *Cryptosomatula* Bruce, 1940.

Subfamily Atomariinae LeConte, 1861

Atomariini J. L. LeConte, 1861: 99 [stem: *Atomari-*]. Type genus: *Atomaria* Stephens, 1829.

Tribe Atomariini LeConte, 1861

Atomariini J. L. LeConte, 1861: 99 [stem: *Atomari-*]. Type genus: *Atomaria* Stephens, 1829.

Ephistemini Casey, 1900: 104 [stem: *Ephistem-*]. Type genus: *Ephistemus* Stephens, 1829.

*Sallтiini Crowson, 1980: 284 [stem: *Sallti-*]. Type genus: *Salltius* Broun, 1893. Comment: unavailable family-group name, proposed after 1930 without description or bibliographic reference to such a description (Art. 13.1).

*Atomaroidini Lyubarsky, 1998: 71 [stem: *Atomaroid-*]. Type genus: *Atomaroides* Lyubarsky, 1989. Comment: unavailable family-group name, proposed after 1930 without description or bibliographic reference to such a description (Art. 13.1).

Tribe Cryptafricini Leschen, 1996

Cryptafricini Leschen, 1996: 623 [stem: *Cryptafric-*]. Type genus: *Cryptafricus* Leschen, 1996.

*Microphagini Lyubarsky, 1998: 39, in key [stem: *Microphag-*]. Type genus: *Microphagus* Leschen, 1996. Comment: unavailable family-group name, proposed after 1930 without description or bibliographic reference to such a description (Art. 13.1).

*Scytomariini Lyubarsky, 1998: 71 [stem: *Scytomari-*]. Type genus: *Scytomaria* Lyubarsky, 1998 [syn. of *Anitamaria* Leschen, 1996]. Comment: unavailable family-group name, proposed after 1930 without description or bibliographic reference to such a description (Art. 13.1).

Tribe Hypocoprini Reitter, 1879

Hypocoprini Reitter, 1879: 74 [stem: *Hypocopr-*]. Type genus: *Hypocoprus* Motschulsky, 1839.

*Alfieriellinae Crowson, 1980: 282 [stem: *Alfieriell-*]. Type genus: *Alfieriella* Wittmer, 1935. Comment: unavailable family-group name, proposed after 1930 without description or bibliographic reference to such a description (Art. 13.1).

*Hypophagina Lyubarsky, 1998: 71 [stem: *Hypophag-*]. Type genus: *Hypophagus* Lyubarsky, 1989. Comment: unavailable family-group name, proposed after 1930 without description or bibliographic reference to such a description (Art. 13.1).

Family Agapythidae Sen Gupta and Crowson, 1969

Agapythinae Sen Gupta and Crowson, 1969b: 579, in key [stem: *Agapyth-*]. Type genus: *Agapytho* Broun, 1921.

Family Priasilphidae Crowson, 1973

Priasilphinae Crowson, 1973a: 56 [stem: *Priasilph-*]. Type genus: *Priasilpha* Broun, 1893.

Family Phloeostichidae Reitter, 1911

Phloeostichini Reitter, 1911: 44 [stem: *Phloeostich-*]. Type genus: *Phloeostichus* W. Redtenbacher, 1842.

Hymaeinae Sen Gupta and Crowson, 1966: 65, in key [stem: *Hymae-*]. Type genus: *Hymaea* Pascoe, 1869.

Family Silvanidae Kirby, 1837

Sylvanidae Kirby, 1837: 110 [stem: *Silvan-*]. Type genus: *Silvanus* Latreille, 1804.

Subfamily Brontinae Blanchard, 1845

Brontites Blanchard, 1845b: 134 [stem: *Bront-*]. Type genus: *Brontes* Fabricius, 1801.

Tribe Brontini Blanchard, 1845

Brontites Blanchard, 1845b: 134 [stem: *Bront-*]. Type genus: *Brontes* Fabricius, 1801. Comment: published before 11 June 1845; original vernacular name available (Art. 11.7.2): first used in latinized form by Erichson (1845 [before 31 June]: 304, as Brontini), treated as available by Lawrence and Newton (1995: 876).

Dendrophagidae Gistel, 1848: [8] [stem: *Dendrophag-*]. Type genus: *Dendrophagus* Schönherr, 1809.

Uleiotaeidae Gistel, 1848: [10] [stem: *Uleiot-*]. Type genus: *Uleiota* Latreille, 1797. Comment: incorrect original stem formation, not in prevailing usage.

Tribe Telephanini LeConte, 1861

Telephanidae J. L. LeConte, 1861: 96 [stem: *Telephan-*]. Type genus: *Telephanus* Erichson, 1845.

Pseudophanini J. L. LeConte, 1861: 96 [stem: *Pseudophan-*]. Type genus: *Pseudophanus* J. L. LeConte, 1859 [syn. of *Cryptamorpha* Wollaston, 1854].

Psammoecini Reitter, 1879: 74 [stem: *Psammoec-*]. Type genus: *Psammoecus* Latreille, 1829.

CRYPTAMORPHINI Casey, 1884: 102 [stem: *Cryptamorph-*]. Type genus: *Cryptamorpha* Wollaston, 1854.

Subfamily SILVANINAE Kirby, 1837

SYLVANIDAE Kirby, 1837: 110 [stem: *Silvan-*]. Type genus: *Silvanus* Latreille, 1804. Comment: incorrect original stem formation, not in prevailing usage.

Family CUCUJIDAE Latreille, 1802

CUCUJIPES Latreille, 1802: 210 [stem: *Cucuj-*]. Type genus: *Cucujus* Fabricius, 1775 [placed on the Official List of Generic Names in Zoology (ICZN 1994a)].

*BIOPHLOCES Motschulsky, 1849: 60 [stem: *Biophloe-*]. Type genus: *Biophloeus* Dejean, 1835 [*nomen oblitum*; this genus name is a senior synonym of the well-established name *Pediacus* Shuckard, 1839 *nomen protectum*; however since *Biophloeus* Dejean has not been used as valid after 1899 to our knowledge, we provide references to support the conservation of *Pediacus* as the valid name for this genus (Art. 23.9.1) (see Appendix 1)]. Comment: original vernacular name unavailable (Art. 11.7.2): not subsequently latinized.

EAROPHILIDAE Gistel, 1856a: 375 [stem: *Earophil-*]. Type genus: *Earophilus* Gistel, 1856 [syn. of *Cucujus* Fabricius, 1775].

Family MYRABOLIIDAE Lawrence and Britton, 1991

MYRABOLIINAE Lawrence and Britton, 1991: 650 [stem: *Myraboli-*]. Type genus: *Myrabolia* Reitter, 1876.

Family CAVOGNATHIDAE Sen Gupta and Crowson, 1966

CAVOGNATHINAE Sen Gupta and Crowson, 1966: 65, in key [stem: *Cavognath-*]. Type genus: *Cavognatha* Crowson, 1964 [syn. of *Taphropiestes* Reitter, 1875].

Family LAMINGTONIIDAE Sen Gupta and Crowson, 1969

LAMINGTONIIDAE Sen Gupta and Crowson, 1969a: 125 [stem: *Lamingtoni-*]. Type genus: *Lamingtonium* Sen Gupta and Crowson, 1969.

Family PASSANDRIDAE Blanchard, 1845

PASSANDRITES Blanchard, 1845b: 134 [stem: *Passandr-*]. Type genus: *Passandra* Dalman, 1817. Comment: published before 11 June 1845; original vernacular name available (Art. 11.7.2): first used in latinized form by Erichson (1845 [before June 31]: 304, as PASSANDRINI), treated as available by Lawrence and Newton (1995: 876).

ANCISTRIINAE Sharp, 1899b: 541 [stem: *Ancistri-*]. Type genus: *Ancistria* Erichson, 1845.

CATOGENINI Grouvelle, 1916: 6 [stem: *Catogen-*]. Type genus: *Catogenus* Westwood, 1830.

Passandrellini Grouvelle, 1916: 6 [stem: *Passandrell-*]. Type genus: *Passandrella* Grouvelle, 1916.

Scalidiini Grouvelle, 1916: 6 [stem: *Scalidi-*]. Type genus: *Scalidia* Erichson, 1845.

Laemotmetini Kessel, 1921: 35 [stem: *Laemotmet-*]. Type genus: *Laemotmetus* Gerstaecker, 1871 [syn. of *Aulonosoma* Motschulsky, 1858].

Family Phalacridae Leach, 1815

Phalacrurida Leach, 1815: 116 [stem: *Phalacr-*]. Type genus: *Phalacrus* Paykull, 1800.

Subfamily Phaenocephalinae Matthews, 1899

Phaenocephalidae A. Matthews, 1899: 205 [stem: *Phaenocephal-*]. Type genus: *Phaenocephalus* Wollaston, 1873.

Subfamily Phalacrinae Leach, 1815

Phalacrurida Leach, 1815: 116 [stem: *Phalacr-*]. Type genus: *Phalacrus* Paykull, 1800. Comment: incorrect original stem formation, not in prevailing usage.

Idiobiidae Gistel, 1856a: 383 [stem: *Idiobi-*]. Type genus: *Idiobius* Gistel, 1856 [syn. of *Olibrus* Erichson, 1845].

Eustilbini Guillebeau, 1892: 149 [stem: *Eustilb-*]. Type genus: *Eustilbus* Sharp, 1888.

Olibrini Guillebeau, 1892: 147 [stem: *Olibr-*]. Type genus: *Olibrus* Erichson, 1845.

Tolyphini Guillebeau, 1892: 147 [stem: *Tolyph-*]. Type genus: *Tolyphus* Erichson, 1845.

Biophytini Guillebeau, 1894: 276, in key [stem: *Biophyt-*]. Type genus: *Biophytus* Guillebeau, 1894.

Megapalpini Guillebeau, 1894: 278, in key [stem: *Megapalp-*]. Type genus: *Megapalpus* Guillebeau, 1893 [preoccupied genus name, not *Megapalpus* Macquart, 1834 [Diptera]; syn. of *Megistopalpus* Guillebeau, 1895]. Comment: permanently invalid (Art. 39): based on preoccupied type genus.

Ochrolitini Guillebeau, 1894: 278, in key [stem: *Ochrolit-*]. Type genus: *Ochrolitus* Sharp, 1889.

Stilbini Jakobson, 1915: 948 [stem: *Stilb-*]. Type genus: *Stilbus* Seidlitz, 1872.

Family Propalticidae Crowson, 1952

Propalticidae Crowson, 1952: 117 [stem: *Propaltic-*]. Type genus: *Propalticus* Sharp, 1879.

Family Laemophloeidae Ganglbauer, 1899

Laemophloeini Ganglbauer, 1899: 606 [stem: *Laemophloe-*]. Type genus: *Laemophloeus* Dejean, 1835.

Nartheciinae Grouvelle, 1908: 453 [stem: *Nartheci-*]. Type genus: *Narthecius* J. L. LeConte, 1861.

Family Tasmosalpingidae Lawrence and Britton, 1991

Tasmosalpinginae Lawrence and Britton, 1991: 651 [stem: *Tasmosalping-*]. Type genus: *Tasmosalpingus* Lea, 1918.

Family Cyclaxyridae Gimmel, Leschen and Ślipiński, 2009

*Cyclaxyrinae Crowson, 1984: 259 [stem: *Cyclaxyr-*]. Type genus: *Cyclaxyra* Broun, 1893 [placed on the Official List of Generic Names in Zoology (ICZN 1988a)]. Comment: unavailable family-group name, proposed after 1930 without description or bibliographic reference to such a description (Art. 13.1).

Cyclaxyridae Gimmel et al., 2009: 513 [stem: *Cyclaxyr-*]. Type genus: *Cyclaxyra* Broun, 1893 [placed on the Official List of Generic Names in Zoology (ICZN 1988a)]. Comment: although this family-group name was used several times before 2009, all previous uses of the name were unavailable (see Gimmel et al. 2009: 512, 513).

Family Kateretidae Kirby, 1837

Catheretidae Kirby, 1837: 107 [stem: *Kateret-*]. Type genus: *Kateretes* Herbst, 1793 [as *Catheretes*, incorrect subsequent spelling of type genus name, not in prevailing usage; *Kateretes* Herbst, 1793 placed on Official List of Generic Names in Zoology (ICZN 1999b)]. Comment: incorrect original stem formation, not in prevailing usage; Kateretidae chosen as correct spelling of the family-group name, given precedence over Brachypteridae Erichson, 1845 and placed on Official List of Family-Group Names in Zoology (ICZN 1999b, as Kateretidae Erichson in Agassiz, 1846); discovery of the older available name proposed by Kirby (1837) removes priority problems with Brachypterinae.

Brachypterinae Erichson, 1845: 125 [stem: *Brachypter-*]. Type genus: *Brachypterus* Kugelann, 1794 [placed on the Official List of Generic Names in Zoology (ICZN 1999b)]. Comment: placed on Official List of Family-Group Names in Zoology (ICZN 1999b).

Cercidae Desmarest, 1851: 291 [stem: *Cerc-*]. Type genus: *Cercus* Latreille, 1797 [syn. of *Kateretes* Herbst, 1793].

Family Nitidulidae Latreille, 1802

Nitidulariae Latreille, 1802: 131 [stem: *Nitidul-*]. Type genus: *Nitidula* Fabricius, 1775. Comment: the classification of Nitidulidae follows Kirejtshuk (2008).

Subfamily Calonecrinae Kirejtshuk, 1982

Calonecrinae Kirejtshuk, 1982: 117 [stem: *Calonecr-*]. Type genus: *Calonecrus* J. Thomson, 1857.

Subfamily Maynipeplinae Kirejtshuk, 1998

Maynipeplinae Kirejtshuk, 1998b: 540 [stem: *Maynipepl-*]. Type genus: *Maynipeplus* Kirejtshuk, 1998.

Subfamily Epuraeinae Kirejtshuk, 1986

Epuraeini Kirejtshuk, 1986: 27 [stem: *Epurae-*]. Type genus: *Epuraea* Erichson, 1843.

Tribe Epuraeini Kirejtshuk, 1986

Epuraeini Kirejtshuk, 1986: 27 [stem: *Epurae-*]. Type genus: *Epuraea* Erichson, 1843.

Tribe Taenioncini Kirejtshuk, 1998

Taenioncini Kirejtshuk, 1998a: 322 [stem: *Taenionc-*]. Type genus: *Taenioncus* Kirejtshuk, 1984.

Subfamily Carpophilinae Erichson, 1842

Carpophilinae Erichson, 1842: 148 [stem: *Carpophil-*]. Type genus: *Carpophilus* Stephens, 1829.

Subfamily Amphicrossinae Kirejtshuk, 1986

Amphicrossini Kirejtshuk, 1986: 28 [stem: *Amphicross-*]. Type genus: *Amphicrossus* Erichson, 1843.

Subfamily Meligethinae Thomson, 1859

Meligethina C. G. Thomson, 1859: 67 [stem: *Meligeth-*]. Type genus: *Meligethes* Stephens, 1830.

Subfamily Nitidulinae Latreille, 1802

Nitidulariae Latreille, 1802: 131 [stem: *Nitidul-*]. Type genus: *Nitidula* Fabricius, 1775.

Tribe Cychramini Gistel, 1848

Cychramidae Gistel, 1848: [4] [stem: *Cychram-*]. Type genus: *Cychramus* Kugelann, 1794.

Tribe Cychramptodini Kirejtshuk and Lawrence, 1992

Cychramptodini Kirejtshuk and Lawrence, 1992: 29, in key [stem: *Cychramptod-*]. Type genus: *Cychramptodes* Reitter, 1878.

Tribe Cyllodini Everts, 1898

Strongylinae Erichson, 1843: 227 [stem: *Strongyl-*]. Type genus: *Strongylus* Herbst, 1792 [preoccupied genus name, not *Strongylus* O. F. Müller, 1780 [Nematoda]; syn. of *Cyllodes* Erichson, 1843]. Comment: permanently invalid (Art. 39): based on preoccupied type genus; Strongylidae Baird, 1853 (type genus *Strongylus* O. F. Müller, 1780) is available in Nematoda.

Cyllodini Everts, 1898: 469, in key [stem: *Cyllod-*]. Type genus: *Cyllodes* Erichson, 1843.

AMBOROTUBINI Leschen and Carlton, 2004: 450 [stem: *Amborotub-*]. Type genus: *Amborotubus* Leschen and Carlton, 2004.

Tribe LAWRENCEROSINI Kirejtshuk, 1991

LAWRENCEROSINI Kirejtshuk, 1991: 863 [stem: *Lawrenceros-*]. Type genus: *Lawrencerosus* Kirejtshuk, 1991.

Tribe MYSTROPINI Murray, 1864

MYSTROPIDAE Murray, 1864: 411 [stem: *Mystrop-*]. Type genus: *Mystrops* Erichson, 1843.

Tribe NITIDULINI Latreille, 1802

NITIDULARIAE Latreille, 1802: 131 [stem: *Nitidul-*]. Type genus: *Nitidula* Fabricius, 1775.

THALYCRINA C. G. Thomson, 1859: 68 [stem: *Thalycr-*]. Type genus: *Thalycra* Erichson, 1843.

POCADIINI Seidlitz, 1872 [Gatt.]: 32 [stem: *Pocadi-*]. Type genus: *Pocadius* Erichson, 1843.

PROMETOPINAE Böving and Craighead, 1931: 37, in key [stem: *Prometopi-*]. Type genus: *Prometopia* Erichson, 1843. Comment: the spelling PROMETOPINAE in the key on page 37 is considered a *lapsus calami* since the correct spelling PROMETOPIINAE was used in the same publication in the conspectus on page 74 and in the caption for plate 35.

ORVOENINI Dajoz, 1980b: 191 [stem: *Orvoeni-*]. Type genus: *Orvoenia* Dajoz, 1980 [syn. of *Megauchenia* W. S. MacLeay, 1825]. Comment: incorrect original stem formation, not in prevailing usage.

Subfamily CILLAEINAE Kirejtshuk and Audisio, 1986

CILLAEINAE Kirejtshuk and Audisio, 1986: 219 [stem: *Cillae-*]. Type genus: *Cillaeus* Laporte, 1835.

Subfamily CRYPTARCHINAE Thomson, 1859

CRYPTARCHINA C. G. Thomson, 1859: 69 [stem: *Cryptarch-*]. Type genus: *Cryptarcha* Shuckard, 1839.

Tribe ARHININI Kirejtshuk, 1987

ARHININI Kirejtshuk, 1987: 63 [stem: *Arhin-*]. Type genus: *Arhina* Murray, 1876.

Tribe CRYPTARCHINI Thomson, 1859

CRYPTARCHINA C. G. Thomson, 1859: 69 [stem: *Cryptarch-*]. Type genus: *Cryptarcha* Shuckard, 1839.

PITYOPHAGINI Reitter, 1891: 163 [stem: *Pityophag-*]. Type genus: *Pityophagus* Shuckard, 1839.

Glischrochilini Iablokoff-Khnzorian, 1966: 314 [stem: *Glischrochil-*]. Type genus: *Glischrochilus* Reitter, 1873.

Tribe Eucalosphaerini Kirejtshuk, 1987

Eucalosphaerini Kirejtshuk, 1987: 80 [stem: *Eucalosphaer-*]. Type genus: *Eucalosphaera* Jelinek, 1978.

Tribe Platyarchini Kirejtshuk, 1998

Platyarchini Kirejtshuk, 1998a: 41, in key [stem: *Platyarch-*]. Type genus: *Platyarcha* Kirejtshuk, 1987.

Subfamily Cybocephalinae Jacquelin du Val, 1858

Cybocéphalites Jacquelin du Val, 1858: 151 [stem: *Cybocephal-*]. Type genus: *Cybocephalus* Erichson, 1844. Comment: original vernacular name available (Art. 11.7.2): first used in latinized form by C. G. Thomson (1862: 123, as Cybocephalidae), generally accepted as in Pakaluk et al. (1994: 230, as Cybocephalinae).

Family Smicripidae Horn, 1880

Smicripini G. H. Horn, 1880d: 268 [stem: *Smicrip-*]. Type genus: *Smicrips* J. L. LeConte, 1878.

Tisiphoninae Sharp, 1900a: 578 [stem: *Tisiphon-*]. Type genus: *Tisiphone* Reitter, 1876 [preoccupied genus name, not *Tisiphone* Hübner, 1816 [Lepidoptera]; syn. of *Smicrips* J. L. LeConte, 1878]. Comment: permanently invalid (Art. 39): based on preoccupied type genus.

Family Bothrideridae Erichson, 1845

Bothriderini Erichson, 1845: 287 [stem: *Bothrider-*]. Type genus: *Bothrideres* Dejean, 1835.

Subfamily Teredinae Seidlitz, 1888

Teredini Seidlitz, 1888 [Gatt.]: 57 [stem: *Tered-*]. Type genus: *Teredus* Dejean, 1835.

Tribe Sosylopsini Dajoz, 1980

Sosylopsini Dajoz, 1980a: 158 [stem: *Sosylops-*]. Type genus: *Sosylopsis* Grouvelle, 1910. Comment: current spelling maintained (Art. 29.5): incorrect stem formation in prevailing usage (should be *Sosylopse-*).

Tribe Sysolini Ślipiński and Pal, 1985

Sysolini Ślipiński and Pal, 1985: 40 [stem: *Sysol-*]. Type genus: *Sysolus* Grouvelle, 1908.

Tribe TEREDINI Seidlitz, 1888

TEREDINI Seidlitz, 1888 [Gatt.]: 57 [stem: *Tered-*]. Type genus: *Teredus* Dejean, 1835.

Subfamily XYLARIOPHILINAE Pal and Lawrence, 1986

XYLARIOPHILINAE Pal and Lawrence, 1986: 208, in key [stem: *Xylariophil-*]. Type genus: *Xylariophilus* Pal and Lawrence, 1986.

Subfamily ANOMMATINAE Ganglbauer, 1899

ANOMMATINI Ganglbauer, 1899: 894 [stem: *Anommat-*]. Type genus: *Anommatus* Wesmael, 1835.

Subfamily BOTHRIDERINAE Erichson, 1845

BOTHRIDERINI Erichson, 1845: 287 [stem: *Bothrider-*]. Type genus: *Bothrideres* Dejean, 1835.

DERETAPHRINI G. H. Horn, 1878: 578 [stem: *Deretaphr-*]. Type genus: *Deretaphrus* Newman, 1842.

DASTARCINI Reitter, 1922a: 39 [stem: *Dastarc-*]. Type genus: *Dastarcus* Walker, 1858.

Family CERYLONIDAE Billberg, 1820

CERYLONIDES Billberg, 1820a: 47 [stem: *Cerylon-*]. Type genus: *Cerylon* Latreille, 1802 [placed on the Official List of Generic Names in Zoology (ICZN 1995c)]. Comment: CERYLONIDAE Billberg, 1820 placed on the Official List of Family-Group Names in Zoology (ICZN 1995c).

Subfamily EUXESTINAE Grouvelle, 1908

EUXESTINAE Grouvelle, 1908: 397, in key [stem: *Euxest-*]. Type genus: *Euxestus* Wollaston, 1858.

PACHYOCHTHESINAE Reitter, 1911: 105 [stem: *Pachyochth-*]. Type genus: *Pachyochthes* Reitter, 1897 [syn. of *Hypodacne* J. L. LeConte, 1875]. Comment: incorrect original stem formation, not in prevailing usage.

METACERYLINI Heinze, 1944: 21 [stem: *Metacerylon-*]. Type genus: *Metacerylon* Grouvelle, 1906. Comment: incorrect original stem formation, not in prevailing usage.

CYCLOXENINI Jeannel and Paulian, 1945: 57, in key [stem: *Cycloxen-*]. Type genus: *Cycloxenus* Arrow, 1925.

TACHYORYCTIDIINI Jeannel and Paulian, 1945: 57, in key [stem: *Tachyoryctidi-*]. Type genus: *Tachyoryctidium* Jeannel and Paulian, 1945.

Subfamily LOEBLIORYLONINAE Ślipiński, 1990

LOEBLIORYLININAE Ślipiński, 1990: 81 [stem: *Loebliorylon-*]. Type genus: *Loebliorylon* Ślipiński, 1990. Comment: incorrect original stem formation, not in prevailing usage.

Subfamily Ostomopsinae Sen Gupta and Crowson, 1973

Ostomopsini Sen Gupta and Crowson, 1973: 400 [stem: *Ostomops-*]. Type genus: *Ostomopsis* Scott, 1922. Comment: current spelling maintained (Art. 29.5): incorrect stem formation in prevailing usage (should be *Ostomopse-*).

Subfamily Murmidiinae Jacquelin du Val, 1858

Murmidiides Jacquelin du Val, 1858: 227 [stem: *Murmidi-*]. Type genus: *Murmidius* Leach, 1822. Comment: original vernacular name available (Art. 11.7.2): first used in latinized form by J. L. LeConte (1861: 78, as Murmidiidae), generally accepted as in Pakaluk et al. (1994: 232, as Murmidiinae).

Subfamily Ceryloninae Billberg, 1820

Cerylonides Billberg, 1820a: 47 [stem: *Cerylon-*]. Type genus: *Cerylon* Latreille, 1802 [placed on the Official List of Generic Names in Zoology (ICZN 1995c)]. Comment: Cerylonidae Billberg, 1820a placed on the Official List of Family-Group Names in Zoology (ICZN 1995c); this family-group name was also used in the same year by Billberg (1820b: 394, as Cerylonides).

Pleosomides Fauvel, 1891: 162 [stem: *Ploeosomat-*]. Type genus: *Ploeosoma* Wollaston, 1854. Comment: incorrect original stem formation, not in prevailing usage.

Lapethinae Sharp, 1894: 445 [stem: *Lapeth-*]. Type genus: *Lapethus* Casey, 1890 [syn. of *Mychocerus* Erichson, 1845].

Mychocerinae Sharp, 1895b: 494 [stem: *Mychocer-*]. Type genus: *Mychocerus* sensu J. L. LeConte, 1869 [syn. of *Mychocerinus* Ślipiński, 1990]. Comment: based on a misidentified type genus.

Aculagnathidae Oke, 1932: 22 [stem: *Aculagnath-*]. Type genus: *Aculagnathus* Oke, 1932.

Dolosidae Dajoz, 1963: 96 [stem: *Dolos-*]. Type genus: *Dolosus* Dajoz, 1963 [subgenus of *Thyroderus* Sharp, 1885].

Family Alexiidae Imhoff, 1856

Alexiadae Imhoff, 1856: [2] 151 [stem: *Alexi-*]. Type genus: *Alexia* Stephens, 1833 [syn. of *Sphaerosoma* Stephens, 1832].

Sphaerosominae Ganglbauer, 1899: 913 [stem: *Sphaerosomat-*]. Type genus: *Sphaerosoma* Stephens, 1832. Comment: incorrect original stem formation, not in prevailing usage.

Family Discolomatidae Horn, 1878

Discolomidae G. H. Horn, 1878: 557 [stem: *Discolomat-*]. Type genus: *Discoloma* Erichson, 1845. Comment: incorrect original stem formation, not in prevailing usage.

Subfamily Notiophyginae Jakobson, 1915

Notiophygidae Jakobson, 1915: 920 [stem: *Notiophyg-*]. Type genus: *Notiophygus* Gory, 1834.

Tribe Dystheamonini John, 1954

Dystheamonini John, 1954: 42 [stem: *Dystheamon-*]. Type genus: *Dystheamon* Grouvelle, 1927.

Tribe Notiophygini Jakobson, 1915

Notiophygidae Jakobson, 1915: 920 [stem: *Notiophyg-*]. Type genus: *Notiophygus* Gory, 1834.

Tribe Pachyplacini John, 1954

Pachyplacini John, 1954: 38 [stem: *Pachyplac-*]. Type genus: *Pachyplacus* John, 1935.

Subfamily Discolomatinae Horn, 1878

Discolomidae G. H. Horn, 1878: 557 [stem: *Discolomat-*]. Type genus: *Discoloma* Erichson, 1845. Comment: incorrect original stem formation, not in prevailing usage.

Subfamily Aphanocephalinae Jakobson, 1904

Aphanocephalidae Jakobson, 1904: 273 [stem: *Aphanocephal-*]. Type genus: *Aphanocephalus* Wollaston, 1873. Comment: name previously attributed to Grouvelle (1912: 198); usage of Aphaenocephalidae by Ganglbauer (1903: 316) in Staphylinoidea is a *lapsus calami* for Phaenocephalidae (now in Phalacridae) according to Lawrence and Newton (1995).

Subfamily Cephalophaninae John, 1954

Cephalophaninae John, 1954: 64 [stem: *Cephalophan-*]. Type genus: *Cephalophanus* John, 1940.

Subfamily Pondonatinae John, 1954

Pondonatinae John, 1954: 19 [stem: *Pondonat-*]. Type genus: *Pondonatus* John, 1954.

Family Endomychidae Leach, 1815

Endomychides Leach, 1815: 116 [stem: *Endomych-*]. Type genus: *Endomychus* Panzer, 1795.

Subfamily Merophysiinae Seidlitz, 1872

Merophysiini Seidlitz, 1872 [Gatt.]: 39 [stem: *Merophysi-*]. Type genus: *Merophysia* Lucas, 1852.

Holoparamecini Seidlitz, 1888 [Gatt.]: 57 [stem: *Holoparamec-*]. Type genus: *Holoparamecus* Curtis, 1833.

Subfamily Pleganophorinae Jacquelin du Val, 1858

Pléganophorides Jacquelin du Val, 1858: 186 [stem: *Pleganophor-*]. Type genus: *Pleganophorus* Hampe, 1855. Comment: original vernacular name available (Art. 11.7.2): first used in latinized form by Gerstaecker (1861: 419, as PLEGANOPHORIDAE), generally accepted as in Pakaluk et al. (1994: 233, as PLEGANOPHORINAE).

TROCHOIDÉITES Chapuis, 1876: 146 [stem: *Trochoide-*]. Type genus: *Trochoideus* Westwood, 1833. Comment: original vernacular name available (Art. 11.7.2): first used in latinized form and generally accepted as in Ganglbauer (1899: 926, as TROCHOIDEINAE); the earlier use of TROCHOÏDEIDAE by Gorham (1874: 185) is not considered to be a proposal for a new family-group taxon but was rather used as a hypothetical example addressed to the editors of the journal.

Subfamily ANAMORPHINAE Strohecker, 1953

ANAMORPHINI Strohecker, 1953: 15, in key [stem: *Anamorph-*]. Type genus: *Anamorphus* J. L. LeConte, 1878.

MYCHOTHENINAE Sasaji, 1978: 8 [stem: *Mychothen-*]. Type genus: *Mychothenus* Strohecker, 1953.

ACRITOSOMATINAE Pakaluk and Ślipiński, 1995: 328 [stem: *Acritosomat-*]. Type genus: *Acritosoma* Pakaluk and Ślipiński, 1995.

Subfamily LEIESTINAE Thomson, 1863

LEIESTINA C. G. Thomson, 1863: 306 [stem: *Leiest-*]. Type genus: *Leiestes* Chevrolat, 1836.

RHANES J. L. LeConte and G. H. Horn, 1883: 120 [stem: *Rhanid-*]. Type genus: *Rhanis* J. L. LeConte, 1854 [preoccupied genus name, not *Rhanis* C. L. Koch, 1846 [Arachnida]; syn. of *Rhanidea* Strohecker, 1953]. Comment: permanently invalid (Art. 39): based on preoccupied type genus; incorrect original stem formation, not in prevailing usage.

PHYMAPHORINA Jakobson, 1915: 961 [stem: *Phymaphor-*]. Type genus: *Phymaphora* Newman, 1838.

Subfamily MYCETAEINAE Jacquelin du Val, 1857

MYCETÉIDES Jacquelin du Val, 1857b: 102 [stem: *Mycetae-*]. Type genus: *Mycetaea* Stephens, 1829. Comment: original vernacular name available (Art. 11.7.2): first used in latinized form by Gerstaecker (1861: 419, as MYCETAEIDAE), generally accepted as in Pakaluk et al. (1994: 233, as MYCETAEINAE); incorrect original stem formation, not in prevailing usage.

Subfamily EUPSILOBIINAE Casey, 1895

EUPSILOBIINI Casey, 1895: 452 [stem: *Eupsilobi-*]. Type genus: *Eupsilobius* Casey, 1895 [syn. of *Eidoreus* Sharp, 1885].

CERASOMMATIDIIDAE Brèthes, 1925: 199 [stem: *Cerasommatidi-*]. Type genus: *Cerasommatidia* Brèthes, 1925.

EIDOREINAE Sasaji, 1986: 235 [stem: *Eidore-*]. Type genus: *Eidoreus* Sharp, 1885.

Subfamily XENOMYCETINAE Strohecker, 1962

XENOMYCETINAE Strohecker, 1962: 801, in key [stem: *Xenomycet-*]. Type genus: *Xenomycetes* G. H. Horn, 1880.

Subfamily DANASCELINAE Tomaszewska, 2000

DANASCELINAE Tomaszewska, 2000: 494, in key [stem: *Danascel-*]. Type genus: *Danascelis* Tomaszewska, 1999. Comment: current spelling maintained (Art. 29.3.1.1): incorrect original stem formation in prevailing usage (should be *Danascelid-*).

Subfamily ENDOMYCHINAE Leach, 1815

ENDOMYCHIDES Leach, 1815: 116 [stem: *Endomych-*]. Type genus: *Endomychus* Panzer, 1795.

*AGARICOPHILES Motschulsky, 1849: 58 [stem: *Agaricophil-*]. Type genus: *Agaricophilus* Motschulsky, 1837. Comment: original vernacular name unavailable (Art. 11.7.2): not subsequently latinized.

CYCLOTOMIDAE Imhoff, 1856: [2] 151 [stem: *Cyclotom-*]. Type genus: *Cyclotoma* Mulsant, 1851.

*AGARICOPHILINAE Lawrence, 1991: 484 [stem: *Agaricophil-*]. Type genus: *Agaricophilus* Motschulsky, 1837. Comment: unavailable family-group name, proposed after 1930 without description or bibliographic reference to such a description (Art. 13.1); proposed as new without reference to AGARICOPHILES Motschulsky, 1849.

Subfamily EPIPOCINAE Gorham, 1873

EPIPOCIDAE Gorham, 1873: 20 [stem: *Epipoc-*]. Type genus: *Epipocus* Germar, 1843.

Subfamily STENOTARSINAE Chapuis, 1876

STÉNOTARSITES Chapuis, 1876: 125 [stem: *Stenotars-*]. Type genus: *Stenotarsus* Perty, 1832. Comment: original vernacular name available (Art. 11.7.2): first used in latinized form by Csiki (1901: 37, as STENOTARSINI), generally accepted as in Shockley et al. (2009: 73, as STENOTARSINAE).

Subfamily LYCOPERDININAE Bromhead, 1838

LYCOPERDINADAE Bromhead, 1838: 419 [stem: *Lycoperdin-*]. Type genus: *Lycoperdina* Latreille, 1807. Comment: name previously attributed to L. Redtenbacher (1844); incorrect original stem formation, not in prevailing usage.

EUMORPHIDAE Gistel, 1848: [10] [stem: *Eumorph-*]. Type genus: *Eumorphus* Weber, 1801.

DAPSINI Gerstaecker, 1858: 170 [stem: *Daps-*]. Type genus: *Dapsa* Latreille, 1829.

CORYNOMALIDAE Gorham, 1873: 14 [stem: *Corynomal-*]. Type genus: *Corynomalus* Gerstaecker, 1857.

AMPHICINI Csiki, 1910: 25 [stem: *Amphic-*]. Type genus: *Amphix* Laporte, 1840.

BECCARIINI Arrow, 1925: 278 [stem: *Beccari-*]. Type genus: *Beccaria* Gorham, 1885 [preoccupied genus name, not *Beccaria* Bourguignat, 1883 [Mollusca]; syn. of *Beccariola* Arrow, 1943]. Comment: permanently invalid (Art. 39): based on preoccupied type genus.

AMPHISTERNINI Strohecker, 1964: 319 [stem: *Amphistern-*]. Type genus: *Amphisternus* Germar, 1843.

Family COCCINELLIDAE Latreille, 1807

COCCINELLIDAE Latreille, 1807: 70 [stem: *Coccinell-*]. Type genus: *Coccinella* Linnaeus, 1758.

Subfamily MICROWEISEINAE Leng, 1920

MICROWEISEINI Leng, 1920: 213 [stem: *Microweise-*]. Type genus: *Microweisea* Cockerell, 1903.

Tribe MICROWEISEINI Leng, 1920

MICROWEISEINI Leng, 1920: 213 [stem: *Microweise-*]. Type genus: *Microweisea* Cockerell, 1903.

Tribe SERANGIINI Pope, 1962

*SERANGIINI Blackwelder, 1945: 450 [stem: *Serangi-*]. Type genus: *Serangium* Blackburn, 1889. Comment: unavailable family-group name, proposed after 1930 without description or bibliographic reference to such a description (Art. 13.1).

SERANGIINI Pope, 1962: 627 [stem: *Serangi-*]. Type genus: *Serangium* Blackburn, 1889.

Tribe SUKUNAHIKONINI Kamiya, 1960

SUKUNAHIKONINI Kamiya, 1960: 24 [stem: *Sukunahikon-*]. Type genus: *Sukunahikona* Kamiya, 1960 [syn. of *Scymnomorphus* Weise, 1897].

SCOTOSCYMNINAE Duverger, 2003: 63 [stem: *Scotoscymn-*]. Type genus: *Scotoscymnus* Weise, 1901. Comment: name proposed to replace SUKUNAHIKONINI Kamiya, 1960 because of the synonymy of the type genus.

Subfamily COCCINELLINAE Latreille, 1807

COCCINELLIDAE Latreille, 1807: 70 [stem: *Coccinell-*]. Type genus: *Coccinella* Linnaeus, 1758.

Tribe ARGENTIPILOSINI Gordon and de Almeida, 1991

ARGENTIPILOSINI Gordon and de Almeida, 1991: 150 [stem: *Argentipilos-*]. Type genus: *Argentipilosa* Gordon and de Almeida, 1991.

Tribe ASPIDIMERINI Mulsant, 1850

ASPIDIMÉRAIRES Mulsant, 1850: 943 [stem: *Aspidimer-*]. Type genus: *Aspidimerus* Mulsant, 1850. Comment: original vernacular name available (Art. 11.7.2):

first used in latinized form by Jakobson (1915: 969, as ASPIDIMERINA), generally accepted as in Pakaluk et al. (1994: 234, as ASPIDIMERINI).

Tribe AZYINI Mulsant, 1850

AZYAIRES Mulsant, 1850: 927 [stem: *Azy-*]. Type genus: *Azya* Mulsant, 1850. Comment: original vernacular name available (Art. 11.7.2): first used in latinized form by Crotch (1874: 279, as AZYAE), generally accepted as in Pakaluk et al. (1994: 233, as AZYINI).

*BUCOLITES Chapuis, 1876: 237 [stem: *Bucol-*]. Type genus: *Bucolus* Mulsant, 1850. Comment: original vernacular name unavailable (Art. 11.7.2): not subsequently latinized.

Tribe BRACHIACANTHINI Mulsant, 1850

BRACHYACANTHAIRES Mulsant, 1850: 520 [stem: *Brachiacanth-*]. Type genus: *Brachiacantha* Chevrolat, 1836 [as *Brachyacantha*, incorrect subsequent spelling of type genus name, not in prevailing usage]. Comment: original vernacular name available (Art. 11.7.2): first used in latinized form by Duverger (1990: 143, as BRACHIACANTHADINI [incorrect stem formation]), generally accepted as in Pakaluk et al. (1994: 249, as BRACHIACANTHINI); incorrect original stem formation, not in prevailing usage.

Tribe CARINODULINI Gordon, Pakaluk and Ślipiński, 1989

CARINODULINI Gordon et al., 1989: 360 [stem: *Carinodul-*]. Type genus: *Carinodula* Gordon et al., 1989.

Tribe CEPHALOSCYMNINI Gordon, 1985

CEPHALOSCYMNINI Gordon, 1985: 66 [stem: *Cephaloscymn-*]. Type genus: *Cephaloscymnus* Crotch, 1872.

Tribe CHILOCORINI Mulsant, 1846

CHILOCORIENS Mulsant, 1846: 166 [stem: *Chilocor-*]. Type genus: *Chilocorus* Leach, 1815. Comment: original vernacular name available (Art. 11.7.2): first used in latinized form by Weise (1885: 239, as CHILOCORINI), generally accepted as in Pakaluk et al. (1994: 234, as CHILOCORINI).

*EXOCHOMAIRES Mulsant, 1850: 465 [stem: *Exochom-*]. Type genus: *Exochomus* Redtenbacher, 1843. Comment: original vernacular name unavailable (Art. 11.7.2): not subsequently latinized.

Tribe CHNOODINI Mulsant, 1850

CHNOODIENS Mulsant, 1850: 907 [stem: *Chnood-*]. Type genus: *Chnoodes* Chevrolat, 1843. Comment: original vernacular name available (Art. 11.7.2): first used in latinized form and generally accepted as in Sicard (1909: 103, as

CHNOODINI); name incorrectly treated as unavailable by Pakaluk et al. (1994: 248).

*SIOLAIRES Mulsant, 1850: 931 [stem: *Siol-*]. Type genus: *Siola* Mulsant, 1850. Comment: original vernacular name unavailable (Art. 11.7.2): not subsequently latinized.

EXOPLECTRIDES Crotch, 1874: xiii [stem: *Exoplectr-*]. Type genus: *Exoplectra* Chevrolat, 1836.

SUMNIINI Hoang, 1982: 138 [stem: *Sumni-*]. Type genus: *Sumnius* Weise, 1892.

Tribe COCCIDULINI Mulsant, 1846

COCCIDULIENS Mulsant, 1846: 266 [stem: *Coccidul-*]. Type genus: *Coccidula* Kugelann, 1798. Comment: original vernacular name available (Art. 11.7.2): first used in latinized form by Gistel (1848: [10], as COCCIDULAEIDAE [incorrect stem formation]), generally accepted as in Pakaluk et al. (1994: 233, as COCCIDULINI).

*RHIZOBIATES Mulsant, 1846: 261 [stem: *Rhyzobi-*]. Type genus: *Rhyzobius* Stephens, 1829 [as *Rhizobius*, unjustified emendation of type genus name by Agassiz (1846b: 327), not in prevailing usage]. Comment: original vernacular name unavailable (Art. 11.7.2): subsequently used in latinized form but not generally attributed to Mulsant (1846); incorrect original stem formation, not in prevailing usage.

RHIZOBIINA C. G. Thomson, 1866: 328 [stem: *Rhyzobi-*]. Type genus: *Rhyzobius* Stephens, 1829 [as *Rhizobius*, unjustified emendation of type genus name by Agassiz (1846b: 327), not in prevailing usage]. Comment: incorrect original stem formation, not in prevailing usage; the name RHIZOBIINAE Buckton, 1883 (type genus *Rhizobius* Burmeister, 1835) is available in Hemiptera.

Tribe COCCINELLINI Latreille, 1807

COCCINELLIDAE Latreille, 1807: 70 [stem: *Coccinell-*]. Type genus: *Coccinella* Linnaeus, 1758.

*ADONIATES Mulsant, 1846: 35 [stem: *Adoni-*]. Type genus: *Adonia* Mulsant, 1846 [syn. of *Hippodamia* Chevrolat, 1836]. Comment: original vernacular name unavailable (Art. 11.7.2): not subsequently latinized.

*HIPPODAMIAIRES Mulsant, 1846: 30 [stem: *Hippodami-*]. Type genus: *Hippodamia* Chevrolat, 1836. Comment: original vernacular name unavailable (Art. 11.7.2): not subsequently latinized.

*MICRASPIAIRES Mulsant, 1846: 162 [stem: *Micraspid-*]. Type genus: *Micraspis* Chevrolat, 1836. Comment: original vernacular name unavailable (Art. 11.7.2): not subsequently latinized; incorrect original stem formation, not in prevailing usage.

*MYSIATES Mulsant, 1846: 125 [stem: *Myzi-*]. Type genus: *Myzia* Mulsant, 1846 [as *Mysia*, alternative original spelling of type genus name; we follow Kovář (2007: 620) in using *Myzia* as the correct spelling for this genus]. Comment:

original vernacular name unavailable (Art. 11.7.2): not subsequently latinized; incorrect original stem formation, not in prevailing usage.

HALYZIAIRES Mulsant, 1846: 123 [stem: *Halyzi-*]. Type genus: *Halyzia* Mulsant, 1846. Comment: original vernacular name available (Art. 11.7.2): first used in latinized form and generally accepted as in Pakaluk et al. (1994: 234, as HALYZIINI).

ALESIAIRES Mulsant, 1850: 343 [stem: *Alesi-*]. Type genus: *Alesia* Mulsant, 1850 [syn. of *Micraspis* Chevrolat, 1836]. Comment: original vernacular name available (Art. 11.7.2): first used in latinized form by Mader (1934: 297, as ALESIINI), generally accepted as in Mader (1954: 93, as ALESIINA).

*CARIAIRES Mulsant, 1850: 228 [stem: *Cari-*]. Type genus: *Caria* Mulsant, 1850 [preoccupied genus name, not *Caria* Hübner, 1823[Lepidoptera]; syn. of *Megalocaria* Crotch, 1874]. Comment: original vernacular name unavailable (Art. 11.7.2): subsequently used in latinized form, e.g., Casey (1899: 72, as CARIINI), but not generally accepted as valid; if name found to be available then permanently invalid (Art. 39): based on preoccupied type genus.

COELOPHORAIRES Mulsant, 1850: 374 [stem: *Coelophor-*]. Type genus: *Coelophora* Mulsant, 1850. Comment: original vernacular name available (Art. 11.7.2): first used in latinized form and generally accepted as in Mader (1954: 93, as COELOPHORINA).

CYDONIAIRES Mulsant, 1850: 429 [stem: *Cydoni-*]. Type genus: *Cydonia* Mulsant, 1850 [syn. of *Cheilomenes* Chevrolat, 1836]. Comment: original vernacular name available (Art. 11.7.2): first used in latinized form and generally accepted as in Mader (1954: 93, as CYDONIINA).

DISCOTOMAIRES Mulsant, 1850: 214 [stem: *Discotom-*]. Type genus: *Discotoma* Mulsant, 1850. Comment: original vernacular name available (Art. 11.7.2): first used in latinized form by Korschevsky (1932: 577, as DISCOTOMINI), generally accepted as in Pakaluk et al. (1994: 234, as DISCOTOMINI).

TYTTHASPIDES Crotch, 1874: xiii [stem: *Tytthaspid-*]. Type genus: *Tytthaspis* Crotch, 1874.

SYNONYCHINI Weise, 1885: 7 [stem: *Synonych-*]. Type genus: *Synonycha* Chevrolat, 1836.

PSYLLOBORINI Casey, 1899: 73 [stem: *Psyllobor-*]. Type genus: *Psyllobora* Chevrolat, 1836.

ANISOSTICTINA Jakobson, 1915: 969 [stem: *Anisostict-*]. Type genus: *Anisosticta* Chevrolat, 1836.

ANISOLEMNIINA Mader, 1954: 93, in key [stem: *Anisolemni-*]. Type genus: *Anisolemnia* Crotch, 1874.

BULAEINI Savoyskaya, 1969: 102 [stem: *Bulae-*]. Type genus: *Bulaea* Mulsant, 1850.

SINGHIKALIINI Miyatake, 1972: 96 [stem: *Singhikali-*]. Type genus: *Singhikalia* Kapur, 1963.

Tribe Cranophorini Mulsant, 1850

Cranophoraires Mulsant, 1850: 939 [stem: *Cranophor-*]. Type genus: *Cranophorus* Mulsant, 1850. Comment: original vernacular name available (Art. 11.7.2): first used in latinized form by Casey (1899: 74, as Cranophorini), generally accepted as in Pakaluk et al. (1994: 233, as Cranophorini).

Tribe Cryptognathini Mulsant, 1850

Cryptognathaires Mulsant, 1850: 496 [stem: *Cryptognath-*]. Type genus: *Cryptognatha* Mulsant, 1850. Comment: original vernacular name available (Art. 11.7.2): first used in latinized form by Gordon (1971: 181, as Cryptognathini), generally accepted as in Pakaluk et al. (1994: 249, as Cryptognathini).

*Pentiliaires Mulsant, 1850: 501 [stem: *Pentili-*]. Type genus: *Pentilia* Mulsant, 1850. Comment: original vernacular name unavailable (Art. 11.7.2): subsequently used in latinized form but not generally attributed to Mulsant (1846).

Pentiliini Casey, 1899: 74 [stem: *Pentili-*]. Type genus: *Pentilia* Mulsant, 1850.

Oeneini Casey, 1899: 74 [stem: *Oene-*]. Type genus: *Oeneis* Mulsant, 1850 [preoccupied genus name, not *Oeneis* Hübner, 1819 [Lepidoptera]; syn. of *Delphastus* Casey, 1899]. Comment: permanently invalid (Art. 39): based on preoccupied type genus; the taxon Oeneini Wheeler, 1903 (type genus *Oeneis* Hübner, 1819) is available in Lepidoptera.

Tribe Cynegetini Thomson, 1866

Cynegetides C. G. Thomson, 1866: 374 [stem: *Cyneget-*]. Type genus: *Cynegetis* Chevrolat, 1836. Comment: current spelling maintained (Art. 29.3.1.1): incorrect stem formation in prevailing usage (should be *Cynegetid-*).

Madaini Gordon, 1975: 206 [stem: *Mad-*]. Type genus: *Mada* Mulsant, 1850. Comment: incorrect original stem formation, not in prevailing usage.

Tribe Diomini Gordon, 1999

Diomini Gordon, 1999: 4 [stem: *Diom-*]. Type genus: *Diomus* Mulsant, 1850.

Tribe Epilachnini Mulsant, 1846

Epilachniens Mulsant, 1846: 190 [stem: *Epilachn-*]. Type genus: *Epilachna* Chevrolat, 1836. Comment: original vernacular name available (Art. 11.7.2): first used in latinized form by Ganglbauer (1898: 947, as Epilachninae), generally accepted as in Pakaluk et al. (1994: 234, as Epilachnini).

*Chnootribaires Mulsant, 1850: 697 [stem: *Chnootrib-*]. Type genus: *Chnootriba* Chevrolat, 1836. Comment: original vernacular name unavailable (Art. 11.7.2): not subsequently latinized.

Subcoccinellini Jakobson, 1915: 968 [stem: *Subcoccinell-*]. Type genus: *Subcoccinella* Huber, 1841.

Tribe Epivertini Pang and Mao, 1979

Epivertini Pang and Mao, 1979: 158 [stem: *Epivert-*]. Type genus: *Epiverta* Dieke, 1947.

Tribe Eremochilini Gordon and Vanderberg, 1987

Eremochilini Gordon and Vanderberg, 1987: 6 [stem: *Eremochil-*]. Type genus: *Eremochilus* Weise, 1912.

Tribe Hyperaspidini Mulsant, 1846

Hypéraspiens Mulsant, 1846: 177 [stem: *Hyperaspid-*]. Type genus: *Hyperaspis* Chevrolat, 1836. Comment: original vernacular name available (Art. 11.7.2): first used in latinized form by Weise (1885: 5, as Hyperaspini), generally accepted as in Pakaluk et al. (1994: 233, as Hyperaspini); incorrect original stem formation, not in prevailing usage.

*Thalassaires Mulsant, 1850: 505 [stem: *Thalass-*]. Type genus: *Thalassa* Mulsant, 1850. Comment: original vernacular name unavailable (Art. 11.7.2): not subsequently latinized.

*Tiphysaires Mulsant, 1850: 516 [stem: *Tiphys-*]. Type genus: *Tiphysa* Mulsant, 1850. Comment: original vernacular name unavailable (Art. 11.7.2): not subsequently latinized.

Tribe Limnichopharini Miyatake, 1994

Limnichopharini Miyatake, 1994: 269 [stem: *Limnichophar-*]. Type genus: *Limnichopharus* Miyatake, 1994.

Tribe Monocorynini Miyatake, 1988

Monocorynini Miyatake, 1988: 28 [stem: *Monocoryn-*]. Type genus: *Monocoryna* Gorham, 1885.

Tribe Noviini Mulsant, 1846

Noviaires Mulsant, 1846: 213 [stem: *Novi-*]. Type genus: *Novius* Mulsant, 1846 [the original spelling used by Mulsant was *Nomius* (p. 213), however, the spelling was changed to *Novius* by Mulsant in the "Addenda et errata" of the same work and is therefore considered a justified emendation (Art. 19.2)]. Comment: the original spelling of this family-group name was Nomiaires (based on the new genus *Nomius* Mulsant, 1846), the spelling of the family-group and genus-group names were changed to Noviaires and *Novius* in the "Addenda et errata" of the same work, Noviaires is therefore treated as the correct original spelling of this family-group name (Art. 19.2); original vernacular name available (Art. 11.7.2): first used in latinized form by Ganglbauer (1899: 977, as Noviini), generally accepted as in Pakaluk et al. (1994: 233, as Noviini).

*Rodoliaires Mulsant, 1850: 901 [stem: *Rodoli-*]. Type genus: *Rodolia* Mulsant, 1850. Comment: original vernacular name unavailable (Art. 11.7.2): not subsequently latinized.

Tribe Ortaliini Mulsant, 1850

Ortaliens Mulsant, 1850: 892 [stem: *Ortali-*]. Type genus: *Ortalia* Mulsant, 1850. Comment: original vernacular name available (Art. 11.7.2): first used in latinized form by Crotch (1874: 274, as Ortaliae), generally accepted as in Pakaluk et al. (1994: 234, as Ortaliini).

Tribe Oryssomini Gordon, 1974

Oryssomini Gordon, 1974: 146 [stem: *Oryssom-*]. Type genus: *Oryssomus* Mulsant, 1850.

Tribe Platynaspini Mulsant, 1846

Platynaspiaires Mulsant, 1846: 215 [stem: *Platynasp-*]. Type genus: *Platynaspis* Redtenbacher, 1843. Comment: original vernacular name available (Art. 11.7.2): first used in latinized form by Casey (1899: 109, as Platynaspini), generally accepted as in Pakaluk et al. (1994: 234, as Platynaspini); current spelling maintained (Art. 29.3.1.1): incorrect stem formation in prevailing usage (should be *Platynaspid-*).

Tribe Plotinini Miyatake, 1994

Plotinini Miyatake, 1994: 279 [stem: *Plotin-*]. Type genus: *Plotina* Lewis, 1896.

Tribe Poriini Mulsant, 1850

Poriens Mulsant, 1850: 884 [stem: *Pori-*]. Type genus: *Poria* Mulsant, 1850. Comment: original vernacular name available (Art. 11.7.2): first used in latinized form and generally accepted as in Gorham (1894: 206, as Poriides [treated as Latin]).

Tribe Scymnillini Casey, 1899

Scymnillini Casey, 1899: 74 [stem: *Scymnill-*]. Type genus: *Scymnillus* G. H. Horn, 1895.

Zilini Gordon, 1985: 74 [stem: *Zil-*]. Type genus: *Zilus* Mulsant, 1850. Comment: name proposed to replace Scymnillini Casey, 1899 because of the synonymy of the type genus at that time.

Tribe Scymnini Mulsant, 1846

Scymniaires Mulsant, 1846: 189 [stem: *Scymn-*]. Type genus: *Scymnus* Kugelann, 1794. Comment: original vernacular name available (Art. 11.7.2): first used in latinized form by Weise (1885: 5, as Scymnini), generally accepted as in

Pakaluk et al. (1994: 233, as SCYMNINI); incorrect original stem formation, not in prevailing usage.

*CRYPTOLAEMAIRES Mulsant, 1853: 267 [stem: *Cryptolaem-*]. Type genus: *Cryptolaemus* Mulsant, 1853. Comment: original vernacular name unavailable (Art. 11.7.2): not subsequently latinized.

Tribe SELVADIINI Gordon, 1985

SELVADIINI Gordon, 1985: 347 [stem: *Selvadi-*]. Type genus: *Selvadius* Casey, 1899.

Tribe SHIROZUELLINI Sasaji, 1967

SHIROZUELLINI Sasaji, 1967: 23 [stem: *Shirozuell-*]. Type genus: *Shirozuella* Sasaji, 1967.

GHANIINI Ahmad, 1973: 449 [stem: *Ghani-*]. Type genus: *Ghanius* Ahmad, 1973.

Tribe STETHORINI Dobzhansky, 1924

STETHORINI Dobzhansky, 1924: 20 [stem: *Stethor-*]. Type genus: *Stethorus* Weise, 1885.

Tribe STICHOLOTIDINI Weise, 1901

PHARINI Casey, 1899: 74 [stem: *Phar-*]. Type genus: *Pharus* Mulsant, 1850 [preoccupied genus name, not *Pharus* Gray, 1840 [Mollusca]; syn. of *Pharoscymnus* Bedel, 1906]. Comment: permanently invalid (Art. 39): based on preoccupied type genus; this family-group name was also proposed in the same year by Ganglbauer (1899: 954, as PHARINI), priority could not be established between those two works; the name PHARINAE Adams and Adams, 1856 in Mollusca (type genus *Pharus* Gray, 1840) is available.

STICHOLOTINI Weise, 1901: 430 [stem: *Sticholotid-*]. Type genus: *Sticholotis* Crotch, 1874. Comment: incorrect original stem formation, not in prevailing usage; correction of stem by Gordon (1977: 186).

CLANINI Weise, 1901: 430 [stem: *Clan-*]. Type genus: *Clanis* Mulsant, 1850 [preoccupied genus name, not *Clanis* Hübner, 1819 [Lepidoptera]; syn. of *Jauravia* Motschulsky, 1858]. Comment: permanently invalid (Art. 39): based on preoccupied type genus.

COELOPTERINI Weise, 1906: column 369 [stem: *Coelopter-*]. Type genus: *Coelopterus* Mulsant and Rey, 1852. Comment: name previously attributed to Della Beffa (1912: 171, as COLEOPTERINI [incorrect stem formation]) in the literature.

Tribe TELSIMIINI Casey, 1899

TELSIMIINI Casey, 1899: 74 [stem: *Telsimi-*]. Type genus: *Telsimia* Casey, 1899.

Tribe TETRABRACHINI Kapur, 1948

LITHOPHILIDAE Imhoff, 1856: [2] 151 [stem: *Lithophil-*]. Type genus: *Lithophilus* Fröhlich, 1799 [preoccupied genus name, not *Lithophilus* Schneider, 1791

[Coleoptera: Carabidae]; syn. of *Tetrabrachys* Kapur, 1948]. Comment: permanently invalid (Art. 39): based on preoccupied type genus.

Tetrabrachinae Kapur, 1948: 320 [stem: *Tetrabrach-*]. Type genus: *Tetrabrachys* Kapur, 1948. Comment: replacement name for Lithophilinae Imhoff, 1856 because of the homonymy of the type genus; current spelling maintained (Art. 29.5): incorrect stem formation in prevailing usage (should be *Tetrabrache-*).

Family Corylophidae LeConte, 1852

Corylophi J. L. LeConte, 1852b: 141 [stem: *Coryloph-*]. Type genus: *Corylophus* Stephens, 1833.

Subfamily Periptyctinae Ślipiński, Lawrence and Tomaszewska, 2001

Periptyctinae Ślipiński et al., 2001: 312 [stem: *Periptyct-*]. Type genus: *Periptyctus* Blackburn, 1825.

Subfamily Corylophinae LeConte, 1852

Corylophi J. L. LeConte, 1852b: 141 [stem: *Coryloph-*]. Type genus: *Corylophus* Stephens, 1833.

Tribe Aenigmaticini Casey, 1900

Aenigmaticini Casey, 1900: 61 [stem: *Aenigmatic-*]. Type genus: *Aenigmaticum* A. Matthews, 1888.

Tribe Cleidostethini Bowestead, Booth, Ślipiński and Lawrence, 2001

Cleidostethini Bowestead et al., 2001: 323 [stem: *Cleidosteth-*]. Type genus: *Cleidostethus* Arrow, 1929.

Tribe Corylophini LeConte, 1852

Corylophi J. L. LeConte, 1852b: 141 [stem: *Coryloph-*]. Type genus: *Corylophus* Stephens, 1833.

Tribe Foadiini Ślipiński, Tomaszewska and Lawrence, 2009

Foadiini Ślipiński et al., 2009: 422 [stem: *Foadi-*]. Type genus: *Foadia* Pakaluk, 1985.

Tribe Orthoperini Jacquelin du Val, 1857

Orthopérites Jacquelin du Val, 1857b: 100 [stem: *Orthoper-*]. Type genus: *Orthoperus* Stephens, 1829. Comment: original vernacular name available (Art. 11.7.2): first used in latinized form by C. G. Thomson (1859: 63, as Orthoperidae), generally accepted as in Ślipiński et al. (2009: 428, as Orthoperini).

Tribe Parmulini Poey, 1854

Clypeastres L. Redtenbacher, 1845: 122 [stem: *Clypeaster-*]. Type genus: *Clypeaster* Dejean, 1821 [preoccupied genus name, not *Clypeaster* Lamarck, 1801

[Echinodermata]; syn. of *Clypastraea* Haldeman, 1842]. Comment: permanently invalid (Art. 39): based on preoccupied type genus; incorrect original stem formation, not in prevailing usage; Clypeasteridae L. Agassiz, 1835 (type genus *Clypeaster* Lamarck, 1801) is available in Echinodermata.

Parmulini Poey, 1854: 323 [stem: *Parmul-*]. Type genus: *Parmulus* Gundlach, 1854 [syn. of *Clypastraea* Haldeman, 1842].

Saciina A. Matthews, 1888: 103 [stem: *Saci-*]. Type genus: *Sacium* J. L. LeConte, 1852 [syn. of *Clypastraea* Haldeman, 1842].

Arthrolipinae Böving and Craighead, 1931: 36, in key [stem: *Arthrolip-*]. Type genus: *Arthrolips* Wollaston, 1854.

Tribe Peltinodini Paulian, 1950

Peltinoditae Paulian, 1950: 19 [stem: *Peltinod-*]. Type genus: *Peltinodes* Paulian, 1950 [syn. of *Holopsis* Broun, 1883]. Comment: precedence (Peltinodini Paulian, 1950 vs Corylophodini Paulian, 1950) given to taxon originally proposed at the higher rank (Art. 24.1).

Corylophodini Paulian, 1950: 21, in key [stem: *Corylophod-*]. Type genus: *Corylophodes* A. Matthews, 1885 [syn. of *Holopsis* Broun, 1883].

Tribe Rypobiini Paulian, 1950

Rhypobiini Paulian, 1950: 48 [stem: *Rypobi-*]. Type genus: *Rypobius* J. L. LeConte, 1852 [as *Rhypobius*, unjustified emendation of type genus name by Gemminger and Harold (1876: 3818), not in prevailing usage]. Comment: incorrect original stem formation, not in prevailing usage.

Gloeosomatini Bowestead, 1999: 128 [stem: *Gloeosomat-*]. Type genus: *Gloeosoma* Wollaston, 1854.

Tribe Sericoderini Matthews, 1888

Sericoderina A. Matthews, 1888: 103 [stem: *Sericoder-*]. Type genus: *Sericoderus* Stephens, 1829.

Tribe Teplinini Pakaluk, Ślipiński and Lawrence, 1994

Peltinini Paulian, 1950: 21, in key [stem: *Peltin-*]. Type genus: *Peltinus* Mulsant, 1861 [preoccupied genus name, not *Peltinus* Rafinesque, 1815 [Coleoptera: Trogossitidae]; syn. of *Teplinus* Pakaluk et al., 1994]. Comment: permanently invalid (Art. 39): based on preoccupied type genus.

Teplinini Pakaluk et al., 1994: 251 [stem: *Teplin-*]. Type genus: *Teplinus* Pakaluk et al., 1994. Comment: replacement name for Peltinini Paulian, 1950 because of the homonymy of the type genus.

Family Akalyptoischiidae Lord, Hartley, Lawrence, McHugh and Miller, 2010

Akalyptoischiidae Lord et al., 2010: 761 [stem: *Akalyptoischi-*]. Type genus: *Akalyptoischion* Andrews, 1976.

Family Latrididae Erichson, 1842

Lathridien Erichson, 1842: 122 [stem: *Latridi-*]. Type genus: *Latridius* Herbst, 1793 [as *Lathridius*, unjustified emendation of type genus name by Illiger (1802), not in prevailing usage]. Comment: the younger name Latridiidae has been used widely for this family in recent literature although Pakaluk et al. (1994: 252) and a small number of subsequent authors, e.g., Míka (2000), Lassau et al. (2005), have used the older name Corticariidae; an application was recently submitted to the Commission in order to conserve usage of the well-established name Latridiidae (Bousquet et al. 2010; see Appendix 6).

Subfamily Latridiinae Erichson, 1842

Lathridien Erichson, 1842: 122 [stem: *Latridi-*]. Type genus: *Latridius* Herbst, 1793 [as *Lathridius*, unjustified emendation of type genus name by Illiger (1802), not in prevailing usage]. Comment: original vernacular name available (Art. 11.7.2): first used in latinized form by L. Redtenbacher (1845: 123, as Lathridii), generally accepted as in Lawrence and Newton (1995: 886, as Latridiidae); incorrect original stem formation, not in prevailing usage.

Subfamily Corticariinae Curtis, 1829

Corticaridae Curtis, 1829: pl. 283 [stem: *Corticari-*]. Type genus: *Corticaria* Marsham, 1802. Comment: incorrect original stem formation, not in prevailing usage; although this is the oldest name for the family, an application was recently submitted by Bousquet et al. (2010) to conserve usage of the well-established name Latridiidae (see Appendix 6).

*Melanophthalmidae Arnett, 1962b: 835 [stem: *Melanophthalm-*]. Type genus: *Melanophthalma* Motschulsky, 1866. Comment: family-group name unavailable (Art. 11.6): originally published as synonym and not made available subsequently; name listed by Arnett as a synonym of Latridiidae and attributed to "auct.", we could find an earlier usage of this name.

†Subfamily Tetrameropseinae Kirejtshuk and Azar, 2008

Tetrameropsinae Kirejtshuk and Azar, 2008: 36 [stem: *Tetrameropse-*]. Type genus: *Tetrameropsis* Kirejtshuk and Azar, 2008. Comment: incorrect original stem formation, not in prevailing usage.

Superfamily Tenebrionoidea Latreille, 1802

Tenebrionites Latreille, 1802: 165 [stem: *Tenebrion-*]. Type genus: *Tenebrio* Linnaeus, 1758. Comment: First Revisers found (Tenebrionoidea Latreille, 1802 vs Mordelloidea Latreille, 1802) are Alonso-Zarazaga and Mansilla-Castrillo (1988: 19).

Family Mycetophagidae Leach, 1815

Mycetophagida Leach, 1815: 110 [stem: *Mycetophag-*]. Type genus: *Mycetophagus* Fabricius, 1792.

Subfamily Esarcinae Reitter, 1882

Esarcini Reitter, 1882a: 115 [stem: *Esarc-*]. Type genus: *Esarcus* Reiche, 1864.

Subfamily Mycetophaginae Leach, 1815

Mycetophagida Leach, 1815: 110 [stem: *Mycetophag-*]. Type genus: *Mycetophagus* Fabricius, 1792.

Tribe Mycetophagini Leach, 1815

Mycetophagida Leach, 1815: 110 [stem: *Mycetophag-*]. Type genus: *Mycetophagus* Fabricius, 1792.

Tritomidae Crotch, 1873a: 78 [stem: *Tritom-*]. Type genus: *Tritoma* Geoffroy, 1762 [senior homonym of *Tritoma* Fabricius, 1775 [Coleoptera: Erotylidae] but placed on the Official Index of Rejected and Invalid Generic Names in Zoology (ICZN 1994a); syn. of *Mycetophagus* Fabricius, 1792]. Comment: permanently invalid (Art. 39): based on suppressed type genus; Tritomidae was also used the same year by Crotch (1873b: 42).

Triphyllini Harold, 1880: 757 [stem: *Triphyll-*]. Type genus: *Triphyllus* Dejean, 1821.

Tribe Typhaeini Thomson, 1863

Typhaeina C. G. Thomson, 1863: 241 [stem: *Typhae-*]. Type genus: *Typhaea* Stephens, 1829.

Typhaeini Nikitsky, 1993: 165 [stem: *Typhae-*]. Type genus: *Typhaea* Stephens, 1829. Comment: family-group name proposed as new without reference to Typhaeina C. G. Thomson, 1863.

Subfamily Bergininae Leng, 1920

Bergini Leng, 1920: 246 [stem: *Bergin-*]. Type genus: *Berginus* Erichson, 1846. Comment: incorrect original stem formation, not in prevailing usage.

Family Archeocrypticidae Kaszab, 1964

Archeocrypticini Kaszab, 1964: 361 [stem: *Archeocryptic-*]. Type genus: *Archeocrypticus* Kaszab, 1964.

Family Pterogeniidae Crowson, 1953

Pterogeniidae Crowson, 1953: 38 [stem: *Pterogeni-*]. Type genus: *Pterogenius* Candèze, 1861.

Family Ciidae Leach, 1819

Cisidae Leach, 1819: 206 [stem: *Ci-*]. Type genus: *Cis* Latreille, 1797. Comment: incorrect original stem formation, not in prevailing usage; Gistel (1848: [6], 1856a: 368) used the name Microtrocteidae for a family that included only the genus *Cis* Latreille, we could not find any genus name on which Microtrocteidae could be based on therefore this family-group name is unavailable.

Subfamily Sphindociinae Lawrence, 1974

Sphindociinae Lawrence, 1974b: 9 [stem: *Sphindoci-*]. Type genus: *Sphindocis* Fall, 1917. Comment: published 28 June 1974; this family-group name was also used in the same year by Lawrence (1974a [22 January]: 24, as Sphindociinae) but the older name is unavailable because it was proposed after 1930 without description or bibliographic reference to such a description (Art. 13.1).

Subfamily Ciinae Leach, 1819

Cisidae Leach, 1819: 206 [stem: *Ci-*]. Type genus: *Cis* Latreille, 1797.

Tribe Ciini Leach, 1819

Cisidae Leach, 1819: 206 [stem: *Ci-*]. Type genus: *Cis* Latreille, 1797. Comment: incorrect original stem formation, not in prevailing usage.

Tribe Orophiini Thomson, 1863

Orophiina C. G. Thomson, 1863: 195 [stem: *Orophi-*]. Type genus: *Orophius* L. Redtenbacher, 1848 [syn. of *Octotemnus* Mellié, 1847].

Octotemnidae Reitter, 1878: 21 [stem: *Octotemn-*]. Type genus: *Octotemnus* Mellié, 1847.

Rhopalodontini Everts, 1898: 517, in key [stem: *Ropalodont-*]. Type genus: *Ropalodontus* Mellié, 1847 [as *Rhopalodontus*, incorrect subsequent spelling of type genus name, not in prevailing usage]. Comment: the fossil Reptilia name Rhopalodontidae Seeley, 1894 (type genus *Rhopalodon* Fischer von Waldheim, 1841) is available, therefore we recomment using the spelling *Ropalodont-* for the ciid name in order to avoid homonymy problems; incorrect original stem formation, not in prevailing usage.

Tribe Xylographellini Kawanabe and Miyatake, 1996

Xylographellini Kawanabe and Miyatake, 1996: 125 [stem: *Xylographell-*]. Type genus: *Xylographella* Miyatake, 1985.

Subtribe Syncosmetina Lopes-Andrade, 2008

Syncosmetina Lopes-Andrade, 2008: 40 [stem: *Syncosmet-*]. Type genus: *Syncosmetus* Sharp, 1891.

Subtribe Xylographellina Kawanabe and Miyatake, 1996

Xylographellini Kawanabe and Miyatake, 1996: 125 [stem: *Xylographell-*]. Type genus: *Xylographella* Miyatake, 1985.

Family Tetratomidae Billberg, 1820

Tetratomaedes Billberg, 1820a: 34 [stem: *Tetratom-*]. Type genus: *Tetratoma* Fabricius, 1790.

Subfamily Tetratominae Billberg, 1820

Tetratomaedes Billberg, 1820a: 34 [stem: *Tetratom-*]. Type genus: *Tetratoma* Fabricius, 1790.

Subfamily Piseninae Miyatake, 1960

Pisenini Miyatake, 1960: 124 [stem: *Pisen-*]. Type genus: *Pisenus* Casey, 1900.

Subfamily Penthinae Lacordaire, 1859

Penthides Lacordaire, 1859: 456 [stem: *Penth-*]. Type genus: *Penthe* Newman, 1838. Comment: original vernacular name available (Art. 11.7.2): first used in latinized form by G. H. Horn (1888: 43, as Penthini), generally accepted as in Young and Pollock (2002: 415, as Penthinae).

Subfamily Hallomeninae Gistel, 1848

Hallomenidae Gistel, 1848: [11] [stem: *Hallomen-*]. Type genus: *Hallomenus* Panzer, 1793.

*Dryalates Mulsant, 1856b: 44 [stem: *Dryal-*]. Type genus: *Dryala* Mulsant, 1856 [syn. of *Hallomenus* Panzer, 1793]. Comment: original vernacular name unavailable (Art. 11.7.2): not subsequently latinized.

Mycétomiens Mulsant, 1856b: 103 [stem: *Mycetomat-*]. Type genus: *Mycetoma* Dejean, 1834. Comment: original vernacular name available (Art. 11.7.2): first used in latinized form and generally accepted as in Desbrochers des Loges (1900: 16, as Mycetomini); incorrect original stem formation, not in prevailing usage.

Subfamily Eustrophinae Gistel, 1848

Eustrophidae Gistel, 1848: [10] [stem: *Eustroph-*]. Type genus: *Eustrophus* Illiger, 1802.

Tribe Eustrophini Gistel, 1848

Eustrophidae Gistel, 1848: [10] [stem: *Eustroph-*]. Type genus: *Eustrophus* Illiger, 1802.

Tribe Holostrophini Nikitsky, 1998

Holostrophini Nikitsky, 1998: 39 [stem: *Holostroph-*]. Type genus: *Holostrophus* G. H. Horn, 1888.

Family Melandryidae Leach, 1815

Melyandrida Leach, 1815: 104 [stem: *Melandry-*]. Type genus: *Melandrya* Fabricius, 1801.

Subfamily Melandryinae Leach, 1815

Melyandrida Leach, 1815: 104 [stem: *Melandry-*]. Type genus: *Melandrya* Fabricius, 1801.

Tribe Anisoxiellini Nikitsky, 2007

Anisoxiellini Nikitsky, 2007: 58 [stem: *Anisoxiell-*]. Type genus: *Anisoxiella* Nikitsky, 1989.

Tribe Dircaeini Kirby, 1837

Dircaeidae Kirby, 1837: 240 [stem: *Dircae-*]. Type genus: *Dircaea* Fabricius, 1798.

Tribe Hypulini Gistel, 1848

Hypulidae Gistel, 1848: [11] [stem: *Hypul-*]. Type genus: *Hypulus* Paykull, 1798.

Maroliini Portevin, 1934: 57, in key [stem: *Maroli-*]. Type genus: *Marolia* Mulsant, 1856.

Tribe Melandryini Leach, 1815

Melyandrida Leach, 1815: 104 [stem: *Melandry-*]. Type genus: *Melandrya* Fabricius, 1801. Comment: incorrect original stem formation, not in prevailing usage.

*Phryganophiles Motschulsky, 1849: 58 [stem: *Phryganophil-*]. Type genus: *Phryganophilus* Sahlberg, 1833. Comment: original vernacular name unavailable (Art. 11.7.2): not subsequently latinized.

Hylepnigalionidae Gistel, 1856a: 384 [stem: *Hylepnigalion-*]. Type genus: *Hylepnigalio* Gistel, 1856 [this genus originally included the species *caraboides* Linnaeus and *caniculatus* (without author name); the type species of *Hylepnigalio* Gistel, 1856 is here considered to be *Chrysomela caraboides* Linnaeus, 1760 by monotypy; **syn. nov.** of *Melandrya* Fabricius, 1801]. Comment: **syn. nov.**

Tribe Orchesiini Mulsant, 1856

Orchésiens Mulsant, 1856b: 27 [stem: *Orchesi-*]. Type genus: *Orchesia* Latreille, 1807. Comment: original vernacular name available (Art. 11.7.2): first used in latinized form by C. G. Thomson (1859: 119, as Orchesiina), generally accepted as in Nikitsky and Pollock (2008: 68, as Orchesiini).

Tribe Serropalpini Latreille, 1829

Serropalpides Latreille, 1829b: 43 [stem: *Serropalp-*]. Type genus: *Serropalpus* Hellenius, 1786.

Tribe Xylitini Thomson, 1864

Xylitina C. G. Thomson, 1864: 316 [stem: *Xylit-*]. Type genus: *Xylita* Paykull, 1798.

Tribe Zilorini Desbrochers des Loges, 1900

Zilorini Desbrochers des Loges, 1900: 2, in key [stem: *Zilor-*]. Type genus: *Zilora* Mulsant, 1856. Comment: spelled Ziloriini in description on page 2 (in key), but correctly spelled Zilorini on p. 17 of the same work.

Zilorini Nikitsky, 2007: 60 [stem: *Zilor-*]. Type genus: *Zilora* Mulsant, 1856. Comment: family-group name proposed as new without reference to Zilorini Desbrochers des Loges, 1900.

Subfamily Osphyinae Mulsant, 1856 (1839)

Nothidae Shuckard, 1839b: 51 [stem: *Noth-*]. Type genus: *Nothus* A. G. Olivier, 1811 [syn. of *Osphya* Illiger, 1807]. Comment: Osphyinae Mulsant, 1856 conserved over this name (Art. 40.2) (see Lawrence and Newton 1995: 888).

*Osphies Motschulsky, 1849: 59 [stem: *Osphy-*]. Type genus: *Osphya* Illiger, 1807. Comment: original vernacular name unavailable (Art. 11.7.2): subsequently used in latinized form but not generally attributed to Motschulsky (1849); incorrect original stem formation, not in prevailing usage.

Osphyens Mulsant, 1856b: 108 [stem: *Osphy-*]. Type genus: *Osphya* Illiger, 1807. Comment: original vernacular name available (Art. 11.7.2): first used in latinized form by Seidlitz (1875 [Gatt.]: 103, as Osphyini), generally accepted as in Pollock (2002: 421, as Osphyinae); name conserved over the older Nothinae Shuckard, 1839 (Art. 40.2) (see Lawrence and Newton 1995: 888).

Conopalpiens Mulsant, 1856b: 105 [stem: *Conopalp-*]. Type genus: *Conopalpus* Gyllenhal, 1810. Comment: original vernacular name available (Art. 11.7.2): first used in latinized form by C. G. Thomson (1864: 313, as Conopalpina), generally accepted as in Hansen (1996: 171, as Conopalpini).

Family Mordellidae Latreille, 1802

Mordellonae Latreille, 1802: 183 [stem: *Mordell-*]. Type genus: *Mordella* Linnaeus, 1758.

†Praemordellinae Ščegoleva-Barovskaya, 1929: 27 [stem: *Praemordell-*]. Type genus: *Praemordella* Ščegoleva-Barovskaya, 1929.

†Liaoximordellidae Wang, 1993: 87 [stem: *Liaoximordell-*]. Type genus: *Liaoximordella* Wang, 1993.

Subfamily Ctenidiinae Franciscolo, 1951

Ctenidiinae Franciscolo, 1951: 56 [stem: *Ctenidi-*]. Type genus: *Ctenidia* Laporte, 1840.

Subfamily Mordellinae Latreille, 1802

Mordellonae Latreille, 1802: 183 [stem: *Mordell-*]. Type genus: *Mordella* Linnaeus, 1758.

Tribe Conaliini Ermisch, 1956

Conaliini Ermisch, 1956: 273 [stem: *Conali-*]. Type genus: *Conalia* Mulsant and Rey, 1858.

Tribe Mordellini Latreille, 1802

Mordellonae Latreille, 1802: 183 [stem: *Mordell-*]. Type genus: *Mordella* Linnaeus, 1758.

Tribe Mordellistenini Ermisch, 1941

Mordellistenini Ermisch, 1941: 714 [stem: *Mordellisten-*]. Type genus: *Mordellistena* A. Costa, 1854.

Tribe Reynoldsiellini Franciscolo, 1957

Reynoldsiellini Franciscolo, 1957: 237 [stem: *Reynoldsiell-*]. Type genus: *Reynoldsiella* Ray, 1930.

Tribe Stenaliini Franciscolo, 1955

Stenaliini Franciscolo, 1955: 177 [stem: *Stenali-*]. Type genus: *Stenalia* Mulsant, 1856.

Family Ripiphoridae Gemminger, 1870 (1855)

Rhipidophoridae Gemminger, 1870: 2117 [stem: *Ripiphor-*]. Type genus: *Ripiphorus* Bosc, 1791 [as *Rhipidophorus*, unjustified emendation of *Ripiphorus* Bosc, 1791 by Perty (1831: xix), not in prevailing usage]. Comment: name conserved over the older Myoditidae Gerstaecker, 1855 (Art. 40.2) (see Lawrence and Newton 1995: 889, as "Myoditini Costa, 1853"); First Reviser (Ripiphoridae Gemminger, 1870 (1855) vs Ptilophoridae Gerstaecker, 1855 vs Ripidiidae Gerstaecker, 1855) not determined, prevailing usage maintained.

Subfamily Ptilophorinae Gerstaecker, 1855

*Ptilophores Motschulsky, 1849: 59 [stem: *Ptilophor-*]. Type genus: *Ptilophorus* Dejean, 1834. Comment: original vernacular name unavailable (Art. 11.7.2): subsequently used in latinized form but not generally attributed to Motschulsky (1849).

Ptilophorini Gerstaecker, 1855: 2 [stem: *Ptilophor-*]. Type genus: *Ptilophorus* Dejean, 1834.

Évaniocérides Lacordaire, 1859: 618 [stem: *Evaniocer-*]. Type genus: *Evaniocera* Guérin-Méneville, 1835 [syn. of *Ptilophorus* Dejean, 1834]. Comment: original vernacular name available (Art. 11.7.2): first used in latinized form by J. L. LeConte

(1862: 276, as EVANIOCERINI), generally accepted as in J. L. LeConte and G. H. Horn (1883: xxxvii, as EVANIOCERINI).

Subfamily PELECOTOMINAE Seidlitz, 1875

PELECOTOMINI Seidlitz, 1875 [Gatt.]: 104 [stem: *Pelecotom-*]. Type genus: *Pelecotoma* Fischer von Waldheim, 1808.

MICHOLAEMINAE Viana, 1971: 69 [stem: *Micholaem-*]. Type genus: *Micholaemus* Viana, 1971.

Subfamily HEMIRHIPIDIINAE Heller, 1921

HEMIRHIPIDIINI Heller, 1921: 168 [stem: *Hemirhipidi-*]. Type genus: *Hemirhipidius* Heller, 1921 [syn. of *Nephrites* Shuckard, 1838].

NEPHRITINAE Selander, 1957: 101 [stem: *Nephrit-*]. Type genus: *Nephrites* Shuckard, 1838. Comment: name proposed to replace HEMIRHIPIDIINI Heller because of the synonymy of the type genus, HEMIRHIPIDIINAE is in prevailing usage (see Lawrence and Newton, 1995: 889).

Subfamily RIPIDIINAE Gerstaecker, 1855

RHIPIDIINI Gerstaecker, 1855: 14 [stem: *Ripidi-*]. Type genus: *Ripidius* Thunberg, 1806 [as *Rhipidius*, unjustified emendation of type genus name by Agassiz (1846b: 327), not in prevailing usage].

Tribe EORHIPIDIINI Iablokoff-Khnzorian, 1986

EORHIPIDIINI Iablokoff-Khnzorian, 1986: 92 [stem: *Eorhipidi-*]. Type genus: *Eorhipidius* Iablokoff-Khnzorian, 1986.

Tribe RIPIDIINI Gerstaecker, 1855

RHIPIDIINI Gerstaecker, 1855: 14 [stem: *Ripidi-*]. Type genus: *Ripidius* Thunberg, 1806 [as *Rhipidius*, unjustified emendation of type genus name by Agassiz (1846b: 327), not in prevailing usage]. Comment: incorrect original stem formation, not in prevailing usage; spelling of family-group and type genus names based on Krell (1996).

Subfamily RIPIPHORINAE Gemminger, 1870 (1855)

RHIPIDOPHORIDAE Gemminger, 1870: 2117 [stem: *Ripiphor-*]. Type genus: *Ripiphorus* Bosc, 1791 [as *Rhipidophorus*, unjustified emendation of *Ripiphorus* Bosc, 1791 by Perty (1831: xix), not in prevailing usage]. Comment: name conserved over the older MYODITINAE Gerstaecker, 1855 (Art. 40.2) (see Lawrence and Newton 1995: 889, as "MYODITINI Costa, 1853").

Tribe MACROSIAGONINI Heyden, 1908

RHIPIPHORITES Laporte, 1840b: 261 [stem: *Ripiphor-*]. Type genus: *Ripiphorus* sensu Fabricius, 1792 [as *Rhipiphorus*, incorrect subsequent spelling of the type genus name, not in prevailing usage; not *Ripiphorus* Bosc, 1791; syn. of *Metoecus* Dejean, 1834]. Comment: original vernacular name available (Art. 11.7.2): first used in latinized form and generally accepted as in Gerstaecker (1855: 17, as Rhipiphorini [incorrect stem formation]); based on a misidentified type genus; name treated here as invalid until an application is submitted to the Commission to suppress it for the Principles of Priority and Homonymy (Art. 65.2.1); also see Ripiphorini Gemminger, 1870.

Macrosiagonini L. Heyden, 1908: 45 [stem: *Macrosiagon-*]. Type genus: *Macrosiagon* Hentz, 1830. Comment: although this is not the oldest name for the tribe, we recommend that an application be sent to the Commission to suppress Ripiphorini Laporte, 1840 because it is based on a misidentified type genus (Art. 65.2.1).

Tribe Ripiphorini Gemminger, 1870 (1855)

*Mioditini A. Costa, 1853: 2 [stem: *Myodit-*]. Type genus: *Myodites* Latreille, 1829 [syn. of *Ripiphorus* Bosc, 1791]. Comment: original vernacular name unavailable (Art. 11.7.2): subsequently used in latinized form but not generally treated as valid attributed to A. Costa, 1853; incorrect original stem formation, not in prevailing usage.

Myoditini Gerstaecker, 1855: 15 [stem: *Myodit-*]. Type genus: *Myodites* Latreille, 1829 [syn. of *Ripiphorus* Bosc, 1791]. Comment : usage of younger name Ripiphorini Gemminger, 1870 conserved over this name (Art. 40.2) (see Lawrence and Newton 1995: 889, as "Myoditini Costa, 1853").

Rhipidophoridae Gemminger, 1870: 2117 [stem: *Ripiphor-*]. Type genus: *Ripiphorus* Bosc, 1791 [as *Rhipidophorus*, unjustified emendation of type genus name by Perty (1831: xix), not in prevailing usage]. Comment: name conserved over the older Myoditini Gerstaecker, 1855 (Art. 40.2) (see Lawrence and Newton 1995: 889, as "Myoditini Costa, 1853"); incorrect original stem formation, not in prevailing usage; as mentioned by Lawrence and Newton (1995: 889) a number of authors used Ripiphoridae before Gemminger, however, their family-group names were based on *Ripiphorus* sensu Fabricius, 1792, which is a synonym of *Metoecus* Dejean, 1834; spelling of family-group and type genus names based on Krell (1996); we recommend that an application be submitted to the Commission to suppress Ripiphorites Laporte, 1840 for the Principles of Priority and Homonymy (Art. 65.2.1).

Family Zopheridae Solier, 1834

Zophérites Solier, 1834: 505 [stem: *Zopher-*]. Type genus: *Zopherus* Gray, 1832. Comment: usage of this name conserved over Colydiidae Billberg, 1820 (Art. 35.5).

Subfamily Colydiinae Billberg, 1820

Colydiides Billberg, 1820b: 394 [stem: *Colydi-*]. Type genus: *Colydium* Fabricius, 1792 [placed on the Official List of Generic Names in Zoology (ICZN 1995c)].

Comment: name placed on Official List of Family-Group Names in Zoology (as COLYDIIDAE Erichson, 1842) and given precedence over ORTHOCERINAE Blanchard, 1845 and CERYLONIDAE Billberg, 1820 whenever their type genera are placed in the same family-group taxon (ICZN 1995c); usage of the younger name ZOPHERIDAE Solier, 1834 conserved over this name (Art. 35.5).

Tribe ACROPINI Sharp, 1894

ACROPINAE Sharp, 1894: 444 [stem: *Acrop-*]. Type genus: *Acropis* H. C. C. Burmeister, 1840. Comment: current spelling maintained (Art. 29.3.1.1): incorrect stem formation in prevailing usage (should be *Acropid-*).

Tribe ADIMERINI Sharp, 1894

ADIMERIDAE Sharp, 1894: 441 [stem: *Adimer-*]. Type genus: *Adimerus* Sharp, 1894.

MONOEDIDAE Schaeffer, 1911: 114 [stem: *Monoed-*]. Type genus: *Monoedus* G. H. Horn, 1882.

Tribe COLYDIINI Billberg, 1820

COLYDIIDES Billberg, 1820b: 394 [stem: *Colydi-*]. Type genus: *Colydium* Fabricius, 1792 [placed on the Official List of Generic Names in Zoology (ICZN 1995c)].

Tribe GEMPYLODINI Sharp, 1893

GEMPYLODINI Sharp, 1893: 256 [stem: *Gempylod-*]. Type genus: *Gempylodes* Pascoe, 1863.

Tribe NEMATIDIINI Horn, 1878

NEMATIDII G. H. Horn, 1878: 573 [stem: *Nematidi-*]. Type genus: *Nematidium* Erichson, 1845.

Tribe ORTHOCERINI Blanchard, 1845

SARROTRIIDES Billberg, 1820a: 9 [stem: *Sarrotri-*]. Type genus: *Sarrotrium* Illiger, 1798 [syn. of *Orthocerus* Latreille, 1797]. Comment: the same name was also published in the same year by Billberg (1820b: 390); SARROTRIIDAE Billberg, 1820 placed on the Official Index of Rejected and Invalid Family-Group Names in Zoology (ICZN 1995c).

ORTHOCÉRITES Blanchard, 1845b: 29 [stem: *Orthocer-*]. Type genus: *Orthocerus* Latreille, 1797 [placed on the Official List of Generic Names in Zoology (ICZN 1995c)]. Comment: ORTHOCERINI Blanchard, 1845 placed on Official List of Family-group names in Zoology (ICZN 1995c).

EUGLOCHIDAE Gistel, 1856a: 382 [stem: *Euglochin-*]. Type genus: *Euglochis* Gistel, 1856 [syn. of *Orthocerus* Latreille, 1797]. Comment: published 18 February 1856; this family-group name was also used in the same year by Gistel (1856b

["31 December"]: 180, as EUGLOCHIDA); incorrect original stem formation, not in prevailing usage.

Tribe RHAGODERINI Horn, 1878

RHAGODERINI G. H. Horn, 1878: 557 [stem: *Rhagoder-*]. Type genus: *Rhagodera* Mannerheim, 1843.

Tribe RHOPALOCERINI Reitter, 1911

APISTINI Ganglbauer, 1899: 873 [stem: *Apist-*]. Type genus: *Apistus* Motschulsky, 1840 [*Apistus* is an unjustified emendation of type genus name, originally spelled *Apeistus*, by Agassiz (1846b: 28), in prevailing usage; the emended name is a junior homonym of *Apistus* Cuvier, 1829 [Pisces]; *Apeistus* Motschulsky, 1840 and *Apistus* Agassiz, 1846 were placed on Official Index of Rejected and Invalid Generic Names in Zoology (ICZN 1986b); syn. of *Rhopalocerus* Redtenbacher, 1842]. Comment: permanently invalid (Art. 39): based on preoccupied type genus; placed on Official Index of Rejected and Invalid Family-Group Names in Zoology (ICZN 1986b).

RHOPALOCERINI Reitter, 1911: 108 [stem: *Rhopalocer-*]. Type genus: *Rhopalocerus* Redtenbacher, 1842 [placed on the Official List of Generic Names in Zoology (ICZN 1986b)]. Comment: placed on the Official List of Family-Group Names in Zoology (ICZN 1986b).

Tribe SYNCHITINI Erichson, 1845

SYNCHITINI Erichson, 1845: 254 [stem: *Synchit-*]. Type genus: *Synchita* Hellwig, 1792. Comment: published before 31 June 1845; this family-group name was also used in the same year by L. Redtenbacher (1845 [before September]: 123, as SYNCHITAE).

DITOMIDAE Imhoff, 1856: [2] 159 [stem: *Bitom-*]. Type genus: *Bitoma* Herbst, 1793 [as *Ditoma*, unjustified emendation of type genus name by Illiger (1807), not in prevailing usage]. Comment: incorrect original stem formation, not in prevailing usage; stem of the family-group name corrected according to Art. 35.4.1 and also to remove from potential homonymy with the older family-group name originally proposed as DITOMICI Bonelli, 1810 (type genus *Ditomus* Bonelli, 1810) in the family CARABIDAE.

COXELINI Seidlitz, 1872 [Gatt.]: 38 [stem: *Coxel-*]. Type genus: *Coxelus* Dejean, 1821.

LANGELANDIINA Fowler, 1889: 192 [stem: *Langelandi-*]. Type genus: *Langelandia* Aubé, 1843.

MEGATAPHRINI Casey, 1890: 309 [stem: *Megataphr-*]. Type genus: *Megataphrus* Casey, 1890.

TARPHIINAE Sharp, 1894: 444 [stem: *Tarphi-*]. Type genus: *Tarphius* Erichson, 1845.

CORTICINI Ganglbauer, 1899: 870 [stem: *Cortic-*]. Type genus: *Corticus* Latreille, 1829.

DIODESMINI Reitter, 1911: 110 [stem: *Diodesmat-*]. Type genus: *Diodesma* Latreille, 1829. Comment: incorrect original stem formation, not in prevailing usage.

TRACHYPHOLINI Grouvelle, 1911: 121 [stem: *Trachypholid-*]. Type genus: *Trachypholis* Erichson, 1845. Comment: incorrect original stem formation, not in prevailing usage.

ENDOPHLOEINI Reitter, 1922a: 17 [stem: *Endophloe-*]. Type genus: *Endophloeus* Dejean, 1834.

PRIOLOMINI Dajoz, 1980a: 127 [stem: *Priolom-*]. Type genus: *Priolomus* Erichson, 1845.

Subfamily ZOPHERINAE Solier, 1834

ZOPHÉRITES Solier, 1834: 505 [stem: *Zopher-*]. Type genus: *Zopherus* Gray, 1832.

Tribe LATOMETINI Ślipiński and Lawrence, 1999

LATOMETINI Ślipiński and Lawrence, 1999: 9, in key [stem: *Latomet-*]. Type genus: *Latometus* Erichson, 1842.

Tribe MONOMMATINI Blanchard, 1845

MONOMMITES Blanchard, 1845b: 16 [stem: *Monommat-*]. Type genus: *Monomma* Klug, 1833. Comment: original vernacular name available (Art. 11.7.2): first used in latinized form by J. L. LeConte (1862: 246, as MONOMMIDAE [incorrect stem formation]), generally accepted as in Lawrence and Newton (1995: 890, as MONOMMATIDAE); incorrect original stem formation, not in prevailing usage.

Tribe PHELLOPSINI Ślipiński and Lawrence, 1999

PHELLOPSINI Ślipiński and Lawrence, 1999: 10, in key [stem: *Phellops-*]. Type genus: *Phellopsis* J. L. LeConte, 1862. Comment: current spelling maintained (Art. 29.5): incorrect stem formation in prevailing usage (should be *Phellopse-*).

Tribe PYCNOMERINI Erichson, 1845

PYCNOMERINI Erichson, 1845: 290 [stem: *Pycnomer-*]. Type genus: *Pycnomerus* Erichson, 1842.

Tribe USECHINI Horn, 1867

USECHINI G. H. Horn, 1867a: 294 [stem: *Usech-*]. Type genus: *Usechus* Motschulsky, 1845.

Tribe ZOPHERINI Solier, 1834

ZOPHÉRITES Solier, 1834: 505 [stem: *Zopher-*]. Type genus: *Zopherus* Gray, 1833. Comment: original vernacular name available (Art. 11.7.2): first used in latinized form by Agassiz (1846b: 392, as ZOPHEROIDAE), generally accepted as in Lawrence and Newton (1995: 891, as ZOPHERIDAE).

Nosodermini Casey, 1907: 280 [stem: *Nosodermat-*]. Type genus: *Nosoderma* Solier, 1841 [preoccupied genus name, not *Nosoderma* Guérin-Méneville, 1838 [Coleoptera: Zopheridae: Zopherinae: Zopherini]; syn. of *Verodes* Casey, 1907; see Foley and Ivie (2007) for comments on type genus]. Comment: permanently invalid (Art. 39): based on preoccupied type genus; all subsequent uses of Nosodermini were based on the junior homonym *Nosoderma* Solier, 1841 (see Foley and Ivie 2007); incorrect original stem formation, not in prevailing usage.

Zopherosini Casey, 1907: 522 [stem: *Zopherose-*]. Type genus: *Zopherosis* A. White, 1859. Comment: incorrect original stem formation, not in prevailing usage.

Family Ulodidae Pascoe, 1869

Ulodinae Pascoe, 1869a: 31 [stem: *Ulod-*]. Type genus: *Ulodes* Erichson, 1842.

Merycidae Crowson, 1953: 37 [stem: *Meryc-*]. Type genus: *Meryx* Latreille, 1802.

Family Promecheilidae Lacordaire, 1859

Proméchilides Lacordaire, 1859: 698 [stem: *Promecheil-*]. Type genus: *Promecheilus* Solier, 1851 [as *Promechilus*, unjustified emendation of type genus name by Lacordaire (1859: 700), not in prevailing usage]. Comment: original vernacular name available (Art. 11.7.2): first used in latinized form by Handlirsch (1925: 594, as Promecheilinae), generally accepted as in Lawrence et al. (2010: 563, as Promecheilidae); incorrect original stem formation, not in prevailing usage.

Perimylopidae St. George, 1939: 212 [stem: *Perimylop-*]. Type genus: *Perimylops* Müller, 1884.

*Parahelopinae Watt, 1975: 424 [stem: *Parahelop-*]. Type genus: *Parahelops* C. O. Waterhouse, 1876. Comment: unavailable family-group name, proposed after 1930 without description or bibliographic reference to such description (Art. 13.1); this name was used several times subsequently, e.g., Elgueta and Arriagada (1989: 37), Lawrence and Britton (1991: 665), Lawrence (1994: 339), but this taxon name remains unavailable (see Lawrence and Newton 1995: 892).

Family Chalcodryidae Watt, 1974

Chalcodryidae Watt, 1974: 24 [stem: *Chalcodry-*]. Type genus: *Chalcodrya* Redtenbacher, 1868.

Family Trachelostenidae Lacordaire, 1859

Trachélosténides Lacordaire, 1859: 567 [stem: *Trachelosten-*]. Type genus: *Trachelostenus* Solier, 1851. Comment: original vernacular name available (Art. 11.7.2): first used in latinized form by Seidlitz (1898: 319, as Trachelostenini), generally accepted as in Lawrence and Newton (1995: 892 as Trachelostenidae).

Family Tenebrionidae Latreille, 1802

Tenebrionites Latreille, 1802: 165 [stem: *Tenebrion-*]. Type genus: *Tenebrio* Linnaeus, 1758. Comment: First Reviser (Tenebrionidae Latreille, 1802 vs Pimeliidae Latreille, 1802 vs Diaperidae Latreille, 1802) not determined, current usage maintained.

Subfamily Zolodininae Watt, 1975

Zolodininae Watt, 1975: 401 [stem: *Zolodin-*]. Type genus: *Zolodinus* Blanchard, 1853.

Subfamily Lagriinae Latreille, 1825 (1820)

Lagriariae Latreille, 1825: 381 [stem: *Lagri-*]. Type genus: *Lagria* Fabricius, 1775. Comment: use of younger family-group name conserved over Lachninae Billberg, 1820 (Art. 40.2) (see Bouchard et al. 2005); the older name Cossyphinae Latreille, 1802 is not given precedence over this name because the current placement of Cossyphini in this subfamily is uncertain (see Bouchard et al. 2005).

Tribe Adeliini Kirby, 1828

Adeliadae Kirby, 1828: 525 [stem: *Adeli-*]. Type genus: *Adelium* Kirby, 1819. Comment: family-group name attributed to Hope (1840a: 188) in recent literature; this is a senior homonym of Adeliini Viereck, 1918 (type genus *Adelius* Haliday, 1833) which is considered a valid tribe in the Hymenoptera: Braconidae: Cheloninae; this case is to be referred to the Commission to remove the homonymy (Art. 55.3.1).

*Apatèlates Mulsant and Rey, 1859: 87 [stem: *Apatel-*]. Type genus: *Apatelus* Mulsant and Rey, 1859. Comment: original vernacular name unavailable (Art. 11.7.2): subsequently used in latinized form, e.g., Heyne and Taschenberg (1907: 208, as Apatelini), but not generally accepted as valid; if evidence is found in the future that would lead to the treatment of Mulsant and Rey's name as available, this would threaten the younger name originally proposed as Apatelinae Grote, 1883 (type genus *Apatele* Hübner, 1822) in Lepidoptera which has been used as valid in recent literature.

Tribe Belopini Reitter, 1917

Calcariens Mulsant, 1854: 268 [stem: *Calcar-*]. Type genus: *Calcar* Dejean, 1821 [preoccupied genus name, not *Calcar* de Montfort, 1810 [Mollusca]; syn. of *Centorus* Mulsant, 1854]. Comment: original vernacular name available (Art. 11.7.2): first used in latinized form and generally accepted as in Seidlitz (1895: 647, as Calcarina); permanently invalid (Art. 39): based on preoccupied type genus.

Belopinae Reitter, 1917: 59 [stem: *Belop-*]. Type genus: *Belopus* Gebien, 1911. Comment: current spelling maintained (Art. 29.5): incorrect stem formation in prevailing usage (should be *Belopod-*).

Tribe Chaerodini Doyen, Matthews and Lawrence, 1990

Chaerodini Doyen et al., 1990: 239 [stem: *Chaerod-*]. Type genus: *Chaerodes* A. White, 1846.

Tribe Cossyphini Latreille, 1802

Cossyphores Latreille, 1802: 164 [stem: *Cossyph-*]. Type genus: *Cossyphus* A. G. Olivier, 1791. Comment: not given precedence over Lagriinae Latreille, 1825 because its current placement is uncertain (see Bouchard et al. 2005).

Tribe Goniaderini Lacordaire, 1859

Goniadérides Lacordaire, 1859: 390 [stem: *Goniader-*]. Type genus: *Goniadera* Perty, 1832. Comment: original vernacular name available (Art. 11.7.2): first used in latinized form by Seidlitz (1895: 614, as Goniaderina), generally accepted as in Gebien (1911: 467, as Goniaderinae).

*Phobéliides Lacordaire, 1859: 393 [stem: *Phobeli-*]. Type genus: *Phobelius* Blanchard, 1845. Comment: original vernacular name unavailable (Art. 11.7.2): subsequently used in latinized form, e.g., Heyne and Taschenberg (1907: 212, as Phobeliini), but not generally accepted as valid.

Eschatoporini Blaisdell, 1906: 78 [stem: *Eschatopori-*]. Type genus: *Eschatoporis* Blaisdell, 1906. Comment: incorrect original stem formation, not in prevailing usage.

Phobeliina Ardoin, 1961: 33 [stem: *Phobeli-*]. Type genus: *Phobelius* Blanchard, 1845.

*Anaedini Skopin, 1964: 7 [stem: *Anaed-*]. Type genus: *Anaedus* Blanchard, 1845. Comment: unavailable family-group name, proposed after 1930 without description or bibliographic reference to such a description (Art. 13.1).

Tribe Laenini Seidlitz, 1895

Laenina Seidlitz, 1895: 669 [stem: *Laen-*]. Type genus: *Laena* Dejean, 1821.

Tribe Lagriini Latreille, 1825 (1820)

Lagriariae Latreille, 1825: 381 [stem: *Lagri-*]. Type genus: *Lagria* Fabricius, 1775. Comment: use of younger family-group name conserved over Lachnini Billberg, 1820 (Art. 40.2) (see Bouchard et al. 2005).

Subtribe Lagriina Latreille, 1825 (1820)

*Lagrien Oken, 1817: 1180 [stem: *Lagri-*]. Type genus: *Lagria* Fabricius, 1775. Comment: original vernacular name unavailable (Art. 11.7.2): subsequently used in latinized form but not generally attributed to Oken (1817).

Lachnaedes Billberg, 1820a: 34 [stem: *Lachn-*]. Type genus: *Lachna* Billberg, 1820 [syn. of *Lagria* Fabricius, 1775]. Comment: the family-group names Lachnidae/-inae/-ini Herrich-Schaeffer 1854 (based on *Lachnus* H. C.

C. Burmeister 1835) are currently used as valid in Hemiptera, LACHNINAE (incorrectly attributed to Passerini (1863)) was placed on the Official List of Family-Group Names in Zoology (ICZN 1956); this case is to be referred to the Commission to remove the homonymy (Art. 55.3.1).

LAGRIARIAE Latreille, 1825: 381 [stem: *Lagri-*]. Type genus: *Lagria* Fabricius, 1775. Comment: use of younger family-group name conserved over LACHNINA Billberg, 1820 (Art. 40.2) (see Bouchard et al. 2005).

LOUBACANTINI Bonadona, 1959: 1034 [stem: *Loubacant-*]. Type genus: *Loubacantus* Bonadona, 1959 [syn. of *Entypodera* Gerstaecker, 1871]. Comment: originally proposed in ANTHICIDAE, transfer to LAGRIINAE by Bonadona (1984).

Subtribe STATIRINA Blanchard, 1845

STATYRITES Blanchard, 1845b: 39 [stem: *Statir-*]. Type genus: *Statira* Lepeletier and Audinet-Serville, 1828 [as *Statyra*, incorrect subsequent spelling of type genus name, not in prevailing usage]. Comment: original vernacular name available (Art. 11.7.2): first used in latinized form by J. L. LeConte (1862: 246, as STATYRINI), generally accepted as in Bouchard et al. (2005: 501, as STATYRINA); incorrect original stem formation, not in prevailing usage.

*HYSTÉRARTHRIDES Lacordaire, 1869: 231 [stem: *Hysterarthr-*]. Type genus: *Hysterarthron* J. Thomson, 1864. Comment: original vernacular name unavailable (Art. 11.7.2): subsequently used in latinized form, e.g., Heyne and Taschenberg (1907: 240, as HYSTERARTHRINI), but not generally accepted as valid; type genus transferred from CERAMBYCIDAE by Ritsema (1892: 54).

Tribe LUPROPINI Ardoin, 1958

LUPROPSINI Ardoin, 1958: 59 [stem: *Luprop-*]. Type genus: *Luprops* Hope, 1833. Comment: incorrect original stem formation, not in prevailing usage.

Tribe PYCNOCERINI Lacordaire, 1859 *nomen protectum*

CHIROSCELIDAE Hope, 1840a: 127 [stem: *Chiroscelid-*]. Type genus: *Chiroscelis* Lamarck, 1804. Comment: published before 30 September 1840; Bouchard et al. (2005: 524) treated this name and CHIROSCELITES Laporte (1840b [before 26 December]: 216) as *nomina oblita*; incorrect original stem formation, not in prevailing usage.

PYCNOCÉRIDES Lacordaire, 1859: 399 [stem: *Pycnocer-*]. Type genus: *Pycnocerus* Westwood, 1841. Comment: *nomen protectum* (see Bouchard et al. 2005: 524); original vernacular name available (Art. 11.7.2): first used in latinized form by Kolbe (1887: 50, as PYCNOCERIDAE), generally accepted as in Bouchard et al. (2005: 501, as PYCNOCERINI).

Prioscelina Skopin, 1964: 30, in key [stem: *Prioscelid-*]. Type genus: *Prioscelis* Hope, 1840. Comment: incorrect original stem formation, not in prevailing usage.

Subfamily Nilioninae Oken, 1843

Nilioniden Oken, 1843: 484 [stem: *Nilion-*]. Type genus: *Nilio* Latreille, 1802 [incorrect subsequent spelling of *Nilion* by Latreille (1804b: 333), incorrect subsequent spelling in prevailing usage, treated as correct original spelling (Art. 33.3.1)]. Comment: original vernacular name available (Art. 11.7.2): first used in latinized form and generally accepted as in Tulk (1847: 614, as Niliondae [incorrect stem formation]); family-group name previously attributed to Lacordaire (1859: 518).

Subfamily Phrenapatinae Solier, 1834

Phrépatides Solier, 1834: 488 [stem: *Phrenapat-*]. Type genus: *Phrenapates* Gray, 1832.

Tribe Archaeoglenini Watt, 1975

Archaeoglenini Watt, 1975: 412 [stem: *Archaeoglen-*]. Type genus: *Archaeoglenes* Broun, 1893.

Tribe Penetini Lacordaire, 1859

Pénétides Lacordaire, 1859: 318 [stem: *Penet-*]. Type genus: *Peneta* Lacordaire, 1859. Comment: original vernacular name available (Art. 11.7.2): first used in latinized form by Seidlitz (1894: 545, as Penetina), generally accepted as in Bouchard et al. (2005: 501, as Penetini).

Phthorini Boddy, 1965: 144 [stem: *Phtor-*]. Type genus: *Phtora* Mulsant, 1854 [as *Phthora*, incorrect subsequent spelling of type genus name, not in prevailing usage; *Phtora* is a junior homonym of *Phtora* Germar, 1836 [Coleoptera: Tenebrionidae: Diaperinae]; syn. of *Clamoris* Gozis, 1886]. Comment: permanently invalid (Art. 39): based on preoccupied type genus.

Tribe Phrenapatini Solier, 1834

Phrépatides Solier, 1834: 488 [stem: *Phrenapat-*]. Type genus: *Phrenapates* Gray, 1832 [as *Phrepates*, incorrect subsequent spelling of type genus name, not in prevailing usage]. Comment: original vernacular name available (Art. 11.7.2): first used in latinized form by Agassiz (1846b: 287, as Phrenapatoidae), generally accepted as in Bouchard et al. (2005: 501, as Phrenapatinae/-ini); incorrect original stem formation, not in prevailing usage.

Subfamily Pimeliinae Latreille, 1802

Pimeliariae Latreille, 1802: 166 [stem: *Pimeli-*]. Type genus: *Pimelia* Fabricius, 1775.

Tribe ADELOSTOMINI Solier, 1834

ADÉLOSTOMITES Solier, 1834: 502 [stem: *Adelostom-*]. Type genus: *Adelostoma* Duponchel, 1827. Comment: original vernacular name available (Art. 11.7.2): first used in latinized form by Agassiz (1846b: 8, as ADELOSTOMOIDAE), generally accepted as in Bouchard et al. (2005: 501, as ADELOSTOMINI); current spelling maintained (Art. 29.5): incorrect stem formation in prevailing usage (should be *Adelostomat-*).

*EURYCHORITES Solier, 1837b: 153 [stem: *Eurychor-*]. Type genus: *Eurychora* Thunberg, 1791. Comment: original vernacular name unavailable (Art. 11.7.2): subsequently used in latinized form but not generally attributed to Solier (1837).

EURYCHORIDAE Hope, 1840a: 121 [stem: *Eurychor-*]. Type genus: *Eurychora* Thunberg, 1791.

Tribe ADESMIINI Lacordaire, 1859 *nomen protectum*

*MACROPODITES Solier, 1834: 501 [stem: *Macropod-*]. Type genus: *Macropoda* Solier, 1835. Comment: unavailable family-group name, not based on an available genus name.

*MACROPODITES Solier, 1835b: 509 [stem: *Macropod-*]. Type genus: *Macropoda* Solier, 1835. Comment: original vernacular name unavailable (Art. 11.7.2): subsequently used in latinized form but not generally accepted as valid.

*ADESMIITES Blanchard, 1845b: 4 [stem: *Adesmi-*]. Type genus: *Adesmia* Fischer von Waldheim, 1822. Comment: original vernacular name unavailable (Art. 11.7.2): subsequently used in latinized form but not generally attributed to Blanchard (1845b).

MACROPODOIDAE Agassiz, 1846b: 221 [stem: *Macropod-*]. Type genus: *Macropoda* Solier, 1835. Comment: *nomen oblitum* (Appendix 1); this name is a junior homonym of MACROPODIDAE Gray, 1821 in Mammalia (type genus *Macropus* Shaw and Nodder, 1790) which is on the Official List of Family-Group Names in Zoology.

MEGAGENIANOS Solier, 1851: 124 [stem: *Megageni-*]. Type genus: *Megagenius* Solier, 1835. Comment: original vernacular name available (Art. 11.7.2): first used in latinized form by Kolbe (1887: 50, as MEGAGENIIDAE), generally accepted as in Bouchard et al. (2007: 386, as MEGAGENIINI); *nomen oblitum*, ADESMIINI Lacordaire, 1859 conserved over this name by Bouchard et al. (2007: 386).

ADESMIIDES Lacordaire, 1859: 22 [stem: *Adesmi-*]. Type genus: *Adesmia* Fischer von Waldheim, 1822. Comment: *nomen protectum* (Appendix 1); original vernacular name available (Art. 11.7.2): first used in latinized form by Baudi di Selve (1875: 22, as ADESMIDAE [incorrect stem formation]), generally accepted as in Bouchard et al. (2005: 502); also ADESMIINI Lacordaire, 1859 conserved over MEGAGENIINI Solier, 1851 by Bouchard et al. (2007: 386).

Épiphysides Lacordaire, 1859: 29 [stem: *Epiphys-*]. Type genus: *Epiphysa* Dejean, 1834. Comment: original vernacular name available (Art. 11.7.2): first used in latinized form by J. L. LeConte (1862: 213, as Epiphysini), generally accepted as in Gebien (1910a: 82, as Epiphysinae).

Tribe Akidini Billberg, 1820

Acides Billberg, 1820a: 32 [stem: *Akid-*]. Type genus: *Akis* Herbst, 1799 [as *Acis*, unjustified emendation of type genus by Billberg (1820a), not in prevailing usage]. Comment: incorrect original stem formation, not in prevailing usage.

Tribe Anepsiini LeConte, 1862

Anepsiini J. L. LeConte, 1862: 215 [stem: *Anepsi-*]. Type genus: *Anepsius* J. L. LeConte, 1851.

Batuliini G. H. Horn, 1870: 270 [stem: *Batuli-*]. Type genus: *Batulius* J. L. LeConte, 1851.

Anchommini G. H. Horn, 1878: 558 [stem: *Anchommat-*]. Type genus: *Anchomma* J. L. LeConte, 1858 [genus originally proposed in Colydiidae]. Comment: incorrect original stem formation, not in prevailing usage.

Tribe Asidini Fleming, 1821

Asidadae Fleming, 1821: 51 [stem: *Asid-*]. Type genus: *Asida* Latreille, 1802.

*Machlides Lacordaire, 1859: 155 [stem: *Machl-*]. Type genus: *Machla* Herbst, 1799 [earlier usage of the type genus name by Lichtenstein (1796) was suppressed for nomenclatural purposes (ICZN 1995e); this genus is currently known under the name *Pseudomachla* Wilke, 1921]. Comment: original vernacular name unavailable (Art. 11.7.2): subsequently used in latinized form and attributed to the original author but not used as valid then (Bouchard et al. 2005: 506).

Astroti G. H. Horn, 1870: 289 [stem: *Astrot-*]. Type genus: *Astrotus* J. L. LeConte, 1858.

Craniotini J. L. LeConte and G. H. Horn, 1883: 361 [stem: *Craniot-*]. Type genus: *Craniotus* J. L. LeConte, 1851.

Machlini Chatanay, 1914: 1 [stem: *Machl-*]. Type genus: *Machla* Herbst, 1799 [earlier usage of the type genus name by Lichtenstein (1796) was suppressed for nomenclatural purposes (ICZN 1995e); this genus is currently known under the name *Pseudomachla* Wilke, 1921].

Parecatini Chatanay, 1914: 2 [stem: *Parecat-*]. Type genus: *Parecatus* Fairmaire, 1900.

Tribe Boromorphini Skopin, 1978

Boromorphini Skopin, 1978: 228, in key [stem: *Boromorph-*]. Type genus: *Boromorphus* Wollaston, 1854.

Tribe Branchini LeConte, 1862

Branchini J. L. LeConte, 1862: 222 [stem: *Branch-*]. Type genus: *Branchus* J. L. LeConte, 1862.

Tribe Caenocrypticini Koch, 1958

Caenocrypticini Koch, 1958: 121 [stem: *Caenocryptic-*]. Type genus: *Caenocrypticus* Gebien, 1920.

Tribe Ceratanisini Gebien, 1937

Apolitina Seidlitz, 1895: 666 [stem: *Apolit-*]. Type genus: *Apolites* Jacquelin du Val, 1861 [preoccupied genus name, not *Apolites* Sundevall, 1835 [Aves]; syn. of *Idastrandiella* Strand, 1929]. Comment: permanently invalid (Art. 39): based on preoccupied type genus.

Anisocerini Reitter, 1906: 477 [stem: *Anisocer-*]. Type genus: *Anisocerus* Faldermann, 1837 [preoccupied genus name, not *Anisocerus* Audinet-Serville, 1835 [Coleoptera: Cerambycidae]; syn. of *Ceratanisus* Gemminger, 1870]. Comment: junior homonym of Anisocerini J. Thomson, 1860 [Coleoptera: Cerambycidae]; permanently invalid (Art. 39): based on preoccupied type genus.

Ceratanisini Gebien, 1937: 791 [stem: *Ceratanis-*]. Type genus: *Ceratanisus* Gemminger, 1870.

Tribe Cnemeplatiini Jacquelin du Val, 1861

Cnémémplatiites Jacquelin du Val, 1861: 286 [stem: *Cnemeplati-*]. Type genus: *Cnemeplatia* A. Costa, 1847.

Subtribe Actizetina Watt, 1992

Actizetina Watt, 1992: 297 [stem: *Actizet-*]. Type genus: *Actizeta* Pascoe, 1875.

Subtribe Cnemeplatiina Jacquelin du Val, 1861

*Autocérides Lacordaire, 1859: 279 [stem: *Autocer-*]. Type genus: *Autocera* Wollaston, 1857 [syn. of *Cnemeplatia* A. Costa, 1847]. Comment: original vernacular name unavailable (Art. 11.7.2): subsequently used in latinized form, e.g., Heyne and Taschenberg (1907: 208, as Autocerini), but not generally accepted as valid.

Cnéméplatiites Jacquelin du Val, 1861: 286 [stem: *Cnemeplati-*]. Type genus: *Cnemeplatia* A. Costa, 1847. Comment: Bouchard et al. (2005: 504) considered Jacquelin du Val's name as unavailable based on the requirements of Art. 11.7.2, however, the recent usage of «Cnemeplatiini Jacquelin du Val, 1861" as a valid taxon by Löbl et al. (2008: 140) made this name available.

Cnemeplatiini Csiki, 1953: 117 [stem: *Cnemeplati-*]. Type genus: *Cnemeplatia* A. Costa, 1847. Comment: proposed as new without reference to Cnéméplatiites Jacquelin du Val, 1861.

Subtribe RONDONIELLINA Ferrer and Moragues, 2000

RONDONIELLINA Ferrer and Moragues, 2000: 100 [stem: *Rondoniell-*]. Type genus: *Rondoniella* Kaszab, 1970.

Subtribe THORICTOSOMATINA Watt, 1992

THORICTOSOMATINA Watt, 1992: 296 [stem: *Thorictosomat-*]. Type genus: *Thorictosoma* Lea, 1919.

Tribe CNEMODININI Gebien, 1910

CNEMODINI G. H. Horn, 1870: 266 [stem: *Cnemodont-*]. Type genus: *Cnemodus* G. H. Horn, 1870 [preoccupied genus name, not *Cnemodus* Herrich-Schaeffer, 1850 [Hemiptera]; syn. of *Cnemodinus* Cockerell, 1906]. Comment: permanently invalid (Art. 39): based on preoccupied type genus; incorrect original stem formation, not in prevailing usage.

CNEMODININAE Gebien, 1910a: 4 [stem: *Cnemodin-*]. Type genus: *Cnemodinus* Cockerell, 1906.

Tribe CONIONTINI Waterhouse, 1858

CONIONTIDAE G. R. Waterhouse, 1858: 59 [stem: *Coniont-*]. Type genus: *Coniontis* Eschscholtz, 1829. Comment: family-group name previously attributed to Lacordaire (1859)/Schaum (1859) (see Bouchard et al. 2005: 502).

COELINI Casey, 1907: 500 [stem: *Coel-*]. Type genus: *Coelus* Eschscholtz, 1829.

EUSATTI Doyen, 1984: 11 [stem: *Eusatt-*]. Type genus: *Eusattus* J. L. LeConte, 1851.

Tribe COSSYPHODINI Wasmann, 1899

COSSYPHODIDAE Wasmann, 1899: 161 [stem: *Cossyphod-*]. Type genus: *Cossyphodes* Westwood, 1851. Comment: downgraded from a subfamily of TENEBRIONIDAE to a tribe of PIMELIINAE by Matthews et al. (2010).

Subtribe COSSYPHODINA Wasmann, 1899

COSSYPHODIDAE Wasmann, 1899: 161 [stem: *Cossyphod-*]. Type genus: *Cossyphodes* Westwood, 1851.

Subtribe COSSYPHODITINA Basilewsky, 1950

COSSYPHODITINAE Basilewsky, 1950a: 187 [stem: *Cossyphodit-*]. Type genus: *Cossyphodites* Brauns, 1901.

Subtribe ESEMEPHINA Steiner, 1980

ESEMEPHINI Steiner, 1980: 391 [stem: *Esemeph-*]. Type genus: *Esemephe* Steiner, 1980.

Subtribe PARAMELLONINA Andreae, 1961

PARAMELLONINAE Andreae, 1961: 200 [stem: *Paramellon-*]. Type genus: *Paramellon* C. O. Waterhouse, 1882. Comment: current spelling maintained (Art. 29.5): incorrect stem formation in prevailing usage (should be *Paramellont-*).

Tribe CRYPTOCHILINI Solier, 1841

CRYPTOCHILITES Solier, 1841: 248 [stem: *Cryptochil-*]. Type genus: *Cryptochile* Latreille, 1829.

Subtribe CALOGNATHINA Lacordaire, 1859

CALOGNATHIDES Lacordaire, 1859: 85 [stem: *Calognath-*]. Type genus: *Calognathus* Guérin-Méneville, 1836. Comment: original vernacular name available (Art. 11.7.2): first used in latinized form by Kolbe (1887: 50, as CALOGNATHIDAE), generally accepted as in Gebien (1910a: 117, as CALOGNATHINAE).

Subtribe CRYPTOCHILINA Solier, 1841

CRYPTOCHILITES Solier, 1841: 248 [stem: *Cryptochil-*]. Type genus: *Cryptochile* Latreille, 1829. Comment: original vernacular name available (Art. 11.7.2): first used in latinized form by Kolbe (1887: 50, as CRYPTOCHILIDAE), generally accepted as in Bouchard et al. (2005: 502, as CRYPTOCHILINI).

Subtribe HOMEBIINA Endrödy-Younga, 1989

HOMEBIINA Endrödy-Younga, 1989: 124 [stem: *Homebi-*]. Type genus: *Homebius* Endrödy-Younga, 1989.

Subtribe HORATOMINA Koch, 1955

HORATOMINA Koch, 1955: 14 [stem: *Horatom-*]. Type genus: *Horatoma* Solier, 1840.

Subtribe VANSONIINA Koch, 1955

VANSONINI Koch, 1955: 12 [stem: *Vansoni-*]. Type genus: *Vansonium* Koch, 1950. Comment: incorrect original stem formation, not in prevailing usage.

Tribe CRYPTOGLOSSINI LeConte, 1862 *nomen protectum*

CENTRIOPTÉRIDES Lacordaire, 1859: 134 [stem: *Centriopter-*]. Type genus: *Centrioptera* Mannerheim, 1843. Comment: original vernacular name available (Art. 11.7.2): first used in latinized form by J. L. LeConte (1862: 220, as CENTRIOPTERAE), generally accepted as in Bouchard et al. (2005: 502, as CENTRIOPTERINI); *nomen oblitum* (see Aalbu 2006: 57).

Cryptoglossini J. L. LeConte, 1862: 220 [stem: *Cryptogloss-*]. Type genus: *Cryptoglossa* Solier, 1837. Comment: *nomen protectum* (see Aalbu 2006: 57).

Tribe Edrotini Lacordaire, 1859

Édrotides Lacordaire, 1859: 31 [stem: *Edrot-*]. Type genus: *Edrotes* J. L. LeConte, 1851. Comment: original vernacular name available (Art. 11.7.2): first used in latinized form by Casey (1907: 279, as Edrotini), generally accepted as in Bouchard et al. (2005: 502, as Edrotini).

Triorophi J. L. LeConte and G. H. Horn, 1883: 362 [stem: *Trioroph-*]. Type genus: *Triorophus* J. L. LeConte, 1851.

Auchmobii J. L. LeConte and G. H. Horn, 1883: 362 [stem: *Auchmobi-*]. Type genus: *Auchmobius* J. L. LeConte, 1851.

Trimytini Casey, 1907: 278 [stem: *Trimytid-*]. Type genus: *Trimytis* Leconte, 1851. Comment: incorrect original stem formation, not in prevailing usage.

Eurymetoponini Casey, 1907: 278 [stem: *Eurymetop-*]. Type genus: *Eurymetopon* Eschscholtz, 1831. Comment: incorrect original stem formation, not in prevailing usage.

Trientomini Casey, 1907: 278 [stem: *Trientom-*]. Type genus: *Trientoma* Solier, 1835.

Tribe Elenophorini Solier, 1837

Elénophorites Solier, 1837a: 638 [stem: *Elenophor-*]. Type genus: *Elenophorus* Dejean, 1821 [syn. of *Leptoderis* Billberg, 1820]. Comment: original vernacular name available (Art. 11.7.2): first used in latinized form by Schaum (1859: 66, as Elenophoridae), generally accepted as in Bouchard et al. (2005: 502, as Elenophorini).

Tribe Epitragini Blanchard, 1845 *nomen protectum*

Lygophilia Rafinesque, 1815: 113 [stem: *Lygophil-*]. Type genus: *Lygophilus* Rafinesque, 1815 [syn. of *Epitragus* Latreille, 1802]. Comment: *nomen oblitum* (see Bouchard et al. 2007: 386).

*Épitragites Solier, 1834: 490 [stem: *Epitrag-*]. Type genus: *Epitragus* Latreille, 1802. Comment: original vernacular name unavailable (Art. 11.7.2): subsequently used in latinized form but not generally attributed to Solier (1834).

Épitragites Blanchard, 1845b: 16 [stem: *Epitrag-*]. Type genus: *Epitragus* Latreille, 1802. Comment: original vernacular name available (Art. 11.7.2): first used in latinized form by Erichson (1847a: 117, as Epitragii), generally accepted as in Bouchard et al. (2005: 502, as Epitragini); *nomen protectum* (see Bouchard et al. 2007: 386).

Tribe Erodiini Billberg, 1820 *nomen protectum*

Cephaceria Rafinesque, 1815: 113 [stem: *Cephacer-*]. Type genus: *Cephacerus* Rafinesque, 1815 [syn. of *Erodius* Fabricius, 1775]. Comment: *nomen oblitum* (see Bouchard et al. 2007: 386).

Erodiides Billberg, 1820a: 32 [stem: *Erodi-*]. Type genus: *Erodius* Fabricius, 1775. Comment: *nomen protectum* (see Bouchard et al. 2007: 386); this family-group name was also used in the same year by Billberg (1820b: 392, as Erodiides).

*Estenogenianos Solier, 1851: 138 [stem: *Stenogeni-*]. Type genus: *Stenogenius* Solier, 1834 [unavailable genus name, proposed in synonymy and not subsequently made available]. Comment: family-group name unavailable (Art. 11.7.1.1): not based on an available genus name at the time; incorrect original stem formation, not in prevailing usage.

*Arthrodeiden Koch, 1943: 483 [stem: *Arthrode-*]. Type genus: *Arthrodeis* Solier, 1834. Comment: original vernacular name unavailable (Art. 11.7.2): proposed after 1899.

Tribe Evaniosomini Lacordaire, 1859

Évaniosomides Lacordaire, 1859: 73 [stem: *Evaniosom-*]. Type genus: *Evaniosomus* Guérin-Méneville, 1834. Comment: original vernacular name available (Art. 11.7.2): first used in latinized form and generally accepted as in Gebien (1910a: 22, as Evaniosominae).

Tribe Falsomycterini Gebien, 1910

Falsomycterinae Gebien, 1910b: 177 [stem: *Falsomycter-*]. Type genus: *Falsomycterus* Pic, 1907.

Tribe Idisiini Medvedev, 1973

Idisiini G. S. Medvedev, 1973: 644 [stem: *Idisi-*]. Type genus: *Idisia* Pascoe, 1866.

Tribe Klewariini Gebien, 1910

Klewariinae Gebien, 1910a: 36 [stem: *Klewari-*]. Type genus: *Klewaria* Reitter, 1910.

Tribe Kuhitangiini Medvedev, 1962

Kuhitangiinae G. S. Medvedev, 1962: 1184 [stem: *Kuhitangi-*]. Type genus: *Kuhitangia* G. S. Medvedev, 1962.

Tribe Lachnogyini Seidlitz, 1894

Lachnogyini Seidlitz, 1894: 490 [stem: *Lachnogy-*]. Type genus: *Lachnogya* Ménétriés, 1849. Comment: subtribal classification according to G. S. Medvedev (2006).

Subtribe Lachnodactylina Reitter, 1904

Lachnodactylina Reitter, 1904: 182 [stem: *Lachnodactyl-*]. Type genus: *Lachnodactylus* Seidlitz, 1898.

Subtribe Lachnogyina Seidlitz, 1894

Lachnogyini Seidlitz, 1894: 490 [stem: *Lachnogy-*]. Type genus: *Lachnogya* Ménétriés, 1849.

Subtribe Netuschiliina Ferrer and Yvinec, 2004

Netuschiliina Ferrer and Yvinec, 2004: 48 [stem: *Netuschili-*]. Type genus: *Netuschilia* Reitter, 1904. Comment: also incorrectly spelled as Netuschilina in the original publication on page 48 (First Revisers are Bouchard et al. 2005: 504).

Tribe Leptodini Lacordaire, 1859

Leptodides Lacordaire, 1859: 108 [stem: *Leptod-*]. Type genus: *Leptodes* Dejean, 1834. Comment: original vernacular name available (Art. 11.7.2): first used in latinized form by Baudi di Selve (1875: 74, as Leptodidae), generally accepted as in Gebien (1910a: 91, as Leptodinae).

Tribe Nycteliini Solier, 1834

Nyctélites Solier, 1834: 502 [stem: *Nycteli-*]. Type genus: *Nyctelia* Latreille, 1825. Comment: original vernacular name available (Art. 11.7.2): first used in latinized form by Hope (1840a: 127, as Nyctelidae [incorrect stem formation]), generally accepted as in Bouchard et al. (2005: 502, as Nycteliini); incorrect original stem formation, not in prevailing usage.

Tribe Nyctoporini Lacordaire, 1859

Nyctoporides Lacordaire, 1859: 130 [stem: *Nyctopor-*]. Type genus: *Nyctoporis* Eschscholtz, 1831. Comment: original vernacular name available (Art. 11.7.2): first used in latinized form by J. L. LeConte (1862: 219, as Nyctoporini), generally accepted as in Bouchard et al. (2005: 502, as Nyctoporini); current spelling maintained (Art. 29.5): incorrect stem formation in prevailing usage (should be *Nyctoporid-*).

Tribe Phrynocarenini Gebien, 1928

Phrynocareninae Gebien, 1928: 105 [stem: *Phrynocaren-*]. Type genus: *Phrynocarenum* Gebien, 1928.

Tribe Physogasterini Lacordaire, 1859

Physogastérides Lacordaire, 1859: 206 [stem: *Physogaster-*]. Type genus: *Physogaster* Lacordaire, 1830 [this genus name has been credited to Guérin-Méneville (1834: 2) in the literature, however the description of *Physogaster mendocinus* by Lacordaire (1830a: 276) makes the genus name available from that author and year]. Comment: original vernacular name available (Art. 11.7.2): first used in latinized form by H. C. C. Burmeister (1875: 488, as Physogasteridae), generally accepted as in Bouchard et al. (2005: 502, as Physogasterini).

Tribe Pimeliini Latreille, 1802

Pimeliariae Latreille, 1802: 166 [stem: *Pimeli-*]. Type genus: *Pimelia* Fabricius, 1775.

Pimidinia Rafinesque, 1815: 113 [stem: *Pimidi-*]. Type genus: *Pimidia* Rafinesque, 1815 [syn. of *Pimelia* Fabricius, 1775]. Comment: incorrect original stem formation, not in prevailing usage.

*Platyopes Motschulsky, 1849: 58 [stem: *Platyop-*]. Type genus: *Platyope* Fischer von Waldheim, 1820. Comment: original vernacular name unavailable (Art. 11.7.2): subsequently used in latinized form but not generally attributed to Motschulsky (1849).

Platyopidae Semenov, 1893a: 260 [stem: *Platyop-*]. Type genus: *Platyope* Fischer von Waldheim, 1820. Comment: family-group name proposed as new without reference to *Platyopes* Motschulsky, 1849; Platyopidae Huene, 1931 has been used in amphibians (type genus *Platyops* Twelvetrees, 1880) but this name is permanently invalid since it is based on a preoccupied genus name.

Leucolaephusini Pierre, 1961: 558 [stem: *Leucolaeph-*]. Type genus: *Leucolaephus* Lucas, 1859. Comment: incorrect original stem formation, not in prevailing usage.

Tribe Praociini Eschscholtz, 1829

Praocidae Eschscholtz, 1829b: 5 [stem: *Praoci-*]. Type genus: *Praocis* Eschscholtz, 1829. Comment: incorrect original stem formation, not in prevailing usage.

Tribe Sepidiini Eschscholtz, 1829

Sepidiae Eschscholtz, 1829b: 4 [stem: *Sepidi-*]. Type genus: *Sepidium* Fabricius, 1775.

Subtribe Hypomelina Koch, 1955

Hypomelina Koch, 1955: 36 [stem: *Hypomel-*]. Type genus: *Hypomelus* Solier, 1843.

Subtribe Molurina Solier, 1834

Molurites Solier, 1834: 505 [stem: *Molur-*]. Type genus: *Moluris* Latreille, 1802. Comment: original vernacular name available (Art. 11.7.2): first used in latinized form by Schaum (1859: 68, as Moluridae), generally accepted as in Bouchard et al. (2005: 502, as Molurina); current spelling maintained (Art. 29.3.1.1): incorrect stem formation in prevailing usage (should be *Molurid-*).

*Psammodoiden Koch, 1953a: 138 [stem: *Psammod-*]. Type genus: *Psammodes* Kirby, 1819. Comment: original vernacular name unavailable (Art. 11.7.2): proposed after 1899; Psammodidae has been used in the Scarabaeidae literature but this is based on an incorrect stem formation (type genus *Psammodius* Fallén, 1807, stem *Psammodi-*).

*Phrynocoloiden Koch, 1953a: 138 [stem: *Phrynocol-*]. Type genus: *Phrynocolus* Lacordaire, 1859. Comment: original vernacular name unavailable (Art. 11.7.2): proposed after 1899.

Subtribe Oxurina Koch, 1955

Oxurina Koch, 1955: 34 [stem: *Oxur-*]. Type genus: *Oxura* Kirby, 1819.

Subtribe Phanerotomeina Koch, 1958

*Phanerotomoide Koch, 1953a: 138 [stem: *Phanerotom-*]. Type genus: *Phanerotoma* Solier, 1843 [preoccupied genus name, not *Phanerotoma* Wesmael, 1838 [Hymenoptera]; syn. of *Phanerotomea* Koch, 1957]. Comment: family-group name unavailable (Art. 11.7.1.1): original name not proposed as a noun.

Phanerotomina Koch, 1955: 37 [stem: *Phanerotom-*]. Type genus: *Phanerotoma* Solier, 1843 [preoccupied genus name, not *Phanerotoma* Wesmael, 1838 [Hymenoptera]; syn. of *Phanerotomea* Koch, 1957]. Comment: permanently invalid (Art. 39): based on preoccupied type genus; Phanerotomini Baker, 1925 (type genus *Phanerotoma* Wesmael, 1838) is available in Hymenoptera: Braconidae.

Phanerotomeina Koch, 1958: 58 [stem: *Phanerotome-*]. Type genus: *Phanerotomea* Koch, 1958. Comment: replacement name for Phanerotomina Koch, 1955 because of the homonymy of the type genus.

Subtribe Sepidiina Eschscholtz, 1829

Sepidiae Eschscholtz, 1829b: 4 [stem: *Sepidi-*]. Type genus: *Sepidium* Fabricius, 1775.

Subtribe Trachynotina Koch, 1955

*Trachynotides Brullé, 1832: 189 [stem: *Trachynot-*]. Type genus: *Trachynotus* Latreille, 1828. Comment: original vernacular name unavailable (Art. 11.7.2): subsequently used in latinized form but not generally attributed to Brullé (1832).

Trachynotina Koch, 1955: 34, in key [stem: *Trachynot-*]. Type genus: *Trachynotus* Latreille, 1828. Comment: Trachynotidae Schrammen, 1924 in Porifera is permanently invalid because it was based on the junior homonym *Trachynotus* Schrammen, 1924; the fish name Trachynotinae Gill, 1861 (type genus *Trachynotus* Agassiz, 1846, an unjustified emendation of *Trachinotus* Lacepède, 1801 and also a junior homonym of *Trachynotus* Latreille, 1828) was later corrected to Trachinotinae, which is used as valid today; Trachynotoidae Förster, 1869 has also been used in ichneumonid literature but this name is permanently invalid since it was based on the junior homonym *Trachynotus* Gravenhorst, 1829.

Tribe STENOSINI Schaum, 1859 (1834)

TAGÉNITES Solier, 1834: 503 [stem: *Tageni-*]. Type genus: *Tagenia* Latreille, 1802 [syn. of *Stenosis* Herbst, 1799]. Comment: original vernacular name available (Art. 11.7.2): first used in latinized form by Hope (1840a: 127, as TAGENIDAE), generally accepted as in G. R. Waterhouse (1845: 30, as TAGENIIDAE [incorrect stem formation]; use of younger name STENOSINI conserved (Art. 40.2) (see Bouchard et al. 2005: 523); incorrect original stem formation, not in prevailing usage.

STENOSIDAE Schaum, 1859: 66 [stem: *Stenos-*]. Type genus: *Stenosis* Herbst, 1799. Comment: published before end of February 1859; this family-group name was also used in the same year by Lacordaire (1859 [before 27 June]: 101, as STÉNOSIDES); use of family-group name conserved over TAGENINI Solier, 1834 (Art. 40.2) (see Bouchard et al. 2005: 523); current spelling maintained (Art. 29.5): incorrect stem formation in prevailing usage (should be *Stenose-*).

PLATAMODINA Reitter, 1900: 82 [stem: *Platamod-*]. Type genus: *Platamodes* Ménétriés, 1849.

TYPHLUSECHINI Casey, 1907: 281 [stem: *Typhlusech-*]. Type genus: *Typhlusechus* Linell, 1897.

ARAEOSCHIZINI Casey, 1907: 484 [stem: *Araeoschiz-*]. Type genus: *Araeoschizus* J. L. LeConte, 1851.

DICHILLINA Reitter, 1916: 137, in key [stem: *Dichill-*]. Type genus: *Dichillus* Jacquelin du Val, 1860.

HARVENGINA Ferrer, 2004: 370 [stem: *Harvengi-*]. Type genus: *Harvengia* Ferrer, 2004. Comment: incorrect original stem formation, not in prevailing usage.

Tribe TENTYRIINI Eschscholtz, 1831

TENTYRIDAE Eschscholtz, 1831: 4 [stem: *Tentyri-*]. Type genus: *Tentyria* Latreille, 1802 [placed on the Official List of Generic Names in Zoology (ICZN 2010c)]. Comment: incorrect original stem formation, not in prevailing usage.

GNATHOSIIDES Lacordaire, 1859: 33 [stem: *Gnathosi-*]. Type genus: *Gnathosia* Fischer von Waldheim, 1821. Comment: original vernacular name available (Art. 11.7.2): first used in latinized form by J. L. LeConte (1862: 213, as GNATHOSIINI), generally accepted as in Skopin (1979: 170, as GNATHOSIINA).

*HYPÉROPIDES Lacordaire, 1859: 60 [stem: *Hyperop-*]. Type genus: *Hyperops* Eschscholtz, 1831. Comment: original vernacular name unavailable (Art. 11.7.2): not subsequently latinized.

CAPNISINI Casey, 1907: 279 [stem: *Capnis-*]. Type genus: *Capnisa* Dejean, 1836 [syn. of *Gnathosia* Fisher von Waldheim, 1821].

PACHYCERINA Skopin, 1979: 170, in key [stem: *Pachycer-*]. Type genus: *Pachycera* Eschscholtz, 1831 [preoccupied genus name, not *Pachycera* Billberg, 1820 [Hemiptera]; syn. of *Oedenocera* Reiche, 1862]. Comment: permanently invalid (Art. 39): based on preoccupied type genus.

Himatismina Skopin, 1979: 170, in key [stem: *Himatism-*]. Type genus: *Himatismus* Erichson, 1843.

Tribe Thinobatini Lacordaire, 1859

Thinobatides Lacordaire, 1859: 63 [stem: *Thinobat-*]. Type genus: *Thinobatis* Eschscholtz, 1831. Comment: original vernacular name available (Art. 11.7.2): first used in latinized form by J. L. LeConte (1862: 214, as Thinobatini), generally accepted as in Bouchard et al. (2005: 502, as Thinobatini); current spelling maintained (Art. 29.3.1.1): incorrect stem formation in prevailing usage (should be *Thinobatid-*).

Tribe Trilobocarini Lacordaire, 1859

Tribolocarides Lacordaire, 1859: 69 [stem: *Trilobocar-*]. Type genus: *Trilobocara* Solier, 1851 [as *Tribolocara*, incorrect subsequent spelling of type genus name, not in prevailing usage]. Comment: original vernacular name available (Art. 11.7.2): first used in latinized form and generally accepted as in Kolbe (1887: 51, as Tribolocarini); incorrect stem formation, not in prevailing usage.

Salaxini Casey, 1907: 282 [stem: *Salac-*]. Type genus: *Salax* Guérin-Méneville, 1834. Comment: incorrect original stem formation, not in prevailing usage.

Tribe Vacronini Gebien, 1910

Vacroninae Gebien, 1910a: 118 [stem: *Vacron-*]. Type genus: *Vacronus* Casey, 1907 [syn. of *Alaephus* G. H. Horn, 1870].

*Eupsophulites Kwieton, 1982: 96 [stem: *Eupsophul-*]. Type genus: *Eupsophulus* Cockerell, 1906. Comment: unavailable family-group name, proposed after 1930 without description or bibliographic reference to such a description (Art. 13.1).

Tribe Zophosini Solier, 1834

Zophosites Solier, 1834: 597 [stem: *Zophos-*]. Type genus: *Zophosis* Latreille, 1802. Comment: original vernacular name available (Art. 11.7.2): first used in latinized form by Stein (1868: 78, as Zophosini), generally accepted as in Bouchard et al. (2005: 501, as Zophosini); current spelling maintained (Art. 29.5): incorrect stem formation in prevailing usage (should be *Zophose-*).

*Onychosites A. Deyrolle, 1867: 79 [stem: *Onychose-*]. Type genus: *Onychosis* Deyrolle, 1867. Comment: original vernacular name unavailable (Art. 11.7.2): not subsequently latinized; incorrect original stem formation, not in prevailing usage.

*Cardiosites A. Deyrolle, 1867: 79 [stem: *Cardiose-*]. Type genus: *Cardiosis* Deyrolle, 1867. Comment: original vernacular name unavailable (Art. 11.7.2): not subsequently latinized; incorrect original stem formation, not in prevailing usage.

Dactylocalcarini Gebien, 1938a: 46 [stem: *Dactylocalcar-*]. Type genus: *Dactylocalcar* Gebien, 1938.

Subfamily TENEBRIONINAE Latreille, 1802

TENEBRIONITES Latreille, 1802: 165 [stem: *Tenebrion-*]. Type genus: *Tenebrio* Linnaeus, 1758. Comment: First Reviser (TENEBRIONINAE Latreille, 1802 vs HELOPINAE Latreille, 1802) not determined, current usage maintained

Tribe ACROPTERONINI Doyen, 1989

ACROPTERONINI Doyen, 1989: 288 [stem: *Acropteron-*]. Type genus: *Acropteron* Perty, 1832. Comment: current spelling maintained (Art. 29.5): incorrect stem formation in prevailing usage (should be *Acropter-*).

Tribe ALPHITOBIINI Reitter, 1917

ALPHITOBIINI Reitter, 1917: 58 [stem: *Alphitobi-*]. Type genus: *Alphitobius* Stephens, 1829 [placed on the Official List of Generic Names in Zoology (ICZN 1975)].

Tribe AMARYGMINI Gistel, 1848

AMARYGMIIDAE Gistel, 1848: [10] [stem: *Amarygm-*]. Type genus: *Amarygmus* Dalman, 1823. Comment: incorrect original stem formation, not in prevailing usage.

MÉGACANTHIDES Lacordaire, 1859: 467 [stem: *Megacanth-*]. Type genus: *Megacantha* Westwood, 1843. Comment: original vernacular name available (Art. 11.7.2): first used in latinized form and generally accepted as in Quedenfeldt (1885: 20, as MEGACANTHIDAE).

MÉRACANTHIDES Lacordaire, 1859: 464 [stem: *Meracanth-*]. Type genus: *Meracantha* Kirby, 1837. Comment: original vernacular name available (Art. 11.7.2): first used in latinized form by J. L. LeConte (1862: 240, as MERACANTHINI), generally accepted as in Gebien (1911: 567, as MERACANTHINAE).

MEGACANTHINA Ardoin, 1962: 960 [stem: *Megacanth-*]. Type genus: *Megacantha* Westwood, 1843. Comment: family-group name proposed as new without reference to MÉGACANTHIDES Lacordaire, 1859.

Tribe AMPHIDORINI LeConte, 1862

*NYCTERINOIDES Solier, 1851: 210 [stem: *Nycterin-*]. Type genus: *Nycterinus* Eschscholtz, 1829. Comment: original vernacular name unavailable (Art. 11.7.2): subsequently used in latinized form but not generally attributed to Solier (1851).

*EMBAPHIONIDES Lacordaire, 1859: 151 [stem: *Embaphi-*]. Type genus: *Embaphion* Say, 1824. Comment: original vernacular name unavailable (Art. 11.7.2): subsequently used in latinized form, e.g., D. B. Thomas (2005: 549, as EMBAPHIONINI), but not generally attributed to Lacordaire (1859); EMBAPHIONINI, as used by D. B. Thomas (2005) is unavailable because it was proposed after 1930 without description or bibliographic reference to such a description (Art. 13.1); incorrect original stem formation, not in prevailing usage.

Amphidorae J. L. LeConte, 1862: 239 [stem: *Amphidor-*]. Type genus: *Amphidora* Eschscholtz, 1829.

Eleodiini Blaisdell, 1909: 27 [stem: *Eleod-*]. Type genus: *Eleodes* Eschscholtz, 1829. Comment: incorrect original stem formation, not in prevailing usage.

Eleodopsinae Blaisdell, 1939: 51 [stem: *Eleodopse-*]. Type genus: *Eleodopsis* Blaisdell, 1939 [syn. of *Eleodes* Eschscholtz, 1829]. Comment: incorrect original stem formation, not in prevailing usage.

Lariversiina La Rivers, 1948: 98 [stem: *Lariversi-*]. Type genus: *Lariversius* Blaisdell, 1947.

Trogloderina La Rivers, 1948: 98 [stem: *Trogloder-*]. Type genus: *Trogloderus* J. L. LeConte, 1879.

*Nycterini Doyen et al., 1990: 244 [stem: *Nycterin-*]. Type genus: *Nycterinus* Eschscholtz, 1829. Comment: unavailable family-group name, proposed after 1930 without description or bibliographic reference to such a description (Art. 13.1); incorrect original stem formation, not in prevailing usage.

Tribe Apocryphini Lacordaire, 1859

Apocryphides Lacordaire, 1859: 432 [stem: *Apocryph-*]. Type genus: *Apocrypha* Eschscholtz, 1831. Comment: original vernacular name available (Art. 11.7.2): first used in latinized form by J. L. LeConte (1862: 217, as Apocryphini), generally accepted as in Gebien (1911: 503, as Apocryphinae).

Diplocyrthini Escalera, 1914b: 355 [stem: *Diplocyrt-*]. Type genus: *Diplocyrtus* Quendenfeldt, 1887 [as *Diplocyrthus*, incorrect subsequent spelling of type genus name, not in prevailing usage]. Comment: incorrect original stem formation, not in prevailing usage.

Tribe Blaptini Leach, 1815

Blapsida Leach, 1815: 101 [stem: *Blapt-*]. Type genus: *Blaps* Fabricius, 1775.

Subtribe Blaptina Leach, 1815

Blapsida Leach, 1815: 101 [stem: *Blapt-*]. Type genus: *Blaps* Fabricius, 1775. Comment: although the original stem formation (*Blaps-*) was correct, the stem *Blapt-* has been used since it was first used by C. G. Thomson (1859: 114, as Blaptidae) and is also used here; an application to the Commission is needed to conserve the current spelling.

Subtribe Gnaptorina Medvedev, 2001

Gnaptorina G. S. Medvedev, 2001: 29 [stem: *Gnaptor-*]. Type genus: *Gnaptor* Brullé, 1832.

Subtribe Gnaptorinina Medvedev, 2001

Gnaptorinina G. S. Medvedev, 2001: 31 [stem: *Gnaptorin-*]. Type genus: *Gnaptorina* Reitter, 1887.

Subtribe Prosodina Skopin, 1960

Prosodina Skopin, 1960: 48 [stem: *Prosod-*]. Type genus: *Prosodes* Eschscholtz, 1829.

Subtribe Remipedellina Semenov, 1907

Remipedellini Semenov, 1907a: 259 [stem: *Remipedell-*]. Type genus: *Remipedella* Semenov, 1907. Comment: this family-group name was also used in the same year by Semenov (1907b: 176, as Remipedellini); priority for the two publications could not be established although Remipedellini Semenov, 1907b could not be considered available since the type genus *Remipedella* was made available only in Semenov (1907a).

Tribe Bolitophagini Kirby, 1837 *nomen protectum*

Eledonaedes Billberg, 1820b: 392 [stem: *Eledon-*]. Type genus: *Eledona* Latreille, 1797. Comment: *nomen oblitum* (see Appendix 1); the name Eledoninae (type genus *Eledone* Leach, 1817) is used as valid in cephalopods (earliest usage in cephalopods found is Eledonidae Rochebrune, 1884).

Bolitophagidae Kirby, 1837: 236 [stem: *Bolitophag-*]. Type genus: *Bolitophagus* Illiger, 1798. Comment: *nomen protectum* (see Appendix 1).

Rhipidandri J. L. LeConte, 1862: 236 [stem: *Rhipidandr-*]. Type genus: *Rhipidandrus* J. L. LeConte, 1862.

Eutomides Lacordaire, 1865: 369 [stem: *Eutom-*]. Type genus: *Eutomus* Lacordaire, 1865 [syn. of *Rhipidandrus* J. L. LeConte, 1862]. Comment: original vernacular name available (Art. 11.7.2): first used in latinized form and generally accepted as in Ferrari (1867: 3, as Eutomides [treated as Latin]); name treated as unavailable by Bouchard et al. (2005: 508) but recently considered available by Alonso-Zarazaga and Lyal (2009: 7); taxon originally described in Curculionidae: Scolytinae.

Tribe Centronopini Doyen, 1989

Centronopini Doyen, 1989: 284 [stem: *Centronop-*]. Type genus: *Centronopus* Solier, 1848. Comment: current spelling maintained (Art. 29.5): incorrect stem formation in prevailing usage (should be *Centronopod-*).

Tribe Cerenopini Horn, 1870

Cerenopi G. H. Horn, 1870: 325 [stem: *Cerenop-*]. Type genus: *Cerenopus* J. L. LeConte, 1851. Comment: current spelling maintained (Art. 29.5): incorrect stem formation in prevailing usage (should be *Cerenopod-*).

Tribe Dissonomini Medvedev, 1968

*Hétérophilates Mulsant and Rey, 1859: 6 [stem: *Heterophyl-*]. Type genus: *Heterophylus* Mulsant and Rey, 1859 [preoccupied genus name, not *Heterophylus* Klug, 1833 [Coleoptera: Tenebrionidae: Diaperinae]; syn. of *Dis-*

sonomus Jacquelin du Val, 1861]. Comment: original vernacular name unavailable (Art. 11.7.2): not subsequently latinized; if found to be available then permanently invalid (Art. 39): based on preoccupied type genus; incorrect original stem formation, not in prevailing usage.

*Dissonomites Jacquelin du Val, 1861: 280 [stem: *Dissonom-*]. Type genus: *Dissonomus* Jacquelin du Val, 1861. Comment: original vernacular name unavailable (Art. 11.7.2): subsequently used in latinized form but not generally attributed to Jacquelin du Val (1861).

Dissonomini G. S. Medvedev, 1968: 211 [stem: *Dissonom-*]. Type genus: *Dissonomus* Jacquelin du Val, 1861.

Tribe Eulabini Horn, 1870

Eulabes G. H. Horn, 1870: 323 [stem: *Eulab-*]. Type genus: *Eulabis* Eschscholtz, 1829.

Tribe Falsocossyphini Ferrer, 2006

Falsocossyphini Ferrer, 2006: 77 [stem: *Falsocossyph-*]. Type genus: *Falsocossyphus* Pic, 1916.

Tribe Heleini Fleming, 1821

Heleadae Fleming, 1821: 51 [stem: *Hele-*]. Type genus: *Helea* Latreille, 1816.

Subtribe Asphalina Matthews and Lawrence, 2005

Asphalina E. G. Matthews and Lawrence, 2005: 544 [stem: *Asphal-*]. Type genus: *Asphalus* Pascoe, 1868.

Subtribe Cyphaleina Lacordaire, 1859

Cyphaléides Lacordaire, 1859: 407 [stem: *Cyphale-*]. Type genus: *Cyphaleus* Westwood, 1841. Comment: original vernacular name available (Art. 11.7.2): first used in latinized form by Pascoe (1866b: 470, as Cyphaleinae), generally accepted as in Bouchard et al. (2005: 502, as Cyphaleina); precedence (Cyphaleini Lacordaire, 1859 vs Nyctozoilini Lacordaire, 1859) given to taxon originally proposed at the higher rank (Art. 24.1).

Nyctozoïlides Lacordaire, 1859: 349 [stem: *Nyctozoil-*]. Type genus: *Nyctozoilus* Guérin-Méneville, 1830. Comment: original vernacular name available (Art. 11.7.2): first used in latinized form by Bates (1872b: 98, as Nyctozoilides [treated as Latin]), generally accepted as in Carter (1911: 138, as Nyctozoilides [treated as Latin]).

Subtribe Heleina Fleming, 1821

Heleadae Fleming, 1821: 51 [stem: *Hele-*]. Type genus: *Helea* Latreille, 1804. Comment: incorrect original stem formation, not in prevailing usage.

Briseinae Carter, 1924: 33 [stem: *Bris-*]. Type genus: *Brises* Pascoe, 1869. Comment: incorrect original stem formation, not in prevailing usage.

Tribe HELOPINI Latreille, 1802

HELOPII Latreille, 1802: 176 [stem: *Helop-*]. Type genus: *Helops* Fabricius, 1775 [placed on the Official List of Generic Names in Zoology (ICZN 2009d)].

Subtribe HELOPINA Latreille, 1802

HELOPII Latreille, 1802: 176 [stem: *Helop-*]. Type genus: *Helops* Fabricius, 1775 [placed on the Official List of Generic Names in Zoology (ICZN 2009d)].

HYPULIA Rafinesque, 1815: 114 [stem: *Hypul-*]. Type genus: *Hypulus* Rafinesque, 1815 [syn. of *Helops* Fabricius, 1775].

*ENOPLOPITES Solier, 1848: 155 [stem: *Enoplopod-*]. Type genus: *Enoplopus* Solier, 1848. Comment: original vernacular name unavailable (Art. 11.7.2): subsequently used in latinized form but not generally attributed to Solier (1848); incorrect original stem formation, not in prevailing usage.

*HÉDYPHANES Motschulsky, 1849: 57 [stem: *Hedyphan-*]. Type genus: *Hedyphanes* Fischer von Waldheim, 1820. Comment: original vernacular name unavailable (Art. 11.7.2): subsequently used in latinized form but not generally attributed to Motschulsky (1849).

ENOPLOPINI Reitter, 1917: 62 [stem: *Enoplopod-*]. Type genus: *Enoplopus* Solier, 1848. Comment: incorrect original stem formation, not in prevailing usage.

NEPHODINI Reitter, 1917: 63 [stem: *Nephod-*]. Type genus: *Nephodes* Blanchard, 1845.

HEDYPHANINA Reitter, 1922b: 6 [stem: *Hedyphan-*]. Type genus: *Hedyphanes* Fischer von Waldheim, 1820.

*STENOTRICHINI Blaisdell, 1939: 50 [stem: *Stenotrich-*]. Type genus: *Stenotrichus* J. L. LeConte, 1862 [syn. of *Helops* Fabricius, 1775]. Comment: unavailable family-group name, proposed after 1930 without description or bibliographic reference to such a description (Art. 13.1).

Subtribe CYLINDRINOTINA Español, 1956

*XANTHOMINI Antoine, 1949: 162 [stem: *Xanthom-*]. Type genus: *Xanthomus* Mulsant, 1854. Comment: unavailable family-group name, proposed after 1930 without description or bibliographic reference to such a description (Art. 13.1).

*ECTROMOPSINI Antoine, 1949: 162 [stem: *Ectromopse-*]. Type genus: *Ectromopsis* Antoine, 1949. Comment: unavailable family-group name, proposed after 1930 without description or bibliographic reference to such a description (Art. 13.1); incorrect original stem formation, not in prevailing usage.

CYLINDRONOTINI Español, 1956: 84 [stem: *Cylindrinot-*]. Type genus: *Cylindrinotus* Faldermann, 1837 [as *Cylindronotus*, unjustified emendation of type genus name by Agassiz (1846b: 111), not in prevailing usage]. Comment: incorrect original stem formation, not in prevailing usage; correction of stem by Bouchard et al. (2005: 509).

Tribe Helopinini Lacordaire, 1859

Hélopinides Lacordaire, 1859: 457 [stem: *Helopin-*]. Type genus: *Helopinus* Solier, 1848.

Subtribe Aptilina Koch, 1958

Aptilina Koch, 1958: 139 [stem: *Aptil-*]. Type genus: *Aptila* Fåhraeus, 1870.

Subtribe Helopinina Lacordaire, 1859

Hélopinides Lacordaire, 1859: 457 [stem: *Helopin-*]. Type genus: *Helopinus* Solier, 1848. Comment: original vernacular name available (Art. 11.7.2): first used in latinized form by Bertkau (1875: 356, as Helopinini), generally accepted as in Gebien (1911: 563, as Helopininae).

Drosochrini Koch, 1958: 133 [stem: *Drosochr-*]. Type genus: *Drosochrus* Erichson, 1843. Comment: unnecessary replacement name for Helopinini Lacordaire, 1859.

Subtribe Micrantereina Reitter, 1917

Micrantereini Reitter, 1917: 60 [stem: *Micrantere-*]. Type genus: *Micrantereus* Solier, 1848.

Subtribe Oncosomina Koch, 1958

Oncosomina Koch, 1958: 134 [stem: *Oncosom-*]. Type genus: *Oncosoma* Westwood, 1843 [unjustified emendation of *Ogcosoma* by Agassiz (1846b: 257), in prevailing usage, treated as justified emendation (Art. 33.2.3.1)]. Comment: current spelling maintained (Art. 29.5): incorrect original stem formation in prevailing usage (should be *Oncosomat-*).

Tribe Melanimonini Seidlitz, 1894 (1854)

Microzoumates Mulsant, 1854: 176 [stem: *Microzo-*]. Type genus: *Microzoum* Dejean, 1834 [syn. of *Melanimon* Steven 1829]. Comment: original vernacular name available (Art. 11.7.2): first used in latinized form by Seidlitz (1889 [Gatt.]: 128, as Microzoina), generally accepted as in Bouchard et al. (2005: 523, as Microzoini); use of Melanimonini Seidlitz, 1894 conserved (Art. 40.2) (see Bouchard et al. 2005: 523); incorrect original stem formation, not in prevailing usage.

Melanimonina Seidlitz, 1894: 449 [stem: *Melanimon-*]. Type genus: *Melanimon* Steven, 1829. Comment: use of younger family-group name conserved (Art. 40.2) (see Bouchard et al. 2005).

Tribe Opatrini Brullé, 1832

Opatrites Brullé, 1832: 213 [stem: *Opatr-*]. Type genus: *Opatrum* Fabricius, 1775.

Subtribe Heterocheirina Koch, 1956

Heterocheirini Koch, 1956: 43 [stem: *Heterocheir-*]. Type genus: *Heterocheira* Lacordaire, 1859.

Subtribe Heterotarsina Blanchard, 1845

Hétérotarsites Blanchard, 1845b: 14 [stem: *Heterotars-*]. Type genus: *Heterotarsus* Latreille, 1829. Comment: original vernacular name available (Art. 11.7.2): first used in latinized form by J. L. LeConte (1862: 231, as Heterotarsini), generally accepted as in Bouchard et al. (2005: 502, as Heterotarsina).

Subtribe Opatrina Brullé, 1832

Opatrites Brullé, 1832: 213 [stem: *Opatr-*]. Type genus: *Opatrum* Fabricius, 1775. Comment: original vernacular name available (Art. 11.7.2): first used in latinized form by Shuckard (1839b: 49, as Opatridae), generally accepted as in Bouchard et al. (2005: 502, as Opatrini).

*Gonocéphalites Solier, 1834: 498 [stem: *Gonocephal-*]. Type genus: *Gonocephalum* Solier, 1834. Comment: original vernacular name unavailable (Art. 11.7.2): subsequently used in latinized form but not generally attributed to Solier (1834).

*Blapstinoides Solier, 1851: 231 [stem: *Blapstin-*]. Type genus: *Blapstinus* Dejean, 1821. Comment: original vernacular name unavailable (Art. 11.7.2): subsequently used in latinized form but not generally attributed to Solier (1851).

Blapstinites Mulsant and Rey, 1853: 258 [stem: *Blapstin-*]. Type genus: *Blapstinus* Dejean, 1821. Comment: original vernacular name available (Art. 11.7.2): first used in latinized form and generally accepted as in J. L. LeConte (1862: 227, as Blapstini).

Stizopides Lacordaire, 1859: 258 [stem: *Stizopod-*]. Type genus: *Stizopus* Erichson, 1843. Comment: original vernacular name available (Art. 11.7.2): first used in latinized form by Heyne and Taschenberg (1907: 207, as Stizopini), generally accepted as in Gebien (1938b: 72 [393], as Stizopini); incorrect original stem formation, not in prevailing usage.

Sclérides Lacordaire, 1859: 263 [stem: *Scler-*]. Type genus: *Scleron* Hope, 1840 [syn. of *Sclerum* Dejean, 1834]. Comment: original vernacular name available (Art. 11.7.2): first used in latinized form by Seidlitz (1890 [Gatt.]: 129, as Sclerina), generally accepted as in Seidlitz (1894: 415, as Sclerina).

*Penthicaires Mulsant and Rey, 1859: 5 [stem: *Penthic-*]. Type genus: *Penthicus* Faldermann, 1836. Comment: original vernacular name unavailable (Art. 11.7.2): subsequently used in latinized form, e.g., Heyne and Taschenberg (1907: 208, as Penthicini, under the valid tribe name Opatrini), but not generally accepted as valid.

*Caediaires Mulsant and Rey, 1859: 124 [stem: *Caedi-*]. Type genus: *Caedius* Blanchard, 1845. Comment: original vernacular name unavailable (Art. 11.7.2): subsequently used in latinized form, e.g., Heyne and Taschenberg (1907: 208, as Caediini, under the valid tribe name Opatrini), but not generally accepted as valid.

*Blacodaires Mulsant and Rey, 1859: 93 [stem: *Blacod-*]. Type genus: *Blacodes* Blanchard, 1845 [syn. of *Blenosia* Laporte, 1840]. Comment: original vernacular name unavailable (Art. 11.7.2): subsequently used in latinized form, e.g., Heyne and Taschenberg (1907: 208, as Blacodini, under the valid tribe name Opatrini), but not generally accepted as valid.

*Clitobiates Mulsant and Rey, 1859: 141 [stem: *Clitobi-*]. Type genus: *Clitobius* Mulsant and Rey, 1859. Comment: original vernacular name unavailable (Art. 11.7.2): subsequently used in latinized form, e.g., Heyne and Taschenberg (1907: 208, as Clitobiini, under the valid tribe name Opatrini), but not generally accepted as valid.

Gonocéphalates Mulsant and Revelière, 1861: 154 [stem: *Gonocephal-*]. Type genus: *Gonocephalum* Solier, 1834. Comment: original vernacular name available (Art. 11.7.2): first used in latinized form and generally accepted as in Gerstaecker (1861: 483, as Gonocephalidae).

*Dilamites Jacquelin du Val, 1861: 279 [stem: *Dilam-*]. Type genus: *Dilamus* Jacquelin du Val, 1861. Comment: original vernacular name unavailable (Art. 11.7.2): not subsequently latinized.

Ammobiini Desbrochers des Loges, 1902b: 1 [stem: *Ammobi-*]. Type genus: *Ammobius* Guérin-Méneville, 1844.

Emmallina Koch, 1956: 51 [stem: *Emmal-*]. Type genus: *Emmalus* Erichson, 1843 [as *Emmallus*, unjustified emendation of type genus name, not in prevailing usage]. Comment: incorrect original stem formation, not in prevailing usage.

Stenolamina Koch, 1956: 49 [stem: *Stenolam-*]. Type genus: *Stenolamus* Gebien, 1920.

Subtribe Neopachypterina Bouchard, Löbl and Merkl, 2007

Pachyptérates Mulsant and Rey, 1859: 83 [stem: *Pachypter-*]. Type genus: *Pachypterus* Lucas, 1846 [preoccupied genus name, not *Pachypterus* Swainson, 1839 [Pisces]; syn. of *Neopachypterus* Bouchard et al., 2007]. Comment: original vernacular name available (Art. 11.7.2): first used in latinized form and generally accepted as in Desbrochers des Loges (1901: 146, as Pachypterini [actually spelled Pachnephorini but this name was corrected to Pachypterini later by Desbroches des Loges (1902a: 85)]; permanently invalid (Art. 39): based on preoccupied type genus.

Pachypterini G. S. Medvedev, 1968: 247 [stem: *Pachypter-*]. Type genus: *Pachypterus* Lucas, 1846 [preoccupied genus name, not *Pachypterus* Swainson, 1839 [Pisces]; syn. of *Neopachypterus* Bouchard et al., 2007]. Comment: family-group name proposed as new without reference to Pachyptérates Mulsant and Rey, 1859; permanently invalid (Art. 39): based on preoccupied type genus.

Neopachypterina Bouchard et al., 2007: 386 [stem: *Neopachypter-*]. Type genus: *Neopachypterus* Bouchard et al., 2007. Comment: replacement name for Pachyptérates Mulsant and Rey, 1859 and Pachypterini G. S. Medvedev, 1968 because of the homonymy of the type genus.

Tribe Palorini Matthews, 2003

*Palorinae E. G. Matthews, 2003a: 50 [stem: *Palor-*]. Type genus: *Palorus* Mulsant, 1854. Comment: unavailable name (Art. 16): not explicitly indicated as new.

Palorinae E. G. Matthews, 2003b: 7 [stem: *Palor-*]. Type genus: *Palorus* Mulsant, 1854. Comment: validation of Palorinae Matthews (2003a).

Tribe Pedinini Eschscholtz, 1829

Pediniden Eschscholtz, 1829b: 4 [stem: *Pedin-*]. Type genus: *Pedinus* Latreille, 1797.

Subtribe Dendarina Mulsant and Rey, 1854

Pandarites Mulsant and Rey, 1854: 153 [stem: *Pandar-*]. Type genus: *Dendarus* Dejean, 1821 [as *Pandarus*, unjustified emendation of type genus name, not in prevailing usage]. Comment: original vernacular name available (Art. 11.7.2): first used in latinized form by Gerstaecker (1854: 226, as Pandaridae [incorrect stem formation]), and generally accepted as in Seidlitz (1889[Gatt.]: 126, as Dendarina).

*Omocratates Mulsant and Rey, 1854: 266, in key [stem: *Omocrat-*]. Type genus: *Omocrates* Mulsant, 1854 [preoccupied genus name, not *Omocrates* H. C. C. Burmeister, 1844 [Coleoptera: Scarabaeidae]; syn. of *Phylan* Dejean, 1821]. Comment: original vernacular name unavailable (Art. 11.7.2): not subsequently latinized; if found to be available then permanently invalid (Art. 39): based on preoccupied type genus.

*Micrositates Mulsant and Rey, 1854: 274 [stem: *Microsit-*]. Type genus: *Micrositus* Mulsant and Rey, 1854. Comment: original vernacular name unavailable (Art. 11.7.2): not subsequently latinized.

*Isocérates Mulsant and Rey, 1854: 188 [stem: *Isocer-*]. Type genus: *Isocerus* Dejean, 1821 [preoccupied genus name, not *Isocerus* Illiger, 1802 [Coleoptera: Cerambycidae]; syn. of *Neoisocerus* Bouchard et al., 2005].

Comment: original vernacular name unavailable (Art. 11.7.2): not subsequently latinized; if found to be available then permanently invalid (Art. 39): based on preoccupied type genus.

*HÉLIOPATHAIRES Mulsant and Rey, 1854: 265 [stem: *Heliopat-*]. Type genus: *Heliopates* Dejean, 1834 [as *Heliopathes*, incorrect subsequent spelling of type genus name, not in prevailing usage]. Comment: original vernacular name unavailable (Art. 11.7.2): not subsequently latinized; incorrect original stem formation, not in prevailing usage.

PHYLACIDES Lacordaire, 1859: 270 [stem: *Phylac-*]. Type genus: *Phylax* Brullé, 1832 [syn. of *Dendarus* Dejean, 1821]. Comment: original vernacular name available (Art. 11.7.2): first used in latinized form and generally accepted as in Seidlitz (1893: 220, as PHYLACINA).

BIOPLANINA A. N. Reichardt, 1936: 24 [stem: *Bioplanet-*]. Type genus: *Bioplanes* Mulsant, 1854. Comment: incorrect original stem formation, not in prevailing usage.

Subtribe EURYNOTINA Mulsant and Rey, 1854

EURYNOTAIRES Mulsant and Rey, 1854: 156 [stem: *Eurynot-*]. Type genus: *Eurynotus* Kirby, 1819. Comment: original vernacular name available (Art. 11.7.2): first used in latinized form by Koch (1956: 25, as EURYNOTINA), generally accepted as in Bouchard et al. (2005: 502, as EURYNOTINA).

PSECTROPINI Kaszab, 1941: 33 [stem: *Psectropod-*]. Type genus: *Psectropus* sensu Kaszab, 1941 [not *Psectropus* Solier, 1848; syn. of *Schyzochelus* Koch, 1954]. Comment: based on misidentified type genus; incorrect original stem formation, not in prevailing usage.

ONCOTINI Koch, 1953c: 267, in key [stem: *Oncot-*]. Type genus: *Oncotus* Blanchard, 1845.

SCHYZOSCHELINA Koch, 1956: 25 [stem: *Schyzoschel-*]. Type genus: *Schyzoschelus* Koch, 1954.

Subtribe LEICHENINA Mulsant, 1854

LEICHENAIRES Mulsant, 1854: 179 [stem: *Leichen-*]. Type genus: *Leichenum* Blanchard, 1845. Comment: original vernacular name available (Art. 11.7.2): first used in latinized form by Casey (1890: 391, as LEICHENINI), generally accepted as in Bouchard et al. (2005: 502, as LEICHENINA).

Subtribe LOENSINA Koch, 1956

*LOENSINI Koch, 1955: 1 [stem: *Loens-*]. Type genus: *Loensus* Lucas, 1920. Comment: unavailable family-group name, proposed after 1930 without description or bibliographic reference to such a description (Art. 13.1).

LOENSINI Koch, 1956: 402 [stem: *Loens-*]. Type genus: *Loensus* Lucas, 1920.

Subtribe Melambiina Mulsant and Rey, 1854

Melambiates Mulsant and Rey, 1854: 267 [stem: *Melambi-*]. Type genus: *Melambius* Mulsant and Rey, 1854. Comment: original vernacular name available (Art. 11.7.2): first used in latinized form by Español (1945: 226, as Melambina [incorrect stem formation]), generally accepted as in Bouchard et al. (2005: 502, as Melambiina).

Litoborinae Antoine, 1941: 19-21 [stem: *Litobor-*]. Type genus: *Litoborus* Mulsant and Rey, 1854.

Zadenina Koch, 1956: 279, in key [stem: *Zaden-*]. Type genus: *Zadenos* Laporte, 1840.

Subtribe Pedinina Eschscholtz, 1829

Pediniden Eschscholtz, 1829b: 4 [stem: *Pedin-*]. Type genus: *Pedinus* Latreille, 1797. Comment: original vernacular name available (Art. 11.7.2): first used in latinized form by G. R. Waterhouse (1845: 32, as Pedinidae), generally accepted as in Bouchard et al. (2005: 502, as Pedinini).

Subtribe Platynotina Mulsant and Rey, 1853

*Hétéroscélites Solier, 1836: 502 [stem: *Heteroscelid-*]. Type genus: *Heteroscelis* Latreille, 1829 [preoccupied genus name, not *Heteroscelis* Latreille, 1829 [Hemiptera]; syn. of *Anomalipus* Guérin-Méneville, 1831]. Comment: original vernacular name unavailable (Art. 11.7.2): not subsequently latinized; if found to be available then permanently invalid (Art. 39): based on preoccupied type genus; incorrect original stem formation, not in prevailing usage.

Platynotaires Mulsant and Rey, 1853: 263 [stem: *Platynot-*]. Type genus: *Platynotus* Fabricius, 1801. Comment: original vernacular name available (Art. 11.7.2): first used in latinized form by J. L. LeConte (1862: 226, as Platynotini), generally accepted as in Bouchard et al. (2005: 502, as Platynotina); First Reviser (Platynotina Mulsant and Rey, 1853 vs Trigonopodina Mulsant and Rey, 1853) not determined, current usage maintained.

Trigonopaires Mulsant and Rey, 1853: 104 [stem: *Trigonopod-*]. Type genus: *Trigonopus* Mulsant and Rey, 1853. Comment: original vernacular name available (Art. 11.7.2): first used in latinized form and generally accepted as in Aalbu (2006: 70, as Trigonopina); incorrect original stem formation, not in prevailing usage.

Gonopides Lacordaire, 1859: 255 [stem: *Gonopod-*]. Type genus: *Gonopus* Latreille, 1829. Comment: original vernacular name available (Art. 11.7.2): first used in latinized form by Heyne and Taschenberg (1907: 207, as Gonopini), generally accepted as in Gebien (1938a: 90, as Gonopini); incorrect original stem formation, not in prevailing usage.

Platynotini Koch, 1953c: 268, in key [stem: *Platynot-*]. Type genus: *Platynotus* Fabricius, 1801. Comment: family-group name proposed as new without reference to Platynotaires Mulsant and Rey, 1853.

Anomalipina Koch, 1954: 427 [stem: *Anomalipod-*]. Type genus: *Anomalipus* Guérin-Méneville, 1831 [unjustified emendation of *Anomalipes* Guérin-Méneville, 1831 by Lacordaire, 1859; unjustified emendation in prevailing usage, treated as justified emendation (Art. 33.2.3.1)]. Comment: incorrect original stem formation, not in prevailing usage.

*anchophthalmoid Koch, 1956: 71, in key [stem: *Anchophthalm-*]. Type genus: *Anchophthalmus* Gerstaecker, 1854. Comment: family-group name unavailable (Art. 11.7.1.1): original name not proposed as a noun.

*selinoid Koch, 1956: 70, in key [stem: *Selin-*]. Type genus: *Selinus* Mulsant and Rey, 1853. Comment: family-group name unavailable (Art. 11.7.1.1): original name not proposed as a noun; Selinini Jeannel, 1948 (type genus *Selina* Motschulsky, 1858) is currently used as a valid in Carabidae.

*melanocratoid Iwan, 1996: 379, 380-381 [stem: *Melanocrat-*]. Type genus: *Melanocratus* Fairmaire, 1895. Comment: family-group name unavailable (Art. 11.7.1.1): original name not proposed as a noun.

Subtribe Pythiopina Koch, 1953

Pythiopini Koch, 1953b: 245 [stem: *Pythiop-*]. Type genus: *Pythiopus* Koch, 1953. Comment: current spelling maintained (Art. 29.5): incorrect stem formation in prevailing usage (should be *Pythiopod-*).

Tribe Platyscelidini Lacordaire, 1859

Platyscélides Lacordaire, 1859: 229 [stem: *Platyscelid-*]. Type genus: *Platyscelis* Latreille, 1818 [placed on the Official List of Generic Names in Zoology (ICZN 1993e)]. Comment: original vernacular name available (Art. 11.7.2): first used in latinized form by Seidlitz (1889 [Gatt.]: 126, as Platyscelina [incorrect stem formation]), generally accepted as in Bouchard et al. (2005: 502, as Platyscelidini); incorrect original stem formation, not in prevailing usage; correction of stem by Egorov (1990: 401) in order to avoid homonymy with Platyscelidae Bate, 1862 in Crustacea.

Tribe Praeugenini De Moor, 1970

*Praogenini Ferreira, 1965: 312 [stem: *Praeugen-*]. Type genus: *Praeugena* Laporte, 1840 [as *Praogena*, unjustified emendation of type genus name, not in prevailing usage]. Comment: unavailable family-group name, proposed after 1930 without description or bibliographic reference to such a description (Art. 13.1); incorrect original stem formation, not in prevailing usage.

Praeugenina De Moor, 1970: 4 [stem: *Praeugen-*]. Type genus: *Praeugena* Laporte, 1840.

Tribe RHYSOPAUSSINI Wasmann, 1896

RHYSOPAUSSIDAE Wasmann, 1896: 613 [stem: *Rhysopauss-*]. Type genus: *Rhysopaussus* Wasmann, 1896.

Tribe SCAURINI Billberg, 1820

SCAURIDES Billberg, 1820a: 31 [stem: *Scaur-*]. Type genus: *Scaurus* Fabricius, 1775. Comment: this family-group name was also used in the same year by Billberg (1820b: 392, as SCAURIDES).

Tribe SCOTOBIINI Solier, 1838

SCOTOBITES Solier, 1838: 39 [stem: *Scotobi-*]. Type genus: *Scotobius* Germar, 1824. Comment: original vernacular name available (Art. 11.7.2): first used in latinized form by Agassiz (1846b: 336, as SCOTOBIOIDAE), generally accepted as in Bouchard et al. (2005: 502, as SCOTOBIINI); incorrect original stem formation, not in prevailing usage.

Tribe TENEBRIONINI Latreille, 1802

TENEBRIONITES Latreille, 1802: 165 [stem: *Tenebrion-*]. Type genus: *Tenebrio* Linnaeus, 1758.

BIUINI Skopin, 1978: 224, in key [stem: *Bi-*]. Type genus: *Bius* Dejean, 1834. Comment: incorrect original stem formation, not in prevailing usage.

Tribe TITAENINI Fauvel, 1905

TITAENINI Fauvel, 1905b: 209 [stem: *Titaen-*]. Type genus: *Titaena* Erichson, 1842.

Tribe TOXICINI Oken, 1843

TOXICIDEN Oken, 1843: 484 [stem: *Toxic-*]. Type genus: *Toxicum* Latreille, 1802.

Subtribe EUDYSANTINA Bouchard, Lawrence, Davies and Newton, 2005

DYSANTINAE Gebien, 1922: 289 [stem: *Dysant-*]. Type genus: *Dysantes* Pascoe, 1871 [preoccupied genus name, not *Dysantes* Foerster, 1868 [Hymenoptera]; syn. of *Eudysantes* Bouchard et al., 2005]. Comment: permanently invalid (Art. 39): based on preoccupied type genus.

EUDYSANTINA Bouchard et al., 2005: 508 [stem: *Eudysant-*]. Type genus: *Eudysantes* Bouchard et al., 2005. Comment: replacement name for DYSANTINA Gebien, 1922 because of the homonymy of the type genus.

Subtribe NYCTEROPINA Lacordaire, 1859

NYCTÉROPIDES Lacordaire, 1859: 388 [stem: *Nycterop-*]. Type genus: *Nycteropus* Klug, 1833. Comment: original vernacular name available (Art. 11.7.2): first used in latinized form by Heyne and Taschenberg (1907: 211, as NYCTEROPINI), generally accepted as in Bouchard et al. (2005: 502, as NYCTEROPINA).

Subtribe Toxicina Oken, 1843

Toxiciden Oken, 1843: 484 [stem: *Toxic-*]. Type genus: *Toxicum* Latreille, 1802. Comment: original vernacular name available (Art. 11.7.2): first used in latinized form and generally accepted as in Tulk (1847: 614, as Toxicidae); family-group name previously attributed to Lacordaire (1859).

Anthraciini Houlbert, 1922b: 185, in key [stem: *Anthraci-*]. Type genus: *Anthracias* Dejean, 1834 [syn. of *Cryphaeus* Klug, 1833].

Tribe Triboliini Gistel, 1848

Triboliidae Gistel, 1848: [4] [stem: *Triboli-*]. Type genus: *Tribolium* W. S. MacLeay, 1825 [placed on the Official List of Generic Names in Zoology (ICZN 1988e)].

Tribe Ulomini Blanchard, 1845

Ulomites Blanchard, 1845b: 16 [stem: *Ulom-*]. Type genus: *Uloma* Dejean, 1821 [placed on the Official List of Generic Names in Zoology (ICZN 1975)]. Comment: original vernacular name available (Art. 11.7.2): first used in latinized form by Blanchard (1853: 164, as Ulomitae), generally accepted as in Bouchard et al. (2005: 502, as Ulomini); current spelling maintained (Art. 29.5): incorrect stem formation in prevailing usage (should be *Ulomat-*).

*Oligocaroides Solier, 1851: 225 [stem: *Oligocar-*]. Type genus: *Oligocara* Solier, 1848. Comment: original vernacular name unavailable (Art. 11.7.2): not subsequently latinized.

Alégoriides Lacordaire, 1859: 325 [stem: *Alegori-*]. Type genus: *Alegoria* Laporte, 1840. Comment: original vernacular name available (Art. 11.7.2): first used in latinized form and generally accepted as in Seidlitz (1894: 545, as Alegoriina).

Neopsectropinae Kaszab, 1941: 30 [stem: *Neopsectropod-*]. Type genus: *Neopsectropus* Kaszab, 1941. Comment: incorrect original stem formation, not in prevailing usage.

Subfamily Alleculinae Laporte, 1840

Alléculites Laporte, 1840b: 242 [stem: *Allecul-*]. Type genus: *Allecula* Fabricius, 1801. Comment: given precedence over the older names Xystropodinae Solier, 1835 and Cteniopodinae Solier, 1835 (Art. 35.5) (see Bouchard et al. 2005).

Tribe Alleculini Laporte, 1840

Alléculites Laporte, 1840b: 242 [stem: *Allecul-*]. Type genus: *Allecula* Fabricius, 1801. Comment: given precedence over the older name Xystropodini Solier, 1835 (Art. 35.5) (see Bouchard et al. 2005).

Subtribe Alleculina Laporte, 1840

Alléculites Laporte, 1840b: 242 [stem: *Allecul-*]. Type genus: *Allecula* Fabricius, 1801. Comment: original vernacular name available (Art. 11.7.2):

first used in latinized form by Packard (1874: iii, as ALLECULIDAE), generally accepted as in Bouchard et al. (2005: 502, as ALLECULINA).

*CYLINDROTHORIDES Lacordaire, 1859: 494 [stem: *Cylindrothor-*]. Type genus: *Cylindrothorus* Solier, 1844. Comment: original vernacular name unavailable (Art. 11.7.2): not subsequently latinized.

UPINELLAE J. L. LeConte, 1866b: 137 [stem: *Upinell-*]. Type genus: *Upinella* Mulsant, 1856.

Subtribe GONODERINA Seidlitz, 1896

GONODERINA Seidlitz, 1896: 83 [stem: *Gonoder-*]. Type genus: *Gonodera* Mulsant, 1856.

PSEUDOCISTELINI Portevin, 1934: 39, in key [stem: *Pseudocistel-*]. Type genus: *Pseudocistela* Crotch, 1873.

Subtribe MYCETOCHARINA Gistel, 1848

MYCETOCHARISIDAE Gistel, 1848: [10] [stem: *Mycetochar-*]. Type genus: *Mycetochara* Berthold, 1827 [as *Mycetochares*, incorrect subsequent spelling not in prevailing usage]. Comment: incorrect original stem formation, not in prevailing usage.

Subtribe XYSTROPODINA Solier, 1835

XYSTROPIDES Solier, 1835a: 229 [stem: *Xystropod-*]. Type genus: *Xystropus* Solier, 1835. Comment: original vernacular name available (Art. 11.7.2): first used in latinized form by Agassiz (1846b: 390, as XYSTROPODIDAE), generally accepted as in Bouchard et al. (2005: 502, as XYSTROPODINA); incorrect original stem formation, not in prevailing usage.

LYSTRONYCHIDES Lacordaire, 1859: 512 [stem: *Lystronych-*]. Type genus: *Lystronychus* Latreille, 1829 [incorrect subsequent spelling of *Lystronichus*, incorrect spelling in current usage, treated as correct original spelling (Art. 33.3.1)]. Comment: original vernacular name available (Art. 11.7.2): first used in latinized form and generally accepted as in J. L. LeConte and G. H. Horn (1883: 390, as LYSTRONYCHI).

Tribe CTENIOPODINI Solier, 1835

CISTELENIAE Latreille, 1802: 188 [stem: *Cistel-*]. Type genus: *Cistela* Fabricius, 1775 [preoccupied genus name, not *Cistela* Geoffroy, 1762 [Coleoptera: BYRRHIDAE]; syn. of *Cteniopus* Solier, 1835; Geoffroy's generic name was suppressed for the purposes of the Principle of Priority but not for the Principle of Homonymy (ICZN 1994a)]. Comment: permanently invalid (Art. 39): based on preoccupied type genus.

CTÉNIOPITES Solier, 1835a: 245 [stem: *Cteniopod-*]. Type genus: *Cteniopus* Solier, 1835. Comment: original vernacular name available (Art. 11.7.2): first used in latinized form by Agassiz (1846b: 107, as CTENIOPODIDAE), generally accepted as in Bouchard et al. (2005: 502, as CTENIOPODINI); incorrect original stem formation, not in prevailing usage.

OMOPHLIENS Mulsant, 1856a: 65 [stem: *Omophl-*]. Type genus: *Omophlus* Dejean, 1834. Comment: original vernacular name available (Art. 11.7.2): first used in latinized form by Seidlitz (1896: 28, as OMOPHLINI), generally accepted as in Ogloblin and Znojko (1950: 9, as OMOPHLINAE).

PHIBALIDAE Gistel, 1856a: 384 [stem: *Phibal-*]. Type genus: *Phibalus* Gistel, 1856 [the genus *Phibalus* Gistel, 1856 originally included only one species therefore the type species is *Chrysomela pubescens* Linnaeus, 1758 by monotypy; this genus has not been used subsequently to our knowledge; **syn. nov.** of *Omophlus* Dejean, 1834]. Comment: **syn. nov.**

PETRIIDAE Semenov, 1893b: 359 [stem: *Petri-*]. Type genus: *Petria* Semenov, 1894.

PODONTINI Portevin, 1934: 45, in key [stem: *Podont-*]. Type genus: *Podonta* Solier, 1835.

Subfamily DIAPERINAE Latreille, 1802

DIAPERIALAE Latreille, 1802: 161 [stem: *Diaper-*]. Type genus: *Diaperis* Geoffroy, 1762 [placed on the Official List of Generic Names in Zoology (ICZN 1994a)].

Tribe CRYPTICINI Brullé, 1832

CRYPTICITES Brullé, 1832: 190 [stem: *Cryptic-*]. Type genus: *Crypticus* Latreille, 1816. Comment: original vernacular name available (Art. 11.7.2): first used in latinized form by C. G. Thomson (1859: 114, as CRYPTICINA), generally accepted as in Bouchard et al. (2005: 503, as CRYPTICINI).

OOCHROTINI Desbrochers des Loges, 1901: 143, in key [stem: *Oochrot-*]. Type genus: *Oochrotus* Lucas, 1852.

Tribe DIAPERINI Latreille, 1802

DIAPERIALAE Latreille, 1802: 161 [stem: *Diaper-*]. Type genus: *Diaperis* Geoffroy, 1762 [placed on the Official List of Generic Names in Zoology (ICZN 1994a)].

Subtribe ADELININA LeConte, 1862

*SITOPHAGIENS Mulsant, 1854: 263 [stem: *Sitophag-*]. Type genus: *Sitophagus* Mulsant, 1854. Comment: original vernacular name unavailable (Art. 11.7.2): not subsequently latinized.

ALPHITOPHAGIDA Gistel, 1856b: 185 [stem: *Alphitophag-*]. Type genus: *Alphitophagus* Stephens, 1832. Comment: although this is the oldest name for the subtribe, we recommend that an application be submitted to the Commission to conserve usage of the name ADELININA J. L. LeConte, 1862.

*GNATHOCÉRITES Jacquelin du Val, 1861: 304 [stem: *Gnatocer-*]. Type genus: *Gnatocerus* Thunberg, 1814 [as *Gnathocerus*, incorrect subsequent spelling of type genus name, not in prevailing usage; *Gnatocerus* Thunberg, 1814 placed on the Official List of Generic Names in Zoology (ICZN 1975)]. Comment: original vernacular name unavailable (Art. 11.7.2): subsequently used in latinized form but not generally attributed to Jacquelin du Val (1861); incorrect original stem formation, not in prevailing usage; GNATHOCERIDAE Schoch, 1894 is available in SCARABAEIDAE.

ADELININI J. L. LeConte, 1862: 237 [stem: *Adelin-*]. Type genus: *Adelina* Dejean, 1835. Comment: this well-established family-group name is threatened by the discovery of the older name ALPHITOPHAGIDA Gistel, 1856; although using Reversal of Precedence (Art. 23.9.1) to conserve usage of ADELININA J. L. LeConte, 1862 would be preferable, this can not be done because we cannot find 25 references in which this name has been used in the last 50 years; current usage is maintained here and an application should be submitted to the Commission in order to conserve ADELININA J. L. LeConte, 1862.

SCHEDAROSINI Reitter, 1876b: 42 [stem: *Schedaros-*]. Type genus: *Schedarosus* Reitter, 1876 [syn. of *Adelina* Dejean, 1835]. Comment: originally placed in the family CUCUJIDAE.

DOLIEMINI Reitter, 1917: 58 [stem: *Doliem-*]. Type genus: *Doliema* Pascoe, 1860 [syn. of *Adelina* Dejean, 1835].

GNATHOCERINI Skopin, 1978: 228, in key [stem: *Gnatocer-*]. Type genus: *Gnatocerus* Thunberg, 1814 [as *Gnathocerus*, incorrect subsequent spelling of type genus name, not in prevailing usage; *Gnatocerus* Thunberg, 1814 placed on the Official List of Generic Names in Zoology (ICZN 1975)]. Comment: incorrect original stem formation, not in prevailing usage; GNATHOCERIDAE Schoch, 1894 is available in SCARABAEIDAE.

Subtribe DIAPERINA Latreille, 1802

DIAPERIALAE Latreille, 1802: 161 [stem: *Diaper-*]. Type genus: *Diaperis* Geoffroy, 1762 [placed on the on Official List of Generic Names in Zoology (ICZN 1994a)].

PENTAPHYLLAIRES Mulsant, 1854: 196 [stem: *Pentaphyll-*]. Type genus: *Pentaphyllus* Dejean, 1821. Comment: original vernacular name available (Art. 11.7.2): first used in latinized form by J. L. LeConte (1862: 236, as PENTAPHYLLI), generally accepted as in Silfverberg (1992: 61, as PENTAPHYLLINI); the name PENTAPHYLLINAE Schindewolf, 1942 (type genus *Pentaphyllum* de Koninck, 1872) is used as valid in Anthozoa; this case is to be referred to the Commission to remove the homonymy (Art. 55.3.1).

PLATYDEMINAE Reitter, 1917: 61 [stem: *Platydem-*]. Type genus: *Platydema* Laporte and Brullé, 1831.

*HOPLOCEPHALINI Kwieton, 1982: 98 [stem: *Hoplocephal-*]. Type genus: *Hoplocephala* Laporte and Brullé, 1831 [based on unjustified emendation of type genus name; originally proposed as *Oplocephala*; emended name in prevailing usage, treated as a justified emendation (Art. 33.2.3.1); syn. of *Neomida* Latreille, 1829]. Comment: unavailable family-group name, proposed after 1930 without description or bibliographic reference to such a description (Art. 13.1).

Tribe Ectychini Doyen, Matthews and Lawrence, 1990

Ectychini Doyen et al., 1990: 247 [stem: *Ectych-*]. Type genus: *Ectyche* Pascoe, 1869.

Tribe Gnathidiini Gebien, 1921

Gnathidiinae Gebien, 1921: 41 [stem: *Gnathidi-*]. Type genus: *Gnathidium* Gebien, 1921.

Subtribe Anopidiina Jeannel and Paulian, 1945

Anopidiini Jeannel and Paulian, 1945: 62 [stem: *Anopidi-*]. Type genus: *Anopidium* Jeannel and Paulian, 1945.

Subtribe Gnathidiina Gebien, 1921

Gnathidiinae Gebien, 1921: 41 [stem: *Gnathidi-*]. Type genus: *Gnathidium* Gebien, 1921.

Tribe Hyociini Medvedev and Lawrence, 1982

Hyocini G. S. Medvedev and Lawrence, 1982: 548 [stem: *Hyoci-*]. Type genus: *Hyocis* Pascoe, 1866.

Subtribe Brittonina Medvedev and Lawrence, 1986

Brittonina G. S. Medvedev and Lawrence, 1986: 574 [stem: *Britton-*]. Type genus: *Brittona* G. S. Medvedev and Lawrence, 1986.

Subtribe Hyociina Medvedev and Lawrence, 1982

Hyocini G. S. Medvedev and Lawrence, 1982: 548 [stem: *Hyoci-*]. Type genus: *Hyocis* Pascoe, 1866. Comment: incorrect original stem formation, not in prevailing usage.

Subtribe Uptonina Medvedev and Lawrence, 1986

Uptonina G. S. Medvedev and Lawrence, 1986: 581 [stem: *Upton-*]. Type genus: *Uptona* G. S. Medvedev and Lawrence, 1986.

Tribe Hypophlaeini Billberg, 1820

Hypophlaeides Billberg, 1820a: 33 [stem: *Hypophlae-*]. Type genus: *Hypophlaeus* Fabricius, 1790 [syn. of *Corticeus* Piller and Mitterpacher, 1783].

CORTICEINI Boddy, 1965: 144 [stem: *Cortice-*]. Type genus: *Corticeus* Piller and Mitterpacher, 1783.

Tribe LEIOCHRININI Lewis, 1894

LEIOCHRININAE Lewis, 1894: 390 [stem: *Leiochrin-*]. Type genus: *Leiochrinus* Westwood, 1883.

Tribe MYRMECHIXENINI Jacquelin du Val, 1858

MYRMÉCHIXÉNITES Jacquelin du Val, 1858: 223 [stem: *Myrmechixen-*]. Type genus: *Myrmechixenus* Chevrolat, 1835. Comment: original vernacular name available (Art. 11.7.2): first used in latinized form by Marseul (1863: 111, as MYRMEKIXENI [incorrect stem formation]), generally accepted as in Bouchard et al. (2005: 503, as MYRMECHIXENINI); originally placed in the family MYCETAEIDAE (syn. of ENDOMYCHIDAE).

Tribe PHALERIINI Blanchard, 1845

PHALÉRIIDES Blanchard, 1845b: 29 [stem: *Phaleri-*]. Type genus: *Phaleria* Latreille, 1802 [placed on the Official List of Generic Names in Zoology (ICZN 1975)]. Comment: original vernacular name available (Art. 11.7.2): first used in latinized form by Gistel (1848: [10], as PHALERIADAE [incorrect stem formation]), generally accepted as in Bouchard et al. (2005: 503, as PHALERIINI).

SEPEDONASTIDAE Gistel, 1856a: 382 [stem: *Sepedonast-*]. Type genus: *Sepedonastes* Gistel, 1856 [syn. of *Phaleria* Latreille, 1802].

CATAPHRONETINI Reitter, 1917: 57 [stem: *Cataphronet-*]. Type genus: *Cataphronetis* Lucas, 1846 [syn. of *Phtora* Germar, 1836].

Tribe SCAPHIDEMINI Reitter, 1922

SCAPHIDEMINI Reitter, 1922b: 2 [stem: *Scaphidem-*]. Type genus: *Scaphidema* Redtenbacher, 1849.

Tribe TRACHYSCELINI Blanchard, 1845

TRACHYSCÉLIDES Blanchard, 1845b: 28 [stem: *Trachyscel-*]. Type genus: *Trachyscelis* Latreille, 1809. Comment: original vernacular name available (Art. 11.7.2): first used in latinized form by G. R. Waterhouse (1858: 60, as TRACHYSCELIDAE), generally accepted as in Bouchard et al. (2005: 503, as TRACHYSCELINI); current spelling maintained (Art. 29.5): incorrect stem formation in prevailing usage (should be *Trachyscelid-*); conservation of the original stem avoids potential homonymy with family-group names based on the type genus *Trachyscelida* G. H. Horn, 1893 in CHRYSOMELIDAE.

Subfamily STENOCHIINAE Kirby, 1837

STENOCHIADAE Kirby, 1837: 238 [stem: *Stenochi-*]. Type genus: *Stenochia* Kirby, 1819.

Tribe Cnodalonini Oken, 1843

Cnodaliden Oken, 1843: 484 [stem: *Cnodalon-*]. Type genus: *Cnodalon* Latreille, 1797. Comment: original vernacular name available (Art. 11.7.2): first used in latinized form and generally accepted as in Tulk (1847: 614, as Cnodalidae); family-group name previously attributed to Gistel (1856a: 382); the correct stem based on *Cnodalon* is *Cnodal-*, therefore the original stem was correctly formed; however, since this family-group name has not been attributed to Oken (1843) in the past and the accepted stem has been *Cnodalon-* to this date, we have kept the currently accepted stem; an application to the Commission is necessary to conserve *Cnodalon-* as the correct stem for this taxon.

Coelometopidae Schaum, 1859: 71 [stem: *Coelometop-*]. Type genus: *Coelometopus* Solier, 1848. Comment: published before February 31, 1859; this family-group name was also used in the same year by Lacordaire (1859 [before 27 June]: 358, as Coelométopides).

Upidae C. G. Thomson, 1859: 116 [stem: *Upid-*]. Type genus: *Upis* Fabricius, 1792. Comment: incorrect original stem formation, not in prevailing usage.

Eutélides Lacordaire, 1859: 354 [stem: *Eutel-*]. Type genus: *Eutelus* Solier, 1843 [preoccupied genus name, not *Eutelus* Walker, 1834 [Hymenoptera]; syn. of *Nodotelus* Koch, 1950]. Comment: original vernacular name available (Art. 11.7.2): first used in latinized form by Pascoe (1866b: 469, as Eutelinae), generally accepted as in Gebien (1911: 431, as Eutelinae); permanently invalid (Art. 39): based on preoccupied type genus; Eutelini Ashmead, 1904 (type genus *Eutelus* Walker, 1834) is available in Hymenoptera.

Misolampides Lacordaire, 1859: 440 [stem: *Misolamp-*]. Type genus: *Misolampus* Latreille, 1807. Comment: original vernacular name available (Art. 11.7.2): first used in latinized form by and generally accepted as in F. Bates (1873: 373, as Misolampides [treated as Latin]).

Catapiestides Lacordaire, 1859: 381 [stem: *Catapiest-*]. Type genus: *Catapiestus* Perty, 1831. Comment: original vernacular name available (Art. 11.7.2): first used in latinized form and generally accepted as in Seidlitz (1895: 614, as Catapiestina).

*Camarides Desmarest, 1860: 165 [stem: *Camari-*]. Type genus: *Camaria* Lepeletier and Audinet-Serville, 1828. Comment: original vernacular name unavailable (Art. 11.7.2): subsequently used in latinized form but not generally attributed to Desmarest (1860).

Polypleuri J. L. LeConte, 1862: 229 [stem: *Polypleur-*]. Type genus: *Polypleurus* Eschscholtz, 1831.

Misolampidiini Reitter, 1917: 60, in key [stem: *Misolampidi-*]. Type genus: *Misolampidius* Solsky, 1876.

Menephilini Reitter, 1920: 15 [stem: *Menephil-*]. Type genus: *Menephilus* Mulsant, 1854.

Stenophanini Reitter, 1922b: 2, 3 [stem: *Stenophan-*]. Type genus: *Stenophanes* Solsky, 1876.

HEGEMONINI Reitter, 1922b: 3 [stem: *Hegemon-*]. Type genus: *Hegemona* Laporte, 1840.

CAMARIINAE Anonymous, 1924: 164 [stem: *Camari-*]. Type genus: *Camaria* Lepeletier and Audinet-Serville, 1828. Comment: this was a translation of the vernacular name "CAMARIINEN" used by Gebien (1919); Gebien's family-group name is not available since it was proposed in a vernacular form after 1899 (Art. 11.7.2); for comments about authorship of Camariinae see Evenhuis et al. (2008: 4).

NODOTELINI Koch, 1950: 67 [stem: *Nodotel-*]. Type genus: *Nodotelus* Koch, 1950. Comment: replacement name for EUTELINI Lacordaire, 1859 because of the homonymy of the type genus.

*THESILEINI Kaszab, 1982: 29 [stem: *Thesile-*]. Type genus: *Thesilea* Haag-Rutenberg, 1878. Comment: unavailable family-group name, proposed after 1930 without description or bibliographic reference to such a description (Art. 13.1).

Tribe STENOCHIINI Kirby, 1837

STENOCHIADAE Kirby, 1837: 238 [stem: *Stenochi-*]. Type genus: *Stenochia* Kirby, 1819. Comment: incorrect original stem formation, not in prevailing usage.

STRONGYLIIDES Lacordaire, 1859: 478 [stem: *Strongyli-*]. Type genus: *Strongylium* Kirby, 1819 [name conserved over *Strongylium* Ditmar, 1809 [Myxomycota] (see Bouchard et al. 2005)]. Comment: original vernacular name available (Art. 11.7.2): first used in latinized form by J. L. LeConte (1862: 241, as STRONGYLIINI), generally accepted as in Aalbu et al. (2002: 483, as STRONGYLIINI).

Tribe TALANINI Champion, 1887 (1883)

DIGNAMPTINI J. L. LeConte and G. H. Horn, 1883: 385 [stem: *Dignampt-*]. Type genus: *Dignamptus* J. L. LeConte, 1878 [syn. of *Talanus* Jacquelin du Val, 1857].

TALANIDES Champion, 1887: 321 [stem: *Talan-*]. Type genus: *Talanus* Jacquelin du Val, 1857. Comment: family-group name conserved over DIGNAMPTINI J. L. LeConte and G. H. Horn, 1883 (Art. 40.2) (see Bouchard et al. 2005).

TENEBRIONIDAE *incertae sedis*

BETSCHIINI Dajoz, 1980a: 135 [stem: *Betschi-*]. Type genus: *Betschia* Dajoz, 1980. Comment: originally placed in the family COLYDIIDAE, the type genus was transferred to TENEBRIONIDAE by Ivie and Ślipiński (1990: 18); placement uncertain although probably close to GNATHIDIINI (Ivie pers. comm. 2008).

Family PROSTOMIDAE Thomson, 1859

*MÉGAGNATHES Motschulsky, 1849: 60 [stem: *Megagnath-*]. Type genus: *Megagnathus* Dejean, 1821 [preoccupied genus name, not *Megagnathus* Billberg, 1820

[Coleoptera: CERAMBYCIDAE]; syn. of *Prostomis* Latreille, 1819]. Comment: original vernacular name unavailable (Art. 11.7.2): not subsequently latinized; if found to be available then permanently invalid (Art. 39): based on preoccupied type genus.

PROSTOMIDAE C. G. Thomson, 1859: 84 [stem: *Prostom-*]. Type genus: *Prostomis* Latreille, 1819. Comment: current spelling maintained (Art. 29.3.1.1): incorrect stem formation in prevailing usage (should be *Prostomid-*).

Family SYNCHROIDAE Lacordaire, 1859

SYNCHROÏDES Lacordaire, 1859: 544 [stem: *Synchro-*]. Type genus: *Synchroa* Newman, 1838. Comment: original vernacular name available (Art. 11.7.2): first used in latinized form by J. L. LeConte (1862: 249, as SYNCHROAE), generally accepted as in Lawrence and Newton (1995: 894, as SYNCHROIDAE).

MALLODRYINI G. H. Horn, 1888: 43 [stem: *Mallodry-*]. Type genus: *Mallodrya* G. H. Horn, 1888.

Family STENOTRACHELIDAE Thomson, 1859

STENOTRACHELIDAE C. G. Thomson, 1859: 124 [stem: *Stenotrachel-*]. Type genus: *Stenotrachelus* Berthold, 1827.

Subfamily STENOTRACHELINAE Thomson, 1859

STENOTRACHELIDAE C. G. Thomson, 1859: 124 [stem: *Stenotrachel-*]. Type genus: *Stenotrachelus* Berthold, 1827.

Subfamily CEPHALOINAE LeConte, 1862

*CEPHALAONIDES Motschulsky, 1860: 140 [stem: *Cephalo-*]. Type genus: *Cephaloon* Newman, 1838 [as *Cephalaon*, incorrect subsequent spelling of type genus name corrected by Motschulsky to *Cephaloon* on pp. 140 in same work]. Comment: original vernacular name unavailable (Art. 11.7.2): subsequently used in latinized form but not generally attributed to Motschulsky (1860).

CEPHALOIDAE J. L. LeConte, 1862: 259 [stem: *Cephalo-*]. Type genus: *Cephaloon* Newman, 1838.

Subfamily NEMATOPLINAE LeConte, 1862

NEMATOPLI J. L. LeConte, 1862: 264 [stem: *Nematopl-*]. Type genus: *Nematoplus* J. L. LeConte, 1855.

Subfamily STOLIINAE Nikitsky, 1985

STOLIINAE Nikitsky, 1985: 32, in key [stem: *Stoli-*]. Type genus: *Stolius* Lewis, 1895.

Family OEDEMERIDAE Latreille, 1810

OEDEMERITES Latreille, 1810: 216 [stem: *Oedemer-*]. Type genus: *Oedemera* A. G. Olivier, 1789.

Subfamily Polypriinae Lawrence, 2005

*Polypriidae Triplehorn and N. F. Johnson, 2005: 441 [stem: *Polypri-*]. Type genus: *Polypria* Chevrolat, 1874. Comment: unavailable family-group name, name published after 1999 and not explicitly indicated as new (Art. 16.1).

Polypriinae Lawrence, 2005: 671, in key [stem: *Polypri-*]. Type genus: *Polypria* Chevrolat, 1874.

Subfamily Calopodinae Costa, 1852 *nomen protectum*

Sparedriidae Gistel, 1848: [11] [stem: *Sparedr-*]. Type genus: *Sparedrus* Dejean, 1821. Comment: *nomen oblitum* (see Appendix 1); incorrect original stem formation, not in prevailing usage.

Calopini A. Costa, 1852: 4 [stem: *Calopod-*]. Type genus: *Calopus* Fabricius, 1775. Comment: original vernacular name available (Art. 11.7.2): first used in latinized form by Semenov (1894: 449, as Calopodidae), generally accepted as in Švihla (2008: 353, as Calopodinae); *nomen protectum* (see Appendix 1); incorrect original stem formation, not in prevailing usage.

Catachirotidae Gistel, 1856a: 384 [stem: *Catachirotet-*]. Type genus: *Catachirotes* Gistel, 1856 [syn. of *Calopus* Fabricius, 1775]. Comment: incorrect original stem formation, not in prevailing usage.

Subfamily Oedemerinae Latreille, 1810

Oedemerites Latreille, 1810: 216 [stem: *Oedemer-*]. Type genus: *Oedemera* A. G. Olivier, 1789.

Tribe Asclerini Gistel, 1848

Ascleraeidae Gistel, 1848: [11] [stem: *Ascler-*]. Type genus: *Asclera* Stephens, 1839. Comment: incorrect original stem formation, not in prevailing usage.

Ganglbaueriidae Semenov, 1894: 450 [stem: *Ganglebaueri-*]. Type genus: *Ganglbaueria* Semenov, 1891.

*Hypasclerini Macnamara, 1971: 164 [stem: *Hypascler-*]. Type genus: *Hypasclera* Kirsch, 1866. Comment: unavailable family-group name, proposed after 1930 without description or bibliographic reference to such a description (Art. 13.1).

*Hypasclerini Švihla, 1986: 161 [stem: *Hypascler-*]. Type genus: *Hypasclera* Kirsch, 1866. Comment: family-group name unavailable (Art. 11.6): originally published as synonym and not made available subsequently.

*Danacerinae Švihla, 1986: 161 [stem: *Danerc-*]. Type genus: *Danerces* Westwood, 1875. Comment: family-group name unavailable (Art. 11.6): originally published as synonym and not made available subsequently; incorrect original stem formation, not in prevailing usage.

*Oxacini Švihla, 1986: 161 [stem: *Oxacid-*]. Type genus: *Oxacis* J. L. LeConte, 1866. Comment: family-group name unavailable (Art. 11.6): originally published as synonym and not made available subsequently; incorrect original stem formation, not in prevailing usage.

Tribe DityLini Mulsant, 1858

*Dityles Motschulsky, 1849: 59 [stem: *Dityl-*]. Type genus: *Ditylus* Fischer von Waldheim, 1817. Comment: original vernacular name unavailable (Art. 11.7.2): subsequently used in latinized form but not generally attributed to Motschulsky (1849).

Ditylates Mulsant, 1858: 100 [stem: *Dityl-*]. Type genus: *Ditylus* Fischer von Waldheim, 1817. Comment: original vernacular name available (Art. 11.7.2): first used in latinized form by Semenov (1894: 450, as Ditylidae), generally accepted as in Kriska (2002: 517, as Ditylini).

Tribe Nacerdini Mulsant, 1858

Nacerdates Mulsant, 1858: 104 [stem: *Nacerd-*]. Type genus: *Nacerdes* Dejean, 1834. Comment: original vernacular name available (Art. 11.7.2): first used in latinized form by Semenov (1894: 450, as Nacerdidae), generally accepted as in Hansen (1996: 175, as Nacerdini).

Tribe Oedemerini Latreille, 1810

Oedemerites Latreille, 1810: 216 [stem: *Oedemer-*]. Type genus: *Oedemera* A. G. Olivier, 1789.

Oncomeradae Gistel, 1856a: 384 [stem: *Oncomer-*]. Type genus: *Oncomera* Stephens, 1829 [subgenus of *Oedemera* A. G. Olivier, 1789; placed on the Official List of Generic Names in Zoology (ICZN 1988f)]. Comment: incorrect original stem formation, not in prevailing usage.

*Oncomerinini Švihla, 1986: 161 [stem: *Oncomerin-*]. Type genus: *Oncomerina* Seidlitz, 1899 [syn. of *Oncomera* Stephens, 1829]. Comment: family-group name unavailable (Art. 11.6): originally published as synonym and not made available subsequently.

Tribe Stenostomatini Mulsant, 1858

Sténostomates Mulsant, 1858: 228 [stem: *Stenostomat-*]. Type genus: *Stenostoma* Latreille, 1810. Comment: original vernacular name available (Art. 11.7.2): first used in latinized form by Semenov (1894: 450, as Stenostomatidae), generally accepted as in Švihla (2008: 369, as Stenostomini [incorrect stem formation]); incorrect original stem formation, not in prevailing usage.

Family Meloidae Gyllenhal, 1810

Melooides Gyllenhal, 1810: 481 [stem: *Melo-*]. Type genus: *Meloe* Linnaeus, 1758 [placed on the Official List of Generic Names in Zoology (ICZN 1999c)]. Comment: given precedence over the older name Horiidae Latreille, 1802 and placed on the Official List of Family-Group Names in Zoology (ICZN 1999c).

Subfamily Eleticinae Wellman, 1910

Eleticides Wellman, 1910: 221 [stem: *Eletic-*]. Type genus: *Eletica* Dejean, 1834. Comment: First Reviser (Eleticinae Wellman, 1910 vs Derideinae Wellman, 1910) not determined, current usage maintained.

Tribe Derideini Wellman, 1910

Derideides Wellman, 1910: 222 [stem: *Deride-*]. Type genus: *Deridea* Westwood, 1875.

Tribe Eleticini Wellman, 1910

Eleticides Wellman, 1910: 221 [stem: *Eletic-*]. Type genus: *Eletica* Dejean, 1834.

Subtribe Eleticina Wellman, 1910

Eleticides Wellman, 1910: 221 [stem: *Eletic-*]. Type genus: *Eletica* Dejean, 1834.

Subtribe Eospastina Selander, 1966

Eospastina Selander, 1966: 474 [stem: *Eospast-*]. Type genus: *Eospasta* Selander, 1966.

Tribe Ertlianini Selander, 1966

Ertliini Kaszab, 1959: 71 [stem: *Ertli-*]. Type genus: *Ertlia* Borchmann, 1942 [preoccupied genus name, not *Ertlia* Aurivillius, 1907 [Coleoptera: Cerambycidae]; syn. of *Morphozonitis* Pic, 1922]. Comment: permanently invalid (Art. 39): based on preoccupied type genus.

Ertlianina Selander, 1966: 478 [stem: *Ertlian-*]. Type genus: *Ertliana* Selander, 1964 [syn. of *Morphozonitis* Pic, 1922]. Comment: replacement name for Ertliini Kaszab, 1959 because of the homonymy of the type genus; although Selander (1991: 78) used Morphozonitina Kaszab, 1969 as the valid name for this taxon (because of the synonymy of *Ertliana* Selander with *Morphozonitis* Pic) the Principle of Priority applies here and therefore the oldest available family-group name Ertlianini should be used as valid instead.

Morphozonitini Kaszab, 1969: 242 [stem: *Morphozonitid-*]. Type genus: *Morphozonitis* Pic, 1922. Comment: replacement name for Ertliini Kaszab, 1959 because of the homonymy of the type genus and Ertlianina Selander, 1966 because of the synonymy of the type genus; incorrect original stem formation, not in prevailing usage.

Tribe Spasticini Kaszab, 1959

Spasticina Kaszab, 1959: 72 [stem: *Spastic-*]. Type genus: *Spastica* Lacordaire, 1859.

Subtribe Anthicoxenina Selander, 1966

Anthicoxenina Selander, 1966: 470 [stem: *Anthicoxen-*]. Type genus: *Anthicoxenus* Fairmaire and Germain, 1860.

Subtribe Protomeloina Abdullah, 1965

Protomeloini M. Abdullah, 1965b: 43 [stem: *Protomelo-*]. Type genus: *Protomeloe* M. Abdullah, 1965.

Subtribe Spasticina Kaszab, 1959

Spasticina Kaszab, 1959: 72 [stem: *Spastic-*]. Type genus: *Spastica* Lacordaire, 1859.

Subtribe Xenospastina Selander, 1966

Xenospastina Selander, 1966: 460, in key [stem: *Xenospast-*]. Type genus: *Xenospasta* Selander, 1966.

Subfamily Meloinae Gyllenhal, 1810

Melooides Gyllenhal, 1810: 481 [stem: *Melo-*]. Type genus: *Meloe* Linnaeus, 1758 [placed on the Official List of Generic Names in Zoology (ICZN 1999c)].

Tribe Cerocomini Leach, 1815

Cerocomatida Leach, 1815: 105 [stem: *Cerocom-*]. Type genus: *Cerocoma* Geoffroy, 1762 [placed on the Official List of Generic Names in Zoology (ICZN 1994a)].

Tribe Epicautini Parker and Böving, 1924

Macrobases J. L. LeConte, 1862: 272 [stem: *Macrobase-*]. Type genus: *Macrobasis* J. L. LeConte, 1862 [subgenus of *Epicauta* Dejean, 1834]. Comment: incorrect original stem formation, not in prevailing usage.

Apterospastides Wellman, 1910: 221 [stem: *Apterospast-*]. Type genus: *Apterospasta* J. L. LeConte, 1862 [syn. of *Macrobasis* J. L. LeConte, 1862].

Epicautini Parker and Böving, 1924: 25 [stem: *Epicaut-*]. Type genus: *Epicauta* Dejean, 1834. Comment: family-group name previously attributed to Denier (1935); this name is threatened by the two older names Macrobaseini J. L. LeConte, 1862 and Apterospastini Wellman, 1910; to maintain nomenclatural stability, we believe that an application should be submitted to the Commission to conserve usage of the well-established name Epicautini.

*Henoini Böving and Craighead, 1931: 79 [stem: *Heno-*]. Type genus: *Henous* Haldeman, 1852 [syn. of *Epicauta* Dejean, 1834]. Comment: unavailable family-group name, proposed after 1930 without description or bibliographic reference to such a description (Art. 13.1).

Tribe Eupomphini LeConte, 1862

Eupomphae J. L. LeConte, 1862: 274 [stem: *Eupomph-*]. Type genus: *Eupompha* J. L. LeConte, 1858. Comment: First Reviser found (Eupomphini J. L. LeConte, 1862 vs Phodagini J. L. LeConte, 1862) is Selander (1991: 78, as Eupomphina).

Phodagae J. L. LeConte, 1862: 274 [stem: *Phodag-*]. Type genus: *Phodaga* J. L. LeConte, 1858.

Calospastides Wellman, 1910: 221 [stem: *Calospast-*]. Type genus: *Calospasta* J. L. LeConte, 1862 [syn. of *Eupompha* J. L. LeConte, 1858].

Gynaecomelodides Wellman, 1910: 221 [stem: *Gynaecomelo-*]. Type genus: *Gynaecomeloe* Wellman, 1910 [syn. of *Cordylospasta* G. H. Horn, 1875]. Comment: incorrect original stem formation, not in prevailing usage.

Cordylospastides Wellman, 1910: 221 [stem: *Cordylospast-*]. Type genus: *Cordylospasta* G. H. Horn, 1875.

Cysteodemides Wellman, 1910: 221 [stem: *Cysteodem-*]. Type genus: *Cysteodemus* J. L. LeConte, 1851.

Tegroderini L. S. Dillon, 1952: 373 [stem: *Tegroder-*]. Type genus: *Tegrodera* J. L. LeConte, 1851.

Megetrina Kaszab, 1959: 80 [stem: *Megetr-*]. Type genus: *Megetra* J. L. LeConte, 1859.

Tribe Lyttini Solier, 1851

Cantharidiae Latreille, 1802: 185 [stem: *Canthar-*]. Type genus: *Cantharis* sensu Latreille, 1802 [not *Cantharis* Linnaeus, 1758; syn. of *Lytta* Fabricius, 1775]. Comment: based on a misidentified type genus, name treated here as invalid until an application is submitted to the Commission to suppress it for the Principles of Priority and Homonymy (Art. 65.2.1); also see Cantharidae Imhoff, 1856.

*Lyttes Motschulsky, 1849: 59 [stem: *Lytt-*]. Type genus: *Lytta* Fabricius, 1775. Comment: original vernacular name unavailable (Art. 11.7.2): subsequently used in latinized form but not generally attributed to Motschulsky (1849).

Lyttoides Solier, 1851: 278 [stem: *Lytt-*]. Type genus: *Lytta* Fabricius, 1775. Comment: original vernacular name available (Art. 11.7.2): first used in latinized form by Gistel (1856a: 385, as Lyttadae [incorrect stem formation]), generally accepted as in Selander (1991: 78, as Lyttina); although this is not the oldest name for the tribe, we recommend that an application be submitted to the Commission to suppress Cantharini Latreille, 1802 because it is based on a misidentified type genus (Art. 65.2.1).

*Alosimates Mulsant, 1857: 358 [stem: *Alosim-*]. Type genus: *Alosimus* Mulsant, 1857. Comment: original vernacular name unavailable (Art. 11.7.2): subsequently used in latinized form, e.g., Selander (1991: 65, as Alosimina), but not generally accepted as valid; Selander's name is also unavailable, it was proposed after 1930 without description or bibliographic reference to such a description (Art. 13.1).

*Lydini Wellman, 1908: 424 [stem: *Lyd-*]. Type genus: *Lydus* Dejean, 1821. Comment: family-group name unavailable (Art. 11.6): originally published as synonym and not made available subsequently; Wellman correctly pointed out that the family-group for a taxon based on *Lydus* should be Lydini, however he did not want to use this family-group name since the name Lydidae Newman, 1834 (type genus *Lyda* Fabricius, 1805) is available for a group of Hymenoptera: Symphyta.

Sybarides Wellman, 1910: 221 [stem: *Sybare-*]. Type genus: *Sybaris* Stephens, 1832. Comment: incorrect original stem formation, not in prevailing usage.

Lydina Kaszab, 1959: 82 [stem: *Lyd-*]. Type genus: *Lydus* Dejean, 1821. Comment: Kaszab proposed the subtribe name Lydina without reference to the comment by Wellman (1908) and his family-group name has been used as valid in Meloidae since (e.g., Turco et al. 2006); the name Lydidae Newman, 1834 (type genus *Lyda* Fabricius, 1805) is available for a group of Hymenoptera: Symphyta; this case is to be referred to the Commission to remove the homonymy (Art. 55.3.1).

Prolyttini Selander, 1960: 25 [stem: *Prolytt-*]. Type genus: *Prolytta* Kaszab, 1959.

Tribe Meloini Gyllenhal, 1810

Melooides Gyllenhal, 1810: 481 [stem: *Melo-*]. Type genus: *Meloe* Linnaeus, 1758 [placed on the Official List of Generic Names in Zoology (ICZN 1999c)].

Proscarabaeidae Gistel, 1848: [11] [stem: *Proscarabae-*]. Type genus: *Proscarabaeus* Schrank, 1781 [syn. of *Meloe* Linnaeus, 1758].

Tribe Mylabrini Rafinesque, 1815

Mylabria Rafinesque, 1815: 114 [stem: *Mylabr-*]. Type genus: *Mylabris* Fabricius, 1775 [placed on the Official List of Generic Names in Zoology (ICZN 1995a)]. Comment: family-group name previously attributed to Billberg (1820a); Brullé (1832: 190) uses the family-group name "Corynites" for a group containing the genera *Sarrotrium* Illiger, 1798, *Orthocerus* Latreille, 1797, *Chiroscelis* Lamarck, 1804, *Toxicum* Latreille, 1802 and *Boros* Herbst, 1797, although Brullé could have based his family-group name on *Coryna* Billberg, 1813 (which is a junior homonym and a synonym of *Hycleus* Latreille, 1817 in Mylabrini) we cannot be sure and therefore we have not considered this a separate family-group name.

Zonabrini Wellman, 1908: 424 [stem: *Zonabr-*]. Type genus: *Zonabris* Harold, 1879 [syn. of *Mylabris* Fabricius, 1775].

Calydina Kaszab, 1960: 126 [stem: *Calyd-*]. Type genus: *Calydus* Reitter, 1896.

Tribe Pyrotini MacSwain, 1956

Pyrotini MacSwain, 1956: 67 [stem: *Pyrot-*]. Type genus: *Pyrota* Dejean, 1834.

Subfamily TETRAONYCINAE Böving and Craighead, 1931

TETRAONYCIDAE Böving and Craighead, 1931: 59, in key [stem: *Tetraonyc-*]. Type genus: *Tetraonyx* Latreille, 1809. Comment: current spelling maintained (Art. 29.5): incorrect stem formation in prevailing usage (should be *Tetraonych-*).

MELOETYPHLINI Borgmeier, 1937: 248 [stem: *Meloetyphl-*]. Type genus: *Meloetyphlus* C. O. Waterhouse, 1872.

Subfamily NEMOGNATHINAE Laporte, 1840

NEMOGNATHITES Laporte, 1840b: 280 [stem: *Nemognath-*]. Type genus: *Nemognatha* Illiger, 1807 [as *Nemognathus*, unjustified emendation of type genus name by Latreille (1829b), not in prevailing usage; *Nemognatha* Illiger, 1807 placed on the Official List of Generic Names in Zoology (ICZN 1999c)]. Comment: NEMOGNATHINAE Laporte, 1840 placed on the Official List of Family-Group Names in Zoology and given precedence over HORIINAE Latreille, 1802 (ICZN 1999c).

Tribe HORIINI Latreille, 1802

HORIALES Latreille, 1802: 182 [stem: *Hori-*]. Type genus: *Horia* Fabricius, 1787 [placed on the Official List of Generic Names in Zoology (ICZN 1999c)]. Comment: HORIIDAE Latreille, 1802 placed on the Official List of Family-Group Names in Zoology (ICZN 1999c); MELOIDAE Gyllenhal, 1810 and NEMOGNATHINAE Laporte, 1840 given precedence over this name (ICZN 1999c).

CISSITINI Kaszab, 1966: 187 [stem: *Cissit-*]. Type genus: *Cissites* Latreille, 1804. Comment: we accept the character about parameres given by Kaszab (1966: 187) to characterize this tribe as sufficient to fulfill the requirements of Art. 13.1; Selander (1991: 66) treated this name as unavailable.

Tribe NEMOGNATHINI Laporte, 1840

NEMOGNATHITES Laporte, 1840b: 280 [stem: *Nemognath-*]. Type genus: *Nemognatha* Illiger, 1807 [as *Nemognathus*, unjustified emendation of type genus name by Latreille (1829b), not in prevailing usage; *Nemognatha* Illiger, 1807 placed on the Official List of Generic Names in Zoology (ICZN 1999c)].

Subtribe NEMOGNATHINA Laporte, 1840

NEMOGNATHITES Laporte, 1840b: 280 [stem: *Nemognath-*]. Type genus: *Nemognatha* Illiger, 1807 [as *Nemognathus*, unjustified emendation of type genus name by Latreille (1829b), not in prevailing usage; *Nemognatha* Illiger, 1807 placed on the Official List of Generic Names in Zoology (ICZN 1999c)]. Comment: original vernacular name available (Art. 11.7.2): first used in latinized form by J. L. LeConte (1862: 270, as NEMOGNATHINI), generally accepted as in Hansen (1996: 176, as NEMOGNATHINAE).

HORNII Dugès, 1889: 213 [stem: *Horni-*]. Type genus: *Hornia* Riley, 1877.

Tricraniides Wellman, 1910: 222 [stem: *Tricrani-*]. Type genus: *Tricrania* J. L. LeConte, 1860.

Subtribe Zonitidina Mulsant, 1857

*Zonites Motschulsky, 1849: 59 [stem: *Zonitid-*]. Type genus: *Zonitis* Fabricius, 1775 [placed on the Official List of Generic Names in Zoology (ICZN 1999c)]. Comment: original vernacular name unavailable (Art. 11.7.2): subsequently used in latinized form but not generally attributed to Motschulsky (1849); incorrect original stem formation, not in prevailing usage.

Zonitaires Mulsant, 1857: 322 [stem: *Zonitid-*]. Type genus: *Zonitis* Fabricius, 1775 [placed on the Official List of Generic Names in Zoology (ICZN 1999c)]. Comment: incorrect original stem formation, not in prevailing usage; Zonitidinae Mulsant, 1857 placed on the Official List of Family-Group Names in Zoology (ICZN 1999c); this family-group name is a senior homonym of Zonitidae Mörch, 1864 in Mollusca (based on *Zonites* Montfort, 1810) which is used as valid in recent literature (see Bologna and Pinto 1997); stem of family-group name changed to *Zonitid-* by the Commission (ICZN 1999c) to remove from homonymy with Mörch's name.

*Leptopalpina Kaszab, 1959: 102, Fig. 97 [stem: *Leptopalp-*]. Type genus: *Leptopalpus* Guérin-Méneville, 1834. Comment: unavailable family-group name, proposed after 1930 without description or bibliographic reference to such a description (Art. 13.1).

Leptopalpina Kaszab, 1963: 138 [stem: *Leptopalp-*]. Type genus: *Leptopalpus* Guérin-Méneville, 1834.

*Palaestrina Kaszab, 1969: 244 [stem: *Palaestr-*]. Type genus: *Palaestra* Laporte, 1840. Comment: unavailable family-group name, proposed after 1930 without description or bibliographic reference to such a description (Art. 13.1).

Subtribe Sitarina Mulsant, 1857

*Apales Motschulsky, 1849: 59 [stem: *Apal-*]. Type genus: *Apalus* Fabricius, 1775. Comment: original vernacular name unavailable (Art. 11.7.2): subsequently used in latinized form but not generally attributed to Motschulsky (1849).

Sitarates Mulsant, 1857: 393 [stem: *Sitar-*]. Type genus: *Sitaris* Latreille, 1802. Comment: original vernacular name available (Art. 11.7.2): first used in latinized form by Dugès (1886: 580, as Sitarini), generally accepted as in Selander (1991: 79, as Sitarina).

Apalides Parker and Böving, 1924: 33 [stem: *Apal-*]. Type genus: *Apalus* Fabricius, 1775.

Stenoriides Parker and Böving, 1924: 32 [stem: *Stenori-*]. Type genus: *Stenoria* Mulsant, 1857.

Tribe STENODERINI Selander, 1991

*STENODERINI Selander, 1964: 1057 [stem: *Stenoder-*]. Type genus: *Stenodera* Eschscholtz, 1818. Comment: unavailable family-group name, proposed after 1930 without description or bibliographic reference to such a description (Art. 13.1).

STENODERINI Selander, 1991: 77 [stem: *Stenoder-*]. Type genus: *Stenodera* Eschscholtz, 1818. Comment: junior homonym of STENODERINI Pascoe, 1867 (type genus *Stenoderus* Dejean, 1821) currently used as valid in CERAMBYCIDAE; this case is to be referred to the Commission to remove the homonymy (Art. 55.3.1).

Family MYCTERIDAE Oken, 1843

MYCTERIDEN Oken, 1843: 484 [stem: *Mycter-*]. Type genus: *Mycterus* Clairville, 1798.

Subfamily MYCTERINAE Oken, 1843

RHINOMACERIDAE Fleming, 1821: 50 [stem: *Rhinomacr-*]. Type genus: *Rhinomacer* Fabricius, 1781 [placed on the Official Index of Rejected and Invalid Generic Names in Zoology (ICZN 2005c); syn. of *Mycterus* Clairville, 1798]. Comment: permanently invalid (Art. 39): based on suppressed type genus; incorrect original stem formation, not in prevailing usage.

MYCTERIDEN Oken, 1843: 484 [stem: *Mycter-*]. Type genus: *Mycterus* Clairville, 1798. Comment: original vernacular name available (Art. 11.7.2): first used in latinized form and generally accepted as in Tulk (1847: 614, as MYCTERIDAE); name previously attributed to Blanchard (1845b: 97).

ARTAXIDAE Gistel, 1848: [8] [stem: *Artax-*]. Type genus: *Artaxus* Gistel, 1848 [syn. of *Mycterus* Clairville, 1798].

Subfamily EURYPINAE Thomson, 1860

EURYPITAE J. Thomson, 1860b: 63 [stem: *Euryp-*]. Type genus: *Eurypus* Kirby, 1819. Comment: the correct stem based on the type genus *Eurypus* is *Eurypod-*, however, we prefer to conserve the incorrect original stem formation for this name (Art. 29.5) to avoid homoymy with EURYPODINI Gahan, 1906 (type genus *Eurypoda* W. Saunders, 1853) currently used as valid in CERAMBYCIDAE.

LACCONOTINI J. L. LeConte, 1862: 254 [stem: *Lacconot-*]. Type genus: *Lacconotus* J. L. LeConte, 1862.

BATOBIINA Seidlitz, 1917a: 88 [stem: *Batobi-*]. Type genus: *Batobius* Fairmaire and Germar, 1863.

THISIINA Seidlitz, 1917a: 91 [stem: *Thisi-*]. Type genus: *Thisias* Champion, 1889.

*STILPNONOTINAE Blackwelder, 1945: 503 [stem: *Stilpnonot-*]. Type genus: *Stilpnonotus* Gray, 1832. Comment: unavailable family-group name, proposed after 1930 without description or bibliographic reference to such a description (Art. 13.1).

Subfamily Hemipeplinae Lacordaire, 1854

Hémipéplides Lacordaire, 1854b: 404 [stem: *Hemipepl-*]. Type genus: *Hemipeplus* Latreille, 1829. Comment: original vernacular name available (Art. 11.7.2): first used in latinized form by J. L. LeConte (1861: 96, as Hemipeplidae), generally accepted as in Pollock (2002: 532, as Hemipeplinae).

Family Boridae Thomson, 1859

Boridae C. G. Thomson, 1859: 117 [stem: *Bor-*]. Type genus: *Boros* Herbst, 1797.

Subfamily Borinae Thomson, 1859

Boridae C. G. Thomson, 1859: 117 [stem: *Bor-*]. Type genus: *Boros* Herbst, 1797.

Subfamily Synercticinae Lawrence and Pollock, 1994

Synercticinae Lawrence and Pollock, 1994: 36, in key [stem: *Synerctic-*]. Type genus: *Synercticus* Newman, 1842.

Family Trictenotomidae Blanchard, 1845

Tricténotomides Blanchard, 1845b: 137 [stem: *Trictenotom-*]. Type genus: *Trictenotoma* Gray, 1832. Comment: original vernacular name available (Art. 11.7.2): first used in latinized form by F. Smith (1851: 18, as Trictenotomidae), generally accepted as in Lawrence and Newton (1995: 897, as Trictenotomidae).

Family Pythidae Solier, 1834

Pythites Solier, 1834: 496 [stem: *Pyth-*]. Type genus: *Pytho* Latreille, 1797. Comment: original vernacular name available (Art. 11.7.2): first used in latinized form by C. G. Thomson (1859: 123, as Pythonidae [incorrect stem formation]), generally accepted as in Hansen (1996: 176, as Pythidae).

Enoptisidae Gistel, 1848: [11] [stem: *Enopt-*]. Type genus: *Enoptes* Gistel, 1848 [syn. of *Pytho* Latreille, 1797]. Comment: incorrect original stem formation, not in prevailing usage.

Ischyomiides Champion, 1886: 258 [stem: *Ischyomi-*]. Type genus: *Ischyomius* Chevrolat, 1878.

Osphyoplesiini Reitter, 1917: 59, in key [stem: *Osphyoplesi-*]. Type genus: *Osphyoplesius* A. Winkler, 1915.

Anaplopinae M. Abdullah, 1966: 147 [stem: *Anaplopod-*]. Type genus: *Anaplopus* Blackburn, 1890. Comment: incorrect original stem formation, not in prevailing usage.

Family Pyrochroidae Latreille, 1806

Pyrochroides Latreille, 1806: 199 [stem: *Pyrochro-*]. Type genus: *Pyrochroa* Geoffroy, 1762 [placed on the Official List of Generic Names in Zoology (ICZN 1994a)].

Subfamily Tydessinae Nikitsky, 1986

Tydessini Nikitsky, 1986: 1188 [stem: *Tydess-*]. Type genus: *Tydessa* Peacock, 1982.

Subfamily Pilipalpinae Abdullah, 1964

Pilipalpini M. Abdullah, 1964: 4 [stem: *Pilipalp-*]. Type genus: *Pilipalpus* Fairmaire, 1876.

Techmessinae Paulus, 1972b: 84, in key [stem: *Techmess-*]. Type genus: *Techmessa* F. Bates, 1874.

Subfamily Pedilinae Lacordaire, 1859

Pédilides Lacordaire, 1859: 574 [stem: *Pedil-*]. Type genus: *Pedilus* Fischer von Waldheim, 1820. Comment: published before 27 June 1859; original vernacular name available (Art. 11.7.2): first used in latinized form by J. L. LeConte (1859a [December]: 46, as Pedilidae), generally accepted as in Lawrence and Newton (1995: 898, as Pedilinae).

Neopedilini M. Abdullah, 1969a: 354 [stem: *Neopedil-*]. Type genus: *Neopedilus* M. Abdullah, 1969 [syn. of *Pedilus* Fischer von Waldheim, 1820].

Subfamily Pyrochroinae Latreille, 1806

Pyrochroides Latreille, 1806: 199 [stem: *Pyrochro-*]. Type genus: *Pyrochroa* Geoffroy, 1762 [placed on the Official List of Generic Names in Zoology (ICZN 1994a)].

Anthomanisidae Gistel, 1848: [11] [stem: *Anthoman-*]. Type genus: *Anthomanes* Gistel, 1848 [syn. of *Pyrochroa* Geoffroy, 1762]. Comment: incorrect original stem formation, not in prevailing usage.

Subfamily Agnathinae Lacordaire, 1859

Agnathides Lacordaire, 1859: 531 [stem: *Agnath-*]. Type genus: *Agnathus* Germar, 1818. Comment: original vernacular name available (Art. 11.7.2): first used in latinized form by Seidlitz (1875 [Gatt.]: 100, as Agnathini), generally accepted as in Young (2002: 542, as Agnathinae).

Cononotini J. L. LeConte, 1862: 256 [stem: *Cononot-*]. Type genus: *Cononotus* J. L. LeConte, 1851.

Family Salpingidae Leach, 1815

Salpingides Leach, 1815: 106 [stem: *Salping-*]. Type genus: *Salpingus* Illiger, 1802.

Subfamily Othniinae LeConte, 1861

Othniidae J. L. LeConte, 1861: 102 [stem: *Othni-*]. Type genus: *Othnius* J. L. LeConte, 1861 [syn. of *Elacatis* Pascoe, 1860].

Elacatidae Cockerell, 1906: 242 [stem: *Elacat-*]. Type genus: *Elacatis* Pascoe, 1860. Comment: junior homonym of Elacatinae Gill, 1861 (type genus *Elacate* Cuvier, 1831) proposed in Pisces; this case is to be referred to the Commission to remove the homonymy (Art. 55.3.1).

Subfamily Prostominiinae Grouvelle, 1914

Prostominini Grouvelle, 1914: 152 [stem: *Prostomini-*]. Type genus: *Prostominia* Reitter, 1889. Comment: incorrect original stem formation, not in prevailing usage.

*Trogocryptinae Crowson, 1953: 51 [stem: *Trogocrypt-*]. Type genus: *Trogocryptus* Sharp, 1900. Comment: unavailable family-group name, proposed after 1930 without description or bibliographic reference to such a description (Art. 13.1); this name has been used subsequently, e.g., Lawrence (1977: 43, 1980: 307), Lawrence and Newton (1995: 900), but it has not been made available.

Subfamily Ageninae Horn, 1878

Agleni G. H. Horn, 1878: 573 [stem: *Aglen-*]. Type genus: *Aglenus* Erichson, 1845.

Subfamily Inopeplinae Grouvelle, 1908

Inopeplini Grouvelle, 1908: 459, in key [stem: *Inopepl-*]. Type genus: *Inopeplus* F. Smith, 1851.

Subfamily Dacoderinae LeConte, 1862

Dacoderini J. L. LeConte, 1862: 216 [stem: *Dacoder-*]. Type genus: *Dacoderus* J. L. LeConte, 1858.

Tretothoracidae Lea, 1910: 210 [stem: *Tretothorac-*]. Type genus: *Tretothorax* Lea, 1910.

Subfamily Aegialitinae LeConte, 1862

Aegialitidae J. L. LeConte, 1862: 241 [stem: *Aegialit-*]. Type genus: *Aegialites* Mannerheim, 1853.

Eurystethidae Seidlitz, 1916: 127 [stem: *Eurysteth-*]. Type genus: *Eurystethes* Seidlitz, 1916 [syn. of *Aegialites* Mannerheim, 1853].

Subfamily Salpinginae Leach, 1815

Salpingides Leach, 1815: 106 [stem: *Salping-*]. Type genus: *Salpingus* Illiger, 1802.

*Rhinosimites Solier, 1834: 496 [stem: *Rhinosim-*]. Type genus: *Rhinosimus* Latreille, 1802. Comment: original vernacular name unavailable (Art. 11.7.2): subsequently used in latinized form but not generally attributed to Solier (1834).

Rhinosimidae Hope, 1840a: 134 [stem: *Rhinosim-*]. Type genus: *Rhinosimus* Latreille, 1802.

Lissodemina Seidlitz, 1917b: 422 [stem: *Lissodem-*]. Type genus: *Lissodema* Curtis, 1833.

Istrisiini Nikitsky, 1992: 483, in key [stem: *Istrisi-*]. Type genus: *Istrisia* Lewis, 1895.

Family Anthicidae Latreille, 1819

Anthicites Latreille, 1819: 363 [stem: *Anthic-*]. Type genus: *Anthicus* Paykull, 1798.

Subfamily Eurygeniinae LeConte, 1862

Eurygenii J. L. LeConte, 1862: 264 [stem: *Eurygeni-*]. Type genus: *Eurygenius* LaFerté-Sénectère, 1849.

Tribe Eurygeniini LeConte, 1862

Eurygenii J. L. LeConte, 1862: 264 [stem: *Eurygeni-*]. Type genus: *Eurygenius* LaFerté-Sénectère, 1849.

Tribe Ictistygnini Borchmann, 1936

Ictistygninae Borchmann, 1936: 534 [stem: *Ictistygn-*]. Type genus: *Ictistygna* Pascoe, 1866.

Tribe Mitraelabrini Abdullah, 1969

Mitraelabrini M. Abdullah, 1969a: 350 [stem: *Mitraelabr-*]. Type genus: *Mitraelabrus* Solier, 1851.

Subfamily Macratriinae LeConte, 1862

Macratriini J. L. LeConte, 1862: 265 [stem: *Macratri-*]. Type genus: *Macratria* Newman, 1838.

Tribe Macratriini LeConte, 1862

Macratriini J. L. LeConte, 1862: 265 [stem: *Macratri-*]. Type genus: *Macratria* Newman, 1838.

†Tribe Camelomorphini Kirejtshuk and Azar, 2008

Camelomorphini Kirejtshuk and Azar, 2008: 40 [stem: *Camelomorph-*]. Type genus: *Camelomorpha* Kirejtshuk and Azar, 2008.

Subfamily Steropinae Jacquelin du Val, 1863

Stéropites Jacquelin du Val, 1863: 365 [stem: *Sterop-*]. Type genus: *Steropes* Steven, 1806. Comment: original vernacular name available (Art. 11.7.2): first used in latinized form by Jakobson (1915: 1022, as Steropina), generally accepted as in Lawrence and Newton (1995: 901, as Steropinae); the older name Steropinae Dana, 1854 is available in copepods (type genus *Sterope* Goodsir, 1845) although it was recently treated as a *nomen oblitum* by Huys (2009: 28); this case is to be referred to the Commission to remove the homonymy (Art. 55.3.1).

Subfamily Copobaeninae Abdullah, 1969

Copobaeninae M. Abdullah, 1969a: 334 [stem: *Copobaen-*]. Type genus: *Copobaenus* Fairmaire and Germain, 1863.

Subfamily Lemodinae Lawrence and Britton, 1991

*Lemodinae Lawrence, 1977: 43 [stem: *Lemod-*]. Type genus: *Lemodes* Boheman, 1858. Comment: unavailable family-group name, proposed after 1930 without description or bibliographic reference to such a description (Art. 13.1).

Lemodinae Lawrence and Britton, 1991: 603, in key [stem: *Lemod-*]. Type genus: *Lemodes* Boheman, 1858.

Subfamily Tomoderinae Bonadona, 1961

Tomoderini Bonadona, 1961: 11 [stem: *Tomoder-*]. Type genus: *Tomoderus* LaFerté-Sénectère, 1849.

Subfamily Anthicinae Latreille, 1819

Anthicites Latreille, 1819: 363 [stem: *Anthic-*]. Type genus: *Anthicus* Paykull, 1798.

Tribe Anthicini Latreille, 1819

Anthicites Latreille, 1819: 363 [stem: *Anthic-*]. Type genus: *Anthicus* Paykull, 1798.

Tribe Endomiini Kaszab, 1956

Endomiini Kaszab, 1956: 45, in key [stem: *Endomi-*]. Type genus: *Endomia* Laporte, 1840.

Tribe Formicomini Bonadona, 1974

Formicomini Bonadona, 1974: 110, in key [stem: *Formicom-*]. Type genus: *Formicomus* LaFerté-Sénectère, 1849.

Tribe Microhoriini Bonadona, 1974

Microhorini Bonadona, 1974: 110, in key [stem: *Microhori-*]. Type genus: *Microhoria* Chevrolat, 1877. Comment: incorrect original stem formation, not in prevailing usage.

Liparoderini Bonadona, 1990: 20 [stem: *Liparoder-*]. Type genus: *Liparoderus* LaFerté-Sénectère, 1849.

Subfamily Notoxinae Stephens, 1829

Notoxidae Stephens, 1829b: 254 [stem: *Notox-*]. Type genus: *Notoxus* Geoffroy, 1762 [placed on the Official List of Generic Names in Zoology (ICZN 1994a)]. Comment: an application needs to be submitted to the Commission to suppress Notoxii Sturm, 1826 (based on the misidentified type genus *Notoxus* sensu Fabricius, 1775) for the Principles of Priority and Homonymy (Art. 65.2.1) to conserve this name as valid.

Family Aderidae Csiki, 1909

Aderidae Csiki, 1909a: 6 [stem: *Ader-*]. Type genus: *Aderus* Stephens, 1829 [placed on the Official List of Generic Names in Zoology (ICZN 1989e)]. Comment: name given precedence over Euglenesidae Seidlitz, 1875 (ICZN 1989e).

†Circaeidae Iablokoff-Khnzorian, 1961: 209 [stem: *Circae-*]. Type genus: *Circaeus* Iablokoff-Khnzorian, 1961.

Tribe Aderini Csiki, 1909

Aderidae Csiki, 1909a: 6 [stem: *Ader-*]. Type genus: *Aderus* Stephens, 1829 [placed on the Official List of Generic Names in Zoology (ICZN 1989e)].

Comment: name given precedence over Euglenesidae Seidlitz, 1875 and placed on the Official List of Family-Group Names in Zoology (ICZN 1989e, as Aderidae A. Winkler (1927)).

Subtribe Aderina Csiki, 1909

Xylophilidae Shuckard, 1839b: 47 [stem: *Xylophil-*]. Type genus: *Xylophilus* Latreille, 1825 [preoccupied genus name, not *Xylophilus* Mannerheim, 1823 [Coleoptera: Eucnemidae]; syn. of *Aderus* Stephen, 1829]. Comment: permanently invalid (Art. 39): based on preoccupied type genus.

Hylophilidae Pic, 1900: 754 [stem: *Hylophil-*]. Type genus: *Hylophilus* Berthold, 1827 [preoccupied genus name, not *Hylophilus* Temminck, 1822 [Aves]; syn. of *Aderus* Stephens, 1829]. Comment: permanently invalid (Art. 39): based on preoccupied type genus.

Aderidae Csiki, 1909a: 6 [stem: *Ader-*]. Type genus: *Aderus* Stephens, 1829 [placed on the Official List of Generic Names in Zoology (ICZN 1989e)]. Comment: placed on the Official List of Family-Group Names in Zoology (ICZN 1989e, as Aderidae A. Winkler (1927)).

Subtribe Cnopina Báguena Corella, 1948

Cnopina Báguena Corella, 1948: 57 [stem: *Cnop-*]. Type genus: *Cnopus* Champion, 1893.

Subtribe Gompeliina Bouchard, 2011

Olotelina Báguena Corella, 1948: 58 [stem: *Olotel-*]. Type genus: *Olotelus* Mulsant and Rey, 1866 [preoccupied genus name, not *Olotelus* Solier, 1851 [Coleoptera: Elateridae]; syn. of *Gompelia* Alonso-Zarazaga, 2010]. Comment: permanently invalid (Art. 39): based on preoccupied type genus.

Gompeliina Bouchard, 2011 **nom. nov.** [stem: *Gompeli-*]. Type genus: *Gompelia* Alonso-Zarazaga, 2010. Comment: replacement name for Olotelina Báguena Corella, 1948 because of the homonymy of the type genus.

Subtribe Syzetoninina Báguena Corella, 1948

Syzetoninini Báguena Corella, 1948: 41 [stem: *Syzetonin-*]. Type genus: *Syzetoninus* Blackburn, 1891.

Tribe Emelinini Báguena Corella, 1948

Emelinini Báguena Corella, 1948: 36 [stem: *Emelin-*]. Type genus: *Emelinus* Casey, 1895.

Tribe Euglenesini Seidlitz, 1875

Euglenini Seidlitz, 1875 [Gatt.]: 106 [stem: *Euglenes-*]. Type genus: *Euglenes* Westwood, 1830 [placed on the Official List of Generic Names in Zoology (ICZN 1989e)].

Subtribe Euglenesina Seidlitz, 1875

Euglenini Seidlitz, 1875 [Gatt.]: 106 [stem: *Euglenes-*]. Type genus: *Euglenes* Westwood, 1830 [placed on the Official List of Generic Names in Zoology (ICZN 1989e)]. Comment: the younger name Aderidae Csiki, 1909 (listed as Aderidae A. Winkler, 1927) was given precedence for the family name, *Euglenes-* was ruled to be the correct stem of this family-group name and Euglenesidae Seidlitz, 1875 was placed on the Official List of Family-Group Names in Zoology (ICZN 1989e); Euglenidae Seidlitz, 1875 was placed on the Official Index of Rejected and Invalid Family-Group Names in Zoology (ICZN 1989e).

Subtribe Pseudolotelina Báguena Corella, 1948

Pseudolotelina Báguena Corella, 1948: 40 [stem: *Pseudolotel-*]. Type genus: *Pseudolotelus* Pic, 1901.

Tribe Phytobaenini Báguena Corella, 1948

Phytobaenini Báguena Corella, 1948: 30 [stem: *Phytobaen-*]. Type genus: *Phytobaenus* Sahlberg, 1834.

Family Scraptiidae Gistel, 1848

Scraptiaeidae Gistel, 1848: [11] [stem: *Scrapti-*]. Type genus: *Scraptia* Latreille, 1807.

Subfamily Scraptiinae Gistel, 1848

Scraptiaeidae Gistel, 1848: [11] [stem: *Scrapti-*]. Type genus: *Scraptia* Latreille, 1807. Comment: incorrect original stem formation, not in prevailing usage.

Tribe Allopodini Franciscolo, 1964

Allopodini Franciscolo, 1964: 176 [stem: *Allopod-*]. Type genus: *Allopoda* J. L. LeConte, 1866.

Tribe Scraptiini Gistel, 1848

Scraptiaeidae Gistel, 1848: [11] [stem: *Scrapti-*]. Type genus: *Scraptia* Latreille, 1807. Comment: family-group name previously attributed to Mulsant (1856)/ Gistel (1856); incorrect original stem formation, not in prevailing usage.

Subfamily Anaspidinae Mulsant, 1856

Anaspiens Mulsant, 1856c: 85 [stem: *Anaspid-*]. Type genus: *Anaspis* Geoffroy, 1762 [placed on the Official List of Generic Names in Zoology (ICZN 1984a)].

Tribe Anaspidini Mulsant, 1856

Anaspiens Mulsant, 1856c: 85 [stem: *Anaspid-*]. Type genus: *Anaspis* Geoffroy, 1762 [placed on the Official List of Generic Names in Zoology (ICZN 1984a)]. Comment: original vernacular name available (Art. 11.7.2): first used in latinized form by C. G. Thomson (1859: 121, as ANASPINA [incorrect stem formation]), generally accepted as in Lawrence and Newton (1995: 903, as ANASPIDINAE); incorrect original stem formation, not in prevailing usage.

Tribe ANASPIMORDINI Franciscolo, 1954

ANASPIMORDINI Franciscolo, 1954: 65, in key [stem: *Anaspimord-*]. Type genus: *Anaspimorda* Ermisch, 1950.

Tribe MENUTHIANASPIDINI Franciscolo, 1972

*CYRTOSCRAPTIINI Franciscolo, 1964: 176 [stem: *Cyrtoscrapti-*]. Type genus: *Cyrtoscraptia* Franciscolo, 1964 [unavailable genus name, proposed after 1930 without description or bibliographic reference to such a description (Art. 13.1); syn. of *Menuthianaspis* Franciscolo, 1972]. Comment: unavailable family-group name, proposed after 1930 without description or bibliographic reference to such a description (Art. 13.1), based on unavailable type genus (Art. 12.2.4).

MENUTHIANASPIDINI Franciscolo, 1972: 123 [stem: *Menuthianaspid-*]. Type genus: *Menuthianaspis* Franciscolo, 1972.

Tribe PENTARIINI Franciscolo, 1954

PENTARIINI Franciscolo, 1954: 64, in key [stem: *Pentari-*]. Type genus: *Pentaria* Mulsant, 1856.

Subfamily LAGRIOIDINAE Abdullah and Abdullah, 1968

LAGRIOIDINI M. Abdullah and A. Abdullah, 1968: 73 [stem: *Lagrioid-*]. Type genus: *Lagrioida* Fairmaire and Germain, 1860.

Subfamily AFREMINAE Levey, 1985

AFREMINAE Levey, 1985: 420 [stem: *Afrem-*]. Type genus: *Afremus* Levey, 1985.

Subfamily ISCHALIINAE Blair, 1920

ISCHALIINAE Blair, 1920: 134 [stem: *Ischali-*]. Type genus: *Ischalia* Pascoe, 1860.

Superfamily CHRYSOMELOIDEA Latreille, 1802

CHRYSOMELINAE Latreille, 1802: 220 [stem: *Chrysomel-*]. Type genus: *Chrysomela* Linnaeus, 1758 [placed on the Official List of Generic Names in Zoology (ICZN 1984c)]. Comment: Švácha and Danilevsky (1987: 17) recognized VESPERINAE, OXYPELTINAE and DISTENIINAE at the family level, and later Švácha et al. (1997: 361) placed ANOPLODERMATINAE and PHILINAE as subfamilies of VESPERIDAE; although the classification is based primarily on larval characters, it is also supported by adult features, as pointed out by Švácha et al. (1997), as well as by some earlier workers, e.g., Gahan (1906), Linsley (1961), Crowson (1981), Saito (1990); this classification will also be used in the upcoming volume 3 of the "Handbook of

Zoology: Coleoptera, beetles" series; First Reviser (CHRYSOMELOIDEA Latreille, 1802 vs MEGALOPODOIDEA Latreille, 1802 vs CERAMBYCOIDEA Latreille, 1802) not determined, current usage maintained.

Family OXYPELTIDAE Lacordaire, 1868

OXYPELTIDES Lacordaire, 1868: 461 [stem: *Oxypelt-*]. Type genus: *Oxypeltus* Blanchard, 1851. Comment: original vernacular name available (Art. 11.7.2): first used in latinized form and generally accepted as in Aurivillius (1912: 254, as OXYPELTINI).

*CHELODERIDOS Germain, 1900: 86 [stem: *Cheloder-*]. Type genus: *Cheloderus* Gray, 1832. Comment: original vernacular name unavailable (Art. 11.7.2): proposed after 1899.

Family VESPERIDAE Mulsant, 1839

VESPÉRAIRES Mulsant, 1839: 214 [stem: *Vesper-*]. Type genus: *Vesperus* Dejean, 1821.

Subfamily PHILINAE Thomson, 1861

PHILITAE J. Thomson, 1861: 297 [stem: *Phil-*]. Type genus: *Philus* Saunders, 1853.

Subfamily VESPERINAE Mulsant, 1839

VESPÉRAIRES Mulsant, 1839: 214 [stem: *Vesper-*]. Type genus: *Vesperus* Dejean, 1821. Comment: original vernacular name available (Art. 11.7.2): first used in latinized form by Heyne and Taschenberg (1907: 240, as VESPERINI), generally accepted as in Villiers (1978: 67, as VESPERINI).

Subfamily ANOPLODERMATINAE Guérin-Méneville, 1840

ANOPLODERMIENS Guérin-Méneville, 1840: 276 [stem: *Anoplodermat-*]. Type genus: *Anoploderma* Guérin-Méneville, 1840.

Tribe ANOPLODERMATINI Guérin-Méneville, 1840

ANOPLODERMIENS Guérin-Méneville, 1840: 276 [stem: *Anoplodermat-*]. Type genus: *Anoploderma* Guérin-Méneville, 1840. Comment: original vernacular name available (Art. 11.7.2): first used in latinized form by J. Thomson (1861: 277, as ANOPLODERMITAE [incorrect stem formation]), generally accepted as in Monné (1994c: 9, as ANOPLODERMATINAE); incorrect original stem formation, not in prevailing usage.

CHERROCRIINAE Prosen, 1960: 90 [stem: *Cherrocri-*]. Type genus: *Cherrocrius* Berg, 1898.

Tribe HYPOCEPHALINI Blanchard, 1845

HYPOCÉPHALIENS Blanchard, 1845b: 135 [stem: *Hypocephal-*]. Type genus: *Hypocephalus* Desmarest, 1832. Comment: original vernacular name available

(Art. 11.7.2): first used in latinized form by Imhoff (1856 [2]: 170, as Hypocephalidae), generally accepted as in Lameere (1913: 94, as Hypocephali).

Tribe Mysteriini Prosen, 1960

Mysterinae Prosen, 1960: 90 [stem: *Mysteri-*]. Type genus: *Mysteria* J. Thomson, 1860. Comment: incorrect original stem formation, not in prevailing usage.

Family Disteniidae Thomson, 1861

Distenitae J. Thomson, 1861: 181 [stem: *Disteni-*]. Type genus: *Distenia* Lepeletier and Audinet-Serville, 1828.

Tribe Cyrtonopini Gressitt, 1940

Cyrtonopini Gressitt, 1940: 27 [stem: *Cyrtonop-*]. Type genus: *Cyrtonops* A. White, 1853.

Tribe Disteniini Thomson, 1861

*Cométites Blanchard, 1845b: 163 [stem: *Comet-*]. Type genus: *Cometes* Lepeletier and Audinet-Serville, 1828. Comment: original vernacular name unavailable (Art. 11.7.2): not subsequently latinized; Monné and Santos-Silva (2008: 261) considered this name a *nomen oblitum*, this action was unnecessary since Cométites Blanchard is not an available name.

Distenitae J. Thomson, 1861: 181 [stem: *Disteni-*]. Type genus: *Distenia* Lepeletier and Audinet-Serville, 1828. Comment: incorrect original stem formation, not in prevailing usage.

Tribe Dynamostini Lacordaire, 1868

Dynamostides Lacordaire, 1868: 196 [stem: *Dynamost-*]. Type genus: *Dynamostes* Pascoe, 1857. Comment: original vernacular name available (Art. 11.7.2): first used in latinized form by Heyne and Taschenberg (1907: 238, as Dynamostini), generally accepted as in Santos-Silva and Martins (2004: 145, as Dynamostini).

Tribe Heteropalpini Villiers, 1961

Heteropalpini Villiers, 1961: 385, in key [stem: *Heteropalp-*]. Type genus: *Heteropalpus* Buquet, 1843.

Family Cerambycidae Latreille, 1802

Cerambicini Latreille, 1802: 211 [stem: *Cerambyc-*]. Type genus: *Cerambyx* Linnaeus, 1758. Comment: First Reviser (Cerambycidae Latreille, 1802 vs Prionidae Latreille, 1802 vs Lepturidae Latreille, 1802) not determined, current usage maintained.

Subfamily Parandrinae Blanchard, 1845

Parandrides Blanchard, 1845b: 134 [stem: *Parandr-*]. Type genus: *Parandra* Latreille, 1802.

Tribe Erichsoniini Thomson, 1861

Erichsonitae J. Thomson, 1861: 274 [stem: *Erichsoni-*]. Type genus: *Erichsonia* Westwood, 1849. Comment: incorrect original stem formation, not in prevailing usage.

Tribe Parandrini Blanchard, 1845

Parandrides Blanchard, 1845b: 134 [stem: *Parandr-*]. Type genus: *Parandra* Latreille, 1802. Comment: original vernacular name available (Art. 11.7.2): first used in latinized form by Gistel (1848: [8], as Parandraeidae [incorrect stem formation]), generally accepted as in Lameere (1913: 4, as Parandrae).

Subfamily Prioninae Latreille, 1802

Prionii Latreille, 1802: 212 [stem: *Prion-*]. Type genus: *Prionus* Geoffroy, 1762 [placed on the Official List of Generic Names in Zoology (ICZN 1994a)].

Tribe Acanthophorini Thomson, 1864

Acanthophoritae J. Thomson, 1864: 289 [stem: *Acanthophor-*]. Type genus: *Acanthophorus* Audinet-Serville, 1832.

Tribe Aegosomatini Thomson, 1861

Aegosomitae J. Thomson, 1861: 308 [stem: *Aegosomat-*]. Type genus: *Aegosoma* Audinet-Serville, 1832. Comment: incorrect original stem formation, not in prevailing usage.

*Catypnides Lacordaire, 1868: 62 [stem: *Catypn-*]. Type genus: *Catypnes* Pascoe, 1864. Comment: original vernacular name unavailable (Art. 11.7.2): not subsequently latinized.

Jamwoninae Kolbe, 1897: 294 [stem: *Jamwon-*]. Type genus: *Jamwonus* Harold, 1879.

Megopides Lameere, 1912: 181 [stem: *Megopid-*]. Type genus: *Megopis* Audinet-Serville, 1832. Comment: incorrect original stem formation, not in prevailing usage.

Tribe Anacolini Thomson, 1857

Anacolites J. Thomson, 1857b: 10 [stem: *Anacol-*]. Type genus: *Anacolus* Berthold, 1827. Comment: original vernacular name available (Art. 11.7.2): first used in latinized form by J. Thomson (1861: 302, as Anacolitae), generally accepted as in Galileo (1987: 482, as Anacolini).

*Poecilosomides Lacordaire, 1868: 185 [stem: *Poekilosomat-*]. Type genus: *Poekilosoma* Audinet-Serville, 1832 [as *Poecilosoma*, unjustified emendation of type genus name by Agassiz (1846b: 301), not in prevailing usage; unjustified emendation also junior homonym of *Poecilosoma* Hübner, 1819]. Comment: original vernacular name unavailable (Art. 11.7.2): subsequently used in latinized form but not generally attributed to Lacordaire (1868); incorrect original stem formation, not in prevailing usage.

Poecilosomi H. W. Bates, 1869: 49 [stem: *Poekilosomat-*]. Type genus: *Poekilosoma* Audinet-Serville, 1832 [as *Poecilosoma*, unjustified emendation of type genus name by Agassiz (1846b: 301), not in prevailing usage; unjustified emendation also junior homonym of *Poecilosoma* Hübner, 1819]. Comment: incorrect original stem formation, not in prevailing usage.

Erythraeninae H. W. Bates, 1875: 52 [stem: *Erythraen-*]. Type genus: *Erythraenus* H. W. Bates, 1875.

*Sobarines Lameere, 1901: 320 [stem: *Sobar-*]. Type genus: *Sobarus* Harold, 1879 [preoccupied genus name, not *Sobarus* Loew, 1855 [Diptera]; a replacement name is needed]. Comment: original vernacular name unavailable (Art. 11.7.2): proposed after 1899.

*Délochiliens Lameere, 1912: 57 [stem: *Delocheil-*]. Type genus: *Delocheilus* J. Thomson, 1860 [as *Delochilus*, unjustified emendation of type genus name by Gemminger (1872: 2777), not in prevailing usage]. Comment: original vernacular name unavailable (Art. 11.7.2): proposed after 1899.

Sobari Lameere, 1913: 85 [stem: *Sobar-*]. Type genus: *Sobarus* Harold, 1879 [preoccupied genus name, not *Sobarus* Loew, 1855 [Diptera]; a replacement name is needed]. Comment: permanently invalid (Art. 39): based on preoccupied type genus.

Delochili Lameere, 1913: 85 [stem: *Delocheil-*]. Type genus: *Delocheilus* J. Thomson, 1860 [as *Delochilus*, unjustified emendation of type genus name by Gemminger and Harold (1872), not in prevailing usage]. Comment: incorrect original stem formation, not in prevailing usage.

Tribe Cacoscelini Thomson, 1861

*Notophysites Blanchard, 1845b: 138 [stem: *Notophyse-*]. Type genus: *Notophysis* Audinet-Serville, 1832. Comment: original vernacular name unavailable (Art. 11.7.2): subsequently used in latinized form but not generally attributed to Blanchard (1845b); incorrect original stem formation, not in prevailing usage.

Cacoscelitae J. Thomson, 1861: 325 [stem: *Cacoscel-*]. Type genus: *Cacosceles* Newman, 1838.

Colpodérides Lacordaire, 1868: 133 [stem: *Colpoder-*]. Type genus: *Colpoderus* Audinet-Serville, 1832 [syn. of *Notophysis* Audinet-Serville, 1832]. Comment: original vernacular name available (Art. 11.7.2): first used in latinized form and generally accepted as in Pascoe (1869b: 673, as Colpoderinae); this name was incorrectly credited to Pascoe, 1869 by Bousquet et al. (2009: 115).

Nothophysini Lameere, 1903a: 4, in key [stem: *Notophyse-*]. Type genus: *Notophysis* Audinet-Serville, 1832 [as *Nothophysis*, unjustified emendation of type genus name by Scudder (1882: 226), not in prevailing usage]. Comment: incorrect original stem formation, not in prevailing usage.

Tribe Callipogonini Thomson, 1861

Callipogonitae J. Thomson, 1861: 323 [stem: *Callipogon-*]. Type genus: *Callipogon* Audinet-Serville, 1832.

Anacanthitae J. Thomson, 1864: 285 [stem: *Anacanth-*]. Type genus: *Anacanthus* Audinet-Serville, 1832 [preoccupied genus name, not *Anacanthus* Gray, 1831 [Pisces]; syn. of *Chorenta* Gistel, 1848]. Comment: permanently invalid (Art. 39): based on preoccupied type genus.

Enoploceritae J. Thomson, 1864: 290 [stem: *Enoplocer-*]. Type genus: *Enoplocerus* Audinet-Serville, 1832.

Orthomegitae J. Thomson, 1864: 294 [stem: *Orthomegal-*]. Type genus: *Orthomegas* Audinet-Serville, 1832. Comment: incorrect original stem formation, not in prevailing usage.

Ctenoscelitae J. Thomson, 1864: 295 [stem: *Ctenoscelid-*]. Type genus: *Ctenoscelis* Audinet-Serville, 1832. Comment: incorrect original stem formation, not in prevailing usage.

Tribe Calocomini Galileo and Martins, 1993

Calocomini Galileo and Martins, 1993: 81 [stem: *Calocom-*]. Type genus: *Calocomus* Audinet-Serville, 1832.

Tribe Cantharocnemini Thomson, 1861

Cantharocnemitae J. Thomson, 1861: 274 [stem: *Cantharocnem-*]. Type genus: *Cantharocnemis* Audinet-Serville, 1832. Comment: current spelling maintained (Art. 29.3.1.1): incorrect original stem formation in prevailing usage (should be *Cantharocnemid-*).

*Scéléocanthides Lacordaire, 1868: 34 [stem: *Sceleocanth-*]. Type genus: *Sceleocantha* Newman, 1840. Comment: original vernacular name unavailable (Art. 11.7.2): subsequently used in latinized form, e.g., Heyne and Taschenberg (1907: 236, as Sceleocanthini), but not generally accepted as valid.

Tribe Ergatini Fairmaire, 1864

Ergatites Fairmaire, 1864: 117 [stem: *Ergat-*]. Type genus: *Ergates* Audinet-Serville, 1832. Comment: original vernacular name available (Art. 11.7.2): first used in latinized form by J. L. LeConte (1873: 286, as Ergatini), generally accepted as in Linsley (1962: 24, as Ergatini).

Tribe Eurypodini Gahan, 1906 (1868)

Zaracides Lacordaire, 1868: 131 [stem: *Zarac-*]. Type genus: *Zarax* Pascoe, 1867 [syn. of *Eurypoda* W. Saunders, 1853]. Comment: original vernacular name available (Art. 11.7.2): first used in latinized form and generally accepted as in

Pascoe (1869b: 672, as ZARACINAE); usage of the younger name EURYPODINI Gahan, 1906 conserved over this name (Art. 40.2).

EURYPODINI Gahan, 1906: 27 [stem: *Eurypod-*]. Type genus: *Eurypoda* W. Saunders, 1853. Comment: name proposed to replace ZARACINI Lacordaire, 1868 because of the synonymy of the type genus; use of family-group name conserved over ZARACINI Lacordaire, 1868 (Art. 40.2); also see comments under EURYPINAE Thomson, 1860 in MYCTERIDAE.

Tribe HOPLIDERINI Thomson, 1864

HOPLIDERITAE J. Thomson, 1864: 290 [stem: *Hoplider-*]. Type genus: *Hoplideres* Audinet-Serville, 1832.

Tribe MACRODONTIINI Thomson, 1861

MACRODONTITAE J. Thomson, 1861: 324 [stem: *Macrodonti-*]. Type genus: *Macrodontia* Lacordaire, 1830. Comment: incorrect original stem formation, not in prevailing usage.

ACANTHINODERITAE J. Thomson, 1864: 294 [stem: *Acanthinoder-*]. Type genus: *Acanthinodera* Hope, 1833.

*ANCISTROTIDES Lacordaire, 1868: 81 [stem: *Ancistrot-*]. Type genus: *Ancistrotus* Audinet-Serville, 1832. Comment: original vernacular name unavailable (Art. 11.7.2): subsequently used in latinized form but not generally attributed to Lacordaire (1868).

*ACALODEGMITES J. Thomson, 1877: 261 [stem: *Acalodegm-*]. Type genus: *Acalodegma* Thomson, 1877. Comment: original vernacular name unavailable (Art. 11.7.2): not subsequently latinized.

ANCISTROTINI Lameere, 1919: 90 [stem: *Ancistrot-*]. Type genus: *Ancistrotus* Audinet-Serville, 1832.

Tribe MACROTOMINI Thomson, 1861

MACROTOMITAE J. Thomson, 1861: 312 [stem: *Macrotom-*]. Type genus: *Macrotoma* Audinet-Serville, 1832.

Subtribe ARCHETYPINA Lameere, 1912

ARCHETYPI Lameere, 1912: 180 [stem: *Archetyp-*]. Type genus: *Archetypus* J. Thomson, 1861.

Subtribe BASITOXINA Lameere, 1912

*MÉCOSARTHRINES Lameere, 1903b: 307 [stem: *Mecosarthr-*]. Type genus: *Mecosarthron* Buquet, 1840. Comment: original vernacular name unavailable (Art. 11.7.2): proposed after 1899.

BASITOXI Lameere, 1912: 180 [stem: *Basitox-*]. Type genus: *Basitoxus* Audinet-Serville, 1832.

MECOSARTHRINI Melzer, 1919: 35 [stem: *Mecosarthr-*]. Type genus: *Mecosarthron* Buquet, 1840.

Subtribe MACROTOMINA Thomson, 1861

MACROTOMITAE J. Thomson, 1861: 312 [stem: *Macrotom-*]. Type genus: *Macrotoma* Audinet-Serville, 1832 [*Macrotoma* Audinet-Serville, 1832 [July] is a junior homonym of *Macrotoma* Laporte, 1832 [April], which is a synonym of *Longina* Wiedemann, 1830 in Diptera; Heffern et al. (2006) applied the Reversal of Precedence (Art. 23.9) to qualify *Macrotoma* Audinet-Serville as *nomen protectum*].

*AULACOPIDES Lacordaire, 1868: 101 [stem: *Aulacopod-*]. Type genus: *Aulacopus* Audinet-Serville, 1832. Comment: original vernacular name unavailable (Art. 11.7.2): subsequently used in latinized form but not generally attributed to Lacordaire (1868); incorrect original stem formation, not in prevailing usage.

AULACOPINAE Kolbe, 1897: 295 [stem: *Aulacopod-*]. Type genus: *Aulacopus* Audinet-Serville, 1832. Comment: incorrect original stem formation, not in prevailing usage.

*CNÉMOPLITIENS Lameere, 1904: 1 [stem: *Cnemoplit-*]. Type genus: *Cnemoplites* Newman, 1842. Comment: original vernacular name unavailable (Art. 11.7.2): proposed after 1899.

CNEMOPLITINAE Schröder, 1905: vii [stem: *Cnemoplit-*]. Type genus: *Cnemoplites* Newman, 1842. Comment: see Bousquet et al. (2009: 17) for comments about the authorship of this name.

PRINOBIINI Vives, 2000: 84 [stem: *Prinobi-*]. Type genus: *Prinobius* Mulsant, 1842. Comment: replacement name for MACROTOMINI Thomson, 1861 because of the homonymy of the type genus.

Subtribe PLATYGNATHINA Gilmour, 1954

PLATYGNATHINA Gilmour, 1954: 33 [stem: *Platygnath-*]. Type genus: *Platygnathus* Audinet-Serville, 1832. Comment: as pointed out by Bousquet et al. (2009: 18) Gilmour (1954: 33) did not provide a description of his new taxon but his new name is available because it was used as valid before 2000, e.g., Ferreira and Veiga Ferreira (1959a: 34, as "PLATYGNATHINA Gilmour, 1954") and was not rejected by an author who, after 1960 and before 2000, expressly applied Article 13 of the then current editions of the Code (Article 13.2.1).

PLATYGNATHINI Quentin and Villiers, 1975: 25 [stem: *Platygnath-*]. Type genus: *Platygnathus* Audinet-Serville, 1832. Comment: family-group name proposed as new without reference to PLATYGNATHINA Gilmour, 1954.

Subtribe Xixuthrina Lameere, 1912

Xixuthri Lameere, 1912: 181 [stem: *Xixuthr-*]. Type genus: *Xixuthrus* J. Thomson, 1864.

Tribe Mallaspini Thomson, 1861

Mallaspitae J. Thomson, 1861: 302 [stem: *Mallasp-*]. Type genus: *Mallaspis* Audinet-Serville, 1832. Comment: current spelling maintained (Art. 29.3.1.1): incorrect original stem formation in prevailing usage (should be *Mallaspid-*).

*Pyrodides Lacordaire, 1868: 174 [stem: *Pyrod-*]. Type genus: *Pyrodes* Audinet-Serville, 1832. Comment: original vernacular name unavailable (Art. 11.7.2): subsequently used in latinized form but not generally attributed to Lacordaire (1868).

Pyrodini Harold, 1879: 165 [stem: *Pyrod-*]. Type genus: *Pyrodes* Audinet-Serville, 1832.

Tribe Mallodonini Thomson, 1861

Mallodonitae J. Thomson, 1861: 318 [stem: *Mallodon-*]. Type genus: *Mallodon* Lacordaire, 1830. Comment: current spelling maintained (Art. 29.5): incorrect original stem formation in prevailing usage (should be *Mallodont-*); we treat this group as a valid tribe as in Drumont and Komiya (2010: 91).

*Sténodontines Lameere, 1902: 66 [stem: *Stenodont-*]. Type genus: *Stenodontes* Audinet-Serville, 1832. Comment: original vernacular name unavailable (Art. 11.7.2): proposed after 1899.

Stenodontini Lameere, 1903a: 54 [stem: *Stenodont-*]. Type genus: *Stenodontes* Audinet-Serville, 1832. Comment: homonym of Stenodontina Schmiedeknecht, 1903 (type genus *Stenodontus* Berthoumieu, 1897) in Hymenoptera; this case is to be referred to the Commission to remove the homonymy (Art. 55.3.1).

Tribe Meroscelisini Thomson, 1861

Meroscelisitae J. Thomson, 1861: 299 [stem: *Meroscelis-*]. Type genus: *Meroscelisus* Audinet-Serville, 1832.

*Tragosomites Fairmaire, 1864: 119 [stem: *Tragosomat-*]. Type genus: *Tragosoma* Audinet-Serville, 1832. Comment: original vernacular name unavailable (Art. 11.7.2): subsequently used in latinized form but not generally attributed to Fairmaire (1864); incorrect original stem formation, not in prevailing usage.

Tragosomitae J. Thomson, 1864: 286 [stem: *Tragosomat-*]. Type genus: *Tragosoma* Audinet-Serville, 1832. Comment: incorrect original stem formation, not in prevailing usage.

Clostérides Lacordaire, 1868: 149 [stem: *Closter-*]. Type genus: *Closterus* Audinet-Serville, 1832. Comment: original vernacular name available (Art. 11.7.2): first used in latinized form by Pascoe (1869b: 676, as Closterinae), generally accepted as in Lameere (1913: 81, as Closteri).

*Monodesmides Lacordaire, 1868: 157 [stem: *Monodesm-*]. Type genus: *Monodesmus* Audinet-Serville, 1832. Comment: original vernacular name unavailable (Art. 11.7.2): subsequently used in latinized form but not generally attributed to Lacordaire (1868).

Monodesminae Gahan, 1890: 299 [stem: *Monodesm-*]. Type genus: *Monodesmus* Audinet-Serville, 1832.

Luluina Gilmour, 1956: 222 [stem: *Lulu-*]. Type genus: *Lulua* Burgeon, 1931.

Tribe Prionini Latreille, 1802

Prionii Latreille, 1802: 212 [stem: *Prion-*]. Type genus: *Prionus* Geoffroy, 1762 [placed on the Official List of Generic Names in Zoology (ICZN 1994a)].

*Prioceria Rafinesque, 1815: 116 [stem: *Priocerat-*]. Type genus: *Prioceras* Rafinesque, 1815 [as pointed out by Bousquet et al. (2009: 19) Rafinesque (1815: 116) listed the genus "*Prioceras* R. sp. do." following the genus "*Prionus* Fabr." in his subfamily Prioceria, the abbreviations after the genus name *Prioceras* mean that this is a new genus by the author "R[afinesque]." and that the new genus includes some species that were included in *Prionus* previously "sp. do."; Rafinesque did not list which species he included in his new genus *Prioceras* and we are not aware of any subsequent validation of this name; *Prioceras* Rafinesque is also listed as a *nomen nudum* in Neave (1940: 889) and Sherborn (1929: 5148)]. Comment: family-group name unavailable (Art. 11.7.1.1): not based on an available genus name at the time; incorrect original stem formation, not in prevailing usage.

*Cyrtognathites Blanchard, 1845b: 138 [stem: *Cyrtognath-*]. Type genus: *Cyrtognathus* Faldermann, 1835 [subgenus of *Dorysthenes* Vigors, 1826]. Comment: original vernacular name unavailable (Art. 11.7.2): subsequently used in latinized form but not generally attributed to Blanchard (1845b).

*Psalidognathites Blanchard, 1845b: 138 [stem: *Psalidognath-*]. Type genus: *Psalidognathus* Gray, 1832. Comment: original vernacular name unavailable (Art. 11.7.2): subsequently used in latinized form but not generally attributed to Blanchard (1845b).

Cyrthognathitae J. Thomson, 1861: 328 [stem: *Cyrtognath-*]. Type genus: *Cyrtognathus* Faldermann, 1835 [as *Cyrthognathus*, incorrect subsequent spelling of type genus name, not in prevailing usage; subgenus of *Dorysthenes* Vigors, 1826]. Comment: incorrect original stem formation, not in prevailing usage.

Psalidognathitae J. Thomson, 1861: 331 [stem: *Psalidognath-*]. Type genus: *Psalidognathus* Gray, 1832.

Prionommitae J. Thomson, 1861: 327 [stem: *Prionommat-*]. Type genus: *Prionomma* A. White, 1853. Comment: incorrect original stem formation, not in prevailing usage.

Orthosomitae J. Thomson, 1864: 284 [stem: *Orthosomat-*]. Type genus: *Orthosoma* Audinet-Serville, 1832. Comment: incorrect original stem formation, not in prevailing usage.

Pithoclitae J. Thomson, 1864: 291 [stem: *Pithocl-*]. Type genus: *Pithocles* J. Thomson, 1864 [syn. of *Derobrachus* Audinet-Serville, 1832].

DEROBRACHITAE J. Thomson, 1864: 291 [stem: *Derobrach-*]. Type genus: *Derobrachus* Audinet-Serville, 1832.

TITANITAE J. Thomson, 1864: 292 [stem: *Titan-*]. Type genus: *Titanus* Audinet-Serville, 1832.

AULACOCERITAE J. Thomson, 1864: 292 [stem: *Aulacocer-*]. Type genus: *Aulacocerus* A. White, 1853.

*PSALIDOCOPTIDES Lacordaire, 1868: 38 [stem: *Psalidocopt-*]. Type genus: *Psalidocoptus* A. White, 1856. Comment: original vernacular name unavailable (Art. 11.7.2): not subsequently latinized.

*POLYARTHRIDES Lacordaire, 1868: 44 [stem: *Polyarthr-*]. Type genus: *Polyarthron* Audinet-Serville, 1832. Comment: original vernacular name unavailable (Art. 11.7.2): subsequently used in latinized form but not generally attributed to Lacordaire (1868).

*MICROPSALIDES Lacordaire, 1868: 42 [stem: *Micropsalid-*]. Type genus: *Micropsalis* Burmeister, 1865 [preoccupied genus name, not *Micropsalis* Meyer, 1859 [Crustacea]; syn. of *Apterocaulus* Fairmaire, 1864]. Comment: original vernacular name unavailable (Art. 11.7.2): subsequently used in latinized form but not generally attributed to Lacordaire (1868); incorrect original stem formation, not in prevailing usage.

POLYARTHRINI Gounelle, 1911: 326 [stem: *Polyarthr-*]. Type genus: *Polyarthron* Audinet-Serville, 1832. Comment: POLYARTHRIDAE Daday, 1893 (type genus *Polyarthra* Ehrenberg, 1834) is available in Rotifera; this case is to be referred to the Commission to remove the homonymy (Art. 55.3.1).

MICROPSALINI Gounelle, 1911: 326 [stem: *Micropsalid-*]. Type genus: *Micropsalis* Burmeister, 1865 [preoccupied genus name, not *Micropsalis* Meyer, 1859 [Crustacea]; syn. of *Apterocaulus* Fairmaire, 1864]. Comment: permanently invalid (Art. 39): based on preoccupied type genus; incorrect original stem formation, not in prevailing usage.

Tribe REMPHANINI Lacordaire, 1868

REMPHANIDES Lacordaire, 1868: 103 [stem: *Remphan-*]. Type genus: *Remphan* G. R. Waterhouse, 1835. Comment: original vernacular name available (Art. 11.7.2): first used in latinized form by Pascoe (1869b: 667, as REMPHANINAE), generally accepted as in Drumont and Komiya (2010: 95, as REMPHANINI); this name was incorrectly credited to Pascoe, 1869 by Bousquet et al. (2009: 18); we treat this group as a valid tribe as in Drumont and Komiya (2010: 95).

RHAPHIPODI Lameere, 1912: 181 [stem: *Rhaphipod-*]. Type genus: *Rhaphipodus* Audinet-Serville, 1832.

Tribe SOLENOPTERINI Lacordaire, 1868

SOLÉNOPTÉRIDES Lacordaire, 1868: 180 [stem: *Solenopter-*]. Type genus: *Solenoptera* Audinet-Serville, 1832. Comment: original vernacular name available

(Art. 11.7.2): first used in latinized form by J. L. LeConte (1873: 286, as SOLENOPTERINI), generally accepted as in Monné (1995c: 37, as SOLENOPTERINI).

*DÉRANCISTRINES Lameere, 1909: 1 [stem: *Derancistr-*]. Type genus: *Derancistrus* Audinet-Serville, 1832. Comment: original vernacular name unavailable (Art. 11.7.2): proposed after 1899.

DERANCISTRINI Lameere, 1912: 181 [stem: *Derancistr-*]. Type genus: *Derancistrus* Audinet-Serville, 1832.

Tribe TERETICINI Lameere, 1913

*TÉRÉTICIENS Lameere, 1912: 72 [stem: *Teretic-*]. Type genus: *Tereticus* C. O. Waterhouse, 1879. Comment: original vernacular name unavailable (Art. 11.7.2): proposed after 1899.

TERETICI Lameere, 1913: 87 [stem: *Teretic-*]. Type genus: *Tereticus* C. O. Waterhouse, 1879.

Tribe VESPEROCTENINI Vives, 2005

VESPEROCTENINI Vives, 2005: 438 [stem: *Vesperocten-*]. Type genus: *Vesperoctenus* H. W. Bates, 1891.

Subfamily LEPTURINAE Latreille, 1802

LEPTURETAE Latreille, 1802: 218 [stem: *Leptur-*]. Type genus: *Leptura* Linnaeus, 1758.

Tribe DESMOCERINI Blanchard, 1845

DESMOCÉRITES Blanchard, 1845b: 163 [stem: *Desmocer-*]. Type genus: *Desmocerus* Dejean, 1821. Comment: original vernacular name available (Art. 11.7.2): first used in latinized form by J. Thomson (1861: 159, as DESMOCERITAE), generally accepted as in Monné (1995b: 1, as DESMOCERINI).

Tribe ENCYCLOPINI LeConte, 1873

ENCYCLOPINI J. L. LeConte, 1873: 326 [stem: *Enclyclop-*]. Type genus: *Encyclops* Newman, 1838.

Tribe LEPTURINI Latreille, 1802

LEPTURETAE Latreille, 1802: 218 [stem: *Leptur-*]. Type genus: *Leptura* Linnaeus, 1758.

*GRAMMOPTÉRATES Mulsant, 1863b: 569 [stem: *Grammopter-*]. Type genus: *Grammoptera* Audinet-Serville, 1835. Comment: original vernacular name unavailable (Art. 11.7.2): not subsequently latinized.

*STRANGALINI Zagajkevich, 1991: 96 [stem: *Strangali-*]. Type genus: *Strangalia* Audinet-Serville, 1835. Comment: family-group name unavailable (Art. 11.6): originally published as synonym and not made available subsequently; incorrect original stem formation, not in prevailing usage.

Tribe Oxymirini Danilevsky, 1997

Oxymirini Danilevsky, 1997: 8 [stem: *Oxymir-*]. Type genus: *Oxymirus* Mulsant, 1862.

Tribe Rhagiini Kirby, 1837

Rhagiadae Kirby, 1837: 178 [stem: *Rhagi-*]. Type genus: *Rhagium* Fabricius, 1775.

*Toxotaires Mulsant, 1839: 230 [stem: *Toxot-*]. Type genus: *Toxotus* Dejean, 1821 [syn. of *Stenocorus* Geoffroy, 1762]. Comment: original vernacular name unavailable (Art. 11.7.2): subsequently used in latinized form but not generally attributed to Mulsant (1839).

*Pachytes Motschulsky, 1849: 60 [stem: *Pachyt-*]. Type genus: *Pachyta* Dejean, 1821. Comment: original vernacular name unavailable (Art. 11.7.2): subsequently used in latinized form but not generally attributed to Motschulsky (1849).

Stenocoritae J. Thomson, 1861: 156 [stem: *Stenocor-*]. Type genus: *Stenocorus* Geoffroy, 1762 [placed on the Official List of Generic Names in Zoology (ICZN 1994a)]. Comment: an application will need to be submitted to the Commission to suppress Stenocoridae Hope, 1834 (based on the misidentified type genus *Stenocorus* sensu Hope, 1834) for the Principles of Priority and Homonymy (Art. 65.2.1) if this name is to be used as valid.

Toxoti J. L. LeConte and G. H. Horn, 1883: 313 [stem: *Toxot-*]. Type genus: *Toxotus* Dejean, 1821 [syn. of *Stenocorus* Geoffroy, 1762]. Comment: junior homonym of Toxotidae Günther, 1860 (type genus *Toxotes* Cuvier and Cloquet, 1816) currently used as valid in Pisces; this case is to be referred to the Commission to remove the homonymy (Art. 55.3.1).

Pachytini Portevin, 1934: 119, in key [stem: *Pachyt-*]. Type genus: *Pachyta* Dejean, 1821.

*Enoploderini Danilevsky, 1997: 9 [stem: *Enoploder-*]. Type genus: *Enoploderes* Faldermann, 1837. Comment: unavailable family-group name, proposed after 1930 without description or bibliographic reference to such a description (Art. 13.1).

Tribe Rhamnusiini Sama, 2009

*Rhamnusiini Danilevsky, 1997: 9 [stem: *Rhamnusi-*]. Type genus: *Rhamnusium* Latreille, 1829. Comment: unavailable family-group name, proposed after 1930 without description or bibliographic reference to such a description (Art. 13.1).

Rhamnusiini Sama, 2009b: 383 [stem: *Rhamnusi-*]. Type genus: *Rhamnusium* Latreille, 1829.

Tribe Teledapini Pascoe, 1871

Teledapinae Pascoe, 1871: 268 [stem: *Teledap-*]. Type genus: *Teledapus* Pascoe, 1871.

Tribe Sachalinobiini Danilevsky, 2010

Sachalinobiini Danilevsky, 2010: 43 [stem: *Sachalinobi-*]. Type genus: *Sachalinobia* Jakobson, 1899.

Tribe Xylosteini Reitter, 1913

Xylosteina Reitter, 1913a: 5 [stem: *Xyloste-*]. Type genus: *Xylosteus* Frivaldszky, 1838.

Subfamily Spondylidinae Audinet-Serville, 1832

Spondylii Audinet-Serville, 1832: 123 [stem: *Spondylid-*]. Type genus: *Spondylis* Fabricius, 1775.

Tribe Anisarthrini Mamaev and Danilevsky, 1973

*Anisarthrites Fairmaire, 1864: 124 [stem: *Anisarthr-*]. Type genus: *Anisarthron* Dejean, 1835. Comment: original vernacular name unavailable (Art. 11.7.2): subsequently used in latinized form but not generally attributed to Fairmaire (1864).

Anisarthronini Mamaev and Danilevsky, 1973: 1260, in key [stem: *Anisarthr-*]. Type genus: *Anisarthron* Dejean, 1835. Comment: family-group name proposed as new without reference to Anisarthrites Fairmaire, 1864; incorrect original stem formation, not in prevailing usage.

Tribe Asemini Thomson, 1861

Asemitae J. Thomson, 1861: 259 [stem: *Asem-*]. Type genus: *Asemum* Eschscholtz, 1830.

*Criomorphates Mulsant, 1863a: 421 [stem: *Criomorph-*]. Type genus: *Criomorphus* Mulsant, 1839 [preoccupied genus name, not *Criomorphus* Curtis, 1831 [Hemiptera]; placed on the Official Index of Rejected and Invalid Generic Names in Zoology (ICZN 1988b); syn. of *Tetropium* Kirby, 1837]. Comment: original vernacular name unavailable (Art. 11.7.2): subsequently used in latinized form but not generally attributed to Mulsant (1863).

*Criocéphalites Fairmaire, 1864: 125 [stem: *Criocephal-*]. Type genus: *Criocephalus* Mulsant, 1839 [syn. of *Arhopalus* Audinet-Serville, 1834]. Comment: original vernacular name unavailable (Art. 11.7.2): subsequently used in latinized form but not generally attributed to Fairmaire (1864).

Tetropiina Seidlitz, 1891 [Gatt.]: 179 [stem: *Tetropi-*]. Type genus: *Tetropium* Kirby, 1837 [placed on the Official List of Generic Names in Zoology (ICZN 1988b)].

Criocephalinae Sharp, 1905: 147 [stem: *Criocephal-*]. Type genus: *Criocephalus* Mulsant, 1839 [syn. of *Arhopalus* Audinet-Serville, 1834].

Criomorphini Portevin, 1927: 20, in key [stem: *Criomorph-*]. Type genus: *Criomorphus* Mulsant, 1839 [preoccupied genus name, not *Criomorphus* Curtis, 1831 [Hemiptera]; placed on the Official Index of Rejected and Invalid

Generic Names in Zoology (ICZN 1988b); syn. of *Tetropium* Kirby, 1837]. Comment: permanently invalid (Art. 39): based on preoccupied type genus.

NOTHORHININI Zagajkevich, 1991: 110 [stem: *Nothorhin-*]. Type genus: *Nothorhina* Redtenbacher, 1845. Comment: placed in synonymy with ASEMINI by Vives and Alonso-Zarazaga (2000: 569).

Tribe ATIMIINI LeConte, 1873

ATIMIINI J. L. LeConte, 1873: 322 [stem: *Atimi-*]. Type genus: *Atimia* Haldeman, 1847.

Tribe SAPHANINI Gistel, 1848

SAPHANIDAE Gistel, 1848: [8] [stem: *Saphan-*]. Type genus: *Saphanus* Audinet-Serville, 1834.

MICHTHYSOMINI J. L. LeConte, 1873: 332 [stem: *Michthisomat-*]. Type genus: *Michthisoma* J. L. LeConte, 1850 [as *Michthysoma*, incorrect subsequent spelling of type genus name, not in prevailing usage]. Comment: incorrect original stem formation, not in prevailing usage.

Tribe SPONDYLIDINI Audinet-Serville, 1832

SPONDYLII Audinet-Serville, 1832: 123 [stem: *Spondylid-*]. Type genus: *Spondylis* Fabricius, 1775. Comment: incorrect original stem formation, not in prevailing usage.

Subfamily NECYDALINAE Latreille, 1825

NECYDALIDES Latreille, 1825: 401 [stem: *Necydal-*]. Type genus: *Necydalis* Linnaeus, 1758.

Subfamily DORCASOMINAE Lacordaire, 1868

DORCASOMIDES Lacordaire, 1868: 456 [stem: *Dorcasom-*]. Type genus: *Dorcasomus* Audinet-Serville, 1834. Comment: original vernacular name available (Art. 11.7.2): first used in latinized form by Kolbe (1897: 299, as DORCASOMINAE), generally accepted as in Aurivillius (1912: 251, as DORCASOMINI).

Subfamily APATOPHYSEINAE Lacordaire, 1869

APATOPHYSIDES Lacordaire, 1869: 234 [stem: *Apatophyse-*]. Type genus: *Apatophysis* Chevrolat, 1860. Comment: original vernacular name available (Art. 11.7.2): first used in latinized form by Heyne and Taschenberg (1907: 240, as APATOPHYSINI), generally accepted as in Švácha and Danilevsky (1988: 125, as APATOPHYSEINAE); incorrect original stem formation, not in prevailing usage; valid status given here as in Danilevsky (2010: 48).

Subfamily CERAMBYCINAE Latreille, 1802

CERAMBICINI Latreille, 1802: 211 [stem: *Cerambyc-*]. Type genus: *Cerambyx* Linnaeus, 1758.

Tribe Acangassuini Galileo and Martins, 2001

Acangassuini Galileo and Martins, 2001: 95 [stem: *Acangassu-*]. Type genus: *Acangassu* Galileo and Martins, 2001. Comment: this name was incorrectly treated as unavailable by Bousquet et al. (2009).

Tribe Achrysonini Lacordaire, 1868

Achrysonides Lacordaire, 1868: 231 [stem: *Achryson-*]. Type genus: *Achryson* Audinet-Serville, 1833. Comment: original vernacular name available (Art. 11.7.2): first used in latinized form by Murray (1870: 169, as Achrysonidae), generally accepted as in Aurivillius (1912: 39, as Achrysonini); current spelling maintained (Art. 29.5): incorrect original stem formation in prevailing usage (should be *Achrys-*).

Tribe Agallissini LeConte, 1873

Agallissini J. L. LeConte, 1873: 321 [stem: *Agalliss-*]. Type genus: *Agallissus* Dalman, 1823.

Tribe Alanizini Di Iorio, 2003

Alanizini Di Iorio, 2003: 1 [stem: *Alaniz-*]. Type genus: *Alanizus* Di Iorio, 2003.

Tribe Anaglyptini Lacordaire, 1868 *nomen protectum*

Anaglyptides Lacordaire, 1868: 404 [stem: *Anaglypt-*]. Type genus: *Anaglyptus* Mulsant, 1839. Comment: original vernacular name available (Art. 11.7.2): first used in latinized form by J. L. LeConte (1873: 319, as Anaglypti), generally accepted as in Linsley (1964: 173, as Anaglyptini); junior homonym of Anaglyptidae Gistel, 1848 (type genus *Anaglyptes* Gistel, 1848; syn. of *Chalcophora* Dejean, 1833) in Coleoptera: Buprestidae; this name was treated as a *nomen protectum* by Bousquet et al. (2009: 41).

Tribe Aphanasiini Lacordaire, 1868

Aphanasiides Lacordaire, 1868: 367 [stem: *Aphanasi-*]. Type genus: *Aphanasium* Dejean, 1835. Comment: original vernacular name available (Art. 11.7.2): first used in latinized form and generally accepted as in Aurivillius (1912: 139, as Aphanasiini).

Tribe Aphneopini Lacordaire, 1868

Aphnéopides Lacordaire, 1868: 421 [stem: *Aphneop-*]. Type genus: *Aphneope* Pascoe, 1863. Comment: original vernacular name available (Art. 11.7.2): first used in latinized form and generally accepted as in Aurivillius (1912: 155, as Aphneopini).

Tribe Auxesini Lepesme and Breuning, 1952

*Auxésides Lacordaire, 1872: 463 [stem: *Auxes-*]. Type genus: *Auxesis* J. Thomson, 1858. Comment: original vernacular name unavailable (Art. 11.7.2): subsequently used in latinized form but not generally attributed to Lacordaire (1872).

Auxesina Lepesme and Breuning, 1952: 140 [stem: *Auxes-*]. Type genus: *Auxesis* J. Thomson, 1858. Comment: current spelling maintained (Art. 29.5): incorrect original stem formation in prevailing usage (should be *Auxese-*).

Psathyrini Quentin, 1954: 103, in key [stem: *Psathyr-*]. Type genus: *Psathyrus* J. Thomson, 1857.

Tribe Basipterini Fragoso, Monné and Campos Seabra, 1987

Basipterini Fragoso et al., 1987: 201 [stem: *Basipter-*]. Type genus: *Basiptera* J. Thomson, 1864.

Tribe Bimiini Lacordaire, 1868

Bimiides Lacordaire, 1868: 464 [stem: *Bimi-*]. Type genus: *Bimia* A. White, 1850. Comment: original vernacular name available (Art. 11.7.2): first used in latinized form and generally accepted as in Aurivillius (1912: 254, as Bimiini).

Sibyllini Cerda, 1973: 115 [stem: *Sibyll-*]. Type genus: *Sibylla* J. Thomson, 1858 [preoccupied genus name, not *Sibylla* Stål, 1856 [Orthoptera]; the valid name for this genus has recently been given as *Sybilla* J. Thomson, 1864 by Monné and Bezark (2009: 30) which is an incorrect subsequent spelling of *Sibylla* J. Thomson, 1858; syn. of *Zehra* Özdikmen, 2008]. Comment: permanently invalid (Art. 39): based on preoccupied type genus.

Tribe Bothriospilini Lane, 1950

Bothriospilinae Lane, 1950: 370 [stem: *Bothriospil-*]. Type genus: *Bothriospila* Aurivillius, 1923.

Tribe Brachypteromatini Sama, 2008

Brachypteromini Sama, 2008: 229 [stem: *Brachypteromat-*]. Type genus: *Brachypteroma* Heyden, 1863. Comment: incorrect original stem formation, not in prevailing usage.

Tribe Callichromatini Swainson, 1840

Callichrominae Swainson, 1840: 293 [stem: *Callichromat-*]. Type genus: *Callichroma* Latreille, 1816. Comment: incorrect original stem formation, not in prevailing usage; this family-group name was incorrectly credited to Swainson and Shuckard (1840) by Bousquet et al. (2009: 42).

Terambidae Gistel, 1848: [8] [stem: *Teramb-*]. Type genus: *Terambus* Gistel, 1848 [syn. of *Aromia* Audinet-Serville, 1834].

Tribe CALLIDIINI Kirby, 1837

CALLIDIADAE Kirby, 1837: 170 [stem: *Callidi-*]. Type genus: *Callidium* Fabricius, 1775 [for comments on problems with the authorship and type species of this type genus see Bousquet et al. (2009: 42)].

*PHYMATODATES Mulsant, 1863a: 397 [stem: *Phymatod-*]. Type genus: *Phymatodes* Mulsant, 1839 [placed on the Official List of Generic Names in Zoology (ICZN 1989a)]. Comment: original vernacular name unavailable (Art. 11.7.2): not subsequently latinized.

Tribe CALLIDIOPINI Lacordaire, 1868

CALLIDIOPSIDES Lacordaire, 1868: 340 [stem: *Callidiop-*]. Type genus: *Callidiopis* A. White, 1855 [as *Callidiopsis*, unjustified emendation of type-genus by Lacordaire (1868: 356), not in prevailing usage]. Comment: original vernacular name available (Art. 11.7.2): first used in latinized form by Pascoe (1869b: 535, as CALLIDIOPSINAE [incorrect stem formation]), generally accepted as in Aurivillius (1912: 115, as CALLIDIOPINI); current spelling maintained (Art. 29.3.1.1): incorrect stem formation in prevailing usage (should be *Callidiopid-*).

Tribe CERAMBYCINI Latreille, 1802

CERAMBICINI Latreille, 1802: 211 [stem: *Cerambyc-*]. Type genus: *Cerambyx* Linnaeus, 1758.

Subtribe CERAMBYCINA Latreille, 1802

CERAMBICINI Latreille, 1802: 211 [stem: *Cerambyc-*]. Type genus: *Cerambyx* Linnaeus, 1758. Comment: Gistel (1856a: 375) used the name CERATAMBYCIDAE but this was based on the incorrect subsequent spelling of the type genus name (incorrectly given as *Caratambyx* in the same work on page 375 but later corrected to *Ceratambyx* by Gistel (1856b: 13)) and was therefore not proposed as a new family-group name.

Subtribe SPHALLOTRICHINA Martins and Monné, 2005

SPHALLOTRICHINA Martins and Monné, 2005: 2 [stem: *Sphallotrich-*]. Type genus: *Sphallotrichus* Fragoso, 1982. Comment: this name was incorrectly treated as unavailable by Bousquet et al. (2009).

Tribe CERTALLINI Fairmaire, 1864

CARTALLITES Fairmaire, 1864: 149 [stem: *Certall-*]. Type genus: *Certallum* Dejean, 1821 [as *Cartallum*, unjustified emendation of type genus name by Audinet-Serville (1834: 94), not in prevailing usage]. Comment: published early September 1864; original vernacular name available (Art. 11.7.2): first used in a latinized form by Sama (1990: 287, as CERTALLINI), generally accepted as in

Vives (2000: 155, as CERTALLINI); incorrect original stem formation, not in prevailing usage.

PYTHEITAE J. Thomson, 1864: 153 [stem: *Pythe-*]. Type genus: *Pytheus* Newman, 1840. Comment: published before October 1864.

*ERIONISPITES Chapuis, 1875: 301 [stem: *Erionisp-*]. Type genus: *Erionispa* Chapuis, 1875 [syn. of *Pytheus* Newman, 1840]. Comment: original vernacular name unavailable (Art. 11.7.2): subsequently used in latinized form but not generally attributed to Chapuis (1875); ERIONISPIDAE was used as valid by Ienistea (1986: 31) but it was not attributed to Chapuis (1875); Ienistea's name is also unavailable, it was proposed after 1930 without description or bibliographic reference to such a description (Art. 13.1); name written with an accent as ÉRIONISPITES on page 301 of the original publication, but without accent in the key and for the description; type genus transferred from CHRYSOMELIDAE by Lameere (1885) and treated as a synonym of *Pytheus* Newman, 1840.

Tribe CHLIDONINI Waterhouse, 1879

CHLIDONINAE C. O. Waterhouse, 1879b: 320 [stem: *Chlidon-*]. Type genus: *Chlidones* C. O. Waterhouse, 1879.

Tribe CLEOMENINI Lacordaire, 1868

CLÉOMÉNIDES Lacordaire, 1868: 405 [stem: *Cleomen-*]. Type genus: *Cleomenes* J. Thomson, 1864. Comment: original vernacular name available (Art. 11.7.2): first used in latinized form by Pascoe (1869b: 554, as CLEOMENINAE), generally accepted as in Adlbauer et al. (2010: 163, as CLEOMENINI); this name was incorrectly credited to Pascoe (1869) by Bousquet et al. (2009: 55); we treat this group as a valid tribe as in Adlbauer et al. (2010: 163).

Tribe CLYTINI Mulsant, 1839

CLYTAIRES Mulsant, 1839: 70 [stem: *Clyt-*]. Type genus: *Clytus* Laicharting, 1784. Comment: original vernacular name available (Art. 11.7.2): first used in latinized form by Gistel (1848: [8], as CLYTIIDAE [incorrect stem formation]), generally accepted as in Aurivillius (1912: 358, as CLYTINI).

NEOCLYTITAE J. Thomson, 1861: 219 [stem: *Neoclyt-*]. Type genus: *Neoclytus* J. Thomson, 1861.

CYLLENITAE J. Thomson, 1864: 184 [stem: *Cyllen-*]. Type genus: *Cyllene* Newman, 1840 [preoccupied genus name, not *Cyllene* Gray, 1834 [Mollusca]; syn. of *Megacyllene* Casey, 1912]. Comment: permanently invalid (Art. 39): based on preoccupied type genus.

Tribe COMPSOCERINI Thomson, 1864

COMPSOCERITAE J. Thomson, 1864: 260 [stem: *Compsocer-*]. Type genus: *Compsocerus* Audinet-Serville, 1834 [for comments on problems with the authorship and type species of this type genus see Bousquet et al. (2009: 44)].

Tribe Coptommatini Lacordaire, 1869

Coptommides Lacordaire, 1869: 221 [stem: *Coptommat-*]. Type genus: *Coptomma* Newman, 1840. Comment: original vernacular name available (Art. 11.7.2): first used in latinized form and generally accepted as in Aurivillius (1912: 488, as Coptommatini); First Reviser (Coptommatini Lacordaire, 1869 vs Navomorphini Lacordaire, 1869) not determined, current usage maintained.

Navomorphides Lacordaire, 1869: 223 [stem: *Navomorph-*]. Type genus: *Navomorpha* A. White, 1855 [syn. of *Coptomma* Newman, 1840]. Comment: original vernacular name available (Art. 11.7.2): first used in latinized form and generally accepted as in Aurivillius (1912: 488, as Navomorphini).

Tribe Curiini LeConte, 1873

Curii J. L. LeConte, 1873: 304 [stem: *Curi-*]. Type genus: *Curius* Newman, 1840.

Tribe Deilini Fairmaire, 1864

*Déilates Mulsant, 1863b: 190 [stem: *Deil-*]. Type genus: *Deilus* Audinet-Serville, 1834. Comment: original vernacular name unavailable (Art. 11.7.2): subsequently used in latinized form but not generally attributed to Mulsant (1863).

Déilates Fairmaire, 1864: 154 [stem: *Deil-*]. Type genus: *Deilus* Audinet-Serville, 1834. Comment: original vernacular name available (Art. 11.7.2): first used in latinized form and generally accepted as in Aurivillius (1912: 294, as Deilini).

Tribe Dejanirini Lacordaire, 1868

Déjanirides Lacordaire, 1868: 460 [stem: *Dejanir-*]. Type genus: *Dejanira* J. Thomson, 1864. Comment: original vernacular name available (Art. 11.7.2): first used in latinized form by Pascoe (1869b: 561, as Dejanirinae), generally accepted as in Aurivillius (1912: 253, as Dejanirini).

Tribe Diorini Lane, 1950

Diorinae Lane, 1950: 373 [stem: *Dior-*]. Type genus: *Diorus* A. White, 1853.

Tribe Distichocerini Pascoe, 1867

*Distichocérites Blanchard, 1845b: 144 [stem: *Distichocer-*]. Type genus: *Distichocera* Kirby, 1819. Comment: original vernacular name unavailable (Art. 11.7.2): subsequently used in latinized form but not generally attributed to Blanchard (1845b).

Distichocerinae Pascoe, 1867a: 125 [stem: *Distichocer-*]. Type genus: *Distichocera* Kirby, 1819.

Tribe Dodecosini Aurivillius, 1912

Dodecosini Aurivillius, 1912: 132 [stem: *Dodecos-*]. Type genus: *Dodecosis* H. W. Bates, 1867. Comment: current spelling maintained (Art. 29.5): incorrect original stem formation in prevailing usage (should be *Dodecose-*).

OLEXANDRELLAEINI Zajciw, 1960: 605 [stem: *Olexandrell-*]. Type genus: *Olexandrella* Zajciw, 1959. Comment: incorrect original stem formation, not in prevailing usage.

Tribe DRYOBIINI Arnett, 1962 *nomen protectum*

DRYOBIINI Arnett, 1962c: 861 [stem: *Dryobi-*]. Type genus: *Dryobius* J. L. LeConte, 1850. Comment: *nomen protectum* (see Bousquet et al. 2009: 45); this is a junior homonym of DRYOBIADAE Gistel, 1856a: 368 (type genus *Dryobia* Gistel, 1856) which is a junior synonym of DRYOPHILIDAE Gistel, 1848.

Tribe EBURIINI Blanchard, 1845

ÉBURIITES Blanchard, 1845b: 145 [stem: *Eburi-*]. Type genus: *Eburia* Lacordaire, 1830. Comment: original vernacular name available (Art. 11.7.2): first used in latinized form by J. Thomson (1861: 237, as EBURITAE), generally accepted as in Monné (1993a: 20, as EBURIINI).

Tribe ECTENESSINI Martins, 1998

ECTENESSINI Martins, 1998: 82 [stem: *Ecteness-*]. Type genus: *Ectenessa* H. W. Bates, 1885.

Tribe ELAPHIDIINI Thomson, 1864

ELAPHIDIONITAE J. Thomson, 1864: 235 [stem: *Elaphidi-*]. Type genus: *Elaphidion* Audinet-Serville, 1834. Comment: incorrect original stem formation, not in prevailing usage; Ivie (1985: 303) pointed out that the correct stem based on *Elaphidion* is *Elaphidi-*; both ELAPHIDIINI and ELAPHIDIONINI have been used in recent literature, we prefer to use the correct spelling of the stem here.

SPHÉRIONIDES Lacordaire, 1868: 312 [stem: *Sphaeri-*]. Type genus: *Sphaerion* Audinet-Serville, 1834. Comment: original vernacular name available (Art. 11.7.2): first used in latinized form by Heyne and Taschenberg (1907: 239, as SPHAERIONINI), generally accepted as in Aurivillius (1912: 96, as SPHAERIONINI); incorrect original stem formation, not in prevailing usage; junior homonym of SPHAERIIDAE Deshayes, 1855 (type genus *Sphaerium* Scopoli, 1877) in Mollusca and SPHAERINA Erichson, 1845 [incorrect original spelling] (type genus *Sphaerius* Waltl, 1838) in Myxophaga, the stem of the beetle family-group name was recently emended to *Sphaerius-* (ICZN 2000); this case is to be referred to the Commission to remove the homonymy (Art. 55.3.1).

STENOSPHENINI J. L. LeConte, 1873: 316 [stem: *Stenosphen-*]. Type genus: *Stenosphenus* Haldeman, 1847.

Tribe ELIGMODERMINI Lacordaire, 1868

ÉLIGMODERMIDES Lacordaire, 1868: 337 [stem: *Eligmoderm-*]. Type genus: *Eligmoderma* J. Thomson, 1864. Comment: original vernacular name available (Art. 11.7.2): first used in latinized form and generally accepted as in Aurivillius

(1912: 114, as Eligmodermini); current spelling maintained (Art. 29.5): incorrect original stem formation in prevailing usage (should be *Eligmodermat-*).

Tribe Erlandiini Aurivillius, 1912

Erlandiini Aurivillius, 1912: 12 [stem: *Erlandi-*]. Type genus: *Erlandia* Aurivillius, 1904.

Tribe Eroschemini Lacordaire, 1868

Éroschémides Lacordaire, 1868: 515 [stem: *Eroschem-*]. Type genus: *Eroschema* Pascoe, 1859. Comment: original vernacular name available (Art. 11.7.2): first used in latinized form by H. W. Bates (1872a: 185, as Erocheminae), generally accepted as in Aurivillius (1912: 287, as Eroschemini); current spelling maintained (Art. 29.5): incorrect original stem formation in prevailing usage (should be *Eroschemat-*).

Tribe Eumichthini Linsley, 1940

Eumichthini Linsley, 1940: 368 [stem: *Eumichth-*]. Type genus: *Eumichthus* J. L. LeConte, 1873.

Tribe Gahaniini Quentin and Villiers, 1969

Gahaniini Quentin and Villiers, 1969: 615, in key [stem: *Gahani-*]. Type genus: *Gahania* Distant, 1907.

Tribe Glaucytini Lacordaire, 1868

Glaucytides Lacordaire, 1868: 405 [stem: *Glaucyt-*]. Type genus: *Glaucytes* J. Thomson, 1857. Comment: original vernacular name available (Art. 11.7.2): first used in latinized form by Pascoe (1869b: 650, as Glaucytinae), generally accepted as in Aurivillius (1912: 435, as Glaucytini); current spelling maintained (Art. 29.5): incorrect stem formation in prevailing usage (should be *Glaucytet-*).

Tribe Graciliini Mulsant, 1839

Graciliaires Mulsant, 1839: 99 [stem: *Gracili-*]. Type genus: *Gracilia* Audinet-Serville, 1834. Comment: original vernacular name available (Art. 11.7.2): first used in latinized form by J. L. LeConte (1873: 300, as Graciliinae), generally accepted as in Monné (1993b: 9, as Graciliini).

Tribe Hesperophanini Mulsant, 1839

Hespérophanaires Mulsant, 1839: 61 [stem: *Hesperophan-*]. Type genus: *Hesperophanes* Dejean, 1835.

Subtribe Daramina Sama, 2008

Daramina Sama, 2008: 224 [stem: *Daram-*]. Type genus: *Daramus* Fairmaire, 1892.

Subtribe Hesperophanina Mulsant, 1839

Hespérophanaires Mulsant, 1839: 61 [stem: *Hesperophan-*]. Type genus: *Hesperophanes* Dejean, 1835 [see Vives and Alonso-Zarazaga (2000: 657) and Bousquet et al. (2009: 47) for a discussion of problems with the type species of this genus]. Comment: original vernacular name available (Art. 11.7.2): first used in latinized form by Pascoe (1869: 523, as Hesperophaninae), generally accepted as in Monné (1993a: 1, as Hesperophanini).

Cerasphoritae J. Thomson, 1861: 234 [stem: *Cerasphor-*]. Type genus: *Cerasphorus* Audinet-Serville, 1834.

Tribe Hesthesini Pascoe, 1867

Hesthesinae Pascoe, 1867a: 127 [stem: *Hesthes-*]. Type genus: *Hesthesis* Newman, 1840. Comment: current spelling maintained (Art. 29.5): incorrect original stem formation in prevailing usage (should be *Hesthese-*).

Tribe Heteropsini Lacordaire, 1868 *nomen protectum*

Dichophyiaeidae Gistel, 1848: [8] [stem: *Dichophyi-*]. Type genus: *Dichophyia* Gistel, 1848. Comment: *nomen oblitum* (see Bousquet et al. 2009: 48). Comment: incorrect original stem formation, not in prevailing usage.

Hétéropsides Lacordaire, 1868: 405 [stem: *Heterops-*]. Type genus: *Heterops* Blanchard, 1842. Comment: *nomen protectum* (see Bousquet et al. 2009: 48); original vernacular name available (Art. 11.7.2): first used in latinized form by H. W. Bates (1872a: 179, as Heteropsinae), generally accepted as in Aurivillius (1912: 438, as Heteropsini); current spelling maintained (Art. 29.5): incorrect stem formation in prevailing usage (should be *Heterop-*); this name was incorrectly attributed to Lacordaire (1869) by Bousquet et al. (2009: 48).

Tribe Hexoplini Martins, 2006

Hexoplonini Martins, 2006: 22 [stem: *Hexopl-*]. Type genus: *Hexoplon* J. Thomson, 1864. Comment: incorrect original stem formation, not in prevailing usage; this name was incorrectly treated as unavailable by Bousquet et al. (2009).

Tribe Holopleurini Chemsak and Linsley, 1974

Holopleurini Chemsak and Linsley, 1974: 183 [stem: *Holopleur-*]. Type genus: *Holopleura* J. L. LeConte, 1873.

Tribe Holopterini Lacordaire, 1868

Holoptérides Lacordaire, 1868: 393 [stem: *Holopter-*]. Type genus: *Holopterus* Blanchard, 1851. Comment: original vernacular name available (Art. 11.7.2): first used in latinized form by Lucas (1920: xxiii, as Holopterini), generally accepted as in Aurivillius (1912: 148, as Holopterini); this tribe was transferred to the subfamily Lepturinae by Vitali (2002: 32) but placed back into the Cerambycinae by Monné (2005a: 301).

Tribe Hyboderini Linsley, 1940

Hyboderini Linsley, 1940: 371 [stem: *Hyboder-*]. Type genus: *Hybodera* J. L. LeConte, 1873.

Tribe Hylotrupini Zagajkevich, 1991

Hylotrupini Zagajkevich, 1991: 67 [stem: *Hylotrup-*]. Type genus: *Hylotrupes* Audinet-Serville, 1834.

Tribe Ibidionini Thomson, 1861

Ibidionitae J. Thomson, 1861: 199 [stem: *Ibidion-*]. Type genus: *Ibidion* Gory, 1833.

Subtribe Compsina Martins and Galileo, 2007

Compsina Martins and Galileo, 2007: 6, in key [stem: *Comps-*]. Type genus: *Compsa* Perty, 1832. Comment: this name was incorrectly treated as unavailable by Bousquet et al. (2009); junior homonym of Compsini Pierce, 1913 (type genus *Compsus* Schönherr, 1823) in Curculionidae; this case is to be referred to the Commission to remove the homonymy (Art. 55.3.1).

Subtribe Ibidionina Thomson, 1861

Ibidionitae J. Thomson, 1861: 199 [stem: *Ibidion-*]. Type genus: *Ibidion* Gory, 1833. Comment: current spelling maintained (Art. 29.5): incorrect stem formation in prevailing usage (should be *Ibidi-*).

*Sydacini Martins, 2003a: 204 [stem: *Sydac-*]. Type genus: *Sydax* Lacordaire, 1868. Comment: name unavailable (Art. 16.1): name not indicated as intentionally new; this taxon was originally described by Martins (1997a: 8-9) but not named.

Subtribe Tropidina Martins and Galileo, 2007

Tropidina Martins and Galileo, 2007: 7 [stem: *Tropid-*]. Type genus: *Tropidion* J. Thomson, 1867. Comment: this name was incorrectly treated as unavailable by Bousquet et al. (2009); incorrect original spelling maintained (should be *Tropidi-*) in order to avoid homonymy with Tropidiini Hull, 1949 (type genus *Tropidia* Meigen, 1822) available in Diptera: Syrphidae.

Tribe Ideratini Martins and Napp, 2009

Ideratini Martins and Napp, 2009: 216 [stem: *Iderat-*]. Type genus: *Ideratus* J. Thomson, 1864.

Tribe Lissonotini Swainson, 1840

Lissonotinae Swainson, 1840: 289 [stem: *Lissonot-*]. Type genus: *Lissonotus* Dalman, 1817. Comment: this family-group name was incorrectly credited to

Swainson and Shuckard (1840) by Bousquet et al. (2009: 49); the junior homonym LISSONOTINI Förster, 1869 (type genus *Lissonota* Gravenhorst, 1829) is available in Hymenoptera: ICHNEUMONIDAE; this case is to be referred to the Commission to remove the homonymy (Art. 55.3.1).

Tribe LUSCOSMODICINI Martins, 2003

LUSCOSMODICINI Martins, 2003b: 30 [stem: *Luscosmodic-*]. Type genus: *Luscosmodicum* Martins, 1970. Comment: this name was incorrectly treated as unavailable by Bousquet et al. (2009).

Tribe LYGRINI Sama, 2008

LYGRINI Sama, 2008: 222 [stem: *Lygr-*]. Type genus: *Lygrus* Fåhraeus, 1872.

Tribe MACRONINI Lacordaire, 1868

ENCHAPTERITAE J. Thomson, 1861: 151 [stem: *Enchopter-*]. Type genus: *Enchoptera* Saunders, 1850 [as *Enchaptera*, incorrect subsequent spelling type genus name, not in prevailing usage]. Comment: incorrect original stem formation, not in prevailing usage; this name has precedence over MACRONINI Lacordaire, 1868 but has not been used as a valid name after 1899 to our knowledge, although we were unable to find 25 references to conserve usage of MACRONINI (Art. 23.9.2), we believe the name MACRONINI should be conserved for this group and an application should be submitted to the Commission (Art. 23.9.3).

MACRONIDES Lacordaire, 1868: 414 [stem: *Macron-*]. Type genus: *Macrones* Newman, 1841. Comment: original vernacular name available (Art. 11.7.2): first used in latinized form by Pascoe (1871: 268, as MACRONINAE), generally accepted as in Aurivillius (1912: 153, as MACRONINI); although this is not the oldest name for the tribe, we recommend that an application be submitted to the Commission to conserve usage of the well-established name MACRONINI Lacordaire, 1868.

Tribe MEGACOELINI Quentin and Villiers, 1969

MEGACOELINI Quentin and Villiers, 1969: 615, in key [stem: *Megacoel-*]. Type genus: *Megacoelus* Lacordaire, 1868.

Tribe METHIINI Thomson, 1860

METHIITAE J. Thomson, 1860a: 127 [stem: *Methi-*]. Type genus: *Methia* Newman, 1842.

Tribe MOLORCHINI Gistel, 1848

MOLORCHIDAE Gistel, 1848: [9] [stem: *Molorch-*]. Type genus: *Molorchus* Fabricius, 1792 [see Bousquet (2008: 620) for a discussion of the type species]. Comment: name previously attributed to Mulsant (1862) but was also used by Marseul (1857a: 166, as MOLORCHIDAE) prior to Mulsant's name.

Tribe Mythodini Lacordaire, 1868

Mythodides Lacordaire, 1868: 418 [stem: *Mythod-*]. Type genus: *Mythodes* J. Thomson, 1864. Comment: original vernacular name available (Art. 11.7.2): first used in latinized form by Pascoe (1871: 268, as Mythodinae), generally accepted as in Aurivillius (1912: 154, as Mythodini).

Tribe Necydalopsini Lacordaire, 1868

Nécydalopsides Lacordaire, 1868: 493 [stem: *Necydalops-*]. Type genus: *Necydalopsis* Blanchard, 1851. Comment: original vernacular name available (Art. 11.7.2): first used in latinized form and generally accepted as in Aurivillius (1912: 275, as Necydalopsini); current spelling maintained (Art. 29.5): incorrect original stem formation in prevailing usage (should be *Necydalopse-*).

Tribe Neocorini Martins, 2005

Neocorini Martins, 2005: 240 [stem: *Neocor-*]. Type genus: *Neocorus* J. Thomson, 1864. Comment: this name was incorrectly treated as unavailable by Bousquet et al. (2009).

Tribe Neostenini Lacordaire, 1868

Néosténides Lacordaire, 1868: 363 [stem: *Neosten-*]. Type genus: *Neostenus* Pascoe, 1857. Comment: original vernacular name available (Art. 11.7.2): first used in latinized form by Pascoe (1871: 268, as Neosteninae), generally accepted as in Aurivillius (1912: 138, as Neostenini).

Tribe Obriini Mulsant, 1839

Obriaires Mulsant, 1839: 95 [stem: *Obri-*]. Type genus: *Obrium* Dejean, 1821. Comment: original vernacular name available (Art. 11.7.2): first used in latinized form by Gistel (1848: [9], as Obriidae), generally accepted as in Monné (1993a: 11, as Obriini).

Tribe Ochyrini Pascoe, 1871

Ochyrinae Pascoe, 1871: 273 [stem: *Ochyr-*]. Type genus: *Ochyra* Pascoe, 1871.

Tribe Oedenoderini Aurivillius, 1912

Oedenoderini Aurivillius, 1912: 358 [stem: *Oedenoder-*]. Type genus: *Oedenoderus* Chevrolat, 1858.

Tribe Oemini Lacordaire, 1868

Oemides Lacordaire, 1868: 216 [stem: *Oem-*]. Type genus: *Oeme* Newman, 1840.

Subtribe Methioidina Martins, 1997

Methioidina Martins, 1997a: 119 [stem: *Methioid-*]. Type genus: *Methioides* Chemsak and Linsley, 1967.

Subtribe Oemina Lacordaire, 1868

*Malacoptérites Blanchard, 1845b: 147 [stem: *Malacopter-*]. Type genus: *Malacopterus* Audinet-Serville, 1833. Comment: original vernacular name unavailable (Art. 11.7.2): not subsequently latinized.

Oemides Lacordaire, 1868: 216 [stem: *Oem-*]. Type genus: *Oeme* Newman, 1840. Comment: original vernacular name available (Art. 11.7.2): first used in latinized form by Pascoe (1869b: 498, as Oeminae), generally accepted as in Aurivillius (1912: 26, as Oemini).

Tribe Opsimini LeConte, 1873

Opsimi J. L. LeConte, 1873: 294 [stem: *Opsim-*]. Type genus: *Opsimus* Mannerheim, 1843.

Tribe Oxycoleini Martins and Galileo, 2003

Oxycoleini Martins and Galileo, 2003: 52 [stem: *Oxycole-*]. Type genus: *Oxycoleus* Lacordaire, 1868. Comment: this name was incorrectly treated as unavailable by Bousquet et al. (2009).

Tribe Paraholopterini Martins, 1997

Paraholopterini Martins, 1997b: 201 [stem: *Paraholopter-*]. Type genus: *Paraholopterus* Cerda and Cekalovic, 1987.

Tribe Phalotini Lacordaire, 1868

Phalotides Lacordaire, 1868: 495 [stem: *Phalot-*]. Type genus: *Phalota* Pascoe, 1863. Comment: original vernacular name available (Art. 11.7.2): first used in latinized form and generally accepted as in Aurivillius (1912: 276, as Phalotini).

Tribe Phlyctaenodini Lacordaire, 1868

Phlycténodides Lacordaire, 1868: 370 [stem: *Phlyctaenod-*]. Type genus: *Phlyctaenodes* Newman, 1840. Comment: original vernacular name available (Art. 11.7.2): first used in latinized form by Aurivillius (1912: 140, as Phlyctaenodini), generally accepted as in Monné (1993b: 20, as Phlyctaenodini); incorrect original stem formation, not in prevailing usage.

Tribe Phoracanthini Newman, 1840

Stenocoridae Hope, 1834: 106 [stem: *Stenocor-*]. Type genus: *Stenocorus* sensu Hope, 1835 [not *Stenocorus* Geoffroy, 1762; syn. of *Phoracantha* Newman, 1840]. Comment: based on a misidentified type genus, name treated here as invalid until an application is submitted to the Commission to suppress it for the Principles of Priority and Homonymy (Art. 65.2.1); also see Stenocoritae J. Thomson, 1861 in Lepturinae: Rhagiini.

Phoracanthidae Newman, 1840: 2 [stem: *Phoracanth-*]. Type genus: *Phoracantha* Newman, 1840.

Tribe Phyllarthriini Lepesme and Breuning, 1956

Phyllarthriini Lepesme and Breuning, 1956: 287 [stem: *Phyllarthri-*]. Type genus: *Phyllarthrius* Hope, 1843.

Tribe Piesarthriini McKeown, 1947

Piesarthrini McKeown, 1947: 55 [stem: *Piesarthri-*]. Type genus: *Piesarthrius* Hope, 1834. Comment: name proposed after 1930 without description or bibliographic reference to such a description (Art. 13.1), however available because it was used as valid before 2000 as in Gressitt (1959: 84, as Piesarthini) and was not rejected by an author who, between 1961 and 1999, applied Article 13 of the then current edition of the Code (see Art. 13.2.1); incorrect original stem formation, not in prevailing usage.

Tribe Piezocerini Lacordaire, 1868

Piézocérides Lacordaire, 1868: 324 [stem: *Piezocer-*]. Type genus: *Piezocera* Audinet-Serville, 1834.

Subtribe Haruspicina Martins, 1976

Haruspicina Martins, 1976: 199 [stem: *Haruspic-*]. Type genus: *Haruspex* J. Thomson, 1864.

Subtribe Piezocerina Lacordaire, 1868

Piézocérides Lacordaire, 1868: 324 [stem: *Piezocer-*]. Type genus: *Piezocera* Audinet-Serville, 1834. Comment: original vernacular name available (Art. 11.7.2): first used in latinized form and generally accepted as in Aurivillius (1912: 102, as Piezocerini).

Zelliboriinae Lane, 1951: 5 [stem: *Zellibori-*]. Type genus: *Zelliboria* Lane, 1951.

Tribe Platyarthrini Bates, 1870

*Coelarthrides Lacordaire, 1868: 405 [stem: *Caelomarth-*]. Type genus: *Caelomarthron* J. Thomson, 1860 [as *Coelarthron*, unjustified emendation of type genus name by Lacordaire (1869: 142), not in prevailing usage; syn. of *Platyarthron* Guérin-Méneville, 1844]. Comment: original vernacular name unavailable (Art. 11.7.2): subsequently used in latinized form but not generally accepted as valid; subsequent usage of Coelarthrinae by Lucas (1920: 17) and Caelarthrinae by Ferreira and Veiga Ferreira (1959b: 331) did not validate this name because Lacordaire's taxon was listed as a synonym of Platyarthrini H. W. Bates; Coelarthridae was used as valid by Ienistea (1986: 30) but it was not attributed to Lacordaire (1868); Ienistea's name is also unavailable, it was proposed after 1930 without description or bibliographic reference to such a description (Art. 13.1); incorrect original stem formation, not in prevailing usage.

PLATYARTHRINAE H. W. Bates, 1870: 419 [stem: *Platyarthr-*]. Type genus: *Platyarthron* Guérin-Méneville, 1844.

Tribe PLECTOGASTERINI Quentin and Villiers, 1969

PLECTOGASTERINI Quentin and Villiers, 1969: 615, in key [stem: *Plectogaster-*]. Type genus: *Plectogaster* C. O. Waterhouse, 1881. Comment: Bousquet et al. (2009: 52) erroneously listed PLECTOGASTRINI as the original spelling.

Tribe PLECTROMERINI Nearns and Braham, 2008

PLECTROMERINI Nearns and Braham, 2008: 19 [stem: *Plectromer-*]. Type genus: *Plectromerus* Haldeman, 1847 [see Bousquet et al. (2009: 53) about problems with the type species of this genus].

Tribe PLEIARTHROCERINI Lane, 1950

PLEIARTHROCERINAE Lane, 1950: 371 [stem: *Pleiarthrocer-*]. Type genus: *Pleiarthrocerus* Bruch, 1915.

Tribe PROTAXINI Gahan, 1906

PROTAXINI Gahan, 1906: 92 [stem: *Protax-*]. Type genus: *Protaxis* Gahan, 1906. Comment: current spelling maintained (Art. 29.5): incorrect original stem formation in prevailing usage (should be *Protaxe-*).

Tribe PROTHEMINI Lacordaire, 1868

PROTHÉMIDES Lacordaire, 1868: 524 [stem: *Prothem-*]. Type genus: *Prothema* Pascoe, 1856. Comment: original vernacular name available (Art. 11.7.2): first used in latinized form by Pascoe (1869b: 578, as PROTHEMINAE), generally accepted as in Aurivillius (1912: 291, as PROTHEMINI).

Tribe PSEBIINI Lacordaire, 1868

*LEPTIDÉITES Fairmaire, 1864: 148 [stem: *Leptide-*]. Type genus: *Leptidea* Mulsant, 1839 [preoccupied genus name, not *Leptidea* Billberg, 1820 [Lepidoptera]; syn. of *Nathrius* Bréthes, 1916]. Comment: original vernacular name unavailable (Art. 11.7.2): subsequently used in latinized form but not generally attributed to Fairmaire (1864).

PSÉBIIDES Lacordaire, 1868: 479 [stem: *Psebi-*]. Type genus: *Psebium* Pascoe, 1864. Comment: original vernacular name available (Art. 11.7.2): first used in latinized form by Kolbe (1897: 299, as PSEBIINAE), generally accepted as in Aurivillius (1912: 261, as PSEBIINI).

LEPTIDEINA Reitter, 1913a: 24 [stem: *Leptide-*]. Type genus: *Leptidea* Mulsant, 1839 [preoccupied genus name, not *Leptidea* Billberg, 1820 [Lepidoptera]; syn. of *Nathrius* Bréthes, 1916]. Comment: permanently invalid (Art. 39): based on preoccupied type genus; LEPTIDEINI Verity, 1947 (type genus *Leptidea* Billberg, 1820) is available in Lepidoptera.

CAMBAIINAE Lane, 1951: 12 [stem: *Cambai-*]. Type genus: *Cambaia* Lane, 1951 [syn. of *Paraleptidea* Gounelle, 1913].

NATHRIINI Arnett, 1962c: 860 [stem: *Nathri-*]. Type genus: *Nathrius* Brèthes, 1916.

Tribe PSEUDOCEPHALINI Aurivillius, 1912 (1861)

AMETROCEPHALITAE J. Thomson, 1861: 256 [stem: *Ametrocephal-*]. Type genus: *Ametrocephala* Blanchard, 1851 [syn. of *Pseudocephalus* Newman, 1842]. Comment: use of younger name PSEUDOCEPHALINI Aurivillius, 1912 conserved over this name (Art. 40.2).

PSEUDOCEPHALINI Aurivillius, 1912: 154 [stem: *Pseudocephal-*]. Type genus: *Pseudocephalus* Newman, 1842. Comment: name proposed to replace AMETROCEPHALINI J. Thomson, 1861 because of the synonymy of the type genus; usage of this name conserved over AMETROCEPHALINI Thomson, 1861 (Art. 40.2).

Tribe PSEUDOLEPTURINI Thomson, 1861

PSEUDOLEPTURITAE J. Thomson, 1861: 146 [stem: *Pseudoleptur-*]. Type genus: *Pseudoleptura* J. Thomson, 1861. Comment: we treat this group as a valid tribe as in Adlbauer et al. (2010: 200).

ERYTHRINAE Pascoe, 1866a: 227 [stem: *Erythr-*]. Type genus: *Erythrus* A. White, 1853.

Tribe PSILOMORPHINI Lacordaire, 1868

PSILOMORPHIDES Lacordaire, 1868: 392 [stem: *Psilomorph-*]. Type genus: *Psilomorpha* Saunders, 1850. Comment: original vernacular name available (Art. 11.7.2): first used in latinized form and generally accepted as in Aurivillius (1912: 148, as PSILOMORPHINI).

Tribe PTEROPLATINI Thomson, 1861

PTEROPLATITAE J. Thomson, 1861: 254 [stem: *Pteroplat-*]. Type genus: *Pteroplatus* Buquet, 1840.

Tribe PYRESTINI Lacordaire, 1868

PYRESTHIDES Lacordaire, 1868: 518 [stem: *Pyrest-*]. Type genus: *Pyrestes* Pascoe, 1857 [as *Pyresthes*, incorrect subsequent spelling of type genus name, not in prevailing usage]. Comment: original vernacular name available (Art. 11.7.2): first used in latinized form by Aurivillius (1912: 288, as PYRESTINI), generally accepted as in Bousquet et al. (2009: 54, as PYRESTINI); incorrect original stem formation, not in prevailing usage.

Tribe RHAGIOMORPHINI Newman, 1841

RHAGIOMORPHIDAE Newman, 1841: 34 [stem: *Rhagiomorph-*]. Type genus: *Rhagiomorpha* Newman, 1840.

Tribe Rhinotragini Thomson, 1861

Rhinotragitae J. Thomson, 1861: 177 [stem: *Rhinotrag-*]. Type genus: *Rhinotragus* Germar, 1824.

Tribe Rhopalophorini Blanchard, 1845

Rhopalophorites Blanchard, 1845b: 152 [stem: *Rhopalophor-*]. Type genus: *Rhopalophora* Audinet-Serville, 1834. Comment: original vernacular name available (Art. 11.7.2): first used in latinized form by Blanchard (1853: 268, as Rhopalophoritae), generally accepted as in Monné (1994a: 1, as Rhopalophorini).

Tribe Rosaliini Fairmaire, 1864

Rosaliites Fairmaire, 1864: 137 [stem: *Rosali-*]. Type genus: *Rosalia* Audinet-Serville, 1834. Comment: original vernacular name available (Art. 11.7.2): first used in latinized form by J. L. LeConte (1873: 310, as Rosaliini), generally accepted as in Linsley (1964: 4, as Rosaliini).

Tribe Sestyrini Lacordaire, 1868

Sestyrides Lacordaire, 1868: 405 [stem: *Sestyr-*]. Type genus: *Sestyra* Pascoe, 1867. Comment: original vernacular name available (Art. 11.7.2): first used in latinized form by Pascoe (1869b: 643, as Sestyrinae), generally accepted as in Aurivillius (1912: 424, as Sestyrini).

Tribe Smodicini Lacordaire, 1868

Smodicides Lacordaire, 1868: 405 [stem: *Smodic-*]. Type genus: *Smodicum* Haldeman, 1847. Comment: original vernacular name available (Art. 11.7.2): first used in latinized form by J. L. LeConte (1873: 294, as Smodici), generally accepted as in Aurivillius (1912: 12, as Smodicini); this name was incorrectly attributed to Lacordaire (1869) by Bousquet et al. (2009: 55).

Tribe Spintheriini Lacordaire, 1869

Spinthériides Lacordaire, 1869: 219 [stem: *Spintheri-*]. Type genus: *Spintheria* J. Thomson, 1861. Comment: original vernacular name available (Art. 11.7.2): first used in latinized form and generally accepted as in Aurivillius (1912: 487, as Spintheriini).

Tribe Stenhomalini Miroshnikov, 1989

Stenhomalini Miroshnikov, 1989: 742 [stem: *Stenhomal-*]. Type genus: *Stenhomalus* A. White, 1855.

Tribe Stenoderini Pascoe, 1867

*Sténodérites Blanchard, 1845b: 163 [stem: *Stenoder-*]. Type genus: *Stenoderus* Dejean, 1821. Comment: original vernacular name unavailable (Art. 11.7.2):

subsequently used in latinized form but not generally attributed to Blanchard (1845b).

SYLLITAE J. Thomson, 1864: 138 [stem: *Syllit-*]. Type genus: *Syllitus* Pascoe, 1859. Comment: this name has precedence over STENODERINI Pascoe, 1867 but has not been used as a valid name after 1899 to our knowledge, unfortunately, we were unable to provide 25 references to conserve usage of STENODERINI (Art. 23.9.2) although we believe the name STENODERINI should be conserved for this group until an application to the ICZN is submitted (Art. 23.9.3).

STENODERINAE Pascoe, 1867b: 311 [stem: *Stenoder-*]. Type genus: *Stenoderus* Dejean, 1821. Comment: senior homonym of STENODERINI Selander, 1991 (type genus *Stenodera* Eschscholtz, 1818) currently used as valid in MELOIDAE; this case is to be referred to the Commission to remove the homonymy (Art. 55.3.1); also, this is not the oldest available name for this tribe, see comments under SYLLITAE Thomson, 1864.

*PTÉROSTÉNIDES Lacordaire, 1868: 410 [stem: *Pterosten-*]. Type genus: *Pterostenus* Laporte, 1840 [syn. of *Stenoderus* Dejean, 1821]. Comment: original vernacular name unavailable (Art. 11.7.2): subsequently used in latinized form but not generally accepted as valid; subsequent usage of PTEROSTENINAE by Aurivillius (1912: 150) and Lucas (1920: 53) did not validate this name because Lacordaire's taxon was listed as a synonym of STENODERINI.

CALLIPRASONINI McKeown, 1947: 71 [stem: *Callipras-*]. Type genus: *Calliprason* A. White, 1843. Comment: name proposed after 1930 without description or bibliographic reference to such a description (Art. 13.1), however available because it was used as valid before 2000 as in Gressitt (1959: 148, as CALLIPRASONINI) and was not rejected by an author who, between 1961 and 1999, applied Article 13 of the then current edition of the Code (see Art. 13.2.1); incorrect original stem formation, not in prevailing usage.

Tribe STENOPTERINI Gistel, 1848

STENOPTERIDAE Gistel, 1848: [9] [stem: *Stenopter-*]. Type genus: *Stenopterus* Illiger, 1804.

Tribe STRONGYLURINI Lacordaire, 1868

STRONGYLURIDES Lacordaire, 1868: 379 [stem: *Strongylur-*]. Type genus: *Strongylurus* Hope, 1834. Comment: original vernacular name available (Art. 11.7.2): first used in latinized form by Pascoe (1869b: 548, as STRONGYLURINAE), generally accepted as in Aurivillius (1912: 144, as STRONGYLURINI).

Tribe TESSAROMMATINI Lacordaire, 1868

TESSAROMMIDES Lacordaire, 1868: 378 [stem: *Tessarommat-*]. Type genus: *Tessaromma* Newman, 1840. Comment: original vernacular name available (Art. 11.7.2): first used in latinized form and generally accepted as in Aurivillius (1912: 143, as TESSAROMMATINI); name also spelled TESSÉROMMIDES in original publication on page 204 (key); incorrect original stem formation, not in prevailing usage.

Tribe THRANIINI Gahan, 1906

THRANIINI Gahan, 1906: 236 [stem: *Thrani-*]. Type genus: *Thranius* Pascoe, 1859.

Tribe THYRSIINI Marinoni and Napp, 1984

THYRSIINI Marinoni and Napp, 1984: 44 [stem: *Thyrsi-*]. Type genus: *Thyrsia* Dalman, 1819.

Tribe TILLOMORPHINI Lacordaire, 1868

TILLOMORPHIDES Lacordaire, 1868: 405 [stem: *Tillomorph-*]. Type genus: *Tillomorpha* Blanchard, 1851. Comment: original vernacular name available (Art. 11.7.2): first used in latinized form by Pascoe (1869b: 554, as TILLOMORPHINAE), generally accepted as in Adlbauer et al. (2010: 206, as TILLOMORPHINI); this name was incorrectly credited to Pascoe, 1869 by Bousquet et al. (2009: 56).

EPIPEDOCERINI Gahan, 1906: 305 [stem: *Epipedocer-*]. Type genus: *Epipedocera* Chevrolat, 1863.

Tribe TORNEUTINI Thomson, 1861

TORNEUTITAE J. Thomson, 1861: 272 [stem: *Torneut-*]. Type genus: *Torneutes* Reich, 1838.

THAUMASIDAE J. Thomson, 1864: 313 [stem: *Thaumas-*]. Type genus: *Thaumasus* Reiche, 1853.

Tribe TRACHYDERINI Dupont, 1836

TRACHYDÉRIDES Dupont, 1836: 1 [stem: *Trachyder-*]. Type genus: *Trachyderes* Dalman, 1817.

Subtribe ANCYLOCERINA Thomson, 1864

ANCYLOCERITAE J. Thomson, 1864: 210 [stem: *Ancylocer-*]. Type genus: *Ancylocera* Audinet-Serville, 1834.

Subtribe TRACHYDERINA Dupont, 1836

TRACHYDÉRIDES Dupont, 1836: 1 [stem: *Trachyder-*]. Type genus: *Trachyderes* Dalman, 1817. Comment: original vernacular name available (Art. 11.7.2): first used in latinized form by J. Thomson (1861: 209, as TRACHYDERITAE), generally accepted as in Monné (1994b: 16, as TRACHYDERINI).

PURPURICENITAE J. Thomson, 1861: 203 [stem: *Purpuricen-*]. Type genus: *Purpuricenus* Dejean, 1821.

TYLOSITAE J. Thomson, 1861: 205 [stem: *Tylose-*]. Type genus: *Tylosis* J. L. LeConte, 1850. Comment: incorrect original stem formation, not in prevailing usage.

Sphaenothecitae J. Thomson, 1861: 212 [stem: *Sphaenothec-*]. Type genus: *Sphaenothecus* Dupont, 1838.

Megaderitae J. Thomson, 1861: 213 [stem: *Megader-*]. Type genus: *Megaderus* Dejean, 1821.

Eriphitae J. Thomson, 1864: 200 [stem: *Eriph-*]. Type genus: *Eriphus* Audinet-Serville, 1834.

Pteracanthitae J. Thomson, 1864: 255 [stem: *Pteracanth-*]. Type genus: *Pteracantha* Newman, 1838.

Metopocoïlitae J. Thomson, 1864: 255 [stem: *Metopocoil-*]. Type genus: *Metopocoilus* Audinet-Serville, 1832.

Sternacanthitae J. Thomson, 1864: 259 [stem: *Sternacanth-*]. Type genus: *Sternacanthus* Audinet-Serville, 1832.

Tropidosomitae J. Thomson, 1864: 256 [stem: *Tropidosomat-*]. Type genus: *Tropidosoma* Perty, 1832. Comment: incorrect original stem formation, not in prevailing usage.

Poecilopéplides Lacordaire, 1868: 404 [stem: *Poecilopepl-*]. Type genus: *Poecilopeplus* Dejean, 1835. Comment: original vernacular name available (Art. 11.7.2): first used in latinized form and generally accepted as in Aurivillius (1912: 449, as Poecilopeplini).

*Dorcacérides Lacordaire, 1868: 404 [stem: *Dorcacer-*]. Type genus: *Dorcacerus* Dejean, 1821. Comment: original vernacular name unavailable (Art. 11.7.2): subsequently used in latinized form but not generally attributed to Lacordaire (1868).

Sténaspides Lacordaire, 1868: 404 [stem: *Stenaspid-*]. Type genus: *Stenaspis* Audinet-Serville, 1834. Comment: original vernacular name available (Art. 11.7.2): first used in latinized form by Pascoe (1869b: 653, as Stenaspidinae), generally accepted as in Aurivillius (1912: 457, as Stenaspini); incorrect original stem formation, not in prevailing usage.

Paristémiides Lacordaire, 1868: 404 [stem: *Paristemi-*]. Type genus: *Paristemia* Westwood, 1841. Comment: original vernacular name available (Art. 11.7.2): first used in latinized form and generally accepted as in J. L. LeConte (1873: 309, as Paristemiini).

Dorcacerinae H. W. Bates, 1870: 430 [stem: *Dorcacer-*]. Type genus: *Dorcacerus* Dejean, 1821. Comment: Aurivillius (1912: 476) used the spelling Dorcadocerini but this was based on *Docadocerus*, which is an incorrect subsequent spelling of the type genus name.

Tribe Tragocerini Pascoe, 1867

Tragocerinae Pascoe, 1867a: 125 [stem: *Tragocer-*]. Type genus: *Tragocerus* Latreille, 1829.

Tribe Trichomesiini Aurivillius, 1912

Trichomesiini Aurivillius, 1912: 276 [stem: *Trichomesi-*]. Type genus: *Trichomesia* Pascoe, 1859.

Tribe Tropocalymmatini Lacordaire, 1868

Tropocalymmides Lacordaire, 1868: 408 [stem: *Tropocalymmat-*]. Type genus: *Tropocalymma* J. Thomson, 1864. Comment: original vernacular name available (Art. 11.7.2): first used in latinized form and generally accepted as in Aurivillius (1912: 150, as Tropocalymmatini); incorrect original stem formation, not in prevailing usage.

Tribe Typhocesini Lacordaire, 1868

Typhocésides Lacordaire, 1868: 539 [stem: *Typhoces-*]. Type genus: *Typhocesis* Pascoe, 1863. Comment: original vernacular name available (Art. 11.7.2): first used in latinized form and generally accepted as in Aurivillius (1912: 296, as Typhocesini).

Tribe Unxiini Napp, 2007

Unxiini Napp, 2007: 312 [stem: *Unxi-*]. Type genus: *Unxia* J. Thomson, 1861.

Tribe Uracanthini Blanchard, 1853

*Uracantitas Blanchard, 1851b: 475 [stem: *Uracanth-*]. Type genus: *Uracanthus* Hope, 1833 [*Uracanthus* is an incorrect subsequent spelling of *Uracantha* Hope, 1833 (p. 64), first used by Hope (1834: 108), in prevailing usage and so deemed to be the correct original spelling (Art. 33.3.1)]. Comment: original vernacular name unavailable (Art. 11.7.2): subsequently used in latinized form but not generally attributed to Blanchard (1851b); incorrect original stem formation, not in prevailing usage.

Uracanthitae Blanchard, 1853: 264 [stem: *Uracanth-*]. Type genus: *Uracanthus* Hope, 1833 [*Uracanthus* is an incorrect subsequent spelling of *Uracantha* Hope, 1833 (p. 64), first used by Hope (1834: 108), in prevailing usage and so deemed to be the correct original spelling (Art. 33.3.1)].

Rhinophthalmitae J. Thomson, 1861: 152 [stem: *Rhinophthalm-*]. Type genus: *Rhinophthalmus* J. Thomson, 1861.

Tribe Vesperellini Sama, 2008

Vesperellini Sama, 2008: 227 [stem: *Vesperell-*]. Type genus: *Vesperella* Dayrem, 1933.

Tribe Xystrocerini Blanchard, 1845

Xystrocérites Blanchard, 1845b: 147 [stem: *Xystrocer-*]. Type genus: *Xystrocera* Audinet-Serville, 1834. Comment: original vernacular name available (Art. 11.7.2): first used in latinized form by J. Thomson (1861: 249, as

Xystroceritae), generally accepted as in Martins and Carvalho (1984: 214, as Xystrocerini).

Subfamily Lamiinae Latreille, 1825

Lamiariae Latreille, 1825: 401 [stem: *Lami-*]. Type genus: *Lamia* Fabricius, 1775.

Tribe Acanthocinini Blanchard, 1845

*Aedilaires Mulsant, 1839: 142 [stem: *Aedil-*]. Type genus: *Aedilis* Audinet-Serville, 1835. Comment: original vernacular name unavailable (Art. 11.7.2): not subsequently latinized.

Acanthocinites Blanchard, 1845b: 154 [stem: *Acanthocin-*]. Type genus: *Acanthocinus* Dejean, 1821. Comment: original vernacular name available (Art. 11.7.2): first used in latinized form by Blanchard (1853: 289, as Acanthocinitae), generally accepted as in Monné (1995a: 1, as Acanthocinini).

Trypanidiitae J. Thomson, 1860a: 7 [stem: *Trypanidi-*]. Type genus: *Trypanidius* Blanchard, 1846.

Dectitae J. Thomson, 1860a: 127 [stem: *Dect-*]. Type genus: *Dectes* J. L. LeConte, 1852.

*Astynomaires Mulsant, 1863b: 286 [stem: *Astynom-*]. Type genus: *Astynomus* Dejean, 1835. Comment: original vernacular name unavailable (Art. 11.7.2): subsequently used in latinized form but not generally attributed to Mulsant (1863b).

Lagocheirinae H. W. Bates, 1863: 100 [stem: *Lagocheir-*]. Type genus: *Lagocheirus* Dejean, 1835.

Liopi J. L. LeConte, 1873: 338 [stem: *Leiopod-*]. Type genus: *Leiopus* Audinet-Serville, 1835 [as *Liopus*, unjustified emendation of type genus name by Agassiz (1846b: 204), not in prevailing usage]. Comment: incorrect original stem formation, not in prevailing usage.

Graphisurini Leng, 1920: 283 [stem: *Graphisur-*]. Type genus: *Graphisurus* Kirby, 1837.

Astynomini Portevin, 1927: 39 [stem: *Astynom-*]. Type genus: *Astynomus* Dejean, 1835.

Tribe Acanthoderini Thomson, 1860

Acanthoderitae J. Thomson, 1860a: 5 [stem: *Acanthoder-*]. Type genus: *Acanthoderes* Audinet-Serville, 1835. Comment: First Reviser (Acanthoderini J. Thomson, 1860 vs Dryoctenini J. Thomson, 1860 vs Oreoderini J. Thomson, 1860) not determined, current usage maintained.

Dryoctenitae J. Thomson, 1860a: 29 [stem: *Dryocten-*]. Type genus: *Dryoctenes* Audinet-Serville, 1835.

Oreoderitae J. Thomson, 1860a: 29 [stem: *Oreoder-*]. Type genus: *Oreodera* Audinet-Serville, 1835.

Hoplosiae J. L. LeConte and G. H. Horn, 1883: 326 [stem: *Oplosi-*]. Type genus: *Oplosia* Mulsant, 1863 [as *Hoplosia*, unjustified emendation of type genus name by Fairmaire (1864), not in prevailing usage]. Comment: incorrect original stem formation, not in prevailing usage.

Tribe Acmocerini Thomson, 1864

Acmoceritae J. Thomson, 1864: 57 [stem: *Acmocer-*]. Type genus: *Acmocera* Dejean, 1835.

Tribe Acridocephalini Dillon and Dillon, 1959

Acridocephalidi E. S. Dillon and L. S. Dillon, 1959a: 49 [stem: *Acridocephal-*]. Type genus: *Acridocephala* Chevrolat, 1855.

Tribe Acrocinini Swainson, 1840

Acrocininae Swainson, 1840: 287 [stem: *Acrocin-*]. Type genus: *Acrocinus* Illiger, 1806. Comment: this family-group name was incorrectly credited to Swainson and Shuckard (1840) by Bousquet et al. (2009: 24).

Tribe Aderpasini Breuning and Teocchi, 1978

Aderpasini Breuning and Teocchi, 1978: 142 [stem: *Aderpas-*]. Type genus: *Aderpas* J. Thomson, 1864.

Tribe Aerenicini Lacordaire, 1872

Aerénicides Lacordaire, 1872: 897 [stem: *Aerenic-*]. Type genus: *Aerenica* Dejean, 1835. Comment: original vernacular name available (Art. 11.7.2): first used in latinized form by Harold (1872: 248, as Aerenicinae), generally accepted as in Aurivillius (1923: 596, as Aerenicini).

Tribe Agapanthiini Mulsant, 1839

Agapanthaires Mulsant, 1839: 172 [stem: *Agapanthi-*]. Type genus: *Agapanthia* Audinet-Serville, 1835. Comment: original vernacular name available (Art. 11.7.2): first used in latinized form by H. W. Bates (1884: 255, as Agapanthinae [incorrect stem formation]), generally accepted as in Aurivillius (1923: 458, as Agapanthiini); incorrect original stem formation, not in prevailing usage.

Hippopsitae J. Thomson, 1860a: 123 [stem: *Hippopse-*]. Type genus: *Hippopsis* Lepeletier and Audinet-Serville, 1828. Comment: incorrect original stem formation, not in prevailing usage.

Nemotragitae J. Thomson, 1864: 93 [stem: *Nemotrag-*]. Type genus: *Nemotragus* Westwood, 1843.

Anauxesitae J. Thomson, 1864: 94 [stem: *Anauxese-*]. Type genus: *Anauxesis* J. Thomson, 1857. Comment: incorrect original stem formation, not in prevailing usage.

Aprosopitae J. Thomson, 1864: 95 [stem: *Aprosop-*]. Type genus: *Aprosopus* Guérin-Méneville, 1844.

Aegoprepinae Pascoe, 1871: 277 [stem: *Aegoprep-*]. Type genus: *Aegoprepes* Pascoe, 1871.

Pachypézides Lacordaire, 1872: 691 [stem: *Pachypez-*]. Type genus: *Pachypeza* Audinet-Serville, 1835. Comment: original vernacular name available (Art. 11.7.2): first used in latinized form and generally accepted as in L. S. Dillon and E. S. Dillon (1945: 12, as Pachypezini).

Spalacopsides Lacordaire, 1872: 701 [stem: *Spalacopse-*]. Type genus: *Spalacopsis* Newman, 1842. Comment: original vernacular name available (Art. 11.7.2): first used in latinized form by Gahan (1890: 325, as Spalacopsinae), generally accepted as in Aurivillius (1923: 360, as Spalacopsini); incorrect original stem formation, not in prevailing usage.

Didymonychini Aurivillius, 1922b: 31 [stem: *Didymonych-*]. Type genus: *Didymonycha* Aurivillius, 1922 [syn. of *Amillarus* J. Thomson, 1857].

*Amillarinen Aurivillius, 1926a: 22 [stem: *Amillar-*]. Type genus: *Amillarus* J. Thomson, 1857. Comment: original vernacular name unavailable (Art. 11.7.2): proposed after 1899.

Hippopsiconini L. S. Dillon and E. S. Dillon, 1945: 11 [stem: *Hippopsicon-*]. Type genus: *Hippopsicon* J. Thomson, 1858.

Tribe Amphoecini Breuning, 1951

Amphoecini Breuning, 1951a: 5 [stem: *Amphoec-*]. Type genus: *Amphoecus* Montrouzier, 1861.

Tribe Ancitini Aurivillius, 1917

Ancitini Aurivillius, 1917: 28 [stem: *Ancit-*]. Type genus: *Ancita* J. Thomson, 1864.

Tribe Ancylonotini Lacordaire, 1869

Ancylonotides Lacordaire, 1869: 391 [stem: *Ancylonot-*]. Type genus: *Ancylonotus* Dejean, 1835. Comment: original vernacular name available (Art. 11.7.2): first used in latinized form by Pascoe (1871: 268, as Ancylonotinae), generally accepted as in Aurivillius (1922a: 152, as Ancylonotini).

Tribe Anisocerini Thomson, 1860

Anisoceritae J. Thomson, 1860a: 31 [stem: *Anisocer-*]. Type genus: *Anisocerus* Lacordaire, 1830.

Onychoceritae J. Thomson, 1864: 19 [stem: *Onychocer-*]. Type genus: *Onychocerus* Lacordaire, 1830.

Platysternides Lacordaire, 1872: 729 [stem: *Platystern-*]. Type genus: *Platysternus* Dejean, 1835. Comment: original vernacular name available (Art. 11.7.2): first used in latinized form and generally accepted as in Aurivillius (1923: 371, as Platysternini); junior homonym of the turtle family Platysternidae Gray, 1869 (type genus *Platysternon* Gray, 1831) which is currently used as valid; this case is to be referred to the Commission to remove the homonymy (Art. 55.3.1).

Tribe Apomecynini Thomson, 1860

Apomecynitae J. Thomson, 1860a: 66 [stem: *Apomecyn-*]. Type genus: *Apomecyna* Dejean, 1821. Comment: name incorrectly spelled Apomecinitae on page 3 but correct spelling used on pages 42, 66 and 68 of the original publication.

Adétides Lacordaire, 1872: 592 [stem: *Adet-*]. Type genus: *Adetus* J. L. LeConte, 1852. Comment: original vernacular name available (Art. 11.7.2): first used in latinized form and generally accepted as in Aurivillius (1922a: 288, as Adetini).

*Agennopsides Lacordaire, 1872: 592 [stem: *Agennopse-*]. Type genus: *Agennopsis* J. Thomson, 1857 [syn. of *Adetus* J. L. LeConte, 1852]. Comment: original vernacular name unavailable (Art. 11.7.2): not subsequently latinized; incorrect original stem formation, not in prevailing usage.

Ptéricoptides Lacordaire, 1872: 601 [stem: *Ptericopt-*]. Type genus: *Ptericoptus* Lacordaire, 1830. Comment: original vernacular name available (Art. 11.7.2): first used in latinized form by Gestro (1876: 140, as Ptericoptini), generally accepted as in Aurivillius (1922a: 294, as Ptericoptini).

Ectatosiides Lacordaire, 1872: 708 [stem: *Ectatosi-*]. Type genus: *Ectatosia* Pascoe, 1857. Comment: original vernacular name available (Art. 11.7.2): first used in latinized form and generally accepted as in Aurivillius (1923: 363, as Ectatosiini).

Ischiolonchides Lacordaire, 1872: 709 [stem: *Ischiolonch-*]. Type genus: *Ischioloncha* J. Thomson, 1860. Comment: original vernacular name available (Art. 11.7.2): first used in latinized form and generally accepted as in Aurivillius (1923: 364, as Ischiolonchini).

Tribe Astathini Thomson, 1864

*Tétraophthalmites Blanchard, 1845b: 160 [stem: *Tetraophthalm-*]. Type genus: *Tetraophthalmus* Dejean, 1835. Comment: original vernacular name unavailable (Art. 11.7.2): subsequently used in latinized form, e.g., Sama (2010: 56, as Tetraophthalmini) but not generally accepted as valid.

Astathitae J. Thomson, 1864: 117 [stem: *Astath-*]. Type genus: *Astathes* Newman, 1842 [syn. of *Tetraophthalmus* Dejean, 1835]. Comment: published before October 1864; this family-group name was also used in the same year by Pascoe (1864 [before 3 October]: 8, as Asthateinae [incorrect stem formation]); Pascoe (1864: 81, 85) refers to the publication by J. Thomson (1864) which is further evidence that the publication by Thomson was published first; this tribe name was incorrectly credited to Pascoe (1864) by Bousquet et al. (2009: 26).

Tribe Batocerini Thomson, 1864

Batoceritae J. Thomson, 1864: 74 [stem: *Batocer-*]. Type genus: *Batocera* Dejean, 1835.

Tribe Calliini Thomson, 1864

Callitae J. Thomson, 1864: 123 [stem: *Calli-*]. Type genus: *Callia* Audinet-Serville, 1835. Comment: incorrect original stem formation, not in prevailing usage.

Gryllicides Lacordaire, 1872: 902 [stem: *Gryllic-*]. Type genus: *Gryllica* J. Thomson, 1860. Comment: original vernacular name available (Art. 11.7.2): first used in latinized form by H. W. Bates (1874a: 234, as Gryllicinae), generally accepted as in Aurivillius (1923: 604, as Gryllicini).

Tribe Ceroplesini Thomson, 1860

Ceroplesitae J. Thomson, 1860a: 95 [stem: *Ceroples-*]. Type genus: *Ceroplesis* Audinet-Serville, 1835 [placed on the Official List of Generic Names in Zoology (ICZN 1986c)].

Subtribe Ceroplesina Thomson, 1860

Ceroplesitae J. Thomson, 1860a: 95 [stem: *Ceroples-*]. Type genus: *Ceroplesis* Audinet-Serville, 1835 [placed on the Official List of Generic Names in Zoology (ICZN 1986c)].

Subtribe Crossotina Thomson, 1864

Crossotitae J. Thomson, 1864: 64 [stem: *Crossot-*]. Type genus: *Crossotus* Audinet-Serville, 1835. Comment: downgraded to subtribe by Sama (2008: 235).

Écyroschémides Lacordaire, 1872: 503 [stem: *Ecyroschemat-*]. Type genus: *Ecyroschema* J. Thomson, 1864. Comment: original vernacular name available (Art. 11.7.2): first used in latinized form and generally accepted as in Aurivillius (1922a: 241, as Ecyroschemini); incorrect original stem formation, not in prevailing usage.

Hécyridides Lacordaire, 1872: 517 [stem: *Hecyrid-*]. Type genus: *Hecyrida* J. Thomson, 1860 [syn. of *Hecyra* J. Thomson, 1857]. Comment: original vernacular name available (Art. 11.7.2): first used in latinized form by Kolbe (1897: 318, as Hecyridinae), generally accepted as in Aurivillius (1922a: 243, as Hecyrini).

Corynofreinae Aurivillius, 1911a: 37 [stem: *Corynofre-*]. Type genus: *Corynofrea* Aurivillius, 1911.

Tribe Cloniocerini Lacordaire, 1872

Cloniocérides Lacordaire, 1872: 590 [stem: *Cloniocer-*]. Type genus: *Cloniocerus* Dejean, 1835. Comment: original vernacular name available (Art. 11.7.2):

first used in latinized form and generally accepted as in Aurivillius (1922a: 287, as CLONIOCERINI).

Tribe COLOBOTHEINI Thomson, 1860

COLOBOTHEITAE J. Thomson, 1860a: 18 [stem: *Colobothe-*]. Type genus: *Colobothea* Lepeletier and Audinet-Serville, 1825 [*Colobothea* is an incorrect subsequent spelling of the original spelling *Colobotea*, in prevailing usage and so deemed to be the correct original spelling (Art. 33.3.1)].

Tribe COMPSOSOMATINI Thomson, 1857

COMPSOSOMITES J. Thomson, 1857b: 70 [stem: *Compsosomat-*]. Type genus: *Compsosoma* Lacordaire, 1830. Comment: original vernacular name available (Art. 11.7.2): first used in latinized form by J. Thomson (1860a: 34, as COMPSOSOMITAE), generally accepted as in Aurivillius (1923: 336, as COMPSOSOMATINI); incorrect original stem formation, not in prevailing usage.

AERENEITES J. Thomson, 1868: 92 [stem: *Aerene-*]. Type genus: *Aerenea* J. Thomson, 1857 [*Aerenea* is an incorrect subsequent spelling of the original spelling *Aerenaea*, introduced by J. Thomson (1860a: 34), in prevailing usage and so deemed to be the correct original spelling (Art. 33.3.1)]. Comment: original vernacular name available (Art. 11.7.2): first used in latinized form by Gounelle (1908: 7, as AERENEINAE), generally accepted as in Aurivillius (1923: 338, as AERENEINI).

Tribe CYRTININI Thomson, 1864

CYRTINITAE J. Thomson, 1864: 41 [stem: *Cyrtin-*]. Type genus: *Cyrtinus* J. L. LeConte, 1852.

ACANTHOMEROSTERNOPLONINI Tippmann, 1956: 10 [stem: *Acanthomerosternopl-*]. Type genus: *Acanthomerosternoplon* Tippmann, 1955 [syn. of *Omosarotes* Pascoe, 1860]. Comment: incorrect original stem formation, not in prevailing usage.

SCOPADINI Villiers, 1980: 587 [stem: *Scopad-*]. Type genus: *Scopadus* Pascoe, 1857.

Tribe DESMIPHORINI Thomson, 1860

DESMIPHORITAE J. Thomson, 1860a: 74 [stem: *Desmiphor-*]. Type genus: *Desmiphora* Audinet-Serville, 1835.

*ANAESTHÉTITES Fairmaire, 1864: 166 [stem: *Anaesthet-*]. Type genus: *Anaesthetis* Dejean, 1835. Comment: original vernacular name unavailable (Art. 11.7.2): not subsequently latinized.

MÉTONIDES Lacordaire, 1869: 387 [stem: *Metont-*]. Type genus: *Meton* Pascoe, 1859. Comment: original vernacular name available (Art. 11.7.2): first used in latinized form and generally accepted as in Aurivillius (1922a: 150, as METONINI); incorrect original stem formation, not in prevailing usage.

HEBESECINAE Pascoe, 1871: 277 [stem: *Hebesecid-*]. Type genus: *Hebesecis* Pascoe, 1865. Comment: incorrect original stem formation, not in prevailing usage.

Amymomides Lacordaire, 1872: 468 [stem: *Amymom-*]. Type genus: *Amymoma* Pascoe, 1866 [preoccupied genus name, not *Amymoma* Latreille, 1797 [Crustacea]; syn. of *Neoamymoma* Marinoni, 1977]. Comment: original vernacular name available (Art. 11.7.2): first used in latinized form and generally accepted as in Aurivillius (1922a: 218, as Amymomini); permanently invalid (Art. 39): based on preoccupied type genus.

Crinotarsides Lacordaire, 1872: 475 [stem: *Crinotars-*]. Type genus: *Crinotarsus* Blanchard, 1853. Comment: original vernacular name available (Art. 11.7.2): first used in latinized form and generally accepted as in Aurivillius (1922a: 229, as Crinotarsini).

Épicastides Lacordaire, 1872: 490 [stem: *Epicast-*]. Type genus: *Epicasta* J. Thomson, 1864. Comment: original vernacular name available (Art. 11.7.2): first used in latinized form and generally accepted as in Aurivillius (1922a: 237, as Epicastini).

Apodasyides Lacordaire, 1872: 623 [stem: *Apodasy-*]. Type genus: *Apodasya* Pascoe, 1863 [the genus name *Chaetosoma* Chevrolat, 1843: 367 is a senior objective synonym of *Apodasya* Pascoe and should be used as valid based on the Principle of Priority; *Chaetosoma* Chevrolat, 1843 is also a senior homonym of the well-established *Chaetosoma* Westwood, 1851, which is the type genus of Chaetosomatidae Crowson, 1952 used as valid in Cucujoidea; because the discovery of the available name *Chaetosoma* Chevrolat, 1843 causes problems for well-established names in Cucujoidea and Cerambycidae, an application was recently submitted by Bousquet and Bouchard (2010) to suppress it for the Principles of Priority and Homonymy (see Appendix 6)]. Comment: original vernacular name available (Art. 11.7.2): first used in latinized form by Kolbe (1897: 321, as Apodasinae [incorrect stem formation]), generally accepted as in Aurivillius (1922a: 305, as Apodasyini).

Nédinides Lacordaire, 1872: 635 [stem: *Nedin-*]. Type genus: *Nedine* J. Thomson, 1864. Comment: original vernacular name available (Art. 11.7.2): first used in latinized form and generally accepted as in Aurivillius (1922a: 317, as Nedinini).

Estolides Lacordaire, 1872: 636 [stem: *Estol-*]. Type genus: *Estola* Fairmaire and Germain, 1859. Comment: original vernacular name available (Art. 11.7.2): first used in latinized form by J. L. LeConte (1873: 340, as Estolae), generally accepted as in Aurivillius (1922a: 317, as Estolini).

Psenocerini J. L. LeConte, 1873: 333 [stem: *Psenocer-*]. Type genus: *Psenocerus* J. L. LeConte, 1852.

Eupogonii J. L. LeConte, 1873: 342 [stem: *Eupogoni-*]. Type genus: *Eupogonius* J. L. LeConte, 1852.

Essisini Aurivillius, 1917: 44 [stem: *Essis-*]. Type genus: *Essisus* Pascoe, 1866.

Velorini Aurivillius, 1917: 32 [stem: *Velor-*]. Type genus: *Velora* J. Thomson, 1864.

Tribe Dorcadionini Swainson, 1840

Dorcadioninae Swainson, 1840: 290 [stem: *Dorcadi-*]. Type genus: *Dorcadion* Dalman, 1817. Comment: this family-group name was incorrectly credited to Swainson and Shuckard (1840) by Bousquet et al. (2009: 29); current spelling maintained (Art. 29.5): incorrect original stem formation in prevailing usage (should be *Dorcadi-*; see Vives and Alonso-Zarazaga 2000: 659); tribe placed in synonymy with Lamiini by Sama (2008: 233).

Dorcadodiidae Gistel, 1856a: 376 [stem: *Dorcadodi-*]. Type genus: *Dorcadodium* Gistel, 1856 [this name is a senior synonym of *Carinatodorcadion* Breuning, 1943 (see Vives and Alonso-Zarazaga 2000: 659 for type species designation of *Dorcadodium* Gistel); we could not find enough references to treat *Dorcadodium* Gistel, 1856 as a *nomen oblitum*, however we believe that an application should be submitted to the Commission to preserve usage of *Carinatodorcadion* Breuning, 1943].

Tribe Dorcaschematini Thomson, 1860

Dorcaschemitae J. Thomson, 1860a: 107 [stem: *Dorcaschemat-*]. Type genus: *Dorcaschema* Haldeman, 1847. Comment: incorrect original stem formation, not in prevailing usage; name spelled Dorchaschemitae on page 104 but spelled Dorcaschemitae on pages 4 and 107 of the original publication.

Protonarthronitae J. Thomson, 1864: 57 [stem: *Protonarthr-*]. Type genus: *Protonarthron* J. Thomson, 1858. Comment: incorrect original stem formation, not in prevailing usage.

Tribe Elytracanthinini Bousquet, 2009

Elytracanthinae Lane, 1955: 281 [stem: *Elytracanth-*]. Type genus: *Elytracantha* Lane, 1955 [preoccupied type genus, not *Elytracantha* Kleine, 1915 [Coleoptera: Brentidae]; syn. of *Elytracanthina* Monné, 2005]. Comment: permanently invalid (Art. 39): based on preoccupied type genus.

Elytracanthinini Bousquet, 2009: 30 [stem: *Elytracanthin-*]. Type genus: *Elytracanthina* Monné, 2005. Comment: replacement name for Elytracanthinae Lane, 1955 because of the homonymy of the type genus.

Tribe Enicodini Thomson, 1864

Enicoditae J. Thomson, 1864: 36 [stem: *Enicod-*]. Type genus: *Enicodes* Gray, 1832. Comment: First Reviser (Enicodini J. Thomson, 1864 vs Nemaschematini J. Thomson, 1864 vs Leptonotini J. Thomson, 1864) not determined, current usage maintained.

Nemaschemitae J. Thomson, 1864: 36 [stem: *Nemaschemat-*]. Type genus: *Nemaschema* J. Thomson, 1861. Comment: incorrect original stem formation, not in prevailing usage.

Leptonotitae J. Thomson, 1864: 36 [stem: *Leptonot-*]. Type genus: *Leptonota* J. Thomson, 1861.

*Énotides Lacordaire, 1872: 487 [stem: *Enotet-*]. Type genus: *Enotes* J. Thomson, 1864. Comment: original vernacular name unavailable (Art. 11.7.2): not subsequently latinized; incorrect original stem formation, not in prevailing usage.

Tribe Eupromerini Galileo and Martins, 1995

Eupromerini Galileo and Martins, 1995: 132 [stem: *Eupromer-*]. Type genus: *Eupromera* Westwood, 1845.

Tribe Forsteriini Tippmann, 1960

Hebestolitae J. Thomson, 1864: 107 [stem: *Hebestol-*]. Type genus: *Hebestola* Blanchard, 1851 [preoccupied genus name, not *Hebestola* Haldeman, 1847 [Coleoptera: Cerambycidae: Lamiinae: Monochamini]; syn. of *Neohebestola* Marinoni, 1977]. Comment: permanently invalid (Art. 39): based on preoccupied type genus.

Forsteriini Tippmann, 1960: 210 [stem: *Forsteri-*]. Type genus: *Forsteria* Tippmann, 1960 [syn. of *Falsamblesthis* Breuning, 1959].

Falsamblesthiini Gilmour, 1961: 131 [stem: *Falsamblesth-*]. Type genus: *Falsamblesthis* Breuning, 1959. Comment: name proposed to replace Forsteriini Tippmann, 1960 because of the synonymy of the type genus; however, because the name was proposed after 1960, it cannot be maintained (Art. 40.2); incorrect original stem formation, not in prevailing usage.

Tribe Gnomini Thomson, 1860

Gnomitae J. Thomson, 1860a: 105 [stem: *Gnom-*]. Type genus: *Gnoma* Fabricius, 1801.

Tribe Gyaritini Breuning, 1950

Gyaritini Breuning, 1950c: 27 [stem: *Gyarit-*]. Type genus: *Gyaritus* Pascoe, 1858.

Tribe Heliolini Breuning, 1951

Heliolini Breuning, 1951a: 8 [stem: *Heliol-*]. Type genus: *Heliolus* Fauvel, 1907.

Tribe Hemilophini Thomson, 1868 *nomen protectum*

Amphionychitae J. Thomson, 1860a: 63 [stem: *Amphionych-*]. Type genus: *Amphionycha* Dejean, 1835 [syn. of *Adesmus* Lepeletier and Audinet-Serville, 1825]. Comment: *nomen oblitum* (see Bousquet et al. 2009: 31).

Hemilophitae J. Thomson, 1868: 189 [stem: *Hemiloph-*]. Type genus: *Hemilophus* Audinet-Serville, 1835. Comment: *nomen protectum* (see Bousquet et al. 2009: 31).

Itesini Lepesme, 1943: 137 [stem: *It-*]. Type genus: *Ites* C. O. Waterhouse, 1880. Comment: incorrect original stem formation, not in prevailing usage; the tribe Itini Reitter, 1913 (type genus *Ita* Tournier, 1878), which is based on the same stem, is currently used as valid in Coleoptera: Curculionidae.

Tribe HOMONOEINI Thomson, 1864

HOMONAEITAE J. Thomson, 1864: 35 [stem: *Homonoe-*]. Type genus: *Homonoea* Newman, 1842 [as *Homonaea*, incorrect subsequent spelling of type genus name, not in prevailing usage]. Comment: incorrect original stem formation, not in prevailing usage.

BUMÉTOPIDES Lacordaire, 1872: 477 [stem: *Bumetopi-*]. Type genus: *Bumetopia* Pascoe, 1858. Comment: original vernacular name available (Art. 11.7.2): first used in latinized form by Gestro (1876: 140, as BUMETOPINI [incorrect stem formation]), generally accepted as in Aurivillius (1922a: 231, as BUMETOPINI [incorrect stem formation]); incorrect original stem formation, not in prevailing usage.

Tribe HYBORHABDINI Aurivillius, 1911

HYBORHABDINAE Aurivillius, 1911b: 22 [stem: *Hyborhabd-*]. Type genus: *Hyborhabdus* Aurivillius, 1911.

Tribe LAMIINI Latreille, 1825

LAMIARIAE Latreille, 1825: 401 [stem: *Lami-*]. Type genus: *Lamia* Fabricius, 1775.

PACHYSTOLAEIDAE Gistel, 1848: [9] [stem: *Pachystol-*]. Type genus: *Pachystola* Dejean, 1835 [syn. of *Lamia* Fabricius, 1775]. Comment: incorrect original stem formation, not in prevailing usage.

PHRISSOMITAE J. Thomson, 1860a: 22 [stem: *Phrissomat-*]. Type genus: *Phrissoma* Dejean, 1835. Comment: incorrect original stem formation, not in prevailing usage; J. Thomson originally spelled the family-group name PHRYSSOMITAE on page 25, however the spelling PHRISSOMITAE was used on pages 2 and 22 in the same publication; tribe placed in synonymy with LAMIINI by Sama (2008: 233).

MORIMITAE J. Thomson, 1864: 77 [stem: *Morim-*]. Type genus: *Morimus* Brullé, 1832.

POTEMNEMINI Aurivillius, 1922a: 117 [stem: *Potemnem-*]. Type genus: *Potemnemus* J. Thomson, 1864.

Tribe LATICRANIINI Lane, 1959

LATICRANIINAE Lane, 1959: 312 [stem: *Laticrani-*]. Type genus: *Laticranium* Lane, 1959.

Tribe MAUESIINI Lane, 1956

MAUESINAE Lane, 1956: 19 [stem: *Mauesi-*]. Type genus: *Mauesia* Lane, 1956. Comment: incorrect original stem formation, not in prevailing usage.

Tribe MEGABASINI Thomson, 1860

MEGABASITAE J. Thomson, 1860a: 30 [stem: *Megabas-*]. Type genus: *Megabasis* Audinet-Serville, 1835. Comment: current spelling maintained (Art. 29.5): incorrect original stem formation in prevailing usage (should be *Megabase-*).

Tribe Mesosini Mulsant, 1839

Mésosaires Mulsant, 1839: 165 [stem: *Mesos-*]. Type genus: *Mesosa* Latreille, 1829. Comment: original vernacular name available (Art. 11.7.2): first used in latinized form by Gistel (1848: [9], as Mesosaeidae [incorrect stem formation]), generally accepted as in López-Pérez (2006: 59, as Mesosini).

Tribe Microcymaturini Breuning and Teocchi, 1985

Microcymaturini Breuning and Teocchi, 1985: 155 [stem: *Microcymatur-*]. Type genus: *Microcymatura* Breuning, 1950.

Tribe Moneilemini Thomson, 1864

Moneilemitae J. Thomson, 1864: 43 [stem: *Moneilem-*]. Type genus: *Moneilema* Say, 1824. Comment: current spelling maintained (Art. 29.5): incorrect original stem formation in prevailing usage (should be *Moneilemat-*).

Tribe Monochamini Gistel, 1848

Monohammidae Gistel, 1848: [9] [stem: *Monocham-*]. Type genus: *Monochamus* Dejean, 1821 [as *Monohammus*, unjustified emendation of type genus name by Dejean (1835: 340), not in prevailing usage]. Comment: incorrect original stem formation, not in prevailing usage.

Taeniotitae J. Thomson, 1864: 76 [stem: *Taeniot-*]. Type genus: *Taeniotes* Audinet-Serville, 1835.

Agnitae J. Thomson, 1864: 83 [stem: *Agni-*]. Type genus: *Agnia* Newman, 1842. Comment: incorrect original stem formation, not in prevailing usage.

Geranitae J. Thomson, 1864: 93 [stem: *Gerani-*]. Type genus: *Gerania* Audinet-Serville, 1835. Comment: incorrect original stem formation, not in prevailing usage.

Ptychodes J. L. LeConte, 1873: 335 [stem: *Ptychod-*]. Type genus: *Ptychodes* Audinet-Serville, 1835.

Goes J. L. LeConte, 1873: 335 [stem: *Goet-*]. Type genus: *Goes* J. L. LeConte, 1852. Comment: incorrect original stem formation, not in prevailing usage.

Docohammidi E. S. Dillon and L. S. Dillon, 1959b: 7 [stem: *Docohamm-*]. Type genus: *Docohammus* Aurivillius, 1908.

Tribe Morimonellini Lobanov, Danilevsky and Murzin, 1981

Morimonellini Lobanov et al., 1981: 790 [stem: *Morimonell-*]. Type genus: *Morimonella* Podaný, 1979. Comment: description by indication (distinguishing characters given in Podany (1979: 43, as Oligorchini)).

Tribe Morimopsini Lacordaire, 1869

Morimopsides Lacordaire, 1869: 289 [stem: *Morimops-*]. Type genus: *Morimopsis* J. Thomson, 1857. Comment: original vernacular name available (Art. 11.7.2): first used in latinized form and generally accepted as in Aurivillius (1922a: 64, as Morimopsini); current spelling maintained (Art. 29.5): incorrect original stem formation in prevailing usage (should be *Morimopse-*).

Tribe Nyctimeniini Gressitt, 1951

Nyctimenitae J. Thomson, 1864: 94 [stem: *Nyctimen-*]. Type genus: *Nyctimene* J. Thomson, 1857 [preoccupied genus name, not *Nyctimene* Borkenhausen, 1797 [Mammalia]; syn. of *Nyctimenius* Gressitt, 1951]. Comment: permanently invalid (Art. 39): based on preoccupied type genus.

Nyctimeniini Gressitt, 1951: 629 [stem: *Nyctimeni-*]. Type genus: *Nyctimenius* Gressitt, 1951. Comment: replacement name for Nyctimenini J. Thomson, 1864 because of the homonymy of the type genus.

Tribe Obereini Thomson, 1864

Obereitae J. Thomson, 1864: 117 [stem: *Obere-*]. Type genus: *Oberea* Dejean, 1835. Comment: published before October 1864; this family-group name was also used in the same year by Pascoe (1864 [before 3 October]: 8, as Obereinae); Pascoe (1864: 81, 85) refers to the publication by J. Thomson (1864) which is further evidence that the publication by Thomson was published first; this tribe name was incorrectly credited to Pascoe (1864) by Bousquet et al. (2009: 33).

Tribe Oculariini Breuning, 1950

Oculariini Breuning, 1950a: 263 [stem: *Oculari-*]. Type genus: *Ocularia* Jordan, 1894.

Tribe Onciderini Thomson, 1860

Oncideritae J. Thomson, 1860a: 38 [stem: *Oncider-*]. Type genus: *Oncideres* Lacordaire, 1830 [*Oncideres* is an incorrect subsequent spelling of *Oncyderes* Lacordaire, 1830, introduced by Audinet-Serville (1835: 67), in prevailing usage and attributed to Lacordaire (1830b), e.g., Monné (2005b: 571), and so deemed to be the correct original spelling (Art. 33.3.1)]. Comment: First Reviser (Onciderini J. Thomson, 1860 vs Hypsiomatini J. Thomson, 1860) not determined, current usage maintained.

Hypsiomitae J. Thomson, 1860a: 109 [stem: *Hypsiomat-*]. Type genus: *Hypsioma* Audinet-Serville, 1835. Comment: incorrect original stem formation, not in prevailing usage.

Hypselominae Pascoe, 1864: 7 [stem: *Hypselom-*]. Type genus: *Hypselomus* Perty, 1832.

Tribe Oncideropsidini Aurivillius, 1922

Oncideropsidini Aurivillius, 1922c: 165 [stem: *Oncideropsid-*]. Type genus: *Oncideropsis* Aurivillius, 1922. Comment: current spelling maintained (Art. 29.5): incorrect original stem formation in prevailing usage (should be *Oncideropse-*).

Tribe Onocephalini Thomson, 1860

Onocephalitae J. Thomson, 1860a: 120 [stem: *Onocephal-*]. Type genus: *Onocephala* J. Thomson, 1857.

Tribe Onychogleneini Aurivillius, 1923

Onychogleneini Aurivillius, 1923: 513 [stem: *Onychoglene-*]. Type genus: *Onychoglenea* Aurivillius, 1922.

Tribe Parmenini Mulsant, 1839

Parménaires Mulsant, 1839: 118 [stem: *Parmen-*]. Type genus: *Parmena* Dejean, 1821. Comment: original vernacular name available (Art. 11.7.2): first used in latinized form by J. Thomson (1864: 38, as Parmenitae), generally accepted as in Villiers (1978: 449, as Parmenini).

Hexathricitae J. Thomson, 1864: 38 [stem: *Hexatrich-*]. Type genus: *Hexatricha* A. White, 1846 [as *Hexathrica*, incorrect subsequent spelling of type genus name, not in prevailing usage]. Comment: the spelling Hexarthricitae (page 38) was corrected to Hexathricitae on page 339 and in the errata (page 483) of the same work; the corrected spelling represents an incorrect original stem formation, not in prevailing usage.

*Dorcadidides Lacordaire, 1869: 257 [stem: *Dorcadid-*]. Type genus: *Dorcadida* A. White, 1846. Comment: original vernacular name unavailable (Art. 11.7.2): subsequently used in latinized form, e.g., Heyne and Taschenberg (1907: 241, as Dorcadidini), but not generally accepted as valid.

Tribe Petrognathini Blanchard, 1845

Pétrognathites Blanchard, 1845b: 160 [stem: *Petrognath-*]. Type genus: *Petrognatha* Leach, 1819. Comment: original vernacular name available (Art. 11.7.2): first used in latinized form by Kolbe (1897: 316, as Petrognathinae), generally accepted as in Aurivillius (1922a: 205, as Petrognathini).

*Omacanthides Lacordaire, 1872: 447 [stem: *Omacanth-*]. Type genus: *Omacantha* Audinet-Serville, 1835 [syn. of *Petrognatha* Leach, 1819]. Comment: original vernacular name unavailable (Art. 11.7.2): subsequently used in latinized form, e.g., Heyne and Taschenberg (1907: 243, as Omacanthini), but not generally accepted as valid; Omacanthidae was used as valid by Ienistea (1986: 30) but it was not attributed to Lacordaire (1872); Ienistea's name is also unavailable, it was proposed after 1930 without description or bibliographic reference to such a description (Art. 13.1).

Tribe Phacellini Lacordaire, 1872

Phacellides Lacordaire, 1872: 664 [stem: *Phacell-*]. Type genus: *Phacellus* Dejean, 1835. Comment: original vernacular name available (Art. 11.7.2): first used in latinized form by Gounelle (1908: 7, as Phacellinae), generally accepted as in Aurivillius (1923: 339, as Phacellini).

Tribe PHANTASINI Kolbe, 1897

*PHANTASIDES Lacordaire, 1869: 285 [stem: *Phantas-*]. Type genus: *Phantasis* J. Thomson, 1860. Comment: original vernacular name unavailable (Art. 11.7.2): subsequently used in latinized form but not generally attributed to Lacordaire (1869).

PHANTASINAE Kolbe, 1897: 306 [stem: *Phantas-*]. Type genus: *Phantasis* J. Thomson, 1860. Comment: current spelling maintained (Art. 29.5): incorrect original stem formation in prevailing usage (should be *Phantase-*).

PHANTASINI Hunt and Breuning, 1957: 51 [stem: *Phantas-*]. Type genus: *Phantasis* J. Thomson, 1860. Comment: family group name proposed as new without reference to PHANTASINAE Kolbe, 1897; current spelling maintained (Art. 29.5): incorrect original stem formation in prevailing usage (should be *Phantase-*).

Tribe PHRYNETINI Thomson, 1864

PHRYNETITAE J. Thomson, 1864: 71 [stem: *Phrynet-*]. Type genus: *Phryneta* Dejean, 1835.

Tribe PHYMASTERNINI Teocchi, 1989

PHYMASTERNINI Teocchi, 1989: 4 [stem: *Phymastern-*]. Type genus: *Phymasterna* Laporte, 1840 [for comments on problems with the authorship and type species of this type genus see Bousquet et al. (2009: 35)].

Tribe PHYTOECIINI Mulsant, 1839

PHYTOECIAIRES Mulsant, 1839: 191 [stem: *Phytoeci-*]. Type genus: *Phytoecia* Dejean, 1835. Comment: original vernacular name available (Art. 11.7.2): first used in latinized form by Pascoe (1864: 8, as PHYTOECIINAE), generally accepted as in Villiers (1978: 521, as PHYTOECIINI).

Tribe POGONOCHERINI Mulsant, 1839

POGONOCHÉRAIRES Mulsant, 1839: 151 [stem: *Pogonocher-*]. Type genus: *Pogonocherus* Dejean, 1821. Comment: original vernacular name available (Art. 11.7.2): first used in latinized form by J. L. LeConte (1873: 340, as POGONOCHERINI), generally accepted as in Villiers (1978: 465, as POGONOCHERINI).

*EXOCENTRITES Fairmaire, 1864: 157 [stem: *Exocentr-*]. Type genus: *Exocentrus* Dejean, 1835. Comment: published early September 1864; original vernacular name unavailable (Art. 11.7.2): subsequently used in latinized form but not generally attributed to Fairmaire (1864).

EXOCENTRINAE Pascoe, 1864: 7 [stem: *Exocentr-*]. Type genus: *Exocentrus* Dejean, 1835. Comment: published 3 October 1864.

ZAPLOI J. L. LeConte and G. H. Horn, 1883: 327 [stem: *Zaplo-*]. Type genus: *Zaplous* J. L. LeConte, 1878.

Tribe Polyrhaphidini Thomson, 1860

Polyrhaphitae J. Thomson, 1860a: 30 [stem: *Polyrhaphid-*]. Type genus: *Polyrhaphis* Audinet-Serville, 1835. Comment: incorrect original stem formation, not in prevailing usage.

Tribe Pretiliini Martins and Galileo, 1990

Pretiliini Martins and Galileo, 1990: 705, in key [stem: *Pretili-*]. Type genus: *Pretilia* H. W. Bates, 1866.

Tribe Proctocerini Aurivillius, 1922

*Cliniides Lacordaire, 1872: 424 [stem: *Clini-*]. Type genus: *Clinia* J. Thomson, 1857 [syn. of *Proctocera* Chevrolat, 1855]. Comment: original vernacular name unavailable (Art. 11.7.2): subsequently used in latinized form, e.g., Breuning (1950d: 411, as Cliniini), but not generally accepted as valid; Cliniidae was used as valid by Ienistea (1986: 30) but it was not attributed to Lacordaire (1872); Ienistea's name is also unavailable, it was proposed after 1930 without description or bibliographic reference to such a description (Art. 13.1).

Proctocerini Aurivillius, 1922a: 182 [stem: *Proctocer-*]. Type genus: *Proctocera* Chevrolat, 1855.

Tribe Prosopocerini Thomson, 1864

Prosopoceritae J. Thomson, 1864: 72 [stem: *Prosopocer-*]. Type genus: *Prosopocera* Blanchard, 1845.

Tribe Pteropliini Thomson, 1860

Pteropliitae J. Thomson, 1860a: 73 [stem: *Pteropli-*]. Type genus: *Pteroplius* Lacordaire, 1830 [as *Pteroplia*, incorrect subsequent spelling of type genus name, not in prevailing usage; *Pteroplius* is an incorrect subsequent spelling of *Pterhoplius* Lacordaire, 1830, introduced by Audinet-Serville (1835: 65), in prevailing usage and attributed to Lacordaire (1830b), e.g., Monné (2005b: 295), and so deemed to be the correct original spelling (Art. 33.3.1)].

Abrynitae J. Thomson, 1864: 44 [stem: *Abryn-*]. Type genus: *Abryna* Newman, 1842.

Protorhopalitae J. Thomson, 1864: 69 [stem: *Protorhopal-*]. Type genus: *Protorhopala* J. Thomson, 1860.

Niphoninae Pascoe, 1864: 56 [stem: *Niphon-*]. Type genus: *Niphona* Mulsant, 1839.

Ataxiides Lacordaire, 1872: 597 [stem: *Ataxi-*]. Type genus: *Ataxia* Haldeman, 1847. Comment: original vernacular name available (Art. 11.7.2): first used in latinized form by J. L. LeConte (1873: 344, as Ataxiini), generally accepted as in Aurivillius (1922a: 291, as Ataxiini).

Emphytoeciides Lacordaire, 1872: 713 [stem: *Emphytoeci-*]. Type genus: *Emphytoecia* Fairmaire and Germain, 1860. Comment: original vernacular name available (Art. 11.7.2): first used in latinized form and generally accepted as in Aurivillius (1923: 365, as EMPHYTOECIINI).

BAROEIDES Lacordaire, 1872: 439 [stem: *Barae-*]. Type genus: *Baraeus* J. Thomson, 1858 [as *Baroeus*, incorrect subsequent spelling of type genus name, not in prevailing usage]. Comment: also spelled BARÉIDES on page 414 of the same work; original vernacular name available (Art. 11.7.2): first used in latinized form and generally accepted as in Aurivillius (1922a: 206, as BARAEINI).

ATOSSIDES Lacordaire, 1872: 496 [stem: *Atoss-*]. Type genus: *Atossa* J. Thomson, 1864 [syn. of *Grammoechus* J. Thomson, 1864]. Comment: original vernacular name available (Art. 11.7.2): first used in latinized form and generally accepted as in Aurivillius (1922a: 149, as ATOSSINI).

METAGNOMINI Aurivillius, 1925: 13 [stem: *Metagnom-*]. Type genus: *Metagnoma* Aurivillius, 1925.

Tribe RHODOPININI Gressitt, 1951

RHODOPIDES Lacordaire, 1872: 450 [stem: *Rhodopid-*]. Type genus: *Rhodopis* J. Thomson, 1857 [preoccupied genus name, not *Rhodopis* Reichenbach, 1854 [Aves]; syn. of *Rhodopina* Gressit, 1951]. Comment: original vernacular name available (Art. 11.7.2): first used in latinized form and generally accepted as in Aurivillius (1922a: 210, as RHODOPINI); incorrect original stem formation, not in prevailing usage; permanently invalid (Art. 39): based on preoccupied type genus.

RHODOPININI Gressitt, 1951: 439 [stem: *Rhodopin-*]. Type genus: *Rhodopina* Gressitt, 1951. Comment: replacement name for RHODOPINI Lacordaire, 1872 because of the homonymy of the type genus.

Tribe SAPERDINI Mulsant, 1839

SAPERDAIRES Mulsant, 1839: 181 [stem: *Saperd-*]. Type genus: *Saperda* Fabricius, 1775 [placed on the Official List of Generic Names in Zoology (ICZN 1980)]. Comment: original vernacular name available (Art. 11.7.2): first used in latinized form by Blanchard (1853: 299, as SAPERDITAE), generally accepted as in Aurivillius (1923: 468, as SAPERDINI).

GLENEÏTAE J. Thomson, 1864: 123 [stem: *Glene-*]. Type genus: *Glenea* Newman, 1842 [for comments on problems with the authorship and type species of this type genus see Bousquet et al. (2009: 37)].

Tribe STENOBIINI Breuning, 1950

STENOBIINI Breuning, 1950b: 305 [stem: *Stenobi-*]. Type genus: *Stenobia* Lacordaire, 1872.

Tribe STERNOTOMINI Thomson, 1860

*STELLOGNATHITES Blanchard, 1845b: 158 [stem: *Stellognath-*]. Type genus: *Stellognatha* Dejean, 1835. Comment: original vernacular name unavailable (Art. 11.7.2): not subsequently latinized.

STERNOTOMITAE J. Thomson, 1860a: 87 [stem: *Sternotom-*]. Type genus: *Sternotomis* Percheron, 1836 [*Sternodonta* Dejean, 1835, a senior synonym of *Sternotomis* Percheron, 1836, is a *nomen oblitum* and *Sternotomis* is a *nomen protectum* following Sama (2009a: 24)]. Comment: current spelling maintained (Art. 29.3.1.1): incorrect original stem formation in prevailing usage (should be *Sternotomid-*).

Tribe TAPEININI Thomson, 1857

TAPEINITES J. Thomson, 1857b: 41 [stem: *Tapein-*]. Type genus: *Tapeina* Lepeletier and Audinet-Serville, 1828. Comment: original vernacular name available (Art. 11.7.2): first used in latinized form by J. Thomson (1860a: 73, as TAPEINITAE), generally accepted as in Aurivillius (1922a: 236, as TAPEININI).

Tribe TETRAOPINI Thomson, 1860

TETRAOPESITAE J. Thomson, 1860a: 66 [stem: *Tetraop-*]. Type genus: *Tetraopes* Schönherr, 1817. Comment: incorrect original stem formation, not in prevailing usage.

Tribe TETRAULAXINI Breuning and Teocchi, 1977

TETRAULAXINI Breuning and Teocchi, 1977: 881 [stem: *Tetraulax-*]. Type genus: *Tetraulax* Jordan, 1903. Comment: current spelling maintained (Art. 29.5): incorrect original stem formation in prevailing usage (should be *Tetraulac-*).

Tribe TETROPINI Portevin, 1927

*POLYOPSIATES Mulsant, 1863b: 340 [stem: *Polyopsi-*]. Type genus: *Polyopsia* Mulsant, 1839 [syn. of *Tetrops* Stephens, 1829]. Comment: original vernacular name unavailable (Art. 11.7.2): not subsequently latinized.

*TÉTROPIDES Planet, 1924: 326 [stem: *Tetrop-*]. Type genus: *Tetrops* Stephens, 1829 [for comments on problems with the authorship and type species of this type genus see Bousquet et al. (2009: 38)]. Comment: original vernacular name unavailable (Art. 11.7.2): proposed after 1899; the argument of Sama (2008: 240) that Planet's name is available, even if published after 1900 in a vernacular form, because it was used as valid in a latinized form and credited to Planet (1924) by Vives (2000: 508) is incorrect (see Bousquet et al. 2009: 38).

TETROPINI Portevin, 1927: 39, in key [stem: *Tetrop-*]. Type genus: *Tetrops* Stephens, 1829 [for comments on problems with the authorship and type species of this type genus see Bousquet et al. (2009: 38)].

Tribe THEOCRINI Lacordaire, 1872

THÉOCRIDES Lacordaire, 1872: 494 [stem: *Theocr-*]. Type genus: *Theocris* J. Thomson, 1858. Comment: original vernacular name available (Art. 11.7.2): first used in latinized form by Gahan (1890: 322, as THEOCRINAE), generally accepted as in Aurivillius (1922a: 238, as THEOCRIDINI [incorrect stem formation]).

Tribe Tmesisternini Blanchard, 1853

Tmesisternitae Blanchard, 1853: 274 [stem: *Tmesistern-*]. Type genus: *Tmesisternus* Latreille, 1829.

Spingnothitae J. Thomson, 1864: 31 [stem: *Sphingnot-*]. Type genus: *Sphingnotus* Perroud, 1855 [as *Spingnothus*, incorrect subsequent spelling of type genus name, not in prevailing usage]. Comment: incorrect original stem formation, not in prevailing usage.

Ichthyosomitae J. Thomson, 1864: 33 [stem: *Ichthyosom-*]. Type genus: *Ichthyosomus* Boisduval, 1835.

*Arsysiides Lacordaire, 1872: 479 [stem: *Arsysi-*]. Type genus: *Arsysia* Pascoe, 1867 [syn. of *Trigonoptera* Perroud, 1855]. Comment: original vernacular name unavailable (Art. 11.7.2): subsequently used in latinized form, e.g., Gestro (1876: 140, as Arsysiini), but not generally accepted as valid; Arsysiidae was used as valid by Ienistea (1986: 30) but it was not attributed to Lacordaire (1872); Ienistea's name is also unavailable, it was proposed after 1930 without description or bibliographic reference to such a description (Art. 13.1).

Trigonopterini Aurivillius, 1922a: 229 [stem: *Trigonopter-*]. Type genus: *Trigonoptera* Perroud, 1855. Comment: replacement name for Arsysiides Lacordaire, 1872 because of the synonymy of the type genus.

Tribe Tragocephalini Thomson, 1857

Tragocephalites J. Thomson, 1857b: 26 [stem: *Tragocephal-*]. Type genus: *Tragocephala* Dejean, 1835. Comment: original vernacular name available (Art. 11.7.2): first used in latinized form by J. Thomson (1860a: 88, as Tragocephalitae), generally accepted as in Aurivillius (1922a: 171, as Tragocephalini).

Tribe Xenicotelini Matsushita, 1933

Xenicotelini Matsushita, 1933: 346 [stem: *Xenicotel-*]. Type genus: *Xenicotela* H. W. Bates, 1884.

Tribe Xenofreini Aurivillius, 1923

Xenofreini Aurivillius, 1923: 375 [stem: *Xenofre-*]. Type genus: *Xenofrea* H. W. Bates, 1885. Comment: this name has been attributed to H. W. Bates (1885: 373) in the literature but we did not find such a name in the Biologia Centrali-Americana or other publications by the same author.

Tribe Xenoleini Lacordaire, 1872

Xénoléides Lacordaire, 1872: 460 [stem: *Xenole-*]. Type genus: *Xenolea* J. Thomson, 1864. Comment: original vernacular name available (Art. 11.7.2): first used in latinized form and generally accepted as in Aurivillius (1922a: 216, as Xenoleini).

Tribe Xylorhizini Lacordaire, 1872

Xylorhizides Lacordaire, 1872: 443 [stem: *Xylorhiz-*]. Type genus: *Xylorhiza* Dejean, 1835. Comment: original vernacular name available (Art. 11.7.2): first used in latinized form by Kolbe (1897: 316, as Xylorrhizinae [incorrect stem formation]), generally accepted as in Aurivillius (1922a: 208, as Xylorhizini).

Tribe Zygocerini Thomson, 1864

Zygoceritae J. Thomson, 1864: 87 [stem: *Zygocer-*]. Type genus: *Zygocera* Erichson, 1842.

Disterninae Pascoe, 1871: 268 [stem: *Distern-*]. Type genus: *Disterna* J. Thomson, 1864.

Family Megalopodidae Latreille, 1802

Megalopides Latreille, 1802: 227 [stem: *Megalopod-*]. Type genus: *Megalopus* Fabricius, 1801.

Subfamily Megalopodinae Latreille, 1802

Megalopides Latreille, 1802: 227 [stem: *Megalopod-*]. Type genus: *Megalopus* Fabricius, 1801. Comment: incorrect original stem formation, not in prevailing usage.

Subfamily Palophaginae Kuschel and May, 1990

Palophaginae Kuschel and May, 1990: 699 [stem: *Palophag-*]. Type genus: *Palophagus* Kuschel, 1990.

Subfamily Zeugophorinae Böving and Craighead, 1931

Zeugophorinae Böving and Craighead, 1931: 63, in key [stem: *Zeugophor-*]. Type genus: *Zeugophora* Kunze, 1818 [placed on the Official List of Generic Names in Zoology (ICZN 1986a)].

Family Orsodacnidae Thomson, 1859

Orsodachnidae C. G. Thomson, 1859: 154 [stem: *Orsodacn-*]. Type genus: *Orsodacne* Latreille, 1802 [placed on the Official List of Generic Names in Zoology (ICZN 2002a)].

Subfamily Orsodacninae Thomson, 1859

Orsodachnidae C. G. Thomson, 1859: 154 [stem: *Orsodacn-*]. Type genus: *Orsodacne* Latreille, 1802 [placed on the Official List of Generic Names in Zoology (ICZN 2002a)]. Comment: incorrect original stem formation, not in prevailing usage.

Subfamily Aulacoscelidinae Chapuis, 1874

Aulacoscélites Chapuis, 1874: 54 [stem: *Aulacoscelid-*]. Type genus: *Aulacoscelis* Duponchel et Chevrolat, 1842. Comment: original vernacular name available

(Art. 11.7.2): first used in latinized form by Heyne and Taschenberg (1907: 245, as AULACOSCELINAE), generally accepted as in Monrós (1959a: 18, as AULACOSCELIDINAE); incorrect original stem formation, not in prevailing usage; a recent application by Santiago-Blay (2008) to conserve usage of the stem *Aulacoscel-* was not approved by the Commission (ICZN 2010b).

Family CHRYSOMELIDAE Latreille, 1802

CHRYSOMELINAE Latreille, 1802: 220 [stem: *Chrysomel-*]. Type genus: *Chrysomela* Linnaeus, 1758 [placed on the Official List of Generic Names in Zoology (ICZN 1984c)]. Comment: First Reviser (CHRYSOMELIDAE Latreille, 1802 vs BRUCHIDAE Latreille, 1802 vs GALERUCIDAE Latreille, 1802) not determined, current usage maintained.

Subfamily SAGRINAE Leach, 1815

SAGRIDA Leach, 1815: 113 [stem: *Sagr-*]. Type genus: *Sagra* Fabricius, 1792.

Tribe CARPOPHAGINI Chapuis, 1874

CARPOPHAGITES Chapuis, 1874: 36 [stem: *Carpophag-*]. Type genus: *Carpophagus* W. S. MacLeay, 1827. Comment: original vernacular name available (Art. 11.7.2): first used in latinized form by Jacoby (1903: 4, as CARPOPHAGINAE), generally accepted as in Clavareau (1913a: 6, as CARPOPHAGINI); the older name CARPOPHAGINAE Selby, 1835 (type genus *Carpophaga* Selby, 1835) is available in Aves (see Bock 1994: 139); this case is to be referred to the Commission to remove the homonymy (Art. 55.3.1).

Tribe DIAPHANOPSIDINI Monrós, 1958

DIAPHANOPSIDINI Monrós, 1958a: 7 [stem: *Diaphanopsid-*]. Type genus: *Diaphanops* Schönherr, 1845. Comment: name proposed after 1930 without description or bibliographic reference to such a description (Art. 13.1), however available because it was used as valid before 2000 as in Seeno and Wilcox (1982: 17, 18) and was not rejected by an author who, between 1961 and 1999, applied Article 13 of the then current edition of the Code (see Art. 13.2.1); current spelling maintained (Art. 29.5): incorrect stem formation in prevailing usage (should be *Diaphanop-*).

Tribe MEGAMERINI Chapuis, 1874

MÉGAMÉRITES Chapuis, 1874: 30 [stem: *Megamer-*]. Type genus: *Megamerus* W. S. MacLeay, 1827. Comment: original vernacular name available (Art. 11.7.2): first used in latinized form by Jacoby (1903: 2, as MEGAMERINAE), generally accepted as in Clavareau (1913a: 4, as MEGAMERINI); First Reviser (MEGAMERINI Chapuis, 1874 vs AMETALLINI Chapuis, 1874 vs MECYNODERINI Chapuis, 1874) not determined, current usage maintained.

Amétallites Chapuis, 1874: 46 [stem: *Ametall-*]. Type genus: *Ametalla* Hope, 1840. Comment: original vernacular name available (Art. 11.7.2): first used in latinized form by Jacoby (1903: 1, as Ametallinae), generally accepted as in Clavareau (1913a: 12, as Ametallini).

Mécynodérites Chapuis, 1874: 44 [stem: *Mecynoder-*]. Type genus: *Mecynodera* Hope, 1840. Comment: original vernacular name available (Art. 11.7.2): first used in latinized form by Jacoby (1903: 1, as Mecynoderinae), generally accepted as in Clavareau (1913a: 11, as Mecynoderini).

Tribe Sagrini Leach, 1815

Sagrida Leach, 1815: 113 [stem: *Sagr-*]. Type genus: *Sagra* Fabricius, 1792.

Subfamily Bruchinae Latreille, 1802

Bruchelae Latreille, 1802: 192 [stem: *Bruch-*]. Type genus: *Bruchus* Linnaeus, 1767 [placed on the Official List of Generic Names in Zoology and older name *Bruchus* Geoffroy, 1762 placed on the Official Index of Rejected and Invalid Generic Names in Zoology (ICZN 1995a)]. Comment: Bruchidae Latreille, 1802 placed on the Official List of Family-Group Names in Zoology (ICZN 1995a).

Tribe Amblycerini Bridwell, 1932

Amblycerinae Bridwell, 1932: 103, in key [stem: *Amblycer-*]. Type genus: *Amblycerus* Thunberg, 1815.

Subtribe Amblycerina Bridwell, 1932

Spermophagidae Crotch, 1873b: 93 [stem: *Spermophag-*]. Type genus: *Spermophagus* sensu G. H. Horn, 1873 [not *Spermophagus* Schönherr, 1833; syn. of *Amblycerus* Thunberg, 1815]. Comment: based on a misidentified type genus, name treated here as invalid until an application is submitted to the Commission to suppress it for the Principles of Priority and Homonymy (Art. 65.2.1); also see Spermophagini Borowiec, 1987 below.

Amblycerinae Bridwell, 1932: 103, in key [stem: *Amblycer-*]. Type genus: *Amblycerus* Thunberg, 1815. Comment: although this is the oldest name for the tribe, we recommend that an application be sent to the Commission in order to conserve usage of Amblycerina Bridwell, 1932 because the older name Spermophagina Crotch, 1873 is based on a misidentified type genus (Art. 65.2.1); MAAZ and CHCL will submit an application to the Commission in order to conserve the current concept of *Amblycerus* Thunberg, 1815 following the discovery of an overlooked type species designation that would alter the concept of the genus (also see Alonso-Zarazaga and Lyal (1999: 8, 23)).

Subtribe Spermophagina Borowiec, 1987

Spermophagini Borowiec, 1987: 27 [stem: *Spermophag-*]. Type genus: *Spermophagus* Schönherr, 1833. Comment: family-group name proposed as new without reference to SPERMOPHAGIDAE Crotch, 1873; an application needs to be submitted to the Commission to suppress SPERMOPHAGIDAE Crotch, 1873 (based on the misidentified type genus *Spermophagus* sensu Horn, 1873) for the Principles of Priority and Homonymy (Art. 65.2.1) to conserve this name as valid.

Tribe BRUCHINI Latreille, 1802

BRUCHELAE Latreille, 1802: 192 [stem: *Bruch-*]. Type genus: *Bruchus* Linnaeus, 1767 [placed on the Official List of Generic Names in Zoology and older name *Bruchus* Geoffroy, 1762 placed on the Official Index of Rejected and Invalid Generic Names in Zoology (ICZN 1995a)]. Comment: BRUCHIDAE Latreille, 1802 placed on the Official List of Family-Group Names in Zoology (ICZN 1995a).

Subtribe ACANTHOSCELIDINA Bridwell, 1946

ACANTHOSCELIDINI Bridwell, 1946: 54, in key [stem: *Acanthoscelid-*]. Type genus: *Acanthoscelides* Schilsky, 1905. Comment: First Reviser (ACANTHOSCELIDINA Bridwell, 1946 vs BRUCHIDIINA Bridwell, 1946) not determined, current usage maintained; description of family-group name not unequivocal (Art. 13.1.1) but name treated as available (Art. 13.2.1); Bridwell (1946: 53) mentioned that he wanted to establish the two new tribes BRUCHIDIINI and ACANTHOSCELIDINI but he followed by saying that it was "premature to attempt a diagnosis" of these tribes; both tribal names were included in his key (p. 54) but their identification was not unequivocal; Bradley (1947: 37) noted that the new tribes BRUCHIDIINI and ACANTHOSCELIDINI were not diagnosed adequately by Bridwell (1946), however he did not cite Article 13 of the then current edition of the Code of Zoological nomenclature to treat those names as unavailable; since BRUCHIDIINI and ACANTHOSCELIDINI have been used as valid and attributed to Bridwell (1946) by subsequent authors, e.g., Bottimer (1968: 1028 and 1015 respectively), we have also treated those names as available here.

BRUCHIDIINI Bridwell, 1946: 54, in key [stem: *Bruchidi-*]. Type genus: *Bruchidius* Schilsky, 1905. Comment: description of family-group name not unequivocal (Art. 13.1.1) but name treated as available (Art. 13.2.1); see comment under ACANTHOSCELIDINI Bridwell, 1946.

Subtribe BRUCHINA Latreille, 1802

BRUCHELAE Latreille, 1802: 192 [stem: *Bruch-*]. Type genus: *Bruchus* Linnaeus, 1767 [placed on the Official List of Generic Names in Zoology and older name *Bruchus* Geoffroy, 1762 placed on the Official Index of Rejected and Invalid Generic Names in Zoology (ICZN 1995a)]. Comment: BRUCHIDAE Latreille, 1802 placed on the Official List of Family-Group Names in Zoology (ICZN 1995a).

Lariidae Bedel, 1901: 341 [stem: *Lari-*]. Type genus: *Laria* Scopoli, 1763 [placed on the Official Index of Rejected and Invalid Generic Names in Zoology (ICZN 1995a); syn. of *Bruchus* Linnaeus, 1767]. Comment: permanently invalid (Art. 39): based on suppressed type genus.

Subtribe Megacerina Bridwell, 1946

Megacerini Bridwell, 1946: 54, in key [stem: *Megacer-*]. Type genus: *Megacerus* Fahraeus, 1839. Comment: Megacerini Viret, 1961 (type genus *Megaceros* Owen, 1844) is used as valid in Mammalia: Cervidae; this case is to be referred to the Commission to remove the homonymy (Art. 55.3.1).

Tribe Eubaptini Bridwell, 1932

Eubaptinae Bridwell, 1932: 103, in key [stem: *Eubapt-*]. Type genus: *Eubaptus* Lacordaire, 1845.

Tribe Kytorhinini Bridwell, 1932

Kytorhininae Bridwell, 1932: 103, in key [stem: *Kytorhin-*]. Type genus: *Kytorhinus* Fischer von Waldheim, 1809.

Tribe Pachymerini Bridwell, 1929

Pachymerinae Bridwell, 1929: 142 [stem: *Pachymer-*]. Type genus: *Pachymerus* Thunberg, 1805. Comment: precedence (Pachymerini Bridwell, 1929 vs Caryedini Brimwell, 1929 vs Caryopemini Brimwell, 1929) given to taxon originally proposed at the higher rank (Art. 24.1).

Subtribe Caryedontina Bridwell, 1929

Caryedini Bridwell, 1929: 143 [stem: *Caryedont-*]. Type genus: *Caryedon* Schönherr, 1823. Comment: incorrect original stem formation, not in prevailing usage; correction of stem by Decelle (1966).

Subtribe Caryopemina Bridwell, 1929

Caryopemini Bridwell, 1929: 143 [stem: *Caryopem-*]. Type genus: *Caryopemon* Jekel, 1855. Comment: current spelling maintained (Art. 29.5): incorrect stem formation in prevailing usage (should be *Caryopemon-*).

Subtribe Pachymerina Bridwell, 1929

Pachymerinae Bridwell, 1929: 142 [stem: *Pachymer-*]. Type genus: *Pachymerus* Thunberg, 1805.

Tribe Rhaebini Blanchard, 1845

Rhaebites Blanchard, 1845b: 180 [stem: *Rhaeb-*]. Type genus: *Rhaebus* Fischer von Waldheim, 1824. Comment: original vernacular name available (Art.

11.7.2): first used in latinized form by Kraatz (1879: 278, as RHAEBINI), generally accepted as in Anton (2010: 353, as RHAEBINI); Blanchard (1845b: 193) also used the spelling RHOEBITES.

Subfamily DONACIINAE Kirby, 1837

DONACIADAE Kirby, 1837: 222 [stem: *Donaci-*]. Type genus: *Donacia* Fabricius, 1775.

Tribe DONACIINI Kirby, 1837

DONACIADAE Kirby, 1837: 222 [stem: *Donaci-*]. Type genus: *Donacia* Fabricius, 1775.

DONACOCIADAE Gistel, 1856a: 377 [stem: *Donacoci-*]. Type genus: *Donacocia* Gistel, 1856 [although this genus has been considered a senior syn. of *Plateumaris* C. G. Thomson, 1859 by some authors, Löbl and Silfverberg (2010: 64) recently chose "*brevicornis* Ahrens, 1810" as the type species for this genus; syn. of *Donacia* Fabricius, 1775]. Comment: incorrect original stem formation, not in prevailing usage.

Tribe HAEMONIINI Chen, 1941

HAEMONIINI Chen, 1941: 8 [stem: *Haemoni-*]. Type genus: *Haemonia* Dejean, 1821 [syn. of *Macroplea* Smouelle, 1819].

Tribe PLATEUMARINI Böving, 1922

PLATEUMARINI Böving, 1922: 50 [stem: *Plateumar-*]. Type genus: *Plateumaris* C. G. Thomson, 1859. Comment: current spelling maintained (Art. 29.3.1.1): incorrect stem formation in prevailing usage (should be *Plateumarid-*).

PLATEUMARINI Askevold, 1990: 633 [stem: *Plateumar-*]. Type genus: *Plateumaris* C. G. Thomson, 1859. Comment: family-group name proposed as new without reference to PLATEUMARINI Böving, 1922; current spelling maintained (Art. 29.3.1.1): incorrect stem formation in prevailing usage (should be *Plateumarid-*).

Subfamily CRIOCERINAE Latreille, 1804

CRIOCERIDES Latreille, 1804c: 159 [stem: *Criocer-*]. Type genus: *Crioceris* Geoffroy, 1762 [placed on the Official List of Generic Names in Zoology (ICZN 1994a)].

Tribe CRIOCERINI Latreille, 1804

CRIOCERIDES Latreille, 1804c: 159 [stem: *Criocer-*]. Type genus: *Crioceris* Geoffroy, 1762 [placed on the Official List of Generic Names in Zoology (ICZN 1994a)]. Comment: published 7 March 1804; this family-group name was also used in the same year by Latreille (1804d [between 19 August and 17 September]: 324, as CRIOCERIDES); current spelling maintained (Art. 29.3.1.1): incorrect stem formation in prevailing usage (should be *Criocerid-*).

Tribe Lemini Gyllenhal, 1813

Lemoideae Gyllenhal, 1813: 632 [stem: *Lem-*]. Type genus: *Lema* Fabricius, 1798 [placed on the Official List of Generic Names in Zoology (ICZN 1970a)].

Tribe Pseudocriocerini Heinze, 1962

Pseudocriocerini Heinze, 1962: 198, in key [stem: *Pseudocriocer-*]. Type genus: *Pseudocrioceris* Pic, 1916. Comment: current spelling maintained (Art. 29.3.1.1): incorrect stem formation in prevailing usage (should be *Pseudocriocerid-*).

Subfamily Cassidinae Gyllenhal, 1813

Cassideae Gyllenhal, 1813: 434 [stem: *Cassid-*]. Type genus: *Cassida* Linnaeus, 1758 [placed on the Official List of Generic Names in Zoology (ICZN 1974b)]. Comment: name placed on the Official List of Family-Group Names in Zoology (ICZN 1974b, as Cassidinae Stephens, 1831); First Reviser found (Cassidinae Gyllenhal, 1813 vs Hispinae Gyllenhal, 1813) is Shuckard (1839b: 68).

Tribe Alurnini Chapuis, 1875

Alurnites Chapuis, 1875: 292 [stem: *Alurn-*]. Type genus: *Alurnus* Fabricius, 1775. Comment: original vernacular name available (Art. 11.7.2): first used in latinized form by Weise (1910: 69, as Alurnini), generally accepted as in Staines (2002: 742, as Alurnini).

*Sphaeropalpites Chapuis, 1875: 359 [stem: *Sphaeropalp-*]. Type genus: *Sphaeropalpus* Guérin-Méneville, 1844 [syn. of *Platyauchenia* Sturm, 1843]. Comment: original vernacular name unavailable (Art. 11.7.2): not subsequently latinized.

Platyaucheniitae Spaeth, 1929b: 113 [stem: *Platyaucheni-*]. Type genus: *Platyauchenia* Sturm, 1843.

Tribe Anisoderini Chapuis, 1875

Anisodérites Chapuis, 1875: 294 [stem: *Anisoder-*]. Type genus: *Anisodera* Chevrolat, 1836. Comment: original vernacular name available (Art. 11.7.2): first used in latinized form by Weise (1911: 40, as Anisoderini), generally accepted as in Würmli (1975: 10, as Anisoderini).

*Lasiochilini Gressitt, 1950: 62 [stem: *Lasiochil-*]. Type genus: *Lasiochila* Weise, 1916. Comment: unavailable family-group name, proposed after 1930 without description or bibliographic reference to such a description (Art. 13.1); the author apparently used this family-group name in error for Anisoderini (see Gressitt 1953: 126); Lasiochilinae Carayon, 1972 (type genus *Lasiochilus* Reuter, 1871) has been used as valid in Hemiptera.

Tribe Aproidini Weise, 1911

Aprioidini Weise, 1911: 41 [stem: *Aproid-*]. Type genus: *Aproida* Pascoe, 1863.

Tribe Arescini Chapuis, 1875

Arescites Chapuis, 1875: 298 [stem: *Aresc-*]. Type genus: *Arescus* Perty, 1832. Comment: original vernacular name available (Art. 11.7.2): first used in latinized form by Weise (1910: 69, as Arescini), generally accepted as in Staines (2002: 741, as Arescini).

Tribe Aspidimorphini Chapuis, 1875

Aspidimorphites Chapuis, 1875: 406 [stem: *Aspidimorph-*]. Type genus: *Aspidimorpha* Hope, 1840. Comment: original vernacular name available (Art. 11.7.2): first used in latinized form by Spaeth (1914: (129), as Aspidormophitae [incorrect stem formation]), generally accepted as in Hincks (1952: 336, as Aspidomorphini).

Tribe Basiprionotini Gressitt, 1952 (1914)

*Prioptérites Chapuis, 1875: 367 [stem: *Priopter-*]. Type genus: *Prioptera* Hope, 1840 [syn. of *Basiprionota* Chevrolat, 1836]. Comment: original vernacular name unavailable (Art. 11.7.2): subsequently used in latinized form but not generally attributed to Chapuis (1875).

Riopteritae Spaeth, 1914: (129) [stem: *Priopter-*]. Type genus: *Prioptera* Hope, 1840 [syn. of *Basiprionota* Chevrolat, 1836]. Comment: usage of Basiprionotini Gressitt, 1952 conserved over this name (Art. 40.2).

Basiprionotini Gressitt, 1952: 444 [stem: *Basiprionot-*]. Type genus: *Basiprionota* Chevrolat, 1836. Comment: published 8 December 1952; name proposed to replace Priopterini because of synonymy of the type genus; this family-group name was also proposed in the same year by Hincks (1952 [31 December]: 329, in key, as Basiprionotini); usage of this name over the older name Priopterini Spaeth, 1929 conserved (Art. 40.2).

Epistictinini Hincks, 1952: 329, in key [stem: *Epistictin-*]. Type genus: *Epistictina* Hincks, 1950.

Tribe Botryonopini Chapuis, 1875

Botryonopites Chapuis, 1875: 291 [stem: *Botryonop-*]. Type genus: *Botryonopa* Guérin-Méneville, 1840 [*Botryonopa* is a incorrect subsequent spelling of the original name *Bothryonopa*, in prevailing usage and so deemed to be the correct original spelling (Art. 33.3.1); for correct author and year of type genus see Staines (2010: 172)]. Comment: original vernacular name available (Art. 11.7.2): first used in latinized form by Weise (1911: 39, as Botryonopini), generally accepted as in Würmli (1975: 8, as Botryonopini).

Tribe Callispini Chapuis, 1875

Callispites Chapuis, 1875: 269 [stem: *Callisp-*]. Type genus: *Callispa* Baly, 1859. Comment: original vernacular name available (Art. 11.7.2): first used in latinized form by Weise (1911: 41, as Callispini), generally accepted as in Würmli (1975: 12, as Callispini).

*Hispodontites Chapuis, 1875: 284 [stem: *Hispodont-*]. Type genus: *Hispodonta* Baly, 1859. Comment: original vernacular name unavailable (Art. 11.7.2): not subsequently latinized.

Tribe Callohispini Uhmann, 1960

Callohispini Uhmann, 1960: 60 [stem: *Callohisp-*]. Type genus: *Callohispa* Uhmann, 1960.

Tribe Cassidini Gyllenhal, 1813

Cassideae Gyllenhal, 1813: 434 [stem: *Cassid-*]. Type genus: *Cassida* Linnaeus, 1758 [placed on the Official List of Generic Names in Zoology (ICZN 1974b)]. Comment: name placed on the Official List of Family-Group Names in Zoology (ICZN 1974b, as Cassidinae Stephens, 1831).

Evaspidotidae Gistel, 1856a: 381 [stem: *Evaspist-*]. Type genus: *Evaspistes* Gistel, 1856 [syn. of *Cassida* Linnaeus, 1758]. Comment: incorrect original stem formation, not in prevailing usage.

*Chiridites Chapuis, 1875: 405 [stem: *Chirid-*]. Type genus: *Chirida* Chapuis, 1875. Comment: original vernacular name unavailable (Art. 11.7.2): not subsequently latinized.

*Hybosites Chapuis, 1875: 380 [stem: *Hybos-*]. Type genus: *Hybosa* Boheman, 1855 [*Hybosa* has been credited to Duponchel and Chevrolat, 1842, e.g., Seeno and Wilcox (1982: 178), but the name is not available from that reference since there is no description nor available species included; this genus was first made available by Boheman (1855: 1)]. Comment: original vernacular name unavailable (Art. 11.7.2): not subsequently latinized.

Basiptites Chapuis, 1875: 379 [stem: *Basipt-*]. Type genus: *Basipta* Chevrolat, 1842. Comment: original vernacular name available (Art. 11.7.2): first used in latinized form and generally accepted as in Hincks (1952: 337, as Basiptini).

Coptocyclitae Spaeth and Reitter, 1926: 7, in key [stem: *Coptocycl-*]. Type genus: *Coptocycla* Chevrolat, 1836. Comment: this family-group name was also used in the same year by Spaeth (1926: 1, as Coptocyclitae) but we could not establish priority.

Charidotitae Spaeth, 1942: 40 [stem: *Charidotid-*]. Type genus: *Charidotis* Boheman, 1854. Comment: name proposed after 1930 without description or bibliographic reference to such a description (Art. 13.1), however available because it was used as valid before 2000 as in Hincks (1952: 343, as Charidotini) and was not rejected by an author who, between 1961 and 1999, applied Article 13 of the then current edition of the Code (see Art. 13.2.1).

Tribe CEPHALOLEIINI Chapuis, 1875

CÉPHALOLÉITES Chapuis, 1875: 277 [stem: *Cephalolei-*]. Type genus: *Cephaloleia* Chevrolat, 1836. Comment: original vernacular name available (Art. 11.7.2): first used in latinized form by Weise (1910: 69, as CEPHALOLEIINI), generally accepted as in Borowiec (1995: 553, as CEPHALOLEIINI); incorrect original stem formation, not in prevailing usage.

Tribe CHALEPINI Weise, 1910

CHALEPINI Weise, 1910: 69 [stem: *Chalep-*]. Type genus: *Chalepus* Thunberg, 1805. Comment: Chalepini Weise, 1910 is a junior homonym of CHALEPIDAE H. C. C. Burmeister, 1847 (type genus *Chalepus* W. S. MacLeay, 1819) in SCARABAEIDAE; the older scarab family-group name is permanently invalid because it is based on a preoccupied type genus; an application to the Commission is needed to conserve usage of CHALEPINI Weise, 1910.

Tribe COELAENOMENODERINI Weise, 1911

COELAENOMENODERINI Weise, 1911: 51 [stem: *Coelaenomenoder-*]. Type genus: *Coelaenomenodera* Blanchard, 1845.

PHARANGISPINI Uhmann, 1940: 122 [stem: *Pharangisp-*]. Type genus: *Pharangispa* Maulik, 1929. Comment: proposed after 1930 without description but available because accompanied by a bibliographic reference (Maulik 1929: 233) to a description (Art. 13.1).

Tribe CRYPTONYCHINI Chapuis, 1875

CRYPTONYCHITES Chapuis, 1875: 286 [stem: *Cryptonych-*]. Type genus: *Cryptonychus* Gyllenhal, 1817. Comment: original vernacular name available (Art. 11.7.2): first used in latinized form by Weise (1911: 45, as CRYPTONYCHINI), generally accepted as in Würmli (1975: 21, as CRYPTONYCHINI).

Tribe DELOCRANIINI Spaeth, 1929

DELOCRANIITAE Spaeth, 1929b: 113 [stem: *Delocrani-*]. Type genus: *Delocrania* Guérin-Méneville, 1844.

Tribe DORYNOTINI Monrós and Viana, 1949 (1923)

*BATONOTITES Chapuis, 1875: 377 [stem: *Batonot-*]. Type genus: *Batonota* Hope, 1839 [syn. of *Dorynota* Chevrolat, 1836]. Comment: original vernacular name unavailable (Art. 11.7.2): subsequently used in latinized form but not generally attributed to Chapuis (1875).

BATONOTITAE Spaeth, 1923: 66 [stem: *Batonot-*]. Type genus: *Batonota* Hope, 1839 [syn. of *Dorynota* Chevrolat, 1836]. Comment: usage of DORYNOTINI Monrós and Viana, 1949 conserved over this name (Art. 40.2).

DORYNOTINI Monrós and Viana, 1949: 392 [stem: *Dorynot-*]. Type genus: *Dorynota* Chevrolat, 1836. Comment: name proposed to replace "BATONOTITES" because of synonymy of the type genus; usage of younger name conserved over BATONOTINI Spaeth, 1923 (Art. 40.2).

Tribe EUGENYSINI Hincks, 1952

*CALASPIDEITAE Spaeth, 1942: 18 [stem: *Calaspide-*]. Type genus: *Calaspidea* Hope, 1840 [syn. of *Eugenysa* Chevrolat, 1836]. Comment: unavailable family-group name, proposed after 1930 without description or bibliographic reference to such a description (Art. 13.1).

EUGENYSINI Hincks, 1952: 329, in key [stem: *Eugenys-*]. Type genus: *Eugenysa* Chevrolat, 1836.

Tribe EURISPINI Chapuis, 1875

EURISPITES Chapuis, 1875: 264 [stem: *Eurisp-*]. Type genus: *Eurispa* Baly, 1859. Comment: original vernacular name available (Art. 11.7.2): first used in latinized form by Weise (1911: 45, as EURISPINI), generally accepted as in Uhmann (1960: 61, as EURISPINI).

Tribe EXOTHISPINI Weise, 1911

EXOTHISPINI Weise, 1911: 51 [stem: *Exothisp-*]. Type genus: *Exothispa* Kolbe, 1897.

Tribe GONIOCHENIINI Spaeth, 1942

GONIOCHENIITAE Spaeth, 1942: 17 [stem: *Goniocheni-*]. Type genus: *Goniochenia* Weise, 1896. Comment: name proposed after 1930 without description or bibliographic reference to such a description (Art. 13.1), however available because it was used as valid before 2000 as in Viana (1964: 217, as GONIOCHENIICI) and was not rejected by an author who, between 1961 and 1999, applied Article 13 of the then current edition of the Code (see Art. 13.2.1).

Tribe GONOPHORINI Chapuis, 1875

GONOPHORITES Chapuis, 1875: 303 [stem: *Gonophor-*]. Type genus: *Gonophora* Chevrolat, 1836. Comment: original vernacular name available (Art. 11.7.2): first used in latinized form by Weise (1911: 54, as GONOPHORINI), generally accepted as in Würmli (1975: 47, as GONOPHORINI).

*WALLACÉITES Chapuis, 1875: 281 [stem: *Wallace-*]. Type genus: *Wallacea* Baly, 1859 [this genus is a senior homonym of *Wallacea* Doleschall, 1859 (see Woodley 2002: 410)]. Comment: original vernacular name unavailable (Art. 11.7.2): subsequently used in latinized form but not generally attributed to Chapuis (1875); WALLACEIAD[A]E [incorrect stem formation] was used as valid by Ienistea (1986: 31) but it was not attributed to Chapuis (1875); Ienistea's name is also unavailable, it was proposed after 1930 without description or bibliographic reference to such a description (Art. 13.1).

Tribe Hemisphaerotini Monrós and Viana, 1951 (1929)

Porphyraspitae Spaeth, 1929a: 29 [stem: *Porphyraspid-*]. Type genus: *Porphyraspis* Hope, 1840 [syn. of *Hemisphaerota* Chevrolat, 1836]. Comment: this family-group name was also used in the same year by Spaeth (1929b: 113); usage of younger name Hemisphaerotini Monrós and Viana, 1951 conserved over this name (Art. 40.2); incorrect original stem formation, not in prevailing usage.

Hemisphaerotini Monrós and Viana, 1951: 370 [stem: *Hemisphaerot-*]. Type genus: *Hemisphaerota* Chevrolat, 1836. Comment: usage of this name conserved over Porphyraspidini Spaeth, 1929 (Art. 40.2).

Tribe Hispini Gyllenhal, 1813

Hispoideae Gyllenhal, 1813: 448 [stem: *Hisp-*]. Type genus: *Hispa* Linnaeus, 1767.

*Monochirites Chapuis, 1875: 330 [stem: *Monochir-*]. Type genus: *Monochirus* Chapuis, 1875 [preoccupied genus name, not *Monochirus* Rafinesque, 1814 [Pisces]; syn. of *Hispellinus* Weise, 1898]. Comment: original vernacular name unavailable (Art. 11.7.2): subsequently used in latinized form but not generally attributed to Chapuis (1875).

*Trichispites Chapuis, 1875: 331 [stem: *Trichisp-*]. Type genus: *Trichispa* Chapuis, 1875. Comment: original vernacular name unavailable (Art. 11.7.2): not subsequently latinized.

Monochirini Weise, 1905: 317 [stem: *Monochir-*]. Type genus: *Monochirus* Chapuis, 1875 [preoccupied genus name, not *Monochirus* Rafinesque, 1814 [Pisces]; syn. of *Hispellinus* Weise, 1897]. Comment: permanently invalid (Art. 39): based on preoccupied type genus.

Tribe Hispoleptini Chapuis, 1875

Hispoleptites Chapuis, 1875: 283 [stem: *Hispolept-*]. Type genus: *Hispopleptis* Baly, 1864. Comment: original vernacular name available (Art. 11.7.2): first used in latinized form by Uhmann (1940: 121, as Hispoleptini), generally accepted as in Staines (2002: 777, as Hispoleptini).

Tribe Hybosispini Weise, 1910

Hybosispini Weise, 1910: 69 [stem: *Hybosisp-*]. Type genus: *Hybosispa* Weise, 1910.

Tribe Imatidiini Hope, 1840

Imatidiidae Hope, 1840a: 152 [stem: *Imatidi-*]. Type genus: *Imatidium* Fabricius, 1801. Comment: name previously attributed to Chapuis (1875).

Tribe Ischyrosonychini Chapuis, 1875

Ischyrosonychites Chapuis, 1875: 382 [stem: *Ischyrosonych-*]. Type genus: *Ischyrosonyx* Sturm, 1843 [syn. of *Eurypedus* Gistel, 1834]. Comment: original vernacular name available (Art. 11.7.2): first used in latinized form by Spaeth (1942: 31, as Ischyronycitae [incorrect stem formation]), generally accepted as in Borowiec (1995: 556, as Ischyrosonychini).

*Physonotitae Spaeth, 1942: 32 [stem: *Physonot-*]. Type genus: *Physonota* Boheman, 1854. Comment: unavailable family-group name, proposed after 1930 without description or bibliographic reference to such a description (Art. 13.1).

Asterizini Hincks, 1952: 330, in key [stem: *Asteriz-*]. Type genus: *Asteriza* Chevrolat, 1836. Comment: Hincks (1952: 336) and Seeno and Wilcox (1982: 171) mention "Asterizitae Spaeth", we could not find any earlier usage of Asterizitae by Spaeth.

Tribe Leptispini Fairmaire, 1868

Leptispites Fairmaire, 1868: 258 [stem: *Leptisp-*]. Type genus: *Leptispa* Baly, 1859. Comment: original vernacular name available (Art. 11.7.2): first used in latinized form by Weise (1911: 44, as Lepthispini [incorrect stem formation]), generally accepted as in Würmli (1975: 18, as Leptispini).

Tribe Mesomphaliini Hope, 1840

Mesomphalidae Hope, 1840a: 160 [stem: *Mesomphali-*]. Type genus: *Mesomphalia* Hope, 1839. Comment: incorrect original stem formation, not in prevailing usage.

*Chélymorphites Chapuis, 1875: 402 [stem: *Chelymorph-*]. Type genus: *Chelymorpha* Chevrolat, 1836. Comment: original vernacular name unavailable (Art. 11.7.2): not subsequently latinized.

*Elytrogonites Chapuis, 1875: 403 [stem: *Elytrogon-*]. Type genus: *Elytrogona* Chevrolat, 1836. Comment: original vernacular name unavailable (Art. 11.7.2): not subsequently latinized.

*Omoplatites Chapuis, 1875: 397 [stem: *Omoplat-*]. Type genus: *Omoplata* Hope, 1840 [syn. of *Echoma* Chevrolat, 1836]. Comment: original vernacular name unavailable (Art. 11.7.2): not subsequently latinized.

Stolaini Hincks, 1952: 329, in key [stem: *Stolad-*]. Type genus: *Stolas* Billberg, 1820. Comment: incorrect original stem formation, not in prevailing usage.

Tribe Nothosacanthini Gressitt, 1952 (1929)

*Hoplionotites Chapuis, 1875: 357 [stem: *Hoplionot-*]. Type genus: *Hoplionota* Hope, 1840 [syn. of *Notosacantha* Chevrolat, 1836]. Comment: original vernacular name unavailable (Art. 11.7.2): subsequently used in latinized form but not generally attributed to Chapuis (1875).

HOPLIONOTITAE Spaeth, 1929b: 113 [stem: *Hoplionot-*]. Type genus: *Hoplionota* Hope, 1840 [syn. of *Notosacantha* Chevrolat, 1836]. Comment: usage of younger name NOTHOSACANTHINI Gressitt, 1952 conserved over this name (Art. 40.2).

NOTOSACANTHINA Gressitt, 1952: 444 [stem: *Notosacanth-*]. Type genus: *Notosacantha* Chevrolat, 1836. Comment: published 8 December 1952; name proposed to replace HOPLIONOTINI Spaeth, 1929 because of the synonymy of the type genus; this family-group name was also proposed in the same year by Hincks (1952 [31 December]: 328, in key, as NOTOSACANTHINI); usage of this name conserved over HOPLIONOTINI Spaeth, 1929 (Art. 40.2).

Tribe OEDIOPALPINI Monrós and Viana, 1947 (1910)

AMPLIPALPINI Weise, 1910: 69 [stem: *Amplipalp-*]. Type genus: *Amplipalpa* Harold, 1875 [syn. of *Oediopalpa* Baly, 1859]. Comment: younger name OEDIOPALPINI Monrós and Viana, 1947 conserved over this name (Art. 40.2).

OEDIOPALPINI Monrós and Viana, 1947: 150, in key [stem: *Oediopalp-*]. Type genus: *Oediopalpa* Baly, 1859. Comment: name proposed to replace AMPLIPALPINI Weise, 1910 because of synonymy of the type genus; usage of this name conserved over the older name AMPLIPALPINI Weise, 1910 (Art. 40.2).

Tribe OMOCERINI Hincks, 1952 (1923)

*TAUROMITES Chapuis, 1875: 372 [stem: *Taurom-*]. Type genus: *Tauroma* Hope, 1839 [syn. of *Omocerus* Chevrolat, 1835]. Comment: original vernacular name unavailable (Art. 11.7.2): subsequently used in latinized form but not generally attributed to Chapuis (1875).

TAUROMITAE Spaeth, 1923: 66 [stem: *Taurom-*]. Type genus: *Tauroma* Hope, 1839 [syn. of *Omocerus* Chevrolat, 1835]. Comment: younger name OMOCERINI Hincks, 1952 conserved over this name (Art. 40.2).

OMOCERINI Hincks, 1952: 329, in key [stem: *Omocer-*]. Type genus: *Omocerus* Chevrolat, 1835. Comment: name proposed to replace TAUROMINI Spaeth, 1923 because of synonymy of the type genus; usage of younger name conserved over TAUROMINI Spaeth, 1923 (Art. 40.2).

Tribe ONCOCEPHALINI Chapuis, 1875

ONCOCÉPHALITES Chapuis, 1875: 308 [stem: *Oncocephal-*]. Type genus: *Onchocephala* Guérin-Méneville, 1844 [this genus name was first made available by Guérin-Méneville (1844: 280-281) although most authors credit Chevrolat (1847: 110, spelled *Oncocephalus*) as the correct author and date of publication of this genus]. Comment: original vernacular name available (Art. 11.7.2): first used in latinized form by Weise (1911: 50, as ONCOCEPHALINI), generally accepted as in Świętojańska et al. (2006: 49, as ONCOCEPHALINI); current spelling maintained (Art. 29.5): incorrect stem formation in prevailing usage (should be *Onchocephal-*).

Choeridionini Weise, 1911: 49 [stem: *Chaeridion-*]. Type genus: *Chaeridiona* Baly, 1869 [as *Choeridiona*, incorrect subsequent spelling of type genus name, not in prevailing usage]. Comment: incorrect original stem formation, not in prevailing usage.

Tribe Promecothecini Chapuis, 1875

Promecothécites Chapuis, 1875: 300 [stem: *Promecothec-*]. Type genus: *Promecotheca* Chevrolat, 1847. Comment: original vernacular name available (Art. 11.7.2): first used in latinized form by Weise (1911: 53, as Promecothecini), generally accepted as in Gressitt (1950: 81, as Promecothecini).

Tribe Prosopodontini Weise, 1910

Prosopodontini Weise, 1910: 69 [stem: *Prosopodont-*]. Type genus: *Prosopodonta* Baly, 1885.

Tribe Sceloenoplini Uhmann, 1930

*Céphalodontites Chapuis, 1875: 313 [stem: *Cephalodont-*]. Type genus: *Cephalodonta* Chevrolat, 1842 [this name was recently attributed to "Chevrolat, 1843" (e.g., Staines (2002: 748)) but it was made available for the first time by Chevrolat (1842: 272); syn. of *Sceloenopla* Chevrolat, 1836]. Comment: original vernacular name unavailable (Art. 11.7.2): subsequently used in latinized form but not generally attributed to Chapuis (1875).

Cephalodontinae Gestro, 1906: 548 [stem: *Cephalodont-*]. Type genus: *Cephalodonta* Chevrolat, 1842 [this name was recently attributed to "Chevrolat, 1843" (e.g., Staines (2002: 748)) but it was made available for the first time by Chevrolat (1842: 272); syn. of *Sceloenopla* Chevrolat, 1836]. Comment: usage of the younger name Sceloenoplini Uhmann, 1930 conserved over this name (Art. 40.2).

Sceloenoplini Uhmann, 1930: 238 [stem: *Sceloenopl-*]. Type genus: *Sceloenopla* Chevrolat, 1836. Comment: although this is not the oldest name for the tribe, we recommend that an application be sent to the Commission in order to conserve usage of Sceloenoplini Uhmann, 1930.

Tribe Spilophorini Chapuis, 1875

Spilophorites Chapuis, 1875: 364 [stem: *Spilophor-*]. Type genus: *Spilophora* Boheman, 1850. Comment: original vernacular name available (Art. 11.7.2): first used in latinized form by Spaeth (1929a: 29, as Spilophoritae), generally accepted as in Borowiec (1995: 553, as Spilophorini); this name is a senior homonym of Spilophorina Krikken, 1984 (type genus *Spilophorus* Schaum, 1848) currently used as valid in Scarabaeidae; this case is to be referred to the Commission to remove the homonymy (Art. 55.3.1).

Tribe Uroplatini Weise, 1910

*Octotomites Chapuis, 1875: 310 [stem: *Octotom-*]. Type genus: *Octotoma* Dejean, 1836. Comment: original vernacular name unavailable (Art. 11.7.2): subsequently used in latinized form but not generally attributed to Chapuis (1875); Octotomidae was used as valid by Ienistea (1986: 31) but it was not attributed to Chapuis (1875); Ienistea's name is also unavailable, it was proposed after 1930 without description or bibliographic reference to such a description (Art. 13.1).

Uroplatini Weise, 1910: 69 [stem: *Uroplat-*]. Type genus: *Uroplata* Chevrolat, 1836 ["*Uroplata* Chevrolat, 1835" placed on the Official List of Generic Names in Zoology (ICZN 1985f)]. Comment: name placed on the Official List of Family-Group Names in Zoology (ICZN 1985f, as Uroplatini Leng, 1920).

Subfamily Chrysomelinae Latreille, 1802

Chrysomelinae Latreille, 1802: 220 [stem: *Chrysomel-*]. Type genus: *Chrysomela* Linnaeus, 1758 [placed on the Official List of Generic Names in Zoology (ICZN 1984c)]. Comment: see Kippenberg (2010b) for an alternative tribal classification within this subfamily.

Tribe Chrysomelini Latreille, 1802

Chrysomelinae Latreille, 1802: 220 [stem: *Chrysomel-*]. Type genus: *Chrysomela* Linnaeus, 1758 [placed on the Official List of Generic Names in Zoology (ICZN 1984c)].

Eleiaeidae Gistel, 1848: [9] [stem: *Elei-*]. Type genus: *Eleia* Gistel, 1848 [syn. of *Chrysomela* Linnaeus, 1758]. Comment: incorrect original stem formation, not in prevailing usage.

Prasocurisidae Gistel, 1848: [10] [stem: *Prasocur-*]. Type genus: *Prasocuris* Latreille, 1802. Comment: incorrect original stem formation, not in prevailing usage.

Chloëmeladae Gistel, 1856a: 379 [stem: *Chloemel-*]. Type genus: *Chloemela* Gistel, 1856 [syn. of *Chrysomela* Linnaeus, 1758]. Comment: incorrect original stem formation, not in prevailing usage.

Doryphores Motschulsky, 1860: 181 [stem: *Doryphor-*]. Type genus: *Doryphora* Illiger, 1807. Comment: original vernacular name available (Art. 11.7.2): first used in latinized form by Gerstaecker (1862: 405, as Doryphorae), generally accepted as in Riley et al. (2002: 639, as Doryphorina).

Gonioctènes Motschulsky, 1860: 179 [stem: *Goniocten-*]. Type genus: *Gonioctena* Chevrolat, 1836. Comment: original vernacular name available (Art. 11.7.2): first used in latinized form by Gerstaecker (1862: 405, as Gonioctenae), generally accepted as in Warchalowski (1994: 100, as Gonioctenina).

Linaeines Motschulsky, 1860: 196 [stem: *Lin-*]. Type genus: *Lina* Latreille, 1829 [syn. of *Chrysomela* Linnaeus, 1758]. Comment: original vernacular name available (Art. 11.7.2): first used in latinized form and generally accepted as in Gerstaecker (1862: 406, as Linae).

Ovosomes Motschulsky, 1860: 211 [stem: *Ovosomat-*]. Type genus: *Ovosoma* Motschulsky, 1860. Comment: original vernacular name available (Art. 11.7.2): first used in latinized form and generally accepted as in Gerstaecker (1862: 406, as Ovosomae); incorrect original stem formation, not in prevailing usage.

Paropsines Motschulsky, 1860: 192 [stem: *Paropse-*]. Type genus: *Paropsis* A. G. Olivier, 1807. Comment: original vernacular name available (Art. 11.7.2): first used in latinized form by Gerstaecker (1862: 405, as Paropsinae), generally accepted as in Riley et al. (2002: 639, as Paropsina); incorrect original stem formation, not in prevailing usage.

Phrathorines Motschulsky, 1860: 216 [stem: *Phrator-*]. Type genus: *Phratora* Chevrolat, 1836. Comment: original vernacular name available (Art. 11.7.2): first used in latinized form by Gerstaecker (1862: 406, as Phratorinae), generally accepted as in Warchalowski (1994: 139, as Phratorina); incorrect original stem formation, not in prevailing usage.

*Australicites Chapuis, 1874: 428 [stem: *Australic-*]. Type genus: *Australica* Chevrolat, 1836 [this name is sometimes treated as a *nomen nudum* (Reid 2006: 55), however it is available since four available species names were included by Chevrolat (1836)]. Comment: original vernacular name unavailable (Art. 11.7.2): not subsequently latinized.

*Clidonotites Chapuis, 1874: 414 [stem: *Clidonot-*]. Type genus: *Clidonotus* Chapuis, 1874. Comment: original vernacular name unavailable (Art. 11.7.2): subsequently used in latinized form but not generally attributed to Chapuis (1874); Clidonotidae was used as valid by Ienistea (1986: 31) but it was not attributed to Chapuis (1874); Ienistea's name is also unavailable, it was proposed after 1930 without description or bibliographic reference to such a description (Art. 13.1).

*Colaspidémites Chapuis, 1874: 364 [stem: *Colaspidem-*]. Type genus: *Colaspidema* Laporte, 1833. Comment: original vernacular name unavailable (Art. 11.7.2): subsequently used in latinized form but not generally attributed to Chapuis (1874); Colaspidemidae was used as valid by Ienistea (1986: 31) but it was not attributed to Chapuis (1874); Ienistea's name is also unavailable, it was proposed after 1930 without description or bibliographic reference to such a description (Art. 13.1).

*Cyrtonites Chapuis, 1874: 364 [stem: *Cyrton-*]. Type genus: *Cyrtonus* Latreille, 1829. Comment: original vernacular name unavailable (Art. 11.7.2): not subsequently latinized.

*Elytrosphoerites Chapuis, 1874: 406 [stem: *Elytrosphoer-*]. Type genus: *Elytrosphoera* Blanchard, 1845. Comment: original vernacular name unavailable (Art. 11.7.2): not subsequently latinized; name incorrectly spelled Elytrosphaerites in the index of the same work (page 449).

Entomoscélites Chapuis, 1874: 418 [stem: *Entomoscelid-*]. Type genus: *Entomoscelis* Chevrolat, 1836. Comment: original vernacular name available (Art. 11.7.2): first used in latinized form by J. L. LeConte and G. H. Horn (1883: 344, as Entomoscelides [treated as Latin]), generally accepted as in Warchalowski (1994: 154, as Entomoscelina); incorrect original stem formation, not in prevailing usage.

*Lycariites Chapuis, 1874: 420 [stem: *Lycari-*]. Type genus: *Lycaria* Stål, 1857. Comment: original vernacular name unavailable (Art. 11.7.2): not subsequently latinized.

Phyllocharites Chapuis, 1874: 422 [stem: *Phyllocharit-*]. Type genus: *Phyllocharis* Dalman, 1824. Comment: original vernacular name available (Art. 11.7.2): first used in latinized form and generally accepted as in Weise (1915: 436, as Phyllocharini); incorrect original stem formation, not in prevailing usage.

*Pyxites Chapuis, 1874: 438 [stem: *Pyxid-*]. Type genus: *Pyxis* Chevrolat, 1843. Comment: original vernacular name unavailable (Art. 11.7.2): not subsequently latinized; incorrect original stem formation, not in prevailing usage.

Phyllodectae J. L. LeConte and G. H. Horn, 1883: 344 [stem: *Phyllodect-*]. Type genus: *Phyllodecta* Kirby, 1837 [subgenus of *Phratora* Chevrolat, 1836].

Dicranosternini Weise, 1915: 436 [stem: *Dicranostern-*]. Type genus: *Dicranosterna* Motschulsky, 1860.

Phaedonini Weise, 1915: 435 [stem: *Phaedon-*]. Type genus: *Phaedon* Latreille, 1829.

Zygogrammini Weise, 1915: 435 [stem: *Zygogrammat-*]. Type genus: *Zygogramma* Chevrolat, 1836. Comment: incorrect original stem formation, not in prevailing usage.

Chrysolinina Chen, 1936: 64, in key [stem: *Chrysolin-*]. Type genus: *Chrysolina* Motschulsky, 1860 [placed on the Official List of Generic Names in Zoology (ICZN 1984c)].

Barymelini Bechyné, 1948: 295 [stem: *Barymel-*]. Type genus: *Barymela* Weise, 1910.

Doryphorini Bechyné, 1950a: 115, in key [stem: *Doryphor-*]. Type genus: *Doryphora* Illiger, 1807. Comment: family-group name proposed as new without reference to Doryphores Motschulsky, 1860.

Monarditini Bechyné, 1953: 85 [stem: *Monardit-*]. Type genus: *Monardita* Bechyné, 1948.

Oreinini Bechyné, 1958: 218, nota [stem: *Orein-*]. Type genus: *Oreina* Chevrolat, 1836. Comment: the older name Oreinini Bleeker, 1863 (type genus *Oreinus* M'Clelland, 1839) is available in Pisces; this case is to be referred to the Commission to remove the homonymy (Art. 55.3.1).

*Phytodectini Bechyné, 1980: 58 [stem: *Phytodect-*]. Type genus: *Phytodecta* Kirby, 1837 [syn. of *Gonioctena* Chevrolat, 1836]. Comment: unavailable family-group name, proposed after 1930 without description or bibliographic reference to such a description (Art. 13.1).

Hispostomini Daccordi, 1981: 181, in key [stem: *Hispostomat-*]. Type genus: *Hispostoma* Weise, 1907. Comment: incorrect original stem formation, not in prevailing usage.

Sphaeratrixini Daccordi, 1981: 181, in key [stem: *Sphaeratrix-*]. Type genus: *Sphaeratrix* Gistel, 1848. Comment: the correct stem based on *Sphaeratrix* is unclear since Gistel did not specify the etymology of his genus therefore we accept the original stem as correct.

*Gastrophysina Steinhausen, 2001: 47 [stem: *Gastrophys-*]. Type genus: *Gastrophysa* Chevrolat, 1836. Comment: unavailable family-group name, proposed after 1999 without explicit intention (Art. 16.1).

Gastrophysina Kippenberg, 2010a: 68 [stem: *Gastrophys-*]. Type genus: *Gastrophysa* Chevrolat, 1836.

Tribe Timarchini Motschulsky, 1860

Timarchaeines Motschulsky, 1860: 187 [stem: *Timarch-*]. Type genus: *Timarcha* Samouelle, 1819. Comment: original vernacular name available (Art. 11.7.2): first used in latinized form by Gerstaecker (1862: 405, as Timarchae), generally accepted as in Burakowski et al. (1990: 115, as Timarchini); incorrect original stem formation, not in prevailing usage.

Subfamily Galerucinae Latreille, 1802

Galerucae Latreille, 1802: 228 [stem: *Galeruc-*]. Type genus: *Galeruca* Geoffroy, 1762 [placed on the Official List of Generic Names in Zoology (ICZN 1994a)].

Tribe Alticini Newman, 1834

Halticites Newman, 1834: 421 [stem: *Altic-*]. Type genus: *Altica* Geoffroy, 1762 [as *Haltica*, unjustified emendation of type genus name by Illiger (1801), not in prevailing usage; *Altica* placed on the Official List of Generic Names in Zoology (ICZN 1994a)]. Comment: incorrect original stem formation, not in prevailing usage.

Hesperidae Swainson, 1840: 310 [stem: *Hesper-*]. Type genus: *Hespera* Weise, 1889.

Longitarses Maehler, 1850: 7 [stem: *Longitars-*]. Type genus: *Longitarsus* Latreille, 1829.

Plectroscelides C. G. Thomson, 1866: 213 [stem: *Plectroscelid-*]. Type genus: *Plectroscelis* Chevrolat, 1836. Comment: incorrect original stem formation, not in prevailing usage.

*Acrocryptites Chapuis, 1875: 36 [stem: *Acrocrypt-*]. Type genus: *Acrocrypta* Baly, 1862. Comment: original vernacular name unavailable (Art. 11.7.2): subsequently used in latinized form but not generally attributed to Chapuis (1875); Acrocryptidae was used as valid by Ienistea (1986: 31) but it was not attributed to Chapuis (1875); Ienistea's name is also unavailable, it was proposed after 1930 without description or bibliographic reference to such a description (Art. 13.1).

*Amphimélites Chapuis, 1875: 34 [stem: *Amphimel-*]. Type genus: *Amphimela* Chapuis, 1875. Comment: original vernacular name unavailable (Art. 11.7.2): not subsequently used in latinized form and treated as valid.

Aphthonites Chapuis, 1875: 69 [stem: *Aphthon-*]. Type genus: *Aphthona* Chevrolat, 1836. Comment: original vernacular name available (Art. 11.7.2): first used in latinized form by J. L. LeConte and G. H. Horn (1883: 352, as Aphthonae), generally accepted as in Bechyné (1968: 1708, as Aphthonini).

Arsipodites Chapuis, 1875: 37 [stem: *Arsipod-*]. Type genus: *Arsipoda* Erichson, 1842. Comment: original vernacular name available (Art. 11.7.2): first used in latinized form by J. L. LeConte and G. H. Horn (1883: 353, as Arsipodes [treated as Latin]), generally accepted as in Bechyné and Špringlová de Bechyné (1973: 26, as Arsipodini).

Aspicélites Chapuis, 1875: 75 [stem: *Aspicel-*]. Type genus: *Aspicela* Dejean, 1836. Comment: original vernacular name available (Art. 11.7.2): first used in latinized form and generally accepted as in G. H. Horn (1889: 166, as Aspicelae).

Blépharidites Chapuis, 1875: 26 [stem: *Blepharid-*]. Type genus: *Blepharida* Chevrolat, 1836. Comment: original vernacular name available (Art. 11.7.2): first used in latinized form by J. L. LeConte and G. H. Horn (1883: 350, as Blepharidae [incorrect stem formation]), generally accepted as in Bechyné (1968: 1724, as Blepharidini).

Crépidodérites Chapuis, 1875: 51 [stem: *Crepidoder-*]. Type genus: *Crepidodera* Chevrolat, 1836. Comment: original vernacular name available (Art. 11.7.2): first used in latinized form and generally accepted as in J. L. LeConte and G. H. Horn (1883: 350, as Crepidoderae).

*Diamphidiites Chapuis, 1875: 24 [stem: *Diamphidi-*]. Type genus: *Diamphidia* Gerstaecker, 1855. Comment: original vernacular name unavailable (Art. 11.7.2): subsequently used in latinized form but not generally attributed to Chapuis (1875); Diamphidiini and Diamphidiidae were subsequently used as valid by Bechyné (1980: 58) and Ienistea (1986: 31) but those names were not attributed to Chapuis (1875); Bechyné's and Ienistea's names are also unavailable, they were proposed after 1930 without description or bibliographic reference to such a description (Art. 13.1).

Diboliites Chapuis, 1875: 137 [stem: *Diboli-*]. Type genus: *Dibolia* Latreille, 1829. Comment: original vernacular name available (Art. 11.7.2): first used in latinized form by J. L. LeConte and G. H. Horn (1883: 350, as Diboliae), generally accepted as in Bechyné (1997: 206, as Diboliini).

*Élithiites Chapuis, 1875: 21 [stem: *Elithi-*]. Type genus: *Elithia* Chapuis, 1875. Comment: original vernacular name unavailable (Art. 11.7.2): subsequently used in latinized form but not generally attributed to Chapuis (1875); this name was originally spelled ELITHIITES in the key but ÉLITHIIDES in description (both p. 21); ELITHIIDAE was used as valid by Ienistea (1986: 31) but it was not attributed to Chapuis (1875); Ienistea's name is also unavailable, it was proposed after 1930 without description or bibliographic reference to such a description (Art. 13.1).

LACTICITES Chapuis, 1875: 123 [stem: *Lactic-*]. Type genus: *Lactica* Erichson, 1847. Comment: original vernacular name available (Art. 11.7.2): first used in latinized form and generally accepted as in J. L. LeConte and G. H. Horn (1883: 350, as LACTICAE).

MNIOPHILITES Chapuis, 1875: 129 [stem: *Mniophil-*]. Type genus: *Mniophila* Stephens, 1831. Comment: original vernacular name available (Art. 11.7.2): first used in latinized form and generally accepted as in J. L. LeConte and G. H. Horn (1883: 350, as MNIOPHILAE).

MONOPLATITES Chapuis, 1875: 91 [stem: *Monoplat-*]. Type genus: *Monoplatus* Chevrolat, 1846 [this genus is usually credited to Clark, 1860, e.g., Seeno and Wilcox (1982: 141) and Linzmeier and Konstantinov (2009), but the name was first made available with a short description by Chevrolat (1846: 333); syn. of *Sphaeronychus* Dejean, 1836]. Comment: original vernacular name available (Art. 11.7.2): first used in latinized form by J. L. LeConte and G. H. Horn (1883: 350, as MONOPLATI), generally accepted as in Bechyné and Špringlová de Bechyné (1975: 132, as MONOPLATINI); treated as a valid tribe and given precedence over SPHAERONYCHINI (Art. 40.1) by Linzmeier and Konstantinov (2009: 657).

*NONARTHRITES Chapuis, 1875: 141 [stem: *Nonarthr-*]. Type genus: *Nonarthra* Baly, 1862. Comment: original vernacular name unavailable (Art. 11.7.2): subsequently used in latinized form but not generally attributed to Chapuis (1875); NONARTHRIDAE was used as valid by Ienistea (1986: 31) but it was not attributed to Chapuis (1875); Ienistea's name is also unavailable, it was proposed after 1930 without description or bibliographic reference to such a description (Art. 13.1).

OEDIONYCHITES Chapuis, 1875: 81 [stem: *Oedionych-*]. Type genus: *Oedionychis* Latreille, 1829. Comment: original vernacular name available (Art. 11.7.2): first used in latinized form by J. L. LeConte and G. H. Horn (1883: 349, as OEDIONYCHES [treated as Latin]), generally accepted as in Bechyné (1968: 1703, as OEDIONYCHINI).

OXYGONITES Chapuis, 1875: 43 [stem: *Oxygon-*]. Type genus: *Oxygona* Chevrolat, 1847 [this genus has been attributed to "Chevrolat, 1837" (see Seeno and Wilcox 1982: 132) but the name is not available from Dejean's second and third editions of his catalogue; it was first made available by Chevrolat (1847: 368)]. Comment: original vernacular name available (Art. 11.7.2): first used in latinized form and generally accepted as in Jacoby (1884a: 34, as OXYGONINAE).

*PROCALITES Chapuis, 1875: 175 [stem: *Procal-*]. Type genus: *Procalus* Clark, 1865. Comment: original vernacular name unavailable (Art. 11.7.2): subsequently used in latinized form but not generally attributed to Chapuis (1875); Procalidae was used as valid by Ienistea (1986: 31) but it was not attributed to Chapuis (1875); Ienistea's name is also unavailable, it was proposed after 1930 without description or bibliographic reference to such a description (Art. 13.1).

Psylliodites Chapuis, 1875: 140 [stem: *Psylliod-*]. Type genus: *Psylliodes* Latreille, 1829. Comment: original vernacular name available (Art. 11.7.2): first used in latinized form and generally accepted as in J. L. LeConte and G. H. Horn (1883: 349, as Psylliodes [treated as Latin]).

Chaetocnemae J. L. LeConte and G. H. Horn, 1883: 354 [stem: *Chaetocnem-*]. Type genus: *Chaetocnema* Stephens, 1831.

Disonychae J. L. LeConte and G. H. Horn, 1883: 351 [stem: *Disonych-*]. Type genus: *Disonycha* Chevrolat, 1836.

Euplectrosceles G. H. Horn, 1889: 167, in key [stem: *Euplectroscelid-*]. Type genus: *Euplectroscelis* Crotch, 1873. Comment: incorrect original stem formation, not in prevailing usage.

Pseudolampses G. H. Horn, 1889: 166, in key [stem: *Pseudolampse-*]. Type genus: *Pseudolampsis* Horn, 1889. Comment: incorrect original stem formation, not in prevailing usage.

Systenae G. H. Horn, 1889: 167, in key [stem: *Systen-*]. Type genus: *Systena* Chevrolat, 1836.

Luperalticini Leng, 1920: 301 [stem: *Luperaltic-*]. Type genus: *Luperaltica* Crotch, 1873.

Octogonotini Weise, 1921: 151 [stem: *Octogonot-*]. Type genus: *Octogonotes* Drapiez, 1820.

Serraticollini B. E. White, 1942: 17 [stem: *Serraticoll-*]. Type genus: *Serraticollis* B. E. White, 1942.

Sphaeronychini Bechyné and Špringlová de Bechyné, 1960: 7 [stem: *Sphaeronych-*]. Type genus: *Sphaeronychus* Dejean, 1836 [*Sphaeronychus* is an incorrect subsequent spelling of *Sphraeronychus* Dejean, 1836, in prevailing usage, treated as correct original spelling (Art. 33.3.1)]. Comment: this taxon was treated as a valid tribe (which included Monoplatini) by Furth (2007: 90).

Disonychina Bechyné and Špringlová de Bechyné, 1966: 142, in key [stem: *Disonych-*]. Type genus: *Disonycha* Chevrolat, 1836. Comment: family-group name proposed as new without reference to Disonychae J. L. LeConte and G. H. Horn, 1883.

Cacoscelini Bechyné, 1968: 1715 [stem: *Cacoscelid-*]. Type genus: *Cacoscelis* Chevrolat, 1836. Comment: Cacoscelini J. Thomson, 1861 (type genus *Cacosceles* Newman, 1838) is available in Cerambycidae and used as valid; the correct stem *Cacoscelid-* should be used for the alticine name in the future in order to avoid homonymy with the cerambycid name.

*Hermaeophagina Bechyné, 1968: 1698 [stem: *Hermaeophag-*]. Type genus: *Hermaeophaga* Foudras, 1859. Comment: unavailable family-group name, proposed after 1930 without description or bibliographic reference to such a description (Art. 13.1).

LONGITARSINI Bechyné, 1968: 1708 [stem: *Longitars-*]. Type genus: *Longitarsus* Latreille, 1829. Comment: family-group name proposed as new without reference to LONGITARSES Maehler, 1850.

LYPNEINA Bechyné, 1968: 1717 [stem: *Lypne-*]. Type genus: *Lypnea* Baly, 1876.

PODAGRICINI Bechyné, 1968: 1718 [stem: *Podagric-*]. Type genus: *Podagrica* Chevrolat, 1836.

*SPHAERODERMINI Bechyné, 1968: 1702 [stem: *Sphaerodermat-*]. Type genus: *Sphaeroderma* Stephens, 1831. Comment: unavailable family-group name, proposed after 1930 without description or bibliographic reference to such a description (Art. 13.1); incorrect original stem formation, not in prevailing usage.

WITTMERALTICINA Bechyné, 1968: 1717 [stem: *Wittmeraltic-*]. Type genus: *Wittmeraltica* Bechyné, 1956.

PSILAPHINA Bechyné and Špringlová de Bechyné, 1973: 27, in key [stem: *Psilaph-*]. Type genus: *Psilapha* Clark, 1865.

DIPHAULACINI Bechyné and Špringlová de Bechyné, 1975: 116 [stem: *Diphaulac-*]. Type genus: *Diphaulaca* Chevrolat, 1836.

MANOBIINI Bechyné and Špringlová de Bechyné, 1975: 132, in key [stem: *Manobi-*]. Type genus: *Manobia* Jacoby, 1885.

MARCAPATIINI Bechyné and Špringlová de Bechyné, 1975: 131 [stem: *Marcapati-*]. Type genus: *Marcapatia* Bechyné, 1958.

MONOMACRINA Bechyné and Špringlová de Bechyné, 1975: 27 [stem: *Monomacr-*]. Type genus: *Monomacra* Chevrolat, 1836.

SYSTENINI Bechyné and Špringlová de Bechyné, 1975: 132, in key [stem: *Systen-*]. Type genus: *Systena* Chevrolat, 1836. Comment: family-group name proposed as new without reference to SYSTENAE G. H. Horn, 1889.

*PHYGASIINI Bechyné, 1980: 58 [stem: *Phygasi-*]. Type genus: *Phygasia* Dejean, 1836. Comment: unavailable family-group name, proposed after 1930 without description or bibliographic reference to such a description (Art. 13.1).

*OCNOSCELINA Seeno and Wilcox, 1982: 127 [stem: *Ocnoscelid-*]. Type genus: *Ocnoscelis* Erichson, 1847. Comment: unavailable family-group name, proposed after 1930 without description or bibliographic reference to such a description (Art. 13.1); incorrect original stem formation, not in prevailing usage.

HERMAEOPHAGINA Bechyné, 1997: 116 [stem: *Hermaeophag-*]. Type genus: *Hermaeophaga* Foudras, 1859.

OCNOSCELINA Bechyné, 1997: 124 [stem: *Ocnoscelid-*]. Type genus: *Ocnoscelis* Erichson, 1847. Comment: incorrect original stem formation, not in prevailing usage.

Tribe DECARTHROCERINI Laboissière, 1937

DECARTHROCERINA Laboissière, 1937: 29 [stem: *Decarthrocer-*]. Type genus: *Decarthrocera* Laboissière, 1937.

Tribe Galerucini Latreille, 1802

Galerucae Latreille, 1802: 228 [stem: *Galeruc-*]. Type genus: *Galeruca* Geoffroy, 1762 [placed on the Official List of Generic Names in Zoology (ICZN 1994a)].

Coelomérites Chapuis, 1875: 196 [stem: *Coelomer-*]. Type genus: *Coelomera* Chevrolat, 1836. Comment: original vernacular name available (Art. 11.7.2): first used in latinized form by Leng (1920: 296, as Coelomerini), generally accepted as in Riley et al. (2002: 639, as Coelomerites [treated as Latin]).

Atysites Chapuis, 1875: 192 [stem: *Atys-*]. Type genus: *Atysa* Baly, 1864. Comment: original vernacular name available (Art. 11.7.2): first used in latinized form by Leng (1920: 296, as Atysini), generally accepted as in Riley et al. (2002: 639, as Atysites [treated as Latin]).

Apophyliites Chapuis, 1875: 182 [stem: *Apophyli-*]. Type genus: *Apophylia* Duponchel and Chevrolat, 1841 [this genus name has been attributed to "J. Thomson, 1858", e.g., Bezděk (2003: 71), however it was made available for the first time with a description by Duponchel and Chevrolat (1841: 31)]. Comment: original vernacular name available (Art. 11.7.2): first used in latinized form by Weise (1923: 124, as Apophydiini [incorrect stem formation]), generally accepted as in Seeno and Wilcox (1982: 96, as Apophyliites [treated as Latin]).

Rupiliites Chapuis, 1875: 213 [stem: *Rupili-*]. Type genus: *Rupilia* Clark, 1864. Comment: original vernacular name available (Art. 11.7.2): first used in latinized form and generally accepted as in Jacoby (1899: 83, as Rupilinae [incorrect stem formation]).

Schematizites Chapuis, 1875: 195 [stem: *Schematiz-*]. Type genus: *Schematiza* Chevrolat, 1836. Comment: original vernacular name available (Art. 11.7.2): first used in latinized form by Wilcox (1971: 101, as Schematizites [treated as Latin]), generally accepted as in Riley et al. (2002: 639, as Schematizites [treated as Latin]).

Mombasicites Laboissière, 1922: 230 [stem: *Mombasic-*]. Type genus: *Mombasica* Fairmaire, 1888.

Chorini Weise, 1923: 124 [stem: *Chorin-*]. Type genus: *Chorina* Baly, 1866. Comment: incorrect original stem formation, not in prevailing usage; correction of stem by Weise (1924: 179).

Leptosonychini Weise, 1924: 180 [stem: *Leptosonych-*]. Type genus: *Leptosonyx* Weise, 1885.

Theonina Laboissière, 1934: 7, in key [stem: *Theon-*]. Type genus: *Theone* Gistel, 1857.

Tribe Hylaspini Chapuis, 1875

Hylaspites Chapuis, 1875: 237 [stem: *Hylasp-*]. Type genus: *Hylaspes* Baly, 1865. Comment: original vernacular name available (Art. 11.7.2): first used in latinized form by Laboissière (1934: 8, as Hylaspina), generally accepted as in

Silfverberg (1990: 120, as HYLASPINI); First Reviser found (HYLASPINI Chapuis, 1875 vs AGELASTICINI Chapuis, 1875) is Silfverberg (1990: 120).

SERMYLITES Chapuis, 1875: 224 [stem: *Sermyl-*]. Type genus: *Sermyla* Chapuis, 1875 [preoccupied genus name, not *Sermyla* Walker 1854 [Lepidoptera], not *Sermyla* Adams, 1854 [Mollusca]; syn. of *Sermylassa* Reitter, 1912]. Comment: original vernacular name available (Art. 11.7.2): first used in latinized form by Jacoby (1884a: 64, as SERMYLINAE), generally accepted as in Seeno and Wilcox (2002: 104, as SERMYLITES [treated as Latin]); permanently invalid (Art. 39): based on preoccupied type genus.

ANTIPHITES Chapuis, 1875: 232 [stem: *Antiph-*]. Type genus: *Antipha* Baly, 1865 [preoccupied genus name, not *Antipha* Walker, 1855 [Lepidoptera]; syn. of *Dercetina* Gressitt and Kimoto, 1963]. Comment: original vernacular name available (Art. 11.7.2): first used in latinized form and generally accepted as in Seeno and Wilcox (1982: 103, as ANTIPHITES [treated as Latin]); permanently invalid (Art. 39): based on preoccupied type genus.

AGELASTICITES Chapuis, 1875: 167 [stem: *Agelastic-*]. Type genus: *Agelastica* Chevrolat, 1836. Comment: original vernacular name available (Art. 11.7.2): first used in latinized form by Weise (1903a: 209, as AGELASTICITES [treated as Latin]), generally accepted as in Burakowski et al. (1991: 23, as AGELASTICINI).

AGELASINI Leng, 1920: 298 [stem: *Agelas-*]. Type genus: *Agelasa* sensu Horn, 1893 [not *Agelasa* Motschulsky, 1860; syn. of *Sermylassa* Reitter, 1912]. Comment: based on a misidentified type genus.

BONESIITES Laboissière, 1926: 91 [stem: *Bonesi-*]. Type genus: *Bonesia* Baly, 1865.

CAPULINI Ogloblin, 1936: 341 [stem: *Capul-*]. Type genus: *Capula* Jakobson, 1925.

GALLERUCIDINI Gressitt and Kimoto, 1963: 390, in key [stem: *Gallerucid-*]. Type genus: *Gallerucida* Motschulsky, 1861.

SERMYLASSINI Mroczkowski, 1991: 22 [stem: *Sermylass-*]. Type genus: *Sermylassa* Reitter, 1912. Comment: replacement name for SERMYLITES Chapuis, 1875 and AGELASINI Leng, 1920 because of the homonymy of the type genus.

Tribe LUPERINI Gistel, 1848

LUPERIIDAE Gistel, 1848: [9] [stem: *Luper-*]. Type genus: *Luperus* Geoffroy, 1762 [placed on the Official List of Generic Names in Zoology (ICZN 1984a)]. Comment: name previously attributed to Chapuis (1875) in the literature; also see Beenen (2010: 464) for an alternative classification within this tribe; incorrect original stem formation, not in prevailing usage.

AULACOPHORITES Chapuis, 1875: 158 [stem: *Aulacophor-*]. Type genus: *Aulacophora* Chevrolat, 1836. Comment: original vernacular name available (Art. 11.7.2): first used in latinized form by Laboissière (1934: 8, as AULACOPHORINA), generally accepted as in Seeno and Wilcox (1982: 97, as AULACOPHORITES [treated as Latin]).

Diabroticites Chapuis, 1875: 165 [stem: *Diabrotic-*]. Type genus: *Diabrotica* Chevrolat, 1836. Comment: original vernacular name available (Art. 11.7.2): first used in latinized form by Leng (1920: 297, as Diabroticini), generally accepted as in Riley et al. (2002: 639, as Diabroticina).

Cérotomites Chapuis, 1875: 229 [stem: *Cerotom-*]. Type genus: *Cerotoma* Chevrolat, 1836. Comment: original vernacular name available (Art. 11.7.2): first used in latinized form by Leng (1920: 298, as Cerotomini), generally accepted as in Riley et al. (2002: 639, as Cerotomites [treated as Latin]).

Platyxanthites Chapuis, 1875: 243 [stem: *Platyxanth-*]. Type genus: *Platyxantha* Baly, 1864. Comment: original vernacular name available (Art. 11.7.2): first used in latinized form and generally accepted as in Jacoby (1884b: 226, as Platyxanthinae).

*Théopéites Chapuis, 1875: 241 [stem: *Theope-*]. Type genus: *Theopea* Baly, 1864. Comment: original vernacular name unavailable (Art. 11.7.2): subsequently used in latinized form but not generally attributed to Chapuis (1875); Theopeidae was used as valid by Ienistea (1986: 31) but it was not attributed to Chapuis (1875); Ienistea's name is also unavailable, it was proposed after 1930 without description or bibliographic reference to such a description (Art. 13.1); Theopeina Clench, 1955 (type genus *Theope* Doubleday, 1847) is available in Lepidoptera.

Scelidites Chapuis, 1875: 184 [stem: *Scelid-*]. Type genus: *Scelida* Chapuis, 1875. Comment: original vernacular name available (Art. 11.7.2): first used in latinized form by Jacoby (1884a: 62, as Scelidinae), generally accepted as in Riley et al. (2002: 639, as Scelidites [treated as Latin]).

Phyllobroticites Chapuis, 1875: 163 [stem: *Phyllobrotic-*]. Type genus: *Phyllobrotica* Chevrolat, 1836. Comment: original vernacular name available (Art. 11.7.2): first used in latinized form by Leng (1920: 297, as Phyllobroticini), generally accepted as in Riley et al. (2002: 639, as Phyllobroticites [treated as Latin]).

*Agetocérites Chapuis, 1875: 177 [stem: *Agetocer-*]. Type genus: *Agetocera* Hope, 1831 [*Agetocera* is an unjustified emendation of the original genus name *Aegelocerus* Hope, 1831 by Hope (1840a: 170); because *Agetocera* is in prevailing usage and has been attributed to the original author and date, e.g. Bezděk (2010a: 73, 2010c: 464) it is treated here as a justified emendation (Art. 33.2.3.1)]. Comment: original vernacular name unavailable (Art. 11.7.2): subsequently used in latinized form but not generally attributed to Chapuis (1875); Agetoceridae was used as valid by Ienistea (1986: 31) but it was not attributed to Chapuis (1875); Ienistea's name is also unavailable, it was proposed after 1930 without description or bibliographic reference to such a description (Art. 13.1).

Mimastrites Chapuis, 1875: 178 [stem: *Mimastr-*]. Type genus: *Mimastra* Baly, 1865. Comment: original vernacular name available (Art. 11.7.2): first used in latinized form by Jacoby (1884a: 67, as Mimastrinae), generally accepted as in Weise (1903b: 334, as Mimastrites [treated as Latin]).

Ornithognathites Chapuis, 1875: 176 [stem: *Ornithognath-*]. Type genus: *Ornithognathus* J. Thomson, 1858. Comment: original vernacular name available (Art. 11.7.2): first used in latinized form by Harold (1877: 110, as Ornithognathinae), generally accepted as in Seeno and Wilcox (1982: 97, as Ornithognathites [treated as Latin]).

Monoleptites Chapuis, 1875: 234 [stem: *Monolept-*]. Type genus: *Monolepta* Chevrolat, 1836. Comment: original vernacular name available (Art. 11.7.2): first used in latinized form by Jacoby (1884a: 61, as Monoleptinae), generally accepted as in Riley et al. (2002: 639, as Monoleptites [treated as Latin]).

Cerophysites Chapuis, 1875: 181 [stem: *Cerophys-*]. Type genus: *Cerophysa* Chevrolat, 1836. Comment: original vernacular name available (Art. 11.7.2): first used in latinized form and generally accepted as in Seeno and Wilcox (1982: 97, as Cerophysides [treated as Latin]).

Phyllecthrites G. H. Horn, 1893: 60, in key [stem: *Phyllecthr-*]. Type genus: *Phyllecthris* Dejean, 1836.

Rhaphidopalpini Weise, 1902: 140 [stem: *Raphidopalp-*]. Type genus: *Raphidopalpa* Chevrolat, 1836 [as *Rhaphidopalpa*, unjustified emendation of type genus name by Rosenhauer (1856: 325), not in prevailing usage; syn. of *Aulacophora* Chevrolat, 1836]. Comment: incorrect original stem formation, not in prevailing usage.

Androlyperini Leng, 1920: 298 [stem: *Androlyper-*]. Type genus: *Androlyperus* Crotch, 1873.

Idacanthites Laboissière, 1921: 63 [stem: *Idacanth-*]. Type genus: *Idacantha* Fairmaire, 1869.

Hyperacanthites Laboissière, 1924: 142 [stem: *Hyperacanth-*]. Type genus: *Hyperacantha* Chapuis, 1879.

Megalognathites Laboissière, 1926: 185 [stem: *Megalognath-*]. Type genus: *Megalognatha* Baly, 1878.

*Exosomites Wilcox, 1965: 93 [stem: *Exosomat-*]. Type genus: *Exosoma* Jacoby, 1903. Comment: unavailable family-group name, proposed after 1930 without description or bibliographic reference to such a description (Art. 13.1); incorrect original stem formation, not in prevailing usage.

*Trachyscelidites Wilcox, 1972: 430 [stem: *Trachyscelid-*]. Type genus: *Trachyscelida* G. H. Horn, 1893. Comment: unavailable family-group name, proposed after 1930 without description or bibliographic reference to such a description (Art. 13.1).

*Adoxiites Wilcox, 1973: 433 [stem: *Adoxi-*]. Type genus: *Adoxia* Broun, 1880. Comment: unavailable family-group name, proposed after 1930 without description or bibliographic reference to such a description (Art. 13.1).

*Eumeleptites Wilcox, 1973: 447 [stem: *Eumelept-*]. Type genus: *Eumelepta* Jacoby, 1892. Comment: unavailable family-group name, proposed after 1930 without description or bibliographic reference to such a description (Art. 13.1).

*XENODITES Wilcox, 1973: 604 [stem: *Xenod-*]. Type genus: *Xenoda* Baly, 1877. Comment: unavailable family-group name, proposed after 1930 without description or bibliographic reference to such a description (Art. 13.1).

*DORYSCITES Wilcox, 1973: 606 [stem: *Dorysc-*]. Type genus: *Doryscus* Jacoby, 1887. Comment: unavailable family-group name, proposed after 1930 without description or bibliographic reference to such a description (Art. 13.1).

Tribe METACYCLINI Chapuis, 1875

MÉTACYCLITES Chapuis, 1875: 212 [stem: *Metacycl-*]. Type genus: *Metacycla* Baly, 1861. Comment: original vernacular name available (Art. 11.7.2): first used in latinized form by Leng (1920: 298, as METACYCLINI), generally accepted as in Riley et al. (2002: 639, as METACYCLINI).

STICTOCEMITES Laboissière, 1922: 220 [stem: *Stictocem-*]. Type genus: *Stictocema* Jacoby, 1906.

*EXORINI Wilcox, 1965: 179 [stem: *Exor-*]. Type genus: *Exora* Chevrolat, 1836. Comment: unavailable family-group name, proposed after 1930 without description or bibliographic reference to such a description (Art. 13.1).

Tribe OIDINI Laboissière, 1921 (1875)

ADORIITES Chapuis, 1875: 155 [stem: *Adori-*]. Type genus: *Adorium* Fabricius, 1801 [syn. of *Oides* Weber, 1801]. Comment: original vernacular name available (Art. 11.7.2): first used in latinized form and generally accepted as in Lea (1925: 16, as ADORIITES [treated as Latin]); usage of younger name OIDINI Laboissière, 1921 conserved (Art. 40.2).

OÏDITES Laboissière, 1921: 38 [stem: *Oid-*]. Type genus: *Oides* Weber, 1801. Comment: usage of this name conserved over ADORIINI Chapuis, 1875 (Art. 40.2).

Subfamily LAMPROSOMATINAE Lacordaire, 1848

LAMPROSOMIDEAE Lacordaire, 1848: 559 [stem: *Lamprosomat-*]. Type genus: *Lamprosoma* Kirby, 1819.

Tribe LAMPROSOMATINI Lacordaire, 1848

LAMPROSOMIDEAE Lacordaire, 1848: 559 [stem: *Lamprosomat-*]. Type genus: *Lamprosoma* Kirby, 1819. Comment: incorrect original stem formation, not in prevailing usage.

Tribe NEOCHLAMYSINI Monrós, 1959

NEOCHLAMYSINI Monrós, 1959b: 29 [stem: *Neochlamys-*]. Type genus: *Neochlamys* Jacoby, 1882. Comment: current spelling maintained (Art. 29.5): incorrect stem formation in prevailing usage (should be *Neochlamyd-*).

Tribe Sphaerocharini Chapuis, 1874

Sphoerocharides Chapuis, 1874: 206 [stem: *Sphaerochar-*]. Type genus: *Sphaerocharis* Lacordaire, 1848 [as *Sphoerocharis*, incorrect subsequent spelling of type genus name, not in prevailing usage]. Comment: original vernacular name available (Art. 11.7.2): first used in latinized form by Heyne and Taschenberg (1907: 248, as Sphaerocharini), generally accepted as in Monrós (1956: 32, as Sphaerocharini); current spelling maintained (Art. 29.5): incorrect stem formation in prevailing usage (should be *Sphaerocharit-*).

Subfamily Cryptocephalinae Gyllenhal, 1813

Cryptocephaloideae Gyllenhal, 1813: 582 [stem: *Cryptocephal-*]. Type genus: *Cryptocephalus* Geoffroy, 1762 [placed on the Official List of Generic Names in Zoology (ICZN 1994a)].

Tribe Clytrini Kirby, 1837

Clythridae Kirby, 1837: 207 [stem: *Clytr-*]. Type genus: *Clytra* Laicharting, 1781 [as *Clythra*, incorrect subsequent spelling of type genus name, not in prevailing usage]. Comment: incorrect original stem formation, not in prevailing usage.

Megalostomideae Lacordaire, 1848: 486 [stem: *Megalostomid-*]. Type genus: *Megalostomis* Chevrolat, 1836. Comment: incorrect original stem formation, not in prevailing usage.

Babideae Lacordaire, 1848: 394 [stem: *Babi-*]. Type genus: *Babia* Chevrolat, 1836. Comment: incorrect original stem formation, not in prevailing usage.

Ischiopachites Chapuis, 1874: 151 [stem: *Ischiopache-*]. Type genus: *Ischiopachys* Chevrolat, 1836. Comment: original vernacular name available (Art. 11.7.2): first used in latinized form by Jacoby and Clavareau (1906: 75, as Ischiopachitae), generally accepted as in Riley et al. (2002: 639, as Ischiopachina); incorrect original stem formation, not in prevailing usage.

Melolonthini Bedel, 1891: 106 [stem: *Melolonth-*]. Type genus: *Melolontha* Geoffroy, 1762 [preoccupied genus name, not *Melolontha* Fabricius, 1775 [Coleoptera: Scarabaeidae]; placed on the Official Index of Rejected and Invalid Generic Names in Zoology (ICZN 1994a); syn. of *Clytra* Laicharting, 1781]. Comment: permanently invalid (Art. 39): based on preoccupied type genus.

Eoclytrini Monrós, 1958b: 35 [stem: *Eoclytr-*]. Type genus: *Eoclytra* Monrós, 1958.

Arateini Moldenke, 1981: 85, in key [stem: *Arate-*]. Type genus: *Aratea* Lacordaire, 1848.

Tribe Cryptocephalini Gyllenhal, 1813

Cryptocephaloideae Gyllenhal, 1813: 582 [stem: *Cryptocephal-*]. Type genus: *Cryptocephalus* Geoffroy, 1762 [placed on the Official List of Generic Names in Zoology (ICZN 1994a)].

Subtribe ACHAENOPINA Chapuis, 1874

ACHAENOPITES Chapuis, 1874: 171 [stem: *Achaenop-*]. Type genus: *Achaenops* Suffrian, 1857. Comment: original vernacular name available (Art. 11.7.2): first used in latinized form and generally accepted as in Clavareau (1913b: 109, as ACHAENOPINI).

Subtribe CRYPTOCEPHALINA Gyllenhal, 1813

CRYPTOCEPHALOIDEAE Gyllenhal, 1813: 582 [stem: *Cryptocephal-*]. Type genus: *Cryptocephalus* Geoffroy, 1762 [placed on the Official List of Generic Names in Zoology (ICZN 1994a)].

Subtribe MONACHULINA Leng, 1920

*MONACHIDEN Suffrian, 1863: 79 [stem: *Monach-*]. Type genus: *Monachus* Chevrolat, 1836 [preoccupied genus name, not *Monachus* Fleming, 1822 [Mammalia], not *Monachus* Kaup, 1829 [Aves]; syn. of *Lexiphanes* Gistel, 1848]. Comment: original vernacular name unavailable (Art. 11.7.2): subsequently used in latinized form but not generally attributed to Suffrian (1863).

MONACHITES Chapuis, 1874: 172 [stem: *Monach-*]. Type genus: *Monachus* Chevrolat, 1836 [preoccupied genus name, not *Monachus* Fleming, 1822 [Mammalia], not *Monachus* Kaup, 1829 [Aves]; syn. of *Lexiphanes* Gistel, 1848]. Comment: original vernacular name available (Art. 11.7.2): first used in latinized form by J. L. LeConte and G. H. Horn (1883: 342, as MONACHI), generally accepted as in Liu (1935: 286, as MONACHINI); permanently invalid (Art. 39): based on preoccupied type genus; MONACHINAE Trouessart, 1897 (type genus *Monachus* Fleming, 1822) is available in Mammalia.

MONACHULINI Leng, 1920: 290 [stem: *Monachul-*]. Type genus: *Monachulus* Leng, 1918 [syn. of *Lexiphanes* Gistel, 1848].

Subtribe PACHYBRACHINA Chapuis, 1874

*PACHYBRACHIDEN Suffrian, 1863: 79 [stem: *Pachybrach-*]. Type genus: *Pachybrachis* Chevrolat, 1836. Comment: original vernacular name unavailable (Art. 11.7.2): subsequently used in latinized form but not generally attributed to Suffrian (1863).

PACHYBRACHITES Chapuis, 1874: 163 [stem: *Pachybrach-*]. Type genus: *Pachybrachis* Chevrolat, 1836. Comment: original vernacular name available (Art. 11.7.2): first used in latinized form by J. L. LeConte and G. H. Horn (1883: 342, as PACHYBRACHI), generally accepted as in Burakowski et al. (1990: 63, as PACHYBRACHINI).

Subtribe STYLOSOMINA Chapuis, 1874

STYLOSOMITES Chapuis, 1874: 162 [stem: *Stylosom-*]. Type genus: *Stylosomus* Suffrian, 1848. Comment: original vernacular name available (Art. 11.7.2):

first used in latinized form by Sharp (1876: 99, as Stylosomites [treated as Latin]), generally accepted as in Liu (1935: 292, as Stylosomini).

Tribe Fulcidacini Jakobson, 1924

*Chlamytes Blanchard, 1845b: 186 [stem: *Chlamyd-*]. Type genus: *Chlamys* Knoch, 1801 [preoccupied genus name, not *Chlamys* Bolten, 1798 [Mollusca]; syn. of *Chlamisus* Rafinesque, 1815]. Comment: original vernacular name unavailable (Art. 11.7.2): subsequently used in latinized form but not generally attributed to Blanchard (1845b); incorrect original stem formation, not in prevailing usage.

Chlamydeae Lacordaire, 1848: 636 [stem: *Chlamyd-*]. Type genus: *Chlamys* Knoch, 1801 [preoccupied genus name, not *Chlamys* Bolten, 1798 [Mollusca]; syn. of *Chlamisus* Rafinesque, 1815]. Comment: incorrect original stem formation, not in prevailing usage; permanently invalid (Art. 39): based on preoccupied type genus.

Fulcidacina Jakobson, 1924: 239 [stem: *Fulcidac-*]. Type genus: *Fulcidax* Clavareau, 1913 [for comments on the authorship of this genus see Bezděk (2010b: 76); syn. of *Poropleura* Lacordaire, 1848]. Comment: we use Fulcidacini Jakobson, 1924 as valid for this taxon as in Bezděk (2010b: 76).

Chlamisinae Gressitt, 1946: 84 [stem: *Chlamis-*]. Type genus: *Chlamisus* Rafinesque, 1815. Comment: replacement name for Chlamydinae Lacordaire, 1848 because of the homonymy of the type genus.

Subfamily Eumolpinae Hope, 1840

Eumolpidae Hope, 1840a: 162 [stem: *Eumolp-*]. Type genus: *Eumolpus* Kugelann, 1798 [an application to suppress *Eumolpus* Kugelann, 1798 and conserve *Eumolpus* Weber, 1801 was recently submitted to the Commission by Moseyko et al. (2010)].

Tribe Bromiini Baly, 1865 (1863)

Adoxinae Baly, 1863: 146 [stem: *Adox-*]. Type genus: *Adoxus* Kirby, 1837 [syn. of *Bromius* Chevrolat, 1836]. Comment: use of younger name Bromiini Baly, 1865 conserved over this name (Art. 40.2).

Heteraspinae Baly, 1863: 147 [stem: *Heteraspid-*]. Type genus: *Heteraspis* Chevrolat, 1836 [this genus (incorrectly attributed to "Chevrolat, 1837") has been treated as a synonym of *Scelodonta* Westwood, 1837 in the literature; *Heteraspis* was in fact made available by Chevrolat (1836: 413) and is a senior synonym of *Scelodonta* Westwood, 1837 (also see Löbl 2010: 83)]. Comment: although this is the oldest name for the tribe, we recommend that an application be sent to the Commission in order to conserve usage of Bromiini Baly, 1865; incorrect original stem formation, not in prevailing usage.

Bromiinae Baly, 1865: 438 [stem: *Bromi-*]. Type genus: *Bromius* Chevrolat, 1836 [an application to conserve *Bromius* Chevrolat, 1836, threatened by the older name *Eumolpus* Kugelann, 1798, was recently submitted to the Commission by Moseyko et al. (2010)]. Comment: use of family-group name conserved over Adoxiini Baly, 1863 (Art. 40.2).

Myochroinae Baly, 1865: 433 [stem: *Myochro-*]. Type genus: *Myochrous* Erichson, 1847.

Leprotites Chapuis, 1874: 268 [stem: *Leprotet-*]. Type genus: *Leprotes* Baly, 1863 [syn. of *Fidia* Motschulsky, 1860]. Comment: original vernacular name available (Art. 11.7.2): first used in latinized form by Lefèvre (1885: 71, as Leprotitae), generally accepted as in Riley et al. (2002: 639, as Leprotites [treated as Latin]); incorrect original stem formation, not in prevailing usage.

Pseudocolaspites Chapuis, 1874: 287 [stem: *Pseudocolaspid-*]. Type genus: *Pseudocolaspis* Laporte, 1833. Comment: original vernacular name available (Art. 11.7.2): first used in latinized form and generally accepted as in Lefèvre (1885: 84, as Pseudocolaspitae); incorrect original stem formation, not in prevailing usage.

Scelodontites Chapuis, 1874: 266 [stem: *Scelodont-*]. Type genus: *Scelodonta* Westwood, 1838 [syn. of *Heteraspis* Chevrolat, 1836]. Comment: original vernacular name available (Art. 11.7.2): first used in latinized form by Lefèvre (1885: 67, as Scelodontitae), generally accepted as in Riley et al. (2002: 639, as Scelodontites [treated as Latin]).

Tomyrites Chapuis, 1874: 264 [stem: *Tomyri-*]. Type genus: *Tomyris* Chapuis, 1874 [preoccupied genus name, not *Tomyris* Eichald, 1831 [Reptilia]; syn. of *Eboo* Reid, 1993]. Comment: original vernacular name available (Art. 11.7.2): first used in latinized form and generally accepted as in Lefèvre (1885: 64, as Tomyritae); permanently invalid (Art. 39): based on preoccupied type genus; incorrect original stem formation, not in prevailing usage.

*Goniopleurites Chapuis, 1875: 247 [stem: *Goniopleur-*]. Type genus: *Goniopleura* Westwood, 1832. Comment: original vernacular name unavailable (Art. 11.7.2): subsequently used in latinized form but not generally attributed to Chapuis (1875); Goniopleuridae was used as valid by Ienistea (1986: 31) but it was not attributed to Chapuis (1875); Ienistea's name is also unavailable, it was proposed after 1930 without description or bibliographic reference to such a description (Art. 13.1); usage of the name Goniopleuridae dates back at least to the 1850's in Trilobita, however it is based on the junior homonym *Goniopleura* Hawle and Corda 1847 and the trilobite family-group name is therefore permanently invalid (Art. 39).

Eubrachini Jacoby, 1908: 432 [stem: *Eubrach-*]. Type genus: *Eubrachis* sensu Baly, 1878 [not *Eubrachis* Dejean, 1836; syn. of *Macrocoma* Chapuis, 1874]. Comment: based on misidentified type genus.

Nerissini Kuntzen, 1912: 42 [stem: *Neriss-*]. Type genus: *Nerissus* Chapuis, 1874.

Cynoini Clavareau, 1914: 109 [stem: *Cynoi-*]. Type genus: *Cyno* T. A. Marshall, 1865. Comment: incorrect original stem formation, not in prevailing usage.

Odontionopini Clavareau, 1914: 63 [stem: *Odontionop-*]. Type genus: *Odontionopa* Erichson, 1842 [preoccupied genus name, not *Odontionopa* Chevrolat,

1836 [Coleoptera: Chrysomelidae: Eumolpinae: Euryopini]; syn. of *Eboo* Reid, 1993]. Comment: permanently invalid (Art. 39): based on preoccupied type genus.

Trichochryseini Clavareau, 1914: 82 [stem: *Trichochryse-*]. Type genus: *Trichochrysea* Baly, 1861.

Lypesthini Chûjô, 1956: 87 [stem: *Lypesthet-*]. Type genus: *Lypesthes* Baly, 1863 [conservation of usage of this name over *Fidia* Motschulsky, 1860 rejected by Commission (ICZN 2009c); syn. of *Fidia* Motschulsky, 1860]. Comment: name proposed to replace Leprotini Chapuis, 1874 because of the synonymy of the type genus; incorrect original stem formation, not in prevailing usage.

Ebooina Reid, 1993: 61 [stem: *Eboo-*]. Type genus: *Eboo* Reid, 1993. Comment: replacement name for Tomyrites Chapuis, 1874 because of the homonymy of the type genus.

Tribe Caryonodini Bechyné, 1951

Caryonodini Bechyné, 1951: 265 [stem: *Caryonod-*]. Type genus: *Caryonoda* Bechyné, 1951.

Tribe Cubispini Monrós, 1954

Cubispini Monrós, 1954: 26 [stem: *Cubisp-*]. Type genus: *Cubispa* Barber, 1946.

Tribe Eumolpini Hope, 1840

Eumolpidae Hope, 1840a: 162 [stem: *Eumolp-*]. Type genus: *Eumolpus* Kugelann, 1798 [an application to suppress *Eumolpus* Kugelann, 1798 and conserve *Eumolpus* Weber, 1801 was recently submitted to the Commission by Moseyko et al. (2010)]. Comment: First Reviser (Eumolpini Hope, 1840 vs Colaspidini Hope, 1840) not determined, current usage maintained.

Colaspidae Hope, 1840a: 163 [stem: *Colaspid-*]. Type genus: *Colaspis* Fabricius, 1801. Comment: incorrect original stem formation, not in prevailing usage.

Corynodina T. A. Marshall, 1865: 29 [stem: *Corynod-*]. Type genus: *Corynodes* Hope, 1840 [syn. of *Platycorynus* Chevrolat, 1836].

Chalcophanites Chapuis, 1874: 256 [stem: *Chalcophan-*]. Type genus: *Chalcophana* Chevrolat, 1836. Comment: original vernacular name available (Art. 11.7.2): first used in latinized form and generally accepted as in Lefèvre (1885: 50, as Chalcophanitae).

Edusites Chapuis, 1874: 306 [stem: *Edus-*]. Type genus: *Edusa* Chapuis, 1874 [preoccupied genus name, not *Edusa* Gistel, 1848 [Tunicata], not *Edusa* Albers, 1860 [Gastropoda] and not *Edusa* Martens 1860 [Mollusca]; syn. of *Edusella* Chapuis, 1874]. Comment: original vernacular name available (Art. 11.7.2): first used in latinized form and generally accepted as in Lefèvre (1885: 111, as Edusitae); permanently invalid (Art. 39): based on preoccupied type genus.

Endocéphalites Chapuis, 1874: 343 [stem: *Endocephal-*]. Type genus: *Endocephalus* Chevrolat, 1836. Comment: original vernacular name available (Art.

11.7.2): first used in latinized form and generally accepted as in Lefèvre (1885: 154, as ENDOCEPHALITAE).

IPHIMÉITES Chapuis, 1874: 230 [stem: *Iphime-*]. Type genus: *Iphimeis* Baly, 1864. Comment: original vernacular name available (Art. 11.7.2): first used in latinized form by Lefèvre (1885: 12, as IPHIMEITAE), generally accepted as in Riley et al. (2002: 639, as IPHIMEITES [treated as Latin]).

CHRYSODINITAE Lefèvre, 1885: 5 [stem: *Chrysodin-*]. Type genus: *Chrysodina* Baly, 1864 [syn. of *Spintherophyta* Dejean, 1836].

EDUSELLINI Clavareau, 1914: 121 [stem: *Edusell-*]. Type genus: *Edusella* Chapuis, 1874.

COLASPOIDINI Chen, 1940: 488 [stem: *Colaspoid-*]. Type genus: *Colaspoides* Laporte, 1833.

Tribe EURYOPINI Chapuis, 1874

EURYOPITES Chapuis, 1874: 302 [stem: *Euryop-*]. Type genus: *Euryope* Dalman, 1824. Comment: original vernacular name available (Art. 11.7.2): first used in latinized form and generally accepted as in Lefèvre (1885: 108, as EURYOPITAE).

ODONTIONOPITES Lefèvre, 1876: 301 [stem: *Odontionop-*]. Type genus: *Odontionopa* Chevrolat, 1836. Comment: original vernacular name available (Art. 11.7.2): first used in latinized form and generally accepted as in Lefèvre (1885: 65, as ODONTIONOPITAE).

PRASOIDEINI Clavareau, 1914: 65 [stem: *Prasoide-*]. Type genus: *Prasoidea* Weise, 1907 [syn. of *Odontionopa* Chevrolat, 1836].

*COLASPOSOMINI Bechyné, 1957: 7, 8 [stem: *Colasposomat-*]. Type genus: *Colasposoma* Laporte, 1833. Comment: unavailable family-group name, proposed after 1930 without description or bibliographic reference to such a description (Art. 13.1); incorrect original stem formation, not in prevailing usage.

COLASPOSOMINI Špringlová de Bechyné, 1960: 3 [stem: *Colasposomat-*]. Type genus: *Colasposoma* Laporte, 1833. Comment: incorrect original stem formation, not in prevailing usage.

Tribe HABROPHORINI Bechyné and Špringlová de Bechyné, 1969

HABROPHORINI Bechyné and Špringlová de Bechyné, 1969: 75 [stem: *Habrophor-*]. Type genus: *Habrophora* Erichson, 1847.

Tribe HEMYDACNINI Bechyné, 1951

HEMYDACNINI Bechyné, 1951: 91 [stem: *Hemydacn-*]. Type genus: *Hemydacne* Jacoby, 1897.

Tribe MEGASCELIDINI Chapuis, 1874

MÉGASCÉLIDES Chapuis, 1874: 82 [stem: *Megascelid-*]. Type genus: *Megascelis* Sturm, 1826. Comment: original vernacular name available (Art. 11.7.2): first used in latinized form by Jacoby and Clavareau (1905: 1, as Megascelidae [incorrect stem formation]), generally accepted as in Riley et al. (2002: 639, as Megascelidini); incorrect original stem formation, not in prevailing usage.

Tribe Merodini Chapuis, 1874

Merodites Chapuis, 1874: 327 [stem: *Merod-*]. Type genus: *Meroda* Baly, 1860. Comment: original vernacular name available (Art. 11.7.2): first used in latinized form and generally accepted as in Lefèvre (1885: 128, as Meroditae).

Tribe Pygomolpini Bechyné, 1949

Pygomolpini Bechyné, 1949: 532 [stem: *Pygomolp-*]. Type genus: *Pygomolpus* Bechyné, 1949.

Tribe Rosiroiini Bechyné, 1950

Rosiroiini Bechyné, 1950b: 148 [stem: *Rosiroi-*]. Type genus: *Rosiroia* Bechyné, 1950.

Tribe Typophorini Baly, 1865

Typophorinae Baly, 1865: 433 [stem: *Typophor-*]. Type genus: *Typophorus* Chevrolat, 1836.

Callisinites Chapuis, 1874: 263 [stem: *Callisin-*]. Type genus: *Callisina* Baly, 1860. Comment: original vernacular name available (Art. 11.7.2): first used in latinized form and generally accepted as in Lefèvre (1885: 63, as Callisinitae).

Métachromites Chapuis, 1874: 295 [stem: *Metachromat-*]. Type genus: *Metachroma* Chevrolat, 1836. Comment: original vernacular name available (Art. 11.7.2): first used in latinized form by Lefèvre (1885: 92, as Metachromitae), generally accepted as in Riley et al. (2002: 639, as Metachromites [treated as Latin]); incorrect original stem formation, not in prevailing usage.

Nodostomites Chapuis, 1874: 261 [stem: *Nodostomat-*]. Type genus: *Nodostoma* Motschulsky, 1860 [syn. of *Basilepta* Baly, 1860]. Comment: original vernacular name available (Art. 11.7.2): first used in latinized form and generally accepted as in Lefèvre (1885: 56, as Nodostomitae); incorrect original stem formation, not in prevailing usage.

Pagriites Lefèvre, 1884: lxvii [stem: *Pagri-*]. Type genus: *Pagria* Lefèvre, 1884. Comment: original vernacular name available (Art. 11.7.2): first used in latinized form and generally accepted as in Lefèvre (1885: 62, as Pagriitae).

Cheirideitae Lefèvre, 1885: 67 [stem: *Cheiride-*]. Type genus: *Cheiridea* Baly, 1878.

Nodinini Chen, 1940: 487 [stem: *Nodin-*]. Type genus: *Nodina* Motschulsky, 1858.

Basileptini Chûjô, 1956: 9 [stem: *Basilept-*]. Type genus: *Basilepta* Baly, 1860. Comment: name proposed to replace Nodostomini Chapuis, 1874 because of the synonymy of the type genus.

Nodini Selman, 1965: 145 [stem: *Nod-*]. Type genus: *Noda* Chevrolat, 1836 [preoccupied genus name, not *Noda* Schellenberg 1803 [Diptera]; syn. of *Brachypnoea* Gistel, 1848]. Comment: permanently invalid (Art. 39): based on preoccupied type genus.

Eumolpinae *incertae sedis*

*Eupalini Verma et al., 2005: 164 [stem: *Eupal-*]. Type genus: *Eupales* Lefèvre, 1885 [although *Floricola* Gistel, 1848 is an older available name for this genus, almost all authors have used *Eupales* Lefèvre, 1885 in the last 125 years; we use *Eupales* Lefèvre, 1885 as valid here pending a vote on its conservation following a recent submission to the Commission by Jolivet and Verma (Bulletin of Zoological Nomenclature 2009: 204; also see Appendix 6)]. Comment: unavailable family-group name, proposed after 1999 without explicit intention (Art. 16.1); included here as "*incertae sedis*" as was done in the most recent volume of the Catalogue of Palaearctic Coleoptera (see Löbl 2010: 83); notice of a new application for the conservation of *Eupales* Lefèvre, 1885 and Eupalini Verma et al. submitted by Jolivet and Verma was recently published in the Bulletin of Zoological Nomenclature (2009: 204; also see Appendix 6).

Subfamily Spilopyrinae Chapuis, 1874

Spilopyrites Chapuis, 1874: 259 [stem: *Spilopyr-*]. Type genus: *Spilopyra* Baly, 1860. Comment: original vernacular name available (Art. 11.7.2): first used in latinized form by Lefèvre (1885: 56, as Spilopyritae), generally accepted as in Reid (2000: 852, as Spilopyrinae); First Reviser found (Spilopyrinae Chapuis, 1874 vs Stenomelinae Chapuis, 1874) is Reid (2000: 852).

Stenomèlites Chapuis, 1874: 421 [stem: *Stenomel-*]. Type genus: *Stenomela* Erichson, 1847. Comment: original vernacular name available (Art. 11.7.2): first used in latinized form and generally accepted as in Monrós (1958c: 146, as Stenomelini).

Hornibiini Crowson, 1946: 79, in key [stem: *Hornibi-*]. Type genus: *Hornibius* Fairmaire, 1888 [syn. of *Hornius* Fairmaire, 1885].

Subfamily Synetinae LeConte and Horn, 1883

Synetae J. L. LeConte and G. H. Horn, 1883: 338 [stem: *Synet-*]. Type genus: *Syneta* Dejean, 1835.

Synetidae Edwards, 1953: 29 [stem: *Synet-*]. Type genus: *Syneta* Dejean, 1835. Comment: family-group name proposed as new without reference to Synetae J. L. LeConte and G. H. Horn, 1883.

†Subfamily Protoscelidinae Medvedev, 1968

Protoscelinae L. N. Medvedev, 1968: 155 [stem: *Protoscelid-*]. Type genus: *Protoscelis* L. N. Medvedev, 1968. Comment: incorrect original stem formation, not in prevailing usage.

Superfamily Curculionoidea Latreille, 1802

Curculionites Latreille, 1802: 195 [stem: *Curculion-*]. Type genus: *Curculio* Linnaeus, 1758.

Family Nemonychidae Bedel, 1882

Nemonychidae Bedel, 1882: 3, in key [stem: *Nemonych-*]. Type genus: *Nemonyx* Redtenbacher, 1845 [placed on the Official List of Generic Names in Zoology (ICZN 2005c)]. Comment: given precedence over Cimberididae Gozis, 1882 and placed on the Official List of Family-Group Names in Zoology (ICZN 2005c).

Subfamily Nemonychinae Bedel, 1882

Nemonychidae Bedel, 1882: 3, in key [stem: *Nemonych-*]. Type genus: *Nemonyx* Redtenbacher, 1845 [placed on the Official List of Generic Names in Zoology (ICZN 2005c)]. Comment: placed on the Official List of Family-Group Names in Zoology (ICZN 2005c).

Subfamily Cimberidinae Gozis, 1882

Cimberidae Gozis, 1882: 58 [stem: *Cimberid-*]. Type genus: *Cimberis* Gozis, 1881 [placed on the Official List of Generic Names in Zoology (ICZN 2005c)]. Comment: Cimberidae Gozis, 1882 placed on the Official Index of Rejected and Invalid Family-Group Names in Zoology and Cimberididae Gozis, 1882 placed on the Official List of Family-Group Names in Zoology (ICZN 2005c).

Tribe Cimberidini Gozis, 1882

Rhinomacerides Schönherr, 1823: column 1136 [stem: *Rhinomacr-*]. Type genus: *Rhinomacer* sensu A. G. Olivier, 1807 [not *Rhinomacer* Fabricius, 1781; syn. of *Cimberis* Gozis, 1881]. Comment: placed on the Official Index of Rejected and Invalid Family-Group Names (ICZN 2005c).

Cimberidae Gozis, 1882: 58 [stem: *Cimberid-*]. Type genus: *Cimberis* Gozis, 1881 [placed on the Official List of Generic Names in Zoology (ICZN 2005c)]. Comment: Cimberidae Gozis, 1882 placed on the Official Index of Rejected and Invalid Family-Group Names in Zoology and Cimberididae Gozis, 1882 placed on the Official List of Family-Group Names in Zoology (ICZN 2005c); incorrect original stem formation, not in prevailing usage.

*Neocimberini O'Brien and Wibmer, 1982: 3 [stem: *Neocimberid-*]. Type genus: *Neocimberis* O'Brien and Wibmer, 1982 [placed on the Official Index of Rejected and Invalid Generic Names in Zoology (ICZN 2005c)]. Comment: unavailable family-group name, proposed after 1930 without description or bibliographic reference to such a description (Art. 13.1); incorrect original stem formation, not in prevailing usage.

Tribe Doydirhynchini Pierce, 1916

Doydirhynchoidea Pierce, 1916: 463, in key [stem: *Doydirhynch-*]. Type genus: *Doydirhynchus* Dejean, 1821.

†Tribe Kuschelomacrini Riedel, 2010

Kuschelomacerini Riedel, 2010: 31 [stem: *Kuschelomacr-*]. Type genus: *Kuschelomacer* Riedel, 2010. Comment: incorrect original stem formation, not in prevailing usage.

Subfamily Rhinorhynchinae Voss, 1922

Rhinorhynchini Voss, 1922: 17 [stem: *Rhinorhynch-*]. Type genus: *Rhinorhynchus* Sharp, 1882.

Tribe Mecomacerini Kuschel, 1994

Mecomacerini Kuschel, 1994: 576, in key [stem: *Mecomacer-*]. Type genus: *Mecomacer* Kuschel, 1954. Comment: current spelling maintained (Art. 29.5): incorrect original stem formation in prevailing usage (should be *Mecomacr-*).

Subtribe Brarina Legalov, 2009

Brarina Legalov, 2009b: 208 [stem: *Brar-*]. Type genus: *Brarus* Kuschel, 1997.

Subtribe Mecomacerina Kuschel, 1994

*Mecomacerini May, 1993: 15, in key [stem: *Mecomacer-*]. Type genus: *Mecomacer* Kuschel, 1954. Comment: unavailable family-group name, description shared with Rhynchitoplesiini, not unequivocal (Art. 13.1.1) (see Alonso-Zarazaga and Lyal 1999: 27).

*Rhynchitoplesiini May, 1993: 15, in key [stem: *Rhynchitoplesi-*]. Type genus: *Rhynchitoplesius* Voss, 1952. Comment: unavailable family-group name, description shared with Mecomacerini, not unequivocal (Art. 13.1.1) (see Alonso-Zarazaga and Lyal 1999: 27).

Mecomacerini Kuschel, 1994: 576, in key [stem: *Mecomacer-*]. Type genus: *Mecomacer* Kuschel, 1954. Comment: current spelling maintained (Art. 29.5): incorrect original stem formation in prevailing usage (should be *Mecomacr-*).

Tribe Rhinorhynchini Voss, 1922

Rhinorhynchini Voss, 1922: 17 [stem: *Rhinorhynch-*]. Type genus: *Rhinorhynchus* Sharp, 1882.

Rhynchitomacerini May, 1993: 15, in key [stem: *Rhynchitomacr-*]. Type genus: *Rhynchitomacer* Voss, 1937. Comment: incorrect original stem formation, not in prevailing usage.

†Subfamily Slonikinae Zherikhin, 1977

Slonikinae Zherikhin, 1977: 179 [stem: *Slonik-*]. Type genus: *Slonik* Zherikhin, 1977.

†Tribe Slonikini Zherikhin, 1977

Slonikinae Zherikhin, 1977: 179 [stem: *Slonik-*]. Type genus: *Slonik* Zherikhin, 1977.

†Tribe Ulyaniscini Legalov, 2009

Ulyaniscini Legalov, 2009a: 128 [stem: *Ulyanisc-*]. Type genus: *Ulyanisca* Gratschev, 1998.

†Subfamily Eccoptarthrinae Arnoldi, 1977

Eccoptarthrini Arnoldi, 1977: 169 [stem: *Eccoptarthr-*]. Type genus: *Eccoptarthrus* Arnoldi, 1977.

Mesophyletinae Poinar, 2006: 879 [stem: *Mesophylet-*]. Type genus: *Mesophyletis* Poinar, 2006. Comment: Poinar (2008: 262) unnecessarily revalidated this name on the basis of Art. 16.2, the criteria of availability were met in the original description in 2006 where the name of the type genus was cited.

†Subfamily Brenthorrhininae Arnoldi, 1977

Brenthorrhininae Arnoldi, 1977: 172 [stem: *Brenthorrhin-*]. Type genus: *Brenthorrhinus* Arnoldi, 1977. Comment: placement according to Legalov (2009c: 289).

†Tribe Brenthorrhinini Arnoldi, 1977

Brenthorrhininae Arnoldi, 1977: 172 [stem: *Brenthorrhin-*]. Type genus: *Brenthorrhinus* Arnoldi, 1977.

†Tribe Brenthorrhinoidini Legalov, 2003

Brenthorrhinoidini Legalov, 2003: 88 [stem: *Brenthorrhinoid-*]. Type genus: *Brenthorrhinoides* Gratshev and Zherikhin, 1996.

†Subfamily Distenorrhininae Arnoldi, 1977

Distenorrhinini Arnoldi, 1977: 170 [stem: *Distenorrhin-*]. Type genus: *Distenorrhinus* Arnoldi, 1977. Comment: placement according to Legalov (2009c: 290).

†Subfamily Eobelinae Arnoldi, 1977

Eobelidae Arnoldi, 1977: 144 [stem: *Eobel-*]. Type genus: *Eobelus* Arnoldi, 1977. Comment: precedence (Oxycorynoidinae Arnoldi, 1977 vs Nanophydinae Arnoldi, 1977 vs Eobelinae Arnoldi, 1977) given to taxon originally proposed at the higher rank (Art. 24.1).

†Tribe Eobelini Arnoldi, 1977

Eobelidae Arnoldi, 1977: 144 [stem: *Eobel-*]. Type genus: *Eobelus* Arnoldi, 1977. Comment: precedence (Eobelini Arnoldi, 1977 vs Procurculionini Arnoldi, 1977) given to taxon originally proposed at the higher rank (Art. 24.1).

†Subtribe Eobelina Arnoldi, 1977

Eobelidae Arnoldi, 1977: 144 [stem: *Eobel-*]. Type genus: *Eobelus* Arnoldi, 1977.

†Subtribe Procurculionina Arnoldi, 1977

Procurculionini Arnoldi, 1977: 157 [stem: *Procurculion-*]. Type genus: *Procurculio* Arnoldi, 1977. Comment: First Reviser (Eccoptothoracina Arnoldi, 1977 vs Procurculionina Arnoldi, 1977) not determined, current usage maintained.

Eccoptothoracini Arnoldi, 1977: 158 [stem: *Eccoptothorac-*]. Type genus: *Eccoptothorax* Arnoldi, 1977.

†Tribe Karataucarini Legalov, 2009

Karataucarini Legalov, 2009c: 288 [stem: *Karataucar-*]. Type genus: *Karataucar* Legalov, 2009.

†Tribe Nanophydini Arnoldi, 1977

Nanophydinae Arnoldi, 1977: 173 [stem: *Nanophyd-*]. Type genus: *Nanophydes* Arnoldi, 1977.

†Tribe Oxycorynoidini Arnoldi, 1977

Oxycorynoidinae Arnoldi, 1977: 159 [stem: *Oxycorynoid-*]. Type genus: *Oxycorynoides* Arnoldi, 1977.

†Tribe Probelini Legalov, 2009

Probelini Legalov, 2009c: 287 [stem: *Probel-*]. Type genus: *Probelus* Arnoldi, 1977.

†Subfamily Paleocartinae Legalov, 2003

Paleocartini Legalov, 2003: 78 [stem: *Paleocart-*]. Type genus: *Paleocartus* Legalov, 2003.

†Tribe Nebrenthorrhinini Legalov, 2007

Nebrenthorrhinina Legalov, 2007: 34 [stem: *Nebrenthorrhin-*]. Type genus: *Nebrenthorrhinus* Legalov, 2003. Comment: placement according to Legalov (2009c: 288).

†Tribe Paleocartini Legalov, 2003

Paleocartini Legalov, 2003: 78 [stem: *Paleocart-*]. Type genus: *Paleocartus* Legalov, 2003. Comment: placement according to Legalov (2009c: 288).

†Subfamily Metrioxenoidinae Legalov, 2009

Metrioxenoidinae Legalov, 2009c: 288 [stem: *Metrioxenoid-*]. Type genus: *Metrioxenoides* Gratshev et al., 1998.

†Subfamily Cretonemonychinae Gratshev and Legalov, 2009

Cretonemonychinae Gratshev and Legalov, 2009: 412 [stem: *Cretonemonych-*]. Type genus: *Cretonemonyx* Gratshev and Legalov, 2009.

†Subfamily Selengarhynchinae Gratshev and Legalov, 2009

Selengarhynchinae Gratshev and Legalov, 2009: 414 [stem: *Selengarhynch-*]. Type genus: *Selengarhynchus* Gratshev and Legalov, 2009.

Family Anthribidae Billberg, 1820

Anthribides Billberg, 1820a: 39 [stem: *Anthrib-*]. Type genus: *Anthribus* Geoffroy, 1762 [placed on the Official List of Generic Names in Zoology (ICZN 1994a)]. Comment: placed on the Official List of Family-Group Names in Zoology and given priority over Choragidae Kirby, 1819 (ICZN 1994b).

Subfamily Anthribinae Billberg, 1820

Anthribides Billberg, 1820a: 39 [stem: *Anthrib-*]. Type genus: *Anthribus* Geoffroy, 1762 [placed on the Official List of Generic Names in Zoology (ICZN 1994a)]. Comment: placed on the Official List of Family-Group Names in Zoology and given priority over Choragidae Kirby, 1819 (ICZN 1994b).

Tribe Anthribini Billberg, 1820

Anthribides Billberg, 1820a: 39 [stem: *Anthrib-*]. Type genus: *Anthribus* Geoffroy, 1762 [placed on the Official List of Generic Names in Zoology (ICZN 1994a)]. Comment: placed on the Official List of Family-Group Names in Zoology and given priority over Choragidae Kirby, 1819 (ICZN 1994b).

Brachytarsina C. G. Thomson, 1859: 128 [stem: *Brachytars-*]. Type genus: *Brachytarsus* Schönherr, 1823 [syn. of *Anthribus* Geoffroy, 1762].

Brachytarsini Pierce, 1930: 22, in key [stem: *Brachytars-*]. Type genus: *Brachytarsus* Schönherr, 1823 [syn. of *Anthribus* Geoffroy, 1762]. Comment: family-group name proposed as new without reference to Brachytarsina C. G. Thomson, 1859.

Tribe Basitropini Lacordaire, 1865

Basitropides Lacordaire, 1865: 566 [stem: *Basitrop-*]. Type genus: *Basitropis* Jekel, 1855. Comment: original vernacular name available (Art. 11.7.2): first used in latinized form by Stein (1868: 115, as Basitropini), generally accepted as in Alonso-Zarazaga and Lyal (1999: 28, as Basitropidini [incorrect stem formation]); Alonso-Zarazaga and Lyal (1999: 28) used the spelling Basitropidini but the same authors subsequently (2002: 4) reverted to Basitropini; precedence (Basitropini Lacordaire, 1865 vs Eugonini Lacordaire, 1865) given to taxon originally proposed at the higher rank (Art. 24.1).

Eugonides Lacordaire, 1865: 569 [stem: *Eugon-*]. Type genus: *Eugonus* Schönherr, 1833. Comment: original vernacular name available (Art. 11.7.2): first used in latinized form and generally accepted as in Sharp (1873: 32, as Eugonides [treated as Latin]); synonymy with Basitropini by Alonso-Zarazaga and Lyal (2002: 4).

Phaenithonini Pierce, 1930: 4, in key [stem: *Phaenithon-*]. Type genus: *Phaenithon* Schönherr, 1826.

Tribe Corrhecerini Lacordaire, 1865

Corrhécérides Lacordaire, 1865: 547 [stem: *Corrhecer-*]. Type genus: *Corrhecerus* Schönherr, 1826. Comment: original vernacular name available (Art. 11.7.2): first used in latinized form by Morimoto (1981: 78, as Corrhecerini), generally accepted as in Alonso-Zarazaga and Lyal (1999: 29, as Corrhecerini).

Nessiarini Morimoto, 1972: 38, in key [stem: *Nessiar-*]. Type genus: *Nessiara* Pascoe, 1860.

Tribe Cratoparini LeConte, 1876

Cratopares J. L. LeConte, 1876: 403 [stem: *Cratopar-*]. Type genus: *Cratoparis* Dejean, 1834 [syn. of *Euparius* Schönherr, 1823]. Comment: current spelling maintained (Art. 29.5): incorrect stem formation in prevailing usage (should be *Cratoparent-*).

Eupariini B. D. Valentine, 1960: 49, in key [stem: *Eupari-*]. Type genus: *Euparius* Schönherr, 1823. Comment: junior homonym of Eupariini A. Schmidt, 1910 (type genus *Euparia* Lepeletier and Serville, 1828) in Scarabaeidae.

†Tribe Cretanthribini Legalov, 2009

Cretanthribini Legalov, 2009c: 291 [stem: *Cretanthrib-*]. Type genus: *Cretanthribus* Legalov, 2009.

Tribe Decataphanini Lacordaire, 1865

Décataphanides Lacordaire, 1865: 556 [stem: *Decataphan-*]. Type genus: *Decataphanes* Labram and Imhoff, 1840. Comment: original vernacular name available (Art. 11.7.2): first used in latinized form by Kolbe (1897: 289, as Decatophaninae [incorrect stem formation]), generally accepted as in Alonso-Zarazaga and Lyal (1999: 30, as Decataphanini).

Tribe Discotenini Lacordaire, 1865

Discoténides Lacordaire, 1865: 500 [stem: *Discoten-*]. Type genus: *Discotenes* Labram and Imhoff, 1841. Comment: original vernacular name available (Art. 11.7.2): first used in latinized form by Pierce (1930: 4, as Discotenini), generally accepted as in Alonso-Zarazaga and Lyal (1999: 30, as Discotenini).

Tribe Ecelonerini Lacordaire, 1865

Ecélonérides Lacordaire, 1865: 562 [stem: *Eceloner-*]. Type genus: *Ecelonerus* Schönherr, 1839. Comment: original vernacular name available (Art. 11.7.2): first used in latinized form by Morimoto (1972: 38, as Ecelonerini), generally accepted as in Alonso-Zarazaga and Lyal (1999: 30, as Ecelonerini).

Tribe Ischnocerini Lacordaire, 1865

Ischnocérides Lacordaire, 1865: 504 [stem: *Ischnocer-*]. Type genus: *Ischnocerus* Schönherr, 1839. Comment: original vernacular name available (Art. 11.7.2): first used in latinized form by J. L. LeConte (1876: 393, as Ischnoceri), generally accepted as in Alonso-Zarazaga and Lyal (2002: 5, as Ischnocerini); the Hymenoptera name Ischnocerini Clement, 1938 (type genus *Ischnoceros* Gravenhorst, 1829) is apparently unavailable.

Meconemini Pierce, 1930: 4, in key [stem: *Meconem-*]. Type genus: *Meconemus* Labram and Imhoff, 1838.

Tribe Gymnognathini Valentine, 1960

Gymnognathini B. D. Valentine, 1960: 48, in key [stem: *Gymnognath-*]. Type genus: *Gymnognathus* Schönherr, 1826.

Tribe Jordanthribini Morimoto, 1980

Jordanthribini Morimoto, 1980: 16 [stem: *Jordanthrib-*]. Type genus: *Jordanthribus* Zimmerman, 1938.

Tribe Mauiini Valentine, 1990

Mauiini B. D. Valentine, 1990: 235, in key [stem: *Maui-*]. Type genus: *Mauia* Blackburn, 1885.

Tribe Mecocerini Lacordaire, 1865

Mécocérides Lacordaire, 1865: 493 [stem: *Mecocer-*]. Type genus: *Mecocerus* Schönherr, 1833 [placed on the Official List of Generic Names in Zoology (ICZN 1972); syn. of *Acanthothorax* Gaede, 1832]. Comment: original vernacular name available (Art. 11.7.2): first used in latinized form by Kolbe (1897: 288, as Mecocerinae), generally accepted as in Alonso-Zarazaga and Lyal (1999: 31, as Mecocerini).

Phloeophilides Lacordaire, 1865: 517 [stem: *Phloeophil-*]. Type genus: *Phloeophilus* Schönherr, 1833. Comment: original vernacular name available (Art.

11.7.2): first used in latinized form and generally accepted as in Olliff (1891: 75, as PHLOEOPHILIDES [treated as Latin]); the older name PHLOEOPHILINI Kiesenwetter, 1863 [CLEROIDEA] was based on an unjustified emendation, the correct stem for the cleroid family-group name is *Phloiophil-*.

CAPPADOCINI Alonso-Zarazaga and Lyal, 1999: 28 [stem: *Cappadoc-*]. Type genus: *Cappadox* Alonso-Zarazaga and Lyal, 1999 [syn. of *Phloeophilus* Schönherr, 1833].

Tribe MYCTEINI Morimoto, 1972

MYCTEINI Morimoto, 1972: 38, in key [stem: *Mycte-*]. Type genus: *Mycteis* Pascoe, 1860.

Tribe OZOTOMERINI Morimoto, 1972

OZOTOMERINI Morimoto, 1972: 37, in key [stem: *Ozotomer-*]. Type genus: *Ozotomerus* Perroud, 1853.

Tribe PIESOCORYNINI Valentine, 1960

PIESOCORYNINI B. D. Valentine, 1960: 49, in key [stem: *Piesocoryn-*]. Type genus: *Piesocorynus* Dejean, 1834.

Tribe PLATYRHININI Bedel, 1882

PLATYRRHINIDAE Bedel, 1882: 3, in key [stem: *Platyrhin-*]. Type genus: *Platyrhinus* Clairville, 1798. Comment: incorrect original stem formation, not in prevailing usage.

Tribe PLATYSTOMINI Pierce, 1916

ANTHOTRIBIDAE Gistel, 1856a: 375 [stem: *Anthotrib-*]. Type genus: *Anthotribus* Gistel, 1856 [preoccupied genus name, not *Anthotribus* Hoffmann, 1803 [Coleoptera: ANTHRIBIDAE: ANTHRIBINAE: ANTHRIBINI]; syn. of *Platystomos* Schneider, 1791]. Comment: permanently invalid (Art. 39): based on preoccupied type genus.

PLATYSTOMOIDEA Pierce, 1916: 463, in key [stem: *Platystom-*]. Type genus: *Platystomos* Schneider, 1791. Comment: the type genus for this family-group name was in fact proposed as a replacement name for *Anthribus* sensu Fabricius, 1790, however it was never used in that sense subsequently (Alonso-Zarazaga and Lyal 1999: 33); an application will be submitted by MAAZ and CHCL to conserve the current concept of *Platystomos* Schneider, 1791 and its associated family-group name.

Tribe PROSCOPORHININI Lacordaire, 1865

PROSCOPORHINIDES Lacordaire, 1865: 544 [stem: *Proscoporhin-*]. Type genus: *Proscoporhinus* Montrouzier, 1861. Comment: original vernacular name available (Art. 11.7.2): first used in latinized form by Ienistea (1986: 32, as PROSOPORHINIDAE [incorrect stem formation]), generally accepted as in Alonso-Zarazaga and Lyal (1999: 33, as PROSCOPORHININI).

Tribe Ptychoderini Jekel, 1855

Ptychoderidae Jekel, 1855: 70 [stem: *Ptychoder-*]. Type genus: *Ptychoderes* Schönherr, 1823.

*Phloeotragides Lacordaire, 1865: 486 [stem: *Phloeotrag-*]. Type genus: *Phloeotragus* Schönherr, 1823. Comment: original vernacular name unavailable (Art. 11.7.2): subsequently used in latinized form but not generally attributed to Lacordaire (1865) and used as valid.

Phloeotraginae Kolbe, 1897: 288 [stem: *Phloeotrag-*]. Type genus: *Phloeotragus* Schönherr, 1823.

Tribe Sintorini Lacordaire, 1865

Sintorides Lacordaire, 1865: 510 [stem: *Sintor-*]. Type genus: *Sintor* Schönherr, 1839. Comment: original vernacular name available (Art. 11.7.2): first used in latinized form by Morimoto (1972: 38, as Sintorini), generally accepted as in Alonso-Zarazaga and Lyal (1999: 34, as Sintorini).

Tribe Stenocerini Kolbe, 1895

Stenocerinarum Kolbe, 1895: 381 [stem: *Stenocer-*]. Type genus: *Stenocerus* Schönherr, 1826. Comment: Stenocerinarum is the genitive form of Stenocerinae and was recognized as such by Alonso-Zarazaga and Lyal (1999: 34).

Allandrini Pierce, 1930: 18 [stem: *Allandr-*]. Type genus: *Allandrus* J. L. LeConte, 1876.

Tribe Tophoderini Lacordaire, 1865

Tophodérides Lacordaire, 1865: 499 [stem: *Tophoder-*]. Type genus: *Tophoderes* Dejean, 1834. Comment: original vernacular name available (Art. 11.7.2): first used in latinized form by Kolbe (1897: 288, as Tophoderinae), generally accepted as in Alonso-Zarazaga and Lyal (1999: 34, as Tophoderini).

Tribe Trigonorhinini Valentine, 1999

Trigonorhinini B. D. Valentine, 1999: 287, in key [stem: *Trigonorhin-*]. Type genus: *Trigonorhinus* Wollaston, 1861.

Tribe Tropiderini Lacordaire, 1865

Tropidérides Lacordaire, 1865: 484 [stem: *Tropider-*]. Type genus: *Tropideres* Schönherr, 1823. Comment: original vernacular name available (Art. 11.7.2): first used in latinized form by Stein (1868: 115, as Tropiderini), generally accepted as in Alonso-Zarazaga and Lyal (1999: 35, as Tropiderini); precedence (Tropiderini Lacordaire, 1865 vs Acorynini Lacordaire, 1865) given to taxon originally proposed at the higher rank (Art. 24.1).

Acorynides Lacordaire, 1865: 512 [stem: *Acoryn-*]. Type genus: *Acorynus* Schönherr, 1833. Comment: original vernacular name available (Art. 11.7.2): first used in latinized form by Kolbe (1897: 289, as Acoryninae), generally accepted as in Morimoto (1980: 22, as Acorynini).

Eurymycterini Pierce, 1930: 15 [stem: *Eurymycter-*]. Type genus: *Eurymycter* J. L. LeConte, 1876.

Tribe Xenocerini Lacordaire, 1865

Xénocerides Lacordaire, 1865: 558 [stem: *Xenocer-*]. Type genus: *Xenocerus* Schönherr, 1833. Comment: original vernacular name available (Art. 11.7.2): first used in latinized form by Heyne and Taschenberg (1907: 235, as Xenocerini), generally accepted as in Alonso-Zarazaga and Lyal (1999: 35, as Xenocerini).

Tribe Xylinadini Lacordaire, 1865

Xylinadides Lacordaire, 1865: 560 [stem: *Xylinad-*]. Type genus: *Xylinades* Latreille, 1828 [syn. of *Xylinada* Berthold, 1827]. Comment: original vernacular name available (Art. 11.7.2): first used in latinized form by Kolbe (1897: 289, as Xylinadinae), generally accepted as in Alonso-Zarazaga and Lyal (1999: 36, as Xylinadini).

Tribe Zygaenodini Lacordaire, 1865

Zygénodides Lacordaire, 1865: 542 [stem: *Zygaenod-*]. Type genus: *Zygaenodes* Pascoe, 1859 [syn. of *Exechesops* Schönherr, 1847]. Comment: original vernacular name available (Art. 11.7.2): first used in latinized form by Kolbe (1897: 289, as Zygaenodinae), generally accepted as in Alonso-Zarazaga and Lyal (1999: 36, as Zygaenodini); incorrect original stem formation, not in prevailing usage.

Hormisci J. L. LeConte, 1876: 396 [stem: *Ormisc-*]. Type genus: *Ormiscus* G. R. Waterhouse, 1845 [as *Hormiscus*, unjustified emendation of type genus name by Agassiz (1846b: 263), not in prevailing usage]. Comment: incorrect original stem formation, not in prevailing usage.

Subfamily Choraginae Kirby, 1819

Choragidae Kirby, 1819: 447 [stem: *Chorag-*]. Type genus: *Choragus* Kirby, 1819 [placed on the Official List of Generic Names in Zoology (ICZN 1994b)]. Comment: placed on the Official List of Family-Group Names in Zoology and not given priority over Anthribidae Billberg, 1820 (ICZN 1994b).

Tribe Apolectini Lacordaire, 1865

Apolectides Lacordaire, 1865: 554 [stem: *Apolect-*]. Type genus: *Apolecta* Pascoe, 1859. Comment: original vernacular name available (Art. 11.7.2): first used in latinized form by Kolbe (1897: 289, as Apolectinae), generally accepted as in Alonso-Zarazaga and Lyal (1999: 28, as Apolectini); Apolectidae Jordan, 1923 (type genus *Apolectus* Cuvier, 1832) is available in Pisces although it is based on a preoccupied genus name and therefore permanently invalid; this case is to be referred to the Commission to remove the homonymy (Art. 55.3.1).

Tribe Araecerini Lacordaire, 1865

Araeocérides Lacordaire, 1865: 588 [stem: *Araecer-*]. Type genus: *Araecerus* Schönherr, 1823 [as *Araeocerus*, unjustified emendation of type genus name by Schönherr (1839), not in prevailing usage]. Comment: original vernacular name available (Art. 11.7.2): first used in latinized form by Stein (1868: 115, as Araeocerini), generally accepted as in Alonso-Zarazaga and Lyal (1999: 37, as Araecerini); incorrect original stem formation, not in prevailing usage.

Tribe Cisanthribini Zimmerman, 1994

Cisanthribini Zimmerman, 1994a: 232 [stem: *Cisanthrib-*]. Type genus: *Cisanthribus* Zimmerman, 1938.

Tribe Choragini Kirby, 1819

Choragidae Kirby, 1819: 447 [stem: *Chorag-*]. Type genus: *Choragus* Kirby, 1819 [placed on the Official List of Generic Names in Zoology (ICZN 1994b)]. Comment: not given priority over Anthribidae Billberg, 1820, placed on the Official List of Family-Group Names in Zoology (ICZN 1994b).

Tribe Valenfriesiini Alonso-Zarazaga and Lyal, 1999

Notioxénides Lacordaire, 1865: 593 [stem: *Notioxen-*]. Type genus: *Notioxenus* Wollaston, 1861 [preoccupied genus name, not *Notioxenus* Motschulsky, 1858 [Coleoptera: Carabidae]; syn. of *Valenfriesia* Alonso-Zarazaga and Lyal, 1999]. Comment: original vernacular name available (Art. 11.7.2): first used in latinized form by Pierce (1930: 31, as Notioxenini), generally accepted as in Morimoto (1978a: 18, as Notioxenini); permanently invalid (Art. 39): based on preoccupied type genus.

Valenfriesiini Alonso-Zarazaga and Lyal, 1999: 38 [stem: *Valenfriesi-*]. Type genus: *Valenfriesia* Alonso-Zarazaga and Lyal, 1999. Comment: replacement name for Notioxenini Lacordaire, 1865 because of the homonymy of the type genus.

Tribe Xenorchestini Lacordaire, 1865

Xénorchestides Lacordaire, 1865: 595 [stem: *Xenorchest-*]. Type genus: *Xenorchestes* Wollaston, 1854. Comment: original vernacular name available (Art. 11.7.2): first used in latinized form by J. L. LeConte (1876: 408, as Xenorchestini), generally accepted as in Alonso-Zarazaga and Lyal (1999: 38, as Xenorchestini).

Homoeoderides Wollaston, 1870: 23 [stem: *Homoeoder-*]. Type genus: *Homoeodera* Wollaston, 1870.

Subfamily Urodontinae Thomson, 1859

Urodontides C. G. Thomson, 1859: 128 [stem: *Urodont-*]. Type genus: *Urodon* Schönherr, 1823 [syn. of *Bruchela* Dejean, 1821].

Bruchelidae Pierce, 1916: 463 [stem: *Bruchel-*]. Type genus: *Bruchela* Dejean, 1821.

†Family Ulyanidae Zherikhin, 1993

Ulyanidae Zherikhin, 1993: 26 [stem: *Ulyan-*]. Type genus: *Ulyana* Zherikhin, 1993.

Family Belidae Schönherr, 1826

Belides Schönherr, 1826: 73 [stem: *Bel-*]. Type genus: *Belus* Schönherr, 1823 [syn. of *Rhinotia* Kirby, 1819].

Subfamily Belinae Schönherr, 1826

Belides Schönherr, 1826: 73 [stem: *Bel-*]. Type genus: *Belus* Schönherr, 1823 [syn. of *Rhinotia* Kirby, 1819].

Tribe Agnesiotidini Zimmerman, 1994

Agnesiotidini Zimmerman, 1994a: 258 [stem: *Agnesiotid-*]. Type genus: *Agnesiotis* Pascoe, 1870.

Tribe Belini Schönherr, 1826

Belides Schönherr, 1826: 73 [stem: *Bel-*]. Type genus: *Belus* Schönherr, 1823 [syn. of *Rhinotia* Kirby, 1819].

Subtribe Belina Schönherr, 1826

Belides Schönherr, 1826: 73 [stem: *Bel-*]. Type genus: *Belus* Schönherr, 1823 [syn. of *Rhinotia* Kirby, 1819].

Subtribe Homalocerina Legalov, 2009

Homalocerina Legalov, 2009d: 308 [stem: *Homalocer-*]. Type genus: *Homalocerus* Schönherr, 1839.

Tribe Pachyurini Kuschel, 1959

Pachyurini Kuschel, 1959a: 253, in key [stem: *Pachyur-*]. Type genus: *Pachyura* Hope, 1834.

Subfamily Oxycoryninae Schönherr, 1840

Oxycorynides Schönherr, 1840: 581 [stem: *Oxycoryn-*]. Type genus: *Oxycorynus* Chevrolat, 1832.

Tribe Aglycyderini Wollaston, 1864

Aglycyderidae Wollaston, 1864: 384 [stem: *Aglycyder-*]. Type genus: *Aglycyderes* Westwood, 1864.

Proterhinides Fauvel, 1891: 154 [stem: *Proterhin-*]. Type genus: *Proterhinus* Sharp, 1878. Comment: original vernacular name available (Art. 11.7.2): first used in latinized form and generally accepted as in Sharp (1899a: 298, as Proterhinidae).

Platycephalitae Paulian, 1944: 118 [stem: *Platycephal-*]. Type genus: *Platycephala* Montrouzier, 1861 [preoccupied genus name, not *Platycephala* Fallén, 1820 [Diptera]; syn. of *Aralius* Kuschel, 1990]. Comment: permanently invalid (Art. 39): based on preoccupied type genus; also Platycephalidae Gill, 1872 (type genus *Platycephalus* Bloch, 1795) is available in Pisces.

Tribe Alloxycorynini Legalov, 2009

Alloxycorynini Legalov, 2009d: 313 [stem: *Alloxycoryn-*]. Type genus: *Alloxycorynus* Voss, 1957.

Tribe Distenorrhinoidini Legalov, 2009

Distenorrhinoidini Legalov, 2009d: 313 [stem: *Distenorrhinoid-*]. Type genus: *Distenorrhinoides* Gratshev and Zherikhin, 2000.

Tribe Metrioxenini Voss, 1953

Metrioxenini Voss, 1953: 124, in key [stem: *Metrioxen-*]. Type genus: *Metrioxena* Pascoe, 1870.

Subtribe Afrocorynina Voss, 1957

Afrocorynini Voss, 1957: 102, in key [stem: *Afrocoryn-*]. Type genus: *Afrocorynus* G. A. K. Marshall, 1955.

Hispodini Voss, 1957: 102, in key [stem: *Hispod-*]. Type genus: *Hispodes* G. A. K. Marshall, 1955.

Subtribe Metrioxenina Voss, 1953

Metrioxenini Voss, 1953: 124, in key [stem: *Metrioxen-*]. Type genus: *Metrioxena* Pascoe, 1870.

Subtribe Zherichinixenina Legalov, 2009

Zherichinixenina Legalov, 2009d: 310 [stem: *Zherichinixen-*]. Type genus: *Zherichinixena* Legalov, 2009.

Tribe Oxycorynini Schönherr, 1840

Oxycorynides Schönherr, 1840: 581 [stem: *Oxycoryn-*]. Type genus: *Oxycorynus* Chevrolat, 1832.

Subtribe Allocorynina Sharp, 1890

Allocoryninae Sharp, 1890: 45 [stem: *Allocoryn-*]. Type genus: *Allocorynus* Sharp, 1890 [syn. of *Rhopalotria* Chevrolat, 1878].

Subtribe Oxycorynina Schönherr, 1840

Oxycorynides Schönherr, 1840: 581 [stem: *Oxycoryn-*]. Type genus: *Oxycorynus* Chevrolat, 1832.

Subtribe Oxycraspedina Marvaldi and Oberprieler, 2006

Oxycraspedina Marvaldi and Oberprieler, 2006: 460 [stem: *Oxycrasped-*]. Type genus: *Oxycraspedus* Kuschel, 1955.

Family Caridae Thompson, 1992

Carinae Thompson, 1992: 882 [stem: *Car-*]. Type genus: *Car* Blackburn, 1897.

Subfamily Carinae Thompson, 1992

Carinae Thompson, 1992: 882 [stem: *Car-*]. Type genus: *Car* Blackburn, 1897.

Caridae Zimmerman, 1994a: 499 [stem: *Car-*]. Type genus: *Car* Blackburn, 1897. Comment: family-group name proposed as new without reference to Carinae Thompson, 1992.

Carinae Kuschel, 1995: 18 [stem: *Car-*]. Type genus: *Car* Blackburn, 1897. Comment: family-group name proposed as new without reference to Carinae Thompson, 1992 or Caridae Zimmerman, 1994.

Subfamily Chilecarinae Legalov, 2009

Chilecarini Legalov, 2009a: 125 [stem: *Chilecar-*]. Type genus: *Chilecar* Kuschel, 1992.

Tribe Carodini Legalov, 2009

Carodesina Legalov, 2009a: 126 [stem: *Carod-*]. Type genus: *Carodes* Zimmerman, 1994. Comment: incorrect original stem formation, not in prevailing usage.

Tribe Chilecarini Legalov, 2009

Chilecarini Legalov, 2009a: 125 [stem: *Chilecar-*]. Type genus: *Chilecar* Kuschel, 1992.

†Subfamily Baissorhynchinae Zherikhin, 1993

Baissorhynchini Zherikhin, 1993: 30 [stem: *Baissorhynch-*]. Type genus: *Baissorhynchus* Zherikhin, 1977.

†Abrocarina Legalov, 2009c: 291 [stem: *Abrocar-*]. Type genus: *Abrocar* Liu and Ren, 2006.

Family Attelabidae Billberg, 1820

Attelabides Billberg, 1820a: 39 [stem: *Attelab-*]. Type genus: *Attelabus* Linnaeus, 1758 [placed on the Official List of Generic Names in Zoology (ICZN 1983a)].

Subfamily Attelabinae Billberg, 1820

Attelabides Billberg, 1820a: 39 [stem: *Attelab-*]. Type genus: *Attelabus* Linnaeus, 1758 [placed on the Official List of Generic Names in Zoology (ICZN 1983a)].

Tribe Attelabini Billberg, 1820

Attelabides Billberg, 1820a: 39 [stem: *Attelab-*]. Type genus: *Attelabus* Linnaeus, 1758 [placed on the Official List of Generic Names in Zoology (ICZN 1983a)].

Subtribe Attelabina Billberg, 1820

Attelabides Billberg, 1820a: 39 [stem: *Attelab-*]. Type genus: *Attelabus* Linnaeus, 1758 [placed on the Official List of Generic Names in Zoology (ICZN 1983a)]. Comment: this family-group name was also used in the same year by Billberg (1820b: 393, as Attelabides).

Subtribe Euscelina Voss, 1925

Euscelina Voss, 1925: 32 [stem: *Euscel-*]. Type genus: *Euscelus* Schönherr, 1833.

Alleuscelina Legalov, 2003: 411 [stem: *Alleuscel-*]. Type genus: *Alleuscelus* Voss, 1937. Comment: proposed as a subtribe of Euscelini.

Clinolabina Legalov, 2003: 410 [stem: *Clinolab-*]. Type genus: *Clinolabus* Jekel, 1860. Comment: proposed as a subtribe of Euscelini.

Subtribe Euscelophilina Voss, 1925

Euscelophilina Voss, 1925: 29 [stem: *Euscelophil-*]. Type genus: *Euscelophilus* Voss, 1925.

Subtribe Henicolabina Legalov, 2007

Henicolabina Legalov, 2007: 282 [stem: *Henicolab-*]. Type genus: *Henicolabus* Voss, 1925.

Subtribe Himatolabina Legalov, 2003

Himatolabina Legalov, 2003: 424 [stem: *Himatolab-*]. Type genus: *Himatolabus* Jekel, 1860. Comment: proposed as a subtribe of Hybolabini.

Subtribe Hybolabina Voss, 1925

Hybolabina Voss, 1925: 191 [stem: *Hybolab-*]. Type genus: *Hybolabus* Jekel, 1860.

Subtribe Isolabina Legalov, 2007

Isolabina Legalov, 2007: 282 [stem: *Isolab-*]. Type genus: *Isolabus* Voss, 1925.

Subtribe Lagenoderina Voss, 1925

Lagenoderina Voss, 1925: 206 [stem: *Lagenoder-*]. Type genus: *Lagenoderus* A. White, 1841.

Subtribe Lamprolabina Voss, 1925

Lamprolabina Voss, 1925: 213 [stem: *Lamprolab-*]. Type genus: *Lamprolabus* Jekel, 1860.

Subtribe Metocalolabina Legalov, 2003

Metocalolabina Legalov, 2003: 433 [stem: *Metocalolab-*]. Type genus: *Metocalolabus* Legalov, 2003.

Subtribe Omolabina Legalov, 2003

Omolabina Legalov, 2003: 426 [stem: *Omolab-*]. Type genus: *Omolabus* Jekel, 1860. Comment: proposed as a subtribe of Hybolabini.

Subtribe Paramecolabina Legalov, 2003

Paramecolabina Legalov, 2003: 439 [stem: *Paramecolab-*]. Type genus: *Paramecolabus* Jekel, 1860.

Subtribe Phialodina Legalov, 2003

Phialodina Legalov, 2003: 437 [stem: *Phialod-*]. Type genus: *Phialodes* Roelofs, 1874.

Subtribe Phymatolabina Voss, 1925

Phymatolabina Voss, 1925: 199 [stem: *Phymatolab-*]. Type genus: *Phymatolabus* Jekel, 1860.

Subtribe Phymatopsinina Legalov, 2003

Phymatopsinina Legalov, 2003: 456 [stem: *Phymatopsin-*]. Type genus: *Phymatopsinus* Voss, 1925. Comment: proposed as a subtribe of Lagenoderini.

Subtribe Pleurolabina Legalov, 2003

Pleurolabina Legalov, 2003: 460 [stem: *Pleurolab-*]. Type genus: *Pleurolabus* Jekel, 1860. Comment: proposed as a subtribe of Lagenoderini.

Tribe Euopini Voss, 1925

Euopsini Voss, 1925: 291 [stem: *Euop-*]. Type genus: *Euops* Schönherr, 1839. Comment: incorrect original stem formation, not in prevailing usage.

Archeuopsina Legalov, 2003: 359 [stem: *Archeuop-*]. Type genus: *Archeuops* Legalov, 2003. Comment: proposed as a subtribe of Euopini; incorrect original stem formation, not in prevailing usage.

SUNIOPSINA Legalov, 2003: 364 [stem: *Suniop-*]. Type genus: *Suniops* Voss, 1928. Comment: proposed as a subtribe of EUOPINI; incorrect original stem formation, not in prevailing usage.

SYNAPTOPSINA Legalov, 2003: 368 [stem: *Synaptop-*]. Type genus: *Synaptops* Jekel, 1860. Comment: proposed as a subtribe of EUOPINI; incorrect original stem formation, not in prevailing usage.

LJUDMILININA Legalov, 2007: 219 [stem: *Ljudmilini-*]. Type genus: *Ljudmilinius* Legalov, 2003. Comment: proposed as a subtribe of EUOPINI; incorrect original stem formation, not in prevailing usage.

PARASYNAPTOPSISINA Legalov, 2007: 227 [stem: *Parasynaptopse-*]. Type genus: *Parasynaptopsis* Legalov, 2003. Comment: proposed as a subtribe of EUOPINI; incorrect original stem formation, not in prevailing usage.

RIEDELININA Legalov, 2007: 218 [stem: *Riedelini-*]. Type genus: *Riedelinius* Legalov, 2003. Comment: proposed as a subtribe of EUOPINI; incorrect original stem formation, not in prevailing usage.

SAWADAEUOPSINA Legalov, 2007: 241 [stem: *Sawadaeuop-*]. Type genus: *Sawadaeuops* Legalov, 2003. Comment: proposed as a subtribe of EUOPINI; incorrect original stem formation, not in prevailing usage.

Tribe PILOLABINI Voss, 1925

PILOLABINI Voss, 1925: 19 [stem: *Pilolab-*]. Type genus: *Pilolabus* Jekel, 1860.

Subfamily APODERINAE Jekel, 1860

APODERIDAE Jekel, 1860: 180 [stem: *Apoder-*]. Type genus: *Apoderus* A. G. Olivier, 1807.

Tribe APODERINI Jekel, 1860

APODERIDAE Jekel, 1860: 180 [stem: *Apoder-*]. Type genus: *Apoderus* A. G. Olivier, 1807.

ANISONYCHINA Legalov, 2003: 555 [stem: *Anisonych-*]. Type genus: *Anisonychus* Voss, 1927. Comment: junior homonym of ANISONYCHIDAE H. C. C. Burmeister, 1844 (type genus *Anisonyx* Latreille, 1807) available in SCARABAEIDAE; this case is to be referred to the Commission to remove the homonymy (Art. 55.3.1).

CENTROCORYNINA Legalov, 2003: 557 [stem: *Centrocoryn-*]. Type genus: *Centrocorynus* Jekel, 1860.

CYCNOTRACHELINA Legalov, 2003: 568 [stem: *Cycnotrachel-*]. Type genus: *Cycnotrachelus* Jekel, 1860.

OPANASSENKOVIINA Legalov, 2003: 554 [stem: *Opanassenkovi-*]. Type genus: *Opanassenkovius* Legalov, 2003.

PSEUDOCYCNOTRACHELINA Legalov, 2003: 525 [stem: *Pseudocycnotrachel-*]. Type genus: *Pseudocycnotrachelus* Legalov, 2003.

Tribe CLITOSTYLINI Voss, 1929

CLITOSTYLINI Voss, 1929a: 192 [stem: *Clitostyl-*]. Type genus: *Clitostylus* Voss, 1929 [syn. of *Trachelismus* Motschulsky, 1870].

Subtribe ALLAPODERINA Legalov, 2003

ALLAPODERINA Legalov, 2003: 467 [stem: *Allapoder-*]. Type genus: *Allapoderus* Voss, 1927.

Subtribe CLITOSTYLINA Voss, 1929

CLITOSTYLINI Voss, 1929a: 192 [stem: *Clitostyl-*]. Type genus: *Clitostylus* Voss, 1929 [syn. of *Trachelismus* Motschulsky, 1870].

Subtribe PSEUDOPHRYSINA Legalov, 2003

PSEUDOPHRYSINA Legalov, 2003: 476 [stem: *Pseudophrys-*]. Type genus: *Pseudophrysus* Legalov, 2003.

Tribe HOPLAPODERINI Voss, 1926

HOPLAPODERINI Voss, 1926: 14 [stem: *Hoplapoder-*]. Type genus: *Hoplapoderus* Jekel, 1860.

Subtribe AFROAPODERINA Legalov, 2003

AFROAPODERINA Legalov, 2003: 482 [stem: *Afroapoder-*]. Type genus: *Afroapoderus* Legalov, 2003.

Subtribe HOPLAPODERINA Voss, 1926

HOPLAPODERINI Voss, 1926: 14 [stem: *Hoplapoder-*]. Type genus: *Hoplapoderus* Jekel, 1860.

Subtribe PARATOMAPODERINA Legalov, 2003

PARATOMAPODERINA Legalov, 2003: 488 [stem: *Paratomapoder-*]. Type genus: *Paratomapoderus* Voss, 1926.

Tribe TRACHELOPHORINI Voss, 1926

TRACHELOPHORINI Voss, 1926: 14, in key [stem: *Trachelophor-*]. Type genus: *Trachelophorus* Jekel, 1860.

Subfamily RHYNCHITINAE Gistel, 1848

RHYNCHITISIDAE Gistel, 1848: [8] [stem: *Rhynchit-*]. Type genus: *Rhynchites* Schneider, 1791.

Tribe Auletini Desbrochers des Loges, 1908

Auletinidae Desbrochers des Loges, 1908: 10, in key [stem: *Aulet-*]. Type genus: *Auletes* Schönherr, 1826. Comment: incorrect original stem formation, not in prevailing usage.

Subtribe Auletina Desbrochers des Loges, 1908

Auletinidae Desbrochers des Loges, 1908: 10, in key [stem: *Aulet-*]. Type genus: *Auletes* Schönherr, 1826. Comment: incorrect original stem formation, not in prevailing usage.

Auletini Pierce, 1913: 365, in key [stem: *Aulet-*]. Type genus: *Auletes* Schönherr, 1826. Comment: family-group name proposed as new without reference to Auletinidae Desbrochers des Loges, 1908.

Subtribe Auletobiina Legalov, 2001

Auletobiina Legalov, 2001: 37 [stem: *Auletobi-*]. Type genus: *Auletobius* Desbrochers des Loges, 1869.

Subtribe Guineauletina Legalov, 2003

Guineauletina Legalov, 2003: 108 [stem: *Guineaulet-*]. Type genus: *Guineauletes* Legalov, 2003.

Subtribe Mandelschtamiina Legalov, 2003

Mandelschtamiina Legalov, 2003: 105 [stem: *Mandelschtami-*]. Type genus: *Mandelschtamius* Legalov, 2003.

Subtribe Pseudauletina Voss, 1933

Pseudauletina Voss, 1933b: 110, in key [stem: *Pseudaulet-*]. Type genus: *Pseudauletes* Voss, 1922.

Subtribe Pseudomesauletina Legalov, 2003

Pseudomesauletina Legalov, 2003: 113 [stem: *Pseudomesaulet-*]. Type genus: *Pseudomesauletes* Legalov, 2001.

Tribe Auletorhinini Voss, 1935

Auletorhinini Voss, 1935b: 509 [stem: *Auletorhin-*]. Type genus: *Auletorhinus* Voss, 1935.

Tribe Byctiscini Voss, 1923

Byctiscini Voss, 1923: 510 [stem: *Byctisc-*]. Type genus: *Byctiscus* C. G. Thomson, 1859.

Subtribe Byctiscina Voss, 1923

Byctiscini Voss, 1923: 510 [stem: *Byctisc-*]. Type genus: *Byctiscus* C. G. Thomson, 1859.

Subtribe Listrobyctiscina Legalov, 2003

Listrobyctiscina Legalov, 2003: 337 [stem: *Listrobyctisc-*]. Type genus: *Listrobyctiscus* Voss, 1923.

Subtribe Svetlanaebyctiscina Legalov, 2003

Svetlanaebyctiscina Legalov, 2003: 323 [stem: *Svetlanaebyctisc-*]. Type genus: *Svetlanaebyctiscus* Legalov, 2001.

Tribe Cesauletini Legalov, 2003

Cesauletini Legalov, 2003: 134 [stem: *Cesaulet-*]. Type genus: *Cesauletes* Hamilton, 1983. Comment: proposed as a tribe of the supertribe Rhynchititae.

Tribe Deporaini Voss, 1929

Deporaini Voss, 1929b: 28 [stem: *Depora-*]. Type genus: *Deporaus* Samouelle, 1819.

Subtribe Chonostropheina Morimoto, 1962

Chonostropheina Morimoto, 1962a: 30, in key [stem: *Chonostrophe-*]. Type genus: *Chonostropheus* Prell, 1924.

Subtribe Deporaina Voss, 1929

Deporaini Voss, 1929b: 28 [stem: *Depora-*]. Type genus: *Deporaus* Samouelle, 1819.

Tribe Minurini Legalov, 2003

Minurini Legalov, 2003: 133 [stem: *Minur-*]. Type genus: *Minurus* G. R. Waterhouse, 1842. Comment: proposed as a tribe of the supertribe Rhynchititae.

Tribe Rhinocartini Voss, 1931

Rhinocartini Voss, 1931: 162, in key [stem: *Rhinocart-*]. Type genus: *Rhinocartus* Voss, 1922.

*Rhynchitallini Voss, 1960: 415 [stem: *Rhynchitall-*]. Type genus: *Rhynchitallus* Voss, 1960. Comment: unavailable family-group name, proposed after 1930 without description or bibliographic reference to such a description (Art. 13.1).

Proteugnamptini Legalov, 2003: 80 [stem: *Proteugnamptini*]. Type genus: *Proteugnamptus* Voss, 1939. Comment: proposed as a tribe in supertribe Rhinocartitae.

Auletanini Legalov, 2003: 84 [stem: *Auletan-*]. Type genus: *Auletanus* Voss, 1922. Comment: proposed as a tribe in the supertribe Rhinocartitae.

†Sanyrevilleina Legalov, 2003: 85 [stem: *Sanyreville-*]. Type genus: *Sanyrevilleus* Gratshev and Zherikhin, 2000. Comment: proposed as a subtribe of Auletanini; raised to tribal level by Legalov (2007).

Proteugnamptini Legalov, 2003: 80 [stem: *Proteugnampt-*]. Type genus: *Proteugnamptus* Voss, 1939. Comment: proposed as a tribe in the supertribe Rhinocartitae.

Vossicartini Legalov, 2003: 79 [stem: *Vossicart-*]. Type genus: *Vossicartus* Legalov, 2003.

Eosalacina Legalov, 2007: 30 [stem: *Eosalac-*]. Type genus: *Eosalacus* Legalov, 2007. Comment: proposed as a subtribe of Proteugnamptini.

Parauletanina Legalov, 2007: 31 [stem: *Parauletan-*]. Type genus: *Parauletanus* Legalov, 2007. Comment: proposed as a subtribe of Sanyrevilleini.

Tribe Rhynchitini Gistel, 1848

Rhynchitisidae Gistel, 1848: [8] [stem: *Rhynchit-*]. Type genus: *Rhynchites* Schneider, 1791. Comment: incorrect original stem formation, not in prevailing usage.

Subtribe Acritorrhynchitina Legalov, 2007

Acritorrhynchitina Legalov, 2007: 70 [stem: *Acritorrhynchit-*]. Type genus: *Acritorrhynchites* Voss, 1941.

Subtribe Anisomerinina Legalov, 2003

Anisomerinina Legalov, 2003: 223 [stem: *Anisomerin-*]. Type genus: *Anisomerinus* Voss, 1933.

Subtribe Eugnamptina Voss, 1930

Eugnamptina Voss, 1930: 67 [stem: *Eugnampt-*]. Type genus: *Eugnamptus* Schönherr, 1839.

Subtribe Lasiorhynchitina Legalov, 2003

Lasiorhynchitina Legalov, 2003: 202 [stem: *Lasiorhynchit-*]. Type genus: *Lasiorhynchites* Jekel, 1860.

Subtribe Perrhynchitina Legalov, 2003

Perrhynchitina Legalov, 2003: 218 [stem: *Perrhynchit-*]. Type genus: *Perrhynchites* Voss, 1953.

Subtribe Rhynchitallina Legalov, 2003

Rhynchitallina Legalov, 2003: 226 [stem: *Rhynchitall-*]. Type genus: *Rhynchitallus* Voss, 1960.

Subtribe RHYNCHITINA Gistel, 1848

RHYNCHITISIDAE Gistel, 1848: [8] [stem: *Rhynchit-*]. Type genus: *Rhynchites* Schneider, 1791. Comment: incorrect original stem formation, not in prevailing usage.

RHYNCHITINI Pierce, 1913: 365, in key [stem: *Rhynchit-*]. Type genus: *Rhynchites* Schneider, 1791. Comment: family-group name proposed as new without reference to RHYNCHITISIDAE Gistel, 1848.

Subtribe TEMNOCERINA Legalov, 2003

TEMNOCERINA Legalov, 2003: 207 [stem: *Temnocer-*]. Type genus: *Temnocerus* Thunberg, 1815.

Subfamily ISOTHEINAE Scudder, 1893

ISOTHEINAE Scudder, 1893: 16 [stem: *Isothe-*]. Type genus: *Isothea* Scudder, 1893. Comment: precedence (ISOTHEINAE Scudder, 1893 vs TOXORHYNCHINAE Scudder, 1893) given to taxon originally proposed at the higher rank (Art. 24.1).

Tribe ISOTHEINI Scudder, 1893

ISOTHEINAE Scudder, 1893: 16 [stem: *Isothe-*]. Type genus: *Isothea* Scudder, 1893.

Subtribe DEPASOPHILINA Legalov, 2003

DEPASOPHILINA Legalov, 2003: 166 [stem: *Depasophil-*]. Type genus: *Depasophilus* Voss, 1922.

†Subtribe ISOTHEINA Scudder, 1893

ISOTHEINAE Scudder, 1893: 16 [stem: *Isothe-*]. Type genus: *Isothea* Scudder, 1893.

†Tribe TOXORHYNCHINI Scudder, 1893

TOXORHYNCHINI Scudder, 1893: 23 [stem: *Toxorhynch-*]. Type genus: *Toxorhynchus* Scudder, 1893.

Subfamily PTEROCOLINAE Lacordaire, 1865

PTÉROCOLIDES Lacordaire, 1865: 190 [stem: *Pterocol-*]. Type genus: *Pterocolus* Schönherr, 1833 [preoccupied genus name, not *Pterocolus* Say, 1831 [Coleoptera: ATTELABIDAE: PTEROCOLINAE]; syn. of *Pterocolus* Say, 1831]. Comment: original vernacular name available (Art. 11.7.2): first used in latinized form by Pascoe (1870b: 437, as PTEROCOLINAE), generally accepted as in Alonso-Zarazaga and Lyal (1999: 44, as PTEROCOLINAE); permanently invalid (Art. 39): based on preoccupied type genus; although strictly speaking the type genus is a junior homonym, and thus cannot be used as valid, to preserve stability, we are considering this as a valid name here; an application to the Commission is necessary to officially conserve usage of this family-group name.

Family BRENTIDAE Billberg, 1820

BRENTHIDES Billberg, 1820a: 40 [stem: *Brent-*]. Type genus: *Brentus* Fabricius, 1787 [as *Brenthus*, unjustified emendation of type genus name by Illiger (1801), not in prevailing usage].

Subfamily BRENTINAE Billberg, 1820

BRENTHIDES Billberg, 1820a: 40 [stem: *Brent-*]. Type genus: *Brentus* Fabricius, 1787 [as *Brenthus*, unjustified emendation of type genus name by Illiger (1801), not in prevailing usage].

Tribe BRENTINI Billberg, 1820

BRENTHIDES Billberg, 1820a: 40 [stem: *Brent-*]. Type genus: *Brentus* Fabricius, 1787 [as *Brenthus*, unjustified emendation of type genus name by Illiger (1801), not in prevailing usage].

Subtribe ARRHENODINA Lacordaire, 1865

ARRHÉNODIDES Lacordaire, 1865: 425 [stem: *Arrhenod-*]. Type genus: *Arrenodes* sensu Lacordaire, 1865 [as *Arrhenodes*, incorrect subsequent spelling of type genus name, not in prevailing usage; not *Arrenodes* Schönherr 1823; syn. of *Estenorhinus* Lacordaire, 1865]. Comment: based on misidentified type genus; MAAZ and CHCL will submit an application to designate as type genus *Arrenodes* Schönherr, 1823 and correct the name of the tribe to ARRENODINI; First Reviser (ARRHENODINA Lacordaire, 1865 vs BELOPHERINA Lacordaire, 1865 vs BELORHYNCHINA Lacordaire, 1865 vs EUTRACHELINA Lacordaire, 1865) not determined, current usage maintained.

BÉLOPHÉRIDES Lacordaire, 1865: 433 [stem: *Belopher-*]. Type genus: *Belopherus* Schönherr, 1833 [syn. of *Belorhynchus* Berthold, 1827]. Comment: original vernacular name available (Art. 11.7.2): first used in latinized form by Pascoe (1872: 317, as BELOPHERINAE), generally accepted as in Sharp (1895a: 53, as BELOPHERINA).

BÉLORHYNCHIDES Lacordaire, 1865: 437 [stem: *Belorhynch-*]. Type genus: *Belorhynchus* Berthold, 1827. Comment: original vernacular name available (Art. 11.7.2): first used in latinized form generally accepted as in Sharp (1895a: 63, as BELORHYNCHINA); placed in synonymy with ARRHENODINI by Alonso-Zarazaga and Lyal (2002: 10).

EUTRACHÉLIDES Lacordaire, 1865: 438 [stem: *Eutrachel-*]. Type genus: *Eutrachelus* Berthold, 1827. Comment: original vernacular name available (Art. 11.7.2): first used in latinized form by Senna (1892: 179, as EUTRACHELINAE), generally accepted as in Sharp (1900b: 385, as EUTRACHELIDES [treated as Latin]).

ORYCHODI Senna, 1895: 213 [stem: *Orychod-*]. Type genus: *Orychodes* Pascoe, 1862.

EUPSALINI Muizon, 1960: 170 [stem: *Eupsalid-*]. Type genus: *Eupsalis* sensu Kleine, 1927 [not *Eupsalis* Lacordaire, 1865; syn. of *Orfilaia* Haedo Rosii, 1955]. Comment: based on a misidentified type genus; incorrect original stem formation, not in prevailing usage.

Subtribe BRENTINA Billberg, 1820

BRENTHIDES Billberg, 1820a: 40 [stem: *Brent-*]. Type genus: *Brentus* Fabricius, 1787 [as *Brenthus*, unjustified emendation of type genus name by Illiger (1801), not in prevailing usage].

Subtribe EREMOXENINA Semenov, 1892

AMORPHOCEPHALIDES Power, 1879: 478 [stem: *Amorphocephal-*]. Type genus: *Amorphocephalus* Schönherr, 1840 [preoccupied genus name, not *Amorphocephalus* Bowdich, 1825 [Pisces]; syn. of *Amorphocephala* Damoiseau, 1966]. Comment: permanently invalid (Art. 39): based on preoccupied type genus.

EREMOXENIDAE Semenov, 1892: 440 [stem: *Eremoxen-*]. Type genus: *Eremoxenus* Semenov, 1892.

PAUSSOBRENTHINI Gestro, 1919: 270 [stem: *Paussobrenth-*]. Type genus: *Paussobrenthus* Gestro, 1919.

Tribe CYLADINI Schönherr, 1823

CYLADES Schönherr, 1823: column 1137 [stem: *Cylad-*]. Type genus: *Cylas* Latreille, 1802.

Tribe CYPHAGOGINI Kolbe, 1892

CYPHAGOGINAE Kolbe, 1892: 162 [stem: *Cyphagog-*]. Type genus: *Cyphagogus* Parry, 1849.

Subtribe ATOPOBRENTINA Damoiseau, 1965

ATOPOBRENTINI Damoiseau, 1965: 1 [stem: *Atopobrent-*]. Type genus: *Atopobrentus* Damoiseau, 1965.

Subtribe CYPHAGOGINA Kolbe, 1892

CYPHAGOGINAE Kolbe, 1892: 162 [stem: *Cyphagog-*]. Type genus: *Cyphagogus* Parry, 1849.

*CALODROMINEN Kolbe, 1916: 50 [stem: *Calodrom-*]. Type genus: *Calodromus* Guérin-Méneville, 1832. Comment: original vernacular name unavailable (Art. 11.7.2): proposed after 1899.

CALODROMINI Kleine, 1922a: 148 [stem: *Calodrom-*]. Type genus: *Calodromus* Guérin-Méneville, 1832.

†Subtribe Dominibrentina Poinar, 2009

Dominibrentini Poinar, 2009: 51 [stem: *Dominibrent-*]. Type genus: *Dominibrentus* Poinar, 2009.

Subtribe Hoplopisthiina Senna and Calabresi, 1919

Hoplopisthi Senna and Calabresi, 1919: 65 [stem: *Hoplopisthi-*]. Type genus: *Hoplopisthius* Senna, 1892. Comment: incorrect original stem formation, not in prevailing usage.

Subtribe Stereodermina Sharp, 1895

Stereodermina Sharp, 1895a: 7 [stem: *Stereoderm-*]. Type genus: *Stereodermus* Lacordaire, 1865.

Tribe Pholidochlamydini Damoiseau, 1962

Pholidochlamydinae Damoiseau, 1962: 26 [stem: *Pholidochlamyd-*]. Type genus: *Pholidochlamys* Lacordaire, 1865. Comment: the original spelling Pholidochlomydinae on page 26 is considered a *lapsus calami* since the the type genus is spelled correctly throughout and the family-group name is spelled correctly on page 18.

Tribe Taphroderini Lacordaire, 1865

Taphrodérides Lacordaire, 1865: 406 [stem: *Taphroder-*]. Type genus: *Taphroderes* Schönherr, 1823. Comment: original vernacular name available (Art. 11.7.2): first used in latinized form by Pascoe (1872: 317, as Taphroderinae), generally accepted as in Alonso-Zarazaga and Lyal (1999: 52, as Taphroderinae).

Ischnomérides Lacordaire, 1865: 414 [stem: *Ischnomer-*]. Type genus: *Ischnomerus* Schönherr, 1840 [preoccupied genus name, not *Ischnomerus* Labram and Imhoff, 1838 [Coleoptera: Curculionidae: Trachelizini]; syn. of *Aulacoderes* Chevrolat, 1839]. Comment: original vernacular name available (Art. 11.7.2): first used in latinized form and generally accepted as in Schoenfeldt (1908: 2, as Ischnomeridae); permanently invalid (Art. 39): based on preoccupied type genus.

Tribe Trachelizini Lacordaire, 1865

Trachélizides Lacordaire, 1865: 417 [stem: *Tracheliz-*]. Type genus: *Trachelizus* Dejean, 1834. Comment: First Reviser (Trachelizini Lacordaire, 1865 vs Ithystenini Lacordaire, 1865) not determined, current usage maintained.

Subtribe Acratina Alonso-Zarazaga, Lyal, Bartolozzi and Sforzi, 1999

Némocéphalides Lacordaire, 1865: 459 [stem: *Nemocephal-*]. Type genus: *Nemocephalus* sensu Lacordaire, 1865 [not *Nemocephalus* Guérin-Méneville, 1827; syn. of *Neacratus* Alonso-Zarazaga et al., 1999]. Comment: original vernacular name available (Art. 11.7.2): first used in latinized form and generally accepted as in Kolbe (1883: 79, as NEMOCEPHALI); based on a misidentified type genus, name treated here as invalid until an application is submitted to the Commission to suppress it for the Principle of Priority (Art. 65.2.1).

ACRATINI Alonso-Zarazaga et al., 1999: 53 [stem: *Acrat-*]. Type genus: *Acratus* Lacordaire, 1865.

Subtribe ITHYSTENINA Lacordaire, 1865

ITHYSTÉNIDES Lacordaire, 1865: 464 [stem: *Ithysten-*]. Type genus: *Ithystenus* Pascoe, 1862. Comment: original vernacular name available (Art. 11.7.2): first used in latinized form by Pascoe (1872: 317, as ITHYSTENINAE), generally accepted as in Alonso-Zarazaga and Lyal (1999: 54, as ITHYSTENINI).

ISCHNOMERI Kolbe, 1883: 74 [stem: *Ischnomer-*]. Type genus: *Ischnomerus* Labram and Imhoff, 1838.

LEPTORRHYNCHIDAE Schoenfeldt, 1908: 69 [stem: *Leptorhynch-*]. Type genus: *Leptorhynchus* Guérin-Méneville, 1838 [as *Leptorrhynchus*, incorrect subsequent spelling of type genus name, not in prevailing usage; preoccupied genus name, not *Leptorhynchus* Clift, 1828 [Reptilia]; syn. of *Ithystenus* Pascoe, 1862]. Comment: permanently invalid (Art. 39): based on preoccupied type genus; incorrect original stem formation, not in prevailing usage.

OZODOCERINI Jakobson, 1911b: 142 [stem: *Ozodecer-*]. Type genus: *Ozodecerus* Chevrolat, 1839. Comment: incorrect original stem formation, not in prevailing usage.

Subtribe MICROTRACHELIZINA Zimmerman, 1994

MICROTRACHELIZINA Zimmerman, 1994b: 182 [stem: *Microtracheliz-*]. Type genus: *Microtrachelizus* Senna, 1893.

Subtribe PSEUDOCEOCEPHALINA Kleine, 1922

CÉOCÉPHALIDES Lacordaire, 1865: 444 [stem: *Ceocephal-*]. Type genus: *Ceocephalus* sensu Lacordaire, 1865 [not *Ceocephalus* Guérin-Méneville, 1833; *Ceocephalus* sensu Lacordaire, 1865 is a mixture of *Orphanobrentus* Damoiseau, 1962 and *Pseudoceocephalus* Kleine, 1920 (see Alonso-Zarazaga and Lyal 1999: 55)]. Comment: original vernacular name available (Art. 11.7.2): first used in latinized form and generally accepted as in Sharp (1900b: 385, as CEOCEPHALIDES [treated as Latin]); based on a misidentified type genus, name treated here as invalid until an application is submitted to the Commission to suppress it for the Principle of Priority (Art. 65.2.1).

UROPTERINI Jakobson, 1911b: 142 [stem: *Uropter-*]. Type genus: *Uroptera* Berthold, 1827.

PSEUDOCEOCEPHALIDAE Kleine, 1922b: 225 [stem: *Pseudoceocephal-*]. Type genus: *Pseudoceocephalus* Kleine, 1920. Comment: this family-group name is preceeded by two older names: CEOCEPHALINA Lacordaire, 1865, which is based on a misidentified type genus, and the newly discovered name UROPTERINA Jakobson, 1911; we recommend that an application be submitted to the Commission to conserve usage of the well-established name PSEUDOCEOCEPHALINA Kleine, 1922.

Subtribe RHYTICEPHALINA Kleine, 1922

RHYTICEPHALINI Kleine, 1922c: 163 [stem: *Rhyticephal-*]. Type genus: *Rhyticephalus* Chevrolat, 1839.

Subtribe TRACHELIZINA Lacordaire, 1865

TRACHÉLIZIDES Lacordaire, 1865: 417 [stem: *Tracheliz-*]. Type genus: *Trachelizus* Dejean, 1834. Comment: original vernacular name available (Art. 11.7.2): first used in latinized form by Pascoe (1872: 317, as TRACHELIZINAE), generally accepted as in Alonso-Zarazaga and Lyal (1999: 53, as TRACHELIZINI); First Reviser (TRACHELIZINA Lacordaire, 1865 vs HEPHEBOCERINA Lacordaire, 1865) not determined, current usage maintained.

HÉPHÉBOCÉRIDES Lacordaire, 1865: 415 [stem: *Hephebocer-*]. Type genus: *Hephebocerus* Schönherr, 1840. Comment: original vernacular name available (Art. 11.7.2): first used in latinized form and generally accepted as in Pascoe (1872: 317, as EPHEBOCERINAE [incorrect stem formation]).

ANCHISTENINI Schedl, 1961: 198 [stem: *Anchiste-*]. Type genus: *Anchisteus* Kolbe, 1883. Comment: incorrect original stem formation, not in prevailing usage.

Subtribe TYCHAEINA Schoenfeldt, 1908

TYCHAEIDAE Schoenfeldt, 1908: 48 [stem: *Tychae-*]. Type genus: *Tychaeus* Fischer von Waldheim, 1823 [syn. of *Nemorhinus* Schönherr, 1823].

Tribe ULOCERINI Schönherr, 1823

ULOCERIDES Schönherr, 1823: column 1137 [stem: *Ulocer-*]. Type genus: *Ulocerus* Schönherr, 1823.

Subfamily EURHYNCHINAE Lacordaire, 1863

EURHYNCHIDES Lacordaire, 1863: 527 [stem: *Eurhynch-*]. Type genus: *Eurhynchus* Kirby, 1828 [placed on the Official List of Generic Names in Zoology (ICZN 1985e)].

†Tribe AXELRODIELLINI Legalov, 2009

AXEIRODIELLINI Legalov, 2009c: 292 [stem: *Axelrodiell-*]. Type genus: *Axelrodiellus* Zherikhin and Gratshev, 2004. Comment: incorrect original stem formation, not in prevailing usage.

Tribe Eurhynchini Lacordaire, 1863

Eurhynchides Lacordaire, 1863: 527 [stem: *Eurhynch-*]. Type genus: *Eurhynchus* Kirby, 1828 [placed on the Official List of Generic Names in Zoology (ICZN 1985e)]. Comment: original vernacular name available (Art. 11.7.2): first used in latinized form by Pascoe (1870b: 436, as Eurhynchinae), generally accepted as in Alonso-Zarazaga and Lyal (1999: 56, as Eurhynchidae); Eurhynchinae Lacordaire, 1863 placed on the Official List of Family-Group Names in Zoology (ICZN 1985e).

Eurhinini Kissinger, 1968: 10 [stem: *Eurhin-*]. Type genus: *Eurhinus* Kirby, 1819 [preoccupied genus name, not *Eurhinus* Illiger, 1807 [Coleoptera: Curculionidae: Baridini]; syn. of *Eurhynchus* Kirby, 1828]. Comment: replacement name for Eurhynchides Lacordaire, 1863 because of the homonymy of the type genus; junior homonym of Eurhinina Lacordaire, 1865 [Curculionidae: Baridinae]; permanently invalid (Art. 39): based on preoccupied type genus.

Subfamily Apioninae Schönherr, 1823

Apionides Schönherr, 1823: column 1136 [stem: *Apion-*]. Type genus: *Apion* Herbst, 1797. Comment: First Reviser (Apioninae Schönherr, 1823 vs Antliarhininae Schönherr, 1823) not determined, current usage maintained.

Supertribe Apionitae Schönherr, 1823

Apionides Schönherr, 1823: column 1136 [stem: *Apion-*]. Type genus: *Apion* Herbst, 1797.

Tribe Apionini Schönherr, 1823

Apionides Schönherr, 1823: column 1136 [stem: *Apion-*]. Type genus: *Apion* Herbst, 1797.

Subtribe Apionina Schönherr, 1823

Apionides Schönherr, 1823: column 1136 [stem: *Apion-*]. Type genus: *Apion* Herbst, 1797. Comment: current spelling maintained (Art. 29.5): incorrect stem formation in prevailing usage (should be *Api-*).

Oxeostomatidae Gistel, 1856a: 374 [stem: *Oxeostom-*]. Type genus: *Oxeostomum* Gistel, 1856 [syn. of *Apion* Herbst, 1797]. Comment: incorrect original stem formation, not in prevailing usage.

Subtribe Aplemonina Kissinger, 1968

Aplemonini Kissinger, 1968: 16 [stem: *Aplemon-*]. Type genus: *Aplemonus* Schönherr, 1847.

Subtribe Aspidapiina Alonso-Zarazaga, 1990

Aspidapiini Alonso-Zarazaga, 1990: 52 [stem: *Aspidapi-*]. Type genus: *Aspidapion* Schilsky, 1901.

Subtribe Catapiina Alonso-Zarazaga, 1990

Catapiina Alonso-Zarazaga, 1990: 114 [stem: *Catapi-*]. Type genus: *Catapion* Schilsky, 1906.

Subtribe Ceratapiina Alonso-Zarazaga, 1990

Ceratapiini Alonso-Zarazaga, 1990: 42 [stem: *Ceratapi-*]. Type genus: *Ceratapion* Schilsky, 1901.

Subtribe Exapiina Alonso-Zarazaga, 1990

Exapiini Alonso-Zarazaga, 1990: 75 [stem: *Exapi-*]. Type genus: *Exapion* Bedel, 1887.

Subtribe Ixapiina Alonso-Zarazaga, 1990

Ixapiini Alonso-Zarazaga, 1990: 71 [stem: *Ixapi-*]. Type genus: *Ixapion* Roudier and Tempère, 1973.

Subtribe Kalcapiina Alonso-Zarazaga, 1990

Kalcapiini Alonso-Zarazaga, 1990: 55 [stem: *Kalcapi-*]. Type genus: *Kalcapion* Schilsky, 1906.

Subtribe Malvapiina Alonso-Zarazaga, 1990

Malvapiini Alonso-Zarazaga, 1990: 65 [stem: *Malvapi-*]. Type genus: *Malvapion* Hoffmann, 1958.

Subtribe Metapiina Alonso-Zarazaga, 1990

Metapiini Alonso-Zarazaga, 1990: 62 [stem: *Metapi-*]. Type genus: *Metapion* Schilsky, 1906.

Subtribe Oxystomatina Alonso-Zarazaga, 1990

Oxystomatini Alonso-Zarazaga, 1990: 111 [stem: *Oxystomat-*]. Type genus: *Oxystoma* Duméril, 1805.

Subtribe Piezotrachelina Voss, 1959

Piezotrachelini Voss, 1959: 51 [stem: *Piezotrachel-*]. Type genus: *Piezotrachelus* Schönherr, 1839.

Subtribe Prototrichapiina Wanat, 1995

Prototrichapiini Wanat, 1995: 19 [stem: *Prototrichapi-*]. Type genus: *Prototrichapion* Voss, 1959.

Subtribe Synapiina Alonso-Zarazaga, 1990

Synapiina Alonso-Zarazaga, 1990: 118 [stem: *Synapi-*]. Type genus: *Synapion* Schilsky, 1902.

Subtribe TRICHAPIINA Alonso-Zarazaga, 1990

TRICHAPIINA Alonso-Zarazaga, 1990: 116 [stem: *Trichapi-*]. Type genus: *Trichapion* Wagner, 1912.

Tribe CHILAPIINI Wanat, 2001

CHILAPIITAE Wanat, 2001: 366 [stem: *Chilapi-*]. Type genus: *Chilapion* Kissinger, 1968.

Tribe NOTERAPIINI Kissinger, 2004

NOTERAPIONINI Kissinger, 2004: 243 [stem: *Noterapi-*]. Type genus: *Noterapion* Kissinger, 2002. Comment: incorrect original stem formation, not in prevailing usage.

Tribe PODAPIINI Wanat, 2001

PODAPIITAE Wanat, 2001: 366 [stem: *Podapi-*]. Type genus: *Podapion* Riley, 1883.

Tribe RHINORHYNCHIDIINI Zimmerman, 1994

RHINORHYNCHIDIINAE Zimmerman, 1994b: 337 [stem: *Rhinorhynchidi-*]. Type genus: *Rhinorhynchidius* Voss, 1922.

Supertribe ANTLIARHINITAE Schönherr, 1823

ANTLIARHINIDES Schönherr, 1823: column 1137 [stem: *Antliarhin-*]. Type genus: *Antliarhis* Billberg, 1820 [as *Antliarhinus*, incorrect subsequent spelling of type genus name, not in prevailing usage].

Supertribe CYBEBITAE Lacordaire, 1863

CYBÉBIDES Lacordaire, 1863: 539 [stem: *Cybeb-*]. Type genus: *Cybebus* Schönherr, 1839. Comment: original vernacular name available (Art. 11.7.2): first used in latinized form by Pascoe (1870b: 436, as CYBEBINAE), generally accepted as in Alonso-Zarazaga and Lyal (1999: 58, as CYBEBINI).

Supertribe MECOLENITAE Wanat, 2001

MECOLENINAE Wanat, 2001: 363 [stem: *Mecolen-*]. Type genus: *Mecolenus* Schönherr, 1847.

Supertribe MYRMACICELITAE Zimmerman, 1994

MYRMACICELINAE Zimmerman, 1994b: 349 [stem: *Myrmacicel-*]. Type genus: *Myrmacicelus* Chevrolat, 1833.

Tribe LISPOTHERIINI Wanat, 2001

LISPOTHERIINI Wanat, 2001: 365 [stem: *Lispotheri-*]. Type genus: *Lispotherium* Faust, 1891.

Tribe Myrmacicelini Zimmerman, 1994

Myrmacicelinae Zimmerman, 1994b: 349 [stem: *Myrmacicel-*]. Type genus: *Myrmacicelus* Chevrolat, 1833.

Supertribe Rhadinocybitae Alonso-Zarazaga, 1992

Rhadinocybini Alonso-Zarazaga, 1992: 193 [stem: *Rhadinocyb-*]. Type genus: *Rhadinocyba* Faust, 1889.

Tribe Notapionini Zimmerman, 1994

Notapionini Zimmerman, 1994b: 317 [stem: *Notapion-*]. Type genus: *Notapion* Zimmerman, 1994. Comment: current spelling maintained (Art. 29.5): incorrect stem formation in prevailing usage (should be *Notapi-*).

Tribe Rhadinocybini Alonso-Zarazaga, 1992

Rhadinocybini Alonso-Zarazaga, 1992: 193 [stem: *Rhadinocyb-*]. Type genus: *Rhadinocyba* Faust, 1889.

Supertribe Tanaitae Schönherr, 1839

Tanaonides Schönherr, 1839: 447 [stem: *Tana-*]. Type genus: *Tanaos* Schönherr, 1826. Comment: incorrect original stem formation, not in prevailing usage.

Subfamily Ithycerinae Schönherr, 1823

Ithycerides Schönherr, 1823: column 1136 [stem: *Ithycer-*]. Type genus: *Ithycerus* Schönherr, 1823.

Pachyrhinchidae Kirby, 1837: 203 [stem: *Pachyrhynch-*]. Type genus: *Pachyrhynchus* Kirby, 1837 [preoccupied genus name, not *Pachyrhynchus* Germar, 1824 [Coleoptera: Curculionidae: Otiorhynchini]; syn. of *Ithycerus* Schönherr, 1823]. Comment: permanently invalid (Art. 39): based on preoccupied type genus; incorrect original stem formation, not in prevailing usage.

†Gobicarini Legalov, 2009c: 291 [stem: *Gobicar-*]. Type genus: *Gobicar* Gratshev and Zherikhin, 1999.

Subfamily Microcerinae Lacordaire, 1863

Microcérides Lacordaire, 1863: 20 [stem: *Microcer-*]. Type genus: *Microcerus* Schönherr, 1833. Comment: original vernacular name available (Art. 11.7.2): first used in latinized form by Pascoe (1870b: 436, as Microcerinae), generally accepted as in Alonso-Zarazaga and Lyal (1999: 63, as Microcerinae).

*Épisides Lacordaire, 1863: 22 [stem: *Epis-*]. Type genus: *Episus* Billberg, 1820. Comment: original vernacular name unavailable (Art. 11.7.2): subsequently used in latinized form, e.g., Heyne and Taschenberg (1907: 224, as Episini), but not generally accepted as valid.

Subfamily Nanophyinae Gistel, 1848

Nanophyeidae Gistel, 1848: [7] [stem: *Nanophy-*]. Type genus: *Nanophyes* Schönherr, 1837 [placed on the Official List of Generic Names in Zoology (ICZN 1989b)].

Tribe Corimaliini Alonso-Zarazaga, 1989

Corimaliini Alonso-Zarazaga, 1989: 225 [stem: *Corimali-*]. Type genus: *Corimalia* Gozis, 1885.

Tribe Nanophyini Gistel, 1848

Nanophyeidae Gistel, 1848: [7] [stem: *Nanophy-*]. Type genus: *Nanophyes* Schönherr, 1837 [placed on the Official List of Generic Names in Zoology (ICZN 1989b)]. Comment: incorrect original stem formation, not in prevailing usage.

Family Dryophthoridae Schönherr, 1825

Dryophthorides Schönherr, 1825: column 588 [stem: *Dryophthor-*]. Type genus: *Dryophthorus* Germar, 1824 [placed on the Official List of Generic Names in Zoology (ICZN 1987c)].

Subfamily Dryophthorinae Schönherr, 1825

Dryophthorides Schönherr, 1825: column 588 [stem: *Dryophthor-*]. Type genus: *Dryophthorus* Germar, 1824 [placed on the Official List of Generic Names in Zoology (ICZN 1987c)].

Subfamily Cryptodermatinae Bovie, 1908

Oxyrhynchides Schönherr, 1823: column 1137 [stem: *Oxyrhynch-*]. Type genus: *Oxyrhynchus* Schönherr, 1823 [preoccupied genus name, not *Oxyrhynchus* Leach, 1818 [Pisces]; syn. of *Cryptoderma* Ritsema, 1885]. Comment: permanently invalid (Art. 39): based on preoccupied type genus.

Cryptoderminae Bovie, 1908: 1 [stem: *Cryptodermat-*]. Type genus: *Cryptoderma* Ritsema, 1885. Comment: replacement name for Oxyrhynchinae Schönherr, 1823 because of the homonymy of the type genus; incorrect original stem formation, not in prevailing usage.

Subfamily Orthognathinae Lacordaire, 1865

Orthognathides Lacordaire, 1865: 311 [stem: *Orthognath-*]. Type genus: *Orthognathus* Schönherr, 1837.

Tribe Orthognathini Lacordaire, 1865

Orthognathides Lacordaire, 1865: 311 [stem: *Orthognath-*]. Type genus: *Orthognathus* Schönherr, 1837. Comment: original vernacular name available (Art. 11.7.2): first used in latinized form by Heyne and Taschenberg (1907: 233, as Orthognathini), generally accepted as in Alonso-Zarazaga and Lyal (1999: 64, as Orthognathini).

SIPALIDES Lacordaire, 1865: 310 [stem: *Sipal-*]. Type genus: *Sipalus* Schönherr, 1825 [preoccupied genus name, not *Sipalus* Fischer, 1813 [Mammalia]; syn. of *Sipalinus* G. A. K. Marshall, 1943]. Comment: original vernacular name available (Art. 11.7.2): first used in latinized form by Pascoe (1870b: 437, as SIPALINAE), generally accepted as in Heyne and Taschenberg (1907: 233, as SIPALINI); permanently invalid (Art. 39): based on preoccupied type genus.

*SIPALININAE G. A. K. Marshall, 1953: 117 [stem: *Sipalin-*]. Type genus: *Sipalinus* G. A. K. Marshall, 1943. Comment: unavailable family-group name, proposed after 1930 without description or bibliographic reference to such a description (Art. 13.1).

SIPALININAE Zimmerman, 1993: 47, in key [stem: *Sipalin-*]. Type genus: *Sipalinus* G. A. K. Marshall, 1943.

Tribe RHINOSTOMINI LeConte, 1874

RHINIDAE J. L. LeConte, 1874b: 466 [stem: *Rhin-*]. Type genus: *Rhina* Latreille, 1806 [preoccupied genus name, not *Rhina* Schneider, 1801 [Pisces]; syn. of *Rhinostomus* Rafinesque, 1815]. Comment: permanently invalid (Art. 39): based on preoccupied type genus.

RHINOSTOMINI Kuschel, 1995: 24, in key [stem: *Rhinostom-*]. Type genus: *Rhinostomus* Rafinesque, 1815 [placed on the Official List of Generic Names in Zoology (ICZN 1955a)].

Subfamily RHYNCHOPHORINAE Schönherr, 1833

RHYNCHOPHORIDES Schönherr, 1833: 26 [stem: *Rhynchophor-*]. Type genus: *Rhynchophorus* Herbst, 1795.

Tribe DIOCALANDRINI Zimmerman, 1993

DIOCALANDRINI Zimmerman, 1993: 99 [stem: *Diocalandr-*]. Type genus: *Diocalandra* Faust, 1894.

Tribe LITOSOMINI Lacordaire, 1865

CALANDRINA C. G. Thomson, 1859: 145 [stem: *Calandr-*]. Type genus: *Calandra* Gistel, 1848 [placed on the Official Index of Invalid and Rejected Generic Names in Zoology (ICZN 1959b); preoccupied genus name, not *Calandra* Clairville, 1798 [Coleoptera: CURCULIONIDAE: RHYNCHOPHORINAE: SPHENOPHORINI]; syn. of *Sitophilus* Schönherr, 1837]. Comment: permanently invalid (Art. 39): based on suppressed type genus.

LITOSOMIDES Lacordaire, 1865: 303 [stem: *Litosom-*]. Type genus: *Litosomus* Lacordaire, 1865 [syn. of *Toxorhinus* Lacordaire, 1865]. Comment: original vernacular name available (Art. 11.7.2): first used in latinized form by Faust (1895: 224, as LITOSOMINORUM), generally accepted as in Alonso-Zarazaga and Lyal (1999: 65, as LITOSOMINI).

Sitophili Csiki, 1936: 68 [stem: *Sitophil-*]. Type genus: *Sitophilus* Schönherr, 1837 [placed on the Official List of Generic Names in Zoology (ICZN 1959b)]. Comment: name proposed to replace Calandrina C. G. Thomson, 1865 because of the synonymy of the type genus (*Calandra* auctorum is a synonym of *Sitophilus* Schönherr, 1837).

Tribe Ommatolampini Lacordaire, 1865

Ommatolampides Lacordaire, 1865: 276 [stem: *Ommatolamp-*]. Type genus: *Ommatolampes* Schönherr, 1837. Comment: original vernacular name available (Art. 11.7.2): first used in latinized form by Heyne and Taschenberg (1907: 233, as Ommatolampini), generally accepted as in Alonso-Zarazaga and Lyal (1999: 66, as Ommatolampini).

Tribe Polytini Zimmerman, 1993

Polytini Zimmerman, 1993: 94 [stem: *Polyt-*]. Type genus: *Polytus* Faust, 1894.

Tribe Rhynchophorini Schönherr, 1833

Rhynchophorides Schönherr, 1833: 26 [stem: *Rhynchophor-*]. Type genus: *Rhynchophorus* Herbst, 1795.

Tribe Sphenophorini Lacordaire, 1865

Calandraedes Billberg, 1820a: 40 [stem: *Calandr-*]. Type genus: *Calandra* Clairville, 1798 [placed on the Official Index of Invalid and Rejected Generic Names in Zoology (ICZN 1959b)]. Comment: permanently invalid (Art. 39): based on suppressed type genus; several family-group names based on *Calandra* Clairville, 1798 were placed on the Official Index of Rejected and Invalid Family-Group Names in Zoology (ICZN 1959b).

Sphénophorides Lacordaire, 1865: 286 [stem: *Sphenophor-*]. Type genus: *Sphenophorus* Schönherr, 1837 [placed on the Official List of Generic Names in Zoology (ICZN 1959b)]. Comment: original vernacular name available (Art. 11.7.2): first used in latinized form by J. L. LeConte (1876: 330, as Sphenophorini), generally accepted as in Alonso-Zarazaga and Lyal (1999: 66, as Sphenophorini); First Reviser (Sphenophorini Lacordaire, 1865 vs Sphenocorynini Lacordaire, 1865) not determined, current usage maintained.

Sphénocorynides Lacordaire, 1865: 279 [stem: *Sphenocoryn-*]. Type genus: *Sphenocorynes* Schönherr, 1837. Comment: original vernacular name available (Art. 11.7.2): first used in latinized form by Aurivillius (1886: 97, as Sphenocoryninae), generally accepted as in Csiki (1936: 20, as Sphenocoryni).

Oxyopisthinae Kolbe, 1899: 5 [stem: *Oxyopisth-*]. Type genus: *Oxyopisthen* sensu Lacordaire, 1865 [not *Oxyopisthen* J. Thomson, 1858; syn. of *Korotyaevius* Alonso-Zarazaga and Lyal, 1999]. Comment: based on misidentified type genus.

Subfamily Stromboscerinae Lacordaire, 1865

Stromboscérides Lacordaire, 1865: 306 [stem: *Stromboscer-*]. Type genus: *Stromboscerus* Schönherr, 1837. Comment: original vernacular name available (Art. 11.7.2): first used in latinized form by Pascoe (1870b: 437, as Strombocerinae [incorrect stem formation]), generally accepted as in Alonso-Zarazaga and Lyal (1999: 68, as Stromboscerinae).

Family Brachyceridae Billberg, 1820

Brachycerides Billberg, 1820a: 39 [stem: *Brachycer-*]. Type genus: *Brachycerus* A. G. Olivier, 1789.

Subfamily Brachycerinae Billberg, 1820

Brachycerides Billberg, 1820a: 39 [stem: *Brachycer-*]. Type genus: *Brachycerus* A. G. Olivier, 1789. Comment: the tribal classification used here follows Oberprieler (2010).

Tribe Brachycerini Billberg, 1820

Brachycerides Billberg, 1820a: 39 [stem: *Brachycer-*]. Type genus: *Brachycerus* A. G. Olivier, 1789.

Protomantiinae Aurivillius, 1886: 21, in key [stem: *Protomanti-*]. Type genus: *Protomantis* Schönherr, 1840.

Tribe Byrsopini Germar, 1829

Cryptopsides Schönherr, 1826: 65 [stem: *Cryptop-*]. Type genus: *Cryptops* Schönherr, 1823 [preoccupied genus name, not *Cryptops* Leach, 1814 [Chilopoda]; syn. of *Byrsops* Germar, 1829]. Comment: permanently invalid (Art. 39): based on preoccupied type genus.

Byrsoptides Germar, 1829: 358 [stem: *Byrsop-*]. Type genus: *Byrsops* Germar, 1829. Comment: incorrect original stem formation, not in prevailing usage.

Byrsopsides Schönherr, 1833: 14 [stem: *Byrsop-*]. Type genus: *Byrsops* Schönherr, 1833 [preoccupied genus name, not *Byrsops* Germar, 1829 [Coleoptera: Curculionidae: Byrsopini]; syn. of *Byrsops* Germar, 1829]. Comment: permanently invalid (Art. 39): based on preoccupied type genus.

Brotheinae G. A. K. Marshall, 1907: 89 [stem: *Brothe-*]. Type genus: *Brotheus* Stephens, 1829. Comment: junior homonym of Brotheini Simon, 1879 in Scorpiones (type genus *Brotheas* C. L. Koch, 1837); this case is to be referred to the Commission to remove the homonymy (Art. 55.3.1).

Brotheusini Alonso-Zarazaga and Lyal, 2006: 24 [stem: *Brotheus-*]. Type genus: *Brotheus* Stephens, 1829. Comment: unnecessary replacement name for Brotheinae Marshall, 1907 because of the homonymy of the family-group name; Brotheinae/-ini/-ina Simon, 1879 (type genus *Brotheas* C. L. Koch, 1837) has been used as valid in Scorpiones in recent literature, e.g., Prendini and W. C. Wheeler (2005); the stem used by Alonso-Zarazaga and Lyal (2006) was chosen in accordance with Art. 29.6.

Subfamily Cryptolarynginae Schalkwyk, 1966

Cryptopharynginae G. A. K. Marshall, 1957: 18 [stem: *Cryptopharyng-*]. Type genus: *Cryptopharynx* G. A. K. Marshall, 1957 [preoccupied genus name, not *Cryptopharynx* Kahl, 1928 [Protozoa]; syn. of *Cryptolarynx* Schalkwyk, 1966]. Comment: permanently invalid (Art. 39): based on preoccupied type genus.

Cryptolarynginae Schalkwyk, 1966: 745 [stem: *Cryptolaryng-*]. Type genus: *Cryptolarynx* Schalkwyk, 1966. Comment: replacement name for Cryptopharynginae G. A. K. Marshall, 1957 because of the homonymy of the type genus.

Periegini Legalov, 2003: 68 [stem: *Perieg-*]. Type genus: *Perieges* Schönherr, 1842.

Subfamily Erirhininae Schönherr, 1825

Erirhinides Schönherr, 1825: column 582 [stem: *Erirhin-*]. Type genus: *Erirhinus* Schönherr, 1825 [syn. of *Notaris* Germar, 1817].

Tribe Aonychini Zimmerman, 1993

Aonychini Zimmerman, 1993: 136, in key [stem: *Aonych-*]. Type genus: *Aonychus* Schönherr, 1844. Comment: junior homonym of Aonychini Davis, 1978 in Mammalia (type genus *Aonyx* Lesson, 1827); this case is to be referred to the Commission to remove the homonymy (Art. 55.3.1).

Aonychusini Zimmerman, 1999: 72 [stem: *Aonychus-*]. Type genus: *Aonychus* Schönherr, 1844. Comment: unnecessary replacement name for Aonychini Zimmerman, 1993 because of the homonymy of the family-group name; the stem used by Zimmerman (1999) was chosen in accordance with Art. 29.6.

Tribe Arthrostenini Reitter, 1913

Arthrostenini Reitter, 1913b: 57, in key [stem: *Arthrosten-*]. Type genus: *Arthrostenus* Schönherr, 1826.

†Tribe Cretuliini Legalov, 2009

Cretuliini Legalov, 2009c: 293 [stem: *Cretuli-*]. Type genus: *Cretulio* Zherikhin, 1993.

Tribe Erirhinini Schönherr, 1825

Erirhinides Schönherr, 1825: column 582 [stem: *Erirhin-*]. Type genus: *Erirhinus* Schönherr, 1825 [syn. of *Notaris* Germar, 1817].

Rhynchaenides Latreille, 1828: 599 [stem: *Rhynchaen-*]. Type genus: *Rhynchaenus* sensu Latreille, 1828 [not *Rhynchaenus* Clairville, 1798; syn. of *Notaris* Germar, 1817]. Comment: based on a misidentified type genus; an application will need to be submitted to the Commission to suppress this name for the Principles of Priority and Homonymy (Art. 65.2.1) if RHYNCAHENIDAE Blanchard, 1853 in CURCULIONINAE: RHAMPHINI: RHAMPHINA is to be used as valid in the future.

NOTARINI Zumpt, 1929: 216 [stem: *Notar-*]. Type genus: *Notaris* sensu Tournier, 1874 [not *Notaris* Germar, 1817; syn. of *Tournotaris* Alonso-Zarazaga and Lyal, 1999]. Comment: based on a misidentified type genus.

NOTODERMINA Voss, 1952: 196, in note [stem: *Notoderm-*]. Type genus: *Notodermus* Desbrochers des Loges, 1875 [syn. of *Procas* Stephens, 1831].

Tribe HIMASTHLOPHALLINI Zherikhin, 1991

HIMASTHLOPHALLINI Zherikhin, 1991: 31 [stem: *Himasthlophall-*]. Type genus: *Himasthlophallus* Egorov and Zherikhin, 1991.

Tribe STENOPELMINI LeConte, 1876

STENOPELMI J. L. LeConte, 1876: 179 [stem: *Stenopelm-*]. Type genus: *Stenopelmus* Schönherr, 1835.

LISSORHOPTRINAE Böving and Craighead, 1931: 67, in key [stem: *Lissorhoptr-*]. Type genus: *Lissorhoptrus* J. L. LeConte, 1876.

Tribe TADIINI Zimmerman, 1993

TADIINI Zimmerman, 1993: 136, in key [stem: *Tadi-*]. Type genus: *Tadius* Pascoe, 1885.

Tribe TANYSPHYRINI Gistel, 1848

TANYSPHYRIDAE Gistel, 1848: [8] [stem: *Tanysphyr-*]. Type genus: *Tanysphyrus* Germar, 1817.

BRACHYPI J. L. LeConte, 1876: 180 [stem: *Brachypod-*]. Type genus: *Brachypus* Schönherr, 1826 [preoccupied genus name, not *Brachypus* Meyer, 1814 [Aves]; syn. of *Brachygyius* G. A. K. Marshall, 1939]. Comment: permanently invalid (Art. 39): based on preoccupied type genus; incorrect original stem formation, not in prevailing usage.

Subfamily OCLADIINAE Lacordaire, 1865

OCLADIIDES Lacordaire, 1865: 79 [stem: *Ocladi-*]. Type genus: *Ocladius* Schönherr, 1825.

Tribe DESMIDOPHORINI Morimoto, 1962

DESMIDOPHORINAE Morimoto, 1962a: 32, in key [stem: *Desmidophor-*]. Type genus: *Desmidophorus* Dejean, 1835.

Tribe OCLADIINI Lacordaire, 1865

OCLADIIDES Lacordaire, 1865: 79 [stem: *Ocladi-*]. Type genus: *Ocladius* Schönherr, 1825. Comment: original vernacular name available (Art. 11.7.2): first used in latinized form by Hustache (1925: 9, as OCLADIINI), generally accepted as in Alonso-Zarazaga and Lyal (1999: 63, as OCLADIINI).

Subfamily RAYMONDIONYMINAE Reitter, 1913

RAYMONDIONYMINI Reitter, 1913b: 58, in key [stem: *Raymondionym-*]. Type genus: *Raymondionymus* Wollaston, 1873.

Tribe MYRTONYMINI Kuschel, 1990

MYRTONYMINA Kuschel, 1990: 80 [stem: *Myrtonym-*]. Type genus: *Myrtonymus* Kuschel, 1990.

Tribe RAYMONDIONYMINI Reitter, 1913

RAYMONDIONYMINI Reitter, 1913b: 58, in key [stem: *Raymondionym-*]. Type genus: *Raymondionymus* Wollaston, 1873.

Family CURCULIONIDAE Latreille, 1802

CURCULIONITES Latreille, 1802: 195 [stem: *Curculion-*]. Type genus: *Curculio* Linnaeus, 1758.

Subfamily CURCULIONINAE Latreille, 1802

CURCULIONITES Latreille, 1802: 195 [stem: *Curculion-*]. Type genus: *Curculio* Linnaeus, 1758.

Tribe ACALYPTINI Thomson, 1859

ACALYPTINA C. G. Thomson, 1859: 143 [stem: *Acalypt-*]. Type genus: *Acalyptus* Schönherr, 1833.

Subtribe ACALYPTINA Thomson, 1859

ACALYPTINA C. G. Thomson, 1859: 143 [stem: *Acalypt-*]. Type genus: *Acalyptus* Schönherr, 1833.

Subtribe DERELOMINA Lacordaire, 1865

DÉRÉLOMIDES Lacordaire, 1865: 9 [stem: *Derelom-*]. Type genus: *Derelomus* Schönherr, 1825. Comment: original vernacular name available (Art. 11.7.2): first used in latinized form by Marseul (1866: 109, as DERELOMIDAE), generally accepted as in Alonso-Zarazaga and Lyal (1999: 77, as DERELOMINI).

HOPLORRHININA Champion, 1903: 277 [stem: *Hoplorhin-*]. Type genus: *Hoplorhinus* Chevrolat, 1878 [as *Hoplorrhinus*, incorrect subsequent spelling of type genus name, not in prevailing usage; syn. of *Celetes* Schönherr, 1836]. Comment: incorrect original stem formation, not in prevailing usage.

CELETINI Voss, 1954: 344 [stem: *Celet-*]. Type genus: *Celetes* Schönherr, 1836. Comment: description by indication (distinguishing characters given in

Champion (1903: 277, as HOPLORRHININA)); name proposed to replace HOPLORHININI Champion, 1903 because of synonymy of the type genus; CELETINI is not in prevailing usage (Art. 40.2.1).

Subtribe NOTOLOMINA Franz, 2006

NOTOLOMINA N. M. Franz, 2006: 276 [stem: *Notolom-*]. Type genus: *Notolomus* J. L. LeConte, 1876.

Subtribe PHYLLOTROGINA Franz, 2006

PHYLLOTROGINA N. M. Franz, 2006: 276 [stem: *Phyllotrog-*]. Type genus: *Phyllotrox* Schönherr, 1843.

Subtribe STAMINODEINA Franz, 2006

STAMINODEINA N. M. Franz, 2006: 277 [stem: *Staminode-*]. Type genus: *Staminodeus* Franz, 2001.

Tribe ACENTRUSINI Alonso-Zarazaga, 2005

ACENTRINA Seidlitz, 1890: 608 [stem: *Acentr-*]. Type genus: *Acentrus* Schönherr, 1845 [preoccupied genus name, not *Acentrus* Duponchel, 1839 [Coleoptera: CURCULIONIDAE: CURCULIONINAE: ACENTRUSINI]; syn. of *Acentrus* Duponchel, 1839]. Comment: permanently invalid (Art. 39): based on preoccupied type genus.

ACENTRUSINI Alonso-Zarazaga, 2005: 92 [stem: *Acentrus-*]. Type genus: *Acentrus* Duponchel, 1839. Comment: replacement name for ACENTRINI Seidlitz, 1890 because of the homonymy of the family-group name; the stem used by Alonso-Zarazaga (2005) was chosen in accordance with Art. 29.6.

Tribe ANCYLOCNEMIDINI Voss, 1962

ANCYLOCNEMINI Voss, 1962: 339 [stem: *Ancylocnemid-*]. Type genus: *Ancylocnemis* G. A. K. Marshall, 1920. Comment: incorrect original stem formation, not in prevailing usage.

Tribe ANTHONOMINI Thomson, 1859

ANTHONOMINA C. G. Thomson, 1859: 144 [stem: *Anthonom-*]. Type genus: *Anthonomus* Germar, 1817.

*BRACHONYDES Tournier, 1874: 68, in key [stem: *Brachonych-*]. Type genus: *Brachonyx* Schönherr, 1825. Comment: original vernacular name unavailable (Art. 11.7.2): subsequently used in latinized form but not generally attributed to Tournier (1874); Alonso-Zarazaga and Lyal (1999: 74) attributed this name to the correct author but did not use it as a valid taxon; incorrect original stem formation, not in prevailing usage.

LONCOPHORINI Pierce, 1916: 467 [stem: *Loncophor-*]. Type genus: *Loncophorus* Chevrolat, 1832.

Bradybatini Pierce, 1919: 26, in key [stem: *Bradybat-*]. Type genus: *Bradybatus* Germar, 1824.

Brachonychina Voss, 1944: 38, in key [stem: *Brachonych-*]. Type genus: *Brachonyx* Schönherr, 1825.

Tachypterellina Voss, 1944: 38, in key [stem: *Tachypterell-*]. Type genus: *Tachypterellus* Fall and Cockerell, 1907 [syn. of *Anthonomus* Germar, 1817].

Tribe Camarotini Schönherr, 1833

Camarotides Schönherr, 1833: 4 [stem: *Camarot-*]. Type genus: *Camarotus* Germar, 1817.

Subtribe Camarotina Schönherr, 1833

Camarotides Schönherr, 1833: 4 [stem: *Camarot-*]. Type genus: *Camarotus* Germar, 1817.

Subtribe Prionomerina Lacordaire, 1863

Prionomérides Lacordaire, 1863: 598 [stem: *Prionomer-*]. Type genus: *Prionomerus* Schönherr, 1835 [syn. of *Odontopus* Say, 1831]. Comment: original vernacular name available (Art. 11.7.2): first used in latinized form by Pascoe (1870b: 436, as Prionomerinae), generally accepted as in Alonso-Zarazaga and Lyal (1999: 75, as Prionomerina).

Tribe Ceratopodini Lacordaire, 1863

Cératopides Lacordaire, 1863: 589 [stem: *Ceratopod-*]. Type genus: *Ceratopus* Schönherr, 1843. Comment: original vernacular name available (Art. 11.7.2): first used in latinized form by Pascoe (1870b: 436, as Ceratopinae [incorrect stem formation]), generally accepted as in Alonso-Zarazaga and Lyal (1999: 76, as Ceratopodini); incorrect original stem formation, not in prevailing usage.

Tribe Cionini Schönherr, 1825

Cionides Schönherr, 1825: column 587 [stem: *Cion-*]. Type genus: *Cionus* Clairville, 1798.

Tribe Cranopoeini Kuschel, 2009

Cranopoeini Kuschel, 2009: 44 [stem: *Cranopoe-*]. Type genus: *Cranopoeus* G. A. K. Marshall, 1931.

Tribe Cryptoplini Lacordaire, 1863

Cryptoplides Lacordaire, 1863: 486 [stem: *Cryptopl-*]. Type genus: *Cryptoplus* Erichson, 1842. Comment: original vernacular name available (Art. 11.7.2): first used in latinized form by J. L. LeConte (1876: 175, as Cryptopli), generally accepted as in Alonso-Zarazaga and Lyal (1999: 76, as Cryptoplini).

Haplonycides Lacordaire, 1865: 16 [stem: *Haplonych-*]. Type genus: *Haplonyx* Schönherr, 1836. Comment: original vernacular name available (Art. 11.7.2): first used in latinized form by Pascoe (1870b: 437, as Haplonychinae), generally accepted as in Schenkling and G. A. K. Marshall (1936: 1, as Haplonychinae); incorrect original stem formation, not in prevailing usage; this name is a junior homonym of Haplonychini Burmeister, 1855 (type genus *Haplonycha* Dejean, 1836) proposed in Scarabaeidae and now a synonym of Liparetrini H. C. C. Burmeister, 1855; this case is to be referred to the Commission to remove the homonymy (Art. 55.3.1) if either of those names is to be used as valid in the future.

Tribe Curculionini Latreille, 1802

Curculionites Latreille, 1802: 195 [stem: *Curculion-*]. Type genus: *Curculio* Linnaeus, 1758.

Subtribe Curculionina Latreille, 1802

Curculionites Latreille, 1802: 195 [stem: *Curculion-*]. Type genus: *Curculio* Linnaeus, 1758.

Balaninidae Gistel, 1848: [7] [stem: *Balanin-*]. Type genus: *Balaninus* Germar, 1817 [syn. of *Curculio* Linnaeus, 1758].

Subtribe Pseudobalaninina Heller, 1925

Pseudobalaninina Heller, 1925b: 94, in key [stem: *Pseudobalanin-*]. Type genus: *Pseudobalaninus* Faust, 1889.

Subtribe Timolina Heller, 1925

Timolina Heller, 1925b: 94, in key [stem: *Timol-*]. Type genus: *Timola* Pascoe, 1886.

Tribe Diabathrariini Lacordaire, 1863

Diabathrariides Lacordaire, 1863: 407 [stem: *Diabathrari-*]. Type genus: *Diabathrarius* Schönherr, 1840. Comment: original vernacular name available (Art. 11.7.2): first used in latinized form by Pascoe (1870b: 436, as Diabathrariinae), generally accepted as in Alonso-Zarazaga and Lyal (1999: 141, as Diabathrariini); transfer from Cyclominae by Oberprieler (2010).

Tribe Ellescini Thomson, 1859

Elleschina C. G. Thomson, 1859: 143 [stem: *Ellesc-*]. Type genus: *Ellescus* Dejean, 1821 [as *Elleschus*, unjustified emendation of type genus name by Schönherr (1837), not in prevailing usage].

Subtribe Dorytomina Bedel, 1886

Dorytomini Bedel, 1886: 280 [stem: *Dorytom-*]. Type genus: *Dorytomus* Germar, 1817.

Subtribe ELLESCINA Thomson, 1859

ELLESCHINA C. G. Thomson, 1859: 143 [stem: *Ellesc-*]. Type genus: *Ellescus* Dejean, 1821 [as *Elleschus*, unjustified emendation of type genus name by Schönherr (1837), not in prevailing usage]. Comment: incorrect original stem formation, not in prevailing usage.

Tribe ERODISCINI Lacordaire, 1863

ERODISCIDES Lacordaire, 1863: 566 [stem: *Erodisc-*]. Type genus: *Erodiscus* sensu Lacordaire, 1863 [Lacordaire excluded the type species of *Erodiscus* Schönherr, 1825 to place it in *Toxeutes* Schönherr, 1843; the five species that Lacordaire included in *Erodiscus* are now placed in *Sicoderus* Vanin, 1986, *Prosicoderus* Vanin, 1986 and *Pimelerodius* Vanin, 1986]. Comment: based on a misidentified type genus; MAAZ and CHCL will submit an application for the conservation of this name by designating *Erodiscus* Schönherr, 1825 as its type genus.

Tribe EUGNOMINI Lacordaire, 1863

EUGNOMIDES Lacordaire, 1863: 499 [stem: *Eugnom-*]. Type genus: *Eugnomus* Schönherr, 1847.

Subtribe EUGNOMINA Lacordaire, 1863

EUGNOMIDES Lacordaire, 1863: 499 [stem: *Eugnom-*]. Type genus: *Eugnomus* Schönherr, 1847. Comment: original vernacular name available (Art. 11.7.2): first used in latinized form by J. L. LeConte (1876: 174, as EUGNOMI), generally accepted as in Alonso-Zarazaga and Lyal (1999: 79, as EUGNOMINI); First Reviser found (EUGNOMINA Lacordaire, 1863 vs SCOLOPTERINA Lacordaire, 1863) is Voss (1937a: 37).

SCOLOPTÉRIDES Lacordaire, 1863: 565 [stem: *Scolopter-*]. Type genus: *Scolopterus* A. White, 1846. Comment: original vernacular name available (Art. 11.7.2): first used in latinized form by Pascoe (1870b: 436, as SCOLOPTERINAE), generally accepted as in Broun (1880: 472 as SCOLOPTERIDAE).

PACTOLINA Voss, 1936: 112, in key [stem: *Pactol-*]. Type genus: *Pactola* Pascoe, 1876.

STEPHANORHYNCHINA Voss, 1936: 112 [stem: *Stephanorhynch-*]. Type genus: *Stephanorhynchus* A. White, 1846.

Subtribe MERIPHINA Marshall, 1937

MERIPHINAE G. A. K. Marshall, 1937a: 329 [stem: *Meriph-*]. Type genus: *Meriphus* Erichson, 1842.

Tribe GONIPTERINI Lacordaire, 1863

GONIPTÉRIDES Lacordaire, 1863: 391 [stem: *Gonipter-*]. Type genus: *Gonipterus* Schönherr, 1833. Comment: original vernacular name available (Art. 11.7.2): first used in latinized form by Pascoe (1870b: 436, as GONIPTERINAE), generally accepted as in Alonso-Zarazaga and Lyal (1999: 141, as GONIPTERINI); transfer from CYCLOMINAE by Oberprieler (2010).

Tribe MECININI Gistel, 1848

MECINIDAE Gistel, 1848: [7] [stem: *Mecin-*]. Type genus: *Mecinus* Germar, 1821.

GYMNETRINA C. G. Thomson, 1859: 143 [stem: *Gymnetr-*]. Type genus: *Gymnetron* Schönherr, 1825. Comment: junior homonym of GYMNETRINAE Swainson, 1839 in Pisces (type genus *Gymnetrus* Bloch, 1795).

MIARIDES Tournier, 1874: 66 [stem: *Miar-*]. Type genus: *Miarus* Schönherr, 1826. Comment: original vernacular name available (Art. 11.7.2): first used in latinized form and generally accepted as in Pierce (1919: 30, as MIARINAE).

MIARINI Zherikhin, 1991: 123 [stem: *Miar-*]. Type genus: *Miarus* Schönherr, 1826. Comment: family-group name proposed as new without reference to MIARIDES Tournier, 1874.

Tribe NERTHOPINI Lacordaire, 1865

NERTHOPIDES Lacordaire, 1865: 19 [stem: *Nerthop-*]. Type genus: *Nerthops* Schönherr, 1826. Comment: original vernacular name available (Art. 11.7.2): first used in latinized form by Pascoe (1870b: 437, as NERTHOPINAE), generally accepted as in Alonso-Zarazaga and Lyal (1999: 80, as NERTHOPINI); precedence (NERTHOPINI Lacordaire, 1865 vs ACALLOPISTINI Lacordaire, 1865) given to taxon originally proposed at the higher rank (Art. 24.1).

*MICROSTYLIDES Lacordaire, 1865: 20 [stem: *Microstyl-*]. Type genus: *Microstylus* Schönherr, 1847. Comment: original vernacular name unavailable (Art. 11.7.2): subsequently used in latinized form but not generally attributed to Lacordaire (1865); subsequent usage of MICROSTYLINAE by Hoffmann (1968: 27) and MICROSTYLIDAE by Ienistea (1986: 33) did not validate this name because these authors did not attribute their names to Lacordaire, 1865; MICROSTYLINAE Hoffmann, 1968: 27 and MICROSTYLIDAE Ienistea, 1986: 33 are also unavailable because they were proposed after 1930 without description or bibliographic reference to such a description (Art. 13.1).

ACALLOPISTIDES Lacordaire, 1865: 22 [stem: *Acallopist-*]. Type genus: *Acallopistus* Schönherr, 1826. Comment: original vernacular name available (Art. 11.7.2): first used in latinized form by Hustache (1932: 371, as ACALLOPISTINAE), generally accepted as in Hustache (1934: 3, as ACALLOPISTINI).

ARCHOLABINAE Voss, 1929a: 200 [stem: *Archolab-*]. Type genus: *Archolabus* Voss, 1929.

Tribe OTIDOCEPHALINI Lacordaire, 1863

OTIDOCÉPHALIDES Lacordaire, 1863: 568 [stem: *Otidocephal-*]. Type genus: *Otidocephalus* Chevrolat, 1832. Comment: original vernacular name available (Art. 11.7.2): first used in latinized form by Pascoe (1870b: 436, as OTIDOCEPHALINAE), generally accepted as in Alonso-Zarazaga and Lyal (1999: 80, as OTIDOCEPHALINI).

Tribe Piazorhinini Lacordaire, 1863

Piazorhinides Lacordaire, 1863: 601 [stem: *Piazorhin-*]. Type genus: *Piazorhinus* Schönherr, 1835. Comment: original vernacular name available (Art. 11.7.2): first used in latinized form by Kirsch (1875: 162, as Piazorhinidae), generally accepted as in Alonso-Zarazaga and Lyal (1999: 81, as Piazorhinini).

Tribe Prionobrachiini Hustache, 1938

Prionobrachiina Hustache, 1938: 90, in key [stem: *Prionobrachi-*]. Type genus: *Prionobrachium* Faust, 1894.

Tribe Pyropini Lacordaire, 1865

Pyropides Lacordaire, 1865: 187 [stem: *Pyrop-*]. Type genus: *Pyropus* Schönherr, 1836. Comment: original vernacular name available (Art. 11.7.2): first used in latinized form by Pascoe (1870b: 437, as Pyropinae), generally accepted as in Alonso-Zarazaga and Lyal (1999: 81, as Pyropini).

Tribe Rhamphini Rafinesque, 1815

Ramphoria Rafinesque, 1815: 115 [stem: *Rhamph-*]. Type genus: *Rhamphus* Clairville, 1798 [as *Ramphorus*, incorrect subsequent spelling of type genus name, not in prevailing usage]. Comment: incorrect original stem formation, not in prevailing usage.

Subtribe Dinorhopalina Voss, 1936

Dinorrhopalina Voss, 1936: 112, in key [stem: *Dinorhopal-*]. Type genus: *Dinorhopala* Pascoe, 1860 [as *Dinorrhopala*, incorrect subsequent spelling of type genus name, not in prevailing usage]. Comment: incorrect original stem formation, not in prevailing usage.

Subtribe Ixalmina Voss, 1936

Ixalmina Voss, 1936: 112, in key [stem: *Ixalm-*]. Type genus: *Ixalma* Pascoe, 1871.

Subtribe Rhamphina Rafinesque, 1815

Ramphoria Rafinesque, 1815: 115 [stem: *Rhamph-*]. Type genus: *Rhamphus* Clairville, 1798 [as *Ramphorus*, incorrect subsequent spelling of type genus name, not in prevailing usage]. Comment: incorrect original stem formation, not in prevailing usage.

Rhamphides Schönherr, 1823: column 1136 [stem: *Rhamph-*]. Type genus: *Rhamphus* Clairville, 1798. Comment: family-group name proposed as new without reference to Ramphoria Rafinesque, 1815.

Orchestides Latreille, 1828: 601 [stem: *Orchest-*]. Type genus: *Orchestes* Illiger, 1798.

RHYNCHENIDAE Blanchard, 1853: 240 [stem: *Rhynchaen-*]. Type genus: *Rhynchaenus* Clairville, 1798. Comment: incorrect original stem formation, not in prevailing usage; an application will need to be submitted to the Commission to suppress RHYNCHAENIDES Latreille, 1828 (based on the misidentified type genus *Rhynchaenus* sensu Latreille, 1828) for the Principles of Priority and Homonymy (Art. 65.2.1) if this name is to be used as valid.

Subtribe TACHYGONINA Lacordaire, 1865

TACHYGONIDES Lacordaire, 1865: 167 [stem: *Tachygon-*]. Type genus: *Tachygonus* Schönherr, 1833 [syn. of *Tachygonus* Guérin-Méneville, 1833]. Comment: original vernacular name available (Art. 11.7.2): first used in latinized form by Pascoe (1870b: 437, as TACHYGONINAE), generally accepted as in Alonso-Zarazaga and Lyal (1999: 82, as TACHYGONINA); the type genus for Lacordaire's TACHYGONIDES is *Tachygonus* Schönherr, 1833; an application will be submitted by MAAZ and CHCL to the Commission to designate *Tachygonus* Guérin-Méneville, [1833] as the type genus of this family-group name.

Tribe SMICRONYCHINI Seidlitz, 1891 *nomen protectum*

DESMORHINES J. L. LeConte, 1876: 167 [stem: *Desmorin-*]. Type genus: *Desmoris* J. L. LeConte, 1876 [syn. of *Smicronyx* Schönherr, 1843]. Comment: *nomen oblitum* (see Appendix 1); incorrect original stem formation, not in prevailing usage.

SMICRONYCINA Seidlitz, 1891 [Gatt.]: 162 [stem: *Smicronych-*]. Type genus: *Smicronyx* Schönherr, 1843. Comment: *nomen protectum* (see Appendix 1); incorrect original stem formation, not in prevailing usage.

Tribe SPHAERIOPOEINI Kuschel, 2003

SPHAERIOPOEINI Kuschel, 2003: 63 [stem: *Sphaeriopoe-*]. Type genus: *Sphaeriopoeus* Kuschel, 2003.

Tribe STOREINI Lacordaire, 1863

STORÉIDES Lacordaire, 1863: 494 [stem: *Store-*]. Type genus: *Storeus* Schönherr, 1843. Comment: original vernacular name available (Art. 11.7.2): first used in latinized form by Kirsch (1877: 170, as STOREIDAE), generally accepted as in Alonso-Zarazaga and Lyal (1999: 82, as STOREINI).

HYBOMORPHIDES Lacordaire, 1865: 141 [stem: *Hybomorph-*]. Type genus: *Hybomorphus* Saunders and Jekel, 1855. Comment: original vernacular name available (Art. 11.7.2): first used in latinized form and generally accepted as in Hustache (1936: 261, as HYBOMORPHINA).

Tribe Styphlini Jekel, 1861

Styphlidae Jekel, 1861: 274 [stem: *Styphl-*]. Type genus: *Styphlus* Schönherr, 1826.

*Orthochaetini A. Winkler, 1932: 1544 [stem: *Orthochaet-*]. Type genus: *Orthochaetes* Germar, 1824. Comment: unavailable family-group name, proposed after 1930 without description or bibliographic reference to such a description (Art. 13.1).

Orthocaetina Morimoto, 1962a: 56, in key [stem: *Orthochaet-*]. Type genus: *Orthochaetes* Germar, 1824. Comment: incorrect original stem formation, not in prevailing usage.

Tribe Tychiini Gistel, 1848

Tychiidae Gistel, 1848: [7] [stem: *Tychi-*]. Type genus: *Tychius* Germar, 1817.

Subtribe Demimaeina Voss, 1937

Demimaeini Voss, 1937b: 144 [stem: *Demimae-*]. Type genus: *Demimaea* Pascoe, 1870.

Subtribe Lignyodina Bedel, 1884

Lignyodini Bedel, 1884a: 66, in key [stem: *Lignyod-*]. Type genus: *Lignyodes* Dejean, 1835.

Subtribe Ochyromerina Voss, 1935

Ochyromerina Voss, 1935a: 228 [stem: *Ochyromer-*]. Type genus: *Ochyromera* Pascoe, 1874.

Endaeini Voss, 1958: 101, in key [stem: *Endae-*]. Type genus: *Endaeus* Schönherr, 1826.

Subtribe Tychiina Gistel, 1848

Tychiidae Gistel, 1848: [7] [stem: *Tychi-*]. Type genus: *Tychius* Germar, 1817.

Miccotrogidae Gistel, 1856a: 371 [stem: *Miccotrog-*]. Type genus: *Miccotrogus* Schönherr, 1825 [syn. of *Tychius* Germar, 1817].

Sibynidae Marseul, 1863: 239 [stem: *Sibini-*]. Type genus: *Sibinia* Germar, 1817 [as *Sibynes*, unjustified emendation of type genus name by Schönherr (1825), not in prevailing usage]. Comment: incorrect original stem formation, not in prevailing usage.

Tribe Ulomascini Lacordaire, 1865

Ulomascides Lacordaire, 1865: 184 [stem: *Ulomasc-*]. Type genus: *Ulomascus* Fairmaire, 1848. Comment: original vernacular name available (Art. 11.7.2): first used in latinized form by Pascoe (1870b: 437, as Ulomascinae), generally accepted as in Alonso-Zarazaga and Lyal (1999: 86, as Ulomascini).

Tribe Viticiini Morimoto, 1983

Viticiinae Morimoto, 1983: 55 [stem: *Vitici-*]. Type genus: *Viticis* Lea, 1930. Comment: transfer from Cyclominae by Oberprieler (2010).

Subfamily Bagoinae Thomson, 1859 *nomen protectum*

Lypriidae Gistel, 1848: [7] [stem: *Lypr-*]. Type genus: *Lyprus* Schönherr, 1826 [syn. of *Bagous* Germar, 1817]. Comment: as pointed out by Alonso-Zarazaga and Lyal (2002: 17) this is the oldest available name for the subfamily; Lyprinae Gistel, 1848 was recently considered a *nomen oblitum* by Colonnelli (2003: 7) however the necessary supporting references were not provided; we hereby consider Lyprinae Gistel 1848 as a *nomen oblitum* (see supporting references in Appendix 1); incorrect original stem formation, not in prevailing usage.

Bagoina C. G. Thomson, 1859: 135 [stem: *Bago-*]. Type genus: *Bagous* Germar, 1817. Comment: *nomen protectum* (see Appendix 1).

Hydronomides Lacordaire, 1863: 483 [stem: *Hydronom-*]. Type genus: *Hydronomus* Schönherr, 1825. Comment: original vernacular name available (Art. 11.7.2): first used in latinized form by J. L. LeConte (1876: 182, as Hydronomi), generally accepted as in Zimmerman (1993: 150, as Hydronomini).

Pseudobagoini Sharp, 1917: 26 [stem: *Pseudobago-*]. Type genus: *Pseudobagous* Sharp, 1917.

Subfamily Baridinae Schönherr, 1836

Barídides Schönherr, 1836: 636 [stem: *Barid-*]. Type genus: *Baris* Germar, 1817 [as *Baridius*, unjustified emendation of type genus name by Schönherr (1825), not in prevailing usage].

Tribe Ambatini Lacordaire, 1863

Ambatides Lacordaire, 1863: 512 [stem: *Ambat-*]. Type genus: *Ambates* Schönherr, 1836. Comment: original vernacular name available (Art. 11.7.2): first used in latinized form by Pascoe (1870b: 436, as Ambatinae), generally accepted as in Alonso-Zarazaga and Lyal (1999: 93, as Ambatini).

*Pteracanthides Chevrolat, 1878: 341 [stem: *Pteracanth-*]. Type genus: *Pteracanthus* Dejean, 1835. Comment: original vernacular name unavailable (Art. 11.7.2): not subsequently latinized; a family-group name with the same stem, Pteracanthina J. Thomson, 1864 (type genus *Pteracantha* Newman, 1838), is available in Coleoptera: Cerambycidae.

Tribe Anopsilini Bondar, 1942

Anopsilini Bondar, 1942: 253 [stem: *Anopsil-*]. Type genus: *Anopsilus* Kirsch, 1870.

Tribe Apostasimerini Schönherr, 1844

Apostasimerides Schönherr, 1844: 1 [stem: *Apostasimer-*]. Type genus: *Apostasimerus* Schönherr, 1844. Comment: see Prena (2009: 33) for details on authorship and taxa included in this tribe.

Subtribe Apostasimerina Schönherr, 1844

*Apostasimerides Schönherr, 1836: 557 [stem: *Apostasimer-*]. Type genus: *Apostasimerus* Schönherr, 1844. Comment: family-group name unavailable (Art. 11.7.1.1): not based on an available genus name at the time.

Apostasimerides Schönherr, 1844: 1 [stem: *Apostasimer-*]. Type genus: *Apostasimerus* Schönherr, 1844.

Subtribe Madopterina Lacordaire, 1865

Madoptérides Lacordaire, 1865: 243 [stem: *Madopter-*]. Type genus: *Madopterus* Schönherr, 1836. Comment: original vernacular name available (Art. 11.7.2): first used in latinized form by Kirsch (1870: 218, as Madopteridae), generally accepted as in Alonso-Zarazaga and Lyal (1999: 96, as Madopterini).

Subtribe Thaliabaridina Bondar, 1943

Thaliabarini Bondar, 1943b: 131 [stem: *Thaliabarid-*]. Type genus: *Thaliabaris* Bondar, 1943. Comment: incorrect original stem formation, not in prevailing usage.

Subtribe Torcina Bondar, 1943

Torcocina Bondar, 1943a: 53 [stem: *Torc-*]. Type genus: *Torcus* Casey, 1922. Comment: incorrect original stem formation, not in prevailing usage.

Subtribe Zygobaridina Pierce, 1907

Centrinides Jekel, 1865: 548 [stem: *Centrin-*]. Type genus: *Centrinus* Schönherr, 1825. Comment: junior homonym of Centrininae Swainson, 1839 in Pisces (type genus *Centrina* Cuvier 1817); this case is to be referred to the Commission to remove the homonymy (Art. 55.3.1).

Zygobari Pierce, 1907: 381 [stem: *Zygobarid-*]. Type genus: *Zygobaris* J. L. LeConte, 1876. Comment: incorrect original stem formation, not in prevailing usage; Alonso-Zarazaga and Lyal (1999: 97) and subsequent workers have used this name as valid instead of the junior homonym Centrinina Jekel, 1865, we continue to use Zygobaridina as valid here.

Limnobarini Casey, 1922: 269 [stem: *Limnobarid-*]. Type genus: *Limnobaris* Bedel, 1885. Comment: incorrect original stem formation, not in prevailing usage.

Tribe Baridini Schönherr, 1836

Barididae Schönherr, 1836: 636 [stem: *Barid-*]. Type genus: *Baris* Germar, 1817 [as *Baridius*, unjustified emendation of type genus name by Schönherr (1825), not in prevailing usage].

Subtribe Baridina Schönherr, 1836

Barididae Schönherr, 1836: 636 [stem: *Barid-*]. Type genus: *Baris* Germar, 1817 [as *Baridius*, unjustified emendation of type genus name by Schönherr (1825), not in prevailing usage].

Subtribe Coelonertina Casey, 1922

Coelonertini Casey, 1922: 94 [stem: *Coelonert-*]. Type genus: *Coelonertus* Solari and Solari, 1906.

Subtribe Coleomerina Casey, 1922

Coleomerini Casey, 1922: 84 [stem: *Coleomer-*]. Type genus: *Coleomerus* Schönherr, 1836.

Subtribe Diorymerina Jekel, 1865

Diorymerides Jekel, 1865: 548 [stem: *Diorymer-*]. Type genus: *Diorymerus* Schönherr, 1825.

Subtribe Eurhinina Lacordaire, 1865

Eurhinides Lacordaire, 1865: 220 [stem: *Eurhin-*]. Type genus: *Eurhinus* Illiger, 1807 [placed on the Official List of Generic Names in Zoology (ICZN 1985e)]. Comment: original vernacular name available (Art. 11.7.2): first used in latinized form by Pierce (1915: 15, as Eurhininae), generally accepted as in Alonso-Zarazaga and Lyal (1999: 92, as Eurhinina); Eurhinini Lacordaire, 1865 placed on the Official List of Family-Group Names in Zoology (ICZN 1985e).

Tribe Madarini Jekel, 1865

Madarides Jekel, 1865: 550 [stem: *Madar-*]. Type genus: *Madarus* Schönherr, 1825. Comment: Madarini, published on 25 January 1865, has priority over Barymerini and Leptoschoinini both published by Lacordaire "before 12 December 1865".

Subtribe Barymerina Lacordaire, 1865

Barymérides Lacordaire, 1865: 259 [stem: *Barymer-*]. Type genus: *Barymerus* Lacordaire, 1865. Comment: original vernacular name available (Art. 11.7.2): first used in latinized form by Faust (1896: 127, as Barymerinorum), generally accepted as in Alonso-Zarazaga and Lyal (1999: 94, as Barymerina).

Sonnetiini Casey, 1922: 321 [stem: *Sonneti-*]. Type genus: *Sonnetius* Casey, 1922.

Subtribe Eutoxina Champion, 1908

Eutoxides Champion, 1908: 360 [stem: *Eutox-*]. Type genus: *Eutoxus* Schönherr, 1844.

Subtribe Leptoschoinina Lacordaire, 1865

Leptoschoïnides Lacordaire, 1865: 236 [stem: *Leptoschoin-*]. Type genus: *Leptoschoinus* Dejean, 1836. Comment: original vernacular name available (Art. 11.7.2): first used in latinized form by Faust (1896: 126, as Leptoschoenorum [incorrect stem formation]), generally accepted as in Alonso-Zarazaga and Lyal (1999: 94, as Leptoschoinina).

Subtribe Madarina Jekel, 1865

Madarides Jekel, 1865: 550 [stem: *Madar-*]. Type genus: *Madarus* Schönherr, 1825.

Subtribe Tonesiina Alonso-Zarazaga and Lyal, 1999

Lytériides Lacordaire, 1865: 249 [stem: *Lyteri-*]. Type genus: *Lyterius* sensu Lacordaire, 1865 [not *Lyterius* Schönherr, 1844; syn. of *Tonesia* Casey, 1922]. Comment: based on a misidentified type genus, name treated here as invalid until an application is submitted to the Commission to suppress it for the Principle of Priority (Art. 65.2.1).

Tonesiina Alonso-Zarazaga and Lyal, 1999: 95 [stem: *Tonesi-*]. Type genus: *Tonesia* Casey, 1922. Comment: although this is not the oldest name for the subtribe, we recommend that an application be submitted to the Commission to suppress Lyteriini Lacordaire, 1865 because it is based on a misidentified type genus (Art. 65.2.1).

Tribe Neosharpiini Hoffmann, 1956

Neosharpini Hoffmann, 1956: 246 [stem: *Neosharpi-*]. Type genus: *Neosharpia* Hoffmann, 1956. Comment: incorrect original stem formation, not in prevailing usage.

Tribe Nertinini Voss, 1954

*Nertides Lacordaire, 1865: 238 [stem: *Nert-*]. Type genus: *Nertus* Schönherr, 1844 [preoccupied genus name, not *Nertus* Boie, 1828 [Aves]; syn. of *Nertinus* G. A. K. Marshall, 1943]. Comment: original vernacular name unavailable (Art. 11.7.2): not subsequently latinized and attributed to Lacordaire (1865); if found to be available then permanently invalid (Art. 39): based on preoccupied type genus.

NERTINI Bondar, 1949: 223 [stem: *Nert-*]. Type genus: *Nertus* Schönherr, 1844 [preoccupied genus name, not *Nertus* Boie 1828 [Aves]; syn. of *Nertinus* G. A. K. Marshall, 1943]. Comment: family-group name proposed as new without reference to NERTIDES Lacordaire, 1865; permanently invalid (Art. 39): based on preoccupied type genus.

NERTININA Voss, 1954: 323 [stem: *Nertin-*]. Type genus: *Nertinus* G. A. K. Marshall, 1943.

Tribe OPTATINI Champion, 1907

OPTATIDES Champion, 1907: 185 [stem: *Optat-*]. Type genus: *Optatus* Pascoe, 1889.

Tribe PANTOTELINI Lacordaire, 1865

PANTOTÉLIDES Lacordaire, 1865: 212 [stem: *Pantotel-*]. Type genus: *Pantoteles* Schönherr, 1845.

Subtribe CYRIONYCHINA Casey, 1922

CYRIONICHINI Casey, 1922: 10 [stem: *Cyrionych-*]. Type genus: *Cyrionyx* Faust, 1896. Comment: incorrect original stem formation, not in prevailing usage.

Subtribe PANTOTELINA Lacordaire, 1865

PANTOTÉLIDES Lacordaire, 1865: 212 [stem: *Pantotel-*]. Type genus: *Pantoteles* Schönherr, 1845. Comment: original vernacular name available (Art. 11.7.2): first used in latinized form by Pascoe (1870b: 437, as PANTOTELINAE), generally accepted as in Alonso-Zarazaga and Lyal (1999: 102, as PANTOTELINI).

Tribe PERIDINETINI Lacordaire, 1865

PÉRIDINÉTIDES Lacordaire, 1865: 209 [stem: *Peridinet-*]. Type genus: *Peridinetus* Schönherr, 1837. Comment: original vernacular name available (Art. 11.7.2): first used in latinized form by Pascoe (1886: cliii, as PERIDINETINAE), generally accepted as in Alonso-Zarazaga and Lyal (1999: 102, as PERIDINETINI).

Subfamily CEUTORHYNCHINAE Gistel, 1848

CENTORHYNCHIDAE Gistel, 1848: [7] [stem: *Ceutorhynch-*]. Type genus: *Ceutorhynchus* Germar, 1824 [as *Centorhynchus*, incorrect subsequent spelling of type genus name, not in prevailing usage; placed on the Official List of Generic Names in Zoology (ICZN 1989c)]. Comment: incorrect original stem formation, not in prevailing usage; First Revisers found (CEUTORHYNCHINAE Gistel, 1848 vs PHYTOBIINAE Gistel, 1848) are Alonso-Zarazaga and Lyal (2002).

Tribe Ceutorhynchini Gistel, 1848

Centorhynchidae Gistel, 1848: [7] [stem: *Ceutorhynch-*]. Type genus: *Ceutorhynchus* Germar, 1824 [as *Centorhynchus*, incorrect subsequent spelling of type genus name, not in prevailing usage; placed on the Official List of Generic Names in Zoology (ICZN 1989c)]. Comment: incorrect original stem formation, not in prevailing usage.

Isorhynchides Lacordaire, 1865: 172 [stem: *Isorhynch-*]. Type genus: *Isorhynchus* Schönherr, 1833. Comment: original vernacular name available (Art. 11.7.2): first used in latinized form by Pascoe (1870b: 437, as Isorhynchinae), generally accepted as in Hustache (1934: 72, as Isorhynchini).

Coeliodes J. L. LeConte, 1876: 268 [stem: *Coeliod-*]. Type genus: *Coeliodes* Schönherr, 1837 [placed on the Official List of Generic Names in Zoology (ICZN 1989c)].

Poophagidae Schultze, 1902: 226 [stem: *Poophag-*]. Type genus: *Poophagus* Schönherr, 1837.

Amalina Wagner, 1936: 166 [stem: *Amal-*]. Type genus: *Amalus* Schönherr, 1825.

Phrydiuchina Wagner, 1938: 163 [stem: *Phrydiuch-*]. Type genus: *Phrydiuchus* Gozis, 1885.

Oxyonyxina Hoffmann, 1957b: 220 [stem: *Oxyonych-*]. Type genus: *Oxyonyx* Faust, 1885. Comment: incorrect original stem formation, not in prevailing usage.

Tribe Cnemogonini Colonnelli, 1979

Cnemogonini Colonnelli, 1979: 2 [stem: *Cnemogon-*]. Type genus: *Cnemogonus* J. L. LeConte, 1876.

Tribe Egriini Pajni and Kohli, 1982

Egriini Pajni and Kohli, 1982: 339, in key [stem: *Egri-*]. Type genus: *Egrius* Pascoe, 1865.

Egriini Colonnelli, 1984: 209 [stem: *Egri-*]. Type genus: *Egrius* Pascoe, 1865. Comment: family-group name proposed as new without reference to Egriini Pajni and Kohli, 1982.

Tribe Hypohypurini Colonnelli, 2004

Hypohypurini Colonnelli, 2004: 10 [stem: *Hypohypur-*]. Type genus: *Hypohypurus* Hustache, 1920.

Tribe Hypurini Schultze, 1902

Hypuridae Schultze, 1902: 209 [stem: *Hypur-*]. Type genus: *Hypurus* Rey, 1882.

Tribe Lioxyonychini Colonnelli, 1984

Lioxyonyxini Colonnelli, 1984: 209 [stem: *Lioxyonych-*]. Type genus: *Lioxyonyx* Hustache, 1933. Comment: incorrect original stem formation, not in prevailing usage.

Tribe Mecysmoderini Wagner, 1938

Mecysmoderina Wagner, 1938: 170 [stem: *Mecysmoder-*]. Type genus: *Mecysmoderes* Schönherr, 1837.

Tribe Mononychini LeConte, 1876

Mononychi J. L. LeConte, 1876: 267 [stem: *Mononych-*]. Type genus: *Mononychus* Germar, 1824 [placed on the Official List of Generic Names in Zoology (ICZN 1989c)].

Tribe Phytobiini Gistel, 1848

Phytobiidae Gistel, 1848: [7] [stem: *Phytobi-*]. Type genus: *Phytobius* Schönherr, 1833 [placed on the Official List of Generic Names in Zoology (ICZN 2001)].

Rhinoncides C. G. Thomson, 1865: 231 [stem: *Rhinonc-*]. Type genus: *Rhinoncus* Schönherr, 1825 [placed on the Official List of Generic Names in Zoology (ICZN 1989c)].

Tribe Scleropterini Schultze, 1902

Scleropteridae Schultze, 1902: 209 [stem: *Scleropter-*]. Type genus: *Scleropterus* Schönherr, 1825. Comment: the family-group name Scléropterides was used earlier by Solier (1834) in the Heteromera, however, Solier's name was not based on a genus name and is therefore unavailable; Agassiz (1846a: 146) incorrectly mentions *Scleropterus* as the type genus of Solier's name but this is impossible because the family-group and genus-group names belonged to different orders; Agassiz (1846b: 335) also latinized Solier's name as Scleropteroidae.

Subfamily Conoderinae Schönherr, 1833

Conoderides Schönherr, 1833: 26 [stem: *Conoder-*]. Type genus: *Conoderes* Schönherr, 1833.

Tribe Arachnopodini Lacordaire, 1865

Arachnopides Lacordaire, 1865: 159 [stem: *Arachnopod-*]. Type genus: *Arachnopus* Guérin-Méneville, 1838 [syn. of *Arachnobas* Boisduval, 1835]. Comment: original vernacular name available (Art. 11.7.2): first used in latinized form by Heyne and Taschenberg (1907: 232, as Arachnopini [incorrect stem formation]), generally accepted as in Alonso-Zarazaga and Lyal (1999: 109, as Arachnopodini); incorrect original stem formation, not in prevailing usage.

Tribe Campyloscelini Schönherr, 1845

Campyloscelides Schönherr, 1845: 197 [stem: *Campyloscel-*]. Type genus: *Campyloscelus* Schönherr, 1845.

Subtribe Campyloscelina Schönherr, 1845

Campyloscelides Schönherr, 1845: 197 [stem: *Campyloscel-*]. Type genus: *Campyloscelus* Schönherr, 1845.

Subtribe Corynemerina Hustache, 1929

Corynemerinae Hustache, 1929: 531 [stem: *Corynemer-*]. Type genus: *Corynemerus* Fåhraeus, 1871.

Subtribe Phaenomerina Faust, 1898

Phaenomerina Faust, 1898: 76, in key [stem: *Phaenomer-*]. Type genus: *Phaenomerus* Schönherr, 1836 [placed on the Official List of Generic Names in Zoology (ICZN 1962)]. Comment: the stem of the senior homonym based on the Scarabaeidae genus *Phaenomeris* was emended to *Phaeromerid-* to remove the homonymy (ICZN 1962); available for family-group taxon below subfamily level and placed on the Official List of Family-Group Names in Zoology (ICZN 1962).

Tribe Conoderini Schönherr, 1833

Conoderides Schönherr, 1833: 26 [stem: *Conoder-*]. Type genus: *Conoderes* Schönherr, 1833. Comment: the name Conoderinae Fleutiaux, 1919 (type genus *Conoderus* Eschscholtz, 1829) is available in Elateridae; this case is to be referred to the Commission to remove the homonymy (Art. 55.3.1).

Conophorides Schönherr, 1837: 719 [stem: *Conophor-*]. Type genus: *Conophorus* Schönherr, 1837 [preoccupied genus name, not *Conophorus* Meigen, 1803 [Diptera]; syn. of *Conoderes* Schönherr, 1833]. Comment: permanently invalid (Art. 39): based on preoccupied type genus.

Tribe Coryssomerini Thomson, 1859

Coryssomerina C. G. Thomson, 1859: 137 [stem: *Coryssomer-*]. Type genus: *Coryssomerus* Schönherr, 1825.

Synophthalmides Lacordaire, 1863: 514 [stem: *Synophthalm-*]. Type genus: *Synophthalmus* Lacordaire, 1863. Comment: original vernacular name available (Art. 11.7.2): first used in latinized form by Hustache (1932: 373, as Synophthalmini), generally accepted as in Hustache (1934: 49, as Synophthalmini).

Metialmini Hustache, 1932: 374, in key [stem: *Metialm-*]. Type genus: *Metialma* Pascoe, 1871.

Tribe Coryssopodini Lacordaire, 1865

Coryssopides Lacordaire, 1865: 163 [stem: *Coryssopod-*]. Type genus: *Coryssopus* Schönherr, 1826. Comment: original vernacular name available (Art. 11.7.2): first used in latinized form by Heyne and Taschenberg (1907: 232, as Coryssopini [incorrect stem formation]), generally accepted as in Alonso-Zarazaga and Lyal (1999: 110, as CORYSSOPODINI); incorrect original stem formation, not in prevailing usage.

*SYMPIÉZOPIDES Lacordaire, 1865: 166 [stem: *Sympiezopod-*]. Type genus: *Sympiezopus* Schönherr, 1837. Comment: original vernacular name unavailable (Art. 11.7.2): subsequently used in latinized form, e.g. Faust (1886b: 367, as SYMPIEZOPINORUM [incorrect stem formation]), but not generally attributed to Lacordaire (1865); incorrect original stem formation, not in prevailing usage.

Tribe LECHRIOPINI Lacordaire, 1865

LÉCHRIOPIDES Lacordaire, 1865: 149 [stem: *Lechriop-*]. Type genus: *Lechriops* Schönherr, 1825. Comment: original vernacular name available (Art. 11.7.2): first used in latinized form by Heyne and Taschenberg (1907: 232, as LECHRIOPINI), generally accepted as in Alonso-Zarazaga and Lyal (1999: 111, as LECHRIOPINI).

COPTURIDAE Desbrochers des Loges, 1892: 68 [stem: *Coptur-*]. Type genus: *Copturus* Schönherr, 1825.

Tribe LOBOTRACHELINI Lacordaire, 1865

LOBOTRACHÉLIDES Lacordaire, 1865: 172 [stem: *Lobotrachel-*]. Type genus: *Lobotrachelus* Schönherr, 1837. Comment: original vernacular name available (Art. 11.7.2): first used in latinized form by Aurivillius (1910: 433, as LOBOTRACHELINAE), generally accepted as in Alonso-Zarazaga and Lyal (1999: 111, as LOBOTRACHELINI).

Tribe MECOPINI Lacordaire, 1865

MÉCOPIDES Lacordaire, 1865: 156 [stem: *Mecop-*]. Type genus: *Mecopus* Schönherr, 1825. Comment: original vernacular name available (Art. 11.7.2): first used in latinized form by Heyne and Taschenberg (1907: 232, as MECOPINI), generally accepted as in Alonso-Zarazaga and Lyal (1999: 111, as MECOPINI); current spelling maintained (Art. 29.5): incorrect original stem formation in prevailing usage (should be *Mecopod-*); if the stem is corrected, it would threaten stability of a widely used name in Orthoptera (see Alonso-Zarazaga and Lyal 1999: 111).

Tribe MENEMACHINI Lacordaire, 1865

MÉNÉMACHIDES Lacordaire, 1865: 27 [stem: *Menemach-*]. Type genus: *Menemachus* Schönherr, 1843. Comment: original vernacular name available (Art. 11.7.2): first used in latinized form by Pascoe (1870b: 437, as MNEMACHINAE [incorrect stem formation]), generally accepted as in Alonso-Zarazaga and Lyal (1999: 112, as MENEMACHINI).

Tribe OTHIPPIINI Morimoto, 1962

OTHIPPIINI Morimoto, 1962a: 47, in key [stem: *Othippi-*]. Type genus: *Othippia* Pascoe, 1874.

Tribe Peloropodini Hustache, 1932

Peleropini Hustache, 1932: 372, in key [stem: *Peloropod-*]. Type genus: *Peloropus* Schönherr, 1836. Comment: incorrect original stem formation, not in prevailing usage.

Tribe Piazurini Lacordaire, 1865

Piazurides Lacordaire, 1865: 144 [stem: *Piazur-*]. Type genus: *Piazurus* Schönherr, 1825. Comment: original vernacular name available (Art. 11.7.2): first used in latinized form by Faust (1899: 100, as Piazurinorum), generally accepted as in Alonso-Zarazaga and Lyal (1999: 113, as Piazurini).

Tribe Sphadasmini Lacordaire, 1865

Sphadasmides Lacordaire, 1865: 161 [stem: *Sphadasm-*]. Type genus: *Sphadasmus* Schönherr, 1844. Comment: original vernacular name available (Art. 11.7.2): first used in latinized form by Faust (1899: 102, as Sphadasminorum), generally accepted as in Alonso-Zarazaga and Lyal (1999: 114, as Sphadasmini).

Tribe Trichodocerini Champion, 1906

Trichodocerides Champion, 1906: 713 [stem: *Trichodocer-*]. Type genus: *Trichodocerus* Chevrolat, 1879.

Tribe Zygopini Lacordaire, 1865

Zygopides Lacordaire, 1865: 142 [stem: *Zygop-*]. Type genus: *Zygops* Schönherr, 1825 [placed on the Official List of Generic Names in Zoology (ICZN 1987e)]. Comment: original vernacular name available (Art. 11.7.2): first used in latinized form by Pascoe (1870b: 437, as Zygopinae), generally accepted as in Alonso-Zarazaga and Lyal (1999: 114, as Zygopini).

Eccoptinae Pierce, 1919: 30, in key [stem: *Eccopt-*]. Type genus: *Eccoptus* Dejean, 1821 [placed on the Official Index of Rejected and Invalid Generic Names in Zoology (ICZN 1987e); syn. of *Zygops* Schönherr, 1825]. Comment: permanently invalid (Art. 39): based on suppressed type genus.

Cylindrocopturini Böving, 1927b: 156, in key [stem: *Cylindrocoptur-*]. Type genus: *Cylindrocopturus* Heller, 1895.

Subfamily Cossoninae Schönherr, 1825

Cossonides Schönherr, 1825: column 587 [stem: *Cosson-*]. Type genus: *Cossonus* Clairville, 1798.

Tribe Acamptini LeConte, 1876

Acampti J. L. LeConte, 1876: 238 [stem: *Acampt-*]. Type genus: *Acamptus* J. L. LeConte, 1876.

Tribe Acanthinomerini Voss, 1972

Acanthinomerini Voss, 1972: 382 [stem: *Acanthinomer-*]. Type genus: *Acanthinomerus* Boheman, 1859.

Tribe Allomorphini Folwaczny, 1973

Allomorphini Folwaczny, 1973: 68, in key [stem: *Allomorph-*]. Type genus: *Allomorphus* Folwaczny, 1968.

Tribe Aphyllurini Voss, 1955

Aphyllurina Voss, 1955a: 276 [stem: *Aphyllur-*]. Type genus: *Aphyllura* Reitter, 1884.

Tribe Araucariini Kuschel, 1966

Araucariini Kuschel, 1966: 4 [stem: *Araucari-*]. Type genus: *Araucarius* Kuschel, 1966.

Tribe Choerorhinini Folwaczny, 1973

Choerorhini Folwaczny, 1973: 69, in key [stem: *Choerorhin-*]. Type genus: *Choerorhinus* Fairmaire, 1858. Comment: incorrect original stem formation, not in prevailing usage.

Tribe Cossonini Schönherr, 1825

Cossonides Schönherr, 1825: column 587 [stem: *Cosson-*]. Type genus: *Cossonus* Clairville, 1798.

Tribe Cryptommatini Voss, 1972

Cryptommatini Voss, 1972: 373 [stem: *Cryptommat-*]. Type genus: *Cryptommata* Wollaston, 1877.

Tribe Dryotribini LeConte, 1876

Dryotribi J. L. LeConte, 1876: 335 [stem: *Dryotrib-*]. Type genus: *Dryotribus* G. H. Horn, 1873.

Cotasterinen Faust, 1886a: 28 [stem: *Cotaster-*]. Type genus: *Cotaster* Motschulsky, 1851. Comment: also as Cotasteriden on p. 31; original vernacular name available (Art. 11.7.2): first used in latinized form by Sainte-Claire Deville (1923: 63, as Cotastrini [incorrect stem formation]), generally accepted as in Hansen (1996: 209, as Cotasterini).

Tribe Microxylobiini Voss, 1972

Microxylobiini Voss, 1972: 431 [stem: *Microxylobi-*]. Type genus: *Microxylobius* Chevrolat, 1836.

Tribe Nesiobiini Alonso-Zarazaga and Lyal, 1999

Nesiotinae Voss, 1972: 338 [stem: *Nesiot-*]. Type genus: *Nesiotes* Wollaston, 1861 [preoccupied genus name, not *Nesiotes* Martens, 1860 [Mollusca]; syn. of *Nesiobius* G. A. K. Marshall, 1943]. Comment: permanently invalid (Art. 39): based on preoccupied type genus.

Nesiobiini Alonso-Zarazaga and Lyal, 1999: 119 [stem: *Nesiobi-*]. Type genus: *Nesiobius* G. A. K. Marshall, 1943. Comment: replacement name for Nesiotinae Voss, 1872 because of the homonymy of the type genus.

Tribe Neumatorini Folwaczny, 1973

Neumatorini Folwaczny, 1973: 69, in key [stem: *Neumator-*]. Type genus: *Neumatora* Normand, 1920.

Tribe Onychiini Chapuis, 1869

Onychiidae Chapuis, 1869: 48 [stem: *Onychi-*]. Type genus: *Onychius* Chapuis, 1869. Comment: see Alonso-Zarazaga and Lyal (2009: 9) for comments on placement.

Tribe Onycholipini Wollaston, 1873

Onycholipides Wollaston, 1873: 454 [stem: *Onycholip-*]. Type genus: *Onycholips* Wollaston, 1861.

Stenoscelides Wollaston, 1877: 82, in key [stem: *Stenoscelid-*]. Type genus: *Stenoscelis* Wollaston, 1861.

Stereocorynina Voss, 1955b: 232 [stem: *Stereocoryn-*]. Type genus: *Stereocorynes* Wollaston, 1873.

Stereocorynini Roudier, 1958: 212 [stem: *Stereocoryn-*]. Type genus: *Stereocorynes* Wollaston, 1873. Comment: family-group name proposed as new without reference to Stereocorynina Voss, 1955.

Tribe Pentarthrini Lacordaire, 1865

Pentarthrides Lacordaire, 1865: 323 [stem: *Pentarthr-*]. Type genus: *Pentarthrum* Wollaston, 1854. Comment: original vernacular name available (Art. 11.7.2): first used in latinized form by Heyne and Taschenberg (1907: 234, as Pentarthrini), generally accepted as in Alonso-Zarazaga and Lyal (1999: 120, as Pentarthrini).

Tribe Proecini Voss, 1956

Proecina Voss, 1956a: 139 [stem: *Proec-*]. Type genus: *Proeces* Schönherr, 1837.

Phylloplatypodini Kato, 1998: 77 [stem: *Phylloplatypod-*]. Type genus: *Phylloplatypus* Kato, 1998.

Tribe Pseudapotrepini Champion, 1909

Pseudapotrepides Champion, 1909: 21 [stem: *Pseudapotrep-*]. Type genus: *Pseudapotrepus* Champion, 1909.

Tribe Rhyncolini Gistel, 1848

Rhynchalidae Gistel, 1848: [6] [stem: *Rhyncol-*]. Type genus: *Rhyncolus* Germar, 1817 [as *Rhyncholus*, incorrect subsequent spelling of type genus name, not in prevailing usage; *Rhyncolus* Germar, 1817 placed on the Official List of Generic Names in Zoology (ICZN 1991b)].

Subtribe Phloeophagina Voss, 1955

Phloeophagina Voss, 1955b: 228 [stem: *Phloeophag-*]. Type genus: *Phloeophagus* Schönherr, 1837.

Subtribe Pseudomimina Voss, 1939

Pseudomimini Voss, 1939: 65 [stem: *Pseudomim-*]. Type genus: *Pseudomimus* Hartmann, 1904.

Subtribe Rhyncolina Gistel, 1848

Rhynchalidae Gistel, 1848: [6] [stem: *Rhyncol-*]. Type genus: *Rhyncolus* Germar, 1817 [as *Rhyncholus*, incorrect subsequent spelling of type genus name, not in prevailing usage; *Rhyncolus* Germar, 1817 placed on the Official List of Generic Names in Zoology (ICZN 1991b)]. Comment: incorrect original stem formation, not in prevailing usage.

Eremotini Voss, 1939: 65 [stem: *Eremot-*]. Type genus: *Eremotes* Wollaston, 1861.

Cotasterosomini Konishi, 1962: 2 [stem: *Cotasterosomat-*]. Type genus: *Cotasterosoma* Konishi, 1962. Comment: incorrect original stem formation, not in prevailing usage.

Himatinini Konishi, 1962: 7 [stem: *Himatin-*]. Type genus: *Himatinum* Cockerell, 1906 [syn. of *Himatium* Wollaston, 1873].

Tribe Tapiromimini Voss, 1972

Tapiromimini Voss, 1972: 380 [stem: *Tapiromim-*]. Type genus: *Tapiromimus* Wollaston, 1877.

Subfamily Cryptorhynchinae Schönherr, 1825

Cryptorhynchides Schönherr, 1825: column 585 [stem: *Cryptorhynch-*]. Type genus: *Cryptorhynchus* Illiger, 1807 [placed on the Official List of Generic Names in Zoology (ICZN 1967b)]. Comment: Cryptorhynchinae Schönherr, 1825 placed on the Official List of Family-Group Names in Zoology (ICZN 1967b).

Tribe Aedemonini Faust, 1898

Acdemonini Faust, 1898: 34 [stem: *Aedemon-*]. Type genus: *Aedemonus* Schönherr, 1837 [as *Acdemonus*, incorrect subsequent spelling of type genus name, not in prevailing usage]. Comment: incorrect original stem formation, not in prevailing usage.

Mecistocerini Morimoto, 1978b: 122 [stem: *Mechistocer-*]. Type genus: *Mechistocerus* Fauvel, 1863 [as *Mecistocerus*, unjustified emendation of type genus name by Pascoe (1870b), not in prevailing usage]. Comment: incorrect original stem formation, not in prevailing usage; correction of stem by Morimoto (1987).

Derbyiellina Zimmerman, 1994a: 645, in key [stem: *Derbyiell-*]. Type genus: *Derbyiella* Lea, 1907.

Tribe Camptorhinini Lacordaire, 1865

Camptorhinides Lacordaire, 1865: 86 [stem: *Camptorhin-*]. Type genus: *Camptorhinus* Schönherr, 1825. Comment: original vernacular name available (Art. 11.7.2): first used in latinized form by Pascoe (1885: 254, as Camptorhininae), generally accepted as in Alonso-Zarazaga and Lyal (1999: 136, as Camptorhinini).

Tribe Cryptorhynchini Schönherr, 1825

Cryptorhynchides Schönherr, 1825: column 585 [stem: *Cryptorhynch-*]. Type genus: *Cryptorhynchus* Illiger, 1807 [placed on the Official List of Generic Names in Zoology (ICZN 1967b)]. Comment: Cryptorhynchinae Schönherr, 1825 placed on the Official List of Family-Group Names in Zoology (ICZN 1967b).

Subtribe Cryptorhynchina Schönherr, 1825

Cryptorhynchides Schönherr, 1825: column 585 [stem: *Cryptorhynch-*]. Type genus: *Cryptorhynchus* Illiger, 1807 [placed on the Official List of Generic Names in Zoology (ICZN 1967b)]. Comment: Cryptorhynchinae Schönherr, 1825 placed on the Official List of Family-Group Names in Zoology (ICZN 1967b).

Cryptorhynchidiini Bradley, 1930: 277, in key [stem: *Cryptorhynchidi-*]. Type genus: *Cryptorhynchidius* Pierce, 1919 [syn. of *Cryptorhynchus* Illiger, 1807].

Subtribe Mecistostylina Lacordaire, 1865

Mécistostylides Lacordaire, 1865: 131 [stem: *Mecistostyl-*]. Type genus: *Mecistostylus* Lacordaire, 1865. Comment: original vernacular name available (Art. 11.7.2): first used in latinized form by G. A. K. Marshall (1916: 10, as Mecistostylini), generally accepted as in Alonso-Zarazaga and Lyal (1999: 129, as Mecistostylina).

Cylindrocosyninae Pascoe, 1875: 55 [stem: *Cylindrocoryn-*]. Type genus: *Cylindrocorynus* Schönherr, 1837. Comment: incorrect original stem formation, not in prevailing usage.

Subtribe Tylodina Lacordaire, 1865

Tylodides Lacordaire, 1865: 90 [stem: *Tylod-*]. Type genus: *Tylodes* Sahlberg, 1823. Comment: original vernacular name available (Art. 11.7.2): first used in latinized form by Pascoe (1885: 255, as Tylodinae), generally accepted as in Alonso-Zarazaga and Lyal (1999: 129, as Tylodina).

Tribe Gasterocercini Zherikhin, 1991

Gasterocercini Zherikhin, 1991: 92 [stem: *Gasterocerc-*]. Type genus: *Gasterocercus* Laporte and Brullé, 1828.

Tribe Psepholacini Lacordaire, 1865

Psépholacides Lacordaire, 1865: 72 [stem: *Psepholac-*]. Type genus: *Psepholax* Lacordaire, 1865. Comment: original vernacular name available (Art. 11.7.2): first used in latinized form by Hustache (1936: 56, as Psepholacina), generally accepted as in Alonso-Zarazaga and Lyal (1999: 136, as Psepholacini); First Reviser (Psepholacini Lacordaire, 1865 vs Strongylopterini Lacordaire, 1865 vs Sympiezoscelini Lacordaire, 1865) not determined, current usage maintained.

Strongyloptérides Lacordaire, 1865: 73 [stem: *Strongylopter-*]. Type genus: *Strongylopterus* Schönherr, 1837. Comment: original vernacular name available (Art. 11.7.2): first used in latinized form and generally accepted as in Hustache (1936: 60, as Strongylopteridina [incorrect stem formation]).

Sympiézoscélides Lacordaire, 1865: 138 [stem: *Sympiezoscel-*]. Type genus: *Sympiezoscelus* G. R. Waterhouse, 1853. Comment: original vernacular name available (Art. 11.7.2): first used in latinized form and generally accepted as in Pascoe (1874: 96, as Sympiezoscelides [treated as Latin]).

Tribe Sophrorhinini Lacordaire, 1865

Sophrorhinides Lacordaire, 1865: 81 [stem: *Sophrorhin-*]. Type genus: *Sophrorhinus* Rouzet, 1855. Comment: original vernacular name available (Art. 11.7.2): first used in latinized form by C. O. Waterhouse (1879a: 310, as Sophororhinae [incorrect stem formation]), generally accepted as in Alonso-Zarazaga and Lyal (1999: 137, as Sophororhinini).

Tribe Torneumatini Bedel, 1884

Torneumatini Bedel, 1884b: 136, in note [stem: *Torneumat-*]. Type genus: *Torneuma* Wollaston, 1860.

Subfamily Cyclominae Schönherr, 1826

Cyclomides Schönherr, 1826: 185 [stem: *Cyclom-*]. Type genus: *Cyclomus* Schönherr, 1826 [syn. of *Epicthonius* Schönherr, 1826]. Comment: the classification of this subfamily follows Oberprieler (2010).

Tribe Amycterini Waterhouse, 1854

Amycteridae G. R. Waterhouse, 1854: 75, in note [stem: *Amycter-*]. Type genus: *Amycterus* Schönherr, 1823.

EUOMIDES Lacordaire, 1863: 315 [stem: *Euom-*]. Type genus: *Euomus* Schönherr, 1847. Comment: original vernacular name available (Art. 11.7.2): first used in latinized form by W. J. MacLeay (1866: 319, as EUOMIDAE), generally accepted as in Schenkling and G. A. K. Marshall (1931a: 27, as EUOMINI).

PSALIDURIDAE Pierce, 1914: 350 [stem: *Phalidur-*]. Type genus: *Phalidura* Fischer von Waldheim, 1823 [as *Psalidura,* incorrect subsequent spelling of type genus name, not in prevailing usage; syn. of *Amycterus* Schönherr, 1823]. Comment: incorrect original stem formation, not in prevailing usage.

*ACANTHOLOPHINI Schenkling and G. A. K. Marshall, 1931a: Amyct. 6 [stem: *Acantholoph-*]. Type genus: *Acantholophus* Boisduval, 1835. Comment: unavailable family-group name, proposed after 1930 without description or bibliographic reference to such a description (Art. 13.1).

Tribe ATERPINI Lacordaire, 1863

ATERPIDES Lacordaire, 1863: 410 [stem: *Aterp-*]. Type genus: *Aterpus* Schönherr, 1826 [syn. of *Aades* Schönherr, 1823]. Comment: precedence (ATERPINI Lacordaire, 1863 vs RHADINOSOMINI Lacordaire, 1863) given to taxon originally proposed at the higher rank (Art. 24.1).

Subtribe ATERPINA Lacordaire, 1863 *nomen protectum*

HELIOMENEIDAE Gistel, 1848: [8] [stem: *Heliomen-*]. Type genus: *Heliomene* Gistel, 1848 [syn. of *Chrysolopus* Germar, 1817]. Comment: as pointed out by Alonso-Zarazaga and Lyal (2002: 22) this is the oldest available name for the subtribe name; HELIOMENEINI Gistel, 1848 was recently considered a *nomen oblitum* by Colonnelli (2003: 7) however the necessary supporting references were not provided; we hereby consider HELIOMENEINI Gistel, 1848 as a *nomen oblitum* (see supporting references in Appendix 1); incorrect original stem formation, not in prevailing usage.

ATERPIDES Lacordaire, 1863: 410 [stem: *Aterp-*]. Type genus: *Aterpus* Schönherr, 1826 [syn. of *Aades* Schönherr, 1823]. Comment: *nomen protectum* (see Appendix 1); original vernacular name available (Art. 11.7.2): first used in latinized form by Pascoe (1870b: 436, as ATERPINAE), generally accepted as in Alonso-Zarazaga and Lyal (1999: 140, as ATERPINI).

*PÉLORORHINIDES Lacordaire, 1863: 415 [stem: *Pelororhin-*]. Type genus: *Pelororhinus* Schönherr, 1834. Comment: original vernacular name unavailable (Art. 11.7.2): subsequently used in latinized form, e.g., Heyne and Taschenberg (1907: 228, as PELORORHININI), Morrone (1997: 101, as PELORHIRHINI [incorrect stem formation]), but not generally accepted as valid; PELORHINIDAE [incorrect stem formation] was used as valid by Ienistea (1986: 33) but it was not attributed to Lacordaire (1863); Ienistea's name is also unavailable, it was proposed after 1930 without description or bibliographic reference to such a description (Art. 13.1).

LOPHOTIDES Jekel, 1865: 546 [stem: *Lophot-*]. Type genus: *Lophotus* Schönherr, 1834 [preoccupied genus name, not *Lophotus* Giorna, 1809 [Pisces]; syn. of *Aegorhinus* Erichson, 1834]. Comment: permanently invalid (Art. 39): based on preoccupied type genus.

RHINARIDES Jekel, 1865: 546 [stem: *Rhinari-*]. Type genus: *Rhinaria* Kirby, 1819. Comment: incorrect original stem formation, not in prevailing usage.

Subtribe RHADINOSOMINA Lacordaire, 1863

RHADINOSOMIDES Lacordaire, 1863: 63 [stem: *Rhadinosom-*]. Type genus: *Rhadinosomus* Schönherr, 1840. Comment: original vernacular name available (Art. 11.7.2): first used in latinized form by Pascoe (1875: 55, as RHADINOSOMINAE), generally accepted as in Alonso-Zarazaga and Lyal (1999: 141, as RHADINOSOMINA).

RHADINOSOMINI Pierce, 1913: 405, in key [stem: *Rhadinosom-*]. Type genus: *Rhadinosomus* Schönherr, 1840. Comment: family-group name proposed as new without reference to RHADINOSOMIDES Lacordaire, 1863.

Tribe CYCLOMINI Schönherr, 1826

CYCLOMIDES Schönherr, 1826: 185 [stem: *Cyclom-*]. Type genus: *Cyclomus* Schönherr, 1826 [syn. of *Epicthonius* Schönherr, 1826].

SOMATODIDES Lacordaire, 1863: 319 [stem: *Somatod-*]. Type genus: *Somatodes* Schönherr, 1840 [placed on the Official List of Generic Names in Zoology (ICZN 1994c)]. Comment: original vernacular name available (Art. 11.7.2): first used in latinized form by Pascoe (1870b: 436, as SOMATODINAE), generally accepted as in Alonso-Zarazaga and Lyal (1999: 144, as SOMATODINI); SOMATODINAE Lacordaire, 1863 placed on the Official List of Family-Group Names in Zoology (ICZN 1994c).

Tribe DICHOTRACHELINI Hoffmann, 1957

DICHOTRACHELINI Hoffmann, 1957a: 60, in note [stem: *Dichotrachel-*]. Type genus: *Dichotrachelus* Stierlin, 1853.

Tribe HIPPORHININI Lacordaire, 1863

HIPPORHINIDES Lacordaire, 1863: 323 [stem: *Hipporhin-*]. Type genus: *Hipporhinus* Schönherr, 1823. Comment: original vernacular name available (Art. 11.7.2): first used in latinized form and generally accepted as in Pascoe (1870b: 436, as HIPPORHININAE).

GRONOPINI Bedel, 1884a: 68, in key [stem: *Gronop-*]. Type genus: *Gronops* Schönherr, 1823.

GRONOPINA Louw and Oberprieler, 1998: 24 [stem: *Gronop-*]. Type genus: *Gronops* Schönherr, 1823. Comment: family-group name proposed as new without reference to GRONOPINI Bedel, 1884.

Tribe Listroderini LeConte, 1876

Listroderi J. L. LeConte, 1876: 124 [stem: *Listroder-*]. Type genus: *Listroderes* Schönherr, 1826.

Palaechtini C. Brinck, 1948: 43 [stem: *Palaechth-*]. Type genus: *Palaechthus* C. O. Waterhouse, 1884 [as *Palaechtus*, incorrect subsequent spelling of type genus name, not in prevailing usage]. Comment: incorrect original stem formation, not in prevailing usage.

Tribe Notiomimetini Wollaston, 1873

Notiomimetides Wollaston, 1873: 440 [stem: *Notiomimet-*]. Type genus: *Notiomimetes* Wollaston, 1873.

Apheli Csiki, 1936: 109 [stem: *Aphel-*]. Type genus: *Aphela* Pascoe, 1865. Comment: description by indication (distinguishing characters given in Wollaston (1873: 440)).

Tribe Rhythirrinini Lacordaire, 1863

Rhytirhinides Lacordaire, 1863: 296 [stem: *Rhythirrin-*]. Type genus: *Rhythirrinus* Schönherr, 1823 [as *Rhytirhinus*, incorrect subsequent spelling of type genus name, not in prevailing usage]. Comment: original vernacular name available (Art. 11.7.2): first used in latinized form by Seidlitz (1875 [Gatt.]: 112, as Rhytirhinini [incorrect stem formation]), generally accepted as in Alonso-Zarazaga and Lyal (1999: 141, as Rhythirrinini); incorrect original stem formation, not in prevailing usage; First Reviser found (Rhythirrinini Lacordaire, 1863 vs Rhyparosomini Lacordaire, 1863 vs Eupagini Lacordaire, 1863) is Kuschel (1971: 251).

Rhyparosomides Lacordaire, 1863: 327 [stem: *Rhyparosom-*]. Type genus: *Rhyparosomus* Schönherr, 1842. Comment: original vernacular name available (Art. 11.7.2): first used in latinized form by Stein (1868: 100, as Rhyparosomini), generally accepted as in Heyne and Taschenberg (1907: 228, as Rhyparosomini).

Eupagides Lacordaire, 1863: 328 [stem: *Eupag-*]. Type genus: *Eupages* Schönherr, 1834. Comment: original vernacular name available (Art. 11.7.2): first used in latinized form by Péringuey (1888: 168, as Eupagidae), generally accepted as in Schenkling and G. A. K. Marshall (1929: 47, as Eupagini).

Rhytidorhinides Wollaston, 1865: 307 [stem: *Rhytidorhin-*]. Type genus: *Rhytidorhinus* Wollaston, 1864 [preoccupied genus name, not *Rhytidorhinus* Agassiz, 1846 [Coleoptera: Curculionidae: Cyclominae]; syn. of *Rhythirrinus* Schönherr, 1823]. Comment: permanently invalid (Art. 39): based on preoccupied type genus.

Subfamily Entiminae Schönherr, 1823

Entimides Schönherr, 1823: column 1138 [stem: *Entim-*]. Type genus: *Entimus* Germar, 1817. Comment: First Revisers found (Entiminae Schönherr, 1823 vs

LORDOPINAE Schönherr, 1823 vs POLYDRUSINAE Schönherr, 1823) are Alonso-Zarazaga and Lyal (1999: 144).

Tribe AGRAPHINI Horn, 1876

AGRAPHI G. H. Horn, 1876: 58 [stem: *Agraph-*]. Type genus: *Agraphus* Say, 1831.

Tribe ALOPHINI LeConte, 1874

ALOPHINI J. L. LeConte, 1874b: 461 [stem: *Aloph-*]. Type genus: *Alophus* Schönherr, 1826 [syn. of *Graptus* Schönherr, 1823].

Tribe ANOMOPHTHALMINI Morrone, 1998

ANOMOPHTHALMINA Morrone, 1998: 86 [stem: *Anomophthalm-*]. Type genus: *Anomophthalmus* Fairmaire, 1884.

Tribe ANYPOTACTINI Champion, 1911

ANYPOTACTINA Champion, 1911: 215 [stem: *Anypotact-*]. Type genus: *Anypotactus* Schönherr, 1840.

Tribe BLOSYRINI Lacordaire, 1863

BLOSYRIDES Lacordaire, 1863: 27 [stem: *Blosyr-*]. Type genus: *Blosyrus* Schönherr, 1823. Comment: original vernacular name available (Art. 11.7.2): first used in latinized form by Péringuey (1888: 144, as BLOSYRIDAE), generally accepted as in Alonso-Zarazaga and Lyal (1999: 146, as BLOSYRINI).

Tribe BRACHYDERINI Schönherr, 1826

BRACHYDERIDES Schönherr, 1826: 94 [stem: *Brachyder-*]. Type genus: *Brachyderes* Schönherr, 1823 [placed on the Official List of Generic Names in Zoology (ICZN 1987a)]. Comment: name placed on the Official List of Family-Group Names in Zoology (ICZN 1987a, as BRACHYDERINAE Schönherr, 1837).

THYLACITIDAE Kirby, 1837: 202 [stem: *Thylacit-*]. Type genus: *Thylacites* Germar, 1817 [placed on the Official Index of Invalid and Rejected Generic Names in Zoology (ICZN 1987a)]. Comment: placed on the Official Index of Rejected and Invalid Family-Group Names in Zoology (ICZN 1987a).

STROPHOSOMIDAE Gistel, 1848: [8] [stem: *Strophosom-*]. Type genus: *Strophosomum* Gistel, 1856 [syn. of *Strophosoma* Billberg, 1820].

Tribe CELEUTHETINI Lacordaire, 1863

CÉLEUTHÉTIDES Lacordaire, 1863: 145 [stem: *Celeuthet-*]. Type genus: *Celeuthetes* Schönherr, 1842.

Subtribe CELEUTHETINA Lacordaire, 1863

CÉLEUTHÉTIDES Lacordaire, 1863: 145 [stem: *Celeuthet-*]. Type genus: *Celeuthetes* Schönherr, 1842. Comment: original vernacular name available

(Art. 11.7.2): first used in latinized form by Faust (1891: 166, as CELEUTHETINAE), generally accepted as in Alonso-Zarazaga and Lyal (1999: 147, as CELEUTHETINI).

COPTORRHYNCHINA Voss, 1940: 279 [stem: *Coptorhynch-*]. Type genus: *Coptorhynchus* sensu Faust, 1897 [not *Coptorhynchus* Guérin-Méneville, 1841; syn. of *Resites* Alonso-Zarazaga and Lyal, 1999]. Comment: incorrect original stem formation, not in prevailing usage; based on a misidentified type genus.

Subtribe ISOPTERINA Morimoto and Kojima, 2001

ISOPTERINA Morimoto and Kojima, 2001: 274 [stem: *Isopter-*]. Type genus: *Isopterus* Faust, 1895.

Tribe CNEORHININI Lacordaire, 1863

CNÉORHINIDES Lacordaire, 1863: 31 [stem: *Cneorhin-*]. Type genus: *Cneorhinus* Schönherr, 1823. Comment: original vernacular name available (Art. 11.7.2): first used in latinized form by Seidlitz (1875 [Gatt.]: 111, as CNEORHININI), generally accepted as in Alonso-Zarazaga and Lyal (1999: 148, as CNEORHININI).

PHILOPEDINI Bedel, 1883: 32, in key [stem: *Philoped-*]. Type genus: *Philopedon* Schönherr, 1826.

DERMATODINI Emden, 1936: 76, in key [stem: *Dermatod-*]. Type genus: *Dermatodes* Schönherr, 1840.

STIGMATRACHELINI Richard, 1983: 8 [stem: *Stigmatrachel-*]. Type genus: *Stigmatrachelus* Schönherr, 1840.

Tribe CRATOPODINI Hustache, 1919

CRATOPINI Hustache, 1919: 473 [stem: *Cratopod-*]. Type genus: *Cratopus* Schönherr, 1823. Comment: incorrect original stem formation, not in prevailing usage.

Tribe CYLYDRORHININI Lacordaire, 1863

CYLINDRORHINIDES Lacordaire, 1863: 339 [stem: *Cylydrorhin-*]. Type genus: *Cylydrorhinus* Guérin-Méneville, 1838 [as *Cylindrorhinus*, incorrect subsequent spelling of type genus name, not in prevailing usage]. Comment: original vernacular name available (Art. 11.7.2): first used in latinized form by Pascoe (1870b: 436, as CYLINDRORHININAE [incorrect stem formation]), generally accepted as in Alonso-Zarazaga and Lyal (1999: 150, as CYLYDRORHININI); incorrect original stem formation, not in prevailing usage.

Tribe CYPHICERINI Lacordaire, 1863

CYPHICÉRIDES Lacordaire, 1863: 220 [stem: *Cyphicer-*]. Type genus: *Cyphicerus* Schönherr, 1823. Comment: First Reviser (CYPHICERINI Lacordaire, 1863 vs PHYTOSCAPHINI Lacordaire, 1863) not determined, current usage maintained.

Subtribe Acanthotrachelina Marshall, 1944

Acanthotrachelini G. A. K. Marshall, 1944: 76, in key [stem: *Acanthotrachel-*]. Type genus: *Acanthotrachelus* Schönherr, 1842.

Subtribe Cyphicerina Lacordaire, 1863

Cyphicérides Lacordaire, 1863: 220 [stem: *Cyphicer-*]. Type genus: *Cyphicerus* Schönherr, 1823. Comment: original vernacular name available (Art. 11.7.2): first used in latinized form by Faust (1888: 284, as Cyphicerinae), generally accepted as in Alonso-Zarazaga and Lyal (1999: 151, as Cyphicerini).

Corigetini Faust, 1885: 167, in key [stem: *Coriget-*]. Type genus: *Corigetus* Desbrochers des Loges, 1872.

Subtribe Mylacorrhinina Reitter, 1913

Mylacorrhynchina Reitter, 1913b: 12, in key [stem: *Mylacorrhin-*]. Type genus: *Mylacorrhina* Reitter, 1913 [syn. of *Altonomus* Desbrochers des Loges, 1907]. Comment: incorrect original stem formation, not in prevailing usage.

Subtribe Myllocerina Pierce, 1913

Myllocerini Pierce, 1913: 421, in key [stem: *Myllocer-*]. Type genus: *Myllocerus* Schönherr, 1826.

Ptochini Reitter, 1913b: 8, in key [stem: *Ptoch-*]. Type genus: *Ptochus* Schönherr, 1826 [placed on the Official List of Generic Names in Zoology (ICZN 1990d)].

Subtribe Phytoscaphina Lacordaire, 1863

Phytoscaphides Lacordaire, 1863: 229 [stem: *Phytoscaph-*]. Type genus: *Phytoscaphus* Schönherr, 1826. Comment: original vernacular name available (Art. 11.7.2): first used in latinized form by J. L. LeConte (1869: 382, as Phytoscaphi), generally accepted as in Alonso-Zarazaga and Lyal (1999: 154, as Phytoscaphina).

*Oxyophthalmina Voss, 1933a: 30 [stem: *Oxyophthalm-*]. Type genus: *Oxyophthalmus* Hochhuth, 1847. Comment: unavailable family-group name, proposed after 1930 without description or bibliographic reference to such a description (Art. 13.1).

Tribe Ectemnorhinini Lacordaire, 1863

Ectemnorhinides Lacordaire, 1863: 562 [stem: *Ectemnorhin-*]. Type genus: *Ectemnorhinus* G. R. Waterhouse, 1853. Comment: original vernacular name available (Art. 11.7.2): first used in latinized form by Pascoe (1870b: 436, as Ectemnorhinae), generally accepted as in Alonso-Zarazaga and Lyal (1999: 155, as Ectemnorhinini).

Canonopsini Dreux and Voisin, 1989: 112 [stem: *Canonopse-*]. Type genus: *Canonopsis* C. O. Waterhouse, 1875. Comment: incorrect original stem formation, not in prevailing usage.

Tribe Elytrurini Marshall, 1956

Elytrurini G. A. K. Marshall, 1956: 5 [stem: *Elytrur-*]. Type genus: *Elytrurus* Boisduval, 1835.

Tribe Embrithini Marshall, 1942

Embrithini G. A. K. Marshall, 1942: 3 [stem: *Embrith-*]. Type genus: *Embrithes* Schönherr, 1842.

Tribe Entimini Schönherr, 1823

Entimides Schönherr, 1823: column 1138 [stem: *Entim-*]. Type genus: *Entimus* Germar, 1817.

Tribe Episomini Lacordaire, 1863

Episomides Lacordaire, 1863: 175 [stem: *Episom-*]. Type genus: *Episomus* Schönherr, 1823. Comment: original vernacular name available (Art. 11.7.2): first used in latinized form by Faust (1894: 185, as Episomini), generally accepted as in Alonso-Zarazaga and Lyal (1999: 156, as Episomini).

Episomini Pierce, 1913: 421, in key [stem: *Episom-*]. Type genus: *Episomus* Schönherr, 1823. Comment: family-group name proposed as new without reference to Episomides Lacordaire, 1863.

Tribe Eudiagogini LeConte, 1874

Promécopides Lacordaire, 1863: 384 [stem: *Promecop-*]. Type genus: *Promecops* Schönherr, 1823 [preoccupied genus name, not *Promecops* Sahlberg, 1823 [Coleoptera: Curculionidae: Entiminae: Eudiagogini]; syn. of *Promecops* Sahlberg, 1823]. Comment: original vernacular name available (Art. 11.7.2): first used in latinized form and generally accepted as in Pascoe (1870b: 436, as Promecopinae); permanently invalid (Art. 39): based on preoccupied type genus.

Eudiagogini J. L. LeConte, 1874b: 454 [stem: *Eudiagog-*]. Type genus: *Eudiagogus* Schönherr, 1840. Comment: First Reviser (Eudiagogini J. L. LeConte, 1874 vs Bathyrini J. L. LeConte, 1874) not determined, current usage maintained.

Bathyrini J. L. LeConte, 1874b: 461 [stem: *Bathyrin-*]. Type genus: *Bathyris* J. L. LeConte, 1874 [syn. of *Colecerus* Schönherr, 1840]. Comment: incorrect original stem formation, not in prevailing usage.

Coleocerini Jekel, 1875: 144 [stem: *Colecer-*]. Type genus: *Colecerus* Schönherr, 1840 [as *Coleocerus*, incorrect subsequent spelling of type genus name, not in prevailing usage]. Comment: incorrect original stem formation, not in prevailing usage.

Tribe Eupholini Günther, 1943

*Eupholini Schenkling and G. A. K. Marshall, 1931b: Lept. 62 [stem: *Euphol-*]. Type genus: *Eupholus* Guérin-Méneville, 1838 [preoccupied genus name, not *Eupholus* Boisduval, 1835 [Coleoptera: Curculionidae: Entiminae]; syn. of *Eupholus* Boisduval, 1835]. Comment: unavailable family-group name, proposed after 1930 without description or bibliographic reference to such a description (Art. 13.1).

Eupholini Günther, 1943: 19 [stem: *Euphol-*]. Type genus: *Eupholus* Guérin-Méneville, 1838 [preoccupied genus name, not *Eupholus* Boisduval, 1835 [Coleoptera: Curculionidae: Entiminae]; syn. of *Eupholus* Boisduval, 1835]. Comment: although this name should be treated as permanently invalid because it is based on a preoccupied type genus (Art. 39), an application will be submitted to the Commission by MAAZ and CHCL to conserve Eupholini Günther, 1943 and designate *Eupholus* Boisduval, 1835 as its type genus.

Eupholini Alonso-Zarazaga and Lyal, 1999: 157 [stem: *Euphol-*]. Type genus: *Eupholus* Boisduval, 1835. Comment: proposed as a new taxon; junior homonym of Eupholini Günther, 1943.

Tribe Eustylini Lacordaire, 1863

Eustylides Lacordaire, 1863: 205 [stem: *Eustyl-*]. Type genus: *Eustylus* Schönherr, 1842. Comment: original vernacular name available (Art. 11.7.2): first used in latinized form by Faust (1886b: 338, as Eustylinorum), generally accepted as in Alonso-Zarazaga and Lyal (1999: 157, as Eustylini).

Exophthalmini G. H. Horn, 1876: 100 [stem: *Exophthalm-*]. Type genus: *Exophthalmus* Schönherr, 1823.

Eustylini Pierce, 1913: 421, in key [stem: *Eustyl-*]. Type genus: *Eustylus* Schönherr, 1842. Comment: family-group name proposed as new without reference to Eustylides Lacordaire, 1863.

Compsi Pierce, 1913: 406, in key [stem: *Comps-*]. Type genus: *Compsus* Schönherr, 1823. Comment: the younger name Compsina Martins and Galileo, 2007 (type genus *Compsa* Perty, 1832) in Cerambycidae is available; this case is to be referred to the Commission to remove the homonymy (Art. 55.3.1).

Exophthalmodina Voss, 1954: 199, in key [stem: *Exophthalmod-*]. Type genus: *Exophthalmodes* Pierce, 1916 [syn. of *Exophthalmus* Schönherr, 1823].

Tribe Geonemini Gistel, 1848

Geonemidae Gistel, 1848: [8] [stem: *Geonem-*]. Type genus: *Geonemus* Schönherr, 1833 [placed on the Official List of Generic Names in Zoology (ICZN 1988d)].

Barynotides Lacordaire, 1863: 37 [stem: *Barynot-*]. Type genus: *Barynotus* Germar, 1817. Comment: original vernacular name available (Art. 11.7.2): first used in latinized form by G. H. Horn (1876: 22, as Barynoti), generally accepted as in Dalla Torre et al. (1936: 32, as Barynotini).

Trigonoscutae J. L. LeConte, 1874b: 456 [stem: *Trigonoscut-*]. Type genus: *Trigonoscuta* Motschulsky, 1853.

Epicaeri G. H. Horn, 1876: 18 [stem: *Epicaer-*]. Type genus: *Epicaerus* Schönherr, 1834.

Calyptilli G. H. Horn, 1876: 26 [stem: *Calyptill-*]. Type genus: *Calyptillus* G. H. Horn, 1876.

Omilei G. H. Horn, 1876: 101 [stem: *Omile-*]. Type genus: *Omileus* G. H. Horn, 1876.

Trigonoscutini Pierce, 1913: 405, in key [stem: *Trigonoscut-*]. Type genus: *Trigonoscuta* Motschulsky, 1853. Comment: family-group name proposed as new without reference to Trigonoscutae J. L. LeConte, 1874.

Menoetiini Pierce, 1913: 373, in key [stem: *Menoeti-*]. Type genus: *Menoetius* Dejean, 1821 [placed on the Official Index of Rejected and Invalid Generic Names in Zoology (ICZN 1987f)]. Comment: permanently invalid (Art. 39): based on suppressed type genus.

Calyptillini Pierce, 1913: 405, in key [stem: *Calyptill-*]. Type genus: *Calyptillus* G. H. Horn, 1876. Comment: family-group name proposed as new without reference to Calyptilli G. H. Horn, 1876.

Tribe Holcorhinini Desbrochers des Loges, 1898

Holcorhinidae Desbrochers des Loges, 1898: 5 [stem: *Holcorhin-*]. Type genus: *Holcorhinus* Schönherr, 1826.

Cyclopterini Reitter, 1913b: 9, in key [stem: *Cyclopter-*]. Type genus: *Cyclopterus* Marseul, 1871 [preoccupied genus name, not *Cyclopterus* Linnaeus, 1758 [Pisces]; syn. of *Nucterocephalus* Desbrochers des Loges, 1897]. Comment: permanently invalid (Art. 39): based on preoccupied type genus; Cyclopteridae Bonaparte, 1831 (type genus *Cyclopterus* Linnaeus, 1758) is available in Pisces.

Tribe Hormorini Horn, 1876

Hormori G. H. Horn, 1876: 23 [stem: *Hormor-*]. Type genus: *Hormorus* G. H. Horn, 1876.

Tribe Laparocerini Lacordaire, 1863

Laparocérides Lacordaire, 1863: 196 [stem: *Laparocer-*]. Type genus: *Laparocerus* Schönherr, 1834. Comment: original vernacular name available (Art. 11.7.2): first used in latinized form by Heyne and Taschenberg (1907: 226, as Laparocerini), generally accepted as in Alonso-Zarazaga and Lyal (1999: 160, as Laparocerini).

LAPAROCERINI Pierce, 1913: 421, in key [stem: *Laparocer-*]. Type genus: *Laparocerus* Schönherr, 1834. Comment: family-group name proposed as new without reference to LAPAROCÉRIDES Lacordaire, 1863.

Tribe LEPTOSTETHINI Lacordaire, 1863

LEPTOSTÉTHIDES Lacordaire, 1863: 258 [stem: *Leptosteth-*]. Type genus: *Leptostethus* G. R. Waterhouse, 1853. Comment: original vernacular name available (Art. 11.7.2): first used in latinized form by Heyne and Taschenberg (1907: 226, as LEPTOSTETHINI), generally accepted as in Alonso-Zarazaga and Lyal (1999: 161, as LEPTOSTETHINI).

Tribe LORDOPINI Schönherr, 1823

LORDOPIDES Schönherr, 1823: column 1142 [stem: *Lordop-*]. Type genus: *Lordops* Schönherr, 1823.

HYPSONOTIDAE Jekel, 1853: I.II.3 [stem: *Hypsonot-*]. Type genus: *Hypsonotus* Germar, 1824.

ALOCORHINI Jekel, 1856: 9bis, in key [stem: *Alocorhin-*]. Type genus: *Alocorhinus* Sahlberg, 1823.

ELYTROXYSI Jekel, 1856: 9bis, in key [stem: *Elytroxe-*]. Type genus: *Elytroxys* Jekel, 1856. Comment: incorrect original stem formation, not in prevailing usage.

EURYLOBI Jekel, 1856: 9bis, in key [stem: *Eurylob-*]. Type genus: *Eurylobus* Schönherr, 1826.

MERODONTI Jekel, 1856: 9bis, in key [stem: *Merodont-*]. Type genus: *Merodontus* Jekel, 1856.

TOMORHINI Jekel, 1856: 9bis, in key [stem: *Tomorhin-*]. Type genus: *Tomorhinus* Jekel, 1856.

Tribe MESOSTYLINI Reitter, 1913

MESOSTYLINI Reitter, 1913b: 8, in key [stem: *Mesostyl-*]. Type genus: *Mesostylus* Faust, 1894.

Tribe MYORHININI Marseul, 1863

MYORHINIDAE Marseul, 1863: 223 [stem: *Myorhin-*]. Type genus: *Myorhinus* Schönherr, 1826 [syn. of *Apsis* Germar, 1820]. Comment: original vernacular name available (Art. 11.7.2): first used in latinized form by Kiesenwetter (1864: 265, as MYORHINIDAE), generally accepted as in Alonso-Zarazaga and Lyal (1999: 162, as MYORHININI).

Tribe NASTINI Reitter, 1913

NASTINI Reitter, 1913b: 9, in key [stem: *Nast-*]. Type genus: *Nastus* Schönherr, 1842.

Tribe Naupactini Gistel, 1848 *nomen protectum*

Iphiides Schönherr, 1823: column 1139 [stem: *Iphi-*]. Type genus: *Iphius* Schönherr, 1823 [syn. of *Alceis* Billberg, 1820]. Comment: *nomen oblitum* (see Appendix 1).

Naupactidae Gistel, 1848: [8] [stem: *Naupact-*]. Type genus: *Naupactus* Dejean, 1821. Comment: *nomen protectum* (see Appendix 1).

Cyphides Lacordaire, 1863: 107 [stem: *Cyph-*]. Type genus: *Cyphus* Germar, 1824 [syn. of *Cyrtomon* Schönherr, 1823]. Comment: original vernacular name available (Art. 11.7.2): first used in latinized form by J. L. LeConte (1874b: 458, as Cyphi), generally accepted as in J. L. LeConte and G. H. Horn (1883: 453, as Cyphi).

Macrostyles J. L. LeConte, 1874b: 457 [stem: *Macrostyl-*]. Type genus: *Macrostylus* Boheman, 1840.

Symmathetes J. L. LeConte, 1874b: 458 [stem: *Symmathet-*]. Type genus: *Symmathetes* Schönherr, 1847 [syn. of *Pantomorus* Schönherr, 1840].

Artipi G. H. Horn, 1876: 91 [stem: *Artipod-*]. Type genus: *Artipus* Schönherr, 1823 [preoccupied genus name, not *Artipus* Sahlberg, 1823 [Coleoptera: Curculionidae: Entiminae: Naupactini]; syn. of *Artipus* Sahlberg, 1823]. Comment: permanently invalid (Art. 39): based on preoccupied type genus.

Platyomina Champion, 1911: 282 [stem: *Platyom-*]. Type genus: *Platyomus* Schönherr, 1823 [preoccupied genus name, not *Platyomus* Sahlberg, 1823 [Coleoptera: Curculionidae: Entiminae: Naupactini]; syn. of *Platyomus* Sahlberg, 1823]. Comment: permanently invalid (Art. 39): based on preoccupied type genus.

Alceidini Pierce, 1913: 404, in key [stem: *Alceent-*]. Type genus: *Alceis* Billberg, 1820. Comment: incorrect original stem formation, not in prevailing usage.

Glaphyrometopi Pierce, 1913: 406, in key [stem: *Glaphyrometop-*]. Type genus: *Glaphyrometopus* Pierce, 1913.

Pseudocyphi Pierce, 1913: 406, in key [stem: *Pseudocyph-*]. Type genus: *Pseudocyphus* Schaeffer, 1905 [syn. of *Platyomus* Sahlberg, 1823].

Neocyphini Hustache, 1919: 476 [stem: *Neocyph-*]. Type genus: *Neocyphus* Bedel, 1883 [syn. of *Cyrtomon* Schönherr, 1823].

Platyomina Voss, 1954: 199, in key [stem: *Platyom-*]. Type genus: *Platyomus* Sahlberg, 1823. Comment: family-group name proposed as new without reference to Platyomina Champion, 1911.

Canephorotomina Voss, 1954: 206, in key [stem: *Canephorotom-*]. Type genus: *Canephorotomus* Voss, 1954 [syn. of *Amitrus* Schönherr, 1840].

Pantomorina Voss, 1954: 207, in key [stem: *Pantomor-*]. Type genus: *Pantomorus* Schönherr, 1840.

Plectrophorina Voss, 1954: 207, in key [stem: *Plectrophor-*]. Type genus: *Plectrophorus* Schönherr, 1826 [preoccupied genus name, not *Plectrophorus* Férussac, 1819 [Mollusca]; syn. of *Plectrophoroides* Wibmer and O'Brien, 1986]. Comment: permanently invalid (Art. 39): based on preoccupied type genus.

Tribe Nothognathini Marshall, 1916

Nothognathides G. A. K. Marshall, 1916: 204, in key [stem: *Nothognath-*]. Type genus: *Nothognathus* G. A. K. Marshall, 1916.

Tribe Omiini Shuckard, 1839

Omiadae Shuckard, 1839b: 60 [stem: *Omi-*]. Type genus: *Omias* Germar, 1817. Comment: senior homonym of Omiini Beck, 1996 in Lepidoptera (type genus *Omia* Hübner, 1821); this case is to be referred to the Commission to remove the homonymy (Art. 55.3.1).

Mylacini Reitter, 1913b: 9, in key [stem: *Mylac-*]. Type genus: *Mylacus* Boheman, 1843 [syn. of *Omias* Germar, 1817].

Tribe Oosomini Lacordaire, 1863

*Oosomides Schönherr, 1823: column 1145 [stem: *Oosom-*]. Type genus: *Oosomus* Schönherr, 1826. Comment: family-group name unavailable (Art. 11.7.1.1): not based on an available genus name at the time.

Oosomides Lacordaire, 1863: 164 [stem: *Oosom-*]. Type genus: *Oosomus* Schönherr, 1826. Comment: original vernacular name available (Art. 11.7.2): first used in latinized form by Kolbe (1891: 26, as Oosominae), generally accepted as in Alonso-Zarazaga and Lyal (1999: 166, as Oosomini).

Oosomini Pierce, 1913: 421, in key [stem: *Oosom-*]. Type genus: *Oosomus* Schönherr, 1826. Comment: family-group name proposed as new without reference to Oosomides Lacordaire, 1863.

Tribe Ophryastini Lacordaire, 1863

Ophryastides Lacordaire, 1863: 256 [stem: *Ophryast-*]. Type genus: *Ophryastes* Germar, 1829. Comment: original vernacular name available (Art. 11.7.2): first used in latinized form by J. L. LeConte (1874b: 454, as Ophryastini), generally accepted as in Alonso-Zarazaga and Lyal (1999: 167, as Ophryastini).

Tribe Ophtalmorrhynchini Hoffmann, 1965

Ophtalmorynchini Hoffmann, 1965: 1411 [stem: *Ophtalmorrhynch-*]. Type genus: *Ophtalmorrhynchus* Hoffmann, 1965. Comment: incorrect original stem formation, not in prevailing usage.

Tribe Otiorhynchini Schönherr, 1826

Loborhynchides Schönherr, 1823: column 1144 [stem: *Loborhynch-*]. Type genus: *Loborhynchus* Schönherr, 1823 [placed on the Official Index of Rejected and Invalid Generic Names in Zoology (ICZN 1972); see Lyal and Alonso-Zarazaga (2010) and Appendix 6; syn. of *Otiorhynchus* Germar, 1822]. Comment: Loborhynchinae Schönherr, 1823 placed on the Official Index of Rejected and Invalid Family-Group Names in Zoology (ICZN 1972).

Otiorhynchides Schönherr, 1826: 203 [stem: *Otiorhynch-*]. Type genus: *Otiorhynchus* Germar, 1822 [placed on the Official List of Generic Names in Zoology (ICZN 1972, as *Otiorhynchus* Germar, 1824); see Lyal and Alonso-Zarazaga (2010) and Appendix 6]. Comment: name placed on the Official List of Family-Group Names in Zoology (ICZN 1972, as Otiorhynchinae Schönherr, 1826).

Brachyrrhinidae Bedel, 1883: 30 [stem: *Brachyrhin-*]. Type genus: *Brachyrhinus* Latreille, 1802 [as *Brachyrrhinus*, incorrect original spelling of type genus name, not in prevailing usage; *Brachyrhinus* Latreille, 1802 placed on the Official Index of Rejected and Invalid Generic Names in Zoology (ICZN 1972); syn. of *Cryphiphorus* Stierlin, 1883]. Comment: name placed on the Official Index of Rejected and Invalid Family-Group Names in Zoology (ICZN 1972).

Tribe Ottistirini Heller, 1925

Ottistirini Heller, 1925a: 56, in key [stem: *Ottistir-*]. Type genus: *Ottistira* Pascoe, 1872.

Tribe Pachyrhynchini Schönherr, 1826

Somatodides Schönherr, 1823: column 1139 [stem: *Somatod-*]. Type genus: *Somatodes* Schönherr, 1823 [placed on the Official Index of Rejected and Invalid Generic Names in Zoology (ICZN 1994c); syn. of *Pachyrhynchus* Germar, 1824]. Comment: name placed on the Official Index of Rejected and Invalid Family-Group Names in Zoology as Somatodini Schönherr, 1823 (ICZN 1994c).

Pachyrhynchides Schönherr, 1826: 88 [stem: *Pachyrhynch-*]. Type genus: *Pachyrhynchus* Germar, 1824 [placed on the Official List of Generic Names in Zoology (ICZN 1970b)]. Comment: name placed on the Official List of Family-Group Names in Zoology as Pachyrhynchini Schönherr, 1826 (ICZN 1970b).

Tribe Peritelini Lacordaire, 1863

Péritélides Lacordaire, 1863: 178 [stem: *Peritel-*]. Type genus: *Peritelus* Germar, 1824. Comment: original vernacular name available (Art. 11.7.2): first used in latinized form by J. L. LeConte (1874b: 455, as Periteli), generally accepted as in Alonso-Zarazaga and Lyal (1999: 171, as Peritelini).

Simoini Pierce, 1913: 421, in key [stem: *Simon-*]. Type genus: *Simo* Dejean, 1821. Comment: incorrect original stem formation, not in prevailing usage.

Paraptochi Pierce, 1913: 423, in key [stem: *Paraptoch-*]. Type genus: *Paraptochus* Seidlitz, 1868.

Homorythmini Hoffmann, 1950: 154 [stem: *Homorhythm-*]. Type genus: *Homorhythmus* Bedel, 1883 [as *Homorythmus*, incorrect subsequent spelling of type genus name, not in prevailing usage; syn. of *Simo* Dejean, 1821]. Comment: incorrect original stem formation, not in prevailing usage.

Tribe Phyllobiini Schönherr, 1826

Phyllobides Schönherr, 1826: 178 [stem: *Phyllobi-*]. Type genus: *Phyllobius* Germar, 1824 [placed on the Official List of Generic Names in Zoology (ICZN 1981c)]. Comment: Phyllobiini Schönherr, 1826 placed on the Official List of Family-Group Names in Zoology (ICZN 1981c); incorrect original stem formation, not in prevailing usage.

Evotini J. L. LeConte, 1874b: 454 [stem: *Evot-*]. Type genus: *Evotus* J. L. LeConte, 1874.

Aphrasti J. L. LeConte, 1874b: 458 [stem: *Aphrast-*]. Type genus: *Aphrastus* Say, 1831.

Metacinopinae Reitter, 1913b: 6, in key [stem: *Metacinop-*]. Type genus: *Metacinops* Kraatz, 1862.

Tribe Polycatini Marshall, 1956

Polycatini G. A. K. Marshall, 1956: 4 [stem: *Polycat-*]. Type genus: *Polycatus* Heller, 1913.

Tribe Polydrusini Schönherr, 1823

Polydrosides Schönherr, 1823: column 1144 [stem: *Polydrus-*]. Type genus: *Polydrusus* Germar, 1817 [as *Polydrosus*, unjustified emendation of type genus name by Schönherr (1823), not in prevailing usage; syn. of *Polydrusus* Germar, 1817; *Polydrusus* Germar, 1817 placed on the Official List of Generic Names in Zoology (ICZN 1981c)]. Comment: placed on the Official List of Family-Group Names in Zoology (ICZN 1981c, as Polydrosini Schönherr, 1823), however, since this family-group name was based on an unjustified emendation of the type genus name, it was corrected (see Art. 32.5.3.2) to Polydrusini by Alonso-Zarazaga and Lyal (1999: 174).

Phyllomanisidae Gistel, 1848: [8] [stem: *Phylloman-*]. Type genus: *Phyllomanes* Gistel, 1848 [preoccupied genus name, not *Phyllomanes* Cabanis, 1847 [Aves]; syn. of *Polydrusus* Germar, 1817]. Comment: permanently invalid (Art. 39): based on preoccupied type genus; incorrect original stem formation, not in prevailing usage.

Liophloeidae Gistel, 1848: [8] [stem: *Liophloe-*]. Type genus: *Liophloeus* Germar, 1817.

Oligocyidae Gistel, 1856a: 373 [stem: *Oligoce-*]. Type genus: *Oligocys* Gistel, 1856 [syn. of *Liophloeus* Germar, 1817]. Comment: incorrect original stem formation, not in prevailing usage.

Scythropides Lacordaire, 1863: 380 [stem: *Scythrop-*]. Type genus: *Scythropus* Schönherr, 1826 [syn. of *Pachyrhinus* Schönherr, 1823]. Comment: original vernacular name available (Art. 11.7.2): first used in latinized form by Marseul (1863: 224, as Scytropidae [incorrect stem formation]), generally accepted as in Hustache (1919: 509, as Scythropini).

AUCHMERESTHINAE Reitter, 1913b: 6, in key [stem: *Auchmeresthet-*]. Type genus: *Auchmeresthes* Kraatz, 1862. Comment: incorrect original stem formation, not in prevailing usage.

Tribe PREMNOTRYPINI Kuschel, 1956

PREMNOTRYPINI Kuschel, 1956: 187 [stem: *Premnotryp-*]. Type genus: *Premnotrypes* Pierce, 1914.

†Tribe PRISTORHYNCHINI Heer, 1847

PRISTORHYNCHIDEN Heer, 1847: 190 [stem: *Pristorhynch-*]. Type genus: *Pristorhynchus* Heer, 1847. Comment: original vernacular name available (Art. 11.7.2): first used in latinized form by Scudder (1893: 29, as PRISTORHYNCHINI), generally accepted as in Alonso-Zarazaga and Lyal (1999: 176, as PRISTORHYNCHINI).

Tribe PRYPNINI Lacordaire, 1863

PRYPNIDES Lacordaire, 1863: 135 [stem: *Prypn-*]. Type genus: *Prypnus* Schönherr, 1823. Comment: original vernacular name available (Art. 11.7.2): first used in latinized form by Heyne and Taschenberg (1907: 225, as PRYPNINI), generally accepted as in Alonso-Zarazaga and Lyal (1999: 176, as PRYPNINI).

Tribe PSALLIDIINI Lacordaire, 1863

PSALIDIIDES Lacordaire, 1863: 138 [stem: *Psallidi-*]. Type genus: *Psallidium* Herbst, 1795 [as *Psalidium*, incorrect subsequent spelling of type genus name, not in prevailing usage]. Comment: original vernacular name available (Art. 11.7.2): first used in latinized form by Everts (1903: 547, as PSALIDIINI [incorrect stem formation]), generally accepted as in Alonso-Zarazaga and Lyal (1999: 176, as PSALLIDIINI); incorrect original stem formation, not in prevailing usage.

Tribe RHYNCOGONINI Sharp, 1919

RHYNCOGONIDES Sharp, 1919a: 77 [stem: *Rhyncogon-*]. Type genus: *Rhyncogonus* Sharp, 1885.

Tribe SCIAPHILINI Sharp, 1891

SCIAPHILINA Sharp, 1891: 167 [stem: *Sciaphil-*]. Type genus: *Sciaphilus* Schönherr, 1823.

SCIAPHILINI Pierce, 1913: 405, in key [stem: *Sciaphil-*]. Type genus: *Sciaphilus* Schönherr, 1823. Comment: family-group name proposed as new without reference to SCIAPHILINA Sharp, 1891.

Tribe SITONINI Gistel, 1848

SITONISIDAE Gistel, 1848: [8] [stem: *Siton-*]. Type genus: *Sitones* Schönherr, 1840 [syn. of *Sitona* Germar, 1817]. Comment: incorrect original stem formation, not in prevailing usage.

Tribe Tanymecini Lacordaire, 1863

Tanymécides Lacordaire, 1863: 82 [stem: *Tanymec-*]. Type genus: *Tanymecus* Germar, 1817.

Subtribe Piazomiina Reitter, 1913

Piazomiini Reitter, 1913b: 28, in key [stem: *Piazomi-*]. Type genus: *Piazomias* Schönherr, 1840.

Subtribe Tainophthalmina Desbrochers des Loges, 1873

Tainophthalmidae Desbrochers des Loges, 1873: 426 [stem: *Tainophthalm-*]. Type genus: *Tainophthalmus* Desbrochers des Loges, 1873.

Amomphi J. L. LeConte, 1874b: 455 [stem: *Amomph-*]. Type genus: *Amomphus* Schönherr, 1848 [syn. of *Aspidiotes* Schönherr, 1847].

Subtribe Tanymecina Lacordaire, 1863

Tanymécides Lacordaire, 1863: 82 [stem: *Tanymec-*]. Type genus: *Tanymecus* Germar, 1817. Comment: original vernacular name available (Art. 11.7.2): first used in latinized form by J. L. LeConte (1874b: 454, as Tanymecini), generally accepted as in Alonso-Zarazaga and Lyal (1999: 178, as Tanymecini).

Hadromerides Jekel, 1865: 545 [stem: *Hadromer-*]. Type genus: *Hadromerus* Schönherr, 1834 [preoccupied genus name, not *Hadromerus* Schönherr, 1823 [Coleoptera: Curculionidae: Entiminae]; syn. of *Hadromeropsis* Pierce, 1913]. Comment: permanently invalid (Art. 39): based on preoccupied type genus.

Siderodactylides Jekel, 1865: 551, in note [stem: *Siderodactyl-*]. Type genus: *Siderodactylus* Schönherr, 1834 [syn. of *Hadromerus* Schönherr, 1823].

Pachnaei J. L. LeConte, 1874b: 457 [stem: *Pachnae-*]. Type genus: *Pachnaeus* Schönherr, 1826.

Minyomeri G. H. Horn, 1876: 17 [stem: *Minyomer-*]. Type genus: *Minyomerus* G. H. Horn, 1876.

Pandeleteini Pierce, 1913: 399, in key [stem: *Pandeletei-*]. Type genus: *Pandeleteius* Schönherr, 1834. Comment: incorrect original stem formation, not in prevailing usage.

Tribe Tanyrhynchini Schönherr, 1826

Tanyrhynchides Schönherr, 1826: 212 [stem: *Tanyrhynch-*]. Type genus: *Tanyrhynchus* Schönherr, 1826.

Éremnides Lacordaire, 1863: 220 [stem: *Eremn-*]. Type genus: *Eremnus* Schönherr, 1823. Comment: original vernacular name available (Art. 11.7.2): first used in latinized form by J. L. LeConte (1869: 380, as Eremnini), generally accepted as in Schenkling and G. A. K. Marshall (1931b: Erem. 13, as Eremnini).

Tribe Thecesternini Lacordaire, 1863

Thécesternides Lacordaire, 1863: 306 [stem: *Thecestern-*]. Type genus: *Thecesternus* Say, 1831. Comment: original vernacular name available (Art. 11.7.2): first used in latinized form by Sharp (1891: 86, as Thecesterninae), generally accepted as in Alonso-Zarazaga and Lyal (1999: 182, as Thecesternini).

Tribe Trachyphloeini Gistel, 1848

Trachyphloeidae Gistel, 1848: [7] [stem: *Trachyphloe-*]. Type genus: *Trachyphloeus* Germar, 1817.

Subtribe Trachyphilina Voss, 1948

Trachyphilina Voss, 1948: 73, in key [stem: *Trachyphil-*]. Type genus: *Trachyphilus* Faust, 1887.

Subtribe Trachyphloeina Gistel, 1848

Trachyphloeidae Gistel, 1848: [7] [stem: *Trachyphloe-*]. Type genus: *Trachyphloeus* Germar, 1817.

Phyllastolidae Gistel, 1856a: 372 [stem: *Phyllastol-*]. Type genus: *Phyllastolus* Gistel, 1856 [syn. of *Trachyphloeus* Germar, 1817].

Cathormiocerini Reitter, 1913b: 8, in key [stem: *Cathormiocer-*]. Type genus: *Cathormiocerus* Schönherr, 1842.

Trachyphloeini Pierce, 1913: 421, in key [stem: *Trachyphloe-*]. Type genus: *Trachyphloeus* Germar, 1817. Comment: family-group name proposed as new without reference to Trachyphloeidae Gistel, 1848.

Pseudocneorrhinini Kôno, 1930: 163, in key [stem: *Pseudocneorhin-*]. Type genus: *Pseudocneorhinus* Roelofs, 1873 [as *Pseudocneorrhinus*, incorrect subsequent spelling of type genus name, not in prevailing usage]. Comment: incorrect original stem formation, not in prevailing usage.

Callirhopalini Voss, 1956b: 23 [stem: *Callirhopal-*]. Type genus: *Callirhopalus* Hochhuth, 1851.

Tribe Tropiphorini Marseul, 1863

Tropiphoridae Marseul, 1863: 220 [stem: *Tropiphor-*]. Type genus: *Tropiphorus* Schönherr, 1842 [placed on the Official List of Generic Names in Zoology (ICZN 1988c)]. Comment: Tropiphorini, published on 15 June 1863, is given priority over Strangaliodini, Byrsopagini, Pantopoeini and Synaptonychini which were published by Lacordaire "before 10 August 1863".

Leptopsides Lacordaire, 1863: 232 [stem: *Leptop-*]. Type genus: *Leptops* Schönherr, 1834 [preoccupied genus name, not *Leptops* Rafinesque, 1820 [Pisces]; syn. of *Leptopius* Oke, 1951]. Comment: original vernacular name available (Art. 11.7.2): first used in latinized form by Pascoe (1870a: 181, as Leptopinae), generally accepted as in Heyne and Taschenberg (1907: 226, as Leptopsini [incorrect stem formation]); permanently invalid (Art. 39):

based on preoccupied type genus; incorrect original stem formation, not in prevailing usage.

Strangaliodides Lacordaire, 1863: 234 [stem: *Strangaliod-*]. Type genus: *Strangaliodes* Schönherr, 1842. Comment: original vernacular name available (Art. 11.7.2): first used in latinized form by G. H. Horn (1876: 37, as Strangaliodes [treated as Latin]), generally accepted as in Voss (1954: 237, as Strangaliodini).

Byrsopagides Lacordaire, 1863: 337 [stem: *Byrsopag-*]. Type genus: *Byrsopages* Schönherr, 1842. Comment: original vernacular name available (Art. 11.7.2): first used in latinized form by Heyne and Taschenberg (1907: 228, as Byrsopagini), generally accepted as in Schenkling and G. A. K. Marshall (1929: 57, as Byrsopagini).

Pantopéides Lacordaire, 1863: 346 [stem: *Pantopoe-*]. Type genus: *Pantopoeus* Schönherr, 1842 [syn. of *Perperus* Schönherr, 1842]. Comment: original vernacular name available (Art. 11.7.2): first used in latinized form and generally accepted as in Faust (1892a: 179, as Pantopeinarum [incorrect stem formation]); incorrect original stem formation, not in prevailing usage.

Synaptonycides Lacordaire, 1863: 372 [stem: *Synaptonych-*]. Type genus: *Synaptonyx* G. R. Waterhouse, 1853. Comment: original vernacular name available (Art. 11.7.2): first used in latinized form and generally accepted as in Wollaston (1877: 158, as Synaptonychides [treated as Latin]); incorrect original stem formation, not in prevailing usage.

Rhigopsini J. L. LeConte, 1874b: 454 [stem: *Rhigopse-*]. Type genus: *Rhigopsis* J. L. LeConte, 1874. Comment: incorrect original stem formation, not in prevailing usage, the comment stating the contrary in Alonso-Zarazaga and Lyal (2002: 28) is incorrect.

Dyslobini J. L. LeConte, 1874b: 454 [stem: *Dyslob-*]. Type genus: *Dyslobus* J. L. LeConte, 1869.

Phyxeles G. H. Horn, 1876: 56 [stem: *Phyxelid-*]. Type genus: *Phyxelis* Schönherr, 1842. Comment: incorrect original stem formation, not in prevailing usage.

Dirotognathini G. H. Horn, 1876: 79 [stem: *Dirotognath-*]. Type genus: *Dirotognathus* G. H. Horn, 1876.

Synirmini Bedel, 1883: 32, in key [stem: *Synirm-*]. Type genus: *Synirmus* Bedel, 1883 [syn. of *Tropiphorus* Schönherr, 1842].

Stenocorynini McKeown, 1939: 408 [stem: *Stenocoryn-*]. Type genus: *Stenocorynus* Schönherr, 1842. Comment: replacement name for Leptopsides Lacordaire, 1863 because of the homonymy of the type genus.

Leptopiinae Oke, 1951: 24 [stem: *Leptopi-*]. Type genus: *Leptopius* Oke, 1951. Comment: replacement name for Leptopsides Lacordaire, 1863 because of the homonymy of the type genus.

Leptosinae G. A. K. Marshall, 1952: 264 [stem: *Leptos-*]. Type genus: *Leptosus* G. A. K. Marshall, 1952 [syn. of *Leptopius* Oke, 1951]. Comment: replacement name for Leptopsides Lacordaire, 1863 because of the homonymy of the type genus.

Tribe Typhlorhinini Kuschel, 1954

Typhlorhinini Kuschel, 1954: 287 [stem: *Typhlorhin-*]. Type genus: *Typhlorhinus* Kuschel, 1954 [syn. of *Hapactorrhynchus* Richard, 1953].

Subfamily Hyperinae Marseul, 1863 (1848)

Hyperidae Marseul, 1863: 224 [stem: *Hyper-*]. Type genus: *Hypera* Germar, 1817. Comment: use of family-group name conserved over Phytonominae Gistel, 1848 (Art. 40.2).

Tribe Cepurini Capiomont, 1867

*Haplopides Lacordaire, 1863: 394 [stem: *Haplopod-*]. Type genus: *Haplopus* Schönherr, 1840 [preoccupied genus name, not *Haplopus* Burmeister, 1838 [Orthoptera]; syn. of *Haplopodus* G. A. K. Marshall, 1946]. Comment: original vernacular name unavailable (Art. 11.7.2): subsequently used in latinized form, e.g., Heyne and Taschenberg (1907: 228, as Haplopini) but not generally accepted as valid; Haplopidae was used as valid by Ienistea (1986: 33) but it was not attributed to Lacordaire (1863); Ienistea's name is also unavailable, it was proposed after 1930 without description or bibliographic reference to such a description (Art. 13.1); if found to be available then permanently invalid (Art. 39): based on preoccupied type genus; incorrect original stem formation, not in prevailing usage.

Cépurides Capiomont, 1867: 438 [stem: *Cepur-*]. Type genus: *Cepurus* Schönherr, 1834. Comment: original vernacular name available (Art. 11.7.2): first used in latinized form by Csiki (1934: 3, as Cepurini), generally accepted as in Alonso-Zarazaga and Lyal (1999: 189, as Cepurini).

Tribe Hyperini Marseul, 1863 (1848)

Phytonomidae Gistel, 1848: [8] [stem: *Phytonom-*]. Type genus: *Phytonomus* Schönherr, 1823 [syn. of *Hypera* Germar, 1817]. Comment: use of the younger names Hyperinae/-ini Marseul, 1863 are conserved over this name (Art. 40.2).

Hyperidae Marseul, 1863: 224 [stem: *Hyper-*]. Type genus: *Hypera* Germar, 1817. Comment: published 15 June 1863; use of family-group name conserved over Phytonomini Gistel, 1848 (Art. 40.2); this family-group name was also used in the same year by Lacordaire (1863 [before 10 August]: 395, as Hypérides).

Coniatina Legalov, 2007: 401 [stem: *Coniat-*]. Type genus: *Coniatus* Germar, 1817.

Macrotarrhusina Legalov, 2007: 401 [stem: *Macrotarrh-*]. Type genus: *Macrotarrhus* Bedel, 1906. Comment: incorrect original stem formation, not in prevailing usage.

Subfamily Lixinae Schönherr, 1823

Lixides Schönherr, 1823: column 1146 [stem: *Lix-*]. Type genus: *Lixus* Fabricius, 1801.

Tribe Cleonini Schönherr, 1826 *nomen protectum*

Geomorides Schönherr, 1823: column 1141 [stem: *Geomor-*]. Type genus: *Geomorus* Schönherr, 1823 [syn. of *Cleonis* Dejean, 1821]. Comment: as pointed out by Alonso-Zarazaga and Lyal (1999: 190) this is the oldest available name for the tribe name, however, it has not been used as valid after 1899 and therefore we treat it here as a *nomen oblitum* (see Appendix 1).

Cleonides Schönherr, 1826: 145 [stem: *Cleon-*]. Type genus: *Cleonus* Schönherr, 1826 [syn. of *Cleonis* Dejean, 1821]. Comment: *nomen protectum* (see Appendix 1).

Cleonidae Kirby, 1837: 198 [stem: *Cleon-*]. Type genus: *Cleonis* Dejean, 1821.

Xerobiadae Gistel, 1856a: 373 [stem: *Xerobi-*]. Type genus: *Xerobia* Gistel, 1856 [syn. of *Cleonis* Dejean, 1821]. Comment: incorrect original stem formation, not in prevailing usage.

*Bothynodérides Chevrolat, 1872: 16 [stem: *Bothynoder-*]. Type genus: *Bothynoderes* sensu Chevrolat, 1872 [not *Bothynoderes* Schönherr, 1823; syn. of *Asproparthenis* Gozis, 1886]. Comment: original vernacular name unavailable (Art. 11.7.2): not subsequently latinized.

*Conorhynchides Chevrolat, 1872: 17 [stem: *Conorhynch-*]. Type genus: *Conorhynchus* Motschulsky, 1860. Comment: original vernacular name unavailable (Art. 11.7.2): subsequently used in latinized form, e.g., Handlirsch (1925: 687 [as Cneorhynchides; *lapsus calami* for Conorhynchides [treated as Latin]), but not generally accepted as valid; Conorhynchinae Gill, 1861 (type genus *Conorhynchus* Gill, 1861, unjustified emendation of *Conorynchus* Nozeman, 1758 and junior homonym of *Conorhynchus* Motschulsky, 1860) is available in Pisces; the Pisces family-group name is permanently invalid since it it based on a preoccupied type genus (Art. 39).

*Cossinodérides Chevrolat, 1872: 18 [stem: *Cossinoder-*]. Type genus: *Cossinoderus* Chevrolat, 1872 [syn. of *Porocleonus* Motschulsky, 1860]. Comment: original vernacular name unavailable (Art. 11.7.2): not subsequently latinized.

Epirhynchini Petri, 1914: 3, in key [stem: *Epirrhynch-*]. Type genus: *Epirrhynchus* Schönherr, 1823 [as *Epirhynchus*, incorrect subsequent spelling of type genus name, not in prevailing usage]. Comment: incorrect original stem formation, not in prevailing usage.

Tribe Lixini Schönherr, 1823

Lixides Schönherr, 1823: column 1146 [stem: *Lix-*]. Type genus: *Lixus* Fabricius, 1801.

*Lariniden Suffrian, 1848: 62 [stem: *Larin-*]. Type genus: *Larinus* Germar, 1824 [preoccupied genus name, not *Larinus* Dejean, 1821 [Coleoptera: Curculionidae: Lixinae: Lixini]; syn. of *Larinus* Dejean, 1821]. Comment: original vernacular name unavailable (Art. 11.7.2): subsequently used in latinized form but not generally attributed to Suffrian (1848); if found to be available then permanently invalid (Art. 39): based on preoccupied type genus.

LARINIDAE Gistel, 1848: [7] [stem: *Larin-*]. Type genus: *Larinus* Germar, 1824 [preoccupied genus name, not *Larinus* Dejean, 1821 [Coleoptera: CURCULIONIDAE: LIXINAE: LIXINI]; syn. of *Larinus* Dejean, 1821]. Comment: permanently invalid (Art. 39): based on preoccupied type genus.

PHYLLONOMEIDAE Gistel, 1856a: 372 [stem: *Phyllonome-*]. Type genus: *Phyllonomeus* Gistel, 1856.

LARININI Petri, 1914: 4, in key [stem: *Larin-*]. Type genus: *Larinus* Germar, 1824 [preoccupied genus name, not *Larinus* Dejean, 1821 [Coleoptera: CURCULIONIDAE: LIXINAE: LIXINI]; syn. of *Larinus* Dejean, 1821]. Comment: family-group name proposed as new without reference to LARINIDAE Gistel, 1848; permanently invalid (Art. 39): based on preoccupied type genus.

Tribe RHINOCYLLINI Lacordaire, 1863

RHINOCYLLIDES Lacordaire, 1863: 433 [stem: *Rhinocyll-*]. Type genus: *Rhinocyllus* Germar, 1817. Comment: original vernacular name available (Art. 11.7.2): first used in latinized form by Faust (1904: 182, as RHINOCYLLINAE), generally accepted as in Alonso-Zarazaga and Lyal (1999: 193, as RHINOCYLLINI).

Subfamily MESOPTILIINAE Lacordaire, 1863

MÉSOPTILIDES Lacordaire, 1863: 563 [stem: *Mesoptili-*]. Type genus: *Mesoptilius* Labram and Imhoff, 1845.

Tribe CARCILIINI Pierce, 1916

CARCILIINAE Pierce, 1916: 465 [stem: *Carcili-*]. Type genus: *Carcilia* Roelofs, 1874.

Tribe LAEMOSACCINI Lacordaire, 1865

LÉMOSACIDES Lacordaire, 1865: 12 [stem: *Laemosacc-*]. Type genus: *Laemosaccus* Schönherr, 1823. Comment: original vernacular name available (Art. 11.7.2): first used in latinized form by Pascoe (1870b: 437, as LAEMOSACCINAE), generally accepted as in Alonso-Zarazaga and Lyal (1999: 194, as LAEMOSACCINI); incorrect original stem formation, not in prevailing usage.

Tribe MAGDALIDINI Pascoe, 1870 *nomen protectum*

THAMNOPHILIDES Schönherr, 1823: column 1136 [stem: *Thamnophil-*]. Type genus: *Thamnophilus* Schönherr, 1823 [preoccupied genus name, not *Thamnophilus* Vieillot, 1816 [Aves]; syn. of *Magdalis* Germar, 1817]. Comment: permanently invalid (Art. 39): based on preoccupied type genus.

SCARDAMYCTISIDAE Gistel, 1848: [7] [stem: *Scardamyct-*]. Type genus: *Scardamyctes* Gistel, 1848 [syn. of *Magdalis* Germar, 1817]. Comment: SCARDAMYCTINI Gistel, 1848 was recently considered a *nomen oblitum* by Colonnelli (2003: 7) however the necessary supporting references were not provided; we hereby consider SCARDAMYCTINI Gistel, 1848 as a *nomen oblitum* (see supporting references in Appendix 1); incorrect original stem formation, not in prevailing usage.

MAGDALINIDAE Gistel, 1856a: 371 [stem: *Magdalin-*]. Type genus: *Magdalinus* Germar, 1843 [placed on the Official Index of Rejected and Invalid Generic Names in Zoology by the Commission (ICZN 1955a); syn. of *Magdalis* Germar, 1817]. Comment: permanently invalid (Art. 39): based on suppressed type genus.

MAGDALINAE Pascoe, 1870b: 436 [stem: *Magdalid-*]. Type genus: *Magdalis* Germar, 1817 [placed on the Official List of Generic Names in Zoology (ICZN 1955a)]. Comment: *nomen protectum* (see Appendix 1); incorrect original stem formation, not in prevailing usage.

RHININAE Pierce, 1916: 465 [stem: *Rhin-*]. Type genus: *Rhina* Latreille, 1802 [preoccupied genus name, not *Rhina* Schneider, 1801 [Pisces]; syn. of *Panus* Schönherr, 1823]. Comment: permanently invalid (Art. 39): based on preoccupied type genus.

Tribe MESOPTILIINI Lacordaire, 1863

MÉSOPTILIDES Lacordaire, 1863: 563 [stem: *Mesoptili-*]. Type genus: *Mesoptilius* Labram and Imhoff, 1845. Comment: original vernacular name available (Art. 11.7.2): first used in latinized form by Pascoe (1870b: 436, as MESOPTILINAE [incorrect stem formation]), generally accepted as in Alonso-Zarazaga and Lyal (1999: 193, as MESOPTILIINI); incorrect original stem formation, not in prevailing usage.

CNEMIDOPHORINI Hustache, 1937: 199, in key [stem: *Cnemidophor-*]. Type genus: *Cnemidophorus* Schönherr, 1835 [preoccupied genus name, not *Cnemidophorus* Wagler, 1830 [Reptilia]; syn. of *Cnemidontus* Schenkling, 1935]. Comment: permanently invalid (Art. 39): based on preoccupied type genus.

CNEMIDONTINI Kuschel, 1955: 271 [stem: *Cnemidont-*]. Type genus: *Cnemidontus* Schenkling, 1935. Comment: replacement name for CNEMIDOPHORINI Hustache, 1937 because of the homonymy of the type genus.

Subfamily MOLYTINAE Schönherr, 1823

MOLYTIDES Schönherr, 1823: column 1142 [stem: *Molyt-*]. Type genus: *Molytes* Schönherr, 1823 [syn. of *Liparus* A. G. Olivier, 1807].

Tribe ANOPLINI Bedel, 1884

ANOPLINI Bedel, 1884a: 67, in key [stem: *Anopl-*]. Type genus: *Anoplus* Germar, 1820.

Tribe AMALACTINI Lacordaire, 1863

AMALACTIDES Lacordaire, 1863: 506 [stem: *Amalact-*]. Type genus: *Amalactus* Schönherr, 1835. Comment: original vernacular name available (Art. 11.7.2):

first used in latinized form by Pascoe (1870b: 436, as AMALACTINAE), generally accepted as in Alonso-Zarazaga and Lyal (1999: 196, as AMALACTINI).

Tribe AMINYOPINI Voss, 1956

AMINYOPINI Voss, 1956b: 30, in key [stem: *Aminyop-*]. Type genus: *Aminyops* Voss, 1956.

NIPHADINI Voss, 1963: 5 [stem: *Niphad-*]. Type genus: *Niphades* Pascoe, 1871.

NIPHADONOTHINA Voss, 1965: 344, in key [stem: *Niphadonoth-*]. Type genus: *Niphadonothus* Voss, 1965.

Tribe AMORPHOCERINI Voss, 1939

AMORPHOCERINI Voss, 1939: 65 [stem: *Amorphocer-*]. Type genus: *Amorphocerus* Schönherr, 1826.

Tribe ANCHONINI Imhoff, 1856

ANCHONIDAE Imhoff, 1856: [2] 213 [stem: *Anchon-*]. Type genus: *Anchonus* Schönherr, 1825.

Tribe BRACHYCEROPSEINI Aurivillius, 1926

BRACHYCEROPSEINAE Aurivillius, 1926b: 2 [stem: *Brachyceropse-*]. Type genus: *Brachyceropsis* Aurivillius, 1926. Comment: Alonso-Zarazaga and Lyal (2002: 18) incorrectly changed their previous opinion in Alonso-Zarazaga and Lyal (1999: 103) regarding the correct stem based on this genus name, the correct stem is *Brachyceropse-*.

Tribe CHOLINI Schönherr, 1825

CHOLIDES Schönherr, 1825: column 584 [stem: *Chol-*]. Type genus: *Cholus* Germar, 1824 [placed on the Official List of Generic Names in Zoology (ICZN 1987d)].

Subtribe CHOLINA Schönherr, 1825

CHOLIDES Schönherr, 1825: column 584 [stem: *Chol-*]. Type genus: *Cholus* Germar, 1824 [placed on the Official List of Generic Names in Zoology (ICZN 1987d)].

AMERININAE Pierce, 1919: 30, in key [stem: *Amerin-*]. Type genus: *Ameris* Dejean, 1821.

Subtribe CHOLOMINA Vaurie, 1974

CHOLOMINI Vaurie, 1974: 4 [stem: *Cholom-*]. Type genus: *Cholomus* Roelofs, 1880.

Subtribe RHINASTINA Vaurie, 1973

RHINASTINI Vaurie, 1973: 4, in key [stem: *Rhinast-*]. Type genus: *Rhinastus* Schönherr, 1825.

Tribe Cleogonini Gistel, 1848

Cleogonidae Gistel, 1848: [7] [stem: *Cleogon-*]. Type genus: *Cleogonus* Schönherr, 1825.

Tribe Conotrachelini Jekel, 1865

Conotrachelides Jekel, 1865: 550 [stem: *Conotrachel-*]. Type genus: *Conotrachelus* Dejean, 1835.

Echinaspini Blatchley, 1922: 121 [stem: *Echinaspid-*]. Type genus: *Echinaspis* Blatchley, 1922 [preoccupied genus name, not *Echinaspis* Haeckel, 1881 [Protista]; syn. of *Microhyus* J. L. LeConte, 1876]. Comment: permanently invalid (Art. 39): based on preoccupied type genus; incorrect original stem formation, not in prevailing usage.

Tribe Cycloterini Lacordaire, 1863

Cyclotérides Lacordaire, 1863: 365 [stem: *Cycloter-*]. Type genus: *Cycloteres* Schönherr, 1843.

Subtribe Cycloterina Lacordaire, 1863

Cyclotérides Lacordaire, 1863: 365 [stem: *Cycloter-*]. Type genus: *Cycloteres* Schönherr, 1843. Comment: original vernacular name available (Art. 11.7.2): first used in latinized form by Heyne and Taschenberg (1907: 228, as Cycloterini), generally accepted as in Alonso-Zarazaga and Lyal (1999: 200, as Cycloterini).

Subtribe Thrombosternina Voss, 1965

Thrombosternina Voss, 1965: 344, in key [stem: *Thrombostern-*]. Type genus: *Thrombosternus* G. A. K. Marshall, 1955.

Tribe Dinomorphini Lacordaire, 1863

Dinomorphides Lacordaire, 1863: 291 [stem: *Dinomorph-*]. Type genus: *Dinomorphus* Perty, 1832. Comment: original vernacular name available (Art. 11.7.2): first used in latinized form by Pascoe (1870b: 436, as Dinomorphinae), generally accepted as in Alonso-Zarazaga and Lyal (1999: 201, as Dinomorphini).

Tribe Emphyastini Lacordaire, 1863

Emphiastides Lacordaire, 1863: 510 [stem: *Emphyast-*]. Type genus: *Emphyastes* Mannerheim, 1852 [as *Emphiastes*, incorrect subsequent spelling of type genus name, not in prevailing usage]. Comment: original vernacular name available (Art. 11.7.2): first used in latinized form by J. L. LeConte (1876: 137, as Emphyastini), generally accepted as in Alonso-Zarazaga and Lyal (1999: 143, as Emphyastina); incorrect original stem formation, not in prevailing usage.

PHYCOCOETES J. L. LeConte, 1876: 189 [stem: *Phycocoet-*]. Type genus: *Phycocoetes* J. L. LeConte, 1876 [preoccupied genus name, not *Phycocoetes* Agassiz, 1846 [Pisces]; syn. of *Thalasselephas* Egorov and Korotyaev, 1976]. Comment: permanently invalid (Art. 39): based on preoccupied type genus.

THALASSELEPHANTINI Alonso-Zarazaga and Lyal, 1999: 208 [stem: *Thalasselephant-*]. Type genus: *Thalasselephas* Egorov and Korotyaev, 1976. Comment: replacement name for PHYCOCOETINI J. L. LeConte, 1876 because of the homonymy of the type genus.

Tribe EUDERINI Lacordaire, 1865

EUDÉRIDES Lacordaire, 1865: 18 [stem: *Euder-*]. Type genus: *Euderes* Schönherr, 1825. Comment: original vernacular name available (Art. 11.7.2): first used in latinized form by Pascoe (1870b: 437, as EUDERINAE), generally accepted as in Alonso-Zarazaga and Lyal (1999: 201, as EUDERINI); the junior homonym EUDERINI Erdös, 1956 (type genus *Euderus* Haliday, 1844) is available in Hymenoptera: EULOPHIDAE; the name ENTIINAE Hedqvist, 1974 is now used as valid in Hymenoptera instead of EUDERINI Erdös, 1956 (see Hansson and Straka 2009: 272).

Tribe GALLOISIINI Morimoto, 1962

GALLOISIINAE Morimoto, 1962b: 375 [stem: *Galloisi-*]. Type genus: *Galloisia* Hustache, 1920.

Tribe GUIOPERINI Lacordaire, 1865

GUIOPÉRIDES Lacordaire, 1865: 78 [stem: *Guioper-*]. Type genus: *Guioperus* Perty, 1832. Comment: original vernacular name available (Art. 11.7.2): first used in latinized form by Heyne and Taschenberg (1907: 232, as GUIOPERINI), generally accepted as in Alonso-Zarazaga and Lyal (1999: 201, as GUIOPERINI).

Tribe HYLOBIINI Kirby, 1837

HYLOBIDAE Kirby, 1837: 196 [stem: *Hylobi-*]. Type genus: *Hylobius* Germar, 1817.

Subtribe EPISTROPHINA Marshall, 1932

EPISTROPHINA G. A. K. Marshall, 1932: 346, in key [stem: *Epistroph-*]. Type genus: *Epistrophus* Kirsch, 1868 [preoccupied genus name, not *Epistrophus* Gistel, 1834 [Coleoptera: ELATERIDAE]; *Epistrophus* Kirsch, 1868 is currently considered a valid genus name]. Comment: as pointed out by Alonso-Zarazaga and Lyal (1999: 202) this family-group name should be treated as permanently invalid (Art. 39) because it is based on a preoccupied type genus, however, an application will be submitted by MAAZ and CHCL to the Commission for the conservation of the genus *Epistrophus* Kirsch, 1868 over its unused senior homonym; meanwhile current usage is maintained.

Subtribe Hylobiina Kirby, 1837

Hylobidae Kirby, 1837: 196 [stem: *Hylobi-*]. Type genus: *Hylobius* Germar, 1817. Comment: incorrect original stem formation, not in prevailing usage.

Heilipinae Faust, 1892b: 202, in key [stem: *Heilipod-*]. Type genus: *Heilipus* Germar, 1824. Comment: incorrect original stem formation, not in prevailing usage.

Syphorbina G. A. K. Marshall, 1932: 346, in key [stem: *Syphorb-*]. Type genus: *Syphorbus* Pascoe, 1881.

Tribe Ithyporini Lacordaire, 1865

Ithyporides Lacordaire, 1865: 50 [stem: *Ithypor-*]. Type genus: *Ithyporus* Schönherr, 1836. Comment: precedence (Ithyporini Lacordaire, 1865 vs Sclerocardiini Lacordaire, 1865) given to taxon originally proposed at the higher rank (Art. 24.1).

Subtribe Colobodina Voss, 1958

Colobodini Voss, 1958: 50, in key [stem: *Colobod-*]. Type genus: *Colobodes* Schönherr, 1837.

Subtribe Ithyporina Lacordaire, 1865

Ithyporides Lacordaire, 1865: 50 [stem: *Ithypor-*]. Type genus: *Ithyporus* Schönherr, 1836. Comment: original vernacular name available (Art. 11.7.2): first used in latinized form by Kirsch (1870: 192, as Ithyporidae), generally accepted as in Alonso-Zarazaga and Lyal (1999: 203, as Ithyporini).

Subtribe Sclerocardiina Lacordaire, 1865

Sclérocardiides Lacordaire, 1865: 317 [stem: *Sclerocardi-*]. Type genus: *Sclerocardius* Schönherr, 1847. Comment: original vernacular name available (Art. 11.7.2): first used in latinized form by Heyne and Taschenberg (1907: 233, as Sclerocardini [incorrect stem formation]), generally accepted as in Alonso-Zarazaga and Lyal (1999: 204, as Sclerocardiina).

Tribe Itini Reitter, 1913

Itini Reitter, 1913b: 58, in key [stem: *It-*]. Type genus: *Ita* Tournier, 1878.

Tribe Juanorhinini Aurivillius, 1931

Juanorhinini Aurivillius, 1931: 465 [stem: *Juanorhin-*]. Type genus: *Juanorhinus* Aurivillius, 1926.

Tribe Lepyrini Kirby, 1837

Lepyridae Kirby, 1837: 197 [stem: *Lepyr-*]. Type genus: *Lepyrus* Germar, 1817.

Tribe Lithinini Lacordaire, 1863

Lithinides Lacordaire, 1863: 349 [stem: *Lithin-*]. Type genus: *Lithinus* Klug, 1833.

Subtribe Lithinina Lacordaire, 1863

Lithinides Lacordaire, 1863: 349 [stem: *Lithin-*]. Type genus: *Lithinus* Klug, 1833. Comment: original vernacular name available (Art. 11.7.2): first used in latinized form by Pascoe (1870b: 436, as Lithininae), generally accepted as in Alonso-Zarazaga and Lyal (1999: 204, as Lithinini).

Subtribe Rhytidophloeina Voss, 1963

Rhytidophloeini Voss, 1963: 3 [stem: *Rhytidophloe-*]. Type genus: *Rhytidophloeus* Schönherr, 1842.

Tribe Lymantini Lacordaire, 1865

Lymantides Lacordaire, 1865: 328 [stem: *Lymant-*]. Type genus: *Lymantes* Schönherr, 1837. Comment: original vernacular name available (Art. 11.7.2): first used in latinized form by Heyne and Taschenberg (1907: 234, as Lymantini), generally accepted as in Alonso-Zarazaga and Lyal (1999: 205, as Lymantini).

Ithaurinae Kuschel, 1959b: 509 [stem: *Ithaur-*]. Type genus: *Ithaura* Pascoe, 1871.

Caecossonina Osella, 1980: 369 [stem: *Caecosson-*]. Type genus: *Caecossonus* Gilbert, 1955.

Tribe Mecysolobini Reitter, 1913

Alcidides Jekel, 1865: 547 [stem: *Alcid-*]. Type genus: *Alcides* Schönherr, 1825 [preoccupied genus name, not *Alcides* Hübner, 1822 [Lepidoptera]; syn. of *Sternuchopsis* Heller, 1918]. Comment: permanently invalid (Art. 39): based on preoccupied type genus.

Mecyslobini Reitter, 1913b: 33, in key [stem: *Mecysolob-*]. Type genus: *Mecysolobus* Reitter, 1905 [as *Mecyslobus*, incorrect subsequent spelling of type genus name, not in prevailing usage]. Comment: incorrect original stem formation, not in prevailing usage.

Alcidodinae G. A. K. Marshall, 1939: 582 [stem: *Alcidod-*]. Type genus: *Alcidodes* G. A. K. Marshall, 1939 [syn. of *Sternuchopsis* Heller, 1918]. Comment: replacement name for Alcidinae Jekel, 1865 because of the homonymy of the type genus.

Tribe Metatygini Pascoe, 1888

Metatyginae Pascoe, 1888: 409 [stem: *Metatyg-*]. Type genus: *Metatyges* Pascoe, 1865 [syn. of *Omophorus* Schönherr, 1835].

Omophorinae G. A. K. Marshall, 1917: 195 [stem: *Omophor-*]. Type genus: *Omophorus* Schönherr, 1835.

METATYGINAE Pierce, 1919: 30, in key [stem: *Metatyg-*]. Type genus: *Metatyges* Pascoe, 1865 [syn. of *Omophorus* Schönherr, 1835]. Comment: family-group name proposed as new without reference to METATYGINAE Pascoe, 1888.

STERNECHOSOMINI Voss, 1958: 43, in key [stem: *Sternechosom-*]. Type genus: *Sternechosomus* Voss, 1958.

Tribe MOLYTINI Schönherr, 1823

MOLYTIDES Schönherr, 1823: column 1142 [stem: *Molyt-*]. Type genus: *Molytes* Schönherr, 1823 [syn. of *Liparus* A. G. Olivier, 1807].

Subtribe LEIOSOMATINA Reitter, 1913

LIOSOMINA Reitter, 1913b: 52, in key [stem: *Leiosomat-*]. Type genus: *Leiosoma* Stephens, 1829 [as *Liosoma*, unjustified emendation of type genus name by Agassiz (1846b: 204), not in prevailing usage]. Comment: incorrect original stem formation, not in prevailing usage.

Subtribe MOLYTINA Schönherr, 1823

MOLYTIDES Schönherr, 1823: column 1142 [stem: *Molyt-*]. Type genus: *Molytes* Schönherr, 1823 [syn. of *Liparus* A. G. Olivier, 1807].

LIPARIDES Latreille, 1828: 597 [stem: *Lipar-*]. Type genus: *Liparus* A. G. Olivier, 1807.

Subtribe PLINTHINA Lacordaire, 1863

PLINTHIDES Lacordaire, 1863: 359 [stem: *Plinth-*]. Type genus: *Plinthus* sensu Westwood, 1838 [not *Plinthus* Germar, 1817; syn. of *Mitoplinthus* Reitter, 1897 but see Alonso-Zarazaga and Lyal (2010) and Appendix 6]. Comment: based on a misidentified type genus; an application was recently submitted to the Commission by Alonso-Zarazaga and Lyal (2010) to keep the name PLINTHINA in its usual sense, with *Plinthus* Germar, 1817 as its type genus (see Appendix 6).

MINYOPIDAE Marseul, 1863: 221 [stem: *Minyop-*]. Type genus: *Minyops* Schönherr, 1823.

Subtribe TYPODERINA Voss, 1965

TYPODERINA Voss, 1965: 343, in key [stem: *Typoder-*]. Type genus: *Typoderus* G. A. K. Marshall, 1953.

Tribe NETTARHININI Lacordaire, 1865

NETTARHINIDES Lacordaire, 1865: 76 [stem: *Nettarhin-*]. Type genus: *Nettarhinus* Schönherr, 1826. Comment: original vernacular name available (Art. 11.7.2): first used in latinized form by Hustache (1936: 62, as NETTARHININA), generally accepted as in Alonso-Zarazaga and Lyal (1999: 206, as NETTARHININI).

Tribe PACHOLENINI Lacordaire, 1863

PACHOLÉNIDES Lacordaire, 1863: 443 [stem: *Pacholen-*]. Type genus: *Pacholenus* Schönherr, 1826. Comment: original vernacular name available (Art. 11.7.2): first used in latinized form by Heyne and Taschenberg (1907: 229, as PACHOLENINI), generally accepted as in Alonso-Zarazaga and Lyal (1999: 206, as PACHOLENINI).

Tribe PAIPALESOMINI Marshall, 1932

PAIPALESOMINI G. A. K. Marshall, 1932: 345, in key [stem: *Paipalesom-*]. Type genus: *Paipalesomus* Schönherr, 1847 [syn. of *Peribleptus* Schönherr, 1843].

Tribe PETALOCHILINI Lacordaire, 1863

PÉTALOCHILIDES Lacordaire, 1863: 517 [stem: *Petalochil-*]. Type genus: *Petalochilus* Schönherr, 1836. Comment: original vernacular name available (Art. 11.7.2): first used in latinized form by Pascoe (1870b: 436, as PETALOCHILINAE), generally accepted as in Alonso-Zarazaga and Lyal (1999: 206, as PETALOCHILINI).

ÉPIPÉDIDES Lacordaire, 1865: 186 [stem: *Epiped-*]. Type genus: *Epipedus* Schönherr, 1842 [preoccupied genus name, not *Epipedus* Spinola, 1837 [Hemiptera]; syn. of *Epipedophyes* G. A. K. Marshall, 1946]. Comment: original vernacular name available (Art. 11.7.2): first used in latinized form and generally accepted as in Pascoe (1870b: 437, as EPIPEDINAE); permanently invalid (Art. 39): based on preoccupied type genus.

HORMOPINI J. L. LeConte, 1876: 320 [stem: *Hormop-*]. Type genus: *Hormops* J. L. LeConte, 1876.

SCHOENHERRIELLINAE Viana, 1952: 231 [stem: *Schoenherriell-*]. Type genus: *Schoenherriella* Viana, 1952 [syn. of *Epipedophyes* G. A. K. Marshall, 1946]. Comment: replacement name for ÉPIPÉDIDES Lacordaire, 1865 because of the homonymy of the type genus.

EPIPEDOPHYINAE Kuschel, 1955: 270 [stem: *Epipedophy-*]. Type genus: *Epipedophyes* G. A. K. Marshall, 1946. Comment: replacement name for ÉPIPÉDIDES Lacordaire, 1865 because of the homonymy of the type genus.

Tribe PHOENICOBATINI Champion, 1914

PHOENICOBATINA Champion, 1914: 416 [stem: *Phoenicobat-*]. Type genus: *Phoenicobates* Champion, 1914.

Tribe PHRYNIXINI Kuschel, 1964

PHRYNIXINAE Kuschel, 1964: 472 [stem: *Phrynix-*]. Type genus: *Phrynixus* Pascoe, 1875.

Tribe PISSODINI Gistel, 1848

PISSODISIDAE Gistel, 1848: [7] [stem: *Pissod-*]. Type genus: *Pissodes* Germar, 1817.

Subtribe Cotasteromimina Morimoto, 1962

Cotasteromimini Morimoto, 1962a: 61, in key [stem: *Cotasteromim-*]. Type genus: *Cotasteromimus* Chûjô and Voss, 1960.

Subtribe Orthorhinina Jekel, 1865

Orthorhinides Jekel, 1865: 548 [stem: *Orthorhin-*]. Type genus: *Orthorhinus* Schönherr, 1825.

Subtribe Pissodina Gistel, 1848

Pissodisidae Gistel, 1848: [7] [stem: *Pissod-*]. Type genus: *Pissodes* Germar, 1817. Comment: incorrect original stem formation, not in prevailing usage.

Angianides Sharp, 1919b: 151 [stem: *Angian-*]. Type genus: *Angianus* Sharp, 1919 [syn. of *Vanapa* Pouillaude, 1915].

Tribe Sternechini Lacordaire, 1863

Sternéchides Lacordaire, 1863: 447 [stem: *Sternech-*]. Type genus: *Sternechus* Schönherr, 1826. Comment: original vernacular name available (Art. 11.7.2): first used in latinized form by Champion (1902: 113, as Sternechina), generally accepted as in Alonso-Zarazaga and Lyal (1999: 208, as Sternechini).

Tribe Styanacini Chûjô and Voss, 1960

Styanacinae Chûjô and Voss, 1960: 2 [stem: *Styanac-*]. Type genus: *Styanax* Pascoe, 1871.

Tribe Trachodini Gistel, 1848

Trachodisidae Gistel, 1848: [7] [stem: *Trachod-*]. Type genus: *Trachodes* Germar, 1824. Comment: incorrect original stem formation, not in prevailing usage.

Blastophiladae Gistel, 1856a: 370 [stem: *Blastophil-*]. Type genus: *Blastophila* Gistel, 1856 [syn. of *Trachodes* Germar, 1824]. Comment: incorrect original stem formation, not in prevailing usage.

Acicnémides Lacordaire, 1865: 31 [stem: *Acicnemid-*]. Type genus: *Acicnemis* Fairmaire, 1849. Comment: original vernacular name available (Art. 11.7.2): first used in latinized form by Klima (1935: Acic. 1, as Acicnemidinae), generally accepted as in Alonso-Zarazaga and Lyal (1999: 196, as Acicnemidini); incorrect original stem formation, not in prevailing usage.

Tribe Trigonocolini Lacordaire, 1863

Trigonocolides Lacordaire, 1863: 592 [stem: *Trigonocol-*]. Type genus: *Trigonocolus* Lacordaire, 1863. Comment: original vernacular name available (Art. 11.7.2): first used in latinized form by Pascoe (1870b: 436, as Trigonocolinae), generally accepted as in Alonso-Zarazaga and Lyal (1999: 208, as Trigonocolini).

MEGARHININAE Faust, 1888: 284 [stem: *Megarhin-*]. Type genus: *Megarhinus* Schönherr, 1835 [preoccupied genus name, not *Megarhinus* Rafinesque, 1820 [Pisces]; syn. of *Trigonocolus* Lacordaire, 1863]. Comment: permanently invalid (Art. 39): based on preoccupied type genus.

Tribe TRYPETIDINI Lacordaire, 1865

TRYPÉTIDES Lacordaire, 1865: 177 [stem: *Trypetid-*]. Type genus: *Trypetes* Schönherr, 1836 [placed on the Official List of Generic Names in Zoology (ICZN 1974a)]. Comment: original vernacular name available (Art. 11.7.2): first used in latinized form by Pascoe (1870b: 437, as TRYPETINAE [incorrect stem formation]), generally accepted as in Alonso-Zarazaga and Lyal (1999: 209, as TRYPETIDINI); incorrect original stem formation, not in prevailing usage; stem based on *Trypetes* Schönherr, 1836 ruled to be *Trypetid-* and TRYPETIDINAE placed on the Official List of Family-Group Names in Zoology (ICZN 1974a, as TRYPETIDINAE Pierce, 1919).

TRYPETESIDAE Heller, 1916: 348, in note [stem: *Trypetes-*]. Type genus: *Trypetes* Schönherr, 1836 [placed on the Official List of Generic Names in Zoology (ICZN 1974a)]. Comment: emendation for TRYPETIDAE Lacordaire, 1865 to avoid homonymy with TRYPETIDAE Loew, 1861 (type genus *Trypeta* Meigen, 1803) in Diptera; corrected spelling not adopted since it creates homonymy problem with TRYPETESIDAE Stebbing, 1910 (type genus *Trypetesa* Norman, 1903) in Crustacea.

Subfamily OROBITIDINAE Thomson, 1859

OROBITINA C. G. Thomson, 1859: 138 [stem: *Orobitid-*]. Type genus: *Orobitis* Germar, 1817. Comment: incorrect original stem formation, not in prevailing usage; correction of stem by Alonso-Zarazaga and Lyal (1999: 210).

Subfamily XIPHASPIDINAE Marshall, 1920

XIPHASPIDINAE G. A. K. Marshall, 1920: 393 [stem: *Xiphaspid-*]. Type genus: *Xiphaspis* G. A. K. Marshall, 1920.

Subfamily SCOLYTINAE Latreille, 1804

SCOLITARII Latreille, 1804c: 156 [stem: *Scolyt-*]. Type genus: *Scolytus* Geoffroy, 1762 [placed on the Official List of Generic Names in Zoology (ICZN 1963b)]. Comment: name placed on the Official List of Family-Group Names in Zoology (ICZN 1963b, as SCOLYTIDAE Westwood, 1838).

Tribe AMPHISCOLYTINI Mandelshtam and Beaver, 2003

AMPHISCOLYTINI Mandelshtam and Beaver, 2003: 2 [stem: *Amphiscolyt-*]. Type genus: *Amphiscolytus* Mandelshtam and Beaver, 2003.

Tribe BOTHROSTERNINI Blandford, 1896

BOTHROSTERNINI Blandford, 1896: 120 [stem: *Bothrostern-*]. Type genus: *Bothrosternus* Eichhoff, 1868.

Tribe CACTOPININI Chamberlin, 1939

CACTOPINAE Chamberlin, 1939: 243 [stem: *Cactopin-*]. Type genus: *Cactopinus* Schwartz, 1899. Comment: incorrect original stem formation, not in prevailing usage; correction of stem by Wood (1978: 113).

Tribe CARPHODICTICINI Wood, 1971

CARPHODICTICINI Wood, 1971: 19 [stem: *Carphodictic-*]. Type genus: *Carphodicticus* Wood, 1971.

Tribe COPTONOTINI Chapuis, 1869

COPTONOTIDAE Chapuis, 1869: 11 [stem: *Coptonot-*]. Type genus: *Coptonotus* Chapuis, 1869.

Tribe CORTHYLINI LeConte, 1876

CORTHYLI J. L. LeConte, 1876: 347 [stem: *Corthyl-*]. Type genus: *Corthylus* Erichson, 1836.

Subtribe CORTHYLINA LeConte, 1876

CORTHYLI J. L. LeConte, 1876: 347 [stem: *Corthyl-*]. Type genus: *Corthylus* Erichson, 1836.

AMPHICRANIDAE Eichhoff, 1878: 460 [stem: *Amphicran-*]. Type genus: *Amphicranus* Erichson, 1836.

XYLEBORIPINA Reitter, 1913c: 102 [stem: *Xyleborip-*]. Type genus: *Xyleborips* Reitter, 1913. Comment: incorrectly spelled XYLOBORIPINA on page 31 of the same work.

GNATHOTRICHINA Balachowsky, 1949: 241 [stem: *Gnathotrich-*]. Type genus: *Gnathotrichus* Eichhoff, 1869.

Subtribe PITYOPHTHORINA Eichhoff, 1878

PITYOPHTHORIDAE Eichhoff, 1878: 173 [stem: *Pityophthor-*]. Type genus: *Pityophthorus* Eichhoff, 1864. Comment: First Revisers found (PITYOPHTHORINA Eichhoff, 1878 vs ARAPTINA Eichhoff, 1878) are Wood and Bright (1992: 949).

ARAPTIDAE Eichhoff, 1878: 305 [stem: *Arapt-*]. Type genus: *Araptus* Eichhoff, 1872.

Tribe CRYPHALINI Lindemann, 1877

CRYPHALOIDEAE Lindemann, 1877: 165 [stem: *Cryphal-*]. Type genus: *Cryphalus* Erichson, 1836.

TRYPOPHLOEINAE Nüsslin, 1911: 375 [stem: *Trypophloe-*]. Type genus: *Trypophloeus* Fairmaire, 1868.

ERNOPORINAE Nüsslin, 1911: 375 [stem: *Ernopor-*]. Type genus: *Ernoporus* C. G. Thomson, 1859.

*EIDOPHERINAE Murayama, 1954: 200 [stem: *Eidophel-*]. Type genus: *Eidophelus* Eichhoff, 1875. Comment: unavailable (Art. 13.1): proposed after 1930 without description or bibliographic reference to such a description; incorrect original stem formation, not in prevailing usage; correction of stem by Wood (1978: 114).

Tribe CRYPTURGINI LeConte, 1876

CRYPTURGI J. L. LeConte, 1876: 387 [stem: *Crypturg-*]. Type genus: *Crypturgus* Erichson, 1836.

†Tribe CYLINDROBROTINI Kirejtshuk, Azar, Beaver, Mandelshtam and Nel, 2009

CYLINDROBROTINI Kirejtshuk et al., 2009: 103 [stem: *Cylindrobrot-*]. Type genus: *Cylindrobrotus* Kirejtshuk et al., 2009.

Tribe DIAMERINI Hagedorn, 1909

DIAMERINAE Hagedorn, 1909: 163 [stem: *Diamer-*]. Type genus: *Diamerus* Erichson, 1836.

STROMBOPHORINI Schedl, 1959: 16 [stem: *Strombophor-*]. Type genus: *Strombophorus* Hagedorn, 1909. Comment: name proposed after 1930 without description or bibliographic reference to such a description (Art. 13.1), however available because it was used as valid before 2000 as in Ferreira (1966: 664, as STROMBOPHORINI) and was not rejected by an author who, between 1961 and 1999, applied Article 13 of the then current edition of the Code (see Art. 13.2.1).

SPHAEROTRYPINI Murayama, 1963: 66 [stem: *Sphaerotryp-*]. Type genus: *Sphaerotrypes* Blandford, 1894.

Tribe DRYOCOETINI Lindemann, 1877

DRYOCOETOIDEAE Lindemann, 1877: 165 [stem: *Dryocoet-*]. Type genus: *Dryocoetes* Eichhoff, 1864 [placed on the Official List of Generic Names in Zoology (ICZN 1979b)].

THAMNURGINAE Nüsslin, 1911: 377 [stem: *Thamnurg-*]. Type genus: *Thamnurgus* Eichhoff, 1864.

TAPHRORYCHINI Reitter, 1913c: 29 [stem: *Taphrorych-*]. Type genus: *Taphrorychus* Eichhoff, 1878.

Tribe HEXACOLINI Eichhoff, 1878

CTENOPHORIDAE Chapuis, 1869: 49 [stem: *Ctenophor-*]. Type genus: *Ctenophorus* Chapuis, 1869 [preoccupied genus name, not *Ctenophorus* Fitzinger, 1843 [Reptilia]; syn. of *Scolytodes* Ferrari, 1867]. Comment: permanently invalid (Art. 39): based on preoccupied type genus.

HEXACOLIDAE Eichhoff, 1878: 306 [stem: *Hexacol-*]. Type genus: *Hexacolus* Eichhoff, 1868. Comment: First Revisers found (PROBLECHILINI Eichhoff, 1878 vs HEXACOLINI Eichhoff, 1878) are Alonso-Zarazaga and Lyal (2009: 58).

Problechilidae Eichhoff, 1878: 167 [stem: *Problechil-*]. Type genus: *Problechilus* Eichhoff, 1878.

Erineophilides Hopkins, 1902: 37 [stem: *Erineophil-*]. Type genus: *Erineophilus* Hopkins, 1902.

Tribe Hylastini LeConte, 1876

Hylastes J. L. LeConte, 1876: 387 [stem: *Hylast-*]. Type genus: *Hylastes* Erichson, 1836.

Hylurgopina Balachowsky, 1949: 122 [stem: *Hylurgop-*]. Type genus: *Hylurgops* J. L. LeConte, 1876.

Tribe Hylesinini Erichson, 1836

Hylesinen Erichson, 1836: 46 [stem: *Hylesin-*]. Type genus: *Hylesinus* Fabricius, 1801. Comment: original vernacular name available (Art. 11.7.2): first used in latinized form by Shuckard (1839b: 63, as Hylesinidae), generally accepted as in Alonso-Zarazaga and Lyal (2009: 63, as Hylesinini); incorrect original stem formation, not in prevailing usage.

Phloeotrupides Lacordaire, 1865: 370 [stem: *Phloeotrup-*]. Type genus: *Phloeotrupes* Erichson, 1836. Comment: original vernacular name available (Art. 11.7.2): first used in latinized form and generally accepted as in Chapuis (1869: 11, as Phloeotrupidae).

Phloeobori Blandford, 1893: 426 [stem: *Phloeobor-*]. Type genus: *Phloeoborus* Erichson, 1836.

Dactylipalpi Blandford, 1893: 426 [stem: *Dactylipalp-*]. Type genus: *Dactylipalpus* Chapuis, 1869.

Alniphagini Murayama, 1963: 29 [stem: *Alniphag-*]. Type genus: *Alniphagus* Swaine, 1918.

Tribe Hylurgini Gistel, 1848

Hylurgidae Gistel, 1848: [6] [stem: *Hylurg-*]. Type genus: *Hylurgus* Latreille, 1806. Comment: status resurrected by Alonso-Zarazaga and Lyal (2009: 69).

Dendroctonides Nüsslin, 1912: 278 [stem: *Dendrocton-*]. Type genus: *Dendroctonus* Erichson, 1836 [placed on the Official List of Generic Names in Zoology (ICZN 1963a)].

Xylechinides Nüsslin, 1912: 281 [stem: *Xylechin-*]. Type genus: *Xylechinus* Chapuis, 1869.

Tomicini Wood, 1978: 118, in key [stem: *Tomic-*]. Type genus: *Tomicus* Latreille, 1802 [placed on the Official List of Generic Names in Zoology (ICZN 1963a)]. Comment: the older name Tomicidae Shuckard, 1839 was based on misidentified type genus (see Alonso-Zarazaga and Lyal 2009); an application will be submitted by MAAZ and CHCL to the Commission to suppress Tomicidae Shuckard, 1839 for the Principles of Priority and Homonymy (Art. 65.2.1).

Tribe Hyorrhynchini Hopkins, 1915

Hyorrhynchinae Hopkins, 1915: 225, in key [stem: *Hyorrhynch-*]. Type genus: *Hyorrhynchus* Blandford, 1894.

*Sueinae Murayama, 1959: 26 [stem: *Sue-*]. Type genus: *Sueus* Murayama, 1951. Comment: unavailable (Art. 13.1): proposed after 1930 without description or bibliographic reference to such a description.

Hyorrhynchini Murayama, 1963: 62 [stem: *Hyorrhynch-*]. Type genus: *Hyorrhynchus* Blandford, 1894. Comment: family-group name proposed as new without reference to Hyorrhynchinae Hopkins, 1915.

Sueinae Murayama, 1963: 4 [stem: *Sue-*]. Type genus: *Sueus* Murayama, 1951.

Tribe Hypoborini Nüsslin, 1911

Hypoborinae Nüsslin, 1911: 376 [stem: *Hypobor-*]. Type genus: *Hypoborus* Erichson, 1836.

Chaetophloeini Schedl, 1966: 361 [stem: *Chaetophloe-*]. Type genus: *Chaetophloeus* J. L. LeConte, 1876.

Tribe Ipini Bedel, 1888

Tomicidae Shuckard, 1839b: 64 [stem: *Tomic-*]. Type genus: *Tomicus* sensu Shuckard, 1839 [not *Tomicus* Latreille, 1802; syn. of *Ips* DeGeer, 1775]. Comment: based on a misidentified type genus (see Alonso-Zarazaga and Lyal 2009); see also Tomicini Wood, 1978; an application will be submitted by MAAZ and CHCL to the Commission to suppress Tomicidae Shuckard, 1839 for the Principles of Priority and Homonymy (Art. 65.2.1).

Ipini Bedel, 1888: 386 [stem: *Ip-*]. Type genus: *Ips* DeGeer, 1775. Comment: an application needs to be submitted to the Commission to suppress Ipini Latreille, 1802 (based on the misidentified type genus *Ips* sensu Latreille, 1802) for the Principles of Priority and Homonymy (Art. 65.2.1) to conserve this name as valid.

Pityogenina Balachowsky, 1949: 244 [stem: *Pityogen-*]. Type genus: *Pityogenes* Bedel, 1888.

Tribe Micracidini LeConte, 1876

Micracides J. L. LeConte, 1876: 367 [stem: *Micracid-*]. Type genus: *Micracis* J. L. LeConte, 1868. Comment: incorrect original stem formation, not in prevailing usage; stem corrected by Alonso-Zarazaga and Lyal (2009: 81).

Hylocuridae Eichhoff, 1878: 298 [stem: *Hylocur-*]. Type genus: *Hylocurus* Eichhoff, 1872.

Tribe Phloeosinini Nüsslin, 1912

Phloeosinides Nüsslin, 1912: 289 [stem: *Phloeosin-*]. Type genus: *Phloeosinus* Chapuis, 1869 [placed on the Official List of Generic Names in Zoology (ICZN 1981a)].

*Dendrosinini Nunberg, 1967: 314 [stem: *Dendrosin-*]. Type genus: *Dendrosinus* Chapuis, 1869. Comment: unavailable (Art. 13.1): proposed after 1930 without description or bibliographic reference to such a description.

Tribe Phloeotribini Chapuis, 1869

Phloeotribidae Chapuis, 1869: 42 [stem: *Phloeotrib-*]. Type genus: *Phloeotribus* Latreille, 1797 [placed on the Official List of Generic Names in Zoology (ICZN 1979a)].

Phthorophloeides Nüsslin, 1912: 289 [stem: *Phthorophloe-*]. Type genus: *Phthorophloeus* Rey, 1885.

Tribe Phrixosomatini Wood, 1978

Phrixosomini Wood, 1978: 118, in key [stem: *Phrixosomat-*]. Type genus: *Phrixosoma* Blandford, 1897. Comment: incorrect original stem formation, not in prevailing usage; correction of stem by Alonso-Zarazaga and Lyal (2009: 91).

Tribe Polygraphini Chapuis, 1869

Polygraphidae Chapuis, 1869: 48 [stem: *Polygraph-*]. Type genus: *Polygraphus* Erichson, 1836.

Carphoborinae Nüsslin, 1911: 376 [stem: *Carphobor-*]. Type genus: *Carphoborus* Eichhoff, 1864.

Tribe Premnobiini Browne, 1962

Premnobiini Browne, 1962: 80 [stem: *Premnobi-*]. Type genus: *Premnobius* Eichhoff, 1878. Comment: family-group name available (Art. 13.1.2): description by indication (distinguishing characters given in Browne (1961)).

Tribe Scolytini Latreille, 1804

Scolitarii Latreille, 1804c: 156 [stem: *Scolyt-*]. Type genus: *Scolytus* Geoffroy, 1762 [placed on the Official List of Generic Names in Zoology (ICZN 1963b)]. Comment: name placed on the Official List of Family-Group Names in Zoology (ICZN 1963b, as Scolytidae Westwood, 1838); incorrect original stem formation, not in prevailing usage.

Eccoptogastridae Gistel, 1848: [6] [stem: *Eccoptogastr-*]. Type genus: *Ekkoptogaster* Herbst, 1793 [as *Eccoptogaster*, incorrect subsequent spelling of type genus name by Gyllenhal (1813), not in prevailing usage; *Ekkoptogaster* Herbst, 1793 and *Eccoptogaster* Erichson, 1836 placed on the Official Index of Rejected and Invalid Genus-Group Names in Zoology (ICZN 1963b); syn. of *Scolytus* Geoffroy, 1762]. Comment: permanently invalid (Art. 39): based on suppressed type genus; Eccoptogasterinae Trédl, 1907 and Eccoptogasterini Reitter, 1906 placed on the Index of Rejected and Invalid Family-Group names in Zoology (ICZN 1963b).

CAMPTOCÉRIDES Lacordaire, 1865: 366 [stem: *Camptocer-*]. Type genus: *Camptocerus* Dejean, 1821. Comment: original vernacular name available (Art. 11.7.2): first used in latinized form and generally accepted as in Chapuis (1869: 49, as CAMPTOCERIDAE).

MINULINI Reitter, 1913c: 12 [stem: *Minul-*]. Type genus: *Minulus* Eggers, 1912.

Tribe SCOLYTOPLATYPODINI Blandford, 1893

SCOLYTOPLATYPINI Blandford, 1893: 428 [stem: *Scolytoplatypod-*]. Type genus: *Scolytoplatypus* C. Schaufuss, 1891. Comment: incorrect original stem formation, not in prevailing usage; correction of stem by Blanford (1895: 84).

TAENIOCERINI Blandford, 1893: 428 [stem: *Taeniocer-*]. Type genus: *Taeniocerus* Blandford, 1893 [preoccupied genus name, not *Taeniocerus* Kaup, 1871 [Coleoptera: PASSALIDAE]; syn. of *Scolytoplatypus* C. Schaufuss, 1891]. Comment: permanently invalid (Art. 39): based on preoccupied type genus.

SPONGOCERINAE Hagedorn, 1909: 163 [stem: *Spongocer-*]. Type genus: *Spongocerus* Blandford, 1893.

Tribe XYLEBORINI LeConte, 1876

XYLEBORI J. L. LeConte, 1876: 358 [stem: *Xylebor-*]. Type genus: *Xyleborus* Eichhoff, 1864 [placed on the Official List of Generic Names in Zoology (ICZN 1968a)].

WEBBINAE Hopkins, 1915: 224, in key [stem: *Webbi-*]. Type genus: *Webbia* Hopkins, 1915. Comment: incorrect original stem formation, not in prevailing usage.

*ECCOPTOPTERINA Browne, 1961: 49 [stem: *Eccoptopter-*]. Type genus: *Eccoptopterus* Motschulsky, 1863. Comment: unavailable (Art. 13.1): proposed after 1930 without description or bibliographic reference to such a description.

Tribe XYLOCTONINI Eichhoff, 1878

XYLOCTONIDAE Eichhoff, 1878: 171 [stem: *Xylocton-*]. Type genus: *Xyloctonus* Eichhoff, 1872.

Tribe XYLOTERINI LeConte, 1876

XYLOTERI J. L. LeConte, 1876: 356 [stem: *Xyloter-*]. Type genus: *Xyloterus* Erichson, 1836.

*TRYPODENDRINAE Trédl, 1907: 70 [stem: *Trypodendr-*]. Type genus: *Trypodendron* Stephens, 1830. Comment: family-group name unavailable (Art. 11.7.1.1): not based on a genus used as valid at the time.

TRYPODENDRINA Nunberg, 1954: 16 [stem: *Trypodendr-*]. Type genus: *Trypodendron* Stephens, 1830.

Subfamily PLATYPODINAE Shuckard, 1839

PLATYPODIDAE Shuckard, 1839b: 64 [stem: *Platypod-*]. Type genus: *Platypus* Herbst, 1793.

Tribe Mecopelmini Thompson, 1992

*Mecopelmini Wood, 1966: 45 [stem: *Mecopelm-*]. Type genus: *Mecopelmus* Blackman, 1944. Comment: unavailable (Art. 13.1): proposed after 1930 without description or bibliographic reference to such a description.

Mecopelminae Thompson, 1992: 873, in key [stem: *Mecopelm-*]. Type genus: *Mecopelmus* Blackman, 1944.

Tribe Platypodini Shuckard, 1839

Platypodidae Shuckard, 1839b: 64 [stem: *Platypod-*]. Type genus: *Platypus* Herbst, 1793.

Crossotarsariae H. Strohmeyer, 1914: 19 [stem: *Crossotars-*]. Type genus: *Crossotarsus* Chapuis, 1865.

Tribe Schedlariini Wood and Bright, 1992

Chapuisiides Blandford, 1895: 89 [stem: *Chapuisi-*]. Type genus: *Chapuisia* Dugès, 1886 [preoccupied genus name, not *Chapuisia* Duvivier, 1885 [Coleoptera: Chrysomelidae]; syn. of *Schedlarius* Wood, 1957]. Comment: permanently invalid (Art. 39): based on preoccupied type genus.

Schedlarini Wood and Bright, 1992: 1087 [stem: *Schedlari-*]. Type genus: *Schedlarius* Wood, 1957. Comment: replacement name for Chapuisiides Blandford, 1895 because of the homonymy of the type genus; incorrect original stem formation, not in prevailing usage; correction of stem and transfer to Platypodinae by Alonso-Zarazaga and Lyal (2009: 17).

Tribe Tesserocerini Strohmeyer, 1914

Tesserocerinae H. Strohmeyer, 1914: 19 [stem: *Tesserocer-*]. Type genus: *Tesserocerus* Saunders, 1836. Comment: precedence (Tesserocerini H. Strohmeyer, 1914 vs Diapodini Strohmeyer, 1914) given to taxon originally proposed at the higher rank (Art. 24.1).

Subtribe Diapodina Strohmeyer, 1914

Diapodariae H. Strohmeyer, 1914: 19 [stem: *Diapod-*]. Type genus: *Diapus* Chapuis, 1865.

Genyocerinae Hopkins, 1915: 225, in key [stem: *Genyocer-*]. Type genus: *Genyocerus* Motschulsky, 1858.

Subtribe Tesserocerina Strohmeyer, 1914

Tesserocerinae H. Strohmeyer, 1914: 19 [stem: *Tesserocer-*]. Type genus: *Tesserocerus* Saunders, 1836. Comment: precedence (Tesserocerina H. Strohmeyer, 1914 vs Cenocephalina Strohmeyer, 1914) given to taxon originally proposed at the higher rank (Art. 24.1).

Symmerariae H. Strohmeyer, 1914: 19 [stem: *Symmer-*]. Type genus: *Symmerus* Chapuis, 1865 [preoccupied genus name, not *Symmerus* Walker,

1848 [Diptera]; syn. of *Chaetastus* Nunberg, 1953]. Comment: permanently invalid (Art. 39): based on preoccupied type genus.

CENOCEPHALARIAE H. Strohmeyer, 1914: 19 [stem: *Cenocephal-*]. Type genus: *Cenocephalus* Chapuis, 1865.

PLATYTARSILIDAE Schedl, 1939: 387 [stem: *Platytarsul-*]. Type genus: *Platytarsulus* Schedl, 1935. Comment: incorrect original stem formation, not in prevailing usage; correction of stem by Wood and Bright (1992: 1084).

PERIOMATINI Schedl, 1939: 397, in key [stem: *Periommat-*]. Type genus: *Periommatus* Chapuis, 1865 [as *Periomatus*, incorrect subsequent spelling of type genus name, not in prevailing usage]. Comment: incorrect original stem formation, not in prevailing usage; correction of stem by Schedl (1972: 249).

PLATYPICERINAE Nunberg, 1953: 44 [stem: *Platypicer-*]. Type genus: *Platypicerus* Nunberg, 1953.

Coleoptera *incertae sedis*

HOMOEOPLASTIDAE Gistel, 1856a: 360. Type genus: *Homoeoplastus* Gistel, 1856 [Gistel includes three species in his new genus, none of which are described in his work nor attributed to authors as having been described previously; we therefore consider *Homoeoplastus* unavailable]. Comment: family-group name unavailable (Art. 11.7.1.1): not based on an available genus name.

PLOCASTEIDAE Gistel, 1856a: 365. Type genus: *Plocastes* Gistel, 1856 [the only species originally included in *Plocastes* is *scaraboides*; Gistel does not mention the author of the species name, nor does he describe the taxon, we therefore consider *Plocastes* as unavailable]. Comment: family-group name unavailable (Art. 11.7.1.1): not based on an available genus name.

SERRATOPALPIDAE Gistel, 1856a: 384. Type genus: *Serratopalpus* Gistel, 1856 [*Serratopalpus helwigii* was the only species originally included in *Serratopalpus*; this new species was not described in Gistel's original paper therefore the genus name is considered unavailable]. Comment: family-group name unavailable (Art. 11.7.1.1): not based on an available genus name.

Acknowledgments

We would like to sincerely thank the following people for their assistance with nomenclatural and/or bibliographical issues during this project: R. Aalbu (Tenebrionidae), A. Bennett (Hymenoptera), C. Chaboo (Chrysomelidae), C. Bellamy (Buprestidae), L. Bocák (various), M. Bologna (Meloidae), P. Bouchet (Gastropoda), M. Brancucci (publication dates), M. Buck (Hymenoptera), P. Cate (Elateridae), D. Chandler (Anthicidae), C. Costa (Elateridae), J. Cracraft (Aves), J. Cumming (Diptera), H. Douglas (Elateridae), M. Engel (Hymenoptera), W. Eschmeyer (Pisces), N. Evenhuis (Diptera), C. Ferraris (Pisces), G. Flores (literature), R. Foottit (Hemiptera), G. Gibson (Hymenoptera), B. Gill (Scarabaeoidea), M. Gimmel (Cyclaxyridae), V. Grebennikov (various), V. Gusarov (Staphylinidae), A. Hamilton (Hemiptera), L. Herman (Staphylinidae), M. Ivie (Tenebrionoidea), D. Iwan (Coleoptera), E. Jendek (Buprestoidea), P. Johnson (Elateridae), P. Jolivet (Chrysomelidae), S. Kazantsev (literature), I. Kitching (Lepidoptera), J. Kolibáč (Cleridae), H. Labrique (literature), D. Lafontaine (Lepidoptera), S. Laplante (various), M. LeCroy (Aves), S. Laplante (Coleoptera), A. Legalov (Curculionoidea), L. LeSage (Chrysomelidae), R. Leschen (various), I. Löbl (various), O. Lonsdale (Diptera), E. Matthews (Tenebrionidae), O. Merkl (Tenebrionidae), E. Michel (ICZN Cases), J. Muona (Eucnemidae, Elateridae), J. Nieto Nafría (Hemiptera), R. Oberprieler (Curculionoidea), J. O'Hara (Diptera), W. Opitz (Cleridae), D. Pavićević (Staphylinidae), K. Philips (Ptinidae), J. Pinto (Meloidae), G. Poinar (Curculionoidea), D. Pollock (Tenebrionoidea), R. Pyle (Pisces), D. Reeder (Mammalia), F. Riedel (Mollusca), C. Roper (Cephalopoda), C. Schmidt (Lepidoptera), N. Simmons (Mammalia), J. Skevington (Diptera), A. Smetana (various), W. Steiner (Tenebrionidae), M. Thayer (stems, Staphyliniformia), C. Thompson (Diptera), F. Vaz-de-Mello (Scarabaeidae), T. Virro (Rotifera), R. Wharton (Hymenoptera) and Q. Yu (Nematoda). M. Thayer provided data on the correct stem of Coleoptera genera. S. Gamman and P. Madaire, the library staff at Agriculture and Agri-Food Canada (Ottawa), are thanked for their assistance with acquisition of important literature. A. Newton partial grant support for catalog database construction: United States National Science Foundation grants 8814449 (Field Museum Coleoptera collection inventory), 0118749 (south temperate Staphylinidae catalog) and 0715705 (world Staphylinini catalog).

References

As mentioned in the *Format of the catalogue and conventions used* section, we have tried to find the most accurate date of publication (given in square brackets) for works cited in the manuscript. As outlined under Art. 21 of the Code, the date to be adopted as the date of publication of a work (and of a contained name or nomenclatural act) is to be determined as follows: if a date of publication is specified in a work it is to be adopted as correct in the absence of evidence to the contrary, if a specific date of publication is not mentioned in the work then the date to be adopted is the last day of the month (when month and year, but not day, are specified) or the last day of the year (when only the year is specified). When the date of publication was not specified within the original work or its containing volume, we tried to obtain the time frame of its appearance as follows: a date cited within an article (such as in the Preface or the submitted or approval date, etc.) indicates that the article was published after that date, while dates of receipt at various libraries, societies, bibliographers and reviewers provide the latest possible date of its publication. To conserve space in discussions on dates of publication, we have used only the first three letters of each month and the following abbreviations:

Acad. Sci. France = Académie des Sciences de France (Comptes Rendus Hebdomadaires des Séances)
Allgem. Bibliogr. Deutschl. = Allgemeine Bibliographie für Deutschland
Allgem. Bibliogr. Monatl. Verz. = Allgemeine Bibliographie. Monatliches Verzeichniss der wichtigern neuen Erscheinungen der deutschen und ausländischen Literatur
Allgem. Lit. Zeit. = Allgemeine Literatur-Zeitung
Ann. Mag. Nat. Hist. = The Annals and Magazine of Natural History
Ann. Sci. Phys. Nat. Agr. Ind. Lyon = Annales des Sciences Physiques et Naturelles, d'Agriculture et d'Industrie (Lyon)
Ann. Soc. Agr. Hist. Nat. Arts Utiles Lyon = Annales de la Société d'Agriculture, Histoire Naturelle et Arts Utiles de Lyon
Ann. Soc. Ent. Belg. = Annales de la Société Entomologique de Belgique
Ann. Soc. Ent. Belg., Comptes-Rendus = Comptes-Rendus des Séances de la Société Entomologique de Belgique
Ann. Soc. Ent. France = Annales de la Société Entomologique de France
Ann. Soc. Ent. France: Bull. Ent. = Annales de la Société Entomologique de France, Bulletin Entomologique
Ann. Soc. Ent. France: Bull. Bibliogr. = Annales de la Société Entomologique de France, Bulletin Bibliographique
Ann. Soc. Imp. Agr. Hist. Nat. Arts Utiles Lyon = Annales de la Société d'Agriculture, Histoire Naturelle et Arts Utiles de Lyon
Ann. Soc. Linn. Lyon = Annales de la Société Linnéenne de Lyon
Annuaire Ent. = Annuaire Entomologique (Caen, par A. Fauvel)

Arch. Nat. Hist. = Archives of Natural History
Arch. Naturg. = Archiv für Naturgeschichte
Berl. Ent. Zeitschr. = Berliner Entomologische Zeitschrift
Bibliogr. France = Bibliographie de la France
Bibliogr. Ital. = Bibliografia Italiana
Biblioth. Hist.-nat. = Bibliotheca Historico-naturalis Physico-chemica et Mathematica
Brit. Critic = The British Critic
Bull. Acad. Imp. Sci. St.-Pétersbourg = Bulletin de l'Académie Impériale des Sciences de St.-Pétersbourg
Bull. Bibliogr. Sci. Phys. = Bulletin Bibliographique des Sciences Médicales, Naturelles et Physiques
Bull. Sci. Nat. Géol. = Bulletin des Sciences Naturelles et de Géologie
Bull. Soc. Ent. France = Bulletin de la Société Entomologique de France
Bull. Soc. Imp. Nat. Moscou = Bulletin de la Société Impériale des Naturalistes de Moscou
Bull. Soc. Linn. Norm. = Bulletin de la Société Linnéenne de Normandie
Cat. Ann. Libr. France = Catalogue Anuel de la Librairie Française (Paris, par C. Reinwald)
Compte-Rendu Soc. Roy. Belg. Géogr. = Compte-Rendu des Actes de la Société Royale Belge de Géographie
Deut. Ent. Zeits. = Deutsche Entomologische Zeitschrift
Edinb. Rev. = The Edinburgh Review, or Critical Journal
Ent. Bericht. = Entomologische Berichten uitgegeven door de Nederlandsche Entomologische Vereeniging
Ent. Litteraturbl. = Entomologische Litteraturblätter, Repertorium der neuesten Arbeiten auf dem Gesammtgebiet der Entomologie (by Friedländer)
Ent. Mon. Mag. = The Entomologist's Monthly Magazine
Ent. Mag. = The Entomological Magazine
Ent. Nachr. = Entomologische Nachrichten
Ent. News = Entomological News
Ent. Record = The Entomologist's Record and Journal of Variation
Ent. Zeit. Stettin = Entomologische Zeitung (Stettin)
Foreign Quart. Rev. = The Foreign Quarterly Review
Kongel. Danske Vidensk. Selsk. Skrifter = Det Kongelige Danske Videnskabernes Selskabs Skrifter
Intell.-Blatt Allgem. Lit. = Intelligenz-Blatt der Allgemeine Literatur-Zeitung
Intell.-Blatt Serap. = Intelligenz-Blatt zum Serapeum
Jour. Nat. Hist. = Journal of Natural History
Jour. Proc. Ent. Soc. London = Journal of the Proceedings of the Entomological Society of London
Jour. Gén. Litt. France = Journal Général de la Littérature de France
Leipz. Repert. = Leipziger Repertorium der deutschen und ausländischen Literatur
Literar. Zeit. = Literarische Zeitung

Literar. Centrbl. = Literarisches Centralblatt für Deutschland
Literary Gazette = The Literary Gazette, and Journal of the Belles Lettres
London Rev. = The London Review, and Literary Journal
Mag. Nat. Hist. = The Magazine of Natural History and Journal of Zoology, Botany, Mineralogy, Geology, and Meteorology
Mem. R. Accad. Sci. Torino = Memorie della Reale Accademia delle Scienze di Torino
Mém. Acad. Sci. Belles-lettres Lyon. Cl. Sci. = Mémoires de l'Académie des Sciences, Belles-lettres et Arts de Lyon: Classe des Sciences
Mém. Prés. Acad. Imp. Sci. St. Pétersbourg = Mémoires Présentés à l'Académie Impériale des Sciences de St. Pétersbourg par Divers Savans et Lus dans ses Assemblées
Mém. Soc. Roy. Sci. Liége = Mémoires de la Société Royale des Sciences de Liége
Monthly Cat. U. S. Govt. Publ. = Monthly Catalogue of United States Government Publications
Nederl. Ent. Ver. = Nederlandsche Entomologische Vereeniging
New Zeal. Parl. Debates = New Zealand, Parliamentary Debates
Neues Allgem. Repert. = Neues Allgemeines Repertorium der neuesten in- und ausländischen Literatur
Öfvers. K. Vetensk.-Akad. Förh. = Öfversigt af Kongliga Vetenskaps-Akademiens Förhandlingar
Opusc. Ent. = Opuscules Entomologiques
Petites Nouv. Ent. = Petites Nouvelles Entomologiques
Proc. Acad. Nat. Sci. Philad. = Proceedings of the Academy of Natural Sciences of Philadelphia
Proc. Amer. Philos. Soc. = Proceedings of the American Philosophical Society
Proc. Ent. Soc. London = Proceedings of the Entomological Society of London
Proc.-Verb. Soc. Agr. Hist. Nat. Arts Utiles Lyon = Procès-Verbaux de la Société d'Agriculture, Histoire Naturelle et Arts Utiles de Lyon
Rép. Trav. Soc. Stat. Mars. = Répertoire des Travaux de la Société de Statistique de Marseille
Rev. Ent. = Revue Entomologique (Silbermann)
Rev. d'Ent. = Revue d'Entomologie (Caen)
Rev. Gén. Sci. = Revue Générale des Sciences Pures et Appliquées
Rev. Zool. = Revue Zoologique
Sitzber. Ges. Naturf. Freunde Berlin = Sitzungsberichte der Gesellschaft Naturforschender Freunde zu Berlin
Sitzber. Königl. Bayer. Akad. Wiss. Münch. = Sitzungsberichte der Königl. Bayerischen Akademie der Wissenschaften zu München
Tijdschr. Ent. = Tijdschrift voor Entomologie
Trans. Ent. Soc. London = The Transactions of the Entomological Society of London
Trans. Roy. Soc. New Zeal. = Transactions of the Royal Society of New Zealand
Verh. Naturf. Ver. Brünn = Verhandlungen des Naturforschenden Vereines in Brünn
Verh. Zool.-Bot. Ges. Wien = Verhandlungen der Kaiserlich-Königlichen Zoologisch-Botanischen Gesellschaft in Wien

Wien. Ent. Zeit. = Wiener Entomologische Zeitung
Wien. Ent. Monatschr. = Wiener Entomologische Monatschrift
Zeitschr. Ver. Naturb. = Zeitschrift des Vereines der Naturbeobachter, Wien
Zool. Anz. = Zoologischer Anzeiger
Zool. Anz., Bibliogr. Zool. = Bibliographia Zoologica, diario Zoologischer Anzeiger adnexa
Zool. Jahresber. = Zoologischer Jahresbericht
Zool. Record = The Zoological Record
Zoologist = The Zoologist: a Popular Miscellany of Natural History
BMNH = The Natural History Museum [formerly British Museum (Natural History)], London, United Kingdom
CNC = Canadian National Collection of Insects, Arachnids and Nematodes, Ottawa, Canada
CUL = Cornell University Library, Ithaca, USA
FMNH = Field Museum of Natural History, Chicago, USA
MCZ = Harvard University, Museum of Comparative Zoology, Cambridge, USA
NRCL = National Resources Canada Library (Earth Sciences), Ottawa, Canada
USNM = National Museum of Natural History, [formerly, United States National Museum], Washington D.C., USA

Aalbu RL (2006) 2006, where are we at: assessing the current state of Tenebrionidae systematics on a global scale (Coleoptera: Tenebrionidae). Cahiers Scientifiques. Centre de Conservation et d'Étude des Collections (Lyon) 10: 55–70. [printed Sep 2006 (p. 148)]

Aalbu RL, Triplehorn CA, Campbell JM, Brown KW, Somerby RE, Thomas DB (2002) Family 106. Tenebrionidae Latreille, 1802 [pp. 463–509]. In: Arnett RH, Jr., Thomas MC, Skelley PE, Frank JH (Eds) American beetles. Volume 2. Polyphaga: Scarabaeoidea through Curculionoidea. CRC Press, Boca Raton, xiv + 861 pp. [19 Jun 2002 (CRC website)]

Abdullah M (1964) A revision of the genus *Pilipalpus* (Coleoptera, Anthicidae: Pedilinae). Beiträge zur Entomologie 14: 3–9. [Mar 1964 (fasc. title page)]

Abdullah M (1965) *Protomeloe crowsoni* a new species of a new tribe (Protomeloini) of the blister beetles (Coleoptera, Meloidae), with remarks on a postulated new pheromone (cantharidin). Entomologisk Tidskrift 86 (1): 43–48. [30 Jul 1965 (wrapper)]

Abdullah M (1966) The taxonomic position of the Australian *Anaplopus tuberculatus*, with a proposed new sub-family (Anaplopinae) of the Tenebrionidae, and including remarks on the family status of the Merycidae (Coleoptera). Entomological News 77 (5): 143–147. [mailed 11 May 1966 (verso of vol. title page)]

Abdullah M (1969a) The natural classification of the family Anthicidae with some ecological and ethological observations (Coleoptera). Deutsche Entomologische Zeitschrift (Neue Folge) 16 (4/5): 323–366. [20 Nov 1969]

Abdullah M (1969b) Conspectus of the current classification of Coleoptera with synonymies. Beiträge zur Entomologie 19 (3/6): 683–685.

Abdullah M (1974) Notes on Sphaeriformia Abdullah, particularly Torridincolidae Steffan (Col. Myxophaga). Annales de la Société Entomologique de France (Nouvelle Série) 10: 959–962.

Abdullah M, Abdullah A (1968) The taxonomic position of *Lagrioida* with a proposed new tribe of the Eurygeniinae (Col., Anthicidae). The Entomologist's Monthly Magazine 104: 73–74. [13 Nov 1968]

Abdullah M, Quadri N-N (1968) The Micropeplidae and Staphylinidae (I. Steninae, Euaesthetinae and Oxyporinae), Coleoptera of Pakistan with descriptions of a new tribe, genus and three species from Karachi. Pakistan Journal of Scientific and Industrial Research 11 (3): 303–312. [Jul 1968 fasc.]

Abeille de Perrin E (1890) Malachiidae. Malachides d'Europe et pays voisins. Annales de la Société Entomologique de France (6) 10 (2): 181–260, 3 pls. [26 Nov 1890 (wrapper); continued in (3) and (4), both 1891]

Abeille de Perrin E (1891) Malachiidae. Malachides d'Europe et pays voisins [continued]. Annales de la Société Entomologique de France (6) 10 (3): 331–420, (4): 567–680. [25 Feb 1891 (fasc. 3 wrapper); Jun 1891 (fasc. 4 wrapper)]

Absolon K (1913) Über *Antrophilon primitivum* nov. gen. n. sp., eine blinde Bathysciine (Coleoptera cavernicola Silphidae) aus dem südillyrischen Faunengebiete. (Aus der zoologischen Abteilung am Landesmuseum in Brünn.). Coleopterologische Rundschau 2 (6/7): 100–109.

Achard J (1914) Un Scaphidiide nouveau de Birmanie (Col.). Bulletin de la Société Entomologique de France 1914: 394–396. [8 July 1914 (Séance)]

Achard J (1924) Essai d'une subdivision nouvelle de la famille des Scaphidiidae. Annales de la Société Entomologique de Belgique 64: 25–31.

Acloque A (1896) Faune de France contenant la description de toutes les espèces indigènes disposées en tableaux analytiques et illustrée de figures représentant les types caractéristiques des genres et des sous-genres. Avec une préface par Edmond Perrier. Coléoptères. J.-B. Baillière et Fils, Paris, 466 pp. [8 Jan 1896 (Bull. Soc. Ent. France 1896: 20); 11 Apr 1896 (Bibliogr. France 1896: 209)]

Ádám L (1987) Staphylinidae of the Kiskunság National Park (Coleoptera) [pp. 126–168]. In: Mahunka S (Ed) The fauna of the Kiskunság National Park, Vol 2. Akadémiai Kiadó, Budapest.

Ádám L (1994) A check-list of the Hungarian Scarabaeoidea with the description of ten new taxa (Coleoptera). Folia Entomologica Hungarica (Series nova) 55: 5–17. [30 May 1994 (title page)]

Ádám L (1996) A check-list of the Hungarian caraboid beetles (Coleoptera). Folia Entomologica Hungarica 57: 5–64.

Ádám L (2001) [systematic content] In: A sátoraljaújhelyi Kazinczy Ferenc Múzeum füzetei V, Sátoraljaújhely. ANAdatok a Zempléni-hegység, a Hernád-völgy, a Bodrogköz, a Rétköz és a Taktaköz holyvafaunájához (Coleoptera). In: Ádám L, Hegyessy G (Eds) A sátoraljaújhelyi Kazinczy Ferenc Múzeum füzetei V, Sátoraljaújhely, 249 pp. [in Hungarian, with English summary]. [15 Nov 2001]

Ádám L (2003) Faunisztikai adatok a Karpat-medencebö l (Coleoptera: Scarabaeoidea). Folia Historico Naturalia Musei Matraensis 27: 101–136.

Adlbauer K, Danilevsky ML, Hubweber L, Löbl I, Morati J, Rapuzzi P, Sama G, Smetana A, Weigel A (2010) Family Cerambycidae (excluding subfamily Prioninae, subfamily Apatophyseinae, and tribe Dorcadionini) [pp. 84–86, 95–241, 264–334]. In: Löbl I, Smetana A (Eds) Catalogue of Palaearctic Coleoptera. Volume 6. Chrysomeloidea. Apollo Books, Stenstrup, 924 pp. [publ. 22 Feb 2010 (verso of title page)]

Agassiz L (1846a) Nomenclator Zoologicus. Fasciculus XI. Continens Coleoptera. Nomina systematica. Generum Coleopterorum, tam viventium quam fossilium, secundum ordinem alphabeticum disposita, adjectis auctoribus, libris in quibus reperiuntur, anno editionis, etymologia et familiis ad quas pertinent. Jent & Gassmann, Soloduri [= Solothurn, Switzerland], xii + 170 pp.

Agassiz L (1846b) Nomenclatoris zoologici index universalis, continens nomina systematica classium, ordinum, familiarum et generum animalium omnium, tam viventium quam fossilium, secundum ordinem alphabeticum unicum disposita, adjectis homonymiis plantarum, nec non variis adnotationibus et emendationibus. Jent & Gassmann, Soloduri [= Solothurn, Switzerland], viii + 393 pp. [1846 (title page); Dec 1845 (date of preface); 29 Dec 1846 (Evenhuis 1997a: 50)]

Agnarsson I, Kunter M (2007) Taxonomy in a changing world: seeking solutions for a science in crisis. Systematic Biology 56(3): 531–539. [Jun 2007]

Ahmad R (1973) A new tribe of the family Coccinellidae (Coleoptera). Bulletin of Entomological Research 62 (3): 449–452. [publ. 26 Feb 1973 (vol. contents)]

Alexeev AV (1994) Jurassic and Lower Cretaceous Buprestidae (Coleoptera) from Eurasia. Paleontological Journal 27 (1A): 9–34. [publ. Jan 1994 (wrapper); 1993 (on paper)]

Alluaud C (1916) Contributions à l'étude des carabiques d'Afrique et de Madagascar [Col.] XII. Observations sur le genre *Melanodes* Chaud. et descriptions de deux espèces nouvelles. Bulletin de la Société Entomologique de France 1916 (15): 226–230. [publ. 13 Nov 1916 (p. [351])]

Alluaud C (1922) Coléoptères. Cicindélides et carabides. Voyage de M le Baron Maurice de Rothschild en Ethiopie et en Afrique orientale anglaise (1904–1905) Résultats scientifiques Animaux articulés 2. Laboratoire d'entomologie (Muséum national d'histoire naturelle), Paris, 483–519.

Alluaud C (1930a) Note sur les scaritides de Madagascar. Afra. Cahiers d'Entomologie 1: 10–23.

Alluaud C (1930b) Étude sur le groupe des sphodrochléniens (Col. Carabidae - Chlaeniitae - Rhopalomelini). Revue de Zoologie et de Botanique Africaines 19: 105–122.

Alluaud C (1934) Mission J. de Lépiney au Soudan Français 1933–34 (Sixième note). Bulletin de la Société des Sciences Naturelles du Maroc 14 (1/3): 30–31. [31 Mar 1934 (vol. title page); printed 15 Oct 1934 (verso) 6 Nov (rec'd at BMNH)]

Alonso-Zarazaga MA (1989) Revision of the supraspecific taxa in the Palaearctic Apionidae Schoenherr, 1823. 1. Introduction and subfamily Nanophyinae Seidlitz, 1891 (Coleoptera, Curculionoidea). Fragmenta Entomologica 21: 205–262.

Alonso-Zarazaga MA (1990) Revision of the supraspecific taxa in the Palaearctic Apionidae Schoenherr, 1823 (Coleoptera, Curculionoidea). 2. Subfamily Apioninae Schoenherr, 1823: introduction, keys and descriptions. Graellsia 46: 19–156. [publ. 31 Dec 1990, date misprinted as 1991 on some pages]

Alonso-Zarazaga MA (1992) Rhadinocybini trib. n. in the Apioninae Schoenherr, 1823 (Coleoptera, Apionidae). Graellsia 48: 193. [31 Dec 1992 (Alonso-Zarazaga and Lyal 1999: 212)]

Alonso-Zarazaga MA (2005) Diagnosis preliminares de nuevos táxones de Curculionidae (Coleoptera). Boletín de la Sociedad Entomológica Aragonesa 37: 89–93. [31 Oct 2005 (wrapper)]

Alonso-Zarazaga MA (2007) Comment on the proposed fixation of the feminine gender of the genus and the form of derivation of family-group names based on *Trachys* Fabricius, 1801 (Insecta, Coleoptera) [case 3335]. Bulletin of Zoological Nomenclature 64 (3): 187. [Sep 2007 issue]

Alonso-Zarazaga MA, Lyal CHC (1999) A world catalogue of families and genera of Curculionoidea (Insecta: Coleoptera) (excepting Scolytidae and Platypodidae). Entomopraxis, S. C. P., Barcelona, 315 pp. [publ. 27 Dec 1999 (verso of title page)]

Alonso-Zarazaga MA, Lyal CHC (2002) Addenda and corrigenda to 'A world catalogue of families and genera of Curculionoidea (Insecta: Coleoptera)'. Zootaxa 63: 1–37. [22 Aug 2002]

Alonso-Zarazaga MA, Lyal CHC (2006) [new taxa] In: Lyal CHC, Alonso-Zarazaga MA: Addenda and corrigenda to 'A world catalogue of families and genera of Curculionoidea (Insecta: Coleoptera)'. 2. Zootaxa 1202: 21–31. [publ. 15 May 2006 (footnote p. 21)]

Alonso-Zarazaga MA, Lyal CHC (2009) A catalogue of family and genus group names in Scolytinae and Platypodinae with nomenclatural remarks (Coleoptera: Curculionidae). Zootaxa No. 2258: 1–134. [publ. 8 Oct 2009 (article title page)]

Alonso-Zarazaga MA, Lyal CHC (2010) Case 3530. Plinthini Lacordaire, 1863 (Insecta, Coleoptera): proposed conservation and *Plinthus* Germar, 1817: proposed conservation by designation of *Curculio megerlei* Panzer, [1803] as the type species. Bulletin of Zoological Nomenclature 67 (4): 292–299.

Alonso-Zarazaga MA, Lyal CHC, Bartolozzi L, Sforzi A (1999) [new taxa] In: Alonso-Zarazaga MA, Lyal CHC: A world catalogue of families and genera of Curculionoidea (Insecta: Coleoptera) (excepting Scolytidae and Platypodidae). Entomopraxis, S. C. P., Barcelona, 315 pp. [27 Dec 1999 (verso of title page)]

Alonso-Zarazaga MA, Mansilla-Castrillo O (1988) Clave artificial de las familias ibero-baleares y macaronésicas del Orden Coleoptera L., 1758. Claves para la identificación de la Fauna española, 20. Cátedra de Entomología, Facultad de Biología, Universidad Complutense, Madrid. 67 pp.

Anderson WH (1949) Larval description and transfer of *Thaumaphrastus karanisensis* from Colydiidae to a new subfamily of Dermestidae (Coleoptera). Bulletin of the Brooklyn Entomological Society 44: 121–127. [Oct 1949]

Andreae H (1961) Chapter II. Coleoptera: Cossyphodidae. In: Hanström B, Brinck P, Rudebeck G (Eds) South African animal life Results of the Lund University Expedition in 1950–1951. Vol. VIII. Almqvist & Wiksell, Stockholm, 198–216.

Andrewes HE (1929) Coleoptera. Carabidae. Vol. I. Carabinae. In: Stephenson J (Ed) The fauna of British India, including Ceylon and Burma. Taylor & Francis, London, xviii + 431pp. + 9pls. [before 24 Jul 1929 (Bull. Soc. Ent. France 1929: 222)]

Andrewes HE (1933) On the types of Oriental Carabidae described by V. de Motschulsky. (Part II). The Transactions of the Royal Entomological Society of London 81: 1–19. [30 Jun 1933]

Anonymous (1924) In: Imperial Bureau of Entomology (Ed) XI. Insecta. Zoological Record [1923]. Pp. 1–347.

Antoine M (1941) Notes d'entomologie marocaine. XXXII. Les Litoborinae du Maroc (Col. Teneb.). Bulletin de la Société des Sciences Naturelles du Maroc 21: 19–52.

Antoine M (1949) Notes d'entomologie marocaine. XLIV. Matériaux pour l'étude des Helopinae du Maroc (Col. Tenebrionides). Bulletin de la Société des Sciences Naturelles du Maroc 25–27 [1945–1947]: 123–162.

Antoine M (1959) Coleopteres Carabiques du Maroc. Troisième partie. Mémoires de la Société des Sciences Naturelles et Physiques du Maroc (Nouvelle Série, Zoologie) 6: 315–465.

Anton K-W (2010) Family Chrysomelidae. Subfamily Bruchinae Latreille, 1802 [pp. 339–353]. In: Löbl I, Smetana A (Eds) Catalogue of Palaearctic Coleoptera. Volume 6. Chrysomeloidea. Apollo Books, Stenstrup, 924 pp. [publ. 22 Feb 2010 (verso of title page)]

Ardoin P (1958) Contribution à l'étude des ténébrionides malgaches. Le genre *Enicmosoma* Gebien. Bulletin de l'Académie Malgache 35 [1957]: 59–77, pls. 1–3.

Ardoin P (1961) Contribution à l'étude des ténébrionides malgaches. Deux nouveaux genres d'Adeliini malgaches. Bulletin de l'Académie Malgache 37 [1959]: 31–38. [1961 (Zool. Record)]

Ardoin P (1962) Essai de révision des Amarygmini africains (Première partie). Bulletin de l'Institut Français d'Afrique Noire 24 (4): 955–1029. [28 Dec 1982 (recorded at CNC)]

Arnett RH, Jr. (1961) Part II Suborder Myxophaga, suborder Polyphaga (part) series Staphyliniformia (part) Hydrophiloidea, Staphylinoidea [pp. (2 unn. +) 211–368]. In: The beetles of the United States (a manual for identification) [original loose-leaf edition]. Catholic University of America Press, Washington D.C., xii + 1112 pp. [+ unn. pp.]. [issued 31 May 1961 (verso of title page for Part III)]

Arnett RH, Jr. (1962a) Part IV Suborder Polyphaga (cont.) series Elateriformia (concl.) Cantharoidea, Series Bostrychiformia Dermestoidea, Bostrychoidea, Series Cucujiformia Cleroidea, Lymexylonoidea, Meloidea [pp. (2 unn. +) 527–644]. In: The beetles of the United States (a manual for identification) [original loose-leaf edition]. Catholic University of America Press, Washington D.C., xii + 1112 pp. [+ unn. pp.]. [issued 8 Jun 1962 (verso of title page for Part V)]

Arnett RH, Jr. (1962b) Part V Suborder Polyphaga (cont.) series Cucujiformia (cont.) Tenebrionoidea, Cucujoidea [pp. (2 unn. +) 645–850]. In: The beetles of the United States (a manual for identification) [original loose-leaf edition]. Catholic University of America

Press, Washington D.C., xii + 1112 pp. [+ unn. pp.]. [issued 20 Aug 1962 (verso of title page for Part VI)]

Arnett RH, Jr. (1962c) Part VI Suborder Polyphaga (concl.) series Cucujiformia (concl.) Chrysomeloidea, Curculionoidea [pp. (2 unn. +) 851–1048]. In: The beetles of the United States (a manual for identification) [original loose-leaf edition]. Catholic University of America Press, Washington D.C., xii + 1112 pp. [+ unn. pp.]. [issued 27 Dec 1962 (footnote p. 1049; evidently the General Index was issued in 1963)]

Arnett RH, Jr. (1967) The systematic position of *Melanactes* and *Pseudomelanactes* (Coleoptera, Elateridae). Entomological News 78 (4): 110–111. [mailed 4 Apr 1967 (verso of vol. title page)]

Arnett RH, Jr., Thomas MC (Eds) (2000) American beetles. Volume 1. Archostemata, Myxophaga, Adephaga, Polyphaga: Staphyliniformia. CRC Press, Boca Raton, xvi + 443 pp. [28 Dec 2000 (CRC Press)]

Arnett RH, Jr., Thomas MC, Skelley PE, Frank JH (Eds) (2002) American beetles. Volume 2. Polyphaga: Scarabaeoidea through Curculionoidea. CRC Press, Boca Raton, xiv + 861 pp.

Arnoldi LV (1977) Rhynchophora and Family Eobelidae [pp. 142–175]. In: Arnoldi LV, Zherikhin VV, Nikritin LM, Ponomarenko AG (Eds): Mezozoyskie Zhestkokrylye. Trudy Paleontologicheskogo Instituta Akademiya Nauk SSSR 161: 1–204. [English translation: Mesozoic Coleoptera; Smithsonian Institution Libraries and N.S.F., Washington D.C.; xii + 285 pp., 1992] [after 5 Sep 1977 (approved to print)]

Arrow GJ (1904) Sound-production in the lamellicorn beetles. The Transactions of the Entomological Society of London 52 (4): 709–750, pl. 26. [23 Dec 1904 (verso of vol. title page)]

Arrow GJ (1911) Upon the Dynamopinae, a new subfamily of Lamellicorn beetles. The Annals and Magazine of Natural History (8) 7 (42): 610–612. [1 Jun 1911 (Evenhuis 2003: 39)]

Arrow GJ (1917) Coleoptera, Lamellicornia Part II. (Rutelinae, Desmonycinae, and Euchirinae). In: Shipley AE, Marshall GAK (Eds) The fauna of British India, including Ceylon and Burma. Taylor and Francis, London, xiii + 387 pp. + 5 pls.

Arrow GJ (1925) Coleoptera. Clavicornia. Erotylidae, Languriidae, and Endomychidae. In: Shipley AE, Scott H (Eds) The fauna of British India, including Ceylon and Burma. Taylor & Francis, London, xvi + 416 pp. + 1 pl., 1 map.

Arrow GJ (1929) A new genus of silphid Coleoptera from Persia. Zoologischer Anzeiger 82: 96–99. [10 Jun 1929 (recorded at BMNH)]

Arrow GJ (1931) Coleoptera Lamellicornia Part III. (Coprinae). In: Stephenson J (Ed) The fauna of British India, including Ceylon and Burma. Taylor and Francis, London, xii + 428 pp. + 10 pls. [Dec 1931 (title page)]

Arrow GJ (1937) Pars 156: Scarabaeidae: Dynastinae. In: Schenkling S (Ed) Coleopterorum Catalogus. Volumen XXI. Scarabaeidae III. W. Junk, Berlin, 124 pp. [16 Aug 1937 (verso of vol. title page)]

Arrow GJ (1940) A nomenclatural note (Coleopt.). Proceedings of the Royal Entomological Society of London (Series B. Taxonomy) 9 (1): 16. [Jan 1940]

Arrow GJ (1950) Coleoptera Lamellicornia Lucanidae and Passalidae Vol. IV. In: Sewell RBS (Ed) The fauna of India including Pakistan, Ceylon, Burma and Malaya. Taylor & Francis,

London, xi + 274 + [1] pp. + 23 pls. [1949 on title page but ad from April 1950 bound in back]

Askevold IS (1990) Reconstructed phylogeny and reclassification of the genera of Donaciinae (Coleoptera: Chrysomelidae). Quaestiones Entomologicae 26: 601–664.

Assing V (2000) A taxonomic and phylogenetic revision of Maorothiini trib. n. from the New Zealand subregion (Coleoptera: Staphylinidae, Staphylininae). Beiträge zur Entomologie 50 (1): 3–64. [11 Apr 2000 (top of article)]

Aubé C (1836) [Pp. 1–64] In: Iconographie et histoire naturelle des Coléoptères d'Europe. Tome cinquième. Hydrocanthares. Méquignon-Marvis Père et Fils, Paris, xi + 415 pp. [Nov / early Dec 1836 (Guignot 1932: 548)]

Audinet-Serville JG (1832) Nouvelle classification de la famille des longicornes. Annales de la Société Entomologique de France 1 (2): 118–201.

Audinet-Serville JG (1834) Nouvelle classification de la famille des longicornes (suite). Annales de la Société Entomologique de France 3 (1): 5–110.

Audinet-Serville JG (1835) Nouvelle classification de la famille des longicornes (suite). Annales de la Société Entomologique de France 4: 5–100, 197–228.

Aurivillius C (1886a) Conspectus generum et specierum Brachyceridarum. Öfversigt af slägten och arter inom familjen Brachyceridae bland Curculioniderna. Öfversigt af Kongliga Vetenskaps-Akademiens Förhandlingar 42 [1885] (7): 5–24. [1886 (wrapper; p. 161); 20 Apr 1886 (recorded at BMNH)]

Aurivillius C (1886b) Ett nytt egendomligt slägte bland Curculioniderna. Entomologisk Tidskrift 7 (2): 95–97.

Aurivillius C (1910) Coleoptera. Curculionidae. In: Sjöstedt Y (Ed) Wissenschaftliche Ergebnisse der Schwedischen Zoologischen Expedition nach dem Kilimandjaro, dem Meru und den umgebenden Massaisteppen Deutsch-Ostafrikas 1905–1906. Vol: 1 Part: 7. P. Palmquists Aktiebolag, Stockholm, 403–435.

Aurivillius C (1911a) Neue oder wenig bekannte Coleoptera Longicornia. 11. Arkiv för Zoologi 7 (3): 1–44. [25 Jan 1911 (verso of vol. title page)]

Aurivillius C (1911b) Neue oder wenig bekannte Coleoptera Longicornia. 12. Arkiv för Zoologi 7 (19): 1–41. [8 Dec 1911 (Musgrave 1932: 8); publ. 23 Dec 1911 (verso of vol. title page)]

Aurivillius C (1912) Pars 39: Cerambycidae: Cerambycinae. In: Schenkling S (Ed) Coleopterorum Catalogus. Volumen XXII. Cerambycidae I. W. Junk, Berlin, 574 pp. [26 Feb 1912 (wrapper); 28 Jan 1912 (contents)]

Aurivillius C (1917) Results of Dr. E. Mjöberg's Swedish Scientific Expeditions to Australia 1910–1913. 12. Cerambycidae. Arkiv för Zoologi 10 (23): 1–50, 3 pls. [28 Mar 1917 (verso of vol. title page)]

Aurivillius C (1922a) Pars 73: Cerambycidae. Lamiinae I [pp. 1–322]. In: Schenkling S (Ed) Coleopterorum Catalogus. Volumen XXIII. Cerambycidae II. W. Junk, Berlin, 704 pp. [1921 (title page); 15 Jan 1922 (wrapper)]

Aurivillius C (1922b) Neue oder wenig bekannte Coleoptera Longicornia. Arkiv för Zoologi 14 (18): 1–32. [26 Jul 1922 (verso of vol. title page)]

Aurivillius C (1922c) Neue Cerambyciden aus der Sammlung G. van Roon. Tijdschrift voor Entomologie 65: 160–173. [31 Dec 1922]

Aurivillius C (1923) Pars 74: Cerambycidae: Lamiinae II [pp. 323–704]. In: Schenkling S (Ed) Coleopterorum Catalogus. Volumen XXIII. Cerambycidae II. W. Junk, Berlin, 704 pp. [2 Jul 1923 (wrapper)]

Aurivillius C (1925) Neue oder wenig bekannte Coleoptera Longicornia. 20. Arkiv för Zoologi 17 A (12): 1–21. [23 Sep 1925 (verso of vol. title page)]

Aurivillius C (1926a) Neue oder wenig bekannte Coleoptera Longicornia. 21. Arkiv för Zoologi 18 A (9): 1–22. [printed 17 Nov 1925 (p. 21), but publ. 5 Jun 1926 (verso of vol. title page); as 1926 in ZooRec]

Aurivillius C (1926b) Sammlungen der Schwedischen Elgon-Expedition im Jahre 1920. 8. Curculioniden. Arkiv för Zoologi 18 A (23): 1–34. [printed 9 Jun 1926 (p. 34) but publ. 13 Oct 1926 (verso of vol. title page)]

Aurivillius C (1931) Coleoptera-Curculionidae von Juan Fernandez und der Oster-Insel [pp. 461–478, pls. 15–16]. In: Skottsberg C (Ed) The Natural History of Juan Fernandez and Easter Island. Vol. III Zoology [1921–1940]. Almqvist & Wiksell, Uppsala, [2] + 688 pp. + 20 pls. [printed 2 Feb 1926 (p. 478); as 1931 (Zool. Record; Neave 1939)]

Baehr M (1982) *Atranopsis* n. gen., eine neue Laufkäfer-Gattung der Agonini aus Syrien (Insecta: Coleoptera: Carabidae). Senckenbergiana Biologica 62: 261–266.

Báguena Corella L (1948) Estudio sobre los Aderidae (Coleópteros Heterómeros). Xylophildae, Hylophilidae sive Euglenidae. Primer ensayo para una nueva ordenación de la familia y estudio de las especies africanas, particularmente de las de los Territorios Españoles del Golfo de Guinea. Instituto de Estudios Africanos, Madrid, xv + 547 + [1] pp.

Balachowsky A (1949) Coléoptères Scolytides. Faune de France. Vol. 50. 320 pp. [30 Jun 1949; 2e trimestre 1949 (dépôt légal, p. 320)]

Balkenohl M (2003) Subfamily Scaritinae Bonelli, 1810 [pp. 219–234]. In: Löbl I, Smetana A (Eds) Catalogue of Palaearctic Coleoptera. Volume 1. Archostemata -Myxophaga - Adephaga. Apollo Books, Stenstrup, 819 pp. [publ. 30 May 2003 (verso of title page)]

Ball GE (1979) Conspectus of carabid classification: history, holomorphology, and higher taxa [pp. 63–111]. In: Erwin TL, Ball GE, Whitehead DR, Halpern AL (Eds) Carabid beetles: their evolution, natural history, and classification. Proceedings of the First International Symposium of Carabidology, Smithsonian Institution, Washington, D C, August 21, 23, and 25, 1976. Dr. W. Junk bv Publishers, The Hague, x + 644 pp. [1979 (copyright)]

Ball GE (1983) Evaluation of the Baron Maximilien de Chaudoir's contribution to classification of cymindine Lebiini and Masoreimorphi (Coleoptera: Carabidae). The Coleopterists Bulletin 36 [1982] (3): 513–530. [mailed 25 Apr 1983 (wrapper)]

Ball GE (1992) The tribe Licinini (Coleoptera: Carabidae): a review of the genus-groups and of the species of selected genera. Journal of the New York Entomological Society 100 (2): 325–380. [mailed 1 Apr 1992 (inside wrapper)]

Ball GE, Shpeley D (2002) Ginemini, *Ginema thomasi*, new tribe, new genus and new species, from Amazonian Bolivia (Coleoptera: Carabidae: Harpalinae). Transactions of the American Entomological Society 128 (1): 75–98. [Mar 2002 issue; Apr 2002 (recorded at CNC)]

Balthasar V (1961) Zwei neue Gattungen der Familie Scarabaeidae (Col.). (107. Beitrag zur Kenntnis der Scarabaeidae.). Entomologische Arbeiten aus dem Museum G. Frey 12 (1): 174–181. [publ. 1 Apr 1961 (vol. Inhalt)]

Baly JS (1863) An attempt at a classification of the Eumolpidae. The Journal of Entomology, Descriptive and Geographical 2 (10): 143–163. [Nov 1863 issue]

Baly JS (1865) Attempt at a classification of the Eumolpidae [continued.]. The Journal of Entomology, Descriptive and Geographical 2 (13): 433–442. [Nov 1865 issue]

Bänninger M (1927) Die Ozaenini (Col. Carab.). 10. Beitrag zur Kenntnis der Carabinae. Deutsche Entomologische Zeitschrift 1927 (3): 177–216 [incl. Nachtrag]. [1 Dec 1927 (p. iii)]

Bänninger M (1938) Monographie der Subtribus Scaritina (Col. Carab.). I, II. Deutsche Entomologische Zeitschrift 1937 (3/4): 81–160. [Feb 1938 (contents)]

Barber HS (1913) The remarkable life-history of a new family (Micromalthidae) of beetles. Proceedings of the Biological Society of Washington 26 (47): 185–190, pl. 4. [8 Aug 1913 (top p. 183)]

Barber HS (1924) New Ptiliidae related to the smallest known beetle. Proceedings of the Entomological Society of Washington 26 (6): 167–178, pls. 7–8. [30 Jun 1924 (verso of vol. title page)]

Basilewsky P (1946) Monographie du genre *Tefflus* Latreille (Col., Carabidae, Panagaeinae). Étude morphologique, systématique et géographique. Novitates Entomologicae Supplementum 5: 1–56.

Basilewsky P (1950a) Descriptions de deux coléoptères myrmécophiles nouveaux du Katanga représentants d'une famille encore inconnue en Afrique centrale (Col. Cossyphodidae). Revue de Zoologie et de Botanique Africaines 43: 182–187.

Basilewsky P (1950b) Études sur les chlaéniens d'Afrique (Col. Carabidae). Bulletin & Annales de la Société Entomologique de Belgique 86 (1/2): 40–54. [5 Mar 1950 (wrapper)]

Basilewsky P (1951a) *Leleupidia luvubuana*, nov. gen. et nov. sp. (Col. Carabidae). Revue de Zoologie et de Botanique Africaines 44: 175–179.

Basilewsky P (1951b) Révision générale des Harpalinae d'Afrique et de Madagascar (Coleoptera Carabidae). Deuxième partie. Annales du Musée Royal du Congo Belge (Série in 8°: Sciences Zoologiques) 9: 1–333 + 6 pls. [25 Sep 1951 (recorded at BMNH)]

Basilewsky P (1953a) Carabidae (Coleoptera Adephaga). Exploration du Parc national de l'Upemba. Fascicle 10. Institut des Parcs Nationaux du Congo Belge, Bruxelles, 252 pp. + 10 pls.

Basilewsky P (1953b) Révision des Leleupidiini (Col. Carabidae Zuphiinae). Revue de Zoologie et de Botanique Africaines 47: 263–281.

Basilewsky P (1963) Révision des Promecognathinae d'Afrique (Coleoptera, Carabidae). Annals of the Transvaal Museum 24 (4): 305-319. [issued Aug 1963 (wrapper)]

Basilewsky P (1973a) Insectes coléoptères. Carabidae Scaritinae. Faune de Madagascar 37: 1–322.

Basilewsky P (1973b) Description d'un coléoptère Carabidae nouveau du sud-ouest africain, représentant d'une sous-famille inédite. Bulletin & Annales de la Societe Royale Belge d'Entomologie 109 (6/12): 224–231. [31 Dec 1973 (wrapper)]

Basilewsky P (1980) Les Reicheiina de l'Afrique du Sud (Coleoptera: Carabidae). Entomologia Generalis 6 (2/4): 293–302. [publ. 28 Nov 1980 (Contents p. v)]

Basilewsky P (1984) Essai d'une classification supragénérique naturelle des carabides lébiens d'Afrique et de Madagascar (Coleoptera Carabidae Lebiinae). Revue de Zoologie Africaine 98 (3): 525–559. [28 Sep 1984 (article header); 29 Sep (wrapper)]

Basilewsky P (1985) Insectes coléoptères. Carabidae Platyninae. Faune de Madagascar 64: 1–543.

Basilewsky P, Grundmann E (1955) Contributions à l'étude systématique des chléniens (Coleoptera Carabidae). Bulletin & Annales de la Société Royale d'Entomologie de Belgique 91 (7/8): 199–206. [31 Aug 1955 (wrapper)]

Bates F (1873) Descriptions of new genera and species of Tenebrionidae from Australia, New Caledonia and Norfolk Island. The Transactions of the Entomological Society of London (4) 1873 (3): 347–380. [5 Aug 1873 (G. Wheeler 1912)]

Bates HW (1863) Contributions to an insect fauna of the Amazon Valley. Coleoptera: Longicornes. The Annals and Magazine of Natural History (3) 12 (68): 100–109. [1 Aug 1863 (Evenhuis 2003: 23)]

Bates HW (1869) Contributions to an insect fauna of the Amazon Valley (Coleoptera, Prionides). The Transactions of the Royal Entomological Society of London 1869: 37–58. [Apr 1869 (footer p. 37)]

Bates HW (1870) Contributions to an insect fauna of the Amazon Valley (Coleoptera, Cerambycidae). The Transactions of the Entomological Society of London (4) 1870 (3): 243–335, (4): 391–444. [(3) 29 Aug 1870; (4) 20 Dec 1870 (G. Wheeler 1912)]

Bates HW (1871) Notes on Carabidae, and descriptions of new species (No. 5). The Entomologist's Monthly Magazine 8 [1871–72]: 29–34.

Bates HW (1872a) On the longicorn Coleoptera of Chontales, Nicaragua. The Transactions of the Entomological Society of London (4) 1872 (3): 163–238. [16 Aug 1872 (G. Wheeler 1912)]

Bates HW (1872b) Notes on Heteromera, and descriptions of new genera and species. The Entomologist's Monthly Magazine 9 (10): 97–99. [before 5 Oct 1872 (Ann. Soc. Ent. Belg. 15 Comptes-Rendus: cx)]

Bates HW (1873) On the geodephagous Coleoptera of Japan. The Transactions of the Entomological Society of London (4) 1873 (2): 219–322. [20 May 1873 (G. Wheeler 1912: 757)]

Bates HW (1874a) Supplement to the longicorn Coleoptera of Chontales, Nicaragua. The Transactions of the Entomological Society of London (4) 1874 (2): 219–235. [27 Apr 1874 (G. Wheeler 1912)]

Bates HW (1874b) On the geodephagous Coleoptera of New Zealand. The Annals and Magazine of Natural History (4) 13 (75): 233–246, (76): 270–277. [1 Mar, 1 Apr 1874 (Evenhuis 2003: 26)]

Bates HW (1875) New genera and species of Prionidae (longicorn Coleoptera). The Entomologist's Monthly Magazine 12 [1875–76]: 47–53. [Jul 1875 (page top)]

Bates HW (1882) Adephaga [part, pp. 41–152, pls. iii-v]. In: Godman FD, Salvin O (Eds) Biologia Centrali-Americana. Insecta. Coleoptera. Vol. I. Part 1. Taylor and Francis, London, x + 316 pp. + 13 pls. [Feb 1882 (pp. 41–72), Apr (73–88), Jun (89–112), Aug (113–136), Oct 1882 (137–152) (signature footers)]

Bates HW (1883) Adephaga [part, pp. 153–256, pls. vi-xii]. In: Godman FD, Salvin O (Eds) Biologia Centrali-Americana. Insecta. Coleoptera. Vol. I. Part 1. Taylor and Francis, London, x + 316 pp. + 13 pls. [Jan 1883 (pp. 153–168), Mar (169–216), Sep (217–240), Dec 1883 (241–256) (signature footers)]

Bates HW (1884) Longicorn beetles of Japan. Additions, chiefly from the later collections of Mr. George Lewis; and notes on the synonymy, distribution, and habits of the previously known species. The Journal of the Linnean Society Zoology 18 (106): 205–262, pls. 1–2. [19 Dec 1884 (verso of vol. title page)]

Bates HW (1885) Longicornia. Supplement [part, pp. 249-436, pls. 17-25]. In: Godman FD, Salvin O (Eds) Biologia Centrali-Americana. Insecta. Coleoptera. Vol. V. Taylor and Francis, London, xii + 525 pp. + 26 pls. [Jan 1885 (pp. 249-272), Mar (273-312), Apr (313-328), May (329-352), Jul (353-384), Aug (385-408), Oct 1885 (409-436) (signature footers)]

Bates HW (1886) On the geodephagous Coleoptera collected by Mr. George Lewis in Ceylon. The Annals and Magazine of Natural History (5) 17 (97): 68-81, (98): 143-156, (99): 199-212. [1 Jan, 1 Feb, 1 Mar 1886 (Evenhuis 2003: 30)]

Bates HW (1887) Lamellicornia [part, pp. 25-160, pls. 2-8]. In: Godman FD, Salvin O (Eds) Biologia Centrali-Americana. Insecta. Coleoptera. Vol. II. Part 2. Pecticornia and Lamellicornia. Taylor and Francis, London, xii + 432 pp. + 24 pls. [May 1887 (pp. 25-64), Jun (65-80), Aug (81-88), Sep (89-104), Oct (105-120), Nov (121-136), Dec 1887 (137-160) (signature footers)]

Bates HW (1888) Rutelidae [pp. 216-296, pls. 12-15], Dynastidae [part, pp. 296-336, pls. 16-17]. In: Godman FD, Salvin O (Eds) Biologia Centrali-Americana. Insecta. Coleoptera. Vol. II. Part 2. Pectinicornia and Lamellicornia. Taylor and Francis, London, xii + 432 pp. + 24 pls. [Apr 1888 (p. 216), Jun (217-248), Aug (249-296), Oct (297-312), Dec 1888 (313-336) (signature footers)]

Bates HW (1891a) Addenda + Introduction [pp. 1–6] + Coleoptera [pp. 7–39]. In: Supplementary appendix to travels amongst the Great Andes of the Equator by Edward Whymper. J. Murray, London, xxii + [1, Addenda] + 147 pp.

Bates HW (1891b) List of the Carabidae (ord. Coleoptera) obtained by Père Cardon in Chota-Nagpore. Bulletin ou Comptes-Rendus des Séances de la Société Entomologique de Belgique 1891: cccxxiv-cccxxxix.

Bates HW (1892) Viaggio di Leonardo Fea in Birmania e regioni vicine. XLIV. List of the Carabidae. Annali del Museo Civico di Storia Naturale di Genova (2) 12 [32]: 267–428, pl. 4. [1892 (title page; Proc. Ent. Soc. London 1892: xxiii)]

Baudet-Lafarge J (1809) Essai sur l'entomologie du Département du Puy-de-Dôme. Monographie des Lamelli-antennes. Landriot, Clermont, 86 pp.

Baudi di Selve F (1875) Europaea et circummediterraneae Faunae Tenebrionidum specierum, quae Comes Dejean in suo Catalogo, editio 3ª consignavit, ex ejusdem collectione in R. Taurinensi Musaeo asservata, cum auctorum hodierne recepta determinatione collatio. Deutsche Entomologische Zeitschrift 19 (1): 17–120. [Jan 1875 (Inhalt p. 3)]

Bechyné J (1948) Notes sur les chrysomélides de l'Amérique du Sud (Col.). Revista de Entomologia (Rio de Janeiro) 19 (1/2): 295–312. [1 Jun 1948 (bottom p. iv)]
Bechyné J (1949) Liste provisoire des eumolpides de la République Argentine et observations diverses sur les eumolpides de l'Amérique du Sud (Col., Chrysomeloidea). Acta Zoologica Lilloana 8: 457–535. [printed 31 Dec 1949 (p. 609)]
Bechyné J (1950a) Notes sur les Chrysomélides de l'Amérique du Sud. II. (Col.). Revista de Entomologia (Rio de Janeiro) 21: 115–156.
Bechyné J (1950b) Eumolpides nouveaux de Madagascar (Col. Phytophaga). Bulletin Mensuel de la Société Linnéenne de Lyon 19 (7): 148–151. [Sep 1950 (top of reprint)]
Bechyné J (1951a) Contribution à la connaissance des eumolpides Malgaches. Bulletin Mensuel de la Société Linnéenne de Lyon 20: 91–94. [Apr 1951]
Bechyné J (1951b) Liste provisoire des eumolpides de Bolivie et observations diverses sur les espèces de l'Amérique du Sud (Col. Phytophaga). Entomologische Arbeiten aus dem Museum G. Frey 2 (2): 227–352. [publ. 31 Dec 1951 (vol. Inhalt)]
Bechyné J (1953) Chrysomelidae (Coleoptera Phytophaga). Exploration du Parc National Albert Mission G F de Witte (1933–1935). Fasc. 79 (5). Institut des Parcs Nationaux du Congo Belge Bruxelles, pp. 83–86.
Bechyné J (1957) Contribution à l'étude des Chrysomeloidea des Iles Mascareignes. I. Eumolpidae. The Mauritius Institute Bulletin 5 (1): 7–21.
Bechyné J (1958) Sur la position systématique de deux Chrysomélides (Col. Phytophaga). Revue Française d'Entomologie 25: 218–220.
Bechyné J (1968) Contribution à la faune du Congo (Brazzaville). Mission A. Villiers et A. Descarpentries. LXXXI. Coléoptères Alticidae. Bulletin de l'Institut Fondamental d'Afrique Noire (Série A: Sciences Naturelles) 30 (4): 1687–1728. [Oct 1968 issue]
Bechyné J (1980) El jeannelismo y la evolución: concepto de las leyes orgánicas sin excepción. Grafindustrial Aragua, Maracay, [10] + 181pp.
Bechyné J (1997) Evaluación de los datos sobre los Phytophaga dañinos en Venezuela (Coleoptera). Parte I. [edited by V. Savini]. Boletin de Entomologia Venezolana Serie Monografias 1: [1] + i-viii, 1–278. [printed Oct 1997 (endleaf)]
Bechyné J, Špringlová de Bechyné B (1960) Beiträge zur Kenntnis der Salvadorenischen Chrysomeloidea (Col. Phytoph.). Pesquisas (Zoologia) 4 (No. 6): 5–73. [submitted post 20 Jul 1960 (p. 70)]
Bechyné J, Špringlová de Bechyné B (1966) Evidenz der bisher bekannten Phenrica-Aitca. (Col. Phytophaga, Alticidae). Entomologisk Tidskrift 87 (3/4): 142–170. [printed 31 Dec 1966 (verso of title page)]
Bechyné J, Špringlová de Bechyné B (1969) La posición sistematica de Megascelis Chevrolat (Col. Phytophaga). Revista de la Facultad de Agronomía, Universidad Central de Venezuela 5 (3): 65–76. [Dec 1969 (date of reprint)]
Bechyné J, Špringlová de Bechyné B (1973) Notas sobre algunos phytophaga de origen paleantartico (Coleoptera). Revista Chilena de Entomología 7: 25–30. [printed 15 Jul 1973; recorded by CNC 23 Dec]

Bechyné J, Špringlová de Bechyné B (1975) Notas sobre la serie filetica de *Monomacra* y sus formas convergentes (Col.: Phytophaga: Alticidae). Revista de la Facultad de Agronomía, Universidad Central de Venezuela 8 (4): 25–140. [Dec 1975 (reprint)]

Bedel L (1879) [Livraison 2, pp. 33–80]. In: Faune des Coléoptères du bassin de la Seine. Vol. 1. Annales de la Société Entomologique de France (5) 9 (publication hors série): 360 pp., 1 pl. [27 Aug 1879]

Bedel L (1881) [pp. i-xxiv, 257–360]. In: Faune des Coléoptères du bassin de la Seine. Vol. 1. Annales de la Société Entomologique de France (6) 1 (publication hors série): 360 pp., 1 pl.

Bedel L (1882) Faune des Coléoptères du bassin de la Seine. Tome VI. Rhynchophora. Annales de la Société Entomologique de France (6) 2 (3, publication hors série): 1–16 (the entire work has 442 pp. + [2] + 1 pl.). [Nov 1882 (Alonso-Zarazaga and Lyal 1999: 213); 1882 (footer p. 1)]

Bedel L (1883) Faune des Coléoptères du bassin de la Seine. Tome VI. Rhynchophora [cont.]. Annales de la Société Entomologique de France (6) 2 (4, publication hors série): 17–32 + pl. 1. [May 1883 (Alonso-Zarazaga and Lyal 1999: 212); 1883 (footer p. 17)]

Bedel L (1884a) Faune des Coléoptères du bassin de la Seine. Tome VI. Rhynchophora [cont.]. Annales de la Société Entomologique de France (6) 3 (3, publication hors série): 65–80. [Dec 1883 (Alonso-Zarazaga and Lyal 1999: 212); 1884 (footer p. 65); 31 Jan 1884 (Table des Matières of 3ième trim.)]

Bedel L (1884b) Faune des Coléoptères du bassin de la Seine. Tome VI. Rhynchophora [cont.]. Annales de la Société Entomologique de France (6) 4 (publication hors série): 129–144. [Oct 1884 (feuille 9); 1884 (footer p. 129)]

Bedel L (1886) Faune des Coléoptères du bassin de la Seine. Tome VI. Rhynchophora [cont.]. Annales de la Société Entomologique de France (6) 6 (3, publication hors série): 249–280. [30 Dec 1886 (Alonso-Zarazaga and Lyal 1999: 212); 1886 (footer p. 249, 265)]

Bedel L (1888) Faune des Coléoptères du bassin de la Seine. Tome VI. Rhynchophora [cont.]. Annales de la Société Entomologique de France (6) 7 ([publication hors série]): 385–442. [pp. 385-end publ. in 1888]

Bedel L (1891) Faune des Coléoptères du bassin de la Seine. Tome V. Annales de la Société Entomologique de France 60 ([publication hors série]): 105–136 (the entire work has 423 pp.). [1891 (footnote p. 105)]

Bedel L (1900) Exploration scientifique de la Tunisie. Catalogue raisonné des Coléoptères de Tunisie. Première partie. Cicindelidae-Staphylinidae. Paris, xiv + 130 pp.

Bedel L (1901) Faune des Coléoptères du bassin de la Seine. Tome V. Annales de la Société Entomologique de France 70 ([publication hors série]): 309–423. [1901 (footer p. 309, etc.)]

Bedel L (1910) Sur l'*Ophionea Chaudoiri* Bohem. et sur quelques genres du groupe des *Colliuris* De Geer [Col. Carabidae]. Bulletin de la Société Entomologique de France 1910 (4): 71–72. [publ. 7 Mar 1910 (p. [433])]

Bedel L (1921) Faune des coléoptères du bassin de la Seine. Vol. 4. Fasc. 2. Sous-ordre Serricornia (Sternoxia). Annales de la Société Entomologique de France 90 ([publication hors série]): 165–362.

Beenen R (2010) subfamily Galerucinae Latreille, 1802 [pp. 443–491]. In: Löbl I, Smetana A (Eds) Catalogue of Palaearctic Coleoptera. Volume 6. Chrysomeloidea. Apollo Books, Stenstrup, 924 pp. [publ. 22 Feb 2010 (verso of title page)]

Bell RT (1998) Where do the Rhysodini (Coleoptera) belong? [pp. 261–272]. In: Ball GE, Casale A, Vigna Taglianti A (Eds) Phylogeny and classification of Caraboidea. (Coleoptera, Adephaga). Proceedings of a symposium, XX International Congress of Entomology. Atti / Museo Regionale de Scienze Naturali, Torino, 543 pp.

Bell RT (2003) Family Rhysodidae Laporte, 1840 [p. 78]. In: Löbl I, Smetana A (Eds) Catalogue of Palaearctic Coleoptera. Volume 1. Archostemata - Myxophaga - Adephaga. Apollo Books, Stenstrup, 819 pp. [publ. 30 May 2003 (verso of title page)]

Bell RT, Bell JR (1978) Rhysodini of the world part I. A new classification of the tribe, and a synopsis of *Omoglymmius* subgenus *Nitiglymmius*, new subgenus (Coleoptera: Carabidae or Rhysodidae). Quaestiones Entomologicae 14: 43–88.

Bell RT, Bell JR (1987) A new subtribe, genus and species of Rhysodini from South Africa (Coleoptera: Carabidae or Rhysodidae). Journal of the Entomological Society of Southern Africa 50: 287–290.

Bell RT, Bell JR (1991) The Rhysodini of Australia (Insecta: Coleoptera: Carabidae or Rhysodidae). Annals of the Carnegie Museum 60: 179–210.

Bellamy CL (1987) [new taxa] In: Bellamy CL, d'Hotman D, Holm E: A review of the Afrotropical Buprestinae with the description of a new tribe, genera and species (Coleoptera: Buprestidae). Journal of the Entomological Society of Southern Africa 50: 217–239.

Bellamy CL (1995) A new subtribe for the monotypic Southern African genus *Bulis* Laporte and Gory (Coleoptera: Buprestidae). Annals of the Transvaal Museum 36 (11): 171–175. [Mar 1995 (footer p. 171)]

Bellamy CL (1996) Further consideration of the subtribe Thomassetiina Bellamy: a new species, new records and placement in the contemporary classification (Coleoptera: Buprestidae). Annals of the Transvaal Museum 36 (16): 215–222. [Mar 1996 (footer p. 222)]

Bellamy CL (2002) Coleoptera Buprestoidea [pp. 1–485]. In: Houston WWK (Ed) Zoological Catalogue of Australia. Vol. 29.5. CSIRO Publishing, Melbourne, xii + 491 pp. + 4 pls. [copyright 2002]

Bellamy CL (2003) An illustrated summary of the higher classification of the superfamily Buprestoidea (Coleoptera). Folia Heyrovskyana Supplementum 10: 1–197.

Bellamy CL (2006) A new subtribe for *Trichinorhipis* Barr, 1948 (Coleoptera: Buprestidae). The Pan-Pacific Entomologist 82 (2): 139–143. [24 Sep 2006 (inside wrapper)]

Bellamy CL (2008a) A world catalogue and bibliography of the jewel beetles (Coleoptera: Buprestoidea). Volume 1. Introduction; fossil taxa; Schizopodidae; Buprestidae: Julodinae - Chrysochroinae: Poecilonotini. Pensoft, Sofia, pp. 1–625. [printed May 2008 (verso of title page)]

Bellamy CL (2008b) A world catalogue and bibliography of the jewel beetles (Coleoptera: Buprestoidea). Volume 2. Chrysochroinae: Sphenopterini through Buprestinae: Stigmoderini Pensoft, Sofia, pp. 631–1260. [printed Jun 2008 (verso of title page)]

Bellamy CL (2008c) A world catalogue and bibliography of the jewel beetles (Coleoptera: Buprestoidea). Volume 3. Buprestinae: Pterobothrini through Agrilinae: Rhaeboscelina. Pensoft, Sofia, pp. 1263–1931. [printed Aug 2008 (verso of title page)]

Bellamy CL (2008d) A world catalogue and bibliography of the jewel beetles (Coleoptera: Buprestoidea). Volume 4. Agrilinae: Agrilina through Trachyini. Pensoft, Sofia, pp. 1935–2684. [printed Sep 2008 (verso of title page)]

Bellamy CL (2009) A world catalogue and bibliography of the jewel beetles (Coleoptera: Buprestoidea). Volume 5. Appendices, bibliography, indices. Pensoft, Sofia, pp. 2689–3264. [printed Feb 2009 (verso of title page)]

Bellés X (1985) Sistemática, filogenia y biogeografía de la subfamilia Gibbiinae (Coleoptera, Ptinidae). Treballs del Museu de Zoologia Supplement No. 3: 3–94.

Benesh B (1955a) Some notes on Neotropical stagbeetles (Coleoptera: Lucanidae). Entomological News 66: 97–104. [Apr issue; mailed 31 Mar 1955 (verso of vol. contents)]

Benesh B (1955b) Some further notes on the stagbeetles, with especial reference to Figulinae (Coleoptera: Lucanidae). Transactions of the American Entomological Society 81 (2): 59–76. [7 Sep 1955 (inside back wrapper)]

Benesh B (1960) Coleopterorum Catalogus (Supplementum). Editio secunda. Pars 8. Lucanidae. W. Junk, The Hague, 178 pp.

Benick L (1920) Über einige brasilianische Aulacotrachelinen und Steninen (Col., Staph.). Översikt av Finska Vetenskaps-Societetens Förhandlingar 62 (A) (4): 1–6. [1920 (reprint; Neave 1939: 359, etc.); 1921 (fasc. wrapper)]

Bernhauer M (1908) Beiträge zur Kenntnis der paläarktischen Staphyliniden-Fauna. I. Münchener Koleopterologische Zeitschrift 3 (3): 320–335. [23 Mar 1908 (wrapper)]

Bernhauer M (1910) Beitrag zur Kenntnis der Staphyliniden-Fauna von Zentralamerika. Verhandlungen der Kaiserlich-Königlichen Zoologisch-Botanischen Gesellschaft in Wien 60 (7/8): 350–393. [20 Oct 1910 (vol. title page)]

Bernhauer M (1911) Zur Staphylinidenfauna Ostindiens und der Sundainseln (3. Beitrag). Entomologische Blätter 7 (2/3, 4): 55–62, 86–93.

Bernhauer M (1915) Zur Staphylinidenfauna des indo-malayischen Gebietes, insbesonders des Himalaya (10. Beitrag). Coleopterologische Rundschau 4 (3): 49–60. [before 6 Mar 1915]

Bernhauer M (1931) Eine neue Subtribus der Quediini (Staphylinidae). Koleopterologische Rundschau 17 84.

Bernhauer M, Scheerpeltz O (1926) Pars 82: Staphylinidae VI [pp. 499–988]. In: Schenkling S (Ed) Coleopterorum Catalogus. Volumen V. Staphylinidae. W. Junk, Berlin, 988 pp. [4 May 1926 (verso of vol. title page)]

Bernhauer M, Schubert K (1910) Pars 19: Staphylinidae I [pp. 1–86]. In: Schenkling S (Ed) Coleopterorum Catalogus. Volumen V. Staphylinidae. W. Junk, Berlin, 988 pp. [30 Aug 1910 (verso of vol. title page)]

Bernhauer M, Schubert K (1911) Pars 29: Staphylinidae II [pp. 87–190]. In: Schenkling S (Ed) Coleopterorum Catalogus. Volumen V. Staphylinidae. W. Junk, Berlin, 988 pp. [10 Apr 1911 (verso of vol. title page)]

Bernhauer M, Schubert K (1912) Pars 40: Staphylinidae III [pp. 191–288]. In: Schenkling S (Ed) Coleopterorum Catalogus. Volumen V. Staphylinidae. W. Junk, Berlin, 988 pp. [28 Mar 1912 (verso of vol. title page)]

Bertkau P (1875) Bericht über die wissenschaftlichen Leistungen auf dem Gebiete der Entomologie während der Jahre 1873 und 1874. Archiv für Naturgeschichte 40 (2): 253–400. [1874 (fasc. title page); cites articles publ. in 1875]

Bertkau P (1878) Bericht über die wissenschaftlichen Leistungen im Gebiete der Arthropoden während die Jahre 1877–78. Archiv für Naturgeschichte 44 (2): 219–562.

Bertkau P (1890) Bericht über die wissenschaftlichen Leistungen im Gebiete der Entomologie während des Jahres 1879. Archiv für Naturgeschichte 56 (Band 2 Heft 2): 1–318. [publ. Aug 1890 (Band 2 Inhalt p. iv)]

Bertkau P (1891) Bericht über die wissenschaftlichen Leistungen im Gebiete der Entomologie während des Jahres 1890. Archiv für Naturgeschichte 57 (Band 2 Heft 2): 1–343. [publ. Dec 1891 (Band 2 Inhalt p. iv)]

Bertrand H (1954) Les Insectes aquatiques d'Europe (genres: larves, nymphes, imagos), Vol. 1. Encyclopédie Entomologique 30. P. Lechevalier, Paris, 556 pp. [17 Apr 1954]

Bertrand H (1967) Clé de détermination des genres de larves d'hydrophilides holarctiques. Cahiers des Naturalistes: Bulletin des Naturalistes Parisiens (Nouvelle Série) 22: 85–88.

Besuchet C (1986) Synonymes et homonyme nouveaux de quelques genres de Psélaphides (Coleoptera). Revue Suisse de Zoologie 93: 257–264.

Besuchet C (1991) Révolution chez les Clavigerinae (Coleoptera, Pselaphidae). Revue Suisse de Zoologie 98: 499–515.

Beutel RG, Leschen RAB (2005) Part 38. Coleoptera, beetles. Volume 1: Morphology and Systematics (Archostemata, Adephaga, Myxophaga, Polyphaga partim). In: Kristensen NP, Beutel RG (Eds) Handbook of zoology A natural history of the phyla of the animal kingdom Volume IV - Arthropoda: Insecta. Walter De Gruyter, Berlin, xi + 567 pp.

Bezděk J (2003) Studies on Asiatic *Apophylia*. Part 1: new synonyms, lectotype designations, redescriptions, descriptions of new species and notes (Coleoptera: Chrysomelidae: Galerucinae). Genus 14: 69–102. [15 Apr 2003]

Bezděk J (2010a) New acts and comments. Chrysomelidae: Galerucinae [p. 73]. In: Löbl I, Smetana A (Eds) Catalogue of Palaearctic Coleoptera. Volume 6. Chrysomeloidea. Apollo Books, Stenstrup, 924 pp. [publ. 22 Feb 2010 (verso of title page)]

Bezděk J (2010b) New acts and comments [p. 76]. In: Löbl I, Smetana A (Eds) Catalogue of Palaearctic Coleoptera. Volume 6. Chrysomeloidea. Apollo Books, Stenstrup, 924 pp. [publ. 22 Feb 2010 (verso of title page)]

Bezděk J (2010c) subfamily Galerucinae Latreille, 1802 [pp. 443–491]. In: Löbl I, Smetana A (Eds) Catalogue of Palaearctic Coleoptera. Volume 6. Chrysomeloidea. Apollo Books, Stenstrup, 924 pp. [publ. 22 Feb 2010 (verso of title page)]

Bickhardt H (1913) H. Sauter's Formosa-Ausbeute. Histeridae II. (Col.). (16. Beitrag zur Kenntnis der Histeriden). Entomologische Mitteilungen 2 (6): 166–177. [1 Jun 1913 (wrapper); 12 Jun (recorded at USNM)]

Bickhardt H (1914) Das System der Histeriden (Vorläufige Mitteilung) (22. Beitrag zur Kenntnis der Histeriden). Entomologische Blätter 10 (9/12): 305–308. [31 Oct 1914]

Bierig A (1934) Géneros y especies nuevas de estafilínidos (Col.) Cubanos. Memorias de la Sociedad Cubana de Historia Natural Felipe Poey 8: 213–223, pl. 11. [25 Oct 1934 (page headers)]

Bierig A (1939) *Litozoon* y *Xenaster* (Col.), 2 géneros nuevos de Staphylinidae. Revista Chilena de Historia Natural 42: 176–180.

Bierig A (1943) Algunos Estaphylinidae (Col.) nuevos de Costa Rica. Revista Chilena de Historia Natural 45: 154–163.

Billberg GJ (1820a) Enumeratio Insectorum in Museo Gust. Joh. Billberg. Gadel, Stockholm, 138 pp. [«31 Dec 1820» (Evenhuis 1997a: 93)]

Billberg GJ (1820b) Novae insectorum species, descriptae. Mémoires de l'Académie Impériale des Sciences de St-Pétersbourg 7 [1815–1816]: 381–395. [1820 (title page); Feb 1820 (art. written p. 559)]

Bílý S (1974) A revision of the genus *Coomaniella* Bourgoin of the Coomaniellini tribe nov. (Coleoptera, Buprestidae). Acta Entomologica Bohemoslovaca 71 (1): 30–41 + pl. 1. [22 Feb 1974 (contents)]

Bílý S (2000) A new concept of Anthaxiini (Coleoptera: Buprestidae). Folia Heyrovskyana 8: 109–114.

Bílý S, Kubáň V, Volkovitsh MG (2009) A study on the tribe Poecilonotini, with a revision of the subtribe Nesotrinchina subtrib. n. and the description of a new genus and species from Papua New Guinea (Coleoptera: Buprestidae: Chrysochroinae). Acta Entomologica Musei Nationalis Pragae 49 (2): 729–767. [15 Dec 2009 (top of article + wrapper)]

Bílý S, Volkovitsh MG (1996) Revision, reclassification and larval morphology of the genus *Paratassa* (Coleoptera: Buprestidae: Paratassini tribus n.). Acta Societatis Zoologicae Bohemicae 60: 325–346. [27 Dec 1996 (issue title page)]

Biström O, Nilsson AN, Wewalka G (1997) A systematic review of the tribes Hyphydrini Sharp and Pachydrini n. trib. (Coleoptera, Dytiscidae). Entomologica Fennica 8 (2): 57–82. [8 Jul 1987 (top of p. 57)]

Blackburn T (1892) Notes on Australian Coleoptera, with descriptions of new species. The Proceedings of the Linnean Society of New South Wales (2) 7 [1893]: 65–151. [1893 (vol. title page); issued 1 Sep 1892 (p. iii)]

Blackwelder RE (1942) Notes on the classification of the staphylinid beetles of the groups Lispini and Osoriinae. Proceedings of the United States National Museum 92: 75–90. [publ. 7 Apr 1942 (p. iii)]

Blackwelder RE (1943) Monograph of the West Indian beetles of the family Staphylinidae. Bulletin of the United States National Museum 182: viii + 658 pp. [Feb 1943 (Monthly Cat. U. S. Govt. Publ. No. 577: 212)]

Blackwelder RE (1944) Checklist of the coleopterous insects of Mexico, Central America, the West Indies, and South America. Part 1. Bulletin of the United States National Museum 185: pp. i-xii + 1–188. [7 Mar 1944 (verso of title page, 1982 ed.)]

Blackwelder RE (1945) Checklist of the coleopterous insects of Mexico, Central America, the West Indies, and South America. Part 3. Bulletin of the United States National Museum 185: 343–550. [21 May 1945 (verso of title page, 1982 ed.)]

Blackwelder RE (1949) Bibliographia. Studies on the dates of publication of books on Coleoptera. The Coleopterists' Bulletin 3: 42–46, 76, 92–94.

Blackwelder RE (1957) Checklist of the coleopterous insects of Mexico, Central America, the West Indies, and South America. Bulletin of the United States National Museum No. 185 (part 6): i-vi, 927–1492. [publ. 15 May 1957 (verso of vol. Contents)]

Blair KG (1920) Notes on the coleopterous genus *Ischalia* Pascoe (fam. Pyrochroidae), with descriptions of two new species from the Philippine Islands. The Entomologist's Monthly Magazine 56: 133–135. [21 Jun 1920 (recorded at USNM)]

Blaisdell FE (1906) New Californian Coleoptera. Entomological News and Proceedings of the Entomological Section of the Academy of Natural Sciences of Philadelphia 17 (3): 71–80. [Mar 1906 issue]

Blaisdell FE (1909) A monographic revision of the Coleoptera belonging to the tenebrionide tribe Eleodiini inhabiting the United States, Lower California, and adjacent islands. Bulletin of the United States National Museum 63: 1–524, pls. 1–13. [issued 24 Jun 1909 (p. ii)]

Blaisdell FE (1939) Studies in the relationships of the subfamilies and tribes of the Tenebrionidae based on the primary genital characters, also descriptions of new species (Coleoptera). Transactions of the American Entomological Society 65 (1): 43–60. [issued 14 Apr 1939 (vol. contents)]

Blanchard CÉ (1845a) Histoire naturelle des insectes leurs moeurs, leurs métamorphoses et leur classification ou traité élémentaire d'entomologie. Tome premier. F. Savy, Paris, v + 398 pp., pls. 1–10. [11 Jun 1845 (Ann. Soc. Ent. France (2) 3: Bull. Ent.: xliii); 28 Jun 1845 (Bibliogr. France 1845: 345)]

Blanchard CÉ (1845b) Histoire naturelle des insectes leurs moeurs, leurs métamorphoses et leur classification ou traité élémentaire d'entomologie. Tome second. F. Savy, Paris, 524 pp., pls. 11–20. [11 Jun 1845 (Ann. Soc. Ent. France (2) 3: Bull. Ent.: xliii), 28 Jun 1845 (Bibliogr. France 1845)]

Blanchard CÉ (1850) [Partie 1, pp. i-iv, 1–128]. In: Ordre des Coleoptères. Tome I. In: Milne-Edwards H, Blanchard CE, Lucas H (Eds). Muséum d'Histoire Naturelle de Paris Catalogue de la Collection Entomologique Classe des Insectes. Gide & Baudry, Paris, iv + 128 pp. [1850 (title page); after 25 Apr 1850 (date of Introduction)]

Blanchard CÉ (1851a) [Partie 2, pp. 129–240]. In: Ordre des Coleoptères. Tome I. In: Milne-Edwards H, Blanchard CE, Lucas H (Eds). Muséum d'Histoire Naturelle de Paris Catalogue de la Collection Entomologique Classe des Insectes. Gide & Baudry, Paris, iv + 240 pp. [1850 (title page); 25 Apr 1850 (date of Introduction)]

Blanchard CÉ (1851b) Coleópteros: Tetramerés, Trimeros, Dimeros [pp. 285–564, pls. 22–32]. In: Gay C (Ed) Historia física y política de Chile Zoología Tomo quinto. 563 pp. Privately published, Paris.

Blanchard CÉ (1853) Description des insectes. In: Hombros J, Jacquinot H (Eds). Voyage au Pole Sud et dans l'Océanie sur les corvettes l'Astrolabe et la Zélée; exécuté par ordre du Roi pendant les années 1837–1838–1839–1840, sous le commandement de M J Dumont-d'Urville, Capitaine de vaisseau; publié par ordre du gouvernement, sous la direction supérieure de M Jacquinot, Capitaine de vaisseau, commandant de la Zélée Zoologie par MM

Hombron et Jacquinot Tome quatrième. Gide & J. Baudry, Paris, [5] + 422 pp. + 19 pls. [1853 (title page); 3 March 1854, «might have been published at the end of 1853» (Clark and Crosnier 2000: 414)]

Blandford WFH (1893) The Scolyto-platypini, a new subfamily of Scolytidae. The Transactions of the Entomological Society of London 1893 (4): 425–442 + pl. xiv. [29 Dec 1893 (G. Wheeler 1912)]

Blandford WFH (1895) Scolytidae [pp. 81–96]. In: Godman FD, Salvin O (Eds) Biologia Centrali-Americana. Insecta. Coleoptera. Vol. IV. Part 6. Taylor and Francis, London, 396 pp. [Dec 1895 (footer, pp. 81, 89)]

Blandford WFH (1896) Scolytidae [continued.; pp. 113–120]. In: Godman FD, Salvin O (Eds) Biologia Centrali-Americana. Insecta. Coleoptera. Vol. IV. Part 6. Taylor and Francis, London, 396 pp. [Mar 1896 (footer, p. 113)]

Blatchley WS (1922) Notes on the Rhynchophora of Eastern North America, with characterizations of new genera and descriptions of new species. (Continued). Journal of the New York Entomological Society 30 (3): 113–127. [Sep issue; 27 Oct 1922 (recorded at USNM); issued 8 Nov 1922 (inside fasc. 4 wrapper)]

Bocák L, Bocáková M (1989) New tribe Lyropaeini, with a description of a new species of *Lyropaeus* (Coleoptera, Lycidae). Polskie Pismo Entomologiczne 58 (4 [1988]): 717–723. [28 Feb 1989]

Bocák L, Bocáková M (1990) Revision of the supergeneric classification of the family Lycidae (Coleoptera). Polskie Pismo Entomologiczne 59 (4 [1989]): 623–676.

Bocák L, Bocáková M (1992) Notes on some genera of the family Lycidae (Insecta: Coleoptera). Entomologica Basiliensia 15: 255–260.

Bocák L, Bocáková M (2008) Phylogeny and classification of the family Lycidae (Insecta: Coleoptera). Annales Zoologici (Warszawa) 58 (4): 695–720. [publ. 30 Dec 2008 (inside wrapper)]

Bocák L, Brlik M (2008) Revision of the family Omalisidae (Coleoptera, Elateroidea). Insect Systematics & Evolution 39 (2): 189–212. [Jun 2008 (top of article)]

Bocáková M (2005) Phylogeny and classification of the tribe Calopterini (Coleoptera, Lycidae). Insect Systematics & Evolution 35: 437–447.

Bock WJ (1994) History and nomenclature of avian family-group names. Bulletin of the American Museum of Natural History 222: 1–281 + 3 [unn.; erratum]. [issued 28 Sep 1994 (title page)]

Boddy DW (1965) Family Tenebrionidae. Pp. 130–189. *Part IV: Macrodactyles, Palpicornes, and Heteromera*. University of Washington Press, Seattle, viii + 268 pp.

Boheman CH (1855) Monographia Cassididarum. Tomus tertius. Officina Nordstedtiana, Holmiae, 543 pp. + pl. 7.

Bollow H (1938) Monographie der palaearktischen Dryopidae, mit Berüchsichtigung der eventuell transgredierenden Arten. (Col.). Mitteilungen der Münchner Entomologischen Gesellschaft 28 (1): 147–187. [20 Aug 1938 (Inhalt)]

Bollow H (1941) Monographie der palaearktischen Dryopidae, mit Berüchsichtigung der eventuell transgredierenden Arten. (Col.). (Fortsetzung). Mitteilungen der Münchner Entomologischen Gesellschaft 31 (1): 1–88, pl. 1. [15 Jan 1941 (Inhalt)]

Bologna MA, Pinto JD (1997) Case 2924. Meloidae Gyllenhal, 1810 and Nemognathinae Castelnau, 1840 (Insecta, Coleoptera): proposed precedence over Horiidae Latreille, 1802. Bulletin of Zoological Nomenclature 54 (4): 226–230 [18 Dec 1997]

Bonadona P (1959) Anthicides récoltés par J. Cantaloube au Cameroun (Coleoptera). Bulletin de l'Institut Français d'Afrique Noire (Série A) 21 (3): 1033–1046. [Jul 1959 (p. [1146])]

Bonadona P (1961) Les Tomoderini de l'Afrique noire et de la région Malgache (Coleoptera Anthicidae). Annales du Musée Royal de l'Afrique Centrale, Tervuren (Série 8vo: Sciences Zoologiques) 91: 1–78. [Jan 1961]

Bonadona P (1974) La classification des Anthicidae de la faune de France (Coleoptera). L'Entomologiste 30 (3): 101–111. [Jun fasc.; dépôt légal 2e trim. 1974; CNC recorded 23 Aug 1974]

Bonadona P (1984) Anthicides nouveaux ou peu connus d'Afrique Noire (Coleoptera, Anthicidae). Revue de Zoologie Africaine 98 (3): 469–504. [publ. 28 Sep 1984 (reprint)]

Bonadona P (1990) Les Anthicidae (Coleoptera) de la faune de France (septième partie). Bulletin Mensuel de la Société Linnéenne de Lyon (Nouvelle Série) 59 (1): 9–24. [Jan 1990 (fasc. title page); 10 Jan (recorded at BMNH)]

Bondar G (1942) Notas entomológicas da Baía. X. Revista de Entomologia (Rio de Janeiro) 13 (3): 225–274. [30 Nov 1942 (publ. date p. 480]

Bondar G (1943a) Notas entomológicas da Baía. XI. Revista de Entomologia (Rio de Janeiro) 14 (1/2): 33–84. [publ. 14 Aug 1943 (p. 336); 31 Jul 1943 (Alonso-Zarazaga and Lyal 1999: 214)]

Bondar G (1943b) Notas entomológicas da Baía. XII. Revista de Entomologia (Rio de Janeiro) 14 (1/2): 85–134. [publ. 14 Aug 1943 (p. 336); 31 Jul 1943 (Alonso-Zarazaga and Lyal 1999: 214)]

Bondar G (1949) Notas entomológicas da Baía. XXI. Revista de Entomologia (Rio de Janeiro) 20 (1/3): 173–228. [31 Aug 1949 (p. 668)]

Bonelli FA (1810) Observations entomologiques. Première partie (cicindélètes et portion des carabiques) [with the "Tabula synoptica exhibens genera carabicorum in sectiones et stirpes disposita"]. [s.n.]. Galletti, F., Turin, 58 pp., 1 table. [reissued in 1812, without the «Tabula Synoptica», in Mem. R. Accad. Sci. Torino 18: 21–78]

Bonelli FA (1813) Observations entomologiques. Deuxième partie. Turin, 52 pp. [This paper was reprinted in Mem. R. Accad. Sci. Torino, 20, 433–484, 1813 (Madge 1989; listed before journal version)]

Borchmann F (1936) 204me fascicule. Coleoptera [Heteromera]. Fam. Lagriidae. In: Wytsman PA (Ed) Genera Insectorum. Vol. XXXV. L. Desmet-Verteneuil, Bruxelles, 561 pp. + 9 pls. [31 Dec 1934 (date on manuscr.); 1936 (wrapper); 16 Nov 1936 (Evenhuis 1994: 60); 1937 (title page)]

Bordoni A (1974) Contributo alla conoscenza sistematica e faunistica degli Staphylinidae italiani - III - Specie italiane del genere *Lithocharis* Boisd. & Lac. (Col. Staphylinidae). XLI contributo alla conoscenza degli Staphylinidae. Redia 55: 321–329. [publ. 21 Dec 1974 (reprint wrapper)]

Bordoni A (1975) Morfologia cefalica e addominale della sottotribu Medina nov. e del genere Medon Stephens in particolare e suoi rapporti con la sistematica (Col. Staphylinidae). Redia 56: 417–446. [1975 (date on separate, not in journal)]

Bordoni A (1980) Studi sui Paederinae - VII - sopra alcuni *Hypomedon* e *Luzea* poco noti (Col. Staphylinidae). 64° contributo alla conoscenza degli Staphylinidae. Redia 63: 169–171. [9 Dec 1980 (reprint wrapper)]

Borgmeier T (1934) Uma nova tribu da subfamilia Aleocharinae (Col. Staph.). Revista de Entomologia (Rio de Janeiro) 4 (4): 451–454. [31 Oct 1934 (p. 534)]

Borgmeier T (1937) Uma nova especie de *Meloetyphlus* Waterhouse (1872), e a hypothese da myrmecophilia deste genero (Col. Meloidae). Revista de Entomologia (Rio de Janeiro) 7 (2/3): 247–255.

Borowiec L (1987) The genera of seed-beetles (Coleoptera, Bruchidae). Rodzaje strąkowców (Coleoptera, Bruchidae). Polskie Pismo Entomologiczne 57 (1): 3–207. [30 Mar 1987 (top of article)]

Borowiec L (1995) Tribal classification of the cassidoid Hispinae (Coleoptera: Chrysomelidae) [pp. 541–558]. In: Pakaluk J, Ślipiński SA (Eds) Biology, phylogeny and classification of Coleoptera: papers celebrating the 80th birthday of Roy A Crowson. Muzeum i Instytut Zoologii PAN, Warszawa, x + 1092 pp. in 2 vols. [31 Mar 1995 (teste S. A. Ślipiński)]

Borowski J, Węgrzynowicz P (2009) *Apate* Fabricius, 1775 (Bostrichidae: coleoptera): a protected name. Annales Zoologici (Warszawa) 59 (2): 189–191.

Bottimer LJ (1968) Notes on Bruchidae of America north of Mexico with a list of world genera. The Canadian Entomologist 100 (10): 1009–1049. [mailed 22 Oct 1968 (p. 1232)]

Bouchard P, Grebennikov VV, Smith ABT, Douglas H (2009) Biodiversity of Coleoptera [pp. 265–301]. In: Foottit RG, Adler PH (Eds). Insect biodiversity: science and society. Blackwell Publishing, Oxford, 656 pp,

Bouchard P, Lawrence JF, Davies A, Newton AF (2005) Synoptic classification of the world Tenebrionidae (Insecta: Coleoptera) with a review of family-group names. Annales Zoologici (Warszawa) 55 (4): 499–530. [31 Dec 2005]

Bouchard P, Löbl I, Merkl O (2007) Nomenclatural notes on tenebrionid beetles of the Palaearctic Region (Insecta: Coleoptera). Annales Zoologici (Warszawa) 57 (3): 385–394.

Boucher S (2006) Évolution et phylogénie des Coléoptères Passalidae (Scarabaeoidea). Les taxons du groupe famille. La tribu néotropicale des Proculini et son complexe *Veturius*. Annales de la Société Entomologique de France (Nouvelle Série) 41 [2005] (3/4): 239–604. [issued 27 Feb 2006 (back of wrapper)]

Bouchet P, Rocroi J-P (2005) Classification and nomenclator of gastropod families. Malacologia 47: 1–397.

Bouchet P, Rocroi J-P (2010) Part 1. Nomenclator of bivalve names of the family-group and above. In: Bouchet P, Rocroi J-P, Bieler R, Carter JG, Coan EV. Nomenclator of Bivalve Families with a Classification of Bivalve Families. Malacologia 52 (2): 1–184.

Boucomont A (1911) Contribution à la classification des Geotrypidae (Col.). Annales de la Société Entomologique de France 79 [1910–1911] (3): 333–350. [22 Feb 1911 (wrapper)]

Bousquet Y (2003) Subfamily Loricerinae [pp. 98–99]; tribe Abacetini [pp. 346–347] [and others]. In: Löbl I, Smetana A (Eds) Catalogue of Palaearctic Coleoptera. Volume 1. Archostemata -Myxophaga - Adephaga. Apollo Books, Stenstrup, 819 pp. [publ. 30 May 2003 (verso of title page)]

Bousquet Y (2008) Nomenclatural and bibliographic notes on Cerambycidae (Coleoptera). The Coleopterists Bulletin 61 [2007] (4): 616–631. [mailed 10 Jan 2008 (inside wrapper)]

Bousquet Y (2009) [new taxon] In: Bousquet Y, Heffern DJ, Bouchard P, Nearns EH. Catalogue of family-group names in Cerambycidae (Coleoptera). Zootaxa 2321: 1–80. [22 Dec 2009]

Bousquet Y, Bouchard P (2010) Case 3513. *Chaetosoma* Westwood, 1851, *Apodasya* Pascoe, 1863 and Chaetosomatidae Crowson, 1952 (Insecta, Coleoptera): proposed conservation. Bulletin of Zoological Nomenclature 67 (2): 140–144. [Jun 2010 issue]

Bousquet Y, Bouchard P, Lord NP (2010) Case 3517. Latridiidae Erichson, 1842 (Insecta, Coleoptera): proposed precedence over Corticariidae Curtis, 1829; and *Corticaria* Marsham, 1802: proposed conservation of usage by designation of *Corticaria ferruginea* Marsham, 1802 as the type species. Bulletin of Zoological Nomenclature 67 (2): 145–150. [Jun 2010 issue]

Bousquet Y, Heffern DJ, Bouchard P, Nearns EH (2009) Catalogue of family-group names in Cerambycidae (Coleoptera). Zootaxa 2321: 1–80. [22 Dec 2009]

Bousquet Y, Laplante S (1999) Les Coléoptères Histérides du Québec. Fabreries, Supplément 8: 1–190. [printed Feb 1999 (and recorded at CNC)]

Bousquet Y, Larochelle A (1993) Catalogue of the Geadephaga (Coleoptera: Trachypachidae, Rhysodidae, Carabidae including Cicindelini) of America north of Mexico. Memoirs of the Entomological Society of Canada 167: 1–397.

Bovie A (1908) 70me fascicule. Coleoptera. Fam. Curculionidae. Subfam. Cryptoderminae. In: Wytsman PA (Ed) Genera Insectorum. Vol. XI. P. Wytsman, Bruxelles, 3 pp. + 1 pl. [1908 (title page); 15 Apr 1908 (date on manuscr.); 23 Jun 1908 (Evenhuis 1994: 54)]

Böving AG (1922) Taxonomy and morphology of the larval stages of *Scobicia declivis* Leconte. In: Burke HE, Hartman RD, Snyder TE: The lead-cable borer or "short-circuit beetle" in California. Appendix. United States Department of Agriculture Technical Bulletin 1107: 49–54. [4 Dec 1922 (top of article)]

Böving AG (1927a) The larva of *Nevermannia dorcatomoides* Fisher with comments on the classification of the Anobiidae according to their larvae (Coleoptera: Anobiidae). Proceedings of the Entomological Society of Washington 29 (3): 51–62. [publ. 21 Apr 1927 (p. 72)]

Böving AG (1927b) Immature stages of *Eumycterus* (?) *saccharidis* Barber, with comments on the classification of the tribe Barini (Coleoptera: Curculionidae). Proceedings of the Entomological Society of Washington 29 (7): 151–159. [16 Nov 1927 (footer p. 164)]

Böving AG, Craighead FC (1931) An illustrated synopsis of the principal larval forms of the order Coleoptera. Entomologica Americana (New Series) 11 [1930] (1): 1–80, (2): 81–160, (3): 161–256, (4): 257–351, pls. 1–125. [(1): publ. 14 Nov 1931; (2): 7 Dec; (3): 9 Dec; (4): 21 Dec 1931 (wrappers)]

Bowestead S (1999) A revision of the Corylophidae (Coleoptera) of the West Palaearctic region. Instrumenta Biodiversitatis 3: 1–203.

Bowestead S, Booth RG, Ślipiński SA, Lawrence JF (2001) The genus *Cleidostethus* Arrow, 1929 reappraisal and transfer from Coccinellidae to Corylophidae (Coleoptera: Cucujoidea). Annales Zoologici (Warszawa) 51 (3): 319–323. [28 Sep 2001 (inside wrapper)]

Bowman JR (1934) The Pselaphidae of North America. Privately published, Pittsburgh, Pennsylvania, [5] + 149 pp. [24 Jul 1934]

Bradley JC (1930) A manual of the genera of beetles of America north of Mexico. Keys for the determination of the families, subfamilies, tribes, and genera of Coleoptera with a systematic list of the genera and higher groups Daw, Illston & Co., Ithaca, x + 360 pp.

Bradley JC (1947) Contributions to our knowledge of the Mylabridae, seu Bruchidae (Coleoptera) with especial reference to the fauna of northeastern America. Psyche 53 [1946] (3/4): 33–42. [mailed 14 Feb 1947 (verso of vol. title page)]

Brancucci M (1980) Morphologie comparée, évolution et systématique des Cantharidae (Insecta: Coleoptera). Entomologica Basiliensia 5: 215–388. [10 Dec 1980 (Inhalt)]

Branden C, Van den (1885) Catalogue des Coléoptères carnassiers aquatiques (Haliplidae, Amphizoïdae, Pelobiidae et Dytiscidae). Annales de la Société Entomologique de Belgique 29 (1): 5–118. [May-Jun 1885 (Compte-Rendu Soc. Roy. Belg. Géogr.)]

Brenske E (1898) Die *Serica*-Arten der Erde. Monographisch bearbeitet. Berliner Entomologische Zeitschrift 42 [1897] (3/4): 345–438. [1898 (volume title page); mid Jul 1898 (wrapper)]

Brèthes J (1925) Coccinellides du British Muséum (avec une nouvelle famille de Coléoptères). Anales del Museo Nacional de Historia Natural Bernardino Rivadavia 33 [1923–25]: 195–214. [20 Nov 1925 (Newton); Dec 1925 (date of vol. introduction)]

Breuning S (1950a) Révision des “Oculariini” [pp. 263–270]. In: Lepesme P (Ed). Longicornia. Études et notes sur les longicornes. Volume I. Paul Lechevalier, Paris, 603 pp. [dépôt légal 1er trimestre 1951 (p. [604]); 1950 (Zool. Record; Neave 1966: 20, etc.)]

Breuning S (1950b) Révision des “Stenobiini” [pp. 305–315]. In: Lepesme P (Ed) Longicornia Études et notes sur les longicornes Volume I. Paul Lechevalier, Paris, 603 pp. [dépôt légal 1er trimestre 1951 (p. [604]); 1950 (Zool. Record; Neave 1966: 20, etc.)]

Breuning S (1950c) Considérations préliminaires sur la classification des lamiaires [pp. 25–28]. In: Lepesme P (Ed) Longicornia. Études et notes sur les longicornes. Volume I. Paul Lechevalier, Paris, 603 pp. [dépôt légal 1er trimestre 1951 (p. [604]); 1950 (Zool. Record; Neave 1966: 20, etc.)]

Breuning S (1950d) Révision des “Proctocerini” [pp. 411–414]. In: Lepesme P (Ed). Longicornia. Études et notes sur les longicornes. Volume I. Paul Lechevalier, Paris, 603 pp. [dépôt légal 1er trimestre 1951 (p. [604]); 1950 (Zool. Record; Neave 1966: 20, etc.)]

Breuning S (1951a) Notes systématiques sur les longicornes de Nouvelle-Calédonie. - I. Bulletin / Institut Royal des Sciences Naturelles de Belgique 27 (32): 1–24. [Jun 1951 (top of article)]

Breuning S, Teocchi P (1977) Regroupement des genres *Brachyolene* Auriv. et *Tetraulax* Jord. dans la tribu des Tetraulaxini, nov. (Coleoptera Cerambycidae Lamiinae). Bulletin de l'Institut Fondamental d'Afrique Noire (Série A. Sciences Naturelles) 38 [1976] (4): 881–891. [publ. 31 Oct 1977 (endleaf)]

Breuning S, Teocchi P (1978) Création de la tribu des Aderpasini, nov. Révision et bionomie des espèces des genres *Aderpas* Thoms. et *Ancylonotopsis* Br. (Coleoptera Cerambycidae Lamiinae). Bulletin de l'Institut Fondamental d'Afrique Noire (Série A. Sciences Naturelles) 39 [1977] (1): 142–168. [publ. 21 Jan 1978 (p. 228)]

Breuning S, Teocchi P (1985) Note concernant les tribus Pachystolini Auriv., Petrognathini Blanch., Xylorhizini Lac. et Microcymaturini nov. (Coleoptera Cerambycidae Lamiinae). Bulletin de l'Institut Fondamental d'Afrique Noire (Série A. Sciences Naturelles) 44 [1982] (1/2): 153–159. [publ. 31 Jul 1985 (endleaf)]

Bridwell JC (1929) A preliminary generic arrangement of the palm bruchids and allies (Coleoptera) with descriptions of new species. Proceedings of the Entomological Society of Washington 31 (8): 141–160. [publ. 26 Dec 1929 (p. 168)]

Bridwell JC (1932) The subfamilies of the Bruchidae (Coleoptera). Proceedings of the Entomological Society of Washington 34 (6): 100–106. [15 Jul 1932 (p. 108)]

Bridwell JC (1946) The genera of beetles of the family Bruchidae in America north of Mexico. Journal of the Washington Academy of Science 36 (2): 52–57.

Brinck C (1948) Coleoptera of Tristan da Cunha. Results of the Norwegian Scientific Expedition to Tristan da Cunha 1937–1938. No. 17. Norske Videnskaps-Akademi I Oslo, Oslo, 121 pp. + 1 pl. [printed Dec 1948 (last page)]

Brinck P (1956) A revision of the Gyrinidae (Coleoptera) of the Ethiopian region. I. Acta Universitatis Lundensis (Nova Series. Andra Avdelningen) 51 (14): 1–140. [printed 10 Sep 1955 (contents); 1956 (endleaf)]

Britton EB (1955) Coleoptera: Scarabaeidae: Melolonthinae and Dynastinae from the Monte Bello Islands, 1952. Proceedings of the Linnean Society of London 165 [1952–53]: 124–126.

Britton EB (1957) A revision of the Australian chafers (Coleoptera: Scarabaeidae: Melolonthinae). Volume 1. British Museum (Natural History), London, viii + 185 pp. + 42 pls. [issued 17 Jun 1957 (title page)]

Britton EB (1971) A new intertidal beetle (Coleoptera: Limnichidae) from the Great Barrier Reef. Journal of Entomology (Series B, Taxonomy) 40: 83–91. [4 Jun 1971 (recorded at CNC)]

Britton EB (1978) A revision of the Australian chafers (Coleoptera: Scarabaeidae: Melolonthinae). Vol. 2. Tribe Melolonthini. Australian Journal of Zoology (Supplementary Series) No. 60: 1–150. [7 Jun 1978 (title page)]

Bromhead EF (1838) Remarks on zoological classification. The Magazine of Natural History (New series) 2 (No. 20): 412–419. [Aug 1838 issue; 25 Nov 1838 (date of vol. Preface)]

Bronn HG (1848) Handbuch einer Geschichte der Natur. Dritter Band. Erster Abtheilung zweite Hälfte. III. Theil: Organisches Leben (Fortsetzung). Index palaeontologicus oder liebersicht der bis jezt bekannten fossilen Organismen. A. Nomenclator palaeontologicus, in alphabetischer Ordnung. N - Z. In: Bischoff GW, Blum JR, Bronn HG, Leonhard KC, von, Leuckart FS, Voigt FS (Eds) Naturgeschichte der drei Reiche, zur allgemeinen Belehrung Fünfzehnter Band. E. Schweizerbart, Stuttgart, pp. 777–1381 + [1]. [1848 (title page)]

Broun T (1880) Manual of the New Zealand Coleoptera [Part I]. Colonial Museum & Geological Survey Department, Wellington, xix + 651 pp. [Jun 1880 (date of vol. Preface p. iv); cited in reading 21 Jun 1880 (Trans. Roy. Soc. New Zeal. 13: 280); 12 Aug 1880 (New Zeal. Parl. Debates 37: 293)]

Broun T (1893) Manual of the New Zealand Coleoptera. Parts V., VI., VII. New Zealand Institute, Wellington, xvii + pp. 975–1504. [14 Jan 1893 (date of Preface); 28 Sep 1893 (Nature 48: 536)]

Brown FM (1964) The dates of publication of the first ten volumes of the Transactions of the American Entomological Society. Transactions of the American Entomological Society 89: 313–321.

Browne FG (1961) The generic characters, habits and taxonomic status of *Premnobius* Eichh. (Coleopt., Scolytidae). Report of the West African Timber Borer Research Unit 4: 45–51. [Jun 1961 (p. [1])]

Browne FG (1962) Two new genera of the Scolytidae (Coleoptera). Report of the West African Timber Borer Research Unit 5: 75–80. [Sep 1962 (p. [2]); 22 Oct 1962 (recorded at BMNH; fide Alonso-Zarazaga and Lyal 2009: 116)]

Bruce N (1951) Cryptophagidae (Coleoptera Polyphaga). Exploration du Parc National Albert, Mission G F de Witte (1933–35). Fascicule 75. Institut des Parcs Nationaux du Congo Belge, Bruxelles, 26 pp., 2 pls.

Bruch C (1933a) Coleópteros mirmecófilos de Misiones (Staph. Pselaph. Hister.). Revista de Entomologia (Rio de Janeiro) 3 (1): 12–37. [22 Mar 1933 (p. iv)]

Bruch C (1933b) Nuevos estafilínidos ecitófilos de Tucumán. Revista de Entomologia (Rio de Janeiro) 3 (2): 205–213. [20 Jul 1933 (p. iv)]

Bruch C (1937) Género y especies nuevos de un estafilínido mirmecófilo (Col. Staph. Aleocharinae). Revista de Entomologia (Rio de Janeiro) 7: 353–356. [11 Oct 1937 (p. 542)]

Brues CT, Melander AL (1932) Classification of insects. A key to the known families of insects and other terrestrial arthropods. Bulletin of the Museum of Comparative Zoology 73: 1–672. [publ. Jan 1932 (p. [2])]

Brullé GA (1832) [Pp. 1–240]. In: Bory de Saint-Vincent JBGM: Expédition scientifique de Morée. Section des sciences physiques. Tome III. - 1.re partie. Zoologie. Deuxième section. - Des animaux articulés. In: Bory de Saint-Vincent JBGM (Ed) F.G. Levrault, Paris, [1] + 400 + [2 (errata)] pp. [publ. dates: Sherborn and Woodward 1901: 335]

Brullé GA (1834) Coléoptères. I. Histoire naturelle des insectes, traitant de leur organisation et de leurs moeurs en général, par MV Audouin, et comprenant leur classification et la description des espèces, par MA Brullé Le tout accompagné de planches gravées sur acier, d'après des peintures exécutées pour cette édition sur la collection du Muséum de Paris Tome IV. F.D. Pillot, Paris, viii + 479 pp. + 16 pls. [text issued in two livraisons: livraison 1 (pp. 1–240) 2 Aug 1834 (Bibliogr. France 1834); livraison 2 (pp. 241–479) Feb 1835 (Bull. Bibliogr. Sci. Phys.), 1835 (Musgrave 1932: 8); we accept 1834 as the publication date of livrason 2 until further evidence is obtained]

Brullé GA (1835) Coléoptères. II. Histoire naturelle des insectes, traitant de leur organisation et de leurs moeurs en général, par MV Audouin, et comprenant leur classification et la description des espèces, par MA Brullé Le tout accompagné de planches gravées sur acier, d'après des peintures exécutées pour cette édition sur la collection du Muséum de Paris. Tome V. F.D. Pillot, Paris, 436 pp., 16 pls.

Burakowski B, Mroczkowski M, Stefańska J (1985) Chrząszcze Coleoptera, Buprestoidea, Elateroidea and Cantharoidea. Katalog Fauny Polski. Część XXIII, tom 10 [Nr 40]. Panstwowe Wydawnictwo Naukowe, Warszawa, 401 pp. + 1 map. [20 Sep 1985 (recorded at CNC)]

Burakowski B, Mroczkowski M, Stefańska J (1986) Chrząszcze Coleoptera, Dermestoidea, Bostrichoidea, Cleroidea i Lymexyloidea. Katalog Fauny Polski. Część XXIII, tom 11 [Nr 42]. Panstwowe Wydawnictwo Naukowe, Warszawa, 242 pp. [printed May 1986 (verso of title page); 5 Sep (recorded at CNC)]

Burakowski B, Mroczkowski M, Stefańska J (1990) Stonkowate - Chrysomelidae, część 1. Katalog Fauny Polski. Część XXIII, tom 16. Chrząszcze Coleoptera. Panstwowe Wydawnictwo Naukowe, Warszawa, 279 pp.

Burakowski B, Mroczkowski M, Stefańska J (1991) Stonkowate - Chrysomelidae, część 2. Katalog Fauny Polski. Część XXIII, tom 17. Chrząszcze Coleoptera. Panstwowe Wydawnictwo Naukowe, Warszawa, 227 pp. [printed Aug 1991 (verso of title page)]

Burgeon L (1935) Catalogues raisonnés de la faune entomologique du Congo Belge. Coléoptères - carabides (première partie). Annales du Musée Royal du Congo Belge (Série in 8°: Sciences Zoologiques) Série III, Section II, Tome II (fascicule 3): 131–257. [Aug 1935 (title page)]

Burmeister F (1939) I. Band: Adephaga. I. Familiengruppe: Caraboidea. Biologie, Ökologie und Verbreitung der europäischen Käfer auf systematischer Grundlage. H. Goecke, Krefeld, 307 pp.

Burmeister HCC (1842) Dritter Band. Besondere Entomologie, Fortsetzung. Coleoptera Lamellicornia Melitophila. Handbuch der Entomologie. T.C.F. Enslin, Berlin, xxii + 828 + [1] pp. [Sep 1842 (date of Vorrede); 28 Dec 1842 (Literar. Zeit.: 1171)]

Burmeister HCC (1844) Besondere Entomologie, Fortsetzung. Erste Abtheilung. Coleoptera Lamellicornia Anthobia et Phyllophaga systellochela. Handbuch der Entomologie. Vierter Band. T.C.F. Enslin, Berlin, xii + 587 + [1] pp.

Burmeister HCC (1846) III. Coleoptera. 7. Lamellicornia. 27. Copris [4 unn. pp.]. In: Genera quaedam insectorum. Iconibus illustravit et descripsit. Volumen I. Continet tabulas XL. A. Burmeister, Berolini, viii + [2] + [120] pp. + 40 pls.

Burmeister HCC (1847) Besondere Entomologie. Fortsetzung. Coleoptera Lamellicornia Xylophila et Pectinicornia. Handbuch der Entomologie. Fünfter Band. T.C.F. Enslin, Berlin, viii + 584 pp.

Burmeister HCC (1855) Besondere Entomologie, Fortsetzung. Zweite Abtheilung. Coleoptera Lamellicornia Phyllophaga chaenochela. Handbuch der Entomologie. Vierter Band. T.C.F. Enslin, Berlin, x + 569 + [1] pp.

Burmeister HCC (1873) Lamellicornia Argentina. Entomologische Zeitung (Stettin) 34 (10/12): 403–417. [early Nov 1873 (end of Heft)]

Burmeister HCC (1875) Melanosoma Argentina. Entomologische Zeitung (Stettin) 36 (10/12): 457–500. [Oct/Dec 1875 issue]

Burmeister HCC, Schaum H (1840) Kritische Revision der Lamellicornia melitophila. Erstes Stück. Zeitschrift für die Entomologie (Germar) 2 (2): 353–420. [1840 (Heft title page); Sep 1840 (Foreign Quart. Rev.: 233)]

Burmeister HCC, Schaum H (1841) Kritische Revision der Lamellicornia melitophila. Zweites Stück. Zeitschrift für die Entomologie (Germar) 3 (1/2): 226–282. [Jun 1841 (Rev. Zool. 4: 201)]

Buysson H, du (1893) Faune Gallo-Rhénane ou species des insectes qui habitent la France, la Belgique, la Hollande, le Luxembourg, la Prusse Rhénane, le Nassau et le Valais avec tableaux synoptiques et planches gravées. Coléoptères. Tome Cinquième. Élatérides (suite). Revue d'Entomologie (Suppl.) 12: 9–72 [Feb-Nov 1893 (first page of each sheet)]

Calder AA (1978) The neotropical genus Anaissus (Elateridae) and its relationship to the Pyrophorinae. Systematic Entomology 3 (4): 295–306. [15 Nov 1978 (recorded at CNC)]

Cameron M (1917a) On a new group of Staphylinidae. The Entomologist's Monthly Magazine 53 (6): 123–125. [Jun 1917; 26 Jun (recorded at USNM)]

Cameron M (1917b) Synonymic note on the group Arpediopsini. The Entomologist's Monthly Magazine 53 (12): 277. [Dec 1917 issue; 3 Jan 1918 (recorded at USNM)]

Cameron M (1918) New Oriental Staphylinidae (3). The Entomologist's Monthly Magazine 54 (9): 214–219. [19 Sep 1918 (recorded at CNC)]

Cameron M (1919) New species of Staphylinidae from Singapore. Part II. The Transactions of the Entomological Society of London 1918 (3/4): 231–246. [29 Mar 1919 (verso of vol. title page)]

Cameron M (1926) Note on the genus *Plagiarthrina* Keys. The Entomologist's Monthly Magazine 62 (8): 184. [16 Aug 1926 (recorded at CNC)]

Cameron M (1933) Description of a new genus of Staphylinidae from India. The Entomologist's Monthly Magazine 69 (5): 103. [10 May 1933 (recorded at CNC)]

Cameron M (1939) Coleoptera, Staphylinidae. In: Sewell RBS (Ed) The fauna of British India, including Ceylon and Burma. Vol. IV. Parts I & II. Taylor & Francis, London, xviii + 691 pp., pls. 1–3. [11 Aug 1939 (title page)]

Cameron M (1944) Description of a new genus of Euaesthetinae (Col., Staph.). The Annals and Magazine of Natural History (11) 11 (73): 68–70. [21 Feb 1944 (Evenhuis 2003: 50)]

Cameron M (1959) New species of Staphylinidae (Col.) from Angola (IV). Publicações Culturais da Companhia de Diamantes de Angola 48: 113–120. [28 Dec 1959 (date of separate)]

Candèze ECA (1857) Monographie des Élatérides. Tome premier. H. Dessain, Liége, viii + 400 pp., 7 pls. [Apr 1857 (date of preface); May 1857 (title page); 29 Jun 1857 (Acad. Sci. France)]

Candèze ECA (1859) Monographie des Élatérides. Tome second. H. Dessain, Liége, 543 pp. + 7 pls. [May 1859 (title page)]

Candèze ECA (1860) Monographie des Élatérides. Tome troisième. H. Dessain, Liége, 512 pp. + 5 pls. [Jul 1860 (title page)]

Candèze ECA (1863) Monographie des Élatérides. Tome quatrième. H. Dessain, Liége, 534 pp. + 6 pls. [May 1863 (title page)]

Candèze ECA (1874) Révision de la Monographie de Élatérides. Premier fascicule. Mémoires de la Société Royale des Sciences de Liège (2) 4: viii+218. [Nov 1874 (vol. title page)]

Candèze ECA (1892) Insectes du Bengale. 24e mémoire. Deuxième note sur les élatérides du Chota-Nagpore. Annales de la Société Entomologique de Belgique 36 (11): 480–495. [before 14 Dec 1892 (Ann. Soc. Ent. France 61: Bull. Ent.: cclxiv)]

Capiomont G (1867) Révision de la tribu des Hypérides, Lacordaire, et en particulier des genres *Hypera* Germ., *Limobius*, Schönh. et *Coniatus* (Germ.) Schönh. renfermant la description de plusieurs genres nouveaux et de 85 espèces nouvelles. Annales de la Société Entomologique de France (4) 7 (3): 417–456 + pl. 11–12.

Carne PB (1957) A systematic revision of the Australian Dynastinae (Coleoptera: Scarabaeidae). Commonwealth Scientific and Industrial Research Organization, Melbourne, 284 pp. [Dec 1957 (title page)]

Carpenter FM (1992) Treatise on invertebrate paleontology. Part R, Arthropoda 4, Vols. 3, 4: Superclass Hexapoda. Geological Society of America and University of Kansas, Lawrence, xxi + 655 pp.

Carter HJ (1911) Revision of the Nyctozoilides - genera and species (fam. Tenebrionidae). Annals of the Queensland Museum 10: 136–166.

Carter HJ (1924) Australian Coleoptera - notes and new species No. iii. The Proceedings of the Linnean Society of New South Wales 49: 19–45.

Carus JV, Engelmann W (1861) Bibliotheca Zoologica. Verzeichniss der Schriften über Zoologie, welche in den periodischen Werken enthalten und vom Jahre 1846–1860 selbständig erschienen sind. Mit Einschluss der allgemein-naturgeschichtlichen, periodischen und palaeontologischen Schriften. Zweiter Band. W. Engelmann, Leipzig, pp. i-xxiv + 951–2144. [Aug 1861 (date of Vorwort); 15 Nov 1861]

Casale A, Laneyrie R (1982) Trechodinae et Trechinae du monde. Tableau des sous-familles, tribus, séries phylétiques, genres et catalogue général des espèces. Mémoires de Biospéologie 9: 1–226. [Oct 1982 (wrapper)]

Casey TL (1884) Revision of the Cucujidae of America north of Mexico. Transactions of the American Entomological Society and Proceedings of the Entomological Section of the Academy of Sciences 11: 69–112 + pls. I-VIII. [Feb 1884 (footer p. 81, 105)]

Casey TL (1890) Coleopterological notices. II. Annals of the New York Academy of Sciences 5: 307–504, pl. iv.

Casey TL (1893) Coleopterological notices, V. Annals of the New York Academy of Sciences 7: 281–606, pl. 1. [Oct 1893 (pp. 281–392), Nov (393–536), Dec 1893 (537–606) (signature footers); 1893 (Zool. Record); before 24 Jan 1894 (Ann. Soc. Ent. France 63: Bull. Ent.: xxxv)]

Casey TL (1895) Coleopterological notices, VI. Annals of the New York Academy of Sciences 8 (6–12): 435–838. [Jul 1895 (pp. 435–514), Aug (515–594), Sep (595–674), Oct (675–722), Nov 1895 (723–838) (signature footers); Nov 1895 (wrapper)]

Casey TL (1897) Coleopterological notices, VII. Annals of the New York Academy of Sciences 9: 285–684. [Feb-Jul 1897 (signature footers); Oct 1897 («Aug»; Ent. News 8: 204)]

Casey TL (1899) A revision of the American Coccinellidae. Journal of the New York Entomological Society 7 (2): 71–169. [Jun 1899 (fasc. title page)]

Casey TL (1900) Review of the American Corylophidae, Cryptophagidae, Tritomidae and Dermestidae, with other studies. Journal of the New York Entomological Society 8 (2): 51–172. [Jun 1900 (fasc. title page); 3 Aug 1900 (recorded at USNM)]
Casey TL (1905) A revision of the American Paederini. Transactions of the Academy of Science of St Louis 15 (2): 17–248. [issued 4 Apr 1905 (wrapper)]
Casey TL (1906) Observations on the staphylinid groups Aleocharinae and Xantholinini, chiefly of America. Transactions of the Academy of Science of St Louis 16 (6): 125–434. [issued 22 Nov l906 (p. 434)]
Casey TL (1907) A revision of the American components of the tenebrionid subfamily Tentyriinae. Proceedings of the Washington Academy of Sciences 9: 275–522. [18 Oct 1907 (reprint wrapper)]
Casey TL (1910) New species of the staphylinid tribe Myrmedoniini [pp. 1–183]. In: Memoirs on the Coleoptera I. New Era Printing Co., Lancaster, Pennsylvania, 205 pp. [issued 24 Sep 1910 (p. [206])]
Casey TL (1912) Descriptive catalogue of the American Byrrhidae [pp. 1–69]. In: Memoirs on the Coleoptera III. The New Era Printing Company, Lancaster, 386 pp. [issued 20 Mar 1912 (p. 386)]
Casey TL (1914a) Some observations on the Carabidae including a new subfamily [pp. 25–44]. In: Memoirs on the Coleoptera V. The New Era Printing Company, Lancaster, 387 pp. [issued 28 Nov 1914 (p. [388])]
Casey TL (1914b) A revision of the Nearctic Harpalinae [pp. 45–305]. In: Memoirs on the Coleoptera V. The New Era Printing Company, Lancaster, 387 pp. [issued 28 Nov 1914 (p. [388])]
Casey TL (1915a) A review of the American species of Rutelinae, Dynastinae and Cetoniinae. Memoirs on the Coleoptera VI. New Era Printing Co., Lancaster, Pennsylvania, 1–394. [Oct-Nov 1915 (footers pp. 1, 305); issued 27 Nov 1915 (p. 460)]
Casey TL (1915b) Studies in some staphylinid genera of North America [pp. 395–450]. In: Memoirs on the Coleoptera VI. New Era Printing Co., Lancaster, Pennsylvania, 460 pp. [issued 27 Nov 1915 (p. 460)]
Casey TL (1922) Studies in the rhynchophorous subfamily Barinae of the Brazilian fauna. Memoirs on the Coleoptera X. New Era Printing Co., Lancaster, Pennsylvania, 520 pp. [issued 25 Nov 1922 (p. 520)]
Cate PC (2007a) New acts and comments. Elateridae [pp. 33–46]. In: Löbl I, Smetana A (Eds) Catalogue of Palaearctic Coleoptera. Volume 4. Elateroidea - Derodontoidea - Bostrichoidea - Lymexyloidea - Cleroidea - Cucujoidea. Apollo Books, Stenstrup, 935 pp. [publ. 30 Jun 2007 (verso of title page)]
Cate PC (2007b) Family Elateridae Leach, 1815 (excluding Cebrioninae, Lissominae and Subprotelaterinae) [pp. 94–207]. In: Löbl I, Smetana A (Eds) Catalogue of Palaearctic Coleoptera. Volume 4. Elateroidea - Derodontoidea - Bostrichoidea - Lymexyloidea - Cleroidea - Cucujoidea. Apollo Books, Stenstrup, 935 pp. [publ. 30 Jun 2007 (verso of title page)]
Célis M-J (1969) Contributions à l'étude des Clavigerinae de l'Afrique (Coleoptera Pselaphidae). 3. Démembrement des Fustigerini et création de deux tribus nouvelles ainsi que d'un

genre inédit. Revue de Zoologie et de Botanique Africaines 80 (3/4): 415–424. [publ. 31 Dec 1969 (wrapper + p. 429)]

Célis M-J (1970) Contribution à l'étude des Clavigerinae de Madagascar (Coleoptera Pselaphidae). Clavigerinae recueillis par le Professeur Dr. H. Franz et par M. J. Vadon. Remaniements apportés à la systématique des Clavigerinae malgaches. Revue de Zoologie et de Botanique Africaines 82 (3/4): 237–270. [30 Dec 1970 (wrapper + p. 381); recorded at CNC 25 Mar 1971]

Cerda MA (1973) Nueva tribu de Cerambycinae (Coleoptera: Cerambycidae). Revista Chilena de Entomología 7: 115–122. [printed 15 Jul 1973 (last page)]

Cerruti M (1970) Description d'un coléoptère psélaphide inédit de l'Erythrée, représentant d'une sous-tribu nouvelle. Revue de Zoologie et de Botanique Africaines 81 (1/2): 117–124. [31 Mar 1970 (wrapper + p. 433)]

Chamberlin WJ (1939) The bark and timber beetles of North America north of Mexico. The taxonomy, biology and control of 575 species belonging to 72 genera of the super family Scolytoidea. Oregon State College Cooperative Association, Corvallis, vi + 513 pp. + 5 pls. [after Aug 1939 (Preface, p. ii)]

Champion GC (1886) Tenebrionidae [part, pp. 137–264]. In: Godman FD, Salvin O (Eds) Biologia Centrali-Americana. Insecta. Coleoptera. Vol IV. Part 1. Heteromera (part). Taylor & Francis, London, xxxiv + 572 pp. + 23 pls. [Apr 1886 (pp. 137–152), May (153–168), Jun (169–184), Jul (185–208), Aug (209–216), Oct (217–224), Nov (225–240), Dec (241–264) (signature footers)]

Champion GC (1887) Tenebrionidae [part, pp. 265–352]. In: Godman FD, Salvin O (Eds) Biologia Centrali-Americana. Insecta. Coleoptera. Vol IV. Part 1. Heteromera (part). Taylor & Francis, London, xxxiv + 572 pp. + 23 pls. [Jan 1887 (pp. 265–272), Jun (273–296), Aug (297–320), Oct (321–328), Dec 1887 (329–352) (signature footers)]

Champion GC (1894) Elateridae [part, pp. 258–304, pls. 11–13]. In: Godman FD, Salvin O (Eds) Biologia Centrali-Americana. Insecta. Coleoptera. Vol. III. Part 1. Serricornia. Taylor and Francis, London, xvi + 690 pp. + 27 pls. [Oct 1894 (pp. 258–264), Nov 1894 (265–304) (signature footers)]

Champion GC (1895) Elateridae [part, pp. 305–440, pls. 14–19]. In: Godman FD, Salvin O (Eds) Biologia Centrali-Americana. Insecta. Coleoptera. Vol. III. Part 1. Serricornia. Taylor and Francis, London, xvi + 690 pp. + 27 pls. [Jan 1895 (pp. 305–312), Mar (313–336), May (337–360), Jul (361–376), Oct (377–400), Dec 1895 (401–440) (signature footers)]

Champion GC (1896) Elateridae [part, pp. 441–556]. In: Godman FD, Salvin O (Eds) Biologia Centrali-Americana. Insecta. Coleoptera. Vol. III. Part 1. Serricornia. Taylor & Francis, London, xvi + 690 pp. + 27 pls. [Jan 1896 (pp. 441–472), Mar (473–496), May (497–528), Jun (529–552), Oct 1896 (553–556) (signature footers)]

Champion GC (1897) Dascillidae [pp. 586–662]. In: Godman FD, Salvin O (Eds) Biologia Centrali-Americana. Insecta. Coleoptera. Vol. III. Part 1. Serricornia. Taylor & Francis, London, xvi + 690 pp. + 27 pls. [Feb 1887 (pp. 586–608), Mar (609–624), May (625–656), Aug 1887 (657–662) (signature footers)]

Champion GC (1902) [Pp. 1–144]. In: Godman FD, Salvin O (Eds) Biologia Centrali-Americana. Insecta. Coleoptera. Vol. IV. Part 4. Rhynchophora. Curculionidae. Curculioninae (part.). Taylor & Francis, London, [viii] + 750 pp. + 35 pls. [May 1902 (pp. 1–32), Aug (33–64), Oct (65–112), Dec 1902 (113–144) (signature footers)]

Champion GC (1903) [Pp. 145–312]. In: Godman FD, Salvin O (Eds) Biologia Centrali-Americana. Insecta. Coleoptera. Vol. IV. Part 4. Rhynchophora. Curculionidae. Curculioninae (part). Taylor & Francis, London, viii + 750 pp. + 35 pls. [Feb 1903 (pp. 145–176), Apr (177–208), May (209–224), Aug (225–280), Nov 1903 (281–312) (signature footers)]

Champion GC (1906) [Pp. 601–750]. In: Godman FD, Salvin O (Eds) Biologia Centrali-Americana. Insecta. Coleoptera. Vol. IV. Part 4. Rhynchophora. Curculionidae. Curculioninae (part). Taylor & Francis, London, viii + 750 pp. + 35 pls. [Jan 1906 (pp. 601–688), Apr 1906 (689–750) (signature footers)]

Champion GC (1907) [Pp. 137–240]. In: Godman FD, Salvin O (Eds) Biologia Centrali-Americana. Insecta. Coleoptera. Vol. IV. Part 5. Rhynchophora. Curculionidae. Curculioninae (continued). Taylor & Francis, London, viii + 513 pp. + 23 pls. [Feb 1907 (137–176), May (177–200), Jul (201–216), Nov 1907 (217–240) (signature footers)]

Champion GC (1908) [Pp. 241–400]. In: Godman FD, Salvin O (Eds) Biologia Centrali-Americana. Insecta. Coleoptera. Vol. IV. Part 5. Rhynchophora. Curculionidae. Curculioninae (continued). Taylor & Francis, London, viii + 513 pp. + 23 pls. [May 1908 (pp. 241–312), Sep (313–376), Nov 1908 (377–400) (signature footers)]

Champion GC (1909) [Pp. 1–78]. In: Godman FD, Salvin O (Eds) Biologia Centrali-Americana. Insecta. Coleoptera. Vol. IV. Part 7. Rhynchophora. Curculionidae. Curculioninae (concluded). Taylor & Francis, London, vii + 221 pp. + 9 pls. [Oct 1909 (pp. 1–48), Dec 1909 (49–78) (signature footers)]

Champion GC (1911) Series Otiorhynchinae Alatae; Supplement to the Thecesterninae and Otiorhynchinae [pp. 169–354]. In: Godman FD, Salvin O (Eds) Biologia Centrali-Americana. Insecta. Coleoptera. Vol. IV. Part 3. Rhynchophora. Curculionidae. Attelabinae, Pterocolinae, Allocoryninae, Apioninae, Thecesterninae, Otiorhynchinae. Taylor & Francis, London, vi + 354 pp. + 15 pls. [May 1911 (pp. 169–240), Oct (241–312), Dec 1911 (313–354) (signature footers)]

Champion GC (1914) Reports of the Percy Sladen Trust expedition to the Indian Ocean in 1905, under the leadership of Mr J. Stanley Gardiner, M. A. Vol. 5. XIX. Coleoptera, Curculionidae. The Transactions of the Linnean Society of London (2nd Ser. Zoology) 16 (4): 393–497 + pls. 22–24. [Jul 1914 (vol. Contents, wrapper)]

Champion GC (1915) Revision of the Mexican and Central American Telephorinae (fam. Telephoridae), with descriptions of new species. The Transactions of the Entomological Society of London 1915 (1): 16–146. [publ. 26 Jun 1915 (wrapper + verso of title page)]

Champion GC (1923) A revision of the Malayan and Indian species of the melyrid subfamily Carphurinae represented in the Hope collection at Oxford and in the British Museum in London (Coleoptera). The Annals and Magazine of Natural History (9) 12 (67): 1–54, pl. 1. [1 Jul 1923 (Evenhuis 2003: 43)]

Champion GC (1924) On a new subfamily of clavicorn Coleoptera. The Entomologist's Monthly Magazine 60: 25–29. [Feb 1924 (issue), 4 Apr (recorded at CNC)]

Champion GC (1925) Some Indian (and Tibetan) Coleoptera (17). The Entomologist's Monthly Magazine 61: 169–181. [Aug 1925 issue; 17 Aug 1925 (recorded at CNC)]

Chandler DS (2001) Biology, morphology and systematics of the ant-like litter beetles of Australia (Coleoptera: Staphylinidae: Pselaphinae). Memoirs on Entomology, International 15: vii + 560 pp. [30 Apr 2001 (p. ii)]

Chapin EA (1924) Classification of the Philippine components of the coleopterous family Cleridae. The Philippine Journal of Science 25 (2): 159–286, pls. 1–5. [18 Sep 1924 (date of issue, volume contents)]

Chapuis F (1869) Synopsis des Scolytides (Prodrome d'un travail monographique). J. Desoer, Liège, 61 pp. [1869 (title page); 1 Jul 1869 (Petites Nouv. Ent.)]

Chapuis F (1874) Histoire naturelle des Insectes. Genera des coléoptères ou exposé méthodique et critique de tous les genres proposés jusqu'ici dans cet ordre d'insectes. Tome dixième. Famille des phytophages. Librairie Encyclopédique de Roret, Paris, iv + 455 pp.

Chapuis F (1875) Histoire naturelle des insectes. Genera des coléoptères ou exposé méthodique et critique de tous les genres proposés jusqu'ici dans cet ordre d'insectes. Tome onzième. Famille des phytophages. Librairie Encyclopédique de Roret, Paris, 420 pp.

Chapuis F (1876) Histoire naturelle des insectes. Genera des coléoptères ou exposé méthodique et critique de tous les genres proposées jusqu'ici dans cet ordre d'insectes. Tome douzième. Famille des érotyliens, des endomychides et des coccinellides. Librairie Encyclopédique de Roret, Paris, 424 pp. [after Mar 1876 (date of Epilogue); 1 May 1876 (Petites Nouv. Ent.)]

Chatanay J (1914) Nouveaux asidides de Madagascar (Col. Tenebrioniade). Insecta / Revue Illustrée d'Entomologie (Rennes) 4 (37): 1–13. [Jan 1914 issue; before Mar 1914 (recorded at USNM)]

Chatzimanolis S, Cohen IM, Schomann A, Solodovnikov A (2010) Molecular phylogeny of the mega-diverse rove beetle tribe Staphylinini (Insecta, Coleoptera, Staphylinidae). Zoologica Scripta 39: 436–449.

Chaudoir M, de (1845) Notices entomologiques sur le gouvernement et la ville de Kiew. Bulletin de la Société Impériale des Naturalistes de Moscou 18 (3): 158–213. [after 6 Aug 1845 (Julian 25 Jul, approved to print)]

Chaudoir M, de (1846) Note sur le groupe des stomides et description d'un nouveau genre de celui des somoplatides. Bulletin de la Société Impériale des Naturalistes de Moscou 19 (4): 511–542. [after 30 Sep 1846 (Julian 18 Sep, approved to print); before 4 Feb 1847 (Ent. Zeit. Stettin)]

Chaudoir M, de (1848) Mémoire sur la famille des carabiques. 1. partie. Bulletin de la Société Impériale des Naturalistes de Moscou 21 (1): 3–134. [after 23 Feb 1848 (Julian 11 Feb, approved to print); 22 Jun 1848 (separate, Ent. Zeit. (Stettin) 9: 161)]

Chaudoir M, de (1861a) Matériaux pour servir à l'étude des cicindélètes et des carabiques. (1-e partie). Bulletin de la Société Impériale des Naturalistes de Moscou 33 (4 [1860]): 269–337. [after 2 Apr 1861 (Julian 21 Mar, approved to print)]

Chaudoir M, de (1861b) Matériaux pour servir à l'étude des cicindélètes et des carabiques (continuation). Bulletin de la Société Impériale des Naturalistes de Moscou 34 (2): 491–576. [after 31 Jul 1861 (Julian 19 Jul, approved to print, verso of fasc. title page)]

Chaudoir M, de (1863) Matériaux pour servir à l'étude des carabiques. 3-e partie. Bulletin de la Société Impériale des Naturalistes de Moscou 35 [1862] (4): 275–320. [after 28 May 1863 (Julian 16 May, approved to print, Bulletin version) / reissued as separate: Moscou, Imprimerie de l'Université Impériale, pp. 1–46, after 26 Mar 1863 (Julian 14 Mar, approved to print, verso of title page)]

Chaudoir M, de (1869) Mémoire sur les thyréoptérides. Annales de la Société Entomologique de Belgique 12 [1868–69]: 113–162. [Sep 1869 (Abeille 1870: xxxvii)]

Chaudoir M, de (1871a) Essai monographique sur le groupe des pogonides. Annales de la Société Entomologique de Belgique 14 [1870–71]: 21–59. [before 20 Feb 1872 (Acad. Sci. France 74: 558)]

Chaudoir M, de (1871b) Monographie des lébiides. Bulletin de la Société Impériale des Naturalistes de Moscou 43 [1870] (3): 111–255. [1871 (wrapper and title page)]

Chaudoir M, de (1872a) Remarques sur le catalogue de MM. de Harold et Gemminger, Tome I. Bulletin de la Société Impériale des Naturalistes de Moscou 44 [1871] (3/4): 279–287. [1872 (title page fasc. 3/4); separate [pp. 1–9] printed 1872 (p. 9)]

Chaudoir M, de (1872b) Observations sur quelques genres de carabiques, avec la description d'espèces nouvelles. Bulletin de la Société Impériale des Naturalistes de Moscou 45 (2): 382–420. [1872 (wrapper and title page)]

Chaudoir M, de (1873a) Essai monographique sur les drimostomides et les cartocérides et description d'un genre nouveau de morionides. Annales de la Société Entomologique de Belgique 15 ([1871–72]): 5–24. [before 12 Feb 1873 (Ann. Soc. Ent. France (5) 3: Bull. Ent.: xxvi)]

Chaudoir M, de (1873b) Monographie des callidides. Annales de la Société Entomologique de Belgique 15 ([1871–72]): 97–204. [12 Feb 1873 (Ann. Soc. Ent. France (5) 3: Bull. Ent.: xxvi)]

Chaudoir M, de (1874) Matériaux pour servir à l'étude des féroniens. Bulletin de la Société Impériale des Naturalistes de Moscou 48 (1): 1–34. [1874 (wrapper); by Oct 1874 (Nachr. Königl. Ges. Wiss. Göttingen 1874: 615)]

Chaudoir M, de (1876a) Étude monographique des masoréides, des tetragonodérides et du genre *Nematotarsus*. Bulletin de la Société Impériale des Naturalistes de Moscou 51 (3): 1–84. [1876 (wrapper); before 17 Apr 1877 (Sitzber. Ges. Naturf. Freunde Berlin 1877: 140]

Chaudoir M, de (1876b) Monographie des Siagonides. Bulletin de la Société Impériale des Naturalistes de Moscou 50 (1): 62–125. [before 6 Sep 1876 (Proc. Ent. Soc. London 1876: xxv)]

Chaudoir M, de (1877) Genres nouveaux et espèces inédites de la famille des carabiques. Bulletin de la Société Impériale des Naturalistes de Moscou 52 (2): 188–268. [before 20 Nov 1877 (Sitzber. Ges. Naturf. Freunde Berlin 1877: 248)]

Chaudoir M, de (1880) Essai monographique sur les morionides. Bulletin de la Société Impériale des Naturalistes de Moscou 55 (1): 317–384. [before 18 Oct 1880 (Zool. Anz. 3: 507)]

Chemsak JA, Linsley EG (1974) Reclassification, synonymy, and descriptions of some North and Central American Cerambycidae (Coleoptera). The Coleopterists Bulletin 28 (4): 181–184. [mailed 31 Dec 1974 (inside wrapper)]

Chen SH (1936) Catalogue des Chrysomelinae de la Chine, de l'Indochine et du Japon. Notes d'Entomologie Chinoise / Musée Heude 3 (5): 63–102. [18 Aug 1936 (fasc. title page); 3 Nov 1936 (recorded at BMNH)]

Chen SH (1940) Notes on Chinese Eumolpidae. Sinensia 11 (5/6): 483–528.

Chen SH (1941) Notes on Donaciine beetles. Sinensia 12 (1/6): 1–17. [Dec 1941 (fasc. title page)]

Chevrolat LAA (1836) [new taxa] In: Dejean PFMA. Catalogue des coléoptères de la collection de M. le comte Dejean. Deuxième Édition [Livraison 5: pp. 361–443]. Méquignon-Marvis Père et Fils, Paris, 443 pp. [end of 1836 (Madge 1988: 318)]

Chevrolat LAA (1842) Cephalodonta [p. 272]. In: D'Orbigny C (Ed) Dictionnaire universel d'histoire naturelle résumant et complétant tous les faits présentés par les encyclopédies, les ancients dictionnaires scientifiques, les oeuvres complètes de Buffon, et les meilleurs traités spéciaux sur les diverses branches des sciences naturelles; donnant la description des êtres et des divers phénomènes de la nature, l'étymologie et la définition des noms scientifiques, les principales applications des corps organiques et inorganiques, à l'agriculture, à la médecine, aux arts industriels, etc; dirigé par M Charles d'Orbigny, et enrichi d'un magnifique atlas de 288 planches gravées sur acier Tome troisième [Livraison 29]. MM. Renard, Martinet et Cie., Paris, 241–296[?]. [Nov 1842 (Evenhuis 1997b: 575)]

Chevrolat LAA (1843) Chaetosoma [p. 367]. In: D'Orbigny C (Ed) Dictionnaire universel d'histoire naturelle… Tome troisième. [Livraisons 30–35]. MM. Renard, Martinet et Cie., Paris. [2 Jan-29 May 1843 (Evenhuis 1997b: 575)]

Chevrolat LAA (1846) Monoplatus [p. 333]. In: D'Orbigny C (Ed) Dictionnaire universel d'histoire naturelle… Tome huitième [Livraisons 85–92]. MM. Renard, Martinet et Cie., Paris, 640 pp. [21 Sep-14 Dec 1846 (Evenhuis 1997b: 576)]

Chevrolat LAA (1847) Oncocephalus [p. 110], Oxygona [p. 368]. In: D'Orbigny C (Ed) Dictionnaire universel d'histoire naturelle… Tome neuvième [Livraisons 97–106]. MM. Renard, Martinet et Cie., Paris, 648 pp. [Feb-Jul 1847 (Evenhuis 1997b: 576)]

Chevrolat LAA (1872) Révision des Cléonides. Revue et Magasin de Zoologie Pure et Appliquée (2) 23 (1): 16–18. [Jan 1872 (p. 5)]

Chevrolat LAA (1878) Diagnoses de nouvelles espèces de Curculionides du genre *Ambates*. Annales de la Société Entomologique de France (5) 7 [1877] (4): 341–346. [10 Apr 1878 (wrapper)]

Chûjô M (1941) Descriptions of some new Erotylidae (Coleoptera) from the Japanese Empire. Kontyû 15 (1): 10–21. [Jul 1941 issue]

Chûjô M (1956) A taxonomic study on the Chrysomelidae (Insecta: Coleoptera) from Formosa Part VIII. Subfamily Eumolpinae. The Philippine Journal of Science 85 (1): 1–180. [Mar 1956 issue]

Chûjô M (1969) Erotylidae (Insecta: Coleoptera). Fauna Japonica. Academic Press of Japan, Tokyo, xii + 316 pp. + 23 pls. [Mar 1969 (verso of title page)]

Chûjô M, Voss E (1960) Neue Curculioniden-Subfamilie, -Gattungen und -Arten von Japan (Coleoptera, Curculionidae). Memoirs of the Faculty of Liberal Arts & Education, Kagawa University (Part II) 94: 1–17.

Clark PF, Crosnier A (2000) The zoology of the *Voyage au pôle sud et dans l'Océanie sur les corvettes l'Astrolabe et la Zélée exécuté par ordre du roi pendant les années 1837–1838–1839–1840 sous le commandement de M. Dumont-d'Urville* (1842–1854): titles, volumes, plates, text, contents, proposed dates and anecdotal history of the publication. Archives of Natural History 27: 407–435.

Clavareau H (1913a) Pars 51: Chrysomelidae. 1. Sagrinae, 2. Donaciinae, 3. Orsodacninae, 4. Criocerinae. In: Schenkling S (Ed) Coleopterorum Catalogus. Volumen XXIV. Chrysomelidae I. W. Junk, Berlin, 103 pp. [6 Mar 1913 (verso of vol. title page)]

Clavareau H (1913b) Pars 51: Chrysomelidae. 5. Megascelinae, 6. Megalopodinae, 7. Clytrinae, 8. Cryptocephalinae, 9. Chlamydinae, 10. Lamprosominae. In: Schenkling S (Ed) Coleopterorum Catalogus. Volumen XXIV. Chrysomelidae I. W. Junk, Berlin, 278 pp. [28 Aug 1913 (verso of vol. title page)]

Clavareau H (1914) Pars 59: Chrysomelidae 11. Eumolpinae. In: Schenkling S (Ed) Coleopterorum Catalogus. Volumen XXIV. Chrysomelidae I. W. Junk, Berlin, 215 pp. [22 Jun 1914 (verso of vol. title page)]

Cline AR, Shockley FW (2010) Biphyllidae LeConte, 1861 [pp. 306–311]. In: Lawrence JF (Ed) Handbook of Zoology Coleoptera, Beetles Volume 2: Morphology and systematics (Elateroidea, Bostrichiformia partim). De Gruyter, Berlin, New York, xiii + 786 pp. [2010 (copyright)]

Cobos A (1955) Estudio sobre los Ptosimites de Ch. Kerremans (Coleoptera, Buprestidae). Bulletin / Institut Royal des Sciences Naturelles de Belgique 31 (13): 1–24. [Apr 1955 (top of article)]

Cobos A (1956) Cuarta nota sobre Bupréstidos (Ins. Coleoptera) neotropicales. - Sobre la posición sistemática del género *Trigonogenium* Gem. et Harold y diversas descripciones de especies nuevas. Archivos del Instituto de Aclimatación 5: 69–95. [15 Feb 1957 (recorded at CNC)]

Cobos A (1959) Octava nota sobre buprestidos neotropicales. Rectificaciones y descripciones diversas. (Coleoptera, Buprestidae). Bulletin / Institut Royal des Sciences Naturelles de Belgique 35 (2): 1–47. [Jan 1959 (top of article)]

Cobos A (1961) Sobre la posición sistemática del género *Potergus* Bonvouloir y revisión de las categorías supragenéricas de la familia Throscidae (Coleoptera). Bulletin / Institut Royal des Sciences Naturelles de Belgique 37 (35): 1–6. [Nov 1961 (top of reprint)]

Cobos A (1963) Comentarios críticos sobre algunos Stemoxia fósiles del ámbar del Báltico recientemente descritos (Coleoptera). Eos, Revista Española de Entomologia 39 (3/4): 345–355. [31 Dec 1939 (p. 482)]

Cobos A (1965) Materiales para el estudio de la familia Eucnemidae. Primera parte (Coleoptera). Eos, Revista Española de Entomologia 40 [1964] (3/4): 289–435. [28 Feb 1965 (p. 588)]

Cobos A (1968) Decimo-sexta nota sobre bupréstidos neotropicales: sobre el género *Mendizabalia* y su posición sistematica (Coleoptera). Bolletino dell'Associazione Romana di Entomologia 23 (2): 17–20. [publ. 24 Apr 1968 (p. 24)]

Cobos A (1974) Notas sobre bupréstidos neotropicales, XIX. El género *Amorphosternus* H. Deyrolle y afines. Archivos del Instituto de Aclimatación 19: 65–81. [printed Apr 1974 (verso of title page); recorded at CNC 25 Jun 1974]

Cobos A (1975) Dos nuevas tribus de bupréstidos (Col. Buprestidae). Eos, Revista Española de Entomologia 49 (1/4): 87–104. [9 Jan 1975 (wrapper)]

Cobos A (1976) Estudio sobre *Rhaeboscelis* Chevrolat, 1837 y géneros afines (Col. Buprestidae). Eos, Revista Española de Entomologia 50 [1974] (1/4): 19–40. [30 Nov 1976 (wrapper)]

Cobos A (1979a) Rectificaciones sinonímicas de bupréstidos (Col.) a niveles supraespecíficos Cinyrini nov. nom. (= Paraleptodomini Cobos, nom. nudum). Boletín de la Asociacion Española de Entomologia 3 (Dec): 226. [Dec 1979 issue, publ Nov 1979 (p. 3)]

Cobos A (1979b) Revisión de la subfamilia Trachyinae a niveles supraespecíficos (Coleoptera, Buprestidae). Acta Entomologica Bohemoslovaca 76 (6): 414–430. [28 Dec 1979 (inside wrapper)]

Cobos A (1980) Ensayo sobre los géneros de la subfamilia Polycestinae (Coleoptera, Buprestidae). Parte I. Eos, Revista Española de Entomologia 54 [1978] (1/4): 15–94. [9 Apr 1980 (wrapper)]

Cobos A (1986) Fauna Ibérica de coleópteros Buprestidae. Consejo Superior de Investigaciones Científicas, Madrid, xi + 364 pp. + 60 plates.

Cobos A (1988) Revisión del género *Geralius* Harold, 1869 (Coleoptera, Buprestidae). Elytron 1 [1987]: 9–16.

Cobos A (1990) Revision del genero *Dismorpha* Gistel (Coleoptera, Buprestidae). Revista Brasileira de Entomologia 34 (3): 539–559. [15 Nov 1990 (each article)]

Cockerell TDA (1906) Preoccupied generic names of Coleoptera. Entomological News and Proceedings of the Entomological Section of the Academy of Natural Sciences of Philadelphia 17 (7): 240–244. [Sep 1906 issue]

Coiffait H (1957) Diagnoses préliminaires de nouveaux Leptotyphlitae (Deuxième note) [Col., Staphylinidae]. Revue Française d'Entomologie 24 (1): 60–81. [printed 1 Apr 1957 (p. 90)]

Coiffait H (1963) Les Leptotyphlitae (Col. Staphylinidae) du Chili; systématique et biogéographie de la sous-famille [pp. 371–383]. In: Delamare-Deboutteville C, Rapoport E (Eds) Biologie de l'Amérique Australe Volume II Études sur la faune du sol. Centre National de la Recherche Scientifique, Paris, 398 + [1] pp. [printed Aug 1963 (endleaf)].

Coiffait H (1978) Coléoptères Staphylinidae de la région paléarctique occidentale. III. Sous-famille Staphylininae, tribu Quediini; sous-famille Paederinae, tribu Pinophilini. Nouvelle Revue d'Entomologie, Supplément 8 (4): 364 pp.

Coiffait H (1982) Coléoptères Staphylinidae de la région paléarctique occidentale. IV. Sous-famille Paederinae, tribu Paederini 1 (Paederi, Lathrobii). Nouvelle Revue d'Entomologie, Supplément 12 (4): 440 pp.

Coiffait H (1984) Coléoptères Staphylinidae de la région paléarctique occidentale. V. Sous-famille Paederinae, tribu Paederini 2; sous-famille Euaesthetinae. Nouvelle Revue d'Entomologie, Supplément 13 (4): 424 pp.

Colonnelli E (1979) Note su *Panophthalmus* e generi affini, con descrizione di una nuova tribu e di una nuova specie della sottofamiglia Ceutorhynchinae (Coleoptera, Curculionidae). Bollettino dell'Associazione Romana di Entomologia 34 (1/4): 1–9. [15 Nov 1979 (back wrapper)]

Colonnelli E (1984) Notes sur quelques Ceutorhynchinae de l'Afrique tropicale (Coleoptera, Curculionidae). Annales Historico-Naturales Musei Nationalis Hungarici 76: 207–238.

Colonnelli E (2003) A revised checklist of Italian Curculionoidea (Coleoptera). Zootaxa 337: 1–142. [24 Oct 2003]

Colonnelli E (2004) Catalogue of Ceutorhynchinae of the world, with a key to genera (Insecta: Coleoptera: Curculionidae). Argania Editio, Barcelona, 124 pp.

Constantin R (2001) Révision du genre *Pagurodactylus* Gorham, 1900 et des genres voisins d'Afrique australe (Coleoptera, Malachiidae). Entomologica Basiliensia 23: 5–87. [1 Sep 2001 (Inhalt)]

Constantin R, Menier JJ (1987) Étude d'un remarquable Melyridae aptère des Iles Canaries: *Gietella fortunata*, n. gen., n. sp., type d'une sous-famille nouvelle Gietellinae (Coleoptera, Cleroidea). Revue Française d'Entomologie (Nouvelle Série) 9 (2): 53–63. [16 Jun 1987 (inside back wrapper)]

Cooman A (1939) Coléoptères Histeridae d'Extrême-Orient, principalement du Tonkin. Notes d'Entomologie Chinoise / Musée Heude 6 (6): 137–142. [15 Aug 1939 (wrapper)]

Costa A (1852) Coleotteri eteromeri, famiglia degli edemeridei - Oedemeridea. Fauna del Regno di Napoli, ossia enumerazione di tutti gli animali che abitano le diverse regione di questo regno e le acque che le bagnano contenente la descrizione de' nuovi o poco esattamente conosciuti Coleotteri, parte 1ª [1849–1854]. G. Sautto, Napoli, 36 pp., pls. 9–11. [1 Jun 1852 (date of introduction p. 4); issued in 5 parts in 1852 (q.v. Sherborn 1937a: 42)]

Costa A (1853) Coleotteri eteromeri, famiglia de' mordellidei - Mordellidea. Fauna del Regno di Napoli, ossia enumerazione di tutti gli animali che abitano le diverse regione di questo regno e le acque che le bagnano contenente la descrizione de' nuovi o poco esattamente conosciuti Coleotteri, parte 1ª [1849–1854]. G. Sautto, Napoli, 32 pp., pls. 19–22. [Oct 1853 (date of introduction p. 2); issued in 4 parts: 1 Dec 1853 (pp. 1–8), 1854 (9–16), Sep 1854 (17–24), 10 Dec 1853 (25–32) (Sherborn 1937a: 42)]

Costa C (1975) Systematics and evolution of the tribes Pyrophorini and Heligmini, with description of Campyloxeninae, new subfamily (Coleoptera, Elateridae). Arquivos de Zoologia, São Paulo 26 (2): 49–190. [8 Oct 1975 (top of article)]

Costa C, Casari-Chen SA (1993) A review of the Pachyderini (Denticollinae) sensu Stibick, 1979, with the description of Platycrepidiini trib. n. (Coleoptera, Elateridae, Pyrophorinae). Revista Brasileira de Entomologia 37 (1): 61–77. [31 Mar 1993 (top p. 61)]

Costa Lima Ad (1962) Micro-Coleóptero representante da nova subfamília Plaumanniolinae (Col., Ptinidae). Revista Brasileira de Biologia 22 (4): 413–418. [Dec 1962 (fasc. title page); 7 May 1963 (recorded at BMNH)]

Craighead FC (1921) Larva of the North American beetle *Sandalus niger* Knoch. Proceedings of the Entomological Society of Washington 23 (2): 44–48. [11 Mar 1921 (verso of vol. title page)]
Crotch GR (1873a) On the arrangement of the families of Coleoptera. Proceedings of the American Philosophical Society 13: 75–87. [7 Feb 1873 (top of pages)]
Crotch GR (1873b) Check list of the Coleoptera of America, North of Mexico. Naturalists' Agency, Salem, Mass., 136 pp. [after 30 Sep 1873 (Remarks p. 128)]
Crotch GR (1873c) Synopsis of the Erotylidae of Boreal America. Transactions of the American Entomological Society 4: 349–363.
Crotch GR (1873d) Revision of the Dytiscidae of the United States. Transactions of the American Entomological Society 4: 383–424. [May 1873 (footer p. 381)]
Crotch GR (1874) A revision of the coleopterous family Coccinellidae. E. W. Janson, London, xv + 311 pp. [after 1 Sep 1874 (letter p. viii)]
Crotch GR (1876) Revision of the coleopterous family Erotylidae. Cistula Entomologica 1 (13): 377–572. [Feb 1876 (wrapper)]
Crowson RA (1946) A revision of the genera of the Chrysomelid group Sagrinae (Coleoptera). The Transactions of the Royal Entomological Society of London 97: 75–115.
Crowson RA (1950) The classification of the families of British Coleoptera. The Entomologist's Monthly Magazine 86: 149–171, 274–288, 327–344. [30 Jun 1950 (pp. 149–171), 14 Oct (274–288), 21 Nov (327–336), 6 Dec 1950 (337–344) (publ. dates p. ii)]
Crowson RA (1952) The classification of the families of British Coleoptera [continued]. The Entomologist's Monthly Magazine 88: 64–72, 109–132. [25 Mar 1952 (pp. 64–72), 8 May (109–120), 26 Jun 1952 (121–132) (publ. dates p. ii)]
Crowson RA (1953) The classification of the families of British Coleoptera [continued]. The Entomologist's Monthly Magazine 89: 37–59, 181–198, 237–248. [3 Feb 1953 (pp. 37–48), 2 Mar (49–59), 9 Jul (181–192), 15 Aug (193–198), "Sep" (237–240), 21 Oct 1953 (241–248) (publ. dates on reprint)]
Crowson RA (1954) The classification of the families of British Coleoptera (concl. with addenda). The Entomologist's Monthly Magazine 90: 57–63. [publ. 8 Apr 1954 (p. ii)]
Crowson RA (1955) The natural classification of the families of Coleoptera. Nathaniel Lloyd, London, [2] + 187 pp.
Crowson RA (1962) Observations on the beetle family Cupedidae, with descriptions of two new fossil forms and a key to the recent genera. The Annals and Magazine of Natural History (13) 5 (51): 147–157, pls. 3–4. [18 Oct 1962 (Evenhuis 2003: 55)]
Crowson RA (1964) A review of the classification of Cleroidea (Coleoptera), with descriptions of two new genera of Peltidae and several new larval types. The Transactions of the Royal Entomological Society of London 116 (12): 275–327. [22 Dec 1964 (wrapper)]
Crowson RA (1966) Further observations on Peltidae (Coleoptera: Cleroidea), with definitions of a new subfamily and of four new genera. Proceedings of the Royal Entomological Society of London (Series B. Taxonomy) 35 (9/10): 119–127. [25 Oct 1966]
Crowson RA (1971) Observations on the superfamily Dascilloidea (Coleoptera: Polyphaga), with the inclusion of Karumiidae and Rhipiceridae. Zoological Journal of the Linnean Society 59 (1): 11–19. [Feb 1971 (top of article)]

Crowson RA (1972) A review of the classification of Cantharoidea (Coleoptera), with the definition of two new families, Cneoglossidae and Omethidae. Revista de la Universidad de Madrid. Estudios de Entomologia 21 (82): 35–77 [1972 (title page)].

Crowson RA (1973a) Further observations on Phloeostichidae and Cavognathidae, with definitions of new genera from Australia and New Zealand. The Coleopterists Bulletin 27 (2): 54–62. [mailed 30 Jun 1973 (wrapper)]

Crowson RA (1973b) On a new superfamily Artematopoidea of polyphagan beetles, with the definition of two new fossil genera from the Baltic amber. Journal of Natural History 7 (2): 225–238. [13 Feb 1973 (Evenhuis 2003: 58)]

Crowson RA (1975) The evolutionary history of Coleoptera, as documented by fossil and comparative evidence. Atti / Congresso Nazionale Italiano di Entomologia. Volume 10. Tip. Coppini, Firenze, 47–90.

Crowson RA (1979) Observations on Clambidae (Coleoptera), with descriptions of a new genus and species and of several larvae. Revue Suisse de Zoologie 86 (3): 611–623. [Sep 1979]

Crowson RA (1980) On amphipolar distribution patterns in some cool climate groups of Coleoptera. Entomologia Generalis 6 (2/4): 281–292. [28 Nov 1980 (p. v)]

Crowson RA (1981) The biology of the Coleoptera. Academic Press, New York, xii + 802 pp.

Crowson RA (1984) The associations of Coleoptera with Ascomycetes [pp. 256–285]. In: Wheeler QD, Blackwell M (Eds) Fungus-insect relationships: perspectives in ecology and evolution. Columbia University Press, New York, NY, xiii + 514 pp.

Crowson RA (1990) A new genus of Boganiidae (Coleoptera) from Australia, with observations on glandular openings, cycad associations and geographical distribution in the family. Journal of the Australian Entomological Society 29 (2): 91–99. [21 Jun 1990 (wrapper)]

Csiki E (1901) Catalogus Endomychidarum. Természetrajzi Füzetek 24: 53 pp. [sep. pagination, inserted after p. 272 of fasc. 1/2]. [1901 (title page); date not specified with other parts (vol. contents) but preface dated Oct [1901]; issues 3/4 publ. 20 Oct 1901]

Csiki E (1903) Die Cicindeliden Ungarns. Mathematische und naturwissenschaftliche berichte aus Ungarn 18 [1901]: 121–144.

Csiki E (1904) [Referate; note in review of Kerremans 1903 (Wytsman's Genera Insectorum, fasc. 12)]. Münchener Koleoperologische Zeitschrift 2 [1904–06] (2): 132–133. [29 Dec 1904]

Csiki E (1909a) Újabb adatok Magyarország bogárfaunájához. Rovartani Lapok 16 (1): 3–6. [28 Jan 1909 (footer p. 1)]

Csiki E (1909b) 2. had: Staphylinoidea. In: Magyarország Bogárfaunája. II Kötet. 1. Füzet. Magyar Nemzeti Múzeum, Budapest, 80 pp. [10 May 1909 (title page)]

Csiki E (1910) Pars 12: Endomychidae. In: Schenkling S (Ed) Coleopterorum Catalogus. Volumen XVI. W. Junk, Berlin, 68 pp. [5 Apr 1910 (verso of vol. title page)]

Csiki E (1919) Pars 70: Scydmaenidae. In: Schenkling S (Ed) Coleopterorum Catalogus. Volumen VII. W. Junk, Berlin, 106 pp. [4 Dec 1919 (verso of vol. title page)]

Csiki E (1927) Carabidae: Carabinae (Partes 91 et 92). In: Schenkling S (Ed) Coleopterorum Catalogus. Volumen I. Carabidae I. W. Junk, Berlin, 622 pp. [15 Dec 1927 (pars 91: pp. 1–314), 22 Dec 1927 (pars 92: pp. 315–622) (verso of vol. title page)]

Csiki E (1928) Carabidae: Mormolycinae, Harpalinae. (Partes 97 et 98). In: Schenkling S (Ed) Coleopterorum Catalogus. Volumen II. Carabidae II. W. Junk, Berlin, 345 pp. [12 Sep 1928 (pars 97: pp. 1–226), 28 Dec 1928 (pars 98: pp. 227–345) (verso of vol. title page)]

Csiki E (1929) Carabidae: Harpalinae III (Pars 104) [pp. 347–527]. In: Schenkling S (Ed) Coleopterorum Catalogus. Volumen II. Carabidae II. W. Junk, Berlin, 1022 pp. [20 Mar 1929 (verso of vol. title page)]

Csiki E (1932a) Pars 121: Carabidae: Harpalinae VI [pp. 1023–1278]. In: Schenkling S (Ed) Coleopterorum Catalogus. Volumen III. Carabidae III: Harpalinae II. W. Junk, Berlin, pp. 1023–1933. [28 May 1932 (verso of title page)]

Csiki E (1932b) Pars 124. Carabidae: Harpalinae VII [pp. 1279–1598]. In: Schenkling S (Ed) Coleopterorum Catalogus. Volumen III. Carabidae III: Harpalinae II. W. Junk, Berlin, pp. 1023–1933. [7 Nov 1932 (verso of vol. title page)]

Csiki E (1933) Pars 126. Carabidae: Harpalinae VIII, Corrigenda et Addenda [pp. 1599–1679]. In: Schenkling S (Ed) Coleopterorum Catalogus. Volumen III. Carabidae III: Harpalinae II. W. Junk, Berlin, pp. 1023–1933. [26 May 1933 (verso of vol. title page)]

Csiki E (1934) Pars 137: Curculionidae: Subfam. Hyperinae In: Schenkling S (Ed) Coleopterorum Catalogus. Volumen XXVIII. W. Junk, Berlin, 66 pp. [22 Jun 1934 (verso of vol. title page)]

Csiki E (1936) Pars 149: Curculionidae: Rhynchophorinae, Cossoninae. In: Schenkling S (Ed) Coleopterorum Catalogus. Volumen XXX. Curculionidae IV. W. Junk, Berlin, 212 pp. [15 Aug 1936 (verso of vol. title page)]

Csiki E (1953) Über neue und bekannte Coleopteren aus Ungarn und den angrenzenden Ländern. Annales Historico-Naturales Musei Nationalis Hungarici (Series Nova) 3 [1952]: 115–135. [31 Jul 1953 (p. [310])]

Cuccodoro G (2007) Family Lymexylidae Fleming, 1821 [pp. 362–363]. In: Löbl I, Smetana A (Eds) Catalogue of Palaearctic Coleoptera. Volume 4. Elateroidea - Derodontoidea - Bostrichoidea - Lymexyloidea - Cleroidea - Cucujoidea. Apollo Books, Stenstrup, 935 pp. [publ. 30 Jun 2007 (verso of title page)]

Curtis J (1829) British entomology; being illustrations and descriptions of the genera of insects found in Great Britain and Ireland: containing coloured figures from nature of the most rare and beautiful species, and in many instances of the plants upon which they are found. Vol. VI. J. Curtis, London, pls. 242–289. [1 Nov 1829 (dates on plates)]

Curtis J (1830) British entomology; being illustrations and descriptions of the genera of insects found in Great Britain and Ireland: containing coloured figures from nature of the most rare and beautiful species, and in many instances of the plants upon which they are found. Vol. VII. J. Curtis, London, pls. 290–337. [1 Feb 1830 (dates on plates)]

Curtis J (1834) British entomology; being illustrations and descriptions of the genera of insects found in Great Britain and Ireland: containing coloured figures from nature of the most rare and beautiful species, and in many instances of the plants upon which they are found. Vol. XI. L. Reeve & Co., London, pls. 482–529. [1 May 1834 (dates on plates)]

Daccordi M (1981) Nuovi taxa di Chrysomelinae afrotropicali. Bollettino del Museo Civico di Storia Naturale di Verona 7 [1980]: 181–195. [31 Mar 1981 (reprint wrapper)]

Dajoz R (1963) *Dolosus leleupi* n.g., n. sp. et *Dolosus basilewskyi* n. sp., types d'une famille nouvelle de Cucujoidea (Coléoptères). Revue de Zoologie et de Botanique Africaines 67 (1/2): 91–96. [30 Mar 1963 (top of article)]

Dajoz R (1964) Anatomie et importance taxinomique des voies génitales femelles d'origine ectodermique chez les Elateridae (Insectes, coléoptères). Cahiers des Naturalistes: Bulletin des Naturalistes Parisiens (Nouvelle Série) 20: 55–72.

Dajoz R (1973) Description du coléoptère *Chiloea chilensis*, n.g., n.sp., type d'une nouvelle famille: Chiloeidae. Annales de la Société Entomologique de France (Nouvelle Série) 9 (1): 173–179. [Jan-Mar 1973 issue; 21 Jun 1973 (recorded at CNC)]

Dajoz R (1980a) Insectes coléoptères: Colydiidae et Cerylonidae. Faune de Madagascar 54: 1–256.

Dajoz R (1980b) Description de *Orvoenia borneensis*, n. gen., n. sp., coléoptère Colydiidae appartenant à une tribu nouvelle. Revue Française d'Entomologie (Nouvelle Série) 2 (4): 190–192. [dépot légal 4e trimestre 1980 (p. 194)]

Dalla Torre KW, Emden M, van, Emden F, van (1936) Pars 147: Curculionidae: Subfam. Brachyderinae I. In: Schenkling S (Ed) Coleopterorum Catalogus. Volumen XXVII. W. Junk, Berlin, 132 pp. [15 Mar 1936 (verso of vol. title page)]

Damoiseau R (1962) Contribution à la connaissance des Brentides (Coleoptera - Brentidae). 4. - Revision de quelques genres africains et américains. Bulletin / Institut Royal des Sciences Naturelles de Belgique 38 (26): 1–35. [Sep 1962 (top of article)]

Damoiseau R (1965) Contribution à la connaissance des Brentidae (Coleoptera - Curculionoidea). 22. - Révision des Calodrominae palaeotropicaux et description d'espèces nouvelles. Bulletin / Institut Royal des Sciences Naturelles de Belgique 41 (34): 1–28. [Sep 1965 (top of article)]

Danilevsky ML (1997) [new taxa] In: Althoff J, Danilevsky ML: Seznam kozličev (Coleoptera, Cerambycoidea) Evrope. A check-list of longicorn beetles (Coleoptera, Cerambycoidea) of Europe. Slovensko Entomološko Društvo Štefana Michielija, Ljubljana, 64 pp.

Danilevsky ML (2010) New acts and comments. Cerambycidae [pp. 43–49]. In: Löbl I, Smetana A (Eds) Catalogue of Palaearctic Coleoptera. Volume 6. Chrysomeloidea. Apollo Books, Stenstrup, 924 pp. [publ. 22 Feb 2010 (verso of title page)]

Darlington PJ, Jr. (1933) A new tribe of Carabidae (Coleoptera) from western United States. The Pan-Pacific Entomologist 9 (3): 110–114. [Jul issue; mailed 8 Sep 1933 (p. 188)]

Darlington PJ, Jr. (1950) Paussid beetles. Transactions of the American Entomological Society 76 (2): 47–142. [25 Aug 1950 (p. iii)]

Darlington PJ, Jr. (1968) The carabid beetles of New Guinea Part 3. Harpalinae (continued): Perigonini to Pseudomorphini. Bulletin of the Museum of Comparative Zoology at Harvard 137 (1): 1–253. [30 Jul 1968 (wrapper)]

De Moor PP (1970) Monograph of the Praeugenina (Coleoptera: Tenebrionidae, Strongyliini). Transvaal Museum Memoir 17: vii + 203 pp., 8 pls., 8 maps. [31 Jul 1970 (title page)]

Decelle J (1966) *Bruchus serratus* Ol., 1790, espèce-type du genre *Caryedon* Schönherr, 1823. Revue de Zoologie et de Botanique Africaines 74 (1/2): 169–173. [30 Sep 1966 (p. 433)]

Dejean PFMA (1825) Species général des coléoptères, de la collection de M. le Comte Dejean. Tome premier. Crevot, Paris, xxx + 463 pp. [1825 (title page); 10 Sep 1825 (Bibliogr. France 1825)]

Dejean PFMA (1835) Catalogue des coléoptères de la collection de M. le comte Dejean. Deuxième Édition [Livraison 4]. Méquignon-Marvis Père et Fils, Paris, pp. 257–360. [22 Aug 1835 (Bibliogr. France 1835: 514)]

Della Beffa G (1912) Revisioine dei Coccinellidi italiana. Rivista Coleotterologica Italiana 10 (8–11): 145–192, i-ii, i-ii [pl. expl.], pls. 1–2. [Aug-Nov 1912 issues]

Denier PCL (1935) Coleopterorum americanorum familiae Meloidarum. Enumeratio synonymica. Revista de la Sociedad Entomológica Argentina 7: 139–176.

Derksen W, Scheiding-Göllner U (1965) Index Litteraturae Entomologicae. Serie II: Die Welt-Litteratur über die gesamte Entomologie von 1864 bis 1900. Band II F-L. Deutsche Akademie der Landwirtschaftwissenschaften, Berlin, 678 pp.

Desbrochers des Loges J (1873) Notes synonymiques. - Remarques diverses. - Description de Coléoptères nouveaux. Annales de la Société Entomologique de France (5) 2 [1872] (4): 420–432. [23 Apr 1873 (p. cii)]

Desbrochers des Loges J (1892) Révision des Baridiides d'Europe et des contrées limitrophes [continued]. Le Frelon 2 (5/6): 65–69. [Aug/Sep 1892 issue]

Desbrochers des Loges J (1898) Monographie des Curculionides appartenant au groupe des Holcorhinidae. Le Frelon 6 (7/8): 1–20 [special pagination]. [15 Apr 1898 (wrapper)]

Desbrochers des Loges J (1900) Faunule des coléoptères de la France et de la Corse. Melandryidae. Le Frelon 9 (1): 1–16, (2): 17–32, (3): 33–48, (4): 49–64 [, (5): 65–76]. [15 Oct 1900 (fasc. 1), 26 Oct (2), 15 Nov (3), 8 Dec (4), 9 Jan 1901 (5) (wrappers)]

Desbrochers des Loges J (1901) Faunule des coléoptères de la France et de la Corse. Tenebrionidae. Le Frelon 9 (9): 137–144, (10/11): 145–176, (12): 177–193. [20 May 1901 (fasc. 9), 10 Jun (10/11), 17 Jul 1901 (12) (wrappers)]

Desbrochers des Loges J (1902a) Faunule des coléoptères de la France et de la Corse. Tenebrionidae (Suite). Le Frelon 10 [(3): 37–42, (4): 49–64, (5): 65–80], (6): 81–96, (7): 97–104. [30 Sep 1901 (fasc. 3), [pp. 43–48 not printed], 24 Oct (4); 19 Dec 1901 (5), 31 Jan 1902 (6), Mar 1902 (7) (wrappers)]

Desbrochers des Loges J (1902b) Faunule des coléoptères de la France et de la Corse: Tenebrionidae (Suite). Le Frelon 11 (1): 1–16, (2): 17–32, (3): 33–48, (4): 49–56. [27 Aug 1902 (fasc. 1), 15 Sep (2), 17 Oct (3), 3 Nov 1902 (4) (wrappers)]

Desbrochers des Loges J (1908) Faunule des coléoptères de la France et de la Corse. Curculionides de la tribu des attélabides et des rhinomacérides. Le Frelon 16 (1): 1–12, (2/3): 13–36. [7 Mar 1908 (fasc. 1), 4 Apr 1908 (2/3) (wrappers)]

Descarpentries A (1970) Note sur un Buprestide inédit de Madagascar appartenant à une nouvelle tribu (Col.). Bulletin de la Société Entomologique de France 74 [1969] (7/8): 188–192. [publ. 26 Feb 1970 (p. 260)]

Desmarest E (1851) Coléoptères. Cicindélètes, carabiques, dytisciens, hydrophiliens, sylphales et nitidulaires [see note]. Encyclopédie d'histoire naturelle ou traité complet de cette science d'après les travaux des naturalistes les plus éminents de tous les pays et de toutes les époques Buffon, Daubenton, Lacépède, G Cuvier, F Cuvier, Geoffroy Saint-Hilaire, Latreille, de Jussieu, Brongniart, etc, etc Ouvrage résumant les observations des auteurs anciens et comprenant toutes les découvertes modernes jusqu'à nos jours. Marescq & Compagnie, Paris,

[2] + 312 pp., 28 pls. [1re partie, livr. 1–6 and 2e partie, livr. 1–3 publ. by 1851 (Ann. Soc. Ent. France (2) 9: Bull. Ent.: cxxv)]

Desmarest E (1857) Coléoptères. Staphyliniens, pselaphiens, dermestiens, ptiniens, cleriens, malachiens, etc. Deuxième partie. In: Chenu JC (Ed) Encyclopédie d'histoire naturelle ou traité complet de cette science d'après les travaux des naturalistes les plus éminents de tous les pays et de toutes les époques Buffon, Daubenton, Lacépède, G Cuvier, F Cuvier, Geoffroy Saint-Hilaire, Latreille, de Jussieu, Brongniart, etc, etc Ouvrage résumant les observations des auteurs anciens et comprenant toutes les découvertes modernes jusqu'à nos jours. Marescq & Compagnie, Paris, [2] + 312 pp., 44 pls. [before 9 May 1857 (both vols. in Bibliogr. France 1857)]

Desmarest E (1860) Coléoptères. Buprestiens, scarabaéiens, piméliens, curculioniens, scolytiens, chrysoméliens, etc. Troisième partie. In: Chenu JC (Ed) Encyclopédie d'histoire naturelle ou traité complet de cette science d'après les travaux des naturalistes les plus éminents de tous les pays et de toutes les époques Buffon, Daubenton, Lacépède, G Cuvier, F Cuvier, Geoffroy Saint-Hilaire, Latreille, de Jussieu, Brongniart, etc, etc Ouvrage résumant les observations des auteurs anciens et comprenant toutes les découvertes modernes jusqu'à nos jours. Marescq & Compagnie, Paris, [3] + 360 pp., 48 pls. [31 Jan 1860 (date of Préface by E. Desmarest); 11 Jul 1860 (Ann. Soc. Ent. France (3) 8: Bull. Ent.: lxiii; cites 3 vols.)]

Deuve T (1997) *Sinozolus yuae* n. gen., n. sp., premier représentant des Zolinae dans l'hémisphère nord (Coleoptera, Trechidae). Bulletin de la Société Entomologique de France 102 (1): 31–37. [Mar 1997 (dépot légal, recorded at CNC 2 Apr)]

Deuve T (2007) Description d'une quatrième espèce de la famille des Gehringiidae: *Helenaea felixi* n. sp., du Yémen (Coleoptera, Caraboidea). Nouvelle Revue d'Entomologie (Nouvelle Série) 23 [2006] (3): 213–218. [Sep/Oct 2006 issue; publ. 2 Apr 2007 (back wrapper)]

Deuve T, Tian M (2001) *Zolinopatrobus nanlingensis* n. gen., n. sp., de la Chine méridionale, premier représentant d'une nouvelle sous-tribu parmi les Trechidae Patrobinae (Coleoptera, Caraboidea). Bulletin de la Société Entomologique de France 106 (4): 417–422. [Oct 2001 (dépot légal)]

Dewailly P (1950) Coléoptères Melolonthini de Madagascar. Mémoires de l'Institut Scientifique de Madagascar (Série A. Biologie Animale) 4 (2): 209–454. [dépôt légal 4e trim. 1950; 8 Mar 1951 (recorded at BMNH)]

Deyrolle A (1867) Monographie de la tribu des zophosites. Annales de la Société Entomologique de France (4) 7 (2): 73–248, pls. 1–4. [15 Oct 1867 (wrapper; p. cxxix)]

Deyrolle H (1865) Description des buprestides de la Malaisie recueillis par M. Wallace. Annales de la Société Entomologique de Belgique 8 [1864]: 1–272, pls. 1–4. [at least Mar 1865 (p. 280); before 9 Aug 1865 (Ann. Soc. Ent. France (5) 14: 514)]

Di Iorio OR (2003) Taxonomy and systematics of Cerambycidae from Argentina, *Alanizus tortuosus* gen. et sp. n. (Coleoptera, Cerambycidae). Les Cahiers Magellanes 19: 1–8. [publ. 30 Jan 2003 (p. 8)]

Didier R, Séguy E (1953) Catalogue illustré des lucanides du globe. Texte. Encyclopédie entomologique, Sér A Travaux généraux. Vol. 27–28. Paul Lechevalier, Paris, 223 pp. [vol. 27], 112 pls. [vol. 28: atlas]. [printed 29 Jun 1953 (endleaf); 4 Nov 1953 (recorded at BMNH)]

Dieckmann L (1983) Beiträge zur Insektenfauna der DDR: Coleoptera - Curculionidae (Tanymecinae, Leptopiinae, Cleoninae, Tanyrhynchinae, Cossoninae, Raymondionymidae, Bagoinae, Tanysphyrinae). Beiträge zur Entomologie 33 (2): 257–381.

Dillon ES, Dillon LS (1959a) The Monochamini (Cerambycidae) of the Ethiopian faunistic region. V. The subtribe Acridocephalidi. The Coleopterists Bulletin 12 [1958]: 49–58. [27 Jan 1959 (back wrapper)]

Dillon ES, Dillon LS (1959b) The Monochamini (Cerambycidae) of the Ethiopian faunistic region. VI. The subtribe Docohammidi. The Coleopterists Bulletin 13 (1): 7–12. [27 Mar 1959 (vol. contents)]

Dillon LS (1952) The Meloidae (Coleoptera) of Texas. American Midland Naturalist 48 (2): 330–420. [Sep 1952 issue]

Dillon LS, Dillon ES (1945) Revision of the tribe Pachypezini (Coleoptera, Cerambycidae). Bulletin of the Brooklyn Entomological Society 40 (1): 11–27. [mailed 15 May 1945 (wrapper)]

Dobzhansky T (1924) Die weiblichen Generationsorgane der Coccinelliden als Artmerkmal betrachtet (Col.). Entomologische Mitteilungen 13 (1): 18–27.

Dolin VG (1973) Iskopaemye formy zhukov-shchelkunov (Elateridae, Coleoptera) is Nizhney Yury Sredney Asii. [Fossil forms of click-beetles (Elateridae, Coleoptera) from the Lower Jurassic of central Asia] [pp. 71–82]. In: Yaroshenko MF, et al. (Ed) Fauna i biologiya nasekomykh Moldavii [Fauna and biology of Moldavian insects]. Izdatel'stvo «Shtiintsa», Institut Zoologii, Akademiya Nauk Moldavskoy SSR, Kishinev, 188 pp. [in Russian].

Dolin VG (1975a) K sistematike mezozoiskikh zhukov-schchelkunov (Coleoptera, Elateridae) [A contribution to the systematics of Mesozoic click beetles (Coleoptera, Elateridae)]. Paleontologicheskii Zhurnal 9 (4): 51–62 [in Russian; English translation in Paleontological Journal 9: 474–486].

Dolin VG (1975b) Zhilkovanie krylev zhukov-shchelkunov (Coleoptera, Elateridae) i ego znachenie dlya sistematiki semeystva. [Wing venation of click beetles (Coleoptera, Elateridae) and its importance for taxonomy of the family]. Zoologicheskii Zhurnal 54 (11): 1618–1633 [in Russian]. [after 30 Oct 1975 (approved to print, last page)]

Dolin VG (1980) Click beetles (Coleoptera, Elateridae) from the Upper Jurassic of Karatau [pp. 17–81]. In: Dolin VG, Panfilov DV, Ponomarenko AG, Pritykina DN (Eds). Iskopaemye nasekomye mezozoya [Fossil insects of the Mesozoic]. Naukova Dumka, Kiev, [40] + 136 pp. + 21 pls. [in Russian].

Dolin VG (1990) Die Stellung der Gattung *Penia* Cast. im System der Elateridae (Coleoptera) auf Grund der Larvenmerkmale. Mitteilungen der Entomologischen Gesellschaft Basel 40 (1/2): 15–19.

Dolin VG (2000) Znachenie lichinochnykh priznakov i zhilkovaniya krylev v sistematike Elateroidea (Coleoptera) [A role of larval and wing venation characters in the systematics of Elateroidea (Coleoptera)]. Doklad na pyat'desyat vtorom ezhegodnom chtenii pamyati N A Kholodkskogo 1 aprelya 1999 g. Zoologicheskii Institut, St-Petersburg, 49 + [1] pp. [in Russian]. [2000 (title page); after 30 Dec 1999 (approved to print, endleaf)]

Doyen JT (1984) Systematics of *Eusattus* and *Conisattus* (Coleoptera; Tenebrionidae; Coniontini; Eusatti). Occasional Papers of the California Academy of Sciences 141: [3] + 1–104. [19 Sep 1984]

Doyen JT (1989) Reconstitution of Coelometopini, Tenebrionini and related tribes of America north of Colombia (Coleoptera: Tenebrionidae). Journal of the New York Entomological Society 97 (3): 277–304. [mailed 22 Aug 1989 (inside wrapper)]

Doyen JT, Matthews EG, Lawrence JF (1990) Classification and annotated checklist of the Australian genera of Tenebrionidae (Coleoptera). Invertebrate Taxonomy 3 [1989] (3): 229–260. [5 Feb 1990 (verso of vol. 4 title page)]

Dreux P, Voisin JF (1989) Sur la systématique des genres de la sous-famille des Ectemnorrhininae (Coleoptera, Curculionidae). Nouvelle Revue d'Entomologie 6 (2): 111–118. [issued 2 Oct 1989 (back wrapper)]

Drumont A, Komiya Z (2010) Family Cerambycidae. Subfamily Prioninae Latreille, 1802 [pp. 86–95]. In: Löbl I, Smetana A (Eds) Catalogue of Palaearctic Coleoptera. Volume 6. Chrysomeloidea. Apollo Books, Stenstrup, 924 pp. [publ. 22 Feb 2010 (verso of title page)]

Dubois A, Bour R (2010) The distinction between family-series and class-series nomina in zoological nomenclature, with emphasis on the nomina created by Batsch (1788, 1789) and on the higher nomenclature of turtles. Bonn Zoological Bulletin 57 (2): 149–171.

Dugès E (1886) Note pour servir à la classification des Méloïdes du Méxique. Bulletin de la Société Zoologique de France 11: 578–582. [publ. 10 Sep 1886 (p. [lv])]

Dugès E (1889) Description of *Leonia rileyi*, a new meloïd genus near *Hornia*. Insect Life 1 (7): 211–213. [Jan issue 1889]

Duncan FM, Waterhouse CO, Peavot H (1937) On the dates of publication of the Society's 'Proceedings', 1859–1926. Proceedings of the Zoological Society of London 1937 A (1): 71–84.

Duponchel PAJ, Chevrolat LAA (1841) Apophylia [p. 31]. In: D'Orbigny C (Ed). Dictionnaire universel d'histoire naturelle. Tome second [Livraisons 13–16]. MM. Renard, Martinet et Cie., Paris, 256 pp. [1 Feb-8 Nov 1841 (Evenhuis 1997b: 575)]

Dupont H (1836) Monographie des trachydérides. Magasin de Zoologie 6: 1–51 + pls. 141–164.

Dupuis C (1986) Dates de publication de l'«Histoire naturelle générale et particulière des crustacés et des insectes» (1802–1805) par Latreille dans le «Buffon de Sonnini». Annales de la Société Entomologique de France (Nouvelle série) 22: 205–210.

Dupuis P (1912) 126me fascicule. Coleoptera Adephaga. Fam. Carabidae. Subfam. Opisthiinae. In: Wytsman PA (Ed) Genera Insectorum. Vol. XX. P. Wytsman, Bruxelles, 2 pp. + 1 pl. [15 Jan 1912 (date on manuscr.); 25 May 1912 (Evenhuis 1994: 56)]

Duverger C (1990) Contribution à l'étude des Hyperaspinae. 1ère note (Coleoptera, Coccinellidae). Bulletin de la Société Linnéenne de Bordeaux 17 [1989] (3): 143–157. [printed 15 Feb 1990 (endleaf)]

Duverger C (2003) Phylogénie des Coccinellidae. Bulletin de la Société Linnéenne de Bordeaux 31 (2): 57–76. [printed 15 Jul 2003 (p. 124)]

Dybas HS (1955) New feather-wing beetles from termite nests in the American tropics (Coleoptera: Ptiliidae). Fieldiana: Zoology 37 (21): 561–577. [publ. 19 Jun 1955 (p. v)]

Dybas HS (1966) Evidence for parthenogenesis in the featherwing beetles, with a taxonomic review of a new genus, and eight new species (Coleoptera: Ptiliidae). Fieldiana: Zoology 51 (2): 11–52. [28 Nov 1966 (top of article)]

Edwards JG (1953) Species of the genus Syneta of the World (Coleoptera: Chrysomeloidea). Wasmann Journal of Biology 11 (1): 23–82. [spring issue (CNC recorded 13 Jul 1953)]

Egorov LV (1990) On systematics of tenebrionid beetles of the tribe Platyscelidini (Coleoptera, Tenebrionidae). Entomologicheskoe Obozrenie 69 (2): 401–412 [in Russian; English translation in Entomological Review, 69 [1990] (6): 137–150]. [approved to print 18 Jun 1990 (back wrapper)]

Eichelbaum F (1909) Katalog der Staphyliniden-Gattungen, nebst Angabe ihrer Literatur, Synonyme, Artenzahl, geographischen Verbreitung und ihrer bekannten Larvenzustände. Mémoires de la Société Entomologique de Belgique 17: 71–280. [1909 (wrapper); distrib. by 26 Dec 1909 (Ann. Soc. Ent. Belg. 53: 515)]

Eichhoff WJ (1878) Ratio, descriptio, emendatio eorum Tomicinorum qui sunt in Dr medic. Chapuisii et autoris ipsius collectionibus et quos praeterea recognovit. Mémoires de la Société Royale des Sciences de Liège (2) 8 (1): i-iv + 1–531 + pls. i-v. [31 Dec 1878 (Alonso-Zarazaga and Lyal 2009: 118); Dec 1878 (title page)]

El Moursy AA (1961) A tentative classification of and a key to the North American genera of the family Byrrhidae (new sense) and family Syncalyptidae (new status) (Coleoptera, Polyphaga, Byrrhoidea). The Coleopterists Bulletin 15 (1): 9–15. [29 Mar 1961 (contents)]

Elgueta DM, Arriagada SG (1989) Estado actual del conocimiento de los Coleópteros de Chile (Insecta: Coleoptera). Revista Chilena de Entomología 17: 5–60. [printed Dec 1989 (p. 4)]

Emden FI, van (1924) Zur Kenntnis der Sandalidae, II. u. III. Entomologische Blätter 20 (2): 86–99.

Emden FI, van (1931) Zur Kenntnis der Sandalidae XI–XVI. Entomologische Blätter 27 (2): 49–59, (3): 107–116, (4): 145–152, pl. 1. [30 Jun (fasc. 2); 30 Sep (3); 31 Dec 1931 (4) (wrappers)]

Emden FI, van (1933) Die Philippinischen Arten der Untergattung *Callirhipis* (Coleoptera). Zur Kenntnis der Sandalidae 19; zugleich Philippinische Sandalidae 2. The Philippine Journal of Science 51 (3): 331–356 [Jul 1933 (wrapper)].

Emden FI, van (1936) Die Anordnung der Brachyderinae-Gattungen im Coleopterorum Catalogus. Stettiner Entomologische Zeitung 97 (1): 66–99. [25 May 1936 (Inhalt)]

Emden FI, van (1958) New South American Carabidae (Coleoptera) with notes on described species. The Annals and Magazine of Natural History (13) 1 (1): 19–32. [publ. 27 Jun 1958 (wrapper)]

Enderlein G (1909) Die Insekten des antarktischen Gebietes. Deutsche Südpolar-Expedition 1901–1903 im Auftrage des Reichsamtes des Innern. X Band. Zoologie II Band. G. Reimer, Berlin, 361–528 + pls. 40–63. [Jan 1909 (Contents p. 3)]

Endrödi S (1966) Monographie der Dynastinae. Teil I. Entomologische Abhandlungen (Dresden) 33: 1 -457.

Endrödy-Younga S (1989) Restructuring of the tribe Cryptochilini (Coleoptera: Tenebrionidae: Tentyriinae). Annals of the Transvaal Museum 35 (6): 109–145. [Dec 1989 (p. 109)]

Endrödy-Younga S (1997) Active extraction of water-dissolved oxygen and descriptions of new taxa of Torridincolidae (Coleoptera: Myxophaga). Annals of the Transvaal Museum 36 (24): 313–332. [Feb 1997 (footer p. 313)]

Endrödy-Younga S, Crowson RA (1986) Boganiidae, a new beetle family for the African fauna (Coleoptera: Cucujoidea). Annals of the Transvaal Museum 34 (12): 253–273. [Nov issue 1986]

Engel MS (2005) Family-group names for bees (Hymenoptera: Apoidea). American Museum Novitates No. 3476: 1–33, 1 table. [11 May 2005 (top of p. 1)]

Engel MS, Bouchard P (2009) Case 3484. Nomiidae Gozis, 1875 (Insecta, Coleoptera): proposed emendation of spelling to Nomiusidae to remove homonymy with Nomiinae Robertson, 1904 (Insecta, Hymenoptera). Bulletin of Zoological Nomenclature 66 (1): 30–33.

Engel MS, Haas F (2007) Family-group names for earwigs (Dermaptera). American Museum Novitates No. 3567: 1–20, 3 tables. [16 May 2007 (wrapper)]

Engel MS, Krishna K (2004) Family-group names for termites (Isoptera). American Museum Novitates 3432: 1–9. [27 Feb 2004 (wrapper)]

Erichson WF (1832) Genera Dyticeorum. Nietack, Berolini, ii + 48 pp. [1832 (title page); before 6 Nov 1833 (Ann. Soc. Ent. France 2: Bull. Ent.: lxv)]

Erichson WF (1836) Systematische Auseinandersetzung der Familie der Borkenkäfer (Bostrichidae). Archiv für Naturgeschichte 2 (1): 45–65. [article read 8 May 1836 (p. 313); before 10 Aug 1836 (Literar. Zeit. 1836: 639); cites Westwood 1836, so it is later]

Erichson WF (1837) Die Käfer der Mark Brandenburg. Erster Band. Erste Abtheilung [pp. i-viii + 1–384]. F.H. Morin, Berlin, viii + 740 pp. [after 20 Jul (forward p. viii); 15 Sep 1837 (Allgem. Bibliogr. Deutschl. 1837: 560)]

Erichson WF (1838) Bericht über die Leistungen in der Entomologie während des Jahres 1837. Archiv für Naturgeschichte 4 (2): 187–264.

Erichson WF (1839a) Die Käfer der Mark Brandenburg. Erster Band. Zweite Abtheilung [pp. 385–740]. F.H. Morin, Berlin, viii + 740 pp. [12 Apr 1839 (Allgem. Bibliogr. Deutschl.)]

Erichson WF (1839b) Genera et species staphylinorum insectorum coleopterorum familiae [pp. 1–400]. F.H. Morin, Berlin, viii + 954 pp. + 5 pls. [1 Nov 1839 (Allgem. Bibliogr. Deutschl. 4: 648)]

Erichson WF (1840a) [Pp. i-viii, 401–954]. In: Genera et species staphylinorum insectorum coleopterorum familiae. F.H. Morin, Berlin, viii + 954 pp. + 5 pls.

Erichson WF (1840b) Die Pachypoden, eine kleine Gruppe aus der Familie der Melolonthen [pp. 29–44]. In: Entomographien, Untersuchungen in dem Gebiete der Entomologie mit besonderer Benutzung der Königl Sammlung in Berlin Erstes Heft. F. H. Morin, Berlin, x + 180 pp., 2 pls. [1 Aug 1840 (date of Vorwort); Sep 1840 (Ent. Zeit. Stettin 1: 144)]

Erichson WF (1842) Beitrag zur Insecten-Fauna von Vandiemensland, mit besonderer Berücksichtigung der geographischen Verbreitung der Insecten. Archiv für Naturgeschichte 8 (1): 83–287, pls. 4–5 [in German; English translations in Papers and Proceedings of the Royal Society of Tasmania, 1859, 3: 289–338 (translated by Miss S.A. Fogg.)]. [1842 (issue title page); before 22 Jun 1842 (Literar. Zeit. 1842: 605)]

Erichson WF (1843) Versuch einer systematischen Eintheilung der Nitidularien. Zeitschrift für die Entomologie (Germar) 4: 225–361. [1843 (title page); before 5 May 1843 (Leipz. Repert. 18: 211)]

Erichson WF (1844) Einige Nachträge zu meinem Versuch einer systematischen Eintheilung der Nitidularien, (Germar's Zeitschr. IV. S.225). Zeitschrift für die Entomologie (Germar) 5: 438–458. [before 20 Aug 1844 (Ent. Zeit. Stettin)]

Erichson WF (1845) [I., II. Lieferungen, pp. 1–320]. In: Naturgeschichte der Insecten Deutschlands. Erste Abtheilung. Coleoptera. Dritter Band. Nicolai, Berlin, vii + 968 pp. [28 May 1845 (Lief. 1: pp. 1–160) (verso of original wrapper); before 15 Oct 1845 (Lief. 2: pp. 161–321) (Intell.-Blatt Serap. 1845: 149)]

Erichson WF (1846) [III. Lieferung, pp. 321–480]. In: Naturgeschichte der Insecten Deutschlands. Erste Abtheilung. Coleoptera. Dritter Band. Nicolai, Berlin, vii + 968 pp. [4 Sep 1846 (Leipz. Repert. 4: 386)]

Erichson WF (1847a) Conspectus insectorum coleopterorum, quae in Republica Peruana observata sunt. Archiv für Naturgeschichte 13 (Band 1, Heft 1): 67–185. [at least Nov 1847 (article dated p. 352)]

Erichson WF (1847b) [IV., V. Lieferungen, pp. 481–800. In: Naturgeschichte der Insecten Deutschlands. Erste Abtheilung. Coleoptera. Dritter Band. Nicolai, Berlin, vii + 968 pp.

Ermisch K (1941) Tribus Mordellistenini. (Col. Mordell.) (8. Beitrag zur Kenntnis der Mordelliden). Mitteilungen der Münchner Entomologischen Gesellschaft 31: 710–726.

Ermisch K (1956) Mordellidae (31. Beitrag zur Kenntnis der Mordelliden) [pp. 269–321]. In: Horion A (Ed) Faunistik der mitteleuropäischen Käfer. Band V: Heteromera. Entomologische Arbeiten aus dem Museum G. Frey, Sonderband, Tutzing bei München, xv + 336 pp.

Erwin TL (1970) A reclassification of bombardier beetles and a taxonomic revision of the North and Middle American species (Carabidae: Brachinida). Quaestiones Entomologicae 6 (1): 4–215. [20 Feb 1970 (p. 1)]

Erwin TL (1979) Thoughts on the evolutionary history of ground beetles: hypotheses generated from comparative faunal analyses of lowland forest sites in temperate and tropical regions [pp. 539–592]. In: Erwin TL, Ball GE, Whitehead DR, Halpern AL (Eds) Carabid beetles: their evolution, natural history, and classification. Proceedings of the First International Symposium of Carabidology, Smithsonian Institution, Washington, D C, August 21, 23, and 25, 1976. Dr. W. Junk bv Publishers, The Hague, x + 644 pp. [1979 (copyright)]

Erwin TL (1985) The taxon pulse: a general pattern of lineage radiation and extinction among carabid beetles [pp. 437–472]. In: Ball GE (Ed) Taxonomy, phylogeny and zoogeography of beetles and ants. Series Entomologica volume 33. Junk, Dordrecht, xiii + 514 pp. + 2 pls.

Erwin TL (1991) The Ground-Beetles of Central America (Carabidae), Part II: Notiophilini, Loricerini, and Carabini. Smithsonian Contributions to Zoology No. 501: iv + 30 pp.

Erwin TL (1994) Arboreal beetles of tropical forests: the Xystosomi group, subtribe Xystosomina (Coleoptera: Carabidae: Bembidiini). Part I. Character analysis, taxonomy, and distribution. The Canadian Entomologist 126 (3): 549–666. [issued Jul 1994 (wrapper)]

Erwin TL (2007) Xenaroswellanini, *Xenaroswelliana deltaquadrant*, new tribe, new genus, and new species from the Cerrado of Estado de Goiás, Brazil (Insecta: Coleoptera: Carabidae). Proceedings of the California Academy of Sciences (4) 58 (27): 561–568. [28 Dec 2007 (wrapper)]

Erwin TL, Sims LL (1984) Carabid beetles of the West Indies (Insecta: Coleoptera): a synopsis of the genera and checklist of tribes of Caraboidea, and of the West Indian species. Quaestiones Entomologicae 20 (4): 351–466 + 1 pl. [Oct 1984]

Escalera MM, de la (1913) Dos nuevas tribus de Drilidos. Boletín de la Real Sociedad Española de Historia Natural 13 (6/7): 318–322. [includes session of 2 Jul 1913]

Escalera MM, de la (1914a) Rectificación sinonimica. Boletín de la Real Sociedad Española de Historia Natural 14: 349. [includes session of 3 Jun 1914]

Escalera MM, de la (1914b) Los Coleópteros de Marruecos. Trabajos del Museo Nacional de Ciencias Naturales (Serie zoológica) 11: 1–553 [10 Nov 1914 (title page)]

Escalera MM, de la (1927) Los *Allotarsus* Grlls. y géneros afines ibero-africanos (Co. Dasyt.). Eos, Revista Española de Entomologia 3: 5–28.

Eschscholtz F (1829a) Zoologischer Atlas, enthaltend Abbildungen und Beschreibungen neuer Thierarten, während des Flottcapitains von Kotzebue zweiter Reise um die Welt, auf der Russisch-Kaiserlichen Kriegsschlupp Predpriaetië in den Jahren 1823–1826. Erstes Heft. G. Reimer, Berlin, iv + 17 pp. [May 1829 (date of forward)]

Eschscholtz F (1829b) Zoologischer Atlas, enthaltend Abbildungen und Beschreibungen neuer Thierarten, während des Flottcapitains von Kotzebue zweiter Reise um die Welt, auf der Russisch-Kaiserlichen Kriegsschlupp Predpriaetië in den Jahren 1823–1826. Drittes Heft. G. Reimer, Berlin, 18 pp.

Eschscholtz F (1831) Zoologischer Atlas, enthaltend Abbildungen und Beschreibungen neuer Thierarten, während des Flottcapitains von Kotzebue zweiter Reise um die Welt, auf der Russisch-Kaiserlichen Kriegsschlupp Predpriaetië in den Jahren 1823–1826. Viertes Heft. G. Reimer, Berlin, 19 pp.

Español F (1945) Nuevos comentarios sistemáticos sobre la subfamilia Opatrinae Reitt. con la descripción de un nuevo representante del Sáhara Español (Col. Tenebrionidae). Eos, Revista Española de Entomologia 20 [1944] (3/4): 213–232, pls 12–17. [publ. 20 Feb1945 (inside wrapper)]

Español F (1956) Los *Probaticus* de España (Col. Tenebrionidae). Eos, Revista Española de Entomologia 32 (1/4): 83–124. [1956 (wrapper); 13 May 1957 (recorded at CNC)]

Evans AV (1989) Revision of the genus *Sparrmannia* Laporte (Coleoptera: Melolonthidae: Melolonthinae). Journal of the Entomological Society of Southern Africa 52 (1): 11–44.

Evenhuis NL (1994) The publication and dating of P. A. Wytsman's Genera Insectorum. Archives of Natural History 21 (1): 49–66.

Evenhuis NL (1997a) Litteratura taxonomica dipterorum (1758–1930). Volume I. A-K. Backhuys Publishers, Leiden, [3] + pp. 1–426.

Evenhuis NL (1997b) Litteratura taxonomica dipterorum (1758–1930). Volume II. L-Z. Backhuys Publishers, Leiden, pp. 427–871.

Evenhuis NL (2003) Publication and dating of the journals forming the Annals and Magazine of Natural History and the Journal of Natural History. Zootaxa 385: 1–68. [publ. 16 Dec 2003 (footer p. 1)]

Evenhuis NL, Pape T, Pont AC (2008) The problems of subsequent typification in genus-group names and use of the Zoological Record: a study of selected post-1930 Diptera genus-group names without type species designation. Zootaxa 1912: 1–44. [publ. 22 Oct 2008 (footer p. 1)]

Everts JE (1898) Coleoptera Neerlandica. De schildvleugelige Insecten van Nederland en het aangrenzend Gebied. Eerste Deel. M. Nijhoff, Gravenhage, viii + 676 + [1] pp. [Mar 1898 (date of forward); before 11 Jul 1898 (Zool. Anz., Bibliogr. Zool. 3: 313)]

Everts JE (1903) [Tweede stuk, pp. 401–796]. In: Coleoptera Neerlandica. De schildvleugelige Insecten van Nederland en het aangrenzend gebied. Tweede Deel. Martinus Nijhoff, 's Gravenhage, iv + 796 + [2] pp. + 8 pls. [Mar 1903 (date of preface p. iv); issued in 2 parts: 1901 (pp. 1–400 + 8 pls.), 1903 (pp. 401–796)]

Fairmaire L (1864) [Livraisons 121–127, pp. 97–176 + pls. 35–45]. In: Jacquelin du Val C, Fairmaire L: Genera des Coléoptères d'Europe comprenant leur classification en famille naturelle, la description de tous les genres, des tableaux dichotomiques destinés à faciliter l'étude, le catalogue de toutes les espèces, de nombreux dessins au trait de caractères. Tome quatrième. Deyrolle fils, Paris, 295 + [2] pp. + 78 pls. [1864 (Zool. Record: 339); livr. 123: early Sep 1864 (J. Thomson 1864: 336); livr. 121–123: 3 Oct 1864 (Jour. Proc. Ent. Soc. London (3) 2: 44); livr. 124–125: 1 May 1865 (Bull. Soc. Linn. Norm. 10: 159)]

Fairmaire L (1868) [Livraisons 137–144, pp. [241?]-295]. In: Jacquelin du Val, C. & Fairmaire, L.: Genera des Coléoptères d'Europe comprenant leur classification en famille naturelle, la description de tous les genres, des tableaux dichotomiques destinés à faciliter l'étude, le catalogue de toutes les espèces, de nombreux dessins au trait de caractères. Tome quatrième. Deyrolle fils, Paris, 295 + [2] pp. + 78 pls. [livr. 137–144 almost certainly included p. 258 (Leptispites): 7 Dec 1868 (Proc. Ent. Soc. London 1868: xlii)]

Fairmaire L, Laboulbène A (1855) 2e Partie [pp. 181–370]. In: Faune entomologique française ou description des insectes qui se trouvent en France. Coléoptères. Tome Premier. Deyrolle, Paris, xxxv + 665 pp.

Faust J (1885) Neue asiatische Rüsselkäfer (aus Turkestan) III. Deutsche Entomologische Zeitschrift 29 (1): 161–190. [Jun 1885 (footer p. 3)]

Faust J (1886a) Bemerkungen zu einigen europäischen Curculioniden-Gattungen. Stettiner Entomologische Zeitung 47 (1/3): 22–31. [early Feb 1886 (p. 128)]

Faust J (1886b) Neue exotische Rüsselkäfer. Deutsche Entomologische Zeitschrift 30 (2): 337–372. [Dec 1886 (Inhalt p. 3)]

Faust J (1888) Neue Rüsselkäfer aller Länder. Stettiner Entomologische Zeitung 49 (7/9): 284–311. [mid Oct 1888 (p. 316)]

Faust J (1891) Neue Rüsselkäfer aller Länder. (Fortsetzung). Stettiner Entomologische Zeitung 51 [1890] (7/12): 165–195. [Jan 1891 (p. 323)]

Faust J (1892) Curculioniden aus dem Malayischen Archipel. Stettiner Entomologische Zeitung 53 (7/9): 184–228. [at least Dec 1892, or 1893 (part submitted end Nov, p. 268)]

Faust J (1892) Fünf neue Curculioniden von Australien. Stettiner Entomologische Zeitung 53 (7/9): 179–184. [«Ende» Nov 1892 (p. 268)]

Faust J (1894) Viaggio di Leonardo Fea in Birmania e regioni vicine. LX. Curculionidae. [Cont.] Annali del Museo Civico di Storia Naturale di Genova 34: 177–192. [22 May 1894 (complete article pp. 153–370, issued in 12 parts in 1894) (Alonso-Zarazaga and Lyal 1999: 225)]

Faust J (1895) Sechs neue Curculioniden-Gattungen und ein neuer *Glochinorhinus*. Stettiner Entomologische Zeitung 56 (7–9): 212–228. [Sep 1895]

Faust J (1896) Reise von E. Simon in Venezuela. Curculionidae. Pars tertia. Stettiner Entomologische Zeitung 57 (1/6): 33–136.

Faust J (1898) Beitrag zur Kenntniss der Fauna von Kamerun, mit besonderer Berücksichtigung der afrikanischen Menemachiden, Isorhynchiden und Campylosceliden. Deutsche Entomologische Zeitschrift 1898 (1): 17–91. [Jun 1898 (p. 3)]

Faust J (1899) Viaggio di Lamberto Loria nella Papuasia orientale. XXIII. Curculionidae. [Cont.]. Annali del Museo Civico di Storia Naturale di Genova 40: 81–128. [29 May 1899 (complete article pp. 1–130, issued in 7 parts in 1899) (Alonso-Zarazaga and Lyal 1999: 225)]

Faust J (1904) Revision der Gruppe Cléonides vrais. Deutsche Entomologische Zeitschrift 1904 (1): 177–284. [Jan 1904 (vol. Inhalt)]

Fauvel A (1874) Faune gallo-rhénane... 2e supplément aux staphylinides. Tome troisième. [livraison 5]. Le Blanc-Hardel, Caen, 25–46. [Jul 1874 (footer p. 25; Fauvel 1875: 54); Sep 1874 (wrapper) / reissued in: Bull. Soc. Linn. Norm. (2) 8: 319–340 (séance 4 May 1874) 17 Apr 1875 (Bibliogr. France 1875: 206)]

Fauvel A (1875) Bibliographie. Annuaire Entomologique 3: 53–56.

Fauvel A (1888) Catalogue des coléoptères Gallo-Rhénans. Revue d'Entomologie 7: 1–16 [special pagination].

Fauvel A (1889) Catalogue des coléoptères gallo-rhénans. Revue d'Entomologie 8: 17–32 [special pagination].

Fauvel A (1891) Les coléoptères de la Nouvelle-Calédonie et dépendances avec descriptions, notes et synonymies nouvelles (1) (Suite). Revue d'Entomologie 10 (6): 148–182. [Jun 1891 issue (footer p. 145)]

Fauvel A (1895) Staphylinides nouveaux de l'Inde et de la Malaisie. Revue d'Entomologie 14 (6–10): 180–286. [Jun-Oct 1895 issues (footers p. 173, etc.)]

Fauvel A (1899) Sur une sous-famille nouvelle de staphylinides, les Trilobitideidae [pp. 3–4]; Sur une tribu nouvelle de staphylinides (Pygostenini) et descriptions de genres et espèces [pp. 5–22] [and other articles]. In: Raffray A, Fauvel A: Genres et espèces de staphylinides nouveaux d'Afrique. Revue d'Entomologie 18 (1–2): 1–44, pl. 1. [Jan-Feb 1899 issues (footers pp. 1, 29)]

Fauvel A (1900) Sur une tribu nouvelle de staphylinides. Revue d'Entomologie 19 (7/8): 123–124. [Aug 1900 (footer p. 109 but 7/8 issued together)]

Fauvel A (1904) Staphylinides exotiques nouveaux. 2e partie. Revue d'Entomologie 23 (4/5): 76–112. [Apr-May 1904 issues (footers pp. 65, 97 but 4/5 issued together)]

Fauvel A (1905a) Sur une tribu nouvelle de staphylinides. Revue d'Entomologie 24 (5/6): 98–100. [May-Jun 1905 issue (footer pp. 97, 129 but 5/6 issued together)]

Fauvel A (1905b) Faune analytique des coléoptères de la Nouvelle-Calédonie. 3e partie. Revue d'Entomologie 24 (10/11, 12): 209–244. [Oct, Nov, Dec 1905 issues (footer pp. 209, 225, 241 but 10/11 issued together)]

Fenyes A (1918) 173a fascicule. Coleoptera. Fam. Staphylinidae. Subfam. Aleocharinae [pp. 1–110]. In: Wytsman PA (Ed) Genera Insectorum. Vol. XXVII. M. Nijhoff, The Hague, 453 pp. + 7 pls. [31 Dec 1913 (date on manuscr., p. 453); 1918 (wrapper); before 1 Jan 1919 (Leng 1920: v, 387); 18 Jun 1919 (Evenhuis 1994: 58)]

Fenyes A (1921) New genera and species of Aleocharinae with a polytomic synopsis of the tribes. Bulletin of the Museum of Comparative Zoology 65 (2): 17–36. [Oct 1921 (article title page)]

Ferrari JA (1867) Die Forst- und Baumzuchtschädlichen Borkenkäfer (Tomicides Lac.) aus der Familie der Holzverderber (Scolytides Lac.), mit besonderer Berücksichtigung vorzüglich der europäischen Formen, und der Sammlung des k. k. zoologischen Kabinetes in Wien. Carl Gerold's Sohn, Wien, [2] + 96 pp. [Jan 1867 (date of Vorwort); before Aug 1867 (Allgem. Bibliogr. Deutschl. 1867: 118)]

Ferraris CJ, Jr. (2000) Book review. International Code of Zoological Nomenclature. Fourth Edition. Copeia 2000 (3): 907–908.

Ferraris CJ, Jr., de Pinna MCC (1999) Higher-level names for catfishes (Actinopterygii: Ostariophysi: Siluriformes). Proceedings of the California Academy of Sciences 51 (1): 1–17.

Ferreira MC (1953) Monografia dos Escarabaeídeos da África do Sul. Tribo-Scarabaeini. I Parte Sub-tribo Pachysomides. Boletím da Sociedade de Estudos de Moçambique 23 (No. 78): 1–87. [Mar-Apr 1953 issue; 16 Oct 1954 (recorded at BMNH)]

Ferreira MC (1965) Catálogo dos Coleópteros de Moçambique. Revista de Entomologia de Moçambique 6 [1963] (1): 1–531 + [1 corrigenda]. [1965 (recorded at FMNH; cites ref. publ. 1st Dec 1964)]

Ferreira MC (1966) Catálogo dos Coleópteros de Moçambique (Conclusão). Revista de Entomologia de Moçambique 6 [1963] (2): [1] + 533–1008 + 1 map. [printed Jan 1966 (back wrapper)]

Ferreira MC, Veiga Ferreira Gd (1959a) Catálogo dos cerambicídeos da Região Etiópica. I parte - Supertribos Parandrina & Prionina. Memórias do Instituto de Investigação Cientifica de Moçambique 1: 1–76 + [1: corrigendum].

Ferreira MC, Veiga Ferreira Gd (1959b) Catálogo dos cerambicídeos da Região Etiópica. II parte. Supertribos Disteniina, Asemina, Cerambycina, Auxesina & Lepturina. Memórias do Instituto de Investigação Cientifica de Moçambique 1: 77–398 + [1: errata].

Ferrer J (2004) Description d'un nouveau genre de Stenosini du Vietnam (Coleoptera, Tenebrionidae). Nouvelle Revue d'Entomologie (Nouvelle Série) 20 [2003] (4): 367–371. [25 Apr 2004 (top of article)]

Ferrer J (2006) Constitution du groupe indo-africain des Falsocossyphini, tribus nova, et description d'un nouveau genre hypogée du Vietnam. Coleoptera, Tenebrionidae. Cahiers

Scientifiques. Centre de Conservation et d'Étude des Collections (Lyon) 10: 75–83. [printed Sep 2006; dépot légal 3e trim. 2006 (p. 148)]

Ferrer J, Moragues G (2000) Contribution à l'étude des Cnemeplatiini. Description de *Rondoniella bremeri* n. sp. du Laos (Coleoptera, Tenebrionidae). Nouvelle Revue d'Entomologie (Nouvelle Série) 17 (2): 99–105. [paru 15 Dec 2000 (back wrapper)]

Ferrer J, Yvinec JH (2004) Révision de la tribu des Lachnogyini Reitter, 1904 sensu nov. et description d'un nouveau genre et d'une espèce nouvelle du désert de Taklamakan, Chine (Coleoptera: Tenebrionidae, Pimeliinae). Annales de la Société Entomologique de France (Nouvelle Série) 40: 41–49. [paru 18 May 2004 (back wrapper)]

Fischer von Waldheim G (1813) Zoognosia tabulis synopticis illustrata, in usum praelectionum Academiae Imperialis Medico-Chirurgicae Mosquensis edita. Editio tertia, classium, ordinum, generum illustratione perpetua aucta. Volumen primum, tabulas synopticas, generales et comparativas, nec non characterum quorundam explicationem iconographicam continens. N. S. Vsevolozsky, Moscow, xiii + [1] + 465pp.

Fischer von Waldheim G (1821) Genera insectorum systematice exposita et analysi iconographica instructa. Volumen primum. Genera Coleopterorum. Augusti Semen, Mosquae, xii + 104 pp., 1 pl. [12 Apr 1821 (date of preface); bound with Entomographia Imperii Russici, Volumen I]

Fitch EA (1881) Obituary. Etienne Mulsant. The Entomologist 14 (213): 46–47. [Feb 1881]

Flach C (1889) Bestimmungstabelle der Trichopterygidae des europäischen Faunengebietes. Verhandlungen der Kaiserlich-Königlichen Zoologisch-Botanischen Gesellschaft in Wien 39 (4): 481–532, pls. 10–14. [end Dec 1889 (wrapper)]

Fleming J (1821) Insecta. Supplement to the Fourth, Fifth and Sixth Editions of the Encyclopaedia Britannica Volume Fifth [Part 1]. 584 pp. A. Constable and Company, Edinburgh, 41–56, pl. 85. [Jul 1821 (Evenhuis 1997a: 226)]

Fleutiaux E (1892a) Catalogue systématique des Cicindelidae, décrits depuis Linné. H. Vaillant-Carmanne, Liège, 186 pp.

Fleutiaux E (1892b) Note sur les Physodactylini. Annales de la Société Entomologique de France 61 (3): 403–412, pl. 8. [28 Dec 1892 (wrapper)]

Fleutiaux E (1902) Essai d'une classification des Melasinae (Eucneminae des auteurs). Annales de la Société Entomologique de France 70 [1901] (3/4): 636–664. [May 1902 (wrapper)]

Fleutiaux E (1919) Elateridae, Trixagidae et Melasidae. Voyage de Ch Alluaud et R Jeannel en Afrique orientale (1911–1912) Résultats scientifiques Coleoptera. Vol. 13. Lhomme, Paris, pp. 1–119 + pl. 8. [15 Mar 1919 (date cited on reprint)]

Fleutiaux E (1920) Études sur les Melasidae (Coleoptera-Serricornia). Première partie. Annales de la Société Entomologique de Belgique 60: 93–104. [6 Jul 1920 (footer p. 91)]

Fleutiaux E (1923) Les Melasidae du Japon (Coléoptères). Annales de la Société Entomologique de France 91 (4): 291–328. [10 Jun 1923 (wrapper)]

Fleutiaux E (1926) Catalogue raisonné des Melasidae des Iles Philippines. Annales de la Société Entomologique de France 95 (1): 29–90. [31 Mar 1926 (wrapper)]

Fleutiaux E (1927a) Description d'un genre nouveau d'Elateridae (Col.). Bulletin de la Société Entomologique de France 32 (6): 104–105. [25 Apr 1927 (p. 324)]

Fleutiaux E (1927b) Les élatérides de l'indochine française (catalogue raisonné). Première partie. Faune des Colonies Françaises 1: 53–122 + pls. 1–2.

Fleutiaux E (1928) Description d'un genre nouveau d'Elateridae de la sous-famille des Hypolithinae. Bulletin de la Société Entomologique de France 1928 (16): 252–254. [24 Nov 1928 (p. 332)]

Fleutiaux E (1936) Les Elateridae de l'Indochine française (6e partie). Annales de la Société Entomologique de France 105 (3): 279–300. [31 Oct 1936 (p. 388)]

Fleutiaux E (1941) Les élatérides de l'Indo-chine française. Huitième et dernière partie. Annales de la Société Entomologique de France 109 [1940] (1/4): 19–40. [15 Mar 1941 (p. 169)]

Fleutiaux E (1947) Révision des élatérides (coléoptères) de l'Indo-chine française. Notes d'Entomologie Chinoise / Musée Heude 11 (8): 233–420. [20 Dec 1947 (top of article)]

Foley IA, Ivie MA (2007) Determination of the correct authorship and type species of *Nosoderma*, and the impact on the nomenclature of the Zopherini (Coleoptera: Zopheridae). The Coleopterists Bulletin 61 (1): 65–74. [publ. 30 Apr 2007 (p. 74)]

Folkerts GW (1979) *Spanglerogyrus albiventris*, a primitive new genus and species of Gyrinidae (Coleoptera) from Alabama. The Coleopterists Bulletin 33 (1): 1–8. [mailed 30 Apr 1979 (wrapper)]

Folwaczny B (1973) Bestimmungstabelle der paläarktischen Cossoninae (Coleoptera, Curculionidae) ohne die nur in China und Japan vorkommenden Gattungen, nebst Angaben zur Verbreitung. Entomologische Blätter für Biologie und Systematik der Käfer 69 (2): 65–180. [30 Oct 1973 (top of article + vol. title page)]

Forbes WTM (1926) The wing folding patterns of the Coleoptera. Journal of the New York Entomological Society 34 (1): 42–68, (2): 91–139 (incl. pls. 7–18). [publ. 2 Apr 1926 (fasc. 1), 28 Jul 1926 (fasc. 2) (inside wrapper of next issue)]

Fowler WW (1887) The Coleoptera of the British Islands. A descriptive account of the families, genera, and species indigenous to Great Britain and Ireland, with notes as to localities, habitats, etc. Vol. I. Adephaga-Hydrophilidae. L. Reeve & Co., London, xxxii + 269 pp. + 12 pls. [1887 (title page)]

Fowler WW (1889) The Coleoptera of the British Islands. A descriptive account of the families, genera, and species indigenous to Great Britain and Ireland, with notes as to localities, habitats, etc. Vol. III. Clavicornia (Leptinidae-Heteroceridae.). L. Reeve & Co., London, 399 pp. + pls. 71–98. [1889 (title page)]

Fowler WW (1912) Coleoptera, General introduction and Cicindelidae and Paussidae. In: Shipley AE (Ed) The fauna of British India, including Ceylon and Burma. Taylor & Francis, London, xx + 529 pp. [Feb 1912 (verso of title page)]

Fox WJ (1913) Dates of publication [pp. vii-xiv]. In: Nolan EJ (Ed): An index to the scientific contents of the Journal and Proceedings of The Academy of Natural Sciences of Philadelphia, 1812–1912, published in commemoration of the centanary of the Academy, March 21, 1912. The Academy of Natural Sciences, Philadelphia, xiv + 1419 pp.

Fragoso SA, Monné MA, Campos Seabra CA (1987) Preliminary considerations on the higher classification of Cerambycinae (Coleoptera, Cerambycidae), with nomenclatural alterations. Revista Brasileira de Biologia 47 (1/2): 189–202. [distrib. 30 May 1987 (footer p. 189)]

Franciscolo ME (1951) Monografia del genere Pselaphostena mihi. XXIII° contributo alla conoscenza dei Mordellidae (Col. Heteromera). Atti della Società Italiana di Scienze Naturali e del Museo Civico di Storia Naturale in Milano 90 (1): 55–76. [Apr 1951 (fasc. title page); 27 Jun 1951 (recorded at BMNH)]

Franciscolo ME (1954) On two species of Anaspidinae (Coleoptera: Scraptiidae) taken on flowers of *Protea abyssinica* in Natal. (XXVIII. Contribution to the knowledge of Mordellidae and Scraptiidae). Proceedings of the Royal Entomological Society of London (B) 23: 63–73. [Apr 1954 issue]

Franciscolo ME (1955) Mordellidae and Scraptiidae in the collections of the Durban Museum. (XXIX. contribution to the knowledge of Mordellidae and Scraptiidae). Durban Museum Novitates 4 (12): 161–185. [issued 30 Jun 1955 (top of article)]

Franciscolo ME (1957) Coleoptera: Mordellidae, a monograph of the South African genera and species. 1. Morphology, subfamily Ctenidiinae and tribe Stenaliini. [Pp. 207–291]. Vol. IV. Almqvist and Wiksell, Stockholm, 508 pp. [1957 (verso of title page)]

Franciscolo ME (1964) Nota preliminare sulla filogenia degli Scraptiidae (Coleoptera Heteromera). (44° contributo alla conoscenza degli Scraptiidae). Atti della Accademia Nazionale Italiana di Entomologia / Rendiconti 11 [1963]: 175–181. [1964 (title page)]

Franciscolo ME (1972) Su alcuni generi poco noti di Anaspidinae (47° contributo alla conoscenza degli Scraptiidae). Memorie della Società Entomologica Italiana 51: 123–155.

Frania HE, Ball GE (2007) Taxonomy and evolution of species of the genus *Euchroa* Brullé (subgenus *Dyschromus* Chaudoir) of central Mexico and the island of Hispaniola (Coleoptera: Carabidae: Pterostichini: Euchroina). Bulletin of Carnegie Museum of Natural History 38: 1–125. [Dec 2007]

Franz H (1989) 4. Beitrag zur Scydmaenidenfauna von Thailand (Die Scydmaenidenausbeute von D.H. Burckhardt u. I. Löbl). Revue Suisse de Zoologie 96 (1): 33–80. [Mar 1989 (wrapper)]

Franz NM (2006) Towards a phylogenetic system of derelomine flower weevils (Coleoptera: Curculionidae). Systematic Entomology 31 (2): 220–287. [18 Apr 2006 (vol. contents)]

Frey G (1975) Eine neue südamerikanische Gattung und Tribus der Melolonthiden. Entomologische Arbeiten aus dem Museum G. Frey 26: 84–86. [publ. 1 Nov 1975 (Inhalt)]

Fujiyama I (1973) Mesozoic insect fauna of East Asia. Part I. Introduction and Upper Triassic faunas. Bulletin of the National Science Museum, Tokyo 16 (2): 331–386, pls. 1–5. [22 Jun 1973 (top of article)]

Furth DG (2007) A new genus and species of flea beetle (Coleoptera: Chrysomelidae: Alticinae) from the rainforest canopy in Costa Rica. Proceedings of the Entomological Society of Washington 109 (1): 90–101. [mailed 2 Jan 2007 (inside wrapper)]

Gahan CJ (1890) On new Longicornia from Africa and Madagascar. The Transactions of the Entomological Society of London 1890 (2): 297–328, pl. 9. [25 Jun 1890 (G. Wheeler 1912)]

Gahan CJ (1906) Coleoptera.-Vol.I. (Cerambycidae). In: Bingham CT (Ed) The fauna of British India, including Ceylon and Burma. Taylor & Francis, London, xviii + 329 pp.

Galewski K (1994) The description of third-stage larvae of *Hydroporus tartaricus* Lec. and *H. marginatus* (Duft.) (Coleoptera, Dytiscidae) with a key to third-stage larvae of central

European species of Hydroporini. Bulletin of the Polish Academy of Sciences Biological Sciences 42 (1): 95–113. [Dec 1994 (contents)]

Galileo MHM (1987) Sistématica das tribos Meroscelisini e Anacolini (Coleoptera, Cerambycidae, Prioninae) nas Américas. 2. Anacolini. Revista Brasileira de Entomologia 31 (4): 481–705. [28 Dec 1987 (wrapper)]

Galileo MHM, Martins UR (1993) Revisão da tribo Solenopterini (Coleoptera, Cerambycidae, Prioninae). Parte I. Transferência de Poekilosoma A.-Serville, 1832 e Calocomus A.-Serville, 1832; os gêneros Prosternodes Thomson, 1860 e Derancistrodes, gen. n. Revista Brasileira de Entomologia 37 (1): 79–99. [31 Mar 1993 (top of article)]

Galileo MHM, Martins UR (1995) Revisão da tribo Eupromerini, trib. n. (Coleoptera, Cerambycidae, Lamiinae) da região neotropical. Revista Brasileira de Entomologia 39 (1): 131–150. [31 Mar 1995 (top of article)]

Galileo MHM, Martins UR (2001) Novos táxons e notas sobre Cerambycidae (Coleoptera) neotropicais. Iheringia (Série Zoologia) 90: 93–106. [25 May 2001 (footer p. 93)]

Ganglbauer L (1886) Coleoptera. Zoologischer Jahresbericht (II. Abtheilung: Arthropoda) 1885: 204–339. [1886 (title page)]

Ganglbauer L (1891a) Die Käfer von Mitteleuropa. Die Käfer der österreichisch-ungarischen Monarchie, Deutschlands, der Schweiz, sowie des französischen und italienischen Alpengebietes. Erster Band. Familienreihe Caraboidea. C. Gerold's Sohn, Wien, iii + 557 pp. [1892 (title page); ausgegeben am 16 Nov 1891 (footnote 557); 14 Nov (date of forward)]

Ganglbauer L (1891b) Cicindelidae, Carabidae [pp. 1–58]. In: Heyden L, von, Reitter E, Weise J (Eds) Catalogus Coleopterorum Europae, Caucasi et Armeniae Rossicae. Mödling & Caen, Berlin, viii + 420 pp.

Ganglbauer L (1895) Die Käfer von Mitteleuropa. Die Käfer der österreichisch-ungarischen Monarchie, Deutschlands, der Schweiz, sowie des französischen und italienischen Alpengebietes. Zweiter Band. Familienreihe Staphylinoidea. 1. Theil: Staphylinidae, Pselaphidae. Carl Gerold's Sohn, Vienna, vi + 880 + [1] pp. [1895 (title page); Mar 1895 (date of forward); Jun 1895 (Rev. d'Ent. 14: 178)]

Ganglbauer L (1898) Die Käfer von Mitteleuropa. Die Käfer der österreichisch-ungarischen Monarchie, Deutschlands, der Schweiz, sowie des französischen und italienischen Alpengebietes. Dritter Band. Familienreihe Staphylinoidea. II. Theil. Scydmaenidae, Silphidae, Clambidae, Leptinidae, Platypsyllidae, Corylophidae, Sphaeriidae, Trichopterygidae, Hydroscaphidae, Scaphidiidae, Histeridae. Familienreihe Clavicornia. Sphaeritidae, Ostomidae, Byturidae, Nitidulidae, Cucujidae, Erotylidae, Phalacridae, Thorictidae, Lathridiidae, Mycetophagidae, Colydiidae, Endomychidae, Coccinellidae [Lief. 1: pp. 1–408]. Carl Gerold's Sohn, Wien, iii + 1046 pp. [Lief. 1: end 1898 (Sharp in Zool. Record 35: 27); 13 Jan 1899 (rev. in Verh. Zool.-Bot. Ges. Wien 49: 124)]

Ganglbauer L (1899) Die Käfer von Mitteleuropa. Die Käfer der österreichisch-ungarischen Monarchie, Deutschlands, der Schweiz, sowie des französischen und italienischen Alpengebietes. Dritter Band. Familienreihe Staphylinoidea. II. Theil. Scydmaenidae, Silphidae, Clambidae, Leptinidae, Platypsyllidae, Corylophidae, Sphaeriidae, Trichopterygidae, Hydroscaphidae, Scaphidiidae, Histeridae. Familienreihe Clavicornia. Sphaeritidae, Ostomidae, Byturidae, Nitidulidae, Cucujidae, Erotylidae, Phalacridae, Thorictidae, Lathridiidae, Mycetophagidae, Colydiidae, Endomychidae, Coccinellidae [Lief. 2: pp. 409–1046 + iii]. Carl Gerold's Sohn, Wien, iii + 1046 pp. [Lief. 2: 1899 (title page); Jun 1899 (date of forward); 11 Oct 1899 (Bull. Soc. Ent. France 1899: 312)]

Ganglbauer L (1903) Systematisch-koleopterologische Studien. Münchener Koleopterologische Zeitschrift 1 (3): 271–319. [5 Mar 1903 (Inhalt)]

Ganglbauer L (1904) Die Käfer von Mitteleuropa. Die Käfer der österreichisch-ungarischen Monarchie, Deutschlands, der Schweiz, sowie des französischen und italienischen Alpengebietes. Vierter Band, erste Hälfte. Dermestidae, Byrrhidae, Nosodendridae, Georyssidae, Dryopidae, Heteroceridae, Hydrophilidae. Carl Gerold's Sohn, Wien, 286 pp.

García M (2001) Nueva subtribu género y especie de Hydrophilini (Coleoptera; Hydrophilidae) del extremo sur-oriental de Venezuela. Boletín del Centro de Investigaciones Biológicas, Universidad del Zulia 35 (2): 151–160. [printed Jul 2001 (endleaf); 4 Dec 2001 (recorded at BMNH)]

Gardner JCM (1933) Immature stages of Indian Coleoptera (13) (Bostrychidae). Indian Forest Records (Entomology Series) 18 (9): 1–19 + pl. 4.

Ge S-Q, Friedrich F, Beutel RG (2010) On the systematic position and taxonomic rank of the extinct myxophagan †*Haplochelus* (Coleoptera). Insect Systematics & Evolution 41 (4): 329–338. [publ. 1 Nov 2010 (BRILL website)]

Gebien H (1910a) Pars 15. Tenebrionidae I [pp. 1–166]. In: Schenkling S (Ed) Coleopterorum Catalogus Volumen XVIII. W. Junk, Berlin, 742 pp. [25 May 1910 (verso of title page)]

Gebien H (1910b) Pars 22. Tenebrionidae II [pp. 167–354]. In: Schenkling S (Ed) Coleopterorum Catalogus Volumen XVIII. W. Junk, Berlin, 742 pp. [1 Nov 1910 (verso of title page)]

Gebien H (1911) Pars 28. Tenebrionidae III [pp. 355–585]. In: Schenkling S (Ed) Coleopterorum Catalogus Volumen XVIII. W. Junk, Berlin, 742 pp. [24 Mar 1911 (verso of title page)]

Gebien H (1919) Monographie der südamerikanischen Camarien (Coleopt. Heterom.) nebst einer Übersicht über die indischen Gattungen der Camariinen. Archiv für Naturgeschichte (Abteilung A) 83 (3): 25–167. [publ. Jul 1919 (wrapper)]

Gebien H (1921) Die Tenebrioniden Westafrikas. Archiv für Naturgeschichte 86 A (6): 1–256. [Jan 1921; 1920 (title page)]

Gebien H (1922) No. V. Coleoptera, Heteromera: Tenebrionidae. In: The Percy Sladen Trust Expedition to the Indian Ocean in 1905, under the leadership of Mr. J. Stanley Gardiner, M. A. Vol. II. The Transactions of the Linnean Society of London (2nd Ser. Zoology) 18: 261–324.

Gebien H (1928) Über einige Gruppen amerikanischer Tenebrioniden (Col. Heter.). Stettiner Entomologische Zeitung 89 (1, 2): 97–164, 167–234. [Mar, Nov 1928 (Inhalt)]

Gebien H (1937) Katalog der Tenebrioniden (Col. Heteromera). Teil I. Pubblicazioni del Museo Entomologico «Pietro Rossi» 2: 505–883. [1 May 1937]

Gebien H (1938a) Die Tenebrioniden (Coleoptera Heteromera) der Namibwüste in Südwestafrika. Abhandlungen herausgegeben vom Naturwissenschaftlichen Verein zu Bremen 30 [1937–38] (3/4): 20–107.

Gebien H (1938b) Katalog der Tenebrioniden. Teil II. Mitteilungen der Münchner Entomologischen Gesellschaft 28: 49–80, 283–314, 397–428 [370–465].

Gemminger M (1869) [new taxa] In: Gemminger M, Harold E, von: Catalogus coleopterorum hucusque descriptorum synonymicus et systematicus. Tom. IV. Scarabaeidae. E. H. Gummi, Monachii [Munich], pp. 979–1346 + [8 (Index, Corrigenda)].

Gemminger M (1870) [new taxa] In: Gemminger M, Harold E, von: Catalogus coleopterorum hucusque descriptorum synonymicus et systematicus. Tom. VII. Tenebrionidae, Nilionidae, Pythidae, Melandryidae, Lagriidae, Pedilidae, Anthicidae, Pyrochroidae, Mordellidae, Rhipidophoridae, Cantharidae, Oedemeridae. E. H. Gummi, Monachii [Munich], pp. 1801–2179 + [11].

Gemminger M (1872) [new taxa] In: Catalogus coleopterorum hucusque descriptorum synonymicus et systematicus. Tom. IX. Scolytidae, Brenthidae, Anthotribidae, Cerambycidae. E. H. Gummi, Monachii, [1] + 2669–2988 + [12] pp. [pars 1: 1872 (title page); before 2 Dec 1872 (Proc. Ent. Soc. London 1872: xliii); pars 2: before 5 May 1873 (Proc. Ent. Soc. London 1873: xvii)]

Gemminger M, Harold E, von (1868a) Catalogus coleopterorum hucusque descriptorum synonymicus et systematicus. Tom. I. Cicindelidae - Carabidae. E. H. Gummi, Monachii [Munich], xxxvi + 424 + [8 (Index)] pp. [by Aug 1868 (Allgem. Bibliogr. Monatl. Verz. 1868: 118)]

Gemminger M, Harold E, von (1868b) Catalogus Coleopterorum hucusque descriptorum synonymicus et systematicus. Tom. II. Dytiscidae, Gyrinidae, Hydrophilidae, Staphylinidae, Pselaphidae, Gnostidae, Paussidae, Scydmaenidae, Silphidae, Trichopterygidae, Scaphididae. E. H. Gummi, Monachii [München], pp. 425–752 + [6 (Index)]. [1868 (title page); 7 Nov 1868 (Sitzber. Königl. Bayer. Akad. Wiss. Münch. 1868 (2): 488); Dec 1868 (Berl. Ent. Zeitschr. 12: 424)]

Gemminger M, Harold E, von (1876) Catalogus coleopterorum hucusque descriptorum synonymicus et systematicus. Tom. XII. Chrysomelidae (Pars II), Languridae, Erotylidea, Endomychidae, Coccinellidae, Corylophidae, Platypsyllidae. Accedit index Generum Universalis. E. H. Gummi, Monachii, pp. 3479–3822 + [2] + lxxiii pp. [1876 (title page); before 1 Nov 1876 (Ent. Nachr. 2: 176); before 6 Dec (Proc. Ent. Soc. London 1876: xxxvi)]

Génier F (2009) Le genre *Eurysternus* Dalman, 1824 (Scarabaeidae: Scarabaeinae: Oniticellini), revision taxonomique et cles de determination illustrees. Pensoft, Sofia and Moscow, 430 pp.

Genise JF (2004) Ichnotaxonomy and ichnostratigraphy of chambered trace fossils in paleosols attributed to coleopterans, ants and termites. In: McIlroy D (Ed) The application of ichnology to palaeoenvironmental and stratigraphic analysis. Geological Society. Special Publication 228, London, 419–453.

Germain P (1897) Apuntes entomolójicos. Agrupacion de los Taurocerastidae. Anales de la Universidad de Chile 97: 287–300, pl. 1.

Germain P (1900) Apuntes entomolójicos. Agrupacion de los Ametrocephalidos. Anales de la Universidad de Chile 107: 79–104.

Germain P (1903) Apuntes entomolójicos. Anales de la Universidad de Chile 112: 343–365.

Germain P (1911) Catálogo de los Coleopteros Chilenos del Museo Nacional. Boletin del Museo Nacional de Chile 3 (2): 47–73. [1911 (title page)]

Germar EF (Ed) (1818) [Literature review of: Cuvier: Le règne animal (pp. 339–362)]. In: Germar EF (Ed) Magazin der Entomologie. Vierter Band. J. C. Hendel und Sohn, Halle, 464 pp. + 3 pls.

Germar EF (1829) Curculionides [pp. 356–359]. In: Ersch JS, Gruber JG (Eds) Allgemeine Encyclopädie der Wissenschaften und Künste in alphabetischer Folge von genannten Schrifstellern Zwanzigster Theil Cos - Czvittinger. J. F.Gleditsch, Leipzig, 458 pp.

Gerstaecker A (1854) Bericht über die im der Entomologie während des Jahres 1853. Archiv für Naturgeschichte 20 (2): 189–288.

Gerstaecker A (1855) Rhipiphoridum Coleopterorum familiae dispositio systematica. Schlesinger, Berlin, [1] + 36 pp. + 1 pl. [4 Jul 1855 (date of original dissertation)]

Gerstaecker A (1858) Monographie der Endomychiden, einer Familie der Coleopteren. Entomographien. Erster Band. W. Engelmann, Liepzig, xiv + 433 pp. + 3 pls. [Jun 1858 (date of preface)]

Gerstaecker A (1861) Bericht über die wissenschaftlichen Leistungen im Gebiete der Entomologie während der Jahre 1859 und 60. Archiv für Naturgeschichte 26 (2): 357–531.

Gerstaecker A (1862) Bericht über die wissenschaftlichen Leistungen im Gebiete der Entomologie während des Jahres 1861. Coleoptera. Archiv für Naturgeschichte 28 (2): 326–414.

Gestro R (1876) Enumerazione dei Longicorni della tribù dei Tmesisternini, raccolti nella regione Austro-Malese dai Sign. O. Beccari, L. M. d'Albertis e A. A. Bruyn. Annali del Museo Civico di Storia Naturale di Genova 9 [1876–77]: 139–182. [27 Nov 1876 (footer p. 129); 3 Dec (p. 161)]

Gestro R (1892) Viaggio di Leonardo Fea in Birmania e regioni vicine. XLVI. Cenno sui Paussidi. Annali del Museo Civico di Storia Naturale «Giacomo Doria» (2) 12: 705–709.

Gestro R (1906) Materiali per lo studio delle Hispidae. XXVIII. Descrizioni di alcune Hispidae inedite. Annali del Museo Civico di Storia Naturale di Genova (3) 2 [42]: 541–557. [20 Dec 1906 (footer p. 529)]

Gestro R (1919) Contribuzione allo studio degli Insetti mirmecofili. Annali del Museo Civico di Storia Naturale «Giacomo Doria» 48: 270–276. [6 Sep 1919]

Gildenkov MY (2000) Relations of phylogenesis in Oxytelinae subfamily (Coleoptera: Staphylinidae): cladism or evolution taxonomy? [pp. 53–56]. In: Krugler ND (Ed) A tribute to Prof VV Stanchinsky. Third International Proceeding, Smolensk 2000. [In Russian, English abstract]. Smolenskiy Gosudarstvennogo Pedagogicheskiy Universitet, Smolensk.

Gildenkov MY (2003) Novaya sistema podsemeistva Oxytelinae (Coleoptera: Staphylinidae). [A new system of subfamily Oxytelinae (Coleoptera: Staphylinidae)]. Izvestiya Khar'kovskogo Entomologicheskogo Obshchestva 10 [2002] (1/2): 32–38 [in Russian, English summary]. [after 26 Dec 2003 (approved to print, verso of vol. title page)]

Gilmour EF (1954) Notes on a collection of Prioninae (Coleoptera, Cerambycidae) from the Institut royal des Sciences naturelles de Belgique. Bulletin / Institut Royal des Sciences Naturelles de Belgique 30 (No. 24): 1–48.

Gilmour EF (1956) Revision of the "Prioninae" of tropical and South Africa [pp. 1–252]. In: Lepesme P (Ed). Longicornia. Études et notes sur les longicornes. Volume III. Paul Lechevalier, Paris, viii + 789 pp. [4e trim. 1956 (dépôt légal, endleaf)]

Gilmour EF (1961) The tribe Falsamblesthiini (Coleoptera, Cerambycidae, Lamiinae). Entomologische Abhandlungen und Berichte aus dem Staatlichen Museum für Tierkunde in Dresden 26 (17): 131–134. [27 Dec 1961 (top of article)]

Gimmel ML, Leschen RAB, Ślipiński SA (2009) Review of the New Zealand endemic family Cyclaxyridae, new family (Coleoptera: Polyphaga). Acta Entomologica Musei Nationalis Pragae 49 (2): 511–528. [publ. 15 Dec 2009 (inside wrapper)]

Girard C (1971) Les Coléoptères Elateridae de Lamto (Côte d'Ivoire). Bulletin de l'Institut Fondamental d'Afrique Noire (Série A, Sciences Naturelles) 33 (3): 549–650. [paru 1 Sep 1971 (p. 774)]

Girard C (1973) Contribution à l'étude des Elateridae de l'Afrique noire (Col.). Bulletin de la Société Entomologique de France 77 (9/10): 276–283. [14 Apr 1973 (p. 320)]

Gistel JNFX (1848) Faunula monacensis cantharologica. Isis von Oken 1848: (6) [unn. pp."1–3"]; (7): [unn. pp. "4–6"]; (8): [unn. p. "7"]; (9): [unn. pp. "8–9"]; (10): [unn. pp. "10–11"]; (11): [unn. pp. «12–13»] (inserted with fasscicles).

Gistel JNFX [as W.G. Tilesius von Tilenau] (1850) Literarische Umschau. Isis Encyclopädische Zeitschrift; Im Auftrage des Münchner Vereins für Naturkunde No. 5 (1): 74–80.

Gistel JNFX (1856a) Die Mysterien der europäischen Insectenwelt. Ein geheimer Schlüssel für Sammler aller Insecten-Ordnungen und Stände, behufs des Fangs, des Aufenhalts-Orts, der Wohnung, Tag- und Jahreszeit u.s.w., oder autoptische Darstellung der Insectenstaats in seinem Zusammenhange zum Bestehen des Naturhaushaltes überhaupt und insbesondere in seinem Einflusse af die phanerogamische und cryptogamische Pfanzenberöltzerrung Europa's. Zum ersten Male nach 25jährigen eigenen Erfahrungen zusammengestellt und herausgegeben. T. Dannheimer, Kempten, xii + 532 pp. [18 Feb 1856 (Evenhuis 1997a: 305)]

Gistel JNFX (1856b) Pleroma zu den Mysterien der europäischen Insectenwelt. Mit einen systematischen Verzeichniss der Smetterlinge und Käfer Europa's. Durch die neueften Entdechungen bis 1856 bereichert. Schorner, Straubing, 250 pp. [«31 December» 1856]

Golbach R (1970) Semiotinae, neuva subfamilia de Elateridae (Col.). Acta Zoologica Lilloana 25 (24): 317–324. [printed 21 May 1970 (end leaf)]

Golbach R (1984) Novedades de la subfamilia Pyrophorinae (Col. Elateridae.). Acta Zoologica Lilloana 38 (1): 81–85.

Gordon RD (1971) A generic review of the Cryptognathini, new tribe, with a description of a new genus (Coleoptera: Coccinellidae). Acta Zoologica Lilloana 26 (12): 179–196. [printed 28 May 1971 (p. 189); 26 Nov 1971 (recorded at BMNH)]

Gordon RD (1974) A review of the Oryssomini, a new tribe of Neotropical Coccinellidae (Coleoptera). The Coleopterists Bulletin 28 (3): 145–154. [issued 28 Oct 1974 (wrapper)]

Gordon RD (1975) A revision of the Epilachninae of the Western Hemisphere (Coleoptera: Coccinellidae). United States Department of Agriculture Technical Bulletin No. 1493: 1–409. [issued Dec 1975 (title page)]

Gordon RD (1977) Classification and phylogeny of the New World Sticholotidinae (Coccinellidae). The Coleopterists Bulletin 31 (3): 185–228. [mailed 28 Oct 1977 (wrapper)]

Gordon RD (1985) The Coccinellidae (Coleoptera) of America north of Mexico. Journal of the New York Entomological Society 93 (1): 1–912. [mailed 3 May 1985 (inside wrapper)]

Gordon RD (1999) South American Coccinellidae (Coleoptera), Part VI: A systematic revision of the South American Diomini, new tribe (Scymninae). Annales Zoologici (Warszawa) 49 (Supplement 1): 1–219. [publ. 30 Sep 1999 (inside wrapper)]

Gordon RD, Almeida LM, de (1991) Sticholotidinae (Coleoptera, Coccinellidae) update: descriptions of new South America taxa. Revista Brasileira de Entomologia 35 (1): 147–154 [31 Mar 1991 (top of article)]

Gordon RD, Pakaluk J, Ślipiński SA (1989) Carinodulini, a new tribe of Sticholotidinae based upon a new genus and species from Mexico (Coleoptera: Coccinellidae). The Coleopterists Bulletin 43 (4): 359–364. [12 Dec 1989 (inside wrapper)]

Gordon RD, Vandenberg N (1987) Eremochilini, a new tribe of Neotropical Epilachninae (Coleoptera: Coccinellidae). Journal of the New York Entomological Society 95 (1): 5–9. [mailed 30 Jan 1987 (inside wrapper)]

Gorham HS (1873) Endomycici recitati. A catalog of the coleopterous group, Endomycici, with descriptions of new species, and notes. Williams and Norgate, London, [1] + 64 pp. + 1 pl. [1873 (title page); May 1873 (date of preface); before 2 Feb 1874 (Proc. Ent. Soc. London 1874: i)]

Gorham HS (1874) On the Endomycici. The Annals and Magazine of Natural History (4) 13 (74): 185. [1 Feb 1874 (Evenhuis 2003: 26)]

Gorham HS (1876) Notes on the coleopterous family Cleridae, with descriptions of new genera and species. Cistula Entomologica 2 (15): 57–106. [7 Aug 1876 (footer p. 57)]

Gorham HS (1880) Lycidae [part, pp. 1–24, pls. 1–2]. In: Godman FD, Salvin O (Eds) Biologia Centrali-Americana. Insecta. Coleoptera. Vol. III. Part 2. Malacodermata. Taylor & Francis, London, xii + 372 pp. + 13 pls. [Dec 1880 (footer p. 1)]

Gorham HS (1881) Lycidae [part], Lampyridae, Telephoridae, Lymexylonidae, Melyridae [part] [pp. 25–112, pls. 4–6]. In: Godman FD, Salvin O (Eds) Biologia Centrali-Americana. Insecta. Coleoptera. Vol. III. Part 2. Malacodermata. Taylor & Francis, London, xii + 372 pp. + 13 pls. [Feb 1881 (pp. 25–40); Apr (41–64); Jun (65–80); Aug (81–88); Oct (89–104); Dec 1881 (105–112) (signature footers)]

Gorham HS (1894) Coccinellidae [part, pp. 193–208]. In: Godman FD, Salvin O (Eds) Biologia Centrali-Americana. Insecta. Coleoptera. Vol. VII. Erotylidae, Endomychidae, and Cocccinellidae. Taylor & Francis, London, xii + 276 pp. + 13 pls. [Jul 1894 (signature footers)]

Gory H, Laporte [=Castelnau] FLNC, de (1836) Chrysobothrites [1 p.]; Genre *Colobogaster* [pp. 3–17 + 3 pls.]. Histoire naturelle et iconographie des insectes Coléoptères, publiée par monographies séparées Suite aux buprestides Texte, tome II. P. Duménil, Paris, [454] pp. + 94 pls. [1835 or 1836 (Sherborn, notes in BMNH copy)]

Gory H, Laporte [=Castelnau] FLNC, de (1839) Sixième groupe. Anthaxites [1 p.]; Genre *Anthaxia* [pp. 1–37 + 8 pls.]. Histoire naturelle et iconographie des insectes Coléoptères, publiée par monographies séparées Suite aux buprestides Texte, tome II. P. Duménil, Paris, [454] pp. + 94 pls. [1839 (Sherborn, notes in BMNH copy)]

Gounelle É (1908) Cérambycides nouveaux ou peu connus de la région néo-tropicale principalement de la sous-région Brésilienne. 2e mémoire. Annales de la Société Entomologique de France 77 (1): 7–20. [8 Jul 1908 (Ann. Soc. Ent. France 78: 586)]

Gounelle É (1911) [Communications.] Note sur *Halycidocrius Philippii* Berg et *Prionapterus staphylinus* Serv. (Col. Cerambycidae). Bulletin de la Société Entomologique de France 1911 (16): pp. 319–327, pl. 2. [8 Nov 1911 (p. 463)]

Gozis MGP, des (1875) Catalogue des coléoptères de France et de la faune Gallo-Rhénane. [s. n.], Montluçon, France, ii + 106 pp. [Oct 1875 (title page); 10 Nov 1875 (Ann. Soc. Ent. France (5) 5: Bull. Ent.: cxcix)]

Gozis MGP, des (1882) Synopsis du genre *Tropideres* Schoenherr et description d'une espèce nouvelle. Feuille des Jeunes Naturalistes 12 (137): 58–59. [1 Mar 1882 (wrapper)]

Gratshev VG, Legalov AA (2009) New taxa of the family Nemonychidae (Coleoptera) from Jurassic and Early Cretaceous. Euroasian Entomological Journal 8 (4): 411–416 [in Russian].

Gravely FH (1914) An account of the oriental Passalidae (Coleoptera), based primarily on the collection in the Indian Museum. Memoirs of the Indian Museum 3 (4): 177–353, pls. 11–13. [Sep 1914 (title page)]

Gravely FH (1918) A contribution towards the revision of the Passalidae of the world. Memoirs of the Indian Museum 7 (1): 1–144. [Dec 1918 (article title page)]

Grebennikov VV (2009) Discheramocephalini, a new pantropical tribe of featherwing beetles (Coleoptera: Ptiliidae): description of new taxa and phylogenetic analysis. Systematic Entomology 34 (1): 113–136. [7 Jan 2009 (vol. contents)]

Grebennikov VV, Newton AF, Jr. (2009) Good-bye Scydmaenidae, or why the ant-like stone beetles should become megadiverse Staphylinidae sensu latissimo (Coleoptera). European Journal of Entomology 106: 275–301.

Green JW (1948) Two new species of Lampyridae from southern Florida, with a generic revision of the Nearctic fauna (Coleoptera). Transactions of the American Entomological Society 74: 61–73. [14 May 1948 (p. 72)]

Green JW (1949) The Lycidae of the United States and Canada. I. The tribe Lycini (Coleoptera). Transactions of the American Entomological Society 75 (2): 53–70. [29 Jul 1949 (p. 70)]

Gressitt JL (1940) The longicorn beetles of Hainan Island. Coleoptera: Cerambycidae. The Philippine Journal of Science 72 (1/2): 1–239, pls. 1–8. [issued 15 Aug 1940 (vol. contents)]

Gressitt JL (1946) Chinese chrysomelid beetles of the subfamily Chlamisinae. Annals of the Entomological Society of America 39 (1): 84–100. [mailed 29 Mar 1946 (p. 708)]

Gressitt JL (1950) The hispine beetles of China (Coleoptera: Chrysomelidae). Lingnan science journal 23 (1/2): 53–142, pls. 4–8.

Gressitt JL (1951) Longicorn beetles of China. In: Lepesme P (Ed). Longicornia. Études et notes sur les longicornes. Volume II. Paul Lechevalier, Paris, 667 pp. + 22 pls. [printed 4th trim. 1951 (end leaf)]

Gressitt JL (1952) The tortoise beetles of China (Chrysomelidae: Cassidinae). Proceedings of the California Academy of Sciences (4) 27 (17): 433–592. [8 Dec 1952 (wrapper)]

Gressitt JL (1953) Supplement to «The hispine beetles of China» (Coleoptera: Chrysomelidae). The Pan-Pacific Entomologist 29 (3): 121–126. [Jul 1953 issue; mailed 25 Aug (vol. p. iv)]

Gressitt JL (1959) Longicorn beetles from New Guinea, I (Cerambycidae). Pacific Insects 1 (1): 59–171. [mailed 17 Jul 1959 (verso of title page)]

Gressitt JL, Kimoto S (1963) The Chrysomelidae (Coleopt.) of China and Korea. Part 2. Pacific Insects Monograph 1B: 301–1026. [25 Feb 1963 (top of article)]

Griffin FJ (1932a) A note on an almost unknown work on Coleoptera. The Entomologist's Monthly Magazine (3) 18: 65–66. [Mar 1932 issue]

Griffin FJ (1932b) On the dates of publication and contents of the parts of Westwood (J. O.), Introduction to the modern classification of Insects, 1838–1840. Proceedings of the Entomological Society of London 6: 83–84.

Griffin FJ (1939) Erichson and others «Naturgeschichte der Insecten Deutschlands I.» Vols. 1–6 [1845-] 1899. Journal of the Society for the Bibliography of Natural History 1 (7): 218. [Feb 1939 (p. 218 footer)]

Griffith E, Pidgeon E (1832) The class Insecta arranged by the Baron Cuvier, with supplementary additions to each order. And notices of new genera and species by George Gray, Esq. Volume the first. Whittaker, Treacher, and Co., London, 570 pp. + 53 pls.

Grouvelle A (1908) Coléoptères de la région indienne. Rhysodidae, Trogositidae, Nitidulidae, Colydiidae, Cucujidae (1er mémoire). Annales de la Société Entomologique de France 77 (2): 315–336, (3): 337–495, pls. 6–9. [14 Oct 1908, 23 Dec 1908 (Ann. Soc. Ent. France 77: 850)]

Grouvelle A (1911) Trachypholini, Colydiidarum nova tribus. Notes from the Leyden Museum 33 [1910–11] (Note X): 121–168. [29 Apr 1911 (vol. contents)]

Grouvelle A (1912) Étude sur les *Aphanocephalus* et descriptions d'espèces nouvelles. Notes from the Leyden Museum 34 (3/4): 197–224. [15 Nov 1912 (vol. contents p. viii)]

Grouvelle A (1914) The Percy Sladen Trust expedition to the Indian Ocean in 1905, under the leadership of Mr. J. Stanley Gardiner, M.A. Volume VI. Coleoptera: Cucujidae, Cryptophagidae [of the Seychelles]. The Transactions of the Linnean Society of London (2nd Ser. Zoology) 17 (1): 141–156. [Dec 1914 (wrapper, vol. contents)]

Grouvelle A (1916) Études sur les Coléopterès. I. Étude sur les *Passandrella*, *Scalidia* et *Laemotmetus* (Passandridae). Mémoires Entomologiques 1: 5–25. [Apr 1916 (fasc. title page)]

Guéorguiev VB (1974) Sur la classification de la sous-famille Bathysciinae (Catopidae, Coleoptera). Doklady Bolgarskoy Akademii Nauk 27 (6): 839–842.

Guérin-Méneville FE (1834) Matériaux pour une classification des Mélasomes (extraits d'une monographie de cette famille). Magasin de Zoologie 4: 1–39 + pls. 101–118.

Guérin-Méneville FE (1840) Description de deux genres nouveaux de la famille des longicornes. Revue Zoologique 1840: 276–277. [Sep 1840 (footer p. 273)]

Guérin-Méneville FE (1843) Note sur un groupe naturel ou une petite tribu de Coléoptères de la famille des Malacodermes. Revue Zoologique 6: 193–194. [Jul 1843 (top of article)]

Guérin-Méneville FE (1844) Iconographie du règne animal de G. Cuvier, ou représentation d'après nature de l'une des espèces les plus remarquables, et souvent non encore figurées, de chaque genre d'animaux. Avec un texte descriptif mis au courant de la science. Ouvrage pouvant servir d'atlas à tous les traités de zoologie. Insectes. Texte. J. B. Baillière, Paris, 576 pp. [1829–1844].

Guignot F (1932) Triosième partie. Systématique [pp. 189–786] In: Les hydrocanthares de France. Hygrobiidae, Haliplidae, Dytiscidae et Gyrinidae de la France continentale avec notes sur les espèces de la Corse et de l'Afrique du nord française. Les frères Douladoure, Toulouse, 1057 pp. [printed 27 Sep 1932 (footer p. 189)]

Guignot F (1942) Description d'un genre nouveau de Dytiscidae. Bulletin Mensuel de la Société Linnéenne de Lyon 11 (1): 10–13. [Jan 1942; 15 May (recorded at BMNH)]

Guignot F (1948) Vingt-septième note sur les hydrocanthares. Bulletin de la Société Linnéenne de Lyon 1948: 163–171.

Guillebeau F (1892) Revision des phalacrides de la faune paléarctique. Revue d'Entomologie 11 (5): 141–197. [May 1892 (footer p. 141)]

Guillebeau F (1894) Descriptions de quelques espècies de la famille des Phalacridae de la collection de M. Antoine Grouvelle. Annales de la Société Entomologique de France 63 (1): 275–310. [30 Jul 1894 (wrapper)]

Günther K (1943) Vermischter Studien über Rüsselkäfer hauptsächlich aus der Sammlung Hartmann, jetzt im Staatl. Museum für Tierkunde, Dresden. Deutsche Entomologische Zeitschrift «Iris», Ergänzungsband 1943: 10–96. [15 Jul 1943 (title page)]

Gurjeva EL (1973) Novaya triba zhukov-shchelkunov Megapenthini tribus n. (Coleoptera, Elateridae). Zoologicheskii Zhurnal 52: 448–451.

Gurjeva EL (1974) Stroenie grudnogo otdela zhukov-shchelkunov (Coleoptera, Elateridae) i znachenie ego priznakov dlya sistemy semeistva. [The thoracic structure of click beetles (Coleoptera, Elateridae) and the significance of the structural characters for the system of the family]. Entomologicheskoe Obozrenie 53 (1): 96–113 [in Russian; English translation in Entomological Review, 53 (1): 67–79]. [20 Feb 1974 (approved to print, verso of title page); 18 Jun 1974 (recorded at CNC)]

Gyllenhal L (1808) Insecta Svecica, Classis I. Coleoptera sive Eleuterata. Tomus I [Pars 1]. F. J. Leverentz, Scaris [Skara], xii + 572 pp. [27 Apr 1808 (date of preface)]

Gyllenhal L (1810) Insecta Suecica, Classis I. Coleoptera sive Eleuterata. Tomi I Pars II. F. J. Leverentz, Scaris [Skara], xx + 660 pp. [1810 (title page)]

Gyllenhal L (1813) Insecta Suecica, Classis I. Coleoptera sive Eleuterata. Tomi I, Pars III. F. J. Leverentz, Scaris [Skara], 730 + 4 pp. [1813 (title page)]

Habu A (1967) Carabidae Truncatipennes group (Insecta: Coleoptera). Fauna Japonica. Biogeographical Society of Japan, Tokyo, xiv + 338 pp. + 27 pls. [May 1967 (verso of title page)]

Habu A (1975) Carabid beetles Mr. A. Sugimoto taken in Ishigaki ls., Ryukyus, by a black-light trap (Coleoptera, Carabidae). Entomological Review of Japan 28 (1/2): 69–84, pls. 7–9. [Aug 1975 (wrapper)]

Habu A (1978) Carabidae Platynini. Fauna Japonica. Yugaku-Sha, Ltd., Tokyo, viii + 447 pp. + 37 pls. [Mar 1978 (verso of title page)]

Habu A (1982) Revised and supplementary notes on and descriptions of the Truncatipennes group of Japan (I). (Coleoptera, Carabidae). Entomological Review of Japan 36 (2): 85–142. [publ. 30 Jan 1982 (inside wrapper)]

Habu A (1984) Revised and supplementary notes on and descriptions of the Truncatipennes group of Japan (IV) (Coleoptera, Carabidae). Entomological Review of Japan 39 (2): 101–139. [publ. 20 Dec 1984 (inside wrapper)]

Hagedorn M (1909) Zur Systematik der Borkenkäfer. Vorläufige Mitteilung. Entomologische Blätter 5 (8): 162–163. [15 Aug 1909 (p. 157)]

Hall WE (2003) *Limulopteryx*, a new genus of Neotropical featherwing beetle (Coleoptera, Staphylinoidea, Ptiliidae) and comments on pterycine ptiliids. In: Cuccodoro G, Leschen RAB (Eds), Systematics of Coleoptera: papers celebrating the retirement of Ivan Löbl. Memoirs on Entomology, International 17: 85–102. [30 Dec 2003 (verso of vol. title page)]

Hall WE (Ed) (2005) Ptiliidae Erichson 1845 [pp. 251–261]. In: Part 38. Coleoptera, beetles. Volume 1: Morphology and Systematics (Archostemata, Adephaga, Myxophaga, Polyphaga partim). In: Kristensen NP, Beutel RG (Eds) Handbook of Zoology. A Natural History of the Phyla of the Animal Kingdom. Volume IV - Arthropoda: Insecta. Walter De Gruyter, Berlin, xi + 567 pp.

Handlirsch A (1925) Systematische Übersicht [pp. 377–1140]. In: Schröder C (Ed) Handbuch der Entomologie. Band III. Geschichte, Literatur, Technik, Paläontologie, Phylogenie, Systematik. Gustav Fischer, Jena, viii + 1201 pp.

Hanley RS (2002) Phylogeny and higher classification of Hoplandriini (Coleoptera: Staphylinidae: Aleocharinae). Systematic Entomology 27 (3): 301–321. [15 Jul 2002 (vol. contents)]

Hansen M (1991) The hydrophiloid beetles. Phylogeny, classification and a revision of the genera (Coleoptera, Hydrophiloidea). Biologiske Skrifter, Det Kongelige Danske Videnskabernes Selskab 40: 1–367. [publ. Dec 1991 (p. 368)]

Hansen M (1996) Katalog over Danmarks biller; Catalogue of the Coleoptera of Denmark. Entomologiske Meddelelser 64 (1): 1–112, (2): 113–231. [(1): Apr; (2): May 1996]

Hansen M (1997) A new subfamily for a remarkable new genus and species of Hydrophilidae from New Guinea (Coleoptera: Hydrophilidae). Annales Zoologici (Warszawa) 47 (1/2): 107–110. [20 Sep 1997 (Newton)]

Hansen M (2004) Family Georissidae [p. 42]; Hydrophilidae [pp. 44–68]. In: Löbl I, Smetana A (Eds) Catalogue of Palaearctic Coleoptera. Volume 2. Hydrophiloidea - Histeroidea - Staphylinoidea. Apollo Books, Stenstrup, 942 pp. [publ. 31 Dec 2004 (verso of title page)]

Hansson C, Straka J (2009) The name Euderinae (Hymenoptera: Eulophidae) is a junior homonym. Proceedings of the Entomological Society of Washington 111 (1): 272–273. [mailed 27 Jan 2009 (inside wrapper)]

Harold E, von (1867) Zur Kenntniss der Gattung *Canthidium* und ihrer nächsten Verwandten. Coleopterologische Hefte 1: 1–61. [Mar 1867 (date of Vorwort p. v); before Jul 1867 (Ent. Mon. Mag. 4: 43)]

Harold E, von (1868) Die chilensischen Aphodiden. Berliner Entomologische Zeitschrift 11[1867] ((3–4)): 278–282. [Jan 1868 (Inhalt for vol.; wrapper)]

Harold E, von (1872) Literatur. Coleopterologische Hefte 10: 207–254. [before 6 Jan 1873 (Proc. Ent. Soc. London 1872: xlv)]

Harold E, von (1876) Literatur. Coleopterologische Hefte 15: 131–172. [1876 (vol. title page); before 1 Nov 1876 (Ent. Nachr. 2: 176)]

Harold E, von (1877) Coleopterorum species novae. Mitteilungen des Münchener Entomologischen Vereins 1: 97–111.

Harold E, von (1879) Bericht über die von den Herren A. v. Homeyer und P. Pogge in Angola und im Lunda-Reiche gesammelten Coleopteren. Coleopterologische Hefte 16: 1–224, pls. 1–2. [by Aug 1879 (Tijdschr. Ent. 24: xlix)]

Harold E, von (1880) Coleoptera. Zoologischer Jahresbericht (W. Engelmann) 1879 (Zweite Hälfte): 735–790.

Hatch MH (1927) Studies on the carrion beetles of Minnesota, including new species. Technical Bulletin / University of Minnesota Agricultural Experiment Station 48: 1–19. [Jun 1927 (wrapper)]

Hatch MH (1928) Silphidae II (Pars 95). In: Schenkling S (Ed) Coleopterorum Catalogus. Volumen VII. W. Junk, Berlin, pp. 63–244. [9 Jun 1928 (verso of vol. title page)]

Hatch MH (1957) The beetles of the Pacific Northwest. Part II: Staphyliniformia. University of Washington Publications in Biology 16 (2): ix + 384 pp.

Hatch MH (1962) The beetles of the Pacific northwest. Part III: Pselaphidae and Diversicornia I. University of Washington Publications in Biology 16: ix + 503 pp. [1961 overprinted with "1962" (title page)]

Háva J (2010) [new tribe] In: Kirejtshuk AG, Háva J, Nel A: New genus and species of subfamily Trinodinae (Coleoptera, Polyphaga, Dermestidae) from Lowermost Eocene French amber. Novyy rod i vid podsemeystva Trinodinae (Coleoptera, Polyphaga, Dermestidae) iz nizhneeotsenovogo Frantsuzskogo yantarya. Zoosystematica Rossica 19 (1): 54–69. [15 Jun 2010 (p. 54 header)]

Hayashi Y (1993) Studies on the Asian Staphylininae, I (Coleoptera, Staphylinidae). Elytra 21 (2): 281–301. [15 Nov 1993 (wrapper)]

Hayek CMF, von (1973) A reclassification of the subfamily Agrypninae (Coleoptera: Elateridae). Bulletin of the British Museum (Natural History) Entomology Supplement 20: 1–309.

Hayek, CMF, von (1989) A short biography of the entomologist James Thomson and the dates of publication of the Archives Entomologiques, Arcana Naturae, Monographie des Cicindélides, Musée Scientifique and Physis. Archives of Natural History 16: 81–99.

Heer O (1838) Fasciculus primus [pp. i-xii, 1–144]. In: Fauna coleopterorum Helvetica. Pars 1. Orellii, Fuesslinii et Sociorum, Turici, xii + 652 pp. [19 Jul 1838 (date of preface); Oct-Dec 1838 (Foreign Quart. Rev. 22: 455)]

Heer O (1839a) Fasciculus secundus [pp. 145–360]. In: Fauna coleopterorum Helvetica. Pars 1. Orelii, Fuesslini et Sociorum, Turici, xii + 652 pp. [1839 (tile page); 30 Apr 1840 (Intell.-Blatt Allgem. Lit.)]

Heer O (1839b) Die Kaefer der Schweiz, mit besonderer Berücksichtigung ihrer geographischen Verbreitung. Erster Theil. Zweite Lieferung. Petitpierre, Neuchatel, 67 pp. [1839 (title page); see Newton and Thayer (1992: 24)]

Heer O (1841a) [Pp. 361–652]. In: Fauna Coleopterorum Helvetica. Pars I. Orelii, Fuesslini & Sociorum, Turici, xii + 652 pp.

Heer O (1843) Ueber *Trichopteryx* Kirby. Entomologische Zeitung (Stettin) 4 (2): 39–62, pls. 1–2. [Feb 1843 issue]

Heer O (1847) Die Insektenfauna der Tertiärgebilde von Oeningen und von Radoboj in Croatien. Erste Abtheilung: Käfer. Neue Denkschriften der Allgemeinen Schweizerischen Gesellschaft für die Gesammten Naturwissenschaften 8 (5): 1–229 + (1), pls. 1–8.

Heffern DJ, Drumont A, Sama G, Komiya Z, Tavakilian G, Santos-Silva A (2006) Reversal of precedence using the International Code of Zoological Nomenclature to maintain prevailing usage of *Macrotoma* Audinet-Serville, 1832 (Coleoptera, Cerambycidae). Bulletin de la Société Entomologique de France 111 (1): 123–127. [publ. Mar 2006]

Heinze EK (1944) Beiträge zur Kenntnis der Tribus Cerylini und Metacerylini (nov.). (Coleoptera: Colydiidae.). Arbeiten über morphologische und taxonomische Entomologie aus Berlin-Dahlem 11 (1): 19–32. [31 Jul 1944 (vol. title page)]

Heinze EK (1962) Die Criocerinen Africas (Col. Chrysomelidae) (31. Beitrag zur Kenntnis der Criocerinen). Entomologische Arbeiten aus dem Museum G Frey 13 (1): 156–270. [publ. 1 Feb 1962 (vol. Inhalt)]

Heller KM (1916) Die Käfer von Neu-Caledonien und den benachbarten Inselgruppen. In: Sarasin F, Roux J (Eds). Nova Caledonia. A. Zoologie. Vol II. Part: 3, pp. 229–364 + 1 [unn.] + pls. 10–11.

Heller KM (1921) Neue Rhipiphoriden (Col.). Tijdschrift voor Entomologie 63 [1920]: 168–175. [1921 (vol. title page)]

Heller KM (1925a) 4. Bestimmungsschlüssel außereuropäischer Käfer. Curculionidae, Tribus n.: Ottistirini. Wiener Entomologische Zeitung 42 (4–10): 55–91 + pl. I. [25 Nov 1925 (wrapper)]

Heller KM (1925b) Studien zur Systematik der altweltlichen Balaninini (Coleoptera: Curculionidae). Stettiner Entomologische Zeitung 86 (2): 86–134. [Dec1925 (vol. Inhalt)]

Heller KM (1926) Fauna sumatrensis (Beitrag Nr. 29). Rhysodidae et familia nova Jacobsoniidae (prope Rhysodidae? Col.). Supplementa Entomologica 14: 126–128. [20 Dec 1926 (title page)]

Herman LH, Jr. (2001) Catalog of the Staphylinidae (Insecta: Coleoptera). 1758 to the end of the second millennium. Parts I-VII. Bulletin of the American Museum of Natural History 265: 1–4218 (in 7 vols.). [18 Jul 2001]

Hetschko A (1915) Verzeichniss der Schriften von Edmund Reitter (1869 bis 1915). In: Festschrift zum siebzigsten Geburtstag Edmund Reitters am 22. Oktober 1915. Wiener Entomologische Zeitung 34 (8/10): 221–270. [22 Oct 1915 (page footers)]

Heyden L, von (1891) [Dytiscidae-Heteroceridae, Cebrionidae-Oedemeridae: pp. 58–77, 210–270]. In: Heyden L, von, Reitter E, Weise J: Catalogus Coleopterorum Europae, Caucasi et Armeniae Rossicae. R. Friedländer & Sohn, Berlin, viii + 420 pp. [13 May 1891 (Ann. Soc. Ent. France 60: Bull. Ent.: lxxxvii)]

Heyden L, von (1908) Bemerkungen und Berichtigungen zum Catal. Coleopt. Europ. Cauc. et Arm. ross. 1906. Wiener Entomologische Zeitung 27 (1): 43–48. [1 Jan 1908 (issue title page)]

Heyne A, Taschenberg O (1905) [Lief. 19/20, pp. 139–170] In: Exotischen Käfer in Wort und Bild. J. F. Schreiber, Esslingen und München, [1905 (Blackwelder 1957: 1117)]

Heyne A, Taschenberg O (1906) [Lief. 21/22, pp. 171–194] In: Exotischen Käfer in Wort und Bild. J. F. Schreiber, Esslingen und München, [1906 (Blackwelder 1957: 1117)]

Heyne A, Taschenberg O (1907) [Lief. 25/26, pp. 195–262] In: Exotischen Käfer in Wort und Bild. J. F. Schreiber, Esslingen und München, [1907 (Blackwelder 1957: 1117)]

Hieke F (2003) Subtribe Amarina [pp. 547–568]. In: Löbl I, Smetana A (Eds) Catalogue of Palaearctic Coleoptera. Volume 1. Archostemata -Myxophaga - Adephaga. Apollo Books, Stenstrup, 819 pp. [publ. 30 May 2003 (verso of title page)]

Hincks WD (1952) The genera of the Cassidinae (Coleoptera: Chrysomelidae). The Transactions of the Royal Entomological Society of London 103 (10): 327–362. [31 Dec 1952 (wrapper)]

Hinton HE (1936) Lepiceridae - a new name for the Cyathoceridae. *Lepicerinus* - a new name for the scolytid genus *Lepicerus*. The Annals and Magazine of Natural History (10) 17 (100): 472–473. [1 Apr 1936 (Evenhuis 2003: 47)]

Hinton HE (1971) The species of *Dryopomorphus* (Coleoptera, Elmidae). The Entomologist 104 (1303): 293–297.

Hlisnikovský J (1963) Neue Liodidae (Coleoptera) aus Neu-Guinea I. Annales Historico-Naturales Musei Nationalis Hungarici 55: 301–311. [1963 (title page); 2 Apr 1964 (recorded at MCZ)]

Hoang DN (1982) Novaya triba bozhikh korovok podsemeistva Coccidulinae (Coleoptera, Coccinellidae). [A new tribe of Coccidulinae (Coleoptera: Coccinellidae)] [pp. 137–140]. In: Medvedev LN (Ed) Zhivotnyi mir Vietnama [Animal world of Vietnam]. Izdatel'stvo Nauka, Moscow, 270 pp. [in Russian]. [after 20 Jul 1982 (approved to print, verso of title page)]

Hoffmann A (1950) Coléopterères curculionides (Première partie). Faune de France. Vol. 52. P. Lechevalier, Paris, 486 pp. [dépot légal 2e trim. 1950 (p. 486)]

Hoffmann A (1956) Curculionides nouveaux rapportés par la Mission G. Remaudière en Iran. Revue de Pathologie Végétale et d'Entomologie Agricole de France 35 (4): 241–249. [Oct-Dec issue]

Hoffmann A (1957a) Observations critiques sur diverses espèces de la tribu des Rhytirrhinini (Col. Curculionidae). Annales de la Société Entomologique de France 126 (1): 59–69. [5 Dec 1957 (vol. contents)]

Hoffmann A (1957b) Description d'une espèce et de genres nouveaux, appartenant à une nouvelle sous-tribu des Ceuthorrhynchini (Col. Curculionidae). Bulletin de la Société Entomologique de France 61 [1956] (9/10): 218–223. [27 Mar 1957 (p. 248)]

Hoffmann A (1965) Contribution à la faune du Congo (Brazzaville) Mission A. Villiers et A. Descarpentries. XIV. Coléoptères Curculionidés. Bulletin de l'Institut Français d'Afrique Noire (Série A, Sciences Naturelles) 27 (4): 1397–1433. [Oct 1965 (issue title page)]

Hoffmann A (1968) The scientific results of the Hungarian soil zoological expedition to the Brazzaville-Congo. 32. Espèces de la famille Curculionidae (Coleoptera). Opuscula Zoologica (Budapest) 8 (1): 11–29.

Holloway BA (1972) The systematic position of the genus *Diphyllostoma* Fall (Coleoptera: Scarabaeoidea). New Zealand Journal of Science 15 (1): 31–38. [Mar 1972 (top of article)]
Holmgren N (1899) Zur Kenntniss der Begattungstasche der Elateriden, zugleich ein Beitrag der Systematik dieser Familie. Vorläufige Mittheilung. Entomologisk Tidskrift 20 (2/3): 197–203. [publ. 26 Sep 1899 (fasc. contents)]
Holyński R (1984) On the Oriental and Notogaean Mastogeniini LeC. Horn (Coleoptera, Buprestidae). Polskie Pismo Entomologiczne 54 (1): 105–114. [30 Apr 1984 (top of article)]
Holyński R (1988) Remarks on the general classification of Buprestidae Leach as applied to Maoraxiina Hol. Folia Entomologica Hungarica 49: 49–54. [30 Jul 1988 (verso of title page)]
Holyński R (1993) A reassessment of internal classification of the Buprestidae Leach (Coleoptera). Crystal Publications of the Natural Science Foundation at Göd (Series Zoologica) 1: 1–42.
Hong Y (1982) Jiuquan pandi kunchong huashi [Mesozoic fossil insects of Jiuquan Basin in Gansu Province]. Geological Publishing House, Peking, 187 pp., 39 pls. [in Chinese].
Hong Y (1998) A new early Cretaceous beetle family - Magnocoleidae fam. n. (Insecta: Coleoptera) in Hebei Province. Geoscience 12 (1): 40–48 [in Chinese]. [Mar 1998]
Hong Y (2002) [Amber insects of China]. Beijing Science and Technology Press, Beijing, x + 653 pp., 48 pls. [in Chinese].
Hope FW (1833) [reading from:] On the characters of several new genera and species of coleopterous insects. Proceedings of the Zoological Society of London 1: 61–64.
Hope FW (1834) Characters and descriptions of several new genera and species of coleopterous insects. Transactions of the Zoological Society of London 1: 91–112. [Jul 1834 (note in BMNH copy)]
Hope FW (1837) The coleopterist's manual, containing the Lamellicorn insects of Linneus and Fabricius. Henry G. Bohn, London, pp. i-xiii + 15–121 + [1], pls. 1–3. ["31 December" 1837; Hope refers to Kirby (1837) on p. 112; as 1837 (Mag. Nat. Hist. (N. S.) 2 (Apr 1838 issue): 211; before 20 Jan 1838 (Literary Gazette 22: 44)]
Hope FW (1838a) The coleopterist's manual, part the second, containing the predaceous land and water beetles of Linneus and Fabricius. Henry G. Bohn, London, xvi + 168 pp., pls. 1–3. [1838 (title page); 4 Mar 1839 (Jour. Proc. Ent. Soc. London 3: xx)]
Hope FW (1838b) Observations on the lamellicorns of Olivier. The Entomological Magazine 5 (4): 312–326. [Apr 1838 issue]
Hope FW (1839) [reading from:] A monograph on Mr. William Sharp MacLeay's coleopterous genus *Euchlora*. Proceedings of the Zoological Society of London 7: 65–75.
Hope FW (1840a) The coleopterist's manual, part the third, containing various families, genera, and species, of beetles, recorded by Linneus and Fabricius. Also, descriptions of newly discovered and unpublished insects. J. C. Bridgewater and Bowdery & Kerby, London, [5] + 191 pp. + 3 pls. [before 30 Sep 1840 (Jour. Proc. Ent. Soc. London 1840: lxxxix)]
Hope FW (1840b) [reading from:] "Descriptions of some new insects collected in Assam, by William Griffith, Esq., Assistant Surgeon in the Madras Medical Establishment". The Annals and Magazine of Natural History (1) 6 (37): 299–301. [1 Dec 1840 (Evenhuis 2003: 14; footer p. 289)]

Hope FW (1842) [reading from:] A monograph on the coleopterous family Phyllophoridae. Proceedings of the Zoological Society of London 10: 73–79.

Hope FW (1845) A catalogue of the lucanoid Coleoptera, in the collection of the Rev. F. W. Hope, M.A., F.R.S., &c. together with descriptions of the new species therein contained. J. C. Bridgewater, London, 31 pp. [1845 (wrapper)]

Hopkins AD (1902) A new genus of scolytids from Florida. Proceedings of the Entomological Society of Washington 5 [1901–03] (1): 34–38. [17 May 1902 (verso of vol. title page)]

Hopkins AD (1915) Contributions towards a monograph of the Scolytid beetles. II. Preliminary classification of the superfamily Scolytoidea. United States Department of Agriculture, Bureau of Entomology / Technical Series 17 (2): i-vi, 165–232, pls. ix-xvi. [9 Jan 1915 (wrapper and title page); 4 Mar 1915 (recorded at BMNH)]

Horn GH (1867a) On *Usechus lacerta* Motsch. Proceedings of the Entomological Society of Philadelphia 6 [1866–67]: 293–294. [Jan 1867 (footer p. 289)]

Horn GH (1867b) Descriptions of some new genera and species of Central American Coleoptera. Proceedings of the Academy of Natural Sciences of Philadelphia 18 (5 [Dec 1866]): 397–401. [Dec 1866 (page footers; incl. Proc. for 26 Dec); 20 Jul 1867 (recorded at USNM)]

Horn GH (1870) Revision of the Tenebrionidae of America, north of Mexico. Transactions of the American Philosophical Society (New Series) 14 (2): 253–404, pls. 14–15. [1870 (title page)]

Horn GH (1873) Revision of the genera and species of the tribe Hydrobiini. Proceedings of the American Philosophical Society 13: 118–137. [read 21 Feb 1873]

Horn GH (1876) [Otiorhynchidae, pp. 13–112] In: LeConte JL, Horn GH: The Rhynchophora of America North of Mexico. Proceedings of the American Philosophical Society held at Philadelphia for promoting useful knowledge 15 (No. 96): vii-xvi, 1–455. [23 Dec 1876 (date of preface p. ix); Dec 1876 (vol. title page)]

Horn GH (1877) Synopsis of the genera and species of the Staphylinide tribe Tachyporini of the United States. Transactions of the American Entomological Society 6: 81–128, pl. 1. [Jun 1877 (footer p. 81)]

Horn GH (1878) Synopsis of the Colydiidae of the United States. Proceedings of the American Philosophical Society held at Philadelphia for promoting useful knowledge 17: 555–592. [19 Apr 1878 (footnote p. 553: "XVII. 101. 3Q. printed June 1, 1878")]

Horn GH (1880a) Synopsis of the Dascyllidae of the United States. Transactions of the American Entomological Society and Proceedings of the Entomological Section of the Academy of Sciences 8 (1/2): 76–114, pl. 1. [Mar-Apr 1880 (page footers); presented 9 Apr 1880 (F. M. Brown 1964: 319)]

Horn GH (1880b) Synopsis of the Silphidae of the United States with reference to the genera of other countries. Transactions of the American Entomological Society and Proceedings of the Entomological Section of the Academy of Sciences 8 ([2]): 219–319, pls. 5–7. [Jul-Sep 1880 (page footers); presented 10 Sep 1880–8 Oct 1880 (F. M. Brown 1964: 319)]

Horn GH (1880c) Synopsis of the Euphoriae. Proceedings of the American Philosophical Society held at Philadelphia for promoting useful knowledge 18 [1878–80]: 397–408 + pl. 4. [printed 2 Jan 1880 (footer p. 397)]

Horn GH (1880d) Revision of the Nitidulidae of the United States. Transactions of the American Entomological Society and Proceedings of the Entomological Section of the Academy of Sciences 7 (4): 267–336, pl. 3. [Jun-Oct 1879 (page footers); publ. 9 Jan 1880 (F. M. Brown 1964: 317)]

Horn GH (1880e) Contributions to the coleopterology of the United States, no. 3. Transactions of the American Entomological Society and Proceedings of the Entomological Section of the Academy of Sciences 8 (1): 139–218. [May 1880 (footer p. 137); 14 Jun-10 Sep 1880 (F. M. Brown 1964: 319)]

Horn GH (1881) On the genera of Carabidae with special reference to the fauna of boreal America. Transactions of the American Entomological Society and Proceedings of the Entomological Section of the Academy of Sciences 9 (2): [1881–1882] 91–196. [12 Dec 1881 (F. M. Brown 1964: 317)]

Horn GH (1888) Miscellaneous coleopterous studies. Transactions of the American Entomological Society and Proceedings of the Entomological Section of the Academy of Sciences 15 (1): 26–48. [22 May 1888 (received at CUL)]

Horn GH (1889) A synopsis of the Halticini of boreal America. Transactions of the American Entomological Society and Proceedings of the Entomological Section of the Academy of Sciences 16 (3): 163–320, pl. [7 Sep 1889 (received at CUL)]

Horn GH (1890) A revision of the Sphaeridiini inhabiting boreal America. Transactions of the American Entomological Society 17 (3): 279–314, pl. 9. [Aug-Nov 1890 (footers); 28 Feb 1891 (recorded at CUL)]

Horn GH (1893) The Galerucini of boreal America. Transactions of the American Entomological Society 20 (2): 57–136. [Apr 1893 (footer p. 57); 19 Jul 1893 (received at CUL)]

Horn W (1893) Bemerkungen und Nachträge zum "Catalogue systématique des Cicindelidae" par Fleutiaux (1893). Deutsche Entomologische Zeitschrift 1893 (2): 321–347. [Jul 1893 (footer p. 3)]

Horn W (1899) Ueber das System der Cicindeliden. Deutsche Entomologische Zeitschrift 1899: 33–51. [Aug 1899 (footer p. 3)]

Horn W (1906) *Odontochila bennigseni euryoides* (nov. subsp.) und das System der Cicindelini. Deutsche Entomologische Zeitschrift 1906 (1): 85–86. [Feb 1906 (vol. contents)]

Horn W (1907) Brullés "*Odontochila* aus dem baltischen Bernstein" und die Phylogenie der Cicindeliden. (Col.). Deutsche Entomologische Zeitschrift 1907 (5): 461–466. [1 Sep 1907 (Inhalt)]

Houlbert C (1912) Tableaux génériques illustrés des coléoptères de France. Supplément à la Faune entomologique Armoricaine. Imprimerie Oberthur, Rennes, 288 pp.

Houlbert C (1922a) Les coléoptères d'Europe France et régions voisines. Anatomie générale; classification et tableaux génériques illustrés. In: Toulouse D (Ed) Encyclopédie Scientifique. Tome second. G. Doin, Paris, 340 + xii pp. [before 15 Jul 1922 (Rev. Gén. Sci. 33: 407)]

Houlbert C (1922b) Les coléoptères d'Europe France et régions voisines. Anatomie générale; classification et tableaux génériques illustrés. In: Toulouse D (Ed) Encyclopédie Scientifique. Tome troisième. G. Doin, Paris, 297 + xii pp. [before 15 Jul 1922 (Rev. Gén. Sci. 33: 407)]

Houlbert C (1934) Faune entomologique armoricaine. Coléoptères, hydrocarabiques. Bulletin de la Société Scientifique de Bretagne 11: 1–147.

Houston WWK, Weir TA (1992) Melolonthinae [pp. 174–358]. In: Houston WWK (Ed) Zoological Catalogue of Australia. Volume 9. Coleoptera: Scarabaeoidea. Australian Government Publishing Service, Canberra, xi + 544 pp.

Howden HF (1997) Podolasiini Howden, new tribe, and a revision of the included genera, *Podolasia* Harold and *Podostena* Howden, new genus (Coleoptera: Scarabaeidae: Melolonthinae). The Coleopterists Bulletin 51 (3): 223–255. [mailed 19 Sep 1997 (inside wrapper)]

Howden HF (2001) A new tribe of Hybosorinae with a description of a new species of *Callosides* Howden (Coleoptera: Scarabaeidae). The Coleopterists Bulletin 55 (2): 199–204. [mailed 12 Jul 2001 (inside wrapper)]

Howden HF, Gill BD (2000) Tribes of New World Ceratocanthinae, with keys to genera and descriptions of new species (Coleoptera: Scarabaeidae). Sociobiology 35 (2B): 281–329. [28 Mar 2000 (recorded at CNC)]

Howden HF, Martínez A (1963) The new tribe Athyreini and its included genera (Coleoptera: Scarabaeidae, Geotrupinae). The Canadian Entomologist 95 (4): 345–352. [mailed 23 Apr 1963 (p. 560)]

Howden HF, Storey RI (1992) Phylogeny of the Rhyparini and the new tribe Stereomerini, with descriptions of new genera and species (Coleoptera; Scarabaeidae; Aphodiinae). Canadian Journal of Zoology 70 (9): 1810–1823. [issued 13 Oct 1992 (fasc. 10 contents, p. iii)]

Hunt JW, Breuning S (1957) New Lamiinae (Coleoptera, Cerambycidae) from South Africa. Durban Museum Novitates 5 (5): 51–70. [issued 10 Dec 1957 (top of article)]

Hustache A (1919) Synopsis des Curculionides de la Faune Malgache. I. Brachydérides et Otiorrhynchides. Annales de la Société Entomologique de France 87 [1918] (3/4): 441–520. [23 Apr 1919 (p. 549)]

Hustache A (1925) Synopsis des curculionides de Madagascar. Bulletin de l'Académie Malgache (Nouvelle Série) 7 [1924]: 1–582. [25 Nov 1925 (Viette 1993: 180)]

Hustache A (1929) Curculionidae. Voyage de Ch Alluaud et R Jeannel en Afrique Orientale (1911–1912) Résultats scientifiques Insectes Coléoptères XIX. Vol: 30. P. Lechevalier, Paris, 365–562 + pls. 10–11. [1 Aug 1929 (title page)]

Hustache A (1932) Essai d'un tableau synoptique des zygopides d'Afrique et de Madagascar [pp. 369–380]. Livre du Centenaire Société Entomologique de France. Société Entomologique de France, Paris, xii + 729 pp. [printed 30 Jun 1932 (endleaf)]

Hustache A (1934) Pars 136: Curculionidae: Subfam. Zygopinae. In: Schenkling S (Ed) Coleopterorum Catalogus. Volumen XXX. W. Junk, Berlin, 96 pp. [10 Jun 1934 (verso of vol. title page)]

Hustache A (1936) Pars 151: Curculionidae. Cryptorhynchinae. In: Schenkling S (Ed) Coleopterorum Catalogus. Volumen XXIX. W. Junk, Berlin, 317 pp. [15 Dec 1936 (verso of vol. title page)]

Hustache A (1937) Magdalinae (Coleoptera, Curculionidae) de l'Amérique méridionale. The Annals and Magazine of Natural History (10) 19 (110): 198–248. [1 Feb 1937 (Evenhuis 2003: 47)]

Hustache A (1938) Ceratopinae Sud-Américains (Col. Curculionides). Miscellanea Entomologica 39 (11): 89–99. [printed 15 Nov 1938 (p. 89)]

Huys R (2009) Unresolved cases of type fixation, synonymy and homonymy in harpacticoid copepod nomenclature (Crustacea: Copepoda). Zootaxa No. 2183: 1–99. [publ. 6 Aug 2009 (title page)]

Hyslop JA (1917) The phylogeny of the Elateridae based on larval characters. Annals of the Entomological Society of America 10 (3): 241–263. [mailed 13 Sep 1917 (after vol. contents)]

Iablokoff-Khnzorian SM (1961) Circaeidae – Novoe semeystvo zhestkokrylykh iz yantarya (Insecta, Coleoptera). [Circaeidae – new family of Coleoptera from amber (Insecta, Coleoptera)]. Doklady Akademii Nauk SSSR 136 (1): 209–210 [in Russian]. [Jan-Feb 1961 / English translation in: Earth Sciences 136: 198–199, Sep 1962]

Iablokoff-Khnzorian SM (1962) Predstaviteli Sternoxia (Coleoptera) iz baltiyskogo yantarya. [Some Sternoxia (Coleoptera) from Baltic amber]. Paleontologicheskii Zhurnal 1962 (3): 81–89 [in Russian]. [after 7 Aug 1962 (approved to print, verso of title page); 28 Dec 1962 (recorded at BMNH)]

Iablokoff-Khnzorian SM (1966) Dva novykh vida zhestkokrylykh iz Armyanskoi SSR (Insecta, Coleoptera). [Two new species of Coleoptera from the Armenian Republic (Insecta, Coleoptera)]. Doklady Akademii Nauk Armyanskoi SSR 42 (5): 309–314 [in Russian]. [23 Nov 1966 (recorded at BMNH)]

Iablokoff-Khnzorian SM (1977) Über die Phylogenie der Lamellicornia (Insecta, Coleoptera). Entomologische Abhandlungen / Staatlichen Museum für Tierkunde in Dresden 41 [1976–77] (5): 135–200. [Oct 3 1977 (top of article)]

Iablokoff-Khnzorian SM (1983) Notes sur la phylogénie des Cucujoidea et le classement général des coléoptères. Deutsche Entomologische Zeitschrift (Neue Folge) 30 (1/3): 45–68. [27 May 1983 (wrapper)]

Iablokoff-Khnzorian SM (1986) Novyy rod i vid zhukov-veeronostsev iz Tadzhikistana (Coleoptera, Rhipiphoridae). [A new genus and species of the beetle family Rhipiphoridae from Tadzhikistan (Coleoptera, Rhipiphoridae)] [in Russian]. Doklady Akademii Nauk Armyanskoi SSR 82 (2): 89–92.

ICZN (1950) Opinion 243. Designation, under the Plenary Powers, for the nominal genus *Carabus* Linnaeus, 1758 (Class Insecta, Order Coleoptera) of a type species in harmony with current nomenclatorial usage. Opinions and Declarations Rendered by the International Commission on Zoological Nomenclature 5 (4): 45–56.

ICZN (1954) Opinion 280. Emendation to *Hygrobia* of the generic name Hygriobia Latreille, 1804 (Class Insecta, Order Coleoptera). Opinions and Declarations Rendered by the International Commission on Zoological Nomenclature 6: 189–203.

ICZN (1955a) Opinion 345. Rejection of a proposal to use the Plenary Powers for the benefit of the generic name *Rhina* Latreille, [1802–1803] (Class Insecta, Order Coleoptera) and addition of the generic name *Rhina* Schneider, 1801 (Class Elasmobranchii) to the Official List of Generic Names in Zoology; designation under the Plenary Powers for the genera *Rhinostomus* Rafinesque, 1815, and *Magdalis* Germar, 1817 (Class Insecta, Order Coleoptera) of type species in harmony with accustomed usage: validation under the Plenary Powers of the specific name *barbicornis* Latreille, [1803–1804], as published in the combination *Rhina barbicornis* (Class Insecta, Order Coleoptera). Opinions and Declarations Rendered by the International Commission on Zoological Nomenclature 10 (12): 353–388.

ICZN (1955b) Opinion 346. Designation, under the Plenary Powers, for the genus *Geotrupes* Latreille, 1796 (Class Insecta, Order Coleoptera) of a type species in harmony with current usage. Opinions and Declarations Rendered by the International Commission on Zoological Nomenclature 10: 389–408.

ICZN (1955c) Direction 28. Addition to the *Official List of Family-Group Names in Zoology* or, as the case may be, to the *Offcial Index of Rejected and Invalid Family-Group Names in Zoology* of the family-group namesinvolved in the cases dealt with in the *Opinions* included in volume 10 of the *Opinions and Declarations rendered by the International Commission on Zoological Nomenclature*, other than family-group names already dealt with in those *Opinions* or in *Directions* included in that volume. Opinions and Declarations Rendered by the International Commission on Zoological Nomenclature 10 (21): 493–510.

ICZN (1956) Direction 54. Addition to the *Official List of Family-Group Names in Zoology* or, as the case may be, to the *Offcial Index of Rejected and Invalid Family-Group Names in Zoology* of the family-group namesinvolved in the cases dealt with in the *Opinions* included in volume 12 of the *Opinions and Declarations rendered by the International Commission on Zoological Nomenclature*, other than family-group names already dealt with in those *Opinions*. Opinions and Declarations Rendered by the International Commission on Zoological Nomenclature 12 (26): 457–470.

ICZN (1957) Opinion 463. Designation under the plenary powers of a type species in harmony with accustomed usage for the genus *Oxypoda* Mannerheim, 1831 (Class Insecta, Order Coleoptera). Opinions and Declarations Rendered by the International Commission on Zoological Nomenclature 16: 15–24.

ICZN (1959a) Opinion 546. Designation under the plenary powers of a type species in harmony with accustomed usage for the genus *Staphylinus* Linnaeus, 1758 (Class Insecta, Order Coleoptera). Opinions and Declarations Rendered by the International Commission on Zoological Nomenclature 20: 141–151.

ICZN (1959b) Opinion 572. Suppression under the plenary powers of the generic name *Calandra* Clairville & Schellenberg, 1798, and validation under the same powers of the specific name *abbreviatus* Fabricius, 1787, as published in the binomen *Curculio abbreviatus* (Class Insecta, Order Coleoptera) Bulletin of Zoological Nomenclature 17 (3–5): 112–116.

ICZN (1961a) Opinion 599. *Bolitochara* Mannerheim, 1831 (Insecta, Coleoptera): designation of a type-species under the plenary powers. Bulletin of Zoological Nomenclature 18: 238–240. [Aug 1961]

ICZN (1961b) Opinion 600. *Ischnopoda* Stephens, 1835, and *Tachyusa* Erichson, 1837 (Insecta, Coleoptera); designations of type-species under the plenary powers. Bulletin of Zoological Nomenclature 18: 241–243. [Aug 1961]

ICZN (1961c) Opinion 604. *Korynetes* Herbst, (1792), and *Necrobia* Olivier, 1795; added to the Official List (Insecta, Coleoptera). Bulletin of Zoological Nomenclature 18: 252–253.

ICZN (1961d) Opinion 618. *Dytiscus cinereus* Linnaeus, 1758 (Insecta, Coleoptera); designation of a neotype under the Plenary Powers. Bulletin of Zoological Nomenclature 18: 365–368.

ICZN (1961e) Opinion 619. *Acilius* Leach, 1817 (Insecta, Coleoptera); preservation by use of the Plenary Powers to vary the ruling given in Opinion 522. Bulletin of Zoological Nomenclature 18: 369–371.

ICZN (1962) Opinion 621. Phaenomerididae Ohaus, 1913, and Phaenomerina Faust, 1898 (Insecta, Coleoptera); addition to the Official List. Bulletin of Zoological Nomenclature 19: 74–75. [23 Mar 1962]

ICZN (1963a) Opinion 670. *Dendroctonus* Erichson, 1836 (Insecta, Coleoptera): Designation of a type-species under the plenary powers with addition of *Tomicus* Latreille, [1802–1803] to the Official List. Bulletin of Zoological Nomenclature 20: 276–278.

ICZN (1963b) Opinion 683. *Scolytus* Geoffroy, 1762 (Insecta,Coleoptera): validated under the Plenary Powers. Bulletin of Zoological Nomenclature 20: 416–417.

ICZN (1964) Opinion 710. *Enhydrus* Laporte, 1834 (Insecta, Coleoptera): validated under the Plenary Powers. Bulletin of Zoological Nomenclature 21: 242–245.

ICZN (1967a) Opinion 806. *Gymnetis* MacLeay, 1819 (Insecta, Coleoptera): designation of a type-species under the Plenary Powers. Bulletin of Zoological Nomenclature 24: 22–23.

ICZN (1967b) Opinion 808. *Cryptorhynchus* Illiger, 1807 (Insecta, Coleoptera): designation of a type-species under the plenary powers. Bulletin of Zoological Nomenclature 24: 83–84.

ICZN (1968a) Opinion 848. *Xyleborus* Bowdich, 1825 (Insecta, Coleoptera): suppressed under the Plenary Powers. Bulletin of Zoological Nomenclature 25: 18–19.

ICZN (1968b) Opinion 862. *Galerita* Fabricius, 1801 (Insecta, Coleoptera): validation under the Plenary Powers. Bulletin of Zoological Nomenclature 25: 98–99.

ICZN (1969a) Opinion 876. *Proteinus* Latreille, 1796 (Insecta, Coleoptera): designation of a type-species under the plenary powers. Bulletin of Zoological Nomenclature 26: 14–15. [May 1969]

ICZN (1969b) Opinion 887. *Bryaxis* Kugelann, 1794 (Insecta, Coleoptera): designation of a type-species under the plenary powers. Bulletin of Zoological Nomenclature 26: 133–135. [24 Oct 1969]

ICZN (1970a) Opinion 908. *Crioceris* Muller, 1764, and *Lema* Fabricius, 1798 (Insecta, Coleoptera): designation of type-species under the Plenary Powers. Bulletin of Zoological Nomenclature 27: 12–13.

ICZN (1970b) Opinion 928. *Pachyrhynchus* Germar, 1824 (Insecta, Coleoptera): validated under the Plenary Powers. Bulletin of Zoological Nomenclature 27: 93–94.

ICZN (1971) Opinion 966. *Xyletinus* Latreille, 1809 (lnsecta, Coleoptera): designation of a type-species under the Plenary Powers. Bulletin of Zoological Nomenclature 28: 81–82.

ICZN (1972) Opinion 982. *Otiorhynchus* Germar, 1824 (lnsecta, Coleoptera): validated under the Plenary Powers. Bulletin of Zoological Nomenclature 29: 19–24.

ICZN (1974a) Opinion 1005. *Trypeta* Meigen, 1803 (lnsecta: Diptera); *Trypetes* Schoenherr, 1836 (lnsecta: Coleoptera); *Trypetesa* Norman, 1903 (Crustacea: Cirripedia): removal of homonymy in family-group names under the Plenary Powers. Bulletin of Zoological Nomenclature 30: 151–152.

ICZN (1974b) Opinion 1023. Cassidae (Mollusca) and Cassidinae (Insecta): placed on the Official List of Family-Group Names in Zoology. Bulletin of Zoological Nomenclature 31: 127–129.

ICZN (1975) Opinion 1039. *Uloma* Dejean, 1821 and *Phaleria* Latreille, 1802 (Itisecta, Coleoptera): designation of type-species under the Plenary Powers. Bulletin of Zoological Nomenclature 32: 136–138.

ICZN (1976) Opinion 1062. *Anobium* Fabricius, 1775: *Grynobius* Thomson, 1859: *Priobium* Motschulsky, 1845 (Coleoptera): designation of type species under the Plenary Powers. Bulletin of Zoological Nomenclature 33: 31–33.

ICZN (1979a) Opinion 1144. *Phloeotribus* (Coleoptera: Scolytidae) ruled to be a justified emendation of *Phloiotribus* Latreille, 1796. Bulletin of Zoological Nomenclature 36: 132–134.

ICZN (1979b) Opinion 1145. *Dryocoetes* Eichhoff, 1864 (Coleoptera: Scolytidae) conserved under the Plenary Powers. Bulletin of Zoological Nomenclature 36: 149–150.

ICZN (1980) Opinion 1155. *Saperda inornata* Say, 1824 (Insecta: Coleoptera): designation of a neotype by the use of the Plenary Powers. Bulletin of Zoological Nomenclature 37: 89–93.

ICZN (1981a) Opinion 1167. *Phloeosinus* Chapuis, 1869 (Coleoptera, Scolytidae) conserved. Bulletin of Zoological Nomenclature 38: 67–68.

ICZN (1981b) Opinion 1178. *Megasternum* Mulsant, 1844, and *Cryptopleurum* Mulsant, 1844 (Insecta, Coleoptera): type species determined. Bulletin of Zoological Nomenclature 38: 114–116.

ICZN (1981c) Opinion 1179. *Polydrusus* Germar, 1817 and *Phyllobius* Germar, 1824 (Insecta, Coleoptera): conserved in accordance with current usage. Bulletin of Zoological Nomenclature 38: 117–119.

ICZN (1982) Opinion 1221. *Baeocera* Erichson, 1845 (Insecta, Coleoptera): designation of type species. Bulletin of Zoological Nomenclature 39: 175–177. [Sep 1982]

ICZN (1983a) Opinion 1239. *Attelabus* Linnaeus, 1758 (Insecta, Coleoptera): type species designated. Bulletin of Zoological Nomenclature 40: 25–26.

ICZN (1983b) Opinion 1244. *Stethaspis* Hope, 1837 (Coleoptera, Scarabaeidae): designation of type species. Bulletin of Zoological Nomenclature 40: 37–38.

ICZN (1983c) Opinion 1250. *Gyrohypnus* Samouelle, 1819, ex Leach MS, *Xantholinus* Dejean, 1821, ex Dahl, and *Othius* Stephens, 1829, ex Leach MS (Insecta, Coleoptera): type species designated for these genera. Bulletin of Zoological Nomenclature 40: 85–87. [Jul 1983]

ICZN (1984a) Opinion 1273. *Anaspis*, *Luperus*, *Lampyris* and *Clerus* (Insecta, Coleoptera): determination of authorship and fixation of type species. Bulletin of Zoological Nomenclature 41: 28–31.

ICZN (1984b) Opinion 1277. *Ptilium* Gyllenhal, 1827 and *Ptenidium* Erichson, 1845 (Insecta, Coleoptera): conserved. Bulletin of Zoological Nomenclature 41: 212–214. [Nov 1984]

ICZN (1984c) Opinion 1279. *Chrysolina* Motschulsky, 1860 (Insecta, Coleoptera): conserved. Bulletin of Zoological Nomenclature 41: 218–220.

ICZN (1985a) Opinion 1299. *Athyreus* MacLeay, 1819 and *Glyptus* Brulle, 1835 (Insecta, Coleoptera): conserved. Bulletin of Zoological Nomenclature 42: 128–129.

ICZN (1985b) Opinion 1307. *Ptinella* Motschulsky, 1844 and *Nephanes* Thomson, 1859 (Insecta, Coleoptera): conserved. Bulletin of Zoological Nomenclature 42: 148–149. [Jun 1985]

ICZN (1985c) Opinion 1323. *Byrrhus semistriatus* Fabricius, 1794 (Insecta, Coleoptera): conserved. Bulletin of Zoological Nomenclature 42: 182–184.

ICZN (1985d) Opinion 1331. Sphaeriidae Jeffreys, 1862 (1820) (Mollusca, Bivalvia) and Microsporidae Reichardt, 1976 (Insecta, Coleoptera): placed on the Official List. Bulletin of Zoological Nomenclature 42: 230–232.

ICZN (1985e) Opinion 1352. *Eurhinus* Schonherr, 1825 (Insecta, Coleoptera): ruled as a justified emendation of *Eurhin* Illiger, 1807. Bulletin of Zoological Nomenclature 42: 287–290.

ICZN (1985f) Opinion 1359. Uroplat- as the stem of family-group names in Reptilia, Sauria and Insecta, Coleoptera: a ruling to remove the homonymy. Bulletin of Zoological Nomenclature 42: 344–346.

ICZN (1985g) International Code of Zoological Nomenclature. Third Edition. Adopted by the XX General Assembly of the International Union of Biological Sciences International Trust for Zoological Nomenclature, London, xx + 338 pp. [Feb 1985]

ICZN (1986a) Opinion 1382. *Zeugophora* Kunze, 1818 (Insecta, Coleoptera): conserved. Bulletin of Zoological Nomenclature 43: 44–45.

ICZN (1986b) Opinion 1397. *Rhopalocerus* W. Redtenbacher, 1842 (Insecta, Coleoptera, Colydidae): conserved. Bulletin of Zoological Nomenclature 43: 150–151.

ICZN (1986c) Opinion 1407. *Lamia aethiops* Fabricius, 1775 designated as type species of *Ceroplesis* Serville, 1835 (Insecta, Coleoptera). Bulletin of Zoological Nomenclature 43: 243–244.

ICZN (1987a) Opinion 1440. *Brachyderes* Schonherr, 1823 and *Cycloderes* Sahlberg, 1823 (Insecta, Coleoptera): conserved. Bulletin of Zoological Nomenclature 44: 144–145.

ICZN (1987b) Opinion 1441. *Elater bimaculatus* Rossi, 1790 (currently *Drasterius bimaculatus*; Insecta, Coleoptera): specific name conserved. Bulletin of Zoological Nomenclature 44: 146.

ICZN (1987c) Opinion 1448. *Dryophthorus* Germar, 1824 (Insecta, Coleoptera): conserved. Bulletin of Zoological Nomenclature 44: 202.

ICZN (1987d) Opinion 1449. *Cholus* Germar, 1824 (Insecta, Coleoptera): conserved. Bulletin of Zoological Nomenclature 44: 203.

ICZN (1987e) Opinion 1450. *Zygops* Schoenherr, 1825 (Insecta, Coleoptera): conserved. Bulletin of Zoological Nomenclature 44: 204.

ICZN (1987f) Opinion 1451. *Lachnopus* Schoenherr, 1840 (Insecta, Coleoptera): conserved. Bulletin of Zoological Nomenclature 44 (3): 205–206. [25 Sep 1987]

ICZN (1988a) Opinion 1472. *Cyclaxyra* Broun, 1893 (Insecta, Coleoptera): conserved. Bulletin of Zoological Nomenclature 45: 69–70.

ICZN (1988b) Opinion 1473. *Tetropium* Kirby, 1837 (Insecta, Coleoptera): conserved. Bulletin of Zoological Nomenclature 45: 71–72.

ICZN (1988c) Opinion 1474. *Tropiphorus* Schonherr, 1842 (Insecta, Coleoptera): conserved. Bulletin of Zoological Nomenclature 45: 73.

ICZN (1988d) Opinion 1493. *Geonemus* Schoenherr, 1833 (Insecta, Coleoptera): *Curculio flabellipes* Olivier, 1807 designated as type species. Bulletin of Zoological Nomenclature 45: 168–169.

ICZN (1988e) Opinion 1495. *Colydium castaneum* Herbst, 1797 (currently *Tribolium castaneum*; Insecta, Coleoptera): specific name conserved. Bulletin of Zoological Nomenclature 45: 171–172.

ICZN (1988f) Opinion 1506. *Oncomera* Stephens, 1829 (Insecta, Coleoptera): *Dryops femorata* Fabricius, 1792 designated as the type species. Bulletin of Zoological Nomenclature 45: 235.

ICZN (1988g) Opinion 1515. Laridae Rafinesque Schmaltz, 1815 (Aves) and Larini LeConte, 1861 (Insecta, Coleoptera): homonymy removed. Bulletin of Zoological Nomenclature 45: 245–246. [23 Sep 1988]

ICZN (1989a) Opinion 1525. *Phymatodes* Mulsant, 1839 and *Phymatestes* Pascoe, 1867 (Insecta, Coleoptera): conserved. Bulletin of Zoological Nomenclature 46: 65–66.

ICZN (1989b) Opinion 1526. *Nanophyes* Schoenherr, 1838 (Insecta, Coleoptera): conserved. Bulletin of Zoological Nomenclature 46: 67–68.

ICZN (1989c) Opinion 1529. *Ceutorhynchus* Germar, 1824, *Rhinoncus* Schoenherr, 1825 and *Curculio assimilis* Paykull, 1792 (Insecta, Diptera): conserved, and *Curculio assimilis* Paykull, 1792 and *Curculio pericarpius* Linnaeus, 1758 designated as the type species of *Ceutorhynchus* and *Rhinoncus* respectively. Bulletin of Zoological Nomenclature 46: 71–73.

ICZN (1989d) Opinion 1546. *Chelonus* Panzer, 1806 (Insecta, Hymenoptera) and *Anomala* Samouelle, 1819 (Insecta, Coleoptera): names conserved. Bulletin of Zoological Nomenclature 46: 149–150.

ICZN (1989e) Opinion 1549. Euglenidae Stein, 1878 (Protista, Flagellata) and Euglenidae Seidlitz, 1875 (Insecta, Coleoptera): homonymy removed, and Aderidae Winkler, 1927 (Insecta, Coleoptera): given precedence over Euglenesidae Seidlitz, 1875. Bulletin of Zoological Nomenclature 46: 193–194. [29 Sep 1989]

ICZN (1990a) Opinion 1577. *Hydrobius* Leach, 1815 (Insecta, Coleoptera): *Dytiscus fuscipes* Linnaeus, 1758 conserved as type species, and *Berosus* Leach, 1817 (Insecta, Coleoptera):conserved. Bulletin of Zoological Nomenclature 47: 71–72. [Mar 1990]

ICZN (1990b) Opinion 1597. *Coryphium angusticolle* Stephens, 1834 (Insecta, Coleoptera): generic and specific names conserved. Bulletin of Zoological Nomenclature 47: 156–157. [Jun 1990]

ICZN (1990c) Opinion 1598. *Ophonus* Dejean 1821 and *Tachys* Dejean, 1821 (Insecta, Coleoptera): *Carabus sabulicola* Panzer, 1796 and *Tachys scutellaris* Stephens, 1828 designated as the respective type species. Bulletin of Zoological Nomenclature 47: 158–159.

ICZN (1990d) Opinion 1616. *Ptochus* Schonherr, 1826 (Insecta, Coleoptera): *Ptochus porcellus* Boheman in Schonherr, 1834 confirmed as the type species. Bulletin of Zoological Nomenclature 47: 231–232.

ICZN (1991a) Opinion 1631. *Ochthebius* Leach, 1815 (Insecta, Coleoptera): *Elophorus marinus* Paykull, 1798 designated as type species. Bulletin of Zoological Nomenclature 48: 80–81. [Mar 1991]

ICZN (1991b) Opinion 1655. *Curculio viridicollis* Fabricius, 1792 (currently *Phyllobius viridicollis*; Insecta, Coleoptera): specific name conserved, and *Rhyncolus* German, 1817: *Curculio ater* Linnaeus, 1758 designated as the type species. Bulletin of Zoological Nomenclature 48: 268–269.

ICZN (1992) Opinion 1681. *Vatellus* Aube, (1837) (Insecta, Coleoptera): conserved. Bulletin of Zoological Nomenclature 49: 165–166.

ICZN (1993a) Opinion 1722. *Acrolocha* Thomson, 1858 (Insecta, Coleoptera): conserved, and *Coprophilus* Latreille, 1829: *Staphylinus striatulus* Fabricius, 1792 designated as the type species. Bulletin of Zoological Nomenclature 50: 164–165.

ICZN (1993b) Opinion 1724. *Helophorus* Fabricius, 1775 (Insecta, Coleoptera): conserved as correct original spelling. Bulletin of Zoological Nomenclature 50: 167–168. [30 Jun 1993]

ICZN (1993c) Opinion 1726. *Mycetoporus* Mannerheim, 1831 (Insecta, Coleoptera): *Tachinus punctus* Gravenhorst, 1806 designated as the type species; *Ischnosoma* Stephens, 1829 conserved; and *Mycetoporus* given precedence over *Ischnosoma*. Bulletin of Zoological Nomenclature 50: 171–173. [30 Jun 1993]

ICZN (1993d) Opinion 1727. *Schizopus* Le Conte, 1858 (Insecta, Coleoptera): placed on the Official List of Generic Names. Bulletin of Zoological Nomenclature 50: 174–175.

ICZN (1993e) Opinion 1729. *Platyscelis* Latreille, 1818 (Insecta, Coleoptera): *Tenebrio hypolithus* Pallas, 1781 designated as the type species, so conserving *Oodescelis* Motschulsky, 1845. Bulletin of Zoological Nomenclature 50: 177–178.

ICZN (1993f) Opinion 1743. Tachinidae Fleming, 1821 (Insecta, Coleoptera): spelling emended to Tachinusidae to remove homonymy with Tachinidae Robineau-Desvoidy, 1830 (Insecta, Diptera), and Tachyporidae MacLeay, 1825 (Insecta, Coleoptera): given precedence. Bulletin of Zoological Nomenclature 50: 248–250. [30 Sep 1993]

ICZN (1994a) Opinion 1754. *Histoire abrégée des insectes qui se trouvent aux environs de Paris* (Geoffroy, 1762): some generic names conserved (Crustacea, Insecta). Bulletin of Zoological Nomenclature 51: 58–70.

ICZN (1994b) Opinion 1756. Anthribidae Billberg, 1820 (Insecta, Coleoptera): given precedence over Choragidae Kirby, 1819. Bulletin of Zoological Nomenclature 51: 72–73. [Mar 1994]

ICZN (1994c) Opinion 1770. *Pachyrhynchus* Germar, 1824, *Somatodes* Schonherr, 1840 and the specific name of *Pachyrhynchus moniliferus* Germar, 1824 (Insecta, Coleoptera): conserved. Bulletin of Zoological Nomenclature 51: 170–171.

ICZN (1994d) Opinion 1772. Metopiini Raffray, 1904 (Insecta, Coleoptera): spelling emended to Metopiasini, and Metopiini Townsend, 1908 (Insecta, Diptera): spelling emended to Metopiaini, so removing the homonymy with Metopiinae Foerster, [1869] (Insecta, Hym.). Bulletin of Zoological Nomenclature 51: 174–175. [Jun 1994]

ICZN (1994e) Opinion 1784. *Buprestis* Linnaeus, 1758 and *Chrysobothris* Eschscholtz, 1829 (Insecta, Coleoptera): conserved by the designation of *Buprestis octoguttata* Linnaeus, 1758 as the type species of *Buprestis*, and *Chrysobothris* and *Dicerca* Eschscholtz, 1829: conserved as the correct original spellings. Bulletin of Zoological Nomenclature 51: 280–282.

ICZN (1995a) Opinion 1809. *Bruchus* Linnaeus, 1767, *Ptinus* Linnaeus, 1767 and *Mylabris* Fabricius, 1775 (Insecta, Coleoptera): conserved. Bulletin of Zoological Nomenclature 52: 208–210. [Jun 1995]

ICZN (1995b) Opinion 1810. *Cryptophagus* Herbst, 1792, *Dorcatoma* Herbst, 1792, *Rhizophagus* Herbst, 1793 and *Colon* Herbst, 1797 (Insecta, Coleoptera): conserved as the correct original spellings, and *Lyctus bipustulatus* Fabricius, 1792 ruled to be the type species of *Rhizophagus*. Bulletin of Zoological Nomenclature 52 (2): 211–213. [30 Jun 1995 (wrapper)]

ICZN (1995c) Opinion 1811. Colydiidae Erichson, 1842 (Insecta, Coleoptera): given precedence over Cerylonidae Billberg, 1820 and Orthocerini Blanchard, 1845 (1820); and *Cerylon* Latreille, 1802: *Lyctus histeroides* Fabricius, 1792 designated as the type species. Bulletin of Zoological Nomenclature 52: 214–216. [Jun 1995]

ICZN (1995d) Opinion 1812. Elmidae Curtis, 1830 (Insecta, Coleoptera): conserved as the correct original spelling, and the gender of *Elmis* Latreille, 1802 ruled to be feminine. Bulletin of Zoological Nomenclature 52: 217–218. [Jun 1995]

ICZN (1995e) Opinion 1820. A.A.H. Lichtenstein's (1796, 1797) *Catalogus musei zoologici... Sectio Tertia. Continens Insecta* and D.H. Schneider's (1800) *Verzeichniss einer Parthei Insekten...*: suppressed, with conservation of some Lichtenstein (1796) names (Insecta and Arachnida). Bulletin of Zoological Nomenclature 52: 283–285.

ICZN (1996a) Opinion 1825. *Poecilonota* Eschscholtz, 1829, *Palmar* Schaefer, 1949 and *Scintillatrix* Obenberger, 1956 (Insecta, Coleoptera): conserved by the designation of *Buprestis variolosa* Paykull, (1799) as the type species of *Poecilonota* and *B. rutilans* Fabricius, (1777) as the type species of *Scintillatrix*. Bulletin of Zoological Nomenclature 53: 57–59.

ICZN (1996b) Opinion 1826. *Melanophila* Eschscholtz, 1829 and *Phaenops* Dejean, 1833 (Insecta, Coleoptera): conserved by the designation of *Buprestis acuminata* De Geer, 1774 as the type species of *Melanophila*. Bulletin of Zoological Nomenclature 53: 60–61.

ICZN (1996c) Opinion 1851. Xantholinini Erichson, 1839 and Quediini Kraatz, [1857] (Insecta, Coleoptera): given precedence over some senior synonyms; *Quedius* Stephens, 1829: *Staphylinus levicollis* Brullé, 1832 designated as the type species. Bulletin of Zoological Nomenclature 53: 215–217. [Sep 1995]

ICZN (1996d) Opinion 1855. *Agonus* Bloch & Schneider, 1801 (Osteichthyes, Scorpaeniformes): conserved; and Agonidae Kirby, 1837 (Insecta, Coleoptera): spelling emended to Agonumidae, so removing the homonymy with Agonidae Swainson, 1839 (Osteichthyes, Scorpaeniformes). Bulletin of Zoological Nomenclature 53: 223–225.

ICZN (1997) Opinion 1862. *Aspidiphorus* Ziegler in Dejean, 1821 (Insecta, Coleoptera): conserved as the correct original spelling, and Sphindidae Jacquelin du Val, (1861): given precedence over Aspidiphoridae Kiesenwetter, 1877 (1859). Bulletin of Zoological Nomenclature 54: 62–64.

ICZN (1998) Opinion 1891. *Crenitis* Bedel, 1881, *Georissus* Latreille, 1809 and *Oosternum* Sharp, 1882 (Insecta, Coleoptera): conserved. Bulletin of Zoological Nomenclature 55: 58–60. [Mar 1998]

ICZN (1999a) International Code of Zoological Nomenclature, Fourth Edition, adopted by the International Union of Biological Sciences. International Trust for Zoological Nomenclature, London, xxix + 306 pp. [Also available from URL: http://www.iczn.org/iczn/index.jsp]. [Aug 1999; recorded Dec 1999]

ICZN (1999b) Opinion 1916. Brachypterinae Zwick, 1973 (Insecta, Plecoptera): spelling emended to Brachypterainae, so removing the homonymy with Brachypterinae Erichson, (1845) (Insecta, Coleoptera); Kateretidae Erichson in Agassiz, (1846): given precedence over Brachypterinae Erichson. Bulletin of Zoological Nomenclature 56: 82–86. [Mar 1999]

ICZN (1999c) Opinion 1918. Meloidae Gyllenhal, 1810 and Nemognathinae Castelnau, 1840 (Insecta, Coleoptera): given precedence over Horiidae Latreille, 1802. Bulletin of Zoological Nomenclature 56: 89–91. [Mar 1999]

ICZN (2000) Opinion 1957. *Sphaerius* Waltl, 1838 (Insecta, Coleoptera): conserved; and Sphaeriidae Erichson, 1845 (Coleoptera): spelling emended to Sphaeriusidae, so removing the homonymy with Sphaeriidae Deshayes, 1854 (1820) (Mollusca, Bivalvia). Bulletin of Zoological Nomenclature 57: 182–184. [Sep 2000]

ICZN (2001) Opinion 1968. *Phytobius* Schoenherr, 1833 (Insecta, Coleoptera): placed on the Official List. Bulletin of Zoological Nomenclature 58: 70–71.

ICZN (2002a) Opinion 1989. *Orsodacne* Latreille, 1802 (Insecta, Coleoptera): *Chrysomela cerasi* Linnaeus, 1758 designated as the type species. Bulletin of Zoological Nomenclature 59: 55.

ICZN (2002b) Opinion 2008. 30 species-group names originally published as junior primary homonyms in *Buprestis* Linnaeus, 1758 (Insecta, Coleoptera): conserved. Bulletin of Zoological Nomenclature 59: 211–216.

ICZN (2002c) Opinion 2009. *Anthaxia* Eschscholtz, 1829 and *Trichocratomerus* Richter, 1949 (Insecta, Coleoptera): generic names conserved by the designation of *Buprestis nitida* Rossi, 1794 (currently *A. fulgurans* (Schrank, 1789)) as the type species of *Anthaxia*. Bulletin of Zoological Nomenclature 59: 217–218.

ICZN (2003a) Opinion 2034 (Case 3087). *Hydrobia* Hartmann, 1821: conserved...; and Hydrobiina Mulsant, 1844 (Coleoptera): spelling emended to Hydrobiusina, so removing the homonymy with Hydrobiidae Troschel, 1857 (Gastropoda). Bulletin of Zoological Nomenclature 60 (2): 152–154. [30 Jun 2003]

ICZN (2003b) Opinion 2054 (Case 3201). *Scarabaeus punctatus* Villers, 1789 (currently *Pentodon bidens punctatus*; Insecta, Coleoptera): specific name conserved. Bulletin of Zoological Nomenclature 60: 247–248.

ICZN (2004a) Opinion 2066 (Case 3202). *Podalgus* Burmeister, 1847 and *Philoscaptus* Brethes, 1919 (Insecta, Coleoptera): current usage of the names conserved by designation of *Podalgus cuniculus* Burmeister, 1847 as the type species of *Podalgus*. Bulletin of Zoological Nomenclature 61: 61–62.

ICZN (2004b) Opinion 2076 (Case 3193). *Chrysodema* Laporte & Gory, 1835 and *Iridotaenia* Deyrolle, 1864 (Insecta, Coleoptera): usage conserved by the designation of *C. sonnerati* Laporte & Gory, 1835 as the type species of *Chrysodema*. Bulletin of Zoological Nomenclature 61: 128–129.

ICZN (2004c) Opinion 2079 (Case 2926). *Trichia* Hartmann, 1840 (Mollusca, Gastropoda): proposed conservation; and Trichiinae Lozek, 1956 (Gastropoda): proposed emendation of spelling to Trichiainae, so removing the homonymy with Trichiidae Fleming, 1821 (Insecta, Coleoptera) not approved. Bulletin of Zoological Nomenclature 61: 177–181.

ICZN (2004d) Opinion 2084 (Case 3209). *Lesteva* Latreille, 1797 and *Anthophagus* Gravenhorst, 1802 (Insecta, Coleoptera): usage conserved by the designation of *L. punctulata* Latreille, 1804 as the type species of *Lesteva*. Bulletin of Zoological Nomenclature 61: 190–191.

ICZN (2005a) Opinion 2098 (Case 3239). *Geostiba* Thomson, 1858 (Insecta, Coleoptera): conserved. Bulletin of Zoological Nomenclature 62: 43–44.

ICZN (2005b) Opinion 2100 (Case 3258). *Acmaeodera* Eschscholtz, 1829 and *Acmaeoderella* Cobos, 1955 (Insecta, Coleoptera): usage conserved by designation of *Buprestis cylindrica* Fabricius, 1775 as the type species of *Acmaeodera*. Bulletin of Zoological Nomenclature 62: 47–48.

ICZN (2005c) Opinion 2111 (Case 3093). Nemonychidae Bedel, November 1882 (Insecta, Coleoptera): given precedence over Cimberididae Gozis, March 1882; and *Cimberis* Gozis, 1881: usage conserved. Bulletin of Zoological Nomenclature 62 (2): 101–103. [30 Jun 2005]

ICZN (2005d) Opinion 2113 (Case 3256). *Leptusa* Kraatz, 1856 and *Cyllopisalia* Pace, 1982 (Insecta, Coleoptera): conserved. Bulletin of Zoological Nomenclature 62: 106–107.

ICZN (2006a) Opinion 2138 (Case 3097). *Bolboceras* Kirby, 1819 (July) (Insecta, Coleoptera): not conserved; priority maintained for *Odonteus* Samouelle, 1819 (June). Bulletin of Zoological Nomenclature 63: 62–64.

ICZN (2006b) Opinion 2149 (Case 3291). Dromiidae Bonelli, 1810 (Insecta, Coleoptera, Caraboidea): emended to Dromiusidae to remove homonymy with Dromiidae De Haan, 1833 (Crustacea, Decapoda, Brachyura, Dromioidea). Bulletin of Zoological Nomenclature 63 (2): 138–139. [30 Jun 2006]

ICZN (2007) Opinion 2186 (Case 3349). *Gnorimus* Le Peletier de Saint-Fargeau & Serville, 1828, and *Osmoderma* Le Peletier de Saint-Fargeau & Serville, 1828 (Insecta, Coleoptera, Scarabaeidae): conserved. Bulletin of Zoological Nomenclature 64: 265–266.

ICZN (2008a) Opinion 2199 (Case 3369). *Bothynus* Hope, 1837 (Insecta, Coleoptera, Scarabaeidae): usage conserved by designation of *Scarabaeus ascanius* Kirby, 1819 as the type species. Bulletin of Zoological Nomenclature 65: 156–157.

ICZN (2008b) Opinion 2211. (Case 3360). The ichnogenus *Coprinisphaera* Sauer, 1955 (Ichnotaxa, Coprinisphaeridae): generic name given precedence over *Fontanai* Roselli, 1939. Bulletin of Zoological Nomenclature 65 (4): 318–319.

ICZN (2008c) Opinion 2214. (Case 3366). *Cisseis* Gory & Laporte de Castelnau, 1839 and *Curis* Gory & Laporte de Castelnau, 1838 (Insecta, Coleoptera, Buprestidae): generic names not conserved. Bulletin of Zoological Nomenclature 65: 325–326.

ICZN (2009a) Opinion 2222 (Case 3335). *Trachys* Fabricius, 1801 (Insecta, Coleoptera): masculine gender of the genus fixed. Bulletin of Zoological Nomenclature 66: 100–102.

ICZN (2009b) Opinion 2229 (Case 3405). *Trigonostomum* Schmidt, 1852 (Platyhelminthes, Trigonostomidae) and *Trigonostomus* Brenske, 1893 (Coleoptera, Scarabaeidae): generic names conserved. Bulletin of Zoological Nomenclature 66: 189–191.

ICZN (2009c) Opinion 2227 (Case 3375). *Fidia* Baly, 1863 and *Lypesthes* Baly, 1863 (Insecta, Coleoptera): usage not conserved and priority maintained for *Fidia* Motschulsky, 1860. Bulletin of Zoological Nomenclature 66: 198–200. [30 Jun 2009]

ICZN (2009d) Opinion 2237 (Case 3422). *Helops* Fabricius 1775 (Insecta, Coleoptera, Tenebrionidae): usage conserved by designation of *Tenebrio caeruleus* Linnaeus, 1758 as the type species. Bulletin of Zoological Nomenclature 66: 369–370.

ICZN (2010a) Opinion 2241 (Case 3377). *Ataenius* Harold, 1867 (Insecta, Coleoptera): precedence given over *Aphodius* Motschulsky, 1862. Bulletin of Zoological Nomenclature 67 (1): 97–98. [Mar 2010 issue]

ICZN (2010b) Opinion 2242 (Case 3398). Aulacoscelinae Chapuis, 1874 (Insecta, Coleoptera, Orsodacnidae or Chrysomelidae): name not conserved. Bulletin of Zoological Nomenclature 67 (1): 99–101. [Mar 2010 issue]

ICZN (2010c) Opinion 2244 (Case 3443). *Tentyria* Latreille, 1802 (Coleoptera, Tenebrionidae): usage conserved. Bulletin of Zoological Nomenclature 67 (1): 104–105.

Ienistea M-A (1986) A new hierarchical system of Arthropoda, mainly referring to insects. YES (Young Entomologists' Society) Quarterly 3 (2): 13–38. [Spring 1986 issue; 22 May 1986 (recorded at CNC)]

Illiger JKW (1801) Namen der Insekten-Gattungen, ihr Genitiv, ihr grammatisches Geschlecht, ihr Silbenmass, ihre Herleitung; zugleich mit den Deutschen Benennungen. Magazin für Insektenkunde 1 [1801]: 125–155. [1801 (Heft 1/2, pp. 1–260)]

Illiger JKW (1802) Aufzählung der Käfergattungen nach der Zahl der Fussglieder. Magazin für Insektenkunde 1 [1801]: 285–305.

Illiger JKW (1807) Vorschlag zur Aufnahme im Fabricischen Systeme fehlende Käfergattungen. Magazin für Insektenkunde 6: 318–349. [1807 (title page)]

Imhoff L (1856) Versuch einer Einführung in das Studium der Koleoptern. L. Imhoff ["auf Kosten des Verfassers"], Basel, xxxi + [2] + 118 + [2] + 272 + [25] pp. + 25 pls. [before 25 Dec 1856 (Allgem. Bibliogr. Deutschl.: 452)]

Ivie MA (1985) Nomenclatorial notes on West Indian Elaphidiini (Coleoptera: Cerambycidae). Pan-Pacific Entomologist 61: 303–314.

Ivie MA (2002) Family 49. Ptilodactylidae Laporte 1836 [pp. 135–138], 69. Bostrichidae [pp. 233–244]. In: Arnett RH, Jr., Thomas MC, Skelley PE, Frank JH (Eds) American beetles. Volume 2. Polyphaga: Scarabaeoidea through Curculionoidea. CRC Press, Boca Raton, xiv + 861 pp. [19 Jun 2002 (CRC website)]

Ivie MA, Ślipiński SA (1990) Catalog of the genera of world Colydiidae (Coleoptera). Annales Zoologici (Warszawa) 43 (Suppl. 1): 1–32. [15 Nov 1990 (top of article)]

Iwan D (1996) Revision of the genera of the newly-established group of Madagascan melanocratoid Platynotina (Coleoptera: Tenebrionidae: Platynotini). Genus 7 (3): 379–449.

Jäch MA (1994) Case 2861. Elmidae Curtis, 1830 and Elmis Latreille, 1802 (Insecta, Coleoptera): proposed confirmation as correct spelling and of feminine gender respectively. Bulletin of Zoological Nomenclature 51 (1): 25–27. [30 Mar 1994]

Jäch MA (2004) Family Hydraenidae [pp. 102–122]. In: Löbl I, Smetana A (Eds) Catalogue of Palaearctic Coleoptera. Volume 2. Hydrophiloidea - Histeroidea - Staphylinoidea. Apollo Books, Stenstrup, 942 pp. [publ. 31 Dec 2004 (verso of title page)]

Jäch MA, Kodada J, Ciampor F (2006) Family Elmidae Curtis, 1830 [pp. 432–440]. In: Löbl I, Smetana A (Eds) Catalogue of Palaearctic Coleoptera. Volume 3. Scarabaeoidea - Scirtoidea - Dascilloidea - Buprestoidea - Byrrhoidea. Apollo Books, Stenstrup, 690 pp. [publ. 30 Jun 2006 (verso of title page)]

Jacobi A (1906) Grundriss der Zoologie für Forstleute. Ergänzungsband zu Lorey's Handbuch der Forstwissenschaft. H. Laupp, Tübingen, xi + 263 pp.

Jacobson HR, Kistner DH (1991) Cladistic study, taxonomic restructuring, and revision of the myrmecophilous tribe Leptanillophilini with comments on its evolution and host relationships (Coleoptera: Staphylinidae; Hymenoptera: Formicidae). Sociobiology 18: 1–150. [14 Feb 1991 (recorded at FMNH)]

Jacobson HR, Kistner DH, Pasteels JM (1986) Generic revision, phylogenetic classification, and phylogeny of the termitophilous tribe Corotocini (Coleoptera: Staphylinidae). Sociobiology 12 (1): 1–245. [8 Oct 1986 (recorded at CNC)]

Jacoby M (1884a) Descriptions of new genera and species of phytophagous Coleoptera from Sumatra. Notes from the Leyden Museum 6 (1): 9–70. [Jan 1884 (vol. contents)]

Jacoby M (1884b) Descriptions of new genera and species of phytophagous Coleoptera collected by Dr. B. Hagen at Serdang (East Sumatra). Notes from the Leyden Museum 6 (4): 201–230. [Oct 1884 (vol. contents)]

Jacoby M (1899) Some new genera and species of phytophagous Coleoptera from India and Ceylon. The Entomologist 32 (430, 431): 67–70, 80–83. [Mar, Apr 1899 (issue title pages)]

Jacoby M (1903) 14a fascicule. Coleoptera Phytophaga sect. Eupoda fam. Sagridae. In: Wytsman PA (Ed) Genera Insectorum. Vol. II. P. Wytsman, Bruxelles, 11 pp. [15 Apr 1903 (date on manuscr., p. 11); 3 Jun 1903 (Evenhuis 1994: 51)]

Jacoby M (1908) Coleoptera, Chrysomelidae. Vol. I. In: Bingham CT (Ed) The fauna of British India, including Ceylon and Burma. Taylor and Francis, London, xx + 534 pp. [Jan 1908 (date of preface)]

Jacoby M, Clavareau H (1905) 32me fascicule. Coleoptera Phytophaga fam. Megascelidae. In: Wytsman PA (Ed) Genera Insectorum. Vol. VI. P. Wytsman, Bruxelles, 6 pp. + 1 pl. [15 May 1905 (date on manuscr., p. 6); 28 Oct 1905 (Evenhuis 1994: 52)]

Jacoby M, Clavareau H (1906) 49me fascicule. Coleoptera Phytophaga fam. Chrysomelidae subfam. Clytrinae. In: Wytsman PA (Ed) Genera Insectorum. Vol. VIII. P. Wytsman, Bruxelles, 87 pp. + 5 pls. [15 Oct 1906 (date on manuscr., p. 87); 10 Nov 1906 (Evenhuis 1994: 53)]

Jacquelin du Val C (1857a) [livr. 49–57: pp. 41–128] In: Manuel entomologique. Généra des coléoptères d'Europe comprenant leur classification en familles naturelles, la description de tous les genres, des tableaux synoptiques destinés à faciliter l'étude, le catalogue de toutes les espèces, de nombreux dessins au trait de charactères et plus de treize cents types représentant un ou plusieurs insectes de chaque genre dessinés et peints d'après nature avec le plus grand soin par M. Jules Migneaux. Tome deuxième. A. Deyrolle, Paris, 285 + [2] pp. [1857 (Tottenham 1949: 360); before 11 Sep 1858 (Bibliogr. France 1858)]

Jacquelin du Val C (1857b) Glanures entomologiques. Annales de la Société Entomologique de France (3) 5 (1): 85–106. [24 Jun 1857 (wrapper)]

Jacquelin du Val C (1858) [livr. 58–67: pp. 129–232]. In: Manuel entomologique. Généra des coléoptères d'Europe comprenant leur classification en familles naturelles, la description de tous les genres, des tableaux synoptiques destinés à faciliter l'étude, le catalogue de toutes les espèces, de nombreux dessins au trait de charactères et plus de treize cents types représentant un ou plusieurs insectes de chaque genre dessinés et peints d'après nature avec le plus grand soin par M. Jules Migneaux. Tome deuxième. A. Deyrolle, Paris, 285 + [2] pp. [1858 (Sherborn notes on BMNH copy wrappers); 1859 (Reinwald 1860: 109)]

Jacquelin du Val C (1859) [livr. 74–87: pp. 1–136]. In: Manuel entomologique. Généra des coléoptères d'Europe comprenant leur classification en familles naturelles, la description de tous les genres, des tableaux synoptiques destinés à faciliter l'étude, le catalogue de toutes les espèces, de nombreux dessins au trait de charactères et plus de treize cents types représentant un ou plusieurs insectes de chaque genre dessinés et peints d'après nature avec le plus grand soin par M. Jules Migneaux. Tome troisième. A. Deyrolle, Paris, 463 + [1] pp. [1859 (Sherborn notes on BMNH copy wrappers)]

Jacquelin du Val C (1860) [livr. 88–102: pp. 137–272]. In: Manuel entomologique. Généra des coléoptères d'Europe comprenant leur classification en familles naturelles, la description de tous les genres, des tableaux synoptiques destinés à faciliter l'étude, le catalogue de toutes les espèces, de nombreux dessins au trait de charactères et plus de treize cents types représentant un ou plusieurs insectes de chaque genre dessinés et peints d'après nature avec le plus grand soin par M. Jules Migneaux. Tome troisième. A. Deyrolle, Paris, 463 + [1] pp. [1860 (Sherborn notes on BMNH copy wrappers)]

Jacquelin du Val C (1861) [livr. 103–110: pp. 273–352]. In: Manuel entomologique. Généra des coléoptères d'Europe comprenant leur classification en familles naturelles, la descrition de tous les genres, des tableaux synoptiques destinés à faciliter l'étude, le catalogue de toutes les espèces, de nombreux dessins au trait de charactères et plus de treize cents types représentant un ou plusieurs insects de chaque genre dessinés et peints d'après nature avec le plus grand soin par M Jules Migneaux. Tome troisième. A. Deyrolle, Paris, 463 + [1] pp. [1861 (Sherborn notes on BMNH copy wrappers)]

Jacquelin du Val C (1863) [livr. 111–120: pp. 353–464]. In: Manuel entomologique. Généra des coléoptères d'Europe comprenant leur classification en familles naturelles, la descrition de tous les genres, des tableaux synoptiques destinés à faciliter l'étude, le catalogue de toutes les espèces, de nombreux dessins au trait de charactères et plus de treize cents types représentant un ou plusieurs insects de chaque genre dessinés et peints d'après nature avec le plus grand soin par M Jules Migneaux. Tome troisième. A. Deyrolle, Paris, 463 + [1] pp. [1863 (Sherborn notes on BMNH copy wrappers)]

Jakobson GG (1904) Kratkiy obsor' klassifikatsiy zhestkokrylych (Coleoptera) [Aperçu des classifications de l'ordre des Coléoptères]. Russkoe Entomologicheskoe Obozrenie 4 (6 [Dec]): 268–276.

Jakobson GG (1906) [Fasc. 4: pp. 241–320] In: Zhuki Rossii i Zapadnoi Evropy. A. F. Devrien, St. Petersburg, 1024 pp. + 83 pls. [in Russian]. [10 Sep 1906 (Griffin 1932a)]

Jakobson GG (1907) [Fasc. 5: pp. 321–400] In: Zhuki Rossii i Zapadnoi Evropy. A. F. Devrien, St. Petersburg, 1024 pp. + 83 pls. [in Russian]. [1 Jun 1907 (Griffin 1932a)]

Jakobson GG (1908) [Fasc. 6: pp. 401–480] In: Zhuki Rossii i Zapadnoi Evropy. A. F. Devrien, St. Petersburg, 1024 pp. + 83 pls. [in Russian]. [1 Mar 1908 (Griffin 1932a)]

Jakobson GG (1909) [Fasc. 7: pp. 481–560] In: Zhuki Rossii i Zapadnoi Evropy. A. F. Devrien, St. Petersburg, 1024 pp. + 83 pls. [in Russian]. [4 Mar 1909 (Griffin 1932a)]

Jakobson GG (1910) [Fasc. 8: pp. 561–640] In: Zhuki Rossii i Zapadnoi Evropy. A. F. Devrien, St. Petersburg, 1024 pp. + 83 pls. [in Russian]. [20 Jan 1910 (Griffin 1932a)]

Jakobson GG (1911a) [Fasc. 9: pp. 641–720] In: Zhuki Rossii i Zapadnoi Evropy. A. F. Devrien, St. Petersburg, 1024 pp. + 83 pls. [in Russian]. [3 Jun 1911 (Griffin 1932a)]

Jakobson GG (1911b) Kritiko-bibliograficheskiy otdyel'. Revue critico-bibliographique. Russkoe Entomologicheskoe Obozrenie 11 (1): 139–146. [1912 (vol. title page); publ. 29 May 1911 (Julian 16 May, vol. p. xix)]

Jakobson GG (1913) [Fasc. 10: pp. 721–864] In: Zhuki Rossii i Zapadnoi Evropy. A. F. Devrien, St. Petersburg, 1024 pp. + 83 pls. [in Russian]. [8 Apr 1913 (Griffin 1932a)]

Jakobson GG (1915) [Fasc. 11: pp. 865–1024] In: Zhuki Rossii i Zapadnoi Evropy. A. F. Devrien, St. Petersburg, 1024 pp. + 83 pls. [in Russian]. [5 Jul 1915 (Griffin 1932a)]

Jakobson GG (1924) Annotationes synonymicae et systematicae de Coleopteris. Russkoe Entomologicheskoe Obozrenie 18 [1922–24] (4): 237–244.

Janssens A (1946) Contribution à l'étude des coléoptères lamellicornes coprophages, XI. - Table synoptique et essai de classification pratique des Scarabaeidae Laparosticti. Bulletin du Musée Royal d'Histoire Naturelle de Belgique 22 (12): 1–13. [Jun 1946 (top of article)]

Janssens E (1950) Classification et zoogeographie (complément du Coleopterorum Catalogus pars 5, Paussidae). Bulletin / Institut Royal des Sciences Naturelles de Belgique 26 (51): 1–19. [Nov 1950 (top of article)]

Jeanne C (1971) Carabiques de la Peninsule Iberique. (10e note.) Bulletin de la Société Linnéenne de Bordeaux 1 (2): 5–18. [Feb 1971 (fasc. title page); 15 Mar 1971 (recorded at BMNH)]

Jeanne C (1972) Carabiques de la peninsule lberique (14e note). Bulletin de la Société Linnéenne de Bordeaux 2 (5): 99–116. [May 1972 (fasc. title page); 13 Jun 1972 (recorded at BMNH)]

Jeanne C (1973) Sur la classification des bembidiides endogés de la région euro-méditerranéenne (Col. Carabidae, Bembidiinae, Anillini). Nouvelle Revue d'Entomologie 3 (2): 83–102.

Jeannel R (1910) Biospeologica XIV. Essai d'une nouvelle classification des Silphides cavernicoles. Archives de Zoologie Expérimentale et Générale (5) 5: 1–48. [20 Mar 1910 (vol. contents)]

Jeannel R (1911) Biospeologica XIX. Révision des Bathysciinae (Coléoptères Silphides). Morphologie, distribution géographique, systématique. Archives de Zoologie Expérimentale et Générale (5) 7 [47] (1): 1–641, pls. 1–24. [2 May 1911 (vol. contents)]

Jeannel R (1921) Silphidae. Voyage de Ch Alluaud et R Jeannel en Afrique orientale (1911–1912) Résultats scientifiques Coléoptères XVI. L. Lhomme, Paris, 229–242. [1 Nov 1921 (title page)]

Jeannel R (1922) Les Trechinae de France. Annales de la Société Entomologique de France 90 [1921] (2): 161–192, (3/4): 295–345. [(2) 22 Mar 1922, (3/4) 26 Jul 1922 (p. 356)]

Jeannel R (1926) Monographie des Trechinae. Morphologie comparée et distribution géographique d'un groupe de coléoptères. (Première livraison). L'Abeille, Journal d'Entomologie 32 (3): 221–550. [1 Jun 1926 (wrapper)]

Jeannel R (1936) Monographie des Catopidae (Insectes coléoptères). Mémoires du Muséum National d'Histoire Naturelle (Nouvelle Série) 1 (1): 1–433. [Jan 1936 (title page)]

Jeannel R (1937) Les Bembidiides endogés (Col. Carabidae). Monographie d'une lignée gondwanienne. Revue Française d'Entomologie 3 [1936] (4): 241–396. [1 Feb 1937 (p. 399)]

Jeannel R (1938) Les migadopides (Coleoptera Adephaga), une lignée subantarctique. Revue Française d'Entomologie 5 (1): 1–55. [15 Apr 1938 (p. 214)]

Jeannel R (1940) Croisière du Bougainville aux îles australes françaises. III. Coléoptères. Mémoires du Muséum National d'Histoire Naturelle (Nouvelle Série) 14: 63–201. ["Oct 1940"]

Jeannel R (1941) Coléoptères carabiques. Première partie. Faune de France 39: 1–571.

Jeannel R (1942) Coléoptères carabiques. Deuxième partie. Faune de France 40 (2): 573–1173. [15 Feb 1942 (p. 1173)]

Jeannel R (1946) Coléoptères carabiques de la région malgache (première partie). Faune de l'Empire Français VI. Office de la Recherche scientifique coloniale, Paris, 372 pp. [dépot légal 1er trim. 1946 (p. 372)]

Jeannel R (1948a) Révision des *Amaurops* et genres voisins (Pselaphidae). Revue Française d'Entomologie 15 (1): 1–19. [28 Feb 1948 (p. 251)]

Jeannel R (1948b) Coléoptères carabiques de la région malgache (deuxième partie). Faune de l'Empire Français X. Office de la Recherche scientifique coloniale, Paris, pp. 373–765. [printed Jun 1948, dépot légal 2e trim. 1948 (endleaf)]

Jeannel R (1949a) Les psélaphides de l'Afrique Orientale (Coleoptera). Mémoires du Muséum National d'Histoire Naturelle (Nouvelle Série) 29 (1): 1–226. [Jun 1949 (endleaf)]

Jeannel R (1949b) Notes biospéologiques. Les coléoptères cavernicoles de la région des Appalaches. III. Étude systématique. Publications du Muséum National d'Histoire Naturelle No. 12 (4): 37–104.

Jeannel R (1949c) Coléoptères carabiques de la région malgache (troisième partie). Faune de l'Empire Français XI. Office de la Recherche scientifique coloniale, Paris, pp. 767–1146. [1949 (p. 1146)]

Jeannel R (1950a) Coléoptères psélaphides. Faune de France. Vol. 53. Librarie de la Faculté des Sciences, Paris, iii + 421 pp. [dépot légal 2e trim. 1950 (p. 421)]

Jeannel R (1950b) Sur la nomenclature des groupements supergénériques. Bulletin of Zoological Nomenclature 3 (7/9): 164–165. [publ. 21 Apr 1950 (p. 237); 19 Apr 1950 (recorded at BMNH)]

Jeannel R (1951) Psélaphides de l'Angola (coléoptères) recueillis par M. A. de Barros Machado. Publicações Culturais da Companhia de Diamantes de Angola No. 9: 125 pp. [8 May 1951 (title page)]

Jeannel R (1952a) Psélaphides recueillis par N. Leleup au Congo Belge. IV. - Faune de l'Itombwe et de la forêt du Rugege. Annales du Musée Royal du Congo Belge (Série in 8°: Sciences Zoologiques) 11: 1–295. [1952 (title page; Zool. Record)]

Jeannel R (1952b) Psélaphides de Saigon. Revue Française d'Entomologie 19 (2): 69–113. [10 Jun 1952 (p. 132)]

Jeannel R (1954a) Les psélaphides de Madagascar. Mémoires de l'Institut Scientifique de Madagascar (Série E. Entomologie) 4 [1953]: 139–344. [1er trim. 1954]

Jeannel R (1954b) Description d'un psélaphide Australien type d'une tribu nouvelle. Revue Française d'Entomologie 21 (2): 99–102. [22 Jun 1954 (p. 148)]

Jeannel R (1954c) Psélaphides recueillis par N. Leleup au Congo Belge sous les auspices de l'Institut pour la Recherche Scientifique en Afrique Centrale (I.R.S.A.C.). VI-XI. Annales du Musée Royal du Congo Belge (Série in 8°: Sciences Zoologiques) 33: [2] + 7–175. [printed Oct 1954 (endleaf)]

Jeannel R (1957) Révision des petits scaritides endogés voisins de *Reicheia* Saulcy. Revue Française d'Entomologie 24 (2): 129–212. [printed 1 Jul 1957 (p. 212)]

Jeannel R (1958a) Révision des psélaphides du Japon. Mémoires du Muséum National d'Histoire Naturelle (Série A: Zoologie) 18 (1): 1–138. [printed 20 Oct 1958 (p. 138)]

Jeannel R (1958b) Sur quelques Trechitae de l'Amerique du Sud (Coleoptera). Entomologische Arbeiten aus dem Museum G. Frey 9 (3): 721–737.

Jeannel R (1959) Révision des psélaphides de l'Afrique intertropicale. Annales du Musée Royal du Congo Belge (Série in 8°: Sciences Zoologiques) 75: 1–742. [printed Sep 1959 (endleaf)]

Jeannel R (1960) Mission zoologique de l'I.R.S.A.C. en Afrique orientale. (P. Basilewsky et N. Leleup, 1957). V. Coleoptera Carabidae Trechinae. Annales du Musée Royal du Congo Belge (Série in 8°: Sciences Zoologiques) 81: 40–76. [printed Mar 1960 (endleaf)]

Jeannel R (1962a) Les psélaphides de la paléantarctide occidentale [pp. 295–479]. In: Delamare-Deboutteville C, Rapoport E (Eds). Biologie de l'Amérique Australe. Vol. I. Études sur la faune du sol. Centre National de la Recherche Scientifique, Paris, 657 pp. [printed Sep 1962 (endleaf)]

Jeannel R (1962b) Les Silphidae, Liodidae, Camiaridae et Catopidae de la paléantarctide occidentale [pp. 481–525]. In: Delamare-Deboutteville C, Rapoport E (Eds). Biologie de l'Amérique Australe. Vol. I. Études sur la faune du sol. Centre National de la Recherche Scientifique, Paris, 657 pp. [Sep 1962]

Jeannel R (1964) Révision des psélaphides de l'Afrique australe [pp. 23–217]. In: The humicolous fauna of south Africa: Pselaphidae and Catopidae (Coleoptera) (N. Leleup expedition 1960–1961). Transvaal Museum Memoir No. 15: 261 pp. [May 1964 (title page); printed Sep 1964 (footer p. 261)]

Jeannel R (1967) Biogéographie de l'Amérique australe [pp. 401–460]. In: Delamare-Debouteville C, Rapoport E (Eds) Biologie de l'Amérique Australe. Volume III. Études sur la faune du sol; documents biogéographiques Centre National de la Recherche Scientifique, Paris, 724 + [1] pp. [printed Apr 1967 (endleaf)]

Jeannel R, Jarrige J (1949) Biospeologica, LXVIII. Coléoptères staphylinides (première série). Archives de Zoologie Expérimentale et Générale 86 (5): 255–392. [22 Nov 1949 (wrapper)]

Jeannel R, Paulian R (1945) Coléoptères. In: Mission scientifique de l'Omo. Tome VI - fascicule 57. Faune des terriers des Rats-Taupes. IV. Coléoptères. Mémoires du Muséum National d'Histoire Naturelle (Nouvelle Série) 19: 51–147. [Jun 1945; dépot légal 4e trim. 1945 (endleaf)]

Jedlička A (1941) Versuch einer Monographie der pal. Carabiden-Gattungen mit abgestutzten Flügeldecken (Truncatipennen) mit Berücksichtigung der indischen Fauna. Jedlička, A. [Selbstverlag], Prague, 27 pp. + 2 pls. [2 May 1941 (reprint p. 1)]

Jedlička A (1963) Monographie der Truncatipennen aus Ostasien. Lebiinae - Odacanthinae - Brachyninae (Coleoptera, Carabidae). Entomologische Abhandlungen und Berichte aus dem Staatlichen Museum für Tierkunde in Dresden 28 [1962–64] (3): 269–579, incl. 6 pls. [11 Dec 1963 (verso of title page)]

Jekel H (1853) Spécimen Fabricia Entomologica. Receuil d'observations nouvelles sur les insectes; monographies, révisions de groupes et de genres, classifications, synonymies et rectifications, descriptions de genres nouveaux et d'espèces nouvelles, &c[a]. Partie 1., Paris, [Préambule] + 8 pp. [I.II] + 8 pp. [I.IV]. [I.II: Feb 1853; Préambule + I.IV: Apr 1853]

Jekel H (1855) Coleoptera. Fam. Curculionides. Part I. In: Saunders WW (Ed). Insecta Saundersiana; or characters of undescribed insects in the collection of William Wilson Saunders, Esq. Vol. 2 (1). John van Voorst, London, 152 + [1] pp. + 2 pls. [30 Jan 1855 (date of Preface); 5 Mar 1855 (Proc. Ent. Soc. London 1855: 80)

Jekel H (1856) Fabricia Entomologica. Receuil d'observations nouvelles sur les insectes; monographies, révisions de groupes et de genres, classifications, synonymies et rectifications, descriptions de genres nouveaux et d'espèces nouvelles, &a. Première partie. H. Jekel, Paris, 12 pp. + table 9 bis [I]. [part I: Feb 1856]

Jekel H (1860) Coleoptera. Fam. Curculionides. Part. II. In: Saunders WW (Ed). Insecta Saundersiana; or characters of undescribed insects in the collection of William Wilson Saunders, Esq. Vol. 2 (2). John van Voorst, London, 155–244 + 1 table + 1 pl. [Mar 1860 (date of Preface); 7 May 1860 (Proc. Ent. Soc. London 1860: 115)]

Jekel H (1861) Tentamenta Entomologica. Journal of Entomology 1 (4): 263–274. [Dec 1861 issue (p. 207)]

Jekel H (1865) Recherches sur la classification naturelle des curculionides. 1ère partie. Annales de la Société Entomologique de France (4) [1864] 4 (3): 537–566. [25 Jan 1865 (wrapper)]

Jekel H (1875) Coleoptera Jekeliana adjecta Eleutheratorum Bibliotheca. Énumeration systématique & synonymique des coléoptères européens & exotiques composant la collection de Henri Jekel. Observations critiques. Descriptions d'espèces nouvelles. Reproduction et traduction de genres et espèces publiés dans des ouvrages rares Français et étrangers de manière à former insensiblement la bibliothèque du coléoptériste. Livraison II [pp. 101–196]. H. Jekel & E. Deyrolle, Paris, 196 pp. [each description or page bears a date: Apr 1875 (p.

101), Jul 1875 (p. 196); however, it appears to have been issued as a single block. Livraison 1: 1873 (wrapper); Livr. 2: 22 Sep 1875 (Ann. Soc. Ent. France (5) 5: Bull. Ent.: clxxvii)]

Johansen JP (1914) Danmarks rovbiller eller billefam. Staphylinidae's. Danske slaegter og arter. Bianco Lunos, Copenhagen, [3] + 660 pp. [May 1914 (date of preface)]

John H (1954) Familiendiagnose der Notiophygidae (= Discolomidae Col.), ihr Verwandtschaftsverhältnis zu den Colydiidae und Bermerkungen zu einigen systematisch noch nicht eingeordneten Gattungen. Entomologische Blätter für Biologie und Systematik der Käfer 50 (1): 9–75. [1 Nov 1954 (wrapper)]

Johnson PJ (2002a) Lectotype designations for Elateridae (Coleoptera) described by George C. Champion in the Biologia Centrali Americana. Dugesiana 9 (1): 15–46. [mailed 31 Jul 2002 (issue title page)]

Johnson PJ (2002b) 57. Throscidae Laporte 1840 [pp. 158–159]; 58. Elateridae Leach 1815 [pp. 160–173]. In: Arnett RH, Jr., Thomas MC, Skelley PE, Frank JH (Eds) American Beetles. Volume 2. Polyphaga: Scarabaeoidea through Curculionoidea. CRC Press, Boca Raton, Florida, xiv + 861 pp.

Jolivet P, Verma KK (2009) Case 3498: *Eupales* Lefèvre 1885 and Eupalini Verma, Gomez-Zurita, Jolivet & Vig, 2005 (Insecta, Coleoptera, Eumolpinae): proposed conservation. In: New applications to the Commission. Bulletin of Zoological Nomenclature 66 (3): 204.

Joy NH (1932) A practical handbook of British beetles. Vol. I (text), Vol. II (plates). H. F. & G. Witherby, London, xxvii + 622 pp., 194 pp. [Jan 1932]

Kamiya H (1960) A new tribe of Coccinellidae (Coleoptera). Kontyû 28 (1): 22–26, pl. 3.

Kapur AP (1948) The genus *Tetrabrachys* (*Lithophilus*) with notes on its biology and a key to the species (Coleoptera, Coccinellidae). The Transactions of the Royal Entomological Society of London 99 (9): 319–339. [30 Sep 1948 (wrapper)]

Kasap H, Crowson RA (1975) A comparative anatomical study of Elateriformia and Dascilloidea (Coleoptera). Transactions of the Royal Entomological Society of London 126: 441–495. [16 May 1975 (recorded at CNC)]

Kaszab Z (1941) Eine neue Unterfamilie und eine neue Tribus aus der Familie der Tenebrionidae. Entomologische Blätter 37 (1): 29–38. [28 Feb 1941 (Inhalt)]

Kaszab Z (1956) Coleoptera IV. Felemás lábfejízes bogarak III. Heteromera III. Fauna Hungariae 9 (3): 1–108 + 1–6 [index]. [manuscript received 5 Jul 1956 (verso of title page)]

Kaszab Z (1959) Phylogenetische Beziehungen des Flügelgeäders der Meloiden (Coleoptera), nebst Beschreibung neuer Gattungen und Arten. Acta Zoologica Academiae Scientiarum Hungaricae 5: 67–114.

Kaszab Z (1960) Die Arten und die systematische Stellung der Meloiden-Gattung *Calydus* Reitter (Coleoptera). Acta Zoologica Academiae Scientiarum Hungaricae 6: 125–134.

Kaszab Z (1963) Merkmale der Adaptation, Spezialisation, Konvergenz, Korrelation und Progression bei den Meloiden (Coleoptera). Acta Zoologica (Budapest) 9: 135–175.

Kaszab Z (1964) The zoological results of Gy. Topál's collectings in South Argentina. 13. Coleoptera - Tenebrionidae. Annales Historico-Naturales Musei Nationalis Hungarici (Pars Zoologica) 56: 353–387. [1964 (title page, Zool. Record)]

Kaszab Z (1966) *Meloetyphlus fuscatus horni* ssp. n. sowie die systematische Stellung der Gattung *Meloetyphlus* Waterhouse (Coleoptera: Meloidae). Beiträge zur Entomologie 16 (1/2): 183–191. [15 Jul 1966 (recorded at CNC)]

Kaszab Z (1969) The system of the Meloidae (Coleoptera). Memorie della Società Entomologica Italiana 48: 241–248. [30 Dec 1969 (Newton)]

Kaszab Z (1982) Die Tenebrioniden Neukaledoniens und der Loyaute-Inseln (Coleoptera). Folia Entomologica Hungarica 43 (2): 1–294. [1 Dec 1982 (verso of issue title page)]

Kato M (1998) Unique leafmining habit in the bark beetle clade: a new tribe, genus, and species of Platypodidae (Coleoptera) found in the Bonin Islands. Annals of the Entomological Society of America 91 (1): 71–80. [Jan 1998 issue (CNC recorded 20 Feb)]

Kaup J (1868a) Prodromus zu einer Monographie der Passaliden. Coleopterologische Hefte 3: 4–32. [2 Nov 1868 (Proc. Ent. Soc. London 1868: xxxvii)]

Kaup J (1868b) Prodromus zu einer Monographie der Passaliden. Coleopterologische Hefte 4: 1–31. [1868 (title page); 1 Mar 1869 (Proc. Ent. Soc. London 1869: v)]

Kaup J (1869) Prodromus zu einer Monographie der Passaliden. Coleopterologische Hefte 5: 1–40. [1869 (title page); 1 Nov 1869 (Proc. Ent. Soc. London 1869: xix)]

Kaup J (1871) Monographie der Passaliden. Berliner Entomologische Zeitschrift 15 (4 [sep. pagination]): 1–125. [Jan 1871 (wrapper)]

Kavanaugh DH (1996) Phylogenetic relationships of genus *Pelophila* Dejean to other basal grade Carabidae (Coleoptera). Annales Zoologici Fennici 33 (1): 31–37. [14 Jun 1996 (top of article)]

Kavanaugh DH, Nègre J (1983) Notiokasiini - a new tribe of Carabidae (Coleoptera) from southeastern South America. The Coleopterists Bulletin 36 [1982] (4): 549–566. [mailed 26 Apr 1983 (wrapper)]

Kawanabe M, Miyatake M (1996) A redescription of *Xylographella punctata* (Coleoptera, Ciidae), with description of a new tribe. Elytra 24 (1): 125–130. [15 May 1996 (p. ii)]

Kazantsev SV (2003) A generic review of Duliticolinae, new subfamily (Coleoptera: Lycidae). Elytron 16 [2002]: 5–21. [publ. Jul 2003 (verso of title page); 28 Nov 2003 (recorded at BMNH)]

Kazantsev SV (2004a) Phylogeny of the tribe Erotini (Coleoptera, Lycidae), with descriptions of new taxa. Zootaxa 496: 1–48. [27 Apr 2004 (footer p. 1)]

Kazantsev SV (2004b) Morphology of Lycidae with some considerations on evolution of the Coleoptera. Elytron 17/18 [2003–2004]: 73–248. [Oct 2004 (verso of title page); 23 Mar 2005 (recorded at FMNH)]

Kazantsev SV (2007) New firefly taxa from Hispaniola and Puerto Rico (Coleoptera: Lampyridae), with notes on biogeography. Novye taksony svetlyachkov s Ispanioly i Puerto-Riko (Lampyriade: Coleoptera), s zamechaniyami po biogeographii. Russian Entomological Journal 15 [2006] (4): 367–392. [printed Jan 2007 (verso of title page)]

Kazantsev SV (2010a) New taxa of Omalisidae, Drilidae and Omethidae, with a note on systematic position of Thilamaninae (Coleoptera). Novye taksony Omalisidae, Drilidae i Omethidae, s zamechaniem po sistematicheskomu polozheniyu Thilmaninae (Coleoptera). Russian Entomological Journal 19 (1): 51–60. [printed Jun 2010 (inside wrapper)]

Kazantsev SV (2010b) New taxa of Lycidae from Samoa, Fiji and Tonga (Coleoptera: Lycidae). Novye taksony Lycidae iz Samoa, Fidzhi i Tonga (Coleoptera: Lycidae). Russian Entomological Journal 18 (3): 195–199. [publ. 5 Apr 2010 (vol. 19 (1): 76)]

Kazantsev SV (2010c) On the status and systematic position of the genera *Falsolucidota* Pic, 1921 and *Macrolycinella* Pic, 1922, with notes on *Lopheros* LeConte, 1881 (Coleoptera: Lycidae). O statuse i sistematicheskom polozhenii rodov *Falsolucidota* Pic, 1921 i *Macrolycinella* Pic, 1922, s zamechaniyami po *Lopheros* LeConte, 1881 (Coleoptera: Lycidae). Russian Entomological Journal 18 [2009] (4): 277–283. [26 Apr 2010]

Kazantsev SV (2010d) Fireflies of Russia and adjacent territories (Coleoptera: Lampyridae). Russian Entomological Journal 19 (3): 187–208.

Kemner NA (1925) Javanische Termitophilen. I. *Schizelythron javanicum* n. g., n. sp., eine neue physogastre Staphylinide von einem neuen, nicht zu den Aleochariden gehörigen Typus, nebst biologischen Bemerkungen über *Jacobsonella termitobia* Silv. Entomologisk Tidskrift 46 (2): 107–126, pl. 3. [10 Jul 1925 (back wrapper)]

Kerremans C (1892) Note sur les chrysodémides Africaines attribuées au genre *Iridotaenia*. Annales de la Société Entomologique de Belgique 36 (2): 49–55. [9 Mar 1892 (Ann. Soc. Ent. France 61: Bull. Ent.: lxxviii)]

Kerremans C (1893) Essai de groupement des Buprestides. Annales de la Société Entomologique de Belgique 37 (3): 94–122. [séance 4 Mar 1893; 10 May 1893 (Ann. Soc. Ent. France 62: Bull. Ent.: ccv)]

Kerremans C (1903) 12 fascicule. Coleoptera Serricornia. Fam. Buprestidae [pp. 1–48]. In: Wytsman P (Ed) Genera Insectorum Vol. II. P. Wytsman, Bruxelles, 338 pp. + 4 pls. [12a: 1902 (title page); 28 Feb 1903 (Evenhuis 1994: 51)]

Kerzhner IM (1984) Daty publikatsii izdaniya «Trudy Russkogo Entomologicheskogo Obshchestva» i «Horae Societatis Entomologicae Rossicae», 1861–1932. Entomologicheskoe Obozrenie 18 (4): 849–857.

Kessel F (1921) Über die Stellung der Passandridae im System. Archiv für Naturgeschichte (Abteilung A) 87 (6): 33–35. [publ. May 1921 (wrapper)]

Kiesenwetter EAH, von (1852) Beiträge zu einer Monographie der Malthinen. Linnaea Entomologica 7: 239–324, pls. 1–2. [29 Oct 1852 (advert. Ent. Zeitung 12: 362)]

Kiesenwetter EAH, von (1858) Naturgeschichte der Insecten Deutschlands. Erste Abtheilung. Coleoptera. Vierter Band. Zweite Lieferung. Bogen 12–24 [pp. 193–384]. Nicolaische Verlagsbuchhandlung, Berlin, vi + 745 + [1] pp. [1858 (title page)]

Kiesenwetter EAH, von (1860) Naturgeschichte der Insecten Deutschlands. Erste Abtheilung. Coleoptera. Vierter Band. Dritte Lieferung. Bogen 25–36 [pp. 385–568]. Nicolaische Verlagsbuchhandlung, Berlin, vi + 745 + [1] pp. [1860 (title page)]

Kiesenwetter EAH, von (1863) Naturgeschichte der Insecten Deutschlands. Erste Abtheilung. Coleoptera. Vierter Band. Dritte Lieferung. Bogen 37–48 [pp. 569–745, Melyridae, etc.]. Nicolaische Verlagsbuchhandlung, Berlin, vi + 745 + [1] pp. [1863 (wrapper); summer 1863 (date of Vorrede); 19 Nov 1863 (Allgem. Bibliogr. Deutschl.)]

Kiesenwetter EAH, von (1864) Beitrag zur Käferfauna Griechenlands. (Neuntes Stück.) Curculionidae. Berliner Entomologische Zeitschrift 8 (3/4): 239–294 + pls. 3–4. [end Dec 1864 (p. iii); 1865 (Derksen and Scheiding-Göllner 1965: 474)]

Kiesenwetter EAH, von (1877) Naturgeschichte der Insecten Deutschlands. Erste Abteilung. Coleoptera. Fünfter Band. [Erste Hälfte]. Erste Lieferung. Bogen 1 bis 13 [pp. 1–200, Anobiadae, Cioidae, Orophyidae, Aspidiphoridae]. Nicolaische Verlagsbuchhandlung, Berlin, 832 pp. [1877 (wrapper); Jul-Dec 1877 (Griffin 1939: 218); 29 Sep 1877 (Literar. Centrbl. 1877: 1360)]

Kikuta T (1986) On the higher taxa of the stag beetle family Lucanidae. In: Papers on entomology presented to prof. Takeshiko Nakane in commemoration of his retirement. Special Bulletin of the Japanese Society of Coleopterology No. 2: 131–138. [10 Oct 1986 (title page)]

Kippenberg H (2010a) New acts and comments. Chrysomelidae: Chrysomelinae [pp. 67–73]. In: Löbl I, Smetana A (Eds) Catalogue of Palaearctic Coleoptera. Volume 6. Chrysomeloidea. Apollo Books, Stenstrup, 924 pp. [publ. 22 Feb 2010 (verso of title page)]

Kippenberg H (2010b) Subfamily Chrysomelinae Latreille, 1802 [pp. 391–443]. In: Löbl I, Smetana A (Eds) Catalogue of Palaearctic Coleoptera. Volume 6. Chrysomeloidea. Apollo Books, Stenstrup, 924 pp. [publ. 22 Feb 2010 (verso of title page)]

Kirby W (1813) Strepsiptera, a new order of insects proposed; and the characters of the order, with those of its genera, laid down. The Transactions of the Linnean Society of London 11 [1815] (1): 86–123, pls. 8–9. [20 May 1813 (Raphael 1970: 64)]

Kirby W (1819) A century of insects, including several new genera described from his cabinet. The Transactions of the Linnean Society of London 12 [1817] (2): 375–482, pls. 21–23. [2 Jul 1819 (Raphael 1970: 64)]

Kirby W (1823) A description of some insects which appear to exemplify Mr. W. S. MacLeay's doctrine of affinity and analogy. The Transactions of the Linnean Society of London 14 (1): 93–110, pl. 3. [28 May-12 Jun 1823 (Raphael 1970: 75)]

Kirby W (1825) A description of such genera and species of insects, alluded to in the «Introduction to Entomology» of Messrs. Kirby and Spence, as appear not to have been before sufficiently noticed or described. The Transactions of the Linnean Society of London 14 (3): 563–572. [31 May 1825 (Raphael 1970: 75)]

Kirby W (1826) [systematic treatments] In: Kirby W, Spence W: An introduction to entomology: or elements of the natural history of insects: with plates. Vol. IV. Longman, Rees, Orme, Brown and Green, London, iv + 634 pp. + pls. 21–30. [Jan 1826 (Sherborn 1922: lxxiii)]

Kirby W (1827) A description of some new genera and species of petalocerous Coleoptera. Zoological Journal 3: 145–158. [Apr-Jul 1827 (top of article)]

Kirby W (1828) A description of some coleopterous insects in the collection of the Rev. F. W. Hope, F. L. S. Zoological Journal 3 [1827–28]: 520–525. [Jan-Apr 1828 issue]

Kirby W (1837) Part the fourth and last. The insects. In: Richardson J (Ed) Fauna boreali-Americana; or the zoology of the northern parts of British America: containing descriptions of the objects of natural history collected on the late northern land expeditions, under command of Captain Sir John Franklin, RN. J. Fletcher, Norwich, xxxix + 325 + [2] pp., 8 pls. [pls. publ. 1 Jan 1830]. [23 Oct 1837 (Evenhuis 1997b: 646), 6 Nov 1837 (Jour. Proc. Ent. Soc. London 1837: lxvi)]

Kirejtshuk AG (1982) Sistematicheskoe polozhenie roda Calonecrus J. Thomson i zamechaniya po philogenii semeistva zhukov-blestyanok (Coleoptera, Nitidulidae).[Systematic position of the genus *Calonecrus* J. Thomson and notes on the phylogeny of the family Nitidulidae (Coleoptera)]. Entomologicheskoe Obozrenie 61 (1): 117–130 [in Russian; English translation in Entomological Review 61 (1): 109–122]. [after 4 Mar 1982 (approved to print, endleaf); recorded at CNC 28 Jun]

Kirejtshuk AG (1986) Analiz stroeniya genitaliy dlya rekonstruktsii philogenii i obosnovaniya sistemy sem. Nitidulidae (Coleoptera). [An analysis of the genitalia morphology and its use in reconstructing the phylogeny and basis of the system of Nitidulidae (Coleoptera).]. Trudy Vsesoyuznogo Entomologicheskogo Obshchestva 68: 22–28 [in Russian]. [after 17 Apr 1986 (approved to print); 22 Sep 1980 (recorded at CNC)]

Kirejtshuk AG (1987) Obzor zhukov-blestyanok podsem. Cryptarchinae (Coleoptera, Nitidulidae) Indo-Malayskoy Oblasti. [Review of the nitidulid beetles of the subfam. Cryptarchinae (Coleoptera, Nitidulidae) from the Indo-Malayan region.]. Trudy Zoologicheskogo Instituta 170: 62–95 [in Russian].

Kirejtshuk AG (1991) Novye rody i vidy zhukov-blestyanok (Coleoptera, Nitidulidae) iz Avstraliyskoy Oblasti. III. [New genera and species of the nitidulid beetles (Coleoptera, Nitidulidae) from Australian region. III.]. Entomologicheskoe Obozrenie 69: 857–878 [in Russian; English translation in Entomological Review 70 (7): 52–73 (Apr 1992)]. [after 16 Jan 1991 (approved to print, last p. of issue)]

Kirejtshuk AG (1994) Parandrexidae fam. n. - Yurskie zhuki iz infraotryada Cucujiformia (Coleoptera, Polyphaga). [Parandrexidae fam. n. - a Jurassic beetles from the Infrarordo Cucujiformia (Coleoptera, Polyphaga)]. Paleontologicheskii Zhurnal 1994 (1): 57–64, pls. 7–8 [in Russian]. [1994; 22 Dec 1994 (recorded at BMNH)]

Kirejtshuk AG (1998a) Nitidulidae (Coleoptera) of the Himalayas and northern Indochina. Part 1: subfamily Epuraeinae. [Theses Zoologicae vol 28]. Koeltz Scientific Books, Koenigstein, 489 pp. publ. Nov 1998 (verso of title page)

Kirejtshuk AG (1998b) Polozhenie podsem. Maynipeplinae subfam. n. (Coleoptera, Nitidulidae) iz ekvatorialnoi Afriki v sisteme, a takzhe zamechaniya po evolyutsii i strukturnym morifikatsiyam y blestyanok. [The position of the subfamily Maynipeplinae subfam. n. (Coleoptera, Nitidulidae) in the classification and notes on the evolution and structural modifications among sap-beetles]. Entomologicheskoe Obozrenie 77 (3): 540–554 [in Russian; English translation in Entomological Review 78: 793–807]. [after 15 Sep 1998 (approved to print); 1998 (Zool. Record)]

Kirejtshuk AG (2008) A current classification of sap beetles (Coleoptera, Nitidulidae). Zoosystematica Rossica 17 (1): 107–122. [23 Nov 2008 (Zool. Record)]

Kirejtshuk AG, Audisio P (1986) [new taxon] In: Kirejtshuk AG: On polyphyly of the Carpophilinae with description of a new subfamily, Cillaeinae (Coleoptera: Nitidulidae). The Coleopterists Bulletin 40 (3): 217–221. [mailed 8 Oct 1986 (inside wrapper)]

Kirejtshuk AG, Azar D (2008) New taxa of beetles (Insecta, Coleoptera) from Lebanese amber with evolutionary and systematic comments. Alavesia 2: 15–46.

Kirejtshuk AG, Azar D (2009) [new taxa] In: Kirejtshuk AG, Azar D, Tafforeau P, Boistel R, Fernandez V: New beetles of Polyphaga (Coleoptera, Polyphaga) from Lower Cretaceous Lebanese amber. Denisia 26: 119–130.

Kirejtshuk AG, Azar D, Beaver RA, Mandelshtam MY, Nel A (2009) The most ancient bark beetle known: a new tribe, genus and species from Lebanese amber (Coleoptera, Curculionidae, Scolytinae). Systematic Entomology 34 (1): 101–112. [7 Jan 2009 (vol. contents)]

Kirejtshuk AG, Chang H, Ren D, Kun SC (2010) Family Lasiosynidae n. fam., new palaeoendemic Mesozoic family from the infraorder Elateriformia (Colopetera: Polyphaga). Annales de la Société Entomologique de France (Nouvelle Série) 46 (1/2): 67–87. [issued 30 Jun 2010 (back wrapper)]

Kirejtshuk AG, Lawrence JF (1992) Cychramptodini, a new tribe of Nitidulidae (Coleoptera) from Australia. Journal of the Australian Entomological Society 31 (1): 29–46. [28 Feb 1992 (wrapper)]

Kirejtshuk AG, Poinar G, Jr. (2006) Haplochelidae, a new family of Cretaceous beetles (Coleoptera: Myxophaga) from Burmese amber. Proceedings of the Entomological Society of Washington 108 (1): 155–164. [mailed 5 Jan 2006; 20 Jan 2006 (recorded at FMNH)]

Kirejtshuk AG, Ponomarenko AG (1990) Iskopaemye zhuki semeystv Peltidae i Nitidulidae (Coleoptera). [Fossil beetles from families Peltidae and Nitidulidae (Coleoptera).]. Paleontologicheskii Zhurnal 1990 (2): 78–88, pl. 7 [in Russian]. [after 3 May 1990 (approved to print)]

Kirejtshuk AG, Ponomarenko AG (2010) A new coleopterous family Mesocinetidae fam. n. (Coleoptera: Scirtoidea) from Late Mesozoic and notes on fossil remains from Shar-Teg (Upper Jurassic, South-Western Mongolia). Novoe semeystvo zhukov Mesocinetidae fam. n. (Coleoptera: Scirtoidea) iz pozdnego mezozoya i zamechaniya ob iskopaemykh ostatkakh iz Shar-Tega (verkhnyaya yura, yugo-zapadnaya Mongolia). Zoosystematica Rossica 19 (2): 301–325. [30 Dec 2010 (top of article)]

Kirsch T (1870) Beiträge zur Käferfauna von Bogotà. (Fünftes Stück: Phanerognathe Curculionen aus der Gruppe der Apostasimeriden). Berliner Entomologische Zeitschrift 13 [1869] (3/4): 187–224. [Jan 1870 (p. ix)]

Kirsch T (1875) Beiträge zur Kenntniss der Peruanischen Käferfauna auf Dr. Abendroth's Sammlungen basirt. (Viertes Stück.). Deutsche Entomologische Zeitschrift 19 (1): 161–208. [Jan 1875 (vol. Inhalt p. 3)]

Kirsch T (1877) [new taxa] In: Kiesenwetter H, von, Kirsch T: Die Käferfauna der Auckland-Inseln, nach Herm. Krone's Sammlungen beschrieben. Deutsche Entomologische Zeitschrift 21 (1): 153–174. [Apr 1877 (vol. Inhalt p. iii)]

Kishii T (1989) Elaterid-beetles from Mt. Houwau-zan and Gozaishi-kôsen Spa in Yamanashi Prefecture, collected by Mr. Kôichi Hosoda in 1987 and 1988 (Coleoptera: Elateridae) "Notes on Elateridae from Japan and its adjacent area (8).". Bulletin of the Héian High School 33: 1–19.

Kissinger DG (1968) Curculionidae subfamily Apioninae of North and Central America with reviews of the world genera Apioninae and world subgenera of *Apion* Herbst (Coleoptera). Taxonomic Publications, South Lancaster, Massachusetts, vii + 559 pp., incl. 41 pls.

Kissinger DG (2004) Review of Noterapion Kissinger from Chile and Argentina (Coleoptera: Apionidae). Insecta Mundi 16 [2002] (4): 221–246. [17 Mar 2004 (back wrapper)]

Kistner DH (1970a) New termitophilous Staphylinidae (Coleoptera) from Hodotermitidae (Isoptera) nests. Journal of the New York Entomological Society 78 (1): 2–16. [mailed 11 Aug 1970 (inside wrapper)]

Kistner DH (1970b) New termitophiles associated with *Longipeditermes longipes* (Haviland) II. The genera *Compactopedia*, *Emersonilla*, *Hirsitilla*, and *Limulodilla*. Journal of the New York Entomological Society 78 (1): 17–32. [mailed 11 Aug 1970 (inside wrapper)]

Kistner DH (1970c) Australian termitophiles associated with *Microcerotermes* (Isoptera: Amitermitinae) I. A new subtribe, genus, and species (Coleoptera, Staphylinidae) with notes on their behavior. Pacific Insects 12 (1): 9–15. [20 May 1970 (on article); mailed 6 Jun 1970 (vol. p. v)]

Kistner DH (1970d) A new genus, species, and subtribe of termitophilous Staphylinidae from Australia (Coleoptera) with a description of its glands. Pacific Insects 12 (3): 499–506. [30 Oct 1970 (on article); mailed 11 Nov 1970 (vol. p. v)]

Kistner DH (1972) A revision of the termitophilous tribe Feldini (Coleoptera, Staphylinidae) with a numerical analysis of the relationships of the species and genera. Contributions of the American Entomological Institute 8 (4): 1–35.

Kistner DH (1993) Cladistic analysis, taxonomic restructuring and revision of the Old World genera formerly classified as Dorylomimini with comments on their evolution and behavior (Coleoptera: Staphylinidae). Sociobiology 22 (2): 151–383.

Kistner DH (1997) [new taxa] In: Kistner DA, Weissflog A, Rosciszewski K, Maschwitz U: New species, new genera, and new records of myrmecophiles associated with army ants (*Aenictus* sp.) with the description of a new subtribe of Staphylinidae (Coleoptera; Formicidae: Aenictinae). Sociobiology 29: 23–221. [by Mar 1997]

Kistner DH, Pasteels JM (1969) A new tribe, genus, and species of termitophilous Aleocharinae (Coleoptera: Staphylinidae) from south-west Africa with a description of its integumentary glands. Annals of the Entomological Society of America 62: 1189–1202.

Kistner DH, Pasteels JM (1970) Taxonomic revision of the termitophilous subtribe Coptotermoeciina (Coleoptera: Staphylinidae) with a description of some integumentary glands and a numerical analysis of their relationships. Pacific Insects 12 (1): 85–115. [20 May 1970 (on article); mailed 6 Jun 1970 (vol. p. v)]

Kistner DH, Watson JAL (1972) A new tribe of termitophilous Aleocharinae (Coleoptera: Staphylinidae) from Australia. Australian Journal of Zoology (Supplementary Series) 17: 1–24. [8 Dec 1972 (wrapper)]

Kleine R (1922a) Neue Brenthiden aus Sumatra nebst Bemerkungen zu schon bekannten Arten. Deutsche Entomologische Zeitschrift 1922 (1): 148–151. [15 Feb 1922 (p. iii)]

Kleine R (1922b) Neue Brenthiden aus dem Dahlemer Museum. Archiv für Naturgeschichte (Abteilung A) 88 (3): 223–228. [publ. Jun 1922 (wrapper)]

Kleine R (1922c) Neuklassifizierung der Brenthidae. Entomologische Blätter 18 (4): 161–163. [30 Dec 1922]

Kleine R (1926) Coleoptera Lycidae. In: Beaufort LF, de, Pulle AA, Rutten L (Eds) Nova Guinea. Résultats des expéditions scientifiques à la Nouvelle Guinée. Vol. XV Zoologie. Livraison II. E.J. Brill, Leiden, pp. 91–195.

Kleine R (1929) Part 1. Neue indische Lycidae nebst faunistischen Bermerkungen. The Indian Forest Records (Entomology Series) 13 [1928] (6): 221–269 + 5 pls. [1929 (issue title page)]

Kleine R (1933) Pars 128: Lycidae. In: Schenkling S (Ed) Coleopterorum Catalogus. Volumen IX. W. Junk, Berlin, 145 pp. [2 Jun 1933 (verso of vol. title page)]

Klima A (1935) Pars 145: Curculionidae: Subfam. Acicnemidinae. In: Schenkling S (Ed) Coleopterorum Catalogus. Volumen XXIX. W. Junk, Berlin, 10 pp. [18 Jun 1935 (verso of vol. title page)]

Klimaszewski J, Pace R, Center TD (2010) [new tribe] In: Klimaszewski J, Pace R, Center TD, Couture J: A remarkable new species of *Himalusa* Pace from Thailand (Coleoptera, Staphylinidae, Aleocharinae): phytophagous aleocharine beetle with potential for bio-control of skunkvine-related weeds in the United States. ZooKeys 35: 1–12. [2 Feb 2010 (p. 1)]

Klimaszewski J, Peck SB (1986) A review of the cavernicolous Staphylinidae (Coleoptera) of eastern North America: Part I. Aleocharinae. Quaestiones Entomologicae 22 (2): 51–113. [issued May 1986 (back wrapper)]

Koch C (1943) Phylogenetische biogeographishe und systematische Studien über ungflügelte Tenebrioniden (Col. Tenebr.) IV. Mitteilungen der Münchner Entomologischen Gesellschaft 33 (2/3): 479–597. [15 Dec 1943 (p. ii, Inhalt)]

Koch C (1950) Proposed change of African generic names in the family Tenebrionidae (Col.). The Entomologist 83 (3): 66–68. [16 Mar 1950 (p. iv)]

Koch C (1953a) Die Tenebrioniden des südlichen Afrikas XIV. Ueber einige neue Molurini aus dem Ungarischen Naturwissenschaftlichen Museum zu Budapest (Vorarbeiten zu einer Monographie der Molurini, 4). Annales Historico-Naturales Musei Nationalis Hungarici 3 [1952]: 137–181. [1953 (vol. title page)]

Koch C (1953b) The Tenebrionidae of southern Africa. XXI. On some new endemic Opatrinae from the Namib Desert. Annals of the Transvaal Museum 22 (2): 231–252.

Koch C (1953c) The Tenebrionidae of southern Africa. XXVI [error for XXVI]. New Port. East African species collected by Dr. A. J. Barbosa. Revista da Facultade de Ciencas (Lisboa) (2a Série: C. Ciencias Naturais) 3: 239–310 + pl. 1.

Koch C (1954) The Tenebrionidae of southern Africa. XXV. New, forgotten or Palaearctic genera and species of Opatrinae. Annals of the Transvaal Museum 22 (3): 419–476. [25 Nov 1954]

Koch C (1955) Monograph of the Tenebrionidae of southern Africa. Vol. I. (Tentyriinae, Molurini. Trachynotina: *Somaticus* Hope). Transvaal Museum Memoir No. 7: xiv + 242pp. + 24 pls + 2 folding maps.

Koch C (1956) II. Tenebrionidae (Coleoptera, Polyphaga). Opatrinae. First Part: Platynotini, Litoborini and Loensini. Exploration du Parc National de l'Upemba Mission G F de Witte. Fascicule 40. Institut des Parcs Nationaux du Congo Belge, Bruxelles, 472 pp. + 35 pls. [1956 (title page)]

Koch C (1958) Tenebrionidae of Angola. Publicações Culturais da Companhia de Diamantes de Angola 39: 1–231. [submitted Jun 1955 (p. 231); 1958 (title page)]

Kolbe HJ (1880) Natürliches System der carnivoren Coleoptera. Deutsche Entomologische Zeitschrift 24 (2): 258–280. [Oct 1880 (p. iii)]

Kolbe HJ (1883) Über die von J. M. Hildebrandt in Madagaskar gefundenen Brenthiden [reading]. Sitzungsberichte der Gesellschaft Naturforschender Freunde zu Berlin 1883: 74–82. [read 22 May 1883]

Kolbe HJ (1887) Eine zoogeographische Studie. Jahresbericht der Zoologischen Section des Westfälischen Provinzial-Vereins für Wissenschaft und Kunst 15 [1886–87]: 47–54. [1887 (title page); last meeting 24 Mar 1887; before 7 Nov 1887 (Zool. Anz. 10: 570)]

Kolbe HJ (1891) Aufzählung der von Herrn Dr. Hans Meyer im Jahre 1889 im Gebiete des Kilimandscharo- und Ugueno-Gebirges gesammelten Coleopteren Stettiner Entomologische Zeitung 52 (1/3): 18–36. [Mar 1891]

Kolbe HJ (1892) Beiträge zur Kenntnis der Brenthiden. Stettiner Entomologische Zeitung 53 ((4/6)): 162–175. [Apr-Jun issue; article written Sep 1892 (p. 175)]

Kolbe HJ (1895) Coleopteren aus Afrika. Stettiner Entomologische Zeitung 55 [1894] (10/12): 361–397. [incl. article dated Apr 1895 (p. 402)]

Kolbe HJ (1897) Coleopteren - Die Käfer Deutsch-Ost-Afrikas. D. Reimer, Berlin, 368 pp. + 4 pls.

Kolbe HJ (1899) Die Oxyopisthinen, eine neue Gruppe der Curculioniden des tropischen Afrika. Stettiner Entomologische Zeitung 60 (1/2): 3–138. [issued Nov 1899 (p. 162)]

Kolbe HJ (1904) Gattungen und Arten der Valgiden von Sumatra und Borneo. Stettiner Entomologische Zeitung 65 (1): 3–57. [Feb 1904 (p. 215)]

Kolbe HJ (1905) Über die Lebensweise und die geographische Verbreitung der coprophagen Lamellicornier. Zoologische Jahrbücher, Supplement 8: 475–594.

Kolbe HJ (1908) Mein System der Coleopteren. Zeitschrift für Wissenschaftliche Insektenbiologie 4 (4): 116–123, (5): 153–162, (6): 219–226, (7): 246–251, (8): 286–294, (10/11): 389–400. [30 Apr 1908 (fasc. 4); 15 Jun (fasc. 5); 26 Jul (fasc. 6); 29 Aug (fasc. 7); 21 Oct (fasc. 8/9); 28 Nov 1908 (fasc. 10/11) (wrappers)]

Kolbe HJ (1910) Die Coleopterenfauna der Seychellen. Nebst Betrachtungen über die Tiergeographie dieser Inselgruppe. Mitteilungen aus dem Zoologischen Museum in Berlin 5 (1): 1–49. [subm. Feb 1910 (title page); publ. Aug 1910 (fasc. contents)]

Kolbe HJ (1916) Beitrag zur Morphologie und Systematik der Taphroderinen (Familie der Brenthiden) Afrikas. Deutsche Entomologische Zeitschrift 1916 (1): 50–67. [1 Apr 1916 (Inhalt p. i)]

Kolbe HJ (1927a) Über die Carabidomemninen, eine neue Gruppe der primitiven Paussidenstufe. Entomologische Blätter 23 (4): 178–187. [31 Dec 1927 (wrapper)]

Kolbe HJ (1927b) Die Einstammigkeit der Paussiden und die primitiven Gattungen dieser myrmekophilen Coleopterenfamilie. Zoologischer Anzeiger 72: 205–218. [subm. 12 Mar 1927]

Kolibáč J (1992) Revision of Thanerocleridae n. stat. (Coleoptera, Cleroidea). Mitteilungen der Schweizerischen Entomologischen Gesellschaft 65 (3/4): 303–340. [printed Dec 1992 (back wrapper)]

Kolibáč J (1998) Classification of the subfamily Hydnocerinae Spinola, 1844 (Coleoptera: Cleridae). Acta Musei Moraviae Časopis Moravského Musea Scientiae Biologicae 83 (1/2): 127–210. [1998 (copyright)]

Kolibáč J (2004) Metaxinidae fam. n., a new family of Cleroidea (Coleoptera). Entomologica Basiliensia 26: 239–268. [issued 24 Nov 2004 (Inhalt)]

Kolibáč J (2006) A review of the Trogossitidae. Part 2: larval morphology, phylogeny and taxonomy (Coleoptera, Cleroidea). Entomologica Basiliensia et Collectionis Frey 28: 105–153. [issued 22 Dec 2006 (inside wrapper)]

Kolibáč J, Huang F (2008) Taxonomic status of the Mesozoic genera *Anhuistoma* Lin, 1985, *Eotenebroides* Ren, 1995, *Lithostoma* Martynov, 1926, *Palaeoendomychus* Zhang, 1992, and *Sinosoronia* Zhang, 1992 (Coleoptera). Entomologica Basiliensia et Collectionis Frey 30: 135–148. [22 Dec 2008 (inside wrapper)]

Konishi M (1962) Taxonomic studies on the Cossoninae of Japan (Coleoptera: Curculionidae). Part I. Insecta Matsumurana 25 (1): 1–17 + pl. I. [Aug 1962 (verso of title page)]

Kôno H (1930) Kurzrüssler aus dem japanischen Reich. Journal of the Faculty of Agriculture, Hokkaido Imperial University 24 (5): 153–242 + pl. I-II. [issued 26 Mar 1930]

Korschevsky R (1932) Pars 120: Coccinellidae II. In: Schenkling S (Ed) Coleopterorum Catalogus. Volumen XVI. W. Junk, Berlin, pp. 225–659. [20 Feb 1932 (verso of vol. title page)]

Kovář I (2007) Family Coccinellidae Latreille 1807 [pp. 568–631]. In: Löbl I, Smetana A (Eds) Catalogue of Palaearctic Coleoptera. Volume 4. Elateroidea - Derodontoidea - Bostrichoidea - Lymexyloidea - Cleroidea - Cucujoidea. Apollo Books, Stenstrup, 935 pp. [publ. 30 Jun 2007 (verso of title page)]

Kraatz G (1856) Staphylinii. Naturgeschichte der Insekten Deutschlands Erste Abteilung Coleoptera Zweiter Band Erste, Zweite Lieferung Bogen 1–24 [pp 1–376]. Nicolaische Buchhandlung, Berlin, viii + 1080 pp. [20 June 1856 (Zerche 1987: 93); Mar 1856 (date of summary inside back wrapper)]

Kraatz G (1857) Staphylinii. Naturgeschichte der Insekten Deutschlands Erste Abteilung Coleoptera Zweiter Band Dritte, Vierte Lieferung Bogen 25–48 [pp 377–768]. Nicolaische Buchhandlung, Berlin, viii + 1080 pp. [May 1857 (Kraatz 1859c: vii); 30 Nov 1857 (Intell.-Blatt Allgem. Lit.); 1857 (wrapper)]

Kraatz G (1859a) [new taxa] In: Schaum HR: Catalogus Coleopterorum Europae. In Verbindung mit Dr. G. Kraatz und H. v. Kiesenwetter. Nicolaische Verlagsbuchhandlung, Berlin, iv + 121 pp. [1859 (wrapper + title page)]

Kraatz G (1859b) Zur kritischen Kenntniss der in Gay's Historia Fisica y Politica von Solier beschriebenen Staphylinen. Berliner Entomologische Zeitschrift 3 (1): 1–16. [by Apr 1859 (cited in Wiener Ent. Monatschr. 3: 128)]

Kraatz G (1859c) Neuere Literatur. Berliner Entomologische Zeitschrift 3: iv-ix.

Kraatz G (1879) *Rhaebus gebleri* Fischer, oder eine neue *Rhaebus*-Art in Europa einheimisch? (Coleopt., Chrysomelin? Sagrin?). Deutsche Entomologische Zeitschrift 23 (2): 276–278. [Nov 1879 (Inhalt p. 3)]

Kraatz G (1880a) Genera Nova Cetonidarum. Entomologische Monatsblätter 2 (12): 17–30. [Feb 1880 issue]

Kraatz G (1880b) Genera Cetonidarum Australiae. Deutsche Entomologische Zeitschrift 24 (1): 177–214. [May 1880 (Inhalt p. iii)]

Kraatz G (1882a) Ueber die Gattung *Clinteria* Burm. Deutsche Entomologische Zeitschrift 26 (1): 49–51. [Apr 1882 (Inhalt p. iii)]

Kraatz G (1882b) Die africanischen Leucoceliden und die ihnen zunächst verwandten Gattungen der Cetoniden. Deutsche Entomologische Zeitschrift 26 (1): 65–78. [Apr 1882 (Inhalt p. iii)]

Krell F-T (1996) Ripiphoridae oder Rhipiphoridae? (Col., Tenebrionoidea). Entomologische Nachrichten und Berichte 40 (1): 47–48. [publ. 20 Apr 1996 (inside wrapper)]

Krikken J (1984) A new key to the suprageneric taxa in the beetle family Cetoniidae, with annotated lists of the known genera. Zoologische Verhandelingen No. 210: 1–75. [5 Sep 1984 (title page)]

Kriska NL (2002) Family 109. Oedemeridae [pp. 514–519]. In: Arnett RH, Jr., Thomas MC, Skelley PE, Frank JH (Eds) American beetles. Volume 2. Polyphaga: Scarabaeoidea through Curculionoidea. CRC Press, Boca Raton, xiv + 861 pp. [19 Jun 2002 (CRC website)]

Kryzhanovskij OL (1972) On the taxonomy of extra-Palaearctic Histeridae (Coleoptera). Entomologica Scandinavica 3 (1): 19–25. [22 Feb 1972; 19 Apr 1972 (recorded at CNC)]

Kryzhanovskij OL (1976) In: Kryzhanovskij OL, Reichardt AN: Zhuki nadsemeistva Histeroidea (Semeistva Sphaeritidae, Histeridae, Synteliidae). Fauna SSSR Novaya Seriya No. 111. Zhestokokrylye Tom V, vyp. 4. Nauka, Leningrad, 434 pp. [in Russian]. [after 25 Nov 1976 (approved to print, last page)]

Kukalová J (1969) On the systematic position of the supposed Permian beetles, Tshecardocoleidae, with a description of a new collection from Moravia. Sborník Geologických Ved, Rada P Paleontologie 11: 139–162, pls. 1–8.

Kuntzen H (1912) Die Eumolpinensubtribus Nerissini (Coleopt. Chrysomelidae). Archiv für Naturgeschichte (Abteilung A) 78 (2): 42–57. [Jun 1912 (wrapper)]

Kurosawa Y (1990) A revisional study on the genus *Micropistus* Théry and its allies (Coleoptera, Buprestidae). Esakia Special Issue No. 1: 57–64. [20 Apr 1990 (wrapper)]

Kuschel G (1954) Un gorgojo ciego de Otiorhynchinae de Madagascar (Aporte 14 de Col. Curculionoïdea). Revue Française d'Entomologie 21 (4): 286–289. [20 Dec 1954 (footer p. 281)]

Kuschel G (1955) Nuevas sinonimias y anotaciones sobre Curculionoidea (Coleoptera). Revista Chilena de Entomología 4: 261–312. [30 Jun 1955 (top of article)]

Kuschel G (1956) Revisión de los Premnotrypini y adiciones a los Bagoini (Aporte 17 sobre Coleoptera Curculionoidea). Boletín del Museo Nacional de Historia Natural (Santiago) (Contribuciones Entomologicas) 26 (6): 187–235. [printed 14 Mar 1956 (inside back wrapper)]

Kuschel G (1959a) Nemonychidae, Belidae y Oxycorynidae de la fauna Chilena, con algunas consideraciones biogeográficas. Investigaciones Zoologicas Chilenas 5: 229–271. [2 Nov 1959 (vol. title page)]

Kuschel G (1959b) Beiträge zur Kenntnis der Curculioniden von Venezuela und Trinidad-Insel (1. Lieferung). Entomologische Arbeiten aus dem Museum Georg Frey 10 (2): 478–514. [publ. 1 Nov 1959 (vol. Inhalt)]

Kuschel G (1964) Insects of Campbell Island. Coleoptera: Curculionidae of the subantarctic islands of New Zealand. Pacific Insects Monograph 7: 416–493. [15 Jul 1964 (p. 2)]

Kuschel G (1966) A cossonine genus with bark-beetle habits, with remarks on relationships and biogeography (Coleoptera Curculionidae). New Zealand Journal of Science 9 (1): 3–29. [Mar 1966 (p. 3); 1 Jul 1966 (recorded at BMNH)]

Kuschel G (1971) Entomology of the Aucklands and other islands south of New Zealand: Coleoptera: Curculionidae. Pacific Insects Monograph 27: 225–259. [publ. 10 Nov 1971 (title page)]

Kuschel G (1990) Beetles in a suburban environment: a New Zealand case study. The identity and status of Coleoptera in the natural and modified habitats of Lynfield, Auckland (1974–1989). DSIR Plant Protection Report No. 3: 1–119. [Dec 1990 (verso of title page)]

Kuschel G (1994) Appendix. Nemonychidae of Australia, New Guinea and New Caledonia [pp. 563–637]. In: Zimmerman EC (Ed). Australian weevils (Coleoptera: Curculionoidea). Volume I. Orthoceri. Anthribidae to Attelabidae. The primitive weevils. CSIRO Australia, Melbourne, xxxii + 741 pp. [28 Feb 1994 (verso of title page)]

Kuschel G (1995) A phylogenetic classification of Curculionoidea to families and subfamilies [pp. 5–33]. In: Anderson RS, Lyal CHC (Eds) Biology and phylogeny of Curculionoidea: Proceedings of the XVIII International Congress of Entomology, July 1988. Memoir of the Entomological Society of Washington 14. [Nov 1995 (reprint recorded by A. Newton)]

Kuschel G (2003) A ball forming weevil from young *Nothofagus* leaves in Chile (Coleoptera: Curculionidae: Curculioninae: Sphaeriopoeini). Revista Chilena de Entomología 29: 59–65.

Kuschel G (2009) New tribe, new genus and species for an Australasian weevil group with notes and keys (Coleoptera, Curculionidae). Revue Française d'Entomologie (Nouvelle Série) 30 (2/4): 41–66. [publ. 13 May 2009 (back wrapper)]

Kuschel G, May BM (1990) Palophaginae, a new subfamily for leaf-beetles, feeding as adult and larva on araucarian pollen in Australia (Coleoptera: Megalopodidae). Invertebrate Taxonomy 3 (6): 697–719. [publ. 8 Jun 1990 (verso of vol. 4 title page)]

Kuwert AF (1890) Bestimmungs-Tabelle der Hydrophiliden Europas, Westasiens und Nordafrikas. Verhandlungen des Naturforschenden Vereines in Brünn 28 [1889]: 3–121. [1890 (vol. title page)]

Kuwert AF (1891) Systematische Uebersicht der Passaliden-Arten und Gattungen. Deutsche Entomologische Zeitschrift 1891 (1): 161–192. [Jun 1891 (Inhalt p. 3)]

Kuwert AF (1893) Die epiphlöinen Gattungen der Cleriden und einige neue Arten derselben. Annales de la Société Entomologique de Belgique 37 (9): 492–497. [11 Oct 1893 (Ann. Soc. Ent. France 62: Bull. Ent: cclxxxv)]

Kuwert AF (1896) Die Passaliden dichotomisch bearbeitet. Novitates Zoologicae 3 (2): 209–231. [26 Jun 1896 (recorded at BMNH)]

Kwieton E (1982) Revue critique des systèmes récents de la famille des Tenebrionidae (Col.). Acta Musei Nationalis Pragae (Series B) 38 (1/2): 79–100.

La Rivers I (1948) Notes on the Eleodini (Coleoptera: Tenebrionidae). Entomological News 59 (4): 96–101. [mailed 2 Jul 1948 (verso of vol. title page)]

Laboissière V (1921) Étude des Galerucini de la collection du Musée du Congo Belge. Première partie. Revue Zoologique Africaine 9 (1): 33–86. [5 Aug 1921 (footer p. 33); 15 Aug 1921 (wrapper)]

Laboissière V (1922) Revision du groupe des Oidites Africains (Col. Chrysomelidae). Annales de la Société Entomologique de France 90 [1921] (3/4): 193–234. [26 Jul 1922 (p. 356)]

Laboissière V (1924) Étude des Galerucini de la collection du Musée du Congo Belge. Quatrième partie. Revue Zoologique Africaine 12: (2): 135–172, (3): 283–315. [(2) 1 Jun 1924; (3) 1 Aug 1924 (tops of articles)]

Laboissière V (1926) Étude des Galerucini de la collection du Musée du Congo Belge, première partie (suite). Revue Zoologique Africaine 14 (1, 2): 91–122, 185–200. [(1) 15 Aug; (2) 15 Nov 1926 (top of articles)]

Laboissière V (1934) Galerucinae de la faune française (Coléoptères). Annales de la Société Entomologique de France 103 (1): 1–108. [publ. 31 Mar 1934 (wrapper)]

Laboissière V (1937) Nouveaux Gallerucinae paléarctiques. Bulletin de la Société Entomologique de France 42 (2): 27–31. [11 Mar 1937 (p. 312)]

Lacordaire JT (1830a) Mémoire sur les habitudes des insectes coléoptères de l'Amérique méridionale. Annales des Sciences Naturelles 20 (Jun): 185–240, (Jul): 241–291. [dates on signature footers]

Lacordaire JT (1830b) Mémoire sur les habitudes des insectes coléoptères de l'Amérique méridionale (Suite et fin). Annales des Sciences Naturelles 21: 149–194. [Oct 1830 (signature footer p. 113)]

Lacordaire JT (1848) Monographie des coléoptères subpentamères de la famille des phytophages. Tome second. Mémoires de la Société Royale des Sciences de Liège 5: vi + 890 pp. [May 1848 (vol. title page)]

Lacordaire JT (1854a) Histoire naturelle des insectes. Genera des Coléoptères ou exposé méthodique et critique de tous les genres proposés jusqu'ici dans cet ordre d'insectes. Tome premier contenant les familles des cicindélètes, carabiques, dytiscides, gyrinides et palpicornes. Librairie Encyclopédique de Roret, Paris, xx + 486 pp. [1854 (title page; Mém. Soc. Roy. Sci. Liége (2) 3 1873: xxxiii); 8 Feb 1854 (Ann. Soc. Ent. France (3) 2: Bull. Ent.: ix)]

Lacordaire JT (1854b) Histoire naturelle des insectes. Genera des Coléoptères ou exposé méthodique et critique de tous les genres proposés jusqu'ici dans cet ordre d'insectes. Tome deuxième contenant les familles des paussides, staphyliniens, ... hétérocérides. Librairie Encyclopédique de Roret, Paris, 548 pp. [1854 (title page); 2 Oct 1854 (Acad. Sci. France)]

Lacordaire JT (1856) Histoire naturelle des insectes. Genera des Coléoptères ou exposé méthodique et critique de tous les genres proposés jusqu'ici dans cet ordre d'insectes. Tome troisième contenant les familles des pectinicornes et lamellicornes. Librairie Encyclopédique de Roret, Paris, 594 pp. [1856 (title page); 29 Mar 1856 (Bibliogr. France 1856)]

Lacordaire JT (1857) Histoire naturelle des insectes. Genera des Coléoptères ou exposé méthodique et critique de tous les genres proposés jusqu'ici dans cet ordre d'insectes. Tome quatrième contenant les familles des buprestides, throscides, eucnémides, élatérides, cébrionides, cérophytides, rhipicérides, dascyllides, malacodermes, clérides, lyméxylones, cupésides, ptiniores, bostrichides et cissides. Librairie Encyclopédique de Roret, Paris, 579 pp. [1857 (title page); 25 May 1857 (Acad. Sci. France); 27 Jun 1857 (Bibliogr. France 1857: 302)]

Lacordaire JT (1859) Histoire naturelle des insectes. Genera des Coléoptères ou exposé méthodique et critique de tous les genres proposés jusqu'ici dans cet ordre d'insectes. Tome cinquième contenant les familles des ténébrionides, cistélides, nilionides, pythides, mélandryides, lagriides, pédilides, anthicides, pyrochroïdes, mordellides, rhipiphorides, stylopides, meloïdes et oedémérides. Librairie Encyclopédique de Roret, Paris, Première partie (pp. 1–400), Deuxième partie (pp. 401–750). [1859 (title page); 27 Jun 1859 (Acad. Sci. France); 16 Jul 1859 (Bibliogr. France 1859)]

Lacordaire JT (1863) Histoire naturelle des insectes. Genera des Coléoptères ou exposé méthodique et critique de tous les genres proposés jusqu'ici dans cet ordre d'insectes. Tome sixième contenant la famille des curculionides. Librairie Encyclopédique de Roret, Paris, 637 pp. [10 Aug 1863 (Acad. Sci. France); 22 Aug 1863 (Bibliogr. France 1863: 398); 1863 (title page)]

Lacordaire JT (1865) Histoire naturelle des insectes. Genera des Coléoptères ou exposé méthodique et critique de tous les genres proposés jusqu'ici dans cet ordre d'insectes. Tome septième contenant les familles des curculionides (suite), scolytides, brenthides, anthribides et bruchides. Librairie Encyclopédique de Roret, Paris, [2] + 620 pp. [1866 (title page); 12 Dec 1865 (The Bookseller 85: 1016); 1865 (Ann. Soc. Ent. France (4) 5: Bull. Bibliogr.: lxxx; Cat. Ann. Libr. France 8 1865: 126); 20 Jan 1866 (Bibliogr. France 1866)]

Lacordaire JT (1868) Histoire naturelle des insectes. Généra des Coléoptères ou exposé méthodique et critique de tous les genres proposés jusqu'ici dans cet ordre d'insectes. Tome huitième contenant les familles des tricténotomides et des longicornes. Librairie Encyclopédique de Roret, Paris, 552 pp. [1869 (title page); 28 Nov 1868 (Bibliogr. France 1868); Nov 1868 (Zool. Record 1868: 194)]

Lacordaire JT (1869) Histoire naturelle des insectes. Genera des coléoptères ou exposé méthodique et critique de tous les genres proposés jusqu'ici dans cet ordre d'insectes. Tome neuvième. Première partie. Famille des longicornes (suite). Librairie Encyclopédique de Roret, Paris, 409 pp. [1869 (title page)]

Lacordaire JT (1872) Histoire naturelle des insectes. Genera des Coléoptères ou exposé méthodique et critique de tous les genres proposés jusqu'ici dans cet ordre d'insectes. Tome neuvième. Deuxième partie. Famille des longicornes (fin). Librairie Encyclopédique de Roret, Paris, pp. 411–930.

Lacroix JP (1979) Contributions à l'étude des coléoptères lucanides du globe. Deux genres nouveaux et onze espèces inédites (Chiasognathinae, Lucaninae, Chalcodinae, Cladognathinae, Dorcinae). Bulletin & Annales de la Société Royale Belge d'Entomologie 114 (10–12): 249–294. [28 Feb 1979 (wrapper)]

Lacroix M (1989) Insectes coléoptères Melolonthidae (1re partie). Faune de Madagascar. 73 (1). Muséum Nationale d'Histoire Naturelle, Paris, 302 pp. [publ. 28 Dec 1989 (p. [303])]

Lacroix M (1997) Insectes coléoptères Hopliidae (1re partie). Faune de Madagascar. 88 (1). Faune de Madagascar, Paris, 399 pp. [publ. 22 Dec 1997 (p. [400])]

Lacroix M (2007) Pachydeminae du monde. (Scarabaeoidea, Melolonthidae). Genera et catalogue commenté. Collection Hannetons, Paris, 450 pp. + 264 pls.

Lafer GS (1989) I. Podotryad Archostemata; II. Podtryad Adephaga [pp. 66–259]. Tom III Zhestkokrylye, ili zhuki Chast 1. Nauka, Leningrad, 572 pp. [in Russian].

Lafer GS (1996) Fam. Sikhotealiniidae Lafer, fam. n. [pp. 390–396]. In: Ler PA (Ed) Tom III Zhestkokrylye, ili zhuki Chast 3. Dal'Nauka, Vladivostok, 556 pp. [in Russian].

LaFerté-Sénectère F, de (1851) Révision de la tribu des patellimanes de Dejean, coléoptères pentamères de la famille des carabiques. Annales de la Société Entomologique de France (2) 9 (2): 209–294. [23 Jul 1851 (wrapper)]

Lameere A (1885) [note in Assemblée mensuelle du 2 février 1884]. Annales de la Société Entomologique de Belgique / Comptes-Rendus des Séances 28 [1884]: lxxx-lxxxiv. [13 May 1885 (Ann. Soc. Ent. France (6) 5: Bull. Ent.: xcv)]

Lameere A (1900) Notes pour la classification des Coléoptères. Annales de la Société Entomologique de Belgique 44 (9): 355–377, 1 pl. [25 Sep 1900 (wrapper)]

Lameere A (1901) Étude sur la phylogénie des longicornes. Annales de la Société Entomologique de Belgique 45 (11): 314–323. [5 Dec 1901 (wrapper)]

Lameere A (1902) Révision des prionides. Quatrième mémoire - Sténodontines. Mémoires de la Société Entomologique de Belgique 9: 63–110. [1902 (title page)]

Lameere A (1903a) Faune entomologique de l'Afrique tropicale. Longicornes. I. Prioninae. Annales du Musée du Congo (Zoologie - série III) 2 (1): 1–114, i-iii + 3 pls. [Dec 1903 (wrapper)]

Lameere A (1903b) Révision des prionides. Huitième mémoire. - Mécosarthrines. Annales de la Société Entomologique de Belgique 47 (9): 307–320. [30 Sep 1903 (signature footer p. 305)]

Lameere A (1904) Révision des prionides. Septième mémoire. - Macrotomines. Mémoires de la Société Entomologique de Belgique 11: 1–216. [1903 (title page); 27 Jul 1904 (Bull. Soc. Ent. France 1904: 222) 29 Jul 1904 (as "1904", Science (N.S.) 20: 156)]

Lameere A (1909) Révision des prionides. Treizième mémoire. - Dérancistrines. Mémoires de la Société Entomologique de Belgique 17: 1–70. [1909 (wrapper); distrib. by 26 Dec 1909 (Ann. Soc. Ent. Belg. 53: 515)]

Lameere A (1912) Révision des prionides. Vingt-et-unième mémoire. - Anacolines. Mémoires de la Société Entomologique de Belgique 21: 1–188. [15 Nov 1912 (wrapper); 23 Jul 1913 (Bull. Soc. Ent. France 1913: 354)]

Lameere A (1913) Pars 52: Cerambycidae: Prioninae. In: Schenkling S (Ed) Coleopterorum Catalogus. Volumen XXII. Cerambycidae I. W. Junk, Berlin, 108 pp. [20 May 1913 (contents)]

Lameere A (1919) 172me fascicule. Coleoptera Longicornia. Fam. Cerambycidae. Subfam. Prioninae. In: Wytsman PA (Ed) Genera Insectorum. Vol. XXVII. L. Desmet-Verteneuil, Bruxelles, 189 pp. + 8 pls. [15 Mar 1919 (date on manuscr., p. 189); 19 May 1919 (Evenhuis 1994: 58)]

Landin BO (1950) Entomological results from the Swedish expedition 1934 to Burma and British India; Coleoptera: Lamellicornia; Fam. Scarabaeidae collected by René Malaise. Arkiv för Zoologi (2) 1 (2): 3–9, 1 pl. [printed 28 Sep 1949 (p. 9); publ. 1 Apr 1950 (fasc. contents)]

Landin BO (1960) Report no. 5 from the Lund University expedition in 1957 to the Azores and Madeira. The Lamellicorn beetles of the Azores (Coleoptera) with some reflexions on the classification of certain Aphodiini. Boletim do Museu Municipal do Funchal 13: 49–84 + 1 pl. [Aug 1960 (title page)]

Lane F (1950) Cerambicídeos neotrópicos II. Sobre a posição sistemática de alguns gêneros. Arquivos de Zoologia do Estado de São Paulo 7 [1949–51] (5): 363–378. [20 Aug 1950 (top of article)]

Lane F (1951) Cerambycoidea neotropica nova II (Coleoptera). Dusenia 2 (1): 1–20 + pl. 1. [printed 25 Jan 1951 (wrapper)]

Lane F (1955) Cerambycoidea neotropica nova III (Coleoptera). Papéis Avulsos do Departamento de Zoologia 12 (13): 281–296. [16 Nov 1955 (top of article)]

Lane F (1956) Cerambycoidea neotropica nova IV (Coleoptera). Dusenia 7 (1): 1–31. [printed 30 Jan 1956 («1955»; wrapper)]

Lane F (1959) Nova subfamilia de Lamiidae (Coleoptera). Papéis Avulsos do Departamento de Zoologia 13 (26): 311–316. [15 Sep 1959 (top of article)]

Lansberge JW, van (1875a) Observations sur la classification des Lamellicornes coprophages. Annales de la Société Entomologique de Belgique 17 [1874] (2): 177–193. [31 Jan 1875 (wrapper)]

Lansberge JW, van (1875b) Monographie des Onitides. Annales de la Société Entomologique de Belgique 18 (1): 5–148. [17 Apr 1875 (wrapper)]

Laporte [=Castelnau] FLNC, de (1834a) Monographie du groupe des Rhipicérites (Coléoptères Pentamères). Annales de la Société Entomologique de France 3 (2): 225–270. [before 1 Sep 1834 (Jour. Proc. Ent. Soc. London 1: xxix)]

Laporte [=Castelnau] FLNC, de (1834b) Études entomologiques, ou description d'insectes nouveaux, et observations sur leur synonymie [livraison 1]. Méquignon-Marvis, Paris, pp. 1–94, pls. 1–2. [in press 2e trim. 1834 (Ann. Soc. Ent. France 3: Bull. Ent.: xxxvi); in press after 1 Jul 1834 (ibid., p. liv); 9 Aug 1834 (Bibliogr. France 1834); 5 Nov 1834 (Ann. Soc. Ent. France 3: Bull. Ent.: lvii)]

Laporte [=Castelnau] FLNC, de (1835a) Études entomologiques, ou description d'insectes nouveaux, et observations sur la synonymie; première partie. Carnassiers [livraison 2]. Méquignon-Marvis, Paris, pp. 95–159, pls. 3–4. [before 6 May 1835 (Rev. Ent. 3: 157, see 1835b); 3 Jun 1835 (Ann. Soc. Ent. France 4: Bull. Ent.: xliv)]

Laporte [=Castelnau] FLNC, de (1835b) Études entomologiques, ou descriptions d'insectes nouveaux et observations sur la synonymie [continued]. Revue Entomologique 3: 157–181. [6 May 1835 (as livr. 15–16; Ann. Soc. Ent. France 4: Bull. Ent.: xlii)]

Laporte [=Castelnau] FLNC, de (1836) Études entomologiques, ou descriptions d'insectes nouveaux et observations sur la synonymie [continued]. Revue Entomologique 4: 5–60, table. [1836 (title page); 7 Feb 1838 (Ann. Soc. Ent. France 7: Bull. Ent.: iv)]

Laporte [=Castelnau] FLNC, de (1840a) Histoire naturelle des insectes Coléoptères; avec une introduction renfermant l'anatomie et la physiologie des animaux articulés, par M. Brullé. Tome premier. Histoire naturelle des animaux articulés, annelides, crustacés, arachnides, myriapodes et insectes Tome troisième. P. Duménil, Paris, cxxv + 324 pp. + 19 pls. [26 Dec 1840 (Acad. Sci. France)]

Laporte [=Castelnau] FLNC, de (1840b) Histoire naturelle des insectes Coléoptères; avec une introduction renfermant l'anatomie et la physiologie des animaux articulés, par M. Brullé. Tome deuxième. Histoire naturelle des animaux articulés, annelides, crustacés, arachnides, myriapodes et insectes Tome troisième. P. Duménil, Paris, 563 + [1 (Tableau du placement des planches)] pp. + 38pls. [26 Dec 1840 (Acad. Sci. France)]

Lapouge G, de (1927) Tribu des Carabini. Tableaux des sous-tribus. Miscellanea Entomologica 30 (6): 45–48. [May-Jun 1927 (wrapper)]

Lassau SA, Hochuli DF, Cassis G, Reid CAM (2005) Effects of habitat complexity on forest beetle diversity: do functional groups respond consistently? Diversity and Distributions: a journal of conservation biogeography 11 (1): 73–82. [Jan 2005 (wrapper); 6 Feb 2005 (recorded at BMNH)]

Latreille PA (1797) Précis des caractères génériques des insectes, disposés dans un ordre naturel. F. Bourdeaux, Brive, xiv + 201+ [7] pp., 1 feuillet pliée. [before 13 Jan 1797 (Evenhuis 1997b: 437)]

Latreille PA (1802) Histoire naturelle, générale et particulière des crustacés et des insectes. Ouvrage faisant suite à l'histoire naturelle générale et particulière, composée par Leclerc de Buffon, et rédigée par C.S. Sonnini, membre de plusieurs sociétés savantes. Familles naturelles des genres. Tome troisième. F. Dufart, Paris, xii + 13–467 + [1] pp. [An X (title page, =1802); Nov 1802 (Evenhuis 1997)]

Latreille PA (1804a) Histoire naturelle, générale et particulière, des crustacés et des insectes. Ouvrage faisant suite aux oeuvres de Leclerc de Buffon, et partie du cours complet d'histoire naturelle rédigé par C. S. Sonnini, membre de plusieurs sociétés savantes. Tome neuvième. F. Dufart, Paris, 416 pp. [An XII (title page, =1804); betw. 19 Aug-17 Sep 1804 (Dupuis 1986: 209)]

Latreille PA (1804b) Histoire naturelle, générale et particulière, des crustacés et des insectes. Ouvrage faisant suite aux oeuvres de Leclerc de Buffon, et partie du cours complet d'histoire naturelle rédigé par C. S. Sonnini, membre de plusieurs sociétés savantes. Tome dixième. F. Dufart, Paris, 445 pp. [An XII (title page, =1804); betw. 19 Aug-17 Sep 1804 (Dupuis 1986: 209)]

Latreille PA (1804c) Tableau méthodique des insectes [pp. 129–200]. Nouveau dictionnaire d'histoire naturelle, appliquée aux arts, pricipalement à l'agriculture et à l'économie rurale et domestique: par une société de naturalistes et d'agriculteurs Tableaux méthodiques d'histoire naturelle Tome XXIV. Déterville, Paris, 84 + 85 + 238 + 18 + 34 pp. [7 Mar 1804 (Evenhuis 1997a: 198)]

Latreille PA (1804d) Histoire naturelle, générale et particulière, des crustacés et des insectes. Ouvrage faisant suite aux oeuvres de Leclerc de Buffon, et partie du cours complet d'histoire naturelle rédigé par C. S. Sonnini, membre de plusieurs sociétés savantes. Tome onzième. F. Dufart, Paris, 422 pp. + pls. 91–93. [An XII (title page, =1804); betw. 19 Aug-17 Sep 1804 (Dupuis 1986: 209)]

Latreille PA (1806) Genera crustaceorum et insectorum secundum ordinem naturalem in familias disposita, iconibus exemplisque plurimis explicata. Tomus secundus. Amand Koenig, Paris, 280 pp. [1807 (title page); Sep[?] 1806 (Jour. Gén. Litt. France 1806 9 (11): 321)]

Latreille PA (1807) Genera crustaceorum et insectorum secundum ordinem naturalem in familias disposita, iconibus exemplisque plurimis explicata. Tomus tertius. A. Koenig, Paris, 258 pp. [1807 (title page); Apr 1807 (Jour. Gén. Litt. France)]

Latreille PA (1810) Considérations générales sur l'ordre naturel des animaux composant les classes des crustacés, des arachnides, et des insectes; avec un tableau méthodique de leurs genres, disposés en familles. F. Schoell, Paris, 444 pp. [23 May 1810 (Evenhuis 1997b: 440)]

Latreille PA (1816) Les crustacés, les arachnides et les insectes. In: Cuvier G (Ed) Le règne animal distribué d'après son organisation, pour servir de base à l'histoire naturelle des animaux et d'introduction à l'anatomie comparée. Tome III. Déterville, Paris, xxix + 653 pp. [1817 (title page); 2 Dec 1816 (Roux 1976: 31)]

Latreille PA (1819) Trachélides [pp. 362–364]. Nouveau dictionnaire d'histoire naturelle, appliquée aux arts, à l'agriculture, à l'économie rurale et domestique, à la médecine, etc Par une société de naturalistes et d'agriculteurs Nouvelle édition Tome XXXIV. Déterville, Paris, [1] + 578 pp.

Latreille PA (1825) Familles naturelles du règne animal, exposées succinctement et dans un ordre analytique, avec l'indication de leurs genres. J.-B. Baillière, Paris, 570 pp. [pp. 337–352 numbered 237–252, in error]. [16 May 1825 (Evenhuis 1997)]

Latreille PA (1828) Rhynchophores ou porte-bec [pp. 584–603]. In: Bory de Saint-Vincent JBGM (Ed) Dictionnaire classique d'histoire naturelle. Tome quatorzième. Pla-Roy. Rey & Gravier; Baudouin Frères, Paris, 710 pp. [Sep 1828 (title page)]

Latreille PA (1829a) Tome IV. Crustacés, arachnides et partie des insectes. In: Cuvier G (Ed) Le règne animal distribué d'après son organisation, pour servir de base à l'histoire naturelle des animaux et d'introduction à l'anatomie comparée, avec figures dessinées d'après nature. Nouvelle édition, revue et augmentée. Déterville, Paris, xxvii + 584 pp. [1829 (title page); 11 Apr 1829 (Bibliogr. France 1829)]

Latreille PA (1829b) Tome V. Suite et fin des insectes. In: Cuvier G (Ed) Le règne animal distribué d'après son organisation, pour servir de base à l'histoire naturelle des animaux et d'introduction à l'anatomie comparée, avec figures dessinées d'après nature. Nouvelle édition, revue et augmentée. Déterville, Paris, xxiv + 556 pp., pls. 17–20. [1829 (title page); 11 Apr 1829 (Bibliogr. France 1829)]

Latreille PA (1834) Distribution méthodique et naturelle des genres de diverses tribus d'insectes coléoptères, de la famille des serricornes [posthumous]. Annales de la Société Entomologique de France 3 (1): 113–170. [before 1 Sep 1834 (Jour. Proc. Ent. Soc. London 1834: xxix)]

Laurent L (1966) Denticollinae, Pleonominae, Athoomorphinae de la région aethiopienne (Coleoptera, Elateridae). Bulletin de la Société Royale des Sciences de Liège 35 (11/12): 801–821. [presented 17 Nov 1966 (footer p. 801)]

Laurent L (1967) La sous-famille Tetralobinae (Coleoptera, Elateridae). Bulletin & Annales de la Société Royale d'Entomologie de Belgique 103 (3/4): 83–109. [31 May 1967 (wrapper)]

Laurent L (1974) Chapter II. Coleoptera: Elateridae [pp. 12–39]. In: Hanström B, Brinck P, Rudebeck G (Eds) South African animal life. Results of the Lund University expedition in 1950–1951. Vol. XV. Berlings, Lund, 532 pp. [1974 (vol. title page)]

Lawrence JF (1974a) The ciid beetles of California (Coleoptera: Ciidae). Bulletin of the California Insect Survey 17: 1–41. [22 Jan 1974]

Lawrence JF (1974b) The larva of *Sphindocis denticollis* Fall and a new subfamily of Ciidae (Coleoptera: Heteromera). Breviora No. 424: 1–14. [28 Jun 1974 (top of article)]

Lawrence JF (1977) The family Pterogeniidae, with notes on the phylogeny of the Heteromera. The Coleopterists Bulletin 31 (1): 25–56. [18 May 1977 (wrapper)]

Lawrence JF (1980) A new genus of Indo-Australian Gempylodini with notes on the constitution of the Colydiidae (Coleoptera). Journal of the Australian Entomological Society 19 (4): 293–310. [15 Dec 1980 (wrapper)]

Lawrence JF (1988) Rhinorhipidae, a new beetle family from Australia, with comments on the phylogeny of the Elateriformia. Invertebrate Taxonomy 2 (1): 1–53. [14 Oct 1988 (verso of vol. 3 contents)]

Lawrence JF (1991) Order Coleoptera: general discussion, family key, many family treatments [pp. 144–658]. In: Stehr FW (Ed) Immature insects. Vol 2. Kendall / Hunt Publishing Co., Dubuque, Iowa, xvi + 975 pp. [by Apr 1991]

Lawrence JF (1994) The larva of *Sirrhas variegatus*, sp.nov., with notes on the Perimylopidae, Ulodidae (stat. n.), Zopheridae and Chalcodryidae (Coleoptera: Tenebrionoidea). Invertebrate Taxonomy 8 (2): 329–349. [28 Jun 1994 (verso of vol. 9 (1) contents)]

Lawrence JF (2005) *Dasytomima*, a new genus of Australian Oedemeridae and its relationship to *Polypria* Chevrolat (Coleoptera: Tenebrionoidea). Annales Zoologici (Warszawa) 55 (4): 663–676. [31 Dec 2005]

Lawrence JF, Britton EB (1991) Coleoptera (beetles) [pp. 543–683]. The insects of Australia A textbook for students and research workers Second edition Volume II. Melbourne University Press, Melbourne, Australia, vi + 543–1137. [avail. by Dec 1991]

Lawrence JF, Newton AF, Jr. (1995) Families and subfamilies of Coleoptera (with selected genera, notes, references and data on family-group names) [pp. 779–1006]. In: Pakaluk J, Ślipiński SA (Eds) Biology, phylogeny and classification of Coleoptera: Papers celebrating the 80th birthday of Roy A Crowson. Vol II. Muzeum i Instytut Zoologii PAN, Warszawa, x + 1092 pp. in 2 vols. [31 Mar 1995 (teste S. A. Ślipiński)]

Lawrence JF, Pollock DA (1994) Relationships of the Australian genus *Synercticus* Newman (Coleoptera: Boridae). Journal of the Australian Entomological Society 33 (1): 35–42. [28 Feb 1994 (wrapper)]

Lawrence JF, Ślipiński SA, Elgueta DM (2010) Promecheilidae Lacordaire 1859 [pp. 563–567]. In: Part 38. Coleoptera, beetles. Volume 2: Morphology and systematics (Elateroidea, Bostrichiformia, Cucujiformia partim). In: Leschen RAB, Beutel RG, Lawrence JF, Ślipiński SA (Eds) Handbook of zoology. Volume IV. Arthropoda: Insecta. W. de Gruyter, New York and Berlin, xiii + 786 pp. [copyright 2010]

Lawrence JF, Stribling JB (1992) A new genus of Ptilodactylidae (Coleoptera: Elateriformia) from North Queensland, with description of the presumed larva. Journal of the Australian Entomological Society 31 (1): 19–27. [28 Feb 1992 (wrapper)]

Lawrence JF, Yoshitomi H (2007) *Nipponocyphon*, a new genus of Japanese Scirtidae (Coleoptera) and its phylogenetic significance. Elytra 35 (2): 507–527. [3 Nov 2007 (top of article)]

Lea AM (1910) Australian and Tasmanian Coleoptera inhabiting or resorting to the nests of ants, bees, and termites. Proceedings of the Royal Society of Victoria (New Series) 23 (1): 116–230 + pls. 25–27. [issued Aug 1910 (reprint wrapper)]

Lea AM (1925) On some new Australian Chrysomelidae (Coleoptera). Proceedings of the Royal Society of Victoria (new series) 37: 1–17. [issued May 1925 (reprint wrapper)]

Leach WE (1815) Entomology [pp. 57–172]. In: Brewster D (Ed). Brewster's Edinburgh Encyclopedia. Volume IX [part I]. W. Blackwood, J. Waugh, etc., Edinburgh, 764 pp. [Apr 1815 (Sherborn 1937b: 112)]

Leach WE (1817) The zoological miscellany; being descriptions of new, or interesting animals. Illustrated with coloured figures, drawn from nature, by R.P. Nodder. Vol. III. E. Nodder & Sons, London, v + [1] + 151 pp. + pls. 121–150. [Dec 1817 (Evenhuis 1997b: 443)]

Leach WE (1819) [new taxa] In: Samouelle G: The Entomologist's useful compendium; or an introduction to the knowledge of British insects, comprising the best means of obtaining and preserving them, and a description of the apparatus generally used; together with the genera of Linné, and the modern method of arranging the classes Crustacea, Myriapoda, spiders, mites, and insects from their affinities and structure, according to the views of Dr. Leach. Also an explanation of the terms used in entomology; a calendar of the times of appearance, and usual situations of near 3,000 species of British insects; with instructions for collecting and fitting up objects for the microscope. Thomas Boys, London, 496 pp., 12 pls. [May 1819 (Brit. Critic, N. S. 11: 559); Opinion 1722 (ICZN 1993a: 164): authorship was ascribed to Leach for those names attributed to him in Samouelle 1819, based on a note (p. 172)]

LeConte JE (1849) Coleopterous insects [pp. 25–36]. In: Catalogue of the fauna and flora of the State of Georgia [Appendix]. In: White G (Ed) Statistics of the State of Georgia. W. Thorne Williams, Savannah, 624 + 77 [Appendix] pp.

LeConte JL (1847) A descriptive catalogue of the geodephagous Coleoptera inhabiting the United States east of the Rocky Mountains. Annals of the Lyceum of Natural History of New York 4 [1846–1848] (8/9): 355–474 [255–374] [due to a printing error, pages after 233 are numbered 100 too high]. [nos. 8/9: Apr 1847 (wrappers)]

LeConte JL (1852a) Synopsis of the Parnidae of the United States. Proceedings of the Academy of Natural Sciences of Philadelphia 6 (1): 41–45. [Feb 1852 (page tops); before 14 Apr 1852 (Fox 1913: xi)]

LeConte JL (1852b) Remarks upon the Coccinellidae of the United States. Proceedings of the Academy of Natural Sciences of Philadelphia 6 (4): 129–145. [Aug 1852 (page tops); 1 Oct 1852 (Proc. Amer. Philos. Soc.)]

LeConte JL (1853a) Descriptions of twenty new species of Coleoptera inhabiting the United States. Proceedings of the Academy of Natural Sciences of Philadelphia 6 (7): 226–235. [Jan 1853 (page tops); 3 Apr 1853 (Fox 1913: xi)]

LeConte JL (1853b) Notes on the classification of the Carabidae of the United States. Transactions of the American Philosophical Society (New Series) 10: 363–403. [read 18 Mar 1853; 1853 (title page)]

LeConte JL (1856) Synopsis of the Melolonthidae of the United States. Journal of the Academy of Natural Sciences of Philadelphia (2) 3 (3): 225–288. [submitted 14 Oct 1856 (Proc. Acad. Nat. Sci. Philad. 1856: 215); publ. 18 Nov 1856 (Fox 1913: ix)]

LeConte JL (1859a) The Coleoptera of Kansas and eastern New Mexico. Smithsonian Contributions to Knowledge 11 (6): vi + 58 pp. [publ. Dec 1859 (p. 58)]

LeConte JL (1859b) Descriptions of some genera and species of Coleoptera from the vicinity of the southern boundary of the United States of America [livr. 3: pp. 121–128, pls. 12–13]. In: Thomson, J. (Ed) Arcana Naturae ou Recueil d'Histoire Naturelle. 132 pp. + 13 pls. [end of 1859, probably early 1860 (Hayek 1989: 93)]

LeConte JL (1861) Classification of the Coleoptera of North America. Part I. [pp. 1–208]. Smithsonian Miscellaneous Collections 3 (No. 136): xxiv + 214 pp. [May 1861 (vol. contents; p. 286); pp. 209–278 in Mar 1862]

LeConte JL (1862) Classification of the Coleoptera of North America. Part I (cont.). Smithsonian Miscellaneous Collections 3 (No. 136): 209–278. [Mar 1862 (vol. contents; p. 286; index [pp. 279–286] evidently published later)]

LeConte JL (1863) List of the Coleoptera of North America. Part. I. Smithsonian Miscellaneous Collections 6 (no. 140): 1–49. [Mar 1863 (advertisement after title page)]

LeConte JL (1866a) List of the Coleoptera of North America. Part. I [continued]. Smithsonian Miscellaneous Collections 6 (no. 140): 50–70. [Apr 1866 (advertisement after title page)]

LeConte JL (1866b) New species of North American Coleoptera. Prepared for the Smithsonian Institution. Part I. Smithsonian Miscellaneous Collections 6 (no. 167): 87–168. [Apr 1866 (advertisement after title page)]

LeConte JL (1867) Additions to the coleopterous fauna of the United States. No. 1. Proceedings of the Academy of Natural Sciences of Philadelphia 18 (5 [Dec 1866]): 361–394. [list of publ. dated 1 Jan 1867 (fasc. 5); 20 Jul 1867 (Fox 1913: xiii)]

LeConte JL (1869) List of Coleoptera collected in Vancouver's Island by Henry and Joseph Matthews, with descriptions of some new species. The Annals and Magazine of Natural History (4) 4 (24): 369–385. [1 Dec 1869 (Evenhuis 2003: 25)]

LeConte JL (1872) On Platypsyllidae, a new family of Coleoptera. Proceedings of the Scientific Meetings of the Zoological Society of London 1872: 799–804, pl. 68. [5 Nov 1872 (page tops)]

LeConte JL (1873) Classification of the Coleoptera of North America. Part II. Smithsonian Miscellaneous Collections 11 (no. 265): 279–348. [May-Jun 1873 (page footers)]

LeConte JL (1874a) Descriptions of new Coleoptera chiefly from the Pacific slope of North America. Transactions of the American Entomological Society 5 (1): 43–72. [Mar 1874 (footer p. 41)]

LeConte JL (1874b) The classification of the rhynchophorus Coleoptera. The American Naturalist 8 (8): 385–470. [Aug 1874 (p. 449 of issue); 26 Aug 1874 (Ann. Soc. Ent. France 1874: Bull. Ent.: clxx)]

LeConte JL (1876) [Pp. 1–12, 112–455] In: LeConte JL, Horn GH: The Rhynchophora of America North of Mexico. Proceedings of the American Philosophical Society held at Philadelphia for promoting useful knowledge 15 (96): i-xvi + 1–455. [23 Dec 1876 (date of preface p. ix); Dec 1876 (fasc. title page)]

LeConte JL (1881) Synopsis of the Lampyridae of the United States. Transactions of the American Entomological Society 9 [1881–82] (1): 15–72. [Feb, Mar, Apr, May, Jun 1881 (footers p. 9, 17, 33, 41, 65); 14 Oct 1881 (F. M. Brown 1964: 317)]

LeConte JL, Horn GH (1883) Classification of the Coleoptera of North America. Smithsonian Miscellaneous Collections 26 (4 [no. 507]): xxxviii + 567 pp. [Feb 1883 (date of advertisement after title page)]

Lecordier C (1977) Les Stagoninae d'Afrique noire (Col. Carabidae). 1re partie. Annales de la Société Entomologique de France (Nouvelle Série) 13 (4): 625–638. [3ème trim. 1977 (fasc. endleaf)]

Lee C-F, Satô M, Shepard WD, Jäch MA (2007) Phylogeny of Psephenidae (Coleoptera: Byrrhoidea) based on larval, pupal and adult characters. Systematic Entomology 32 (3): 502–538. [17 Apr 2007 (online); 18 Jul 2007 (vol. contents)]

Lefèvre É (1876) Descriptions d'Eumolpides nouveaux ou peu connus. Revue et Magasin de Zoologie Pure et Appliquée (3) 4 (8): 278–311. [before 15 Nov 1876 (Petites Nouv. Ent. 2: 88)]

Lefèvre É (1884) Communications [descriptions de trois genres nouveaux de Coléoptères de la famille des Eumolpides]. Bulletin des Séances de la Société Entomologique de France (6) 4 (2): lxv-lxvii. [8 Oct 1884 (Lefèvre 1895)]

Lefèvre É (1885) Eumolpidarum hucusque cognitarum catalogus, sectionum conspectu systematico, generum sicut et specierum nonnullarum novarum descriptionibus adjunctis. Mémoires de la Société Royale des Sciences de Liège (2) 11 (Nr. 16): 1–172. [Dec 1885 (vol. title page)]

Lefèvre É (1895) Annales de la Société Entomologique de France. Tables générales de 1881 à 1890 inclusivement. Paris, Société Entomologique de France, 239 + [1] pp. [1895 (title page)]

Legalov AA (2001) Revision der holarktischen Auletini (Coleoptera: Attelabidae). Reviziya golarkticheskich Auletini (Coleoptera: Attelabidae). Russian Entomological Journal 10 (1): 33–66. [23 May 2001 (journal website)]

Legalov AA (2003) Taxonomy, classification and phylogeny of the leaf-rolling weevils (Coleoptera: Rhynchitidae, Attelabidae) of the world fauna. Taksonomiya, klassifikatsiya i filogeniya rinkhitid i trubkovertov (Coleoptera: Rhynchitidae, Attelabidae) mirovoi fauny. Kapitel, Novosibirsk, 733 + 350 pp. [printed 13 Nov 2003]

Legalov AA (2007) Leaf-rolling weevils (Coleoptera, Rhynchitidae, Attelabidae) of the world fauna. Leaf-rolling weevils. Agro-Siberia, 523 pp.

Legalov AA (2009a) A review of fossil and recent species of the family Ithyceridae (Coleoptera) from the world fauna. Amurskii Zoologicheskii Zhurnal 1 (2): 117–131 + pls. 1–4.

Legalov AA (2009b) Annotated checklist of fossil and Recent species of the family Nemonychidae (Coleoptera) from the world fauna. Amurskii Zoologicheskii Zhurnal 1 (3): 200–213.

Legalov AA (2009c) Contribution to the knowledge of the Mesozoic Curculionoidea (Coleoptera). Amurskii Zoologicheskii Zhurnal 1 (4): 283–295.

Legalov AA (2009d) Annotated checklist of Recent and fossil species of the family Belidae (Coleoptera) from the world fauna. Amurskii Zoologicheskii Zhurnal 1 (4): 296–324, + pls. V-XI.

Leiler TE (1976) Zur Kenntnis der Entwicklungsstadien und der Lebensweise nord- und mitteleuropäischer Eucnemiden (Col.). Entomologische Blätter für Biologie und Systematik der Käfer 72 (1): 10–50. [25 Jul 1976 (wrapper)]

Leleup N (1969a) Position systématique et nouvelle diagnose du genre *Halorabyxis* Jeannel, endémique des Mascareignes. Définition de *H. vinsoni* Jeannel et description d'une espèce inédite (Coleoptera Pselaphidae Bythininae). Bulletin & Annales de la Société Royale d'Entomologie de Belgique 105 (7/8): 136–145. [15 Sep 1969 (wrapper)]

Leleup N (1969b) Une nouvelle tribu de Psélaphides (Col.) découverte par H. Franz à Madagascar. Bulletin & Annales de la Société Royale d'Entomologie de Belgique 105 (9/10): 281–288. [30 Nov 1969 (wrapper)]

Leleup N (1970) Contributions à l'étude des coléoptères psélaphides de l'Afrique. 7. Une nouvelle sous-tribu de Batrisini de l'Afrique intertropicale. Bulletin & Annales de la Société Royale d'Entomologie de Belgique 105 (11/12): 305–341. [28 Feb 1970 (wrapper)]

Leleup N (1973) Contributions à l'étude des coléoptères psélaphides de l'Afrique.13. Description de trois genres inédits et de quatorze espèces nouvelles des régions intertropicales, nouvelle définition des Ambicocerina et création de la tribu des Barrosellini. Bulletin & Annales de la Société Royale Belge d'Entomologie 109 (1/3): 43–91. [31 Mar 1973 (wrapper)]

Leng CW (1920) Catalogue of the Coleoptera of America, North of Mexico. John D. Sherman, Jr., Mount Vernon, New York, x + 470 pp. [1 Dec 1920 (date of preface)]

Leng CW, Mutchler AJ (1916) Descriptive catalogue of West Indian Cicindelinae. Bulletin of the American Museum of Natural History 35 (36): 681–699.

Leng CW, Mutchler AJ (1922) The Lycidae, Lampyridae and Cantharidae (Telephoridae) of the West Indies. Bulletin of the American Museum of Natural History 46: 413–499. [24 Aug 1922]

Lepesme P (1943) Un remarquable cérambycide nouveau de Guyane. Revue Française d'Entomologie 9 [1942] (3/4): 135–137. [15 Aug 1943 (p. 151)]

Lepesme P, Breuning S (1952) Note préliminaire sur la classification des coléoptères cérambycides. Transactions of the IXth International Congress of Entomology, Amsterdam, August 17–24, 1951 Volume I. Amsterdam, 139–142.

Lepesme P, Breuning S (1956) Une tribu nouvelle des Cerambycinae (Coleoptera Cerambycidae). Revue de Zoologie et de Botanique Africaines 53 (3/4): 287–304. [7 Jul 1956 (p. 523)]

Leschen RAB (1996) Phylogeny and revision of the genera of Cryptophagidae (Coleoptera: Cucujoidea). The University of Kansas Science Bulletin 55 (15): 549–634. [5 Jul 1996 (top of article)]

Leschen RAB (2003) Erotylidae (Insecta: Coleoptera: Cucujoidea): phylogeny and review. Fauna of New Zealand. Ko te Aitanga Pepeke o Aotearoa. Number 47. Manaaki Whenua Press, Lincoln, 108 pp. [5 Jun 2003]

Leschen RAB, Carlton CE (2004) A new tribe, genus and species of nitidulid beetle (Coleoptera: Nitidulidae: Nitidulinae) from Bolivia. The Coleopterists Bulletin 58: 443–451.

Leschen RAB, Beutel RG, Lawrence JF (Eds) (2010) Part 39. Coleoptera, beetles. Volume 2: Morphology and systematics (Elateroidea, Bostrichiformia, Cucujiformia partim). In: Kristensen NP, Beutel RG (Eds) Handbook of zoology. A natural history of the phyla of the animal kingdom. Volume IV - Arthropoda: Insecta. Walter De Gruyter, Berlin, xiii + 786 pp.

Lesne P (1896) Revision des Coléoptères de la famille des Bostrychides. Annales de la Société Entomologique de France 65 (1): 95–127, pls. 8–9. [before 9 Dec 1896 (Bull. Soc. Ent. France 1896: 409)]

Lesne P (1910) Revision des Coléoptères de la famille des Bostrychides. 6e mémoire: Dinapatinae et Apatinae. Annales de la Société Entomologique de France 78 [1909] (4): 471–574, pls. 13–17. [27 Apr 1910 (Annales vol. 79: 540)]

Lesne P (1921a) Les espèces typiques de *Trogoxylon*, (Col. Lyctidae). Position systématique de ce genre. Bulletin de la Société Entomologique de France 1921 (16): 228–231. [24 Nov 1921 (p. [323]))]

Lesne P (1921b) Classification des coléoptères xylophages de la famille des Bostrychides. Compte-rendu de la session Association française pour l'avancement des sciences 44: 285–289. [1921 (title page)]

Lesne P (1934) Note sur un Bostrychide Néo-Zélandais l'*Euderia squamosa* Broun (Coleopt.). Annales de la Société Entomologique de France 103 (3/4): 389–393. [31 Dec 1934 (wrapper)]

Léveillé A (1888) Catalogue de la famille des temnochilides. Annales de la Société Entomologique de France (6) 8 (4): 429–448. [10 Jun 1889 (wrapper)]

Levey B (1978) A new tribe, Epistomentini, of Buprestidae (Coleoptera) with a redefinition of the tribe Chrysochroini. Systematic Entomology 3 (2): 153–158. [Apr issue; 9 May 1978 (recorded at CNC)]

Levey B (1985) Afreminae: a new subfamily of Anthicidae (Coleoptera) from southern Africa. Entomologica Scandinavica 15 [1984] (3): 419–422. [31 Jan 1985 (top p. 299)]

Lewis G (1882) Synteliidae: a family to include *Syntelia* & *Sphaerites*, with a note of a new species of the first genus. The Entomologist's Monthly Magazine 19: 137–138. [Nov 1882 issue]

Lewis G (1894) On the Tenebrionidae of Japan. The Annals and Magazine of Natural History (6) 13 (77): 377–400, (78): 465–484, pl. 13. [1 May, 1 Jun 1894 (Evenhuis 2003: 33)]

Lichtenstein AAH (1796) Catalogus musei zoologici ditissimi Hamburgi, d. III. Februar 1796 auctionis lege distrahendi. Sectio Tertia. Continens Insecta. Schniebes, Hamburg, 13 + 224 pp.

Liebke M (1929) Laufkäfer-Studien VI. Entomologischer Anzeiger 9: 245–247, 261–265, 297–298.

Lindemann K (1877) Monographie der Borkenkaefer Russlands. Die cryphaloiden Tomiciden. Bulletin de la Société Impériale des Naturalistes de Moscou 51 (3 [1876]): 148–169. [before 25 Apr 1877 (Ann. Soc. Ent. France (5) 7 Bull. Bibliogr.: 17)]

Lindroth CH (1956) A revision of the genus *Synuchus* Gyllenhal (Coleoptera: Carabidae) in the widest sense, with notes on *Pristosia* Motschulsky (*Eucalathus* Bates) and *Calathus* Bonelli. The Transactions of the Royal Entomological Society of London 108 (11): 485–576. [31 Dec 1956 (wrapper)]

Lindroth CH (1968) The ground-beetles (Carabidae, excl. Cicindelinae) of Canada and Alaska. Part 5. Opuscula Entomologica Supplementum 33: 649–944.

Linsley EG (1940) A reclassification of the tribe Obriini of LeConte (Coleoptera, Cerambycidae). Journal of the New York Entomological Society 48 (4): 367–377. [Dec issue; 21 Dec 1940 (recorded at CNC)]

Linsley EG (1961) The Cerambycidae of North America. Part I. Introduction. University of California Publications in Entomology 18: i–vi, 1–97 + 35 pls. [issued 29 May 1961 (verso of title page)]

Linsley EG (1962) The Cerambycidae of North America. Part II. Taxonomy and classification of the Parandrinae, Prioninae, Spondylinae, and Aseminae. University of California Publications in Entomology 19: i-v, 1–102 + [1] + 1 pl. [issued 19 Jan 1962 (verso of title page)]

Linsley EG (1964) The Cerambycidae of North America. Part V. Taxonomy and classification of the subfamily Cerambycinae, tribes Callichromini through Ancylocerini. University of California Publications in Entomology 22: i-viii + 1–197 + 1 pl. [issued 19 Mar 1964 (verso of title page)]

Linzmeier AM, Konstantinov AS (2009) A new genus of flea beetles (Coleoptera: Chrysomelidae) from the south of Brazil. Proceedings of the Entomological Society of Washington 111 (3): 656–665.

Liu G (1935) Catalogue of the phytophagous beetles of China [continued]. Lingnan science journal 14 (2): 285–298.

Lobanov AL, Danilevsky ML, Murzin SV (1981) Sistematicheskiy spisok usachey (Coleoptera, Cerambycidae) fauny SSSR. I. [Systematic list of longicorn beetles (Coleoptera, Cerambycidae) of the USSR. I]. Entomologicheskoe Obozrenie 60 (4): 784–803 [in Russian: not translated in Entomological Review]. [after 10 Dec 1981 (approved to print, back wrapper); 13 Apr 1982 (recorded at CNC)]

Löbl I (2010) New acts and comments [p. 83]. In: Löbl I, Smetana A (Eds) Catalogue of Palaearctic Coleoptera. Volume 6. Chrysomeloidea. Apollo Books, Stenstrup, 924 pp. [publ. 22 Feb 2010 (verso of title page)]

Löbl I, Merkl O, Ando K, Bouchard P, Lillig M, Masomuto K, Schawaller W (2008) Family Tenebrionidae [in part, pp. 105–113, 120–127, 139–219, 238–241, 257, 276–277, 297–319, 339–352]. In: Löbl I, Smetana A (Eds) Catalogue of Palaearctic Coleoptera. Volume 5. Tenebrionoidea. Apollo Books, Stenstrup, 670 pp. [publ. 15 Apr 2008 (verso of title page)]

Löbl I, Silfverberg H (2010) New acts and comments. Chrysomelidae: Donaciinae [p. 64]. In: Löbl I, Smetana A (Eds) Catalogue of Palaearctic Coleoptera. Volume 6. Chrysomeloidea. Apollo Books, Stenstrup, 924 pp. [publ. 22 Feb 2010 (verso of title page)]

Löbl I, Smetana A (Eds) (2003) Catalogue of Palaearctic Coleoptera. Volume 1. Archostemata -Myxophaga - Adephaga. Apollo Books, Stenstrup, 819 pp. [publ. 30 May 2003 (verso of title page)]

Löbl I, Smetana A (Eds) (2004) Catalogue of Palaearctic Coleoptera. Volume 2. Hydrophiloidea - Histeroidea - Staphylinoidea. Apollo Books, Stenstrup, 942 pp. [publ. 31 Dec 2004 (verso of title page)]

Löbl I, Smetana A (Eds) (2006) Catalogue of Palaearctic Coleoptera. Volume 3. Scarabaeoidea - Scirtoidea - Dascilloidea - Buprestoidea - Byrrhoidea. Apollo Books, Stenstrup, 690 pp. [publ. 30 Jun 2006 (verso of title page)]

Löbl I, Smetana A (Eds) (2007) Catalogue of Palaearctic Coleoptera. Volume 4. Elateroidea - Derodontoidea - Bostrichoidea - Lymexyloidea - Cleroidea - Cucujoidea. Apollo Books, Stenstrup, Denmark, 935 pp. [30 Jun 2007 (verso of title page)]

Löbl I, Smetana A (2008) Errata [pp. 21–27]. In: Löbl I, Smetana A (Eds) Catalogue of Palaearctic Coleoptera. Volume 5. Tenebrionoidea. Apollo Books, Stenstrup, 670 pp. [publ. 15 Apr 2008 (verso of title page)]

Löbl I, Smetana A (Eds) (2010) Catalogue of Palaearctic Coleoptera. Volume 6. Chrysomeloidea. Apollo Books, Stenstrup, 924 pp. [publ. 22 Feb 2010 (verso of title page)]

Lohse GA (1964) Staphylinidae I (Micropeplinae bis Tachyporinae). In: Freude H, Harde KW, Lohse GA (Eds) Die Käfer Mitteleuropas. Band 4. Goecke & Evers, Krefeld, 264 pp. [1964 (copyright); 22 Nov 1964 (recorded by Smetana)]

Lohse GA (1969) Vorschläge zur Änderung der Aleocharinensystematik (Coleoptera: Staphylinidae) [pp. 169–175]. In: Bericht über die 10 Wanderversammlung deutscher Entomologen, 15 bis 19 September 1965 in Dresden Tagungsberichte Nr 80. Deutsche Akademie der Landwirtschaftswissenschaften zu Berlin, Berlin, 175 pp.

Lohse GA (1974) Hypocyphtinae, Aleocharinae [in part; pp. 7–72, 221–292]. In: Freude H, Harde KW, Lohse GA (Eds) Die Käfer Mitteleuropas Band 5 Staphylinidae II (Hypocyphtinae und Aleocharinae), Pselaphidae. Goecke & Evers, Krefeld, 381 pp. [1974 (copyright)]

Lohse GA (1989) 23. Familie: Staphylinidae [pp. 121–127, 129–240]. In: Lohse GA, Lucht WH (Eds) Die Käfer Mitteleuropas. Band 12. 1. Supplementband mit Katalogteil. Goecke & Evers, Krefeld, 346 pp. [1989 (copyright); 30 Jul 1989 (recorded by Smetana)]

Lopatin IK (1961) [New family of Coleoptera (Sogdiidae, fam. n.) from Tadzhikistan]. Izvestiya Akademii Nauk Tadzhikskoi SSR Otdelenie Biologicheskikh Nauk 1: 121–125 [in Russian].

Lopes-Andrade C (2008) An essay on the tribe Xylographellini (Coleoptera: Tenebrionoidea: Ciidae). Zootaxa 1832: 1–110. [publ. 30 Jul 2008 (footer p. 1)]

López-Pérez JJ (2006) Encontrar rápidamente la clasificación de los Cerambycidae del mundo (Coleoptera: Chrysomeloidea). Boletín de la Sociedad Andaluza de Entomología No. 13 [2005]: 44–66. [accepted 8 Mar 2006 (p. 66)]

Lord NP, Hartley CS, Lawrence JV, McHugh MFW, Miller KB (2010) Phylogenetic analysis of the minute brown scavenger beetles (Coleoptera: Latridiidae), and recognition of a new beetle family, Akalyptoischiidae fam.n. (Coleoptera: Cucujoidea). Systematic Entomology 35 (4): 753–763. [28 Jun 2010 (online); Oct 2010 issue]

Lorenz W (1998a) Nomina carabidarum - a directory of the scientific names of ground beetles. (Insecta, Coleoptera «Geadephaga»: Trachypachidae and Carabidae incl. Paussinae, Cicindelinae, Rhysodinae). First edition. W. Lorenz, Tutzing, Germany, iv + 937 pp.

Lorenz W (1998b) Systematic list of extant ground beetles of the world (Insecta Coleoptera "Geadephaga": Trachypachidae and Carabidae incl. Paussinae, Cicindelinae, Rhysodinae). First edition. W. Lorenz, Tutzing, Germany, [ii] + 502 pp.

Lorenz W (2005) Systematic list of extant ground beetles of the world (Insecta Coleoptera «Geadephaga»: Trachypachidae and Carabidae incl. Paussinae, Cicindelinae, Rhysodinae). Second edition. W. Lorenz, Tutzing, Germany, 530 pp.

Louw S, Oberprieler RG (1998) [new taxon] In: Louw S: Solving the riddle: combining life-history analysis and morphological comparison in weevil phylogenetics. In: Colonnelli E, Louw S, Osella G (Eds) Taxonomy, ecology and distribution of Curculionoidea (Coleoptera: Polyphaga). Proceedings of a Symposium (28 August, 1996, Florence, Italy). XX International Congress of Entomology. Atti del Museo Regionale di Scienze Naturali, Torino: 19–26.

Lucas R (1920) Catalogus alphabeticus generum et subgenerum Coleopterorum orbis terrarum totius (famil., trib., subtr., sect. incl.). Pars I. Archiv für Naturgeschichte (Abteilung A) 84 [1918] (1/5): i-xxxi, 1–696 [also as separate, 1920, Nicolai, Berlin, same pagin.]. [Jan, Feb 1920]

Luna de Carvalho E (1951) Contribution pour un nouveau catalogue de la famille des Paussides (Col. Carab. Isochaeta). Memorias e Estudos do Museu Zoologico da Universidade de Coimbra 207: 1–55 + 2 pls.

Luna de Carvalho E (1961) Révision des Paussides appartenant aux tribus des Pentaplatarthrini et Hylotorini (2e contribution à l'étude monographique des Paussides) (Col. Carabidae Paussinae). Revue de Zoologie et de Botanique Africaines 63 (1/2): 1–19. [30 Mar 1961 (p. 392)]

Luna de Carvalho E (1989) Essai monographique des coléoptères protopaussines et paussines contenant des descriptions et iconographie des taxa actuelles et fossiles avec des clefs dichotomiques de toutes les espèces. Memórias do Instituto de Investigação Científica Tropical (2) 70 [1987]: 1–1026. [printed May 1989 (endleaf)]

Lyal CHC, Alonso-Zarazaga MA (2010) Case 3529. *Otiorhynchus* Germar, 1824 and *Loborhynchus* Schoenherr, 1823 (Insecta, Coleoptera): proposed emendation of entries on the Official List of Generic Names in Zoology. Bulletin of Zoological Nomenclature 67 (4): 289–291.

Lynch Arribálzaga F (1878) Notas sobre dos Athyreitae de Buenos Aires. Naturalista Argentino: El Revista de Historia Natural 1 (5): 145–149.

Lynch Arribálzaga F (1884) Los Estafilinos de Buenos Aires. Boletín de la Academia Nacional de Ciencias en Córdoba 7 (1): 5–256, (3): 257–392. [12 Apr 1884 (date of Introduction); 1884 (title pages of both fasc.); by Dec 1884 (Biblioth. Hist.-nat. 34 (2): 149)]

Lyubarsky GY (1998) Phylogenetics of the beetles of the family Cryptophagidae: a cladistic analysis. Zoologicheskie Issledovania 1: 1–92 [in Russian].

MacDonald WB (1845) [translation of Erichson's review of Mulsant, 1842] In: Report on Zoology, MDCCCXLII [pp. 1–348]. In: Reports on the progress of zoology and botany 1841, 1842. The Ray Society, Edinburgh, viii + 43 + 348 + 104 pp.

Machado A (1992) Monografía de los carábidos de las Islas Canarias (Insecta, Coleoptera). Instituto de Estudios Canarios, La Laguna, 734 pp. [1992 (title page)]

Machatschke JW (1959) Phylogenetische Untersuchungen über die Sericini (sensu Dalla Torre 1912) (Coleoptera: Lamellicornia, Melolonthidae). Beiträge zur Entomologie 9 (7/8): 730–746. [Dec 1959 (top p. 713); recorded at CNC 19 Jul 1960]

MacLeay WJ (1866) New species of Amycteridae. The Transactions of the Entomological Society of New South Wales 1 (5): 319–340. [6 Aug 1866 (read to Society); 1866 (wrapper)]

MacLeay WJ (1887) Miscellanea Entomologica, No. III. The Scaritidae of New Holland. The Proceedings of the Linnean Society of New South Wales (2) 2 [1888] (1): 115–134. [issued 18 May 1887 (vol. 2 contents p. iii)]

MacLeay WS (1819) Horae entomologicae: or essays on the annulose animals. Vol. I. Part I. Containing general observations on the geography, manners, and natural affinities of the insects which compose the genus *Scarabaeus* of Linnaeus; to which are added a few incidental remarks on the genera *Lucanus* and *Hister* of the same author. With an appendix and plates. S. Bagster, London, xxx + [2] + 160 pp., 3 pls. [Nov 1819 (Evenhuis 1997)]

MacLeay WS (1825) Annulosa Javanica, or an attempt to illustrate the natural affinities and analogies of the insects collected in Java by Thomas Horsfield, M.D. F.L. & G.S. and deposited by him in the museum of the honourable East-India Company. Number 1. Kingsbury, Parbury, and Allen, London, xii + 50 pp., 1 pl. [July 1825 (Evenhuis 1997b: 507)]

Macnamara JP (1971) Emendation of taxa of Micronesian Oedemeridae (Coleoptera). Entomological News 82 (6): 164. [publ. 28 Jun 1971 (verso of vol. title page)]

MacSwain JW (1956) A classification of the first instar larvae of the Meloidae (Coleoptera). University of California Publications in Entomology (University of California Press) 12: 1–182 + pls. 1–30. [issued 10 Jul 1956 (verso of title page)]

Mader L (1934) Evidenz der paläarktischen Coccinelliden und ihrer Aberrationen in Wort und Bild. I. Teil: Epilachnini, Coccinellini, Halyziini, Synonychini. Zeitschrift des Vereines der Naturbeobachter, Wien 1924 (Beilag): 289–328. [Apr, May, Aug, Sep 1924 (pp. 289–335 signature footers); but publ. later in Zeitschr. Ver. Naturb. (appearance in Zoo. Rec.); 412 pp. + 64 pls. total]

Mader L (1954) Coccinellidae. III. Teil. Exploration du Parc National Albert Mission G F de Witte (1933–1935). Fasc. 80. Institut des Parcs Nationaux du Congo Belge Bruxelles, 206 pp.

Madge RB (1988) The publication dates of Dejean's catalogues. Archives of Natural History 15: 317–321.

Madge RB (1989) A catalogue of the family-group names in the Geodephaga, 1758–1985 (Coleoptera: Carabidae s. lat.). Entomologica Scandinavica 19: 459–474. [4 May 1989]

Maehler, FJ (1850) Enumeratio coleopterorum circa Heidelbergam indigenarum adjectia synonymis locisque natalibus. Mohr, Heidelbergae. viii + 116 pp.

Maes JM (1992a) Lista de los Lucanidae (Coleoptera) del mundo. Revista Nicaragüense de Entomología 22A: 1–60. [Dec 1992 (Zool. Record)]

Maes JM (1992b) Lista de los Lucanidae (Coleoptera) del mundo. Revista Nicaragüense de Entomología 22B: 61–121. [Dec 1992 (Zool. Record)]

Majer K (1987) Comparative morphology and proposed major taxonomy of the family Melyridae (Insecta, Coleoptera). Polskie Pismo Entomologiczne 56 [1986] (4): 719–859. [28 Feb 1987 (top of article); printed Mar 1987 (verso of title page)]

Majer K (1990) A new tribe Listrini trib. n., including two new genera (Coleoptera, Melyridae). Acta Entomologica Bohemoslovaca 87 (5): 368–384. [25 Oct 1990 (inside wrapper)]

Majer K (1995a) A review of the classification of the Melyridae and related families (Coleoptera, Cleroidea). Entomologica Basiliensia 17 [1994]: 319–390. [30 Apr 1995 (p. 3)]

Majer K (1995b) Revision of the family Mauroniscidae (Insecta: Coleoptera: Cleroidea). Entomologische Abhandlungen (Dresden) 57 (3): 57–89. [1 Dec 1995 (top of article)]

Majer K (2000) [new taxa] In: Kubáň V, Majer K, Kolibáč J: Classification of the tribe Coraebini Bedel, 1921 (Coleoptera, Buprestidae, Agrilinae). Acta Musei Moraviae Časopis Moravského Musea Scientiae Biologicae 85 (2): 185–287. [22 Jun 2001 (recorded at BMNH)]

Majer K (2002) Subfamilial classification of the family Malachiidae (Coleoptera, Cleroidea). Entomologica Basiliensia 24: 179–244. [24 Sep 2002 (Inhalt)]

Mamaev BM (1976) [Morphological types of xylophagous beetle larvae (Coleoptera, Eucnemidae) and their evolutionary importance] [pp. 136–155]. In: Mamaev BM (Ed) [Evolutionary morphology of woodboring larvae]. Nauka, Moscow, 202 pp. [in Russian].

Mamaev BM, Danilevsky ML (1973) Novye dannye po sisteme podsemeistva Aseminae (Coleoptera, Cerambycidae) v svyazi s osobennostyami stroeniya lichinok. [New data on systematic status of the subfamily Aseminae (Coleoptera, Cerambycidae) with reference to the morphology of larvae]. Zoologicheskii Zhurnal 52 (8): 1257–1261 [in Russian]. [after 3 Aug 1973 (approved to print, endleaf); 6 Nov (recorded at CNC)]

Mandelshtam MY, Beaver RA (2003) *Amphiscolytus* - a new genus, and Amphiscolytini - a new tribe of Scolytidae (Coleoptera) for *Dacryophthorus capensis* Schedl. Zootaxa 298: 1–8. [publ. 16 Sep 2003 (footer p. 1)]

Mann WM (1921) Three new myrmecophilous beetles. Proceedings of the United States National Museum 59: 547–552. [publ. 5 Oct 1921 (p. v)]

Mann WM (1924) Myrmecophiles from the western United States and Lower California. Annals of the Entomological Society of America 17 (1): 87–95. [mailed 18 Apr 1924 (p. 476)]

Marinoni RC, Napp DS (1984) Thyrsiini, uma nova tribo para Cerambycinae (Coleoptera, Cerambycidae). Revista Brasileira de Entomologia 28 (1): 39–49. [22 Feb 1984 (wrapper)]

Marseul S-A, de (1857a) Catalogue des coléoptères d'Europe de Marseul, Paris, xvi + 200 pp. [1 Mar 1857 (date of preface); 24 Jun 1857 (Ann. Soc. Ent. France (3) 5: Bull. Ent.: lxxxiii); 26 Sep 1857 (Bibliogr. France 1857)]

Marseul S-A, de (1857b) Essai monographique sur la famille des Histérides (Suite). Annales de la Société Entomologique de France (3) 5 (1): 109–167, (2): 397–428, (3): 429–516, pls. 10–11. [(1): 24 Jun; (2): 9 Sep; (3): 14 Nov 1857 (wrappers)]

Marseul S-A, de (1863) Catalogue des coléoptères d'Europe et du Bassin de la Méditerranée en Afrique & en Asie. Deuxième édition. A. Deyrolle, Paris, 300 pp. [15 Jun 1863 ("Principales abréviations...")]

Marseul S-A, de (1866) Catalogus Coleopterorum Europae et confinium. Friedlander & Sohn, Berlin, [12] + 131 pp. [1866 (date of avertissement); 1866 (Zool. Record)]

Marseul S-A, de (1882) Les entomologistes et leurs écrits. L'Abeille, Mémoires d'Entomologie 20: 20–39.

Marshall GAK (1907) A revision of the genus Synthocus, Schönh., and its allies (Curculionidae). Transactions of the South African Philosophical Society 18 (1): 89–120, pl. 6. [Oct 1907 (contents)]
Marshall GAK (1916) Coleoptera. Rhynchophora: Curculionidae. In: Shipley AE (Ed) The fauna of British India, including Ceylon and Burma. Taylor & Francis, London, xv + 367 pp. [1916 (title page); Oct 1916 (date of preface)]
Marshall GAK (1917) On new species of Indian Curculionidae.- Part III. The Annals and Magazine of Natural History (8) 19 (110): 188–198. [1 Feb 1917 (Evenhuis 2003: 41)]
Marshall GAK (1920) On new species of Curculionidae from Africa. The Annals and Magazine of Natural History (9) 6 (34): 369–398, pl. 13. [1 Oct 1920 (Evenhuis 2003: 42)]
Marshall GAK (1932) Notes on the Hylobiinae (Col., Curc.). The Annals and Magazine of Natural History (10) 9 (52): 341–355. [1 Apr 1932 (Evenhuis 2003: 46)]
Marshall GAK (1937a) New Curculionidae (Col.) from New Zealand. Transactions and proceedings of the Royal Society of New Zealand 67 (3): 316–340 + pl. 45. [issued Dec 1937 (fasc. title page)]
Marshall GAK (1937b) On some Oriental Cossoninae (Col. Curc.). Proceedings of the Royal Entomological Society of London (Series B, Taxonomy) 6 (3): 54–58. [15 Mar 1937 (wrapper)]
Marshall GAK (1939) New tropical African Curculionidae (Col.). The Annals and Magazine of Natural History (11) 3 (18): 561–583. [1 Jun 1939 (Evenhuis 2003: 48)]
Marshall GAK (1942) On some East African Otiorrhynchinae (Col., Curcul.). The Annals and Magazine of Natural History (11) 9 (49): 1–26. [1 Jan 1942 (Evenhuis 2003: 49)]
Marshall GAK (1944) On the genera of the tribe Cyphicerini (Col., Curc.). I. The Annals and Magazine of Natural History (11) 11 (74): 73–98. [18 Mar 1944 (Evenhuis 2003: 50)]
Marshall GAK (1952) Taxonomic notes on Curculionidae (Col.). The Annals and Magazine of Natural History (12) 5 (51): 261–270. [1 Mar 1952 (Evenhuis 2003: 52)]
Marshall GAK (1953) On a collection of Curculionidae (Coleoptera) from Angola. Publicações Culturais da Companhia de Diamantes de Angola 16: 99–119. [separate publ. 3 Sep 1953 (title page of article)]
Marshall GAK (1956) The Otiorrhynchine Curculionidae of the tribe Celeuthetini (Col.). Trustees of the British Museum, London, 134 pp. [issued Jan 1956 (title page)]
Marshall GAK (1957) A new subfamily of Curculionidae (Coleoptera). Proceedings of the Royal Entomological Society of London (Series B. Taxonomy) 26 (1/2): 17–20. [26 Feb 1957 (wrapper)]
Marshall TA (1865) Corynodinorum recensio. Journal of the Proceedings of the Linnean Society of London 8 (1): 24–50.
Martínez A (1968) Insectos nuevos o poco conocidos XIII. Ceratocanthini nom. n. para Acanthocerini (Coleoptera, Scarabaeidae, Troginae). Revista de la Sociedad Entomológica Argentina 30 [1967] (1/4): 9–16. [15 Feb 1968 (footer p. 9)]
Martínez A, Viana MJ (1964) Una nueva subfamilia de coleopteros (Ptinidae; Fabiinae). Neotropica 10 (31): 7–14. [14 Jul 1964]

Martins UR (1976) Sistemática e evolução da tribo Piezocerini (Coleoptera, Cerambycidae). Arquivos de Zoologia, São Paulo 27 ((3/4)): 165–370 + 8 pls. [28 May 1976 (top of article); 1 Jun 1976 (recorded at BMNH)]

Martins UR (1997a) Tribo Oemini [pp. 5–155]. In: Martins UR (Ed). Cerambycidae Sul-Americanos (Coleoptera). Taxonomia. Volume 1. Sociedade Brasileira de Entomologia, São Paulo, [5] + 217 pp. [May 1997 (date of Preface)]

Martins UR (1997b) Tribo Paraholopterini [pp. 201–207]. In: Martins UR (Ed). Cerambycidae Sul-Americanos (Coleoptera). Taxonomia. Volume 1. Sociedade Brasileira de Entomologia, São Paulo, [5] + 217 pp. [May 1997 (date of Preface)]

Martins UR (1998) Tribo Ectenessini [pp. 81–182]. In: Martins UR (Ed). Cerambycidae Sul-Americanos (Coleoptera). Taxonomia. Volume 2. Sociedade Brasileira de Entomologia, São Paulo, vi + 195 pp.

Martins UR (2003a) Tribo Sydacini [pp. 203–213]. In: Martins UR (Ed) Cerambycidae Sul-Americanos (Coleoptera). Taxonomia. Volume 6. Sociedade Brasileira de Entomologia, São Paulo, vii + 232 pp. [Dec 2003 (title page)]

Martins UR (2003b) Tribo Luscosmodicini [pp. 29–33]. In: Martins UR (Ed). Cerambycidae Sul-Americanos (Coleoptera). Taxonomia. Volume 6. Sociedade Brasileira de Entomologia, São Paulo, vii + 232 pp. [Dec 2003 (title page)]

Martins UR (2005) Tribe Neocorini [pp. 239–270]. In: Martins UR (Ed). Cerambycidae (Coleoptera) Sul-Americanos. Taxonomia. Volume 5. Editora da Universidade de São Paulo, São Paulo, v + 284 pp.

Martins UR (2006) Tribo Hexoplonini [pp. 21–211]. In: Martins UR (Ed). Cerambycidae Sul-Americanos (Coleoptera). Taxonomia. Volume 8. Sociedade Brasileira de Entomologia, São Paulo, ii + 234 pp. [Jan 2006 (title page)]

Martins UR, Carvalho SM (1984) Considerações sobre a classificação da tribo Methiini com a revalidação de Xystrocerini Blanchard, 1845 e Oemini Pascoe, 1869 (Coleoptera, Cerambycidae). Papéis Avulsos de Zoologia 35 (20): 209–224. [30 Nov 1984 (top of article)]

Martins UR, Galileo MHM (1990) Lamiinae (Coleoptera, Cerambycidae) com garras tarsais apendiculadas e descrição de Pretiliini, trib. n. Revista Brasileira de Entomologia 34 (4): 703–708. [30 Dec 1990 (top of article)]

Martins UR, Galileo MHM (2003) Tribo Oxycoleini [pp. 51–63]. In: Martins UR (Ed). Cerambycidae Sul-Americanos (Coleoptera). Taxonomia. Volume 6. Sociedade Brasileira de Entomologia, São Paulo, vii + 232 pp. [Dec 2003 (title page)]

Martins UR, Galileo MHM (2007) Tribo Ibidionini subtribo Tropidina [pp. 1–176]. In: Martins UR (Ed). Cerambycidae Sul-Americanos (Coleoptera). Taxonomia. Volume 9. Sociedade Brasileira de Entomologia, São Paulo, vii + 232 pp. [Nov 2007 (title page)]

Martins UR, Monné MA (2005) Tribo Cerambycini [pp. 1–218]. In: Martins UR (Ed). Cerambycidae (Coleoptera) Sul-Americanos. Taxonomia. Volume 5. Editora da Universidade de São Paulo, São Paulo, v + 284 pp.

Martins UR, Napp DS (2009) Tribo Ideratini [215–221]. In: Martins UR (Ed) Cerambycidae Sul-Americanos (Coleoptera). Taxonomia. Volume 10. Sociedade Brasileira de Entomologia, São Paulo. vi + 373 pp.

Martynov AV (1933) Permskie iskopaemye nasekomye Arkhangel'skogo Kraya. Chast' II. Setchatokrylye, vislokrylye i zhuki, s prislozheniem opisaniya dvukh novykh zhukov iz Tikhikh Gor [Permian fossil insects from the Arkhangelsk District. Part II. Neuroptera, Megaloptera and Coleoptera with the description of two new beetles from Tikhie Gory] [in Russian]. Trudy Paleontologicheskogo Instituta Akademiya Nauk SSSR 2: 63–96. [after 17 Jun 1933 (approved to print); 21 Sep 1933 (recorded at NRCL)]

Martynov AV (1937) Permian fossil insects from Kargala and their relationships Trudy Paleontologicheskogo Instituta Akademiya Nauk SSSR 7 (2): 1–92 [in Russian]. [after 27 Jul 1937 (approved to print)]

Marvaldi AE, Oberprieler RG (2006) [new taxa] In: Marvaldi AE, Oberprieler RG, Lyal CHC, Bradbury T, Anderson RS: Phylogeny of the Oxycoryninae sensu lato (Coleoptera: Belidae) and evolution of host-plant associations. Invertebrate Systematics 20 (4): 447–476. [Aug 2006]

Mateu J (1963) Notas sobre tres séries filéticas de Lebiidae (Lichnasthenini Thomson, Singilini Jeannel, Somotrichini nov.) (Coleoptera-Carabidae) y rectificaciones sinonìmicas. Annali del Museo Civico di Storia Naturale "Giacomo Doria" 74 [1963–64]: 122–139. [separate publ. 1 Oct 1963 (vol. contents)]

Mateu J (1992) Du nouveau sur les subtribus de Zuphiini et description d'un nouveau genre et d'une nouvelle espèce de cette tribu. (Coleoptera, Carabidae, Zuphiini). Entomologica Basiliensia 15: 195–204. [31 Dec 1992 (Inhalt)]

Mateu J (2002) Sur un genre nouveau et une espèce cavernicole inédite appartenant à une nouvelle sous-famille de Coléoptères Carabiques Promecognathidae. Revue Française d'Entomologie (Nouvelle Série) 24 (1): 67. [publ. 15 Mar 2002 (inside back wrapper)]

Matsushita M (1933) Beitrag zur Kenntnis der Cerambyciden des Japanischen Reichs. Journal of the Faculty of Agriculture of the Hokkaido Imperial University 34 (2): 157–445. [30 Dec 1933]

Matthews A (1888) Corylophidae [pp. 102–125, pl. 3]. In: Godman FD, Salvin O (Eds) Biologia Centrali-Americana. Insecta. Coleoptera. Vol. II. Part 1. Taylor & Francis, London, xii + 717 pp., 19 pls. [Jan 1888 (footer p. 97)]

Matthews A (1899) A monograph of the coleopterous families Corylophidae and Sphaeriidae. O. E. Janson & Son, London, [8] + 220 + [1] pp. + 9 pls. [posthumous; edited by Philip B. Mason]. [Nov 1899 (date of preface)]

Matthews EG (2003a) The Palorus group - a new subfamily of Tenebrionidae (Insecta, Coleoptera). Spixiana 26 (1): 49–50. [1 Mar 2003 (wrapper)]

Matthews EG (2003b) *Ulomotypus* Broun a member of the new subfamily Palorinae, with remarks on *Aphtora* Bates and *Demtrius* Broun (Coleoptera, Tenebrionidae). New Zealand Entomologist 26: 7–14. [Dec 2003 (contents); Sep 2004 (recorded at CNC)]

Matthews EG, Lawrence JF (2005) New taxa, new synonymy and new generic records for Australian Tenebrionidae (Coleoptera). Annales Zoologici (Warszawa) 55 (4): 531–547. [31 Dec 2005 (verso of fasc. title page)]

Matthews EG, Lawrence JF, Bouchard P, Steiner WE, Ślipiński SA (2010) Tenebrionidae Latreille, 1802 [pp. 574–659]. In: Leschen RAB, Beutel RG, Lawrence JF (Eds) Handbook of Zoology. Coleoptera, Beetles. Volume 2: Morphology and systematics (Elateroidea, Bostrichiformia partim). De Gruyter, Berlin, New York, xiii + 786 pp. [2010 (copyright)]

Matthews EG, Stebnicka ZT (1986) A review of *Dermarziella* Balthasar, with a transfer from Aphodiinae to Scarabaeinae (Coleoptera: Scarabaeidae). Australian Journal of Zoology 34 (3): 449–461. [3 Jun 1986 (contents for issue 4)]

Maulik S (1929) Injurious Hispinae from the Solomon Islands. Bulletin of Entomological Research 20 (2): 233–239. [Aug 1929 (wrapper)]

May BM (1993) Larvae of Curculionoidea (Insecta: Coleoptera): a systematic overview. Fauna of New Zealand. Ko te Aitanga Pepeke o Aotearoa. Number 28. Manaaki Whenua Press, Lincoln, 223 pp. [Jun 1993 (p. 225)]

Mayor A (2007) family Malachiidae Fleming, 1821 [pp. 415–454] [and others]. In: Löbl I, Smetana A (Eds) Catalogue of Palaearctic Coleoptera. Volume 4. Elateroidea - Derodontoidea - Bostrichoidea - Lymexyloidea - Cleroidea - Cucujoidea. Apollo Books, Stenstrup, 935 pp. [publ. 30 Jun 2007 (verso of title page)]

Mazur S (1973) Sphaeritidae i Gniliki - Histeridae. In: Klucze do oznaczania owadów Polski. Czesc XIX. Chrzaszcze - Coleoptera. Zeszyt 11–12. Panstwowe Wydawnictwo Naukowe, Warszawa, 74 pp.

Mazur S (2004) Family Histeridae [pp. 68–102]. In: Löbl I, Smetana A (Eds) Catalogue of Palaearctic Coleoptera. Volume 2. Hydrophiloidea - Histeroidea - Staphylinoidea. Apollo Books, Stenstrup, 942 pp. [publ. 31 Dec 2004 (verso of title page)]

McDermott FA (1964) The taxonomy of the Lampyridae (Coleoptera). Transactions of the American Entomological Society 90 (1): 1–72. [issued 27 Mar 1964 (inside back wrapper)]

McKay IJ (1991) Cretaceous Promecognathinae (Coleoptera: Carabidae): a new genus, phylogenetic reconstruction and zoogeography. Biological Journal of the Linnean Society 44 (1): 1–12. [Sep 1991 issue]

McKenna MC, Bell SK (1997) Classification of mammals above the species level. Columbia University Press, New York, xii + 631 pp. [1997 (Copyright)]

McKeown KC (1939) A note on the synonymy of *Leptops* (Coleoptera: Curculionidae). The Proceedings of the Linnean Society of New South Wales 64: 408. [1939 (vol. title page)]

McKeown KC (1947) Catalogue of the Cerambycidae (Coleoptera) of Australia. Memoirs of the Australian Museum No. 10: 1–190. [2 May 1947 (title page)]

Medvedev GS (1962) Novoe podsemeystvo chernotelok (Coleoptera, Tenebrionidae) iz Turkmenii. Zoologicheskii Zhurnal 41 (8): 1184–1189. [Aug 1962 (inside wrapper)]

Medvedev GS (1968) Zhestkokrylye Tom XIX, vypusk 2. Zhuki-Chernotelki (Tenebrionidae). Podsemeysvo Opatrinae, triby Platynotini, Dendarini, Pedinini, Dissonomini, Pachypterini, Opatrini (chast') i Heterotarsini. [Coleoptera. Darkling beetles (Tenebrionidae). Subfamily Opatrinae]. Fauna SSSR Novaya Seriya. No. 97. Nauka, Leningrad, 284 + [2] pp. [in Russian]. [after 20 Jun 1968 (approved to print, endleaf)]

Medvedev GS (1973) Polozhenie rodov *Leichenum* Dej. i *Idisia* Pask. (Coleoptera, Tenebrionidae) v sisteme i opisanie novogo roda iz severnykh Karakumov[Position of the genera *Leichenum* Dej. and *Idisia* Pasc. (Coleoptera, Tenebrionidae) in the system and a description of a new genus from northern Karakums]. Entomologicheskoe Obozrenie 52 (3):

644–650 [in Russian; English translation in Entomological Review 52: 428–433]. [after 18 Sep 1973 (approved to print, back wrapper)]

Medvedev GS (2001) Evolyutsiya i sistema zhukov-chernotelok triby Blaptini (Coleoptera, Tenebrionidae) [Evolution and system of darkling beetles of the tribe Blaptini (Coleoptera, Tenebrionidae)]. Chteniya pamyati Nikolaya Aleksandrovicha Kholodkovskogo [Readings in memory of N A Kholodkovskii]. Vyp. 53. Russkoe Entomologicheskoe Obshchestvo, St. Petersburg, 332 pp. [in Russian]. [after 29 Dec 2000 (approved to print, endleaf); 2001 (title page)]

Medvedev GS (2006) K sistematike i nomenklature zhukov-chernotelok trib Phalerini, Lachnogyini, Klewariini i Blaptini (Coleoptera, Tenebrionidae) [To the systematics and nomenclature of the tenebrionid-beetle tribes Phaleriini, Lachnogyini, Klewariini Nlaptini (Coleoptera, Tenebrionidae)]. Entomologicheskoe Obozrenie 85 (3): 555–579, 715 [in Russian]. [after 31 Aug 2006 (approved to print, back wrapper)]

Medvedev GS, Lawrence JF (1982) In: Lawrence JF, Medvedev GS: Novaya triba zhukov-chernotelok (Coleoptera, Tenebrionidae) iz Avstralii i ee polozhenie v sisteme [A new tribe of the tenebrionid beetles (Coloeptera, Tenebrionidae) from Australia and its systematic position]. Entomologicheskoe Obozrenie 61 (3): 548–571 [in Russian; English translation in Entomological Review 61(3): 85–107]. [after 9 Sep 1982 (approved to print, back wrapper)]

Medvedev GS, Lawrence JF (1986) Zhuki-chernotelki triby Hyocini (Coleoptera, Tenebrionidae) Avstralii. III. Podtriby Brittonina subtrib. n. i Uptonina subtrib. n. [Tenebrionid beetles of the tribe Hyocini (Coleoptera, Tenebrionidae) of Australia. 3. Subtribes Brittonina subtrib. n. and Uptonina subtrib. n.]. Entomologicheskoe Obozrenie 65 (3): 574–591 [in Russian; English translation in Entomological Review 66 (1)[1987]: 161–179]. [after 26 Aug 1986 (approved to print, back wrapper)]

Medvedev LN (1968) Zhuki-listoedy Jury Karatau (Coleoptera, Chrysomelidae) [Leaf beetles of the Karatau Jurassic (Coleoptera, Chrysomelidae)] [pp. 155–165]. In: Rohdendorf BB (Ed) Yurskie nasekomye Karatau [Jurassic insects of Karatau]. Izdatelstvo "Nauka", Moscow, 252 pp., 25 pls. [in Russian]. [after 26 Jun 1968 (approved to print)]

Medvedev LN, Kazantsev SV (1992) A new subfamily and a new genus of lycid beetles (Coleoptera, Lycidae) from southeast Asia. In: Medvedev LN (Ed) Systematics and ecology of insects of Vietnam. Nauka, Moskva, 55–60 [in Russian].

Medvedev SI (1951) Zhestkokrylye Tom. X. vyp. 1. Plastinchatousye (Scarabaeidae), podsem. Melolonthinae, ch. 1 (khrushchi). Fauna SSSR Novaya Seriya. No. 46. Zoologicheskiy Institut Akademia Nauk SSSR, Moscow, 513 + [2] pp. [after 16 May 1951 (approved to print, endleaf)]

Medvedev SI (1952) Zhestkokrylye Tom. X. vyp. 1. Plastinchatousye (Scarabaeidae), podsem. Melolonthinae, ch. 2 (khrushchi). Fauna SSSR Novaya Seriya No. 52. Zoologicheskiy Institut Akademia Nauk SSSR, Moscow, 274 + [2] pp. [after 30 Jun 1952 (approved to print, endleaf)]

Melzer J (1919) Os longicornios Brazileiros da sub-familia "Prioninae." Tomando em consideração particular as especies do Estado de São Paulo. Revista do Museu Paulista 11: 3–207 + 10 pls. [1 Dec 1919 (date of Prefacio)]

Míka P (2000) Corticariidae (Coleoptera) from southern Thailand; Part 1 - review of the genus *Melanophthalma* with description of some new species. Corticariidae (Coleoptera) jižního Thajska; 1. část. - přehled druhů rodu *Melanophthalma* s popisy několíka nových druhů. Klapalekiana 36 (1/3): 125–143. [29 Aug 2000 (p. 125 header)]

Mikšić R (1976) Monographie der Cetoniinae der paläarktischen und orientalischen Region. Band 1. Allgemeiner Teil. Systematischer Teil: Gymnetini (Taenioderina, Chalcotheina). Forstinstitut in Sarajevo, Sarajevo, [2] + 444 pp. + 10 [unn.] pls. [1976 (title page)]

Miller KB (2000) Cladistic analysis of the tribes of Dytiscinae and the phylogenetic position of the genus *Notaticus* Zimmermann (Coleoptera: Dytiscidae). Insect Systematics & Evolution 31 (2): 165–177. [Sep 2000 (top of article)]

Miller KB (2001) On the phylogeny of the Dytiscidae (Insecta: Coleoptera) with emphasis on the morphology of the female reproductive system. Insect Systematics & Evolution 32 (1): 45–92. [Apr 2001 (top of article)]

Miller KB (2009) On the systematics of Noteridae (Coleoptera: Adephaga: Hydradephaga): phylogeny, description of a new tribe, genus and species, and survey of female genital morphology. Systematics and Biodiversity 7 (2): 191–214. [issued 26 May 2009 (top of article)]

Miroshnikov AI (1989) Novye i maloizvestnye zhuki-drovoseki (Coleoptera, Cerambycidae) s Dalnego Vostoka i sistematicheskoe polozhenie roda *Stenhomalus* White, 1855. [New and little known longhorn beetles (Coleoptera, Cerambycidae) from the Far East and the systematic position of the genus Stenhomalus White, 1855]. Entomologicheskoe Obozrenie 68 (4): 739–747 [in Russian: English translation in Entomological Review 69 (4)[1990]: 58–66]. [after 8 Dec 1989 (approved to print, back wrapper)]

Miwa Y (1931) A study on the lucanid Coleoptera from the Japanese Empire 1. Transactions of the Natural History Society of Formosa 21: 315–325.

Miyatake M (1960) The genus Pisenus Casey and some notes on the family Tetratomidae (Coleoptera). Transactions of the Shikoku Entomological Society 6 [1958–60] (8): 121–134. [30 Jun 1960 (verso of vol. title page)]

Miyatake M (1972) A new Formosan species belonging to the genus Singhikalia Kapur. with proposal of a new tribe (Coleoptera: Coccinellidae). Transactions of the Shikoku Entomological Society 11 [1971–73] (3): 92–98. [30 Jun 1972 (verso of vol. title page)]

Miyatake M (1988) A revision of the genus *Monocoryna* Gorham, with proposal of a new tribe (Coleoptera: Coccinellidae). Transactions of the Shikoku Entomological Society 19 (1/2): 25–46. [30 Nov 1988 (wrapper)]

Miyatake M (1994) Revisional studies on Asian genera of the subfamily Sticholotidinae (Coleoptera: Coccinellidae). Memoirs of the College of Agriculture, Ehime University [Ehime Daigaku Nogakubu Kiyo] 38 (2): 223–293. [Mar 1994 (wrapper)]

Mjöberg E (1917) Results of Dr. E. Mjöberg's Swedish scientific expeditions to Australia 1910–1913. 14. Cetoniidae, Rutelidae, Passalidae, Chysomelidae: subfam. Sagrinae, Cassidinae, Hispinae. Arkiv för Zoologi 11 (3): 1–19. [printed 10 Mar 1917 (p. 19); publ. 9 Oct 1917 (verso of vol. title page)]

Mjöberg E (1925) The mystery of the so-called "trilobite larvae" or "Perty's larvae" definitely solved. Psyche 32 (3): 119–157, pls. 3–4. [Jun 1925 issue (top of article); 11 Aug 1925 (recorded at CNC)]

Moldenke AR (1981) A generic reclassification of the New World Clytrinae (Coleoptera: Chrysomelidae) with a description of new species. Entomologische Arbeiten aus dem Museum Georg Frey 29/30: 75–116. [publ. 1 Dec 1981 (Inhalt)]

Monné MA (1993a) Part II. Subfamily Cerambycinae: tribes Hesperophanini and Eburiini. Catalogue of the Cerambycidae (Coleoptera) of the Western Hemisphere. Sociedade Brasileira de Entomologia, São Paulo, 77 pp.

Monné MA (1993b) Part VI. Subfamily Cerambycinae: tribes Eligmodermini, Callidiopini, Curiini, Graciliini, Obriini, Hyboderini, Eumichthini, Phlyctaenodini, Holopterini, Stenoderini, Pseudocephalini and Bimiini. Catalogue of the Cerambycidae (Coleoptera) of the Western Hemisphere. Sociedade Brasileira de Entomologia, São Paulo, 47 pp.

Monné MA (1994a) Part X. Subfamily Cerambycinae: tribes Rhopalophorini, Heteropsini, Thyrsiini, Agallissini, Platyarthrini, Pteroplatini, Holopleurini and Lissonotini. Catalogue of the Cerambycidae (Coleoptera) of the Western Hemisphere. Sociedade Brasileira de Entomologia, São Paulo, 81 pp.

Monné MA (1994b) Part XI. Subfamily Cerambycinae: tribes Torneutini, Trachyderini and Basipterini. Catalogue of the Cerambycidae (Coleoptera) of the Western Hemisphere. Sociedade Brasileira de Entomologia, São Paulo, 157 pp.

Monné MA (1994c) Part XII. Subfamily Parandrinae, Anoplodermatinae, Spondylinae, Aseminae and Oxypeltinae. Catalogue of the Cerambycidae (Coleoptera) of the Western Hemisphere. Sociedade Brasileira de Entomologia, São Paulo, 56 pp.

Monné MA (1995a) Part XVIII. Subfamily Lamiinae: tribe Acanthocinini. Catalogue of the Cerambycidae (Coleoptera) of the Western Hemisphere. Sociedade Brasileira de Entomologia, São Paulo, 196 pp.

Monné MA (1995b) Part XXI. Subfamily Lepturinae. Catalogue of the Cerambycidae (Coleoptera) of the Western Hemisphere. Sociedade Brasileira de Entomologia, São Paulo, 159 pp.

Monné MA (1995c) Part XXII. Subfamily Prioninae. Catalogue of the Cerambycidae (Coleoptera) of the Western Hemisphere. Sociedade Brasileira de Entomologia, São Paulo, 115 pp.

Monné MA (2005a) Catalogue of the Cerambycidae (Coleoptera) of the Neotropical Region. Part I. Subfamily Cerambycinae. Zootaxa 946: 1–765. [publ. 20 Apr 2005 (footer p. 3)]

Monné MA (2005b) Catalogue of the Cerambycidae (Coleoptera) of the Neotropical Region. Part II. Subfamily Lamiinae. Zootaxa 1023: 1–760.

Monné MA, Bezark LG (2009) Checklist of the Cerambycidae, or longhorned wood-boring beetles, of the Western Hemisphere. BioQuip Publications, Rancho Dominguez, [1] + 482 pp.

Monné MA, Santos-Silva A (2008) Disteniinae Thomson (Insecta, Coleoptera): a protected name. Revista Brasileira de Entomologia 52 (2): 261–262. [Jun 2008 issue]

Monrós F (1954) Notes and synonyms in Chrysomelidae (Coleoptera). Proceedings of the Entomological Society of Washington 56 (1): 23–26. [26 Feb 1954 (p. iv)]

Monrós F (1956) Revision génerica de Lamprosominae con descripción de algunos géneros y especies nuevas (Col., Chrysomelidae). Revista Agronómica del Noroeste Argentino (Tucumán) 2: 25–77.

Monrós F (1958a) Pars 51 (Editio Seconda). Chrysomelidae: Sagrinae. Coleopterorum Catalogus Supplementa. Uitgeverij Dr. W. Junk, 's-Gravenhage, 19 pp. [15 Nov 1958 (title page)]

Monrós F (1958b) Descripcion de una nueva tribus de Clytrinae (Col., Chrysomelidae). Acta Zoologica Lilloana 15: 35–39. [printed 29 Aug 1958 (endleaf)]

Monrós F (1958c) Consideraciones sobre la fauna del Sur de Chile y revision de la tribus Stenomelini (Coleoptera, Chrysomelidae). Acta Zoologica Lilloana 15: 143–153. [printed 29 Aug 1958 (endleaf)]

Monrós F (1959a) Los géneros de Chrysomelidae. Opera Lilloana 3: 1–337.

Monrós F (1959b) Notes on Lamprosomatinae (Chrysomelidae). The Coleopterists Bulletin 12 [1958]: 29–33. [issued 27 Jan 1959 (back wrapper)]

Monrós F, Viana MJ (1947) Revisión sistematica do los Hispidae Argentinos (Insecta, Coleop. Chrysomeloid.). Anales del Museo Argentino de Ciencias Naturales «Bernardino Rivadavia» / Anales 42 [Entomologia No. 162]: 125–324. [10 Dec 1947 (title page)]

Monrós F, Viana MJ (1949) Revision de las especies Argentinas de Dorynotini (Col. Cassidinae). (1a contribución al conocimiento de Cassidinae). Acta Zoologica Lilloana 8: 391–426. [printed 31 Dec 1949]

Monrós F, Viana MJ (1951) Las Cassidinae de la seccion Hemisphaerotina, con revision de las especies Argentinas (Col. Cassidinae). (2a contribucion al conocimiento de Cassidinae). Acta Zoologica Lilloana 11: 367–395. [printed 13 Nov 1951]

Moore BP (1998) The enigmatic tribe Physocrotaphini Chaudoir (Helluodini auct.) (Coleoptera: Carabidae). Phylogeny and classification of Caraboidea (Coleoptera: Adephaga) Proceedings of a symposium (28 August, 1996, Florence, Italy) 20 International Congress of Entomology. 1998. Museo Regionale di Scienze Naturali - Atti, 369–380.

Moore I (1964) A new key to the subfamilies of the Nearctic Staphylinidae and notes on their classification. The Coleopterists Bulletin 18 (3): 83–91. [publ. 24 Sep 1964 (vol. contents)]

Morawitz A (1889) Entomologische Beiträge. Bulletin de l'Académie Impériale des Sciences de St.-Pétersbourg 33 [1890] (1) : 33–82. [read 1 Nov 1888; printed Apr 1889 (wrapper); 27 Dec 1889 (recorded at MCZ); also issued in: Mélanges Biologiques tirés du Bulletin Physico-Mathematique de l'Académie Impériale des Sciences de St Petersbourg 13: 5–54]

Moret P (2005) Los Coleópteros Carabidae del Páramo en los Andes del Ecuador. Sistemática, ecología y biogeografía (Insecta, Coleoptera, Carabidae) [Monografía 2]. Centro de Biodiversidad y Ambiente, Pontificia Universidad Católica del Ecuador, Quito, 306 pp. [Jun 2005 (title page)]

Morimoto K (1962a) Key to families, subfamilies, tribes and genera of the superfamily Curculionoidea of Japan excluding Scolytidae, Platypodidae and Cossoninae. (Comparative morphology, phylogeny and systematics of the superfamily Curculionoidea of Japan. III). Journal of the Faculty of Agriculture, Kyushu University, Fukuoka 12 (1): 21–66. [30 Jun 1962 (Alonso-Zarazaga and Lyal 1999: 244)]

Morimoto K (1962b) Descriptions of a new sub-family, new genera and species of the family Curculionidae of Japan (comparative morphology, phylogeny and systematics of the super-family Curculionoidea of Japan. II). Journal of the Faculty of Agriculture, Kyushu University, Fukuoka 11 (4): 375–409. [4 Mar 1962 (Alonso-Zarazaga and Lyal 1999: 244)]
Morimoto K (1972) A key to the genera of Oriental Anthribidae (Coleoptera). Bulletin of the Government Forest Experiment Station No. 246: 35–54, 17 pls.
Morimoto K (1978a) The family Anthribidae of Japan (Coleoptera). Part 1. Esakia 12: 17–47. [31 Nov 1978 (Kyushu Univ. Repository)]
Morimoto K (1978b) On the genera of Oriental Cryptorhynchinae (Coleoptera: Curculionidae). Esakia 11: 121–143. [issued 31 Mar 1978 (Kyushu Univ. Repository)]
Morimoto K (1980) The family Anthribidae of Japan (Coleoptera). Part 3. Esakia 15: 11–47. [issued 31 Oct 1980 (Kyushu Univ. Repository)]
Morimoto K (1981) The family Anthribidae of Japan (Coleoptera). Part 4. Esakia 17: 53–107. [30 Nov 1981 (Kyushu Univ. Repository)]
Morimoto K (1983) The family Curculionidae of Japan. II. Viticiinae, subfam. n. Esakia 30: 55–62. [15 Dec 1983 (Kyushu Univ. Repository)]
Morimoto K (1987) The family Curculionidae of Japan VI. Tribe Mechistocerini Part 1 (Insecta, Coleoptera). Journal of the Faculty of Agriculture, Kyushu University 31 (4): 321–343. [Mar 1987]
Morimoto K, Kojima H (2001) Isopterina, a new subtribe of the tribe Celeuthetini, with notes on the related taxa (Coleoptera, Curculionidae). Elytra 29 (2): 265–283. [15 Nov 2001 (wrapper)]
Morrone JJ (1997) Nomenclatural notes on the subfamily Cyclominae (Coleoptera: Curculionidae). Acta Entomologica Chilena 21: 101–102.
Morrone JJ (1998) Anomophthalmina subtrib. n., an endemic subtribe of Entimini (Coleoptera: Curculionidae) from western Patagonia. Revista de la Sociedad Entomológica Argentina 57 (1/4): 85–89. [printed Jan 1998 (footer p. 140)]
Morvan DM (2004) Nouvelle tribu et nouvelles espèces de Platyninae. Nouvelles espèces de Tiruka et de Sphodrini de la région orientale. Asia.. Loened Aziad Amprevaned Feuraskelleged C'hwiledig 9: 1–65. [Apr 2004 (wrapper)]
Moseyko AG, Sprecher-Uebersax E, Löbl I (2010) Case 3519. *Eumolpus* Weber, 1801, *Chrysochus* Chevrolat in Dejean, 1836 and *Bromius* Chevrolat in Dejean, 1836 (Insecta, Coleoptera, Chrysomelidae): proposed conservation of usage. Bulletin of Zoological Nomenclature 67 (3): 218–224. [Sep 2010 issue]
Motschulsky V, de (1845) Über die Ptilien Russland's. Bulletin de la Société Impériale des Naturalistes de Moscou 18 (4): 504–539, pls. 9–10. [after 12 Aug 1845 (Julian 30 Jul; approved to print, verso of title page)]
Motschulsky V, de (1849) Coléoptères reçus d'un voyage de M. Handschuch dans le midi de l'Espagne enumerés et suivis de notes. Bulletin de la Société Impériale des Naturalistes de Moscou 22 (3): 52–163. [after 15 Jul 1849 (approved to print, verso of part 2 title page); by Nov 1849 (Ent. Zeit. Stettin 10)]

Motschulsky V, de (1850) Die Käfer Russlands. I. Insecta carabica. W. Gautier, Moscow, iv + xi + 91 pp. [after 5 May 1850 (approved to print, verso of title page); 23 Oct 1850 (Ann. Soc. Ent. France (2) 8: Bull. Ent.: lviii)]

Motschulsky V, de (1852) Malthinides. Études Entomologiques 1: 1–15. [1852 (footer p. 24)]

Motschulsky V, de (1855) Sur les collections coléoptèrologiques de Linné et de Fabricius. Études Entomologiques 4: 25–71. [1855 (vol. title page)]

Motschulsky V, de (1857) Sur les collections coléoptèrologiques de Linné et de Fabricius (continuation). Études Entomologiques 5: 60–75. [1856 (vol. title page); 5 Feb 1857 (Julian 24 Jan 1857: Bull. Soc. Imp. Nat. Moscou 30 (2): 614)]

Motschulsky V, de (1860) Coléoptères rapportés de la Sibérie orientale et notamment des pays situés sur les bords du fleuve Amour par Mm. Schrenck, Maack, Ditmar, Voznessensky, etc. In: Schrenck L (Ed) Reisen und Forschungen im Amur-Lande in den Jahren 1854–1856 im Auftrage der Keiserl Akademie der Wissenschaften zu St Petersburg ausgeführt und in Verbindung mit mehreren Gelehrten herausgegeben Band II Zweite Lieferung Coleopteren. Eggers & Comp., St Petersburg, 79–257 + map. [13 Jun 1860 (Ann. Soc. Ent. France (3) 8: Bull. Ent.: lvi)]

Motschulsky V, de (1862) Entomologie spéciale. Remarques sur la collection d'insectes de V. Motschulsky. Études Entomologiques 11: 15–55. [1862 (vol. title page)]

Mroczkowski M (1991) [new name] In: Burakowski B, Mroczkowski M, Stefańska J: Część XXIII, tom 17. Chrząszcze Coleoptera. Stonkowate - Chrysomelidae, część 2. Katalog Fauny Polski Nr. 50: 1–227. [printed Aug 1991 (verso of title page)]

Muizon J, de (1960) Faune des brenthides d'Afrique. Mémoires de l'Institut Francais d'Afrique Noire 59: 1–256. [printed Aug 1960 (p. 256)]

Mulsant É (1839) 1re livraison. - Longicornes. Histoire naturelle des Coléoptères de France. Maison, Paris, xi + 304 pp. + 3 pls. [1840 (title page); 19 Oct 1839 (Bibliogr. France 1839)]

Mulsant É (1842) Lamellicornes. Histoire naturelle des Coléoptères de France. [Tome 2]. Maison, Paris, viii + 623 pp. + [8, rectif.] + 3 pls. [6 Aug 1842 (Bibliogr. France 1842)]

Mulsant É (1844) Palpicornes. Histoire naturelle des Coléoptères de France. [Tome 3]. Maison, Paris, vii + 197 + [7, errata + suppl.] pp. + 1 pl. [18 May 1844 (Bibliogr. France 1844)]

Mulsant É (1846) Sulcicolles. - Sécuripalpes. Histoire naturelle des Coléoptères de France. [Tome 4]. Maison, Paris, xxiv + [2] + 26 + 280 + [4] pp. + 1 pl. [26 Sep 1846 (Bibliogr. France 1846)]

Mulsant É (1850) Species des Coléoptères trimères sécuripalpes. Annales des Sciences Physiques et Naturelles et d'Industrie (Lyon) (2) 2 [1849–50] (1/2): 1–1104. [1850 (title page; Ann. Soc. Ent. France (2) 8: Bull. Ent.: lxxiii); 19 Dec 1850 (Rép. Trav. Soc. Stat. Mars. 14: 543; 27 Jan 1851 (Proc. Ent. Soc. London 1851: 59) / as separate: 1851, Paris, C. Savy Jeune, xi + 1104 pp.]

Mulsant É (1853) Suplément à la monographie des coléoptères trimères sécuripalpes. Annales de la Société Linnéenne de Lyon (Nouvelle Série) 1 [1852–53]: 129–333. [1853 (title page); 1854? (vol. incl. members appointed to 31 Déc 1853); reissued in: Opusc. Ent. 3 1853: 1–205]

Mulsant É (1854) Latigènes. Histoire naturelle des Coléoptères de France. [Tome 5]. Maison, Paris, x + 396 pp. [2 Oct 1854 (date of dedication)]
Mulsant É (1856a) Pectinipèdes. Histoire naturelle des Coléoptères de France. [Tome 6]. Maison, Paris, [4] + 96 + [12] pp. [1856 (title page); 20 Dec 1856 (Bibliogr. France 1856)]
Mulsant É (1856b) Barbipalpes. Histoire naturelle des Coléoptères de France L. Maison, [4] + 115 pp. + 1 pl. [12 Sep 1856 (date of dedication); 1856 (title page, Bull. Bibliogr. Sci. Phys. 1860: 49) / reissued in: Ann. Soc. Linn. Lyon (N. S.) 3: 193–304, 2 pls., 1856 (vol. title page); both 5 Sep 1857 (Bibliogr. France 1857)]
Mulsant É (1856c) Tribu des Longipèdes. Histoire naturelle des Coléoptères de France. L. Maison, Paris, [4] + 171 + [1] + [4] (suppl.) pp. + 1 pl. [10 Oct 1856 (date of dedication) / reissued in: Ann. Soc. Linn. Lyon (N. S.) 3: 305–471, 1856 (vol. title page); both 5 Sep 1857 (Bibliogr. France 1857)]
Mulsant É (1857) [Histoire naturelle des Coléoptères de France]. Vésicants. Annales de la Société Linnéenne de Lyon (Nouvelle Série) 4: 209–409. [17 Nov 1857 (deceased member noted p. ix, list to 31 Dec) / separate: Paris, Magnin, Blanchard et Cie., [2] + 201 + [9] (suppl.) pp. + 1 pl., 10 Nov. 1857 (date of dedication); both 28 Aug 1858 (Bibliogr. France 1858: 450, 458)]
Mulsant É (1858) [Histoire naturelle des Coléoptères de France]. Angustipennes. Annales de la Société Linnéenne de Lyon (Nouvelle Série) 5: 65–238. [28 Sep 1858 (member deceased noted; list to 31 Dec) / separate: Paris, Magnin, Blanchard & Cie., [4] + 172 + [3] (suppl.) pp., 16 Dec 1858 (date of dedication); 24 Sep 1859 (Bibliogr. France 1859: 445)]
Mulsant É (1862) Mollipennes. Histoire naturelle des Coléoptères de France. Magnin, Blanchard & Cie., Paris, [4] + 440 pp. + 3 pls. [16 May 1862 (date of dedication); 1862 (title page); 13 Jun 1863 (Bibliogr. France 1863) / reissued in: Ann. Soc. Linn. Lyon (N. S.) 9: 57–496, Feb 1863 (title page); 15 Aug 1863 (Bibliogr. France 1863)]
Mulsant É (1863a) [Histoire naturelle des Coléoptères de France]. Longicornes. Annales des Sciences Physiques et Naturelles, d'Agriculture et d'Industrie (Lyon) (Troisième Série) 6 [1862]: 307–466. [explicitely "extrait des Annales de la Société..." (Table Alphabétique) but publ. as follows: Ann. Sci. Phys. Nat. Agr. Ind. Lyon (3) 6: 307–466, (incl. séance 19 Dec 1862; 9 May 1863 [Bibliogr. France 1863]; = sep. pp. 1–162) / separate: Paris: Magnin, Blanchard & Cie., [3] + 590 pp., 1863 (1862–63 [title page]; 2 Jan 1864 (Bibliogr. France 1864) / Ann. Sci. Phys. Nat. Agr. Ind. Lyon 7 [1863]: 97–320 (incl. Tableau to 31 Dec 1863; 1864; = separate pp. 161bis-384); 8 [1864]: 1–208 (incl. Tableau to 31 Dec 1864; 1865; = separate pp. 385–590)]
Mulsant É (1863b) Longicornes [pp. 161bis-384]. Histoire naturelle des Coléoptères de France. Magnin, Blanchard & Cie., Paris, [3] + 590 pp. [reissued in Ann. Sci. Phys. Nat. Agr. Ind. Lyon (3) 6 [1862]: 307–466 (1863); 7 [1863]: 97–320 (1864); 8 [1864]: 1–208 (1865)]
Mulsant É, Revelière E (1861) Description d'un Coléoptère nouveau, constituant un nouveau genre, dans la tribu des Opatates. Annales de la Société Linnéenne de Lyon (Nouvelle Série) 7 [1860]: 153–157. [Feb 1861 (vol. title page); 3 Aug 1861 (Bibliogr. France 1861)]
Mulsant É, Rey C (1853) Essai d'une division des derniers Mélasomes. Mémoires de l'Académie Nationale des Sciences, Belles-Lettres et Arts de Lyon Classe des Sciences (Nouvelle

Série) 2 [1852]: 226–329. [1852 (vol. title page); 14 Jun 1853 (pres. to Acad.); 18 Jul 1853 (article p. 219 pres. to Acad.) / reissued in: Opusc. Ent. 4: 1–104]

Mulsant É, Rey C (1854) Essai d'une division des derniers Mélasomes [cont'd] Pandarites. Mémoires de l'Académie Impériale des Sciences, Belles-lettres et Arts de Lyon Classe des Sciences (Nouvelle Série) 4: 153–332. [1854 (title page); after 14 Feb 1854 (art. submitted p. 18) / reissued in: Opusc. Ent. 5: 1–189 (of 255 pp.), 25 Oct 1854 (date of dedication); from p. 165, each page has one line less text than the original]

Mulsant É, Rey C (1859) Essai d'une division des derniers Mélasomes, famille des parvilabres, quatrième tribu Opatrites. Opuscules Entomologiques 10: 1–160. [1859 (title page; Reinwald 1860: 153); 20 Jun 1859 (date of dedication); 25 Feb 1860 (Bibliogr. France 1860) (before Mémoires version)]

Mulsant É, Rey C (1864a) [Histoire naturelle des Coléoptères de France]. Tribu des Angusticolles; Tribu des Diversipalpes. Annales de la Société Linnéenne de Lyon (Nouvelle Série) 10: 247–380, pls. 1–2; pp. 381–404. [1863–64 (title page); Feb 1863 (error for 1864, wrapper); 17 Sep 1864 (Bibliogr. France 1864) / separate: Paris, Magnin, Blanchard & Co., [4] + 134 pp., 2 pls., [2] + 23 + [1] pp., 1863–64 (title page); 8 Dec and 10 Dec 1863 (dates of dedications); 24 Sep 1864 (Bibliogr. France 1864)]

Mulsant É, Rey C (1864b) Histoire naturelle des Coléoptères de France. Térédiles. F. Savy, Paris, 394 pp., 9 pls. [24 Jul 1864 (date of dedication); 18 Mar 1865 (Bibliogr. France 1865) / reissued in: Ann. Soc. Linn. Lyon 11 [1864]: 289–420, Feb 1865 (title page), 13 May 1865 (Bibliogr. France 1865); and 2 [1865]: 1–284 + 10 pls., 6 Jan 1866 (title page), 28 Apr 1866 (Bibliogr. France 1866)]

Mulsant É, Rey C (1867a) Histoire naturelle des Coléoptères de France. Scuticolles. F. Savy, Paris, [4] + 186 pp. + 2 pls. [1867 (title page); 4 Feb 1867 (date of dedication) / reissued in: Ann. Soc. Linn. Lyon (N. S.) 15 [1867]: 1–188 + 3 pls., 15 Jan 1868 (title page), 17 Oct. 1868 (Bibliogr. France 1868)]

Mulsant É, Rey C (1867b) Histoire naturelle des Coléoptères de France. Vésiculifères. F. Savy, Paris, [2] + 2 + 308 + 4 (numbered 125–128) + [9 (explication des planches)] pp. + 7 pls. [14 Nov 1866 (date of dedication); 1867 (title page); 28 Sep 1867 (Bibliogr. France 1867) / reissued in: Ann. Sci. Phys. Nat. Agr. Ind. Lyon (3) 11 [1867]: 625–943 + 7 pls., 1867 (title page); 17 Jul 1869 (Bibliogr. France 1869)]

Mulsant É, Rey C (1868a) [Histoire naturelle des Coléoptères de France]. Tribu des Floricoles. Annales de la Société Linnéenne de Lyon 15 [1867]: 237–402. [15 Jan 1868 (title page); 17 Oct 1868 (Bibliogr. France 1868) / separate: Paris: Deyrolle, [4] + 315 + [3] + [19] pp. + 19 pls., 10 Oct 1867 (date of dedication); Mar 1868 (title page) / Floricoles (Suite). Ann. Soc. Linn. Lyon (N. S.) 16[1868]: 83–231 + 19 pls., 28 Dec 1868 (title page)]

Mulsant É, Rey C (1868b) Histoire naturelle des Coléoptères de France. Gibbicolles. Deyrolle, Paris, 226 pp., 14 pls. [8 Sep 1868 (date of dedication); Nov 1868 (title page); 1 May 1869 (Bibliogr. France 1869); 1869 (Arch. Naturg. 1870: 102) / reissued in: Ann. Soc. Imp. Agr. Hist. Nat. Arts Utiles Lyon (4) 1: 179–420 + 14 pls., 1869 (title page), 1870 (date stamp); 15 Jul 1871 (Bibliogr. France 1871)]

Mulsant É, Rey C (1869) Histoire naturelle des Coléoptères de France. Piluliformes. Deyrolle, Paris, [4] + 175 + [3] pp. + 2 pls. [1869 (title page); 25 Apr 1869 (date of dedication); 20

Nov 1869 (Bibliogr. France 1869) / reissued in: Ann. Soc. Linn. Lyon (N. S.) 17: 201–378 + [2] + 2 pls., 28 Dec 1869 (title page), 16 Apr 1870 (Bibliogr. France 1870)]

Mulsant É, Rey C (1871a) [Histoire naturelle des Coléoptères de France]. Tribu des Lamellicornes. Annales de la Société d'Agriculture, Histoire Naturelle et Arts Utiles de Lyon (4) 2 [1869]: 241–650. [1870 (title page); before 3 Nov 1871 (Proc.-Verb. Soc. Agr. Hist. Nat. Arts Utiles Lyon 1871: lxxvi) / separate (1871b)]

Mulsant É, Rey C (1871b) Histoire naturelle des Coléoptères de France. Lamellicornes - Pectinicornes [pp. 411–735]. Deyrolle, Paris, [2] + 735 + [1] + 42 + [1] pp. + 3 pls. [Sep 1871 (title page); 8 Sep 1871 (date of dedication); 1 Aug 1872 (Petites Nouv. Ent.) / to p. 410 in 1871a / reissued in: Ann. Soc. Agr. Hist. Nat. Arts Utiles Lyon (4) 3[1870]: 155–480 + 2 pls. (Lamellicornes) + 481–523 + [5, explic.] + 1 pl. (Pectinicornes), 1871 (title page); 12 Apr 1873 (Bibliogr. France 1873)]

Mulsant É, Rey C (1871c) Histoire naturelle des Coléoptères de France. Brévipennes. Aléochariens [Huitième Branche, Bolitocharaires]. Deyrolle Fils, Paris, [4] + 321 + pp. + [5, explic.] + [4, table, erratum] + 5 pls. [1871 (title page); 25 Aug 1871 (date of dedication) / reissued in: Ann. Soc. Linn. Lyon (N. S.) 19 [1872]: 91–411 + [11] + 5 pls., 31 Dec 1872 (title page); 9 Jul 1873 (Ann. Soc. Ent. France (5) 3: Bull. Ent.: cxxxii)]

Mulsant É, Rey C (1872a) Tribu des Uncifères [sep. pp. 1–57]. In: Histoire naturelle des Coléoptères de France. Improsternés - Uncifères - Diversicornes, Spinipèdes. Deyrolle, Paris, [2] + 17 + [1] + 57 + 39 + [1] + 57 + [1] pp. + 2 pls. [Dec 1872 (title page); 8 Dec 1872 (date of dedication); 3 May 1873 (Ann. Soc. Ent. Belg. 16 Comptes-Rendus: lvi) / reissued in: Ann. Soc. Agr. Hist. Nat. Arts Utiles Lyon (4) 4: 79–136, 1872 (title page); 11 Apr 1874 (Ann. Soc. Ent. Belg. 17 Comptes-Rendus: lxiv)]

Mulsant É, Rey C (1872b) Tribu des Diversicornes [sep. pp. 1–39 + 1]. In: Histoire naturelle des Coléoptères de France. Improsternés - Uncifères - Diversicornes, Spinipèdes. Deyrolle, Paris, [2] + 17 + [1] + 57 + 39 + [1] + 57 + [1] pp. + 2 pls. [Dec 1872 (title page); 8 Dec 1872 (date of dedication); 3 May 1873 (Ann. Soc. Ent. Belg. 16 Comptes-Rendus: lvi) / reissued in: Ann. Soc. Agr. Hist. Nat. Arts Utiles Lyon (4) 4: 137–176, 1872 (title page); 11 Apr 1874 (Ann. Soc. Ent. Belg. 17 Comptes-Rendus: lxiv)]

Mulsant É, Rey C (1873a) Histoire naturelle des Coléoptères de France. Brévipennes. Aléochariens [Première branche dinardaires - cinquième branche oligotaires]. Deyrolle Fils, Paris, [4] + 155 pp. + pls. 1–2. [before 25 Dec 1873 (Fauvel letter to Brisout de Barneville in: Annuaire Ent. 1873) / reissued in: Mém. Acad. Sci. Belles-lettres Lyon. Cl. Sci. 20 [1873–74]: 23–175, 2 pls. (without Tableau), after Feb 1874 (p. 341), 11 Jul 1874 (Bibliogr. France 1874: 355)]

Mulsant É, Rey C (1873b) Histoire naturelle des Coléoptères de France. Brévipennes. Aléochariens (Suite) [Tableau... des Aléochariens + Septième Branche, Myrmédoniaires, part 1]. Deyrolle, Paris, 695 pp., pls. 1–5. [Dec 1873 (title page); 18 Dec 1873 (date of dedication); 25 Dec 1873 (Fauvel letter to Brisout de Barneville in Annuaire Ent. 1873); 11 Apr 1874 (Ann. Soc. Ent. Belg. 17 Comptes-Rendus: lxiv) / reissued in: Ann. Soc. Agr. Hist. Nat. Arts Utiles Lyon (4) 6 [1873]: 33–727-[738 incl. explic. + 5 pls.], 1874 (title page), 19 Feb 1875 (Proc.-Verb. Soc. Agr. Hist. Nat. Arts Utiles Lyon 1875: xxx)]

Mulsant É, Rey C (1874) [Histoire naturelle des Coléoptères de France]. Tribu des brévipennes: Famille des aléochariens (suite). Sixième branche: Aléocharaires. Annales de la Société Linnéenne de Lyon (Nouvelle Série) 20 [1873]: 285–447. [1874 (title page); 9 Sep 1874 (Ann. Soc. Ent. France (5) 4: Bull. Ent.: clxxx) / separate: Paris: Deyrolle, [2] + 565 pp. + 5 pls., Dec 1874 (title page); 8 Dec 1874 (date of dedication); 1 Feb 1875 (Petites Nouv. Ent.) / pp. 163–565 reissued in: Ann. Soc. Linn. Lyon (N. S.) 21 [1874]: 1–403 + [5, explic.] + 5 pls., 1875 (title page); 14 Apr 1875 (Ann. Soc. Ent. France (5) 5: Bull. Ent.: lxxix)]

Mulsant É, Rey C (1876) [Histoire naturelle des Coléoptères de France]. Tribu des brévipennes. [Staphyliniens.]. Annales de la Société d'Agriculture, Histoire Naturelle et Arts Utiles de Lyon (4) 8 [1875]: 145–856, [7, explic.] + pls. 1–6. [1876 (title page); 21 Jul 1877 (Bibliogr. France 1877) / separate: Paris: Deyrolle, [2] + 712 pp. + [7, explic.] + 6 pls., May 1877 (title page); 17 May 1877 (date of dedication); 11 Aug 1877 (Bibliogr. France 1877)]

Mulsant É, Rey C (1878a) Histoire naturelle des Coléoptères de France. Brévipennes. Pédériens. - Évesthétiens. Deyrolle, Paris, [2] +338 + [2, tables] + [7, explic.] pp. + 6 pls. [1878 (title page); 10 May 1878 (date of dedication); 15 Nov 1878 (Petites Nouv. Ent.) / reissued in: Ann. Soc. Linn. Lyon (N. S.) 24 [1877]: 1–338 + [2, tables] + [7, explic.], pls. 1–6, Oct 1878 (title page), 13 Nov 1878 (Ann. Soc. Ent. France (5) 8: Bull. Bibliogr.: 48)]

Mulsant É, Rey C (1878b) [Histoire naturelle des Coléoptères de France]. Tribu des brévipennes: Cinquième famille: Oxyporiens. Sixième famille: Oxyteliens. Annales de la Société d'Agriculture, Histoire Naturelle et Arts Utiles de Lyon (4) 10 [1877]: 443–850 + [8, explic.], pls. 1–7. [1878 (title page); 5 Jul 1878 (last art. read at Soc.) / separate: Paris: Deyrolle, [2] + 408 + 8 [explic.] + 7 pls., 1879 (title page); 10 Jan 1879 (date of dedication)]

Mulsant É, Rey C (1880) In: Rey, C. [Histoire naturelle des Coléoptères de France]. Tribu des brévipennes. Onzième famille: Omaliens. Douzième famille: Pholidiens. Annales de la Société Linnéenne de Lyon (Nouvelle Série) 27: 1–430, pls. 1–6. [1880 (title page); incl. list of members for 1881]

Muona J (1979) Staphylinidae [pp. 14–28]. In: Silfverberg H (Ed) Enumeratio Coleopterorum Fennoscandiae et Daniae. Helsingfors Entomologiska Bytesförening, Helsinki, vi + 79 pp. [1979 (title page)]

Muona J (1991a) The Eucnemidae of south-east Asia and the western Pacific - a biogeographical study. Australian Systematic Botany 4 (1): 165–182. [3 Sep 1991 (p. 165)]

Muona J (1991b) A revision of the Indomalesian tribe Galbitini new tribe (Coleoptera, Eucnemidae). Entomologica Scandinavica Supplement 39: 1–67. [1 Nov 1991 (p. 1)]

Muona J (1993) Review of the phylogeny, classification and biology of the family Eucnemidae (Coleoptera). Entomologica Scandinavica Supplement 44: 1–133. [30 Jun 1993 (p. 1)]

Muona J (1994) Eucnemidae Eschscholtz, 1829 (Insecta, Coleoptera); proposed conservation over Melasidae Fleming, 1821. (Case 2938). In: Notices. Bulletin of Zoological Nomenclature 51 (3): 185.

Muona J (2007) Eucnemidae [pp. 81–87]. In: Löbl I, Smetana A (Eds) Calalogue of Palaearctic Coleoptera. Volume 4. Elateroidea - Derodontoidea - Bostrichoidea - Lymexyloidea - Cleroidea - Cucujoidea. Apollo Books, Stenstrup, Denmark, 935 pp. [30 Jun 2007 (verso of title page)]

Muona J, Alaruikka D (2007) Eucnemidae [p. 32]. In: New nomenclatorial acts, and comments. In: Löbl I, Smetana A (Eds) Catalogue of Palaearctic Coleoptera. Volume 4. Elateroidea - Derodontoidea - Bostrichoidea - Lymexyloidea - Cleroidea - Cucujoidea. Apollo Books, Stenstrup, Denmark, 935 pp. [30 Jun 2007 (verso of title page)]

Murayama JJ (1954) Scolytid-Fauna of the northern half of Honshu with a distribution table of all the Scolytid-species described from Japan. Bulletin of the Faculty of Agriculture Yamaguti [Yamaguchi] University 5: 149–212. [Dec 1954 (reprint wrapper)]

Murayama JJ (1959) An introduction to the studies of Japanese Scolytidae. Forestry Leaves 11 (3): 25–32. [1 Sep 1959]

Murayama JJ (1963) Studies in the scolytid fauna of the northern half of the Far East, V. Hylesininae. Shukosha Press, Fukuoka, Japan, iii + [i] + 72 pp. [publ. 31 Oct 1963 (reprint fide R. A. Beaver, pers. comm. to Alonso-Zarazaga)]

Murray A (1858) List of Coleoptera received from Old Calabar, on the west coast of Africa [continued]. The Annals and Magazine of Natural History (3) 2 (11): 340–349. [1 Nov 1858 (Evenhuis 2003: 21)]

Murray A (1864) Monograph of the family of Nitidulariae. Transactions of the Entomological Society of London 24 (3): 211–414 + pls. 32–36.

Murray A (1870) List of Coleoptera received from Old Calabar, on the west coast of Africa [continued]. The Annals and Magazine of Natural History (4) 6 (31): 44–56, pls. 2, 3; (32): 161–176. [(31) 1 Jul 1870; (32) 1 Aug 1870 (Evenhuis 2003: 25)]

Musgrave A (1932) Bibliography of Australian entomology, 1775-l930 with biographical notes on authors and collectors. Royal Zoological Society of New South Wales, Sydney, viii + 380 pp. [publ. Sep 1932 (title page)]

Nagel P (1987) Arealsystemanalyse afrikanischer Fühlerkäfer (Coleoptera, Carabidae, Paussinae). Ein Beitrag zur Rekonstruktion der Landschaftsgenese. Erdwissenschaftliche Forschung Band XXI. Band XXI. F. Steiner, Stuttgart, 233 pp.

Nakane T, Kishii T (1956) On the subfamilies of Elateridae from Japan (Coleoptera). Kontyû 24 (4): 201–206, pls. 22–23. [1 Oct 1956 (verso of title page)]

Naomi S-I (1986) A taxonomic study on the subfamily Megalopininae (Coleoptera, Oxyporidae) of Japan, with descriptions of two new species. Kontyû 54 (2): 344–352. [25 Jun 1986 (wrapper)]

Napp DS (2007) Unxiini, uma nova tribo de Cerambycinae (Coleoptera, Cerambycidae). Revista Brasileira de Entomologia 51 (3): 312–340. [mailed 30 Sep 2007 (contents)]

Naviaux R (1991) Les cicindèles de Thaïlande, étude faunistique (Coleoptera Cicindelidae). Bulletin Mensuel de la Société Linnéenne de Lyon 60 (7): 209–288. [Sep 1991 (fasc. title page); 6 Aug 1991 (recorded at BMNH)]

Nearns EH, Branham MA (2008) Revision and phylogeny of the tribes Curiini LeConte and Plectromerini Nearns and Branham, new tribe (Coleoptera: Cerambycidae: Cerambycinae). Memoirs of the American Entomological Society No. 47: 1–117. [issued 1 Mar 2008 (title page)]

Neave SA (1939) Nomenclator Zoologicus. A list of the names of genera and subgenera in zoology from the tenth edition of Linnaeus 1758 to the end of 1935. Vols. 1–2. Zoological Society of London, London, xiv + 957 + 1025 pp.

Neave SA (1940) Nomenclator Zoologicus. A list of the names of genera and subgenera in zoology from the tenth edition of Linnaeus 1758 to the end of 1935. Vol. III. M-P. Zoological Society of London, London, 1065 pp.

Neave SA (1966) Nomenclator Zoologicus. Vol. VI, 1946–1955. Zoological Society of London, London, ix + 329 pp.

Neboiss A (1956) A check list of Australian Elateridae (Coleoptera). Memoirs of the National Museum of Victoria 22 (2): 1–75. [25 Jul 1956]

Nelson GH (1982) A new tribe, genus, and species of North American Buprestidae with consideration of subfamilial and tribal categories. The Coleopterists Bulletin 35 [1981] (4): 431–450. [mailed 23 Jun 1982 (wrapper)]

Newman E (1834) Attempted division of British insects into natural orders. The Entomological Magazine 2 (4): 379–431. [Oct 1834 issue (note in vol. 1 p. 4 indicates the Ent. Mag. was to be published quarterly on the 1st of the month)]

Newman E (1840) Entomological notes. The Entomologist 1 (1): 1–16, (2): 17–32. [Nov + Dec 1840 issues]

Newman E (1841) Entomological notes (continued). The Entomologist 1 (3): 33–37. [Jan 1841 issue]

Newton AF, Jr. (1988) Fooled by flatness: subfamily shifts in subcortical Staphylinidae (Coleoptera). The Coleopterists Bulletin 42 (3): 255–262. [mailed 19 Sep 1988 (inside wrapper)]

Newton AF, Jr. (1997) Review of Agyrtidae (Coleoptera), with a new genus and species from New Zealand. Annales Zoologici (Warszawa) 47 (1/2): 111–156. [20 Sep 1997]

Newton AF, Jr. (1998) Phylogenetic problems, current classification and generic catalog of world Leiodidae (including Cholevidae) [pp. 41–178]. In: Giachino PM, Peck SB (Eds). Phylogeny and evolution of subterranean and endogean Cholevidae (=Leiodidae Cholevinae). Proceedings of the XXth International Congress of Entomology. 1996. Firenze. Atti / Museo Regionale di Scienze Naturali, Torino, 1998, 295 pp. [Dec 1998 (printed)]

Newton AF, Jr., Thayer MK (1992) Current classification and family-group names in Staphyliniformia (Coleoptera). Fieldiana: Zoology (New Series) No. 67: 1–92. [29 May 1992 (wrapper)]

Newton AF, Jr., Thayer MK (1995) Protopselaphinae new subfamily for Protopselaphus new genus from Malaysia, with a phylogenetic analysis and review of the Omaliine Group of Staphylinidae including Pselaphidae (Coleoptera) [pp. 219–320]. In: Pakaluk J, Ślipiński SA (Eds) Biology, phylogeny and classification of Coleoptera: Papers celebrating the 80th birthday of Roy A Crowson. Muzeum i Instytut Zoologii PAN, Warszawa, x + 1092 pp. in 2 vols. [31 Mar 1995 (teste S. A. Ślipiński)]

Nikitsky NB (1985) Zhestkokrylie semeystv Tetratomidae i Cephaloidae Vostochnoy Palearktiki [Beetles of the families Tetratomidae and Cephaloidae from the eastern Palaearctic]. Sbornik Trudov Zoologicheskogo Muzeya MGU 23: 26–37 [in Russian]. [1985 (wrapper)]

Nikitsky NB (1986) Semeystvo Pilipalpidae stat. n. (Coleoptera, Heteromera), ego sostav i sistematicheskie svyazi [The family Pilipalpidae stat. n. (Coleoptera, Heteromera), its composition and taxonomic relationships]. Zoologicheskii Zhurnal 65 (8): 1178–1189 [in Russian]. [after 22 Jul 1986 (approved to print)]

Nikitsky NB (1992) Sem. Salpingidae [pp. 482–493]. In: Ler PA (Ed) Opredelitel' nasekomykh Dal'nego Vostoka SSSR Tom 3 - Zheskokrylye, ili zhuki Chast 2 [Keys to the identification of insects of the Soviet Far East 3 - Coleoptera, or beetles Part 2]. Nauka, Saint-Petersburg, 704 pp. [in Russian]

Nikitsky NB (1993) Zhuki-griboedy (Coleoptera, Mycetophagidae) fauny Rossii i sopredel'nykh stran. [Fungivorous beetles (Coleoptera, Mycetophagidae) of the fauna of Russia and adjacent countries]. Izd-vo Moskovskogo Universiteta, Moskva, 183 pp.

Nikitsky NB (1998) Generic classification of the beetle family Tetratomidae (Coleoptera, Tenebrionoidea) of the world, with description of new taxa. Series Faunistica No. 9. Pensoft Publishers, Moscow, 80 pp. [printed Jan 1998 (verso of title page)]

Nikitsky NB (2007) Novye taksony zhukov-griboedov i zhukov-tenelyubov (Coleoptera, Mycetophagidae, Melandryidae) [New taxa of the hairy fungus beetles and false darkling beetles (Coleoptera, Mycetophagidae, Melandryidae)]. Byulleten' Moskovskogo Obshchestva Ispytateley Prirody Otdel Biologicheskiy 112 (4): 58–62 [in Russian]. [Jul-Aug 2007 issue]

Nikitsky NB, Lawrence JF, Kirejtshuk AG, Gratshev WG (1994) A new beetle family, Decliniidae fam. n., from the Russian Far East and its taxonomic relationships (Coleoptera Polyphaga). Novoe semeystvo zhestkokrylykh Decliniidae fam. n. s Dalnego Vostoka Rossii i ego taksonomicheskie sbyazi (Coleoptera Polyphaga). Russian Entomological Journal 2 (5/6): 3–10. [printed Jan 1994 (inside wrapper)]

Nikitsky NB, Pollock DA (2008) Family Melandryidae [pp. 64–73]. In: Löbl I, Smetana A (Eds) Catalogue of Palaearctic Coleoptera. Volume 5. Tenebrionoidea. Apollo Books, Stenstrup, 670 pp. [publ. 15 Apr 2008 (verso of title page)]

Nikolajev GV (1970) [Taxonomic position of groups included in the subfamily Geotrupinae (Col., Scarabaeidae)]. Materials of the Second Scientific Session of Young Specialists and PhD Candidates (Thesis of Lectures), Kazakhstan, 31–34 [in Russian].

Nikolajev GV (1990) [The usage of the particularities of wing venation for clarification of the system of Lamellicornia beetles]. In: Medvedev GS, Korotyaev BA (Eds) Advances of Entomology in USSR: Coleoptera Proceedings of the X Meeting of the All-Soviet Entomology Society, Leningrad, September 11–15, 1989. USSR Academy of Sciences Zoology Institute, Leningrad, 98–99 [in Russian].

Nikolajev GV (1992) Taksonomicheskie priznaki i sostav rodov mezozoyskikh plastinchatousykh zhukov (Coleoptera, Scarabaeidae) [Taxonomic characters and generic composition of Mesozoic scarab beetles (Coleoptera, Scarabaeidae)]. Paleontologicheskii Zhurnal 1992 (1): 76–88. [in Russian, English summary] [after 14 Feb 1992 (approved to print)]

Nikolajev GV (1995a) Materialy k sistematike podsemeystva Ochodaeinae (Coleoptera, Scarabaeidae) [New data on the systematics of the subfamily Ochodaeinae (Coleoptera, Scarabaeidae)]. Zoologicheskii Zhurnal 74 (8): 72–82 [in Russian; English translation in Entomological Review 75: 113–123]. [after 11 Jul 1995 (approved to print); Aug 1995 issue]

Nikolajev GV (1995b) Novoe podsemeystvo plastinchatousykh zhukov (Coleoptera, Scarabaeidae) iz nizhnego mela Zabaykalya i ego philogeneticheskie svyazi [A new subfamily of lamellicorn beetles (Coleoptera, Scarabaeidae) from the Lower Cretaceous of Transbaykalia and its phylogenetic relationships]. Paleontologicheskii Zhurnal 1995 (2): 147–151 [in

Russian; English translation in Paleontological Journal 29: 185–192]. [after 19 Apr 1995 (approved to print, p. 160)]

Nikolajev GV (1996) Vidy plastinchatousykh zhukov (Coleoptera, Scarabaeidae) s bol'shim kolichestvom borozdok na nadkryl'yakh iz nizhnemelovogo mestonakhozhdeniya Baysa v Zabaykal'e [Lamellicorn beetle species (Coleoptera, Scarabaeidae) with multistriate elytra, from the Lower Cretaceous Baysa locality, Transbaikalia]. Paleontologicheskii Zhurnal 1996 (2): 91–99. [in Russian; English translation in Paleontological Journal 30: 217–224 (1996)] [after 12 Mar 1996 (approved to print)]

Nikolajev GV (1998) Vidy plastinchatousykh zhukov gruppy pleurosticti (Coleoptera, Scarabaeidae) iz nizhnego mela Zabaykal'ya [Pleurostict lamellicorn beetles (Coleoptera, Scarabaeidae) from the Lower Cretaceous of Transbaikalia]. Paleontologicheskii Zhurnal 1998 (5): 77–84. [in Russian; English translation in Paleontological Journal 32: 513–520 (1998)] [after 6 Aug 1998 (approved to print)]

Nikolajev GV (1999a) O polifilii podsemeystva Penichrolucaninae (Coleoptera, Lucanidae) i vydelenii iz ego sostava monotipichnoy triby Brasilucanini [On the polyphyly of the subfamily Penichrolucaninae (Coleoptera, Lucanidae), with the erection of the new monotypic tribe Brasilucanini]. Tethys Entomological Research 1: 171–172 [in Russian].

Nikolajev GV (1999b) Materialy k sistematike plastinchatousykh zhukov podsemeystva Hybosorinae (Coleoptera, Scarabaeidae) s vydeleniem novoy triby dlya chetyrekh monotipichnykh rodov iz Yuzhnoy Ameriki i opisaniem novykh taksonov iz nizhnego mela Zabaylak'ya [New data on the systematics of the scarab beetles of the subfamily Hybosorinae (Coleoptera, Scarabaeidae): the establishment of the new tribe comprising four monotypic South American genera, and description of some new taxa from the Lower Cretaceous of Transbaikalia]. Tethys Entomological Research 1: 173-182 [in Russian].

Nikolajev GV (2000a) Novoe podsemeystvo plastinchatousykh zhukov (Coleoptera, Scarabaeidae) iz nizhnego mela Zabaykal'ya i Mongolii i ego polozhenie v sisteme nadsemeysvta [A new subfamily of lamellicorn beetles (Coleoptera, Scarabaeoidae) from the Lower Cretaceous of Transbaikalia and Mongolia, and its position within the superfamily]. Paleontologicheskii Zhurnal 2000 (4): 63–66. [in Russian; English summary] [Jul-Aug issue; after 7 Jun 2000 (approved to print)]

Nikolajev GV (2000b) New subfamily of the stag beetles (Coleoptera: Scarabaeoidea: Lucanidae) from the Mesozoic of Mongolia, and its position in the system of the superfamily. Paleontological Journal 34 (suppl. 3): S327-S330. [2000 (copyright)]

Nikolajev GV (2002) O veroyatnoy prinadlezhnosti k podsemeystvu Pleocominae novogo roda plastinchatousykh zhukov iz nizhnego mela Mongolii [To possible assignment of a new genus of lamellicorn beetles from the Lower Cretaceous of Mongolia to the subfamily Pleocominae]. Paleontologicheskii Zhurnal 2002 (3): 51–55 [in Russian, English summary; English translation in Paleontological Journal 36: 279–282 (2002)] [after 8 Apr 2002 (approved to print); Sep 2002]

Nikolajev GV (2003a) Taksonomicheskiy sostav podsemeystva Bolboceratinae (Coleoptera, Scarabaeidae) Palearktiki [The taxonomic composition of the subfamily Bolboceratinae (Coleoptera, Scarabaeidae) from Palaearctic faunistic region]. Tethys Entomological Research 8: 187–206 [in Russian]. [7 Apr 2003 (page top); 8–9 Apr 2003 (inside back wrapper)]

Nikolajev GV (2003b) Opisanie lichinki roda *Eubolbitus* Reitter (Coleoptera, Scarabaeidae, Bolboceratinae). A description of the larva of the genus *Eubolbitus* Reitter (Coleoptera, Scarabaeidae, Bolboceratinae). Evraziatskii Entomologicheskii Zhurnal 1 (2): 207–209 [in Russian]. [after 22 Jan 2003 (approved to print, p. 242)]

Nikolajev GV (2005a) Omorgini (Coleoptera, Scarabaeidae, Troginae) - novaya triba plastinchatousykh zhukov. Omorgini (Coleoptera, Scarabaeidae, Troginae) - a new tribe of scarab beetles. Evraziatskii Entomologicheskii Zhurnal 4 (4): 321–322 [in Russian]. [after 22 Dec 2005 (approved to print, inside back wrapper)]

Nikolajev GV (2005b) Plastinchatousye zhuki podsemeystva Glaphyrinae (Coleoptera, Scarabaeidae) v nizhnem mele Zabaykal'ya [Scaraboid beetles of the subfamily Glaphyrinae (Coleoptera, Scarabaeidae) from the Lower Cretaceous of Transbaikalia]. Zhivotnyy mir Dal'nego Vostoka [Animal World of the Far East] 5: 69–78 [in Russian].

Nikolajev GV (2006) O taksonomicheskom statuse verkhneoligotsenovogo roda *Ceruchites* Statz (Coleoptera, Lucanidae). On the taxonomic status of the Upper Oligocene genus Ceruchites Statz (Coleoptera, Lucanidae). Evraziatskii Entomologicheskii Zhurnal 5 (2): 133–134 [in Russian]. [after 12 Jun 2006 (approved to print, inside back wrapper)]

Nikolajev GV (2007a) Mezozoyskiy etap evolyutsii plastinchatousykh (Insecta: Coleoptera: Scarabaeoidea). Kazak Universiteti, Almaty, 221 pp.

Nikolajev GV (2007b) Novoe podsemeystvo plastinchatousykh zhukov semeystva Hybosoridae (Coleoptera: Scarabaeoidea) iz mezozoya Azii [A new Hybosoridae subfamily (Coleoptera: Scarabaeoidaea) from the Mesozoic of Asia]. Izvestiya Natsionalnoi Akademii Nauk Respubliki Kazakhstan Seriya Biologicheskaya i Meditsinskaya 2 (260): 47–48 [in Russian]. [Mar-Apr 2007 issue]

Nikolajev GV (2008) [Taxonomic structure of the subfamily Aegialiinae (Coleoptera, Scarabaeidae) according to its geographic distribution]. Evraziatskii Entomologicheskii Zhurnal 7 (2): 148–149.beu

Nikolajev GV (2010) [On the systematic position of the new Scarabaeoidea genus (Coleoptera) from the Lower Cretaceous of Transbaikalia]. Tethys Entomological Research 17: 67–72 [in Russian].

Nilsson AN (2003) Family Dytiscidae [pp. 35–78] [and others]. In: Löbl I, Smetana A (Eds) Catalogue of Palaearctic Coleoptera. Volume 1. Archostemata - Myxophaga - Adephaga. Apollo Books, Stenstrup, 819 pp. [publ. 30 May 2003 (verso of title page)]

Nilsson AN (2005) Family Noteridae [pp. 87–153]. Nilsson, A N Vondel, B J van: World catalog of insects Vol 7 Amphizoidae, Aspidytidae, Haliplidae, Noteridae and Paelobiidae (Coleoptera, Adephaga). Apollo Books, Stenstrup, 171 pp. [publ. 24 Jun 2005 (verso of title page)]

Nilsson AN, Roughley RE, Brancucci M (1989) A review of the genus- and family-group names of the family Dytiscidae Leach (Coleoptera). Entomologica Scandinavica 20: 287–316. [1989 (recorded FMNH 29 January 1990)]

Noonan GR (1976) Synopsis of the supra-specific taxa of the tribe Harpalini (Coleoptera: Carabidae). Quaestiones Entomologicae 12 (1): 3–87. [Jan 1976 issue; 2 Feb (recorded at CNC)]

Nordmann A (1837) Symbolae ad monographiam staphylinorum. Academiae Caesareae Scientiarum, Petropoli [St. Petersburg], 167 pp., 2 pls. [1837 (title page)]

Notman H (1920) Staphylinidae from Florida in the collection of the American Museum of Natural History, with descriptions of new genera and species. Bulletin of the American Museum of Natural History 42: 693–732, pl. 39. [15 Dec 1920]

Nunberg M (1953) Nowa podrodzina, rodzaj i gatunek w rodzinie wyrynnikowatych (Platypodidae, Coleoptera). Novoe podsemeystvo, rod i vid v semeystve Platypodidae (Coleoptera). A new subfamily, genus and species of the family Platypodidae (Coleoptera). Annales Musei Zoologici Polonici 15 (5): 43–53 + pls. 9–10. [20 Apr 1953 (top of article); 14 Jul 1953 (recorded at BMNH)]

Nunberg M (1954) Korniki - Scolytidae, Wyrynniki - Platypodidae. Klucze do Oznaczania Owadów Polski 19 (99/100): 1–106. [5 Jun 1954 (sent to press, p. 2)]

Nunberg M (1967) Zur Kenntnis der afrikanischen Borken- und Kernkäfer-Fauna (Coleoptera Scolytidae et Platypodidae). Revue de Zoologie et de Botanique Africaines 76 (3/4): 313–340. [30 Dec 1967 (top p. 313)]

Nüsslin O (1911) Phylogenie und System der Borkenkäfer. (Fortsetzung aus Heft 11). Zeitschrift für Wissenschaftliche Insektenbiologie 7 (12): 372–378. [15 Dec 1911 (wrapper + p. 369)]

Nüsslin O (1912) Zur Phylogenie und Systematik der einheimischen Hylesinen - Die Gattungen. Naturwissenschaftliche Zeitschrift für Forst- und Landwirtschaft 10 (4/5): 267–290. [Apr-May 1912 issue; recorded 2 May 1912 (BMNH, in pencil; fide Alonso-Zarazaga and Lyal 2009: 125)]

Obenberger J (1920) Révision du genre *Bubastes* Lap. & Gory (Col. Buprestidae). Annales de la Société Entomologique de France 89 (1): 89–108. [10 Nov 1920 (p. 440)]

Oberprieler RG (1997) Classification of the African Saturniidae (Lepidoptera) - the quest for natural groups and relationships [pp. 142–155]. In: The Inaugural Conference on African Lepidoptera, Nairobi - Kenya -1–8 May 1997. Metamorphosis. Occasional Supplement No. 3: 142–155. [Dec 1997 (title page)]

Oberprieler RG (2010) A reclassification of the weevil subfamily Cyclominae (Coleoptera: Curculionidae). Zootaxa 2515: 1–35. [publ. 23 Jun 2010 (p. 1 footer)]

Oberprieler RG, Marvaldi AE, Anderson RS (2007) Weevils, weevils, weevils everywhere. Zootaxa 1668: 491–520. [publ. 21 Dec 2007 (footer p. 491)]

O'Brien CW, Wibmer GJ (1982) Annotated checklist of the weevils (Curculionidae sensu lato) of North America, Central America, and the West Indies (Coleoptera: Curculionoidea). Memoirs of the American Entomological Institute 34: i-ix, 1–382. [1982 (title page)]

Ocampo FC (2006) Phylogenetic analysis of the subfamily Anaidinae based on morphological data and revision of the New World tribe Anaidini. In: Ocampo FC (Ed): Phylogenetic analysis of the scarab family Hybosoridae and monographic revision of the New World subfamily Anaidinae (Coleoptera: Scarabaeoidea). Bulletin of the University of Nebraska State Museum 19: 13–177. [issued 1 May 2006 (verso of title page)]

Ogloblin DA (1936) Nasekomye zhestkokrylye. T. XXVI, vyp. 1. Listoedy, Galerucinae. Fauna SSSR. Novaya serie No 8. Izdatelistvo Akademii Nauk SSSR, Moskva, xiv + 455 + [1] pp. [after 19 Dec 1936 (approved to print, verso of title page)]

Ogloblin DA, Znojko DV (1950) Zhestkokrylye. Tom XVIII, vyp. 8. Pyltseedy (Sem. Alleculidae), ch. 2. Podsem. Omophlinae. In: Pavlovskiy EN, Shtakelberg AA (Eds) Fauna SSSR. Novaya seriya No 44. Izdatel'stvo Akademii Nauk SSSR, Moskva, 133 + [2] pp. [in Russian]. [after 5 Oct 1950 (approved to print, endleaf)]

Ohaus F (1912) Révision der Adoretini. (Col. Lamell. Rutelin.). Deutsche Entomologische Zeitschrift 1912 (2): 141–156, (3): 267–282, (4): 411–426, (5): 509–524, (6): 625–643. [(2): 1 Apr; (3): 30 May; (4): 31 Jul; (5): 30 Sep 1912 (Inhalt p. iii)]

Ohaus F (1918) Pars 66: Scarabaeidae: Euchirinae, Phaenomerinae, Rutelinae. In: Schenkling S (Ed) Coleopterorum Catalogus. Volumen XX. Scarabaeidae II. W. Junk, Berlin, 241 pp. [27 Jul 1918 (verso of vol. title page)]

Oke C (1925) New Australian Coleoptera (Part I.). The Victorian Naturalist 42 (1): 6–15. [6 May 1925 (p. 1)]

Oke C (1932) Aculagnathidae. A new family of Coleoptera. Proceedings of the Royal Society of Victoria (New Series) 44 (1): 22–24 + pl. 2. [issued 29 Feb 1932 (vol. title page)]

Oke CG (1951) The Coleoptera of the Russell Grimwade Expedition. Memoirs of the National Museum of Victoria 17: 19–25.

Oken L (1817) Cuviers und Okens Zoologien neben einander gestellt. Isis von Oken 1817 (8 [144–148]): cols. 1145–1176, [4 unn.], 1177–1178, [4 unn.], 1180–1182 [double cols., as 1780–82], [2 unn.], 1183–1185 [double cols., 1185 unn.].

Oken L (1843) Lehrbuch der naturphilosophie. Dritte, neu bearbeitete Auflage. F. Schulthess, Zürich, xii + 523 pp. [1843 (title page)]

Olexa A (1982) Revision der paläarktischen Arten der Gattung *Anapleus* (Coleoptera, Histeridae). Acta Entomologica Bohemoslovaca 79: 37–45. [Jan 1982 (back wrapper)]

Olivier E (1884) Essai d'une révision des espèces européennes et circaméditerranéennes de la famille des Lampyrides. L'Abeille, Journal d'Entomologie 22 (284–285): 1–56, pls. 1–2. [27 Aug 1884 (wrapper)]

Olivier E (1907) 53me fascicule. Coleoptera. Fam. Lampyridae. In: Wytsman PA (Ed) Genera Insectorum. Vol. VIII. Verteneuil & Desmet, Bruxelles, 74 pp. + 3 pls. [15 Jan 1907 (date on manuscr., p. 74); 18 Feb 1907 (Evenhuis 1994: 53)]

Olliff AS (1891) Coleoptera - (Continued) [pp. 58–81]. In: Whymper E (Ed) Supplementary Appendix to Travels amongst the Great Andes of the Equator. John Murray, London, xxii + [1 unn., Addenda] + 147 pp. + 14 pls. [Dec 1891 (The Literary News 12: 363)]

Opitz W (2002) Family 73. Cleridae [pp. 267–280]. In: Arnett RH, Jr., Thomas MC, Skelley PE, Frank JH (Eds) American beetles. Volume 2. Polyphaga: Scarabaeoidea through Curculionoidea. CRC Press, Boca Raton, xiv + 861 pp. [19 Jun 2002 (CRC website)]

Opitz W (2009) Classification, natural history, and evolution of Neorthopleurinae subfam. n. (Coleoptera, Cleridae) Part I. Generic composition of the subfamily and key to genera. Entomologica Basiliensia et Collectionis Frey 31: 135–207. [publ. 28 Dec 2009 (back wrapper)]

Opitz W (2010) Classification, natural history, phylogeny, and subfamily composition of the Cleridae and generic content of the subfamilies (Coleoptera: Cleroidea). Entomologica Basiliensia et Collectionis Frey 32: 31–128. [publ. 20 Dec 2010 (back wrapper)]

Opitz W, Herman LH (2009) Epiphloeinae Kuwert, 1893 and Ichneinae Spinola, 1841: two names for the same subfamily of checkered beetles (Coleoptera: Cleridae). The Coleopterists Bulletin 63 (2): 183–189. [publ. 24 Jun 2009 (p. 189)]

Orchymont Ad (1916) De la place que doivent occuper dans la classification les sous-familles des Sphaeridiinae et des Hydrophilinae [Col. Hydrophilidae]. Bulletin de la Société Entomologique de France 1916 (15): 235–240. [13 Nov 1916 (endleaf)]

Orchymont Ad (1919) Contribution à l'étude des sous-familles des Sphaeridiinae et des Hydrophilinae (Col. Hydrophilidae). Annales de la Société Entomologique de France 88 (1/2): 105–168. [24 Nov 1919 (p. 419)]

Osella G (1980) Nuovi curculionidi ciechi della regione Neotropicale (Coleoptera). Fragmenta Entomologica 15 (2): 365–422. [30 Jul 1980 (vol. contents)]

Outerelo R, Gamarra P (1985) 10. Las familias y géneros de los estafilínidos de la Península Ibérica. Claves para la identificación de la fauna Española. Universidad Complutense, Madrid, 139 pp. [Jul 1985 (wrapper); déposito légal 1986 (verso of title page)]

Özdikmen H (2008) New family group and genus group names for Eucnemidae (Coleoptera). Munis Entomology & Zoology 3 (2): 675–676. [Jun 2008 (title page); 20 Jun 2008 (recorded at CNC)]

Özdikmen H, Darilmaz MC (2010) Case 3514. Enhydrini Régimbart, 1882 (Insecta, Coleoptera): proposed emendation of spelling to Enhydrusini to remove homonymy with Enhydrini Gray, 1825 (Mammalia, Mustelidae). Bulletin of Zoological Nomenclature 67 (4): 285–288.

Pace R (1988) *Sphuridaethes loebli*, nuovo genere e nuova specie di Aleocharinae termitofila del Borneo (Coleoptera, Staphylinidae) (XCIX contributo alla conoscenza delle Aleocharinae). Revue Suisse de Zoologie 95 (4): 979–985. [Dec 1988 (wrapper)]

Pace R (2000a) Aleocharinae di Papua-Nuova Guinea (Coleoptera, Staphylinidae) (156º contributo alla conoscenza delle Aleocharinae). Bulletin de l'Institut Royal des Sciences Naturelles de Belgique Entomologie 70: 109–163. [22 Dec 2000]

Pace R (2000b) Athexenia nilgiriensis n. gen. e n. sp. della tribù Termitodiscini dell'India (Coleoptera, Staphylinidae) (155° contributo alla conoscenza delle Aleocharinae). Nouvelle Revue d'Entomologie (Nouvelle Série) 16 [1999] (4): 335–341. [issued 30 Jun 2000 (back wrapper)]

Pace R (2003) Aleocharinae della Penisola di Malacca (Malaysia) (Coleoptera, Staphylinidae). Bollettino del Museo Regionale di Scienze Naturali, Torino 20 [2002] (1): 33–78. [31 Mar 2003 (top of article)]

Pacheco MF (1964) Sistemática, filogenia y distribución de los heterocéridos de América (Coleoptera: Heteroceridae). Escuela Nacional de Agricultura, Colegio de Post-Graduados, Chapingo, xii + 155 pp. + 55 pls. [Jan 1964 (title page); 1965 (auct.)]

Packard AS (1874) Guide to the study of insects, and a treatise on those injurious and beneficial to crops: for the use of colleges, far-schools, and agriculturists. Fourth edition. Naturalists' Agency, Salem, viii + 715 pp. + 15 pls. [Apr 1874 (date of preface)]

Pajni HR, Kohli SK (1982) Taxonomic studies on Indian Ceutorhynchinae (Coleoptera: Curculionidae). Tribe Mecysmoderini. Oriental Insects 16 (3): 337–372. [31 Dec 1982 (vol. title page)]

Pakaluk J, Ślipiński SA (1995) Acritosomatinae, a new subfamily of Endomychidae for *Acritosoma* new genus based upon two new species from Mexico and Peru (Coleoptera: Cucujoidea). Genus 6 (3/4): 327–335. [15 Dec 1995 (p. 327 header)]

Pakaluk J, Ślipiński SA, Lawrence JF (1994) Current classification and family-group names in Cucujoidea (Coleoptera). Genus 5 (4): 223–268. [28 Dec 1994 (top of reprint)]

Pal TK, Lawrence JF (1986) A new genus and subfamily of mycophagous Bothrideridae (Coleoptera: Cucujoidea) from the Indo-Australian region, with notes on related families. Journal of the Australian Entomological Society 25 (3): 185–210. [29 Aug 1986 (wrapper)]

Palmer AR (1957) Miocene arthropods from the Mojave Desert, California. United States Geological Survey Professional Paper 294G: 237–279.

Pang X-F, Mao JL (1979) [Fauna of economic insects of China. Coleoptera: Coccinellidae (II). Part 14]. Science Press, Beijing, 170 pp. + xvi pls. [in Chinese, title translation taken from subsequent citation].

Park JK, Trac DH, Will K (2006) Carabidae from Vietnam (Coleoptera). Journal of Asia-Pacific Entomology 9 (2): 85–105. [Jun 2006 (wrapper)]

Park O (1951) Cavernicolous pselaphid beetles of Alabama and Tennessee, with observations on the taxonomy of the family. Geological Survey of Alabama / Museum Paper 31: 1–107. [15 Oct 1951 (wrapper)]

Park O (1952) A revisional study of Neotropical pselaphid beetles. Part two. Tribe Euplectini sensu latiore. Chicago Academy of Sciences, Special Publication No. 9 (2): 53–150 [incl. 7 pls.]. [18 Aug 1952 (p. 53)]

Park O (1953) New or little known pselaphid beetles of the United States, with observations on taxonomy and evolution of the family Pselaphidae. Bulletin of the Chicago Academy of Sciences 9 (14): 249–283. [16 Jan 1953 (p. 249)]

Park O (1976) [new taxa] In: Park O, Wagner JA, Sanderson MW: Review of the pselaphid beetles of the West Indies, (Coleoptera: Pselaphidae). Fieldiana: Zoology 68: i-xi, 1–90. [12 Nov 1976 (wrapper)]

Park O, Schuster RO (1955) A new subtribe of pselaphid beetles from California. Chicago Academy of Sciences Natural History Miscellanea No. 148: 1–6. [20 Jun 1955 (p. 1)]

Parker JB, Böving AG (1924) The blister beetle *Tricania sanguinipennis* - biology, descriptions of different stages, and systematic relationship. Proceedings of the United States National Museum 64 (23): 1–40 + 5 pls. [publ. 30 Jul 1924 (p. vii)]

Parry FJS (1864) A catalogue of lucanoid Coleoptera; with illustrations and descriptions of various new and interesting species. The Transactions of the Entomological Society of London (3) 2 (1): 1–113, pls. 1–12. [4 Jul 1864 (G. Wheeler 1912: 756)]

Parry FJS (1870) A revised catalogue of the lucanoid Coleoptera; with remarks on the nomenclature, and descriptions of new species. The Transactions of the Entomological Society of London 1870: 53–118, pls. 1–3. [14 Mar 1870 (G. Wheeler 1912: 757)]

Pascoe FP (1864) Longicornia Malayana; or a descriptive catalogue of the species of the three longicorn families Lamiidae, Cerambycidae and Prionidae, collected by Mr. A. R. Wallace in the Malay Archipelago. The Transactions of the Entomological Society of London (3) 3 (1): 1–96. [3 Oct 1864 (G. Wheeler 1912: 756); 712 pp. in all]

Pascoe FP (1866a) Catalogue of longicorn Coleoptera, collected in the Island of Penang by James Lamb, Esq. (Part I.). Proceedings of the Scientific Meetings of the Zoological Society of London 1866: 222–267, pls. 26–28. [8 May 1866 (page headers)]

Pascoe FP (1866b) Notices of new or little-known genera and species of Coleoptera [concl.]. The Journal of Entomology Descriptive and Geographical 2 (14): 443–493, pls. 18–19. [Jun 1866 issue]

Pascoe FP (1867a) On the Longicornia of Australia, with a list of all the described species, &c. The Journal of the Linnean Society Zoology 9 [1868] (35): 113–142, pls. 3–4. [30 Jan 1867 (wrapper); continued from (34): 80–112 publ. 11 Oct 1866]

Pascoe FP (1867b) Characters of some new genera of the coleopterous family Cerambycidae. The Annals and Magazine of Natural History (3) 19 (113): 307–319. [1 May 1867 (Evenhuis 2003: 24)]

Pascoe FP (1869a) Descriptions of new genera and species of Tenebrionidae from Australia and Tasmania. The Annals and Magazine of Natural History (4) 3 (13): 29–45, pl. 10, (14): 132–153, pl. 11. [(13): 1 Jan, (14): 1 Feb 1869 (Evenhuis 2003: 24)]

Pascoe FP (1869b) Longicornia Malayana; or a descriptive catalogue of the species of the three longicorn families Lamiidae, Cerambycidae and Prionidae, collected by Mr. A. R. Wallace in the Malay Archipelago. The Transactions of the Entomological Society of London (3) 3 (6): 497–552, (7): 553–712. [(6): 26 Jan 1869, (7): 13 Oct 1869 (G. Wheeler 1912: 756); 712 pp. in all]

Pascoe FP (1870a) Descriptions of some genera and species of Australian Curculionidae. Transactions of the Entomological Society of London 1870 (2): 181–212. [16 Jun 1870 (G. Wheeler 1912: 757)]

Pascoe FP (1870b) Contributions towards a knowledge of the Curculionidae. Part I. The Journal of the Linnean Society Zoology 10 (47): 434–458, pls. 17–19. [17 Jan 1870 (verso of vol. title page)]

Pascoe FP (1871) Descriptions of new genera and species of longicorns, including three new subfamilies. The Annals and Magazine of Natural History (4) 8 (46): 268–281, pl. 13. [1 Oct 1871 (Evenhuis 2003: 25)]

Pascoe FP (1872) Notes on Coleoptera, with descriptions of new genera and species. - Part II. The Annals and Magazine of Natural History (4) 10: 317–326, pl. 15. [1 Nov 1872 (Evenhuis 2003: 26)]

Pascoe FP (1874) Contributions towards a knowledge of the Curculionidae. Part IV. Journal of the Linnean Society Zoology 12: 1–99 + pl. i-iv. [11 Feb 1874]

Pascoe FP (1875) Additions to the Australian Curculionidae. Part VIII. The Annals and Magazine of Natural History (4) 16 (91): 55–67, pl. 1. [1 Jul 1875 (Evenhuis 2003: 27)]

Pascoe FP (1885) List of the Curculionidae of the Malay Archipelago collected by Dr. Odoardo Beccari, L. M. D'Albertis, and others. Annali del Museo Civico di Storia Naturale di Genova (2) 2 [22]: 201–332, pls. 1–3. [1885 (title page); 24 Feb 1886 (Ann. Soc. Ent. France (6) 6: Bull. Ent.: xl)]

Pascoe FP (1886) List of Curculionidae found by Mr Van Volxem in the neighbourhood of Rio Janeiro. Bulletin ou Comptes-Rendus des Séances de la Société Entomologique de Belgique (3) No. 76: cli-clvi. [séance 4 Sep 1886; 11 May 1887 (Ann. Soc. Ent. France (6) 7: Bull. Ent.: xciii)]

Pascoe FP (1888) Descriptions of some new genera and species of Curculionidae, mostly Asiatic. - Part V. The Annals and Magazine of Natural History (6) 2 (11): 409–418. [31 Oct 1888 (Evenhuis 2003: 31)]

Paulian R (1933) Revision des Coptodactylini (Col. Lamellicornia). Bulletin de la Société Entomologique de France 38 (5): 67–74. [6 Apr 1933 (p. 344)]

Paulian R (1944) Les Aglycyderidae, une famille relicte (Col.). Revue Française d'Entomologie 10 [1943] (3/4): 113–119. [28 Mar 1944]

Paulian R (1945) III Coléoptères Scarabéides de l'Indochine. Première partie. Faune de l'Empire Français. Librairie Larose, Paris, 225 + [3] pp. + map. [dépôt légal 1er trim. 1945 (last page)]

Paulian R (1950) Les Corylophidae d'Afrique (Coleoptera). Mémoires de l'Institut Francais d'Afrique Noire 12: 1–126.

Paulian R (1959) Un nouveau Coléoptère scarabéide de Madagascar, type d'une nouvelle sous-famille. Le Naturaliste Malgache 10 [1958] (1/2): 39–46. [Jan-Mar 1959 issue; 1er trim. 1959 (dépôt légal, p. 179)]

Paulian R (1984) Les Orphnidae américains (coléoptères, Scarabaeoidea). Annales de la Société Entomologique de France (Nouvelle Série) 20 (1): 65–92. [30 Mar 1984 (inside back wrapper)]

Paulian R, Descarpentries A (1982) I. Systématique [pp. 1–67]. In: Insectes Coléoptères. Cetoniidae Euchroeina. Faune de Madagascar. 57. ORSTOM, Paris, 88 pp. [printed 26 Feb 1982 (p. 88); publ. 24 Feb 1982 (p. [89])]

Paulsen MJ, Hawks DC (2008) Platyceroidini, a new tribe of North American stag beetles (Coleoptera: Lucanidae: Lucaninae). Insecta Mundi 51/58 (58): 1–2. [issued 5 Dec 2008 (fasc. title page)]

Paulus HF (1972a) Die systematische und phylogenetische Stellung der Karumiidae, mit einer Beschreibung von *Escalerina serraticornis* n. sp. aus S-Persien. Senckenbergiana Biologica 53 [1971]: 37–54. [14 Apr 1972 (article header)]

Paulus HF (1972b) Neue Pyrochroidae aus Nepal (Coleoptera, Heteromera), mit einer Diskussion der verwandschaftlichen Verhältnisse der Familie. Zeitschrift der Arbeitsgemeinschaft Österreichischer Entomologen 23 (3): 75–85. [issued Jun 1972 (p. 104)]

Paulus HF (1973) Revision der Familie Byrrhidae I: Zur Systematik und Faunistik der west-paläarktischen Vertreter der Gattung *Curimopsis* Ganglbauer 1902 (Col.: Byrrhidae: Syncalyptinae). Senckenbergiana Biologica 54 (4/6): 353–367. [21 Dec 1973 (p. 353 header)]

Paulus HF (1975) *Penicillophorus ctenotarsus* n.gen. et n. sp. aus Kolumbien, mit einer Beschreibung einer neuen Tribus Penicillophorini der Phengodidae (Col., Polyphaga, Cantharoidea). Zeitschrift der Arbeitsgemeinschaft Österreichischer Entomologen 25 [1973] (3/4): 69–80. [Jan 1975 (p. 132)]

Paulus HF (1981) Insects of Saudi Arabia. Coleoptera: fam. Cebrionidae. Fauna of Saudi Arabia 3: 257–265. [1 Dec 1981 (contents)]

Paykull G (1798) Fauna Suecica: Insecta (Coleoptera). Tomus I Joh. F. Edman, Upsaliae, viii + 358 + [2, errata] pp. [17 Jan 1798 (date of Proëmium); before 20 Jun 1798 (Allgem. Litt. Zeit.)]

Peacock ER (1978) *Trichelodes* (= *Hexanodes*), type of Trichelodini new tribe, and its affinities with the Trinodini (Coleoptera: Dermestidae). New Zealand Journal of Zoology 5 (2): 341–349. [19 Jun 1978 (vol. 5 contents)]

Pederzani F (1995) Keys to the identification of the genera and subgenera of adult Dytiscidae (sensu lato) of the world. (Coleoptera Dytiscidae). Atti della Accademia Roveretana degli Agiati (Serie VII: Classe di Scienze Matematiche Fisiche e Naturali) 4 B: 5–83. [printed Dec 1995; 23 Jan 1996 (recorded at BMNH)]

Pereira FS (1954) O género *Scatonomus* Er. (Coleoptera, Scarabaeidae). Revista Brasileira de Biologia 1: 53–78. [publ. Jan 1954 (vol. contents)]

Pereira FS, Martínez A (1956) Algunas notas sinonimicas en Phanaeini (Col. Scarabaeidae, Coprinae). Revista Brasileira de Entomologia 5: 229–239. [publ. 20 Jun 1956 (vol. indice)]

Péringuey L (1888) Second contribution to the South-African coleopterous fauna. Transactions of the South African Philosophical Society 4 [1884–88]: 67–195 + [4, pl. expl.], pls. 1–4. [read at Soc. 31 Mar 1886 (also in page headers); 1888 (Derksen and Scheiding-Göllner 1965)]

Péringuey L (1896) Descriptive catalogue of the Coleoptera of South Africa. Part II. Cicindelidae supplement. Carabidae. Transactions of the South African Philosophical Society 7 [1893–96]: 99–121, pl. 11, pp. 123–623 + i-xiv [bibliog.], pls. 3–10. [1896 (page header)]

Péringuey L (1902) Descriptive catalogue of the Coleoptera of South Africa. Fam. Scarabaeidae. Sub-families: Rutelinae, Hopliinae. Transactions of the South African Philosophical Society 12 [1901–1902]: 561–896. [issued 21 Aug 1902 (title page)]

Péringuey L (1904) Descriptive catalogue of the Coleoptera of South Africa. Fam. Lucanidae. Fam. Scarabaeidae. Sub-families: Sericinae, Melolonthinae, Trichiinae, Valginae, Cetoniinae. Transactions of the South African Philosophical Society 13: 1–293, pls. 1–4. [issued 19 Aug 1904 (title page)]

Péringuey L (1907) Descriptive catalogue of the Coleoptera of South Africa. Fam. Lucanidae. Fam. Scarabaeidae. Sub-families: Sericinae, Melolonthinae, Trichiinae, Valginae, Cetoniinae. Transactions of the South African Philosophical Society 13: 289–546, pl. 47. [issued 5 Apr 1907 (title page)]

Perkins PD (1980) Aquatic beetles of the family Hydraenidae in the Western Hemisphere: classification, biogeography and inferred phylogeny (Insecta: Coleoptera). Quaestiones Entomologicae 16 (1/2): 5–554 + fig. 1. [Jan 1980 (contents); Jan 1981 (teste Perkins)]

Perkins PD (1994) [new taxa] In: Perkins PD, Balfour-Browne J: A contribution to the taxonomy of aquatic and humicolous beetles of the family Hydraenidae in southern Africa. Fieldiana: Zoology (New Series) No. 77: viii + 1–159. [31 Aug 1994 (wrapper)]

Perkins PD (1997) Life on the effective bubble: exocrine secretion delivery systems (ESDS) and the evolution and classification of beetles in the family Hydraenidae (Insecta: Coleoptera). Annals of Carnegie Museum 66 (2): 89–207. [22 May 1997 (wrapper)]

Perkins PD (2005) A revision of the African hygropetric genus *Coelometopon* Janssens, and description of *Oomtelecopon* new genus (Coleoptera: Hydraenidae). Zootaxa 949: 1–103. [publ. 20 Apr 2005 (p. 3 footer)]

Perkovsky EE (1997a) Neotropospeonellina subtrib. n. iz Venesuely kak klyuch k poznaniyu istorii Leptodirini (Coleoptera, Leiodidae). [Neotropospeonellina subtrib. n. from Venezuela as key to knowledge of the history of Leptodirini (Coleoptera, Leiodidae)] [in Russian]. Vestnik Zoologii 31 (3): 59. [May-Jun 1997 issue]

Perkovsky EE (1997b) Significance of antennal characters, particularly the presence of Hamann's organ on the first segment of a club, for the systematics of the agyrtid-colonid group of Staphylinoidea (Coleoptera: Agyrtidae, Leiodidae, Colonidae). Dopovidi Nacionalnoi Akademii Nauk Ukraini 1997 (11): 167–174 [in Russian, English summary].

Perkovsky EE (2002) Agirtidno-kolonidnya gruppa semeystv stafilinoidov (Coleoptera: Agyrtidae, Leiodidae, Colonidae) mirovoy fauny: sistematika i filogeniya. Avtoreferat dissertatsii na soiskanie uchenoy stepeni doktora biologicheskikh nauk, Intitut Zoologii im I. I. Shmal'gauzena NAN Ukarainy; Paleontologicheskiy Institut RAN, Moskva, Moskva, 44 pp. [26 Apr 2002 (to printer)]

Perreau M (2000) Catalogue des Coléoptères Leiodidae Cholevinae et Platypsyllinae. Mémoires de la Société Entomologique de France No. 4: 1–460. [Jun 2000 (dépôt légal)]

Perreau M (2004) Family Leiodidae [pp. 133–203]. In: Löbl I, Smetana A (Eds) Catalogue of Palaearctic Coleoptera. Volume 2. Hydrophiloidea - Histeroidea - Staphylinoidea. Apollo Books, Stenstrup, 942 pp. [publ. 31 Dec 2004 (verso of title page)]

Perty JAM (1830) [Fasc. I] Insecta Brasiliensia. In: Spix J, Martius C: Delectus animalium articulatorum, quae in itinere per Brasiliam annis MDCCCXVII-MDCCCXX jussu et auspiciis Maximilliani Josephi I. Bavariae regis augustissimi peracto. Frid. Fleischer, Monachii [München], pp. 1–60, pls. 1–12. [see: Sherborne 1922: ci; Isis von Oken 1832: 137, 1212; Isis von Oken 1833: 1164; 44 pages issued as thesis in 1833, added to Delectus publ. as a book]

Perty JAM (1831) Observationes nonnullae in Coleoptera Indiae Orientalis. Dissertatio philosophico-entomologica, quam unacum praemissis thesibus auctoritate et consensu illustris philosophorum ordinis in Academia Ludovico-Maximilianea facultatem legendi rite adepturus. M. Lindauer, Monachii, xxxxiv pp., 1 pl.

Petri K (1914) Beiträge zur Käferfauna. Verhandlungen und Mittheilungen des Siebenbürgischen Vereins für Naturwissenschaften zu Hermannsstadt 65 (Festschrift): 1–23. [after 2 Sep 1914 (title page)]

Peyerimhoff P, de (1914) Nouveaux Coléoptères du Nord-Africain (Dix-huitième note: Récoltes de M. R. de Borde à Biskra). Bulletin de la Société Entomologique de France 1914 (8): 245–251. [13 May 1914 (p. [528])]

Philips TK (2002) Family 70. Anobiidae [pp. 245–260]. In: Arnett RH, Jr., Thomas MC, Skelley PE, Frank JH (Eds) American beetles. Volume 2. Polyphaga: Scarabaeoidea through Curculionoidea. CRC Press, Boca Raton, xiv + 861 pp. [19 Jun 2002 (CRC website)]

Pic M (1900) Hylophilidae, Anthicidae et Pedilidae de l'ile de Sumatra. Annales de la Société Entomologique de France 68 [1899] (4): 754–760. [Jun 1900 (wrapper)]

Pic M (1912) Pars 48: Anobiidae. In: Schenkling S (Ed) Coleopterorum Catalogus. Volumen X. W. Junk, Berlin, 92 pp. [18 Oct 1912 (verso of vol. title page)]

Pic M (1914) Pars 58: Dascillidae, Helodidae, Eucinetidae. In: Schenkling S (Ed) Coleopterorum Catalogus. Volumen X. W. Junk, Berlin, 65 pp. [3 Feb 1914 (verso of vol. title page)]

Pic M (1915) Diagnoses de nouveaux genres et nouvelles espèces de Scaphidiides. L'Échange, Revue Linnéenne 31 (368): 30–32, (369): 35–36, (370); 40, (371): 43–44. [(fasc. 368): Aug, (369): Sep, (370): Oct, (371): Nov 1915]

Pic M (1928) Notes et descriptions. Mélanges Exotico-Entomologiques 51: 1–36. [12 Apr 1928 (title page)]

Pic M (1929) Pars 103: Dasytidae: Melyrinae. In: Schenkling S (Ed) Coleopterorum Catalogus. Volumen X. W. Junk, Berlin, 32 pp. [20 Mar 1929 (verso of vol. title page)]

Pic M (1930) Contribution a l'étude des coléoptères malacodermes (2e article). Annales de la Société Entomologique de France 99 (4): 311–324. [31 Dec 1930 (p. 372, wrapper)]

Pic M (1937) Pars 155: Dasytidae: Dasytinae. In: Schenkling S (Ed) Coleopterorum Catalogus. Volumen X. W. Junk, s-Gravenhage, 130 pp. [12 Aug 1937 (verso of vol. title page)]

Pierce WD (1907) Contribution to the knowledge of Rhynchophora. Entomological News and Proceedings of the Entomological Section of the Academy of Natural Sciences of Philadelphia 18 (9): 379–385. [Nov 1907 issue]

Pierce WD (1913) Miscellaneous contributions to the knowledge of the weevils of the families Attelabidae and Brachyrhinidae. Proceedings of the United States National Museum 45 (No. 1988): 365–426. [publ. 23 May 1913 (p. vi)]

Pierce WD (1914) New potato weevils from Andean South America. Journal of Agricultural Research 1 (4): 347–352, pls. 39–41. [10 Jan 1914 (p. 347 footer)]

Pierce WD (1915) Descriptions of some weevils reared from cotton in Peru. United States Department of Agriculture, Office of the Secretary / Report No. 102: 1–16, 2 pls. [issued 25 Jan 1915 (wrapper)]

Pierce WD (1916) Studies of weevils (Rhynchophora) with descriptions of new genera and species. Proceedings of the United States National Museum 51 (No. 2159): 461–473. [16 Dec 1916 (p. viii)]

Pierce WD (1919) Contributions to our knowledge of the weevils of the superfamily Curculionoidea. Proceedings of the Entomological Society of Washington 21 (2): 21–36. [publ. 26 Feb 1919 (p. 40)]

Pierce WD (1930) Studies of the North American weevils belonging to the superfamily Platystomoidea. Proceedings of the United States National Museum 77 (No. 2840): 1–34, pls. 1–5. [10 Dec 1930 (vol. contents)]

Pierre F (1961) Écologie et distribution géographique des *Storthocnemis* Karsch. Affinités. (Tenebrionidae). XI Internationaler Kongress für Entomologie, Wien 17 bis 25 August 1960 Verhandlungen Band 1, Vienna, 555–558.

Pittino R (1985) Insects of Saudi Arabia. Coleoptera Scarabaeoidea: a revision of the family Aphodiidae. Fauna of Saudi Arabia 6 [1984]: 267–360. [issued 28 Feb 1985 (contents)]

Pittino R, Mariani G (1986) A revision of the Old World species of the genus *Diastictus* Muls. and its allies (*Platytomus* Muls., *Pleurophorus* Muls., *Afrodiastictus* n. gen, *Bordatius* n. gen.)

(Coleoptera, Aphodiidae, Psammodiini). Giornale Italiano di Entomologia 3 (12): 1–165. [30 Apr 1986 (wrapper)]

Planet LM (1924) Histoire naturelle des longicornes de France. Encyclopédie entomologique II (Série A). Paul Lechevalier, Paris, 386 pp. [printed 15 Aug 1924 (endleaf)]

Platia G (2008) Order Coleoptera, family Elateridae [pp. 194–210]. In: Harten A, van (Ed) Arthropod Fauna of the United Arab Emirates. Volume 1. Dar Al Ummah, Abu Dhabi, 754 pp. [publ. 30 Nov 2007 (article, p. 194); 20 Jan 2008 (verso of vol. title page)]

Podany C (1979) *Morimonella bednariki* (Oligorchini). Tribu, genre et espèces nouveaux (Col. Cerambycidae). Bulletin de la Société Entomologique de Mulhouse 1979 (3): 43–45. [28 Sep 1979 (contents); 15 Oct 1979 (recorded at BMNH)]

Poey F (1854) XXV. Conspectus familiarum coleopterorum. Memorias sobre la historia natural de la isla de Cuba, acompañadas de sumarios Latinos y extractos en francés Tomo 1° [pp 302–337]. Barcina, Habana, 463 pp. + 34 pls. [pp. 281–453, pls. 31–44 publ. Jun 1854 (p. 449)]

Poggi RG (2010) Gli "Annali" pubblicati dal Museo Civico di Storia Naturale "Gaicomo Doria" di Genova: storia del periodico ed indice generali dei primi cento volumi (1870–2009). Annali del Museo Civico di Storia Naturale «Giacomo Doria» 101: 1–529. [printed Sep 2010 (endleaf)]

Poggi RG, Nonveiller A, Colla D, Pavićević D, Rada T (2001) Thaumastocephalini, a new tribe of Pselaphinae for *Thaumastocephalus folliculipalpus* n. gen., n. sp., an interesting new troglobious species from central Dalmatia (Croatia) (Coleoptera, Staphylinidae, Pselaphinae). Annali del Museo Civico di Storia Naturale "Giacomo Doria" 94: 1–20. [5 Dec 2001 (reprint wrapper)]

Poinar G, Jr. (2006) *Mesophyletis calhouni* (Mesophyletinae), a new genus, species, and subfamily of early Cretaceous weevils (Coleoptera: Curculionoidea: Eccoptarthridae) in Burmese amber. Proceedings of the Entomological Society of Washington 108 (4): 878–884. [mailed 12 Oct 2006 (inside wrapper)]

Poinar G, Jr. (2008) Type genus for Mesophyletinae, a subfamily of early Cretaceous weevils (Coleoptera: Curculionoidea: Eccoptarthridae) in Burmese amber. Proceedings of the Entomological Society of Washington 110 (1): 262. [Jan 2008 issue; mailed 7 Jan 2008 (inside wrapper)]

Poinar G, Jr. (2009) *Dominibrentus leptus*, n. gen., n. sp. (Curculionoidea, Brentidae, Cyphagoginae, Dominibrentini, n. tribe), a straight-snouted weevil in Dominican amber. Historical Biology 21 (1/2): 51–55. [Mar-Jun 2009 issue]

Poinar G, Jr., Brown AE (2004) A new subfamily of Cretaceous antlike stone beetles (Coleoptera: Scydmaenidae: Hapsomelinae) with an extra leg segment. Proceedings of the Entomological Society of Washington 106 (4): 789–796. [mailed 24 Sep 2004 (inside wrapper)]

Poinar GO, Jr. (2009) *Dominibrentus leptus*, n. gen., n. sp. (Curculionoidea, Brentidae, Cyphagoginae, Dominibrentini, n. tribe), a straight-snouted weevil in Dominican amber. Historical Biology 21 (1/2): 51–55. [Mar-Jun 2009 issue]

Pollock DA (2002) Family 100. Melandridae [pp. 417–422]; 112. Mycteridae [pp. 530–533]. In: Arnett RH, Jr., Thomas MC, Skelley PE, Frank JH (Eds) American beetles. Volume 2.

Polyphaga: Scarabaeoidea through Curculionoidea. CRC Press, Boca Raton, xiv + 861 pp. [publ. 19 Jun 2002 (CRC website)]

Pongrácz A (1935) Die eozäne Insektenfauna des Geiseltales. Nova Acta Academiae Caesareae Leopoldino-Carolinae Naturae Curiosorum (Neue Folge) 2: 485–572, pls. 1–7.

Ponomarenko AG (1961) O sistematicheskom polozhenii *Coptoclava longipoda* Ping (Insecta, Coleoptera) [On the systematic position of *Coptoclava longipoda* Ping (Insecta, Coleoptera)]. Paleontologicheskii Zhurnal 1961 (3): 67–72 [in Russian]. [after 4 Sep 1961 (approved to print)]

Ponomarenko AG (1966) Zhuki semeystva Cupedidae iz nizhnego triasa Sredney Azii [Beetles of the family Cupedidae, Lower Triassic of Soviet Central Asia]. Paleontologicheskii Zhurnal 1966 (4): 47–68, pls. 3–4 [in Russian]. [after 17 Nov 1966 (approved to print)]

Ponomarenko AG (1968) Zhuki-arkhostematy yury Karatau (Coleoptera, Archostemata) [Archostematan beetles from the Jurassic of Karatau (Coleoptera, Archostemata)] [pp. 118–138, pls. 13–16]. In: Rohdendorf BB (Ed). Yurskie nasekomye Karatau [Jurassic insects of Karatau]. Izdatelstvo "Nauka", Moscow, 252 pp., 25 pls. [in Russian]. [after 26 Jun 1968 (approved to print)]

Ponomarenko AG (1969a) Istoricheskoe razvitie zhestkokrylykh -Arkhostemat [Historical development of the Coleoptera - Archostemata]. Trudy Paleontologicheskogo Instituta Akademiya Nauk SSSR 125: 1–240 [in Russian]. [after 18 Sep 1969 (approved to print)]

Ponomarenko AG (1969b) Cretaceous insects from Labrador. 4. A new family of beetles (Coleoptera: Archostemata). Psyche 76 (3): 306–310. [Sep 1969 issue; mailed 31 Dec 1969]

Ponomarenko AG (1977) [new taxa] In: Arnoldi LV, Zherikhin VV, Nikritin LM, Ponomarenko AG (Eds): Mezozoyskie Zhestkokrylye. Trudy Paleontologicheskogo Instituta Akademiya Nauk SSSR 161: 1–204 [English translation 1992: Mesozoic Coleoptera. Smithsonian Institution Libraries and N.S.F., Washington D.C.; xii + 285 pp.]. [after 5 Sep 1977 (approved to print)]

Ponomarenko AG (1985) Zhestkokrylye iz yury Sibiri i zapadnoy Mongolii [Coleoptera from the Jurassic of Siberia and western Mongolia] [pp. 47–87]. In: Rasnitsyn AP (Ed): Yurskie nasekomye Sibiri i Mongolii. Trudy Paleontologicheskogo Instituta Akademiya Nauk SSSR 211: 1–192, 18 pls. [after 20 Nov 1984 (approved to print); 1985 (title page)]

Ponomarenko AG (1994) Two new species of Mesozoic dytiscoid beetles from Asia. Paleontological Journal 27 (1A): 182–191. [Jan 1994 (wrapper); 1993 (in paper)]

Pope RD (1962) A review of the Pharini (Coleoptera: Coccinellidae). The Annals and Magazine of Natural History (13) 4 [1961] (46): 627–640. [publ. 26 Mar 1962 (wrapper)]

Portevin G (1914) Révision du genre *Estadia* Fairm. (Col.). Annales de la Société Entomologique de Belgique 58: 199–201. [1 Aug 1914 (p. 189)]

Portevin G (1926) Les Liodidae de l'Inde [pp. 75–83]. Encyclopédie Entomologique. Série B. Tome I. Coleoptera Tome 1 (1925–26), fasc. 2. Paul Lechevalier, Paris, 191 pp. [15 Jan 1926 (footer p. 63)]

Portevin G (1927) Tableaux dichotomiques pour la détermination des longicornes de France. Encyclopédie Entomologique. Vol. II. Supplément. Paul Lechevalier, Paris, 53 pp. [printed 14 Jul 1927 (endleaf)]

Portevin G (1929) Histoire naturelle des Coléoptères de France. Tome I. Adephaga. Polyphaga: Staphylinoidea. Encyclopédie Entomologique. Série A [Tome XII]. Paul Lechevalier, Paris, xii + 649 pp., pls. 1–5. [19 Nov 1929]

Portevin G (1931) Histoire naturelle des Coléoptères de France. Tome II. Polyphaga: Lamellicornia, Palpicornia Diversicornia. Encyclopédie Entomologique. Série A [Tome XIII]. Paul Lechevalier & Fils, Paris, vi + [1] + 542 pp. + pls. 6–10. [29 Jun 1931]

Portevin G (1934) Histoire naturelle des Coléoptères de France. Tome III. Polyphaga: Heteromera, Phytophaga. Encyclopédie Entomologique. Série A [Tome XVII]. Paul Lechevalier & Fils, Paris, vi + [1] + 374 pp. + pls. 11–15. [by Sep 1934 (Acad. Sci. France 199: 1164)]

Potts RWL (1974) Revision of the Scarabaeidae: Anomalinae 1. The genera occurring in the United States and Canada (Coleoptera). The Pan-Pacific Entomologist 50 (2): 148–154. [mailed 8 Nov 1974 (inside wrapper)]

Pouillaude I (1917) Les Cétonides Malgaches (suite). Insecta / Revue Illustrée d'Entomologie 6 (Apr-Jun): 49–75. [pp. 19–48 publ. Feb 1917 (p. 19 footer); before 28 Mar 1917 (Bull. Soc. Ent. France 1917: 130)]

Power G (1879) Notes pour servir à la Monographie des Brenthides. Annales de la Société Entomologique de France (5) 8 [1878] (4): 477–496. [26 Mar 1879 (wrapper)]

Prena J (2009) Apostasimerini (Coleoptera: Curculionidae: Baridinae). Rectification of authorship, year of publication, rank, and taxa included. The Coleopterists Bulletin 63 (1): 33–34. [mailed 22 Apr 2009 (inside wrapper)]

Prendini L, Wheeler WC (2005) Scorpion higher phylogeny and classification, taxonomic anarchy, and standards for peer review in online publishing. Cladistics 21 (5): 446–494. [Oct 2005 issue; 6 Dec (recorded at CNC)]

Prokin AA, Ren D (2010) New Mesozoic diving beetles (Coleoptera, Dytiscidae) from China. Paleontologicheskii Zhurnal 2010 (5): 47–53 [in Russian; English transl. in: Paleontological Journal 44 (5): 526–533].

Prosen AF (1960) Notas sobre la família Anoplodermatidae (Coleopt. Cerambycoidea). Anales del Instituto de Medicina Regional 5 (2): 87–100.

Puchkov AV, Matalin AV (2003) Subfamily Cicindelinae Latreille, 1802 [pp. 99–118]. In: Löbl I, Smetana A (Eds) Catalogue of Palaearctic Coleoptera. Volume 1. Archostemata - Myxophaga - Adephaga. Apollo Books, Stenstrup, 819 pp. [publ. 30 May 2003 (verso of title page)]

Puthz V (1967) The scientific results of the Hungarian soil zoological expedition to the Brazzaville-Congo. 4. Beitrag zur Kenntnis der Unterfamilie Steninae (Coleoptera: Staphylinidae). Opuscula Zoologica (Budapest) 7 (1): 185–194.

Putzeys JAAH (1867) Révision générales des clivinides. Annales de la Société Entomologique de Belgique 10 [1866]: 1–242. [printed before 29 Sep 1867 (Ann. Soc. Ent. Belg. 11 Comptes-Rendus: xlvi)]

Quedenfeldt G (1885) Verzeichniss der von Herrn Major a. D. von Mechow in Angola und am Quango-Strom 1878–1881 gesammelten Tenebrioniden und Cisteliden. Berliner Entomologische Zeitschrift 29 (1): 1–38, pl. 3. [Jul 1885 (wrapper)]

Quedenfeldt G (1887) Verzeichniss der von Herrn Major a. D. von Mechow in Angola und am Quango-Strom 1878–1881 gesammelten Anthothribiden und Bostrychiden. Berliner Entomologische Zeitschrift 30 [1886] (2): 303–328, pl. 8. [Jan 1887 (fasc. cover)]

Quentin RM (1954) Contribution à l'étude des Coléoptères Cerambycidae. I. - A propos du genre *Combesius* Lepesme. Revue Française d'Entomologie 21 (2): 103–108. [printed 22 Jun 1954 (p. 148)]

Quentin RM, Villiers A (1969) Révision des Plectogasterini, nov. trib. [Col. Cerambycidae Cerambycinae]. Annales de la Société Entomologique de France (Nouvelle Série) 5 (3): 613–646. [dépôt légal 3ième trim. 1969 (p. 774)]

Quentin RM, Villiers A (1975) Insectes Coléoptères Cerambycidae Parandrinae et Prioninae. Faune de Madagascar. 40. Paris, 251 pp. + map. [publ. 29 Aug 1975 (p. [252])]

Raffray A (1890) Étude sur les Psélaphides. V. Tableaux synoptiques.- Notes et synonymie. Revue d'Entomologie 9 (3–6): 81–172. [Mar 1890 (footer p. 77), Apr (p. 93), May (p. 125), Jun (p. 157)]

Raffray A (1900) Psélaphides et Clavigérides de Madagascar. Annales de la Société Entomologique de France 68 [1899] (3): 516–525. [Feb 1900 (wrapper); 28 Feb 1900 (Bull. Soc. Ent. France 1900: 77)]

Raffray A (1904) Genera et catalogue des Psélaphides. Annales de la Société Entomologique de France 72 (4): 484–604, 3 pls. [Sep 1904 (wrapper)]

Raffray A (1908) 64me fascicule. Coleoptera. Fam. Pselaphidae. In: Wytsman PA (Ed) Genera Insectorum. Vol. X. P. Wytsman, Bruxelles, 487 pp. + 9 pls. [15 Feb 1908 (date on manuscr., p. 486); 17 Mar 1908 (recorded at BMNH fide Evenhuis 1994: 54); 8 Apr 1908 (Bull. Soc. Ent. France 1908: 85)]

Raffray A (1917) Sur le genre *Mirus* Saulcy (*Imirus* Reitt.). Bulletin de la Société Entomologique de France 1917 (5): 108–110. [11 Apr 1917 (p. [402])]

Rafinesque CS (1815) Analyse de la nature ou tableau de l'univers et des corps organisés. C. S. Rafinesque, Palermo, 224 pp. [publ. betw. Apr and 21 Jul 1815 (Stafleu and Cowan 1983: 557)]

Rakovič M (1987) A revision of the genus *Odochilus* Harold with remarks on the tribal classification of the subfamily Aphodiinae (Coleoptera, Scarabaeidae). Acta Entomologica Bohemoslovaca 84 (1): 27–44. [30 Jan 1987 (contents p. 4)]

Rakovič M, Král D (1997) New taxa, new combinations and current taxonomic status of tribes and genera of Psammodiinae (Coleoptera: Scarabaeoidea: Aphodiidae). Acta Societatis Zoologicae Bohemicae 61 (3): 233–247. [17 Oct 1997 (verso of vol. title page)]

Raphael S (1970) The publication dates of the Transactions of the Linnean Society of London, Series I, 1791–1875. Biological Journal of the Linnean Society 2: 61–76. [Mar 1970 (top of article)]

Ratcliffe BC (2002) Family 23. Lucanidae [pp. 6–9]. In: Arnett RH, Jr., Thomas MC, Skelley PE, Frank JH (Eds) American beetles. Volume 2. Polyphaga: Scarabaeoidea through Curculionoidea. CRC Press, Boca Raton, xiv + 861 pp. [19 Jun 2002 (CRC website)]

Redtenbacher L (1844) Tentamen dispositionis generum et specierum coleopterorum pseudotrimerorum Archiducatus Austriae. Zeitschrift für die Entomologie (Germar) 5: 113–132. [1844 (reprint of original thesis dated 1844a)]

Redtenbacher L (1845) Die Gattungen der deutschen Kaefer-Fauna nach der analytischen Methode bearbeitet, nebst einem kurz gefassten Leitfaden, zum Studium dieses Zweiges der Entomologie. C. Ueberreuter, Wien, [13] + 177 + [1] pp. + 2 pls. [Jul 1845 (date of preface); Sep 1845 (Ent. Zeit. Stettin 6: 295)]

Reed EC (1874) On the Coleoptera Geodephaga of Chile. Proceedings of the Scientific Meetings of the Zoological Society of London 1874 (1): 48–70, pl. 13. [Jun 1874 (Duncan, Waterhouse and Peavot 1937: 73)]

Rees BE (1943) Classification of the Dermestidae (larder, hide, and carpet beetles) based on larval characters, with a key to the North American genera. United States Department Agriculture / Miscellaneous Publications No. 511: 1–18. [issued Apr 1943]

Régimbart M (1879) Étude sur la classification des Dytiscidae. Annales de la Société Entomologique de France (5) 8 [1878] (4): 447–466, pl. 10. [26 Mar 1879 (wrapper; Ann. Soc. Ent. France (5) 9: Bull. Bibliogr.: 13)]

Régimbart M (1882) Essai monographique de la famille des Gyrinidae. 1re partie. Annales de la Société Entomologique de France (6) 2 (3): 379–400. [27 Dec 1882 (Lefèvre 1895)]

Reichardt AN (1936) Zhuki-chernotelki triby Opatrini palearkticheskoy oblasti [Darkling beetles of the tribe Opatrini (Coleoptera Tenebrionidae) of the Palearctic Region]. Opredeliteley po fauny SSSR [Keys to the fauna of the USSR]. Tom. 19. Zoological Institute of the Russian Academy of Sciences, Moscow 224 pp. [in Russian] [1936 (title page); after 27 Jun 1936 (approved to print)]

Reichardt AN (1941) Nasekomye zhestkokrylye. Sem. Sphaeritidae i Histeridae (Chast 1) [Fam. Sphaeritidae and Histeridae (part 1)]. Fauna SSSR Novaya Seriya. Tom III, wyp. 5. Zoologicheskiy Institut Akademia Nauk SSSR, Moscow, xiii + 419 + 1 pp.

Reichardt H (1974) Monograph of the neotropical Helluonini, with notes and discussions on Old World forms (Coleoptera: Carabidae). Studia Entomologica (nova série) 17 (1/4): 211–302. [publ. 31 Oct 1974 (p. 560)]

Reichardt H (1976) Sphaeriidae in Insecta and Mollusca: Comments on the Secretary's revised proposals. Comment (4). Bulletin of Zoological Nomenclature 32 (4): 203–204. [30 Jan 1976 (p. 193)]

Reiche L (1866) Étude des espèces de mylabrides de la collection de L. Reiche, suivie d'une note sur le genre *Trigonurus* Mulsant et description d'une espèce nouvelle. Annales de la Société Entomologique de France (4) 5 [1865] (4): 627–642. [26 May 1866 (p. xci)]

Reid CAM (1993) *Eboo*, nom. n.: redescription of type species (Coleoptera: Chrysomelidae: Eumolpinae). The Coleopterists Bulletin 47 (1): 61–67. [mailed 30 Mar 1993 (inside wrapper)]

Reid CAM (1999) A new generic synonym in the Australian Lucanidae (Coleoptera). The Coleopterists Bulletin 53 (2): 175–177. [mailed 17 Jun 1999 (inside wrapper)]

Reid CAM (2000) Spilopyrinae Chapuis: a new subfamily in the Chrysomelidae and its systematic placement (Coleoptera). Invertebrate Taxonomy 14 (6): 837–862. [1 Dec 2000 (recorded at FMNH)]

Reid CAM (2006) A taxonomic revision of the Australian Chrysomelinae, with a key to the genera (Coleoptera: Chrysomelidae). Zootaxa 1292: 1–119. [publ. 14 Aug. 2006 (footer p. 1)]

Reinwald C (1860) Catalogue annuel de la librairie française. Deuxième année - 1859. C. Reinwald, paris, viii + 328 pp.

Reitter E (1875) Revision der europäischen Crytophagiden. Deutsche Entomologische Zeitschrift 19 (3): 1–86 [sep. pagin.]. [Mar 1875 (fasc. title page)]

Reitter E (1876a) Systematische Eintheilung der Trogositidae (Familia coleopterorum). Verhandlungen des Naturforschenden Vereines in Brünn 14 [1875]: 3–69, pls. 1–2. [1876 Februar (handwritten in BMNH reprint); 15 Aug 1876 (Petites Nouv. Ent.) / reissued in: Deut. Ent. Zeits. 20 (4): 27–95, pls. 1–2, publ. May 1876 (fasc. title page)]

Reitter E (1876b) Neue Gattungen und Arten aus der Familie der Cucujidae. Coleopterologische Hefte 15: 37–64. [1 Nov 1876 (Ent. Nachr. 2: 176)]

Reitter E (1878) Nachträge und Ergänzungen zur Bearbeitung der Cioiden von H. v. Kiesenwetter. Deutsche Entomologische Zeitschrift 22 (1): 21–30. [May 1878 (Inhalt)]

Reitter E (1879) Bestimmungs-Tabellen der europäischen Coleopteren. I. Enthaltend die Familien: Cucujidae, Telmatophilidae, Tritomidae, Mycetaeidae, Endomychidae, Lyctidae und Sphindidae. Verhandlungen der Kaiserlich-Königlichen Zoologisch-Botanischen Gesellschaft in Wien 29 (1 Halbjahr): 71–100. [1880 (vol. title page); 1 Oct 1879 (Ent. Nachr. 5: 260) / separate: 30 pp., 1880]

Reitter E (1882a) Bestimmungs-Tabellen der europäischen Coleopteren. VI. Enthaltend die Familien: Colydiidae, Rhysodidae, Trogositidae. Verhandlungen des Naturforschenden Vereines in Brünn 20 [1881]: 113–149. [1882 (vol. title page); at least May 1882 (papers cited) Jun 1883 (Tijdschr. Ent.) / separate: 37 pp., 1882]

Reitter E (1882b) Versuch einer systematischen Eintheilung der Clavigeriden und Pselaphiden. Verhandlungen des Naturforschenden Vereines in Brünn 20 [1881]: 177–211. [1882 (vol. title page); at least May 1882 (papers cited)]

Reitter E (1882c) [Clavigeridae, Pselaphidae, Scydmaenidae] In: Naturgeschichte der Insecten Deutschlands. Erste Abtheilung, Coleoptera. Dritter Band, Zweite Abtheilung [Erste Lieferung. Bogen 1 bis 13]. Nicolai'sche Verlags-Buchhandlung, Berlin, vi + 198 pp. [1882 (title page); 15 Oct 1882 (Ent. Nachr.); Jul-Dec 1882 (Griffin 1939: 218)]

Reitter E (1884) Bestimmungs-Tabellen der europäischen Coleopteren. XII. Necrophaga (Platypsyllidae, Leptinidae, Silphidae, Anisotomidae und Clambidae). Reitter, Brünn, pp. 3–122. [separate, end 1884 (Deut. Ent. Zeits. 30: 219); 1884 (Zool. Jahresber. 1884) / reissued in: Verh. Naturf. Ver. Brünn 23 (1): 3–122, 1885 (Heft title page)]

Reitter E (1886) Beitrag zur Systematik der Grotten-Silphiden. Wiener Entomologische Zeitung 5 (9): 313–316. [20 Nov 1886 (wrapper + page footers)]

Reitter E (1887) Bestimmungs-Tabellen der Europäischen Coleopteren. III. Enthaltend die Familien: Scaphididae, Lathridiidae und Dermestidae. II. Auflage. E. Reitter, Mödling, 75 pp. [1886 (title page); 1887 (wrapper; Wien. Ent. Zeit. 34: 227)]

Reitter E (1889a) Zwei neue Coleopteren-Gattungen aus Transkaukasien. Wiener Entomologishe Zeitung 8 (9): 289–292 + pl. 4. [20 Nov 1889 (wrapper; page footers)]

Reitter E (1889b) Bemerkungen und Berichtigungen zu den Clavicornen in der Fauna Baltica 2. Aufl. und Fauna Transsylvanica von Dr. G. Seidlitz. Ein Beitrag zur Lösung strittiger, coleopterologischer Fragen. Deutsche Entomologische Zeitschrift 1889 (2): 289–318. [Nov 1889 (Inhalt)]

Reitter E (1891) [Pselaphiden bis Elateriden (pp. 123–210); Cerambyciden (pp. 337–355)]. In: Reitter E (Ed) Heyden, L,von, Reitter, E and Weise, J: Catalogus Coleopterorum Europae, Caucasi et Armeniae Rossicae. R. Friedländer & Sohn, Berlin, vii + 420 pp. [13 May 1891 (Ann. Soc. Ent. France 60: Bull. Ent.: lxxxvii)]

Reitter E (1894) Bestimmungs-Tabelle der Coleopteren-Familie der Cleriden, des palaearktischen Faunengebietes. XXVIII. Heft. Verhandlungen des Naturforschenden Vereines in Brünn 32 [1893]: 37–89. [1894 (Hetschko 1915: 243) / separate, with different title: Brünn: Reitter, 55 pp. + [1, Index], 1894 (wrapper)]

Reitter E (1900) Bestimmungs-Tabelle der Tenebrioniden-Abtheilungen: Tentyrini und Adelostomini aus Europa und den angrenzenden Ländern [XLII. Heft]. Reitter, Paskau, [1] + pp. 82–197. [separate, 1900 (title page) / reissued in: Verh. Naturf. Ver. Brünn 39 [1900]: 82–197, 1901 (title page)]

Reitter E (1904) Bestimmungs-Tabelle der Tenebrioniden. Unterfamilien: Lachnogyini, Akidini, Pedinini, Opatrini und Trachyscelini aus Europa und den angrenzenden Ländern [LIII. Heft]. Verhandlungen des Naturforschenden Vereines in Brünn 42 [1903]: 25–189. [1904 (vol. title page) / separate, with different title: Brünn: Reitter, pp. 25–189, May 1904 (wrapper); appeared Aug 1904 (Semenov 1904b: 324, footnote)]

Reitter E (1905) Bestimmungs-Tabelle der palaearctischen, mit *Athous* verwandten Elateriden (subtribus Athouina), mit einer Uebersicht der verwandten Coleopteren-Familien: Sternoxia und mit einem Bestimmungsschlüssel der Gattungen der Elateridae [LVI. Heft]. Verhandlungen des Naturforschenden Vereines in Brünn 43 [1904]: 3–122. [1904 (vol. title page); 1905 (Hetschko 1915: 257) / separate, with different title: Brünn: Reitter, 122 pp., 1905 (wrapper)]

Reitter E (1906) [families Trichopterygidae, Tenebrionidae and others] In: Heyden L, von, Reitter E, Weise J: Catalogus Coleopterorum Europae, Caucasi et Armeniae rossicae. Editio secunda. R. Friedländer & Sohn, Berlin, iv pp. + 750 cols. + pp. 751–774. [1906 (title page)]

Reitter E (1909) Fauna Germanica. Die Käfer des Deutschen Reiches. Nach der analytischen Methode bearbeitet. II. Band. K. G. Lutz, Stuttgart, 392 pp., pls. 41–80. [1909 (title page); 1 Jan 1910 (Ent. Bericht. 3 (51): 40); Feb 1910 (Ent. Litteraturbl.)]

Reitter E (1911) Fauna Germanica. Die Käfer des Deutschen Reiches. Nach der analytischen Methode bearbeitet. III. Band. K. G. Lutz, Stuttgart, 436 pp., pls. 81–128. [1911 (title page); 1 Jan 1912 (Ent. Bericht. 3 (63): 224)]

Reitter E (1913a) Fauna Germanica. Die Käfer des Deutschen Reiches. Nach der analytischen Methode bearbeitet. IV. Band. K. G. Lutz, Stuttgart, 236 pp. + pls. 129–152. [1912 (title page); 1913 (Zool. Record; Arch. Naturg. 80B); Feb 1914 (Ent. Litteraturbl.); 1 Mar 1914 (Ent. Bericht. 4 (76): 62)]

Reitter E (1913b) Bestimmungs-Tabellen der europäischen Coleopteren. LXVIII. Heft (68.) enthält Bestimmungs-Schlüssel für die Unterfamilien, Tribus und Gattungen der Curculionidae. (19. Teil.). Bestimmungs-Schlüssel der mir bekannten europäischen Gattungen der Curculionidae, mit Einschluss der mir bekannten Gattungen aus dem palaearctischen Gebiete. 90 pp. [separate, 1912 (wrapper); 1913 (Arch. Naturg. 80B (5): 129) / reissued in: Verh. Naturf. Ver. Brünn 51[1912]: 1–90, 1913 (wrapper); 19 May 1914 (recorded at BMNH)]

Reitter E (1913c) Bestimmungs-Tabelle der Borkenkäfer (Scolytidae) aus Europa und den angrenzenden Ländern [XXXI. Heft]. Wiener Entomologische Zeitung 32 (Beiheft): 116 pp. [15 May 1913 (title page) / separate: Paskau: Reitter, 116 pp., 1913 (wrapper)]

Reitter E (1916) Bestimmungstabelle der Tenebrioniden, enthaltend die Zopherini, Elenophorini, Leptodini, Stenosini und Lachnogyini aus der paläarktischen Fauna [Heft 79]. Wiener Entomologische Zeitung 35 (5/7): 129–171. [12 Sep 1916 (wrapper) / separate, with different title on cover: Paskau: E. Reitter, pp. 129–171, 1916 (wrapper)]

Reitter E (1917) Bestimmungs-Schlüssel für die Unterfamilien und Tribus der paläarktischen Tenebrionidae [81. Heft]. Wiener Entomologische Zeitung 36 (3/5): 51–66, (9/10): 296 [Berichtigung]. [10 Jul (fasc. 3/5), 30 Dec 1917 (9/10) (wrappers) / separate, with different title on cover: Paskau: E. Reitter, pp. 51–66, without Bericht., 1917 (wrapper)]

Reitter E (1920) Bestimmungs-Tabellen der europaeischen Coleopteren. Heft 87. Tenebrionidae. XV. Teil. Belopinae, Borinae, Tenebrioninae und Coelometopinae aus der palaearktischen Fauna [cover]. Bestimmungstabelle der Unterfamilien: Belopinae, Borinae, Tenebrioninae und Coelometopinae der Tenebrioniden (Col.). Emmerich Reitter, Paskau, 24 pp. [1920 (wrapper)]

Reitter E (1922a) Bestimmungs-Tabellen der europaeischen Coleopteren. VI. Heft. Enthaltend die Familien: Colydiidae, Rhysodidae, Ostomidae. Zweite, gänzlilch umgearbeitete und auf die palaearktische Fauna ausgedehnte Auflage. Emmerich Reitter, Troppau, 73 pp. [1922 (wrapper)]

Reitter E (1922b) Bestimmungstabelle der paläarktischen Helopinae (Col. Tenebrionidae). I. Teil [Heft 92]. Wiener Entomologische Zeitung 39 (1/4): 1–44. [30 Mar 1922 (fasc. title page) / separate, with different title on cover: Troppau: E. Reitter, pp. 1–44, 1922 (wrapper) / Tiel II: 39 (5/10): 113–171, 25 Oct 1922 (fasc. title page)]

Rey C (1886) Histoire naturelle des Coléoptères de France (suite) [Tribu des Palpicornes, part 2]. Annales de la Société Linnéenne de Lyon (Nouvelle Série) 32 [1885]: 1–186 + [4, errata and explic. planches], pls. 1–2. [1886 (vol. title page)]

Ribera I, Beutel RG, Balke M, Vogler AP (2002) Discovery of Aspidytidae, a new family of aquatic Coleoptera. Proceedings of the Royal Society of London / Biological Sciences (Series B) 269 (1507): 2351–2356. [22 Nov 2002 (paper); 18 Oct 2002 (online)]

Richard R (1983) Insectes coléoptères. Curculionidae Stigmatrachelini. Faune de Madagascar. 62. ORSTOM, Paris, 194 pp. [publ. 22 Dec 1983 (p. [195])]

Richter AA (1949) Nasekomye zhestkokrylye. Tom XIII, vypusk 2. Zlatki (Buprestidae). Chast' 2. Fauna SSSR Novaya seriya No 37. Izdatel'stvo Akademii Nauk SSSR, Moskva, 255 pp. [in Russian].

Riedel A (2006) Revision of the subgenus Metaeuops Legalov of Euops Schoenherr (Coleoptera, Curculionoidea, Attelabidae) from the Papuan region. Zootaxa 1181: 1–102. [publ. 21 Apr 2006 (footer p. 1)]

Riedel A (2010) A new tribe, genus and species of Nemonychidae from Baltic amber (Coleoptera: Curculionoidea: Nemonychidae: Cimberidinae). Insect Systematics & Evolution 41 (1): 29–38. [17 Jun 2010 (recorded at CNC)]

Riley EG, Clark SM, Flowers RW, Gilbert AJ (2002) Family 124. Chrysomelidae [pp. 617–691]. In: Arnett RH, Jr., Thomas MC, Skelley PE, Frank JH (Eds) American beetles. Volume 2. Polyphaga: Scarabaeoidea through Curculionoidea. CRC Press, Boca Raton, xiv + 861 pp. [19 Jun 2002 (CRC website)]

Ritsema C (1869) [untitled note on *Platypsyllus castoris*]. Petites Nouvelles Entomologiques (1) 1 (10): 23–38. [15 Nov 1869 (top of issue)]

Ritsema C (1884) Bijlagen. I. Lijst van Coleoptera verzameld in de Afdeeling Saleijer. Bijdragen tot de taal-, land- en volkenkunde van Nederlandsch-Indië (4) 8 (3): 492–498. [1884 (title page); at least Aug 1884 / reissued as: Bijdrage tot de Kennis der Coleopteren-Fauna van het eiland Saleijer en vat het naburige eilandje Poeloe-Katela. Tijdschr. Ent. 27 (4): 253–264, by Jan 1885 (Tijdschr. Ent. 29)]

Ritsema C (1892) On two genera described by James Thomson in his "Systema Cerambycidarum". Notes from the Leyden Museum 14 (1/2): 54. [5 Jul 1892 (recorded at MCZ)]

Rivalier E (1971) Remarques sur la tribu des Cicindelini (Col.Cicindelidae) et sa subdivision en sous-tribus. Nouvelle Revue d'Entomologie 1 (2): 135–143. [30 Jun 1971 (fasc. title page)]

Roeschke H (1898) Carabidologische Notizen III. Entomologische Nachrichten 24 (17/18): 283–285. [Sep 1898 (fasc. title page); 3 Oct 1898 (recorded at USNM)]

Rohdendorf BB (1944) A new family of Coleoptera from the Permian of the Urals. Comptes Rendus (Doklady) de l'Academie des Sciences de l'URSS (Nouvelle Série) 44 (6): 252–253. [30 Aug 1944 (top of issue title page)]

Rohdendorf BB (1961) Nadotryad Coleopteroidea. Zhestkokryloobraznye [Superorder Coleopteroidea] [pp. 393–469]. In: Rohdendorf BB, Becker-Migdisova EE, Martynova OM, Sharov AG (Eds): Paleozoyskie nasekomye Kuznetskogo Basseyna [Palaeozoic insects of the Kuznetsk Basin]. Trudy Paleontologicheskogo Instituta Akademiya Nauk SSSR 85: 1–705 [in Russian]. [after 30 Jun 1961 (approved to print)]

Roig-Juñent S (2000) The subtribes and genera of the tribe Broscini (Coleoptera: Carabidae): cladistic analysis, taxonomic treatment, and biogeographical considerations. Bulletin of the American Museum of Natural History 255: 1–90. [issued 25 Sep 2000 (title page)]

Roig-Juñent S, Cicchino AC (2001) *Chaltenia patagonica*, new genus and species belonging to Chalteniina, a new subtribe of Zolini (Coleoptera: Carabidae). The Canadian Entomologist 133 (5): 651–670. [Sep 2001 (wrapper)]

Rosenhauer WG (1856) Thiere Andalusiens nach dem Resultate einer Reise zusammengestellt, nebst den Beschreibungen von 249 neuer oder bis jetzt unbeschriebenen Gattungen und Arten. Theodor Blaesing, Erlangen, viii + 429 pp. + 3 pls. [1856 (title page); Oct 1856 (date of preface); Nov 1856 (Evenhuis 1997b: 668)]

Roudier A (1958) Curculionides de l'archipel de Madère. Revue Française d'Entomologie 25 (3): 199–214. [printed 15 Sep 1958 (p. 240)]

Roughley RE (2000) [new taxa] In: Larson DJ, Alarie Y, Roughley RE: Predaceous diving beetles (Coleoptera: Dytiscidae) of the Nearctic region, with emphasis on the fauna of Canada and Alaska. National Research Coucil of Canada Research Press, Ottawa, xiv + 982 pp. [2000 (title page)]

Rousseau E (1905) 38me fascicule. Coleoptera Adephaga. Fam. Carabidae Subfam. Anthiinae. In: Wytsman PA (Ed) Genera Insectorum. Vol. VI. P. Wytsman, Bruxelles, 19 pp. + 2 pls. [14 Jun 1905 (date on manuscr., p. 19); 1905 (title page); 26 May 1906 (Evenhuis 1994)]

Roux C (1976) On the dating of the first edition of Cuvier's *Règne Animal*. Journal of the Society for the Bibliography of Natural History 8 (1): 31.

Sabrosky CW (1999) Family-group names in Diptera. Myia 10: 1–360.

Sahlberg J (1876) Enumeratio Coleopterorum brachelytrorum Fenniae. Systematisk förteckning öfver de inom Finlands naturalhistoriska område hittills funna Coleoptera Brachelytra... I. Staphylinidae. Acta Societatis pro Fauna et Flora Fennica 1 [1875–77] (3): 1–248. [5 Feb 1876 (title page)]

Sainte-Claire Deville J (1923) Faune des Coléoptères du Bassin de la Seine par Louis Bedel. Supplément aux Rhynchophora. Rédigé d'après les notes de L. Bedel Annales de la Société Entomologique de France 6 bis (1, publication hors série): 1–80 (of 159).

Saito A (1990) Female reproductive organs of cerambycid beetles from Japan and the neighboring areas. I. Philini through Atimiini. Elytra 18: 231–260.

Sama G (1990) Description d'un *Certallum* de l'Afrique du nord (Coleoptera Cerambycidae). L'Entomologiste 46: 287–289. [Dec 1990 (dépôt légal, back wrapper)]

Sama G (2008) Preliminary note on the cerambycid fauna of North Africa with the description of new taxa (Insecta Coleoptera Cerambycidae). Quaderno di Studi e Notizie di Storia Naturale della Romagna 27: 217–245. [Dec 2008 (top of article)]

Sama G (2009a) New nomenclatural acts in Cerambycidae (Coleoptera, Cerambycidae). Entomologia Africana 14: 22–26.

Sama G (2009b) [new taxon] In: Sama G, Sudre J: New nomenclatural acts in Cerambycidae. II. (Coleoptera). Bulletin de la Société Entomologique de France 114 (3): 383–388. [Sep 2009 (back wrapper)]

Sama G (2010) New acts and comments. Cerambycidae [pp. 49–58]. In: Löbl I, Smetana A (Eds) Catalogue of Palaearctic Coleoptera. Volume 6. Chrysomeloidea. Apollo Books, Stenstrup, 924 pp. [publ. 22 Feb 2010 (verso of title page)]

Sánchez-Ruiz M (1996) Catalogo bibliografico de las especies de la familia Elateridae (Coleoptera) de la Peninsula Iberica e Islas Baleares. Documentos Fauna Iberica 2. Museo Nacional de Ciencias Naturales Consejo Superior de Investigaciones Cientificas, Madrid, 265 pp.

Santiago-Blay JA (2008) Case 3398. Aulacoscelinae Chapuis, 1874 (Insecta, Coleoptera, Orsodacnidae or Chrysomelidae): proposed conservation. Bulletin of Zoological Nomenclature 65 (2): 97–105. [Jun 2008 issue]

Santos-Silva A, Martins UR (2004) Notas e descrições em Disteniinae (Coleoptera, Cerambycidae). Revista Brasileira de Zoologia 21 (1): 145–152. [Mar 2004 issue]

Sasaji H (1967) A revision of the Formosan Coccinellidae (I) the subfamily Sticholotinae, with a establishment of a new tribe (Coleoptera). Etizenia 25: 1–28. [10 Dec 1967 (wrapper)]

Sasaji H (1978) Notes on the Japanese Endomychidae, with an establishment of a new subfamily (Coleoptera). Memoirs of the Faculty of Education, Fukui University (Series II, Natural Science) 28: 1–31. [25 Dec 1978]

Sasaji H (1986) Systematic position of the genus *Eidoreus* Sharp (Coleoptera: Clavicornia). In: Papers on entomology presented to Prof. Takehiko Nakane in commemoration of his retirement. Special Bulletin of the Japanese Society of Coleopterology: 229–235. [10 Oct 1986]

Satô M (2006) Family Chelonariidae [pp. 454–455]. In: Löbl I, Smetana A (Eds) Catalogue of Palaearctic Coleoptera. Volume 3. Scarabaeoidea - Scirtoidea - Dascilloidea - Buprestoidea - Byrrhoidea. Apollo Books, Stenstrup, 690 pp. [publ. 30 Jun 2006 (verso of title page)]

Saulcy F, de (1870) [footnote, p. 90] In: Heyden, L. von. Entomologische Reise nach dem südlichen Spanien, der Sierra Guadarrama und Sierra Morena, Portugal und der Cantabrischen Gebirgen. Mit Beschreibungen der neuen Arten von L. v. Heyden und den Mitgliedern des Berliner entomol. Vereins. Berliner Entomologische Zeitschrift 14 (Beiheft): 1–218, pls. 1–2. [1870 (wrapper; Arch. Naturg. for 1870; Berl. Ent. Zeitschr. 14, etc.); 1871 (Evenhuis 1997, without comment)]

Savoyskaya GI (1969) O vydelenii novykh taksonomiceskikh kategoriy koktsinellid (Col., Coccinellidae) [On the formation of new taxonomic categories of coccinellids (Col., Coccinellidae)]. Vestnik Selskokhozyaystvennoy Nauki Kazakhstana, Alma-Ata 12 (9): 101–106. [24 Sep 1969 (to printer)]

Ščegoleva-Barovskaya T (1929) Der erste Vertreter de Familie Mordellidae aus der Jura-formation Turkestans [in Russian, German summary]. Comptes Rendus de l'Academie des Sciences de l'URSS 8 (1): 27–29. [5 Apr 1929 (recorded at BMNH)]

Schaeffer C (1911) New Coleoptera and miscellaneous notes. Journal of the New York Entomological Society 19 (2): 113–126. [Jun 1911 issue]

Schalkwyk AD, van (1966) Change of curculionid (Coleoptera) generic name from *Cryptopharynx* to *Cryptolarynx*. Suid-Afrikaanse Tydskrif vir Landbouwetenskap / South African Journal of Agricultural Science 9 (3): 745. [Sep 1966 issue]

Schauberger E (1934) Zur Kenntnis der paläarktischen Harpalinen (Vierzehnter Beitrag). Koleopterologische Rundschau 20 (3/4): 99–118. [31 Jul 1934 (wrapper)]

Schauberger E (1937) Zur Kenntnis der australisch-melanesischen Harpalinen. (Zweiter Beitrag). Entomologische Rundschau 54 (21): 272–275. [1 Mar 1937 (top of fasc.)]

Schaufuss LW (1872) Tabellen-Entwurf zur Bestimmung der Pselaphiden-Gattungen. Nunquam Otiosus 2 (Lief. 1): 243–248. [24 Jul 1872 (footnote p. 280)]

Schaufuss LW (1882a) Pselaphidarum Monographiae. Annali del Museo Civico di Storia Naturale "Giacomo Doria" 18 [1882–83]: 173–206. [7–8 Mar 1882 (footer pp. 161, 177)]

Schaufuss LW (1882b) Neue Pselaphiden im Museo Civico di Storia Naturale zu Genua. Annali del Museo Civico di Storia Naturale "Giacomo Doria" 18: 349–399. [11 Jul (footer p. 338) -26 Aug 1882 (p. 385); Aug 1882 (Ann. Soc. Ent. France (6) 2: Bull. Ent.: cli)]

Schaufuss LW (1889) Neue Scydmaeniden im Museum Ludwig Salvator. Berliner Entomologische Zeitschrift 33 (1): 1–42. [end Sep 1889 (wrapper)]

Schaufuss LW (1890) System-Schema der Pselaphiden, ein Blick in die Vorzeit, in die Gegenwart und in die Zukunft. Tijdschrift voor Entomologie 33 [1889–90] (2): 101–162, pls. 2–6 and foldout. [Jun 1890 (Nederl. Ent. Ver.)]

Schaum HR (1857a) Beitrag zur Käferfauna Griechenlands. Erstes Stück: Cicindelidae, Carabici, Dytiscidae, Gyrinidae. Berliner Entomologische Zeitschrift 1: 116–158. [1857 (title page); after Aug 1857 (p. 115)]

Schaum HR (1857b) Naturgeschichte der Insecten Deutschlands. Erste Abtheilung. Coleoptera. Erster Band. Erste Hälfte [zweite Lieferung, pp. 191–352, Carabici]. Nicolai, Berlin, vi + 791 pp. [Mar 1857 (p. v)]

Schaum HR (1859) Catalogus Coleopterorum Europae. In Verbindung mit Dr. G. Kraatz und H. v. Kiesenwetter. Nicolaische Verlagsbuchhandlung, Berlin, iv + 121 pp. [1859 (wrapper); Feb 1859 (Wien. Ent. Monatschr. 3: 64, reviewed in Jul, p. 210)]

Schaum HR (1860) Naturgeschichte der Insecten Deutschlands. Erste Abtheilung. Coleoptera. Erster Band. Erste Hälfte [vierter Lieferung, pp. 553–791, Carabici]. Nicolai, Berlin, vi + 791 pp. [printed early 1859 (p. v); Feb 1860 (date of forward)]

Schaum HR (1863) Descriptions of four new genera of Carabidae. The Journal of Entomology, Descriptive and Geographical 2 (8): 74–78, pl. 4. [Apr 1863 issue]

Schedl KE (1939) Die Einteilung und geographische Verbreitung der Platypodidae. 56. Beitrag zur Morphologie und Systematik der Scolytidae und Platypodidae [pp. 377–410]. In: Jordan K, Hering EM (Eds) Verhandlungen, VII. Internationaler Kongress für Entomologie. Band I. Internationalen Kongresse für Entomologie, Weimar, 617 pp., 29 pls. [Apr 1939 (title page)]

Schedl KE (1959) Scolytidae und Platypodidae aus Angola (Coleoptera). 1. Beitrag. Publicações Culturais da Companhia de Diamantes de Angola No. 48: 15-[28]. [separate publ. 1 Dec 1959 (title page)]

Schedl KE (1961) 1. Beitrag zur Systematik afrikanischer Brenthiden (Col.). Entomologische Arbeiten aus dem Museum G Frey Tutzing bei München 12 (1): 185–204. [publ. 1 Apr 1961 (vol. Inhalt)]

Schedl KE (1966) Interessante und neue Scolytoidea aus Afrika. 244. Beitrag zur Morphologie und Systematik der Scolytoidea. Revista de Entomologia de Moçambique 8 (1 [1965]): 349–379. [printed Oct 1966 (back wrapper)]

Schedl KE (1972) *Monographie der Familie Platypodidae. Coleoptera.* W. Junk, Den Haag, v + 322 pp. [publ. 30 Jun 1972]

Scheerpeltz O (1934) Pars 130. Staphylinidae VIII. In: Schenkling S (Ed) Coleopterorum Catalogus. Volumen VI. Staphylinidae II. W. Junk, Berlin, pp. 1501–1881. [2 Jan 1934 (verso of vol. title page)]

Scheerpeltz O (1944) Erster Nachtrag zur Bestimmungstabelle der in der paläarktischen Region durch Arten vertretenen Gattungen der XVII. Fam. Staphylinidae (35. Beitrag zur Kenntnis der paläarktischen Staphyliniden). Koleopterologische Rundschau 30 (4/6): 169–172. [30 Sep 1944 (wrapper)]

Scheerpeltz O (1968) Teil XV fa: Coleoptera-Staphylinidae. Catalogus Faunae Austriae Ein systematisches Verzeichnis aller auf österreichischem Gebiet festgestellten Tierarten. Springer-Verlag, Wien, 279 pp. [1968 (cover)]

Scheerpeltz O (1974) Coleoptera: Staphylinidae (exclus. Subfam. Paederinae, except. pars min.) [pp. 43–394]. In: Hanström B, Brinck P, Rudebeck G (Eds) South African animal life. Results of the Lund University expedition in 1950–1951. Vol. XV. Swedish Natural Science Research Council, Stockholm., 532 pp. [printed 1974 (verso of title page)]

Schenkling S (1910) Pars 23: Cleridae. In: Schenkling S (Ed) Coleopterorum Catalogus. Volumen X. W. Junk, Berlin, pp. 1–174. [29 Nov 1910 (verso of vol. title page)]

Schenkling S (1915) Pars 64: Derodontidae, Lymexylonidae, Micromalthidae. In: Schenkling S (Ed) Coleopterorum Catalogus. Volumen X. W. Junk, Berlin, pp. 3–6, 1–14. [30 Jan 1915 (verso of vol. title page)]

Schenkling S (1921) Pars 72: Scarabaeidae: Cetoniinae. In: Schenkling S (Ed) Coleopterorum Catalogus. Volumen XXI. Scarabaeidae III. W. Junk, Berlin, 431 pp. [20 Jan 1921 (verso of vol. title page)]

Schenkling S (1922) Pars 75: Scarabaeidae: Trichiinae, Valginae. In: Schenkling S (Ed) Coleopterorum Catalogus. Volumen XXI. Scarabaeidae III. W. Junk, Berlin, 58 pp. [1 Apr 1922 (verso of vol. title page)]

Schenkling S (1927) Pars 88: Elateridae I [pp. 265–636]. In: Schenkling S (Ed) Coleopterorum Catalogus. Volumen XI. W. Junk, Berlin, 636 pp. [22 Feb 1927 (verso of vol. title page)]

Schenkling S, Marshall GAK (1929) Pars 106: Curculionidae: Subfam. Byrsopinae, Rhytirrhininae, Thecesterninae, Hipporrhininae, Rhyparosominae In: Schenkling S (Ed) Coleopterorum Catalogus. Volumen XXVIII. W. Junk, Berlin, 62 pp. [15 Aug 1929 (verso of vol. title page)]

Schenkling S, Marshall GAK (1931a) Pars 116: Curculionidae: Dinomorphinae, Somatodinae, Amycterinae, Gonipterinae. In: Schenkling S (Ed) Coleopterorum Catalogus. Volumen XXVIII. Curculionidae II. W. Junk, Berlin, pp. 1–2 (Dinomorphinae, Somatodinae), pp. 1–39 (Amycterinae), pp. 1–11 (Gonipterinae). [30 Mar 1931 (verso of vol. title page)]

Schenkling S, Marshall GAK (1931b) Pars 114. Curculionidae. Eremninae, Leptopinae, Tanyrrhynchinae, Cylindrorrhinae, Thecesterninae (Suppl.), Rhytirrhininae (Suppl.), Rhyparosominae (Suppl.). In: Schenkling S (Ed) Coleopterorum Catalogus. Volumen XXVIII. Curculionidae II. W. Junk, Berlin, pp. 1–39 (Eremninae), 1–83 (Leptopinae), 1–10 (Tanyrrhynchinae), 1–23 (Cylindrorrhinae), 1–4 (Thecesterninae, Rhytirrhininae, Rhyparosominae). [12 Feb 1931 (verso of vol. title page)]

Schenkling S, Marshall GAK (1936) Pars 150: Curculionidae: Subfam. Haplonychinae. In: Schenkling S (Ed) Coleopterorum Catalogus. Volumen XXIX. W. Junk, Berlin, 8 pp. [15 Aug 1936 (verso of vol. title page)]

Schimmel R, Tarnawski D (2009) Monograph of the tribe Quasimusini (Insecta: Coleoptera, Elateridae, Negastriinae). Polish Taxonomical Monographs 17: 1–169.

Schiødte JMC (1848) Mödet den 4de Juni [untitled]. Oversigt over det Kongelige Danske Videnskabernes Selskabs Forhandlinger 1847: 67–69. [1848 (vol. title page)]

Schiødte JMC (1849) Specimen faunae subterraneae. Bidrag til den underjordiske fauna. Bianco Luno, Kjöbenhavn, 39 pp., 4 pls. [1849 (title page) / reissued, same pagin. in: Kongel. Danske Vidensk. Selsk. Skrifter 2, 1851]

Schlechtendal DHR, von (1894) Beiträge zur Kenntnis fossiler Insekten aus dem Braunkohlengebirge von Rott am Siebengebirge. Abhandlungen der Naturforschenden Gesellschaft zu Halle 20: 197–228, pls. 12–14. [28 Aug 1894 (recorded at BMNH)]

Schmidt A (1910a) 110me fascicule. Coleoptera Lamellicornia. Fam. Aphodiidae. Genera Insectorum. Vol. XVII. P. Wytsman, Bruxelles, 155 pp. + 3 pls. [15 Dec 1909 (date on manuscr., p. 155); 1910 (title page); 31 Dec 1910 (Evenhuis 1994: 55)]

Schmidt A (1910b) Pars 20: Aphodiinae. In: Schenkling S (Ed) Coleopterorum Catalogus. Volumen XIX. Junk, W., Berlin, 111 pp. [30 Sep 1910 (verso of vol. title page)]

Schmidt F (1852) Zwei neue Arten von *Leptoderus*. Entomologische Zeitung (Stettin) 13 (11): 381–382. [Nov 1852 issue]

Schmidt J (1885) Tabellen zur Bestimmung der europäischen Histeriden. Berliner Entomologische Zeitschrift 29 (2): 279–330. [Dec 1885 (wrapper)]

Schoch G (1894) Ueber die Systematik der Cetoniden. Mittheilungen der Schweizerischen Entomologischen Gesellschaft 9 (4): 164–225. [Oct 1894 (wrapper)]

Schoch G (1895) Die Genera und Species meiner Cetonidensammlung. I. Teil: Goliathidae, Gymnetidae, Madagassae, Schizorrhinidae. Zürcher & Furrer, Zürich, iii + 63 + [1] pp.

Schoenfeldt H, von (1908) 65me fascicule. Coleoptera. Fam. Brenthidae. In: Wytsman PA (Ed) Genera Insectorum Vol. XI. P. Wytsman, Bruxelles, 88 pp. + 2 pls. [15 Mar 1908 (date on manuscr., p. 88); 23 Jun 1908 (Evenhuis 1994: 54)]

Scholtz CH (1988) [new taxa] In: Scholtz CH, d'Hotman D, Evans AV, Nel A: Phylogeny and systematics of the Ochodaeidae (Insecta: Coleoptera: Scarabaeoidea). Journal of the Entomological Society of Southern Africa 51 (2): 207–240. [1988 (copyright)]

Scholtz CH (1991) Nomenclatural changes in the Ochodaeidae (Coleoptera: Scarabaeoidea). The Coleopterists Bulletin 45 (1): 30. [mailed 21 Mar 1991 (inside wrapper)]

Schönherr CJ (1823) Curculionides. Tabula synoptica familiae curculionidum. Isis von Oken 1823 (10): cols. 1132–1146. [publ. Oct 1823 (Dieckmann 1983: 260)]

Schönherr CJ (1825) Continuatio tabulae synopticae familiae curculionidum. Isis von Oken 1825 (5): cols. 581–588. [May 1825 issue]

Schönherr CJ (1826) Curculionidum dispositio methodica, cum generum characteribus, descriptionibus atque observationibus variis, seu prodromus ad Synonymiae Insectorum, partem IV. Fleischer, Lipsiae, x +[1]+ 338 pp. [1826 (title page); 7 Oct 1825 (date of preface p. x)]

Schönherr CJ (1833) Genera et species curculionidum, cum synonymia hujus familiae. Species novae aut hactenus minus cognitae, descriptionibus a Dom. Leonardo Gyllenhal, C. H. Boheman, et entomologis aliis illustratae. Tomus primus. - Pars prima. Roret, Paris, xv + 381 pp. [8 Aug 1833 (Ann. Soc. Ent. France 2: Bull. Ent.: xlvi)]

Schönherr CJ (1836) Genera et species curculionidum, cum synonymia hujus familiae. Species novae aut hactenus minus cognitae, descriptionibus a Dom. Leonardo Gyllenhal, C. H. Boheman, et entomologis aliis illustratae. Tomus tertius.- Pars secunda. Roret, Paris, pp. 507–858. [16 Mar 1836 (Ann. Soc. Ent. France 5: Bull. Ent.: xx)]

Schönherr CJ (1837) Genera et species curculionidum, cum synonymia hujus familiae. Species novae aut hactenus minus cognitae, descriptionibus a Dom. Leonardo Gyllenhal, C. H. Boheman, et entomologis aliis illustratae. Tomus quartus.- Pars secunda. Roret, Paris, pp. 601–1121 + [1122–1124, Corrigenda]. [1837 (Erichson 1838: 231); before 6 Jan 1838 (Bibliogr. France 1838: 10)]

Schönherr CJ (1839) Genera et species curculionidum, cum synonymia hujus familiae. Species novae aut hactenus minus cognitae, descriptionibus Dom. L. Gyllenhal, C. H. Boheman, O. J. Fahraeus et entomologis aliis illustratae. Tomus quintus. -Pars prima. Roret, Paris, viii + 456 pp. [15 Jun 1839 (Bibliogr. France 1839: 286); 7 Aug 1839 (Ann. Soc. Ent. France 8: Bull. Ent.: xxiii)]

Schönherr CJ (1840) Genera et species curculionidum, cum synonymia hujus familiae. Species novae aut hactenus minus cognitae, descriptionibus Dom. L. Gyllenhal, C. H. Boheman, O. J. Fahraeus et entomologis aliis illustratae. Tomus quintus. - Pars secunda, supplementum continens. Roret, Paris, viii + 465–970 + [4, Corrig.] pp. [18 Mar 1840 (Ann. Soc. Ent. France 9: Bull. Ent.: viii); 21 Mar 1840 (Bibliogr. France 1840: 141)]

Schönherr CJ (1844) Genera et species curculionidum, cum synonymia hujus familiae. Species novae aut hactenus minus cognitae, descriptionibus a Dom. L. Gyllenhal, C. H. Boheman, O. J. Fahraeus et entomologis aliis illustratae. Tomus octavus. - Pars prima. Supplementum continens. Roret, Paris, pp. 1–442 [439–442: Corrig.]. [17 Apr 1844 (Ann. Soc. Ent. France (2) 2: Bull. Ent.: xxxiv)]

Schönherr CJ (1845) Genera et species curculionidum, cum synonymia hujus familiae. Species novae aut hactenus minus cognitae, descriptionibus a Dom. L. Gyllenhal, C. H. Boheman, O. J. Fahraeus et entomologis aliis illustratae. Tomus octavus. - Pars secunda. Supplementum continens. Roret, Paris, [5] + 504 pp. [342–454: Mantissa Curculionidum] + 28 foldouts [Synopsis Geographica]. [12 Mar 1845 (Ann. Soc. Ent. France (2) 3: Bull. Ent.: xiv); 17 May 1845 (Bibliogr. France 1845: 262)]

Schröder C (1905) Inhalts-Verzeichnis [rev. of Lameere, 1903]. Zeitschrift für Wissenschaftliche Insektenbiologie 1: iii-xv.

Schuler L (1970) La classification des Harpalomorphi de France. Les formes étranges (coléoptères carabiques). L'Entomologiste 25 [1969] (5/6): 108–118. [dép. légal 1e trim. 1970 (endleaf)]

Schultze A (1902) Kritisches Verzeichniss der bis jetzt beschriebenen Palaearctischen Ceuthorrhynchinen unter Nachweis der darauf bezüglichen wichtigsten Arbeiten und Angabe ihrer bekannten geographischen Verbreitung sowie einem Vorwort. Deutsche Entomologische Zeitschrift 1902 (1): 193–226. [Jul 1902 (Inhalt)]

Schwarz O (1897) Über die systematische Stellung der Elateriden-Gattungen *Dicronychus* Cast. und *Tarsalgus* Cand. Deutsche Entomologische Zeitschrift 1897 (1): 9–16. [Jul 1897 (Inhalt)]

Schwarz O (1902) Ueber die Elateriden-Gattungen *Protelater* Sharp und *Anaspasis* Cand. und ihre systematische Stellung nebst Beschreibung einer neuen Gattung und Art. Deutsche Entomologische Zeitschrift 1902 (2): 364–366. [Oct 1902 (Inhalt)]

Schwarz O (1906) 46A / 46B fascicule [pp. 1–112 / 113–224]. Coleoptera. Fam. Elateridae. In: Wytsman PA (Ed) Genera Insectorum. Vol. VII. P. Wytsman, Bruxelles, 370 pp. [1906 (title page); both fasc. 10 Nov 1906 (Evenhuis 1994: 53); continued with plates in 46C, 1907]

Schwarz O (1907) 51me fascicule. Coleoptera. Fam. Dicronychidae. In: Wytsman PA (Ed) Genera Insectorum. Vol. VIII. Verteneuil & Desmet, Bruxelles, 5 pp. [15 Jan 1907 (date on manuscr., p. 74); 26 Mar 1907 (Evenhuis 1994: 53)]

Sciaky R (1996) New taxa and new synonyms among Pterostichinae from Asia (Coleoptera Carabidae). Entomofauna Zeitschrift für Entomologie 17 (29): 429–440. [31 Dec 1996]

Scudder SH (1882) Nomenclator Zoologicus. An alphabetical list of all generic names that have been employed by naturalists for recent and fossil animals from the earliest times to the close of the year 1879. In two parts: I. Supplemental list. II. Universal index. Bulletin of the United States National Museum No. 19: xxi + 376 + [1] + 340 pp. [26 Apr 1882 (date of advertisement)]

Scudder SH (1893) Tertiary rhynchophorous Coleoptera of the United States. Monographs of the United States Geological Survey 21: 1–206, pls. 1–12. [1893 (title page)]

Seeno TN, Wilcox JA (1982) Leaf beetle genera (Coleoptera: Chrysomelidae). Entomography 1: 1–221.

Seevers CH (1941) Taxonomic investigations of some termitophilous Staphylinidae of the subfamilies Aleocharinae and Trichopseniinae (new subfamily). Annals of the Entomological Society of America 34 (2): 318–346, pls. 1–3. [mailed 2 Jul 1941 (p. iv)]

Seevers CH (1944) A new subfamily of beetles parasitic on mammals. Staphylinidae, Amblyopininae. Field Museum of Natural History Zoological Series 28 (3): 155–172, pls. 10–12. [17 Mar 1944 (wrapper)]

Seevers CH (1957) A monograph on the termitophilous Staphylinidae (Coleoptera). Fieldiana: Zoology 40: 1–334. [9 Apr 1957 (wrapper)]

Seevers CH (1958) A revision of the Vatesini, a tribe of Neotropical myrmecophiles (Coleoptera, Staphylinidae). Revista Brasileira de Entomologia 8: 181–202. [15 May 1958 (Indice)]

Seevers CH (1965) The systematics, evolution and zoogeography of staphylinid beetles associated with army ants (Coleoptera, Staphylinidae). Fieldiana: Zoology 47 (2): 139–351. [22 Mar 1965 (wrapper)]

Seevers CH (1978) [new taxa] In: Seevers CH, Herman LH: A generic and tribal revision of the North American Aleocharinae (Coleoptera: Staphylinidae). Fieldiana: Zoology 71: vi + 275 pp. [28 Apr 1978 (wrapper)]

Seidlitz G (1872) Fauna Baltica. Die Käfer (Coleoptera) der Ostseeprovinzen Russlands [2. Lieferung. Gattungen, pp. 25–48, Arten, pp. 129–208]. H. Laakmann, Dorpat, xlii + 142 + 560 pp. [Jan 1872 (Vorwort)]

Seidlitz G (1874) Fauna Baltica. Die Käfer (Coleoptera) der Ostseeprovinzen Russlands [3. Lieferung. Gattungen, pp. 49–80, Arten, pp. 209–340]. H. Laakmann, Dorpat, xlii + 142 + 560 pp. [Jan 1874 (date of Vorwort); by Apr 1874 (Berl. Ent. Zeitschr. 18: 240)]

Seidlitz G (1875) Fauna Baltica. Die Käfer (Coleoptera) der Ostseeprovinzen Russlands [4. Lieferung. Gattungen, pp. 81–142, Arten, pp. 341–560]. H. Laakmann, Dorpat, xlii + 142 + 560 pp. [to printer 11 Oct 1875 (verso of title page)]

Seidlitz G (1888) Fauna Baltica. Die Kaefer (Coleoptera) der Deutschen Ostseeprovinzen Russlands. Zweite neu bearbeitete Auflage [3. Lieferung. Familien, pp. xli-xlviii; Gattungen, pp. 49–80; Arten, pp. 225–336]. Hartungsche Verlagsdruckerei, Königsberg, [9] + lvi + 192 + 818 pp. + 1 pl. [Fall 1888 (Vorwort)]

Seidlitz G (1889) Fauna Baltica. Die Kaefer (Coleoptera) der Deutschen Ostseeprovinzen Russlands. Zweite neu bearbeitete Auflage [4. Lieferung. Gattungen, pp. 81–128; Arten, pp. 337–512]. Hartungsche Verlagsdruckerei, Königsberg, [9] + lvi + 192 + 818 pp. + 1 pl. [Spring 1889 (Vorwort)]

Seidlitz G (1890) Fauna Baltica. Die Kaefer (Coleoptera) der Deutschen Ostseeprovinzen Russlands. Zweite neu bearbeitete Auflage [5. Lieferung. Gattungen, pp. 129–160, Arten, pp.

513–608]. Hartungsche Verlagsdruckerei, Königsberg, [9] + lvi + 192 + 818 pp. + 1 pl. [Spring 1890 (Vorwort)]

Seidlitz G (1891) Fauna Baltica. Die Kaefer (Coleoptera) der Deutschen Ostseeprovinzen Russlands. Zweite neu bearbeitete Auflage [6. Lieferung. Vorwort; Familien, pp. xlix-lvi; Gattungen, pp. 161–192; Arten, etc., pp. 609–818]. Hartungsche Verlagsdruckerei, Königsberg, [9] + lvi + 192 + 818 pp. + 1 pl. [Feb 1891 (Vorwort)]

Seidlitz G (1893) Erste Abteilung. Coleoptera. Fünfter Band. [Erste Hälfte]. Zweite Lieferung. Bogen 13a bis 25 [pp. 201–400, Tenebrionidae]. In: Kiesenwetter H, von, Seidlitz G (Eds). Naturgeschichte der Insecten Deutschlands. Nicolaische Verlags-Buchhandlung, Berlin, 882 pp. [Mar 1893 (Vorbemerkung to 1898 issue: xxviii)]

Seidlitz G (1894) Erste Abteilung. Coleoptera. Fünfter Band. [Erste Hälfte]. Dritte Lieferung. Bogen 26 bis 38 [pp. 401–608, Tenebrionidae]. In: Kiesenwetter H, von, Seidlitz G (Eds) Naturgeschichte der Insecten Deutschlands. Nicolaische Verlags-Buchhandlung, Berlin, 882 pp. [May 1894 (Vorbemerkung to 1898 issue: xxviii)]

Seidlitz G (1895) Erste Abteilung. Coleoptera. Fünfter Band. [Erste Hälfte]. Vierte Lieferung. Bogen 39 bis 50 [pp. 609–800, Tenebrionidae]. In: Kiesenwetter H, von, Seidlitz G (Eds) Naturgeschichte der Insecten Deutschlands. Nicolaische Verlags-Buchhandlung, Berlin, 882 pp. [Sep 1896 (Vorbemerkung to 1898 issue: xxviii) but 1895 (title page) and reviewed Jul 1895 (Ent. Nachr. 21: 222)]

Seidlitz G (1896) Erste Abteilung. Coleoptera. Fünfter Band. Zweite Hälfte. Erste Lieferung. Bogen 1 bis 19 [pp. 1–304, Alleculidae]. In: Kiesenwetter H, von, Seidlitz G (Eds) Naturgeschichte der Insecten Deutschlands. Nicolaische Verlags-Buchhandlung, Berlin, 1206 pp. [Nov 1896 (date of preface); before 14 Jan 1897 («1896» Zool. Anz., Bibliogr. Zool. 2: 73)]

Seidlitz G (1898) Erste Abteilung. Coleoptera. Fünfter Band. Zweite Hälfte. Zweite Lieferung. Bogen 20 bis 43 [pp. 305–680, Nachtrag, Lagriidae, Melandryidae]. In: Kiesenwetter H, von, Seidlitz G (Eds) Naturgeschichte der Insecten Deutschlands. Nicolaische Verlags-Buchhandlung, Berlin, 1206 pp. [May 1898 (Introd. on verso of fasc. wrapper); before 22 Jun 1898 (Bull. Soc. Ent. France 1898: 252)]

Seidlitz G (1916) Die letzten Familien der Heteromeren (Col.). Deutsche Entomologische Zeitschrift 1916 (2): 113–128. [1 Jul 1916 (Inhalt)]

Seidlitz G (1917a) Die letzten Familien der Heteromeren (Col.). Deutsche Entomologische Zeitschrift 1917 (1): 65–116. [1 Jul 1917 (Inhalt)]

Seidlitz G (1917b) Die letzten Familien der Heteromeren (Col.) (Fortsetzung). Deutsche Entomologische Zeitschrift 1916 (5/6): 387–498. [1 Feb 1917 (Inhalt)]

Selander RB (1957) The systematic position of the genus *Nephrites* and the phylogenetic relationships of the higher groups of Rhipiphoridae (Coleoptera). Annals of the Entomological Society of America 50 (1): 88–103. [issued 15 Feb 1957 (inside wrapper)]

Selander RB (1960) Bionomics, systematics, and phylogeny of *Lytta*, a genus of blister beetles (Coleoptera, Meloidae). Illinois Biological Monographs No. 28: vi + 295 pp. [distrib. 30 Dec 1960 (verso of title page)]

Selander RB (1964) Sexual behavior in blister beetles (Coleoptera: Meloidae) I. The genus Pyrota. The Canadian Entomologist 96 (8): 1037–1082. [mailed 25 Sep 1964 (p. 1260)]

Selander RB (1966) A classification of the genera and higher taxa of the Meloid subfamily Eleticinae (Coleoptera). The Canadian Entomologist 98 (5): 449–481. [mailed 11 May 1966 (p. 672)]

Selander RB (1991) On the nomenclature and classification of the Meloidae (Coleoptera). Insecta Mundi 5 (2): 65–94. [Jun issue]

Selman BJ (1965) A revision of the Nodini and a key to the genera of Eumolpidae of Africa (Coleoptera: Eumolpidae). Bulletin of the British Museum (Natural History) Entomology 16 (3): 143–174. [issued 23 Aug 1965 (verso of title page)]

Semenov AP (1892) De Brenthidarum genere novo palaearctico. Trudy Russkago Entomologicheskago Obshchestva / Horae Societatis Entomologicae Rossicae 26 [1891–92] (3/4): 438–443. [7 Jul 1892 (Julian 25 Jun, after Index)]

Semenov AP (1893a) Symbolae ad cognitionem Pimeliidarum. Trudy Russkago Entomologicheskago Obshchestva / Horae Societatis Entomologicae Rossicae 27 [1892–93] (1/2): 249–264. [22 Apr 1893 (Julian 10 Apr, after Index)]

Semenov AP (1893b) De Coleopterorum familia nova. Mélanges Biologiques tirés du Bulletin de l'Académie Impériale des Sciences de St-Pétersbourg 13: 359–366. [read 14 Apr 1893; 1894 (Derksen & Scheiding-Göllner 1965); reissued in: Bull. Acad. Imp. Sci. St.-Pétersbourg (N. S.) 3 [35]: 607–614 (printed Aug 1894)]

Semenov AP (1894) Symbolae ad cognitionem Oedemeridarum. Trudy Russkago Entomologicheskago Obshchestva / Horae Societatis Entomologicae Rossicae 28 [1893–94]: 449–474. [printed Nov 1894 (verso of title page)]

Semenov AP (1904a) Analecta coleopterologica. VIII. Russkoe Entomologicheskoe Obozrenie 4 (5): 201–202.

Semenov AP (1904b) [Review of Reitter: Bestimmungs-Tabelle, LIII Heft]. Russkoe Entomologicheskoe Obozrenie 6 (3/4): 324–326.

Semenov AP (1907a) De novo Pimeliinorum genere, quod tribum peculiarem repraesentat (Coleoptera, Tenebrionidae). Russkoe Entomologicheskoe Obozrenie 6 [1906] (3/4): 257–260. [Dec 1906 (footer); 1907 (Zool. Record; Kerzhner, pers. corresp.; fasc. includes events in 1907)]

Semenov AP (1907b) Synopsis generum tribus Platyopinorum (Coleoptera, Tenebrionidae Pimeliini). Horae Societatis Entomologicae Rossicae 38 [1907–1908]: 175–184. [publ. 3 May 1907 (Roman; p. vii); Oct 1908 (verso of vol. title page)]

Semenov AP (1909) [note in Bulletin Entomologique, séance of 8 Dec]. Russkoe Entomologicheskoe Obozrenie 8 [1908]: xxv. [Jan 1909 (footer)]

Semenov AP (1914) Analecta coleopterologica. XVIII. Russkoe Entomologicheskoe Obozrenie 14 [1914] (1): 14–22. [14 Jul 1914 (Gregorian; p. xxii); 1915 (vol. title page)]

Semenov AP, Martynov AB (1925) [new taxa] In: Martynov AB: Ob odnom interesnom iskopaemom zhuke iz yurekikh slantsev severnogo Turkestana. On a new interesting fossil beetle from the jurassic beds in North Turkestan. Russkoe Entomologicheskoe Obozrenie 19 (2): 73–78, pl. 1 [in Russian]. [1925 (fasc. title page)]

Semenov AP, Pjatakova V (1936) Sur un genre irrédit, représentant une nouvelle tribu dans la famille Elateridae (Coleoptera). Nový neznámý rod, zástupce nové skupiny v čeledi Elateridae. Časopis České Společnosti Entomologické 33 (3): 101–103. [15 May 1936 (wrapper)]

Semenov AP, Reichardt AN (1925) Vysoko spetsializovannyy predstavitel novykh roda i triby podsemeystva Aphodiini (Coleoptera, Scarabaeidae). De novo peculiari Aphodiinorum genere, tribum propriam formante (Coleoptera, Scarabaeidae). Revue Russe d'Entomologie 19 (2): 83–88. [1925 (fasc. title page)]

Sen Gupta T (1967) A new subfamily of Languriidae (Coleoptera) based on four genera, with a key to the species of *Toramus*. Proceedings of the Royal Entomological Society of London (Series B. Taxonomy) 36 (11/12): 167–176. [29 Dec 1967]

Sen Gupta T (1968a) Review of the genera of the tribe Loberini (Coleoptera: Languriidae). Breviora / Museum of Comparative Zoology No. 303: 1–27. [31 Dec 1968 (top of article)]

Sen Gupta T (1968b) Revision of the genera of Cladoxenini (= Cladoxeninae Arrow) and Thallisellini trib. n. of the family Languriidae (Coleoptera: Clavicornia). Journal of Natural History 2 (4): 463–475. [1 Oct 1968 (Jour. Nat. Hist. 3: 608)]

Sen Gupta T (1970) On the taxonomy of Erotylidae (Insecta: Coleoptera: Clavicornia), with descriptions of two new larvae. Proceedings of the Zoological Society of Calcutta 22 [1969] (2): 97–107, pls. 5–9. [issued 20 Feb 1970 (wrapper)]

Sen Gupta T (1979) A new subfamily of Merophysiidae (Clavicornia: Coleoptera) and descriptions of two new species of *Gomya* Dajoz and its larva. Revue Suisse de Zoologie 86 (3): 691–698. [Sep 1979 (fasc. title page)]

Sen Gupta T (1988) Review of the genera of the family Rhizophagidae (Clavicornia: Coleoptera) of the world. Memoirs of the Zoological Survey of India 17: 1–58, pls. i-xxiv. [publ. Dec 1988 (verso of title page)]

Sen Gupta T, Crowson RA (1966) A new family of cucujoid beetles, based on six Australian and one New Zealand genera. The Annals and Magazine of Natural History (13) 9 (97/99): 61–85. [1 Sep 1966 (wrapper)]

Sen Gupta T, Crowson RA (1969a) On a new family of Clavicornia (Coleoptera) and a new genus of Languriidae. Proceedings of the Royal Entomological Society of London (Series B. Taxonomy) 38 (9/10): 125–131. [21 Nov 1969 (wrapper)]

Sen Gupta T, Crowson RA (1969b) Further observations on the family Boganiidae, with definition of two new families Cavognathidae and Phloeostichidae. Journal of Natural History 3 (4): 571–590. [1 Oct 1969 (Jour. Nat. Hist. 3: 608)]

Sen Gupta T, Crowson RA (1971) A review of classification of the family Languriidae (Coleoptera: Clavicornia) and the place of Languriidae in the natural system of Clavicornia. Memoirs of the Zoological Survey of India 15 (2): 1–42. [publ. Jun 1971 (wrapper)]

Sen Gupta T, Crowson RA (1973) A review of the classification of Cerylonidae (Coleoptera, Clavicornia). Transactions of the Royal Entomological Society of London 124 (4): 365–446. [12 Nov 1973 (reprint dedication)]

Sen Gupta T, Crowson RA (1979) The coleopteran family Sphindidae. The Entomologist's Monthly Magazine 113 [1977] (3): 177–191. [publ. 12 Feb 1979 (wrapper)]

Senna A (1892) Contributions to the knowledge of the family Brenthidae. VIII. Notes from the Leyden Museum 14 (3/4): 161–186. [Jul 1892 (vol. endleaf); 5 Jul 1892 (recorded at MCZ)]

Senna A (1892) Contributions to the knowledge of the family Brenthidae. VIII. Notes from the Leyden Museum 14 (3/4): 161–186. [Feb 1892]

Senna A (1895) Descriptions of some new Brenthidae. Notes from the Leyden Museum 16 (3/4): 213–226. [15 Mar 1895 (wrapper)]

Senna A, Calabresi E (1919) Contribuzione allo studio dei Brentidi. Revisione del gruppo Hoplopisthi. Bullettino della Società Entomologica Italiana 50 [1918]: 63–77. [20 Dec 1919 (wrapper)]

Sharp D (1873) Description of two new genera and three new species of Anthribidae from New Zealand. The Entomologist's Monthly Magazine 10 (7): 29–32. [Jul 1873 issue; 7 Jul 1873 (Proc. Ent. Soc. London 1873: xxiv)]

Sharp D (1876) Descriptions of some new genera and species of New Zealand Coleoptera. The Entomologist's Monthly Magazine 13: 70–77, 97–102. [Aug-Sep, Oct 1876]

Sharp D (1877) Descriptions of some new forms of aberrant Melolonthini from Australia forming a distinct subtribe (Systellopides). Annali del Museo Civico di Storia Naturale "Giacomo Doria" 9 [1876–77]: 311–320. [7 Mar 1877 (p. 305)]

Sharp D (1880) Avis préliminaire d'une nouvelle classification de la famille des Dytiscidae. Comptes Rendus des Séances de la Société Entomologique de Belgique 23: cxlvii-cli. [13 Oct 1880 (Ann. Soc. Ent. France (5) 10 Bibliogr. Ent.: 40)]

Sharp D (1882) Hydrophilidae [part], Heteroceridae, Parnidae, Georissidae, Cyathoceridae [pp. 113–144, pl. 4]. In: Godman FD, Salvin O (Eds) Biologia Centrali-Americana. Insecta. Coleoptera. Vol. I. Part 2. Taylor & Francis, London, xv + 824 pp. + 19 pls. [Dec 1882 (signature footers)]

Sharp D (1883) Staphylinidae [part, pp. 145–312]. In: Godman FD, Salvin O (Eds) Biologia Centrali-Americana. Insecta. Coleoptera. Vol. I. Part 2. Taylor & Francis, London, xv + 824 pp. + 19 pls. [May (pp. 145–192), Jul (193–240), Sep (241–288), Nov (289–296), Dec 1883 (297–312) (signature footers)]

Sharp D (1884) Staphylinidae [part, pp. 313–392, pls. 8–9]. In: Godman FD, Salvin O (Eds) Biologia Centrali-Americana. Insecta. Coleoptera. Vol. I. Part 2. Taylor & Francis, London, xv + 824 pp. + 19 pls. [Feb (pp. 313–344), Apr (345–376), Jun 1884 (377–392) (signature footers)]

Sharp D (1886a) On New Zealand Coleoptera. With descriptions of new genera and species. Scientific Transactions of the Royal Dublin Society (2) 3 [1883–1887]: 351–456, pls. 12–13. [Nov 1886 (wrapper; contents p. iv)]

Sharp D (1886b) Staphylinidae [part, pp. 537–672]. In: Godman FD, Salvin O (Eds) Biologia Centrali-Americana. Insecta. Coleoptera. Vol. I. Part 2. Taylor & Francis, London, xv + 824 pp. + 19 pls. [Jun (pp. 537–552), Jul (553–576), Aug (577–608),Oct (609–632), Nov (633–648), Dec 1886 (649–672) (signature footers)]

Sharp D (1887) Staphylinidae [part], Supplement, Index [pp. 673–824], Introduction [pp. i-xv]. In: Godman FD, Salvin O (Eds) Biologia Centrali-Americana. Insecta. Coleoptera. Vol. I. Part 2. Taylor & Francis, London, xvi + 824 pp. + 19 pls. [Jan (pp. 673–704), Feb (705–736), Mar (737–744), Sep (745–768), Oct (769–800), Nov 1887 (801–824, i-xvi) (signature footers)]

Sharp D (1889) The Staphylinidae of Japan [continued]. The Annals and Magazine of Natural History (6) 3 (13): 28–44, (14): 108–121, (15): 249–267, (16): 319–334, (17): 406–419, (18): 463–476. [1 Jan (13) - 1 Jun (18) 1889 (Evenhuis 2003: 31)]

Sharp D (1890) [Pp. 41–80, pls. 2–3]. In: Godman FD, Salvin O (Eds) Biologia Centrali-Americana. Insecta. Coleoptera. Vol. 4. Part 3. Rhynchophora. Curculionidae. Attelabinae, Pterocolinae, Allocoryninae, Apioninae, Thecesterninae, Otiorhynchinae. Taylor & Francis, London, vi + 354 pp. + 15 pls. [Nov (pp. 41–64), Dec 1890 (65–80) (signature footers)]

Sharp D (1891) [Pp. 81–168] In: Godman FD, Salvin O (Eds) Biologia Centrali-Americana. Insecta. Coleoptera. Vol. 4. Part 3. Rhynchophora. Curculionidae. Attelabinae, Pterocolinae, Allocoryninae, Apioninae, Thecesterninae, Otiorhynchinae. Taylor & Francis, London, vi + 354 pp. + 15 pls. [Feb (pp. 81–88), Oct (89–136), Nov 1891 (137–168) (signature footers)]

Sharp D (1893) On the genus *Mecedanum*, Erichson. The Entomologist's Monthly Magazine 29: 255–258. [Nov 1893 issue]

Sharp D (1894) Adimeridae, Colydiidae [part] [pp. 441–488]. In: Godman FD, Salvin O (Eds) Biologia Centrali-Americana. Insecta. Coleoptera. Vol. II. Part 1. Taylor & Francis, London, xii + 717 + [1] pp. + 19 pls. [Oct (pp. 441–464), Nov 1894 (465–488) (signature footers)]

Sharp D (1895a) Brenthidae [pp. 1–80, pls. 1–3]. In: Godman FD, Salvin O (Eds) Biologia Centrali-Americana. Insecta. Coleoptera. Vol. IV. Part 6. Taylor & Francis, London, [4] + 396 pp. + 14 pls. [May (pp. 1–32), Jul (33–48), Aug 1895 (49–80) (signature footers)]

Sharp D (1895b) Colydiidae [part, pp. 489–496]. In: Godman FD, Salvin O (Eds) Biologia Centrali-Americana. Insecta. Coleoptera. Vol. II. Part 1. Taylor & Francis, London, xii + 717 + [1] pp. + 19 pls. [Jan 1895 (signature footer)]

Sharp D (1899a) Insects. Part II. Hymenoptera continued (Tubulifera and Aculeata), Coleoptera, Strepsiptera, Lepidoptera, Diptera, Aphaniptera, Thysanoptera, Hemiptera, Anoplura. In: Harmer SF, Shipley AE (Eds) The Cambridge Natural History. Volume VI. Macmillan & Co., London, xii + 626 pp. [1899 (verso of title page of 2nd ed.); before 15 Sep 1899 (Ent. Record 11: 249)]

Sharp D (1899b) Colydiidae [part], Rhysodidae, Cucujidae [part] [pp. 497–560]. In: Godman FD, Salvin O (Eds) Biologia Centrali-Americana. Insecta. Coleoptera. Vol. II. Part 1. Taylor & Francis, London, xii + 717 + [1] pp. + 19 pls. [Jun (pp. 497–520), Aug (521–552), Sep 1899 (553–560) (signature footers)]

Sharp D (1900a) Cucujidae [part], Monotomidae, Cryptophagidae [part] [pp. 561–624]. In: Godman FD, Salvin O (Eds) Biologia Centrali-Americana. Insecta. Coleoptera. Vol. II. Part 1. Taylor & Francis, London, xii + 717 + [1] pp. + 19 pls. [Feb (pp. 561–584), Mar (585–608), Apr 1900 (609–624) (signature footers)]

Sharp D (1900b) On the Insects from New Britain [pp. 381–394 + pl. 35]. In: Willey A (Ed) Zoological results based on material from New Britain, New Guinea, Loyalty Islands and elsewhere, collected during the years 1895, 1896 and 1897. Part IV. University Press, Cambridge. [May 1900 (title page)]

Sharp D (1905) The genus *Criocephalus*. With notes on the habits of *Asemum striatum* and *Criocephalus ferus*. The Transactions of the Entomological Society of London 1905 (1): 145–176, pl. 9. [20 May 1905 (verso of vol. title page)]

Sharp D (1917) Studies in Rhynchophora. 1. Tribe Pseudobagoini. The Entomologist's Monthly Magazine 53 (2): 26–32. [Feb 1917 (wrapper); 23 Feb 1917 (recorded at USNM)]

Sharp D (1919a) Studies in Rhynchophora (Coleoptera). V. The genus *Rhyncogonus*. Proceedings of the Hawaiian Entomological Society 4 [1918] (1): 77–82. [Jun 1919 (wrapper); mailed 15 Jul 1919 (fasc. endleaf)]

Sharp D (1919b) Studies in Rhynchophora. 4. An aberrant new genus and tribe from New Guinea. Entomologist's Monthly Magazine 55 (7): 151–153. [Jul 1919 (wrapper); 22 Jul 1919 (recorded at USNM)]

Sharp D, Muir F (1912) The comparative anatomy of the male genital tube in Coleoptera. The Transactions of the Entomological Society of London 1912 (3): 477–642, pls. 42–78. [24 Dec 1912 (verso of vol. title page)]

Shepard WD (2002) Family 46. Limnichidae [pp. 125–126]. In: Arnett RH, Jr., Thomas MC, Skelley PE, Frank JH (Eds) American beetles. Volume 2. Polyphaga: Scarabaeoidea through Curculionoidea. CRC Press, Boca Raton, xiv + 861 pp. [19 Jun 2002 (CRC website)]

Sherborn CD (1922) [Introduction and bibliography, pp. i-cxxxi + 1–128] In: Index animalium sive index nominum quae ab A.D. MDCCLVIII generibus et speciebus animalium imposita sunt. Societatibus eruditorum adiuvantibus. Sectio secunda a kalendis ianuariis, MDCCCI usque ad finem decembris, MDCCCL. The Trustees of the British Museum, London, cxxxi + 7056 pp.

Sherborn CD (1929) Index animalium sive index nominum quae ab A.D. MDCCLVIII generibus et speciebus animalium imposita sunt. Sectio secunda. A Kalendis Ianuariis, MDCCCI usque ad finem Decembris, MDCCCL. Pars XXI. The Trustees of the British Museum, London, pp. 5139–5348.

Sherborn CD (1934) Dates of publication of catalogues of natural history (post 1850) issued by the British Museum. The Annals and Magazine of Natural History (10) 13: 308–312.

Sherborn CD (1937a) On the dates of publication of Costa (O. G.) and (A.) Fauna del Regno di Napoli, 1829–1886. Journal of the Society for the Bibliography of Natural History 1 [1936–1943] (2): 35–47. [Feb 1937 (fasc. footers)]

Sherborn CD (1937b) Brewster's Edinburgh Encyclopaedia. Issued in 18 vols. from 18- to 1830. Journal of the Society for the Bibliography of Natural History 1[1936–1943]: 112.

Sherborn CD, Woodward BB (1901) Dates of publication of the zoological and botanical portions of some French voyages., etc. The Annals and Magazine of Natural History (7) 8 (44): 161–164, (46): 333–336, (47): 491–494.

Shipp JW (1894) The coprophagous lamellicorns; a revised list of species belonging to the genera *Pachylomerus*, Kirby, and *Ateuchus*, Weber. The Entomologist 27: 254–257. [Sep 1894 issue]

Shockley FW, Tomaszewska KW, McHugh JV (2009) An annotated checklist of the handsome fungus beetles of the world (Coleoptera: Cucujoidea: Endomychidae). Zootaxa No. 1999: 1–113. [4 Feb 2009 (p. 2)]

Shokhin IV (2007) Materialy k faune plastinchatousykh zhukov (Coleoptera, Scarabaeoidea) Yuzhnoy Rossii. Contribution to the fauna of lamellicorn beetles (Coleoptera, Scarabaeoidea) of Southern Russia, with some nomenclatural changes in the family Scarabaeidae.

Kavkaziy Entomologicheskiy Byulleten / Caucasian Entomological Bulletin 3 (2): 105–185.

Shuckard WE (1839a) Elements of British entomology, containing a general introduction to the science, a systematic description of all the genera, and a list of all the species of British insects, with a history of their transformation, habits, economy, and distribution, with outline figures of the families and their larvae and pupae, an explanation of the technical terms, and full directions for collecting. Part I. Hippolyte Baillière, London, 240 pp. [before 3 Jun 1839 (Jour. Proc. Ent. Soc. Lond. 1839: xxv)]

Shuckard WE (1839b) [text and new taxa] In: Spry, W. & Shuckard, W. E.: The British Coleoptera delineated, consisting of figures of all the genera of British beetles, drawn in outline by W. Spry, M.E.S. W. Crofts, London, viii + 83 pp., 86 + 8 pls. [issued in 8 parts in 1839: 3 Jun (1/2); 1 Jul (3); 5 Aug (4); 7 Oct (5/6); 4 Nov (7); 2 Dec (8) 1839 (Jour. Proc. Ent. Soc. London 1839: xxv-xxxiv); 1840 (vol. title page); Jul 1840 (date of preface)]

Sicard A (1909) Révision des coccinellides de la faune Malgache. II. Cocinellidae aphidiphagae. Annales de la Société Entomologique de France 78 (1): 63–165. [Jul 1909 (wrapper); 13 Jul 1909 (Bull. Soc. Ent. France 1909: 221)]

Silfverberg H (1990) The nomenclaturally correct names of some family-groups in Coleoptera. Entomologica Fennica 1 (2): 119–121. [17 Aug 1990 (top of article)]

Silfverberg H (1992) Enumeratio Coleopterorum Fennoscandiae, Daniae et Baltiae. Helsingfors Entomologiska Bytesförening, Helsinki, v + 94 pp. [2 Jun 1992 (recorded at FMNH)]

Skinner H (1892) Entomological literature [review]. Entomological News and Proceedings of the Entomological Section of the Academy of Natural Sciences of Philadelphia 3 (5): 123–125. [mailed 2 May 1892 (p. 164)]

Skopin NG (1960) Materialy po morfologii i ekologii lichinok triby Blaptini (Coleoptera, Tenebrionidae). [Material on the larval morphology and ecology of the tribe Blaptini (Coleoptera, Tenebrionidae)]. Trudy Instituta Zoologii Akademiya Nauk Kazakhskoy SSR (Entomologiya) 11: 36–71, pls. 1–13 [in Russian]. [12 Feb 1960 (author's reprint dedication)]

Skopin NG (1964) Die Larven der Tenebrioniden des Tribus Pycnocerini (Coleoptera Heteromera). Annales du Musée Royal de l'Afrique Centrale, Tervuren (Série 8vo: Sciences Zoologiques) 127: 1–35 + [1], 16 pls. [1964 (title page)]

Skopin NG (1978) Tenebrionidae [pp. 223–266]. In: Klausnitzer B (Ed) Ordnung Coleoptera (larven). W. Junk, The Hague, vi + 378 pp. [1978 (title page)]

Skopin NG (1979) Systematische Stellung der Gattung *Scythis* Schaum, 1865, sowie Revision der Arten (Coleoptera, Tenebrionidae). Annales Historico-Naturales Musei Nationalis Hungarici 71: 169–183. [1979 (title page)]

Ślipiński SA (1990) A monograph of the world Cerylonidae (Coleoptera; Cucujoidea). Part I - Introduction and higher classification. Annali del Museo Civico di Storia Naturale "Giacomo Doria" 88 [1990–91]: 1–273. [5 Sep 1990 (index p. [687])]

Ślipiński SA (1992) Larinotinae - a new subfamily of Trogossitidae (Coleoptera), with notes on the constitution of Trogossitidae and related families of Cleroidea. Revue Suisse de Zoologie 99 (2): 439–463. [printed Jun 1992]

Ślipiński SA, Lawrence JF (1999) Phylogeny and classification of Zopheridae sensu novo (Coleoptera: Tenebrionoidea) with a review of the genera of Zopherinae (excluding Monommatini). Annales Zoologici (Warszawa) 49 (1/2): 1–53. [accepted May 1999]

Ślipiński SA, Lawrence JF, Tomaszewska W (2001) The placement of *Periptyctus* Blackburn in Corylophidae (Coleoptera: Curculionoidea) with descriptions of a new genus and subfamily. Annales Zoologici (Warszawa) 51: 311–317.

Ślipiński SA, Pal TK (1985) Sysolini - new tribe of Bothriderinae (Coleoptera, Colydiidae), with a description of a new species of *Sysolus* from Viet-Nam. Sysolini - nowe plemię w pedrodzinie Bothriderinae (Coleoptera, Colydiidae), z opisem nowego gatunku *Sysolus* z Wietnamu. Polskie Pismo Entomologiczne 55: 39–44. [30 Mar 1985 (top of article)]

Ślipiński SA, Tomaszewska W, Lawrence JF (2009) Phylogeny and classification of Corylophidae (Coleoptera: Cucujoidea) with descriptions of new genera and larvae. Systematic Entomology 34 (3): 409–433. [Jul issue; 22 Jun 2009 (vol. contents)]

Sloane TG (1890) Studies in Australian entomology. No. iv. - New genera and species of Carabidae. The Proceedings of the Linnean Society of New South Wales (2) 5 [1890]: 641–653. [issued 16 Dec 1890 (Contents p. v)]

Sloane TG (1898) On Carabidae from West Australia, sent by Mr. A. M. Lea (with descriptions of new genera and species, synoptic tables, &c.). The Proceedings of the Linnean Society of New South Wales 23 (3): 444–520. [issued 9 Dec 1898 (Contents p. iv)]

Sloane TG (1903) Studies in Australian Entomology. No. xii. New Carabidae (Panageini, Bembidiini, Pogonini, Platysmatini, Platynini, Lebiini, with revisional lists of genera and species, some notes on synonymy, &c.). The Proceedings of the Linnean Society of New South Wales 28 [1903–04] (3): 566–642. [issued 23 Dec 1903 (Contents p. viii)]

Sloane TG (1905) Revisional notes on Australian Carabidae. Part I. Tribes Carabini, Pamborini, Pseudozaenini, Clivini; and the genus *Nebriosoma*. The Proceedings of the Linnean Society of New South Wales 29 [1904–05] (4): 699–733. [issued 10 Apr 1905 (Contents p. ix)]

Sloane TG (1907a) Studies in Australian Entomology. No. xv. New genera and species of Carabidae, with some notes on synonymy (Clivinini, Scaritini, Cunipectini, Trigonotomini and Lebiini). The Proceedings of the Linnean Society of New South Wales 32 [1907–08] (2): 346–381. [issued 20 Aug 1907 (Contents p. vii)]

Sloane TG (1907b) Further Carabidae from German New Guinea and its dependencies. (Col.). Deutsche Entomologische Zeitschrift 1907 (5): 467–474. [1 Sep 1907 (Inhalt p. iii)]

Sloane TG (1920) The Carabidae of Tasmania. The Proceedings of the Linnean Society of New South Wales 45 [1920–21] (1): 113–178. [issued 25 Jun 1920 (Contents p. iii)]

Sloane TG (1923a) The classification of the family Carabidae. The Transactions of the Entomological Society of London 1923 [1923–24] (1/2): 234–250. [publ. 10 Aug 1923 (verso of vol. title page)]

Sloane TG (1923b) Supplementary note to the classification of the Carabidae. The Transactions of the Entomological Society of London 1923 (1/2): 250a-250c. [publ. 10 Aug 1923 (verso of vol. title page)]

Smetana A (1975) Revision of the New World genera of the tribe Omicrini trib. n. of the hydrophilid subfamily Sphaeridiinae (Coleoptera). Studies on Neotropical Fauna 10 (2): 153–182. [Dec 1975 (wrapper)]

Smetana A (1983) The status of the staphylinid genera *Derops* Sharp and *Rimulincola* Sanderson (Coleoptera). Entomologica Scandinavica 14 (3): 269–279. [30 Sep 1983 (top of article)]

Smith ABT (2006) A review of the family-group names for the superfamily Scarabaeoidea (Coleoptera) with corrections to nomenclature and a current classification. Coleopterists Society Monograph 5: 144–204. [29 Dec 2006]

Smith F (1851) List of the coleopterous insects in the collection of the British Museum. Part I [error for V]. - Cucujidae, &c. Trustees of the British Museum, London, [1] + 25 pp. [30 Nov 1851 (date of introduction); 13 Dec 1851 (Sherborn 1934: 310)]

Smrz J (1982) Comparative anatomy of proventriculus and intraelytral structure of the suborder Adephaga (Coleoptera). Acta Universitatis Carolinae - Biologica 1980 (3/4): 213–296. [1982 (verso of fasc. title page)]

Solier AJJ (1834) Essai d'une division des coléoptères hétéromères, et d'une monographie de la famille des Collaptèrides. Annales de la Société Entomologique de France 3 (3): 479–636, pls. 12–15. [by Jan 1835 (as "1834", Ent. Mag.: 526); 2 Feb 1835 (Jour. Proc. Ent. Soc. London 1835: xli)]

Solier AJJ (1835a) Prodrome de la famille des Xystropides. Annales de la Société Entomologique de France 4 (2): 229–248. [before 28 Sep 1835 (Acad. Sci. France 1: 161)]

Solier AJJ (1835b) Essai sur les collaptérides (suite). Annales de la Société Entomologique de France 4 (3): 509–574, pls. xiv-xv. [before 21 Dec 1835 (Acad. Sci. France 1: 517)]

Solier AJJ (1835c) Extrait d'une lettre adressée par M. Solier à M. Audouin relative aux Buprestides. In: Errata et addenda. Annales de la Société Entomologique de France 3 [1834] (4): xcix-c. [before 4 May 1835 (Jour. Proc. Ent. Soc. London 1835: liii)]

Solier AJJ (1836) Essai sur les collaptèrides [sic] (suite). Annales de la Société Entomologique de France 5 (3): 403–512, pls. 11–13. [before 31 Oct 1836 (Acad. Sci. France 3: 521)]

Solier AJJ (1837a) Essai sur les collaptérides (suite). Annales de la Société Entomologique de France 5 [1836] (4): 635–684, pls. 23–24. [before 6 Mar 1837 (Acad. Sci. France 4: 380)]

Solier AJJ (1837b) Essai sur les collaptérides (suite). Annales de la Société Entomologique de France 6 (2): 151–172, 1 pl. [before 9 Oct 1837 (Acad. Sci. France 5 : 527)]

Solier AJJ (1838) Essai sur les collaptèrides [sic] (suite). Annales de la Société Entomologique de France 7 (1): 5–73, pls. 1–3; (2): 159–199, pls. 7–8. [before 30 Jul (fasc. 1), before 19 Nov (fasc. 2) 1838 (Acad. Sci. France 7: 287, 905)]

Solier AJJ (1841) Essai sur les collaptérides (suite). Annales de la Société Entomologique de France 9 [1840] (3): 207–270, pls. 9–10 [pp. 255–270 as 355–370]. [Apr 1841 (wrapper)]

Solier AJJ (1848) Essai sur les collaptérides. Studi Entomologici 1 (2): 149–370, 12 pls. [1848 (title page)]

Solier AJJ (1849) Zoología. Tomo cuarto. Insectos. Orden III. Coleópteros. [pp. 105–380, 414–511, pls. 1–12]. In: Gay C (Ed) Historia física y política de Chile según documentos adquiridos en esta república durante doce años de residencia en ella y publicada bajo los auspicios del supremo gobierno. C. Gay, Paris, 511 pp. [1849 (title page)]

Solier AJJ (1851) Zoología. Tomo quinto. Fauna Chilena. Insectos. Coleópteros [pp. 5–285]. In: Gay C (Ed) Historia física y política de Chile según documentos adquiridos en esta república durante doce años de residencia en ella y publicada bajo los auspicios del supremo gobierno. C. Gay, Paris, 564 pp. + 561–563 [bis, Index]. [1851 (title page)]

Solodovnikov AY, Newton AF, Jr. (2005) Phylogenetic placement of Arrowinini trib. n. within the subfamily Staphylininae (Coleoptera: Staphylinidae), with revision of the relict South African genus *Arrowinus* and description of its larva. Systematic Entomology 30 (3): 398–441. [9 Dec 2004 (online); 18 Jul 2005 (vol. contents)]

Soriano C, Ponomarenko AG, Delclos X (2007) Coptoclavid beetles (Coleoptera: Adephaga) from the Lower Cretaceous of Spain: a new feeding strategy in beetles. Palaeontology 50 (2): 525–536. [Mar 2007 issue; 15 Mar 2007 (online)]

Spaeth F (1914) Über die paläarktischen Cassiden mit besonderer Berücksichtigung jener von Asien. Verhandlungen der Kaiserlich-Königlichen Zoologisch-Botanischen Gesellschaft in Wien 64 (5/6): (128)-(147). [5 Sep 1914 (vol. title page)]

Spaeth F (1923) Ueber *Batonota* Hope (Col. cassid.). Wiener Entomologische Zeitung 40 (1/4): 65–76. [publ. 25 Apr 1923 (wrapper)]

Spaeth F (1926) Monographie der zur Gruppe der Coptocyclitae gehorigen amerikanischen Cassidinen (Col.): I. Die Gattungen mit gekammten Klauen. Supplementa Entomologica 13: 1–108. [15 Sep 1926 (title page)]

Spaeth F (1929a) Ist *Militinaspis cassididoides* eine Cassidinae? Koleopterologische Rundschau 15 (1): 28–30. [15 Jul 1929 (wrapper)]

Spaeth F (1929b) Die Gattung *Hemisphaerota* Spaeth (Coleopt. Chrysom. Cassid.). Koleopterologische Rundschau 15 (2/3): 111–131. [30 Sep 1929 (wrapper)]

Spaeth F (1942) Cassidinae (Col. Chrysom.) [pp. 11–43]. In: Titschack E (Ed) Beiträge zur Fauna Perus. Nach der Ausbeute der Hamburger Süd-Peru Expedition 1936, anderer Sammlungen, wie auch auf Grund von Literaturangaben. Band II. G. Fischer, Jena, vii + 344 pp.

Spaeth F, Reitter E (1926) Bestimmungs-Tabellen der europäischen Coleopteren. 95. Heft. Cassidinae der palaearktischen Region. Emmerich Reitter, Troppau, 68 pp.

Spangler PJ (1999) A new tribe, genus, and species of limnichid beetle, *Wooldridgeus perforatus*, from Malaysia (Coleoptera: Limnichidae: Wooldridgeini). Insecta Mundi 12 [1998] (3/4): 181–187. [1999 (p. 320 cites paper publ. in Jan 1999)]

Spangler PJ, Steiner WE, Jr. (2005) A new aquatic beetle family, Meruidae, from Venezuela (Coleoptera: Adephaga). Systematic Entomology 30 (3): 339–357. [Jul issue; 18 Jul 2005 (vol. contents)]

Speidel W, Naumann CM (2004) A survey of family-group names in noctuid moths (Insecta: Lepidoptera). Systematics and Biodiversity 2 (2): 191–221. [issued Nov 2004]

Spinola M (1841) Monographie des térédiles. Revue Zoologique 1841 (4): 70–76. [Mar 1841 issue]

Spinola M (1844) Essai Monographique sur les Clérites, Insectes Coléoptères. Tome premier. Ponthenier, Gènes, ix + 386 pp. [1844 (title page); "sur le point d'être terminé" (insert after Ann. Soc. Ent. France (2) 2 (3): Bull. Ent.: lxiv, publ. 8 Jan 1845); Apr 1845 (Bibliogr. Ital. (N. S.) 1: 131; first of many notices found); several bibliographies cite as 1845]

Špringlová de Bechyné B (1960) Essai monographique du genre *Eumolpus* (Coleoptera Phytophaga). Institut Royale des Sciences Naturelles de Belgique Mémoires (Deuxième série) 60: 1–79. [distrib. 31 Oct 1960 (title page)]

St. George RA (1939) The larva of *Perimylops antarcticus* Müller and the systematic position of the family Perimylopidae (Coleoptera). Proceedings of the Entomological Society of Washington 41 (7): 207–214. [27 Oct 1939 (verso of vol. title page)]

Stafleu FA, Cowan RS (1983) Taxonomic literature: a selective guide to botanical publications and collections with dates, commentaries and types. 2nd ed. Vol. 4: P - Sak. Regnum Veg. No. 110. Scheltema & Holkema, Bohn, 1214 pp.

Staines CL (2002) The new world tribes and genera of hispines (Coleoptera: Chrysomelidae: Cassidinae). Proceedings of the Entomological Society of Washington 104 (3): 721–784. [mailed 26 Jun 2002 (inside wrapper)]

Staines CL (2010) Some nomenclatural notes on hispines (Coleoptera: Chrysomelidae: Hispinae). Proceedings of the Entomological Society of Washington 112 (1): 172–173. [Jan 2010 issue]

Standfuss K, Kerzhner IM (2004) Publikationsdaten der "Trudy Russkogo Obshchestva" und "Horae Societatis Entomologicae Rossicae", 1861–1932. Daty publikatsii izdaniya "Trudy Russkogo Entomologicheskogo Obshchestva" i "Horae Societatis Entomologicae Rossicae", 1861–1932. Entomofauna Zeitschrift für Entomologie 25 (14): 237–248. [10 Aug 2004 (top of article)]

Stebnicka ZT (1977) A revision of the world species of the tribe Aegialiini (Coleoptera, Scarabaeidae, Aphodiinae). Acta Zoologica Cracoviensia 22 (11): 397–506, pl. 23. [31 Dec 1977 (top of article)]

Stebnicka ZT (1999) *Lomanoxia* Martinez, 1951, and a new tribe Lomanoxiini with notes on comparative morphology (Coleoptera: Scarabaeoidea: Aphodiinae). Acta Zoologica Cracoviensia 42 (2): 279–286. [17 Dec 1999 (top of article)]

Stebnicka ZT (2001) Aphodiinae (Insecta: Coleoptera: Scarabaeidae). Fauna of New Zealand. Ko te Aitanga Pepeke o Aotearoa. Number 42. Manaaki Whenua Press, Lincoln, Canterbury, New Zealand, 61 pp.

Stebnicka ZT, Howden HF (1995) Revision of Australian genera in the tribes Aphodiini, Aegialiini and Proctophanini (Coleoptera: Scarabaeidae: Aphodiinae). Invertebrate Taxonomy 9 (4): 709–766. [13 Oct 1995 (vol. 10 (1) contents)]

Stebnicka ZT, Howden HF (1996) Australian genera and species in the tribes Odontolochini, Psammodiini, Rhyparini, Stereomerini and part of the Eupariini (Coleoptera: Scarabaeoidea: Aphodiinae). Invertebrate Taxonomy 10 (1): 97–170. [1 Mar 1996 (11 (1) contents)]

Steel WO (1950) Notes on the Omaliinae (Col., Staphylinidae). (3) A new tribe and three new genera from New Zealand. The Entomologist's Monthly Magazine 86 (2): 54–64. [14 Mar 1950 (verso of vol. title page)]

Steel WO (1966) A revision of the staphylinid subfamily Proteininae (Coleoptera) I. Transactions of the Royal Entomological Society of London 118: 285–311. [13 Dec 1966]

Steffan AW (1964) Torridincolidae, coleopterorum nova familia e regione aethiopica. Entomologische Zeitschrift 74 (17): 193–200. [1 Sep 1964 (top of article)]

Stein JPEF (1868) Catalogus Coleopterorum Europae. F. Nicolai, Berlin, iv + 149 pp. [1868 (title page)]

Stein JPEF, Weise J (1877) Catalogi Coleopterorum Europae. Editio secunda. Libraria Nicolai, Berlin, iv + 209 pp. [by Dec 1877 (Biblioth. Hist.-nat. 27 (2): 141).]

Steiner WE, Jr. (1980) A new tribe, genus, and species of cossyphodine from Peru (Coleoptera: Tenebrionidae). Proceedings of the Entomological Society of Washington 82 (3): 384–395. [mailed 11 Jul 1980 (inside wrapper)]

Steinhausen WR (2001) Die Puppen mitteleuropäischer Blattkäfer - Eine vorläufige Bestimmungstabelle 1. Teil (Coleoptera: Chrysomelidae). Mitteilungen der Münchner Entomologischen Gesellschaft 91: 35–63. [1 Oct 2001 (top of article)]

Stephens JF (1827) Illustrations of British entomology; or, a synopsis of indigenous insects: containing their generic and specific distinctions; with an account of their metamorphoses, times of appearance, localities, food, and economy, as far as practicable. Mandibulata. Vol. I [pp. 1–76, pls. 1–5]. Baldwin & Cradock, London, iv + 186 + [2] pp. + 9 pls. [1828 (title page); 1 May 1827 (p. 1 footer); 1 Jun (p. 17); 1 Jul (p. 29); 1 Aug (p. 37); 1 Sep (p. 45); 1 Oct (p. 53); 1 Nov (p. 61); 1 Dec (p. 69)]

Stephens JF (1828) Illustrations of British entomology; or, a synopsis of indigenous insects: containing their generic and specific distinctions; with an account of their metamorphoses, times of appearance, localities, food, and economy, as far as practicable. Mandibulata. Vol. II [pp. 1–112, pls. 10–14]. Baldwin & Cradock, London, 200 pp. + pls. 10–15]. [1829 (title page); 1 Jul 1828 (p. 1 footer); 1 Aug (p. 33); 1 Sep (p. 65); 1 Oct (p. 97)]

Stephens JF (1829a) The nomenclature of British insects; being a compendious list of such species as are contained in the Systematic Catalogue of British Insects, and forming a guide to their classification, &c. &c. Baldwin and Cradock, London, [2] + 68 + [1] pp. [1 Jun 1829 (Blackwelder 1949: 92)]

Stephens JF (1829b) A systematic catalogue of British insects: being an attempt to arrange all the hitherto discovered indigenous insects in accordance with their natural affinities. Containing also the references to every English writer on entomology, and to the principal foreign authors. With all the published British genera to the present time. [Part 1, Insecta Mandibulata]. Baldwin and Cradock, London, xxxiv + 416 pp. [15 Jul 1829]

Stibick JNL (1976) A revision of the Hypnoidinae of the world (Col. Elateridae). Part I. Introduction, phylogeny, biogeography. The Hypnoidinae of North and South America. The genera *Berninelsonius* and *Ligmargus*. Eos, Revista Española de Entomologia 51 [1975] (1/4): 143–223. [30 Nov 1976 (wrapper)]

Stibick JNL (1979) Classification of the Elateridae (Coleoptera). Relationships and classification of the subfamilies and tribes. Pacific Insects 20 (2/3): 145–186. [16 May 1979 (top of article)]

Stichel W (1923) Zur Phylogenesis eines geologisch jungen Formenkreises der Käfer, der Ditominen (Carab. Harpal.). Zeitschrift für Wissenschaftliche Insektenbiologie 18 (3/4): 41–50, (5/7): 81–100, (8/9): 145–162, (10/11): 209–242, 2 pls. [1 Apr 1923 (fasc. 3/4); 15 May (5/7); 1 Jul (8/9); 1 Oct (10/11) 1923 (wrappers)]

Straneo SL (1951) Sur la tribu des Metiini (Antarctiini auct.) (Coleoptera Pterostichidae). Revue Française d'Entomologie 18 (2): 56–88. [15 Jul 1951 (p. 49 footer)]

Streubel AV (1839) Ueber die Stellung der Brachyelytren oder Staphylinen im natürlichen System (Ein Beytrag zur Classification der Käfer). Isis von Oken 1839 (2): 126–137. [not before Mar 1839 (submission dates of ads in Heft 2)]

Strohecker HF (1953) 210e fascicule. Coleoptera. Fam. Endomychidae. In: Wytsman PA (Ed) Genera Insectorum. L. Desmet-Verteneuil, Bruxelles, 140 pp. + 5 pls. [1953 (title page); 4 Jun 1953 (Evenhuis 1994: 61)]

Strohecker HF (1962) Key to the genera of the United States [p. 801]. In: Arnett RH, Jr: Part V Suborder Polyphaga (cont.) series Cucujiformia (cont.) Tenebrionoidea, Cucujoidea. Pp. [2 unn.] + 645–850 In: The beetles of the United States (a manual for identification) [original loose-leaf edition]. Catholic University of America Press, Washington D.C., xii + 1112 pp. [+ unn. pp.]. [issued 20 Aug 1962 (verso of title page for Part VI)]

Strohecker HF (1964) A synopsis of the Amphisternini (Coleoptera: Endomychidae). Pacific Insects 6 (2): 319–357. [mailed 31 Aug 1964 (p. v)]

Strohmeyer G (1928) Systematisches und Zoogeographisches über die Cypholobini (Carab. Anthiinae). Ein Beitrag zur Kenntnis der Fauna des afrikanischen Trockenwaldes. Mitteilungen aus dem Zoologischen Museum in Berlin 14 (2): 287–462, pls. 1–17. [1928 (title page); at least Mar 1928 (previous article submitted)]

Strohmeyer H (1914) 163me fascicule. Coleoptera. Fam. Platypodidae. In: Wytsman PA (Ed) Genera Insectorum. Vol. XXV. V. Verteneuil & L. Desmet, Bruxelles, pp. 1–55, pls. 1–12. [15 May 1914 (date on manuscr.); 20 Jun 1914 (Evenhuis 1994: 58)]

Sturm J (1826) Catalog meiner Insecten-Sammlung. Erster Theil. Käfer. J. Sturm, Nürnberg, viii + 207 pp. + 4 pls. [1826 (title page)]

Sturm J (1843) Catalog der Kaefer-Sammlung von Jacob Sturm. J. Sturm, Nürnberg, xii + 386 pp. + 6 pls. [Jan 1843 (date of preface); Aug 1843 (Ent. Zeit. Stettin 4: 255)]

Štys P, Jansson A (1988) Check-list of recent family-group and genus-group names of Nepomorpha (Heteroptera) of the world. Acta Entomologica Fennica 50: 1–44.

Suffrian E (1848) Bemerkungen über einige deutsche Rüsselkäfer. (Fortszetzung). Stettiner Entomologische Zeitung 9 (2): 52–62. [Feb 1848 issue]

Suffrian E (1863) Zur Kenntniss der südamerikanischen Cryptocephalen. Linnaea Entomologica 15: 77–344. [end Oct 1863 (date of Vorwort); 15 Nov 1863 (Intell.-Blatt Serap. 24: 166)]

Summers SV (1874) Catalogue of the Coleoptera from the region of Lake Pontchartrain, La. Bulletin of the Buffalo Society of Natural Sciences 2 [1874–75]: 78–99. [1875 (vol. title page); May 1874 (footer p. 73)]

Švácha P, Danilevsky ML (1987) Cerambycoid larvae of Europe and Soviet Union (Coleoptera, Cerambycoidea). Part I. Acta Universitatis Carolinae - Biologica 30 [1986] (1/2): 1–176. [Oct 1987 (top of article)]

Švácha P, Danilevsky ML (1988) Cerambycoid larvae of Europe and Soviet Union (Coleoptera, Cerambycoidea). Part II. Acta Universitatis Carolinae - Biologica 31 [1987] (3/4): 121–284. [Dec 1988 (top of article)]

Švácha P, Wang J-J, Chen S-C (1997) Larval morphology and biology of *Philus antennatus* and *Heterophilus punctulatus*, and systematic position of the Philinae (Coleoptera: Cerambycidae and Vesperidae). Annales de la Société Entomologique de France (Nouvelle Série) 33 (3): 323–369. [issued 30 Sep 1997 (inside back wrapper)]

Svenson GJ, Branham MA (2007) Case 3402. Photinini LeConte, 1881 (Insecta, Coleoptera) and Photininae Giglio-Tos, 1915 (Insects, Mantodea): proposed resolution of homonymy between family-group names. Bulletin of Zoological Nomenclature 64 (4): 243–251. [20 Dec 2007]

Švihla V (1986) Revision of the generic classification of the Old World Oedemeridae (Coleoptera). Sborník Národního Musea v Praze 41 B [1985] (3/4): 141–238.

Švihla V (2008) Family Oedemeridae [pp. 353–369]. In: Löbl I, Smetana A (Eds) Catalogue of Palaearctic Coleoptera. Volume 5. Tenebrionoidea. Apollo Books, Stenstrup, 670 pp. [publ. 15 Apr 2008 (verso of title page)]

Swainson W (1840) [classification system and marked paragraphs] In: Swainson W, Shuckard WE: On the history and natural arrangement of insects. In: Lardner D (Ed): The Cabinet Cyclopaedia. Natural History. Longman, Orme, Brown, Green & Longmans and Taylor, London, iv + 406 pp. [1840 (title page); Jul 1840 (date of preface); 14 Dec 1840 (Literary Gazette 24: 775); Jan 1841 (reviewed in Entomologist: 38)]

Świętojańska J, Chorzępa M, Ghate H (2006) Description of last instar larva and pupa of *Chaeridiona picea* Baly, 1869 and *Oncocephala quadrilobata* (Guerin, 1844) (Coleoptera: Chrysomelidae: Cassidinae: Oncocephalini) from India. Zootaxa 1341: 49–68. [publ. 23 Oct 2006 (footer p. 49)]

Szymczakowski W (1964) Analyse systématique et zoogéographique des Catopidae (Coleoptera) de la région orientale. Acta Zoologica Cracoviensia 9 (2): 55–289. [30 Apr 1964 (Contents)]

Tangelder IRM, Krikken J (1982) Termitophilous scarabs of the tribe Corythoderini: a taxonomic review (Coleoptera: Aphodiidae). Zoologische Verhandelingen No. 194: 1–114. [20 Jul 1982 (wrapper)]

Telnov D (2004) Check-list of Latvian Beetles (Insecta: Coleoptera). Second edition. Compendium of Latvian Coleoptera. Volume I. Entomological Society of Latvia, Rīga, 114 pp.

Teocchi P (1989) Transfert de *Poimenesperus lugens* White dans le genre *Tragon* Murray et mise en synonymie de *Tragon tragonoides* Lepesme (Coleoptera Cerambycidae Lamiinae). Bulletin de la Société Sciences Nat 61: 4. [Mar 1989 (title page); 28 Mar 1989 (recorded at BMNH)]

Théry A (1929) Description d'un genre nouveau de la famille des Buprestides (Col.) appartenant à la faune chinoise et remarques diverses. Bulletin de la Société Entomologique de France 1929 (4): 57–60. [4 Apr 1929 (p. 336)]

Thomas DB (2005) Blasisdell's formae and homonyms in the genus *Eleodes* Eschscholtz (Coleoptera: Tenebrionidae: Embaphionini). Annales Zoologici (Warszawa) 55 (4): 549–560. [31 Dec 2005 (inside wrapper)]

Thomas MC (2008) Beetles (Coleoptera) [pp. 437–447]. In: Capinera JL (Ed). Encyclopedia of Entomology. 2nd edition. Volume 1 A-C. Springer, Netherlands.

Thompson RT (1992) Observations on the morphology and classification of weevils (Coleoptera, Curculionoidea) with a key to major groups. Journal of Natural History 26 (4): 835–891. [18 Sep 1992 (dates in vol. 27 (1))]

Thomson CG (1857) Skandinaviens Coleoptera, synoptiskt bearbetade. Häftet I. Carabici. Berlingska Boktryckeriet, Lund, ii + 64 pp. [1857 (title page); Dec 1856 (date of preface); by Mar 1858 (Allgem. Bibliogr. Monatl. Verz. 1858: 43)]

Thomson CG (1858) Försök till uppställning af Sveriges Staphyliner. Öfversigt af Kongliga Vetenskaps-Akademiens Förhandlingar 15 (1): 27–40. [11 Jan 1858 (p. 27 footer)]

Thomson CG (1859) Skandinaviens Coleoptera, synoptiskt bearbetade. Tom. I. Berlingska Boktryckeriet, Lund, [3] + 290 pp. [Apr 1859 (date of Förord); before 14 Dec 1859 (Öfvers. K. Vetensk.-Akad. Förh. 16 (10): 448)]

Thomson CG (1860) Skandinaviens Coleoptera, synoptiskt bearbetade. II. Tom. Berlingska Boktryckeriet, Lund, 304 pp. [1860 (title page)]

Thomson CG (1862) Skandinaviens Coleoptera, synoptiskt bearbetade. Tom. IV. Lundbergska Boktryckeriet, Lund, 269 pp. [1862 (title page)]

Thomson CG (1863) Skandinaviens Coleoptera, synoptiskt bearbetade. Tom. V. Lundbergska Boktryckeriet, Lund, 340 pp. [1863 (title page); Sep-Oct 1863 (Allgem. Bibliogr. Monatl. Verz.)]

Thomson CG (1864) Skandinaviens Coleoptera, synoptiskt bearbetade. Tom. VI. Lundbergska Boktryckeriet, Lund, 385 pp. [1864 (title page); 5 Sep 1864 (Jour. Proc. Ent. Soc. London 1864: 38)]

Thomson CG (1865) Skandinaviens Coleoptera, synoptiskt bearbetade. Tom. VII. Lundbergska Boktryckeriet, Lund, 394 pp. [Häftet I: 7 Aug 1865 (Jour. Proc. Ent. Soc. London 1865: 106); II: 1 Jan 1866 (ibid., p. 132)]

Thomson CG (1866) Skandinaviens Coleoptera, synoptiskt bearbetade. Tom. VIII. Lundbergska Boktryckeriet, Lund, 409 + lxxv pp. [1866 (title page)]

Thomson CG (1867) Skandinaviens Coleoptera, synoptiskt bearbetade. Tom. IX [Supplementa]. Lundbergska Boktryckeriet, Lund, 407 pp. [1867 (title page)]

Thomson J (1857a) Monographie des cicindélides ou exposé méthodique et critique des tribus, genres et espèces de cette famille. Tome premier. Ballière, Paris, 66 pp. [25 Feb 1857 (livr.1: pp. xvii, 1–18, pls. 1–3) (Ann. Soc. Ent. France (3) 5: Bull. Ent.: xxiv); 1857 (livr. 2: 19–42, pls. 4–6) (wrapper); 1857 (livr. 3: 43–66, pls. 7–10) (wrapper), 21 Dec 1858 (Proc. Acad. Nat. Sci. Philad. 1858: xxviii)]

Thomson J (1857b) Monographie de la tribu des anacolites, de la famille des longicornes [pp. 7–20 + 3 pls. in livr. 1]. In: Thomson J (Ed) Archives entomologiques ou recueil contenant des illustrations d'insectes nouveaux ou rares. Tome premier. Société Entomologique de France, Paris, 514 + [1] pp. + 1 + 21 pls. [15 Mar 1857 (date of Préface); 25 Mar 1857 (Ann. Soc. Ent. France (3) 5: Bull. Ent.: xxxiii)]

Thomson J (1858) Voyage au Gabon. Deuxième partie. Histoire naturelle proprement dite. Insectes. I. Ordre coléoptères [livr. 13–14: pp. 29–80; livr. 15: 81–239]. In: Thomson J (Ed) Archives entomologiques ou recueil contenant des illustrations d'insectes nouveaux ou rares. Tome deuxième. Société Entomologique de France, Paris, 469 + [2] pp. + 1 + 15 pls. [livr. 13–15: Jun 1858 (Arch. Nat. Hist. 16: 90–91)]

Thomson J (1859) Notice historique sur le genre *Cicindela* suivie de la description de sept espèces nouvelles de Cicindelidae [livr. 2: pp. 85–92]. In: Thomson J (Ed). Arcana Naturae ou Recueil d'Histoire Naturelle. J. B. Ballière et fils, Paris, 132 pp. + 13 pls. [end of 1859, probably early 1860 (Hayek 1989: 93)]

Thomson J (1860a) Essai d'une classification de la famille des Cérambycides et matériaux pour servir a une monographie de cette famille [livr. 1: pp. xvi + 1–128]. J. Thomson, Paris, xvi + 396 pp. + 3 pls. [1 May 1860 (date of dedication)]

Thomson J (1860b) Matériaux pour servir a une monographie nouvelle de la famille des Clérides [pp. 46–67]. In: Thomson J (Ed) Musée Scientifique ou Recueil d'Histoire Naturelle. J. Thomson, Paris, 96 pp., 9 pls. [1860 (title page); 1860 (livr. 1, pp. 1–40) (Bull. Bibliogr. Sci. Phys. 1: 181); 1860 (livr. 2, pp. 41–72) (Arch. Nat. Hist. 16: 94)]

Thomson J (1861) Essai d'une classification de la famille des Cérambycides et matériaux pour servir a une monographie de cette famille [livr. 2–4: pp. 129–396]. J. Thomson, Paris, xvi + 396 pp. + 3 pls. [1861 (wrapper, p. 396 footer); Jul 1861 (Ann. Mag. Nat. Hist. (3) 8: 41–46, Bates discusses Pseudolepturitae, Distenitae [in part 2], and the genera *Cheloderus* and *Oxypeltis* [in part 3]); Aug 1861 (ibid., p. 150, genus *Aegomorphus* [in Supplément, p. 336])]

Thomson J (1864) Systema cerambycidarum ou exposé de tous les genres compris dans la famille des cérambycides et familles limitrophes. [pp. 1–352]. H. Dessain, Liége, 578 pp. [1 Jan 1864 (date of dedication); 1864 (livr. 1–3: Zool. Record 1864: 336); before Oct 1864 (livr. 1–2: pp. 1–272: cited by Pascoe, 1864: 81, 85); Apr 1865 (livr. 3: pp. 273–352: Allgem. Bibliogr. Monatl. Verz. 1865: 44); 27 Dec 1865 (livr. 4: 353–578: Ann. Soc. Ent. France (4) 6: 243)]

Thomson J (1868) Physis recueil d'histoire naturelle. Tome deuxième. Société Entomologique de France, Paris, 208 pp. [15 May 1868 (livr. 4, pp. 5–40); 1 Aug 1868 (livr. 5, pp. 41–100); 15 Oct 1868 (livr. 6, pp. 101–208) (all Bibliogr. France 1868)]

Thomson J (1877) Typi cerambycidarum musei Thomsoniani. Revue et Magasin de Zoologie Pure et Appliquée (3) 5 [40]: 249–279.

Thon T (1827) [review of Dejean (1825–1826); pp. 12–16] In: Entomologisches Archiv. Erster Band, Erstes Heft. T. Thon, Jena, iv + 16 + [2] pp., 2 pls. [Jul-Oct 1827 (Foreign Quart. Rev. 1: 670)]

Thunberg CP (1789) Periculum entomologicum, quo characteres generum insectorum. J. Edman, Upsaliae, 16 pp.

Tillyard RJ (1924) Upper Permian Coleoptera and a new order from the Belmont beds, New South Wales. The Proceedings of the Linnean Society of New South Wales 49 (4): 429–435, pls. 45–46. [issued 29 Dec 1924 (contents)]

Tippmann FF (1956) Eine neue, morphologisch einmalige Lamiine aus Ost-Perú (Coleoptera: Cerambycidae, subfam. Lamiinae). Entomologische Blätter für Biologie und Systematik der Käfer 51 [1955] (1/2): 8–14. [1 Apr 1956 (wrapper)]

Tippmann FF (1960) Studien über neotropische Longicornier III (Coleoptera: Cerambycidae.). Koleopterologische Rundschau 37/38 [1959/60]: 82–217.

Tishechkin AK (2007) Phylogenetic revision of the genus *Mesynodites* (Coleoptera: Histeridae: Hetaeriinae) with descriptions of new tribes, genera and species. Sociobiology 49 (1): 1–167. [19 Jan 2007 (recorded at CNC)]

Tomaszewska W (2000) Morphology, phylogeny and classification of adult Endomychidae (Coleoptera: Cucujoidea). Annales Zoologici (Warszawa) 50 (4): 449–558. [20 Dec 2000 (verso of issue title page)]

Tömösváry Ö (1883) Adatok hazánk Thysanura-Faunájához. Magyar Tudományos Akadémia Mathematikai és Természettudományi Közlemények 18: 119–131, 1 pl. [1883 (Zool. Jahresber. 1: 162)]

Tottenham CE (1949) The generic names of the British Staphylinidae with a check list of the species. The generic names of British insects Part 9. Royal Entomological Society, London, 345–466. [30 Aug 1949]

Tottenham CE (1954) Vol. IV. Part 8(a). Coleoptera. Staphylinidae. Section (a) Piestinae to Euaesthetinae. Handbooks for the identification of British insects. Royal Entomological Society of London, London, pp. 1–79. [30 Nov 1954 (wrapper)]

Tournier H (1874) Matériaux pour servir à la monographie de la tribu des Érirrhinides de la famille des Curculionides (Coléoptères). Annales de la Société Entomologique de Belgique 17 (1): 63–116. [30 Jul 1874 (wrapper)]

Tôyama M (1986) The buprestid genus *Chalcophorella* Kerremans and its related genera (Coleoptera, Buprestidae). In: Ueno SI (Ed) Entomological papers presented to Yoshihiko Kurosawa on the occasion of his retirement, 188–193.

Tôyama M (1987) The systematic positions of some buprestid genera (Coleoptera, Buprestidae). Elytra 15: 1–11.

Trédl R (1907) Nahrungspflanzen und Verbreitungsgebiete der Borkenkäfer Europas. (Schluss). Entomologische Blätter 3 (5): 69–72. [18 May 1907 (p. 65)]

Triplehorn CA, Johnson NF (2005) Borror and DeLong's introduction to the study of insects. Seventh edition. Thomson Brooks/Cole, Belmont, California, [4, liner] + x + 864 pp.

Tschitschérine T (1898) Matériaux pour servir à l'étude des feroniens. IV. Horae Societatis Entomologicae Rossicae 32 (1/2): 1–224. [22 Jul 1898 (répartitions p. v; Standfuss and Kerzhner 2004: 242); 1899 (title page; Madge 1989)]

Tschitschérine T (1899) Sur l'emploi des noms de *Feronia* et de *Platysma* et sur les rapports des *Zabrus* avec les *Amara* (Col.). Bulletin de la Société Entomologique de France 1899 (4): 83–85. [séance 22 Feb 1899]

Tschitschérine T (1900) Mémoire sur la tribu des Harpalini. Horae Societatis Entomologicae Rossicae 34 [1899–1900] (3/4): 335–370. [14 Aug 1900 (Julian 1 Aug, répartitions p. vi)]

Tschitschérine T (1902) Notes sur les Platysmatini de l'Australie. Horae Societatis Entomologicae Rossicae 35 (3/4): 502–534. [3 May 1902 (Julian 20 Apr, p. vi)]

Tschitschérine T (1903) Chaetodactylini, subtribus nova Platysmatinorum (Coleoptera, Carabidae). Horae Societatis Entomologicae Rossicae 36 (1/2): 157–163. [2 Feb 1903 (Julian 20 Jan, p. vi)]

Tulk A (1847) Elements of Physiophilosophy. [English translation of: Oken, L., 1843: Lehrbuch der naturphilosophie. Dritte, neu bearbeitete Auflage]. The Ray Society, London, xix + [1] + 665 pp. [1847 (title page)]

Turco F, Di Giulio A, Bologna MA (2006) First-instar larval morphology in the subtribe Lydina (Coleoptera, Meloidae, Lyttini), with discussion on its systematic value. Deutsche Entomologische Zeitschrift 53 (2): 213–222 [Nov 2006 issue; 13 Oct 2006 (publ. online)]

Uéno S-I (1957) Blind aquatic beetles of Japan, with some accounts of the fauna of Japanese subterranean waters. Archiv für Hydrobiologie 53 (2): 250–296, 3 pls. [Apr 1957 (top of article)]

Uhmann E (1930) Hispinen aus Costa Rica aus der Ausbeute der Herrn. Ferd. Nevermann. 20. Beitrag zur Kenntnis der Hispinen (Col. Chrys.). Folia Zoologica et Hydrobiologica 1 [1929–30] (2): 209–256. [20 May 1930 (wrapper)]

Uhmann E (1940) Die Klauen der Hispinen. 90. Beitrag zur Kenntnis der Hispinen (Col. Chrys.). Zoologischer Anzeiger 130 (5/6): 119–123. [15 May 1940 (wrapper)]

Uhmann E (1960) Hispinae aus Madagascar. 2. Teil. 192. Beitrag zur Kenntnis der Hispinae (Col. Chrysomelidae). Revue Française d'Entomologie 27 (1): 58–91. [printed 12 Mar 1960 (p. 100)]

Valentine BD (1960) The genera of the weevil family Anthribidae North of Mexico (Coleoptera). Transactions of the American Entomological Society 86 (1): 41–85. [issued 17 Jun 1960 (list of papers)]

Valentine BD (1990) The Anthribidae of the Seychelles and Mascarene Islands: taxonomy, keys and a bibliographic catalogue (Coleoptera). Insecta Mundi 3 (3 [Sep 1989]): 233–246. [9 Mar 1990 (4: 132)]

Valentine BD (1999) A review of Nearctic and some related Anthribidae (Coleoptera). Insecta Mundi 12 [1998] (3/4): 251–296. [1999 (p. 320 cites paper publ. in Jan 1999)]

Valentine JM (1987) Some ancient and zoogeographically significant carabid beetles from the south Pacific (Coleoptera: Carabidae), with descriptions of new taxa. Bishop Museum Occasional Papers 27: 73–89. [publ. 19 Feb 1987 (verso of title page)]

Vaurie P (1973) The weevil genera *Homalinotus* and *Ozopherus* of the Neotropical Cholinae (Coleoptera, Curculionidae). Bulletin of the American Museum of Natural History 152 (1): 1–50. [29 Jun 1973 (title page)]

Vaurie P (1974) A new tribe for the South American genera *Cholomus* and *Irenarchus* (Coleoptera, Curculionidae, Cholinae). American Museum Novitates No. 2548: 1–13. [issued 26 Sep 1974 (p. 1 footer)]

Vaz-de-Mello FZ (2008) Synopsis of the new subtribe Scatimina (Coleoptera: Scarabaeidae: Scarabaeinae: Ateuchini), with descriptions of twelve new genera and review of *Genieridium*, new genus. Zootaxa No. 1955: 1–75. [publ. 5 Dec 2008 (footer p. 3)]

Verma KK, Gomez-Zurita J, Jolivet P (2005) Biology of *Eupales ulema* (Germar, 1813), and its taxonomic placement among Eumolpinae (Coleoptera, Chrysomelidae). Nouvelle Revue d'Entomologie (Nouvelle Série) 22 (2): 155–164. [30 Sep 2005 (back wrapper)]

Viana MJ (1952) Nueva subfamilia de Curculionidae, para la Argentina y Bolivia. Subfamilia Schöenherriellinae. (Coleopt.). Revista de la Sociedad Entomológica Argentina 15 (4): 231–236. [31 Dec 1952 (wrapper)]

Viana MJ (1964) Revision sistematica de las especies argentinas de "Goniocheniici" Spaeth y catalogo bibliografico de la tribus (Coleopt., Chrysomelidae, Cassidinae). Revista del Museo Argentino de Ciencias Naturales "Bernardino Rivadavia", Entomologia 1: 217–266.

Viana MJ (1971) Micholaeminae nueva subfamilia de Rhipiphoridae y *Micholaemus gerstaeckeri* nuevo género y especie de la República Argentina (Insecta, Coleoptera). Revista de la Sociedad Entomológica, Argentina 33 (1/4): 69–76. [Dec 1971 (vol. title page)]

Viana MJ, Martínez A (1971) Una nueva subfamilia de Anobiidae (Insecta, Coleoptera). Neotropica 17 (54): 121–126. [1 Dec 1971 (p. 105)]

Vienna P (1974) Gli Histeridae paleartici conservati nella collezione del Museo Civico di Storia Naturale di Milano (Coleoptera). Atti della Società Italiana di Scienze Naturali e del Museo Civico di Storia Naturale di Milano 115 (3/4): 271–284. [15 Dec 1974 (p. 424)]

Viette P (1993) Quelle date de publication pour le volume du Bulletin de l'Académie Malgache, nouvelle série, tome VII, anné 1924? Bulletin de la Société Entomologique de France 98 (2): 180. [Jun 1993 issue; Aug 1993 (dépôt légal, back wrapper)]

Villiers A (1961) Sur la structure des palpes maxillaires de quelques Disteniinae (Coleoptera Cerambycinae) [pp. 382–385]. XI Internationaler Kongreß für Entomologie Wien, 17 bis 25 August 1960 Verhandlungen Band I, 803 pp.

Villiers A (1978) Faune des coleopteres de France 1. Cerambycidae. Encyclopédie Entomologique Série A, Travaux généraux. Vol. 42. 552 pp. [Oct 1978 (Bull. Ent. Soc. France 84: 100)]

Villiers A (1980) Coléoptères Cerambycidae des Antilles françaises III. Lamiinae. Annales de la Société Entomologique de France (Nouvelle Série) 16 (4): 541–598. [31 Dec 1980 (back wrapper)]

Vitali F (2002) Versetzung der Tribus Holopterini Lacordaire, 1869 zur Unterfamilie Lepturinae (Coleoptera, Cerambycidae). Entomofauna 23 (3): 29–33. [publ. 15 Apr 2002]

Vives E (2000) Coleoptera, Cerambycidae. In: Ramos Sánchez A (Ed) Fauna Iberica. Vol. 12. Museo Nacional de Ciencias Naturales, Madrid, 715 pp. + 5 pls. + [1, adenda] [pp. 567–661 are by Vives and Alonso-Zarazaga]. [publ. 20 Dec 2000 (verso of title page)]

Vives E (2005) Révision du genre *Vesperus* Dejean 1821 (Coleoptera: Cerambycidae). Annales de la Société Entomologique de France (Nouvelle Série) 40 [2004] (3/4): 437–457. [10 Aug 2005 (back wrapper)]

Vives E, Alonso-Zarazaga MA (2000) Apéndice 1. Nomenclatura: lista de sinónimos y combinaciones [pp. 567–661]. In: Vives E: Coleoptera, Cerambycidae. In: Ramos Sánchez A (Ed) Fauna Iberica. Vol. 12. Museo Nacional de Ciencias Naturales, Madrid, 715 pp. + 5 pls. + [1, adenda]. [publ. 20 Dec 2000 (verso of title page)]

Vogt H (1967) 53. Familie: Cucujidae [pp. 83–104]. In: Freude H, Harde KW, Lohse GA (Eds). Die Käfer Mitteleuropas. Band 7. Clavicornia. Goecke & Evers, Krefeld, 310 pp.

Volkovitsh MG (2001) The comparative morphology of antennal structures in Buprestidae (Coleoptera): evolutionary trends, taxonomic and phylogenetic implications. Part 1. Acta Musei Moraviae Časopis Moravského Muzea Scientiae Biologicae 86 (1/2): 43–169.

Volkovitsh MG (2008) A revision of the buprestid subtribe Xenopsina subtr. n. with description of new species of the genera *Xenopsis* Saund. and *Sommaia* Toyama (Coleoptera, Buprestidae, Polycestinae) and notes on the systematic position of the subtribe. Entomologicheskoe Obozrenie 87: 627–649.

Voss E (1922) Monographische Bearbeitung der Unterfamilie Rhynchitinae (Curc.). I. Teil, Nemonychini - Auletini. Archiv für Naturgeschichte (Abteilung A) 88 (8): 1–113. [Aug 1922 (Newton); 10 Oct 1922 (recorded at BMNH)]

Voss E (1923) Indo-Malayische Rhynchitinen (Curculionidae). II, Zehnter Beitrag zur Kenntnis der Curculioniden. The Philippine Journal of Science 22 (5): 489–514. [5 May 1923 (vol. contents)]
Voss E (1925) Die Unterfamilien Attelabinae und Apoderinae. (Col. Curc.). (18. Beitrag zur Kenntnis der Curculioniden) (Fortsetzung). Stettiner Entomologische Zeitung 85 [1924] (1/2): 1–78 + pls. 1–3, 191–304. [May 1925 (Inhalt)]
Voss E (1926) Die Unterfamilien Attelabinae und Apoderinae. (Col. Curc.). (18. Beitrag zur Kenntnis der Curculioniden) (Fortsetzung). Stettiner Entomologische Zeitung 87 (1): 1–89 + pls. 4–6. [May 1926 (Inhalt)]
Voss E (1929a) Die Unterfamilien Attelabinae und Apoderinae (Col. Curc.). (18. Beitrag zur Kenntnis der Curculioniden.) (Fortsetzung). Stettiner Entomologische Zeitung 90 (2): 161–242. [Dec 1929 (Inhalt)]
Voss E (1929b) Einige bisher unbeschriebene Rhynchitinen der palaearktischen Region. (Col. Curc.) (27 Beitrag zur Kenntnis der Curculioniden). Entomologische Blätter 25 (1): 24–29. [31 Mar 1929 (wrapper)]
Voss E (1930) Die Attelabiden der Hauserschen Sammlung. (Col. Curc.) (28. Beitrag zur Kenntnis der Curculioniden). Wiener Entomologische Zeitung 47 (2): 65–88. [10 Jun 1930 (wrapper)]
Voss E (1931) Monographie der Rhynchitinen-Tribus Rhinomacerini und Rhinorhynchini. II. Teil der Monographie der Rhynchitinae-Pterocolinae. (36. Beitrag zur Kenntnis der Curculioniden). B. Spezieller Teil I. Unterfamilie Rhynchitinae. Entomologische Blätter 27 (4): 162–167. [31 Dec 1931 (wrapper)]
Voss E (1933a) Vier neue Rüßlerarten aus Szetschwan nebst Bemerkungen zu bekannten Arten. (Col. Curc.). (44. Beitrag zur Kenntnis der Curculioniden). Entomologisches Nachrichtenblatt (Troppau) 7 (1): 27–34. [Jan 1933 (article footers)]
Voss E (1933b) Monographie der Rhynchitinen-Tribus Auletini. III. Teil der Monographie der Rhynchitinae-Pterocolinae. (37. Beitrag zur Kenntnis der Curculioniden). Stettiner Entomologische Zeitung 94 (1): 108–136. [15 May 1933 (wrapper)]
Voss E (1934) Die Cossoninen Afrikas und Madagaskars des Deutschen Entomologischen Instituts der Kaiser-Wilhelm-Gesellschaft. (Schluss). Arbeiten über morphologische und taxonomische Entomologie aus Berlin-Dahlem 1 (3): 186–194. [12 Sep 1934]
Voss E (1935a) Fünf Rüßler aus den Tribus Prionomerini und Anthonomini von Java. (Col. Curc.). (59. Beitrag zur Kenntnis der Curculioniden). Entomologische Blätter 31 (6): 228–233. [31 Dec 1935 (wrapper)]
Voss E (1935b) Einige unbeschriebene Curculioniden aus dem Indomalayischen Archipel. 58. Beitrag zur Kenntnis der Curculioniden. The Philippine Journal of Science 56 (4): 509–522. [Apr 1935]
Voss E (1936) Über die Tribus Scolopterini sowie einige verwandte Gattungen und Gruppen von Neu-Seeland (Coleoptera: Curculionidae). (62. Beitrag zur Kenntnis der Curculioniden) (Schluss). Arbeiten über morphologische und taxonomische Entomologie aus Berlin-Dahlem 3 (2): 110–121. [8 Jun 1936 (vol. title page)]

Voss E (1937a) Nochmals die Tribus Scolopterini, sowie zur Abgrenzung der Subfamilie Eugnominae (Coleoptera: Curculionidae). Arbeiten über Morphologische und Taxonomische Entomologie aus Berlin-Dahlem 4 (1): 37–43. [publ. 25 Jan 1937 (vol. Inhalt)]
Voss E (1937b) Ein weiterer Beitrag zur Kenntnis der Curculioniden Javas. (65. Beitrag zur Kenntnis der Curculioniden). Tijdschrift voor Entomologie 80 (1/2): 127–166. [May 1937 (vol. title page)]
Voss E (1939) Rhynchitinen, Attelabinen und Cossoninen aus dem Kongo-Gebiet (Col., Curc.). Revue de Zoologie et de Botanique Africaines 32 (1): 42–82. [25 Feb 1939 (p. 42); 30 Mar 1939 (recorded at BMNH)]
Voss E (1940) Über Rüsselkäfer der Indomalayischen Subregion, vorwiegend von Java. (Coleoptera: Curculionidae.) Teil II. (89. Beitrag zur Kenntnis der Curculioniden). Arbeiten über morphologische und taxonomische Entomologie aus Berlin-Dahlem 7 (4): 279–287. [20 Dec 1940 (wrapper)]
Voss E (1944) Anthonominen-Studien (Col., Curc.). (97. Beitrag zur Kenntnis der Curculioniden). Stettiner Entomologische Zeitung 105 (1/2): 34–51. [1 Jun 1944 (wrapper)]
Voss E (1948) Über einige von Herrn Dr. Wettstein auf Kreta gesammelte Curculioniden (Col. Curc.). (109. Beitrag zur Kenntnis der Curculioniden). Annalen des Naturhistorischen Museums in Wien 56: 70–76. [Aug 1948 (wrapper)]
Voss E (1952) Mandschurische Rüßler aus dem Museum G. Frey. (121. Beitrag zur Kenntnis der Curculioniden). Mitteilungen der Münchner Entomologischen Gesellschaft 42: 190–205. [15 Nov 1952 (title page); 1 Nov 1952 (top of p. 1)]
Voss E (1953) Einige Rhynchophoren der Bernsteinfauna (Col.). Mitteilungen aus dem Geologischen Staatsinstitut in Hamburg 22: 119–140. [Feb 1953 (top of article)]
Voss E (1954) Curculionidae (Col.) [pp. 193–376]. In: Titschack E (Ed) Beiträge zur Fauna Perus nach der Ausbeute der Hamburger Südperu-Expedition 1936, anderer Sammlungen, wie auch auf Grund von Literaturangaben. Band IV. Wissenschaftliche Bearbeitungen. G. Fischer, Jena, viii + 386 pp. [10 Dec 1954 (recorded at BMNH)]
Voss E (1955a) Bemerkenswerte Ergebnisse einer Revision der Attelabiden des Ungarischen Naturwissenschaftlichen Museums zu Budapest, nebst Bemerkungen zur Cossoninen-Gattung *Aphyllura* Reitt. (Coleoptera). Annales Historico-Naturales Musei Nationalis Hungarici (Series nova) 6: 269–277. [printed 14 Jul 1955 (vol. title page)]
Voss E (1955b) Zur Synonymie und systematischen Stellung europäischer Cossoninen-Gattungen unter Berücksichtigung einiger Gattungen der madeirischen Fauna. (126. Beitrag zur Kenntnis der Curculioniden). Mitteilungen der Münchner Entomologischen Gesellschaft 44/45 [1954–55]: 182–239. [1 Aug 1955 (vol. title page)]
Voss E (1956a) Die von Biró auf Neu-Guinea aufgefundenen Rüsselkäfer (Col.). I. Annales Historico-Naturales Musei Nationalis Hungarici (Series nova) 7: 121–142.
Voss E (1956b) Über einige in Fukien (China) gesammelte Rüssler. V, nebst einer neuen Gattung und Art aus Yunnan (Col. Curc.). Entomologische Blätter für Biologie und Systematik der Käfer 51 [1955] (1/2): 21–45. [1 Apr 1956 (wrapper)]
Voss E (1957) *Archimetrioxena electrica* Voss und ihre Beziehungen zu rezenten Formenkreisen (Col., Curc.). Deutsche Entomologische Zeitschrift (Neue Folge) 4 (1/2): 95–102. [1 Apr 1957 (vol. Inhalt)]

Voss E (1958) Ein Beitrag zur Kenntnis der Curculioniden im Grenzgebiet der Orientalischen zur Paläarktischen Region (Col., Curc.) die von J. Klapperich und Tschung Sen in der Provinz Fukien gesammelten Rüsselkäfer (132. Beitrag zur Kenntnis der Curculioniden). Decheniana Beihefte 5: 1–139. [Jun 1958 (title page); 16 Jun 1958 (date of Vorwort); May 1958 (top of article)]

Voss E (1959) Ein Beitrag zur Kenntnis der Apioniden-Fauna Zentralafrikas (Coleoptera Curculionoidea). Annales du Musée Royal du Congo Belge (Série in 8°. Sciences Zoologiques) 76: 7–119. [printed Sep 1959 (endleaf); 22 Dec 1959 (recorded at BMNH)]

Voss E (1960) Nowy rodzaj i podrodzaj z podroziny Rhynchitinae z Chin (Coleoptera, Curculionidae). Noviy rod i podrod podsemeystva Rhynchitinae iz Kitaya (Coleoptera, Curculionidae). Eine neue Gattung und Untergattung der Subfamilie Rhynchitinae aus China (Coleoptera, Curculionidae). (154. Beitrag zur Kenntnis der Curculioniden). Annales Zoologici (Warszawa) 18 (22): 413–420. [20 Apr 1960 (title page); 2 Nov 1960 (recorded at BMNH)]

Voss E (1962) Attelabidae, Apionidae, Curculionidae (Coleoptera Rhynchophora). Exploration du Parc National de l'Upemba Mission G F de Witte. Fascicule 44. Institut des Parcs Nationaux du Congo Belge, Bruxelles, 380 pp. [publ. 15 Sep 1962 (endleaf)]

Voss E (1963) Neue Curculioniden und Gattungen aus Madagaskar (Col.). (184. Beitrag zur Kenntnis der Curculioniden). Entomologische Mitteilungen aus dem Zoologischen Staatsinstitut u Zoologischen Museum Hamburg 3 (47): 1–6. [30 Dec 1963]

Voss E (1965) Mission zoologique de l'I.R.S.A.C. en Afrique Orientale (P. Basilewsky et N. Leleup, 1957) Résultats scientifiques. Cinquième partie. Coleoptera Curculionidae II (Schluss). Annales du Musée Royal de l'Afrique Centrale, Tervuren (Série in 8°, Zoologie) 138: 293–377. [printed Jun 1965 (verso of title page)]

Voss E (1972) [new taxa] In: Decelle J, Voss E: 35. Fam. Curculionidae In: La faune terrestre de l'ile de Sainte-Helene. Deuxième partie. II. lnsectes. 9. Coleoptera. Annales du Musée Royal de l'Afrique Centrale, Tervuren (Sciences Zoologiques) 192: 306–515. [Mar 1972 (vol. title page)]

Vulcano MA, Martínez A, Pereira FS (1961) Eurysternini, nueva tribu de Scarabaeinae (Col. Scarabaeidae). Actas y Trabajos del Primer Congreso Sudamericano de Zoología (La Plata, 12–24 octubre 1959). Sección IV: Entomología [1960]. Tomo III. pp. 267–271. [printed 8 Sep 1961 (endleaf)]

Wagner H (1936) Vorstudien zu einer monographischen Bearbeitung der Ceuthorhynchinen Mitteleuropas (Col.Curcul.). Entomologische Blätter 32 (4): 161–170. [31 Aug 1936 (wrapper)]

Wagner H (1938) Monographie der paläarktischen Ceuthorrhynchinae (Curcul.). Entomologische Blätter 34 (4): 145–172. [15 Aug 1938 (wrapper)]

Wallace AR (1860) On the zoological geography of the Malay Archipelago. Journal of the Proceedings of the Linnean Society Zoology 4: 172–184. [read 3 Nov 1859]

Wanat M (1995) Systematics and phylogeny of the tribe Ceratapiini (Coleoptera: Curculionoidea: Apionidae). Genus. International Journal of Invertebrate Taxonomy (Supplement) [1994]: 1–406. [10 Feb 1995 (verso of title page)]

Wanat M (2001) Genera of Australo-Pacific Rhadinocybinae and Myrmacicelinae, with biogeography of the Apionidae (Coleoptera: Curculionoidea) and phylogeny of the Brentidae (s. lato). Mantis, Olsztyn, 432 pp. [15 Oct 2001 (verso of title page)]

Wang WI (1993) On Liaoximordellidae fam. n. (Coleoptera, Insecta) from the Jurassic of western Liaoning Province, China. Acta Geologica Sinica [Di Zhi Xue Bao] 67 (1): 86–94 + pl. 1 [in Chinese, English summary]. [Feb 1993 (top of article); 10 May 1993 (recorded at BMNH)]

Warchalowski A (1994) Chrysomelidae. Stonkowate (Insecta: Coleoptera). Część IV (podplemiona: Chrysomelina, Gonioctenina, Phratorina i Entomoscelina oraz podrodzina Galerucinae). Fauna Polski Fauna Poloniae. Tome 16. 279 pp. [1994 (title page)]

Wasmann E (1887) Neue Brasilianische Staphyliniden, bei *Eciton hamatum* gesammelt von Dr. W. Müller. Deutsche Entomologische Zeitschrift 31 (2): 403–416, pl. 5. [Nov 1887 (vol. Inhalt p. i)]

Wasmann E (1894) Kritisches Verzeichniss der Myrmekophilen und Termitophilen Arthropoden. Mit Angabe der Lebensweise und mit Beschreibung neuer Arten. F. L. Dames, Berlin, xv + 231 pp. [1894 (title page)]

Wasmann E (1896) Viaggio di Leonardo Fea in Birmania e regioni vicine, LXXII. Neue Termitophilen und Termiten aus Indien. Annali del Museo Civico di Storia Naturale "Giacomo Doria" (2) 16 [36]: 613–630, pl. 2. [27 Jun 1896 (Poggi 2010: 122)]

Wasmann E (1899) Neue Termitophilen und Myrmecophilen aus Indien. Deutsche Entomologische Zeitschrift 1899 (1): 145–169, pls. 1–2. [end Jul 1899 (p. 37)]

Wasmann E (1904a) Die moderne Biologie und die Entwicklungstheorie. Zweite, vermehrte Auflage. Herder, Freiburg im Breisgau, xii + 323 pp., 4 pls.

Wasmann E (1904b) Zur Kenntniss der Gäste der Treiberameisen und ihrer Wirthe am obern Congo, nach den Sammlungen und Beobachtungen von P. Herm. Kohl C. SS. C. bearbeitet (138. Beitrag zur Kenntnis der Myrmekophilen und Termitophilen.). Zoologische Jahrbücher, Supplement 7: 611–682, pls. 31–33.

Wasmann E (1912) *Mimanomma spectrum*, ein neuer Dorylinengast des extremsten Mimikrytypus. Zoologischer Anzeiger 39 (13/14): 473–481. [30 Apr 1912 (fasc. title page); 7 May 1912 (recorded at BMNH)]

Wasmann E (1916a) Neue dorylophile Staphyliniden Afrikas (Col.). (217. Beitrag zur Kenntnis der Myrmekophilen). Entomologische Mitteilungen 5 ((1/4)): 92–109, (5/8): 134–147, pl. 3. [25 Mar (fasc. 1/4), 15 Jul (5/8) 1916 (wrappers)]

Wasmann E (1916b) Wissenschaftliche Ergebnisse einer Forschungsreise nach Ostindien, ausgeführt im Auftrage der Kgl. Preuß. Akademie der Wissenschaften zu Berlin von H. v. Buttel-Reepen. V. Termitophile und myrmecophile Coleopteren. Gesammelt von Herrn Prof. Dr. v. Buttel-Reepen in den Jahren 1911–1912. Zoologische Jahrbücher, Abteilung für Systematik, Geographie und Biologie der Tiere 39 (2): 169–210, pls. 4, 5. [25 Feb 1916 (vol. Inhalt p. iii)]

Wasmann E (1917) Neue Anpassungstypen bei Dorylinengästen Afrikas (Col., Staphylinidae). (218. Beitrag zur Kenntnis der Myrmekophilen.). Zeitschrift für Wissenschaftliche Zoologie 117 (2): 257–360, pls. 7–10. [19 Jun 1917 (Inhalt p. iii)]

Wasmann E (1918) Myrmekophile und termitophile Coleopteren aus Ostindien, hauptsächlich gesammelt von P. J. Assmuth S. J. II. Scarabaeidae (223. Beitrag zur Kenntnis der Myrmecophilen und Termitophilen.). Wiener Entomologische Zeitung 37 (1/3): 1–23, pls. 1–2. [25 Mar 1918 (wrapper)]

Wasmann E (1920) Wissenschaftliche Ergebnisse der Deutschen Zentral-Africa-Expedition 1907–1908, unter Führung Adolf Friedrichs, Herzogs zu Mecklenburg. Tijdschrift voor Entomologie 62 (3/4): 109–130. [15 Jan 1920 (wrapper)]

Wasmann E (1928) Zur Kenntnis der Carabidomemninen. (278. Beitrag zur Kenntnis der Myrmecophilen). Tijdschrift voor Entomologie 71: 267–276, pl. 6. [31 Dec 1928 (verso of vol. title page)]

Waterhouse CO (1879a) Notice of a small collection of Coleoptera from Jamaica, with descriptions of new species from the West Indies. The Transactions of the Entomological Society of London (4) 1878 (4): 303–311. [6 Jan 1879 (G. Wheeler 1912)]

Waterhouse CO (1879b) Descriptions of new Coleoptera from East Africa and Madagascar. The Transactions of the Entomological Society of London 1879 (4): 319–321. [27 Dec 1879 (G. Wheeler 1912: 758)]

Waterhouse GR (1845) Descriptions of coleopterous insects collected by Charles Darwin, esq., in the Galapagos Islands. The Annals and Magazine of Natural History (1) 16 (102): 19–41. [1 Jul 1845 (Evenhuis 2003: 16)]

Waterhouse GR (1854) Notes on the species of *Amycterus* and allied genera, with descriptions of some new species. The Transactions of the Entomological Society of London (2) 3 (2): 75–80. [28 Oct 1854 (G. Wheeler 1912)]

Waterhouse GR (1858) Catalogue of British Coleoptera. Taylor and Francis, London, 117 pp.

Watt JC (1974) Chalcodryidae: a new family of heteromerous beetles (Coleoptera: Tenebrionoidea). Journal of the Royal Society of New Zealand 4 (1): 19–38. [Mar 1974 (contents of fasc. 2)]

Watt JC (1975) A revised subfamily classification of Tenebrionidae (Coleoptera). New Zealand Journal of Zoology 1 [1974] (4): 381–452. [publ. 27 Jan 1975 (verso of title page of vol. 2, fasc. 1)]

Watt JC (1992) Relationships of Actizeta and Cnemeplatiini (Coleoptera: Tenebrionidae). Systematic Entomology 17 (3): 287–299. [21 Jul 1992 (vol. contents)]

Watts CHS (1978) A revision of the Australian Dytiscidae (Coleoptera). Australian Journal of Zoology (Supplementary Series) 26 (57): 1–166. [24 Apr 1978 (wrapper)]

Weise J (1885) Bestimmungs-Tabellen der europäischen Coleopteren. II. Heft. Coccinellidae. II. Auflage. Mit Berücksicksichtigung der Arten aus dem nördlichen Asien. E. Reitter, Mödling, 83 pp.

Weise J (1901) Coccinelliden aus Ceylon gesammelt von Dr. Horn. Deutsche Entomologische Zeitschrift 44 [1900]: 417–445. [Feb 1901 (p. 3)]

Weise J (1902) Afrikanische Chrysomeliden. Archiv für Naturgeschichte 68 (1 (2)): 119–174.

Weise J (1903a) Afrikanische Chrysomeliden. Archiv für Naturgeschichte 69 (1 (2)): 197–226.

Weise J (1903b) Afrikanische Galerucinen. Deutsche Entomologische Zeitschrift 1903 (2): 321–334. [Aug 1903 (Inhalt p. 3)]

Weise J (1905) Bemerkungen über Hispinen. Deutsche Entomologische Zeitschrift 1905 (2): 317–320. [Sep 1905 (Inhalt)]

Weise J (1906) [families Coccinellidae, Chrysomelidae and Curculionidae] In: Heyden L, von, Reitter E, Weise J: Catalogus Coleopterorum Europae, Caucasi et Armeniae rossicae. Editio secunda. R. Friedländer & Sohn, Berlin, iv pp. + 750 cols. + pp. 751–774. [1906 (title page)]

Weise J (1910) Beitrag zur Kenntnis der amerikanischen Hispinen. Archiv für Naturgeschichte 76 (1): 67–127, pl. 3.

Weise J (1911) Pars 35. Fam. Chrysomelidae. Hispinae. In: Schenkling S (Ed) Coleopterorum Catalogus Volumen XXV. W. Junk, Berlin, 94 pp. [5 Aug 1911 (verso of vol. title page)]

Weise J (1915) Übersicht der Chrysomelini. Deutsche Entomologische Zeitschrift 1915 (4): 434–436. [15 Aug 1915 (Inhalt)]

Weise J (1921) Wissenschaftliche Ergebnisse der schwedischen entomologischen Reise des Herrn Dr. A. Roman in Amazonas 1914–1915. 6. Chrysomelidae. Arkiv för Zoologi 14 (1 (1)): 1–205. [16 Aug 1921 (verso of vol. title page)]

Weise J (1923) Uebersicht der Galerucinen. Wiener Entomologische Zeitung 40 (1/4): 124. [25 Apr 1923 (wrapper)]

Weise J (1924) Pars 35. Fam. Chrysomelidae. Galerucinae. In: Schenkling S (Ed) Coleopterorum Catalogus Volumen XXV. W. Junk, Berlin, 225 pp. [30 May 1924 (verso of vol. title page)]

Wellman C (1908) [nom. n.] In: Sitzung vom 16.III.08. Deutsche Entomologische Zeitschrift 1908 (3): 424. [1 May 1908 (Inhalt)]

Wellman C (1910) On the classification of the Lyttidae (Meloidae s. Cantharidae auctt.). Entomological News and Proceedings of the Entomological Section of the Academy of Natural Sciences of Philadelphia 21 (5): 211–222. [May issue; 2 May 1910 (recorded at USNM)]

Wenzel RL (1944) On the classification of the histerid beetles. Field Museum of Natural History Zoological Series 28 (2): 51–151, pls. 1–9. [19 Jan 1944 (wrapper)]

Werner YL (2006) The case of impact factor versus taxonomy: a proposal. Journal of Natural History 40 (21/22): 1285–1286. [publ. 28 Sep 2006 (footer p. 1285)]

Westcott RL (1968) A new subfamily of blind beetle from Idaho ice caves with notes on its bionomics and evolution (Coleoptera: Leiodidae). Contributions in Science / Los Angeles County Museum of Natural History No. 141: 14 pp. [14 Jun 1968 (title page); 15 Jul 1968 (recorded at BMNH)]

Westwood JO (1838) Synopsis of the genera of British insects [pp. 1–48]. In: An introduction to the modern classification of insects, founded on the natural habits and corresponding organization of the different families. Vol. II. Longman, Orme, Brown, Green & Longmans, London. [7 May (pp. 1–16); 9 Aug (17–32); 5 Nov 1838 (33–48) (Griffin 1932b: 83)]

Wheeler G (1912) On the dates of the publications of the Entomological Society of London. Transactions of the Entomological Society of London 1911 (4): 750–767. [Jan 1912]

Wheeler QD (1986) Revision of the genera of Lymexylidae (Coleoptera: Cucujiformia). Bulletin of the American Museum of Natural History 183 (2): 113–210. [issued 13 Jun 1986 (title page)]

White BE (1942) A new genus and species of Coleoptera (Chrysomelidae) from southwestern United States. Entomological News 53 (1): 16–21. [mailed 23 Jan 1942 (verso of vol. title page)]

White RE (1971) A new subfamily in Anobiidae (Coleoptera). Annals of the Entomological Society of America 64 (6): 1301–1304. [15 Nov 1971 (vol. contents)]

White RE (1974) Type-species for world genera of Anobiidae (Coleoptera). Transactions of the American Entomological Society 99 [1973] (4): 415–475. [issued 8 Jan 1974 (vol. contents)]

White RE (1982) Family: Anobiidae. United States Department of Agriculture / Agriculture Handbook (A catalog of the Coleoptera of America north of Mexico) No. 529–70: xi + 58 + [1, map] pp. [Aug 1982 (title page)]

Whitehead DR, Ball GE (1975) Classification of the Middle American genus *Cyrtolaus* Bates (Coleoptera: Carabidae: Pterostichini). Quaestiones Entomologicae 11: 591–619. [Oct issue; 8 Dec 1975 (recorded at CNC)]

Wilcox JA (1965) A synopsis of the North American Galerucinae (Coleoptera: Chrysomelidae). Bulletin. New York State Museum and Science Service No. 400: iv + 226 pp. [Aug 1965 (wrapper)]

Wilcox JA (1971) Pars 78, Fasc. 1 (Editio Seconda). Chrysomelidae: Galerucinae Oidini, Galerucini, Metacyclini, Sermylini. Coleopterorum Catalogus Supplementa. Uitgeverij Dr. W. Junk, 's-Gravenhage, pp. 1–220. [15 Jun 1971 (wrapper)]

Wilcox JA (1972) Pars 78, Fasc. 2 (Editio Seconda). Chrysomelidae: Galerucinae Luperini: Aulacophorina, Diabroticina. Coleopterorum Catalogus Supplementa. Uitgeverij Dr. W. Junk, 's-Gravenhage, pp. 221–431. [15 Mar 1972 (wrapper)]

Wilcox JA (1973) Pars 78, Fasc. 3 (Editio Seconda). Chrysomelidae: Galerucinae. Luperini: Luperina. Coleopterorum Catalogus Supplementa. Uitgeverij Dr. W. Junk, 's-Gravenhage, pp. 433–664. [20 Aug 1973 (title page)]

Winkler A (1925) Catalogus Coleopterorum regionis palaearcticae. I. A. Caraboidea, B. Palpicornia, Staphylinoidea, C. Diversicornia. Pars 3 [pp. 241–368]. Albert Winkler, Wien, vi + [2] + 1698 pp. [28 Feb 1925 (verso of title page)]

Winkler A (1927) Catalogus Coleopterorum regionis palaearcticae. II. D. Heteromera, Lamellicornia, E. Phytophaga, F. Rhynchophora. [pp. 817–880]. In: Pars 7 [pp. 753–880]. Albert Winkler, Wien, vi + [2] + 1698 pp. [16 Sep 1927 (verso of title page)]

Winkler A (1932) Catalogus Coleopterorum regionis palaearcticae. II. D. Heteromera, Lamellicornia, E. Phytophaga, F. Rhynchophora. Pars 13 [pp. 1521–1702]. Albert Winkler, Wien, vi + [2] + 1698 pp. [10 Dec 1932 (verso of title page)]

Winkler JR (1953) Supplementary remarks to R. Kleine's «Bestimmungs-Tabellen» with the descriptions of new species of east Asiatic Lycidae. Acta Entomologica Musei Nationalis Pragae 28 [1952]: 401–410. [1953 (vol. title page); accepted 2 Dec 1952 (p. 401)]

Winkler JR (1964) A revision of the genus *Dieropsis* Gahan, 1908, type of a new subfamily Dieropsinae n. subf. (Coleoptera: Cleridae). Acta Universitatis Carolinae - Biologica 1964 (3): 305–329. [publ. 20 Sep 1964 (p. 329)]

Winkler JR (1980) A revision of the new subfamily Cleropiestinae subf. n. (Coleoptera: Cleridae). Acta Universitatis Carolinae - Biologica 1978: 437–456. [Dec 1980 (top of p. 437)]

Winkler JR (1982) Subfamilies and neutral terms proposed for groups higher than subfamily in Cleridae (Coleoptera) - purpose, definitions, identification key. Acta Universitatis Carolinae - Biologica 1980: 517–531. [Nov 1982 (top of p. 517)]

Winkler JR (1987) Berendtimiridae fam. n., a new family of fossil beetles from Baltic Amber (Coleoptera, Cantharoidea). Mitteilungen der Münchner Entomologischen Gesellschaft 77: 51–59. [1 Dec 1987 (wrapper)]

Wittmer W (1970) On some Cantharidae (Coleoptera) of the United States. The Coleopterists' Bulletin 24 (2): 42–46. [publ. 25 Jun 1970 (vol. contents)]

Wittmer W (1976) 35. Beitrag zur Kenntnis der neotropischen Fauna. Coleoptera: Fam. Malachiidae. Studia Entomologica 19 (1/4): 421–449. [publ. 30 Dec 1976 (Indice)]

Wolfe GW, Roughley RE (1990) A taxonomic, phylogenetic, and zoogeographic analysis of Laccornis Gozis (Coleoptera: Dytiscidae) with the description of Laccornini, a new tribe of Hydroporinae. Quaestiones Entomologicae 26 (3): 273–354. [issued Nov 1990 (inside wrapper)]

Wollaston TV (1864) Catalogue of the coleopterous insects of the Canaries in the collection of the British Museum. British Museum, London, xiii + 648 pp. [issued 25 Jun 1864 (Sherborn 1934: 310)]

Wollaston TV (1865) Coleoptera Atlantidum, being an enumeration of the Coleopterous Insects of the Madeiras, Salvages and Canaries. Taylor & Francis, London, xlvii + 526 pp. + 140 pp. (Appendix, Indices). [25 Oct 1865 (date of preface); 30 Dec 1865 (London Rev. 11 (287): p. i)]

Wollaston TV (1870) On the Coleoptera of St. Helena [concluded]. The Annals and Magazine of Natural History (4) 5 (25): 18–37. [1 Jan 1870 (Evenhuis 2003: 25)]

Wollaston TV (1873) On the genera of the Cossonidae. The Transactions of the Entomological Society of London 1873 (4): 427–657. [Oct 1873 issue; 13 Oct 1873 (G. Wheeler 1912: 757)]

Wollaston TV (1877) Coleoptera Sanctae-Helenae. J. van Voorst, London, xxv + 256 pp. + 1 pl. [1877 ("a few weeks before his death" [4 Jan 1878]: Proc. Ent. Soc. London 1877: lxiv)]

Wood SL (1966) New records and species of Neotropical Platypodidae (Coleoptera). The Great Basin Naturalist 26 (3/4): 45–70. [31 Dec 1966 (wrapper); 25 Feb 1967 (recorded at BMNH)]

Wood SL (1971) New records and species of Neotropical bark beetles (Scolytidae: Coleoptera) Part V. Brigham Young University Science Bulletin (Biological Series) 15 (3): [2] + 1–54. [Dec 1971 (wrapper + title page); 18 Feb 1972 (recorded at MCZ)]

Wood SL (1971) New records and species of Neotropical bark beetlesw (Scolytidae: Coleoptera) Part V. Brigham Young University Science Bulletin (Biological Series) 15 (3): [2] + 1–54. [Dec 1971 (wrapper)]

Wood SL (1978) A reclassification of the subfamilies and tribes of Scolytidae (Coleoptera). Annales de la Société Entomologique de France (Nouvelle Série) 14 (1): 95–122. [issued 28 Feb 1978 (back wrapper)]

Wood SL, Bright DE (1992) A Catalog of Scolytidae and Platypodidae (Coleoptera), Part 2: Taxonomic Index. Volume B. Great Basin Naturalist Memoirs No. 13: [2] + 835–1553. [Dec 1992 (copyright)]

Woodley NE (2002) *Wallacea* Baly, a valid name in the Hispinae (Coleoptera: Chrysomelidae). The Coleopterists Bulletin 55 [2001] (4): 410. [mailed 2 Jan 2002 (inside wrapper)]

Würmli M (1975) Gattungsmonographie der altweltischen Hispinen (Coleoptera: Chrysomelidae: Hispinae). Entomologische Arbeiten aus dem Museum G Frey 26: 1–83. [publ. 1 Nov 1975 (vol. Inhalt)]

Young DK (2002) Family 115. Pyrochroidae [pp. 540–543]. In: Arnett RH, Jr, Thomas MC, Skelley PE, Frank JH (Eds) American beetles. Volume 2. Polyphaga: Scarabaeoidea through Curculionoidea. CRC Press, Boca Raton, xiv + 861 pp. [19 Jun 2002 (CRC website)]

Young DK, Pollock DA (2002) Family 99. Tetratomidae [pp. 413–416]. In: Arnett RH, Jr, Thomas MC, Skelley PE, Frank JH (Eds) American beetles. Volume 2. Polyphaga: Scarabaeoidea through Curculionoidea. CRC Press, Boca Raton, xiv + 861 pp. [19 Jun 2002 (CRC website)]

Zaballos JP, Jeanne C (1994) Nuevo catalogo de los Carabidos (Coleoptera) de la Peninsula Iberica. Monografias Sociedad Entomológica Aragonesa 1: 1–159.

Zagajkevich IK (1991) Taksonomiya i ekologiya usachey. [Taxonomy and ecology of Cerambycidae]. Naukova Dumka, Kiev, 178 pp. [in Russian].

Zahradník P (2007) family Ptinidae Latreille, 1802 [subfamilies Anobiinae, Dorcatominae, etc., pp. 339–362]. In: Löbl I, Smetana A (Eds) Catalogue of Palaearctic Coleoptera. Volume 4. Elateroidea - Derodontoidea - Bostrichoidea - Lymexyloidea - Cleroidea - Cucujoidea. Apollo Books, Stenstrup, 935 pp. [publ. 30 Jun 2007 (verso of title page)]

Zahradník P (2009) Order Coleoptera, Family Ptinidae [pp. 174–186]. In: van Harten A (Ed.) Arthropod Fauna of the United Arab Emirates. Volume 2. Dar Al Ummah, Abu Dhabi.

Zaitzev FA (1908) Catalogue des Coléoptères aquatiques des familles des Dryopidae, Georyssidae, Cyathoceridae, Heteroceridae et Hydrophilidae. Horae Societatis Entomologicae Rossicae 38 (4): 283–420. [14 Oct 1908 (Julian 1 Oct, p. vii; Kerzhner 1984)]

Zajciw D (1960) Uma tribo, três gêneros e três espécies novas de Longicórneos do Brasil (Col., Cerambycidae). Anais da Academia Brasileira de Ciências 31 [1959] (4): 605–616. [31 Dec 1959 (footer p. 605); 1960 (wrapper)]

Zalat S, Saleh R, Angus R, Kaschef A (2000) Diving beetles (Coleoptera: Dytiscidae and Noteridae) of Egypt. Egyptian Journal of Natural History 2: i-ii, 1–107. [Dec 2000 (vol. title page)]

Zalesskiy GM (1947) O dvykh novykh permskikh zhukakh [On two new Permian beetles]. Doklady Akademii Nauk SSSR (Novaya Seriya) 56 (8): 857–860 [in Russian]. [27 Feb 1948 (recorded at BMNH)]

Zamotajlov AS (2000) Lissopogonini, a new tribe of the subfamily Patrobinae from East Asia (Coleoptera: Carabidae). Zoosystematica Rossica 8 (2): 266. [publ. 16 Mar 2000 (fasc. contents)]

Zamotajlov AS (2002) Opyt razrabotki philogeneticheskoy sistemy zhuzhelits podsemeystva Patrobinae (Coleoptera, Carabidae) [Inferring phylogenetic system of the carabid subfamily Patrobinae (Coleoptera, Carabidae)]. Chteniya Pamyati Nikolaya Aleksandrovicha Kholodkovskogo 55 (1): 1–147 [in Russian]. [to printer 20 Nov 2002 (endleaf)]

Zamotajlov AS, Lafer GS (2001) K poznaniyu zhuzhelits roda Platidiolus Chaudoir, 1878 (Coleoptera, Carabidae) iz Kontinental'noy Azii [Contribution to the knowledge of the carabid genus *Platidiolus* Chaudoir, 1878 (Coleoptera, Carabidae) from Continental Asia]. Entomologicheskoe Obozrenie 80 (2): 411–435, 543 [in Russian; English translation in Entomological Review, 81 [2001] (4), 402–422]. [after 31 May 2001 (approved to print, back wrapper)]

Zerche L (1987) Beitrag zur Kenntnis der Gattung *Thiasophila* Kraatz, 1856 (Coleoptera, Staphylinidae, Aleocharinae). Entomologische Blätter für Biologie und Systematik der Käfer 83 (2/3): 91–114. [31 Dec 1987 (top of article)]

Zerche L (1990) Monographie der paläarktischen Coryphiini (Coleoptera, Staphylinidae, Omaliinae). Akademie der Landwirtschaftswissenschaften der Deutschen Demokratischen Republik, Berlin, 413 pp. [recorded 21 Sep 1990]

Zerche L (2004) Revision der Gattung *Aegialites* Mannerheim (Coleoptera: Salpingidae: Aegialitinae). A revision of the genus Aegialites Mannerheim (Coleoptera: Salpingidae: Aegialitinae). Stuttgarter Beiträge zur Naturkunde (Serie A (Biologie)) 666: 1–116, 4 pls. [5 Aug 2004 (top of article)]

Zherikhin VV (1977) [Family Curculionidae Latreille, 1802] [pp. 178–182, and other parts] In: Arnoldi LV, Zherikhin, VV, Nikritin LM, Ponomarenko AG (Eds): Mezozoyskie Zhestkokrylye. Trudy Paleontologicheskogo Instituta Akademiya Nauk SSSR 161: 1–204 [English translation: Mesozoic Coleoptera; Smithsonian Institution Libraries and N.S.F., Washington D.C.; xii + 285 pp., 1992] [after 5 Sep 1977 (approved to print)].

Zherikhin VV (1991) [new taxa] In: Zherikhin VV, Egorov AB: Zhuki-dolgonosiki (Coleoptera, Curculionidae) Dalnego Vostoka SSSR (obzor podsemeistv s opisaniem novykh taksonov). [Curculionidae (Coleoptera) of the Far East of the USSR (subfamilies with the description of new taxa)]. Akademija Nauk SSSR, Dal'nevostochnoe Otdelenie, Biologo-Pochvenniy Institut, Vladivostok, 1–164 [in Russian]. [1990 (title page); after 18 Dec 1991 (approved to print, p. 164)]

Zherikhin VV (1993) [new taxa] In: Gromov VV, Dmitriev VY, Zherikhin VV, Lebedev EL, Ponomarenko AG, Rasnitsyn AP, Sukacheva ID: Melovye entomofauny Basseyna r. Ul'i (Zapadnoe Priokhot'e) [Cretaceous insect fauna from Uljya-river Basin (West Okhot region)] Trudy Paleontologicheskogo Instituta Akademiya Nauk SSSR 252: 5–60 [in Russian]. [after 28 Jan 1993 (approved to print)]

Zherikhin VV, Gratshev VG (1994) Obrieniidae, fam. n., the oldest Mesozoic weevils (Coleoptera, Curculionoidea). Paleontological Journal 27 (1A): 50–69. [publ. Jan 1994 (wrapper); 1993 (top of article)]

Zimmerman EC (1993) Australian weevils (Coleoptera: Curculionoidea). Volume III. Nanophyidae, Rhynchophoridae, Erirhinidae, Curculionidae: Amycterinae, Literature Consulted. CSIRO, East Melbourne, x + 854 pp. [1 Oct 1993 (verso of title page)]

Zimmerman EC (1994a) Australian weevils (Coleoptera: Curculionoidea). Volume I. Orthoceri, Anthribidae to Attelabidae, the primitive weevils. CSIRO, East Melbourne, xxxii + 741 pp. [28 Feb 1994 (verso of title page)]

Zimmerman EC (1994b) Australian weevils (Coleoptera: Curculionoidea). Volume II. Brentidae, Eurhynchidae, Apionidae, and a chapter on immature stages by Brenda May. CSIRO, East Melbourne, x + 755 pp. [28 Feb 1994 (verso of title page)]

Zimmerman EC (1999) [Aonychusini nom. n.] In: Alonso-Zarazaga MA, Lyal CHC: A world catalogue of families and genera of Curculionoidea (Insecta: Coleoptera) (excepting Scolytidae and Platypodidae). Entomopraxis, S. C. P., Barcelona, 315 pp. [publ. 27 Dec 1999 (verso of title page)]

Zimmermann A (1919) Die Schwimmkäfer des Deutschen Entomologischen Museums in Berlin-Dahlem. Archiv für Naturgeschichte (Abteilung A) 83 [1917] (12): 68–249. [Dec 1919 (wrapper)]

Zimmermann C (1832) Ueber die bisherige Gattung *Amara*. Ein Beitrag zu einer künftigen Monographie der hieher gehörigen Thiere. Faunus. Zeitschrift für Zoologie und Vergleichende Anatomie 1: 5–40. [1832 (title page) ; 1833 (Neues Allg. Repert. für 1833: 100)]

Zumpt F (1929) Revision der Genera *Notaris* Germ., *Lepidonotaris* m., *Thryogenes* Bed., *Grypus* Germ., *Icaris* Tourn. und *Picianus* m. (Col. Curc.). Coleopterologisches Centralblatt 3 (5/6): 213–239. [5 May 1929 (vol. Inhalt)]

Zunino M (1984) Sistematica generica dei Geotrupinae (Coleoptera, Scarabaeoidea: Geotrupidae), filogenesi della sottofamiglia e considerazioni biogeografiche. Bollettino del Museo Regionale di Scienze Naturali, Torino 2 (1): 9–162. [23 Mar 1984 (top of article); printed 15 Jun 1984 (fasc. endleaf)]

Zunino M (1985a) Las relaciones taxonómicas de los Phanaeina (Coleoptera, Scarabaeinae) y sus implicancias biogeográficas (Resumen). Revista Peruana de Entomologia 26 [1983] (1): 21–23. [publ. 20 Nov 1985 (wrapper)]

Zunino M (1985b) Las relaciones taxonomicas de los Phanaeina (Coleoptera, Scarabaeinae) y sus implicaciones biogeograficas. Folia Entomologica Mexicana 64: 101–115. [distrib. 20 Dec 1985 (Contenido)]

Appendix I

Conservation of younger names using Reversal of Precedence (Art. 23.9). Cases listed in alphabetical order by family.

Carabidae

Supporting references for the conservation of Perigonini Horn, 1881 over Trechicini Bates, 1873 (Art. 23.9.2). The taxon name Trechicini Bates, 1873 has not been used as valid after 1899 to our knowledge.

Assmann T (1999) A new anophthalmic genus of Perigonini from the Iberian Peninsula (Insecta, Coleoptera, Carabidae). Spixiana 22 (3): 255–262.

Ball GE, Bousquet Y (2000) Carabidae Latreille, 1810 [pp. 32–132]. In: Arnett RH, Thomas MC (Eds) American beetles. Volume 1. Archostemata, Myxophaga, Adephaga, Polyphaga: Staphyliniformia. CRC Press, Boca Raton, xv + 443 pp.

Basilewsky P (1989) Révision des Perigonini d'Afrique (Coleoptera Carabidae). Revue de Zoologie Africaine 103 (4): 413–452.

Bousquet Y (2003) Tribe Perigonini G.H. Horn, 1881 [pp. 448–449]. In: Löbl I, Smetana A (Eds) Catalogue of Palaearctic Coleoptera. Volume 1. Archostemata - Myxophaga - Adephaga. Apollo Books, Stenstrup, 819 pp.

Bousquet Y, Larochelle A (1993) Catalogue of the Geadephaga (Coleoptera: Trachypachidae, Rhysodidae, Carabidae including Cicindelini) of America north of Mexico. Memoirs of the Entomological Society of Canada 167: 1–397.

Dajoz R (2002) Les coléoptères carabidés et ténébrionidés: écologie et biologie. Tec et Doc, Paris, xi + 522 pp.

Darlington PJ, Jr. (1968) The carabid beetles of New Guinea Part III. Harpalinae (continued): Perigonini to Pseudomorphini. Bulletin of the Museum of Comparative Zoology at Harvard 137 (1): 1–253.

Erwin TL (1985) The taxon pulse: a general pattern of lineage radiation and extinction among carabid beetles [pp. 437–472]. In: Ball GE (Ed) Taxonomy, phylogeny and zoogeography of beetles and ants. Series Entomologica Volume 33. Junk, Dordrecht, xiii + 514 pp. + 2 pls.

Erwin TL, Sims LL (1984) Carabid beetles of the West Indies (Insecta: Coleoptera): a synopsis of the genera and checklist of tribes of Caraboidea, and of the West Indian species. Quaestiones Entomologicae 20 (4): 351–466 + 1 pl.

Hůrka K (1996) Carabidae of the Czech and Slovak republics. Kabourek, Zlin, 565 pp.

Kryzhanovskij OL, Belousov IA, Kabak II, Kataev BM, Makarov KV, Shilenkov VG (1995) A checklist of the ground beetles of Russia and adjacent lands (Insecta, Coleoptera, Carabidae). Pensoft Series Faunistica No. 3. Pensoft Publishers, Sofia and Moscow, 271 pp.

Lafer GS (1989) I. Podotryad Archostemata; II. Podtryad Adephaga [pp. 66–259]. Tom III Zhestkokrylye, ili zhuki Chast 1. Nauka, Leningrad, 572 pp. [in Russian].

Larochelle A, Larivière M-C (2001) Carabidae (Insecta: Coleoptera): catalogue. Fauna of New Zealand. Ko te Aitanga Pepeke o Aotearoa. Number 43, Manaaki Whenua Press, Lincoln, 285 pp.

Larochelle A, Larivière M-C (2007) Carabidae (Insecta: Coleoptera): synopsis of supraspecific taxa. Fauna of New Zealand. Ko te Aitanga Pepeke o Aotearoa. Number 60, Manaaki Whenua Press, Lincoln, 188 pp.

Lindroth CH (1968) The ground-beetles (Carabidae, excl. Cicindelinae) of Canada and Alaska. Part 5. Opuscula Entomologica Supplementum 33: 649–944.

Lorenz W (2005) Systematic list of extant ground beetles of the world (Insecta Coleoptera «Geadephaga»: Trachypachidae and Carabidae incl. Paussinae, Cicindelinae, Rhysodinae). Second Edition. W. Lorenz, Tutzing, [ii] + 530 pp.

Martínez C (2005) Introducción a los escarabajos Carabidae (Coleoptera) de Colombia. Instituto de Investigación de Recursos Biológicos Alexander von Humboldt, Bogota, 546 pp.

Moore BP, Weir TA, Pyke JE (1987) Carabidae [pp. 23–320]. In: Lawrence JF, Moore BP, Pyke JE, Weir TA (Eds) Zoological catalogue of Australia. Volume 4. Coleoptera: Archostemata, Myxophaga and Adephaga. Australian Government Publishing Service, Canberra, viii + 444 pp.

Moret P (2008) Four new species of *Diploharpus* Chaudoir, 1850 from Ecuador (Coleoptera, Carabidae, Perigonini). Memoirs on Biodiversity 1: 201–208.

Nelson RE (1991) First records of *Perigona pallipennis* (LeC.) and *Perigona nigriceps* (Dej.) (Coleoptera: Carabidae: Perigonini) from Maine: easternmost records for the genus in North America. The Coleopterists Bulletin 45 (3): 284–285.

Perrault GG (1985) Etudes sur la faune des Carabidae de Guyane. 2. Une nouvelle espèce de *Perigona* microphthalme (Coleoptera, Carabidae, Perigonini). L'Entomologiste 41: 61–64.

Perrault GG (1988) Notes sur la tribu des Perigonini (Coleoptera Carabidae) avec les descriptions de deux sous-genres et d'une espèce. Entomologische Blätter für Biologie und Systematik der Käfer 84 (1/2): 11–16.

Perrault GG (1992a) Notes sur les types des espèces néotropicales du genre *Perigona* décrites par Bates (Coleoptera, Carabidae, Perigonini). Nouvelle Revue d'Entomologie Nouvelle Série 8 [1991] (4): 398.

Perrault GG (1992b) Etudes sur la faune des Carabidae de Guyane. III. Deux espèces nouvelles de *Diploharpus* (Perigonini). L'Entomologiste 48 (2): 99–103.

Reichardt H (1977) A synopsis of the genera of Neotropical Carabidae (Insecta: Coleoptera). Quaestiones Entomologicae 13: 346–493.

Carabidae

Supporting references for the conservation of Anisodactylina Lacordaire, 1854 over Eurytrichina LeConte, 1848 (Art. 23.9.2). The taxon name Eurytrichina LeConte, 1848 has not been used as valid after 1899 to our knowledge.

Ball GE, Bousquet Y (2000) Carabidae Latreille, 1810 [pp. 32–132]. In: Arnett RH, Thomas MC (Eds) American beetles. Volume 1. Archostemata, Myxophaga, Adephaga, Polyphaga: Staphyliniformia. CRC Press, Boca Raton, xv + 443 pp.

Bousquet Y, Larochelle A (1993) Catalogue of the Geadephaga (Coleoptera: Trachypachidae, Rhysodidae, Carabidae including Cicindelini) of America north of Mexico. Memoirs of the Entomological Society of Canada 167: 1–397.

Bousquet Y, Tchang J-P (1992) Anisodactyline larvae (Coleoptera: Carabidae: Harpalini): descriptions of genus-group taxa of eastern Canada and phylogenetic remarks. The Canadian Entomologist 124 (5): 751–783.

Habu A (1973) Carabidae: Harpalini (Insecta: Coleoptera). Fauna Japonica, Keigaku Publishing Co., Tokyo, xiii + 430 pp. + 24 pls.

Hůrka K (1996) Carabidae of the Czech and Slovak republics. Kabourek, Zlin, 565 pp.

Ito N (1997) Some new taxa of the subtribe Anisodactylina from Asia with a key to the Asian genera (Coleoptera, Carabidae, Harpalini). Japanese Journal of Entomology 65 (4): 816–825.

Ito N (2003a) Subtribe Anisodactylina Lacordaire, 1854 [pp. 360–363]. In: Löbl I, Smetana A (Eds) Catalogue of Palaearctic Coleoptera. Volume 1. Archostemata - Myxophaga - Adephaga. Apollo Books, Stenstrup, 819 pp.

Ito N (2003b) Notes on species of the harpaline subtribe Anisodactylina (Coleoptera, Carabidae) from China. Special Bulletin of the Japanese Society of Coleopterology 6: 79–86.

Kaiser M (2004) Faunistik und Biogeographie der Anisodactylinae und Harpalinae Westfalens (Coleoptera: Carabidae). Abhandlungen aus dem Westfälischen Museum für Naturkunde 66 (3): 1–155.

Kataev BM (2003) A new genus and species of the subtribe Anisodactylina from south-western Australia (Coleoptera: Carabidae: Harpalini). Acta Zoologica Academiae Scientiarum Hungaricae 48 (3): 173–179.

Kataev BM (2002) On some new and little-known species of the Anisodactylina and Harpalina (the Selenophori group) from East Asia and Oriental region (Coleoptera: Carabidae: Harpalini). O nekotorykh novykh i maloizvestnykh vidakh Anisodactylina i Harpalina (gruppa Selenophori) iz vostochnoy Azii i Oriental'nogo regiona (Coleoptera: Carabidae: Harpalini). Russian Entomological Journal 11 (3): 241–252.

Kataev BM, Wrase DW (2001) A new genus and a new species of the subtribe Anisodactylina from Vietnam and remarks on the taxonomic position of *Hiekea picipes* N. Ito 1997 (Coleoptera: Carabidae: Harpalini). Linzer Biologische Beiträge 33: 637–646.

Kryzhanovskij OL, Belousov IA, Kabak II, Kataev BM, Makarov KV, Shilenkov VG (1995) A checklist of the ground beetles of Russia and adjacent lands (Insecta, Coleoptera, Carabidae). Pensoft Series Faunistica No. 3. Pensoft Publishers, Sofia and Moscow, 271 pp.

Larochelle A, Larivière M-C (2001) Carabidae (Insecta: Coleoptera): catalogue. Fauna of New Zealand. Ko te Aitanga Pepeke o Aotearoa. Number 43, Manaaki Whenua Press, Lincoln, 285 pp.

Larochelle A, Larivière M-C (2005) Harpalini (Insecta: Coleoptera: Carabidae: Harpalinae). Fauna of New Zealand. Ko te Aitanga Pepeke o Aotearoa. Number 53, Manaaki Whenua Press, Lincoln, 160 pp.

Larochelle A, Larivière M-C (2007) Carabidae - (Insecta: Coleoptera): synopsis of supraspecific taxa. Fauna of New Zealand. Ko te Aitanga Pepeke o Aotearoa. Number 60, Manaaki Whenua Press, Lincoln, 188 pp.

Lindroth CH (1968) The ground-beetles (Carabidae, excl. Cicindelinae) of Canada and Alaska. Part 5. Opuscula Entomologica Supplementum 33: 649–944.

Machado A (1992) Monografia de los carábidos de las Islas Canarias (Insecta, Coleoptera). Instituto de Estudios Canarios, La Laguna, 734pp.

Martínez C (2005) Introducción a los escarabajos Carabidae (Coleoptera) de Colombia. Instituto de Investigación de Recursos Biológicos Alexander von Humboldt, Bogota, 546 pp.

Martinez-Navarro EM, Galian J, Serrano J (2003) Phylogenetic relationships among subtribes of Harpalini Bonelli 1810 (Coleoptera, Carabidae), and *Harpalus* Latreille 1802 and some related western Palearctic taxa. A preliminary cladistic analysis using morphological and karyotypic characters. Acta Entomológica Ibérica e Macaronésica 1: 59–65.

Martinez-Navarro EM, Galian J, Serrano J (2005) Phylogeny and molecular evolution of the tribe Harpalini (Coleoptera, Carabidae) inferred from mitochondrial cytochrome-oxidase I. Molecular Phylogenetics and Evolution 35: 127–146.

Noonan GR (1973) The anisodactylines (Insecta: Coleoptera: Carabidae: Harpalini): classification, evolution, and zoogeography. Quaestiones Entomologicae 9 (4): 266–480.

Noonan GR (1974) *Allendia*, a new South American genus with notes on its evolutionary relationships to other genera of Anisodactylina (Coleoptera: Carabidae: Harpalini). The Coleopterists Bulletin 28 (4): 219–227.

Noonan GR (1976) Synopsis of the supra-specific taxa of the tribe Harpalini (Coleoptera: Carabidae). Quaestiones Entomologicae 12 (1): 3–87.

Reichardt H (1977) A synopsis of the genera of Neotropical Carabidae (Insecta: Coleoptera). Quaestiones Entomologicae 13 (4): 346–493.

Cucujidae

Supporting references for the conservation of *Pediacus* Shuckard, 1839 over *Biophloeus* Dejean, 1835 (Art. 23.9.2). The genus name *Biophloeus* Dejean, 1835 has not been used as valid after 1899 to our knowledge.

Alexander KNA (1983) Some records of *Pediacus dermestoides* (Fab.) (Col., Cucujidae) from central southern England. Entomologist's Monthly Magazine 119: 219.

Alexander KNA (1987) *Pediacus dermestoides* (Fab.) (Col., Cucujidae) new to Cornwall, and some comments on its apparent increase in abundance in southern Britain. Entomologist's Monthly Magazine 123: 162.

Beutel RG, Pollock DA (2000) Larval head morphology of *Phycosecis litoralis* (Pascoe) (Coleoptera: Phycosecidae) with phylogenetic implications. Invertebrate Taxonomy 14 (6): 825–835.

Böhme J (2005) Band K. Katalog (Faunistische Übersicht). 2. Auflage. Die Käfer Mitteleuropas, Elsevier GmbH, Spektrum A. V., Munich, xii + 515 pp.

Bousquet Y (1991) Family Cucujidae: flat bark beetles [pp. 219–220]. In: Bousquet Y (Ed) Checklist of the beetles of Canada and Alaska. Research Branch Publication 1861/E, Ottawa, vi + 430 pp.

Burakowski B, Mroczkowski M, Stefańska J (1986) Chrząszcze Coleoptera. Cucujoidea, część 1. Katalog Fauny Polski. Część XXIII, tom 12 [Nr 43]. Panstwowe Wydawnictwo Naukowe, Warszawa, 266 pp.

Cooter J (1977) *Pediacus depressus* (Herbst, 1797) (Col.: Cucujidae) new to Scotland. The Entomologist's Record and Journal of Variation 89 (12): 339–340.

Crowson RA (1973) Further observations on Phloeostichidae and Cavognathidae, with definitions of new genera from Australia and New Zealand. The Coleopterists Bulletin 27 (2): 54–62.

Grosso-Silva JM (2002) Registos interessantes de coleópteros (Insecta, Coleoptera) para Portugal (3ª nota). Primeiro registo ibérico de *Pediacus dermestoides* (Fabricius, 1792) (Cucujidae). Boletín de la Sociedad Entomológica Aragonesa 31: 49–54.

Háva J (2000) Faunistic records from the Czech Republic - 95. Coleoptera. Klapalekiana 36 (1/3): 36.

Jansen RP, van de Sande C (2008) The genus *Pediacus* in The Netherlands (Coleoptera: Cucujidae). Entomologische Berichten 68 (1): 17–20.

Lawrence JF (1991) Cucujidae (Cucujoidea) (including Catogenidae, Laemophloeidae, Passandridae, Silvanidae) [pp. 463–488]. In: Stehr FW (Ed) Immature Insects. Volume 2. Kendall / Hunt Publishing Company, Dubuque, xvi + 975 pp.

Lawrence JF, Britton EB (1994) Australian Beetles. Melbourne University Press, Carlton, x + 192 pp. + 16 pls.

Lawrence JF, Newton AF, Jr. (1995) Families and subfamilies of Coleoptera (with selected genera, notes, references and data on family-group names) [pp. 779–1006]. In: Pakaluk J, Ślipiński SA (Eds) Biology, phylogeny, and classification of Coleoptera: papers celebrating the 80th birthday of Roy A Crowson Volume 2. Muzeum i Institut Zologii PAN, Warszawa, x + 1092 pp. in 2 vols.

Majka CG (2008) The flat bark beetles (Coleoptera, Silvanidae, Cucujidae, Laemophloeidae) of Atlantic Canada. ZooKeys 2: 221–238.

Martínez de Murguía L, Lapaza J, Salaberria E, Méndez M, Molino-Olmedo F (2004) Coleópteros saproxílicos (Insecta: Coleoptera) de un hayedo acidófilo en regeneración del norte peninsular. Munibe (Sciencias Naturales) 55: 167–182.

Muona J, Rutanen I (1994) The short-term impact of fire on the beetle fauna in boreal coniferous forest. Annales Zoologici Fennici 31: 109–121.

Nakane T, Ohbayashi K, Nomura S, Kurosawa Y (1984) Iconographia Insectorum Japonicorum, Colore Naturali Edita. Volume 2 (Coleoptera). Hokuryukan, Tokyo, 461 pp.

Nikitskij NB, Belov VV (1979) Novyy I maloizvestnyy vidy Clavicornia (Coleoptera) iz Talysha. New and poorly known species of Clavicornia (Coleoptera) from Talysh. Zoologicheskii Zhurnal 58: 849–854.

Sasaji H (1983) Contribution to the taxonomy of the superfamily Cucujoidea (Coleoptera) of Japan and her adjacent districts, 1. Memoir of the Faculty of Education Fukui University (Series II: Natural Science) 33: 17–52.

Thomas MC (1988) Generic key to the known larvae of the Cucujidae, Passandridae, and Silvanidae of America North of Mexico (Coleoptera). Insecta Mundi 2 (2): 81–89.

Thomas MC (2002) 82. Cucujidae Latreille 1802 [pp. 329–330]. In: Arnett RH, Jr., Thomas MC, Skelley PE, Frank JH (Eds) American beetles. Volume 2. Polyphaga: Scarabaeoidea through Curculionoidea. CRC Press, New York, xiv + 861 pp.

Thomas MC (2004) A revision of *Pediacus* Shuckard (Coleoptera: Cucujidae) for America north of Mexico, with notes on other species. Insecta Mundi 17 (3/4): 157–177.

Vavra J (1993) Nové druhy brouků (Coleoptera) pro území Slovenska. New species of beetles (Coleoptera) for the territory of Slovakia. Klapalekiana 29 (1/2): 61–62.

Węgrzynowicz P (2007) Family Cucujidae Latreille, 1802 [pp. 502–503]. In: Löbl I, Smetana A (Eds) Catalogue of Palaearctic Coleoptera. Volume 4. Apollo Books, Stenstrup, 935 pp.

Curculionidae

Supporting references for the conservation of Smicronychini Seidlitz, 1891 over Desmorini LeConte, 1876 (Art. 23.9.2). The taxon name Desmorini LeConte, 1876 has not been used as valid after 1899 to our knowledge.

Alonso-Zarazaga MA (2002) Lista preliminar de los Coleoptera Curculionoidea del área ibero-balear, con descripción de *Melicius* gen. nov. y nuevas citas. Boletín de la Sociedad Entomológica Aragonesa 31: 9–33.

Alonso-Zarazaga MA, Ugarte San Vicente I, Coello P (2009) Nuevos datos sobre la distribución de *Sharpia rubida* (Rosenhauer, 1856) (Coleoptera, Curculionidae, Smicronychini). Boletín de la Sociedad Entomológica Aragonesa 44: 573–574.

Anderson DM, Cox ML (1997) *Smicronyx* species (Coleoptera: Curculionidae), economically important seed predators of witchweeds (*Striga* spp.) (Scrophulariaceae) in sub-Saharan Africa. Bulletin of Entomological Research 87 (1): 3–17.

Anderson RS (2002) IV. Curculioninae Latreille, 1802 [pp. 732–740]. In: Arnett RH, Jr., Thomas MC, Skelley PE, Frank JH (Eds) American beetles. Volume 2. Polyphaga: Scarabaeoidea through Curculionoidea CRC Press, Boca Raton, xiv + 861 pp.

Angelov PA (1980) Coleoptera, Curculionidae. IV čast Calandrinae 2. [Coleoptera, Curculionidae. Part 4. Calandrinae. 2]. Fauna na B'lgariya 10. B'lgarska Akademia na Naukite, Sofia, 301 pp. [in Bulgarian].

Arnett RH, Jr. (2000) American insects. A handbook of the insects of America north of Mexico. Second edition. CRC Press, Boca Raton, xvii + 1003 pp.

Balsbaugh EU, Aarhus DG (1990) Checklist and new state records of Curculionidae (broad sense) (Coleoptera) for North Dakota. Journal of the Kansas Entomological Society 63(2): 227–236.

Bolu H, Legalov AA (2008) On the Curculionoidea (Coleoptera) fauna of almond (*Amygdalus communis* L.) orchards in south-eastern and eastern Anatolia in Turkey. Baltic Journal of Coleopterology 8(1): 75–85.

Braunert C (2009) Verzeichnis der Rüsselkäfer Luxemburgs (Coleoptera, Curculionoidea) mit Ausnahme der Borkenkäfer (Scolytinae) und Kernkäfer (Platypodinae). Bulletin de la Société des Naturalistes Luxembourgeois 110: 125–142.

Davis SR (2007) Nectarivory in a weevil, *Smicronyx squalidus* (Coleoptera: Curculionidae: Curculioninae), on *Desmanthus* (Fabaceae). Polish Journal of Entomology 76 (3): 221–224.

Franz NM (2006) Towards a phylogenetic system of derelomine flower weevils (Coleoptera: Curculionidae). Systematic Entomology 31 (2): 220–287.

Hoffmann A (1968) The scientific results of the Hungarian Soil Zoological Expedition to the Brazzaville-Congo. 32. Espèces de la famille Curculionidae (Coleoptera). Opuscula Zoologica 8(1): 11–29.

Hyder DE, Oseto CY (2005) Structure of the stridulatory apparatus and analysis of the sound produced by *Smicronyx fulvus* and *Smicronyx sordidus* (Coleoptera, Curculionidae, Erirhininae, Smicronychini). Journal of Morphology 201(1): 69–84.

Li S, Sun L, Oseto CY, Ferris VR (2007) Phylogenetic analyses and a method for rapid molecular diagnosis of two sunflower seed weevils (Coleoptera: Curculionidae). Annals of the Entomological Society of America 100(5): 649–654.

Lyal CHC, Douglas DA, Hine SJ (2006) Morphology and systematic significance of sclerolepidia in the weevils (Coleoptera: Curculionoidea). Systematics and Biodiversity 4(2): 203–241.

Majka C, Anderson RS, McAlpine DF, Webster RP (2007) The weevils (Coleoptera: Curculionoidea) of the Maritime Provinces of Canada, I: New records from New Brunswick. The Canadian Entomologist 139(3): 378–396.

Marvaldi AE, Lanteri AA (2005) Key to higher taxa of South American weevils based on adult characters (Coleoptera, Curculionoidea). Revista Chilena de Historia Natural 78(1): 65–87.

Marvaldi AE, Morrone JJ (2000) Phylogenetic systematics of weevils (Coleoptera: Curculionoidea): a reappraisal based on larval and adult morphology. Insect Systematics & Evolution 31 (1): 43–58.

Morrone JJ (2000) Mexican weevils (Coleoptera: Curculionoidea): a preliminary key to families and subfamilies. Acta Zoologica Mexicana (Nueva Serie) 80: 131–141.

Pešic S (2003) Balkan weevils (Curculionoidea) in The Natural History Museum London (Reitter's collection). Kragujevac Journal of Science 25: 163–172.

Piper GL, Mulford BL (1980) Life history observations on *Smicronyx commixtus* Dietz (Coleoptera: Curculionidae) in southeastern Washington. The Coleopterists Bulletin 34(3): 295–298.

Roseland CR, Bates MB, Oseto CY (1990) Role of a male-produced pheromone of the red sunflower seed weevil (Coleoptera: Curculionidae) in host finding. Environmental Entomology 19(6): 1675–1680.

Salas-Araiza MD, O'Brien CW, Romero-Napoles J (2001) Curculionoidea (Insecta: Coleoptera) from the state of Guanajuato, Mexico. Insecta Mundi 15 (1): 45–57.

Tanner VM (1966) Rhynchophora beetles of the Nevada Test Site. Brigham Young University Science Bulletin (Biological Series) 8(2): 1–35.

Wanat M, Mokrzycki T (2005) A new checklist of the weevils of Poland (Coleoptera: Curculionoidea). Genus 16: 69–117.

Curculionidae

Supporting references for the conservation of Bagoinae C.G. Thomson, 1859 over Lyprinae Gistel 1848 (Art. 23.9.2). The taxon name Lyprinae Gistel 1848 has not been used as valid after 1899 to our knowledge. Note: Lyprinae

Gistel, 1848 was recently considered a *nomen oblitum* by Colonnelli (2003: 7), however the necessary supporting references were not provided.

Alonso-Zarazaga MA, Lyal CHC (1999) A world catalogue of families and genera of Curculionoidea (Insecta: Coleoptera) (excepting Scolytidae and Platypodidae). Entomopraxis, S.C.P., Barcelona, 315 pp.

Alonso-Zarazaga MA, Lyal CHC (2002) Addenda and corrigenda to 'A world catalogue of families and genera of Curculionoidea (Insecta: Coleoptera)'. Zootaxa 63: 1–37.

Amann E, Brandstetter CM, Kapp A (1994) Käfer am Wasser. (Gattungsschlussel der (semi-) aquatischen Käfer Mitteleuropas). Erster Vorarlberger Coleopterologischer Verein, Burs, i, 1–38

Anderson RS (2002) V. Bagoinae Thomson 1859 [p. 740]. In: Arnett RH, Jr., Thomas MC, Skelley PE, Frank JH (Eds) American beetles. Volume 2. Polyphaga: Scarabaeoidea through Curculionoidea. CRC Press, Boca Raton.

Angelov P (1989) Eine neue *Bagous*-Art aus Bulgarien (Insecta, Coleoptera, Curculionidae: Bagoinae). Reichenbachia 27 (1): 69–70.

Caldara R, O'Brien CW (1998) Systematics and evolution of weevils of the genus *Bagous*. VI. Taxonomic treatment of the species of the western Palearctic Region (Coleoptera Curculionidae). Memorie della Società Entomologica Italiana 76 [1997]: 131–347.

Colonnelli E (2003) A revised checklist of Italian Curculionoidea (Coleoptera). Zootaxa 337: 1–142.

DeLoach CJ (1976) Identification and biological notes on the species of *Neochetina* that attack Pontederiaceae in Argentina (Coleoptera: Curculionidae: Bagoini). The Coleopterists Bulletin 29 [1975] (4): 257–265.

Dieckmann L (1983) Beiträge zur Insektenfauna der DDR: Coleoptera - Curculionidae (Tanymecinae, Leptopiinae, Cleoninae, Tanyrhynchinae, Cossoninae, Raymondionyminae, Bagoinae, Tanysphyrinae). Beiträge zur Entomologie 33 (2): 257–381.

Dieckmann L (1990) Revision der mitteleuropäischen Arten der *Bagous collignensis*-Gruppe (Insecta, Coleoptera, Curculionidae: Bagoinae). Reichenbachia 27 (2): 141–145.

Freude H, Harde KW, Lohse GA (1983) Die Käfer Mitteleuropas. Band 11. Goecke & Evers, Krefeld, 342 pp.

Kissinger DG (1966) *Cyrtobagous* Hustache, a genus of weevils new to the United States fauna (Coleoptera: Curculionidae: Bagoini). The Coleopterists' Bulletin 20 (4): 125–127.

Menet D (1997) Les curculionides du nord de la France. 7ème partie: sous-familles des Pissodinae, Anoplinae, Tanysphyrinae, Bagoinae et Cossoninae. Bulletin de la Société Entomologique du Nord de la France 285: 3–10.

Morris MG (1985) *Bagous brevis* Gyllenhal new to Ireland from the Burren, Co. Clare, with a brief review of the Irish Bagoini (Coleoptera: Curculionidae). Irish Naturalists' Journal 21: 400–403.

O'Brien CW (1976) A taxonomic revision of the New World subaquatic genus *Neochetina* (Coleoptera: Curculionidae: Bagoini). Annals of the Entomological Society of America 69 (2): 165–174.

O'Brien CW, Askevold IS (1992) Systematics and evolution of weevils of the genus *Bagous* Germar (Coleoptera: Curculionidae), 1. Species of Australia. Transactions of the American Entomological Society 118 (4): 331–452.

O'Brien CW, Marshall GB (1979) U.S. *Bagous*, bionomic notes, a new species, and a new name (Bagoini, Erirhininae, Curculionidae, Coleoptera). Southwestern Entomologist 4 (2): 141–149.

Pajni HR, Kamal Tewari P (1986) On a new species of genus *Hydronomidius* (Bagoinae: Curculionidae) from India. Journal of the Bombay Natural History Society 82 [1985] (3): 610–613.

Pesarini C (1978) Tabelle per la determinazione dei generi di curculionidi italiani (Coleoptera). Informatore del Giovane Entomologo 19: 1–8.

Smreczyński S (1972) Ryjkowce-Curculionidae. Podrodzina Curculioninae. Klucze do Oznaczania Owadów Polski [Nr 77]. Część XIX. Chrząszcze – Coleoptera, Zeszyt 98d. Panstwowe Wydawnictwo Naukowe, Warszawa, 195 pp.

Sprick P (2000) Eignung einer Insektengruppe für die Fauna-Flora-Habitat-Richtlinie der EU (92/43/EWG, 21. Mai 1992) am Beispiel der Rüsselkäfer-Unterfamilie Bagoinae (Col., Curculionidae) (Beiträge zur Ökologie phytophager Käfer III). Insecta, Zeitschrift für Entomologie und Naturschutz 6: 61–96.

Sprick P (2001) Suitability of an insect group for the habitats directive of the EU: the weevil subfamily Bagoinae (Col., Curculionidae). Contributions to the ecology of phytophagous beetles VII. [pp. 7–40] In: Stüben PE (Ed) Studies on taxonomy, biology and ecology of Curculionoidea. Snudebiller 2. Curculio-Institut, Mönchengladbach [CD ROM].

Stresemann E, Hannemann H-J, Klausnitzer B, Senglaub K (1989) Exkursionsfauna für die Gebiete der DDR und der BRD. Band 2/1. Wirbellose: Insekten - Erster Teil. 8 Aufl. Volk und Wissen Volkseigener Verlag, Berlin, 504 pp.

Warner RE (1970) *Neochetina eichhorniae*, a new species of weevil from waterhyacinth, and biological notes on it and *N. bruchi* (Coleoptera: Curculionidae: Bagoini). Proceedings of the Entomological Society of Washington 72 (4): 487–496.

Zimmerman EC (1994) Australian weevils (Coleoptera: Curculionoidea). Volume 1. Orthoceri Anthribidae to Attelabidae, the primitive weevils. CSIRO, Melbourne, i-xxxii, pp. 1–741.

Curculionidae

Supporting references for the conservation of Aterpina Lacordaire, 1863 over Heliomenina Gistel, 1848 (Art. 23.9.2). The taxon name Heliomenina Gistel, 1848 has not been used as valid after 1899 to our knowledge.

Alonso-Zarazaga MA, Lyal CHC (1999) A world catalogue of families and genera of Curculionoidea (Insecta: Coleoptera) (excepting Scolytidae and Platypodidae). Entomopraxis, S.C.P., Barcelona, 315 pp.

Barratt BIP, Evan AA, Ferguson CM, Barker GM, McNeill MR, Phillips CB (1997) Laboratory nontarget host range of the introduced parasitoids *Microctonus aethiopoides* and *M. hyperodae* (Hymenoptera: Braconidae) compared with field parasitism in New Zealand. Environmental Entomology 26(3): 694–702.

Elgueta M (1975) Una nueva especie de Aterpinae (Coleoptera. Curculionidae). Revista Chilena de Entomología 8 [1974]: 133–134.

Elgueta M (2000) Dos especies nuevas de *Aegorhinus* (Coleoptera: Curculionidae: Aterpini) de Chile. Two new *Aegorhinus* species (Coleoptera: Curculionidae: Aterpinae) from Chile. Acta Entomológica Chilena 24: 7–18.

Escalante T, Linaje M, Illoldi-Rangel P, Rivas M, Estrada P, Neira F, Morrone JJ (2009) Ecological niche models and patterns of richness and endemism of the southern Andean genus *Eurymetopum* (Coleoptera, Cleridae). Revista Brasileira de Entomologia 53(3): 379–385.

Hawkeswood TJ (1991) Review of the history, biology and host plants of the Australian weevil *Chrysolophus spectabilis* (Fabricius) (Coleoptera: Curculionidae: Aterpinae). Spixiana 14(1): 17–25.

Hunsdoerfer AK, Rheinheimer J, Wink M (2009) Towards the phylogeny of the Curculionoidea (Coleoptera): reconstructions from mitochondrial and nuclear ribosomal DNA sequences. Zoologischer Anzeiger 248 (1): 9–31.

Kuschel G (1982) Apionidae and Curculionidae (Coleoptera) from the Poor Knights Islands, New Zealand. Journal of the Royal Society of New Zealand 12(3): 273–282.

Leschen RAB, Lawrence JF, Kuschel G, Thorpe S, Wang Q (2003) Coleoptera genera of New Zealand. New Zealand Entomologist 26: 15–28.

Lyal CHC, Douglas DA, Hine SJ (2006) Morphology and systematic significance of sclerolepidia in the weevils (Coleoptera: Curculionoidea). Systematics and Biodiversity 4(2): 203–241.

Marvaldi AE, Morrone JJ (1999) Immature stages of *Rhyparonotus altarensis* (Olliff) (Coleoptera: Curculionidae: Molytinae), with comments on larval characters in Anchonini and Molytinae. Journal of the New York Entomological Society 106 [1998] (2/3): 95–104.

Marvaldi AE, Morrone JJ (2000) Phylogenetic systematics of weevils (Coleoptera: Curculionoidea): a reappraisal based on larval and adult morphology. Insect Systematics & Evolution 31 (1): 43–58.

May BM (1993) Larvae of Curculionoidea (Insecta: Coleoptera): a systematic overview. Fauna of New Zealand . Ko te Aitanga Pepeke o Aotearoa. Number 28, Manaaki Whenua Press, Lincoln, 225 pp.

Morrone JJ (1998) On Udvardy's Insulantarctica province: a test from the weevils (Coleoptera: Curculionoidea). Journal of Biogeography 25 (5): 947–955.

Morrone JJ (2000) Mexican weevils (Coleoptera: Curculionoidea): a preliminary key to families and subfamilies. Acta Zoologica Mexicana (Nueva Serie) 80: 131–141.

Morrone JJ (2002) Checklist of the species of Cyclominae (Coleoptera: Curculionidae) occurring in America south of the United States. Revista de la Sociedad Entomológica Argentina 61 (1/2): 11–24.

Morrone JJ, Marvaldi AE (1998) *Listroderes abditus* or *Antarctobius abditus*?: a simultaneous analysis of larval and adult characters (Coleoptera: Curculionidae). European Journal of Entomology 95 (3): 429–436.

Morrone JJ, Roig-Junent S (2000) Synopsis and cladistics of the American Aterpini (Coleoptera: Curculionidae, Cyclominae). Entomologica Scandinavica 30 [1999] (4): 417–434.

Oberprieler RG, Marvaldi AE, Anderson RS (2007) Weevils, weevils, weevils everywhere. Zootaxa 1668: 491–520.

Parra LB, Mutis AT, Aguilera AP, Rebolledo RR, Quiroz AC (2009) Estado del conocimiento sobre el cabrito del frambueso (CF), *Aegorhinus superciliosus* (Guérin) (Coleoptera: Curculionidae). Idesia 27(1): 57–65.

Pérez-Schultheiss J, Arriagada AM, Baessolo L (2009) Aterpini (Coleoptera: Curculionidae) del Parque Nacional Isla Guamblín, Archipiélago de los Chonos, Aysén, Chile. Boletín de la Sociedad Entomológica Aragonesa 45: 249–252.

Posadas P, Morrone JJ (2003) Biogeografía histórica de la familia Curculionidae (Coleoptera) en las subregiones Subantártica y Chilena Central. Revista de la Sociedad Entomológica Argentina 62(1/2): 75–84.

Richardson BJ, Oberprieler RG (2007) The diversity of Linnaean communities: a way of detecting invertebrate groups at risk of extinction. Journal of Insect Conservation 11(3): 287–297.

Setliff GP (2007) Annotated checklist of weevils from the Papuan region (Coleoptera, Curculionoidea). Zootaxa 1536: 1–296.

Vergara OE, Jerez V, Parra LE (2006) Diversidad y patrones de distribución de coleópteros en la Región del Biobío, Chile: una aproximación preliminar para la conservación de la diversidad. Revista Chilena de Historia Natural 79: 369–388.

Curculionidae

Supporting references for the conservation of Naupactini Gistel, 1848 over Iphiini Schönherr, 1823 (Art. 23.9.2). The taxon name Iphiini Schönherr, 1823 has not been used as valid after 1899 to our knowledge.

Bordon C (1991) El género *Macrostylus* Boheman (Col. Curc. Brachyderinae, Naupactini) en Venezuela. Acta Biologica Venezuelica 13: 1–50.

Chen Y (1991) [The Chinese *Mesagroicus* Schönherr (Coleoptera: Curculionidae, Naupactini)]. Acta Entomologica Sinica 34 (4): 468–469 [in Chinese].

Coscarón M, del C, Díaz NB, Lanteri AA, Loiacono MS (1991) Importancia taxonómica del revestimiento tegumentario en la tribu Naupactini. I. Género *Cyrtomon* y taxa afines (Coleoptera: Curculionidae). Neotropica 37 (No. 97): 31–54.

Díaz NB, Loiácono MS, Coscarón M, del C, Lanteri AA (1990) Importancia taxonómica de las piezas bucales en la tribu Naupactini. 1. Género *Cyrtomon* Schoenherr y taxa afines (Coleoptera, Curculionidae). Revista Brasileira de Entomologia 34 (4): 861–876.

Díaz NB, Loiácono MS, Lanteri AA, Coscarón M, del C (1991) Importancia taxonómica de la piezas bucales en la tribu Naupactini. II. Las especies del género *Cyrtomon* Schoenherr (Coleoptera: Curculionidae). Neotropica 36 (No. 96) [1990]: 93–99.

Ferragu M, Richard R (1986) Cinq curculionides nouveaux de Madagascar, des Mascareignes et d'Afrique australe (Coleoptera). Revue Française d'Entomologie (Nouvelle Série) 8 (4): 149–155.

Gomez CA, Lanteri AA (2006) Primer registro de *Naupactus ruizi* (Coleoptera: Curculionoidea) asociado con *Pinus ponderosa* (Gymnospermae: Pinaceae) en Patagonia. Revista de la Sociedad Entomológica Argentina 65: 107–109.

Guadalupe del Río M, Lanteri AA (2007) Taxonomic revision of *Melanocyphus* Jekel (Coleoptera: Curculionidae). Studies on Neotropical Fauna and Environment 42: 127–132.

Guadalupe del Río M, Lanteri AA (2008) *Thoraconaupactus*, a new Brazilian genus of broad nosed weevil (Coleoptera: Curculionidae: Entiminae) associated with *Leucaena* (Fabaceae). Entomological News 118 [2007] (5): 459–469.

Guadalupe del Río M, Lanteri AA, Guedes JVC (2006) Taxonomic revision and cladistic analysis of *Teratopactus* Heller (Coleoptera: Curculionidae). Invertebrate Systematics 20 (5): 585–602.

Guedes JVC, Lanteri AA, Parra JRP (2005) Chave de identificação, ocorrência e distribuição dos curculionídeos-das-raízes dos citros em São Paulo e Minas Gerais. Neotropical Entomology 34 (4): 577–584.

Lanteri AA (1981) Estudio comparativo de las estructuras genitales en la tribu Naupactini. I. Los caracteres a nivel genérico de *Naupactus* Schönherr, *Teratopactus* Heller y *Trichonaupactus* Hustache (Coleoptera: Curculionidae). Revista de la Sociedad Entomológica Argentina 40 (1/4): 273–278.

Lanteri AA (1989) Estudio sistemático de los géneros *Trichocyphus* Heller y *Mendozella* Hustache (Coleoptera: Curculionidae). Boletín de la Sociedad de Biología de Concepción 60: 139–147.

Lanteri AA (1990) Revisión sistemática del género *Priocyphus* Hustache 1939 y creación de los géneros *Priocyphopsis* y *Lamprocyphopsis* (Coleoptera, Curculionidae). Revista Brasileira de Entomologia 34 (2): 403–422.

Lanteri AA (1992) Systematics, cladistics and biogeography of a new weevil genus, *Galapaganus* (Coleoptera: Curculionidae) from the Galápagos Islands, and coasts of Ecuador and Perú. Transactions of the American Entomological Society 118 (3): 227–267.

Lanteri AA (1995) Systematic revision of *Ericydeus* Pascoe (Coleoptera: Curculionidae). Entomologica Scandinavica 26 (4): 393–424.

Lanteri AA (2004) New taxonomic and biogeographic information on *Galapaganus femoratus* species group (Coleoptera: Curculionidae: Entiminae). Transactions of the American Entomological Society 130 (2/3): 177–192.

Lanteri AA, Guadalupe del Río M (2003) Revision of the genus *Briarius* (Fischer de Waldheim) (Coleoptera: Curculionidae). Insect Systematics & Evolution 34 (3): 281–294.

Lanteri AA, Guadalupe del Río M (2005) Taxonomic revision of *Thoracocyphus* Emden (Coleoptera: Curculionidae). Insect Systematics & Evolution 35 (4): 449–456.

Lanteri AA, Guadalupe del Río M (2005) Taxonomy of the monotypic genus *Trichaptus* Pascoe (Coleoptera: Curculionidae: Entiminae), a potential weevil mimic of Mutillidae. The Coleopterists Bulletin 59 (1): 47–54.

Lanteri AA, Guadalupe del Río M (2006) Taxonomic revision of the genus *Cyphopsis* Roelofs (Coleoptera, Curculionidae). Deutsche Entomologische Zeitschrift 53 (2): 275–281.

Lanteri AA, Guadalupe del Río M (2006) Taxonomic revision of the monotypic genus *Acyphus* Heller (Coleoptera: Curculionidae) with comments on infraspecific variation. Zootaxa 1312: 59–68.

Lanteri AA, Guedes JVC, Parra JRP (2002) Weevils injurious for roots of citrus in São Paulo state, Brazil. Neotropical Entomology 31 (4): 561–569.

Lanteri AA, Morrone JJ (1991) Cladistic analysis of *Priocyphis* Hustache and related genera (Coleoptera: Curculionidae). Proceedings of the Entomological Society of Washington 93 (2): 278–287.

Lanteri AA, Normark BB (1995) Parthenogenesis in the tribe Naupactini (Coleoptera: Curculionidae). Annals of the Entomological Society of America 88 (6): 722–731.

Marvaldi AE, Loiacono MS (1994) First instar larvae in the tribe Naupactini (Coleoptera, Curculionidae). Revista Brasileira de Entomologia 38 (2): 453–466.

Normark BB (1996) Phylogeny and evolution of parthenogenetic weevils of the *Aramigus tessellatus* species complex (Coleoptera: Curculionidae: Naupactini): evidence from mitochondrial DNA sequences. Evolution 50 (2): 734–745.

Pesarini C (1977) Tabelle per la determinazione dei generi di curculionidi italiani (Coleoptera). 2. L'Informatore del Giovane Entomologo 18: 9–16.

Wibmer GJ, O'Brien CW (1986) Annotated checklist of the weevils (Curculionidae sensu lato) of South America (Coleoptera: Curculionoidea). Memoirs of the American Entomological Institute No. 39: i-xvi, 1–563.

Curculionidae

Supporting references for the conservation of Cleonini Schönherr, 1826 over Geomorini Schönherr, 1823 (Art. 23.9.2). The taxon name Geomorini Schönherr, 1823 has not been used as valid after 1899 to our knowledge.

Alonso Zarazaga MA, Lyal CHC (2009) On the nomenclature of some genera in the Cleonini (Coleoptera, Curculionidae, Lixinae). Graellsia 65(2): 241–242.

Alonso-Zarazaga MA, Lyal CHC (1999) A world catalogue of families and genera of Curculionoidea (Insecta: Coleoptera) (excepting Scolytidae and Platypodidae). Entomopraxis, S.C.P., Barcelona, 315 pp.

Anderson RS (1988) Systematics, phylogeny and biogeography of New World weevils traditionally of the tribe Cleonini (Coleoptera: Curculionidae; Cleoninae). Quaestiones Entomologicae 23 [1987] (4): 431–709.

Anderson RS (1989) New synonymy in North American *Stephanocleonus* Motschulsky (Coleoptera: Curculionidae). The Coleopterists Bulletin 43 (1): 93.

Arzanov YG (2006) *Borisocleonus* gen. n. – novyy rod dolgonosikov triby Cleonini (Coleoptera: Curculionidae: Lixinae). *Borisocleonus* gen. n. - a new genus of weevils from the tribus of Cleonini (Coleoptera: Curculionidae: Lixinae). Russian Entomological Journal 15 (1): 63–66.

Aslam NA (1963) On the genera of Indo-Pakistan Cleoninae and Hylobiinae (Coleoptera: Curculionidae). Bulletin of the British Museum (Natural History) (Entomology) 13 (3): 47–66.

Balsbaugh EU, Aarhus DG (1990) Checklist and new state records of Curculionidae (broad sense) (Coleoptera) for North Dakota. Journal of the Kansas Entomological Society 63(2): 227–236.

Barratt BIP, Evan AA, Ferguson CM, Barker GM, McNeill MR, Phillips CB (1997) Laboratory nontarget host range of the introduced parasitoids *Microctonus aethiopoides* and *M. hyperodae* (Hymenoptera: Braconidae) compared with field parasitism in New Zealand. Environmental Entomology 26(3): 694–702.

Brun LA, Sheppard AW, Carrara A (1994) Host range of *Pachycerus cordiger* (= *P. scabrosus*) (Col.: Curculionidae). Entomophaga 38[1993] (4): 537–539.

Dieckmann L (1983) Beiträge zur Insektenfauna der DDR: Coleoptera Curculionidae (Tanymecinae, Leptopiinae, Cleoninae, Tanyrhynchinae, Cossoninae, Raymondionyminae, Bagoinae, Tanysphyrinae). Beiträge zur Entomologie 33 (2): 257–381.

Gratshev VG, Zherichin VV (2003) The fossil record of weevils and related beetle families (Coleoptera, Curculionoidea). Acta Zoologica Cracoviensia 46 (Supplement. Fossil Insects): 129–138.

Holecová M, Rozek M, Karagyan G (2000) Karyological notes on three weevil species from Armenia (Coleoptera, Curculionidae, Cleoninae). Folia Biologica (Cracow) 48(1–2): 25–27.

Hunsdoerfer AK, Rheinheimer J, Wink M (2009) Towards the phylogeny of the Curculionoidea (Coleoptera): reconstructions from mitochondrial and nuclear ribosomal DNA sequences. Zoologischer Anzeiger 248 (3): 9–31.

Kingsolver JM (1962) Notes on fossil Cleoninae (Coleoptera: Curculionidae). Psyche 69 (1): 47–49.

Louw S (1993) Breeding populations of *Lixus carinerostris* Boheman and *Calodemas prolixus* Faust (Coleoptera: Curculionidae) co-existing on Mesembryanthemaceae. The Coleopterists Bulletin 47(4): 335–339.

Meregalli M (2002) Notes on the genus *Pachycerus* Schoeherr 1823 with description of a new species from Somalia (Coleoptera Curculionidae Lixinae). Tropical Zoology 15 (2): 233–242.

Meregalli M (2004) *Pseudeumecops* n. gen., a new genus of Cleonini from East Africa (Coleoptera, Curculionidae, Lixinae). Italian Journal of Zoology 71 (2): 153–163.

Meregalli M (2008) Taxonomic relationships between *Pachycerus* and *Rhabdorrhynchus* (Coleoptera: Curculionidae: Lixinae), with descriptions of two new species of *Rhabdorrhynchus* from the Arabian Peninsula. Zoological Journal of the Linnean Society 152(1): 25–37.

Meregalli M (2009) Revision of the Indo-African *Pachycerus* Schoenherr, 1823, with a description of four new species (Coleoptera: Curculionidae: Lixinae). Zoological Journal of the Linnean Society 157(2): 295–325.

Perrin HM, Meregalli M (2008) Désignation de lectotypes des espèces de Cleonini, décrites par Gebler et Chevrolat, dans les collections du MNHN à Paris (Coleoptera, Curculionidae, Lixinae). Revue Française d'Entomologie (Nouvelle Série) 29 [2007] (4): 129–148.

Takenouchi Y (1973) A revised study of the chromosomes of twenty-four species of Japanese weevils (Coleoptera: Curculionidae). Genetica 44(4): 621–632.

Volovnik SV (1989) On distribution and ecology of some species of cleonine weevils (Coleoptera: Curculionidae) I. Tribe Cleonini. Entomological Review 68(3): 138–144.[publ. originally in *Entomologicheskoe Obozrenie* 68 (1): 86–92]

Volovnik SV (2005) On parasites and predators of Cleoninae weevils (Col. Curculionidae) in Ukrainian steppe. Anzeiger für Schädlingskunde 67(4): 77–79.

Volovnik SV (2007) On distribution and ecology of some species of cleonines (Coleoptera, Curculionidae): IV. Genus *Lixus* F., subgenus *Eulixus* Reitt. Entomological Review 87 (7): 840–847. [publ. originally in *Entomologicheskoe Obozrenie* 86 (3): 521–531]

Wheeler AG, Jr., Whitehead DR (1985) *Larinus planus* (F.) in North America (Coleoptera: Curculionidae: Cleoninae) and comments on biological control of Canada thistle. Proceedings of the Entomological Society of Washington 87(4): 751–758.

Curculionidae

Supporting references for the conservation of Magdalidini Pascoe, 1870 over Scardamyctini Gistel, 1848 (Art. 23.9.2). The taxon name Scardamyctini Gistel, 1848 has not been used as valid after 1899 to our knowledge.

Alonso-Zarazaga MA (2002) Lista preliminar de los Coleoptera Curculionoidea del área iberobalear, con descripción de *Melicius* gen. nov. y nuevas citas. Boletín de la Sociedad Entomológica Aragonesa 31: 9–33.

Alonso-Zarazaga MA, Lyal CHC (1999) A world catalogue of families and genera of Curculionoidea (Insecta: Coleoptera) (excepting Scolytidae and Platypodidae). Entomopraxis, S.C.P., Barcelona, 315 pp.

Anderson RS (1997) Weevils (Coleoptera: Curculionoidea, excluding Scolytinae and Platypodinae) of the Yukon [pp. 523–562]. In: Danks HV, Downes JA (Eds) Insects of the Yukon. Biological Survey of Canada (Terrestrial Arthropods), Ottawa, 1034 pp.

Anderson RS (2002) XV. Mesoptiliinae Lacordaire, 1863 [p. 786]. In: Arnett RH, Jr., Thomas MC, Skelley PE, Frank JH (Eds) American Beetles. Volume 2. Polyphaga: Scarabaeoidea through Curculionoidea CRC Press, Boca Raton, xiv + 861 pp.

Balsbaugh EU, Aarhus DG (1990) Checklist and new state records of Curculionidae (broad sense) (Coleoptera) for North Dakota. Journal of the Kansas Entomological Society 63(2): 227–236.

Braunert C (2009) Verzeichnis der Rüsselkäfer Luxemburgs (Coleoptera, Curculionoidea) mit Ausnahme der Borkenkäfer (Scolytinae) und Kernkäfer (Platypodinae). Bulletin de la Société des Naturalistes Luxembourgeois 110: 125–142.

Bright DE, Bouchard P (2008) Coleoptera, Curculionidae, Entiminae. Weevils of Canada and Alaska. Volume 2. The Insects and Arachnids of Canada. Part 25, NRC Research Press, Ottawa, xiii + 327 pp.

Burrini AG, Magnano L, Magnano AR, Scala C, Baccetti B (1988) Spermatozoa and phylogeny of Curculionoidea (Coleoptera). International Journal of Insect Morphology and Embryology 17(1): 1–50.

Crowson RA (1984) The associations of Coleoptera with Ascomycetes [pp. 256–285]. In: Wheeler Q, Blackwell M (Eds) Fungus-insect relationships: perspectives in ecology and evolution. Columbia University Press, New York, xiii + 514 pp.
Elgueta M, Arriagada SG (1989) Estado actual del conocimiento de los Coleópteros de Chile (Insecta: Coleoptera). Revista Chilena de Entomología 17: 5–60.
Hollis D (Ed), (1980) Animal identification: a reference guide. Volume 3. Insects. London, British Museum (Natural History) and John Wiley, viii + 160 pp.
Hunsdoerfer AK, Rheinheimer J, Wink M (2009) Towards the phylogeny of the Curculionoidea (Coleoptera): reconstructions from mitochondrial and nuclear ribosomal DNA sequences. Zoologischer Anzeiger 248 (1): 9–31.
Kojima H, Morimoto K (2004) An online checklist and database of the Japanese weevils (Insecta: Coleoptera: Curculionoidea) (excepting Scolytidae and Platypodidae). Bulletin of the Kyushu University Museum 2: 33–147.
Kuschel G (2003) Nemonychidae, Belidae, Brentidae (Insecta: Coleoptera: Curculionoidea). Fauna of New Zealand. Ko te Aitanga Pepeke o Aotearoa. Number 45, Manaaki Whenua Press, Lincoln, 97 pp.
Lu X, Legalov AA, Zhang R (2005) A new subgenus and two new species of *Magdalis* Germar from northern China (Coleoptera: Curculionidae: Magdalinae). The Coleopterists Bulletin 59(3): 369–378.
Majka C, Anderson RS, McAlpine DF, Webster RP (2007) The weevils (Coleoptera: Curculionoidea) of the Maritime Provinces of Canada, I: New records from New Brunswick. The Canadian Entomologist 139(3): 378–396.
Mann F. G. (1960) Regiones biogeográficas de Chile. Investigaciones Zoológicas Chilenas 6: 15–49.
McKenna DD, Sequeira AS, Marvaldi AE, Farrell BD (2009) Temporal lags and overlap in the diversification of weevils and flowering plants. Proceedings of the National Academy of Sciences of the United States of America 106(17): 7083–7088.
Morris MG (1993) A check list of the weevils of Ireland (Coleoptera: Curculionoidea). Entomologist's Gazette 44 (4): 289–296.
Morris MG (2002) True weevils (part I): Coleoptera: Curculionidae (subfamilies Raymondionyminae to Smicronychinae). Handbooks for the Identification of British Insects 5(17b): 1–149.
Morrone JJ, Muñiz R, Asiain J, Márquez J (2002) Lista de las especies de Curculionoidea (Insecta: Coleoptera) depositadas en la colección del Museo de Zoología 'Alfonso L. Herrera,' Facultad de Ciencias, UNAM (MZFC). Acta Zoologica Mexicana (Nueva serie) 87: 147–165.
O'Brien CW, Wibmer GJ (1982) Annotated checklist of the weevils (Curculionidae sensu lato) of North America, Central America, and the West Indies (Coleoptera: Curculionoidea). Memoirs of the American Entomological Institute 34: i-ix, 1–382.
Pešic S (2003) Balkan weevils (Curculionoidea) in The Natural History Museum London: world part. Kragujevac Journal of Science 25: 139–162.

Ribera I (1999) Evolución, filogenia y clasificación de los Coleoptera (Arthropoda: Hexapoda). Boletín de la Sociedad Entomológica Aragonesa 26(Evolución y filogenia de Arthropoda): 435–458.

Wanat M, Mokrzycki T (2005) A new checklist of the weevils of Poland (Coleoptera: Curculionoidea). Genus 16 (1): 69–117.

Elateridae

Supporting references for the conservation of Agrypninae/-ini Candèze, 1857 over Adelocerinae/-ini Gistel, 1848 and Pangaurinae/-ini Gistel, 1856 (Art. 23.9.2). The taxon name Pangaurinae/-ini Gistel, 1856 has not been used as valid after 1899 to our knowledge. The taxon name Adelocerinae/-ini has been used as valid after 1899 but attributed to Buysson, 1893 who proposed the name as new, without reference to Gistel's name.

Casari SA (2002) Review of the genus *Chalcolepidius* Eschscholtz, 1829 (Coleoptera, Elateridae, Agrypninae). Revista Brasileira de Entomologia 46 (3): 263–428.

Casari SA (2004) New species of *Alaus* Eschscholtz, 1829 (Coleoptera: Elateridae, Agrypninae, Hemirhipini). Annales de la Société Entomologique de France (Nouvelle Série) 39 [2003] (4): 315–333.

Casari SA (2006) Larva, pupa and adult of *Aeolus cinctus* Candèze (Coleoptera, Elateridae, Agrypninae). Revista Brasileira de Entomologia 50 (3): 347–351.

Casari SA (2008) Cladistic analysis of Hemirhipini with establishment of *Propalaus* gen. nov. (Coleoptera, Elateridae, Agrypninae). Papéis Avulsos de Zoologia 48 (16): 139–180.

Chakraborty P, Chakraborty S (2000) Agrypninae (Coleoptera: Elateridae) of India: a taxonomic review and checklist. Records of the Zoological Survey of India 98: 71–83.

Chassain J (2005) *Lacon giuglarisi*, espèce nouvelle pour la fauna de Guyane française (Coleoptera, Elateridae, Agrypninae). L'Entomologiste 61 (3): 131–134.

Chassain J (2009) Description d'une nouvelle espèce guyanaise et brésilienne d'élatéride du genre *Coctilelater* (Costa, 1975) (Coleoptera Elateridae Agrypninae Pyrophorini). L'Entomologiste 65 (1): 35–38.

Cobos A (1966) Nuevo género y especie de Agrypninae (Col. Elateridae). Annales de la Société Entomologique de France (Nouvelle Série) 2 (3): 651–653.

Costa C, Cesari-Chen SA (1984) Larvae of Neotropical Coleoptera. VIII. Elateridae: Agrypninae, Elaterinae and Physorhininae. Revista Brasileira de Entomologia 28 (3): 315–328.

Costa C, Rosa SP, Chassain J (2009) A taxonomic revision and revalidation of *Nycterilampus* Montrouzier (Coleoptera: Elateridae, Agrypninae). Zootaxa 2017: 47–57.

Hoffman RL (2007) The distribution of *Conoderus scissus* (Schaeffer) with notes on some taxonomic characters (Coleoptera: Elateridae: Agrypninae). Banisteria 30: 32–34.

Jiang S-H, Chen X-Q, Wu S-J, Meng Z-Y, Li G-J (2009) Molecular phylogenetic analysis of Elateridae (Insecta: Coleoptera) based on 28S rDNA gene fragments. Acta Entomologica Sinica 52 (1): 74–83 [in Chinese, English summary].

Kishii T (1995) A study on the elaterid-beetles of Shibata Collection from Taiwan, I (Coleoptera, Elateridae). On the subfamilies Oxynopterinae and Agrypninae. The Entomological Review of Japan 50 (1): 1–14, pls. 1–4.

Lee S-H, Woo K-S, Lee YI (1987) Taxonomic study on the subfamily Agrypninae (Coleoptera: Elateridae) in Korea. Korean Journal of Plant Protection 26: 1–7.

Ohira H (2002) Notes on the morphological structure of *Agrypnus* species from Japan (I). (Coleoptera: Elateridae, Agrypninae, *Agrypnus*). Miscellaneous Reports of the Hiwa Museum for Natural History 41: 53–67.

Ohira H (2004) Notes on the morphological structure of *Agrypnus* species from Japan (III). (Coleoptera: Elateridae, Agrypninae, *Agrypnus*, *Colaulon* group). Miscellaneous Reports of the Hiwa Museum for Natural History 43: 67–90.

Platia G (2008) Description of three wingless species of *Rismethus* Fleutiaux, 1947 (Coleoptera, Elateridae: Agrypnini) from Western Malaysia. Folia Entomologica Hungarica 69: 179–184.

Rosa SP (2004) Nova espécie do gênero *Ptesimopsia* Costa (Coleoptera, Elateridae, Agrypninae). Revista Brasileira de Entomologia 48 (2): 227–228.

Rosa SP (2004) Revisão do gênero *Opselater* Costa (Coleoptera, Elateridae, Agrypninae). Revista Brasileira de Entomologia 48 (2): 203–219.

Rosa SP (2009) Description of *Coctilelater minimus* sp. nov. (Coleoptera, Elateridae, Agrypninae). Revista Brasileira de Entomologia 53 (1): 36–37.

Smith JW, Enns WR (1977) The click beetle subfamilies Agrypninae, Pyrophorinae, and Melanotinae (Coleoptera: Elateridae) in Missouri - part I. Journal of the Kansas Entomological Society 50 (3): 436–468.

Smith JW, Enns WR (1978) The click beetle subfamilies Agrypninae, Pyrophorinae, and Melanotinae (Coleoptera: Elateridae) in Missouri - part II. Journal of the Kansas Entomological Society 51 (1): 42–74.

Stibick JNL (1979) Classification of the Elateridae (Coleoptera). Relationships and classification of the subfamilies and tribes. Pacific Insects 20 (2/3): 145–186.

Tarnawski D, Buchholz L (2008) Sprezykowate -Elateridae. Część ogolna oraz podrodziny: Agrypninae, Negastriinae i Diminae. Klucze do Oznaczania Owadów Polski [Nr. 172]. Część XIX. Chrząszcze - Coleoptera. Zeszyt 34a. Polskie Towarzystwo Entomologiczne, Warszawa, 125 pp.

Vats LK, Vasu V (1993) Taxonomic importance of internal organs of reproduction in Agrypninae (Coleoptera: Elateridae). Journal of Entomological Research 17 (1): 65–70.

Elateridae

Supporting references for the conservation of Prosternini Gistel, 1856 over Diacanthini Gistel, 1848 (Art. 23.9.2). The taxon name Diacanthini Gistel, 1848 has not been used as valid after 1899 to our knowledge.

Barševskis A (2005) Catalogue of click-beetles (Coleoptera: Elateridae) of Latvia. Proceedings on Taxonomy and Faunistics of Beetles (Coleoptera) dedicated to the 100th birthday of the Latvian entomologist Mihails Stiprais (1905–1990). Pp. 7–28.

Bishop DJ, Majka CG, Bondrup-Nielsen S, Peck SB (2009) Deadwood and saproxylic beetle diversity in naturally disturbed and managed spruce forests in Nova Scotia. ZooKeys 22: 309–340.

Casari SA (2008) A phylogenetic study of the subtribe Dicrepidiina (Elateridae, Elaterinae, Ampedini). Revista Brasileira de Entomologia 52 (2): 182–260.

Johnson PJ (1995) A new genus of Conoderini, with new generic classifications for *Ctenicera sleeperi* Becker and *Ctenicera pilatei* (Champion), and a new species from Jamaica (Coleoptera: Elateridae). The Coleopterists Bulletin 49 (1): 59–71.

Johnson PJ (2002a) 58. Elateridae Leach 1815 [pp. 160–173]. In: Arnett RH, Jr., Thomas MC, Skelley PE, Frank JH (Eds) American beetles. Volume 2. Polyphaga: Scarabaeoidea through Curculionoidea. CRC Press, Boca Raton, xiv + 861 pp.

Johnson PJ (2002b) Lectotype designations for Elateridae (Coleoptera) described by George C. Champion in the Biologia Centrali-Americana. Dugesiana 9 (1): 15–46.

Kalniņš M, Juceviča E, Karpa A, Salmane I, A. P, Telnov D (2007) Bezmugurkaulnieki [pp. 106–151]. In: Pilāts V (Ed) Biologiskā Daudzveidība Gaujas Nacionālajā Parkā Biodiversity in Gauja National Park. Gaujas nacionālā parka administrācija, Sigulda.

Kishii T (1997) A study on the elaterid-beetles of Shibata collection from Taiwan, VI. (Coleoptera: Elateridae). On the subfamily Denticollinae: Tribe Prosternini. The Entomological Review of Japan 52 (1): 15–18.

Kishii T (2004) Two new species of the subfamily Dendrometrinae (Coleoptera: Elateridae) from Japan. The Entomological Review of Japan 59 (1): 61–66.

Kishii T, Jiang S-H (1996) Notes on the Chinese Elateridae (3) (Coleoptera). Entomological Review of Japan 51 (2): 97–102.

Kishii T, Jiang S-H (1999) Notes on the Chinese Elateridae (Coleoptera) (4). Entomological Review of Japan 54 (1): 11–19.

Lundberg B (1995) Catalogus Coleopterorum Sueciae. Naturhistoriska Riksmuseet, Stockholm [unpaginated].

Majka CG, Johnson PJ (2008) The Elateridae (Coleoptera) of the Maritime Provinces of Canada: faunal composition, new records, and taxonomic changes. Zootaxa 1811: 1–33.

Ohira H (2007) Notes on the morphological structure of *Actenicerus* species (Coleoptera: Elateridae, Dendrometrinae, Prosternini) from Japan (8). Nanki Seibutu 49: 131–134.

Ohira H (2008) Notes on the morphological structure of *Actenicerus* species (Coleoptera: Elateridae, Dendrometrinae, Prosternini) from Japan (9). Nanki Seibutu 50: 79–81.

Recalde JI, Sánchez-Ruiz A (2003) Elateridae (Coleoptera) forestales de Navarra (III). Acerca de la presencia del género *Paraphotistus* Kishii, 1966 en la Península Ibérica. Boletín de la Sociedad Entomológica Aragonesa 32: 95–97.

Sánchez-Ruiz A (1996) Catalogo bibliográfico de las especies de la familia Elateridae (Coleoptera) de la Península Iberica e Islas Baleares. Documentos Fauna Ibérica, 2. Museo Nacional de Ciencias Naturales, CSIC, Madrid, 265 pp.

Sánchez-Ruiz A (1998) Familia Elateridae (Insecta: Coleoptera). Catalogus de la Entomofauna Aragonesa 17: 13–17.

Sánchez-Ruiz A (1999) Los Elateridae (Coleoptera) de Los Monegros. Boletín de la Sociedad Entomológica Aragonesa 24 [1998]: 167–168.

Sánchez-Ruiz A, Muñoz J, Blasco-Zumeta J (1998) Nuevos datos para la fauna de Elateridae (Coleoptera) de Aragón. Boletín de la Sociedad Entomológica Aragonesa 22: 13–15.

Schneider MC, Rosa SP, Almeida MC, Costa C, Cella DM (2007) Strategies of karyotype differentiation in Elateridae (Coleoptera, Polyphaga). Micron 38 (6): 590–598.

Silfverberg H (1992) Enumeratio Coleopterorum Fennoscandiae, Daniae et Baltiae. Helsingfors Entomologiska Bytesförening, Helsinki, v + 94 pp.

Telnov D (2004) Check-list of Latvian beetles (Insecta: Coleoptera). Second edition. Compendium of Latvian Coleoptera 1. Entomological Society of Latvia, Rïga, 113 pp.

Telnov D, Barsevskis A, Savich F, Kovalevsky F, Berdnikov S, Doronin M, Cibulskis R, Ratniece D (1997) Check-list of Latvian beetles (Insecta: Coleoptera). Mitteilungen des Internationalen Entomologischen Vereins, Supplement 5: 1–141.

Zapata de la Vega JL, Sánchez-Ruiz A (2002) El género *Selatosomus* Stephens, 1830 en la península Ibérica (Coleoptera: Elateridae: Dendrometrinae: Prosternini). Boletín de la Sociedad Entomológica Aragonesa 30: 101–110.

Monotomidae

Supporting references for the conservation of *Monotoma* Herbst, 1793 over *Monotoma* Panzer, 1792 (Art. 23.9.2). The taxon name *Monotoma* Panzer, 1792 has not been used as valid after 1899 to our knowledge.

Arnett RH, Jr. (1962b) Part V Suborder Polyphaga (cont.) series Cucujiformia (cont.) Tenebrionoidea, Cucujoidea [pp. (2 unn. +) 645–850]. In: The beetles of the United States (a manual for identification) [original loose-leaf edition]. Catholic University of America Press, Washington D.C., xii + 1112 pp. [+ unn. pp.].

Arnett RH, Jr. (2000) American insects. A handbook of the insects of America north of Mexico. Second edition. CRC Press, Boca Raton, xvii + 1003 pp.

Bousquet Y (1991) Family Rhizophagidae rhizophagid beetles [pp. 218–219]. In: Bousquet Y (Ed) Checklist of beetles of Canada and Alaska. Agriculture Canada, Publication 1861/R, 430 pp.

Bousquet Y (2002) Family 79. Monotomidae Laporte 1840 [pp. 319–321]. In: Arnett RH, Thomas MC, Skelley PE, Frank JH (Eds) American beetles. Volume 2. Polyphaga: Scarabaeoidea through Curculionoidea. CRC Press, Boca Raton, xiv + 861 pp.

Bousquet Y (2010) Monotomidae Laporte, 1840 [pp. 319–324]. In: Part 38. Coleoptera, beetles. Volume 2: Morphology and systematics (Elateroidea, Bostrichiformia, Cucujiformia partim). In: Leschen RAB, Beutel RG, Lawrence JF, Ślipiński SA (Eds) Handbook of zoology. Volume IV. Arthropoda: Insecta. W. de Gruyter, New York and Berlin, xiii + 786 pp.

Bousquet Y, Laplante S (2000) Taxonomic review of the Canadian species of the genus *Monotoma* Herbst (Coleoptera: Monotomidae). Proceedings of the Entomological Society of Ontario 130 [1999]: 67–96.

De Marzo L (2005) Una peculiarità anatomica del dotto eiaculatore in *Monotoma* Herbst, 1793 (Insecta Coleoptera Rhizophagidae). Naturalista Siciliano (4) 29 (1/2): 67–70.

Downie NM, Arnett RH, Jr. (1996) The beetles of northeastern North America. Volume II: Polyphaga: Series Bostrichiformia through Curculionoidea. Sandhill Crane Press, Gainesville, pp. i-x, 891–1721.

Hansen M, Kristensen S (1991) To nye danske biller af slaegten *Monotoma* Herbst (Coleoptera, Monotomidae). Entomologiske Meddelelser 59 (2): 41–44.

Holzschuh C, Lohse GA (1981) Eine neue Art der Gattung *Monotoma* Herbst aus Mitteleuropa: *Monotoma gotzi* n. sp. Entomologische Blätter für Biologie und Systematik der Käfer 77 (3): 175–177.

Jelínek J (2007) Family Monotomidae Laporte, 1840 [pp. 491–495]. In: Löbl I, Smetana A (Eds) Catalogue of Palaearctic Coleoptera. Volume 4. Elateroidea - Derodontoidea - Bostrichoidea - Lymexyloidea - Cleroidea - Cucujoidea. Apollo Books, Stenstrup, 935 pp.

Johnson C, Otero JC (2010) Description of a new *Monotoma* (*Monotomina*) species (Coleoptera: Monotomidae) from the Iberian Peninsula. Entomologica Fennica 21 (2): 104–106.

Klimaszewski J, Watt JC (1997) Coleoptera: family-group review and keys to identification. Fauna of New Zealand. Ko te Aitanga Pepeke o Aotearoa. Number 37. Manaaki Whenua Press, Lincoln, 197 pp.

Kuschel G (1979) The genera *Monotoma* Herbst (Rhizophagidae) and *Anommaus* Wesmael (Cerylidae) in New Zealand (Coleoptera). New Zealand Entomologist 7 (1): 44–48.

Majka CG, Bousquet Y (2010) Monotomidae (Coleoptera) of the Maritime Provinces of Canada. Journal of the Acadian Entomological Society 6: 1–8.

Milander G (1992) On the Monotomidae (Coleoptera) of Estonia. Izvestiya Akademii Nauk Estonskoi SSR Seriya Biologicheskaya 41: 66–71.

Nikitsky NB (1986) Zhestkokrylye podsemeystv Monotominae i Thioninae (Coleoptera, Rhizophagidae) dalnego Vostoka SSSR [Beetles of the subfamilies Monotominae and Thioninae (Coleoptera, Rhizophagidae) in the Soviet Far East]. Zoologicheskii Zhurnal 65 (11): 1622–1630 [in Russian].

Pal TK (2000). On five monotomid species deposited in the Museum of Natural History of the Humboldt-University in Berlin (Coleoptera, Cucujoidea). Mitteilungen aus dem Museum für Naturkunde in Berlin, Zoologische Reihe 76 (1): 113–119.

Peacock ER (1977) Vol. V, Part 5(a). Coleoptera Rhizophagidae. Handbooks for the identification of British insects. Royal Entomological Society of London, London, pp. 1–19.

Peck SB (2006) The beetles of the Galápagos Islands, Ecuador: evolution, ecology, and diversity (Insecta: Coleoptera). NRC Research Press, Ottawa, xiii + 313 pp.

Peck SB, Thomas MC (1998) A distributional checklist of the beetles (Coleoptera) of Florida. Arthropods of Florida and neighboring land areas. Volume 16, Florida Department of Agriculture and Consumer Services, Gainesville, viii + 180 pp.

Sen Gupta T (1977) A new genus and species of Rhizophagidae (Clavicornia: Coleoptera) from Sikkim. Oriental Insects 11 (4): 531–536.

Sen Gupta T (1988) Review of the genera of the family Rhizophagidae (Clavicornia: Coleoptera) of the world. Memoirs of the Zoological Survey of India 17: 1–58, pls. i-xxiv.

Silfverberg H (1992) Enumeratio Coleopterorum Fennoscandiae, Daniae et Baltiae. Helsingfors Entomologiska Bytesförening, Helsinki, v + 94 pp.

Ślipiński SA (1981) Klucze do oznaczania owadów Polski [Nr. 119]. Część XIX. Chrząszcze - Coleoptera. Zeszyt 63. Monotomidae. Polskie Towarzystwo Entomologiczne, Warszawa, 11 pp.

Vorst O (2010). Monotomidae [pp. 124–125]. In: Vorst O (Ed) Catalogus van de Nederlandse kevers. Nederlandse Entomologische Vereniging, Amsterdam, 317 pp.

Oedemeridae

Supporting references for the conservation of Calopodinae A. Costa, 1852 over Sparedrinae Gistel, 1848 (Art. 23.9.2). The taxon name Sparedrinae Gistel, 1848 has not been used as valid after 1899 to our knowledge.

Arnett RH, Jr. (1962) Part V. Suborder Polyphaga (cont.). Series Cucujiformia (Cont.). Tenebrionoidea, Cucujoidea [pp. (2 unn. +) 645–850]. In: The beetles of the United States (a manual for identification) [original loose-leaf edition]. The Catholic University of America Press, Washington D.C., xii + 1112 pp. [+ unn. pp.].

Arnett RH, Jr. (1993) American insects. A handbook of the insects of America north of Mexico. The Sandhill Crane Press, Inc., Gainesville, xiii + 850 pp.

Arnett RH, Jr. (2000) American insects. A handbook of the insects of America north of Mexico. Second edition. CRC Press, Boca Raton, xvii + 1003 pp.

Bologna MA (2005) *Zonitis fernancastroi*, a new species for the Italian fauna, and additional records of Meloidae and Oedemeridae (Coleoptera). Bollettino della Società Entomologica Italiana 137 (2): 107–114.

Campbell JM (1991) Family Oedemeridae false blister beetles [p. 266]. In: Bousquet Y (Ed) Checklist of beetles of Canada and Alaska. Agriculture Canada Research Branch Publication 1861/E, vi + 430 pp.

Crowson RA (1967) The natural classification of the families of Coleoptera. E.W. Classey Ltd., Middlesex (England), [1] + 214 pp. [Note: this is a reprint].

Gómez de Dios MÁ, Verdugo Páez A (2008) Nuevo registro de *Sparedrus lencinae* Vázquez, 1988 (Oedemeridae: Calopodinae) de la Comunidad Autónoma de Andalucía (España). Boletín de la Sociedad Entomológica Aragonesa 42: 461–462.

Hatch MH (1965) The beetles of the Pacific Northwest. Part IV: Macrodactyles, Palpicornes, and Heteromera. University of Washington Press, Seattle, viii + 268 pp.

Hudson L (1975) A systematic revision of the New Zealand Oedemeridae (Coleoptera, Insecta). Journal of the Royal Society of New Zealand 5 (3): 227–274.

Hůrka K (2005) Brouci České a Slovenské republiky. Käfer der Tschechischen und Slowakischen Republik. Vit Kabourek, Zlín, 390 pp.

Kaszab Z (1969) 70. Familie: Oedemeridae [pp. 79–92]. In: Freude H, Harde KW, Lohse GA (Eds) Die Käfer mitteleuropas. Band 8. Teredilia, Heteromera, Lamellicornia. Goecke & Evers, Krefeld, 388 pp.

Klausnitzer B (1999) 112. Familie: Oedemeridae [pp. 310–324]. In: Klausnitzer B (Ed) Die Larven der Käfer mitteleuropas. 5 Band. Polyphaga Teil 4. Goecke & Evers, Krefeld, 336 pp.

Kriska NL (2002) 109. Oedemeridae Latreille 1810 [pp. 514–519]. In: Arnett RH, Jr., Thomas MC, Skelley PE, Frank JH (Eds) American beetles. Volume 2. Polyphaga: Scarabaeoidea through Curculionoidea. CRC Press, Boca Raton, xiv + 861 pp.

Kubisz D (2006) Oedemeridae and Scraptiidae of Poland (Coleoptera, Tenebrionoidea). Monografie Faunistyczne 24: 1–165.

Lawrence JL (2005) *Dasytomima*, a new genus of Australian Oedemeridae and its relationship to *Polypria* Chevrolat (Coleoptera: Tenebrionoidea). Annales Zoologici (Warszawa) 55 (4): 663–676.

Lundberg B (1995) Catalogus Coleopterorum Sueciae. Naturhistoriska Riksmuseet, Stockholm [unpaginated].

Švihla V (1986) Revision of the generic classification of the Old World Oedemeridae (Coleoptera). Sborník Národního Muzea v Praze (Řada B - Přírodní Vědy) 41 [1985] (3/4): 141–238.

Švihla V (1993) Oedemeridae [pp. 112–113]. In: Jelínek J (Ed) Check-list of Czechoslovak insects IV (Coleoptera). Seznam československých brouků. Folia Heyrovskyana Supplementum 1, 172 pp.

Švihla V (2008) Family Oedemeridae Latreille, 1810 [pp. 353–369]. In: Löbl I, Smetana A (Eds) Catalogue of Palaearctic Coleoptera. Volume 5. Tenebrionoidea. Apollo Books, Strenstrup, 670 pp.

Telnov D (2004) Check-list of Latvian beetles (Insecta: Coleoptera). Second edition. Compendium of Latvian Coleoptera 1. Entomological Society of Latvia, Rïga, 113 pp.

Telnov D, Barsevskis A, Savich F, Kovalevsky F, Berdnikov S, Doronin M, Cibulskis R, Ratniece D (1997) Check-list of Latvian beetles (Insecta: Coleoptera). Mitteilungen des Internationalen Entomologischen Vereins, Supplement 5: 1–141.

Vázquez XA (1988) A new species of Oedemeridae from the Iberian Peninsula (Coleoptera). Nouvelle Revue d'Entomologie (Nouvelle série) 5 (3): 259–261.

Vázquez XA (1993) Coleoptera, Oedemeridae, Pyrochroidae, Pythidae, Mycteridae. Fauna Iberica, Vol 5, Museo Nacional de Ciencias Naturales, Madrid, 181 pp.

Vázquez XA (2002) European Fauna of Oedemeridae. Argania editio S.C.P., Barcelona, 179 pp.

Zahradník J (1985) Käfer Mittel- und Nordwesteuropas. Ein Bestimmungsbuch für Biologen und Naturfreunde. Paul Parey, Hamburg, 498 pp.

Scarabaeidae

Supporting references for the conservation of *Pachypus* Dejean, 1821 over *Pachypus* Billberg, 1820 (Art. 23.9.2). The genus name *Pachypus* Billberg, 1820 has not been used as valid after 1899 to our knowledge.

Ahrens D (2006) The phylogeny of Sericini and their position within the Scarabaeidae based on morphological characters (Coleoptera: Scarabaeidae). Systematic Entomology 31 (1): 113–144.

Arnone M, Sparacio I (1990) Il *Pachypus caesus* Erichson 1840: brevi note sulla biologia e la distribuzione in Sicilia (Coleoptera Scarabaeoidea). Naturalista Siciliano (4) 14 (1): 63–71.

Ballerio A, Tacchetti M (1990) *Pachypus candidae* (Petagna 1786) (Coleoptera Pachipodidae). Bollettino della Società Entomologica Italiana 121 [1989] (3): 234.

Baraud J (1977) Coléoptères Scarabaeoidea. Faune de l'Europe occidentale: Belgique, France, Grande-Bretagne, Italie, Péninsule Ibérique. Nouvelle Revue d'Entomologie Supplement 7 (1): 1–352.

Baraud J (1980) Nouveaux coléoptères Scarabaeoidea d'Afrique du nord. Nouvelle Revue d'Entomologie 10 (3): 279–284.

Baraud J (1992) Coléoptères Scarabaeoidea d'Europe. Faune de France 78. Société Linnéenne de Lyon, Lyon, ix + 856 pp. + 11 pls.

Bidault J (1999) Voyage entomologique en Haute Corse (Coleoptera, Carabidae, Cerambycidae, Scarabaeidae). Revue de l'Association Roussillonnaise d'Entomologie 8: 21–23.

Cantot P, Cantot C, Cantot S (1988) Sur deux espèces de coléoptères capturées en Corse (Scarabaeidae & Cerambycidae). L' Entomologiste 44 (6): 308.

Crovetti A (1970) Contributi alla conoscenza dei Coleotteri Scarabeidi. 1. Il genere *Pachypus* Serville (Coleoptera, Scarabaeidae, Pachypodinae). Bollettino di Zoologia Agraria e Bachicoltura (Ser. 2) 9 [1968–69]: 134–188.

Crowson RA (1995) Some interesting evolutionary parallels in Coleoptera [pp. 63–85]. In: Pakaluk J, Ślipiński SA (Eds). Biology, phylogeny and classification of Coleoptera: Papers celebrating the 80th birthday of Roy A Crowson. Muzeum i Instytut Zoologii PAN, Warszawa, x + 1092 pp. in 2 vols.

Dupuis F (2005) L'abdomen et les genitalia des femelles de coléoptères Scarabaeoidea (Insecta, Coleoptera). Zoosystema 27 (4): 733–823.

Evans AV, Bellamy CL (1996) An inordinate foundness for beetles. Nevraumont Publishing Company, New York, 208 pp.

Evans G (1975) The life of beetles. George Allen & Unwin Ltd, London, 232 pp.

Král D, Löbl I (2006) Subfamily Pachypodinae Erichson, 1840 [p. 180]. In: Löbl I, Smetana A (Eds). Catalogue of Palaearctic Coleoptera. Volume 3. Scarabaeoidea - Scirtoidea - Dascilloidea - Buprestoidea - Byrrhoidea. Apollo Books, Stenstrup, 690 pp.

Kukalová-Peck J, Lawrence JF (2004) Relationships among coleopteran suborders and major endoneopteran lineages: evidence from hind wing characters. European Journal of Entomology 101 (1): 95–144.

Lapiana F, Sparacio I (2006) I Coleotteri Lamellicorni delle Madonie (Sicilia) (Insecta Coleoptera Lucanoidea et Scarabaeoidea). Naturalista Siciliano (4) 30 (2): 227–292.

Lawrence JF, Newton AF, Jr. (1995) Families and subfamilies of Coleoptera (with selected genera, notes, references and data on family-group names) [pp. 779–1006]. In: Pakaluk J, Ślipiński SA (Eds). Biology, phylogeny and classification of Coleoptera: Papers celebrating the 80th birthday of Roy A Crowson Vol II. Muzeum i Instytut Zoologii PAN, Warszawa, x + 1092 pp. in 2 vols.

Meinecke C-C (1975) Riechsensillen und Systematik der Lamellicornia (Insecta, Coleoptera). Zoomorphologie 82 (1): 1–42.

Ocampo FC, Ruiz-Manzanos E, Marvaldi AE (2010) Systematic revision, cladistics and biogeography of the genus *Neogutierrezia* Martínez (Coleoptera: Scarabaeidae) and its phylogenetic placement in Rutelinae based on structural alignment of 28S rDNA sequences. Invertebrate Systematics 24 (1): 81–111.

Paulian R (1988) Biologie des Coléoptères. Lechevalier, Paris, xi + 719 pp.

Pesarini C (2004) Insetti della fauna Italiana Coleotteri Lamellicorni. Natura (Milan) 93 (2): 1–130.

Sabella G, Sparacio I (2004) Il ruolo dei Parchi Siciliani nella conservazione di taxa di insetti di particolare interesse naturalistico (Insecta Coleoptera et Lepidoptera Rhopalocera). Naturalista Siciliano (4) 28 (1): 477–508.

Scholtz CH, Grebennokov V (2005) 13. Scarabaeoidea Latreille, 1802 [pp. 367–425]. In: Beutel RG, Leschen RAB (Eds). Handbook of zoology. A natural history of the phyla of the animal kingdom. Volume IV - Arthropoda: Insecta. Walter De Gruyter, Berlin, xi + 567 pp.

Smith ABT (2006) A review of the family-group names for the superfamily Scarabaeoidea (Coleoptera) with corrections to nomenclature and a current classification. Coleopterists Society Monograph 5: 144–204.

Sparacio I (2008) Una nuova specie di *Pachypus* Dejean di Sardegna (Coleoptera, Scarabaeoidea, Pachypodidae). Doriana 8 (360): 1–13.

Scarabaeidae

Supporting references for the conservation of *Sparrmannia* Laporte, 1840 over *Leocaeta* Dejean, 1833 and *Cephalotrichia* Hope, 1837 (Art. 23.9.2). The generic names *Leocaeta* Dejean, 1833 and *Cephalotrichia* Hope, 1837 have not been used as valid after 1899 to our knowledge.

Ahrens D (2006) The phylogeny of Sericini and their position within the Scarabaeidae based on morphological characters (Coleoptera: Scarabaeidae). Systematic Entomology 31 (1): 113–144.

Browne DJ, Scholtz CH (1995) Phylogeny of the families of Scarabaeoidea (Coleoptera) based on characters of the hindwing articulation, hindwing base and wing venation. Systematic Entomology 20 (3): 145–173.

Browne DJ, Scholtz CH (1996) The morphology of the hind wing articulation and wing base of the Scarabaeoidea (Coleoptera) with some phylogenetic implications. Bonner Zoologische Monographien 40: 1–200.

Browne DJ, Scholtz CH (1998) Evolution of the scarab hindwing articulation and wing base: a contribution toward the phylogeny of the Scarabaeidae (Scarabaeoidea: Coleoptera). Systematic Entomology 23 (4): 307–326.

Chown SL, Nicolson SW (2004) Insect physiological ecology: mechanisms and patterns. Oxford University Press, Oxford, ix + 243 pp.

Chown SL, Scholtz CH (1993) Temperature regulation in the nocturnal melolonthine *Sparrmannia flava*. Journal of Thermal Biology 18 (1): 25–33.

Coca-Abia M (2007) Phylogenetic relationships of the subfamily Melolonthinae (Coleoptera, Scarabaeidae). Insect Systematics & Evolution 38 (4): 447–472.

d'Hotman D, Scholtz CH (1991) Comparative morphology of the male genitalia of derived groups of Scarabaeoidea (Coleoptera). Elytron 4 [1990]: 3–39.

Evans AV (1988) Three new species of *Glyptoglossa* Brenske (Coleoptera: Scarabaeidae: Melolonthinae). Journal of the Entomological Society of Southern Africa 51 (1): 87–96.

Evans AV (1989) Revision of the genus *Sparrmannia* Laporte (Coleoptera: Melolonthidae: Melolonthinae). Journal of the Entomological Society of Southern Africa 52 (1): 11–44.

Ferreira MC (1966) Catálogo dos Coleópteros de Moçambique (Conclusão). Revista de Entomologia de Moçambique 6 [1963] (2): [1] + 533–1008 + 1 map.

Hardy, A.R. (1978) Placement of the genus *Benedictia* Sanderson (Coleoptera: Scarabaeidae). The Coleopterists Bulletin 32 (1): 67–70.

Katovich K (2008) A generic-level phylogenetic review of the Macrodactylini (Coleoptera: Scarabaeidae: Melolonthinae). Insecta Mundi 23: 1–78.

Lacroix M (2004) Contribution à la connaissance des Pachydeminae africains, IV. Nouvelles espèces (Coleoptera, Melolonthidae). Coléoptères 10 (9): 107–118.

Lacroix M (2007) Pachydeminae du monde: genera et catalogue (Coleoptera, Melolonthidae). Collection Hannetons, Paris, 450 pp.

Nel A, Scholtz CH (1990) Comparative morphology of the mouthparts of adult Scarabaeoidea. Entomology Memoir, Department of Agricultural Development, Republic of South Africa 80: 1–84.

Paulian R (1993) Les Coléoptères: à la conquête de la terre. Sociéte Nouvelle des Éditions Boubée, Paris, 241 + [4] pp.

Picker M, Griffiths C, Weaving A (2002) Field guide to insects of South Africa. Struik Publishers, Cape Town, 440 pp.

Sanmartín I (2003) Dispersal vs. vicariance in the Mediterranean: historical biogeography of the Palearctic Pachydeminae (Coleoptera, Scarabaeoidea). Journal of Biogeography 30 (12): 1883–1897.

Sanmartín I, Martín-Piera F (2003) First phylogenetic analysis of the subfamily Pachydeminae (Coleoptera, Scarabaeoidea, Melolonthidae): the Palearctic Pachydeminae. Journal of Zoological Systematics and Evolutionary Research 41 (1): 2–46.

Scholtz CH (1988) Biology of *Sparrmannia flava* Arrow (Coleoptera: Scarabaeidae: Melolonthinae). The Coleopterists Bulletin 42 (1): 57–62.

Scholtz CH, Chown SL (1995) The evolution of habitat use and diet in the Scarabaeoidea: a phylogenetic approach [pp. 355–374]. In: Pakaluk J, Ślipiński SA (Eds) Biology, phylogeny and classification of Coleoptera: Papers celebrating the 80th birthday of Roy A Crowson. Volume one. Muzeum i Instytut Zoologii PAN, Warszawa, x + 1092 pp. in 2 vols.

Scholtz CH, Grebennikov VV (2005) Scarabaeiformia Crowson, 1960. Chapter 12 [pp. 345–366]. In: Beutel RG, Leschen RAB (Eds). Handbook of zoology. A natural history of the phyla of the animal kingdom. Volume IV - Arthropoda: Insecta. Walter De Gruyter, Berlin, xi + 567 pp.

Scholtz CH, Holm E (1985) Insects of southern Africa. Butterworths, Durban, 502 p. + xii plates.

Smith ABT (2006) A review of the family-group names for the superfamily Scarabaeoidea (Coleoptera) with corrections to nomenclature and a current classification. Coleopterists Society Monograph 5: 144–204.

Tenebrionidae

Supporting references for the conservation of Adesmiini Lacordaire, 1859 over

Macropodini Agassiz, 1846 (Art. 23.9.2). The taxon name Macropodini Agassiz, 1846 has not been used as valid after 1899 to our knowledge and is a junior homonym of the well-established name Macropodidae Gray, 1821 [Mammalia].

Alfieri A (1976) The Coleoptera of Egypt. Mémoires de la Société Entomologique d'Égypte 5: 1–361.

Ardoin P (1972) Liste des espèces de Tenebrionidae (Coleoptera) récoltées au Sudan par les expéditions finlandaises (1962–1964) (Zoological contribution from the Finnish expeditions to the Sudan no. 27). Commentationes Biologicae 49: 1–20.

Bouchard P, Lawrence JF, Davies AE, Newton AF (2005) Synoptic classification of the world Tenebrionidae (Insecta: Coleoptera) with a review of family-group names. Annales Zoologici (Warszawa) 55 (4): 499–530.

Cloudsley Thompson JL (2001) Thermal and water relations of desert beetles. Naturwissenschaften 88: 447–460.

Doyen JT (1994) Cladistic relationships among Pimeliine Tenebrionidae (Coleoptera). Journal of the New York Entomological Society 101 [1993] (4): 443–514.

Duncan FD (2003) The role of the subelytral cavity in respiration in a tenebrionid beetle, *Onymacris multistriata* (Tenebrionidae: Adesmiini). Journal of Insect Physiology 49: 339–346.

Ferreira MC (1965) Catálogo dos Coleópteros de Moçambique. Revista de Entomologia de Moçambique 6 [1963] (1): 1–531 + [1, corrigenda].

Ferreira MC (1967) Catálogo dos Coleópteros de Angola. Revista de Entomologia de Moçambique 8 [1965] (2): 415–1317.

Girard C (1967) Coléoptères ténébrionidés Erodiini, Tentyriini et Adesmiini récoltés par J. Mateu dans le massif de l'Ennedi. Bulletin de l'Institut Français de l'Afrique Noire (Série A) 29 (3): 945–953.

Hanrahan SA, Nicolson SW (1984) The effect of dehydration on the Malpighian tubules of *Onymacris plana* (Coleoptera: Adesmiini). Elektronmikroskopievereniging van Suidelike Afrika Verrigtings 14: 107–108.

Hauffe HC, Pietruszka RD, Seely MK (1988) Observations on the behaviour of *Onymacris laeviceps* Gebien (Coleoptera: Tenebrionidae: Adesmiini) in the central Namib Desert dunes. Journal of the Entomological Society of Southern Africa 51: 183–192.

Kaszab Z (1981) Insects of Saudi Arabia. Coleoptera: fam. Tenebrionidae (part 2). Fauna of Saudi Arabia 3: 276–401.

Koch C (1962) The Tenebrionidae of Southern Africa. XXXI. Comprehensive notes on the tenebrionid fauna of the Namib desert. Annals of the Transvaal Museum 24 (2/3): 61–106.

Kwieton E (1981) Esquisse entomogéographique de l'Algérie et de l'histoire du désert Saharien, à la base des coléoptères Tenebrionidae. Anais da Faculdade de Ciências Universidade do Porto 62: 189–237.

Kwieton E (1982) Revue critique des systèmes récents de la famille des Tenebrionidae (Col.). Sborník Národního Muzea v Praze (Řada B - Přírodní Vědy) 38 (1/2): 79–100.

McClain E, Praetorius RL, Hanrahan SA, Seely MK (1984) Dynamics of the wax bloom of a seasonal Namib Desert tenebrionid, *Cauricara phalangium* (Coleoptera: Adesmiini). Oecologia 63 (3): 314–319.

Penrith M-L (1979) Revision of the western southern African Adesmiini (Coleoptera: Tenebrionidae). Cimbebasia (Series A) 5: 1–94.

Penrith M-L (1986) Relationships in the tribe Adesmiini (Coleoptera: Tenebrionidae) and a revision of the genus *Stenodesia* Reitter. Annals of the Transvaal Museum 34 (13): 275–302.

Roer H (1981) Weitere Untersuchungen zur Anpassung des Namibwüstenkäfers *Onymacris r. rugatipennis* (Haag 1875, Col.: Tenebrionidae, Adesmiini) am das Trockenflussbett des Kuiseb in Südwestafrika. Mitteilungen der Deutschen Gesellschaft für Allgemeine und Angewandte Entomologie 3 (1/3): 218–222.

Roer H (1983) Aktionsraum und Anpassungsphanomene des Dunenkafers *Onymacris laeviceps* Gebien (Col.: Tenebrionidae, Adesmiini) in der Namibwüste. Bonner Zoologische Beiträge 34 (1/3): 357–369.

Roer H (1984) Dispersion dynamics of desert beetles of the genus *Onymacris* (Col., Tenebrionidae, Adesmiini) in the Namib [p. 302]. XVII International Congress of Entomology. Abstract Volume, 954 pp.

Roer H (1985) Dispersion dynamics of tenebrionids of the genus *Onymacris* (Col.: Tenebrionidae, Adesmiini) in the Namib Desert. Journal SWA Scientific Society 39: 65–69.

Roer H (1986) Zur Anpassung des Schwarzkäfers *Onymacris unguicularis* (Haag) (Col.: Tenebrionidae, Adesmiini) an die Nebelzone der Namibwüste. Bonner Zoologische Beiträge 37 (2): 143–154.

Skopin NG (1978) Tenebrionidae [pp. 223–266]. In: Klausnitzer B (Ed) Ordnung Coleoptera (Larven). W. Junk, The Hague, vi + 378 pp.

Veiga Ferreira G, da (1966) Catálogo dos tipos de insectos existentes no Museu Dr. Álvaro de Castro. Revista de Entomologia de Moçambique 7 [1964] (1): 197–216.

Tenebrionidae

Supporting references for the conservation of Bolitophagini Kirby, 1837 over Eledonini Billberg, 1820 (Art. 23.9.2). The taxon name Eledonini Billberg, 1820 has not been used as valid in Tenebrionidae after 1899 to our knowledge. Furthermore, conservation of Bolitophagini avoids homonymy with Eledoninae used as valid in cephalopods.

Aalbu RL (2006) 2006, where are we at: assessing the currents state of Tenebrionidae systematics on a global scale (Coleoptera: Tenebrionidae). Cahiers Scientifiques. Centre de Conservation et d'Étude des Collections (Lyon) 10: 55–70.

Aalbu RL, Flores GE, Triplehorn CA (2002) Tenebrionidae [pp. 499–512]. In: Bousquets JL, Morrone JJ (Eds) Biodiversidad, taxonomía y biogeografía de arthropódos de México: Hacia una síntesis de su conocimiento Volumen III. Universidad Nacional Autonóma de México, México, x + 690 pp.

Aalbu RL, Triplehorn CA, Campbell JM, Brown KW, Somerby RE, Thomas DB (2002) 106. Tenebrionidae Latreille 1802 [pp. 463–509]. In: Arnett RHJ, Thomas MC, Skelley PE, Frank JH (Eds) American beetles. Volume 2. Polyphaga: Scarabaeoidea through Curculionoidea. CRC Press, Boca Raton, xiv + 861 pp.

Bouchard P, Lawrence JF, Davies AE, Newton AF (2005) Synoptic classification of the world Tenebrionidae (Insecta: Coleoptera) with a review of family-group names. Annales Zoologici (Warsaw) 55 (4): 499–530.

Downie NM, Arnett RH, Jr. (1996) The beetles of northeastern North America. Volume 2: Polyphaga: Series Bostrichiformia through Curculionoidea. Sandhill Crane Press, Gainesville, pp. i-x, 891–1721.

Doyen JT, Matthews EG, Lawrence JF (1990) Classification and annotated checklist of the Australian genera of Tenebrionidae (Coleoptera). Invertebrate Taxonomy 3 [1989] (3): 229–260.

Doyen JT, Poinar GO, Jr (1994) Tenebrionidae from Dominican amber (Coleoptera). Entomologica Scandinavica 25 (1): 27–51.

Endrödy-Younga S (1998) *Afrobyrsax capensis* spec. nov., with a tropical West African relationship (Coleoptera: Tenebrionidae: Bolitophagini). Annals of the Transvaal Museum 36 (30): 421–423.

Español F (1985) Los Bolitophaginae de la fauna española (Col., Tenebrionidae). Publicaciones del Departamento de Zoología, Universidad de Barcelona 11: 61–64.

Hornig U (1993) Das Typenmaterial der Tenebrionidae im Staatlichen Museum für Tierkunde Dresden. Teil 1: Tribus Platyscelini, Praocini, Pedinini, Opatrini, Phaleriini, Crypticini, Bolitophagini, Rhipidandrini, Diaperini, Gnathidiini, Leiochrini et Phrenapatini (Insecta, Coleoptera). Entomologische Abhandlungen (Dresden) 55 (2): 153–161.

Jung BH, Kim SY, Kim JI (2007) Taxonomic review of the tribe Bolitophagini in Korea (Coleoptera: Tenebrionidae: Tenebrioninae). Entomological Research 37 (3): 190–196.

Kompantseva TV (1994) Mycetophilous tenebrionid fauna of Bolitophagini and Diaperini (Coleoptera, Tenebrionidae) from Middle Asia. Byulleten' Moskovskogo Obshchestva Ispytatelei Prirody Otdel Biologicheskii 99: 44–47.

Kompantseva TV (1996) Larva of *Rhipidandrus crenipennis* (Motschulsky, 1858), and the position of the genus in Bolitophagini (Coleoptera Tenebrionidae). Lichinka *Rhipidandrus crenipennis* (Motschulsky, 1858), I polozhenie roda v tribe Bolitophagini (Coleoptera Tenebrionidae). Russian Entomological Journal 4 (1/4) [1995]: 55–59.

Löbl I, Ando K, Bouchard P, Iwan D, Lillig M, Masumoto K, Merkl O, Nabozhenko M, Novák V, Petterson R, Schawaller W, Soldati F (2008) Family Tenebrionidae Latreille, 1802 [pp. 105–352]. In: Löbl I, Smetana A (Eds) Catalogue of Palaearctic Coleoptera. Volume 5. Tenebrionoidea. Apollo Books, Stenstrup, 670 pp.

Masumoto K (2006) Tenebrionid beetles (Coleoptera) from the Palau Islands collected by Keiichi Takahashi. Entomological Review of Japan 61 (2): 143–161.

Masumoto K, Akita K (2001) Two new tenebrionid species from Japan. Special Publication of the Japanese Coleopterist Society 1: 247–250.

Matthews EG, Bouchard P (2008) Tenebrionid beetles of Australia: descriptions of tribes, keys to genera, catalogue of species. Australian Biological Resources Study, Canberra, viii + 398 pp.

Miyatake M (1964) Notes on the tribe Bolitophagini of Japan, with the descriptions of four new genera and two new species (Coleoptera: Tenebrionidae). Transactions of the Shikoku Entomological Society 8 (2): 59–84.

Miyatake M (1970) A revision of the genus *Byrsax* Pascoe of Japan, with some notes on the Japanese Bolitophagini (Coleoptera: Tenebrionidae). Transactions of the Shikoku Entomological Society 10 (3/4): 116–126.

Schawaller W (2006) A new species of *Cryphaeus* and new records of other fungus-adapted tenebrionids from South Africa (Coleoptera: Tenebrionidae). Annals of the Transvaal Museum 43: 69–74.

Soldati F, Soldati L, Thieren Y (2009) Discovery of *Eledonoprius serrifrons* (Reitter, 1890) in Corsica, a new species for the fauna of France (Coleoptera, Tenebrionidae, Bolitophagini). Rutilans 12: 33–36.

Steiner WE, Jr. (2004) New distribution records for *Eleates depressus* (Randall) in the southeastern United States (Coleoptera: Tenebrionidae; Bolitophaginae) and notes on these occurrences. Banisteria 24: 52–53.

Tezcan S, Karsavuran Y, Pehlivan E, Keskin B, Ferrer J (2004) Contributions to the knowledge of the Tenebrionidae (Coleoptera) from Turkey Part I. Lagriinae, Pimeliinae, Bolitophaginae, Diaperinae. Turkiye Entomoloji Dergisi 28: 99–114.

Tovar AC, Sáez Bolaño J, Baena M (2008) Nuevas citas de Bolitophagini Kirby, 1837 (Coleoptera, Tenebrionidae) de España. Boletín de la Sociedad Entomológica Aragonesa 42: 361–365.

Throscidae

Supporting references for the conservation of Throscidae Laporte, 1840 over Stereolidae Rafinesque, 1815 (Art. 23.9.2). The taxon name Stereolidae Rafinesque, 1815 has not been used as valid after 1899 to our knowledge.

Allen AA (2001) The identity of *Trixagus elateroides* (Throscidae) of British authors. Coleopterist 10: 37.

Burakowski B (2000) Redescription of *Aulonothroscus laticollis* (Rybinski, 1897) (Coleoptera: Throscidae). Annales Zoologici (Warszawa) 50: 27–34.

Franz H, von (1982) Revision der *Throscus*-Arten (Coleoptera) der Makaronesischen Inseln. Zeitschrift der Arbeitsgemeinschaft Österreichischer Entomologen 34 (1/2): 49–58.

Gruszka A, Tarnawszki D (1994) *Trixagus elateroides* (Heer) (Coleoptera, Throscidae), *Isorhipis melasoides* (Cast.) and *Dirhagus pygmaeus* (F.) (Coleoptera, Eucnemidae) - new records from Poland. Wiadomosci Entomologiczne 13: 256.

Hornig U (2003) Faunistic records from the Czech Republic - 164. Coleoptera: Throscidae. Klapalekiana 39 (1/3): 129.

Hornig U (2005) Fauna der Throscidae der Oberlaustiz (Col.). Entomologische Nachrichten und Berichte 49 (2): 123–126.

Johnson PJ (1997) New distribution records for three Throscidae (Coleoptera) in South Dakota. Prairie Naturalist 29: 51–52.

Johnson PJ (2002) 57. Throscidae Laporte 1840 [pp. 158–159]. In: Arnett RH, Jr., Thomas MC, Skelley PE, Frank JH (Eds) American beetles. Volume 2. Polyphaga: Scarabaeoidea through Curculionoidea. CRC Press, Boca Raton, xiv + 861 pp.

Kistner DH, Abdel-Galil FA (1986) A new genus and species of termitophilous Throscidae from South Africa (Coleoptera). Sociobiology 12 (2): 305–313.

Leseigneur L (1995) Statut actuel des genres *Trixagus* Kugelann, 1794, et *Throscus* Latreille, 1796. Désignation des lectotypes des espèces paléarctiques de H. de Bonvouloir (Coleoptera, Throscidae). Bulletin de la Société Entomologique de France 100 (4): 347–359.

Leseigneur L (1996) *Trixagus atticus* Reitter et *T. minutus* Rey, deux espèces méconnues d'Europe occidentale présentes en France (Coleoptera Throscidae). Bulletin Mensuel de la Société Linnéenne de Lyon 65: 181–192.

Leseigneur L (1997) Réhabilitation de *Trixagus gracilis* Wollaston (Coleoptera, Throscidae). Bulletin de la Société Entomologique de France 102 (2): 137–142.

Leseigneur L (2005a) Description de *Trixagus meybohmi* n. sp. et note sur la morphologie des *Trixagus* du groupe *carinifrons* (Coleoptera, Throscidae). Bulletin de la Société Entomologique de France 110 (1): 89–96.

Leseigneur L (2005b) *Trixagus angelinii* n. sp. (Coleoptera, Throscidae). Bulletin de la Société Entomologique de France 110 (2): 181–184.

Leseigneur L, Gueorguiev B (2006) A contribution to the study of Bulgarian Throscidae (Coleoptera: Elateriformia). Historia Naturalis Bulgarica 17: 43–46.

Muona J (1993) Eucnemidae and Throscidae from Baltic amber (Coleoptera). Entomologische Blätter für Biologie und Systematik der Käfer 89: 15–45.

Muona J (1995) The phylogeny of Elateroidea (Coleoptera), or which tree is best today? Cladistics 11 (4): 317–341.

Muona J (2002) *Trixagus leseigneuri* n. sp. (Coleoptera, Throscidae). Bulletin de la Société Entomologique de France 107 (2): 187–190.

Nikitsky NB, Semenov VB (2001) To the knowledge of the beetles (Coleoptera) of the Moscow region. Byulleten' Moskovskogo Obshchestva Ispytatelei Prirody Otdel Biologicheskii 106: 38–49 [in Russian].

Van-Meer C (1998) *Aulonothroscus laticollis* (Rybinski) (Coleoptera, Throscidae), une espèce nouvelle pour la faune d'Europe occidentale. Bulletin de la Société Linnéenne de Bordeaux 26: 181–183.

Wedmann S (1994) Fossile Vertreter der Eucnemidae und Throscidae (Insecta: Coleoptera) aus der mitteleozanen Messel-Formation. Courier Forschungsinstitut Senckenberg 170: 65–73.

Yensen E (1975) A review of the genus *Pactopus* Leconte (Coleoptera: Throscidae). The Coleopterists Bulletin 29 (2): 87–91.

Yensen E (1975) A revision of the North American species of *Trixagus* Kugelann (Coleoptera: Throscidae). Transactions of the American Entomological Society 101 (1): 125–166.

Yensen E (1981) A new species of *Trixagus* from Panama and a key to New World *Trixagus* (Coleoptera: Throscidae). The Coleopterists Bulletin 34 [1980] (3): 257–261.

Yoshizawa N (2003) Notes on the *Drapetes torigaii* Nakane (Coleoptera, Throscidae) in Nagano, Japan. Gekkan-Mushi 391: 14–15.

Trogossitidae

Supporting references for the conservation of Lophocaterini Crowson, 1964 over Lycoptini Casey, 1890 (Art. 23.9.2). The taxon name Lycoptini Casey, 1890 has not been used as valid after 1899 to our knowledge.

Beutel RG, Leschen RAB (2005) Classification [pp. 11–16]. In: Beutel RG, Leschen RAB (Eds). Handbook of zoology. A natural history of the phyla of the animal kingdom. Volume IV - Arthropoda: Insecta. Walter De Gruyter, Berlin, xi + 567 pp.

Crowson RA (1964) A review of the classification of Cleroidea (Coleoptera), with descriptions of two new genera of Peltidae and of several new larval types. Transactions of the Royal Entomological Society of London 116 (12): 275–327.

Crowson RA (1970) Further observations on Cleroidea (Coleoptera). Proceedings of the Royal Entomological Society of London (Series B) 39 (1/2): 1–20.

Klausnitzer B (1996) Die Larven der Käfer Mitteleuropas. 3. Band. Polyphaga. Teil 2. Goecke & Evers, Krefeld, 335 pp.

Klimaszewski J, Watt JC (1997) Coleoptera: family-group review and keys to identification. Fauna of New Zealand. Ko te Aitanga Pepeke o Aotearoa. Number 37, Manaaki Whenua Press, Lincoln, 199 pp.

Kolibáč J (2006) A review of the Trogossitidae. Part 2: larval morphology, phylogeny and taxonomy (Coleoptera, Cleroidea). Entomologica Basiliensia et Collectionis Frey 28: 105–153.

Kolibáč J (2007) Family Trogossitidae Latreille, 1802 [pp. 364–366]. In: Löbl I, Smetana A (Eds) Catalogue of Palaearctic Coleoptera. Volume 4. Elateroidea - Derodontoidea - Bostrichoidea - Lymexyloidea - Cleroidea - Cucujoidea. Apollo Books, Stenstrup, 935 pp.

Lafer GS (1992) 55. Sem. Peltidae (Lophocateridae) [pp. 82–86]. In: Ler PA (Ed) Opredelitel' nasekomykh Dal'nego Vostoka SSSR Tom 3 - Zheskokrylye, ili zhuki Chast 2 [Keys to the identification of insects of the Soviet Far East 3 - Coleoptera, or beetles Part 2]. Nauka, Saint-Petersburg, 704 pp. [in Russian].

Lawrence JF, Britton EB (1991) Coleoptera (beetles) [pp. 543–683]. In: The insects of Australia: a textbook for student and research workers. Second edition. Volume II. Melbourne University Press, Carlton, pp. i-vi + 543–1137.

Lawrence JF, Newton AFJ (1995) Families and subfamilies of Coleoptera (with selected genera, notes, references and data on family-group names) [pp. 779–1006]. In: Pakaluk J, Ślipiński SA (Eds) Biology, phylogeny, and classification of Coleoptera: papers celebrating the 80th birthday of Roy A Crowson. Museum i Instytut Zoologii PAN, Warszawa, x + 1092 pp. in 2 vols.

Leschen RAB (2000) Beetles feeding on bugs (Coleoptera, Hemiptera): repeated shifts from mycophagous ancestors. Invertebrate Taxonomy 14 (6): 917–929.

Leschen RAB (2002) 72. Trogossitidae Latreille, 1802 [pp. 263–266]. In: Arnett RH, Jr., Thomas MC, Skelley PE, Frank JH (Eds) American beetles. Volume 2. Polyphaga: Scarabaeoidea through Curculionoidea. CRC Press, Boca Raton, xiv + 861 pp.

Leschen RAB, Lawrence JF, Kuschel G, Thorpe S, Wang Q (2003) Coleoptera genera of New Zealand. New Zealand Entomologist 26: 15–28.

Leschen RAB, Lawrence JF, Ślipiński SA (2005) Classification of basal Cucujoidea (Coleoptera: Polyphaga): cladistic analysis, keys and review of new families. Invertebrate Systematics 19 (1): 17–73.

Lucht W (1998) 32.c Familie: Lophocateridae [pp. 207–208]. In: Lucht W, Klausnitzer B (Eds) Die Käfer Mitteleuropas. 4. Supplementband. Goecke & Evers, Krefeld, 398 pp.

Lundberg B (1995) Catalogus Coleopterorum Sueciae. Naturhistoriska Riksmuseet, Stockholm [unpaginated].

Machado A, Oromí P (2000) Elenco de los Coleópteros de las Islas Canarias. Instituto de Estudios Canarios, La Laguna, 306 + [2] pp.

McKenna DD, Farrell BD (2009) Beetles (Coleoptera) [pp. 278–289]. In: Hedges SB, Kumar S (Eds) The timetree of life. Oxford University Press Inc., New York.

Peck SB (2005) A checklist of the beetles of Cuba with data on distributions and bionomics (Insecta: Coleoptera). Arthropods of Florida and neighboring land areas. Volume 18, Florida Department of Agriculture and Consumer Services, Gainesville, vi + 241 pp.

Peck SB, Thomas MC (1998) A distributional checklist of the beetles (Coleoptera) of Florida. Arthropods of Florida and neighboring land areas. Volume 16, Florida Department of Agriculture and Consumer Services, Gainesville, viii + 180 pp.

Ribera I (1999) Evolución, filogenia y clasificación de los Coleoptera (Arthropoda: Hexapoda). Boletín de la Sociedad Entomológica Aragonesa 26: 435–458.

Silfverberg H (1992) Enumeratio Coleopterorum Fennoscandiae, Daniae et Baltiae. Helsingfors Entomologiska Bytesförening, Helsinki, v + 94 pp.

Ślipiński SA (1992) Larinotinae - a new subfamily of Trogossitidae (Coleoptera), with notes on the constitution of Trogossitidae and related families of Cleroidea. Revue Suisse de Zoologie 99 (2): 439–463.

Tait SM, Dahlsten DL, Gill RJ, Doyen JT (1990) Life history of the incense cedar scale, *Xylococculus macrocarpae* (Homoptera: Margarodidae), on incense cedar in California with a description of the larvae of one of its common predators, *Eronyxa expansus* van Dyke (Coleoptera: Trogositidae). Hilgardia 58 (2): 1-19.

Telnov D (2004) Check-list of Latvian beetles (Insecta: Coleoptera). Second edition. Compendum of Latvian Coleoptera 1. Entomological Society of Latvia, Rïga, 113 pp.

Telnov D, Barsevskis A, Savich F, Kovalevsky F, Berdnikov S, Doronin M, Cibulskis R, Ratniece D (1997) Check-list of Latvian beetles (Insecta: Coleoptera). Mitteilungen des Internationalen Entomologischen Vereins, Supplement 5: 1-141.

Appendix 2

Coleoptera family-group name changes required based on the Principle of Priority. The action that has been taken to fix the problem, or recommendation for future work, is mentioned for each case. Cases listed in alphabetical order by family.

Family	Change from:	Change to:	Action	Reference
Aderidae	Aderidae/-ini Csiki, 1909	Euglenesidae/-ini Seidlitz, 1875	Younger name given precedence over older name and placed on the Official List of Family-Group Names in Zoology	ICZN (1989e)
Anthribidae	Anthribidae Billberg, 1820	Choragidae Kirby, 1819	Younger name given precedence over older name and placed on the Official List of Family-Group Names in Zoology	ICZN (1994b)
Brentidae	Acratina Alonso-Zarazaga et al., 1999	Nemocephalina Lacordaire, 1865	An application to the Commission is needed to suppress the older name because it is based on a misidentified type genus (Art. 65.2.1)	
Brentidae	Pseudoceocephalina Kleine, 1922	Uropterina Jakobson, 1911	An application to the Commission is needed to conserve usage of the well-established name Pseudoceocephalina Kleine, 1922	
Brentidae	Pseudoceocephalina Kleine, 1922	Ceocephalina Lacordaire, 1865	An application to the Commission is needed to suppress the older name because it is based on a misidentified type genus (Art. 65.2.1)	
Buprestidae	Chalcophorina Lacordaire, 1857	Anaglyptina Gistel, 1848	Use of younger name conserved (Art. 40.2)	this paper
Callirhipidae	Callirhipidae Emden, 1924	Zenoidae LeConte, 1866	An application to the Commission is needed to conserve usage of the well-established name Callirhipidae Emden, 1924	
Cantharidae	Cantharidae/-inae/-ini Imhoff, 1856	Telephoridae/-nae/-ini Leach, 1815	Use of younger name conserved (Art. 40.2)	Lawrence and Newton (1995)
Carabidae	Anisodactylina Lacordaire, 1854	Eurytrichina LeConte, 1847	Use of younger name conserved (Art. 23.9.2)	this paper
Carabidae	Harpalinae Bonelli, 1810	Graphipterinae Latreille, 1802	Use of younger name conserved (Art. 35.5)	this paper

Family	Change from:	Change to:	Action	Reference
Carabidae	Hexagoniini Horn, 1881	Trigonodactylini Brullé, 1834	Use of younger name conserved (Art. 40.2)	this paper
Carabidae	Iresiina Rivalier, 1971	Euprosopina Horn, 1893	An application to the Commission is needed to conserve usage of the well-established name Iresiina Rivalier, 1971	
Carabidae	Iresiina Rivalier, 1971	Eucalliina Horn, 1893	An application to the Commission is needed to conserve usage of the well-established name Iresiina Rivalier, 1971	
Carabidae	Iresiina Rivalier, 1971	Distipsiderina Horn, 1893	An application to the Commission is needed to conserve usage of the well-established name Iresiina Rivalier, 1971	
Carabidae	Perigonini Horn, 1881	Trechicini Bates, 1873	Use of younger name conserved (Art. 23.9.2)	this paper
Carabidae	Salcediini/-ina Alluaud, 1930	Zelmini/-ina Andrewes, 1929	Use of younger name conserved (Art. 40.2)	this paper
Cerambycidae	Eurypodini Gahan, 1906	Zaracini Lacordaire, 1868	Use of younger name conserved (Art. 40.2)	this paper
Cerambycidae	Hemilophini Thomson, 1868	Amphionychini Thomson, 1860	Older name treated as a *nomen oblitum*	Bousquet et al. (2009)
Cerambycidae	Macronini Lacordaire, 1868	Enchopterini Thomson, 1861	An application to the Commission is needed to conserve usage of the well-established name Macronini Lacordaire, 1868	
Cerambycidae	Pseudocephalini Aurivillius, 1912	Ametrocephalini Thomson, 1861	Use of younger name conserved (Art. 40.2)	Bousquet et al. (2009)
Cerambycidae	Stenoderini Pascoe, 1867	Syllitini Thomson, 1864	An application to the Commission is needed to conserve usage of the well-established name Stenoderini Pascoe, 1867	
Chrysomelidae	Amblycerina Bridwell, 1932	Spermophagina Crotch, 1873	An application to the Commission is needed to suppress the older name because it is based on a misidentified type genus (Art. 65.2.1)	
Chrysomelidae	Basiprionotini Gressitt, 1952	Priopterini Spaeth, 1914	Use of younger name conserved (Art. 40.2)	this paper
Chrysomelidae	Bromiini Baly, 1865	Adoxiini Baly, 1863	Use of younger name conserved (Art. 40.2)	this paper
Chrysomelidae	Bromiini Baly, 1865	Heteraspini Baly, 1863	An application to the Commission is needed to conserve usage of the well-established name Bromiini Baly, 1865	

Family	Change from:	Change to:	Action	Reference
Chrysomelidae	Dorynotini Monrós and Viana, 1949	Batonotini Spaeth, 1923	Use of younger name conserved (Art. 40.2)	this paper
Chrysomelidae	Hemisphaerotini Monrós and Viana, 1951	Porphyraspidini Spaeth, 1929	Use of younger name conserved (Art. 40.2)	this paper
Chrysomelidae	Nothosacanthini Gressitt, 1952	Hoplonotini Spaeth, 1929	Use of younger name conserved (Art. 40.2)	this paper
Chrysomelidae	Oediopalpini Monrós and Viana, 1947	Amplipalpini Weise, 1910	Use of younger name conserved (Art. 40.2)	this paper
Chrysomelidae	Oidini Laboissière, 1921	Adoriini Chapuis, 1875	Use of younger name conserved (Art. 40.2)	this paper
Chrysomelidae	Omocerini Hincks, 1952	Tauromini Spaeth, 1923	Use of younger name conserved (Art. 40.2)	this paper
Chrysomelidae	Sceloenoplini Uhmann, 1930	Cephalodontini Gestro, 1906	An application to the Commission is needed to conserve usage of the well-established name Sceloenoplini Uhmann, 1930	
Cryptophagidae	Ipidae/-inae/-ini Latreille, 1802	Cryptophagidae/-inae/-ini Kirby, 1826	An application to the Commission is needed to suppress the older name because it is based on a misidentified type genus (Art. 65.2.1)	
Curculionidae	Aterpini/-ina Lacordaire, 1863	Heliomeneini/-ina Gistel, 1848	Use of younger name conserved (Art. 23.9.2)	this paper
Curculionidae	Bagoinae Thomson, 1859	Lyprinae Gistel, 1848	Use of younger name conserved (Art. 23.9.2)	this paper
Curculionidae	Cleonini Schönherr, 1826	Geomorini Schönherr, 1823	Use of younger name conserved (Art. 23.9.2)	this paper
Curculionidae	Hyperinae/-ini Lacordaire, 1863	Phytonominae/-ini Gistel, 1848	Use of younger name conserved (Art. 40.2)	this paper
Curculionidae	Magdalidini Pascoe, 1870	Scardamyctini Gistel, 1848	Use of younger name conserved (Art. 23.9.2)	this paper
Curculionidae	Mesoptiliinae Lacordaire, 1863	Scardamyctinae Gistel, 1848	Use of younger name conserved (Art. 35.5)	this paper
Curculionidae	Naupactini Gistel, 1848	Iphiini Schönherr, 1823	Use of younger name conserved (Art. 23.9.2)	this paper
Curculionidae	Smicronychini Seidlitz, 1891	Desmorini LeConte, 1876	Use of younger name conserved (Art. 23.9.2)	this paper
Curculionidae	Tonesiina Alonso-Zarazaga and Lyal	Lyteriina Lacordaire, 1865	An application to the Commission is needed to suppress the older name because it is based on a misidentified type genus (Art. 65.2.1)	
Curculionidae	Zygobaridina Pierce, 1907	Centrinina Jekel, 1865	Use of younger name conserved because the older name is a junior homonym	Alonso-Zarazaga and Lyal (1999)

Family	Change from:	Change to:	Action	Reference
Dascilloidea	Dascilloidea/-idae/-inae/-ini Guérin-Méneville, 1843	Atopoidea/-idae/-inae/-ini Laporte, 1834	Use of younger name conserved (Art. 40.2)	Lawrence and Newton (1995)
Dascilloidea	Dascilloidea Guérin-Méneville, 1843	Rhipiceroidea Latreille, 1834	Use of younger name conserved because relationships between Dascillidae and Rhipiceridae are uncertain	
Derodontoidea	Derodontoidea LeConte, 1861	Nosodendroidea Erichson, 1846	Use of younger name conserved (Art. 35.5)	this paper
Dryopidae	Dryopidae Billberg, 1820	Parnidae Leach, 1817	Use of younger name conserved (Art. 40.2)	Lawrence and Newton (1995)
Elateridae	Agrypninae/-ini Candèze, 1857	Adelocerinae/-ini Gistel, 1848	Use of younger name conserved (Art. 23.9.2)	this paper
Elateridae	Agrypninae/-ini Candèze, 1857	Pangaurinae/-ini Gistel, 1856	Use of younger name conserved (Art. 23.9.2)	this paper
Elateridae	Agrypninae Candèze, 1857	Oophorinae Gistel, 1848	Use of younger name conserved (Art. 35.5)	this paper
Elateridae	Denticollinae/-ini/-ina Stein and Weise, 1877	Campylinae/-ini/-ina Gistel, 1848	Use of younger name conserved (Art. 40.2)	this paper
Elateridae	Hypnoidini Schwarz, 1906	Cryptohypnini Candèze, 1860	Use of younger name conserved (Art. 40.2)	this paper
Elateridae	Melanotini Candèe, 1859	Cratonychini Gistel, 1848	Use of younger name conserved (Art. 40.2)	Sánchez-Ruiz (1996)
Elateridae	Prosternini Gistel, 1856	Diacanthini Gistel, 1848	Use of younger name conserved (Art. 23.9.2)	this paper
Elateridae	Thylacosterninae Fleutiaux, 1920	Pterotarsinae Fleutiaux, 1902	An application to the Commission will be submitted by J. Muona and H. Silfverberg (pers. comm. 2010) to conserve usage of the well-established name Thylacosterninae Fleutiaux, 1920	
Elateroidea/-idae	Elateroidea/-idae Leach, 1815	Cebrionoidea/-idae Latreille, 1802	An application to the Commission will be sent to conserve usage of the well-established names Elateroidea/-idae Leach, 1815	P. J. Johnson (in Lawrence and Newton 1995, pers. comm. 2009)

Family	Change from:	Change to:	Action	Reference
Erotylidae	Encaustini Crotch, 1876	Engidini MacLeay, 1825	An application to the Commission is needed to suppress the older name because it is based on a misidentified type genus (Art. 65.2.1)	
Eucnemidae	Eucnemidae Eschscholtz, 1829	Melasidae Fleming, 1821	An application to the Commission has been sent to conserve usage of the well-established name Eucnemidae Fleming, 1821	see Appendix 6
Gyrinidae	Enhydrini Régimbart, 1882	Dineutini Desmarest, 1851	Use of younger name conserved (Art. 35.5)	this paper
Hydrophilidae	Coelostomatini Heyden, 1891	Cyclonotini Horn, 1890	Use of younger name conserved (Art. 40.2)	Hansen (1991)
Hygrobiidae	Hygrobiidae Régimbart, 1879	Pelobiinae Erichson, 1837	Use of younger name conserved (Art. 40.2)	Lawrence and Newton (1995)
Latridiidae	Latridiidae Erichson, 1842	Corticariidae Curtis, 1829	An application to the Commission was submitted by Bousquet et al. (2010) to conserve usage of the well-established name Latridiidae Erichson, 1842	
Leiodidae	Catopocerinae/-ini Hatch, 1927	Pinodytinae/-ini Horn, 1880	Use of younger name conserved (Art. 40.2)	Newton and Thayer (1992)
Leiodidae	Coloninae Horn, 1880	Myloechinae Thomson, 1859	Use of younger name conserved (Art. 40.2)	Newton and Thayer (1992)
Leiodidae	Leptodirini/-ina Lacordaire, 1854	Stagobiini/-ina Schiødte, 1849	Use of younger name conserved (Art. 40.2)	Newton and Thayer (1992)
Lepiceroidea	Lepiceroidea/-idae Hinton, 1936	Cyathoceroidea/-idae Sharp, 1882	Use of younger name conserved (Art. 40.2)	Lawrence and Newton (1995)
Lymexyloidea	Lymexyloidea/-idae Fleming, 1821	Hylecoetoidea/-idae Germar, 1818	Use of younger name conserved (Art. 35.5)	this paper
Melandryidae	Osphyinae Mulsant, 1856	Nothinae Shuckard, 1839	Use of younger name conserved (Art. 40.2)	Lawrence and Newton (1995)

Family	Change from:	Change to:	Action	Reference
Meloidae	Epicautini Parker and Böving, 1924	Macrobasini LeConte, 1862	An application to the Commission is needed to conserve usage of the well-established name Epicautini Parker and Böving, 1924	
Meloidae	Epicautini Parker and Böving, 1924	Apterospastini Wellman, 1910	An application to the Commission is needed to conserve usage of the well-established name Epicautini Parker and Böving, 1924	
Meloidae	Lyttini Solier, 1851	Cantharini Latreille, 1802	An application to the Commission is needed to suppress the older name because it is based on a misidentified type genus (Art. 65.2.1)	
Meloidae	Meloidae Gyllenhal, 1810	Horiidae Latreille, 1802	Younger name given precedence over older name and placed on the Official List of Family-Group Names in Zoology	ICZN (1999c)
Meloidae	Meloidae/-inae Gyllenhal, 1810	Cantharidae/-inae Latreille, 1802	An application to the Commission is needed to suppress the older name because it is based on a misidentified type genus (Art. 65.2.1)	
Meloidae	Nemognathinae Laporte, 1840	Horiinae Latreille, 1802	Younger name given precedence over older name and placed on the Official List of Family-Group Names in Zoology	ICZN (1999c)
Nemonychidae	Nemonychidae Bedel, 1882	Cimberididae Gozis, 1882	Younger name given precedence over older name and placed on the Official List of Family-Group Names in Zoology	ICZN (2005c)
Oedemeridae	Calopodinae Costa, 1852	Sparedrinae Gistel, 1848	Use of younger name conserved (Art. 23.9.2)	this paper
Psephenidae	Placonychinae Horn, 1880	Eubrianacinae Jakobson, 1913	Use of younger name conserved (Art. 40.2)	Lawrence and Newton (1995)
Ptiliidae	Acrotrichinae Reitter, 1909	Cleopteriinae Gistel, 1856	Use of younger name conserved (Art. 40.2)	Newton and Thayer (1992)
Ptiliidae	Ptinellini Reitter, 1906	Neuglenini Reitter, 1891	Use of younger name conserved (Art. 40.2)	this paper
Ptilodactylidae	Anchytarsinae Champion, 1897	Coloboderinae Erichson, 1847	An application to the Commission is needed to conserve usage of the well-established name Anchytarsinae Champion, 1897	

Family	Change from:	Change to:	Action	Reference
Ripiphoridae	Macrosiagonini Heyden, 1908	Ripiphorini Laporte, 1840	An application to the Commission is needed to suppress the older name because it is based on a misidentified type genus (Art. 65.2.1)	
Ripiphoridae	Ripiphoridae/-inae/-ini Gemminger, 1870	Myoditidae/-inae/-ini Gerstaecker, 1855	Use of younger name conserved (Art. 40.2)	Lawrence and Newton (1995)
Scarabaeidae	Anomalini/-ina Streubel, 1839	Euchlorini/-ina Hope, 1839	Use of younger name conserved (Art. 23.9.2)	Smith (2006)
Scarabaeidae	Cremastocheilini Burmeister and Schaum, 1841	Macromini Burmeister and Schaum, 1840	Use of younger name conserved (Art. 35.5)	Smith (2006)
Scarabaeidae	Eupariini Schmidt, 1910	Ataeniini Harold, 1868	Use of younger name conserved (Art. 23.9.2)	Smith (2006)
Scarabaeidae	Heterosternina Bates, 1888	Macropnina Horn, 1867	Use of younger name conserved (Art. 23.9.2)	Smith (2006)
Scarabaeidae	Oniticellini Kolbe, 1905	Drepanocerina van Lansberge, 1875	Use of younger name conserved (Art. 35.5)	Smith (2006)
Schizophoroidea	Schizophoroidea Ponomarenko, 1968	Schizocoleoidea Rohdendorf, 1961	Use of younger name conserved (Art. 35.5)	this paper
Sphindidae	Sphindidae/-inae Jacquelin Du Val, 1860	Aspidiphoridae/-inae Kiesenwetter, 1877 (1859)	Ruled under the plenary power that the younger name is to be given precedence over the older name	ICZN (1997)
Staphylinidae	Astenina Hatch, 1857	Suniina Sharp, 1886	An application to the Commission is needed to suppress the older name because it is based on a misidentified type genus (Art. 65.2.1)	
Staphylinidae	Athetini/-ina Casey, 1910	Callicerini Jakobson, 1908	An application to the Commission was submitted by Gusarov to conserve usage of the well-established names Athetini/-ina Casey, 1910	
Staphylinidae	Dolicaonina Casey, 1905	Gnathymenina Solier, 1849	An application to the Commission is needed to conserve usage of the well-established name Dolicaonina Casey, 1905	
Staphylinidae	Eusphalerini Hatch, 1957	Anthobiini Portevin, 1929	An application to the Commission is needed to suppress the older name because it is based on a misidentified type genus (Art. 65.2.1)	
Staphylinidae	Goniaceritae/-ini Reitter, 1882	Goniastitae/-ini Schaufuss, 1872	Use of younger name conserved (Art. 40.2)	Newton and Thayer (1992)

Family	Change from:	Change to:	Action	Reference
Staphylinidae	Xantholinini Erichson, 1839	Agrodini Nordmann, 1837	Younger name given precedence over older names and placed on the Official List of Family-Group Names in Zoology	ICZN (1996c)
Staphylinidae	Xantholinini Erichson, 1839	Gyrohypnini Kirby, 1837	Younger name given precedence over older names and placed on the Official List of Family-Group Names in Zoology	ICZN (1996c)
Staphylinidae	Xanthopygina Sharp, 1884	Platycnemina Nordmann 1837	Younger name given precedence over older name and placed on the Official List of Family-Group Names in Zoology	ICZN (1996c)
Tenebrionidae	Adelinina LeConte, 1862	Alphitophagina Gistel, 1856	An application to the Commission is needed to conserve usage of the well-established name Adelinina LeConte, 1862	
Tenebrionidae	Adesmiini Lacordaire, 1859	Megageniini Solier, 1851	Use of younger name conserved (Art. 23.9.2)	Bouchard et al. (2007)
Tenebrionidae	Adesmiini Lacordaire, 1859	Macropodini Agassiz, 1846	Use of younger name conserved (Art. 23.9.2)	this paper
Tenebrionidae	Alleculinae Laporte, 1841	Ctenioopodinae Solier, 1835	Use of younger name conserved (Art. 35.5)	Bouchard et al. (2005)
Tenebrionidae	Alleculinae/-ini Laporte, 1840	Xystropodinae/-ini Solier, 1835	Use of younger name conserved (Art. 35.5)	Bouchard et al. (2005)
Tenebrionidae	Bolitophagini Kirby, 1837	Eledonini Billberg, 1820	Use of younger name conserved (Art. 23.9.2)	this paper
Tenebrionidae	Cryptoglossini LeConte, 1862	Centriopterini Lacordaire, 1859	Use of younger name conserved (Art. 23.9.2)	Aalbu (2006)
Tenebrionidae	Epitragini Blanchard, 1845	Lygophilini Rafinesque, 1815	Use of younger name conserved (Art. 23.9.2)	Bouchard et al. (2007)
Tenebrionidae	Erodiini Billberg, 1820	Cephacerini Rafinesque, 1815	Use of younger name conserved (Art. 23.9.2)	Bouchard et al. (2007)
Tenebrionidae	Lagriinae Latreille, 1825	Cossyphinae Latreille, 1802	Use of younger name conserved because placement of older taxon uncertain	Bouchard et al. (2005)
Tenebrionidae	Lagriinae/-ini/-ina Latreille, 1825	Lachninae/-ini/-ina Billberg, 1820	Use of younger name conserved (Art. 40.2)	Bouchard et al. (2005)

Family	Change from:	Change to:	Action	Reference
Tenebrionidae	Melanimini Seidlitz, 1894	Microzoini Mulsant, 1854	Use of younger name conserved (Art. 40.2)	Bouchard et al. (2005)
Tenebrionidae	Pycnocerini Lacordaire, 1859	Chiroscelini Laporte, 1840	Use of younger name conserved (Art. 23.9.2)	Bouchard et al. (2005)
Tenebrionidae	Stenosini Lacordaire, 1859	Tagenini Solier, 1834	Use of younger name conserved (Art. 40.2)	Bouchard et al. (2005)
Tenebrionidae	Talanini Champion, 1887	Dignamptini LeConte and Horn, 1883	Use of younger name conserved (Art. 40.2)	Bouchard et al. (2005)
Throscidae	Throscidae Laporte, 1840	Stereolidae Rafinesque, 1815	Use of younger name conserved (Art. 23.9.2)	this paper
Trogossitidae	Lophocaterini Crowson, 1964	Lycoptini Casey, 1890	Use of younger name conserved (Art. 23.9.2)	this paper
Zopheridae	Zopheridae Solier, 1834	Colydiidae Billberg, 1820	Use of younger name conserved (Art. 35.5)	this paper

Appendix 3

Coleoptera family-group name changes required based on the Principle of Homonymy. The action that has been taken to fix the problem, or recommendation for future work, is mentioned for each case. Cases listed in alphabetical order by family.

Family	Coleoptera name:	Homonym:	Type genera	Action	Reference
Anthicidae	Notoxinae Stephens, 1829	Notoxini Sturm, 1826 [Coleoptera: Cleridae]	same, older name based on misidentified type genus	An application to the Commission is needed to suppress the older name because it is based on a misidentified type genus (Art. 65.2.1)	
Anthicidae	Steropinae Jacquelin du Val, 1863	Steropinae Dana, 1854 [Copepoda]	different	An application to the Commission is needed to remove the homonymy (Art. 55.3.1)	
Anthribidae	Apolectini Lacordaire, 1865	Apolectidae Jordan, 1923	different	An application to the Commission is needed to remove the homonymy (Art. 55.3.1)	
Anthribidae	Euparini Valentine, 1960	Eupariina Schmidt, 1910 [Coleoptera: Scarabaeidae]	different	An application to the Commission is needed to remove the homonymy (Art. 55.3.1)	
Attelabidae	Anisonychina Legalov, 2003	Anisonychidae Burmeister, 1844 [Coleoptera: Scarabaeidae]	different	An application to the Commission is needed to remove the homonymy (Art. 55.3.1)	
Brachyceridae	Aonychini Zimmerman, 1993	Aonychini Davis, 1978 [Mammalia]	different	An application to the Commission is needed to remove the homonymy (Art. 55.3.1)	
Brachyceridae	Brotheini Marshall, 1907	Brotheinae Simon, 1879 [Arachnida]	different	An application to the Commission is needed to remove the homonymy (Art. 55.3.1)	
Buprestidae	Anaglyptina Gistel, 1848	Anaglyptini Lacordaire, 1868 [Coleoptera: Cerambycidae]	different	Older name treated as a *nomen oblitum* (Art. 23.9.2)	Bousquet et al. (2009)
Cantharidae	Cantharidae Imhoff, 1856	Cantharidae Latreille, 1802	same, older name based on misidentified type genus	An application to the Commission is needed to suppress the older name because it is based on a misidentified type genus (Art. 65.2.1)	

Family	Coleoptera name:	Homonym:	Type genera	Action	Reference
Carabidae	Agonidae Kirby, 1837	Agonidae Swainson, 1839 [Pisces]	different	Stem of beetle name emended and Agoniumidae Kirby, 1837 placed on the Official List of Family-Group Names in Zoology	ICZN (1996d)
Carabidae	Amblytelini Blackburn, 1892	Amblytelinae Viereck, 1918 [Hymenoptera]	different	An application to the Commission is needed to remove the homonymy (Art. 55.3.1)	
Carabidae	Arthropterini Wasmann, 1928	Arthropteridae Jordan, 1923 [Pisces]	same, Coleoptera genus older	An application to the Commission is needed to conserve the Coleoptera name	
Carabidae	Bradybaenina Csiki, 1932	Bradybaenidae Pilsbry, 1939 [Mollusca]	different	An application to the Commission is needed to remove the homonymy (Art. 55.3.1)	
Carabidae	Colpodini Chaudoir, 1872	Colpodidae Poche, 1913 [Protista]	different	An application to the Commission is needed to remove the homonymy (Art. 55.3.1)	
Carabidae	Granigerina Antoine, 1959 [Carabidae: Harpalinae]	Granigerini Bedel, 1900 [Carabidae: Trechinae]	same, older name based on misidentified type genus	An application to the Commission is needed to suppress the older name because it is based on a misidentified type genus (Art. 65.2.1)	
Carabidae	Nomiini Gozis, 1875	Nomiidae Robertson, 1904 [Hymenoptera]	different	An application was submitted to the Commission to emend the stem of the beetle family-group name	Engel and Bouchard (2009)
Cerambycidae	Anaglyptini Lacordaire, 1868	Anaglyptina Gistel, 1848 [Coleoptera: Buprestidae]	different	Older name treated as a *nomen oblitum* (Art. 23.9.2)	Bousquet et al. (2009)
Cerambycidae	Compsina Martins and Galileo, 2007	Compsini Pierce, 1913 [Coleoptera: Curculionidae]	different	An application to the Commission is needed to remove the homonymy (Art. 55.3.1)	
Cerambycidae	Dryobiini Arnett, 1962	Dryobiini Gistel, 1856 [Coleoptera: Ptinidae]	different	Older name treated as a *nomen oblitum*	Bousquet et al. (2009)
Cerambycidae	Lissonotini Swainson, 1840	Lissonotini Förster, 1869 [Hymenoptera]	different	An application to the Commission is needed to remove the homonymy (Art. 55.3.1)	
Cerambycidae	Platysternini Lacordaire, 1872	Platysternidae Gray, 1869 [Reptilia]	different	An application to the Commission is needed to remove the homonymy (Art. 55.3.1)	

Family	Coleoptera name:	Homonym:	Type genera	Action	Reference
Cerambycidae	Polyarthrini Gounelle, 1911	Polyarthridae Daday, 1893 [Rotifera]	different	An application to the Commission is needed to remove the homonymy (Art. 55.3.1)	
Cerambycidae	Sphaeriini Lacordaire, 1868	Sphaeriidae Deshayes, 1855 [Mollusca]	different	An application to the Commission is needed to remove the homonymy (Art. 55.3.1)	
Cerambycidae	Sphaeriini Lacordaire, 1868	Sphaeriidae Erichson, 1845 [Coleoptera: Myxophaga]	different	Stem of older beetle name emended and Sphaeriusidae Erichson, 1845 placed on the Official List of Family-Group Names in Zoology	ICZN (2000)
Cerambycidae	Stenocorini Thomson, 1861 [Cerambycidae: Lepturinae]	Stenocorini Hope, 1834 [Cerambycidae: Cerambycinae]	same, older name based on misidentified type genus	An application to the Commission is needed to suppress the older name because it is based on a misidentified type genus (Art. 65.2.1)	
Cerambycidae	Stenoderini Pascoe, 1867	Stenoderini Selander, 1991 [Coleoptera: Meloidae]	different	An application to the Commission is needed to remove the homonymy (Art. 55.3.1)	
Cerambycidae	Stenodontini Lameere, 1903	Stenodontina Schmiedeknecht, 1903 [Hymenoptera]	different	An application to the Commission is needed to remove the homonymy (Art. 55.3.1)	
Cerambycidae	Toxotini LeConte and Horn, 1883	Toxotidae Günther, 1860 [Pisces]	different	An application to the Commission is needed to remove the homonymy (Art. 55.3.1)	
Chaetosomatidae	Chaetosomatidae Crowson, 1952	Chaetosomatidae Claus, 1872 [Nematoda]	same, Coleoptera genus older	An application to the Commission was submitted by YB and PB to conserve the Coleoptera name	
Chrysomelidae	Carpophagini Chapuis, 1874	Carpophaginae Selby, 1835 [Aves]	different	An application to the Commission is needed to remove the homonymy (Art. 55.3.1)	
Chrysomelidae	Chalepini Weise, 1910 [Chrysomelidae: Cassidinae]	Chalepidae Burmeister, 1847 [Scarabaeidae: Dynastinae]	same, Scarabaeidae name older	An application to the Commission is needed to conserve usage of the well established Chrysomelidae name	
Chrysomelidae	Oreinini Bechyné, 1958	Oreinini Bleeker, 1863 [Pisces]	different	An application to the Commission is needed to remove the homonymy (Art. 55.3.1)	

Family	Coleoptera name:	Homonym:	Type genera	Action	Reference
Chrysomelidae	Megacerini Bridwell, 1946	Megacerini Viret, 1961 [Mammalia]	different	An application to the Commission is needed to remove the homonymy (Art. 55.3.1)	
Chrysomelidae	Spermophagina Borowiec, 1987 [Chrysomelidae: Bruchinae: Amblycerini: Spermophagina]	Spermophagina Crotch, 1873 [Chrysomelidae: Bruchinae: Amblycerini: Amblycerina]	same, older name based on misidentified type genus	An application to the Commission is needed to suppress the older name because it is based on a misidentified type genus (Art. 65.2.1)	
Chrysomelidae	Spilophorini Chapuis, 1875 [Chrysomelidae: Cassidinae]	Spilophorina Krikken, 1984 [Coleoptera: Scarabaeidae]	different	An application to the Commission is needed to remove the homonymy (Art. 55.3.1)	
Cleridae	Trichodini Portevin, 1931	Trichodina Maitland, 1851[Protozoa]	different	An application to the Commission is needed to remove the homonymy (Art. 55.3.1)	
Cryptophagidae	Ipini Latreille, 1802 [Cryptophagidae: Cryptophaginae: Cryptophagini]	Ipini Bedel, 1888 [Curculionidae: Scolytinae]	same, older name based on misidentified type genus	An application to the Commission is needed to suppress the older name because it is based on a misidentified type genus (Art. 65.2.1)	
Curculionidae	Centrinina Jekel, 1865	Centrininae Swainson, 1839 [Pisces]	different	The younger name Zygobaridina is used as valid instead of Centrinina	Alonso-Zarazaga and Lyal (1999)
Curculionidae	Compsini Pierce, 1913	Compsina Martins and Galileo, 2007 [Coleoptera: Cerambycidae]	different	An application to the Commission is needed to remove the homonymy (Art. 55.3.1)	
Curculionidae	Conoderinae Schönherr, 1833	Conoderini Fleutiaux, 1919 [Coleoptera: Elateridae]	different	An application to the Commission is needed to remove the homonymy (Art. 55.3.1)	
Curculionidae	Euderini Lacordaire, 1865	Euderini Erdös, 1956 [Hymenoptera]	different	A junior synonym is used as valid instead of Euderini in Hymenoptera	Hansson and Straka (2009)
Curculionidae	Haplonychini Lacordaire, 1865	Haplonychidae Burmeister, 1855 [Coleoptera: Scarabaeidae]	different	An application to the Commission is needed to remove the homonymy (Art. 55.3.1)	

Family	Coleoptera name:	Homonym:	Type genera	Action	Reference
Curculionidae	Ipini Bedel, 1888 [Curculionidae: Scolytinae]	Ipini Latreille, 1802 [Cryptophagidae: Cryptophaginae: Cryptophagini]	same, older name based on misidentified type genus	An application to the Commission is needed to suppress the older name because it is based on a misidentified type genus (Art. 65.2.1)	
Curculionidae	Ominii Shuckard, 1839	Ominii Beck, 1996 [Lepidoptera]	different	An application to the Commission is needed to remove the homonymy (Art. 55.3.1)	
Curculionidae	Phaenomerina Faust, 1898	Phaenomerini Erichson, 1847 [Coleoptera: Scarabaeidae]	different	Stem of Scarabaeidae name emended and Phaenomerididae (with incorrect author and year "Ohaus, 1913") placed on the Official List of Family-Group Names in Zoology	ICZN (1962)
Curculionidae	Rhynchaenini Blanchard, 1853	Rhynchaenini Latreille, 1828	same, older name based on misidentified type genus	An application to the Commission is needed to suppress the older name because it is based on a misidentified type genus (Art. 65.2.1)	
Curculionidae	Tomicini Wood, 1978 [Curculionidae: Scolytinae: Hylurgini]	Tomicini Shuckard, 1839 [Curculionidae: Scolytinae: Ipini]	same, older name based on misidentified type genus	An application to be submitted to the Commission by MAAZ and CHCL to suppress the older name because it is based on a misidentified type genus (Art. 65.2.1)	
Curculionidae	Trypetini Lacordaire, 1865	Trypetidae Loew, 1861[Diptera]	different	Stem of beetle name emended and Trypetidinae (with incorrect author and year "Pierce, 1919") placed on the Official List of Family-Group Names in Zoology	ICZN (1974a)
Elateridae	Conoderini Fleutiaux, 1919	Conoderinae Schönherr, 1833 [Coleoptera: Curculionidae]	different	An application to the Commission is needed to remove the homonymy (Art. 55.3.1)	
Elateridae	Drasteriini Houlbert, 1912	Drasteriini Wiltshire, 1976 [Lepidoptera]	different	An application to the Commission is needed to remove the homonymy (Art. 55.3.1)	
Elateridae	Ludiini Candèze, 1857	Ludiini Aurivilius, 1904 [Lepidoptera]	different	The Lepidoptera name was replaced by Micragonini Cockerell, 1914	Oberprieler (1997)

Family	Coleoptera name:	Homonym:	Type genera	Action	Reference
Elateroidea	Cantharidae Imhoff, 1856	Cantharidae Latreille, 1802	same, older name based on misidentified type genus	An application to the Commission is needed to suppress the older name because it is based on a misidentified type genus (Art. 65.2.1)	
Elmidae	Larini LeConte, 1861	Laridae Rafinesque-Schmaltz, 1815 [Aves]	different	Stem of beetle name emended and Laraini LeConte, 1861 placed on the Official List of Family-Group Names in Zoology	ICZN (1988g)
Gyrinidae	Enhydrini Régimbart, 1882	Enhydrini Gray, 1825	different	An application was recently submitted to the Commission by Özdikmen and Darilmaz (2010) to remove the homonymy	
Helotidae	Helotidae Chapuis, 1876	Helotidae Adams et al., 1854 [Pisces]	different	An application to the Commission is needed to remove the homonymy (Art. 55.3.1)	
Hydrophilidae	Hydrobiina Mulsant, 1844	Hydrobiidae Troschel 1857 [Mollusca]	different	Stem of beetle name emended and Hydrobiusina Mulsant, 1844 placed on the Official List of Family-Group Names in Zoology	ICZN (2003a)
Lampyridae	Photinini LeConte, 1881	Photininae Giglio-Tos, 1915 [Mantodea]	different	An application to the Commission was submitted by Svenson and Branham (2007) to remove the homonymy (Art. 55.3.1)	
Leiodidae	Anisotomidae Reitter, 1884 [Leiodinae: Agathidiini]	Anisotomidae Erichson, 1845 [Leiodinae: Leiodini]	same, older name based on misidentified type genus	An application to the Commission is needed to suppress the older name because it is based on a misidentified type genus (Art. 65.2.1)	
Leiodidae	Triarthrini Jeannel, 1962	Triarthridae Ulrich, 1930 [Trilobita]	different	An application to the Commission is needed to remove the homonymy (Art. 55.3.1)	
Meloidae	Lydina Kaszab, 1959	Lydidae Newman, 1834 [Hymenoptera]	different	An application to the Commission is needed to remove the homonymy (Art. 55.3.1)	
Meloidae	Stenoderini Selander, 1991	Stenoderini Pascoe, 1867 [Coleoptera: Cerambycidae]	different	An application to the Commission is needed to remove the homonymy (Art. 55.3.1)	

Family	Coleoptera name:	Homonym:	Type genera	Action	Reference
Meloidae	Zonitina Mulsant, 1857	Zonitidae Mörch, 1864 [Mollusca]	different	Stem of beetle name emended and Zonitidinae Mulsant, 1857 placed on the Official List of Family-Group Names in Zoology	ICZN (1999c)
Plastoceridae	Plastoceridae Crowson, 1972	Plastocerini LeConte, 1861 [Coleoptera: Elateridae]	same, older name based on misidentified type genus	An application to the Commission is needed to suppress the older name because it is based on a misidentified type genus (Art. 65.2.1)	
Prionoceridae	Prionoceridae Lacordaire, 1857	Prionocerini Savchenko, 1966 [Diptera]	different	An application to the Commission is needed to remove the homonymy (Art. 55.3.1)	
Ptinidae	Dryobiini Gistel, 1856	Dryobiini Arnett, 1962 [Coleoptera: Cerambycidae]	different	Older name treated as a *nomen oblitum* (Art. 23.9.2)	Bousquet et al. (2009)
Ripiphoridae	Ripiphoridae/-inae/-ini Gemminger, 1870 [Ripiphoridae: Ripiphorinae: Ripiphorini]	Ripiphoridae/-inae/-ini Laporte, 1840 [Ripiphoridae: Ripiphorinae: Macrosiagonini]	same, older name based on misidentified type genus	An application to the Commission is needed to suppress the older name because it is based on a misidentified type genus (Art. 65.2.1)	
Salpingidae	Elacatidae Cockerell, 1906	Elacatinae Gill, 1861 [Pisces]	different	An application to the Commission is needed to remove the homonymy (Art. 55.3.1)	
Scarabaeidae	Anisonychidae Burmeister, 1844	Anisonychini Legalov, 2003 [Coleoptera: Attelabidae]	different	An application to the Commission is needed to remove the homonymy (Art. 55.3.1)	
Scarabaeidae	Chalepidae Burmeister, 1847 [Scarabaeidae: Dynastinae]	Chalepini Weise, 1910 [Chrysomelidae: Cassidinae]	same, Scarabaeidae name older	An application to the Commission is needed to conserve usage of the well established name in Chrysomelidae	
Scarabaeidae	Cryptodontina Lacordaire, 1856	Cryptodontidae Dall, 1895 [Mollusca]	different	An application to the Commission is needed to remove the homonymy (Art. 55.3.1)	
Scarabaeidae	Euchroeina Paulian and Descarpentries, 1982	Euchroeidae Dahlbom, 1854 [Hymenoptera]	different	An application to the Commission is needed to remove the homonymy (Art. 55.3.1)	

Family	Coleoptera name:	Homonym:	Type genera	Action	Reference
Scarabaeidae	Eupariina Schmidt, 1910	Eupariini Valentine, 1960 [Coleoptera: Anthribidae]	different	An application to the Commission is needed to remove the homonymy (Art. 55.3.1)	
Scarabaeidae	Haplonychidae Burmeister, 1855	Haplonychini Lacordaire, 1865 [Coleoptera: Curculionidae]	different	An application to the Commission is needed to remove the homonymy (Art. 55.3.1)	
Scarabaeidae	Macrophyllina Burmeister, 1855	Macrophyllina Gray, 1866 [Mammalia]	different	An application to the Commission is needed to remove the homonymy (Art. 55.3.1)	
Scarabaeidae	Phaenomerini Erichson, 1847	Phaenomerina Faust, 1898 [Coleoptera: Curculionidae]	different	Stem of Scarabaeidae name emended and Phaenomeridini Ohaus, 1913 [in error for Erichson, 1847] placed on the Official List of Family-Group Names in Zoology	ICZN (1962)
Scarabaeidae	Spilophorina Krikken, 1984	Spilophorini Chapuis, 1875 [Coleoptera: Chrysomelidae]	different	An application to the Commission is needed to remove the homonymy (Art. 55.3.1)	
Scarabaeidae	Trichiini Fleming, 1821	Trichiinae Lozek, 1956 [Mollusca]	different	The mollusc name was placed on the Official Index of Rejected and Invalid Family-Group Names in Zoology	ICZN (2004c)
Scarabaeidae	Trigonostomina Ohaus, 1912	Trigonostomidae Graff, 1905 [Platyhelminthes]	different	Stem of beetle name emended and Trigonostomusina Ohaus, 1912 placed on the Official List of Family-Group Names in Zoology	ICZN (2009b)
Sphaeriusidae	Sphaeriidae Erichson, 1845	Sphaeriini Lacordaire, 1868 [Coleoptera: Cerambycidae] and Sphaeriidae Deshayes, 1855 [Mollusca]	different	Stem of older beetle name emended and Sphaeriusidae Erichson, 1845 placed on the Official List of Family-Group Names in Zoology	ICZN (2000)
Staphylinidae	Callicerina Jakobson, 1908	Callicerina Rondani, 1856 [Diptera]	different	An application to the Commission was recently submitted by Gusarov to suppress the Coleoptera name	
Staphylinidae	Metopiini Raffray, 1904	Metopiinae Foerster, 1869 [Hymenoptera]	different	Stem of beetle name emended and Metopiasini Raffray, 1904 placed on the Official List of Family-Group Names in Zoology	ICZN (1994d)

Family	Coleoptera name:	Homonym:	Type genera	Action	Reference
Staphylinidae	Tachinidae Fleming, 1821	Tachinidae Robineau-Desvoidy, 1830 [Diptera]	different	Stem of beetle name emended and Tachinusidae Fleming, 1821 placed on the Official List of Family-Group Names in Zoology	ICZN (1993f)
Staphylinidae	Toxoderini Bernhauer and Schubert, 1911	Toxoderini Saussure, 1869 [Mantodea]	different	An application to the Commission is needed to remove the homonymy (Art. 55.3.1)	
Tenebrionidae	Adeliini Kirby, 1828	Adeliini Viereck, 1918 [Hymenoptera]	different	An application to the Commission is needed to remove the homonymy (Art. 55.3.1)	
Tenebrionidae	Lachnini Billberg, 1820	Lachnini Herrich-Schaeffer 1854 [Hemiptera]	different	An application to the Commission is needed to remove the homonymy (Art. 55.3.1)	
Tenebrionidae	Macropodini Agassiz, 1846	Macropodidae Gray, 1821[Mammalia]	different	Beetle name treated as a *nomen oblitum* (Art. 23.9.2)	this paper
Tenebrionidae	Pentaphyllini Mulsant, 1854	Pentaphyllinae Schindewolf, 1942 [Anthozoa]	different	An application to the Commission is needed to remove the homonymy (Art. 55.3.1)	

Appendix 4

COLEOPTERA names as they appear on the *Official List of Family-Group Names in Zoology*. Cases listed in alphabetical order by family. Names on the *Official List* with the symbol "°" differ from those presented in our catalogue (spelling, authorship and/or year of publication).

Family	Name on *Official List*	Opinion(s) or Direction	Reference
ADERIDAE	°ADERIDAE Winkler, 1927	1549	ICZN 1989e
ADERIDAE	EUGLENESIDAE Seidlitz, 1875	1549	ICZN 1989e
ANTHRIBIDAE	ANTHRIBIDAE Billberg, 1820	1756	ICZN 1994b
ANTHRIBIDAE	CHORAGIDAE Kirby, 1819	1756	ICZN 1994b
BRENTIDAE	°EURHININI Lacordaire, 1866	1352	ICZN 1985e
BRENTIDAE	EURHYNCHINAE Lacordaire, 1863	1352	ICZN 1985e
BUPRESTIDAE	TRACHEIDAE Laporte, 1835	2222	ICZN 2009a
CARABIDAE	AGONUMIDAE Kirby, 1837	1855	ICZN 1996d
CARABIDAE	DROMIUSIDAE Bonelli, 1810	2149	ICZN 2006b
CARABIDAE	°GALERITINI Lacordaire, 1854	862	ICZN 1968b
CERYLONIDAE	CERYLONIDAE Billberg, 1820	1811	ICZN 1995c
CHRYSOMELIDAE	BRUCHIDAE Latreille, 1802	1809	ICZN 1995a
CHRYSOMELIDAE	°CASSIDINAE Stephens, 1831	1023	ICZN 1974b
CHRYSOMELIDAE	°UROPLATINI Leng, 1920	1359	ICZN 1985f
CURCULIONIDAE	°BRACHYDERINAE Schönherr, 1837	1440	ICZN 1987a
CURCULIONIDAE	CRYPTORHYNCHINAE Schönherr, 1825	808	ICZN 1967b
CURCULIONIDAE	OTIORHYNCHINAE Schönherr, 1826	982	ICZN 1972
CURCULIONIDAE	PACHYRHYNCHINI Schönherr, 1826	928	ICZN 1970b
CURCULIONIDAE	PHAENOMERINA Faust, 1898	621	ICZN 1962
CURCULIONIDAE	PHYLLOBIINI Schönherr, 1826	1179	ICZN 1981c
CURCULIONIDAE	°POLYDROSINI Schönherr, 1823	1179	ICZN 1981c
CURCULIONIDAE	°SCOLYTIDAE Westwood, 1838	683	ICZN 1963b
CURCULIONIDAE	SOMATODINAE Lacordaire, 1863	1770	ICZN 1994c
CURCULIONIDAE	°TRYPETIDINAE Pierce, 1919	1005	ICZN 1974a
DYTISCIDAE	°DYTISCIDAE Leach, 1817	619	ICZN 1961e
ELMIDAE	ELMIDAE Curtis, 1830	1812	ICZN 1995d
ELMIDAE	LARAINI LeConte, 1861	1515	ICZN 1988g
GEOTRUPIDAE	°GEOTRUPINI Latreille, 1806	Direction 28	ICZN 1955c
HYDROPHILIDAE	HYDROBIUSINA Mulsant, 1844	2034	ICZN 2003a
KATERETIDAE	BRACHYPTERINAE Erichson, [1845]	1916	ICZN 1999a
KATERETIDAE	°KATERETIDAE Erichson in Agassiz, [1846]	1916	ICZN 1999a
MELOIDAE	HORIIDAE Latreille, 1802	1918	ICZN 1999b
MELOIDAE	MELOIDAE Gyllenhal, 1810	1918	ICZN 1999b
MELOIDAE	NEMOGNATHINAE Castelnau, 1840	1918	ICZN 1999b
MELOIDAE	ZONITIDINAE Mulsant, 1857	1918	ICZN 1999b
NEMONYCHIDAE	CIMBERIDIDAE Gozis, 1882	2111	ICZN 2005c
NEMONYCHIDAE	NEMONYCHIDAE Bedel, 1882	2111	ICZN 2005c

Family	Name on *Official List*	Opinion(s) or Direction	Reference
Orsodacnidae	Aulacoscelidinae Chapuis, 1874	2242	ICZN 2010a
Ptinidae	Ptinidae Latreille, 1802	1809	ICZN 1995a
Scarabaeidae	°Phaenomerididae Ohaus, 1913	621	ICZN 1962
Scarabaeidae	Trigonostomusina Ohaus, 1912	2229	ICZN 2009b
Sphaeriusidae	Microsporidae Crotch, 1873	1331, 1957	ICZN 1985d, 2000
Sphaeriusidae	Sphaeriusidae Erichson, 1845	1957	ICZN 2000
Sphindidae	Aspidiphoridae Kiesenwetter, 1877	1862	ICZN 1997
Sphindidae	°Sphindidae Jacquelin du Val, [1861]	1862	ICZN 1997
Staphylinidae	Agrodini Nordmann, 1837	1851	ICZN 1996c
Staphylinidae	Bolitocharini Thomson, 1859	599	ICZN 1961a
Staphylinidae	Gyrohypnini Kirby, 1837	1851	ICZN 1996c
Staphylinidae	Metopiasini Raffray, 1904	1772	ICZN 1994d
Staphylinidae	Oxypodides Thomson, 1859	463	ICZN 1957
Staphylinidae	Platycnemini Nordmann, 1837	1851	ICZN 1996c
Staphylinidae	Quediini Kraatz, [1857]	1851	ICZN 1996c
Staphylinidae	°Staphylinidae Latreille, [1803-1804]	546	ICZN 1959a
Staphylinidae	Tachinusidae Fleming, 1821	1743	ICZN 1993f
Staphylinidae	Tachyporidae MacLeay, 1825	1743	ICZN 1993f
Staphylinidae	Xantholinini Erichson, 1839	1851	ICZN 1996c
Zopheridae	°Colydiidae Erichson, 1842	1811	ICZN 1995c
Zopheridae	Orthocerini Blanchard, 1845 (1820)	1811	ICZN 1995c
Zopheridae	Rhopalocerini Reitter, 1911	1397	ICZN 1986b

Appendix 5

Coleoptera type genus names as they appear on the *Official List of Generic Names in Zoology*. Cases listed in alphabetical order by family.

Family	Name on *Official List*	Opinion(s)	Reference(s)
Aderidae	*Aderus* Stephens, 1829	1549	ICZN 1989e
Aderidae	*Euglenes* Westwood, 1830	1549	ICZN 1989e
Anthribidae	*Anthribus* Geoffroy, 1762	1754	ICZN 1994a
Anthribidae	*Mecocerus* Schönherr, 1833	982	ICZN 1972
Anthribidae	*Choragus* Kirby, 1819	1756	ICZN 1994b
Attelabidae	*Attelabus* Linnaeus, 1758	1239	ICZN 1983a
Bostrichidae	*Bostrichus* Geoffroy, 1762	1754	ICZN 1994a
Brentidae	*Eurhynchus* Kirby, in Kirby & Spence, 1828	1352	ICZN 1985
Brentidae	*Nanophyes* Schönherr, 1838	1526	ICZN 1989b
Buprestidae	*Acmaeodera* Eschscholtz, 1829	2100	ICZN 2005b
Buprestidae	*Actenodes* Dejean, 1833	2008	ICZN 2002b
Buprestidae	*Anthaxia* Eschscholtz, 1829	2009	ICZN 2002c
Buprestidae	*Astraeus* Laporte & Gory, 1837	795	ICZN 1966
Buprestidae	*Buprestis* Linnaeus, 1758	1784	ICZN 1994e
Buprestidae	*Chrysobothris* Eschscholtz, 1829	1784	ICZN 1994e
Buprestidae	*Chrysodema* Laporte & Gory, 1835	2076	ICZN 2004b
Buprestidae	*Dicerca* Eschscholtz, 1829	1784	ICZN 1994e
Buprestidae	*Iridotaenia* Deyrolle, 1864	2076	ICZN 2004b
Buprestidae	*Melanophila* Eschscholtz, 1829	1826	ICZN 1996b
Buprestidae	*Nascio* Gory & Laporte, 1838	2008	ICZN 2002b
Buprestidae	*Poecilonota* Eschscholtz, 1829	1825	ICZN 1996a
Buprestidae	*Trachys* Fabricius, 1801	2222	ICZN 2009a
Byrrhidae	*Byrrhus* Linnaeus, 1767	1754	ICZN 1994a
Byrrhidae	*Simplocaria* Stephens, 1829	1323	ICZN 1985c
Carabidae	*Agonum* Bonelli, 1810	1855	ICZN 1996d
Carabidae	*Carabus* Linnaeus, 1758	243	ICZN 1950
Carabidae	*Dromius* Bonelli, 1810	2149	ICZN 2006b
Carabidae	*Galerita* Fabricius, 1801	862	ICZN 1968b
Carabidae	*Glyptus* Brullé, 1835	1299	ICZN 1985a
Carabidae	*Ophonus* Dejean, 1821	1598	ICZN 1990c
Carabidae	*Tachys* Dejean, 1821	1598	ICZN 1990c
Cerambycidae	*Ceroplesis* Audinet-Serville, 1835	1407	ICZN 1986
Cerambycidae	*Phymatodes* Mulsant, 1839	1525	ICZN 1989a
Cerambycidae	*Prionus* Geoffroy, 1762	1754	ICZN 1994a
Cerambycidae	*Saperda* Fabricius, 1775	1155	ICZN 1980
Cerambycidae	*Stenocorus* Geoffroy, 1762	1754	ICZN 1994a
Cerambycidae	*Tetropium* Kirby, 1837	1473	ICZN 1988b
Cerylonidae	*Cerylon* Latreille, 1802	1811	ICZN 1995c
Chrysomelidae	*Altica* Geoffroy, 1762	1754	ICZN 1994a
Chrysomelidae	*Bruchus* Linnaeus, 1767	1809	ICZN 1995a

Family	Name on *Official List*	Opinion(s)	Reference(s)
Chrysomelidae	*Cassida* Linnaeus, 1758	1023	ICZN 1974b
Chrysomelidae	*Chrysolina* Motschulsky, 1860	1279	ICZN 1984c
Chrysomelidae	*Chrysomela* Linnaeus, 1758	1279	ICZN 1984c
Chrysomelidae	*Crioceris* Geoffroy, 1762	908, 1754	ICZN 1970a, 1994a
Chrysomelidae	*Cryptocephalus* Geoffroy, 1762	1754	ICZN 1994a
Chrysomelidae	*Galeruca* Geoffroy, 1762	1754	ICZN 1994a
Chrysomelidae	*Lema* Fabricius, 1798	908	ICZN 1970a
Chrysomelidae	*Luperus* Geoffroy, 1762	1273	ICZN 1984a
Chrysomelidae	*Uroplata* Chevrolat, 1835	1359	ICZN 1985
Cleridae	*Clerus* Geoffroy, 1762	1273	ICZN 1984a
Cleridae	*Korynetes* Herbst, (1792)	604	ICZN 1961c
Cleridae	*Necrobia* Olivier, 1795	604	ICZN 1961c
Cleridae	*Notoxus* Geoffroy, 1762	1754	ICZN 1994a
Cleridae	*Trichodes* Herbst, 1792	1273	ICZN 1984a
Coprinisphaeridae	*Coprinisphaera* Sauer, 1955	2211	ICZN 2008b
Cryptophagidae	*Cryptophagus* Herbst, 1792	1810	ICZN 1995b
Cucujidae	*Cucujus* Fabricius, 1775	1754	ICZN 1994a
Curculionidae	*Brachyderes* Schönherr, 1823	1440	ICZN 1987
Curculionidae	*Ceutorhynchus* Germar, 1824	1529	ICZN 1989c
Curculionidae	*Cholus* Germar, 1824	1449	ICZN 1987d
Curculionidae	*Coeliodes* Schönherr, 1837	1529	ICZN 1989c
Curculionidae	*Cryptorhynchus* Illiger, 1807	808	ICZN 1967b
Curculionidae	*Dendroctonus* Erichson, 1836	670	ICZN 1963a
Curculionidae	*Dryocoetes* Eichhoff, 1864	1145	ICZN 1979b
Curculionidae	*Eurhinus* Illiger, 1807	1352	ICZN 1985
Curculionidae	*Geonemus* Schönherr, 1833	1493	ICZN 1988d
Curculionidae	*Magdalis* Germar, 1817	345	ICZN 1955a
Curculionidae	*Mononychus* Germar, 1824	1529	ICZN 1989c
Curculionidae	*Otiorhynchus* Germar, 1824	982	ICZN 1972
Curculionidae	*Pachyrhynchus* Germar, 1824	928	ICZN 1970b
Curculionidae	*Phaenomerus* Schönherr, 1836	621	ICZN 1962
Curculionidae	*Phloeosinus* Chapuis, 1869	1167	ICZN 1981a
Curculionidae	*Phloeotribus* Latreille, 1796	1144	ICZN 1979a
Curculionidae	*Phyllobius* Germar, 1824	1179	ICZN 1981c
Curculionidae	*Phytobius* Schönherr, 1833	1968	ICZN 2001
Curculionidae	*Polydrusus* Germar, 1817	1179	ICZN 1981c
Curculionidae	*Ptochus* Schönherr, 1826	1616	ICZN 1990d
Curculionidae	*Rhinoncus* Schönherr, 1825	1529	ICZN 1989c
Curculionidae	*Rhyncolus* Germar, 1817	1655	ICZN 1991b
Curculionidae	*Scolytus* Geoffroy, 1762	683	ICZN 1963b
Curculionidae	*Somatodes* Schönherr, 1840	1770	ICZN 1994c
Curculionidae	*Tomicus* Latreille, [1802]	670	ICZN 1963a
Curculionidae	*Tropiphorus* Schönherr, 1842	1474	ICZN 1988c
Curculionidae	*Trypetes* Schönherr, 1836	1005	ICZN 1974a
Curculionidae	*Xyleborus* Eichhoff, 1864	848	ICZN 1968a

Family	Name on *Official List*	Opinion(s)	Reference(s)
Curculionidae	*Zygops* Schönherr, 1825	1450	ICZN 1987e
Cyclaxyridae	*Cyclaxyra* Broun, 1893	1472	ICZN 1988a
Dermestidae	*Anthrenus* Geoffroy, 1762	1754	ICZN 1994a
Dryophthoridae	*Dryophthorus* Germar, 1824	1448	ICZN 1987c
Dryophthoridae	*Rhinostomus* Rafinesque, 1815	345	ICZN 1955a
Dryophthoridae	*Sitophilus* Schönherr, 1838	572	ICZN 1959b
Dryophthoridae	*Sphenophorus* Schönherr, 1838	572	ICZN 1959b
Dytiscidae	*Acilius* Leach, 1817	619	ICZN 1961e
Dytiscidae	*Dytiscus* Linnaeus, 1758	619	ICZN 1961e
Dytiscidae	*Graphoderus* Dejean, 1833	618	ICZN 1961d
Dytiscidae	*Vatellus* Aubé, [1837]	1681	ICZN 1992
Elateridae	*Drasterius* Eschscholtz, 1829	1441	ICZN 1987b
Elmidae	*Elmis* Latreille, 1802	1812	ICZN 1995d
Elmidae	*Lara* Le Conte, 1852	1515	ICZN 1988g
Erotylidae	*Tritoma* Fabricius, 1775	1754	ICZN 1994a
Geotrupidae	*Athyreus* MacLeay, 1819	1299	ICZN 1985a
Geotrupidae	*Bolboceras* Kirby, 1819	2138	ICZN 2006a
Geotrupidae	*Geotrupes* Latreille, 1796	346	ICZN 1955b
Geotrupidae	*Odonteus* Samouelle, 1819	2138	ICZN 2006a
Gyrinidae	*Enhydrus* Laporte, 1834	710	ICZN 1964
Gyrinidae	*Gyrinus* Geoffroy, 1762	1754	ICZN 1994a
Hydraenidae	*Ochthebius* Leach, 1815	1631	ICZN 1991a
Hydrophilidae	*Berosus* Leach, 1817	1577	ICZN 1990a
Hydrophilidae	*Georissus* Latreille, 1809	1891	ICZN 1998
Hydrophilidae	*Helochares* Mulsant, 1844	710	ICZN 1964
Hydrophilidae	*Helophorus* Fabricius, 1775	1724	ICZN 1993b
Hydrophilidae	*Hydrobius* Leach, 1815	1577	ICZN 1990a
Hydrophilidae	*Hydrophilus* Geoffroy, 1762	1754	ICZN 1994a
Hydrophilidae	*Megasternum* Mulsant, 1844	1178	ICZN 1981b
Hygrobiidae	*Hygrobia* Latreille, 1804	280	ICZN 1954
Kateretidae	*Brachypterus* Kugelann, 1794	1916	ICZN 1999a
Kateretidae	*Kateretes* Herbst, 1793	1916	ICZN 1999a
Lampyridae	*Lampyris* Geoffroy, 1762	1273	ICZN 1984a
Leiodidae	*Colon* Herbst, 1797	1810	ICZN 1995b
Lucanidae	*Platycerus* Geoffroy, 1762	1754	ICZN 1994a
Megalopodidae	*Zeugophora* Kunze, 1818	1382	ICZN 1986
Meloidae	*Cerocoma* Geoffroy, 1762	1754	ICZN 1994a
Meloidae	*Horia* Fabricius, 1787	1918	ICZN 1999b
Meloidae	*Meloe* Linnaeus, 1758	1918	ICZN 1999b
Meloidae	*Mylabris* Fabricius, 1775	1809	ICZN 1995a
Meloidae	*Nemognatha* Illiger, 1807	1918	ICZN 1999b
Meloidae	*Zonitis* Fabricius, 1775	1918	ICZN 1999b
Monotomidae	*Rhizophagus* Herbst, 1793	1810	ICZN 1995b
Nemonychidae	*Cimberis* Gozis, 1881	2111	ICZN 2005c
Nemonychidae	*Nemonyx* Redtenbacher, 1845	2111	ICZN 2005c

Family	Name on *Official List*	Opinion(s)	Reference(s)
Oedemeridae	*Oncomera* Stephens, 1829	1506	ICZN 1988f
Omalisidae	*Omalisus* Geoffroy, 1762	1754	ICZN 1994a
Orsodacnidae	*Orsodacne* Latreille, 1802	1989	ICZN 2002a
Ptiliidae	*Nephanes* Thomson, 1859	1307	ICZN 1985b
Ptiliidae	*Ptenidium* Erichson, 1845	1277	ICZN 1984b
Ptiliidae	*Ptilium* Gyllenhal, 1827	1277	ICZN 1984b
Ptiliidae	*Ptinella* Motschulsky, 1844	1307	ICZN 1985b
Ptinidae	*Anobium* Fabricius, 1775	1062	ICZN 1976
Ptinidae	*Dorcatoma* Herbst, 1792	1810	ICZN 1995b
Ptinidae	*Ptilinus* Geoffroy, 1762	1754	ICZN 1994a
Ptinidae	*Ptinus* Linnaeus, 1767	1809	ICZN 1995a
Ptinidae	*Xyletinus* Latreille, 1809	966	ICZN 1971
Pyrochroidae	*Pyrochroa* Geoffroy, 1762	1754	ICZN 1994a
Scarabaeidae	*Ataenius* Harold, 1867	2241	ICZN 2010a
Scarabaeidae	*Anomala* Samouelle, 1819	1546	ICZN 1989d
Scarabaeidae	*Bothynus* Hope, 1837	2199	ICZN 2008a
Scarabaeidae	*Copris* Geoffroy, 1762	1754	ICZN 1994a
Scarabaeidae	*Gymnetis* MacLeay, 1819	806	ICZN 1967a
Scarabaeidae	*Melolontha* Fabricius, 1775	1754	ICZN 1994a
Scarabaeidae	*Osmoderma* Lepeletier & Serville, 1828	2186	ICZN 2007
Scarabaeidae	*Pelidnota* MacLeay, 1819	2054	ICZN 2003b
Scarabaeidae	*Pentodon* Hope, 1837	2054	ICZN 2003b
Scarabaeidae	*Phaenomeris* Hope, 1833	621	ICZN 1962
Scarabaeidae	*Podalgus* Burmeister, 1847	2066	ICZN 2004a
Scarabaeidae	*Stethaspis* Hope, 1837	1244	ICZN 1983b
Scarabaeidae	*Trichius* Fabricius, 1775	2079	ICZN 2004c
Scarabaeidae	*Trigonostomus* Brenske, 1893	2229	ICZN 2009b
Schizopodidae	*Schizopus* LeConte, 1858	1727	ICZN 1993d
Scraptiidae	*Anaspis* Geoffroy, 1762	1273	ICZN 1984a
Sphaeriusidae	*Microsporus* Kolenati, 1846	1331	ICZN 1985d
Sphaeriusidae	*Sphaerius* Waltl, 1838	1331, 1957	ICZN 1985d, 2000
Sphindidae	*Aspidiphorus* Ziegler in Dejean, 1821	1862	ICZN 1997
Sphindidae	*Sphindus* Megerle in Dejean, 1821	1862	ICZN 1997
Staphylinidae	*Tachinus* Gravenhorst, 1802	1743	ICZN 1993f
Staphylinidae	*Acrotona* Thomson, 1859	600	ICZN 1961b
Staphylinidae	*Agrodes* Nordmann, 1837	1851	ICZN 1996c
Staphylinidae	*Anthophagus* Gravenhorst, 1802	2084	ICZN 2004d
Staphylinidae	*Atheta* Thomson, 1858	600	ICZN 1961b
Staphylinidae	*Baeocera* Erichson, 1845	1221	ICZN 1982
Staphylinidae	*Bolitochara* Mannerheim, 1831	599	ICZN 1961a
Staphylinidae	*Bryaxis* Kugelann, 1794	887	ICZN 1969b
Staphylinidae	*Bythinus* Leach, 1817	887	ICZN 1969b
Staphylinidae	*Coprophilus* Latreille, 1829	1722	ICZN 1993a
Staphylinidae	*Coryphium* Stephens, 1834	1597	ICZN 1990b
Staphylinidae	*Creophilus* Samouelle, 1819	546	ICZN 1959a

Family	Name on *Official List*	Opinion(s)	Reference(s)
Staphylinidae	*Geostiba* Thomson, 1858	2098	ICZN 2005a
Staphylinidae	*Gyrohypnus* Samouelle, 1819	1250	ICZN 1983c
Staphylinidae	*Ischnopoda* Stephens, 1835	600	ICZN 1961b
Staphylinidae	*Leptusa* Kraatz, 1856	2113	ICZN 2005d
Staphylinidae	*Lesteva* Latreille, 1797	2084	ICZN 2004d
Staphylinidae	*Metopias* Gory, 1832	1772	ICZN 1994d
Staphylinidae	*Mycetoporus* Mannerheim, 1831	1726	ICZN 1993c
Staphylinidae	*Othius* Stephens, 1829	1250	ICZN 1983c
Staphylinidae	*Oxypoda* Mannerheim, 1831	463	ICZN 1957
Staphylinidae	*Platycnemus* Nordmann, 1837	1851	ICZN 1996c
Staphylinidae	*Proteinus* Latreille, 1796	876	ICZN 1969a
Staphylinidae	*Quedius* Stephens, 1829	1851	ICZN 1996c
Staphylinidae	*Staphylinus* Linnaeus, 1758	546	ICZN 1959a
Staphylinidae	*Tachyporus* Gravenhorst, 1802	1743	ICZN 1993f
Staphylinidae	*Tachyusa* Erichson, 1837	600	ICZN 1961b
Staphylinidae	*Xantholinus* Dejean, 1821	1250	ICZN 1983c
Staphylinidae	*Zyras* Stephens, 1835	599	ICZN 1961a
Tenebrionidae	*Alphitobius* Stephens, 1829	1039	ICZN 1975
Tenebrionidae	*Diaperis* Geoffroy, 1762	1754	ICZN 1994a
Tenebrionidae	*Gnatocerus* Thunberg, 1814	1039	ICZN 1975
Tenebrionidae	*Helops* Fabricius, 1775	2237	ICZN 2009d
Tenebrionidae	*Phaleria* Latreille, [1802]	1039	ICZN 1975
Tenebrionidae	*Platyscelis* Latreille, 1818	1729	ICZN 1993e
Tenebrionidae	*Tentyria* Latreille, 1802	2244	ICZN 2010c
Tenebrionidae	*Tribolium* MacLeay, 1825	1495	ICZN 1988e
Tenebrionidae	*Uloma* Dejean, 1821	1039	ICZN 1975
Trogossitidae	*Peltis* Kugelann, 1792	1754	ICZN 1994a
Zopheridae	*Colydium* Fabricius, 1792	1811	ICZN 1995c
Zopheridae	*Orthocerus* Latreille, 1796	1811	ICZN 1995c
Zopheridae	*Rhopalocerus* W. Redtenbacher, 1842	1397	ICZN 1986

Appendix 6

Summary of cases involving Coleoptera family-group names and/or their type genera awaiting a ruling by the Commission. Cases listed in alphabetical order by family.

Carabidae

Engel and Bouchard (2009: Case 3484) reported that the Coleoptera name Nomiidae Gozis, 1875 and the Hymenoptera name Nomiidae Robertson, 1904 are homonyms based on similar but not identical type genera. Since the Hymenoptera name is currently used as valid and the Coleoptera name is not, the Commission was asked to emend the latter to Nomiusidae Gozis, 1875.

Chaetosomatidae

Bousquet and Bouchard (2010) submitted an application to the Commission (Case 3513) to conserve usage of Chaetosomatidae Crowson, 1952 over Chaetosomatidae Claus, 1872 [Nematoda]. In addition, the Commission was asked to conserve usage of *Chaetosoma* Westwood, 1851 and *Apodasya* Pascoe, 1863, both threatened by the discovery of the older name *Chaetosoma* Chevrolat, 1843.

Chrysomelidae

An application to conserve *Eupales* Lefèvre, 1885 and Eupalini Verma et al. 2005 was recently submitted by Jolivet and Verma (Bulletin of Zoological Nomenclature 2009: 204; see notice of "New applications to the Commission"). This Case (number 3498) has not been published yet.

Moseyko et al. (2010) submitted an application to the Commission (Case 3519) to conserve usage of *Eumolpus* Weber, 1801 and *Bromius* Chevrolat, 1836.

Curculionidae

An application to conserve usage of Plinthini Lacordaire, 1863 and to designate *Plinthus* Germar, 1817 as its type genus was recently submitted by Alonso-Zarazaga and Lyal (2010: Case 3530).

An application to emend the entries for *Otiorhynchus* and *Loborhynchus* in the Official List of Generic Names in Zoology was recently submitted to the Commission by Lyal and Alonso-Zarazaga (2010: Case 3529).

Eucnemidae

Muona (1994; see notice of "New applications to the Commission") submitted an application to conserve usage of Eucnemidae Eschscholtz, 1829 over Melasidae Fleming, 1821. This Case (number 2938) was not subsequently published in the Bulletin of Zoological Nomenclature and is apparently not closed (ICZN secretariat pers. comm. 2010). The status of this application is still uncertain.

Gyrinidae

The name Enhydrini Régimbart, 1882 is a junior homonym of the mammalian name Enhydrini Gray, 1825. An application was recently submitted to the Commission by Özdikmen and Darilmaz (2010: Case 3514) to remove the homonymy by changing the stem of the beetle name to Enhydrus-.

Lampyridae

Svenson and Branham (2007: Case 3402) submitted an application to remove the homonymy between Photinini LeConte, 1881 and Photininae Giglio-Tos, 1915 [Mantodea]. Both family-group names are correctly formed and are based on similar but not identical type genera. The Commission was asked to emend the Mantodea name to Photinaini Giglio-Tos, 1915.

Latridiidae

Bousquet et al. (2010) submitted an application to the Commission (Case 3517) to give precedence to Latridiidae Erichson, 1842 over Corticariidae Curtis, 1829 whenever their type genera are placed in the same family-group taxon and to conserve the current usage of *Corticaria* Marsham, threatened by the discovery of an overlooked type species designation.

Staphylinidae

An application to conserve Athetini Casey, 1910 and Athetina Casey, 2010 was recently submitted by Gusarov (Bulletin of Zoological Nomenclature 2010: 270; see notice of "New applications to the Commission"). This Case (number 3537) has not been published yet.

Index

Family-group names are listed in Small caps and genus-group names in *italics*.